Springer Monographs in Mathematics

Springer
New York
Berlin
Heidelberg
Barcelona
Hong Kong
London
Milan
Paris
Singapore
Tokyo

N.P. Landsman

Mathematical Topics Between Classical and Quantum Mechanics

With 15 Illustrations

Springer

N.P. Landsman
Korteweg-de Vries Institute for Mathematics
University of Amsterdam
Plantage Muidergracht 24
Amsterdam 1018 TV
The Netherlands

Mathematics Subject Classification (1991): 81S10, 81PXX, 58FXX, 81RXX, 81TXX

Library of Congress Cataloging-in-Publication Data
Landsman, N.P. (Nicolaas P.)
Mathematical topics between classical and quantum mechanics / N.P. Landsman.
p. cm.
Includes bibliographical references and index.
ISBN 0-387-98318-X (hardcover : alk. paper)
1. Quantum theory—Mathematics. 2. Quantum field theory—Mathematics. 3. Hilbert space. 4. Geometry, Differential. 5. Mathematical physics. I. Title.
QC174.17.M35L36 1998
530.12—dc21 98-18391

Printed on acid-free paper.

Camera-ready copy prepared from the author's LaTeX files.
Printed and bound by Edwards Brothers, Inc., Ann Arbor, MN.
Printed in the United States of America.

9 8 7 6 5 4 3 2 1

ISBN 0-387-98318-X Springer-Verlag New York Berlin Heidelberg SPIN 10636251

I realize that the disappearance of a culture does not signify the disappearance of human value, but simply of certain means of expressing this value, yet the fact remains that I have no sympathy for the current of European civilization and do not understand its goals, if it has any. So I am really writing for friends who are scattered throughout the corners of the globe.

Our civilization is characterized by the word "progress". Progress is its form rather than making progress one of its features. Typically it constructs. It is occupied with building an ever more complicated structure. And even clarity is only sought as a means to this end, not as an end in itself. For me on the contrary clarity, perspicuity are valuable in themselves. I am not interested in constructing a building, so much as in having a perspicuous view of the foundations of typical buildings.

Ludwig Wittgenstein

Preface

Subject Matter

The original title of this book was *Tractatus Classico-Quantummechanicus*, but it was pointed out to the author that this was rather grandiloquent. In any case, the book discusses certain topics in the interface between classical and quantum mechanics. Mathematically, one looks for similarities between Poisson algebras and symplectic geometry on the classical side, and operator algebras and Hilbert spaces on the quantum side. Physically, one tries to understand how a given quantum system is related to its alleged classical counterpart (the classical limit), and vice versa (quantization).

This monograph draws on two traditions: The algebraic formulation of quantum mechanics and quantum field theory, and the geometric theory of classical mechanics. Since the former includes the geometry of state spaces, and even at the operator-algebraic level more and more submerges itself into noncommutative geometry, while the latter is formally part of the theory of Poisson algebras, one should take the words "algebraic" and "geometric" with a grain of salt!

There are three central themes. The first is the relation between constructions involving observables on one side, and pure states on the other. Thus the reader will find a unified treatment of certain aspects of the theory of Poisson algebras, operator algebras, and their state spaces, which is based on this relationship. Roughly speaking, observables relate to each other by an algebraic structure, whereas pure states are tied together by transition probabilities (in both cases topology plays an additional role). The discussion of quantization shows both sides of the coin. One side involves a mapping of functions on the classical phase space into some operator algebra; at the other side one has coherent states, which define a map from the phase space itself into a projective Hilbert space. The duality between these sides is neatly exhibited in what is sometimes called Berezin quantization.

The second theme is the analogy between the C^*-algebra of a Lie groupoid and the Poisson algebra of the corresponding Lie algebroid. For example, the role played by groups and fiber bundles in classical and quantum mechanics may be understood on the basis of this analogy.

Thirdly, we describe the parallel between symplectic reduction in classical mechanics (with Marsden–Weinstein reduction as an important special case) and Rieffel induction (a tool for constructing representations of operator algebras) in quantum mechanics. This provides an interesting example of the mathematical similarities alluded to above, and in addition leads to a powerful strategy for the quantization of constrained systems in physics.

Various examples illustrate the abstract theory: The reader will find particles moving on a curved space in an external gauge field, magnetic monopoles, low-dimensional gauge theories, topological quantum effects, massless particles, and θ-vacua. On the other hand, the reader will not find path integrals, geometric quantization, the WKB-approximation, microlocal analysis, quantum chaos, or quantum groups. The connection between these topics and those treated in this book largely remains to be understood.

Prerequisites, Level, and Organization of the Book

This book should be accessible to mathematicians with a good undergraduate education and some prior knowledge of classical and quantum mechanics, and to theoretical physicists who have not completely abstained from functional analysis. It is assumed that the reader has at least seen the description of classical mechanics in terms of symplectic geometry, and knows the standard Hilbert space description of a quantum-mechanical particle moving in $\mathbb{R}^3$.

The reader should be familiar with the basics of the theory of manifolds, Lie groups, Banach spaces, and Hilbert spaces, say at the level of a first course. The necessary concepts in operator algebras, Riemannian and symplectic geometry, and fiber bundles are developed from scratch, but some previous exposure to these subjects would do no harm.

It is suggested that the reader start by going through the informal Introductory Overview as a whole. The main text is of a technical nature. The various chapters are logically related to each other, but can be read almost independently. To study a given chapter it is usually sufficient to be familiar with the preceding chapters merely at the level of the Introductory Overview. Some technical details will, of course, depend on previous material in a deeper way. One should by all means go through the list of conventions and notation below.

In the interest of clarity and continuity, no credits or references to the literature are given in the main text. These may be found in the Notes, which in addition contain comments and elaborations on the main text. If no reference for a particular result is given, it is either standard or new (we leave this decision to the reader).

The author would be happy if glaring omissions in the notes or references were pointed out to him.

In the Index, entries refer only to the location where an entry is defined and/or occurs for the first time.

Conventions and Notation

Unless explicitly indicated otherwise, or obvious from the context, our conventions are as follows.

General

- The (Roman) chapter number is used only in cross-referencing between different chapters. In such references, numbers in brackets refer to equations and those without refer to paragraphs (e.g., 1.2.3) or to sections (such as 1.2).
- The symbol ■ means "end of proof". The symbol □ stands for "end of incomplete proof".
- The equation $A := B$ means that A is by definition equal to B.
- The abbreviation "iff" means "if and only if".
- An index that occurs twice is summed over, i.e., $a_i a_i := \sum_i a_i a_i$.
- Projections between spaces are denoted by τ; in case of possible confusion we write $\tau_{E \to Q}$ for the pertinent projection from E to Q.
- The symbol $\restriction$ means "restricted to".
- The symbol 1_X stands for the function on X that is identically one.
- We put $0 \in \mathbb{R}^+$ but $0 \notin \mathbb{N}$.

Functional Analysis

- Vector spaces are over $\mathbb{C}$, and functions are $\mathbb{C}$-valued. Vector spaces over $\mathbb{R}$ are denoted by $\mathcal{V}_{\mathbb{R}}$ etc.; spaces of real-valued functions are written, for example, $C^\infty(P, \mathbb{R})$. The only exception to this rule is formed by Lie algebras $\mathfrak{g}$, which are always real except when the complexification $\mathfrak{g}_{\mathbb{C}}$ is explicitly indicated (this occurs only in III.1.10, III.1.11, and IV.3.6).
- The space $C_0(X)$, where X is a locally compact Hausdorff space, consists of all continuous functions on X that vanish at infinity; the space of all compactly supported continuous functions on X is denoted by $C_c(X)$, and the bounded continuous functions form $C_b(X)$. These are usually seen as normed spaces under the **sup-norm**

$$\|f\|_\infty := \sup_{x \in X} |f(x)|.$$

- When X has the discrete topology (relative to which all functions are continuous), we often write $\ell(X)$, $\ell_c(X)$, $\ell^\infty(X)$, $\ell_0(X)$ for $C(X)$, $C_c(X)$, $L^\infty(X)$, and $C_0(X)$.

- The topological dual of a topological vector space V is denoted by V^*; hence the double dual is V^{**}. The action of $\theta \in V^*$ on $v \in V$ is denoted by $\theta(v)$. Multilinear forms α are similarly denoted by $\alpha(v_1, \ldots, v_n)$.
- When confusion might arise otherwise, we write $X \dot{+} Y$ for $X + Y$ in $V_1 \oplus V_2$, where $X \in V_1$ and $Y \in V_2$ (for example, in $V \oplus V$ the expression $X + Y$ would be ambiguous, denoting either $X + Y \dot{+} 0$, where $X + Y \in V \simeq V \oplus 0 \subset V \oplus V$, or $X \dot{+} Y$, or $0 \dot{+} X + Y$).

Hilbert Spaces

- Inner products $(\ ,\ )$ in a Hilbert space $\mathcal{H}$ are linear in the second entry and antilinear in the first.
- If $\mathcal{K}$ is a closed subspace of a Hilbert space $\mathcal{H}$, then $[\mathcal{K}]$ denotes the orthogonal projection onto $\mathcal{K}$. If $\Psi \in \mathcal{H}$, we write $[\Psi]$ for $[\mathbb{C}\Psi]$.
- The symbol $\mathbb{S}\mathcal{H}$ denotes the space of all unit vectors in $\mathcal{H}$. The projective space of $\mathcal{H}$ is called $\mathbb{P}\mathcal{H}$; hence $\mathbb{P}\mathbb{C}^N = \mathbb{C}\mathbb{P}^{N-1}$.
- The symbols $\mathfrak{B}(\mathcal{H})$, $\mathfrak{B}_0(\mathcal{H})$, $\mathfrak{B}_1(\mathcal{H})$, $\mathfrak{B}_2(\mathcal{H})$ stand for the collections of all bounded, compact, trace-class, Hilbert–Schmidt operators on $\mathcal{H}$. The unit operator in $\mathfrak{B}(\mathcal{H})$ is called $\mathbb{I}$. We write $\mathfrak{M}_N(\mathbb{C})$ for $\mathfrak{B}(\mathbb{C}^N)$.
- When A and B are operators on $\mathcal{H}$, the symbol $[A, B]$ stands for the commutator $AB - BA$. We also use $\{A, B\}_\hbar := i[A, B]/\hbar$.
- In the context of the previous item, or more generally when A and B are elements of a Jordan algebra or a C^*-algebra, $A \circ B$ denotes $\frac{1}{2}(AB + BA)$. In all other situations, $\circ$ has its usual meaning of composition; i.e., when f and g are suitable functions, one has $f \circ g(x) := f(g(x))$.
- We say that two Hilbert spaces are **naturally isomorphic** if they are related by a unitary isomorphism whose construction is independent of a choice of basis.
- The Hilbert space $L^2(\mathbb{R}^n)$ is defined with respect to Lebesgue measure.

Our convention for the inner product is the one mainly used in the physics literature. Its motivation, however, is mathematical. Firstly, each $\Psi \in \mathcal{H}$ defines a linear functional on $\mathcal{H}$ by $\Psi(\Phi) := (\Psi, \Phi)$, without the need to change the order. Secondly, the convention is the same as for "inner products" taking values in a C^*-algebra, which for good reasons are always taken to be linear in the second entry; see IV.2.

C^*-Algebras

- The set of self-adjoint elements in a C^*-algebra $\mathfrak{A}$ is called $\mathfrak{A}_{\mathbb{R}}$. Its state space is $\mathcal{S}(\mathfrak{A})$, and its pure state space is $\mathcal{P}(\mathfrak{A})$.
- The unitization of a C^*-algebra $\mathfrak{A}$ is called $\mathfrak{A}_{\mathbb{I}}$.
- States on a C^*-algebra are denoted by ω; pure states are sometimes also called ρ, σ, or ψ. The state space of $\mathfrak{A}$ is called $\mathcal{S}(\mathfrak{A})$; the pure state space is denoted by $\mathcal{P}(\mathfrak{A})$.

- The Gelfand transform $\hat{A}$ of $A \in \mathfrak{A}_{\mathbb{R}}$ is the function on $\mathcal{P}(\mathfrak{A})$ defined by $\hat{A}(\omega) := \omega(A)$. When $\mathfrak{A}$ is commutative, this concept is used for general $A \in \mathfrak{A}$.
- Representations of a C^*-algebra are generically denoted by π.
- The GNS-representation corresponding to a state ω is called π_ω, with canonical cyclic vector Ω_ω.
- Equivalence of representations means unitary equivalence.
- The representation of a C^*-algebra $\mathfrak{A}$ induced (in the sense of Rieffel) by a representation π_χ of a C^*-algebra $\mathfrak{B}$ on a Hilbert space $\mathcal{H}_\chi$ is denoted by $\pi^\chi(\mathfrak{A})$, realized on a Hilbert space $\mathcal{H}^\chi$.
- Transformation group C^*-algebras are called action C^*-algebras.

Group Representations and Actions

- Group representations on a Hilbert space are tacitly assumed to be continuous and unitary.
- The adjoint action of a Lie group G on its Lie algebra $\mathfrak{g}$ is denoted by Ad; the dual coadjoint action on $\mathfrak{g}^*$ is called Co, i.e., $\text{Co}(x) := \text{Ad}^*(x^{-1})$.
- When $H \subset G$ is a closed subgroup, the representation of G induced (in the sense of Mackey) from a representation $U_\chi(H)$ on a Hilbert space $\mathcal{H}_\chi$ is denoted by $U^\chi(G)$, and is realized on a Hilbert space called $\mathcal{H}^\chi$.
- The unitary dual of a group G is denoted by $\hat{G}$.
- Equivalence of group representations means unitary equivalence.

Differential Geometry

- All manifolds (Lie groups included) are assumed to be real, smooth, connected, Hausdorff, finite-dimensional, and paracompact.
- If $\varphi : M \to N$ is a smooth map between two manifolds, the pullback is denoted by φ^*, and the pushforward is φ_* (often called $T\varphi$ or φ' in the literature). In particular, for $g \in C^\infty(N)$ the function $\varphi^* g$ in $C^\infty(M)$ is $g \circ \varphi$.
- We denote a point on a manifold Q by q, with coordinates q^i (in a given chart; $i = 1, \ldots, \dim(Q)$). The dependence of the coordinates on the chart is suppressed in the notation. We write ∂_i for $\partial/\partial q^i$. The point $p_i dq^i$ in the fiber $T_q^* Q$ of the cotangent bundle T^*Q at q then has canonical coordinates (p_i, q^i); we denote this point by (p, q). Similarly, the point $v^i \partial_i$ in the fiber $T_q Q$ of the tangent bundle TQ at q has coordinates (v^i, q_j), and we sometimes label this simply as (v, q).
- The action of $\theta \in T_q^* Q$ on $v \in T_q Q$ is written as $\theta_q(v)$. Similarly for multilinear forms, e.g., $\mathbf{g}_q(v, w)$ stands for a Riemannian inner product of $v, w \in T_q Q$.
- The tangent vector (field) to a curve $c(\cdot)$ is called $\dot{c}(\cdot)$.
- The symbol $\wedge^n(Q)$ stands for the bundle of n-forms over Q. Also, $\wedge_n(Q)$ is the dual vector bundle of $\wedge^n(Q)$, i.e., the bundle of totally antisymmetric contravariant tensors.

- The space of compactly supported smooth sections of a vector bundle E is denoted by $\Gamma(E)$.

Acknowledgements

This book was written between March 1996 and March 1998, based on research starting in 1989. The project was financed by the E.P.S.R.C., the Alexander von Humboldt Stiftung, and the Royal Netherlands Academy of Arts and Sciences. It was carried out at the Department of Applied Mathematics and Theoretical Physics of the University of Cambridge (October 1989–September 1993, October 1994–June 1997), the II. Institut für Theoretische Physik of the University of Hamburg (October 1993–September 1994), the Korteweg–de Vries Institute for Mathematics of the University of Amsterdam (July 1997–March 1998) and the Erwin Schrödinger Institute for Mathematical Physics in Vienna (September–October 1997).

Many mathematical ideas in this book may be traced back to J. von Neumann; key physical insights originated with P.A.M. Dirac. In addition, the author has been inspired by the work of F.A. Berezin, P. Bona, A. Connes, V. Guillemin, R. Haag, K. Hepp, C.J. Isham, G.W. Mackey, J.E. Marsden, B. Mielnik, M.A. Rieffel, F.W. Shultz, J.-M. Souriau, S. Sternberg, A. Weinstein, and P. Xu.

The research of the author's Ph.D. students Mark Robson, Urs Wiedemann, and Ken Wren contributed to this work, as did his collaboration with Noah Linden. Helpful comments, suggestions, and corrections on the manuscript were received from Hendrik Grundling, Brian Hall, Eli Hawkins, Marc Rieffel, Simon Ruijsenaars, Erik Thomas, Gijs Tuynman, and Alan Weinstein.

The author is grateful to Jeremy Butterfield, Robbert Dijkgraaf, Gérard Emch, Klaus Fredenhagen, Chris Isham, Dick Kadison, Daniel Kastler, Jerry Marsden, and John C. Taylor for moral and other forms of support.

The book was written during a happy time, shared with Imke and our cat Pauli.

Klaas Landsman
University of Amsterdam

Contents

Introductory Overview

I. Observables and Pure States

The aim of the first chapter is to give two descriptions of classical and quantum mechanics, each of which enables one to see in a different way what their common properties as well as their striking differences are. The first description focuses on the observables of the theory, whereas the second one is based on the pure states.

Observables

Consider a particle moving in the configuration space $Q = \mathbb{R}^3$. Its phase space is the cotangent bundle $T^*\mathbb{R}^3 \simeq \mathbb{R}^6$, and the collection of classical observables is taken as $\mathfrak{A}_{\mathbb{R}}^0 = C^\infty(T^*\mathbb{R}^3, \mathbb{R})$. This is a real vector space under pointwise addition and scalar multiplication by real numbers.

Ordinary (pointwise) multiplication of $f, g \in \mathfrak{A}_{\mathbb{R}}^0$, which for the moment we write as $f \circ g$, naturally defines a bilinear map on $\mathfrak{A}_{\mathbb{R}}^0$. This map is commutative and associative. In addition, in mechanics a key role is played by the **Poisson bracket**

$$\{f, g\} := \frac{\partial f}{\partial p^i}\frac{\partial g}{\partial q_i} - \frac{\partial f}{\partial q_i}\frac{\partial g}{\partial p^i}.$$

Hence $\mathfrak{A}_{\mathbb{R}}^0$ becomes a real Lie algebra under the Poisson bracket. This bracket is related to $\circ$ by the Leibniz rule, which says that $g \mapsto \{f, g\}$ is a derivation of $\circ$ for all $f \in \mathfrak{A}_{\mathbb{R}}^0$, in that $\{f, g \circ h\} = \{f, g\} \circ h + g \circ \{f, h\}$. Hence one coins the abstract definition of a Poisson bracket on a commutative (but not necessarily associative) algebra as a Lie bracket satisfying the Leibniz rule with respect to the product defining the algebra.

In quantum mechanics the above system is described by an infinite-dimensional space; to avoid complications we shall instead look at an N-level quantum system

($N < \infty$). The set of its observables $\mathfrak{A}_{\mathbb{R}}$ is the real vector space $\mathfrak{M}_N(\mathbb{C})_{\mathbb{R}}$ of Hermitian complex $N \times N$ matrices. A symmetric bilinear product on $\mathfrak{A}_{\mathbb{R}}$ is given by

$$A \circ B := \tfrac{1}{2}(AB + BA).$$

In addition, $\mathfrak{A}_{\mathbb{R}}$ admits a Poisson bracket defined by

$$\{A, B\}_{\hbar} := \frac{i}{\hbar}(AB - BA),$$

where $\hbar \in \mathbb{R}\backslash\{0\}$; in physics $\hbar$ has a specific numerical value, and is known as **Planck's constant**. A difference with the classical case is that $\circ$ now fails to be associative.

The following algebraic structure of the set of observables of classical or quantum mechanics may be extracted from the above considerations. A **Jordan–Lie algebra** $\mathfrak{A}_{\mathbb{R}}$ is a real vector space equipped with two bilinear maps, $\circ$ and $\{\, ,\, \}$ that are commutative and anticommutative, respectively. For each $A \in \mathfrak{A}_{\mathbb{R}}$, the map $B \mapsto \{A, B\}$ is a derivation of the Poisson structure $(\mathfrak{A}_{\mathbb{R}}, \{\, ,\, \})$; this makes $(\mathfrak{A}_{\mathbb{R}}, \{\, ,\, \})$ a real Lie algebra. Also, $B \mapsto \{A, B\}$ is a derivation of the Jordan structure $(\mathfrak{A}_{\mathbb{R}}, \circ)$; this is the Leibniz rule. Finally, the **associator identity**

$$(A \circ B) \circ C - A \circ (B \circ C) = \tfrac{1}{4}\hbar^2\{\{A, C\}, B\}$$

holds, for some constant $\hbar \in \mathbb{R}$. For $\hbar = 0$, in which case the commutative product is associative, one speaks of a **Poisson algebra**; this associativity is an algebraic characterization of classical mechanics.

The identity $(A \circ B) \circ A^2 = A \circ (B \circ A^2)$, where $A^2 := A \circ A$, which makes $(\mathfrak{A}_{\mathbb{R}}, \circ)$ a so-called (real) **Jordan algebra**, is implied by these axioms. A **JB-algebra** is defined as a Jordan algebra for which $\mathfrak{A}_{\mathbb{R}}$ is a Banach space, and the norm and the **Jordan product** $\circ$ are related by certain axioms. We refer to a Jordan–Lie algebra $\mathfrak{A}_{\mathbb{R}}$ for which $(\mathfrak{A}_{\mathbb{R}}, \circ)$ is a JB-algebra as a **JLB-algebra** (for Jordan–Lie–Banach).

A **C^*-algebra** is a complex Banach space equipped with an associative multiplication and an involution *, such that the C^*-axioms

$$\|AB\| \leq \|A\|\,\|B\|, \quad \|A^*A\| = \|A\|^2$$

are satisfied. It can be shown that any C^*-algebra is isomorphic to a norm-closed subalgebra of $\mathfrak{B}(\mathcal{H})$ for some Hilbert space $\mathcal{H}$.

In elementary quantum mechanics one assumes that every (bounded) observable of a given theory corresponds to a (bounded) self-adjoint operator on a Hilbert space $\mathcal{H}$, and vice versa. This assumption may be dropped, in which case the system is said to possess **superselection rules**. The assumption that the observables form the **self-adjoint part** $\mathfrak{A}_{\mathbb{R}} := \{A \in \mathfrak{A} \mid A^* = A\}$ of a C^*-algebra $\mathfrak{A}$ then naturally emerges. A crucial point is now that a JLB-algebra is the self-adjoint part of a C^*-algebra.

The **state space** $\mathcal{S}(\mathfrak{A})$ of a C^*-algebra $\mathfrak{A}$ (with unit $\mathbb{I}$) consists of all linear functionals ω on $\mathfrak{A}$ that are **positive** (that is, $\omega(A^*A) \geq 0$ for all $A \in \mathfrak{A}$) and **normalized** (i.e., $\omega(\mathbb{I}) = 1$). Such states ω are automatically continuous, so that

$\mathcal{S}(\mathfrak{A}) \subset \mathfrak{A}^*$. The space $\mathcal{S}(\mathfrak{A})$ is equipped with the w^*-topology inherited from $\mathfrak{A}^*$. If $\lambda \in (0, 1)$ and $\omega_1, \omega_2 \in \mathcal{S}(\mathfrak{A})$, then $\lambda\omega_1 + (1-\lambda)\omega_2 \in \mathcal{S}(\mathfrak{A})$. Moreover, $\mathcal{S}(\mathfrak{A})$ is a closed subset of the unit ball of $\mathfrak{A}^*$. Hence $\mathcal{S}(\mathfrak{A})$ is a compact convex set.

The state space of $\mathfrak{A} = \mathfrak{M}_n(\mathbb{C})$ consists of the density matrices on $\mathbb{C}^N$. At the opposite extreme, so to speak, one can show that the state space of $\mathfrak{A} = C(X)$ for a compact Hausdorff space X consists of the probability measures on X.

A **representation** of a C^*-algebra $\mathfrak{A}$ is a linear map $\pi : \mathfrak{A} \to \mathfrak{B}(\mathcal{H})$, for some Hilbert space $\mathcal{H}$, such that

$$\pi(AB) = \pi(A)\pi(B); \quad \pi(A^*) = \pi(A)^*.$$

For the JLB-algebra $\mathfrak{A}_{\mathbb{R}}$ this means that $\pi : \mathfrak{A}_{\mathbb{R}} \to \mathfrak{B}(\mathcal{H})_{\mathbb{R}}$ satisfies

$$\pi(\{A, B\}_1) = \{\pi(A), \pi(B)\}_1; \ \pi(A \circ B) = \pi(A) \circ \pi(B);$$

here $\{A, B\}_1 := i(AB - BA)$ and $A \circ B := \frac{1}{2}(AB + BA)$, etc.

There is a remarkable correspondence between states and representations of a C^*-algebra. It is given by the **GNS-construction**. Given a state ω on a C^*-algebra $\mathfrak{A}$, this construction produces a representation π_ω on some Hilbert space $\mathcal{H}_\omega$ containing a unit vector Ω_ω that is cyclic for $\pi_\omega(\mathfrak{A})$ (that is, $\pi_\omega(\mathfrak{A})\Omega_\omega$ is dense in $\mathcal{H}_\omega$). These objects are related by

$$(\Omega_\omega, \pi_\omega(A)\Omega_\omega) = \omega(A) \quad \forall A \in \mathfrak{A}.$$

Conversely, let a vector $\Omega \in \mathcal{H}$ be cyclic for some representation $\pi(\mathfrak{A})$. Then $\omega(A) = (\Omega, \pi(A)\Omega)$ defines a state on $\mathfrak{A}$ whose GNS-representation is equivalent to π.

Pure States

A state is called **pure** if it cannot be written as a convex combination of other states. The set of pure states of a C^*-algebra $\mathfrak{A}$ is denoted by $\mathcal{P}(\mathfrak{A})$; any state ω can be approximated by finite sums $\sum_i p_i\rho_i$, where $\sum_i p_i = 1$ and all ρ_i are pure. The pure state space of $\mathfrak{M}_n(\mathbb{C})$ and $C(X)$ may be identified with the projective space $\mathbb{PC}^N$ and with X, respectively.

It is often convenient to look at $A \in \mathfrak{A}_{\mathbb{R}}$ as a function $\hat{A}$ on $\mathcal{P}(\mathfrak{A})$; this is accomplished by putting $\hat{A}(\rho) = \rho(A)$. The map $A \mapsto \hat{A}$ is the **Gelfand transform**. The ensuing realization of $\mathfrak{A}_{\mathbb{R}}$ as a space of functions on its pure state space is faithful. In this realization $\|A\|$ equals the sup-norm $\|\hat{A}\|_\infty$ of $\hat{A}$ over $\mathcal{P}(\mathfrak{A})$.

A representation π is called **irreducible** if the set $\pi(\mathfrak{A})\Psi$ is dense in $\mathcal{H}$ for every $\Psi \in \mathcal{H}$. The special significance of pure states in the context of the GNS-construction is that the corresponding representations are irreducible.

The pure states of a classical system are the points of its phase space P. A manifold P whose associated space of smooth functions $C^\infty(P, \mathbb{R})$ is equipped with a Poisson bracket (satisfying the Leibniz property with respect to pointwise multiplication) is called a **Poisson manifold**. Each function $h \in C^\infty(P, \mathbb{R})$ then defines a **Hamiltonian vector field** ξ_h by

$$\xi_h f = \{h, f\}.$$

Hence on $P = T^*\mathbb{R}^3$ we have

$$\xi_h = \frac{\partial h}{\partial p_i}\frac{\partial}{\partial q^i} - \frac{\partial h}{\partial q^i}\frac{\partial}{\partial p_i}.$$

Hamilton's equations of motion for a curve $\sigma(t)$ in P are

$$\frac{d\sigma(t)}{dt} = \xi_h(\sigma(t));$$

solutions are called **Hamiltonian flows** or curves.

Poisson manifolds form the main source of Poisson algebras. The example $P = T^*\mathbb{R}^3$ is special in that at each point $\sigma \in P$ the collection of Hamiltonian vector fields ξ_h spans the tangent space $T_\sigma P$. Poisson manifolds with this nondegeneracy property are called **symplectic**. Traditionally, classical mechanics used to be described in terms of symplectic manifolds, but many Poisson manifolds that are not symplectic have turned out to be relevant in physics. A system whose phase space is not symplectic may be said to possess classical superselection rules.

The most important result in the theory of Poisson manifolds is that any such manifold admits a (generally singular) foliation by subspaces on which the ξ_h span the tangent space. These subspaces therefore acquire a symplectic structure, and are accordingly called the **symplectic leaves** of P. Such leaves are characterized by the properties that any two of their points can be connected by a piecewise smooth Hamiltonian curve in P and that any Hamiltonian flow must stay within a given leaf.

The simplest nontrivial illustration is provided by $P = \mathbb{R}^3$, with Poisson bracket given by $\{x, y\} = z$ and its cyclic permutations. The symplectic leaves are the spheres S^2_r of radius r; there is a jump in dimension of the leaves at $r = 0$, rendering the foliation a singular one.

We return to quantum mechanics. Let $H \in \mathfrak{M}_n(\mathbb{C})_{\mathbb{R}}$, and define $\tilde{H} \in C^\infty(\mathbb{C}^N)$ by

$$\tilde{H}(\Psi) := (\Psi, H\Psi).$$

The Hilbert space $\mathcal{H} = \mathbb{C}^N$ (seen as a real manifold) has a natural nondegenerate Poisson structure, characterized by

$$\{\tilde{A}, \tilde{B}\}_\hbar = \frac{i}{\hbar}(\widetilde{AB - BA}).$$

Since a quantum-mechanical state is normalized to unit length and defined only up to a phase, the space of pure states is the projective space $\mathbb{P}\mathcal{H}$, rather than $\mathcal{H}$. Fortunately, the considerations above can be transferred to $\mathbb{P}\mathcal{H}$ almost without modification. In particular, $\tilde{H}$ may be seen as a function $\hat{H}$ on $\mathbb{P}\mathcal{H}$, and the above Poisson structure projects to one on $\mathbb{P}\mathcal{H}$. The Hamiltonian flow of $\hat{H}$ with respect to that structure is then precisely the projection to $\mathbb{P}\mathcal{H}$ of the unitary time evolution on $\mathcal{H}$ that solves the Schrödinger equation with Hamiltonian H.

Given a Poisson manifold P, we define a **representation** of the Poisson algebra $\mathfrak{A}^0_{\mathbb{R}} = C^\infty(P, \mathbb{R})$ on a symplectic manifold S (with associated Poisson bracket $\{\, ,\, \}_S$) as a linear map $\pi : \mathfrak{A}^0_{\mathbb{R}} \to C^\infty(S, \mathbb{R})$ satisfying three properties, of which the two most important are

$$\pi(\{f, g\}) = \{\pi(f), \pi(g)\}_S; \quad \pi(f \circ g) = \pi(f) \circ \pi(g).$$

One recognizes the analogy with the definition of a representation of a JLB-algebra. A representation π of $C^\infty(P, \mathbb{R})$ on S is always associated to a smooth Poisson map $J : S \to P$ through $\pi = J^*$.

A representation $\pi(\mathfrak{A}^0_{\mathbb{R}})$ on S is said to be **irreducible** if at every point $\sigma \in S$ the collection of Hamiltonian vector fields $\{\xi_{\pi(f)}(\sigma), f \in \mathfrak{A}^0_{\mathbb{R}}\}$ spans the (real) tangent space $T_\sigma S$. Interestingly, the notion of irreducibility for representations of JLB-algebras (and therefore of C^*-algebras), looked upon as spaces of functions on their pure state spaces, can be shown to be identical to the one for Poisson algebras.

The pure state space $\mathcal{P}(\mathfrak{A})$ of a C^*-algebra $\mathfrak{A}$ is a Poisson manifold in a certain generalized sense; it is foliated by symplectic leaves of the form $\mathbb{P}\mathcal{H}_\alpha$, where each Hilbert space $\mathcal{H}_\alpha$ corresponds to an equivalence class of irreducible representations of $\mathfrak{A}$. The basic theorem on irreducible representations is the same for Poisson algebras of the type $C^\infty(P, \mathbb{R})$, where P is a finite-dimensional manifold (which we here consider to be the pure state space of $C^\infty(P, \mathbb{R})$), and C^*-algebras (where commutative C^*-algebras are understood to have the zero Poisson structure), where $S = \mathbb{P}\mathcal{H}$ for some Hilbert space $\mathcal{H}$. It is the following: If a symplectic manifold S carries an irreducible representation π of a C^*-algebra or a Poisson algebra $\mathfrak{A}_{\mathbb{R}}$, then S must be isomorphic (as a symplectic manifold) to a symplectic leaf of the space of pure states of $\mathfrak{A}_{\mathbb{R}}$, or to a covering space thereof. Up to isomorphism, $\pi(f)$ is simply the restriction of f to the leaf in question (composed with the covering projection if necessary).

Saying that $\mathbb{P}\mathcal{H}$ equipped with a certain Poisson structure is the pure state space of quantum mechanics clearly does not fully characterize this theory. For by comparison with classical mechanics we know that the observables of quantum mechanics do not comprise all functions in $C^\infty(\mathcal{H}, \mathbb{R})$, but only those of the form $\hat{H}$, where $H \in \mathfrak{M}_n(\mathbb{C})_{\mathbb{R}}$ (or $\mathfrak{B}(\mathcal{H})_{\mathbb{R}}$).

The essential extra ingredient of quantum mechanics is the existence of transition probabilities between pure states. A **transition probability** on a set $\mathcal{P}$ is a function $p : \mathcal{P} \times \mathcal{P} \to [0, 1]$ satisfying $p(\rho, \sigma) = 1 \iff \rho = \sigma$ and $p(\rho, \sigma) = 0 \iff p(\sigma, \rho) = 0$. All transition probabilities in physics are **symmetric** in that $p(\rho, \sigma) = p(\sigma, \rho)$. The transition probabilities of classical mechanics are trivial: $p(\rho, \sigma) = \delta_{\rho\sigma}$. In quantum mechanics, on the other hand, where $\mathcal{P} = \mathbb{P}\mathcal{H}$, the function p assumes the form $p(\varphi, \psi) = |(\Phi, \Psi)|^2$ (where the unit vectors $\Phi, \Psi \in \mathcal{H}$ project to $\varphi, \psi \in \mathbb{P}\mathcal{H}$).

From Pure States to Observables

We have seen that classical mechanics is described by Poisson algebras of observables of the type $\mathfrak{A}_{\mathbb{R}} = C^\infty(P, \mathbb{R})$, where P is a Poisson manifold. The algebra of observables of a quantum-mechanical system (perhaps possessing superselection rules) is the self-adjoint part $\mathfrak{A}_{\mathbb{R}}$ of a C^*-algebra $\mathfrak{A}$, realized as a certain collection of functions on the pure state space $\mathcal{P} = \mathcal{P}(\mathfrak{A})$. This space is a generalized Poisson manifold, which, like its classical counterpart P, is foliated by symplectic leaves. Classical and quantum mechanics share the property of **unitarity**. This means that the Hamiltonian flow $\rho \mapsto \rho(t)$ generated by a given observable preserves the transition probabilities, in that $p(\rho(t), \sigma(t)) = p(\rho, \sigma)$ for all t for which the flow is defined.

A **Poisson space with a transition probability** is, roughly speaking, at the same time a symmetric transition probability space $\mathcal{P}$ and a Poisson manifold, such that the Poisson structure is unitary.

The quantum mechanics of an N-level system, whose algebra of observables is $\mathfrak{M}_n(\mathbb{C})_{\mathbb{R}}$, has the property that its pure state space $\mathcal{P} = \mathbb{P}\mathbb{C}^N$ is irreducible as a transition probability space. In general, a transition probability space is called **irreducible** if it is not the union of two (nonempty) orthogonal subsets. A **sector** C of a transition probability space $\mathcal{P}$ is a subset of $\mathcal{P}$ with the property that $p(\rho, \sigma) = 0$ for all $\rho \in C$ and all $\sigma \in \mathcal{P}\backslash C$. Thus a transition probability space is the disjoint union of its irreducible sectors. In classical mechanics each point of $\mathcal{P}$ is a sector.

The superposition principle of quantum mechanics (which is normally expressed in terms of vectors in a Hilbert space) can be described in the present language. For any subset Q of $\mathcal{P}$ we define the orthoplement

$$Q^{\perp} := \{\sigma \in \mathcal{P} | p(\rho, \sigma) = 0 \,\forall \rho \in Q\}.$$

The possible superpositions of the pure states ρ, σ are then the elements of $\{\rho, \sigma\}^{\perp\perp}$. If ρ and σ lie in different sectors, then clearly $\{\rho, \sigma\}^{\perp\perp} = \{\rho, \sigma\}$.

It turns out that the pure state space of quantum mechanics with (discrete) superselection rules can be characterized (up to technicalities) by the following three properties (or axioms):

- QM1: The pure state space $\mathcal{P}$ is a Poisson space with a transition probability.
- QM2: For each pair (ρ, σ) of points that lie in the same sector of $\mathcal{P}$, $\{\rho, \sigma\}^{\perp\perp}$ is isomorphic to $\mathbb{P}\mathbb{C}^2$ as a transition probability space.
- QM3: The sectors of $(\mathcal{P}, p)$ as a transition probability space coincide with the symplectic leaves of $\mathcal{P}$ as a Poisson space.

Here $\mathbb{P}\mathbb{C}^2$ is understood to be equipped with the usual Hilbert space transition probabilities. The universality of the transition probabilities (and, by implication, of the Poisson structure) of quantum mechanics is notable, as is the third property (which is not shared by classical mechanics).

To characterize classical mechanics, one simply postulates

- CM1: The pure state space $\mathcal{P}$ is a Poisson space with a transition probability.
- CM2: The transition probabilities are $p(\rho, \sigma) = \delta_{\rho\sigma}$.

One can reconstruct the algebra of observables $\mathfrak{A}_{\mathbb{R}}$ from its pure state space, equipped with the structure of a Poisson space with a transition probability. Given a general transition probability space $(\mathcal{P}, p)$, we first define the real vector space $\mathfrak{A}_{\mathbb{R}}(\mathcal{P})$ as a certain subspace of the real Banach space $\ell^{\infty}(\mathcal{P})$. For simplicity we assume that $(\mathcal{P}, p)$ has a finite basis (here a **basis** B of $\mathcal{P}$ is a pairwise orthogonal subset for which $\sum_{\rho \in B} p(\rho, \sigma) = 1$ for all $\sigma \in \mathcal{P}$). The space $\mathfrak{A}_{\mathbb{R}}(\mathcal{P})$ in question then consists of all finite linear combinations $\sum_i \lambda_i p_{\rho_i}$, where $\lambda_i \in \mathbb{R}$, $\rho_i \in \mathcal{P}$, and $p_\rho(\sigma) := p_{\rho\sigma}$. This will be the collection of observables, which are seen to be essentially linear combinations of the transition probabilities.

Axioms QM1 and QM2 imply the existence of a spectral theorem in $\mathfrak{A}_{\mathbb{R}}(\mathcal{P})$, saying that every $A \in \mathfrak{A}_{\mathbb{R}}(\mathcal{P})$ has a spectral resolution $A := \sum_j \lambda_j p_{e_j}$, where the e_j are pairwise orthogonal and the eigenvalues λ_j are real. The spectral theorem equips $\mathfrak{A}_{\mathbb{R}}(\mathcal{P})$ with a squaring map, for given the spectral resolution above one can define A^2 by $A^2 = \sum_j \lambda_j^2 p_{e_j}$. Subsequently, one defines a map $\circ$ on $\mathfrak{A}_{\mathbb{R}}(\mathcal{P})$ by

$$A \circ B := \tfrac{1}{4}((A + B)^2 - (A - B)^2).$$

Axiom QM2 implies that this map is bilinear, so that $\circ$ indeed defines a Jordan product. This product, combined with the sup-norm, turns $\mathfrak{A}_{\mathbb{R}}(\mathcal{P})$ into a JB-algebra; the relevant axioms are satisfied as a consequence of the fact that the Jordan product comes from a spectral resolution. Had the transition probabilities been trivial, this Jordan product would have been pointwise multiplication, implying associativity.

Given a Poisson structure on $\mathcal{P}$, any function h on $\mathcal{P}$ whose restriction to each symplectic leaf is smooth defines a Hamiltonian flow $\sigma \mapsto \sigma(t)$ on $\mathcal{P}$. This defines a one-parameter family of maps $\alpha_t : \mathfrak{A}_{\mathbb{R}}(\mathcal{P}) \to \mathfrak{A}_{\mathbb{R}}(\mathcal{P})$, given by $\alpha_t(f) : \sigma \mapsto f(\sigma(t))$. It is not difficult to show that unitarity (guaranteed by Axiom QM1) implies that α_t is a Jordan homomorphism; that is, $\alpha_t(f \circ g) = \alpha_t(f) \circ \alpha_t(g)$. The derivative of the homomorphism property with respect to t yields the Leibniz rule, since

$$\frac{df(\sigma(t))}{dt} = \{h, f\}(\sigma(t)).$$

Quite unlike the situation in classical mechanics, in quantum mechanics the Poisson structure of the pure state space turns out to be determined by the axioms up to a collection of constants (one for each sector). Suitable rescalings then lead to a single constant $\hbar$. It is remarkable that the curious associator "identity" is satisfied by the ensuing Poisson bracket. Therefore, at the end of the day $(\mathfrak{A}_{\mathbb{R}}(\mathcal{P}), \circ, \{\ ,\ \}, \|\cdot\|)$ becomes a JLB-algebra. This enables one to endow the complexification $\mathfrak{A}_{\mathbb{R}}(\mathcal{P})$ with the properties of a C^*-algebra, of which $\mathfrak{A}_{\mathbb{R}}(\mathcal{P})$ is the self-adjoint part. In analogy with classical mechanics, the algebra of observables $\mathfrak{A}_{\mathbb{R}}(\mathcal{P})$ is realized (even as a Banach space) as a subspace of $\ell^\infty(\mathcal{P}, \mathbb{R})$.

In passing from pure states to algebras of observables one has the correspondences listed in Table 1.

II. Quantization and the Classical Limit

The second chapter relates classical and quantum mechanics to each other. Such a relation is possible on the basis of the structural similarities between the mathe-

Pure state space	*Algebra of observables*
transition probabilities	Jordan product
Poisson structure	Poisson bracket
unitarity	Leibniz rule

TABLE 1. From pure states to algebras of observables

matical description of these theories laid out in Chapter I, and it can be approached from the point of view of either observables or pure states.

Foundations

The problem of quantizing a given classical system is as old as quantum mechanics itself. Initially, the term "quantization" indicated the fact that at a microscopic scale certain physical quantities assume only discrete values, sometimes called quantum numbers. This was found to be true particularly for energy levels of bound states, as well as for, e.g., angular momentum and electrical charge. Such discreteness is easily understood within the Hilbert space formalism of quantum mechanics, where self-adjoint operators may or may not have a discrete spectrum, and is no longer seen as the defining property of a quantum theory.

In the modern literature "quantization" refers to the passage from a classical to a "corresponding" quantum theory. This notion goes back to the time that the correct formalism of quantum mechanics was beginning to be discovered, and from that time to the present day practically all known quantum-mechanical models have been constructed on the basis of some quantization procedure. Nonetheless, Barry Simon wrote:

> *It seems to me that there has been in the literature entirely too much emphasis on quantization (i.e. general methods of obtaining quantum mechanics from classical methods) as opposed to the converse problem of the classical limit of quantum mechanics. This is unfortunate since the latter is an important question for various areas of modern physics while the former is, in my opinion, a chimera.*

In the present book the conception of quantization used in this quotation, which indeed applies to geometric quantization and related approaches, is replaced by a different one: We see quantization as the study of the possible correspondence between a given classical theory, given as a Poisson algebra or a Poisson manifold and perhaps a Hamiltonian, and a given quantum theory, mathematically expressed as a certain algebra of observables or a pure state space, and perhaps a time evolution. For this purpose it is not at all necessary that the quantum theory be formulated in terms of classical structures. On the basis of this understanding quantization and the classical limit are two sides of the same coin.

Early thought on both quantization and the classical limit was guided by Bohr's "correspondence principle", which was a rather vague idea to the effect that quantum mechanics should converge to classical mechanics for $\hbar \to 0$, and also in the limit of large quantum numbers. The second aspect, including its relation to the first, will be studied in Chapter III. The use of the limit $\hbar \to 0$ is sometimes criticized on the argument that $\hbar$ is a constant, but what is meant here is simply that $\hbar$ should be small compared to other relevant quantities of the same dimension; this includes the case where units have been chosen in which $\hbar$ is dimensionless and equal to 1!

In Chapter I classical and quantum mechanics are formulated in such a way that they look structurally similar for any value of $\hbar$. On the observable side one classically has a Poisson algebra $(\mathfrak{A}^0_{\mathbb{R}}, \circ, \{\ ,\ \})$, in which $\circ$ is associative, whereas quantum-mechanically one has the self-adjoint part $(\mathfrak{A}^\hbar_{\mathbb{R}}, \circ_\hbar, \{\ ,\ \}_\hbar)$, $\hbar \neq 0$, of a C^*-algebra. One now needs a proper way of expressing the idea that $(\mathfrak{A}^\hbar_{\mathbb{R}}, \ldots)$ is the quantization of $(\mathfrak{A}^0_{\mathbb{R}}, \ldots)$, and that the latter is the classical limit of the former. For this to be possible in the first place, the quantum algebra of observables $\mathfrak{A}^\hbar_{\mathbb{R}}$ must be defined for all values $\hbar \in I_0$, where I_0 is a certain subset of $\mathbb{R}$ that has 0 as an accumulation point (I_0 may be discrete, e.g., $I_0 = \{1/n, n \in \mathbb{N}\}$, or an interval, such as $I_0 = (0, 1]$; another example would be $I_0 = \mathbb{R}\backslash\{0\}$).

The essence of quantization is now that there should be a family of linear maps $\mathcal{Q}_\hbar : \mathfrak{A}^0_{\mathbb{R}} \to \mathfrak{A}^\hbar_{\mathbb{R}}$, $\hbar \in I_0$; the operator $\mathcal{Q}_\hbar(f)$ is interpreted as the quantum observable corresponding to the classical observable f. A mathematically precise version of Bohr's correspondence principle, at least as far as the algebraic structure is concerned, is then given by the conditions

$$\lim_{\hbar \to 0} \|\mathcal{Q}_\hbar(f) \circ_\hbar \mathcal{Q}_\hbar(g) - \mathcal{Q}_\hbar(fg)\|_\hbar = 0$$

and

$$\lim_{\hbar \to 0} \|\{\mathcal{Q}_\hbar(f), \mathcal{Q}_\hbar(g)\}_\hbar - \mathcal{Q}_\hbar(\{f, g\})\|_\hbar = 0,$$

for all $f, g \in \mathfrak{A}^0_{\mathbb{R}}$; here a possible $\hbar$-dependence of the operations in $\mathfrak{A}^\hbar_{\mathbb{R}}$ has been indicated. Together with the continuity of $\hbar \mapsto \|\mathcal{Q}_\hbar(f)\|_\hbar$ for all $f \in \mathfrak{A}^0_{\mathbb{R}}$, these conditions define what is meant by a **strict quantization**.

From the perspective of pure states the classical theory is characterized by a Poisson manifold $(P, \{\ ,\ \})$. Quantization should relate this to a family of Poisson spaces with a transition probability $(\mathcal{P}_\hbar, p, \{\ ,\ \}_\hbar)$, $\hbar \in I_0$, satisfying the "QM" axioms of Chapter I. This relation is given by a **pure state quantization**, which is a collection of injections $q_\hbar : P \to \mathcal{P}_\hbar$ ($\hbar \in I_0$) that embed the classical pure state space into its quantum counterpart. These maps should satisfy certain conditions motivated by the correspondence principle. One such condition is obviously

$$\lim_{\hbar \to 0} p(q_\hbar(\rho), q_\hbar(\sigma)) = \delta_{\rho\sigma},$$

stating that the quantum-mechanical transition probabilities converge to the classical ones. It is interesting to relate this condition to the one on the Jordan product of observables. Assume that P is discrete; then $\mathfrak{A}^0 = \ell_0(P)$ is generated by functions of the type $p^{\rm cl}_\sigma : \rho \mapsto \delta_{\rho\sigma}$. Given a pure state quantization $q_\hbar$, we can hope to define a strict quantization $\mathcal{Q}^B_\hbar$ of $\mathfrak{A}^0_{\mathbb{R}}$ by linear extension of $\mathcal{Q}^B_\hbar(p^{\rm cl}_\sigma) := p_{q_\hbar(\sigma)}$. The spectral resolution of $f \in \mathfrak{A}^0_{\mathbb{R}}$ is $f = \sum_\sigma f(\sigma) p^{\text{cl}}_\sigma$, so that $\mathcal{Q}^B_\hbar(f) = \sum_\sigma f(\sigma) p_{q_\hbar(\sigma)}$. For small $\hbar$ the right-hand side approximates the spectral representation of $\mathcal{Q}^B_\hbar(f)$, since the $q_\hbar(\sigma)$ become almost orthogonal. Therefore, $\mathcal{Q}^B_\hbar(f)^2$ is approximately $\sum_\sigma f(\sigma)^2 p_{q_\hbar(\sigma)}$, which equals $\mathcal{Q}^B_\hbar(f^2)$. Hence $\mathcal{Q}^B_\hbar(f)^2 \to \mathcal{Q}^B_\hbar(f^2)$ for $\hbar \to 0$, which is equivalent to the condition on the Jordan product.

For nondiscrete P the notion of a pure state quantization has been worked out only when $P = S$ is symplectic and each $\mathcal{P}_\hbar$ is irreducible, being equal to $\mathbb{P}\mathcal{H}_\hbar$ for some Hilbert space $\mathcal{H}_\hbar$. In the cases we consider, the sum over points in P is then replaced by the **Liouville measure** μ_L on S, locally given by $d\mu_L(p, q) := d^n p d^n q/(2\pi)^n$. In addition, a function $c : I_0 \to \mathbb{R}\backslash\{0\}$ appears. The conditions on a pure state quantization $q_\hbar$ are stated in terms of the **Berezin quantization** of $f \in \mathfrak{A}^0_{\mathbb{R}} = C_0(S, \mathbb{R})$. This is an operator $\mathcal{Q}^B_\hbar(f)$ on $\mathcal{H}_\hbar$, defined (for each $\hbar \in I_0$) by its expectation values

$$(\Psi, \mathcal{Q}^B_\hbar(f)\Psi) := c(\hbar) \int_S d\mu_L(\sigma)\, p(q_\hbar(\sigma), \psi) f(\sigma),$$

where $\psi \in \mathbb{P}\mathcal{H}_\hbar$ is the projection of the unit vector Ψ to $\mathbb{P}\mathcal{H}_\hbar$. This expression evidently generalizes $\mathcal{Q}^B_\hbar(f)$ in the previous paragraph. The function c is fixed by imposing the first condition $\mathcal{Q}^B_\hbar(1_S) = \mathbb{I}$. The second requirement on $q_\hbar$ is that in the limit $\hbar \to 0$ the above expression with $\psi = q_\hbar(\rho)$ converge to $f(\rho)$ for all $f \in \mathfrak{A}^0_{\mathbb{R}}$ and all $\rho \in S$. Finally, each $q_\hbar$ should pull the canonical symplectic form on $\mathbb{P}\mathcal{H}_\hbar$ back to the one on S.

Let us assume that each $q_\hbar(\sigma) \in \mathbb{P}\mathcal{H}_\hbar$ is the projection of a unit vector $\Psi^\sigma_\hbar \in \mathcal{H}_\hbar$. The map $W : \mathcal{H}_\hbar \to L^2(S, c(\hbar)\mu_L)$ defined by $W\Psi(\sigma) := (\Psi^\sigma_\hbar, \Psi)$ is then a partial isometry. Defining p to be the projection onto the image of W, and U to be W, seen as a map from $\mathcal{H}_\hbar$ to $pL^2(S, c(\hbar)\mu_L)$, we obtain

$$U\mathcal{Q}^B_\hbar(f)U^{-1} = pfp,$$

where f is seen as a multiplication operator on $L^2(S, c(\hbar)\mu_L)$. In this way, quantum observables act on a subspace of L^2(phase space), rather than on L^2(configuration space), as is more usual in quantization theory, in an extremely elegant fashion.

Quantization on Flat Space

Our main illustration of strict as well as pure state quantization will come from the manifold $S = T^*\mathbb{R}^n$, equipped with its canonical Poisson bracket; this makes S symplectic. This manifold is particularly well structured in being both a cotangent bundle and a Kähler manifold (the latter comprise a class of complex manifolds of which more examples will be encountered in the next chapter). It turns out that a Kähler manifold often admits a strict Berezin quantization, which is derived from a pure state quantization as explained above. The observables on cotangent bundles, on the other hand, are best quantized using a prescription going back to Weyl, which is not directly related to a pure state quantization. The phase space $T^*\mathbb{R}^n$, then, may be quantized either way; Berezin quantization enjoys the advantage of positivity, whereas Weyl quantization has better symmetry properties.

In both methods the **Heisenberg group** $\tilde{H}_n$ plays a central role; this is the connected and simply connected Lie group whose Lie algebra $\mathfrak{h}_n = \mathbb{R}^{2n+1}$ is described by

$$[P_i, Q^j] = -\delta_i^j Z; \quad [P_i, Z] = [Q^j, Z] = 0,$$

in terms of a suitable basis $\{P_i, Q^j, Z\}_{i,j=1,\dots,n}$. The Heisenberg group is nilpotent, and the exponential map $\text{Exp} : \mathfrak{h}_n \to \tilde{H}_n$ is a diffeomorphism. For each $\hbar \neq 0$ there exists an irreducible representation $U_{\frac{1}{\hbar}}$ on $\mathcal{H} = L^2(\mathbb{R}^n)$, given by

$$U_{\frac{1}{\hbar}}\,(\text{Exp}(-uQ + vP + tZ))\,\Psi(x) := e^{-i\left(t+\frac{1}{2}uv-ux\right)/\hbar}\Psi(x - v),$$

where $uQ := u_i Q^i$, etc. Of special significance are the **Weyl operators**

$$U_{\frac{1}{\hbar}}(p, q) := U_{\frac{1}{\hbar}}(p, q, 0) = e^{\frac{i}{\hbar}(pQ_\hbar^S - qP_\hbar^S)},$$

where $Q_\hbar^{S,i} = x^i$ and $P_{\hbar,i}^S = -i\hbar\partial/\partial x^i$ are the **position operator** and **momentum operator** of elementary quantum mechanics.

Both Berezin quantization $\mathcal{Q}_\hbar^B$ and Weyl quantization $\mathcal{Q}_\hbar^W$ are defined for $\hbar \in I_0 = \mathbb{R}\backslash\{0\}$, and map $\mathfrak{A}_\mathbb{R}^0 = C_0(T^*\mathbb{R}^n, \mathbb{R})$ into $\mathfrak{A}^\hbar = \mathfrak{B}_0(L^2(\mathbb{R}^n))_\mathbb{R}$ (the self-adjoint part of the C^*-algebra of compact operators on $L^2(\mathbb{R}^n)$). Both are given by an expression of the form

$$\mathcal{Q}_\hbar(f) := \hbar^{-n} \int_{T^*\mathbb{R}^n} d\mu_L(p, q)\, f(p, q) U_{\frac{1}{\hbar}}(p, q) A U_{\frac{1}{\hbar}}(p, q)^*.$$

For **Weyl quantization** one puts $A = 2^n P$, where P is the (nonpositive) **parity operator** P on $L^2(\mathbb{R}^n)$, defined by $P\Psi(x) := \Psi(-x)$. To obtain Berezin quantization one chooses the positive operator $A = [\Psi_\hbar^0]$, which is the projection onto the (unit) vector

$$\Psi_\hbar^0(x) := (\pi\hbar)^{-n/4} e^{-x^2/(2\hbar)}.$$

The pure state quantization $q_\hbar^B$ associated with Berezin quantization is given by $\mathcal{H}_\hbar = L^2(\mathbb{R}^n)$ for all $\hbar \neq 0$, and $q_\hbar^B(p, q) = \psi_\hbar^{(p,q)}$, where the right-hand side is given by projecting the unit vector

$$\Psi_\hbar^{(p,q)} := U_{\frac{1}{\hbar}}(p, q)\Psi_\hbar^0$$

to $\mathbb{P}L^2(\mathbb{R}^n)$. In terms of $z := (q + ip)/\sqrt{2}$, the transition probabilities between quantized pure states are

$$p(q_\hbar^B(z), q_\hbar^B(w)) = e^{-|z-w|^2/\hbar},$$

which evidently converges to the classical transition probability δ_{zw} as $\hbar \to 0$. The Hilbert space $L^2(T^*\mathbb{R}^n, \mu_L)$ is naturally isomorphic to $L^2(\mathbb{C}^n, \mu_G)$, where μ_G is a suitable Gaussian measure on $\mathbb{C}^n$. The projection p in the preceding section then projects on the subspace of functions on $\mathbb{C}^n$ that are entire in $\bar{z}$. Accordingly, Berezin quantization on flat space assumes the pulchritudinous form of sandwiching a multiplication operator on $L^2(\mathbb{C}^n, \mu_G)$ between two identical projections, whose image is a space of entire functions.

A comparison between classical dynamics and its quantum counterpart is of central importance to the theory of quantization and the classical limit. If the classical Hamiltonian h lies in $\mathfrak{A}^0_{\mathbb{R}}$, this comparison is straightforward. In that case, the quantum Hamiltonian $H_\hbar := \mathcal{Q}_\hbar(h)$ lies in $\mathfrak{A}^\hbar_{\mathbb{R}}$, and Dirac's property implies that for fixed t and for all $f \in \mathfrak{A}^0_{\mathbb{R}}$ one has

$$\lim_{\hbar \to 0} \|\mathcal{Q}_\hbar(\alpha_t^0(f)) - \alpha_t^\hbar(\mathcal{Q}_\hbar(f))\| = 0.$$

Here $\alpha_t^0(f) : \sigma \mapsto f(\sigma(t))$, and $\alpha_t^\hbar(A) := e^{itH_\hbar/\hbar} A e^{-itH_\hbar/\hbar}$.

It so happens that most Hamiltonians on $T^*\mathbb{R}^n$ used in physics are unbounded, so that the above norm-convergence is somewhat unrealistic. A silver lining on this generic unboundedness, however, is the fact that for Hamiltonians that are at most quadratic in the canonical variables (p, q) the excellent equivariance properties of Weyl quantization imply that $\mathcal{Q}_\hbar^W(\alpha_t^0(f)) - \alpha_t^\hbar(\mathcal{Q}_\hbar(f)) = 0$ for any $\hbar$. For $\mathcal{Q}_\hbar^B$ instead of $\mathcal{Q}_\hbar^W$ this equation holds for Hamiltonians that in addition are $O(2n)$-invariant.

Convergence from quantum to classical dynamics for more general unbounded Hamiltonians may be achieved by looking at the time evolution of particular pure states. Most literature on this subject is concerned with the **time-dependent WKB method**, where one assumes that the initial wave function is of the form $\Psi_\hbar(x) = \rho_\hbar(x)\exp(iS(x)/\hbar)$, where S is real and independent of $\hbar$, and $\rho_\hbar$ is a real formal power series in $\hbar$ (of which only the zeroth-order term is relevant in the classical limit). An approximate solution $\Psi_\hbar(x, t)$ to the time-dependent Schrödinger equation is then constructed in terms of a classical trajectory between $x(0)$ and $x(t) = x$, where $x(0)$ is determined by the requirement that the trajectory with initial data $(dS(x_0), x_0)$ indeed arrives at x after time t. Such initial pure states are quite peculiar, since in the classical limit they typically converge to mixed states on $\mathfrak{A}^0$: The support of the mixed state on $C_0(T^*\mathbb{R}^n)$ in question, which is a probability measure on $T^*\mathbb{R}^n$, is the so-called Lagrangian submanifold of $T^*\mathbb{R}^n$ defined by S and ρ_0 (this is the collection of points $(dS(q), q)$, where $q \in \text{supp}\,(\rho_0)$). Moreover, the WKB method works without further ado only if the projected flow defines a diffeomorphism of the configuration space $\mathbb{R}^n$ for all $t' \in [0, t]$.

We concentrate on a different method, which works well if the initial state $\Psi_\hbar$ converges to a pure state in the classical limit. We will specifically look at the (coherent) state $\Psi_\hbar^{(p,q)}$ defined earlier, whose classical limit is the point $(p, q) \in T^*\mathbb{R}^n$. The method is based on Taylor-expanding the quantum Hamiltonian $H = H(P_\hbar^S, Q_\hbar^S)$, which, up to suitable ordering, is obtained by substituting $(P_\hbar^S, Q_\hbar^S)$ for (p, q) in the classical Hamiltonian $h(p, q)$ around the classical trajectory $(p(t), q(t))$. This method works well for classical Hamiltonians of the type

$$h(p, q) = \frac{(p - eA(q))^2}{2m} + V(q),$$

which describe a particle moving in an external potential V and magnetic field $\nabla \times A$. One can prove that for all $f \in \mathfrak{A}^0_{\mathbb{R}}$, and $\mathcal{Q}_\hbar = \mathcal{Q}^W_\hbar$ or $\mathcal{Q}_\hbar = \mathcal{Q}^B_\hbar$, one has

$$\lim_{\hbar \to 0} \left(\Psi_\hbar^{(p,q)}, \left[\mathcal{Q}_\hbar(\alpha_t^0(f)) - \alpha_t^\hbar(\mathcal{Q}_\hbar(f)) \right] \Psi_\hbar^{(p,q)} \right) = 0.$$

Quantization on Riemannian Manifolds

According to the general theory of relativity, gravitational fields are described by a pseudo-Riemannian metric $\mathbf{g}$ on spacetime. To describe the motion of a test particle in a static external gravitational field we therefore assume that space is a Riemannian manifold $(Q, \mathbf{g})$; the corresponding phase space is the cotangent bundle $S = T^*Q$, whose canonical symplectic structure is independent of $\mathbf{g}$. The metric provides an isomorphism between T^*Q and the tangent bundle TQ, and it turns out to be easier to discuss mechanics on TQ. The natural Hamiltonian on TQ is

$$h(v, q) = \tfrac{1}{2}\mathbf{g}_{ij}(q)v^i v^j.$$

The Hamiltonian flow $(v(t), q(t))$ on T^*Q is known as **geodesic motion**, since $q(t)$ is a geodesic on Q; the tangent vector to this geodesic is $v(t)$, which is parallel transported along the geodesic.

Using the geometric structure, it is possible to generalize the Weyl quantization method on the flat space $\mathbb{R}^n$ to any Riemannian manifold. The key to this generalization lies in rewriting Weyl's prescription as

$$\mathcal{Q}^W_\hbar(f)\Psi(x) = \int_{\mathbb{R}^n} d^n y\, K^W_\hbar \lfloor f \rfloor (x, y)\Psi(y),$$

with kernel $K^W_\hbar \lfloor f \rfloor(x, y) = \hbar^{-n} \check{f}((x-y)/\hbar, \tfrac{1}{2}(x+y))$. Here $\check{f}(v, q)$ is the partial Fourier transform of $f(p, q)$ in the fiber direction of T^*Q; this is a function on TQ. We now recognize $\tfrac{1}{2}(x+y)$ as the midpoint of the geodesic connecting x and y, and $(x-y)$ as its tangent vector at this midpoint; the map $(x, y) \mapsto ((x-y), \tfrac{1}{2}(x+y))$ provides a diffeomorphism between $\mathbb{R}^n \times \mathbb{R}^n$ and $T\mathbb{R}^n$.

When $(Q, \mathbf{g})$ is complete (in that the motion generated by h is defined for all times), and in addition has the property that any two points are connected by a unique geodesic, one has $Q \simeq \mathbb{R}^n$ as a manifold. Moreover, the obvious generalization of the geodesic construction above provides a diffeomorphism between $Q \times Q$ and TQ. In general, one has to proceed locally, using the geodesic midpoint construction to obtain a diffeomorphism between a neighborhood of the diagonal embedding $\delta(Q)$ in $Q \times Q$ and the zero section Q in TQ. On a suitable choice of functions in $\mathfrak{A}_0 = C_0(T^*Q)$, this still enables one to generalize the Weyl prescription to obtain a strict quantization map $\mathcal{Q}^W_\hbar$. For suitable (real) f, the operator $\mathcal{Q}^W_\hbar(f)$ is a compact (self-adjoint) operator on $L^2(Q)$ (defined with respect to the canonical Riemannian measure on Q).

The single most important property of $\mathcal{Q}^W_\hbar$ is that it is equivariant under isometries. To explain this, we first note that the group Diff(Q) of diffeomorphisms of Q acts on T^*Q by pullback; call this action ρ^0. Accordingly, each $\varphi \in \text{Diff}(Q)$

defines an automorphism $\alpha^0_\varphi(f) = f \circ \varphi^*$ of $\mathfrak{A}^0$. Furthermore, there is a natural representation $\rho^\hbar$ of $\text{Diff}(Q)$ on $L^2(Q)$, which defines automorphisms $\alpha^\hbar_\varphi$ on $\mathfrak{A} = \mathfrak{B}_0(L^2(Q))$, as explained before in the context of $\mathbb{R}^n$. If, then, φ is an isometry of $(Q, \mathbf{g})$, for all suitable f one has

$$\alpha^\hbar_\varphi\left(\mathcal{Q}^W_\hbar(f)\right) = \mathcal{Q}^W_\hbar(\alpha^0_\varphi(f)).$$

It is possible to extend $\mathcal{Q}^W_\hbar$ to certain unbounded classical observables, in particular to functions that are polynomial in the canonical momenta. The Weyl quantization of the classical Hamiltonian is

$$\mathcal{Q}^W_\hbar(h) = -\tfrac{1}{2}\hbar^2(\Delta - \tfrac{1}{3}\mathbf{R}),$$

containing not only the Laplace–Beltrami operator Δ, but picking up an additional term proportional to the Ricci scalar $\mathbf{R}$. If $(Q, \mathbf{g})$ is complete and $\mathbf{R}$ is bounded, this quantum Hamiltonian is essentially self-adjoint on the domain $C^\infty_c(Q) \subset L^2(Q)$. However, even when these conditions are not met one can prove results on the convergence of quantum to classical dynamics similar to those in the flat case.

III. Groups, Bundles, and Groupoids

In Chapter III we construct Poisson algebras and C^*-algebras from well-known geometric objects, namely Lie groups and their Lie algebras, and principal fiber bundles and their associated "infinitesimal" objects. These Poisson and C^*-algebras turn out to be related by a strict quantization. The theory of Lie groupoids and algebroids then provides a perspective unifying these seemingly diverse classes of examples, as well as providing new ones.

Lie Groups and Lie Algebras

Let $\mathfrak{g}$ be a Lie algebra. The (minus) **Lie–Poisson structure** on the dual $\mathfrak{g}^*$ is given by the Poisson bracket

$$\{\tilde{X}, \tilde{Y}\}_- = -\widetilde{[X, Y]},$$

where each $X \in \mathfrak{g}$ defines a linear function $\tilde{X}(\theta) := \theta(X)$ on $\mathfrak{g}^*$. Physically, the associated Poisson algebra $C^\infty(\mathfrak{g}^*_-, \mathbb{R})$ is the classical algebra of observables of an immobile particle whose only degrees of freedom are "internal". For example, when $\mathfrak{g} = \mathfrak{so}(3)$ it describes a spinning particle, the magnitude of whose spin is not fixed.

In the spirit of Chapter I one may then look for representations of $C^\infty(\mathfrak{g}^*_-, \mathbb{R})$ on a symplectic manifold S. Such a representation corresponds to a Poisson map $J : S \to \mathfrak{g}^*_-$. The representation theory of $C^\infty(\mathfrak{g}^*_-, \mathbb{R})$ is closely related to the existence of $\mathfrak{g}$-actions on S, i.e., homomorphisms $X \mapsto \xi_X$ from $\mathfrak{g}$ into the space of vector fields on S. For given such a representation, one finds a $\mathfrak{g}$-action by $\xi_X := \xi_{J_X}$, with $J_X := J^*\tilde{X}$. Conversely, a $\mathfrak{g}$-action $X \mapsto \xi_{J^*\tilde{X}}$ generated by some smooth map $J : S \to \mathfrak{g}^*_-$ in this way is called **Hamiltonian**, and J is known as a **momentum map** for the action. It does not follow that the momentum map of a Hamiltonian $\mathfrak{g}$-action is a Poisson map with respect to the Poisson bracket displayed above: In general, one has

$$\{J_X, J_Y\}_S = -J_{[X,Y]} - \Gamma(X, Y),$$

where Γ is constant on S, defining a so-called 2-cocycle on $\mathfrak{g}$.

A smooth action of a Lie group G on a manifold S leads to a $\mathfrak{g}$-action through $\xi_X f(\sigma) = df(\mathrm{Exp}(tX)\sigma)/dt|t = 0$; the G-action is said to be Hamiltonian when the associated $\mathfrak{g}$-action is. When $J : S \to \mathfrak{g}^*_-$ is a Poisson map (i.e., $\Gamma = 0$), the G-action and the $\mathfrak{g}$-action are called **strongly Hamiltonian**, and J is said to be **equivariant**.

The **coadjoint action** Co of G on $\mathfrak{g}^*$ is the dual of the adjoint action. The main theorem on the Lie–Poisson structure is that the symplectic leaves of $\mathfrak{g}^*_-$ are precisely the coadjoint orbits. This endows the coadjoint orbits with the **Lie symplectic structure.** For example, the coadjoint orbits in $\mathfrak{so}(3)^*$ are two-spheres; picking an orbit fixes the magnitude of the classical spin. More generally, a coadjoint orbit plays the role of a classical charge. When G is abelian, as in the theory of electromagnetism (where $H = U(1)$), the charge is just a number. The significance of the coadjoint orbits in representation theory is that (up to covering spaces) every irreducible representation $\pi_{\mathcal{O}}$ of the Poisson algebra $C^\infty(\mathfrak{g}^*_-, \mathbb{R})$ is realized on such an orbit $\mathcal{O}$.

In quantum mechanics the focus is on G-actions on a projective Hilbert space $\mathbb{P}\mathcal{H}$. These actions should not merely respect the Poisson structure on $\mathbb{P}\mathcal{H}$, but must in addition preserve the quantum-mechanical transition probabilities. Such G-actions on $\mathbb{P}\mathcal{H}$ turn out to be given by linear unitary G-"actions" U on $\mathcal{H}$ itself, which satisfy $U(x)U(y) = c(x, y)U(xy)$. Here $c : G \times G \to U(1)$ is a so-called **multiplier**, which measures to what extent U differs from a true representation of G. A multiplier on G is the "global" analogue of a 2-cocycle Γ on the Lie algebra $\mathfrak{g}$. Indeed, let J be the momentum map of the associated action on $\mathbb{P}\mathcal{H}$; it is given by

$$J_X(\psi) = i\hbar(\Psi, dU(X)\Psi),$$

where the unit vector Ψ is a lift of $\psi \in \mathbb{P}\mathcal{H}$ to $\mathcal{H}$, and

$$dU(X) := \frac{d}{dt} U(\mathrm{Exp}(tX))_{|t=0}.$$

The presence of a multiplier in the G-action on $\mathcal{H}$ is then reflected by

$$\{J_X, J_Y\}_\hbar = -J_{[X,Y]} - \hbar\Gamma(X, Y).$$

What is the quantum-mechanical counterpart of the Lie–Poisson algebra $C^\infty(\mathfrak{g}^*_-, \mathbb{R})$? This turns out to be the **group C^*-algebra** $C^*(G)$. Here the shift from the "infinitesimal" object $\mathfrak{g}^*$ to the "global" object G in passing from classical to quantum mechanics is typical. To define $C^*(G)$ (for a group whose left and right Haar measures dx coincide, for simplicity) one starts from the convolution operation on, say, $C_c^\infty(G)$, that is,

$$f * g(x) := \int_G dy\, f(xy^{-1})g(y).$$

One adds an involution $f^*(x) := \overline{f(x^{-1})}$ so as to turn $C_c^\infty(G)$ into a *-algebra. In the associativity of convolution and the involutive nature of $f \mapsto f^*$ this algebra reflects the corresponding properties $(xy)x = x(yz)$ and $(xy)^{-1} = y^{-1}x^{-1}$ of the group G itself. One then equips $C_c^\infty(G)$ with an appropriate norm, and closes it so as to obtain the C^*-algebra $C^*(G)$.

The quantum analogue of the correspondence between $\mathfrak{g}$-actions on symplectic manifolds and representations of $C^\infty(\mathfrak{g}^*_-, \mathbb{R})$ is then a basic theorem about $C^*(G)$, stating that there is a bijective correspondence between representations U of G on Hilbert spaces and nondegenerate representations π of $C^*(G)$ as a C^*-algebra, given by $\pi(f) = \int_G dx\, f(x)U(x)$.

When G is compact there is a neat quantum analogue of the decomposition of $\mathfrak{g}^*_-$ as the union of its symplectic leaves (which, as we saw, are just the coadjoint orbits). From the Peter–Weyl theorem, one has the decomposition

$$C^*(G) \simeq \bigoplus_{\gamma \in \hat{G}} \mathfrak{M}_{d_\gamma}(\mathbb{C}),$$

where $\hat{G}$ is the space of all (equivalence classes of) irreducible representations of G, and d_γ is the dimension of a given such representation.

The analogy between the Poisson algebra $C^\infty(\mathfrak{g}^*_-, \mathbb{R})$ and the C^*-algebra $C^*(G)$ is further illustrated by the construction of a strict quantization relating the two. One here chooses $\mathfrak{A}^0 = C_0(\mathfrak{g}^*)$ and $\mathfrak{A}^\hbar = C^*(G)$, and, roughly speaking, defines the quantization map $\mathcal{Q}_\hbar : \mathfrak{A}^0_\mathbb{R} \to \mathfrak{A}^\hbar_\mathbb{R}$ by

$$\mathcal{Q}_\hbar(f)(\mathrm{Exp}(X)) := \int_{\mathfrak{g}^*} \frac{d^n\theta}{(2\pi\hbar)^n}\, e^{\frac{i}{\hbar}\theta(X)} f(\theta).$$

For compact or nilpotent Lie groups one can show that this indeed defines a strict quantization. The nature of this prescription may be illustrated by the fact that in any representation π of $C^*(G)$ one obtains (transgressing the realm of bounded operators)

$$\pi(\mathcal{Q}_\hbar(\tilde{X})) = i\hbar dU(X).$$

Hence from the Lie–Poisson bracket above and the property $[dU(X), dU(Y)] = dU([X, Y])$ one immediately verifies that

$$\frac{i}{\hbar}\left[\pi(\mathcal{Q}_\hbar(\tilde{X})), \pi(\mathcal{Q}_\hbar(\tilde{Y}))\right] = \pi(\mathcal{Q}_\hbar(\{\tilde{X}, \tilde{Y}\}_-)).$$

Rather than $C^\infty(\mathfrak{g}^*_-, \mathbb{R})$ one may try to quantize the Poisson algebra $C^\infty(\mathcal{O}, \mathbb{R})$, where $\mathcal{O}$ is a coadjoint orbit in $\mathfrak{g}^*$. For compact G this is indeed possible, with the interesting feature that Planck's constant is "quantized"; this reflects the compactness of the classical phase space $\mathcal{O}$. One starts from an irreducible representation $U_\gamma(G)$ on a (finite-dimensional) Hilbert space $\mathcal{H}_\gamma$ labeled by a highest weight γ, with highest weight vector Ψ_γ. Let $J : \mathbb{P}\mathcal{H}_\gamma \to \mathfrak{g}^*_-$ (as displayed above) be the momentum map of the G-action on $\mathbb{P}\mathcal{H}$ associated to the representation U on $\mathcal{H}$. One then has to assume that $\mathcal{O}$ contains $J(\psi_\gamma)$. When this is the case, for $k \in \mathbb{Z}$ one defines $\mathcal{H}_\hbar$ for $\hbar = 1/k$ as $\mathcal{H}^{\text{hw}}_{\gamma/\hbar}$ (i.e., the carrier space of the representation with highest weight $k\gamma$), upon which the map $q_\hbar : \mathcal{O} \to \mathbb{P}\mathcal{H}_\hbar$, defined by

$$q_\hbar(\text{Co}(x)\gamma) := \tau_{\mathcal{H}_\hbar \to \mathbb{P}\mathcal{H}_\hbar}(U_{\gamma/\hbar}(x)\Psi_{\gamma/\hbar}),$$

is a pure state quantization of $\mathcal{O}$ on $I_0 := 1/\mathbb{N}$. The associated Berezin quantization $\mathcal{Q}^B_\hbar$ then turns out to be strict. It is also G-equivariant: With $\alpha^0_x(f) := f \circ \text{Co}(x^{-1})$ for arbitrary $x \in G$ and $f \in C^\infty(\mathcal{O})$, one has

$$\mathcal{Q}^B_{1/k}(\alpha^0_x(f)) = U_{k\gamma}(x)\mathcal{Q}^B_{1/k}(f)U_{k\gamma}(x)^*.$$

Internal Symmetries and External Gauge Fields

The description of purely spatial degrees of freedom of a single particle having been given in Chapter II, and the treatment of purely internal variables having just been sketched, the goal is now to combine these.

The appropriate mathematical tool is the theory of **principal fiber bundles**. When Q is a manifold and H a Lie group, a principal H-bundle P is defined by a free H-action on P and by a projection $\tau : \mathsf{P} \to Q$. These must be such that locally $\mathsf{P} \simeq Q \times H$, relative to which the H-action becomes the canonical right action on the second variable, and τ is projection onto the first. In particular, $\mathsf{P}/H \simeq Q$. This setup is the starting point for the classical as well as the quantum theory of a particle that moves on Q and has internal degrees of freedom related to H.

More generally, a bundle over a manifold Q with typical fiber F is a space B with a projection $\tau : \mathsf{B} \to Q$ such that locally $\mathsf{B} \simeq Q \times F$, and τ is projection onto the first variable. Apart from principal bundles, where F is a Lie group, an important class is formed by **vector bundles**, where F is a vector space. A **section** of B is a map $s : Q \to \mathsf{B}$ for which $\tau \circ s = \text{id}$.

A most important concept, used in classical as well as in quantum mechanics, is that of an **associated bundle**: Given a smooth H-action L on some manifold M, the associated bundle $\mathsf{M} = \mathsf{P} \times_H M$ is $(\mathsf{P} \times M)/H$, where the H-action on $\mathsf{P} \times M$ defining the quotient is given by $h : (x, m) \mapsto (xh^{-1}, L_h(m))$. This is a bundle over Q with typical fiber M. The projection $\tau_{\mathsf{M} \to Q}$ is given by $\tau_{\mathsf{M} \to Q}([x, m]_H) = \tau(x)$.

The classical theory is based on the Poisson manifold $(T^*\mathsf{P})/H$. Here the H-action on $T^*\mathsf{P}$ is the pullback of the given action on P; the canonical Poisson bracket on the cotangent bundle $T^*\mathsf{P}$ quotients to one on $(T^*\mathsf{P})/H$, defining its Poisson structure. The symplectic leaves of $(T^*\mathsf{P})/H$ are of the form $J^{-1}(\mathcal{O})/H$, where $J : T^*\mathsf{P} \to \mathfrak{h}^*_-$ is the momentum map of the associated $\mathfrak{h}$-action on $T^*\mathsf{P}$, and $\mathcal{O}$ is a coadjoint orbit in $\mathfrak{h}^*$. The choice of an orbit $\mathcal{O}$ specifies a classical charge; the orbit contains internal degrees of freedom, which in physics couple to an external gauge field. There is a correspondence

$$\pi_{\mathcal{O}}(C^\infty(\mathfrak{h}^*_-, \mathbb{R})) \longleftrightarrow \pi^{\mathcal{O}}(C^\infty((T^*\mathsf{P})/H, \mathbb{R}))$$

between the coadjoint orbits $\mathcal{O}$ in $\mathfrak{h}^*$, here in the guise of the irreducible representations $\pi_{\mathcal{O}}$ of $C^\infty(\mathfrak{h}^*_-, \mathbb{R})$, and the irreducible representations of the Poisson algebra $C^\infty((T^*\mathsf{P})/H, \mathbb{R})$: For each such orbit one obtains an irreducible representation $\pi^{\mathcal{O}}$ on the symplectic manifold $(T^*\mathsf{P})^{\mathcal{O}} := J^{-1}(\mathcal{O})/H$.

A symplectic leaf $(T^*\mathsf{P})^{\mathcal{O}}$ is locally of the form $T^*Q \times \mathcal{O}$. However, to separate the spatial and internal degrees of freedom in an intrinsic fashion, one needs to choose a **connection** on P. This is a decomposition of each tangent space $T_x\mathsf{P}$ into an (intrinsically defined) vertical subspace (which projects to zero under τ), and a complement, called the horizontal tangent space at x. Choosing such a decomposition turns out to be equivalent to the specification of an $\mathfrak{h}$-valued 1-form $\mathbf{A}$ on P, with certain properties. The part of $\mathbf{A}$ that lives on Q (relative to a local factorization $\mathsf{P} \simeq Q \times H$) is the physicist's gauge field or Yang–Mills field.

Let us introduce the manifold

$$\mathsf{P} *_Q T^*Q := \{(x, \sigma) \in \mathsf{P} \times T^*Q \mid \tau_{\mathsf{P}\to Q}(x) = \tau_{T^*Q\to Q}(\sigma)\}.$$

This is a principal H-bundle over T^*Q if one defines its projection to be the one onto the second variable, and its H-action to be essentially the H-action on P. Choosing a connection then leads to the realization of $(T^*\mathsf{P})^{\mathcal{O}}$ as a bundle associated to $\mathsf{P} *_Q T^*Q$ by the coadjoint representation of H on $\mathcal{O}$. In this realization the Poisson bracket on $(T^*\mathsf{P})^{\mathcal{O}}$ depends on $\mathbf{A}$.

The basic tool in the construction of (unbounded) physical observables on the phase space $(T^*\mathsf{P})^{\mathcal{O}}$ is the group $\mathrm{Aut}(\mathsf{P})$ of automorphisms of the bundle P; this group consists of those diffeomorphisms on P that commute with the H-action. Any diffeomorphism on P pulls back to one of $T^*\mathsf{P}$; a bundle automorphism in addition maps $J^{-1}(\mathcal{O})$ into itself, and quotients to a Poisson map on $(T^*\mathsf{P})^{\mathcal{O}}$. The momentum map for this reduced action $\rho_0^{\mathcal{O}}$ of $\mathrm{Aut}(\mathsf{P})$ on $(T^*\mathsf{P})^{\mathcal{O}}$ then gives the classical observables that are linear in the (conventional) momentum. Functions of the configuration variable q are more easily obtained, namely from the natural projection $\tau_{(T^*\mathsf{P})^{\mathcal{O}}\to Q}$.

From the perspective of ("classical") representation theory the symplectic space $(T^*\mathsf{P})^{\mathcal{O}}$ therefore plays a double role: It firstly carries the irreducible representation $\pi^{\mathcal{O}}$ of the Poisson algebra $C^\infty((T^*\mathsf{P})/H, \mathbb{R})$, and secondly it supports the Poisson action $\rho_0^{\mathcal{O}}$ of the group $\mathrm{Aut}(\mathsf{P})$.

To specify a "natural" Hamiltonian $h^{\mathcal{O}}$ on all leaves $(T^*\mathsf{P})^{\mathcal{O}}$ in one go, one needs a Riemannian metric $\mathbf{g}^Q$ on Q and a connection $\mathbf{A}$ on P, as above. In the $\mathbf{A}$-dependent realization of $(T^*\mathsf{P})^{\mathcal{O}}$ as the associated bundle $(\mathsf{P} *_Q T^*Q) \times_H \mathcal{O}$ one then simply puts

$$h^{\mathcal{O}}(p, q, \theta) = \tfrac{1}{2}\mathbf{g}_Q^{\mu\nu}(q)p_\mu p_\nu.$$

For simplicity this has been expressed in local coordinates, but $h^{\mathcal{O}}$ is an intrinsically defined function. In the original definition of $(T^*\mathsf{P})^{\mathcal{O}}$ as a subspace of $(T^*\mathsf{P})/H$ this reads

$$h_A^{\mathcal{O}}(p, q, \theta) = \tfrac{1}{2}\mathbf{g}_Q^{\mu\nu}(q)(p_\mu - \theta_i A_\mu^i(q))(p_\nu - \theta_j A_\nu^j(q)).$$

The associated equations of motion are the so-called **Wong equations**.

We now turn to the associated quantum theory. For compact H an appropriate (complexified) quantum algebra of observables is $\mathfrak{B}_0(L^2(\mathsf{P}))^H$, the C^*-algebra of compact operators on $L^2(\mathsf{P})$ (defined with respect to some H-invariant measure equivalent to the Lebesgue measure) that commute with the representation U_R of H, given by

$$U_R(h)\Psi(x) = \Psi(xh).$$

On the basis of the quantization theory on Riemannian manifolds in Chapter II one shows that $\mathfrak{B}_0(L^2(\mathsf{P}))^H$ is a strict quantization of $C_0((T^*\mathsf{P})/H)$. Noticing that $C_0((T^*\mathsf{P})/H, \mathbb{R}) \simeq C_0(T^*\mathsf{P}, \mathbb{R})^H$, the associated quantization map is simply given by restriction of the Weyl quantization map $\mathcal{Q}_\hbar^W$ on $T^*\mathsf{P}$ (where P has been equipped with an H-invariant Riemannian metric).

From a representation-theoretic viewpoint, the quantum counterpart of $(T^*\mathsf{P})^{\mathcal{O}}$ is a Hilbert space $\mathcal{H}^\chi$, constructed as follows. One starts with a representation $U_\chi(H)$ on a Hilbert space $\mathcal{H}_\chi$, and then considers the vector bundle $\mathsf{H}^\chi := \mathsf{P} \times_H \mathcal{H}_\chi$ associated to P by the representation U_χ; locally $\mathsf{H}^\chi \simeq Q \times \mathcal{H}_\chi$. In contrast with the classical situation, the relevant object is not this associated bundle itself, but rather its space of smooth sections $\Gamma(\mathsf{H}^\chi)$ (with compact support). This space may be realized as a space of maps $\Psi^\chi : \mathsf{P} \to \mathcal{H}_\chi$ satisfying the equivariance condition

$$\Psi^\chi(xh^{-1}) = U_\chi(h)\Psi^\chi(x)$$

for all $x \in \mathsf{P}$ and $h \in H$. Exploiting the fact that the fiber $\mathcal{H}_\chi$ is a Hilbert space, one can equip $\Gamma(\mathsf{H}^\chi)$ with an inner product; its closure is $\mathcal{H}^\chi$.

In analogy to the classical situation, $\mathcal{H}^\chi$ plays a double role in representation theory. Firstly, it carries a representation π^χ of the C^*-algebra $\mathfrak{B}_0(L^2(\mathsf{P}))^H$, given by

$$\pi^\chi(K)\Psi^\chi(x) = \int_{\mathsf{P}} d\mu(y)\, K(x, y)\Psi^\chi(y).$$

This representation is irreducible iff $U_\chi(H)$ is irreducible, so that we obtain a correspondence

$$U_\chi(H) \longleftrightarrow \pi^\chi(\mathfrak{B}_0(L^2(\mathsf{P}))^H)$$

analogous to the classical correspondence $\pi_{\mathcal{O}} \leftrightarrow \pi^{\mathcal{O}}$. Secondly, $\mathcal{H}^\chi$ carries the **induced representation** $U^\chi(\mathrm{Aut}(\mathsf{P}))$, defined by

$$U^\chi(\varphi)\Psi^\chi(x) = \sqrt{\frac{d\nu(\varphi_Q^{-1}(\tau(x)))}{d\nu(\tau(x))}}\, \Psi^\chi(\varphi^{-1}(x)),$$

where ν is a measure on Q that is naturally defined by μ, and φ_Q is the diffeomorphism of Q associated to φ in the obvious way.

The well-known construction of induced representations of a (Lie) group G is a special case: one takes $\mathsf{P} = G$, defined as a bundle over $Q = G/H$ through the canonical right action of $H \subset G$ on G. The left action of G on itself then realizes G as a subgroup of $\mathrm{Aut}(G)$, so that the equation above applies. This is called **Mackey induction**. When ν is G-invariant we obtain

$$U^{\chi}(y)\Psi^{\chi}(x) = \Psi^{\chi}(y^{-1}x).$$

We return to the case of a general bundle P. The correspondence between the classical and the quantum theory is further illustrated by the remarkable equation

$$i\hbar dU^{\chi}(\xi) = \pi^{\chi}(\mathcal{Q}_{\hbar}^{W}(J_{\xi})).$$

Here ξ is an element of the Lie algebra of Aut(P), that is, an H-invariant vector field on P, and J is the momentum map for the Aut(P)-action on $T^*\mathsf{P}$. Since the latter commutes with the H-action, its momentum map is H-invariant, so that each J_{ξ} lies in $C^{\infty}(T^*\mathsf{P}, \mathbb{R})^H$. Extending the definition of $\mathcal{Q}_{\hbar}^{W}$ to suitable unbounded functions, the right-hand side is therefore well-defined. When $U_{\chi}(H)$ actually corresponds to a coadjoint orbit $\mathcal{O}$, one may regard $\pi^{\chi}(\mathcal{Q}_{\hbar}^{W}(J_{\xi}))$ as the quantization of $J_{\xi}^{\mathcal{O}}$, but in the absence of such a correspondence the right-hand side still makes sense as the quantum "ξ-momentum" in the sector χ. In particular, one has

$$\frac{i}{\hbar}[\pi^{\chi}(\mathcal{Q}_{\hbar}^{W}(J_{\xi})), \pi^{\chi}(\mathcal{Q}_{\hbar}^{W}(J_{\eta}))] = \pi^{\chi}\left(\mathcal{Q}_{\hbar}^{W}(\{J_{\xi}, J_{\eta}\})\right).$$

The quantum Hamiltonian on $\mathcal{H}^{\chi}$ defined by Weyl quantization is

$$H_{\hbar}^{\chi} = -\tfrac{1}{2}\hbar^2\left(\Delta_{\chi}^{\mathbf{A}} - \tfrac{1}{3}\mathbf{R}_{Q} + \tfrac{1}{12}\mathbf{F}^2 - C^{\chi}\right).$$

Here $\Delta_{\chi}^{\mathbf{A}}$ is a gauge-covariant Laplacian, and the other terms are geometric objects acting as multiplication operators, all constructed from $\mathbf{g}^{Q}$ and $\mathbf{A}$.

As in Chapter II, the possible convergence of the classical equations of motion generated by $h_{A}^{\mathcal{O}}$ to their quantum counterparts generated by $H_{\hbar}^{\chi}$ can be analyzed. One has to find a suitable analogue of the coherent states $\Psi_{\hbar}^{(p,q)}$ used for this purpose when only spatial degrees of freedom are present. Our discussion on the quantization of the Poisson algebra $C^{\infty}(\mathcal{O}, \mathbb{R})$, where $\mathcal{O}$ is a coadjoint orbit in $\mathfrak{h}^*$, suggests how to proceed. We assume that $\mathcal{O}$ is associated with an irreducible representation $U_{\chi}(H)$, labeled by a highest weight χ, in that $\mathcal{O}$ contains $J(\psi_{\chi})$. For "quantized" $\hbar = 1/k$, $k \in \mathbb{N}$, we then replace $\Psi_{\hbar}^{(p,q)}$ by the unit vectors

$$\Psi_{1/k}^{(p,q,h)} := \Psi_{1/k}^{(p,q)} \otimes U_{k\chi}(h)\Psi_{k\chi}$$

in $L^2(\mathbb{R}^n) \otimes \mathcal{H}_{k\chi}$, where $\Psi_{k\chi}$ is a normalized highest weight vector in $\mathcal{H}_{k\chi}$. The desired convergence may then be shown for $k \to \infty$. The proof makes essential use of the G-equivariance of the Berezin quantization of $C^{\infty}(\mathcal{O}, \mathbb{R})$.

Everything said so far may be explicitly calculated in the simplest nontrivial example, where the bundle $\mathsf{P}(Q, H)$ is $SO(3)(S^2, SO(2))$. This bundle supports a certain canonical connection, which in physics terms describes the field of a magnetic monopole sitting at the origin. The symplectic leaves in $\mathfrak{so}(2) = \mathbb{R}$ are just numbers e, identified with the electric charge of the particle moving on S^2. The symplectic leaves $(T^*SO(3))^e$ are diffeomorphic to T^*S^2, but one still sees the effect of a nonzero charge e in all relevant quantities, such as the momentum map for the reduced $SO(3)$ action on $(T^*SO(3))^e$.

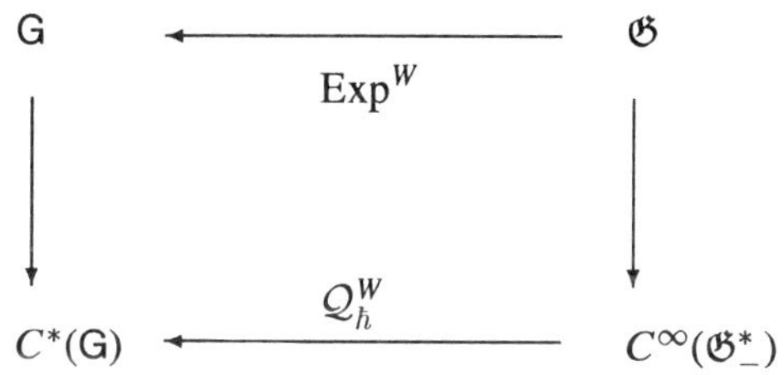

FIGURE 1. Groupoids and algebroids in quantization

The quantum theory is formulated in terms of the line bundles H^n, $n \in \mathbb{Z}$, defined by the irreducible representations $U_n(\alpha) = \exp(-in\alpha)$ of $SO(2)$. These representations are the quantum analogues of the coadjoint orbits e, illustrating the quantization of electrical charge. It should be mentioned that the Hilbert space $\mathcal{H}^n$ is insensitive to the topology of the line bundle H^n; the relevance of this bundle in quantum mechanics lies in the fact that the space of sections $\Gamma(\mathsf{H}^n)$, which does "see" the topology, provides a domain of essential self-adjointness for the basic quantum observables.

Lie Groupoids and Lie Algebroids

Our aim is to explain Figure 1. Let us first look at a case where we already know what all the entries and arrows mean, namely when $\mathsf{G} = G$ is a Lie group and $\mathfrak{G} = \mathfrak{g}$ its Lie algebra. In that case, we have seen that we can canonically associate a Poisson algebra $C^\infty(\mathfrak{g}^*_-, \mathbb{R})$ with $\mathfrak{g}$ and a C^*-algebra $C^*(G)$ with G, in such a way that under favorable circumstances (e.g., when G is compact), $C^*(G)$ is a strict quantization of $C^\infty(\mathfrak{g}^*_-)$. The central ingredient in the construction of the quantization map $\mathcal{Q}_\hbar^W = \mathcal{Q}_\hbar$ was the usual exponential map $\mathrm{Exp}^W = \mathrm{Exp} : \mathfrak{g} \to G$.

The quantization of a system with configuration space Q fits into this diagram as well. We already know three of the four corners: For $\mathfrak{G}$ we would like to read the tangent bundle TQ, with associated Poisson algebra $C^\infty(T^*Q, \mathbb{R})$, and we wish $C^*(\mathsf{G})$ to be the algebra of compact operators $\mathfrak{B}_0(L^2(Q))$. The object G should then be chosen as $Q \times Q$, equipped with structures such that one may firstly construct TQ as an associated infinitesimal object (in analogy to the construction of a Lie algebra from a Lie group), and secondly can define the C^*-algebra $C^*(Q \times Q)$ through the construction of a convolution and an involution on $C_c^\infty(Q \times Q)$.

The appropriate starting point is the concept of a **groupoid**. This is a generalization of a group, in which multiplication is only partially defined. When it is defined, it is associative, and each element has an inverse. For example, one may say that two elements (q_1, q_1') and (q_2, q_2') of $Q \times Q$ can be multiplied iff $q_1' = q_2$, in which case $(q_1, q_1')(q_1', q_2') := (q_1, q_2')$. This reflects the way arrows are composed; one therefore interprets a point $(q, q') \in Q \times Q$ as an arrow from q' to q.

Thus the inverse is $(q, q')^{-1} := (q', q)$. The ensuing object is the **pair groupoid** on Q.

Every groupoid G may be thought of as a collection of arrows connecting points on some space Q, called the **base** of G. The collection of all elements of the form $\gamma\gamma^{-1}, \gamma \in \mathsf{G}$, is naturally isomorphic to the base Q, and this isomorphism leads to an inclusion $\iota : Q \hookrightarrow \mathsf{G}$. Arrows in $\iota(Q)$ evidently start and end at the same point. In a pair groupoid one has $\iota(q) = (q, q)$.

In a group all arrows start and end at the unit e, so that any two elements may be composed. An intermediate possibility is an **action groupoid**. Given a group G and a G-action on a set Q, we look at $G \times Q$ as a collection of arrows between points in Q, in such a way that (x, q) starts at $x^{-1}q$ and ends at q. Accordingly, the product $(x, q)(y, q')$ is defined when $q' = x^{-1}q$, in which case $(x, q)(y, x^{-1}q) := (xy, q)$. The inverse is $(x, q)^{-1} := (x^{-1}, x^{-1}q)$. Hence $\iota(q) = (e, q)$.

When all relevant objects are manifolds and all operations are smooth, one speaks of a **Lie groupoid**. Given a Lie groupoid G, one can turn $C_c^\infty(\mathsf{G})$ into a convolution *-algebra that reflects the basic properties of the groupoid operations. For example, for a Lie group this reproduces the *-algebra we have already encountered. For a pair groupoid one obtains

$$f * g(q_1, q_2) = \int_Q d\nu(q)\, f(q_1, q)g(q, q_2)$$

and $f^*(q_1, q_2) = \overline{f(q_2, q_1)}$. On an action groupoid one has

$$f * g(x, q) = \int_G dy\, f(xy, q)g(y^{-1}, y^{-1}x^{-1}q)$$

and $f^*(x, q) = \overline{f(x^{-1}, x^{-1}q)}$.

One sees that in these two cases the involution is defined by

$$f^*(\gamma) = \overline{f(\gamma^{-1})};$$

this is, in fact, always true. One can put a norm on $C_c^\infty(\mathsf{G})$, and complete it so as to obtain a C^*-algebra $C^*(\mathsf{G})$. For the pair groupoid $Q \times Q$ one then finds $C^*(Q \times Q) = \mathfrak{B}_0(L^2(Q))$.

Given a principal H-bundle P over Q, one may form the "quotient" of the pair groupoid $\mathsf{P} \times \mathsf{P}$ by H, obtaining the **gauge groupoid** $\mathsf{P} \times_H \mathsf{P}$ of the bundle. This is a groupoid with base Q; an arrow $[x, y]_H$ starts at $\tau(y)$ and ends at $\tau(x)$ (where τ is the bundle projection on P). The C^*-algebra $C^*(\mathsf{P} \times_H \mathsf{P})$ of this groupoid turns out to be isomorphic to $\mathfrak{B}_0(L^2(Q)) \otimes C^*(H)$. For compact H this is nothing but the C^*-algebra $\mathfrak{B}_0(L^2(\mathsf{P}))^H$ we have already encountered, and $C^*(\mathsf{P} \times_H \mathsf{P})$ is in every respect the correct generalization of $B_0(L^2(\mathsf{P}))^H$ to the case where H is noncompact. In particular, given a representation $U_\chi(H)$ one may construct an induced representation π^χ of $C^*(\mathsf{P} \times_H \mathsf{P})$, which is irreducible iff U_χ is. This leads to a bijective correspondence

$$U_\chi(H) \longleftrightarrow \pi^\chi(C^*(\mathsf{P} \times_H \mathsf{P}))$$

between the representations of H and the representations of $C^*(\mathsf{P} \times_H \mathsf{P})$.

The C^*-algebra of an action groupoid $G \times Q$ is usually written as $C^*(G, Q)$, and is called an **action C^*-algebra**. The C^*-algebra $C^*(G, Q)$ has the remarkable property that each of its representations corresponds to a **system of imprimitivity** $(U, \tilde{\pi})$, where U is a representation of G, and $\tilde{\pi}$ is a representation of $C_0(Q)$, satisfying the covariance condition

$$U(x)\tilde{\pi}(\tilde{f})U(x)^* = \tilde{\pi}(\alpha_x(\tilde{f})).$$

Here $\alpha_x(\tilde{f}) : q \mapsto \tilde{f}(x^{-1}q)$. This condition is an integrated form of the "canonical" commutation relation

$$\frac{i}{\hbar}[\mathcal{Q}_\hbar(\tilde{X}), \mathcal{Q}_\hbar(\tilde{f})] = \mathcal{Q}_\hbar(\xi_X \tilde{f}),$$

for $\tilde{f} \in C_c^\infty(Q)$ and $X \in \mathfrak{g}$. Here $\mathcal{Q}_\hbar(\tilde{X}) := i\hbar dU(X)$ and $\mathcal{Q}_\hbar(\tilde{f}) := \tilde{\pi}(\tilde{f})$; recall the definition of the linear function $\tilde{X} \in C^\infty(\mathfrak{g}^*, \mathbb{R})$ and of the vector field ξ_X on Q.

Turning to the top right corner in Figure 1, we now describe the "infinitesimal" object $\mathfrak{G}$ associated to a Lie groupoid G, generalizing the concept of a Lie algebra. The **Lie algebroid** of a Lie groupoid with base Q is a vector bundle over Q, which apart from the bundle projection $\tau : \mathfrak{G} \to Q$ enjoys another linear map $\tau_a : \mathfrak{G} \to TQ$, called the **anchor**. In addition, there is a Lie bracket $[\,,\,]$ on the space of sections of $\mathfrak{G}$, which is related to the usual commutator on vector fields on Q through the anchor. These objects are all constructed from G; the bundle $\mathfrak{G}$ itself is built from the geometry of the map $\iota : Q \hookrightarrow \mathsf{G}$ (it is the normal bundle of this inclusion), the Lie bracket is derived from the commutator of left-invariant vector fields on G (much as in the case of a Lie algebra), and the anchor is the derivative of the map from G to Q that assigns to an arrow its starting point.

For example, the Lie algebroid of the pair groupoid $Q \times Q$ is the tangent bundle TQ with the obvious Lie bracket; the anchor is, of course, the identity map. The Lie algebroid of the action groupoid $G \times Q$ is the **action algebroid** $\mathfrak{g} \times Q$, regarded as a trivial bundle over Q. Identifying sections of $\mathfrak{g} \times Q$ with $\mathfrak{g}$-valued functions $X(\cdot)$ on Q, the Lie bracket on constant sections is simply the bracket $[\,,\,]_\mathfrak{g}$ in $\mathfrak{g}$. More generally, one has

$$[X, Y]_{\mathfrak{g}\times Q}(q) = [X(q), Y(q)]_\mathfrak{g} + \xi_Y X(q) - \xi_X Y(q).$$

The anchor comes out as $\tau_a(X, q) = -\xi_X(q)$. Finally, the Lie algebroid of the gauge groupoid $\mathsf{P} \times_H \mathsf{P}$ is $(T\mathsf{P})/H$ as a vector bundle over Q, with commutator inherited from the usual one on vector fields on P.

This brings us to the downward arrow on the right in Figure 1, namely the construction of a Poisson algebra from the Lie algebroid $\mathfrak{G}$. The Poisson manifold in question is the dual bundle $\mathfrak{G}^*$, the Poisson bracket on $C^\infty(\mathfrak{G}^*, \mathbb{R})$ being determined by the special cases

$$\{f, g\}_- = 0;$$
$$\{\tilde{s}, f\}_- = -\tau_a \circ sf;$$
$$\{\tilde{s}_1, \tilde{s}_2\}_- = -\widetilde{[s_1, s_2]}_\mathfrak{G}.$$

Here f and g are functions on the base Q, and $\tilde{s}$ is a linear function on $\mathfrak{G}^*$ defined by a section s of $\mathfrak{G}$ in the obvious way.

For $\mathfrak{G} = TQ$ this is simply the usual Poisson structure on the cotangent bundle T^*Q; we see that this structure ultimately derives from the groupoid operations on $Q \times Q$. The relevant special cases of the bracket on the Poisson algebra $C^\infty(\mathfrak{g}^*_- \times Q, \mathbb{R})$ determined by the action groupoid $G \times Q$ are as follows. Firstly, for functions f, g depending only on Q one has the obvious $\{f, g\}_- = 0$. Secondly, on constant sections (identified with linear functions on $\mathfrak{g}^*$) one has $\{\tilde{X}, \tilde{Y}\}_- = -\widetilde{[X, Y]}$. Finally, the "mixed" bracket is $\{\tilde{X}, \tilde{f}\}_- = \xi_X \tilde{f}$. One sees from these Poisson brackets that a representation of the Poisson algebra $C^\infty(\mathfrak{g}^* \times Q, \mathbb{R})$ on a symplectic manifold is essentially a **classical system of imprimitivity**, being the classical analogue of the system of imprimitivity determined by a representation of the corresponding groupoid C^*-algebra $C^*(G, Q)$.

Now to the top horizontal arrow. It turns out that the exponential map $\mathrm{Exp} : \mathfrak{g} \to G$ on a Lie algebra can be generalized to a map $\mathrm{Exp}^W : \mathfrak{G} \to \mathsf{G}$ from the Lie algebroid $\mathfrak{G}$ into a corresponding Lie groupoid G. This generalized exponential map, however, depends on the choice of a connection (or covariant derivative) on the vector bundle $\mathfrak{G}$ over Q. Since for a Lie algebra the base space of this bundle consists of only one point, there is no need for a connection in this case. In an action Lie groupoid $\mathfrak{g} \times Q$ one does not need a connection either in order to define Exp^W. In terms of $\mathrm{Exp} : \mathfrak{g} \to G$ the map $\mathrm{Exp}^W : \mathfrak{g} \times Q \to G \times Q$ is given by

$$\mathrm{Exp}^W(X, q) = (\mathrm{Exp}(X), \mathrm{Exp}(\tfrac{1}{2}X)q).$$

On a pair Lie algebroid TQ one does need a connection; this is, of course, nothing but an affine connection. The latter leads to an exponential map exp in the sense of affine geometry, in terms of which $\mathrm{Exp}^W : TQ \to Q \times Q$ is

$$\mathrm{Exp}^W(X) = (\exp_{\tau(X)}(-\tfrac{1}{2}X), \exp_{\tau(X)}(\tfrac{1}{2}X)),$$

where $\tau := \tau_{TQ \to Q}$. For example, the affine connection may be the Levi-Civita connection provided by a Riemannian metric on Q.

At last, we are now in a position to define the generalized Weyl quantization map $\mathcal{Q}^W_\hbar : C^\infty(\mathfrak{G}^*, \mathbb{R}) \to C^*(\mathsf{G})_\mathbb{R}$ (restricted to suitable bounded functions); this is the bottom line of Figure 1. In analogy with the prescription for groups, in a rough sketch it is given by

$$\mathcal{Q}^W_\hbar(f)(\mathrm{Exp}^W(X)) = \int \frac{d^n\theta}{(2\pi\hbar)^n} \, e^{\frac{i}{\hbar}\theta(X)} f(\theta),$$

where the integration is over the fiber of $\mathfrak{G}^*$ above $\tau_{\mathfrak{G} \to Q}(X)$.

Using the above formulae for Exp^W, one verifies that this prescription indeed reduces to the Weyl quantization of $C^\infty(T^*Q, \mathbb{R})$ explained in Chapter II, as well as to the quantization of $C^\infty(\mathfrak{g}^*_-, \mathbb{R})$ discussed above.

IV. Reduction and Induction

In symplectic geometry one has the concept of **symplectic reduction**, whose aim is the construction of new symplectic manifolds from old ones; it may be interpreted as a tool in the representation theory of Poisson algebras. There exists an analogous technique in the representation theory of C^*-algebras, called **induction**. The final chapter develops the analogy between reduction on the classical side and induction on the quantum side. In physics these techniques play a central role in the classical and quantum theory of constrained systems.

Reduction

The general concept of symplectic reduction is as follows. Let (S, ω) be a symplectic manifold, and let C be a submanifold of S. The restriction ω_C of the symplectic form ω to C is closed, but not necessarily nondegenerate. Suppose ω_C is degenerate. The tangent bundle TC to C then contains a subbundle $\mathcal{N}_C := TC \cap TC^{\perp}$, where $TC^{\perp}$ consists of all vectors in TC on which ω_C identically vanishes. Under favorable circumstances, the collection of all curves in C that are tangent to $\mathcal{N}_C$ defines a foliation Φ_C of C, whose quotient $\mathsf{S}^C := C/\Phi_C$ is a manifold.

The essential point is now that the **reduced space** S^C is equipped with a symplectic form ω^C, whose pullback to C under the projection from C to S^C is ω_C. This is possible because the "directions of degeneracy" $\mathcal{N}_C$ of ω_C have disappeared in the construction of the reduced space.

In physics the submanifold $C \subset \mathsf{S}$ is defined by constraints on the allowed initial states of a given dynamical system; Gauss's law in electrodynamics is a typical example. Flows along $\mathcal{N}_C$ are often generated by gauge transformations, which do not modify the physical state of the system, and correspond to a redundancy in the description of the system in terms of the degrees of freedom in S. The passage from S to C then implements the constraints, whereas the subsequent step from C to S^C eliminates the gauge redundancy. In any case, one should firmly keep in mind that symplectic reduction is generically a two-step procedure (except when ω_C is nondegenerate, so that C itself is symplectic, and $\mathsf{S}^C = C$).

Suppose that $TC^{\perp}$ is contained in TC, in which case C is called **coisotropic**. The collection of all smooth functions on S that are constant on the leaves of Φ_C is then a Poisson algebra, which in physics is the algebra of **weak observables** $\mathfrak{A}_w^C$ of the system. Each $f \in \mathfrak{A}_w^C$ evidently "reduces" to a well-defined function $\pi^C(f)$ on the reduced space, and the map π^C is a representation of $\mathfrak{A}_w^C$ in $C^\infty(\mathsf{S}^C, \mathbb{R})$.

The following specialization of the above reduction scheme plays a central role in this chapter. Suppose one has a pair of symplectic manifolds (S, ω_S) and (S_ρ, ω_ρ), a Poisson manifold P, and a pair of Poisson morphisms $J : S \to P^-$ and $J_\rho : S_\rho \to P$ (here P^- is P with minus its Poisson bracket). One then takes $\mathsf{S} = S \times S_\rho$, equipped with the symplectic form $\omega := \omega_S + \omega_\rho$. We write Φ_ρ for the null foliation Φ_C. The submanifold

$$C = S *_P S_\rho := \{(\sigma, \alpha) \in S \times S_\rho | J(\sigma) = J_\rho(\alpha)\}$$

is then coisotropic, and leads to a reduced space

$$S_J^\rho := (S *_P S_\rho)/\Phi_\rho.$$

Most physically relevant examples of symplectic reduction are a special case of this construction, which we call **special symplectic reduction**. From the point of view of representation theory the main feature of the construction emerges when one has a second Poisson manifold P_2, and a Poisson map $J_2 : S \to P_2$, such that the pullback $J_2^* C^\infty(P_2, \mathbb{R})$ in $C^\infty(S, \mathbb{R})$ Poisson-commutes with $J^* C^\infty(P, \mathbb{R})$. The map $J^\rho : S_J^\rho \to P_2$, given by

$$J^\rho([\sigma, \alpha]_{\Phi_\rho}) = J_2(\sigma)$$

is then well-defined, and is a Poisson map. Pulling back, one obtains a representation $(J^\rho)^*$ of the Poisson algebra $C^\infty(P_2, \mathbb{R})$ on S_J^ρ, which is said to be **reduced** by the representation J_ρ^* of $C^\infty(P, \mathbb{R})$.

Writing J_1 for J etc., we denote this situation by

$$P_2 \xleftarrow{J_2} S \xrightarrow{J_1} P_1.$$

Denote the set of all $f \in C^\infty(S, \mathbb{R})$ for which $\{f, J_2^* g\} = 0$ for all $g \in C^\infty(P_2, \mathbb{R})$ by $J^* C^\infty(P_2, \mathbb{R})'$. The existence of the manifold S and the maps J_1, J_2 implies that P_1 and P_2 stand in a certain relationship to each other, which is particularly close if $J_2^* C^\infty(P_2, \mathbb{R})' = J_1^* C^\infty(P_1, \mathbb{R})$ as well as $J_1^* C^\infty(P_1, \mathbb{R})' = J_2^* C^\infty(P_2, \mathbb{R})$, and J_1 and J_2 are surjective, with connected and simply connected level sets in S.

If, given P_1 and P_2, one can find S, J_1, and J_2 such that these, and some additional technical conditions are met, one says that P_1 and P_2 are **Morita equivalent**. The **classical imprimitivity theorem** then states that $C^\infty(P_1, \mathbb{R})$ and $C^\infty(P_2, \mathbb{R})$ have equivalent representation theories. Specifically, every representation of $C^\infty(P_2, \mathbb{R})$ is reduced from some representation of $C^\infty(P_1, \mathbb{R})$, and vice versa, and this bijection preserves irreducibility.

The idea of the proof of this theorem is as simple as it is elegant, and is most easily formulated if we use Poisson maps J rather than representations $\pi = J^*$ (one may always pass from one to the other). Given a Poisson map $J_\rho : S_\rho \to P_1$, one constructs the reduced space S_1^ρ by special symplectic reduction. As explained above, this leads to a Poisson map $J_1^\rho : S_1^\rho \to P_2$. One now turns the diagram $P_2 \overset{J_2}{\leftarrow} S \overset{J_1}{\rightarrow} P_1$ around, obtaining $P_1 \overset{J_1}{\leftarrow} S^- \overset{J_2}{\rightarrow} P_2$. Applying special symplectic reduction once again, this time from $J_\sigma := J_1^\rho$, one obtains a reduced space S_2^σ and a Poisson map $J_2^\sigma : S_2^\sigma \to P_1$. Using all the assumptions involved in the Morita equivalence of P_1 and P_2, one then shows that S_2^σ is symplectomorphic to S_ρ, such that J_2^σ is equivalent to J_ρ. This works in the opposite direction as well.

Specializing special symplectic reduction, we now assume that $P = \mathfrak{h}^*$ (where $\mathfrak{h}^*$ is the dual of the Lie algebra $\mathfrak{h}$ of a connected Lie group H), and $J : S \to \mathfrak{h}^*_-$ is an equivariant momentum map coming from a strongly Hamiltonian H-action on S. Moreover, we take S_ρ to be a coadjoint orbit $\mathcal{O}$ in $\mathfrak{h}^*$ (equipped with the Lie symplectic structure), so that J_ρ is simply the inclusion map. The ensuing doubly specialized reduction procedure is called **Marsden–Weinstein**

reduction. The reduced space $S_J^{\mathcal{O}}$ obtained by special symplectic reduction from these data is easily seen to be diffeomorphic to $J^{-1}(\mathcal{O})/H$, which thereby acquires a symplectic structure. The symplectic space $J^{-1}(\mathcal{O})/H$ is called a **Marsden–Weinstein quotient**.

Since H acts on S by Poisson maps, the Poisson bracket on S descends to a Poisson bracket on S/H. However, the latter is not symplectic, unless H is discrete. As a Poisson manifold, S/H is foliated by its symplectic leaves. It turns out that these leaves are precisely the Marsden–Weinstein quotients $J^{-1}(\mathcal{O})/H$. This allows us to see the phase spaces $(T^*\mathsf{P})^{\mathcal{O}}$ of Chapter III in a new light. In particular, when P and H are connected and simply connected, the correspondence $\pi_{\mathcal{O}} \leftrightarrow \pi^{\mathcal{O}}$ between the irreducible representations of $C^\infty(\mathfrak{h}^*_-, \mathbb{R})$ and those of $C^\infty((T^*\mathsf{P})/H, \mathbb{R})$ found in Chapter III comes out as a consequence of the classical imprimitivity theorem.

In a generalization of this construction, which we call **Kazhdan–Kostant–Sternberg reduction**, the inclusion of $\mathcal{O}$ into $\mathfrak{h}^*$ is replaced by a general Poisson map $J_\rho : S_\rho \rightarrow \mathfrak{h}^*_+$, where S_ρ is symplectic. We assume that J_ρ is minus the momentum map of a strongly Hamiltonian H-action on S_ρ. Special symplectic reduction then leads to to a reduced space $(T^*\mathsf{P})^\rho$.

In the special case that $\mathsf{P} = G$ is a Lie group, seen as a principal bundle over $Q = G/H$, we thus obtain a reduced space $(T^*G)^\rho$ for each strongly Hamiltonian H-space S_ρ. This reduced space carries a classical system of imprimitivity. Firstly, the left G-action on T^*G (pulled back from the left action on G) reduces to a G-action on $(T^*G)^\rho$. This yields a Poisson map $J^\rho_{(1)} : (T^*G)^\rho \rightarrow \mathfrak{g}^*_-$. Secondly, since $(T^*G)^\rho$ is a bundle over G/H, one has a map $J^\rho_{(2)} : (T^*G)^\rho \rightarrow G/H$; this is a Poisson map with respect to the zero Poisson structure on G/H. These combine to form a representation of the Poisson algebra $C^\infty(\mathfrak{g}^*_- \times G/H, \mathbb{R})$ of the action algebroid defined by the canonical G-action on G/H.

Without any connectedness assumptions, the **classical transitive imprimitivity theorem** now states that any classical system of imprimitivity for G and G/H, in other words, any representation of the Poisson algebra $C^\infty(\mathfrak{g}^*_- \times G/H, \mathbb{R})$, is equivalent to one of the above form.

So far, we have (tacitly) assumed that S/H and each $J^{-1}(\mathcal{O})/H$ are manifolds. When H is compact this is the case when the H-action on S is free. A fascinating situation arises when one drops this assumption. Without loss of generality, we may restrict ourselves to the case $\mathcal{O} = \{0\}$. It turns out that the reduced space S_J^0 is the (disjoint) union of certain symplectic manifolds $S^0_{[K]}$, each of which corresponds to the conjugacy class $[K]$ of the stabilizer $K \subset H$ of some point in S. The reduced space has a Poisson structure, which restricted to each $S^0_{[K]}$ is equivalent to the symplectic structure of that subspace. Any Hamiltonian flow in $J^{-1}(0)/H$ necessarily stays inside a given subspace $S^0_{[K]}$. In view of the last point, the decomposition of S_J^0 is somewhat reminiscent of the foliation of a Poisson manifold by its symplectic leaves.

Induction

Almost every aspect of special symplectic reduction has a counterpart in the context of Hilbert spaces and C^*-algebras, albeit with subtle changes.

In special symplectic reduction one starts from a Poisson map $J : S \to P^-$. Equivalently, one has a representation $J^* : C^\infty(P^-) \to C^\infty(S, \mathbb{R})$, which may alternatively be regarded as an antirepresentation of $C^\infty(P, \mathbb{R})$. The quantum analogue of the (complexified) Poisson algebra $C^\infty(P)$ is taken to be a C^*-algebra $\mathfrak{B}$; the antirepresentation J^* should then correspond to a right action $\pi_{\mathrm{R}}(\mathfrak{B})$ on a linear space of some sort. These are easy analogies. In the absence of an underlying space P for $\mathfrak{B}$, it is clear that the equivalent classical objects J and J^* should be disentangled in quantum theory.

The quantum counterpart of J in operator theory is a **Hilbert C^*-module** over the C^*-algebra $\mathfrak{B}$. This consists of a complex linear space $\mathcal{E}$, a linear right action π_{R} of $\mathfrak{B}$ on $\mathcal{E}$, and a "$\mathfrak{B}$-valued inner product" $\langle\, ,\, \rangle_{\mathfrak{B}} : \mathcal{E} \times \mathcal{E} \to \mathfrak{B}$.

The sesquilinear form $\langle\, ,\, \rangle_{\mathfrak{B}}$ must firstly satisfy $\langle \Psi, \Phi \rangle_{\mathfrak{B}}^* = \langle \Phi, \Psi \rangle_{\mathfrak{B}}$, generalizing the behavior of an ordinary $\mathbb{C}$-valued inner product under complex conjugation. Secondly, the $\mathfrak{B}$-valued inner product should intertwine π_{R} with the canonical right action of $\mathfrak{B}$ on itself (given by multiplication on the right); in other words, one requires that $\langle \Psi, \pi_{\mathrm{R}}(B)\Phi \rangle_{\mathfrak{B}} = \langle \Psi, \Phi \rangle_{\mathfrak{B}} B$. Furthermore, one imposes positive definiteness, in that $\langle \Psi, \Psi \rangle_{\mathfrak{B}} \geq 0$, with equality iff $\Psi = 0$. It is finally required that $\mathcal{E}$ be complete in the norm $\|\Psi\| := \|\langle \Psi, \Psi \rangle_{\mathfrak{B}}\|^{1/2}$.

For example, $\mathfrak{B}$ is a Hilbert C^*-module over itself, with $\pi_{\mathrm{R}}(B)A := AB$ and $\langle A, B \rangle_{\mathfrak{B}} := A^*B$. Also, a Hilbert space $\mathcal{H}$ is a Hilbert C^*-module over $\mathbb{C}$ in its inner product.

So far, we have stated the first half of the input for "quantum induction". In special symplectic reduction one furthermore has a second Poisson map $J_\rho : S_\rho \to P$, where S_ρ is a symplectic manifold. In quantum theory S_ρ is replaced by a Hilbert space $\mathcal{H}_\chi$. There is no quantum counterpart of J_ρ, but the associated representation $J_\rho^* : C^\infty(P, \mathbb{R}) \to C^\infty(S_\rho, \mathbb{R})$ corresponds to a representation π_χ of $\mathfrak{B}$ on $\mathcal{H}_\chi$.

The construction of the classical reduced space S_J^ρ is replaced by a procedure called **Rieffel induction**. Table 2 presents a summary of the analogy between special symplectic reduction and Rieffel induction, which proceeds as follows. One first equips $\mathcal{E} \otimes \mathcal{H}_\chi$ with a sesquilinear form $(\, ,\,)_0^\chi$, defined by linear extension of

$$(\Psi \otimes v, \Phi \otimes w)_0^\chi := (v, \pi_\chi(\langle \Psi, \Phi \rangle_{\mathfrak{B}})w)_\chi,$$

where $\Psi, \Phi \in \mathcal{E}$ and $v, w \in \mathcal{H}_\chi$. This form is positive semidefinite, because $(\, ,\,)_\chi$ and $\langle\, ,\, \rangle_{\mathfrak{B}}$ are. Then form the quotient of $\mathcal{E} \otimes \mathcal{H}_\chi$ by the null space $\mathcal{N}_\chi$ of $(\, ,\,)_0^\chi$; this is evidently a pre-Hilbert space. The induced space

$$\mathcal{H}^\chi := (\mathcal{E} \otimes \mathcal{H}_\chi / \mathcal{N}_\chi)^-$$

is the completion of $\mathcal{E} \otimes \mathcal{H}_\chi / \mathcal{N}_\chi$ in the inner product inherited from $(\, ,\,)_0^\chi$.

For the quantum counterpart of the reduced representation $(J^\rho)^*(C^\infty(P_2))$ on S_J^ρ in special symplectic reduction, we define the notion of an **adjointable operator**

Special symplectic reduction	*Rieffel induction*
Poisson algebra $C^\infty(P)$	C^*-algebra $\mathfrak{B}$
symplectic manifold S	linear space $\mathcal{E}$
$J^* : C^\infty(P) \to C^\infty(S)$	$\pi_{\mathrm{R}}(\mathfrak{B})$ on $\mathcal{E}$
Poisson map $J : S \to P$	$\mathfrak{B}$-valued inner product $\langle\, ,\rangle_{\mathfrak{B}}$
symplectic manifold S_ρ	Hilbert space $\mathcal{H}_\chi$
representation $J_\rho^* : C^\infty(P) \to C^\infty(S_\rho)$	representation $\pi_\chi : \mathfrak{B} \to \mathfrak{B}(\mathcal{H}_\chi)$
Cartesian product $S \times S_\rho$	tensor product $\mathcal{E} \otimes \mathcal{H}_\chi$
constraint manifolds $S *_P S_\rho$	$\mathcal{E} \otimes \mathcal{H}_\chi$
null foliation Φ_ρ	null space $\mathcal{N}_\chi$
$S_J^\rho = (S *_P S_\rho)/\Phi_\rho$	$\mathcal{H}^\chi = (\mathcal{E} \otimes \mathcal{H}_\chi/\mathcal{N}_\chi)^-$
$J^* C^\infty(P)' \subset C^\infty(S)$	$C^*(\mathcal{E}, \mathfrak{B})$
Poisson algebra $C^\infty(P_2)$	C^*-algebra $\mathfrak{A}$
reduced representation $(J^\rho)^*(C^\infty(P_2))$	induced representation $\pi^\chi(\mathfrak{A})$

TABLE 2. Special symplectic reduction and Rieffel induction

on $\mathcal{E}$. This is an operator A that has an adjoint with respect to the $\mathfrak{B}$-valued inner product; in other words, one has

$$\langle \Psi, A\Phi \rangle_{\mathfrak{B}} = \langle A^*\Psi, \Phi \rangle_{\mathfrak{B}}$$

for some operator A^* on $\mathcal{E}$. The space of all adjointable operators on $\mathcal{E}$ is a C^*-algebra, denoted by $C^*(\mathcal{E}, \mathfrak{B})$. An adjointable operator A has the property that $A \otimes \mathbb{I}_\chi$ maps the null space $\mathcal{N}_\chi$ into itself (here $\mathbb{I}_\chi$ is the unit operator on $\mathcal{H}_\chi$). Hence $A \otimes \mathbb{I}_\chi$ induces an operator on $\mathcal{E} \otimes \mathcal{H}_\chi/\mathcal{N}_\chi$ in a natural way; under suitable boundedness assumptions the latter operator extends to an operator $\pi^\chi(A)$ on $\mathcal{H}^\chi$.

To complete the picture, suppose one has a morphism of a C^*-algebra $\mathfrak{A}$ into $C^*(\mathcal{E}, \mathfrak{B})$. Composing with this morphism, one may (with slight abuse of notation) look at π^χ as a representation of $\mathfrak{A}$ on the induced space $\mathcal{H}^\chi$. This representation is said to be **induced** (in the sense of Rieffel) by the representation $\pi_\chi(\mathfrak{B})$ with respect to the Hilbert C^*-module $\mathcal{E}$ over $\mathfrak{B}$.

With regard to the analogies listed in Table 2, it is remarkable that the constraint manifold of classical mechanics has no quantum counterpart. In other words, in quantum mechanics it is not necessary to impose the constraints (at least in the case that all constraints are first class in the sense of Dirac, which is the case in special symplectic reduction). As opposed to classical reduction, which is a two-step procedure, the construction of the induced space $\mathcal{H}^\chi$ in Rieffel induction has only one step, corresponding to the second step of symplectic reduction.

The physical interpretation of $\mathcal{H}^\chi$ is that it is the physical state space of the system, in which all gauge (and perhaps other unphysical) degrees of freedom have been removed. (Traditional approaches to constrained quantization instead try to mimic the first step of symplectic reduction, imposing the constraints on the Hilbert space of states of the unconstrained system. This has turned out not to work, except in the very simplest examples.)

We now explain the quantum analogue of Morita equivalence of Poisson algebras. In preparation, we need a refinement of the C^*-algebra $C^*(\mathcal{E}, \mathfrak{B})$, which is analogous to the restriction of the C^*-algebra $\mathfrak{B}(\mathcal{H})$ of all bounded operators on a Hilbert space $\mathcal{H}$ to the C^*-algebra $\mathfrak{B}_0(\mathcal{H})$ of all compact operators on $\mathcal{H}$. Namely, we define $C_0^*(\mathcal{E}, \mathfrak{B})$ as the C^*-subalgebra of $C^*(\mathcal{E}, \mathfrak{B})$ that is generated by all operators of the type $T^{\mathfrak{B}}_{\Psi,\Phi}$, where Ψ, $\Phi \in \mathcal{E}$, and

$$T^{\mathfrak{B}}_{\Psi,\Phi} Z := \Psi\langle\Phi, Z\rangle_{\mathfrak{B}}.$$

Two C^*-algebras $\mathfrak{A}$ and $\mathfrak{B}$ are now said to be **Morita equivalent** when there exists a full Hilbert C^*-module $\mathcal{E}$ over $\mathfrak{B}$ under which $\mathfrak{A} \simeq C_0^*(\mathcal{E}, \mathfrak{B})$. We write $\mathfrak{A} \rightleftharpoons \mathcal{E} \rightleftharpoons \mathfrak{B}$. Here a Hilbert C^*-module is called **full** when the collection $\{\langle\Psi, \Phi\rangle_{\mathfrak{B}}\}$, where Ψ, Φ run over $\mathcal{E}$, is dense in $\mathfrak{B}$.

For example, $\mathfrak{B}_0(\mathcal{H})$ is Morita equivalent to $\mathbb{C}$, with $\mathcal{E} = \mathcal{H}$.

As in the classical case, Morita equivalence implies that $\mathfrak{A}$ and $\mathfrak{B}$ have equivalent representation theories, the bijection preserving irreducibility. More precisely, the **quantum imprimitivity theorem** states that every representation of $\mathfrak{A}$ is equivalent to one that is Rieffel-induced from some representation $\pi_\chi(\mathfrak{B})$, and vice versa. The proof uses exactly the same idea as its classical counterpart. The crucial step of turning $P_2 \overset{J_2}{\leftarrow} S \overset{J_1}{\rightarrow} P_1$ around to $P_1 \overset{J_1}{\leftarrow} S^- \overset{J_2}{\rightarrow} P_2$ in the classical proof now works as follows. The conjugate space $\overline{\mathcal{E}}$ is equal to $\mathcal{E}$ as a real vector space, but has the conjugate action of complex scalars. The replacement of S by S^- corresponds to the replacement of $\mathcal{E}$ by $\overline{\mathcal{E}}$. Moreover, the expression

$$\langle\Psi, \Phi\rangle_{C_0^*(\mathcal{E},\mathfrak{B})} := T^{\mathfrak{B}}_{\Psi,\Phi},$$

in combination with the right action $\pi_{\mathrm{R}}(A)\Psi := A^*\Psi$, where $A \in C_0^*(\mathcal{E}, \mathfrak{B})$, defines $\overline{\mathcal{E}}$ as a full Hilbert C^*-module over $C_0^*(\mathcal{E}, \mathfrak{B})$. Similarly, the right action of $\mathfrak{B}$ on $\mathcal{E}$ is turned into a left action on $\overline{\mathcal{E}}$ by acting with the adjoint. Hence $\mathfrak{A} \rightleftharpoons \mathcal{E} \rightleftharpoons \mathfrak{B}$ turns around to $\mathfrak{B} \rightleftharpoons \overline{\mathcal{E}} \rightleftharpoons \mathfrak{A}$.

The theorem is then proved by starting with a representation $\pi_\chi(\mathfrak{B})$, Rieffel-inducing with respect to $\mathfrak{A} \rightleftharpoons \mathcal{E} \rightleftharpoons \mathfrak{B}$ to obtain a representation $\pi^\chi(\mathfrak{A})$, using the latter to construct an induced representation of $\mathfrak{B}$ with respect to $\mathfrak{B} \rightleftharpoons \overline{\mathcal{E}} \rightleftharpoons \mathfrak{A}$, and finally showing that this representation of $\mathfrak{B}$ is equivalent to π_χ. This procedure works in both directions.

In view of the Morita equivalence $\mathfrak{B}_0(\mathcal{H}) \rightleftharpoons \mathcal{H} \rightleftharpoons \mathbb{C}$, an immediate corollary of the quantum imprimitivity theorem is that the C^*-algebra of compact operators has only one irreducible representation.

There is a quantum analogue of Marsden–Weinstein reduction. Instead of a strongly Hamiltonian H-action on a symplectic manifold S, we start from a representation $U(H)$ on a Hilbert space $\mathcal{H}$, and the role of the Poisson algebra $C^\infty(\mathfrak{h}^*_-)$ is now played by the group C^*-algebra $\mathfrak{B} = C^*(H)$. Suppose, for simplicity, that H is compact. We then take $\mathcal{E} = \mathcal{H}$ (actually, $\mathcal{E}$ is a certain completion of $\mathcal{H}$, but we will not bother with this detail), on which $C^*(H)$ acts from the right by

$$\pi_{\mathrm{R}}(f) = \int_H dh\, f(h)U(h)^{-1}.$$

The $C^*(H)$-valued inner product on $\mathcal{H}$ is defined by letting $\langle\Psi, \Phi\rangle_{C^*(H)}$ be the function $h \mapsto (\Psi, U(h)\Phi)$.

We may now proceed with Rieffel induction from some representation π_χ of $C^*(H)$; as we have seen in Chapter III, we may equivalently assume that we have a representation U_χ of H. The form $(\,,\,)_0^\chi$ on $\mathcal{H} \otimes \mathcal{H}_\chi$ reads

$$(\tilde{\Psi}, \tilde{\Phi})_0^\chi = \int_H dh\, (\tilde{\Psi}, U \otimes U_\chi(h)\tilde{\Phi})_{\mathcal{H}\otimes\mathcal{H}_\chi}.$$

Although for compact H this integral may be explicitly computed, we leave the expression as it stands, and remark that it is valid for noncompact groups as well. The only difference with the compact case is that in general $\mathcal{E}$ is a suitably chosen dense subspace of $\mathcal{H}$ (for in the noncompact case the convergence of the H-integration needs attention).

A most interesting instance of "quantum Marsden–Weinstein reduction" arises in the context of a principal H-bundle P. We take $\mathcal{H} = L^2(\mathsf{P})$ (defined with respect to some H-invariant measure), which carries the unitary representation $U(H) := U_R(H)$ naturally constructed from the given right action of H on P. Hence we obtain a right action of $C^*(H)$ on $L^2(\mathsf{P})$, eventually leading to a Hilbert C^*-module $\mathcal{E}$ over $C^*(H)$. The Hilbert space $\mathcal{H}^\chi$ constructed by Rieffel induction from $U_\chi(H)$ is then naturally isomorphic to the space $\mathcal{H}^\chi$ defined earlier in the context of Mackey induction.

We may compute the C^*-algebra $C_0^*(\mathcal{E}, C^*(H))$. Remarkbly, this turns out to be the C^*-algebra $C^*(\mathsf{P} \times_H \mathsf{P})$ of the gauge groupoid of the bundle. It follows that $C^*(\mathsf{P} \times_H \mathsf{P})$ and $C^*(H)$ are Morita equivalent. The bijective correspondence $U_\chi(H) \leftrightarrow \pi^\chi(C^*(\mathsf{P} \times_H \mathsf{P}))$ found in Chapter III then follows from the quantum imprimitivity theorem. Let U_L be the canonical representation of Aut(P) on $L^2(\mathsf{P})$. For each $\varphi \in \text{Aut}(\mathsf{P})$ the operator $U(\varphi)$ commutes with $\pi_{\mathsf{R}}(C^*(H))$, which implies that it is adjointable. The Rieffel-induced representative $\pi^\chi(U(\varphi))$ coincides with the induced representative $U^\chi(\varphi)$ defined in Chapter III.

Specializing to the case where $\mathsf{P} = G$ is a Lie group, and realizing that the action C^*-algebra $C^*(G, G/H)$ is isomorphic to the gauge groupoid C^*-algebra $C^*(G \times_H G)$, we conclude that $C^*(G, G/H)$ and $C^*(H)$ are Morita equivalent. Applied to this situation, the general quantum imprimitivity theorem then implies the **quantum transitive imprimitivity theorem**. To explain what this theorem means, first observe that $\mathcal{H}^\chi$ carries a transitive system of imprimitivity, in which $U(G) = U^\chi(G)$, and $\tilde{\pi}(C_0(G/H))$ is defined by

$$\tilde{\pi}(\tilde{f})\Psi^\chi(x) = \tilde{f}([x]_H)\Psi^\chi(x).$$

The theorem now states that for any system of imprimitivity for G with $Q = G/H$ there exists a representation $U_\chi(H)$ such that the system is equivalent to the one on $\mathcal{H}^\chi$ just defined. This is the exact quantum counterpart of the classical transitive imprimitivity theorem discussed earlier.

Applications in Relativistic Quantum Theory

The most interesting applications of Rieffel's and other induction techniques in physics appear to be to the mathematically rigorous study of gauge field theories. We wish to clarify three aspects of such theories.

Firstly, there is a close (but not universal) relationship between gauge invariance and the masslessness of the field quanta of a quantum gauge theory. A shadow of this relationship may already be seen in the classical theory of massless relativistic particles and fields; infinite-dimensional Marsden–Weinstein reduction will be seen to play a central role.

Secondly, the passage from an unconstrained Yang–Mills theory to its physical sector involves a tremendous reduction in degrees of freedom; when the underlying space is a circle this reduction even leads to a finite-dimensional theory. This will be proved with induction techniques.

Thirdly, it was discovered in the sixties and seventies that the quantization of certain physical systems, notably gauge theories, may involve parameters (beyond $\hbar$) without a classical analogue. These so-called **vacuum angles** or **θ-angles** emerge in a transparent way when one quantizes constrained systems using induced representations.

We start with a description of the coadjoint orbits of the Poincaré group. This group is the semidirect product $P = L \ltimes_\rho \mathcal{M}$ of the Lorentz group $L = SO(3, 1)$ and the additive group $\mathcal{M} = \mathbb{R}^4$ (equipped with the Minkowski metric $\mathrm{diag}\,(1, -1, -1, -1)$, of which L is the connected isometry group). The action ρ of L on $\mathcal{M}$ with respect to which the semidirect product is formed is simply the defining action of $SO(3, 1)$ on $\mathbb{R}^4$. A central role in the description of the coadjoint orbits of such semidirect products is played by the dual action $\rho^*(L)$ on $\mathcal{M}^* \simeq \mathbb{R}^4$.

A nontrivial analysis shows that each coadjoint orbit $\mathcal{O}^P$ in $\mathfrak{p}^*$ is isomorphic (as a symplectic manifold) to a Marsden–Weinstein quotient of the type $(T^*\mathsf{P})^{\mathcal{O}}$, which we have encountered before in a different context. Here we have to take $\mathsf{P} = L$, whereas $\mathcal{O}$ is a coadjoint orbit of the stabilizer $L_{\tilde{p}}$ of some point $\tilde{p}$ in $\mathcal{M}^*$ under the action $\rho^*(L)$. Hence the orbits are fibered over $T^*(L/L_{\tilde{p}})$, with typical fiber $\mathcal{O}$.

The phase space of a massless relativistic particle with positive energy is obtained by choosing $\tilde{p} = (1, 0, 0, -1)$. Its stabilizer is isomorphic to the Euclidean group $E(2) := SO(2) \ltimes_\rho \mathbb{R}^2$ in dimension 2. Hence classical massless particles are further classified by the coadjoint orbits $\mathcal{O}$ of $E(2)$. The dual of the Lie algebra of $E(2)$ is $\mathbb{R}^3$, whose coadjoint orbits are either cylinders $C_r = S^1_r \times \mathbb{R}$ (where the circle S^1_r of radius $r > 0$ lies in the (x, y)-plane, and $\mathbb{R}$ is the z-axis), or points $(0, 0, h)$.

Only the latter are believed to be of physical relevance; the parameter h is called the **helicity** of the particle. For example, a classical photon has helicity 1 or -1, and a classical graviton has helicity ± 2. The phase space $\mathcal{O}^P_{0,+,h}$ of such particles is diffeomorphic to $T^*(L/E(2))$, but the Poisson bracket contains an additional term (beyond the canonical cotangent bundle bracket) proportional to h.

There is an almost complete parallel between the coadjoint orbits and the irreducible representations of P. The only difference lies in the fact that the latter are classified by the irreducible representations of the stabilizers $L_{\tilde{p}}$, rather than by their coadjoint orbits. Relativistic massless quantum particles are therefore classified by the irreducible representations of $E(2)$. The physically relevant representations $U^{0,+,h}$ are again labeled by the helicity h, which in quantum theory assumes only (half-) integral values. The Hilbert space $\mathcal{H}^{0,+,h}$ carrying $U^{0,+,h}$ is $L^2(L/E(2))$.

In a remarkable twist of nature and mathematics, the physically relevant irreducible representations of P describe both quantized particles and classical fields. However, the massless relativistic fields occurring in the Lagrangians and Hamiltonians of classical field theory do not transform under $U^{0,+,h}$, but under a so-called **covariant representation** of P. This is a (generally nonunitary) representation $\mathcal{R}^\lambda$ that is (Mackey) induced from a (nonunitary) representation of L. The lack of unitarity does not matter for classical physics, since the "covariant" action of P on the space of fields should be seen in a symplectic context; it is, indeed, strongly Hamiltonian.

Gauge fields A transform under the covariant **vector representation** $\mathcal{R}^V(P)$, defined by

$$\mathcal{R}^V(\Lambda, v)A^\mu(q) = \Lambda^\mu{}_\nu A^\nu(\Lambda^{-1}(q - v)).$$

In order to reach $U^{0,+,\pm 1}$, as a first step one imposes the infinite number of constraints $\Box A^\mu = 0$ on the space S^V of all gauge fields, and performs symplectic reduction. This leads to a symplectic space $S^{0,+,V}$, whose configuration space part consists of all solutions of the above wave equation whose Cauchy data are square-integrable in a suitable sense.

The second step of the passage from S^V to $\mathcal{H}^{0,+,\pm 1}$, then, involves the **gauge group** $\mathcal{G}$. This is the real Hilbert space of real solutions λ of the wave equation $\Box\lambda = 0$ on $\mathcal{M}$ whose (weak) derivative $\partial\lambda$ (seen as a four-vector with components $\partial_\mu\lambda$) lies in $S^{0,\mathbb{R},V}$. The connection between gauge invariance and masslessness in classical free field theories is now as follows. The gauge group acts on $S^{0,\mathbb{R},V}$ by

$$\lambda : A_\mu \mapsto A_\mu + \partial_\mu\lambda;$$

this action is strongly Hamiltonian, with momentum map J, and the Marsden–Weinstein quotient $J^{-1}(0)/\mathcal{G}$ is $\mathcal{H}^{0,+,1} \oplus \mathcal{H}^{0,+,-1}$. Moreover, the reduction of the covariant action $\mathcal{R}^V(P)$ on S^V to $S^{0,\mathbb{R},V}$ further reduces to an action on $J^{-1}(0)/\mathcal{G}$, which coincides with the representation $U^{0,+,1} \oplus U^{0,+,-1}$.

We now quantize this reduction procedure with the aid of a generalization of the quantum Marsden–Weinstein induction technique, which is suitable for dealing with infinite-dimensional groups. We start as if the gauge group were locally compact, and consider a Hilbert space $\mathcal{H}$ carrying a representation $U(\mathcal{G})$. To construct $\mathcal{H}$ we exploit the fact that S^V, previously looked upon as a symplectic space, may be turned into a Hilbert space $\mathcal{H}^V$. For $\mathcal{H}$ we then take the **bosonic Fock space**

$$\exp(\mathcal{H}^V) := \bigoplus_{l=0}^{\infty} \otimes_s^l \mathcal{H}^V,$$

where $\otimes_s^l \mathcal{H}^V$ is the symmetrized tensor product of l copies of $\mathcal{H}^V$. We define a map $\sqrt{\mathrm{Exp}} : \mathcal{H}^V \to \exp(\mathcal{H}^V)$ by

$$\sqrt{\mathrm{Exp}}(A) := \sum_{l=0}^{\infty} \frac{\otimes^l A}{\sqrt{l!}};$$

it follows that the inner product of two exponential vectors is

$$\left(\sqrt{\mathrm{Exp}}(A), \sqrt{\mathrm{Exp}}(B)\right) = e^{(A,B)_{\mathcal{H}^V}}.$$

Since the exponential vectors are dense in $\mathcal{H}$, the natural representation $U(\mathcal{G})$ we use is characterized by its matrix elements

$$\left(\sqrt{\mathrm{Exp}}(A), U_F(\lambda)\sqrt{\mathrm{Exp}}(B)\right) = e^{(A,B)_{\mathcal{H}^V}} e^{-\frac{1}{2}\|\lambda\|^2_{\mathcal{H}^V}} e^{(A,\partial\lambda)_{\mathcal{H}^V} - (\partial\lambda, B)_{\mathcal{H}^V}}.$$

Mimicking the finite-dimensional situation, we would like to integrate the above function of λ over $\mathcal{G}$ with respect to a $\mathcal{G}$-invariant "Lebesgue" measure. This is impossible, but fortunately one may combine the nonexistent Lebesgue measure on $\mathcal{G}$ with the factor $\exp(-\frac{1}{2}\|\lambda\|^2_{\mathcal{H}^V})$. This combination yields a Gaussian measure μ on a certain completion $\mathcal{G}_c$ of $\mathcal{G}$. We then put

$$\left(\sqrt{\mathrm{Exp}}(A), \sqrt{\mathrm{Exp}}(B)\right)_0^{\mathrm{id}} := e^{(A,B)_{\mathcal{H}^V}} \int_{\mathcal{G}_c} d\mu(\lambda) e^{(A,\partial\lambda)_{\mathcal{H}^V} - (\partial\lambda, B)_{\mathcal{H}^V}}.$$

One may proceed with the construction of the induced space $\mathcal{H}^{\mathrm{id}}$ as usual, obtaining the correct quantum field theory of photons. In particular, the gauge group $\mathcal{G}$ is trivially represented in $\mathcal{H}^{\mathrm{id}}$, and Gauss's law is satisfied.

Following this treatment of the connection between masslessness and gauge invariance in classical and quantum electromagnetism, we turn to a different class of models for a discussion of the remaining two points of interest in gauge theories.

Classical Yang–Mills theory on a circle with structure group H is defined by the configuration space $\mathcal{A}_{\mathbb{R}} = L^2(S^1, \mathfrak{h})$, with phase space

$$S = T^*\mathcal{A}_{\mathbb{R}} \simeq \mathcal{A} = L^2(S^1, \mathfrak{h}_{\mathbb{C}}).$$

Here the inner product in L^2 is defined with respect to an $\mathrm{Ad}(H)$-invariant inner product on $\mathfrak{h}$. The **gauge group** $\mathcal{G}$ of the model is the Sobolev loop group $\mathcal{H}_1(S^1, H)$, consisting of those $g \in C(S^1, H)$ whose (weak) derivative $\dot{g} := g^{-1}dg/d\alpha$ lies in $\mathcal{A}_{\mathbb{R}}$. These definitions guarantee that the action

$$g : A \mapsto A^g := \mathrm{Ad}(g)A + gdg^{-1} = g(A - \dot{g})g^{-1}$$

of $\mathcal{G}$ on $\mathcal{A}_{\mathbb{R}}$ is smooth. This action lifts to a strongly Hamiltonian action on the phase space $\mathcal{A}$, given by the same formula (with A replaced by a complex connection Z).

The point is now that the Marsden–Weinstein quotient $S^0 = J^{-1}(0)/\mathcal{G}$ (where J the momentum map of the above $\mathcal{G}$-action) is symplectomorphic to the finite-dimensional cotangent bundle $T^*(H/\mathrm{Ad}(H))$. The identification of this reduced space with the physical phase space rests on the fact that $J^{-1}(0)$ is the subspace of S on which Gauss's law holds. The key element of the proof of this symplectomorphism is the construction of the **Wilson loop** $\mathcal{W} : \mathcal{A}_{\mathbb{R}} \to H$. We first define $\hat{\mathcal{W}} : L^2(S^1, \mathfrak{h}) \to C([0,1], H)$ as the solution of the differential equation

$$\left(\frac{\partial}{\partial\alpha} + A\right)\hat{\mathcal{W}}_A(\alpha) = 0,$$

with initial condition $\hat{\mathcal{W}}_A(0) = e$ (here $\hat{\mathcal{W}}_A := \hat{\mathcal{W}}(A)$). With $\mathcal{W}(A) := \hat{\mathcal{W}}_A(1)$, one shows that $\mathcal{W}(A^g) = \mathcal{W}(A)$ for all based gauge transformations $g \in \mathcal{G}_e$ (i.e., $g(0) = e$). Hence $\mathcal{W}$ quotients to a map from $\mathcal{A}_{\mathbb{R}}/\mathcal{G}_e$ to H; the peculiar feature of the model is that this map is a diffeomorphism. Moreover, $\mathcal{W}$ complexifies to a map $\mathcal{W}_{\mathbb{C}}$ from the phase space $\mathcal{A}$ to the complexification $H_{\mathbb{C}}$ of H. Since $H_{\mathbb{C}}$ is diffeomorphic to T^*H, the map $\mathcal{W}_{\mathbb{C}}$ restricts and quotients to a symplectomorphism between S^0 and T^*H.

The quantization of this reduction procedure follows the lines of our earlier treatment of photons. The unconstrained phase space S is quantized as the bosonic Fock space $\mathcal{H} := \exp(\mathcal{A})$, on which the gauge group is represented by

$$U_F(g)\sqrt{\mathrm{Exp}}(Z) := e^{-\frac{1}{2}\|\dot{g}\|^2 + (\dot{g}, Z)}\sqrt{\mathrm{Exp}}(Z^g).$$

Hence the matrix element of $U_F(g)$ between two exponential vectors again contains a Gaussian factor $\exp(-\frac{1}{2}\|\dot{g}\|^2)$, which we wish to combine with the nonexistent Haar measure on $\mathcal{G}$. This leads to a version of the well-known **Wiener measure** μ^W, conditioned to the space of continuous loops on H. We may therefore put

$$\left(\sqrt{\mathrm{Exp}}(W), \sqrt{\mathrm{Exp}}(Z)\right)_0^{\mathrm{id}} := \int_{LH} d\mu^W(g)\, e^{(W, Z^g) + (\dot{g}, Z)}.$$

For technical reasons the integral is over $LH = C(S^1, H)$ rather than over the gauge group, which is a sup-dense subspace of LH that happens to have μ^W-measure zero.

The induced space $\mathcal{H}^{\mathrm{id}}$ defined by induction with respect to the above form is naturally isomorphic to the subspace of $L^2(H)$ that is invariant under the representation defined by the adjoint action. In fact, replacing LH in the above integral by the space LH_e of based loops, the induced space $\mathcal{H}_e^{\mathrm{id}}$ is $L^2(H)$ itself. A function $f \in C^\infty(H)$ defines $\mathcal{W}_f \in C^\infty(\mathcal{A})$ by $\mathcal{W}_f(A) := f(\mathcal{W}(A))$. The quantization of the observable $\mathcal{W}_f$ on $L^2(H)$ then comes out to be the multiplication operator f. When f is a class function, this operator has a well-defined restriction to $\mathcal{H}^{\mathrm{id}}$.

The identification of $\mathcal{H}_e^{\mathrm{id}}$ with $L^2(H)$ makes essential use of the **Hall coherent states** $\tilde{\Psi}_\hbar^z$ in $L^2(H)$; this is a recently discovered family of coherent states that is labeled by the points z in $H_{\mathbb{C}}$. The complexified Wilson loop $\mathcal{W}_{\mathbb{C}} : \mathcal{A} \to H_{\mathbb{C}}$ of the classical theory has a quantum counterpart, which (up to normalization) maps $\sqrt{\mathrm{Exp}}(Z)$ to $\tilde{\Psi}_{1/2}^{\mathcal{W}(Z)}$.

We finally turn to vacuum angles. In view of the equalities

$$\pi_0(\mathcal{G}) := \mathcal{G}/\mathcal{G}^0 = \pi_0(LH) = \pi_1(H),$$

where $\mathcal{G}^0$ is the identity component of $\mathcal{G}$, the gauge group is disconnected whenever the first homotopy group of the structure group H is nontrivial. For example, in the abelian case $H = U(1)$ one has $\pi_0(\mathcal{G}) = \pi_1(U(1)) = \mathbb{Z}$. This motivates us to make some general comments on the quantization of Marsden–Weinstein quotients by a disconnected group $\mathcal{G}$.

The space $S^0 = J^{-1}(0)/\mathcal{G}$ may be constructed in two steps: one firstly forms $J^{-1}(0)/\mathcal{G}^0$, which is a symplectic space. Secondly, one quotients the latter by the discrete group $\pi_0(\mathcal{G})$, again obtaining a symplectic space, which is isomorphic to S^0. We isolate the second step. Although on the classical side $\pi_0(\mathcal{G})$ possesses only the trivial coadjoint orbit $\{0\}$, on the quantum side it will have nontrivial irreducible representations, which have no classical counterpart. A **vacuum angle** is an element of the unitary dual $\widehat{\pi_0(\mathcal{G})}$; for Yang–Mills theory on a circle this is the same as $\widehat{\pi_1(H)}$. For $H = U(1)$ one therefore finds $\widehat{\mathcal{G}} = \hat{\mathbb{Z}} = U(1)$, explaining the alternative name **θ-angles**.

Pick a $\theta \in \widehat{\pi_0(\mathcal{G})}$, with corresponding representation $U_\theta(\pi_0(\mathcal{G}))$. Instead of forming the physical state space by induction from the trivial representation of the gauge group, as we have done so far, we have the freedom of inducing from the representation $\tilde{U}_\theta(\mathcal{G})$, derived from U_θ via the canonical projection from $\mathcal{G}$ to $\mathcal{G}/\mathcal{G}^0$. The quantum observables of the gauge-invariant system, such as the physical Hamiltonian, then explicitly depend on θ, since these operators are constructed by an induction procedure that depends on θ. Hence one obtains a different physical theory for each $\theta \in \widehat{\pi_0(\mathcal{G})}$. In other words, the gauge-invariant theory admits **inequivalent quantizations**, classified by θ.

The θ-dependence may be shown quite explicitly in the $U(1)$ gauge theory on a circle. As we have seen, the reduced classical theory of this model describes a particle moving on $U(1)$; the corresponding θ-dependent quantum theory turns out to be a description of the **Aharonov–Bohm effect**.

CHAPTER I

Observables and Pure States

1 The Structure of Algebras of Observables

1.1 Jordan–Lie Algebras and C^*-Algebras

In this section we specify the key algebraic and functional-analytic structures relevant to classical and quantum mechanics. Our main aim is to look at a C^*-algebra from the point of view of its self-adjoint part. From this perspective the relationship between the respective algebraic structures of classical and quantum mechanics is particularly transparent.

Recall that an **algebra** is a vector space with a (not necessarily commutative or associative) bilinear and distributive multiplication $\circ$. We write $A^2 := A \circ A$.

Definition 1.1.1. *A (real)* **Jordan algebra** *is a (real) algebra where*

$$A \circ B = B \circ A; \tag{1.1}$$

$$A \circ (B \circ A^2) = (A \circ B) \circ A^2. \tag{1.2}$$

The simplest motivation for (1.2) is that it is automatically satisfied when the Jordan product $\circ$ comes from an associative product via $A \circ B = \frac{1}{2}(AB + BA)$. However, not all Jordan algebras arise in this way.

Theories of dynamical significance have a second algebraic operation.

Definition 1.1.2. *A* **Jordan–Lie algebra** *is a real vector space $\mathfrak{A}_{\mathbb{R}}$ equipped with two bilinear maps $\circ$ and $\{\ ,\ \}$ (referred to as the* **Jordan product** *and the* **Poisson bracket***, respectively), such that the following conditions are satisfied. Firstly, one has*

$$A \circ B = B \circ A;$$

$$\{A, B\} = -\{B, A\} \tag{1.3}$$

for all $A, B \in \mathfrak{A}_{\mathbb{R}}$. *Secondly, for each* A, *the map* $B \mapsto \{A, B\}$ *is a derivation of* $(\mathfrak{A}_{\mathbb{R}}, \circ)$ *as well as of* $(\mathfrak{A}_{\mathbb{R}}, \{\ ,\ \})$. *This means that the* **Leibniz rule**

$$\{A, B \circ C\} = \{A, B\} \circ C + B \circ \{A, C\} \tag{1.4}$$

as well as the **Jacobi identity**

$$\{A, \{B, C\}\} = \{\{A, B\}, C\} + \{B, \{A, C\}\} \tag{1.5}$$

must hold for all $A, B, C \in \mathfrak{A}_{\mathbb{R}}$. *Finally, for all* $A, B, C \in \mathfrak{A}_{\mathbb{R}}$ *and some* $\hbar^2 \in \mathbb{R}$ *one requires the* **associator identity**

$$(A \circ B) \circ C - A \circ (B \circ C) = \tfrac{1}{4}\hbar^2\{\{A, C\}, B\}. \tag{1.6}$$

A Jordan–Lie algebra in which $\circ$ *is associative is called a* **Poisson algebra**.

It follows from these axioms that $(\mathfrak{A}_{\mathbb{R}}, \circ)$ is a real Jordan algebra, whereas $(\mathfrak{A}_{\mathbb{R}}, \{\ ,\ \})$ is a real Lie algebra. In connection with Jordan–Lie algebras, the terminology (non) associative always refers to the Jordan product.

The following definitions are recorded for later use.

Definition 1.1.3. *A* **Jordan morphism** *between Jordan–Lie algebras* $\mathfrak{A}_{\mathbb{R}}$ *and* $\mathfrak{B}_{\mathbb{R}}$ *is a linear map* $\beta : \mathfrak{A}_{\mathbb{R}} \to \mathfrak{B}_{\mathbb{R}}$ *satisfying* $\beta(A \circ B) = \beta(A) \circ \beta(B)$ *for all* $A, B \in \mathfrak{A}_{\mathbb{R}}$. *Similarly, a* **Poisson morphism** *between such algebras is a linear map satisfying* $\beta(\{A, B\}) = \{\beta(A), \beta(B)\}$. *A map between Jordan–Lie algebras that is simultaneously a Jordan morphism and a Poisson morphism is called simply a* **morphism**. *An invertible (Jordan, Poisson) morphism* $\alpha : \mathfrak{A}_{\mathbb{R}} \to \mathfrak{A}_{\mathbb{R}}$ *is called a* **(Jordan, Poisson) automorphism**, *and an invertible (Jordan, Poisson) morphism* $\alpha : \mathfrak{A}_{\mathbb{R}} \to \mathfrak{B}_{\mathbb{R}}$ *is a (Jordan, Poisson)* **isomorphism**.

We now equip the algebras introduced above with a norm.

Definition 1.1.4. *A* JB**-algebra** *is simultaneously a real Jordan algebra and a Banach space in which for all* $A, B \in \mathfrak{A}_{\mathbb{R}}$ *one has*

$$\|A \circ B\| \leq \|A\|\,\|B\|; \tag{1.7}$$

$$\|A\|^2 \leq \|A^2 + B^2\|. \tag{1.8}$$

The motivation for the axioms of a JB-algebra will emerge in due course. Putting $A = B$ in (1.7) and $B = 0$ in (1.8), one sees that given (1.7), axiom (1.8) is equivalent to the pair

$$\|A^2\| = \|A\|^2; \tag{1.9}$$

$$\|A^2\| \leq \|A^2 + B^2\|. \tag{1.10}$$

Definition 1.1.5. *A* JLB**-algebra** *is a* JB*-algebra* $\mathfrak{A}_{\mathbb{R}}$ *equipped with a Poisson bracket that makes it a Jordan–Lie algebra for some* $\hbar^2 \geq 0$.

A JLB-algebra with $\hbar = 0$ may alternatively be regarded as a Poisson algebra with zero Poisson structure. A Poisson algebra with nonzero Poisson bracket cannot, in general, be normed in such a way that the bracket is defined on the norm-completion of the algebra. For this reason Poisson algebras are usually not studied in the setting of Banach spaces.

A JLB-algebra $\mathfrak{A}_{\mathbb{R}}$ turns out to be the real part of a complex associative algebra $\mathfrak{A}$ of a much-studied type.

An **involution** on a complex algebra is a real-linear map $A \mapsto A^*$ such that for all $A, B \in \mathfrak{A}$ and $\lambda \in \mathbb{C}$ one has

$$A^{**} = A; \tag{1.11}$$

$$(AB)^* = B^* A^*; \tag{1.12}$$

$$(\lambda A)^* = \overline{\lambda} A^*. \tag{1.13}$$

A ***-algebra** is a complex associative algebra with an involution.

Definition 1.1.6. *A* C^***-algebra** *is a complex Banach space* $\mathfrak{A}$ *that is at the same time a* **-algebra, such that for all* $A, B \in \mathfrak{A}$ *one has*

$$\|AB\| \leq \|A\| \; \|B\|; \tag{1.14}$$

$$\|A^* A\| = \|A\|^2. \tag{1.15}$$

Combining the two axioms for a C^*-algebra leads to $\|A\| \leq \|A^*\|$; replacing A by A^* and using (1.11) yields

$$\|A^*\| = \|A\|. \tag{1.16}$$

It can actually be shown that (1.15) implies (1.14), but this highly nontrivial fact distracts from the guiding idea that a C^*-algebra is a specialization of a **Banach algebra**. This is a complex Banach space and an associative algebra, in which all A, B satisfy (1.14). This property guarantees that left and right multiplication are bounded, hence continuous; in fact, multiplication is a jointly continuous operation. For example, the space $\mathfrak{B}(\mathcal{B})$ of all linear maps on a Banach space $\mathcal{B}$ is a Banach algebra under the norm

$$\|A\| := \sup\{\|A\Psi\| \mid \Psi \in \mathcal{B}, \; \|\Psi\| = 1\}. \tag{1.17}$$

The C^*-axioms are motivated by the following example. Consistent with the above terminology, a ***-algebra of bounded operators** on some Hilbert space $\mathcal{H}$ is a collection of bounded operators on $\mathcal{H}$ that is closed under addition, scalar multiplication, operator multiplication, and taking adjoints. Thus the role of the involution is played by the adjoint. Recall the definition of the norm of a bounded operator:

$$\|A\|^2 := \sup\{(A\Psi, A\Psi) \mid \Psi \in \mathcal{H}, \; (\Psi, \Psi) = 1\}. \tag{1.18}$$

Since in a Hilbert space the norm is defined by $\|\Psi\|^2 = (\Psi, \Psi)$, eq. (1.18) is evidently a special case of (1.17). Hence $\|A\Psi\| \leq \|A\| \; \|\Psi\|$, which implies (1.14). Moreover, we estimate

$$\|A\Psi\|^2 = (A\Psi, A\Psi) = (\Psi, A^* A\Psi) \leq \|A^* A\| \; \|\Psi\|^2,$$

so that $\|A\|^2 \leq \|A^*A\|$, which is $\leq \|A^*\| \, \|A\|$ by (1.14). This leads to (1.16) by the argument preceding that equation; the ensuing inequality $\|A^*A\| \leq \|A\|^2$ then implies (1.15).

Definition 1.1.7. *A* **morphism** *between C^*-algebras $\mathfrak{A}, \mathfrak{B}$ is a linear map $\varphi : \mathfrak{A} \to \mathfrak{B}$ such that*

$$\varphi(AB) = \varphi(A)\varphi(B); \tag{1.19}$$

$$\varphi(A^*) = \varphi(A)^* \tag{1.20}$$

for all $A, B \in \mathfrak{A}$. An **isomorphism** *is a bijective morphism. Two C^*-algebras are* **isomorphic** *when there exists an isomorphism between them.*

It is clear that a C^*-algebra morphism between $\mathfrak{A}$ and $\mathfrak{B}$ restricts to a morphism (in the sense of 1.1.3) between the associated JLB-algebras $\mathfrak{A}_{\mathbb{R}}$ and $\mathfrak{B}_{\mathbb{R}}$ (cf. 1.1.9), and vice versa. Morphisms between C^*-algebras have excellent properties; see 1.3.10. For example, an isomorphism is automatically isometric.

Theorem 1.1.8. *A norm-closed *-algebra $\mathfrak{A}$ in $\mathfrak{B}(\mathcal{H})$ is a C^*-algebra (with operator multiplication as the product, etc.). Conversely, any C^*-algebra is isomorphic to a norm-closed *-algebra in $\mathfrak{B}(\mathcal{H})$, for some Hilbert space $\mathcal{H}$.*

The computation following (1.18) establishes the first half. The proof of the converse will be given at the end of 1.5. □

An element A of a *-algebra $\mathfrak{A}$ for which $A^* = A$ is called **self-adjoint**. The **self-adjoint part** $\mathfrak{A}_{\mathbb{R}}$ is the collection of all self-adjoint elements in $\mathfrak{A}$, seen as a real vector space. Since one may write

$$A = A' + iA'' := \tfrac{1}{2}(A + A^*) + i\tfrac{-i}{2}(A - A^*), \tag{1.21}$$

every element of $\mathfrak{A}$ is a linear combination of two self-adjoint elements.

A **commutative C^*-algebra** is a C^*-algebra in which the associative multiplication is commutative. The connection between JB-algebras, Jordan–Lie algebras, and C^*-algebras is as follows.

Theorem 1.1.9. *Let $\mathfrak{A}$ be a C^*-algebra, and choose $\hbar \in \mathbb{R}\backslash\{0\}$. Equipped with the norm inherited from $\mathfrak{A}$, and the operations*

$$A \circ B := \tfrac{1}{2}(AB + BA);$$

$$\{A, B\}_\hbar := \frac{i}{\hbar}[A, B], \tag{1.22}$$

the self-adjoint part $\mathfrak{A}_{\mathbb{R}}$ of $\mathfrak{A}$ is a JLB-algebra. When $\mathfrak{A}$ is noncommutative, the parameter $\hbar$ in (1.6) equals $\hbar$ in (1.22); in particular, one has $\hbar^2 > 0$. When $\mathfrak{A}$ is commutative, $\mathfrak{A}_{\mathbb{R}}$ is a Poisson algebra with zero Poisson bracket.

Conversely, given a JLB-algebra $\mathfrak{A}_{\mathbb{R}}$ for which $\hbar^2 \geq 0$, its complexification $\mathfrak{A}$ is a C^*-algebra under the operations

$$AB := A \circ B - \tfrac{1}{2}i\hbar\{A, B\}; \tag{1.23}$$

$$(A + iB)^* := A - iB; \tag{1.24}$$

$$\|A\| := \|A^*A\|^{\frac{1}{2}}. \tag{1.25}$$

In (1.24) we assume that $A, B \in \mathfrak{A}_{\mathbb{R}}$, and concerning (1.25) we remark that $A^*A \in \mathfrak{A}_{\mathbb{R}}$ for any $A \in \mathfrak{A}$.

To prove the first half of the theorem, first note that by (1.16) the involution in $\mathfrak{A}$ is continuous, so that $\mathfrak{A}_{\mathbb{R}}$ is a closed subspace of $\mathfrak{A}$. The axioms for a JLB-algebra are trivially verified, except (1.10). We defer the proof of this property to the end of 1.4.

In the opposite direction, it is trivially verified that the product (1.23) is associative as a consequence of the properties of a Jordan–Lie algebra. When (1.25) defines a norm, the property (1.15) holds by construction.

When $\mathfrak{A}_{\mathbb{R}}$ is associative, so that $\mathfrak{A}$ is commutative, the norm (1.25) on $\mathfrak{A}$ simply becomes $\|A + iB\| = \|A^2 + B^2\|^{\frac{1}{2}}$, where $A, B \in \mathfrak{A}_{\mathbb{R}}$. All axioms for a norm are then easily derived from (1.9) and (1.10).

The proof that (1.25) is a norm also in the noncommutative case, as well as the proof of (1.14), will be given at the end of 1.4, too. □

One could replace the minus sign on the right-hand side of (1.23) by a plus sign; that choice leads to a C^*-algebra as well, which is anti-isomorphic to the one based on the minus sign.

1.2 Spectrum and Commutative C^*-Algebras

We are going to examine to what extent the closely related notions of spectrum and functional calculus of a (bounded) self-adjoint operator A on a Hilbert space $\mathcal{H}$ generalize to the context of C^*-algebras and JLB-algebras. On the way we discuss the structure of commutative C^*-algebras. In this section we do not use Theorem 1.1.9, except in the commutative case, for the outstanding part of the proof of this theorem will depend on some of the results below.

Recall that the spectrum $\sigma(A)$ of $A \in \mathfrak{B}(\mathcal{H})$ is the set of those $z \in \mathbb{C}$ for which $A - z\mathbb{I}$ has no (bounded two-sided) inverse; when A is self-adjoint, the **spectral radius** $r(A)$ appears in the fundamental equality

$$\|A\| = r(A) := \sup\{|z| \mid z \in \sigma(A)\}. \tag{1.26}$$

Since the presence of a unit is crucial in these definitions, we have to go through the following considerations. A **unit** $\mathbb{I}$ in a JB-algebra $\mathfrak{A}_{\mathbb{R}}$ is an element such that $A \circ I = A$ for all $A \in \mathfrak{A}_{\mathbb{R}}$; a JB-algebra with a unit is called **unital**. When $\mathfrak{A}_{\mathbb{R}}$ is a JLB-algebra, its unit becomes a unit of the C^*-algebra $\mathfrak{A}$, in that $\mathbb{I}A = A\mathbb{I} = A$ for all $A \in \mathfrak{A}$. This follows by putting $B = C = \mathbb{I}$ in (1.4), implying $\{A, \mathbb{I}\} = 0$ for all A, and subsequently applying (1.23). In any case, taking the adjoint of $\mathbb{I}^*\mathbb{I} = \mathbb{I}^*$ yields $\mathbb{I}^*\mathbb{I} = \mathbb{I}$; hence $\mathbb{I}^* = \mathbb{I}$. Also, (1.15) then implies $\|\mathbb{I}\| = 1$.

When a C^*-algebra or a JLB-algebra has no unit, one may add one.

Proposition 1.2.1. *For every C^*-algebra without unit there exists a unique unital C^*-algebra $\mathfrak{A}_{\mathbb{I}}$, called the* **unitization** *of $\mathfrak{A}$, and an isometric (hence injective) morphism $\mathfrak{A} \to \mathfrak{A}_{\mathbb{I}}$, such that $\mathfrak{A}_{\mathbb{I}}/\mathfrak{A} \simeq \mathbb{C}$.*

Let $\mathfrak{B}(\mathfrak{A})$ be the Banach algebra of all bounded linear maps on $\mathfrak{A}$. Whether or not $\mathfrak{A}$ is unital, the map $\rho : \mathfrak{A} \to \mathfrak{B}(\mathfrak{A})$, given by

$$\rho(A)B := AB, \tag{1.27}$$

is isometric. To see this, note that $\|\rho(A)\| \leq \|A\|$ by (1.14), whereas the opposite inequality follows from (1.15).

Now let $\mathfrak{A}$ be a nonunital C^*-algebra, and form $\mathfrak{A}_{\mathbb{I}} := \mathfrak{A} \oplus \mathbb{C}$. Extend ρ to $\mathfrak{A}_{\mathbb{I}}$ by $\rho(A\dot{+}z)B := AB + zB$, so that $\rho(0\dot{+}1) = \mathbb{I}$ (the unit in $\mathfrak{B}(\mathfrak{A})$). Equipped with the norm (1.17) and the algebraic structure of $\mathfrak{B}(\mathfrak{A})$, and with the involution $\rho(A\dot{+}z)^* := \rho(A^*) + \overline{z}\mathbb{I}$, the vector space $\rho(\mathfrak{A}_{\mathbb{I}})$ is easily shown to be a unital C^*-algebra. Since ρ is a vector space isomorphism between $\mathfrak{A}_{\mathbb{I}}$ and $\rho(\mathfrak{A}_{\mathbb{I}})$, one may transfer the C^*-algebraic structure on the latter to $\mathfrak{A}_{\mathbb{I}}$. Restricted to $\mathfrak{A} \subset \mathfrak{A}_{\mathbb{I}}$, one recovers $\mathfrak{A}$ as a C^*-algebra. Uniqueness follows from 1.2.4.4 below. □

Definition 1.2.2. *Let $\mathfrak{A}$ be either a unital C^*-algebra, or the complexification of a unital JLB-algebra $\mathfrak{A}_{\mathbb{R}}$. The* **spectrum** *$\sigma(A)$ of $A \in \mathfrak{A}$ is defined as the set of those $z \in \mathbb{C}$ for which $A - z\mathbb{I}$ has no (two-sided) inverse in $\mathfrak{A}$.*

When $\mathfrak{A}$ is nonunital, one puts $\sigma(A) := \sigma_{\mathbb{I}}(A\dot{+}0)$, where $\sigma_{\mathbb{I}}$ stands for the spectrum in the unitization of $\mathfrak{A}$.

In the nonunital case 0 always lies in $\sigma(A)$, as it is obvious from 1.2.1 that $A \in \mathfrak{A}$ never has an inverse in $\mathfrak{A}_{\mathbb{I}}$. The theory of Banach algebras shows that $\sigma(A)$ is a compact subset of $\mathbb{C}$. For later use we note that

$$\sigma(zA) = z\sigma(A); \tag{1.28}$$

$$\sigma(AB) \cup \{0\} = \sigma(BA) \cup \{0\} \tag{1.29}$$

for all $A, B \in \mathfrak{A}$; the first property is obvious, and the second follows, because for $z \neq 0$ the invertibility of $AB - z$ implies the invertibility of $BA - z$. Namely, one computes that $(BA - z)^{-1} = B(AB - z)^{-1}A - z^{-1}\mathbb{I}$.

For any locally compact Hausdorff space X, we regard the space $C_0(X)$ of all continuous functions on X that vanish at infinity as a Banach space in the sup-norm. A basic fact of topology and analysis is that $C_0(X)$ is complete in this norm. Convergence in the sup-norm is the same as uniform convergence. What's more, it is easily verified that $C_0(X)$ is a commutative C^*-algebra under pointwise addition, multiplication, and complex conjugation (defining the involution). When X is compact, the function 1_X, which is 1 for every x, is the unit $\mathbb{I}$. One checks that the spectrum of $f \in C(X)$ is simply the set of values of f. On $C_0(X)$, with X noncompact, one has to supplement this set with zero.

Theorem 1.2.3. *Let $\mathfrak{A}$ be a commutative C^*-algebra. There exists a locally compact Hausdorff space X for which $\mathfrak{A}$ is isomorphic to $C_0(X)$. When $\mathfrak{A}$ is unital, X is compact, so that $\mathfrak{A} \simeq C(X)$. The space X is unique up to homeomorphism.*

Similarly, an associative JLB-algebra $\mathfrak{A}_{\mathbb{R}}$ is isomorphic to some $C_0(X, \mathbb{R})$, where X is locally compact; when $\mathfrak{A}_{\mathbb{R}}$ has a unit, X is compact.

For simplicity we assume that $\mathfrak{A}$ is unital; if it isn't, one would start by adjoining a unit. The proof is based on a technique that applies to general commutative unital

Banach algebras; we state the main facts without proof. Consider the space $\Delta(\mathfrak{A})$ of all nonzero multiplicative linear functionals ω on $\mathfrak{A}$ (that is, $\omega : \mathfrak{A} \to \mathbb{C}$ satisfies $\omega(AB) = \omega(A)\omega(B)$ for all A, B). Each $\omega \in \Delta(\mathfrak{A})$ is continuous, and satisfies $\|\omega\| = \omega(\mathbb{I}) = 1$. Thence it is easily seen that $\Delta(\mathfrak{A})$ is a closed subspace of $\mathfrak{A}^*$ with the w^*-topology. By the Banach–Alaoglu theorem, $\Delta(\mathfrak{A})$ is therefore a compact Hausdorff space in the relative w^*-topology.

The **Gelfand transform** of $A \in \mathfrak{A}$ is the function $\hat{A}$ on $\Delta(\mathfrak{A})$ defined by

$$\hat{A}(\omega) := \omega(A). \tag{1.30}$$

Since the relative w^*-topology on $\Delta(\mathfrak{A})$ coincides with the weakest topology that makes all functions $\hat{A}$ continuous, it is clear that the Gelfand transform maps $\mathfrak{A}$ into $C(\Delta(\mathfrak{A}))$. It is immediate that the image of $\mathfrak{A}$ in $C(\Delta(\mathfrak{A}))$ separates points. Regarding $C(\Delta(\mathfrak{A}))$ as a commutative Banach algebra in the sup-norm, as explained above, the multiplicativity of each $\omega \in \Delta(\mathfrak{A})$ implies that the Gelfand transform is a homomorphism. The spectrum of $A \in \mathfrak{A}$ coincides with the set of values of $\hat{A}$ on $\Delta(\mathfrak{A})$; in other words,

$$\sigma(A) = \sigma(\hat{A}) = \{\hat{A}(\omega) | \, \omega \in \Delta(\mathfrak{A})\}. \tag{1.31}$$

This implies that

$$\|\hat{A}\|_\infty = r(A), \tag{1.32}$$

where the spectral radius $r(A)$ is defined in (1.26). In any Banach algebra, commutative or not, one has

$$r(A) = \lim_{n\to\infty} \|A^n\|^{1/n}. \tag{1.33}$$

Now assume that $\mathfrak{A}$ is a commutative C^*-algebra; accordingly, regard $C(\Delta(\mathfrak{A}))$ as a commutative C^*-algebra. We first show that $\omega \in \Delta(\mathfrak{A})$ is real on $\mathfrak{A}_\mathbb{R}$. Pick $A \in \mathfrak{A}_\mathbb{R}$, and suppose that $\omega(A) = \alpha + i\beta$, where $\alpha, \beta \in \mathbb{R}$. Since $\omega(\mathbb{I}) = 1$, one has $\omega(B) = i\beta$, where $B := A - \alpha\mathbb{I}$ is self-adjoint. Hence for $t \in \mathbb{R}$ one computes $|\omega(B + it\mathbb{I})|^2 = \beta^2 + 2t\beta + t^2$. On the other hand, using $\|\omega\| = 1$ and (1.15) we estimate $|\omega(B + it\mathbb{I})|^2 \leq \|B\|^2 + t^2$. Hence $\beta^2 + t\beta \leq \|B\|^2$ for all $t \in \mathbb{R}$. For $\beta > 0$ this is impossible. For $\beta < 0$ we repeat the argument with B replaced by $-B$. Hence $\beta = 0$, so that $\omega(A)$ is real when $A = A^*$. Consequently, by (1.30) the function $\hat{A}$ is real-valued. Writing $\varphi(A) := \hat{A}$, condition (1.20) follows.

Secondly, for $A \in \mathfrak{A}_\mathbb{R}$ one combines (1.15) with (1.33) to obtain $\|A\| = r(A)$, which with (1.33) implies $\|\hat{A}\|_\infty = \|A\|$. For general A this equality then follows via (1.15) in both $\mathfrak{A}$ and $C(\Delta(\mathfrak{A}))$, and the fact that $A^*A \in \mathfrak{A}_\mathbb{R}$. Thus the Gelfand transform is isometric, and therefore injective. Finally, surjectivity follows from the Stone–Weierstrass theorem.

The uniqueness of X follows from the fact that when X and Y are compact Hausdorff spaces, the commutative Banach algebras $C(X)$ and $C(Y)$ are isomorphic iff X and Y are homeomorphic; this is equivalent to the statement that $\Delta(C(X))$ is homeomorphic to X. The homeomorphism is given by letting $x \in X$ correspond with $\omega_x \in \Delta(C(X))$, defined by $\omega_x(f) := f(x)$. The assumption that X is compact and Hausdorff, hence normal, is used to prove that the evaluation map $x \mapsto \omega_x$

is injective, whereas the compactness of X implies that the evaluation map is surjective.

The case of a commutative JLB-algebra $\mathfrak{A}_{\mathbb{R}}$ may be reduced to that of a commutative C^*-algebra by complexification; see Theorem 1.1.9. □

Theorem 1.2.4. *Let $\mathfrak{A}$ be either a unital C^*-algebra or the complexification of a unital JLB-algebra $\mathfrak{A}_{\mathbb{R}}$.*

1. *The spectrum $\sigma(A)$ of $A \in \mathfrak{A}_{\mathbb{R}}$ in $\mathfrak{A}$ is equal to its spectrum in the C^*-algebra $C^*(A)$ generated by A and $\mathbb{I}$. In particular, $\sigma(A)$ is a subset of the real line.*
2. *The compact Hausdorff space $\Delta(C^*(A))$ of Theorem 1.2.3 is homeomorphic to $\sigma(A)$, and the C^*-algebra $C^*(A)$ of the preceding item is isomorphic to $C(\sigma(A))$. Under this isomorphism the function $\hat{A} \in C(X)$ is mapped into $\mathrm{id}_{\sigma(A)} : t \mapsto t$.*
3. *The* **continuous functional calculus** *for self-adjoint operators A on a Hilbert space applies verbatim to $\mathfrak{A}_{\mathbb{R}}$: In particular, for each $A \in \mathfrak{A}_{\mathbb{R}}$ and $f \in C(\sigma(A))$ there exists an operator $f(A) \in \mathfrak{A}$ that is the obvious expression when f is polynomial (and in the general case is given by uniformly approximating f by polynomials on the basis of the Stone–Weierstrass theorem), and has the properties*

$$\sigma(f(A)) = f(\sigma(A)); \tag{1.34}$$

$$\|f(A)\| = \|f\|_{\infty}. \tag{1.35}$$

4. *For $A \in \mathfrak{A}_{\mathbb{R}}$ the norm is given by (1.26); for general A one has $\|A\| = \sqrt{r(A^*A)}$. Hence the norm in a C^*-algebra is unique, in that a *-algebra that is a C^*-algebra in some norm admits no other norm in which it is a C^*-algebra.*

If $A = A^*$, then $A_z := A - z$ is normal for any $z \in \mathbb{C}$ (i.e., A_z commutes with its adjoint $A_{\bar{z}}$). Suppose that $z \notin \sigma(A)$, so that A_z is invertible. Consider the commutative C^*-algebra $C^*(A_z, A_z^{-1})$ generated by A_z, its inverse, and the unit. By Theorem 1.2.3 one has $C^*(A_z, A_z^{-1}) \simeq C(X)$, where $X = \Delta(C^*(A_z, A_z^{-1}))$. Since A_z is invertible in $C^*(A_z, A_z^{-1})$, it must be that $\hat{A}_z(x) \neq 0$ for all $x \in X$. It is then elementary that $\hat{A}_z^{-1}$ is a uniform limit of polynomials in A_z, $\hat{A}_{\bar{z}}$, and 1_X. Transferring this back to $C^*(A_z, A_z^{-1})$ by the inverse of the Gelfand transform, it follows that $C^*(A_z, A_z^{-1}) = C^*(A_z) = C^*(A)$. Hence when $A - z$ is invertible in $\mathfrak{A}$ its inverse lies in $C^*(A)$, which implies the first claim in 1.2.4.1.

Consider 1.2.3 and its proof with $\mathfrak{A} = C^*(A)$. We see from (1.31) and the fact that $\hat{A}$ is real-valued for $A \in \mathfrak{A}_{\mathbb{R}}$ that the spectrum of A in $C^*(A)$ is real. When now $\mathfrak{A}$ has the meaning of the present Theorem 1.2.4, the second claim in 1.2.4.1 follows from the first one just proved.

For 1.2.4.2, consider the map $\hat{A} : X \to \mathbb{R}$. It is clear from (1.31) that the image of $\hat{A}$ is $\sigma(A)$. To prove injectivity, assume $\hat{A}(\omega_1) = \hat{A}(\omega_2)$. Then $\omega_1(A) = \omega_2(A)$ by (1.30), whence $\omega_1(A)^n = \omega_2(A)^n$ by multiplicativity of $\omega_i \in \Delta(C^*(A))$. Since the linear span of all polynomials is dense in $C^*(A)$, and the ω_i are continuous, this yields $\omega_1 = \omega_2$. The map $\hat{A}$ is continuous, because $\hat{A} \in C(X)$ by 1.2.3. To prove continuity of the inverse, one checks that for $z \in \sigma(A)$ the functional $\hat{A}^{-1}(z) \in \Delta(C^*(A))$ maps A to z (and hence A^n to z^n, etc.). Finally, given the homeomorphism $\Delta(C^*(A)) \simeq \sigma(A)$, the second isomorphism in $C^*(A) \simeq C(\Delta(C^*(A))) \simeq C(\sigma(A))$ follows from the topological fact that a compact Hausdorff space X is determined by $C(X)$, and vice versa; cf. the proof of 1.2.3.

The existence of the continuous functional calculus should now be obvious. Since $f(\sigma(A))$ is the set of values of f on $\sigma(A)$, (1.34) follows from (1.31), with A replaced by $f(A)$. The fact that for C^*-algebras the Gelfand transform is isometric yields (1.35).

The corresponding statements for a JLB-algebra follow by complexification, using the commutative part of 1.1.9.

The first claim in 1.2.4.4 follows from (1.35) with $f = \text{id}$. The second claim follows from the first, (1.15), and the property $A^*A \in \mathfrak{A}_{\mathbb{R}}$. Hence the norm is determined by the algebraic structure. ∎

For later use we record that for $A \in \mathfrak{A}_{\mathbb{R}}$, Theorem 1.2.4 implies

$$\sigma(A) = 0 \iff A = 0. \tag{1.36}$$

1.3 Positivity, Order, and Morphisms

Recall that a (bounded) operator $A \in \mathfrak{B}(\mathcal{H})$ on a Hilbert space is called **positive** when $(\Psi, A\Psi) \geq 0$ for all $\Psi \in \mathcal{H}$; this property is equivalent to $A^* = A$ and $\sigma(A) \subseteq \mathbb{R}^+$. This notion of positivity induces a partial ordering $\leq$ in $\mathfrak{B}(\mathcal{H})$, in which $A \leq B$ when $B - A \geq 0$. Our aim is to generalize these concepts to C^*-algebras and JLB-algebras.

Definition 1.3.1. *A* **partially ordered vector space** $(\mathfrak{A}_{\mathbb{R}}, \leq)$ *consists of a real vector space* $\mathfrak{A}_{\mathbb{R}}$ *and either one of the following equivalent data:*

- *A* **positive cone** $\mathfrak{A}^+$ *in* $\mathfrak{A}_{\mathbb{R}}$*; this is a subset for which* (i) $A \in \mathfrak{A}^+$ *and* $t \in \mathbb{R}^+$ *implies* $tA \in \mathfrak{A}^+$, (ii) $A, B \in \mathfrak{A}^+$ *implies* $A+B \in \mathfrak{A}^+$, *and* (iii) $\mathfrak{A}^+ \cap -\mathfrak{A}^+ = 0$.
- *A* **linear partial ordering**, *i.e., a partial ordering* $\leq$ *in which* $A \leq B$ *implies* $A + C \leq B + C$ *for all* $C \in \mathfrak{A}_{\mathbb{R}}$ *and* $tA \leq tB$ *for all* $t \in \mathbb{R}^+$.

The equivalence between these two structures is as follows: Given $\mathfrak{A}_{\mathbb{R}}^+$ one defines $A \leq B$ if $B - A \in \mathfrak{A}_{\mathbb{R}}^+$, and given $\leq$ one puts $\mathfrak{A}_{\mathbb{R}}^+ = \{A \in \mathfrak{A}_{\mathbb{R}} \mid 0 \leq A\}$.

Definition 1.3.2. *Let* $\mathfrak{A}_{\mathbb{R}}$ *be a* JLB*-algebra or the self-adjoint part of a* C^**-algebra. An element* $A \in \mathfrak{A}_{\mathbb{R}}$ *is called* **positive** *when its spectrum is positive; i.e.,* $\sigma(A) \subset \mathbb{R}^+$. *We write* $A \geq 0$ *or* $A \in \mathfrak{A}^+$, *where*

$$\mathfrak{A}^+ := \{A \in \mathfrak{A}_{\mathbb{R}} \,|\, \sigma(A) \subset \mathbb{R}^+\}. \tag{1.37}$$

It is immediate from (1.31) that $A \in \mathfrak{A}_{\mathbb{R}}$ is positive iff its Gelfand transform $\hat{A}$ is pointwise positive in $C(\sigma(A))$.

Theorem 1.3.3. *The set (1.37) of all positive elements of a C^*-algebra $\mathfrak{A}$ or a JLB-algebra $\mathfrak{A}_{\mathbb{R}}$ is a positive cone. This cone may alternatively be expressed as*

$$\mathfrak{A}_{\mathbb{R}}^+ = \{A^2 \mid A \in \mathfrak{A}_{\mathbb{R}}\} \tag{1.38}$$

$$= \{B^*B \mid B \in \mathfrak{A}\}. \tag{1.39}$$

Property (i) in 1.3.1 follows from (1.28). Since $\sigma(A) \subseteq [0, r(A)]$, we have $|c - t| \leq c$ for all $t \in \sigma(A)$ and all $c \geq r(A)$. Hence $\sup_{t\in\sigma(A)} |c1_{\sigma(A)} - \hat{A}| \leq c$ by (1.31) and 1.2.4.2, so that $\|c1_{\sigma(A)} - \hat{A}\|_\infty \leq c$. Gelfand transforming back to $C^*(A)$, this implies $\|c\mathbb{I} - A\| \leq c$ for all $c \geq \|A\|$ by 1.2.4.3. Inverting this argument, one sees that if $\|c\mathbb{I} - A\| \leq c$ for some $c \geq \|A\|$, then $\sigma(A) \subset \mathbb{R}^+$. Using this with A replaced by $A + B$ and $c = \|A\| + \|B\|$ leads to property (ii). Finally, when $A \in \mathfrak{A}^+$ and $A \in -\mathfrak{A}^+$, it must be that $\sigma(A) = 0$, hence $A = 0$ by (1.36). This proves property (iii).

If $\sigma(A) \subset \mathbb{R}^+$ and $A = A^*$, then $\sqrt{A} \in \mathfrak{A}_{\mathbb{R}}$ is defined by the continuous functional calculus for $f = \sqrt{\cdot}$ and satisfies $\sqrt{A}^2 = A$. Hence $\mathfrak{A}^+ \subseteq \{A^2 \mid A \in \mathfrak{A}_{\mathbb{R}}\}$. The opposite inclusion follows from (1.34) and 1.2.4.2. This proves (1.38).

The inclusion $\mathfrak{A}^+ \subseteq \{B^*B \mid B \in \mathfrak{A}\}$ is trivial from (1.38).

Lemma 1.3.4. *Every $A \in \mathfrak{A}_{\mathbb{R}}$ has a decomposition $A = A_+ - A_-$, where $A_+, A_- \in \mathfrak{A}^+$ and $A_+A_- = 0$. Moreover, $\|A_\pm\| \leq \|A\|$.*

Apply the continuous functional calculus with $f = \mathrm{id}_{\sigma(A)} = f_+ - f_-$, where $\mathrm{id}_{\sigma(A)}(t) = t$, $f_+(t) = \max\{t, 0\}$, and $f_-(t) = \max\{-t, 0\}$. The bound follows from (1.35) with A replaced by $A_\pm$. ■

Apply this decomposition to $A = B^*B$ (noting that $A = A^*$); it follows that $(A_-)^3 = -(BA_-)^*BA_-$. Since $\sigma(A_-) \subset \mathbb{R}^+$ as A_- is positive, we see from (1.34) with $f(t) = t^3$ that $(A_-)^3 \geq 0$. Hence $-(BA_-)^*BA_- \geq 0$.

Lemma 1.3.5. *If $-C^*C \in \mathfrak{A}^+$ for some $C \in \mathfrak{A}$, then $C = 0$.*

Write $C = D + iE$, where $D, E \in \mathfrak{A}_{\mathbb{R}}$ (cf. (1.21)), so that $C^*C = 2D^2 + 2E^2 - CC^*$. Applying (1.29) with A replaced by C and B by C^*, we see that the assumption $\sigma(C^*C) \subset \mathbb{R}^-$ implies $\sigma(CC^*) \subset \mathbb{R}^-$; since C^*C is the sum of three positive terms, and $\mathfrak{A}^+$ is a positive cone, it follows that $C^*C \in \mathfrak{A}^+$. Hence the starting assumption $\sigma(C^*C) \subset \mathbb{R}^-$ implies $\sigma(C^*C) \subset \mathbb{R}^+$, so that $\sigma(C^*C) = 0$. Hence $C^*C = 0$ by (1.36).

In a C^*-algebra this implies $C = 0$ by (1.15). In a complexified JLB-algebra we replace C by C^* in the above argument, so that $CC^* = 0$ as well as $C^*C = 0$; hence $D^2 + E^2 = 0$, whence $D = E = 0$ by (1.8). ■

The last claim before the lemma therefore implies $BA_- = 0$. As $(A_-)^3 = -(BA_-)^*BA_- = 0$, we see that $(A_-)^3 = 0$, and finally $A_- = 0$ by the continuous functional calculus with $f(t) = t^{1/3}$. Hence $B^*B = A_+$, which lies in $\mathfrak{A}^+$. ■

When $A = A^*$ one checks the validity of

$$-\|A\|\,\mathbb{I} \leq A \leq \|A\|\,\mathbb{I} \tag{1.40}$$

by taking the Gelfand transform of $C^*(A)$. The implication

$$-B \leq A \leq B \implies \|A\| \leq \|B\| \tag{1.41}$$

then follows, because $-B \leq A \leq B$ and (1.40) with A replaced by B yield $-\|B\|\,\mathbb{I} \leq A \leq \|B\|\,\mathbb{I}$, so that $\sigma(A) \subseteq [-\|B\|, \|B\|]$; hence $\|A\| \leq \|B\|$ by (1.26).

An important consequence of (1.39) is the fact that inequalities of the type $A_1 \leq A_2$ for $A_1, A_2 \in \mathfrak{A}_{\mathbb{R}}$ are stable under conjugation by arbitrary elements $B \in \mathfrak{A}$, so that $A_1 \leq A_2$ implies $B^*A_1B \leq B^*A_2B$. This is because $A_1 \leq A_2$ is the same as $A_2 - A_1 \geq 0$; by (1.39) there is an $A_3 \in \mathfrak{A}$ such that $A_2 - A_1 = A_3^*A_3$. But then $(A_3B)^*A_3B \geq 0$, and this is nothing but $B^*A_1B \leq B^*A_2B$. For example, replace A in (1.40) by A^*A, and use (1.15) in a C^*-algebra, or (1.25) in a complexified JLB-algebra. This yields $A^*A \leq \|A\|^2\mathbb{I}$. Applying the above principle to any $A, B \in \mathfrak{A}$ gives

$$B^*A^*AB \leq \|A\|^2 B^*B. \tag{1.42}$$

Definition 1.3.6. *A* **positive map** $\mathcal{Q} : \mathfrak{A} \to \mathfrak{B}$ *between two C^*-algebras is a linear map with the property that $A \geq 0$ in $\mathfrak{A}$ implies $\mathcal{Q}(A) \geq 0$ in $\mathfrak{B}$.*

Proposition 1.3.7. *A positive map between C^*-algebras is $*$-preserving and bounded.*

One infers from 1.3.4 that $\mathcal{Q}(\mathfrak{A}_{\mathbb{R}}) \subseteq \mathfrak{B}_{\mathbb{R}}$; the $\mathbb{C}$-linearity of $\mathcal{Q}$ then proves the first claim.

For the second claim, let us first show that boundedness on $\mathfrak{A}^+$ implies boundedness on $\mathfrak{A}$. Using (1.21) and 1.3.4, we can write $A = A'_+ - A'_- + iA''_+ - iA''_-$, where A'_+ etc. are positive. Since $\|A'\| \leq \|A\|$ and $\|A''\| \leq \|A\|$ by (1.21), we have $\|B\| \leq \|A\|$ for $B = A'_+, A'_-, A''_+,$ or A''_- by 1.3.4. Hence if $\|\mathcal{Q}(B)\| \leq c\|B\|$ for all $B \in \mathfrak{A}^+$ and some $c > 0$, then $\|\mathcal{Q}(A)\| \leq 4c\|A\|$.

Now assume that $\mathcal{Q}$ is not bounded; by the previous argument it is not bounded on $\mathfrak{A}^+$, so that for each $n \in \mathbb{N}$ there is an $A_n \in \mathfrak{A}_1^+$ such that $\|\mathcal{Q}(A_n)\| \geq n^3$ (here $\mathfrak{A}_1^+$ consists of all $A \in \mathfrak{A}^+$ with $\|A\| \leq 1$). The series $\sum_{n=0}^{\infty} n^{-2}A_n$ obviously converges to some $A \in \mathfrak{A}^+$. Since $\mathcal{Q}$ is positive, we have $\mathcal{Q}(A) \geq n^{-2}\mathcal{Q}(A_n) \geq 0$ for each n. Hence by (1.41), $\|\mathcal{Q}(A)\| \geq n^{-2}\|\mathcal{Q}(A_n)\| \geq n$ for all $n \in \mathbb{N}$, which is impossible. Thus $\mathcal{Q}$ is bounded on $\mathfrak{A}^+$, and therefore on $\mathfrak{A}$ by the previous paragraph. ■

Since a morphism $\varphi : \mathfrak{A} \to \mathfrak{B}$ satisfies $\varphi(B^*B) = \varphi(B)^*\varphi(B)$, it is clear from (1.39) that a morphism is a positive map. For later reference we collect this, and other good properties of morphisms. In preparation, we define a **left ideal** in a C^*-algebra $\mathfrak{A}$ as a closed linear subspace $\mathfrak{I} \subset \mathfrak{A}$ such that $A \in \mathfrak{I}$ implies $BA \in \mathfrak{I}$ for all $B \in \mathfrak{A}$. Similarly, a **right ideal** is a closed linear subspace $\mathfrak{I} \subseteq \mathfrak{A}$ such that $A \in \mathfrak{I}$ implies $AB \in \mathfrak{I}$ for all $B \in \mathfrak{A}$. An **ideal** is both a left and a right ideal.

A proper ideal cannot contain a unit $\mathbb{I}$; in order to prove properties of ideals one needs a suitable replacement of a unit.

Definition 1.3.8. *An* **approximate unit** *in a nonunital C^*-algebra $\mathfrak{A}$ consists of a directed set Λ (i.e., a set with a partial order $\leq$ in which for all λ_1, λ_2 there is a $\lambda \geq \lambda_i$, $i = 1, 2$) and a family $\{\mathbb{I}_\lambda\}_{\lambda\in\Lambda}$ of elements of $\mathfrak{A}$ for which $\mathbb{I}_\lambda^* = \mathbb{I}_\lambda$ and $\sigma(\mathbb{I}_\lambda) \subset [0, 1]$ (whence $\|\mathbb{I}_\lambda\| \leq 1$), so that for all $A \in \mathfrak{A}$ one has*

$$\lim_{\lambda\to\infty} \|\mathbb{I}_\lambda A - A\| = \lim_{\lambda\to\infty} \|A\mathbb{I}_\lambda - A\| = 0. \tag{1.43}$$

For example, an approximate unit in $C_0(\mathbb{R})$ may be constructed with $\Lambda = \mathbb{N}$, taking $\mathbb{I}_n$ to be a continuous function that is 1 on $[-n, n]$ and vanishes for $|x| > n + 1$. More generally, it can be shown that every nonunital C^*-algebra $\mathfrak{A}$ has an approximate unit; when $\mathfrak{A}$ is separable, Λ may be chosen countable. The technique of approximate units allows us to prove the main properties of ideals in C^*-algebras.

Theorem 1.3.9. *Let $\mathfrak{J}$ be an ideal in a C^*-algebra $\mathfrak{A}$.*

1. *If $J \in \mathfrak{J}$ then $J^* \in \mathfrak{J}$; in other words, every ideal in a C^*-algebra is self-adjoint.*
2. *The quotient $\mathfrak{A}/\mathfrak{J}$ is a C^*-algebra in the norm $\|\tau(A)\| := \inf_{J\in\mathfrak{J}} \|A + J\|$, the multiplication $\tau(A)\tau(B) := \tau(AB)$, and the involution $\tau(A)^* := \tau(A^*)$.*

Note that the involution in $\mathfrak{A}/\mathfrak{J}$ is well-defined because of 1.3.9.1.

Put $\mathfrak{J}^* := \{A^* | A \in \mathfrak{J}\}$, and note that $\mathfrak{J} \cap \mathfrak{J}^*$ is a C^*-subalgebra of $\mathfrak{A}$. Hence it has an approximate unit $\{\mathbb{I}_\lambda\}$. Pick $J \in \mathfrak{J}$, and use (1.15) and 1.3.8 to estimate

$$\|J^* - J^*\mathbb{I}_\lambda\|^2 \leq \|(J^*J - J^*J\mathbb{I}_\lambda)\| + \|(JJ^* - JJ^*\mathbb{I}_\lambda)\|.$$

Now J^*J and JJ^* both lie in $\mathfrak{J} \cap \mathfrak{J}^*$, so that both terms on the right-hand side vanish for $\lambda \to \infty$. Hence J^* is a norm-limit of elements in $\mathfrak{J}$; since $\mathfrak{J}$ is closed, it follows that $J^* \in \mathfrak{J}$.

We omit the well-known proof that $\mathfrak{A}/\mathfrak{J}$ is a Banach algebra in the given norm and multiplication. To prove the C^*-property (1.15), we first note that

$$\|\tau(A)\| = \lim_{\lambda\to\infty} \|A - A\mathbb{I}_\lambda\|, \tag{1.44}$$

for any $A \in \mathfrak{A}$ and approximate unit $\{\mathbb{I}_\lambda\}$ in $\mathfrak{J}$. To prove this, we first add a unit to $\mathfrak{A}$ if necessary. For any $J \in \mathfrak{J}$ we have $A - A\mathbb{I}_\lambda = (A + J)(\mathbb{I} - \mathbb{I}_\lambda) + J(\mathbb{I}_\lambda - \mathbb{I})$, so that $\|A - A\mathbb{I}_\lambda\| \leq \|A + J\| \, \|\mathbb{I} - \mathbb{I}_\lambda\| + \|J\mathbb{I}_\lambda - J\|$. Since

$$\|\mathbb{I} - \mathbb{I}_\lambda\| \leq 1 \tag{1.45}$$

from 1.3.8 and the proof of 1.3.3, we obtain $\lim_{\lambda\to\infty} \|A - A\mathbb{I}_\lambda\| \leq \|A + J\|$. For each $\epsilon > 0$ we can choose $J \in \mathfrak{J}$ such that $\|\tau(A)\| + \epsilon \geq \|A + J\|$. Using this J in the previous inequality, letting $\epsilon \to 0$, and noting the obvious $\|A - A\mathbb{I}_\lambda\| \geq \|\tau(A)\|$, we obtain (1.44).

Successively using (1.44), (1.15) in $\mathfrak{A}$, (1.45), (1.44) once again, and the definition of the C^*-operations in $\mathfrak{A}/\mathfrak{J}$, we obtain $\|\tau(A)\|^2 \leq \|\tau(A)\tau(A)^*\|$. By the argument preceding 1.1.7, this implies (1.15). □

Theorem 1.3.10. *Let $\varphi : \mathfrak{A} \to \mathfrak{B}$ be a morphism between C^*-algebras.*

1. *The kernel of φ is an ideal in $\mathfrak{A}$. Conversely, every ideal in a C^*-algebra is the kernel of some morphism.*

2. *One has* $\|\varphi\| = 1$*, and therefore* $\|\varphi(A)\| \leq \|A\|$ *for all* $A \in \mathfrak{A}$.
3. *The map* φ *is isometric when it is injective.*
4. *The image* $\varphi(\mathfrak{A})$ *is a* C^**-subalgebra of* $\mathfrak{B}$.
5. *The map* φ *is positive.*

It is clear that $\sigma(\varphi(A^*A)) \subseteq \sigma(A^*A)$, so that the inequality in 1.3.10.2 follows from 1.2.4.4. It follows that φ is continuous, so that $\ker(\varphi)$ is closed. It is then obvious from (1.19) that $\ker(\varphi)$ is an ideal. On the other hand, a given ideal $\mathfrak{J}$ is the kernel of the canonical projection $\tau : \mathfrak{A} \to \mathfrak{A}/\mathfrak{J}$. Now $\mathfrak{A}/\mathfrak{J}$ is a C^*-algebra and $\varphi := \tau$ is a morphism with $\mathfrak{J} = \ker(\varphi)$.

Assume that there is a $B \in \mathfrak{A}$ for which $\|\varphi(B)\| \neq \|B\|$. By (1.15), (1.19), and (1.20) this implies $\|\varphi(B^*B)\| \neq \|B^*B\|$. Put $A := B^*B$, noting that $A^* = A$. By (1.26) we must have $\sigma(A) \neq \sigma(\varphi(A))$. Since $\sigma(\varphi(A)) \subseteq \sigma(A)$ in any case, this implies $\sigma(\varphi(A)) \subset \sigma(A)$. By Urysohn's lemma there is a nonzero $f \in C(\sigma(A))$ that vanishes on $\sigma(\varphi(A))$, so that $f(\varphi(A)) = 0$. By the continuous functional calculus we have $\varphi(f(A)) = 0$, proving 1.3.10.3 by reductio ad absurdum.

Define $\psi : \mathfrak{A}/\ker(\varphi) \to \mathfrak{B}$ by $\psi \circ \tau = \varphi$, with $\tau : \mathfrak{A} \to \mathfrak{A}/\ker(\varphi)$ the canonical projection. Then ψ is an injective morphism, so that it is isometric by 1.3.10.3. Hence $\|\varphi\| = 1$, since $\|\tau\| = 1$. Since $\psi(\mathfrak{A}/\ker(\varphi)) = \varphi(\mathfrak{A})$, it follows that φ has closed range in $\mathfrak{B}$. Since φ is a morphism, this implies that $\varphi(\mathfrak{A})$ is a C^*-algebra in the norm of $\mathfrak{B}$. ∎

1.4 States

We now change our perspective, and pass from observables to states.

Definition 1.4.1. *A* **state** *on a* C^**-algebra* $\mathfrak{A}$ *is a linear map* $\omega : \mathfrak{A} \to \mathbb{C}$ *for which* $\omega(A) \geq 0$ *for all* $A \in \mathfrak{A}^+_{\mathbb{R}}$ **(positivity)** *and* $\|\omega\| = 1$ **(normalization)**. *The* **state space** $\mathcal{S}(\mathfrak{A})$ *of* $\mathfrak{A}$ *is the set of all states on* $\mathfrak{A}$.

For example, on $\mathfrak{A} = \mathfrak{B}(\mathcal{H})$ every unit vector $\Omega \in \mathcal{H}$ defines a state ω by

$$\omega(A) = (\Omega, \pi(A)\Omega). \tag{1.46}$$

This is, indeed, the original notion of a state as used in quantum mechanics.

Combining 1.3.4 with positivity, we see that a state is real-valued on $\mathfrak{A}_{\mathbb{R}}$; in view of (1.21) we then infer that a state is a **Hermitian functional** on $\mathfrak{A}$, in that

$$\omega(A^*) = \overline{\omega(A)} \tag{1.47}$$

for all $A \in \mathfrak{A}$. In particular, a state is determined by its values on $\mathfrak{A}^+$. Combining the positivity of ω with (1.39) one sees that $(A, B)_\omega := \omega(A^*B)$ defines a pre-inner product on $\mathfrak{A}$. Hence from the Cauchy–Schwarz inequality we obtain the useful bound

$$|\omega(A^*B)|^2 \leq \omega(A^*A)\omega(B^*B). \tag{1.48}$$

Proposition 1.4.2. *A linear map* $\omega : \mathfrak{A} \to \mathbb{C}$ *on a unital* C^**-algebra is positive iff* ω *is bounded and* $\|\omega\| = \omega(\mathbb{I})$. *Hence a state* ω *on a unital* C^**-algebra may* *equivalently be characterized as a positive linear functional for which $\omega(\mathbb{I}) = 1$ or as a bounded linear functional for which $\|\omega\| = \omega(\mathbb{I}) = 1$.*

A state ω on a C^*-algebra without unit has a unique extension to a state $\omega_{\mathbb{I}}$ on the unitization $\mathfrak{A}_{\mathbb{I}}$, given by

$$\omega_{\mathbb{I}}(A + \lambda\mathbb{I}) := \omega(A) + \lambda. \tag{1.49}$$

When ω is positive and $A = A^*$ we have, using (1.40), the bound $|\omega(A)| \leq \omega(\mathbb{I})\|A\|$. For general A the same inequality follows from (1.48) with $A = \mathbb{I}$, (1.15), and the bound just derived. The upper bound is reached by $A = \mathbb{I}$.

To prove the converse claim, we first note that the argument after (1.33) may be copied, showing that ω is real on $\mathfrak{A}_{\mathbb{R}}$. Next, we show that $A \geq 0$ implies $\omega(A) \geq 0$. Choose $s > 0$ small enough so that $\|\mathbb{I} - sA\| \leq 1$. For $\omega \neq 0$ one has $\|\mathbb{I} - sA\| \geq |\omega(\mathbb{I} - sA)|/\omega(\mathbb{I})$, so that $|\omega(\mathbb{I}) - s\omega(A)| \leq \omega(\mathbb{I})$. This is only possible when $\omega(A) \geq 0$.

As to the positivity of $\omega_{\mathbb{I}}$, we observe that $|\omega(A - A\mathbb{I}_{\lambda})| \to 0$ for any approximate unit in $\mathfrak{A}$. Using (1.48) with $B = \mathbb{I}_{\lambda}$, this leads to $|\omega(A)|^2 \leq \omega(A^*A)$. Combining this inequality with (1.47), the definition (1.49) leads to $\omega_{\mathbb{I}}((A + \lambda\mathbb{I})^*(A + \lambda\mathbb{I})) \geq |\omega(A) + \overline{\lambda}|^2 \geq 0$. Hence ω is positive by (1.39). ∎

An important feature of a state space $\mathcal{S}(\mathfrak{A})$ is that it is a convex set. (A convex set $\mathcal{C}$ in a vector space $\mathcal{V}$ is a subset of $\mathcal{V}$ such that the convex sum $\lambda v + (1 - \lambda)w$ belongs to $\mathcal{C}$ whenever $v, w \in \mathcal{C}$ and $\lambda \in [0, 1]$. Geometrically, this means that the line segment between any two points in $\mathcal{C}$ lies in $\mathcal{C}$. It follows that a finite sum $\sum_i p_i v_i$ belongs to $\mathcal{C}$ when all $p_i \geq 0$ and $\sum_i p_i = 1$, and all $v_i \in \mathcal{C}$.) In the unital case it is clear that $\mathcal{S}(\mathfrak{A})$ is convex, since both positivity and normalization are clearly preserved under convex sums. In the nonunital case one arrives at this conclusion most simply via (1.49).

Let $\mathcal{S}(\mathfrak{A})$ be the state space of a unital C^*-algebra $\mathfrak{A}$. Each element ω of $\mathcal{S}(\mathfrak{A})$ is continuous, so that $\mathcal{S}(\mathfrak{A}) \subset \mathfrak{A}^*$. Since w^*-limits obviously preserve positivity and normalization, we see that $\mathcal{S}(\mathfrak{A})$ is closed in $\mathfrak{A}^*$ if the latter is equipped with the w^*-topology. Moreover, $\mathcal{S}(\mathfrak{A})$ is a closed subset of the unit ball of $\mathfrak{A}^*$, so that $\mathcal{S}(\mathfrak{A})$ is compact in the relative w^*-topology by the Banach–Alaoglu theorem. It follows that the state space of a unital C^*-algebra is a compact convex set.

The very simplest example is $\mathfrak{A} = \mathbb{C}$, in which case $\mathcal{S}(\mathfrak{A})$ is a point. The next case is $\mathfrak{A} = \mathbb{C} \oplus \mathbb{C} = \mathbb{C}^2$. The dual is $\mathbb{C}^2$ as well, so that each element of $\left(\mathbb{C}^2\right)^*$ is of the form $\omega(\lambda\dot{+}\mu) = c_1\lambda_1 + c_2\lambda_2$. Positive elements of $\mathbb{C} \oplus \mathbb{C}$ are of the form $\lambda\dot{+}\mu$ with $\lambda \geq 0$ and $\mu \geq 0$, so that a positive functional must have $c_1 \geq 0$ and $c_2 \geq 0$. Since $\mathbb{I} = 1\dot{+}1$, normalization yields $c_1 + c_2 = 1$. Identifying 0 with the functional mapping $\lambda\dot{+}\mu$ to λ, and 1 with the one mapping it to μ, we conclude that $\mathcal{S}(\mathbb{C} \oplus \mathbb{C})$ may be identified with the interval $[0, 1]$.

Now consider $\mathfrak{A} = \mathfrak{M}_2(\mathbb{C})$. We identify $\mathfrak{M}_2(\mathbb{C})$ with its dual through the pairing $\omega(A) = \mathrm{Tr}\,\omega A$. It follows that $\mathcal{S}(\mathfrak{A})$ consists of all positive 2×2 matrices ρ with $\mathrm{Tr}\,\rho = 1$; these are the density matrices of quantum mechanics. To identify $\mathcal{S}(\mathfrak{A})$

with a familiar compact convex set, we parametrize

$$\rho = \frac{1}{2}\begin{pmatrix} 1+x & y+iz \\ y-iz & 1-x \end{pmatrix}, \tag{1.50}$$

where $x, y, z \in \mathbb{R}$. The positivity of this matrix then corresponds to the constraint $x^2 + y^2 + z^2 \leq 1$. Hence $\mathcal{S}(\mathfrak{M}_2(\mathbb{C}))$ is the unit ball in $\mathbb{R}^3$.

There are lots of states:

Proposition 1.4.3. *For every $A \in \mathfrak{A}$ and $a \in \sigma(A)$ there is a state ω_a on $\mathfrak{A}$ for which $\omega(A) = a$. When $A = A^*$ there exists a state ω such that $|\omega(A)| = \|A\|$.*

If necessary we add a unit to $\mathfrak{A}$; in the present context this is justified by (1.49). Define a linear map $\tilde{\omega}_a : \mathbb{C}A \oplus \mathbb{C}\mathbb{I} \to \mathbb{C}$ by $\tilde{\omega}_a(\lambda A + \mu\mathbb{I}) := \lambda a + \mu$. Since $a \in \sigma(A)$, one has $\lambda a + \mu \in \sigma(\lambda A + \mu\mathbb{I})$; this easily follows from the definition of $\sigma(A)$. In any Banach algebra one has $r(A) \leq \|A\|$; applying this with A replaced by $\lambda A + \mu\mathbb{I}$ implies $|\tilde{\omega}_a(\lambda A + \mu\mathbb{I})| \leq \|\lambda A + \mu\mathbb{I}\|$. Since $\tilde{\omega}_a(\mathbb{I}) = 1$, it follows that $\|\tilde{\omega}\| = 1$. By the Hahn–Banach theorem there exists an extension ω_a of $\tilde{\omega}$ to $\mathfrak{A}$ of norm 1. Since $\|\omega_a\| = \omega_a(\mathbb{I}) = 1$, this extension is a state, which clearly satisfies $\omega_a(A) = \tilde{\omega}_a(A) = a$.

Since $\sigma(A)$ is closed, there is an $a \in \sigma(A)$ for which $r(A) = |a|$. For this a, and $A = A^*$, one has $|\omega(A)| = |a| = r(A) = \|A\|$ by (1.26); cf. 1.2.4.4. ∎

Corollary 1.4.4. *For all $A \in \mathfrak{A}_{\mathbb{R}}$ one has*

$$\|A\| = \sup\{|\omega(A)| \mid \omega \in \mathcal{S}(\mathfrak{A})\}. \tag{1.51}$$

Hence if $\omega(A) = 0$ for all states $\omega \in \mathcal{S}(\mathfrak{A})$, then $A = 0$.

Our goal is to give a geometric realization of a unital C^*-algebra $\mathfrak{A}$ as a certain function space, somewhat in the spirit of Theorem 1.2.3. A function f on a convex set K is called **affine** if it preserves convexity, that is, if

$$f(\lambda\omega_1 + (1-\lambda)\omega_2) = \lambda f(\omega_1) + (1-\lambda) f(\omega_2) \;\forall \omega_1, \omega_2 \in K, \lambda \in [0, 1]. \tag{1.52}$$

The space $A(K, \mathbb{R})$ of all real-valued affine continuous functions on a compact convex set K has a positive cone $A(K, \mathbb{R})^+$, consisting of all positive functions (cf. 1.3.1). Equivalently, $A(K, \mathbb{R})^+$ has a linear partial ordering, in which $f \leq g$ when $f(\omega) \leq g(\omega)$ for all $\omega \in K$. Also, $A(K, \mathbb{R})^+$ is a Banach space in the sup-norm in the case that K is Hausdorff.

Theorem 1.4.5. *The self-adjoint part $\mathfrak{A}_{\mathbb{R}}$ of a unital C^*-algebra $\mathfrak{A}$ is isomorphic as a partially ordered Banach space to the space $A(\mathcal{S}(\mathfrak{A}), \mathbb{R})$ of all real-valued affine continuous functions on the state space $\mathcal{S}(\mathfrak{A})$ of $\mathfrak{A}$ (equipped with the relative w^*-topology).*

The isomorphism in question is given by (1.30), now seen as a map from $\mathfrak{A}$ to the space of functions on the state space $\mathcal{S}(\mathfrak{A})$. It is immediate that this transform is injective. It is a well-known fact in functional analysis that a Banach space $\mathfrak{A}_{\mathbb{R}}$ may be identified under (1.30) with the subspace of its double dual $\mathfrak{A}_{\mathbb{R}}^{**}$ consisting of all w^*-continuous linear functionals on $\mathfrak{A}^*_{\mathbb{R}}$. Since linear functionals on $\mathfrak{A}^*$ are automatically affine on $\mathcal{S}(\mathfrak{A})$, and since we know that a state is real-valued on $\mathfrak{A}_{\mathbb{R}}$, the transform (1.30) maps $\mathfrak{A}_{\mathbb{R}}$ into $A(\mathcal{S}(\mathfrak{A}), \mathbb{R})$. Reinterpreting (1.51) as the equality $\|A\| = \|\hat{A}\|_\infty$, it follows that the map $A \mapsto \hat{A}$ is isometric.

Without proof we state that any Hermitian functional $\varphi \in \mathfrak{A}^*$ on a C^*-algebra $\mathfrak{A}$ has a unique decomposition $\varphi = t_1\omega_1 - t_2\omega_2$, where $\omega_i \in \mathcal{S}(\mathfrak{A})$ and $t_i \in \mathbb{R}^+$. This implies that an element f of $A(\mathcal{S}(\mathfrak{A}), \mathbb{R})$ has a unique extension to a Hermitian linear functional $\tilde{f}$ on $\mathfrak{A}^*$, which is evidently w^*-continuous, and is therefore given by an element $A \in \mathfrak{A}$ (cf. the preceding paragraph). The function $\hat{A}$ in (1.30) is evidently f, so that the image of $\mathfrak{A}$ under $A \mapsto \hat{A}$ is all of $A(\mathcal{S}(\mathfrak{A}), \mathbb{R})$.

Finally, it is trivial from the pertinent definitions that the transform (1.30) preserves positivity. □

As promised, we now complete the proof of Theorem 1.1.9. We discuss the unital case; the nonunital case may be reduced to this, using 1.2.1.

For the first half it remained to be shown that the norm on the self-adjoint part of a C^*-algebra satisfies (1.10). This follows from the order inequality $A^2 \leq A^2 + B^2$ (which is derived from (1.38) and the linearity of the partial ordering) and (1.41).

In the second half we need to prove that (1.25) defines a norm on $\mathfrak{A}$. Firstly, the property $\|A\| = 0 \Rightarrow A = 0$ follows from Lemma 1.3.5. Secondly, the triangle inequality follows by successively using (1.25) with A replaced by $A + B$, (1.51), (1.48), and again (1.25) and (1.51), this time from right to left. Finally, we use (1.42); taking the norm in $\mathfrak{A}_{\mathbb{R}}$ and using (1.25) yields (1.14). ■

Thus from now on a JLB-algebra and the self-adjoint part of a C^*-algebra will be one and the same object.

1.5 Representations and the GNS-Construction

In the theory of C^*-algebras, Hilbert spaces are most naturally regarded as modules, and the material of this section explains how the usual Hilbert space framework of quantum mechanics emerges from the algebraic setting.

Definition 1.5.1. *A* **representation** *of a C^*-algebra $\mathfrak{A}$ on a Hilbert space $\mathcal{H}$ is a morphism $\pi : \mathfrak{A} \to \mathfrak{B}(\mathcal{H})$.*

From the Jordan–Lie point of view, this means that $\pi : \mathfrak{A}_{\mathbb{R}} \to \mathfrak{B}(\mathcal{H})_{\mathbb{R}}$ is a morphism of Jordan–Lie algebras (cf. 1.1.3); here the Jordan–Lie structure on both spaces is given by (1.22). In view of 1.3.10.2 a representation π is automatically continuous; hence $\|\pi(A)\| \leq \|A\|$ for all $A \in \mathfrak{A}$. When π is faithful, this sharpens to $\|\pi(A)\| = \|A\|$ by 1.3.10.3.

There is a natural equivalence relation in the set of all representations of $\mathfrak{A}$: Two representations π_1, π_2 on Hilbert spaces $\mathcal{H}_1, \mathcal{H}_2$, respectively, are called **equivalent** if there exists a unitary isomorphism $U : \mathcal{H}_1 \to \mathcal{H}_2$ such that $U\pi_1(A)U^* = \pi_2(A)$ for all $A \in \mathfrak{A}$.

The map $\pi(A) = 0$ for all $A \in \mathfrak{A}$ is a representation; more generally, such trivial π may occur as a summand. To exclude this possibility, one says that a representation is **nondegenerate** if 0 is the only vector annihilated by all representatives of $\mathfrak{A}$.

A representation π is called **cyclic** if its carrier space $\mathcal{H}$ contains a **cyclic vector** Ω for π; this means that the closure of $\pi(\mathfrak{A})\Omega$ (which in any case is a closed subspace of $\mathcal{H}$) coincides with $\mathcal{H}$.

Proposition 1.5.2. *Any nondegenerate representation π is a direct sum of cyclic representations.*

The proof uses a lemma that appears in many other proofs as well.

Lemma 1.5.3. *Let $\mathfrak{M}$ be a *-algebra in $\mathfrak{B}(\mathcal{H})$, Ψ a nonzero vector $\mathcal{H}$, and p the projection onto the closure of $\mathfrak{M}\Psi$. Then $p \in \mathfrak{M}'$ (that is, $[p, A] = 0$ for all $A \in \mathfrak{M}$).*

If $A \in \mathfrak{M}$, then $Ap\mathcal{H} \subseteq p\mathcal{H}$ by definition of p. Hence $p^{\perp}Ap = 0$ with $p^{\perp} = \mathbb{I} - p$. When $A = A^*$ this yields $[A, p] = 0$; by (1.21) this is true for all $A \in \mathfrak{M}$. ■

Apply this lemma with $\mathfrak{M} = \pi(\mathfrak{A})$; the assumption of nondegeneracy guarantees that p is nonzero, and the conclusion implies that $A \mapsto p\pi(A)$ defines a cyclic subrepresentation of $\mathfrak{A}$ on $p\mathcal{H}$. This process may be repeated on $p^{\perp}\mathcal{H}$, etc. ■

If π is a nondegenerate representation of a C^*-algebra $\mathfrak{A}$ on $\mathcal{H}$, then any unit vector $\Omega \in \mathcal{H}$ defines a state $\omega \in \mathcal{S}(\mathfrak{A})$, referred to as a **vector state** relative to π, by means of (1.46). Conversely, from any state $\omega \in \mathcal{S}(\mathfrak{A})$ on $\mathfrak{A}$ one can construct a cyclic representation π_ω on a Hilbert space $\mathcal{H}_\omega$ with cyclic vector Ω_ω in the following way. We restrict ourselves to the unital case; the general case follows by adding a unit to $\mathfrak{A}$ and extending ω to $\mathfrak{A}_{\mathbb{I}}$ by (1.49).

Construction 1.5.4.

1. *Given $\omega \in \mathcal{S}(\mathfrak{A})$, define the sesquilinear form $(\,,\,)_0^\omega$ on $\mathfrak{A}$ by*
$$(A, B)_0^\omega := \omega(A^*B). \tag{1.53}$$
Since ω is a state, hence a positive functional, this form is positive semidefinite (this means that $(A, A)_0^\omega \geq 0$ for all A). Its null space
$$\mathcal{N}_\omega = \{A \in \mathfrak{A} \mid \omega(A^*A) = 0\} \tag{1.54}$$
is a left ideal in $\mathfrak{A}$.
2. *The form $(\,,\,)_0^\omega$ projects to an inner product $(\,,\,)_\omega$ on the quotient $\mathfrak{A}/\mathcal{N}_\omega$. If $V_\omega : \mathfrak{A} \to \mathfrak{A}/\mathcal{N}_\omega$ is the canonical projection, then by definition*
$$(V_\omega A, V_\omega B)_\omega := (A, B)_0^\omega. \tag{1.55}$$
The Hilbert space $\mathcal{H}_\omega$ is the closure of $\mathfrak{A}/\mathcal{N}_\omega$ in this inner product.
3. *The representation $\pi_\omega(\mathfrak{A})$ is firstly defined on $\mathfrak{A}/\mathcal{N}_\omega \subset \mathcal{H}_\omega$ by*
$$\pi_\omega(A)V_\omega B := V_\omega AB; \tag{1.56}$$
it follows that π_ω is continuous. Hence $\pi_\omega(A)$ may be defined on all of $\mathcal{H}_\omega$ by continuous extension of (1.56).

4. *The cyclic vector is defined by $\Omega_\omega = V_\omega \mathbb{I}$, so that*

$$(\Omega_\omega, \pi_\omega(A)\Omega_\omega) = \omega(A) \quad \forall A \in \mathfrak{A}. \tag{1.57}$$

We now prove the various claims made here. First note that the null space $\mathcal{N}_\omega$ of $(\, ,\,)_0^\omega$ can be defined in two equivalent ways, since

$$\mathcal{N}_\omega := \{A \in \mathfrak{A} \mid (A, A)_0^\omega = 0\} = \{A \in \mathfrak{A} \mid (A, B)_0^\omega = 0 \,\forall\, B \in \mathfrak{A}\}. \tag{1.58}$$

The equivalence follows from (1.48). With the continuity of ω, the equality (1.58) implies that $\mathcal{N}_\omega$ is a left ideal. Hence π_ω in (1.56) is well-defined on the dense subspace $\mathfrak{A}/\mathcal{N}_\omega \subset \mathcal{H}_\omega$, where it clearly satisfies (1.19), with $\varphi \to \pi_\omega$. Also, (1.20) may be verified from (1.55) and (1.53).

To prove that π_ω is bounded on $\mathfrak{A}/\mathcal{N}_\omega$, we compute $\|\pi_\omega(A)\Psi\|^2$ for $\Psi = V_\omega B$, where $A, B \in \mathfrak{A}$. From (1.55) and (1.53) one has $\|\pi_\omega(A)\Psi\|^2 = \omega(B^*A^*AB)$. By (1.42) and the positivity of ω one has $\omega(B^*A^*AB) \leq \|A\|^2\omega(B^*B)$. But $\omega(B^*B) = \|\Psi\|^2$, so that $\|\pi_\omega(A)\| \leq \|A\|$. Hence π_ω may be extended to all of $\mathcal{H}_\omega$, where (1.55) and (1.53) hold by continuity.

Proposition 1.5.5. *If a representation $\pi(\mathfrak{A})$ on $\mathcal{H}$ is cyclic, then the GNS-representation $\pi_\omega(\mathfrak{A})$ on $\mathcal{H}_\omega$ defined by any vector state Ω (corresponding to a cyclic unit vector $\Omega \in \mathcal{H}$) is equivalent to $\pi(\mathfrak{A})$.*

The operator $U : \mathcal{H}_\omega \to \mathcal{H}$ implementing the equivalence is initially defined on the dense subspace $\pi_\omega(\mathfrak{A})\Omega_\omega$ by $U\pi_\omega(A)\Omega_\omega = \pi(A)\Omega$; this operator is well-defined, for $\pi_\omega(A)\Omega_\omega = 0$ implies $\pi(A)\Omega = 0$ by the GNS-construction. It follows from (1.57) that U is unitary as a map from $\mathcal{H}_\omega$ to $U\mathcal{H}_\omega$, but since Ω is cyclic for π, the image of U is $\mathcal{H}$. Hence U is unitary. One verifies that U intertwines π_ω and π. ■

Corollary 1.5.6. *If the Hilbert spaces $\mathcal{H}_1, \mathcal{H}_2$ of two cyclic representations π_1, π_2 each contain a cyclic vector $\Omega_1 \in \mathcal{H}_1$, $\Omega_2 \in \mathcal{H}_2$ such that*

$$\omega_1(A) := (\Omega_1, \pi_1(A)\Omega_1) = (\Omega_2, \pi_2(A)\Omega_2) =: \omega_2(A)$$

for all $A \in \mathfrak{A}$, then $\pi_1(\mathfrak{A})$ and $\pi_2(\mathfrak{A})$ are equivalent.

By 1.5.5 the representation π_1 is equivalent to the GNS-representation π_{ω_1}, and π_2 is equivalent to π_{ω_2}. On the other hand, π_{ω_1} and π_{ω_2} are induced by the same state $\omega_1 = \omega_2$, so they must coincide. ■

The state ω is called **faithful** when its GNS-representation π_ω is faithful. This is guaranteed when $\mathcal{N}_\omega = 0$, but note that even in that case $\mathcal{H}_\omega$ does not coincide with $\mathfrak{A}$, as the topology on $\mathfrak{A}$ in the operator-norm is finer than the topology of the norm $\|A\|_\omega^2 := (A, A)_0^\omega$.

The GNS-construction leads to a simple proof of Theorem 1.1.8, which uses the following notion.

Definition 1.5.7. *The* **universal representation** *π_u of a C^*-algebra $\mathfrak{A}$ is the direct sum of all its GNS-representations $\pi_\omega, \omega \in \mathcal{S}(\mathfrak{A})$; hence it is defined on the Hilbert space $\mathcal{H}_u = \oplus_{\omega \in \mathcal{S}(\mathfrak{A})} \mathcal{H}_\omega$.*

Theorem 1.1.8 then follows by taking $\mathcal{H} = \mathcal{H}_{\text{u}}$; the desired isomorphism is π_{u}. If $\pi_{\text{u}}(A) = 0$ for some $A \in \mathfrak{A}$, then $\pi_\omega(A)\Omega_\omega = 0$ for all states ω, whence $\|\pi_\omega(A)\Omega_\omega\|^2 = \omega(A^*A) = 0$ by the GNS-construction, so that $A^*A = 0$ by 1.4.3, and finally $A = 0$ by (1.15). Hence π_{u} is faithful, and therefore isometric by 1.3.10.3. ∎

1.6 Examples of C^*-Algebras and State Spaces

In this section we give some elementary examples of C^*-algebras.

Example 1.6.1. *Commutative C^*-algebras*

Let X be a discrete space. Take $\mathfrak{A} := \ell_0(X)$, which is the closure (in the sup-norm) of $\ell_c(X)$. The space $\ell_0(X)$ is a C^*-algebra under pointwise multiplication and complex conjugation; see 1.2. By elementary Banach space theory, the dual of $\mathfrak{A}$ is $\mathfrak{A}^* = \ell^1(X)$ under the pairing $\rho(f) = \operatorname{Tr} \rho f := \sum_{x \in X} \rho(x) f(x)$. The positive cone in $\mathfrak{A}$ or $\mathfrak{A}^*$ consists of the positive functions f or ρ. The state space $\mathcal{S}(\mathfrak{A})$ is the set of those positive functions ρ for which $\operatorname{Tr} \rho = \sum_x \rho(x) = 1$.

For example, for a given $y \in X$, the function $\rho = \delta_y$, defined by $\delta_y(x) := \delta_{xy}$, is a state; one clearly has $\delta_y(f) = f(y)$. Hence by Corollary 1.5.6 the one-dimensional representation π_y, defined on $\mathcal{H}_y = \mathbb{C}$ by $\pi_y(f) := f(y)$, is equivalent to the GNS-representation π_{δ_y} (the pertinent cyclic vector in $\mathbb{C}$ is simply $\Omega = 1$).

A positive normalized function on X defines a faithful state when it is strictly positive on X. The GNS-representation $\pi_\rho(\ell_0(X))$ of a faithful state ρ is equivalent to the representation π on $\mathcal{H} = \ell^2(X)$ (with counting measure) by multiplication operators, i.e., $\pi(f)\Psi(x) := f(x)\Psi(x)$. To see this, we first write the inner product in $\ell^2(X)$ as $(\Psi, \Phi) = \operatorname{Tr} \Psi^*\Phi := \sum_x \overline{\Psi(x)}\Phi(x)$. Then note that since $\operatorname{Tr} \rho = 1$, one has $\rho^{1/2} \in \ell^2(X)$. It is clear from the property $\rho(x) > 0$ for all x that $\rho^{1/2}$ is a cyclic vector for $\pi(\ell_0(X))$, with the property $(\rho^{1/2}, \pi(f)\rho^{1/2}) = \rho(f)$ for all $f \in \ell_0(X)$. The equivalence between π_ρ and π then follows from Corollary 1.5.6.

Adding the fact that the double dual of $\mathfrak{A}$ is $\ell_0(X)^{**} = \ell^\infty(X)$, we summarize the situation by

$$\ell_c(X) \subseteq \ell^1(X) = \ell_0(X)^* \subseteq \ell^2(X) \subseteq \ell_0(X) \subseteq \ell^\infty(X) = \ell_0(X)^{**}. \tag{1.59}$$

When X is finite all inclusions are replaced by equalities; when X is infinite all inclusions are strict.

Now take X to be a locally compact Hausdorff space, and put $\mathfrak{A} := C_0(X)$ with the sup-norm; this is the closure of $C_c(X)$. Recall that a **Radon measure** is a Borel measure that is inner regular with respect to compact sets. By the Riesz representation theorem, $\mathfrak{A}^*$ is the space of all complex Radon measures μ on X with finite total mass $\mu(X)$. With $\mathfrak{A}^+$ consisting of the positive functions in $\mathfrak{A}$, the dual cone $\mathfrak{A}^{*+}$ is the subspace of $\mathfrak{A}^*$ of nonnegative finite Radon measures. The state space $\mathcal{S}(\mathfrak{A}) = M_1^+(X)$ then consists of the probability measures on X. The GNS-representation π_μ of a state $\mu \in \mathcal{S}(\mathfrak{A})$ is realized on $\mathcal{H}_\mu = L^2(X, \mu)$, on which $\pi_\mu(f)$ is f as a multiplication operator.

Example 1.6.2. *Noncommutative C^*-algebras*

When a Hilbert space $\mathcal{H} = \mathbb{C}^N$ is finite-dimensional, the "maximally noncommutative" C^*-algebra of operators on $\mathcal{H}$ is the algebra $\mathfrak{M}_N(\mathbb{C})$ of $N \times N$ matrices. The appropriate generalization of $\mathfrak{M}_N(\mathbb{C})$ to the infinite-dimensional case is as follows. A **projection** in a *-algebra is an element satisfying $p^2 = p^* = p$.

Definition 1.6.3. *Let $\mathcal{H}$ be a Hilbert space. The *-algebra $\mathfrak{B}_f(\mathcal{H})$ of* **finite-rank operators on** *$\mathcal{H}$ is the (finite) linear span of all finite-dimensional projections on $\mathcal{H}$. In other words, an operator $A \in \mathfrak{B}(\mathcal{H})$ lies in $\mathfrak{B}_f(\mathcal{H})$ when $A\mathcal{H} := \{A\Psi | \Psi \in \mathcal{H}\}$ is finite-dimensional.*

The **C^*-algebra $\mathfrak{B}_0(\mathcal{H})$ of compact operators** *on $\mathcal{H}$ is the norm-closure of $\mathfrak{B}_f(\mathcal{H})$ in $\mathfrak{B}(\mathcal{H})$ (with all C^*-algebraic operations borrowed from $\mathfrak{B}(\mathcal{H})$).*

It is clear that $\mathfrak{B}_f(\mathcal{H})$ is a *-algebra, since $p^* = p$ for any projection p. It is obvious that $\mathfrak{B}_f(\mathcal{H})$ is closed under right multiplication by elements of $\mathfrak{B}(\mathcal{H})$; since it is a *-algebra, it is therefore also closed under left multiplication. By continuity of multiplication in $\mathfrak{B}(\mathcal{H})$, it follows that $\mathfrak{B}_0(\mathcal{H})$ is an ideal in $\mathfrak{B}(\mathcal{H})$. It is easily verified that the unit operator $\mathbb{I}$ lies in $\mathfrak{B}_0(\mathcal{H})$ iff $\mathcal{H}$ is finite-dimensional.

We know from the theory of single operators on a Hilbert space that the image of the unit ball in $\mathcal{H}$ under an element $A \in \mathfrak{B}_0(\mathcal{H})$ is compact (in the strong topology on $\mathcal{H}$); this explains the name of $\mathfrak{B}_0(\mathcal{H})$. A self-adjoint operator $A \in \mathfrak{B}(\mathcal{H})$ is compact iff $A = \sum_i a_i[\Psi_i]$ (norm-convergent sum), where each eigenvalue a_i has finite multiplicity, and $\lim_{i\to\infty} |a_i| = 0$ (where the eigenvalues have been ordered so that $a_i \leq a_j$ when $i > j$). In other words, the set of eigenvalues is discrete, and can have only 0 as a possible accumulation point.

We now wish to determine the state space of $\mathfrak{B}_0(\mathcal{H})$. This involves the study of a number of other subspaces of $\mathfrak{B}(\mathcal{H})$, whose definition we recall.

Definition 1.6.4. *The* **Hilbert–Schmidt norm** *$\|A\|_2$ of $A \in \mathfrak{B}(\mathcal{H})$ is defined by*

$$\|A\|_2^2 := \sum_i \|A\mathbf{e}_i\|^2 = \mathrm{Tr}\,(|A|^2), \tag{1.60}$$

where $\{\mathbf{e}_i\}_i$ is an arbitrary basis of $\mathcal{H}$; the right-hand side is independent of the choice of basis. Also, $|A| := \sqrt{A^*A}$ is defined by the continuous functional calculus. The **Hilbert–Schmidt class** *$\mathfrak{B}_2(\mathcal{H})$ consists of all $A \in \mathfrak{B}(\mathcal{H})$ for which $\|A\|_2 < \infty$.*

The **trace norm** *$\|A\|_1$ of $A \in \mathfrak{B}(\mathcal{H})$ is defined by*

$$\|A\|_1 := \||A|^{1/2}\|_2^2 = \mathrm{Tr}\,|A|. \tag{1.61}$$

The **trace class** *$\mathfrak{B}_1(\mathcal{H})$ consists of all $A \in \mathfrak{B}(\mathcal{H})$ for which $\|A\|_1 < \infty$.*

The noncommutative analogue of (1.59) is as follows.

Theorem 1.6.5. *One has the inclusions*

$$\mathfrak{B}_f(\mathcal{H}) \subseteq \mathfrak{B}_1(\mathcal{H}) = \mathfrak{B}_0(\mathcal{H})^* \subseteq \mathfrak{B}_2(\mathcal{H}) \subseteq \mathfrak{B}_0(\mathcal{H}) \subseteq \mathfrak{B}(\mathcal{H}) = \mathfrak{B}_0(\mathcal{H})^{**}, \tag{1.62}$$

where the (isometric) identification of $\mathfrak{B}_1(\mathcal{H})$ with $\mathfrak{B}_0(\mathcal{H})^*$ is made through the pairing

$$\rho(A) = \operatorname{Tr} \rho A, \tag{1.63}$$

and $\mathfrak{B}(\mathcal{H})$ is (isometrically) identified with $\mathfrak{B}_1(\mathcal{H})^* = \mathfrak{B}_0(\mathcal{H})^{**}$ through the pairing $A(\rho) = \operatorname{Tr} \rho A$.

The definition of the norms in (1.6.4) easily leads to $\|A\| \leq \|A\|_i$ for $i = 1, 2$. Since $\mathfrak{B}_0(\mathcal{H})$, $\mathfrak{B}_1(\mathcal{H})$, and $\mathfrak{B}_2(\mathcal{H})$ are the completions of $\mathfrak{B}_f(\mathcal{H})$ in the norms $\|\cdot\|$, $\|\cdot\|_1$, and $\|\cdot\|_2$, respectively, these inequalities imply that $\mathfrak{B}_i(\mathcal{H}) \subseteq \mathfrak{B}_0(\mathcal{H})$ for $i = 1, 2$. Using the characterization of self-adjoint compact operators mentioned above, one then infers from 1.6.4 that $\|A\|_2 \leq \|A\|_1$, so that $\mathfrak{B}_1(\mathcal{H}) \subseteq \mathfrak{B}_2(\mathcal{H})$.

The inclusions $\mathfrak{B}_1(\mathcal{H}) \subseteq \mathfrak{B}_0(\mathcal{H})^*$ and $\mathfrak{B}(\mathcal{H}) \subseteq \mathfrak{B}_1(\mathcal{H})^*$ both follow from the (nontrivial) estimate

$$|\operatorname{Tr} \rho A| \leq \|A\| \, \|\rho\|_1. \tag{1.64}$$

To show that $\mathfrak{B}_0(\mathcal{H})^* \subseteq \mathfrak{B}_1(\mathcal{H})$ one restricts a given element $\hat{\rho} \in \mathfrak{B}_0(\mathcal{H})^*$ to $\mathfrak{B}_2(\mathcal{H})$, on which it is continuous. Now, the operator space $\mathfrak{B}_2(\mathcal{H})$ is a Hilbert space in the inner product $(A, B) := \operatorname{Tr} A^* B$, so that by Riesz–Fischer there must be an operator $\rho \in \mathfrak{B}_2(\mathcal{H})$ such that $\hat{\rho}(A) = \operatorname{Tr} \rho A$ for all $A \in \mathfrak{B}_2(\mathcal{H})$. One then shows that $|\operatorname{Tr} p|\rho|| \leq \|\hat{\rho}\|$ for any finite-dimensional projection p, which implies that $\|\rho\|_1 \leq \|\hat{\rho}\|$, so that $\rho \in \mathfrak{B}_1(\mathcal{H})$. With the opposite inequality from (1.64), this proves that $\mathfrak{B}_1(\mathcal{H}) = \mathfrak{B}_0(\mathcal{H})^*$ isometrically.

To establish the inclusion $\mathfrak{B}_1(\mathcal{H})^* \subseteq \mathfrak{B}(\mathcal{H})$, pick $\hat{A} \in \mathfrak{B}_1(\mathcal{H})^*$, and define a quadratic form Q_A on $\mathcal{H}$ by $Q_A(\Psi, \Phi) := \hat{A}(|\Phi\rangle\langle\Psi|)$. Here the operator $|\Phi\rangle\langle\Psi|$ is defined by $|\Phi\rangle\langle\Psi|\Omega := (\Psi, \Omega)\Phi$. This form is easily seen to be bounded by $\|\hat{A}\|$, so that it is implemented by a bounded operator A, in that $Q_A(\Psi, \Phi) = (\Psi, A\Phi)$. By linear extension to $\mathfrak{B}_f(\mathcal{H})$ and subsequently continuous extension to $\mathfrak{B}_1(\mathcal{H})$, this implies that $\hat{A}(\rho) = \operatorname{Tr} \rho A$, with $\|A\| \leq \|\hat{A}\|$. Since (1.64) implies the opposite inequality, this proves the last claim. □

Corollary 1.6.6. *The state space $\mathcal{S}(\mathfrak{B}_0(\mathcal{H}))$ of the C^*-algebra $\mathfrak{B}_0(\mathcal{H})$ of all compact operators on some Hilbert space $\mathcal{H}$ consists of all density matrices, where a* **density matrix** *is an element $\rho \in \mathfrak{B}_1(\mathcal{H})$ that is positive ($\rho \geq 0$) and has unit trace ($\operatorname{Tr} \rho = 1$), and the corresponding state is defined in (1.63).*

Since $\rho \in \mathfrak{B}_1(\mathcal{H})$ is compact, one may diagonalize it by $\rho = \sum_i p_i[\Psi_i]$. Using $A = [\Psi_i]$, which is positive, the condition $\rho(A) \geq 0$ yields $p_i \geq 0$. Conversely, when all $p_i \geq 0$, the operator ρ is positive. The normalization condition $\|\rho\|_1 = \sum p_i = 1$ completes the characterization of $\mathcal{S}(\mathfrak{B}_0(\mathcal{H}))$. ■

Proposition 1.6.7.

1. *For each unit vector $\Psi \in \mathcal{H}$ the GNS-representation $\pi_\psi(\mathfrak{B}_0(\mathcal{H}))$ corresponding to the density matrix $\rho = [\Psi]$ is equivalent to the defining representation.*
2. *The GNS-representation π_ρ corresponding to a faithful state ρ on $\mathfrak{B}_0(\mathcal{H})$ is equivalent to the representation $\hat{\pi}_\rho(\mathfrak{B}_0(\mathcal{H}))$ on the Hilbert space $\mathfrak{B}_2(\mathcal{H})$ of Hilbert–Schmidt operators given by left multiplication, i.e., $\hat{\pi}_\rho(A)B := AB$.*

The first claim is immediate from the property $\text{Tr}\,[\Psi]A = (\Psi, A\Psi)$ and 1.5.5. For the second, it is obvious from 1.6.4 that for $A \in \mathfrak{B}(\mathcal{H})$ and $B \in \mathfrak{B}_2(\mathcal{H})$ one has $\|AB\|_2 \leq \|A\|\,\|B\|_2$, so that the representation $\hat{\pi}_\rho$ is well-defined. When $\rho \in \mathfrak{B}_1(\mathcal{H})$ and $\rho \geq 0$, then $\rho^{1/2} \in \mathfrak{B}_2(\mathcal{H})$, and it is easily seen that $\rho^{1/2}$ is cyclic for $\hat{\pi}_\rho(\mathfrak{B}_0(\mathcal{H}))$ when $\hat{\rho}$ is faithful. Using the fact that for $A, B \subset \mathfrak{B}_2(\mathcal{H})$ one has $\text{Tr}\,AB = \text{Tr}\,BA$, we compute $(\rho^{1/2}, \hat{\pi}_\rho(A)\rho^{1/2}) = \rho(A)$. The equivalence between π_ρ and $\hat{\pi}_\rho$ then follows from 1.5.6. ■

1.7 Von Neumann Algebras

In this section we state some basic facts about von Neumann algebras (which will be used only as ancillary tools). The **commutant** $\mathfrak{M}'$ of some collection $\mathfrak{M}$ of bounded operators on a Hilbert space is the set of all bounded operators that commute with all elements of $\mathfrak{M}$; the **bicommutant** $\mathfrak{M}''$ is the commutant of $\mathfrak{M}'$. One verifies that $\mathfrak{M}''' = \mathfrak{M}'$. The main result is the so-called **double commutant theorem**, which we will first state in the finite-dimensional case.

Proposition 1.7.1. *Let $\mathcal{H} = \mathbb{C}^n$ be a finite-dimensional Hilbert space, and let $\mathfrak{M}$ be a *-algebra (and hence a C^*-algebra) in $\mathfrak{B}(\mathcal{H}) = \mathfrak{M}_n(\mathbb{C})$ containing $\mathbb{I}$. Then $\mathfrak{M}'' = \mathfrak{M}$.*

Choose some $\Psi \in \mathcal{H}$, form the linear subspace $\mathfrak{M}\Psi$ of $\mathcal{H}$, and consider the projection $p = [\mathfrak{M}\Psi]$ onto this subspace. By Lemma 1.5.3 one has $p \in \mathfrak{M}'$. Hence $A \in \mathfrak{M}''$ commutes with p. Since $\mathbb{I} \in \mathfrak{M}$, we therefore have $\Psi = \mathbb{I}\Psi \in \mathfrak{M}\Psi$, so $\Psi = p\Psi$, and $A\Psi = Ap\Psi = pA\Psi \in \mathfrak{M}\Psi$. Hence $A\Psi = A_0\Psi$ for some $A_0 \in \mathfrak{M}$.

Choose $\Psi_1, \ldots, \Psi_n \in \mathcal{H}$, and regard $\Omega := \Psi_1 \dot{+} \cdots \dot{+} \Psi_n$ as an element of $\mathcal{H}^n := \oplus^n \mathcal{H} \simeq \mathcal{H} \otimes \mathbb{C}^n$ (the direct sum of n copies of $\mathcal{H}$), where Ψ_i lies in the ith copy. Identify $\mathfrak{B}(\mathcal{H}^n)$ with the algebra $\mathfrak{M}_n(\mathfrak{B}(\mathcal{H}))$ of $n \times n$ matrices with entries in $\mathfrak{B}(\mathcal{H})$, and embed $\mathfrak{M}$ in $\mathfrak{M}_n(\mathfrak{B}(\mathcal{H}))$ by $A \mapsto \delta(A) := A\mathbb{I}_n^{\otimes}$, where $\mathbb{I}_n^{\otimes}$ is the unit in $\mathfrak{M}_n(\mathfrak{B}(\mathcal{H}))$; this is the diagonal matrix in $\mathfrak{M}_n(\mathfrak{B}(\mathcal{H}))$ in which all diagonal entries are A.

Now use the first part of the proof, with $\mathcal{H}$, $\mathfrak{M}$, A, and Ψ replaced by $\mathcal{H}^n$, $\delta(\mathfrak{M})$, $\mathbb{A} := \delta(A)$, and Ω, respectively. Hence given $\Psi_1, \ldots, \Psi_n$ and $\delta(A) \in \delta(\mathfrak{M})$ there exists $\mathbb{A}_0 \in \delta(\mathfrak{M})''$ such that $\delta(A)\Omega = \mathbb{A}_0\Omega$. For arbitrary $\mathbb{B} \in \mathfrak{M}_n(\mathfrak{B}(\mathcal{H}))$, compute $([\mathbb{B}, \delta(A)])_{ij} = [B_{ij}, A]$. Hence $\delta(\mathfrak{M})' = \mathfrak{M}_n(\mathfrak{M}')$. It is easy to see that $\mathfrak{M}_n(\mathfrak{M}')' = \mathfrak{M}_n(\mathfrak{M}'')$, so that $\delta(\mathfrak{M})'' = \delta(\mathfrak{M}'')$. Therefore, $\mathbb{A}_0 = \delta(A)_0$ for some $A_0 \in \mathfrak{M}$. Hence $A\Psi_i = A_0\Psi_i$ for all $i = 1, \ldots, n$. Since the Ψ_i were arbitrary, this proves that $A = A_0 \in \mathfrak{M}''$. ■

As it stands, Proposition 1.7.1 is not valid when $\mathfrak{M}_n(\mathbb{C})$ is replaced by $\mathfrak{B}(\mathcal{H})$, where $\dim(\mathcal{H}) = \infty$. To describe the appropriate refinement, we define two locally convex topologies on $\mathfrak{B}(\mathcal{H})$ that are weaker than the norm topology we have been using so far.

The seminorms $p_\Psi(A) := \|A\Psi\|$ define the **strong topology** on $\mathfrak{B}(\mathcal{H})$, so that $A_\lambda \to A$ strongly when $\|(A_\lambda - A)\Psi\| \to 0$ for all $\Psi \in \mathcal{H}$. In the proof of 1.7.2

we will use the fact that a neighborhood basis of A is given by all sets of the form $\{B \in \mathfrak{B}(\mathcal{H}) \mid \|(A - B)\Psi_i\| < \epsilon \text{ for all } i = 1, \ldots, n\}$, where $\epsilon > 0$, $n \in \mathbb{N}$, and $\Psi_1, \ldots, \Psi_n \in \mathcal{H}$.

The **weak topology** on $\mathfrak{B}(\mathcal{H})$ is defined by the seminorms $p_{\Psi,\Phi}(A) := |(\Psi, A\Phi)|$, so that $A_\lambda \to A$ weakly when $|(\Psi, (A_\lambda - A)\Psi)| \to 0$ for all $\Psi \in \mathcal{H}$. The norm topology is stronger than the strong topology, which in turn is stronger than the weak topology.

Theorem 1.7.2. *Let $\mathfrak{M}$ be a *-algebra in $\mathfrak{B}(\mathcal{H})$ containing $\mathbb{I}$. The following are equivalent:*

1. $\mathfrak{M}'' = \mathfrak{M}$.
2. *$\mathfrak{M}$ is closed in the weak operator topology.*
3. *$\mathfrak{M}$ is closed in the strong operator topology.*

It is easily verified from the definition of weak convergence that the commutant $\mathfrak{N}'$ of a *-algebra $\mathfrak{N}$ is always weakly closed. If $\mathfrak{M}'' = \mathfrak{M}$, then $\mathfrak{M} = \mathfrak{N}'$ for $\mathfrak{N} = \mathfrak{M}'$, so that $\mathfrak{M}$ is weakly closed. Hence $1 \Rightarrow 2$. Since the weak topology is weaker than the strong topology, $2 \Rightarrow 3$ is trivial.

To prove $3 \Rightarrow 1$, we adapt the proof of 1.7.1 to the infinite-dimensional situation. Instead of $\mathfrak{M}\Psi$, which may not be closed, we consider its closure $\overline{\mathfrak{M}\Psi}$, so that $p = [\overline{\mathfrak{M}\Psi}]$. Hence $A \in \mathfrak{M}''$ implies $A \in \overline{\mathfrak{M}\Psi}$; in other words, for every $\epsilon > 0$ there is an $A_\epsilon \in \mathfrak{M}$ such that $\|(A - A_\epsilon)\Psi\| < \epsilon$. For $\mathcal{H}^n$ this means that $\|\delta(A - A_\epsilon)\Omega\|^2 < \epsilon^2$. The left-hand side of this inequality equals the sum $\sum_{i=1}^n \|(A - A_\epsilon)\Psi_i\|^2$, so that $\|(A - A_\epsilon)\Psi_i\| < \epsilon$ for all $i = 1, \ldots, n$. It follows that $A_\epsilon \to A$ strongly for $\epsilon \to 0$. Since all $A_\epsilon \in \mathfrak{M}$ and $\mathfrak{M}$ is strongly closed, this implies that $A \in \mathfrak{M}$, so that $\mathfrak{M}'' \subseteq \mathfrak{M}$. Since trivially $\mathfrak{M} \subseteq \mathfrak{M}''$, this proves $3 \Rightarrow 1$. ∎

This theorem is remarkable, for it relates a topological condition ($\mathfrak{M}$ being closed in certain topologies) to an algebraic one ($\mathfrak{M}$ being its own bicommutant). A similar but simpler example of such a theorem states that a linear subspace $\mathcal{K}$ of a Hilbert space is closed iff $\mathcal{K} = \mathcal{K}^{\perp\perp}$ (where $\mathcal{K}^\perp$ is the orthogonal complement of $\mathcal{K}$).

Definition 1.7.3. *A *-algebra $\mathfrak{M}$ (containing the unit operator) of bounded operators on some Hilbert space is called a* **von Neumann algebra** *if it satisfies one (hence all) of the conditions in 1.7.2.*

We know from 1.6.5 that $\mathfrak{B}(\mathcal{H}) = \mathfrak{B}_1(\mathcal{H})^*$; the pertinent w^*-topology on $\mathfrak{B}(\mathcal{H})$ is often called the **σ-weak topology**. This topology is generated by the seminorms $p_\rho(A) := |\mathrm{Tr}\, \rho A|$, and is clearly stronger than the weak topology (but weaker than the norm topology). Hence a von Neumann algebra $\mathfrak{M} \subseteq \mathfrak{B}(\mathcal{H})$ is closed in the (relative) σ-weak topology.

Moreover, a von Neumann algebra $\mathfrak{M}$ is closed in the norm topology (defined by the norm (1.18)) as well, so that it is a C^*-algebra. A state on $\mathfrak{M} \subseteq \mathfrak{B}(\mathcal{H})$ of the form (1.63) for a density matrix ρ (cf. 1.6.6) is called **normal**. The linear span of all normal states in $\mathfrak{M}^*$ is called the **predual** $\mathfrak{M}_*$ of $\mathfrak{M}$. For example, the predual of $\mathfrak{B}(\mathcal{H})$ is $\mathfrak{B}_1(\mathcal{H})$, and more generally one has $\mathfrak{M} = \mathfrak{M}_*^*$ as a Banach space. The set $\mathcal{N}(\mathfrak{M}) := \mathcal{S}(\mathfrak{M}) \cap \mathfrak{M}_*$ of all normal states on $\mathfrak{M}$ is called the **normal state space** of $\mathfrak{M}$.

All von Neumann algebras in this book are of the form $\mathfrak{M} = \pi(\mathfrak{A})''$, where π is a representation of some C^*-algebra $\mathfrak{A}$. In particular, one may take $\pi = \pi_u$; cf. 1.5.7.

Proposition 1.7.4. *The bidual $\mathfrak{A}^{**}$ of a C^*-algebra $\mathfrak{A}$ is isomorphic (as a Banach space) to $\pi_u(\mathfrak{A})''$. Through this isomorphism, $\mathfrak{A}^{**}$ acquires the structure of a von Neumann algebra (and therefore of a C^*-algebra).*

The proof is a highly nontrivial generalization of the proof of 1.6.5. The equality $\mathfrak{B}_0(\mathcal{H})^* = \mathfrak{B}_1(\mathcal{H})$ is now replaced by the fact that $\mathfrak{A}^*$ is the linear span of all functionals of the form $A \mapsto (\Psi, \pi_u(\mathfrak{A})\Phi)$, where $\Psi, \Phi \in \mathcal{H}_u$. This characterization is then used to show that $\mathfrak{A}^*$ is the predual of $\pi_u(\mathfrak{A})''$, so that $\mathfrak{A}^{**} = \pi_u(\mathfrak{A})''$. □

In the context of Theorem 1.4.5, we note that when K is a Hausdorff compact convex set, the bidual of $A(K, \mathbb{R})$ (with sup-norm) is the space $A_b(K, \mathbb{R})$ of all bounded real-valued affine functions on K. Hence for a C^*-algebra $\mathfrak{A}$ one has $\mathfrak{A}^{**}_{\mathbb{R}} \simeq \pi_u(\mathfrak{A})''_{\mathbb{R}} \simeq A_b(\mathcal{S}(\mathfrak{A}), \mathbb{R})$. The predual of $\mathfrak{A}^{**}$ is obviously $\mathfrak{A}^{**}_* = \mathfrak{A}^*$, and the normal state space is $\mathcal{N}(\mathfrak{A}^{**}) = \mathcal{S}(\mathfrak{A})$. More generally, for any von Neumann algebra $\mathfrak{M}$ one has $\mathfrak{M}_{\mathbb{R}} \simeq A_b(\mathcal{N}(\mathfrak{M}), \mathbb{R})$ as partially ordered Banach spaces. This isomorphism maps the σ-weak topology on $\mathfrak{M}_{\mathbb{R}}$ to the topology of pointwise convergence on $A_b(\mathcal{N}(\mathfrak{M}), \mathbb{R})$.

The **center** of a von Neumann algebra $\mathfrak{M}$ is $\mathfrak{M} \cap \mathfrak{M}'$; this is the set of all elements of $\mathfrak{M}$ that commute with every element in the algebra. The following proposition allows one to regard $\pi(\mathfrak{A})''$ as a von Neumann subalgebra of $\mathfrak{A}^{**}$.

Proposition 1.7.5. *If π is a cyclic representation of a C^*-algebra $\mathfrak{A}$, there exists a projection p in the center of $\pi_u(\mathfrak{A})''$ such that $\pi(\mathfrak{A})''$ is isomorphic (as a von Neumann algebra) to $p\pi_u(\mathfrak{A})''$.*

The idea of the proof is that the morphism $\pi \circ \pi_u^{-1}$ from $\pi_u(\mathfrak{A})$ to $\pi(\mathfrak{A})$ is σ-weakly continuous, so that it can be extended to a morphism from $\pi_u(\mathfrak{A})''$ to $\pi(\mathfrak{A})''$. The kernel of this extension is a σ-weakly closed ideal in $\pi_u(\mathfrak{A})''$. It can be shown that a σ-weakly closed ideal in a von Neumann algebra $\mathfrak{M}$ is of the form $q\mathfrak{M}$, where q is a projection in the center of $\mathfrak{M}$. Applying this to the case at hand yields 1.7.5, with $p = \mathbb{I} - q$. □

2 The Structure of Pure State Spaces

2.1 Pure States and Compact Convex Sets

In this section we look at a subspace of the state space on a C^*-algebra, which may be interpreted as a quantum analogue of the phase space of a classical system.

Let us return to 1.4. One observes that the compact convex sets one naturally has in mind have a boundary; this particularly applies to the state spaces of the

C^*-algebras $\mathbb{C}$, $\mathbb{C} \oplus \mathbb{C}$, and $\mathfrak{M}_2(\mathbb{C})$. The intrinsic definition of this boundary is as follows.

Definition 2.1.1. *An* **extreme point** *in a convex set K is a member ω of K that can be decomposed as $\omega = \lambda\omega_1 + (1-\lambda)\omega_2$, where $\lambda \in (0,1)$, iff $\omega_1 = \omega_2 = \omega$. The collection $\partial_e K$ of extreme points in K is called the* **extreme boundary** *of K.*

An extreme point in the state space $K = \mathcal{S}(\mathfrak{A})$ of a C^*-algebra $\mathfrak{A}$ is called a **pure state**. *A state that is not pure is called* **mixed**. *We write $\mathcal{P}(\mathfrak{A})$, or simply $\mathcal{P}$, for $\partial_e\mathcal{S}(\mathfrak{A})$, referred to as the* **pure state space** *of $\mathfrak{A}$.*

Thus the single state on $\mathbb{C}$ is pure, the pure states on $\mathbb{C} \oplus \mathbb{C}$ are the points 0 and 1 in $[0, 1]$, and the pure states on $\mathfrak{M}_2(\mathbb{C})$ are the matrices ρ in (1.50) for which $x^2 + y^2 + z^2 = 1$. These are the projections onto one-dimensional subspaces of $\mathbb{C}^2$, and we see that $\mathcal{P}(\mathfrak{M}_2(\mathbb{C}))$ may be identified with the unit sphere in $\mathbb{R}^3$. More generally, one has

Proposition 2.1.2. *The pure state space of $\mathfrak{B}_0(\mathcal{H})$ consists of all one-dimensional projections, so that any pure state on $\mathfrak{B}_0(\mathcal{H})$ is a vector state (1.46) in $\mathcal{H}$.*

This is immediate from 1.6.6, the spectral theorem applied to a density matrix, and 2.1.1. ■

A useful reformulation of the notion of a pure state is as follows.

Proposition 2.1.3. *A state is pure iff $0 \le \rho \le \omega$ for a positive functional ρ implies $\rho = t\omega$ for some $t \in \mathbb{R}^+$.*

We assume that $\mathfrak{A}$ is unital; if not, use 1.2.1 and (1.49). For $\rho = 0$ or $\rho = \omega$ the claim is obvious. When ω is pure and $0 \le \rho \le \omega$, with $0 \ne \rho \ne \omega$, then $0 < \rho(\mathbb{I}) < 1$, since $\omega - \rho$ is positive; hence $\|\omega - \rho\| = \omega(\mathbb{I}) - \rho(\mathbb{I}) = 1 - \rho(\mathbb{I})$. Hence $\rho(\mathbb{I})$ would imply $\omega = \rho$, whereas $\rho(\mathbb{I}) = 0$ implies $\rho = 0$, contrary to assumption. Hence $\omega_1 := (\omega - \rho)/(1 - \rho(\mathbb{I}))$ and $\omega_2 := \rho/\rho(\mathbb{I})$ are states, and $\omega = \lambda\omega_1 + (1-\lambda)\omega_2$ with $\lambda = 1 - \rho(\mathbb{I})$. Since ω is pure, by 2.1.1 we have $\rho = \rho(\mathbb{I})\omega$.

Conversely, if ω is decomposed as in 2.1.1, then $0 \le \lambda\omega_1 \le \omega$, so that $\lambda\omega_1 = t\omega$ by assumption; normalization gives $t = \lambda$, hence $\omega_1 = \omega = \omega_2$, and ω is pure. ■

Here is another example of a pure state space.

Proposition 2.1.4. *The pure state space of the commutative C^*-algebra $C_0(X)$ (equipped with the relative w^*-topology) is homeomorphic to X.*

The case that X is not compact may be reduced to the compact case by passing from $\mathfrak{A} = C_0(X)$ to $\mathfrak{A}_{\mathbb{I}} = C(\tilde{X})$ (where $\tilde{X}$ is the one-point compactification of X); cf. (2.2) below. In view of the proof of Theorem 1.2.3, we then merely need to prove that any pure state on $C(X)$ is multiplicative, and vice versa; $\mathcal{P}(C(X))$ and $\Delta(C(X))$ are both equipped with the relative w^*-topology.

Let $\omega_x \in \Delta(C(X))$ (cf. the proof of 1.2.3), and suppose a functional ρ satisfies $0 \le \rho \le \omega_x$. Then $\ker(\omega_x) \subseteq \ker(\rho)$, and $\ker(\omega_x)$ is a maximal ideal, so that

$\ker(\omega_x) = \ker(\rho)$. Since two functionals on any vector space are proportional when they have the same kernel, it follows from 2.1.3 that ω_x is pure.

Conversely, let ω be a pure state, and pick a $g \in C(X)$ with $0 \leq g \leq 1_X$. Define a functional ω_g on $C(X)$ by $\omega_g(f) := \omega(fg)$. Since $\omega(f) - \omega_g(f) = \omega(f(1-g))$, and $0 \leq 1 - g \leq 1_X$, one has $0 \leq \omega_g \leq \omega$. Hence $\omega_g = t\omega$ for some $t \in \mathbb{R}^+$ by 2.1.3. Putting $f = 1_X$ yields $t = \omega(g)$. Since any function is a linear combination of functions g for which $0 \leq g \leq 1_X$, it follows that ω is multiplicative. ■

It could be that a given convex set contains no extreme points at all; think of an open convex cone. When K is compact, this possibility is excluded by a basic theorem in functional analysis, which we state without proof. The **convex hull** $\text{co}(V)$ of a subset V of a vector space is defined by

$$\text{co}(V) := \{\lambda v + (1 - \lambda)w \mid v, w \in V, \lambda \in [0, 1]\}. \tag{2.1}$$

Theorem 2.1.5. *A compact convex set K embedded in a locally convex vector space is the closure of the convex hull of its extreme points. In other words, $K = \overline{\text{co}}(\partial_e K)$.*

Although the state space of a C^*-algebra $\mathfrak{A}$ without unit (such as $\mathfrak{B}_0(\mathcal{H})$ or $C_0(X)$) is not compact, Theorem 2.1.5 may nonetheless be used. For the pure state space of $\mathfrak{A}$ may be described in terms of the pure state space $\mathcal{P}(\mathfrak{A}_{\mathbb{I}})$ of its unitization $\mathfrak{A}_{\mathbb{I}}$; cf. 1.2.1. Define a functional ω_∞ by $\omega_\infty(A + \lambda\mathbb{I}) = \lambda$ for all A; this is easily seen to be a pure state on $\mathfrak{A}_{\mathbb{I}}$. Taking (1.49) into account, one obtains a homeomorphism

$$\mathcal{P}(\mathfrak{A}) \simeq \mathcal{P}(\mathfrak{A}_{\mathbb{I}})\backslash\{\omega_\infty\}. \tag{2.2}$$

The extreme boundary $\partial_e K$ of a compact convex set is not necessarily closed, so that the pure state space $\mathcal{P}(\mathfrak{A})$ of a unital C^*-algebra $\mathfrak{A}$, while always a Hausdorff space, is not generally compact. Nonetheless, it is interesting to realize $\mathfrak{A}_{\mathbb{R}}$ as a subspace of $C(\mathfrak{A}_{\mathbb{R}}(\mathcal{P}), \mathbb{R})$, somewhat in the spirit of 1.4.5. To do so, we replace $\Delta(\mathfrak{A})$ in the definition (1.30) of the Gelfand transform by the pure state space of an arbitrary C^*-algebra; cf. 2.1.4.

Definition 2.1.6. *Let $\mathfrak{A}_{\mathbb{R}}$ be the self-adjoint part of a C^*-algebra $\mathfrak{A}$. The* **Gelfand transform** *of $A \in \mathfrak{A}_{\mathbb{R}}$ is the function $\hat{A} : \mathcal{P}(\mathfrak{A}) \to \mathbb{R}$ defined by (1.30). The subspace $\{\hat{A} \mid A \in \mathfrak{A}_{\mathbb{R}}\}$ of $\ell^\infty(\mathcal{P}(\mathfrak{A}), \mathbb{R})$ is denoted by $\hat{\mathfrak{A}}_{\mathbb{R}}$.*

The extension of the Gelfand transform from $\mathfrak{A}_{\mathbb{R}}$ to $\mathfrak{A}$ is useful only for commutative C^*-algebras; in the noncommutative case the first claim below would not hold if $\mathfrak{A}_{\mathbb{R}}$ were replaced by $\mathfrak{A}$.

Theorem 2.1.7. *The Gelfand transform is an isomorphism between $\mathfrak{A}_{\mathbb{R}}$ and $\hat{\mathfrak{A}}_{\mathbb{R}} \subseteq C(\mathcal{P}(\mathfrak{A}), \mathbb{R})$, seen as partially ordered Banach spaces (here the order in $\mathfrak{A}_{\mathbb{R}}$ is defined by 1.3.3 and 1.3.1, whereas the order in $C(\mathcal{P}(\mathfrak{A}), \mathbb{R})$ is defined by the cone of pointwise positive functions).*

The equality $\hat{\mathfrak{A}}_{\mathbb{R}} = C(\mathcal{P}(\mathfrak{A}), \mathbb{R})$ occurs iff $\mathfrak{A}$ is commutative and unital, in which case $\mathcal{P}(\mathfrak{A})$ is closed.

The first claim follows from 1.4.5 and 2.1.5: Any $A \in A(K, \mathbb{R})$ is determined by its values on $\partial_e K$ (for it is affine and continuous). The inclusion $\hat{\mathfrak{A}}_{\mathbb{R}} \subseteq C(\mathcal{P}(\mathfrak{A}), \mathbb{R})$ is immediate from the definition of the relative w^*-topology. The claim about the order is a trivial consequence of the pertinent definition, too. The last claim follows from 1.2.3 and 2.1.4. ■

An alternative proof may be obtained from the following sharpening of Proposition 1.4.3.

Proposition 2.1.8. *For every $A \in \mathfrak{A}_{\mathbb{R}}$ and $a \in \sigma(A)$ there is a pure state ω_a on $\mathfrak{A}$ for which $\omega_a(A) = a$. There exists a pure state ω such that $|\omega(A)| = \|A\|$.*

We extend the state in the proof of 1.4.3 to $C^*(A)$ by multiplicativity and continuity, that is, we put $\tilde{\omega}_a(A^n) = a^n$, etc. It follows from 2.1.4 that this extension is pure. One easily checks that the set of all extensions of $\tilde{\omega}_a$ to $\mathfrak{A}$ (which extensions we know to be states; see the proof of 1.4.3) is a closed convex subset K_a of $\mathcal{S}(\mathfrak{A})$; hence it is a compact convex set. By Theorem 2.1.5, K_a has at least one extreme point ω_a. If ω_a were not an extreme point in $\mathcal{S}(\mathfrak{A})$, it would be decomposable as in 2.1.1. But in that case ω_1 and ω_2 would both coincide on $C^*(A)$ with $\tilde{\omega}_a$, so that ω_a cannot be an extreme point of K_a. ■

In any case, when $\mathfrak{A}$ is noncommutative one would like to characterize $\hat{\mathfrak{A}}_{\mathbb{R}}$ in $C(\mathcal{P}(\mathfrak{A}), \mathbb{R})$. This will be done in Theorem 3.2.1.

2.2 *Pure States and Irreducible Representations*

In this section we start our analysis of irreducible representations of C^*-algebras and their connection to pure states.

Definition 2.2.1. *A representation π of a C^*-algebra $\mathfrak{A}$ on a Hilbert space $\mathcal{H}$ is called* **irreducible** *if a closed subspace of $\mathcal{H}$ that is stable under $\pi(\mathfrak{A})$ is either $\mathcal{H}$ or 0.*

This definition should be familiar from the theory of group representations. The defining representations of $\mathfrak{M}_n(\mathbb{C})$, $\mathfrak{B}_0(\mathcal{H})$, and $\mathfrak{B}(\mathcal{H})$ are evidently irreducible.

Proposition 2.2.2. *Each of the following conditions is equivalent to the irreducibility of $\pi(\mathfrak{A})$:*

1. $\pi(\mathfrak{A})' = \mathbb{C}\mathbb{I}$, *or, equivalently,* $\pi(\mathfrak{A})'' = \mathfrak{B}(\mathcal{H})$ (**Schur's lemma**).
2. *Every nonzero vector Ω in $\mathcal{H}$ is cyclic for $\pi(\mathfrak{A})$ (i.e., $\pi(\mathfrak{A})\Omega$ is dense in $\mathcal{H}$ for all $\Omega \neq 0$).*

The commutant $\pi(\mathfrak{A})'$ is a *-algebra in $\mathfrak{B}(\mathcal{H})$, so when it is nontrivial it must contain a self-adjoint element A that is not a multiple of $\mathbb{I}$. It follows from Theorem 1.7.2 and the spectral theorem that the projections in the spectral resolution of A lie in $\pi(\mathfrak{A})'$ if A does. Hence when $\pi(\mathfrak{A})'$ is nontrivial it contains a nontrivial projection p. But then $p\mathcal{H}$ is stable under $\pi(\mathfrak{A})$, contradicting irreducibility. Hence 2.2.1 $\Rightarrow$ 2.2.2.1.

Conversely, when $\pi(\mathfrak{A})' = \mathbb{C}\mathbb{I}$ and π is reducible, one finds a contradiction because the projection onto the alleged nontrivial stable subspace of $\mathcal{H}$ commutes with $\pi(\mathfrak{A})$.

When there exists a vector $\Psi \in \mathcal{H}$ for which $\pi(\mathfrak{A})\Psi$ is not dense in $\mathcal{H}$, we can form the projection onto the closure of $\pi(\mathfrak{A})\Psi$. By Lemma 1.5.3, with $\mathfrak{M} = \pi(\mathfrak{A})$, this projection lies in $\pi(\mathfrak{A})'$, so that by Schur's lemma π cannot be irreducible. Hence 2.2.1 $\Rightarrow$ 2.2.2.2. The converse is trivial. ■

The connection between representations and states (see 1.5) can be refined when a state is pure.

Theorem 2.2.3. *The GNS-representation $\pi_\omega(\mathfrak{A})$ of a state $\omega \in \mathcal{S}(\mathfrak{A})$ is irreducible iff ω is pure.*

When ω is pure yet $\pi_\omega(\mathfrak{A})$ reducible, there is a nontrivial projection $p \in \pi_\omega(\mathfrak{A})'$ by Schur's lemma. Let Ω_ω be the cyclic vector for π_ω. If $p\Omega_\omega = 0$, then $Ap\Omega_\omega = pA\Omega_\omega = 0$ for all $A \in \mathfrak{A}$, so that $p = 0$, since π_ω is cyclic. Similarly, $p^\perp\Omega_\omega = 0$ is impossible. We may then decompose $\omega = \lambda\psi + (1-\lambda)\psi^\perp$, where ψ and $\psi^\perp$ are states defined as in (1.46), with $\Psi := p\Omega_\omega/\|p\Omega_\omega\|$, $\Psi^\perp := p^\perp\Omega_\omega/\|p^\perp\Omega_\omega\|$, and $\lambda = \|p^\perp\Omega_\omega\|^2$. Hence ω cannot be pure, so that π_ω is irreducible by reductio ad absurdum.

In the opposite direction, suppose π_ω is irreducible, yet ω decomposable as in 2.1.1. Then $\lambda\omega_1 - \omega = (1-\lambda)\omega_2$, which is positive; hence $\lambda\omega_1(A^*A) \leq \omega(A^*A)$ for all $A \in \mathfrak{A}$. By (1.48) this yields $|\lambda\omega_1(A^*B)|^2 \leq \omega(A^*A)\omega(B^*B)$ for all A, B. This makes the quadratic form $\hat{Q}$ on $\pi_\omega(\mathfrak{A})\Omega_\omega$ by $\hat{Q}(\pi_\omega(A)\Omega_\omega, \pi_\omega(B)\Omega_\omega) := \lambda\omega_1(A^*B)$ well-defined. Furthermore, $\hat{Q}$ is bounded with norm 1, so that $\hat{Q}$ can be extended to $\mathcal{H}_\omega$ by continuity. Since ω is a Hermitian functional, one has $\hat{Q}(\Phi, \Psi) = \overline{\hat{Q}(\Psi, \Phi)}$.

Thus there exists a self-adjoint operator Q on $\mathcal{H}_\omega$ such that $\hat{Q}(\Psi, \Phi) = (\Psi, Q\Phi)$ for all $\Psi, \Phi \in \mathcal{H}$. In other words, one has $(\pi_\omega(A)\Omega_\omega, Q\pi_\omega(B)\Omega_\omega) = \lambda\omega_1(A^*B)$. Since π_ω is a representation, one computes that $[Q, \pi_\omega(C)] = 0$ for all $C \in \mathfrak{A}$, so that $Q \in \pi_\omega(\mathfrak{A})'$. Since π_ω is irreducible, one must have $Q = t\mathbb{I}$ for some $t \in \mathbb{R}$. Hence ω_1 is proportional to ω, and therefore equal to ω by normalization, so that ω is pure. ■

Here are some easy consequences of this result, culminating in 2.2.6.

Proposition 2.2.4. *If $(\pi(\mathfrak{A}), \mathcal{H})$ is irreducible, then the GNS-representation $(\pi_\omega(\mathfrak{A}), \mathcal{H}_\omega)$ defined by any vector state ω (corresponding to a unit vector $\Omega \in \mathcal{H}$) is equivalent to $(\pi(\mathfrak{A}), \mathcal{H})$. In particular, any vector state in an irreducible representation is pure.*

Immediate from 2.2.2.2, 1.5.5, and 2.2.3. ■

Corollary 2.2.5. *Every irreducible representation of a C^*-algebra comes from a pure state via the GNS-construction.*

Combine 2.2.4 and 2.2.3. ■

This leads straight to a basic result in the theory of C^*-algebras:

Theorem 2.2.6. *The C^*-algebra $\mathfrak{B}_0(\mathcal{H})$ of all compact operators on some Hilbert space possesses only one irreducible representation, up to equivalence, namely the defining one.*

This is immediate from 2.1.2, 2.2.5, and 2.2.4. ■

Two pure states ρ and σ on a C^*-algebra $\mathfrak{A}$ are said to be **equivalent** if the associated GNS-representations π_ρ and π_σ are equivalent; we write $\rho \sim \sigma$. It is easily verified from the definition of this notion of equivalence that $\sim$ is an equivalence relation in $\mathcal{P}(\mathfrak{A})$. It follows from 2.2.4 that all vector states in $\mathcal{H}_\rho$ are equivalent to ρ. Conversely, any state $\sigma \sim \rho$ is given by a vector state in $\mathcal{H}_\rho$, for if $U : \mathcal{H}_\sigma \rightarrow \mathcal{H}_\rho$ intertwines π_σ and π_ρ, then the vector state defined by $U\Omega_\sigma \in \mathcal{H}_\rho$ coincides with σ. Since the intertwiner is unique by Schur's lemma, one thus obtains a bijection between the equivalence class $[\rho]$ of a given pure state ρ and the set of vector states in $\mathcal{H}_\rho$.

The topological aspects of this bijection will be clarified in 2.5. For now, we are led to a manageable refinement of the the universal representation (cf. 1.5.7), which is still faithful.

Definition 2.2.7. *The* **reduced atomic representation** *π_{ra} of a C^*-algebra $\mathfrak{A}$ is the direct sum over irreducible representations $\pi_{\text{ra}} = \oplus_{\rho\in[\mathcal{P}(\mathfrak{A})]}\pi_\rho$ (on the Hilbert space $\mathcal{H}_{\text{ra}} = \oplus_{\rho\in[\mathcal{P}(\mathfrak{A})]}\mathcal{H}_\rho$), where one includes one representative of each equivalence class in $\mathcal{P}(\mathfrak{A})$.*

The specific choice of pure states in each equivalence class affects the reduced atomic representation only within (unitary) equivalence. Replacing the use of 1.4.3 in the proof of Theorem 1.1.8 by 2.1.8, one infers that π_{ra} is indeed faithful. If ρ and σ are inequivalent pure states, Schur's lemma implies that

$$\pi_{\text{ra}}(\mathfrak{A})'' = \oplus_{\omega\in[\mathcal{P}(\mathfrak{A})]}\mathfrak{B}(\mathcal{H}_\omega). \tag{2.3}$$

If $\mathfrak{A}$ is commutative, so that $\mathfrak{A}_{\mathbb{R}} \simeq C(\mathcal{P}(\mathfrak{A}), \mathbb{R})$ (see 2.1.7), one easily infers that $\pi_{\text{ra}}(\mathfrak{A})'' = \ell^\infty(\mathcal{P}(\mathfrak{A}))$. On the noncommutative side, we infer

Proposition 2.2.8. *Every finite-dimensional C^*-algebra is a direct sum of matrix algebras.*

Since $\mathfrak{A}$ is finite-dimensional and $\mathfrak{A} \simeq \pi_{\text{ra}}(\mathfrak{A})$, the right-hand side must be finite-dimensional. Hence by 1.7.1 and (2.3) one has $\mathfrak{A} \simeq \oplus_{\omega\in[\mathcal{P}(\mathfrak{A})]}\mathfrak{B}(\mathcal{H}_\omega)$, where each $\mathcal{H}_\omega$ is finite-dimensional and the sum is finite. ■

2.3 Poisson Manifolds

We return to Poisson algebras (cf. 1.1.2). The main source of such algebras is the following.

Definition 2.3.1. *A* **Poisson manifold** *is a manifold P equipped with a bilinear operation $\{\,,\,\} : C^\infty(P,\mathbb{R}) \times C^\infty(P,\mathbb{R}) \to C^\infty(P,\mathbb{R})$ with the property that $(C^\infty(P,\mathbb{R}), \circ, \{\,,\,\})$, where $\circ$ is pointwise multiplication, is a Poisson algebra.*

By definition, the map $g \mapsto \{f, g\}$ (for fixed $f \in C^\infty(P,\mathbb{R})$) is a derivation on $C^\infty(P,\mathbb{R})$, and this implies that the Poisson bracket $\{f, g\}(\sigma)$ depends only on the differentials df and dg at $\sigma \in P$. Therefore, there exists a smooth antisymmetric tensor field $B \in \Gamma(\wedge_2(P))$ such that

$$\{f, g\} = B(df, dg). \tag{2.4}$$

The Jacobi identity implies that the **Poisson tensor** B must satisfy

$$\iota_B d\iota_B \alpha = 0 \quad \forall \alpha \in \wedge^3(P). \tag{2.5}$$

(Recall that the insertion ι of $A \in \wedge_n(P)$ into $\beta \in \wedge^{n+m}(P)$ produces an element $\iota_A\beta \in \wedge^m(P)$ defined by $(\iota_A\beta)(C) = \beta(A \wedge C)$ for all $C \in \wedge_m(P)$.) If P is finite-dimensional, this can be conveniently stated in terms of local coordinates $\{\sigma^a\}$: If $B^{ab}(\sigma) = B_\sigma(d\sigma^a \wedge d\sigma^b)$, so that $B^{ba} = -B^{ab}$, then

$$B^{ea}\frac{\partial B^{bc}}{\partial \sigma^e} + B^{ec}\frac{\partial B^{ab}}{\partial \sigma^e} + B^{eb}\frac{\partial B^{ca}}{\partial \sigma^e} = 0. \tag{2.6}$$

Conversely, an element $B \in \Gamma(\wedge_2(P))$ satisfying (2.5) (or 2.6)) defines a Poisson bracket by (2.4).

The Poisson tensor B defines a linear map $B^\sharp : T^*P \to TP$ by

$$(B^\sharp(\alpha))(\beta) := B(\alpha, \beta), \tag{2.7}$$

where α and β lie in the same fiber in T^*P. If $h \in C^\infty(P,\mathbb{R})$, the image $B^\sharp(dh)$ is usually written as ξ_h, and called the **Hamiltonian vector field** of h. Hence

$$\xi_h f = B^\sharp(dh) f = \{h, f\}. \tag{2.8}$$

By virtue of the Jacobi identity, one has

$$[\xi_f, \xi_g] = \xi_{\{f,g\}} \tag{2.9}$$

and

$$L_{\xi_f} B = 0 \tag{2.10}$$

for all $f, g \in C^\infty(P,\mathbb{R})$, where L is the Lie derivative. Hence $f \mapsto \xi_f$ is a homomorphism from $C^\infty(P,\mathbb{R})$ into the subspace of $\Gamma(TP)$ (regarded as a Lie algebra under the commutator) of vector fields preserving the Poisson structure.

If $c : I \to P$ (where $I \subseteq \mathbb{R}$ is some interval containing 0) is a curve in P for which $c(0) = \sigma$, we write $\sigma(t)$ for $c(t)$. Given $h \in C^\infty(P,\mathbb{R})$, Hamilton's equations of motion for such a curve are

$$\frac{d\sigma(t)}{dt} = \xi_h(\sigma(t)). \tag{2.11}$$

A curve satisfying this equation for some h is called a **Hamiltonian curve**. The corresponding flow, given by

$$F_t(\sigma) = \sigma(t), \tag{2.12}$$

is called the **Hamiltonian flow** of h. A trivial consequence of (2.11) and (2.8) is

Proposition 2.3.2. *A function $h \in C^\infty(P, \mathbb{R})$ is constant along the flow trajectories it generates.*

The theory of ordinary differential equations (Picard iterations) guarantees existence and uniqueness of a local solution for each initial value $c(0) \in P$ and t in some compact interval around 0. When the motion exists, one has the property $F_s \circ F_t = F_{s+t}$. Given h and $c(0)$, it may happen that the motion is not defined for all $t \in \mathbb{R}$, in which case the vector field ξ_h is called **incomplete**. If ξ_h has compact support, it is always complete.

Given $h \in C^\infty(P, \mathbb{R})$ with Hamiltonian flow $\sigma(t)$, one constructs a one-parameter family of linear maps $\alpha_t^0 : C^\infty(P, \mathbb{R}) \to C^\infty(P, \mathbb{R})$ by

$$\alpha_t^0(f)(\sigma) := f(\sigma(t)). \tag{2.13}$$

This family is evidently defined only for those t for which the solution of (2.11) is defined for any initial value. One infers from (2.8) and (2.11) that the infinitesimal version of (2.13) is

$$\frac{d\alpha_t^0(f)}{dt} = \{h, \alpha_t^0(f)\}; \tag{2.14}$$

here the derivative is understood pointwise. The following result is a local version of the "infinitesimal" fact (2.10).

Proposition 2.3.3. *If $\alpha_t^0(f)$ satisfies (2.14), then α_t^0 is a morphism (cf. 1.1.3) of $C^\infty(P, \mathbb{R})$ for each t for which it is defined.*

The Leibniz rule and (2.14) imply $d[\alpha_t(fg)]/dt = d[\alpha_t(f)\alpha_t(g)]/dt$; the proposition follows by integrating this relation. ■

If the motion pertinent to h is defined for all $t \in \mathbb{R}$, one obtains a one-parameter group of automorphisms in this way. Equation (2.14) evidently makes sense in any Poisson algebra.

Definition 2.3.4. *An element h of a Poisson algebra is called* **complete** *if the one-parameter family of automorphisms defined by (2.14) is defined for all $t \in \mathbb{R}$.*

For Poisson algebras of the type $C^\infty(P, \mathbb{R})$ this amounts to saying that the flow of ξ_h is complete.

We will frequently need the notion of a **Poisson map** $J : (P_1, B^1) \to (P_2, B^2)$; this is a smooth map such that, in obvious notation,

$$J^*\{f, g\}_2 = \{J^*f, J^*g\}_1 \tag{2.15}$$

for all $f, g \in C^\infty(P_2, \mathbb{R})$. Equivalently,

$$B_\sigma^1(J^*\alpha, J^*\beta) = B_{J(\sigma)}^2(\alpha, \beta) \tag{2.16}$$

for all $\sigma \in P_1$ and all $\alpha, \beta \in T^*_{J(\sigma)}P_2$.

It follows from the Jacobi identity that the flow $F_t : P \to P$ of each Hamiltonian vector field ξ_f is a Poisson map (for all t for which the flow is defined). Moreover, a chasing of the definitions shows the validity of

Proposition 2.3.5. *For any Poisson map $J : P_1 \to P_2$ one has*

$$J_* \xi_{J^* f} = \xi_f \circ J \tag{2.17}$$

for all $f \in C^\infty(P_2, \mathbb{R})$. Moreover, the image of the flow of $\xi_{J^* f}$ under J is the flow of ξ_f.

In the present setting, symplectic spaces are regarded as special instances of Poisson manifolds.

Definition 2.3.6. *A Poisson manifold for which the map $B^\sharp$ is an isomorphism is called* **symplectic**. *If $B_\sharp : TP \to T^*P$ is the inverse of $B^\sharp$, the* **symplectic form** *$\omega \in \Gamma(\wedge^2(P))$ is defined by*

$$\omega(X, Y) := (B_\sharp(X))(Y). \tag{2.18}$$

As a consequence of the Jacobi identity (or (2.5)), ω is closed ($d\omega = 0$). In terms of the symplectic form, the Poisson bracket reads

$$\{f, g\} = -\omega(\xi_f, \xi_g), \tag{2.19}$$

where ξ_f and ξ_g are defined as in (2.8), that is,

$$\xi_f = B_\sharp^{-1}(df), \tag{2.20}$$

and this is equivalent to the connection

$$i_{\xi_f}\omega = df. \tag{2.21}$$

The following characterization of symplectic manifolds follows directly from the definition and the local existence of Hamiltonian flows.

Proposition 2.3.7. *A Poisson manifold is symplectic iff one of the following equivalent properties is satisfied:*

- *The collection of Hamiltonian vector fields $\{\xi_f, f \in C^\infty(P, \mathbb{R})\}$, or, equivalently, the image of $B^\sharp$, spans $T_\sigma P$ at each $\sigma \in P$.*
- *Any two points of P can be connected by a piecewise smooth Hamiltonian curve.*

When P is finite-dimensional, the first condition simply states that at every point the rank of $B^\sharp$ (that is, the dimension of the image of $B^\sharp$ at a given point) equals the dimension of P.

The cotangent bundle T^*Q of any manifold Q is symplectic.

Definition 2.3.8. *The* **canonical symplectic form** *ω on a cotangent bundle T^*Q is given by $\omega = -d\Theta$, where Θ is a one-form on T^*Q defined by*

$$\Theta_\sigma(X) := \sigma_{\tau(\sigma)}(\tau_*(X)), \tag{2.22}$$

where $\tau := \tau_{T^*Q \to Q}$.

In canonical coordinates (p, q) on T^*Q this reads

$$\omega = dq^i \wedge dp_i, \tag{2.23}$$

and the associated Poisson bracket is given by

$$\{f, g\} := \frac{\partial f}{\partial p_i}\frac{\partial g}{\partial q^i} - \frac{\partial f}{\partial q^i}\frac{\partial g}{\partial p_i}. \tag{2.24}$$

A diffeomorphism $J : S_1 \to S_2$ between two symplectic manifolds that is a Poisson map is called a **symplectomorphism**; S_1 and S_2 are **symplectomorphic** when such a map exists.

2.4 The Symplectic Decomposition of a Poisson Manifold

In this section we argue that an arbitrary finite-dimensional Poisson manifold is foliated by symplectic subspaces; this is somewhat analogous to the decomposition of a finite-dimensional C^*-algebra as a direct sum of matrix algebras; cf. 2.2.8. In preparation, we recall some differential geometry.

Definition 2.4.1. *A* **distribution** *D on a manifold P is a subset of the tangent bundle TP such that $D_\sigma := D \cap T_\sigma P$ is a vector space for each $\sigma \in P$. The* **rank** *of D at σ is the dimension of D_σ.*

A distribution is called **smooth** *if for every $\sigma \in P$ and $v \in D_\sigma$ there is a smooth vector field ξ, defined on a neighborhood $\mathcal{N}$ of σ, such that $\xi(\rho) \in D_\rho$ for all $\rho \in \mathcal{N}$, and $\xi(\sigma) = v$. Such a ξ is called a* **local section** *of D.*

A distribution is called **involutive** *if for any pair ξ_1, ξ_2 of local sections one has $[\xi_1, \xi_2](\rho) \in D_\rho$ in their common domain of definition.*

A distribution D on P is **completely integrable** *when each point $\sigma \in P$ lies in an immersed submanifold $S_\sigma \subseteq P$ whose tangent space at σ is D_σ.*

(One sometimes speaks of a **generalized** distribution when the rank of D is not constant on P; we will, instead, speak of a **regular distribution** when the rank is constant.)

Hence a completely integrable distribution defines a foliation of P, whose leaves are the S_σ. The leaves of a completely integrable foliation may have varying dimension. (Such a foliation is sometimes called **singular**; again, we will rather use the adjective **regular** when the leaf dimension is constant.)

For smooth regular distributions the question of complete integrability is settled by the well-known **Frobenius theorem**, which states that D is completely integrable iff it is involutive. In general, one needs a stronger condition (the "singular Frobenius theorem") to arrive at completely integrability, which we state without proof.

Lemma 2.4.2. *A smooth distribution D is integrable iff at each $\sigma \in P$ one can choose local sections $\xi_1, \ldots, \xi_{\text{rank}(D_\sigma)}$ that span D_σ with the property that for an* *arbitrary local section ξ of D (defined around σ) one has*

$$[\xi, \xi_i](\sigma(t)) = \lambda_i^j(t)\xi_j(\sigma(t)) \tag{2.25}$$

for small enough t. Here $\sigma \mapsto \sigma(t)$ is the (local) flow generated by ξ, and the λ_i^j are certain functions of t.

We return to Poisson manifolds. In general, the map $B^\sharp$ may fail to be surjective. The image of $B^\sharp$ defines a distribution D on TP, which is easily seen to be smooth; for it is generated by the Hamiltonian vector fields, each of which is a smooth section of the tangent bundle. The rank of $B^\sharp$ is not necessarily constant, so that D may not be regular.

Definition 2.4.3. *A* **symplectic leaf** *in a Poisson manifold (P, B) is a maximal set of points that are equivalent under the following equivalence relation: $\rho \sim \sigma$ iff ρ and σ can be connected by a piecewise smooth Hamiltonian curve.*

The terminology will be justified shortly. This equivalence relation leads to a decomposition $P = \cup_\alpha S_\alpha$, where each S_α is a symplectic leaf.

Lemma 2.4.4. *The rank of $B^\sharp$ is constant on each symplectic leaf.*

This is simply because the flow of each Hamiltonian vector field ξ_f is a Poisson map, and such maps leave B (and therefore $B^\sharp$) invariant, cf. (2.10) and (2.16). In particular, the pushforward of a Hamiltonian flow F_t maps the image of $B^\sharp$ at some σ into its image at $F_t(\sigma)$. ■

Using 2.4.4, 2.4.2 (with $\xi_i = \xi_{f_i}$ for suitable f_i), and (2.9), one infers that D is completely integrable, and it will become clear shortly that the leaves of the foliation defined by D are just the symplectic leaves of S.

In general, a given symplectic leaf $S_\alpha \subset P$ may not be a submanifold of P. Nonetheless, one may turn S_α into a manifold by a standard procedure of (singular) foliation theory. In the present context, this is accomplished by defining a chart around a given $\sigma \in S_\alpha$ in the following way. Let the rank of $B^\sharp$ at σ be n, and choose functions $f_1, \ldots, f_n$ such that $\{\xi_{f_i}\}_{i=1,\ldots,n}$ spans the image of $B^\sharp$ at σ. There is an $\epsilon > 0$ and an ϵ-ball $O_\epsilon \subset \mathbb{R}^n$ around 0 such that $F : O_\epsilon \to P$, defined by

$$F(t_1, \ldots, t_n) = F_{t_1}^1 \circ \cdots \circ F_{t_n}^n(\sigma),$$

where F_t^i denotes the flow of ξ_{f_i}, is a bijection.

Lemma 2.4.5. *Applying the above procedure for a sufficient number of points $\sigma \in S_\alpha$ leads to an atlas on S_α that is well-defined and independent, up to smooth equivalence, of the choice of the f_i at each point. The dimension of S_α with this manifold structure is the rank of $B^\sharp$.*

The pushforward of each F_t^i, and therefore of $F_{t_1}^1 \circ \cdots \circ F_{t_n}^n$, maps the image of $B^\sharp$ at σ into its image at $F_t^i(\sigma)$. ■

Equipped with this manifold structure and topology, each S_α is an injectively immersed submanifold of P; that is, the inclusion $\iota_\alpha : S_\alpha \hookrightarrow P$ is continuous and of constant rank, equal to the dimension of S_α, at each point.

The singular Frobenius theorem 2.4.2 applies; indeed, the leaf of the pertinent foliation is locally given by $F(O_\epsilon)$.

Lemma 2.4.6. *If $f \in C^\infty(P, \mathbb{R})$ vanishes on S_α, then $\{f, g\}(\sigma) = 0$ for all $\sigma \in S_\alpha$ and all $g \in C^\infty(P, \mathbb{R})$. Therefore, one can define a Poisson bracket $\{\ ,\ \}_\alpha$ on S_α by*

$$\{\iota_\alpha^* f, \iota_\alpha^* g\}_\alpha := \iota_\alpha^*\{f, g\}. \tag{2.26}$$

Each S_α is a symplectic manifold, and each inclusion ι_α is a Poisson map.

If $f = 0$ on S_α, then $\{f, g\} = -\xi_g f = 0$, since ξ_g is tangent to S_α. ■

Thus we arrive at

Theorem 2.4.7. *For each finite-dimensional Poisson manifold P there exists a family $\{S_\alpha\}$ of symplectic manifolds, and injective Poisson immersions $\iota_\alpha : S_\alpha \hookrightarrow P$, such that $P = \cup_\alpha \iota_\alpha(S_\alpha)$ (disjoint union). Each subset $\iota_\alpha(S_\alpha)$ is a symplectic leaf of P as defined in 2.4.3. The value of the Poisson bracket $\{f, g\}$ at some $\sigma \in P$ depends only on the restrictions of f and g to the symplectic leaf through σ.*

In the text preceding the theorem we have made no notational distinction between S_α and $\iota_\alpha(S_\alpha)$. Indeed, if each S_α is a submanifold, one can simply say that $P = \cup_\alpha S_\alpha$ as manifolds.

2.5 (Projective) Hilbert Spaces as Symplectic Manifolds

In this section we look at the geometric structure of $\mathcal{P}(\mathfrak{B}_0(\mathcal{H}))$.

Definition 2.5.1. *The* **projective space** *$\mathbb{P}\mathcal{H}$ of a Hilbert space $\mathcal{H}$ is the space of one-dimensional complex linear subspaces of $\mathcal{H}$. Equivalently, $\mathbb{P}\mathcal{H}$ is the quotient $\mathbb{S}\mathcal{H}/U(1)$ of the unit sphere*

$$\mathbb{S}\mathcal{H} := \{\Psi \in \mathcal{H} \mid (\Psi, \Psi) = 1\} \tag{2.27}$$

by the action of $U(1) \simeq \mathbb{T}$, given by $z : \Psi \mapsto z\Psi$, where $|z| = 1$.

The identification of vector states in $\mathcal{H}$, one-dimensional projections on $\mathcal{H}$, and points of $\mathbb{P}\mathcal{H}$ is immediately clear from this realization. Hence we conclude from 2.1.2 that $\mathbb{P}\mathcal{H} \simeq \mathcal{P}(\mathfrak{B}_0(\mathcal{H}))$ (as collections of linear functionals on $\mathfrak{B}_0(\mathcal{H})$ for the moment), and $\mathbb{P}\mathcal{H} \subseteq \mathcal{P}(\mathfrak{B}(\mathcal{H}))$; when $\mathcal{H}$ is infinite-dimensional $\mathbb{P}\mathcal{H}$ does not nearly exhaust $\mathcal{P}(\mathfrak{B}(\mathcal{H}))$.

The space $\mathbb{P}\mathcal{H}$ can be topologized by restricting the usual (norm) Hilbert space topology on $\mathcal{H}$ to $\mathbb{S}\mathcal{H}$, and quotienting it to $\mathbb{P}\mathcal{H} \simeq \mathbb{S}\mathcal{H}/U(1)$. We will denote the image of $\Psi \in \mathbb{S}\mathcal{H}$ in $\mathbb{P}\mathcal{H}$ under the canonical projection $\tau : \mathbb{S}\mathcal{H} \to \mathbb{P}\mathcal{H}$ by ψ; conversely, given $\psi \in \mathbb{P}\mathcal{H}$, such a $\Psi \in \mathbb{S}\mathcal{H}$ will stand for an arbitrary preimage of ψ (and similarly for φ, Φ, etc.).

We now give $\mathbb{P}\mathcal{H}$ the structure of a real manifold. For $\psi \in \mathbb{P}\mathcal{H}$ and $\Psi \in \mathbb{S}\mathcal{H}$, define a neighborhood $\mathcal{N}_\psi := \{\varphi \in \mathbb{P}\mathcal{H} \mid (\Psi, \Phi) \neq 0\}$; this is indeed an open set in the quotient topology. Then $\mathcal{N}_\psi$ is mapped into $\Psi^\perp \subset \mathcal{H}$ by

$$F_\psi(\varphi) = \frac{\Phi}{(\Psi, \Phi)} - \Psi = \frac{[\Psi^\perp]\Phi}{(\Psi, \Phi)} \tag{2.28}$$

(which depends only on the lift Φ), where $[\Psi^\perp]$ is the projection onto $\Psi^\perp \subset \mathcal{H}$. Clearly, ψ is mapped into the null vector, and the image of this map is open in $\Psi^\perp$. It is easily checked that this map is a homeomorphism between $\mathcal{N}_\psi$ and its image. We now let ψ (more precisely, Ψ) vary over a basis in $\mathcal{H}$, and for each such ψ we construct an (arbitrary) reference isomorphism between $\Psi^\perp$ and a fixed reference Hilbert space $\mathcal{H}'$ with two (real) dimensions less than $\mathcal{H}$. This leads to a collection of charts, making $\mathbb{P}\mathcal{H}$ a Hilbert manifold, modeled on $\mathcal{H}'$ (equipped with the strong topology). We will refer to the topology on $\mathbb{P}\mathcal{H}$ considered so far as its **manifold topology**.

Proposition 2.5.2. *The following topologies on $\mathbb{P}\mathcal{H}$ coincide:*

1. *The manifold topology.*
2. *The w^*-topology relative to $\mathbb{P}\mathcal{H} \subset \mathfrak{B}_0(\mathcal{H})^*$.*
3. *The w^*-topology relative to $\mathbb{P}\mathcal{H} \subset \mathfrak{B}(\mathcal{H})^*$.*

It is quite trivial to verify that the topology on $\mathbb{P}\mathcal{H}$ that is inherited from the strong topology on $\mathcal{H}$ is stronger than the topology in 2.5.2.3, which in turn is stronger than the one of 2.5.2.2. Using the fact that $\mathfrak{B}_0(\mathcal{H})$ is generated by the one-dimensional projections on $\mathcal{H}$, one verifies that the topology in 2.5.2.2 coincides with the one induced by the weak topology on $\mathcal{H}$. Since the strong and the weak Hilbert space topologies coincide on $\mathbb{S}\mathcal{H}$, the equivalence between 2.5.2.2 and 2.5.2.3 follows.

It follows from (2.28) that for arbitrary $\varphi \in N_\psi$, one has

$$\frac{\varphi(A)}{\varphi([\Psi])} = (F_\psi(\varphi), AF_\psi(\varphi)) + (\Psi, AF_\psi(\varphi)) + (F_\psi(\varphi), A\Psi) + (\Psi, A\Psi). \tag{2.29}$$

It is clear from this equation that $F_\psi(\varphi_n) \to F_\psi(\varphi)$ strongly implies $\varphi_n(A) \to \varphi(A)$, so that $\varphi_n \to \varphi$ in the topology of 2.5.2.3. Hence the manifold topology on $\mathbb{P}\mathcal{H}$ is stronger than the topology of 2.5.2.3. Conversely, if $\varphi_n(A) \to \varphi(A)$, then each term on the right-hand side of (2.29) must converge, so that $F_\psi(\varphi_n) \to F_\psi(\varphi)$ weakly and $(F_\psi(\varphi_n), AF_\psi(\varphi_n)) \to (F_\psi(\varphi_n), AF_\psi(\varphi_n))$. Taking $A = \mathbb{I}$, these conditions imply $F_\psi(\varphi_n) \to F_\psi(\varphi)$ strongly, so that the topology of 2.5.2.3 is stronger than the manifold topology. Hence the topologies in 2.5.2.1 and 2.5.2.3 coincide. ∎

Corollary 2.5.3. *The pure state space of the C^*-algebra $\mathfrak{B}_0(\mathcal{H})$ (with relative w^*-topology) is homeomorphic to the projective space $\mathbb{P}\mathcal{H}$ (with manifold topology).*

Theorem 2.5.4. *The pure state space $\mathcal{P}(\mathfrak{A})$ of a C^*-algebra $\mathfrak{A}$ is a disjoint union $\mathcal{P}(\mathfrak{A}) = \cup_\alpha \mathbb{P}\mathcal{H}_\alpha$, where $\mathcal{H}_\alpha$ is isomorphic to the irreducible GNS-representation* *space of an arbitrary state in $\mathbb{P}\mathcal{H}_\alpha$. All states in a given subspace $\mathbb{P}\mathcal{H}_\alpha$ are equivalent, and any two states lying in different such subspaces are inequivalent. The inclusion map of any $\mathbb{P}\mathcal{H}_\alpha$ (equipped with the manifold topology) into $\mathcal{P}(\mathfrak{A})$ (with the w^*-topology) is continuous.*

The set-theoretic part of this claim follows from the comments after the proof of 2.2.6. The topological part is a consequence of (1.57) and the equalities of the topologies in 2.5.2.1 and 2.5.2.3. ■

Of course, the disjoint union in 2.5.4 is meant in a set-theoretic rather than a topological sense (the $\mathbb{P}\mathcal{H}_\alpha$ are not necessarily components of $\mathcal{P}(\mathfrak{A})$).

We now embark on a description of $\mathbb{P}\mathcal{H}$ as a symplectic manifold, starting with the corresponding analysis of $\mathcal{H}$. Regarding $\mathcal{H}$ as a real vector space, we identify the tangent bundle $T\mathcal{H}$ with $\mathcal{H}\times\mathcal{H}$ in the usual way: For any $\Psi\in\mathcal{H}$, an element $\Phi\in\mathcal{H}$ defines a tangent vector $V(\Phi)\in T_\Psi\mathcal{H}$ by

$$V(\Phi)_\Psi f = \frac{df}{dt}(\Psi + t\Phi)_{|t=0}. \tag{2.30}$$

If $V(\Phi)$ is tangent to $\mathbb{S}\mathcal{H}$, the derivative τ_* will project it to an element $v(\Phi)$ of $T\mathbb{P}\mathcal{H}$. This applies to tangent vectors of the form $V(iA\Psi)$, where $A^* = A$ in $\mathfrak{B}(\mathcal{H})$, and Ψ is arbitrary. We observe that for any $\Psi\in\mathcal{H}$ the collection of vectors $\{iA\Psi \mid A\in\mathfrak{B}(\mathcal{H})_\mathbb{R}\}$, while not being equal to $\mathcal{H}$ because of the restriction to $\mathfrak{B}(\mathcal{H})_\mathbb{R}$, contains $\Psi^\perp$. It then follows from the above discussion of the manifold structure of $\mathbb{P}\mathcal{H}$ that for all $\psi\in\mathbb{P}\mathcal{H}$ one has

$$T_\psi\mathbb{P}\mathcal{H} = \{v(iA\Psi) \mid A\in\mathfrak{B}(\mathcal{H})_\mathbb{R}\}. \tag{2.31}$$

We now show that $\mathcal{H}$ and $\mathbb{P}\mathcal{H}$ are both examples of (real) symplectic manifolds (the real structure depends on the choice of a basis). Further to the identification $T\mathcal{H}\simeq\mathcal{H}\times\mathcal{H}$ (see (2.30)) we identify $T^*\mathcal{H}$ with $\mathcal{H}\times\mathcal{H}$: For $\Phi\in\mathcal{H}$ the one-form $O(\Phi)$ is defined by

$$(O(\Phi))(V(\Omega)) := \mathrm{Re}\,(\Phi,\Omega). \tag{2.32}$$

Note that $O(\Phi) = df_\Phi$, where $f_\Phi(\Omega) := \mathrm{Re}\,(\Phi,\Omega)$.

A Poisson tensor on $\mathcal{H}$ may be defined for any $\hbar\in\mathbb{R}\backslash\{0\}$ by

$$B(O(\Phi), O(\Omega)) := -\frac{1}{2\hbar}\mathrm{Im}\,(\Phi,\Omega). \tag{2.33}$$

It follows that

$$B^\sharp(O(\Phi)) = -\frac{1}{2\hbar}V(i\Phi). \tag{2.34}$$

Since this map is evidently invertible, one infers

Proposition 2.5.5. *The Poisson manifold $(\mathcal{H}, B)$ is symplectic. The symplectic form ω is given by*

$$\omega(V(\Phi), V(\Omega)) = 2\hbar\,\mathrm{Im}\,(\Phi,\Omega). \tag{2.35}$$

Let H be a self-adjoint element of $\mathfrak{B}(\mathcal{H})$. Define $\tilde{H} \in C^\infty(\mathcal{H}, \mathbb{R})$ by

$$\tilde{H}(\Psi) := (\Psi, H\Psi). \tag{2.36}$$

The corresponding Hamiltonian vector field is

$$\xi_{\tilde{H}}(\Psi) = -V(\frac{i}{\hbar}H\Psi). \tag{2.37}$$

The (real) linear span of $V(\Psi)$ and all $\xi_{\tilde{H}}(\Psi)$ is $T_\Psi\mathcal{H}$. The Poisson bracket of functions of the type (2.36) is (cf. (1.22))

$$\{\tilde{A}, \tilde{B}\} = \frac{i}{\hbar}(\widetilde{[A, B]}) = \widetilde{\{A, B\}}_\hbar. \tag{2.38}$$

If U is a unitary operator on $\mathcal{H}$, the pullback $U^*\tilde{A}$ equals $\widetilde{U^{-1}AU}$. It then follows from (2.38) and (2.15) that each such U defines a Poisson map.

The **Schrödinger equation** "$H\Psi(t) = i\hbar d\Psi(t)/dt$" of quantum mechanics is nothing but (2.11) with (2.37). The solution of this equation is the Hamiltonian flow generated by H, given by

$$\Psi(t) = e^{-itH/\hbar}\Psi. \tag{2.39}$$

We now pass to $\mathbb{P}\mathcal{H}$. Recall the action of $U(1)$ on $\mathcal{H}$ (cf. 2.5.1); it is easily checked that this is a Poisson map for each $z \in U(1)$. Consider $\mathcal{H}^* := \mathcal{H}\backslash\{0\}$; since each point of $\mathcal{H}^*$ has the same stabilizer (namely $\{e\}$), it follows that $\mathcal{H}^*/U(1)$ is a manifold. Moreover, $\mathcal{H}^*/U(1)$ is a Poisson manifold: If $\tau : \mathcal{H}^* \to \mathcal{H}^*/U(1)$ is the canonical projection, then $\tau_*B(\Psi) = \tau_*B(z\Psi)$ for all z and Ψ, so that we can consistently define a Poisson tensor B_R on $\mathcal{H}^*/U(1)$ at some point $\psi = \tau(\Psi)$ by $B_R(\psi) = \tau_*B(\Psi)$. Equivalently, the Poisson bracket $\{\, ,\, \}$ on $\mathcal{H}^*/U(1)$ is taken to be

$$\tau^*\{f, g\}_R = \{\tau^* f, \tau^* g\}, \tag{2.40}$$

which is well-defined by the same argument. The Jacobi identity and the Leibniz rule follow from the fact that they are satisfied on P.

Although $\mathcal{H}^*/U(1)$ may be infinite-dimensional, the statement of Theorem 2.4.7 actually applies.

Proposition 2.5.6. *The symplectic leaves of the Poisson manifold $\mathcal{H}^*/U(1)$ are the spaces $S_r = \mathcal{H}_r/U(1)$, where $\mathcal{H}_r = \{\Psi \in \mathcal{H} \mid (\Psi, \Psi) = r^2\}$, so that $\mathcal{H}^*/U(1) = \cup_{r>0}\mathcal{H}_r/U(1)$. The projective space $\mathbb{P}\mathcal{H}$ may be identified with S_1. Hence $\mathbb{P}\mathcal{H}$ is symplectic; the symplectic form ω is explicitly given by*

$$\omega_\psi(v(iA\Psi), v(iB\Psi)) = -i\hbar\widehat{[A, B]}(\psi), \tag{2.41}$$

and the corresponding Poisson bracket is

$$\{\hat{A}, \hat{B}\} = \frac{i}{\hbar}\widehat{[A, B]} = \widehat{\{A, B\}}_\hbar, \tag{2.42}$$

cf. (1.22) and (2.38).

See the text below (2.30) for the definition of v. We will show that S_1 is a symplectic leaf of $\mathcal{H}^*/U(1)$; the argument for the other S_r is similar. For each $H \in \mathfrak{B}(\mathcal{H})_{\mathbb{R}}$ we here have introduced the function $\hat{H}$ on $\mathbb{P}\mathcal{H}$ by

$$\hat{H}(\psi) = \hat{H}(\tau(\Psi)) := \tilde{H}(\Psi), \tag{2.43}$$

where $\tilde{H}$ is given by (2.36), and Ψ is now assumed to be a unit vector. Note that

$$\|H\| = \|\hat{H}\|_\infty, \tag{2.44}$$

where the norm on the left-hand side is the operator norm in $\mathfrak{B}(\mathcal{H})$. Indeed, our notation $\hat{H}$ is motivated by the fact that (2.43) is a special case of the Gelfand transform (1.30). It follows directly from the definition of the manifold structure of $\mathbb{P}\mathcal{H}$ that $\hat{H}$ is smooth for each $H \in \mathfrak{B}(\mathcal{H})_{\mathbb{R}}$. Equation (2.37) implies

$$\xi_{\hat{H}}(\tau(\Psi)) = -v\left(\frac{i}{\hbar}H\Psi\right). \tag{2.45}$$

The fact that each S_r is symplectic now follows from Propositions 2.3.7 and 2.2.2, and (2.37) or (2.45). The Poisson bracket (2.42) is derived from (2.38); it is, of course, consistent with (2.19), (2.41), and (2.45).

Finally, the continuity of the inclusion of S_1 into $\mathcal{H}^*/U(1)$ is immediate from Proposition 2.5.2. ∎

It follows from the comment after (2.37) that the Poisson structure is completely determined by the special case (2.42).

If $\mathcal{H} = \mathbb{C}^N$ is finite-dimensional, the symplectic form defined by (2.41) is $\hbar$ times the well-known **Fubini–Study** form on $\mathbb{P}\mathbb{C}^N$.

As on $\mathcal{H}$, each unitary operator U (projected to a map on $\mathbb{P}\mathcal{H}$) is a Poisson map with respect to (2.42). The Schrödinger equation, projected to $\mathbb{P}\mathcal{H}$, is a special case of (2.11): If, in somewhat sloppy notation, $\psi(t)$ is the flow obtained by projecting $\Psi(t)$ (cf. (2.39)) from $\mathbb{S}\mathcal{H}$ to $\mathbb{P}\mathcal{H}$, one has from (2.45)

$$\frac{d\psi(t)}{dt} = \xi_{\hat{H}}(\psi(t)). \tag{2.46}$$

In particular, the flow is complete for any H. As a matter of notation, we write the solution as

$$\psi(t) = e^{-it\hat{H}/\hbar}\psi. \tag{2.47}$$

The right-hand side is by definition the projection of (2.39) to $\mathbb{P}\mathcal{H}$.

Eigenvalues and eigenvectors have a neat description in the present language, too.

Proposition 2.5.7. *A vector $\Psi \in \mathcal{H}$ is an eigenvector of an operator $H \in \mathfrak{B}(\mathcal{H})_{\mathbb{R}}$ iff $\psi = \tau(\Psi)$ is a critical point of $\hat{H}$ (i.e., $d\hat{H}(\psi) = 0$); the corresponding eigenvalue is $\hat{H}(\psi)$.*

This is perhaps obvious from the minimax description of eigenvalues, but here is a direct proof. The property $d\hat{H}(\psi) = 0$ is the same as $X\hat{H}(\psi) = 0$ for all $X \in T_\psi\mathbb{P}\mathcal{H}$. By (2.31) and (2.39), this is equivalent to $(\Psi, (HA - AH)\Psi) = 0$, or $(A\Psi, H\Psi) = \overline{(A\Psi, H\Psi)}$, for all $A \in \mathfrak{B}(\mathcal{H})_{\mathbb{R}}$. Hence $(\Psi, H\Phi) \in \mathbb{R}$ for all $\Phi \in \Psi^{\perp}$, which is possible only if $(\Psi, H\Phi)$ vanishes for all $\Phi \in \Psi^{\perp}$. This implies that Ψ must be an eigenvector of H. ∎

From the symplectic point of view, the two steps in the construction of $\mathbb{P}\mathcal{H}$ appear in reverse order. Firstly, one pulls the symplectic form ω on $\mathcal{H}$ back to $\mathbb{S}\mathcal{H}$; here it is degenerate. Secondly, this degeneracy is removed upon quotienting $\mathbb{S}\mathcal{H}$ by $U(1)$, arriving at $\mathbb{P}\mathcal{H}$ once more. See IV.1.5.

2.6 *Representations of Poisson Algebras*

We look at symplectic manifolds as the classical analogues of modules for Poisson algebras (cf. the opening remark in 1.5).

Definition 2.6.1. *A* **representation of a Poisson algebra** $(\mathfrak{A}_{\mathbb{R}}, \circ, \{\,,\,\})$ *is a linear map* $\pi : \mathfrak{A}_{\mathbb{R}} \to C^{\infty}(S, \mathbb{R})$, *where* S *is a symplectic manifold, satisfying*

$$\pi(f \circ g) = \pi(f)\pi(g);$$
$$\{\pi(f), \pi(g)\}_S = \pi(\{f, g\}) \tag{2.48}$$

(where $\{\,,\,\}_S$ *is the Poisson bracket on* S*), as well as preserving completeness.*

The condition (2.48) says simply that $\pi : \mathfrak{A}_{\mathbb{R}} \to C^{\infty}(S, \mathbb{R})$ is a morphism, assuming that the Jordan product in $C^{\infty}(S, \mathbb{R})$ is represented by pointwise multiplication (cf. 1.1.3). The completeness requirement means that the flow of $\xi_{\pi(h)}$ is defined for all times if h is complete in $\mathfrak{A}_{\mathbb{R}}$, cf. 2.3.4. It is imposed to eliminate constructions of the type $\mathfrak{A}_{\mathbb{R}} = C^{\infty}(P, \mathbb{R})$, $P' \neq P$ open in P, and π being simply restriction to P'.

There is a natural notion of equivalence. Namely, two representations $\pi_1 : \mathfrak{A}_{\mathbb{R}} \to C^{\infty}(S_1, \mathbb{R})$ and $\pi_2 : \mathfrak{A}_{\mathbb{R}} \to C^{\infty}(S_2, \mathbb{R})$ are called **equivalent** if there exists a symplectomorphism $J : S_1 \to S_2$ such that $J^*\pi_2(f) = \pi_1(f)$ for all $f \in \mathfrak{A}_{\mathbb{R}}$.

We can analyze the structure of representations of Poisson algebras of a slightly more general type than $C^{\infty}(P, \mathbb{R})$, where P is a Poisson manifold.

Definition 2.6.2. *A* **Poisson space** P *is a Hausdorff topological space together with a linear subspace* $\mathfrak{A}_{\mathbb{R}} \subset C(P, \mathbb{R})$ *and a collection* S_α *of symplectic manifolds (called the symplectic leaves of* P*), as well as continuous injections* $\iota_\alpha : S_\alpha \hookrightarrow P$, *such that:*

1. $P = \cup_\alpha \iota_\alpha(S_\alpha)$ *(disjoint union).*
2. $\mathfrak{A}_{\mathbb{R}}$ *separates points.*
3. $\mathfrak{A}_{\mathbb{R}} \subseteq C^{\infty}_L(P, \mathbb{R})$, *where* $C^{\infty}_L(P, \mathbb{R})$ *consists of all* $f \in C(P, \mathbb{R})$ *for which* $\iota_\alpha^* f \in C^{\infty}(S_\alpha, \mathbb{R})$ *for each* α.
4. $\mathfrak{A}_{\mathbb{R}}$ *is closed under the Poisson bracket*

$$\{f, g\}(\iota_\alpha(\sigma)) = \{\iota_\alpha^* f, \iota_\alpha^* g\}_\alpha(\sigma). \tag{2.49}$$

If the ambient space P carries additional structure, such as a uniformity or a smooth structure, one can refine the above definition in the obvious way; such refinements will play an important role in later sections.

Definition 2.6.3. *A* **uniform Poisson space** *is a Poisson space P in which the topology is defined by a uniformity on P and that satisfies Definition 2.6.2 with $C(P,\mathbb{R})$ replaced by the space $C_u(P,\mathbb{R})$ of uniformly continuous functions on P.*

Similarly, a **smooth Poisson space** is a Poisson space for which P is a manifold and $C(P,\mathbb{R})$ is replaced by $C^\infty(P,\mathbb{R})$. By Theorem 2.4.7, a smooth Poisson space with $\mathfrak{A}_\mathbb{R} = C^\infty(P,\mathbb{R})$ is nothing but a Poisson manifold. The more general concept of a Poisson space is useful when the symplectic leaves do not fit together to form a manifold. This happens in the context of singular symplectic reduction, cf. IV.1.11. Moreover, we will show in 3.2.2 that the pure state space of a C^*-algebra is a uniform Poisson space. In any case, the object $C_L^\infty(P,\mathbb{R})$ defined in 2.6.2.3 is the function space intrinsically related to a (general, uniform, or smooth) Poisson space P.

Definition 2.6.2 does not entail that $\mathfrak{A}_\mathbb{R}$ is a Poisson algebra under pointwise multiplication as the Jordan product, but an interesting result arises when one makes that assumption. In preparation for this, we remark that the notion of a Poisson map makes sense in the context of Poisson spaces: It is still defined by (2.15).

Proposition 2.6.4. *Let $(P,\mathfrak{A}_\mathbb{R})$ be a locally compact Poisson space for which $\mathfrak{A}_\mathbb{R}$ is a Poisson algebra under pointwise multiplication. If $\pi : \mathfrak{A}_\mathbb{R} \to C^\infty(S,\mathbb{R})$ is a representation of $\mathfrak{A}_\mathbb{R}$ on a finite-dimensional symplectic manifold S, then there exists a continuous map $J : S \to P$ such that $\pi = J^*$.*

For simplicity we show this for compact P, and assume that $\mathfrak{A}_\mathbb{R}$ contains the unit function 1_P. The Stone–Weierstrass theorem then implies that $\mathfrak{A}_\mathbb{R}$ is dense in $C(P,\mathbb{R})$ in the sup-norm. Take a point $\sigma \in S$, and define a linear functional $\tilde{J}_\sigma$ on $\mathfrak{A}_\mathbb{R}$ by $\tilde{J}_\sigma(f) = (\pi(f))(\sigma)$. By the first member of (2.48), this functional is multiplicative. If it were defined on all of $C(P,\mathbb{R})$, we could immediately conclude from this that $\tilde{J}_\sigma$ is continuous; a positivity argument shows that this follows in the present case as well. Hence we extend $\tilde{J}_\sigma$ to all of $C(P,\mathbb{R})$. It follows that $\tilde{J}_\sigma$ defines a pure state, and pure states on $C(P,\mathbb{R})$ correspond to points of P (see 2.1.4). Hence $\tilde{J}_\sigma$ corresponds to a point $J(\sigma)$ in P, and this defines the desired map $J : S \to P$. The continuity of J follows from a technical argument in the theory of commutative C^*-algebras. The second member of (2.48) obviously implies that J is a Poisson map. □

Corollary 2.6.5. *If $\mathfrak{A}_\mathbb{R} = C^\infty(P,\mathbb{R})$ for a Poisson manifold P and $\pi : \mathfrak{A}_\mathbb{R} \to C^\infty(S,\mathbb{R})$ is a representation, then there exists a smooth Poisson map $J : S \to P$ such that $\pi = J^*$.*

The smoothness of J follows from the property $\pi = J^*$. ■

There is a natural notion of irreducibility for representations of Poisson algebras.

Definition 2.6.6. *A representation π of a Poisson algebra $\mathfrak{A}_{\mathbb{R}}$ is called* **irreducible** *if*

$$\{\xi_{\pi(f)}(\sigma) \mid f \in \mathfrak{A}_{\mathbb{R}}\} = T_\sigma S \;\forall \sigma \in S. \tag{2.50}$$

If S is infinite-dimensional, it is understood that one takes the closure of the left-hand side in the definition. The finite-dimensional irreducible representations of a Poisson algebra associated with a locally compact Poisson space (Definition 2.6.2) can be described concretely.

Theorem 2.6.7. *Under the assumptions of Proposition 2.6.4, let π be irreducible. Then S is symplectomorphic to a symplectic leaf S_α of P, or to a covering space thereof.*

In fact, it will follow from the proof below that an irreducible representation space S of a locally compact Poisson space has to be finite-dimensional. The proof of this theorem is based on 2.6.4; we have $\pi = J^*$ for $J : S \to P$. For each $\sigma \in S$, let $S_{\alpha(\sigma)}$ be the symplectic leaf for which $\iota_{\alpha(\sigma)}(S_{\alpha(\sigma)})$ contains $J(\sigma)$ (cf. 2.6.2); we will henceforth identify $\iota_{\alpha(\sigma)}(S_{\alpha(\sigma)})$ and $S_{\alpha(\sigma)}$. By irreducibility, any $X \in T_\sigma S$ can be written as $X = \xi_{\pi(f)}$ for some $f \in \mathfrak{A}_{\mathbb{R}}$. We define a linear map $J_* : T_\sigma S \to T_{J(\sigma)} S_{\alpha(\sigma)}$ by

$$J_* \xi_{\pi(f)}(\sigma) := \xi_f(J(\sigma)). \tag{2.51}$$

The notation is consistent: If P is a manifold, J_* is indeed the pushforward of J, cf. (2.17). The fact that J_* is well-defined follows from its injectivity, which we will now demonstrate. If $J_*\xi_{\pi(f)}(\sigma) = 0$, then $\{f, g\}(J(\sigma)) = 0$ for all $g \in \mathfrak{A}_{\mathbb{R}}$, since J is a Poisson map. But then $\omega^S_\sigma(\xi_{\pi(f)}, \xi_{\pi(g)}) = 0$ for all g, where ω^S is the symplectic form on S. Since ω^S is nondegenerate, this implies $\xi_{\pi(f)} = 0$, which proves injectivity. Now, J_* is evidently surjective as well, because $S_{\alpha(\sigma)}$ is symplectic. Hence J_* is an isomorphism.

Combining this result with Propositions 2.3.5 and 2.3.7, we conclude that $J(S) \subseteq S_{\alpha(\sigma)}$, where S and $S_{\alpha(\sigma)}$ are locally symplectomorphic (since π is a representation). The completeness of π (see Definition 2.6.1) implies that $J(S) = S_{\alpha(\sigma)}$. For if the inclusion $J(S) \subseteq S_{\alpha(\sigma)}$ were proper, we could take a neighborhood $\mathcal{N}$ of a boundary point of $J(S)$ in $S_{\alpha(\sigma)}$, and take $\sigma_1 \in \mathcal{N} \cap J(S)$ and $\sigma_2 \in \mathcal{N}$ but $\sigma_2 \notin J(S)$, such that σ_1 and σ_2 are connected by a Hamiltonian curve tangent to ξ_f (cf. 2.3.7). We then consider the Hamiltonian curve in S tangent to $\xi_{\pi(f)}$ and passing through s_1, where $J(s_1) = \sigma_1$ (s_1 may not be unique). By 2.3.5 this curve is mapped onto the Hamiltonian curve connecting σ_1 and σ_2, but this is impossible because of our assumptions on σ_2. Hence the curve in S in question must suddenly stop somewhere, contradicting the completeness of π.

A similar argument shows that J is a covering projection. For J not to be a covering projection, there must exist a point $\sigma_3 \in S_{\alpha(\sigma)}$, a neighborhood $\mathcal{N}_{\sigma_3}$ of σ_3, and a connected component $J_i^{-1}(\mathcal{N}_{\sigma_3})$ of $J^{-1}(\mathcal{N}_{\sigma_3})$ such that $J(J_i^{-1}(\mathcal{N}_{\sigma_3})) \subset \mathcal{N}_{\sigma_3}$ is a proper inclusion. But in that case we could choose points $\sigma_1 \in J(J_i^{-1}(\mathcal{N}_\sigma))$ and $\sigma_2 \in \mathcal{N}_{\sigma_3}$ but $\sigma_2 \notin J(J_i^{-1}(\mathcal{N}_{\sigma_3}))$, which can be connected by a smooth Hamiltonian curve, tangent to some vector field ξ_g. Let $\sigma_1 = J(s_1)$ for some s_1, and consider the flow of $\xi_{\pi(g)}$ through s_1 in S. Then this flow must either suddenly stop, contradicting the completeness of π, or continue outside $J_i^{-1}(\mathcal{N}_\sigma)$ to a point s_2 for which $J(s_2) = \sigma_2$, contradicting the assumptions on σ_2. Hence J must be a covering projection, and Theorem 2.6.7 is proved. ∎

We now return to C^*-algebras and their pure state spaces. Take a C^*-algebra $\mathfrak{A}$ with pure state space $\mathcal{P} = \mathcal{P}(\mathfrak{A})$ (equipped with the w^*-topology), and identify its self-adjoint part $\mathfrak{A}_{\mathbb{R}}$ with a subspace of $C(\mathcal{P}, \mathbb{R})$ by the Gelfand transform $\hat{A} \in C(\mathcal{P}, \mathbb{R})$; see (1.30). We will occasionally drop the hat on A.

Proposition 2.6.8. *The pure state space $\mathcal{P} = \mathcal{P}(\mathfrak{A})$ of a C^*-algebra $\mathfrak{A}$ (where $\mathfrak{A}_{\mathbb{R}}$ is identified with a subspace of $C(\mathcal{P}, \mathbb{R})$ through the Gelfand transform (1.30)), equipped with the irreducible representation spaces $S_\alpha = \mathbb{P}\mathcal{H}_\alpha$ and the inclusion maps ι_α, is a Poisson space.*

This is a trivial consequence of 2.5.4; note that the Poisson bracket in the sense of 2.6.2 coincides with the one (1.22) originally defined on $\mathfrak{A}_{\mathbb{R}}$. Recall that the choice of each $\mathcal{H}_\alpha$ is arbitrary within unitary equivalence; the Poisson structure on $\mathcal{P}(\mathfrak{A})$ is independent of the particular choices made by the comment following (2.42). The spaces $\mathbb{P}\mathcal{H}_\alpha$ are now seen to be the symplectic leaves of $\mathcal{P}$. ∎

Proposition 2.6.8 recognizes the fact that (the self-adjoint parts of) C^*-algebras fall under the theory of Poisson spaces. This point of view receives further support from a reconsideration of the notion of a representation π of a C^*-algebra on a Hilbert space $\mathcal{H}$ (see 1.5.1). As explained in 2.5, we may identify $\mathfrak{B}(\mathcal{H})_{\mathbb{R}}$ with a subspace of the Poisson algebra $C^\infty(\mathcal{H}, \mathbb{R})$, so that π maps $A \in \mathfrak{A}_{\mathbb{R}}$ to $\widehat{\pi(A)} \in C^\infty(\mathbb{P}\mathcal{H}, \mathbb{R})$ (cf. (2.43)). It follows from 2.5.6 that $\pi : \mathfrak{A}_{\mathbb{R}} \to C^\infty(\mathcal{H}, \mathbb{R})$, thus interpreted, is a Poisson morphism.

Proposition 2.6.9. *A representation π of a C^*-algebra $\mathfrak{A}$ on a Hilbert space $\mathcal{H}$ is irreducible iff for every $\psi \in \mathbb{P}\mathcal{H}$ the set $\{v(i\pi(A)\Psi) \mid A \in \mathfrak{A}_{\mathbb{R}}\}$ of tangent vectors is dense in $T_\psi\mathbb{P}\mathcal{H}$* (**Poisson irreducibility**).

This follows from (2.31) and 2.2.2.2. Note that $T_\Psi\mathcal{H}$ equals $\{V(A\Psi) \mid \Psi \in \mathfrak{B}(\mathcal{H})\}$, but does not equal $\{V(A\Psi) \mid \Psi \in \mathfrak{B}(\mathcal{H})_{\mathbb{R}}\}$; nonetheless, $T_\psi\mathbb{P}\mathcal{H}$ is given by (2.31). This is because the orthogonal complement of $\{V(A\Psi) \mid \Psi \in \mathfrak{B}(\mathcal{H})_{\mathbb{R}}\}$ in $T_\Psi\mathcal{H}$ projects to zero in $T_\psi\mathbb{P}\mathcal{H}$. ∎

Combining 2.6.9 and (2.45), we see that the notions of irreducibility of a representation of a C^*-algebra (Definition 2.2.1) and of a Poisson algebra (Definition 2.6.6) coincide (cf. 2.2.1). Therefore, on the Poisson side there is a close formal similarity between C^*-algebras and Poisson algebras as far as their respective representation theories are concerned. Indeed, combining Theorems 2.5.4 and 2.6.7 and Proposition 2.6.8, we obtain (under the above identifications)

Corollary 2.6.10. *Let $\mathfrak{A}_{\mathbb{R}}$ be either a Poisson algebra defined by a locally compact Poisson space P, or the self-adjoint part of a C^*-algebra with pure state space $\mathcal{P}$. Then, up to equivalence, every irreducible representation of $\mathfrak{A}_{\mathbb{R}}$ is given by the* *restriction of $\mathfrak{A}_{\mathbb{R}}$ to a symplectic leaf of P (or $\mathcal{P}$, respectively), or by this restriction preceded by the covering projection on a covering space of such a leaf.*

Note that $\mathbb{P}\mathcal{H}$ has no nontrivial covering spaces. If the Poisson space P is compact and the Poisson algebra $\mathfrak{A}_{\mathbb{R}}$ contains the unit function, then $\mathfrak{A}_{\mathbb{R}}$ is dense in $C(P, \mathbb{R})$ (cf. the proof of 2.6.4); in that case, P is actually the pure state space of $\mathfrak{A}$ as a C^*-algebra (cf. 2.1). If P is merely locally compact, the same conclusion holds if $\mathfrak{A}_{\mathbb{R}}$ is contained in $C_0(P)$.

The difference between representations of C^*-algebras and Poisson algebras lies on the Jordan side; from the point of view of pure states, the Jordan structure on $\mathfrak{A}_{\mathbb{R}}$ eventually originates from a novel structure on $\mathcal{P}(\mathfrak{A})$.

2.7 Transition Probability Spaces

Here is the structure alluded to at the end of the preceding section.

Definition 2.7.1. *A* **transition probability** *on a set $\mathcal{P}$ is a function*

$$p : \mathcal{P} \times \mathcal{P} \to [0, 1] \tag{2.52}$$

that satisfies

$$p(\rho, \sigma) = 1 \iff \rho = \sigma \tag{2.53}$$

and

$$p(\rho, \sigma) = 0 \iff p(\sigma, \rho) = 0. \tag{2.54}$$

A set with such a transition probability is called a **transition probability space**.

The following set of definitions is natural and self-evident.

Definition 2.7.2. *A family of subsets of a transition probability space $\mathcal{P}$ is called* **orthogonal** *if $p(\rho, \sigma) = 0$ whenever ρ and σ do not lie in the same subset. The space $\mathcal{P}$ is called* **reducible** *if it is the union of two (nonempty) orthogonal subsets; if not, it is said to be* **irreducible**. *A* **component** *C of $\mathcal{P}$ is a subset $C \subset \mathcal{P}$ such that C and $\mathcal{P}\backslash C$ are orthogonal. An irreducible component of $\mathcal{P}$ is called a* **sector**.

Thus any transition probability space is the disjoint union of its sectors.

Certain subsets of $\mathcal{P}$ are of special significance. The **orthoplement** of $Q \subset \mathcal{P}$ is defined by

$$Q^{\perp} = \{\sigma \in \mathcal{P} \mid p(\rho, \sigma) = 0 \,\forall \rho \in Q\}. \tag{2.55}$$

It is immediately obvious that if $R \subseteq S$, then $S^{\perp} \subseteq R^{\perp}$, and that $T \subseteq T^{\perp\perp}$. Putting $R = Q$ and $S = Q^{\perp\perp}$ shows that $Q^{\perp\perp\perp} \subseteq Q^{\perp}$; putting $T = Q^{\perp}$ yields $Q^{\perp} \subseteq Q^{\perp\perp\perp}$. Hence $Q^{\perp} = Q^{\perp\perp\perp}$. Accordingly, the **orthoclosure** of a subset $Q \subseteq \mathcal{P}$ is defined as $Q^{\perp\perp}$, and Q is called **orthoclosed** if $Q = Q^{\perp\perp}$. It follows that $Q^{\perp}$ (and therefore $Q^{\perp\perp}$) is always orthoclosed. Also, one easily sees that any component $C \subseteq \mathcal{P}$ is orthoclosed; this applies in particular to $\mathcal{P}$ itself, and to any sector of $\mathcal{P}$.

Definition 2.7.3. *A* **basis** *of a transition probability space $\mathcal{P}$ is an orthogonal family B of points of $\mathcal{P}$ with the property that*

$$\sum_{\rho \in B} p(\rho, \sigma) = 1 \ \forall \sigma \in \mathcal{P} \tag{2.56}$$

(if B is infinite the sum is defined as the least upper bound of all finite partial sums).

The transition probability space is called **symmetric** if

$$p(\rho, \sigma) = p(\sigma, \rho) \ \ \forall \rho, \sigma \in \mathcal{P}. \tag{2.57}$$

The simplest example of a symmetric transition probability space is obtained by taking any set $\mathcal{P}$, and putting

$$p(\rho, \sigma) := \delta_{\rho\sigma}. \tag{2.58}$$

Proposition 2.7.4. *In a symmetric transition probability space all bases have the same cardinality.*

Let B_1 and B_2 be two bases. If both are finite, (2.56) shows that the cardinality $\text{card}(B_1)$ of B_1 is given by $\sum_{\rho \in B_1} \sum_{\sigma \in B_2} p(\rho, \sigma)$. But then the symmetry of p implies that this must equal $\text{card}(B_2)$. The same calculation shows that it is impossible that B_1 is finite and B_2 infinite (and vice versa). Let both be infinite. For fixed $\sigma \in B_2$, define $R(\sigma) = \{\rho \in B_1 \mid p(\rho, \sigma) > 0\}$. By (2.56), $R(\sigma)$ can be at most countable. Hence the set $\cup_{\sigma \in B_2} R(\sigma)$ has the same cardinality as B_2. On the other hand, this set is contained in B_1, so that $\text{card}(B_2) \leq \text{card}(B_1)$. The symmetry of p leads to the opposite inequality, so that $\text{card}(B_1) = \text{card}(B_2)$. ■

Consequently, one can define the **dimension** of a symmetric transition probability space as the cardinality of any of its bases. If B is a basis, then $B^{\perp\perp} = \mathcal{P}$.

Clearly, any subset of $\mathcal{P}$ is a transition probability space if one simply restricts p to it. Not every orthoclosed subset is necessarily the orthoclosure of a maximal orthogonal subset contained in it, however: There exist examples of orthoclosed subsets that do not have any basis. To exclude pathological cases, we impose the following

Definition 2.7.5. *A transition probability space is* **well-behaved** *if:*

- *It is symmetric.*
- *Every orthoclosed subset Q of $\mathcal{P}$ has the property that any maximal orthogonal subset of Q is a basis of Q.*

In a well-behaved transition probability space any set of the type $Q^{\perp}$ is orthoclosed. Moreover, any orthogonal subset S has the property

$$S^{\perp\perp} = \left\{ \rho \in \mathcal{P} \mid \sum_{\sigma \in S} p(\rho, \sigma) = 1 \right\}, \tag{2.59}$$

since one can complete S with a basis of $S^{\perp}$ to form a basis of $\mathcal{P}$.

For each point ρ in an arbitrary transition probability space $\mathcal{P}$, the function p_ρ on $\mathcal{P}$ is defined by

$$p_\rho(\sigma) := p(\rho, \sigma). \tag{2.60}$$

Proposition 2.7.6. *Let $\mathcal{P}$ be a well-behaved transition probability space. For each orthoclosed subset $Q \subseteq \mathcal{P}$ the function*

$$p_Q := \sum_{i=1}^{\dim(Q)} p_{e_i} \tag{2.61}$$

is independent of the choice of basis $\{e_i\}$ of Q. Moreover,

$$Q = \{\rho \in \mathcal{P} \mid p_Q(\rho) = 1\}. \tag{2.62}$$

Choose a basis $B = \{e_i\} \cup \{u_j\}$ of $\mathcal{P}$ that contains the given basis of Q; clearly, $u_j \in Q^\perp$ for all j. By (2.56), $p_Q = \sum_i p_{e_i} = 1 - \sum_j p_{u_j}$, in which the right-hand side is clearly independent of the choice of basis of Q. We now prove (2.62). If $\rho \in Q$, then $\sum_j p(\rho, u_j) = 0$, so that $p_Q(\rho) = 1$ by (2.56). If $p_Q(\rho) = 1$, then $p(\rho, u_j) = 0$ for all j, so that $\rho \in (\cup_j u_j)^\perp = Q^{\perp\perp} = Q$. ■

2.8 Pure State Spaces as Transition Probability Spaces

This section is devoted to the result that the pure state space of a C^*-algebra is a well-behaved transition probability space. To see this in perspective, we start in the more general context of compact convex sets; cf. 1.4.5 and preceding text. We will routinely omit the hat on the Gelfand transform.

Let K be a compact convex set (in a Hausdorff vector space). An extreme point $\rho \in \partial_e K$ is called **norm-exposed** if there exists some $A \in A_b(K, \mathbb{R})$, with $\|A\| = 1$, such that $\{\omega \in K \mid A(\omega) = 1\} = \rho$. Equivalently, A satisfies $A(\rho) = 1$ and $A(\omega) < 1$ for all $\omega \in K\backslash\{\rho\}$.

Proposition 2.8.1. *Let K be a Hausdorff compact convex set with the property that every extreme point is norm-exposed. Then the formula*

$$p(\rho, \sigma) := \inf\{A(\rho) \mid A \in A_b(K, \mathbb{R}), 0 \leq A \leq 1_K, A(\sigma) = 1\} \tag{2.63}$$

defines a transition probability on the extreme boundary $\partial_e K$ of K. The expression (2.63) is not changed if the infimum is taken over $A(K, \mathbb{R})$ instead of $A_b(K, \mathbb{R})$.

If $\rho = \sigma$, then $p(\rho, \sigma) = 1$ by definition. The converse follows immediately from the extra requirement on K. Condition (2.54) is easily verified if one rewrites (2.63) as

$$p(\sigma, \rho) = 1 - \sup\{A(\sigma) \mid A \in A_b(K, \mathbb{R}), 0 \leq A \leq 1_K, A(\rho) = 0\}. \tag{2.64}$$

The claim that we may minimize over A in $A(K, \mathbb{R})$ follows from the density of $A(K, \mathbb{R})$ in $A_b(K, \mathbb{R})$ in the topology of pointwise convergence. ■

If $\mathfrak{A}$ is a commutative unital C^*-algebra, its self-adjoint part is of the form $\mathfrak{A}_\mathbb{R} = C(\mathcal{P}(\mathfrak{A}), \mathbb{R})$ by Theorem 2.1.7, where the pure state space is a compact

Hausdorff space. Since $\mathfrak{A}_{\mathbb{R}}^{**}$ contains $\ell^\infty(\mathcal{P}(\mathfrak{A}))$, one immediately sees that (2.63) leads to (2.58). If one minimizes (2.63) only over $A(K, \mathbb{R}) \simeq C(\mathcal{P}(\mathfrak{A}), \mathbb{R})$, the same result follows from Urysohn's lemma; since $\mathcal{P}(\mathfrak{A})$ is compact and Hausdorff, it is normal.

Theorem 2.8.2. *The pure state space $\mathcal{P}(\mathfrak{A})$ of a C^*-algebra $\mathfrak{A}$ is a well-behaved transition probability space under (2.63). The transition probabilities are explicitly given by $p(\rho, \sigma) = 0$ if ρ and σ are inequivalent, and*

$$p(\rho, \sigma) = |(\Omega_\rho, \Omega_\sigma)|^2 \tag{2.65}$$

if ρ and σ are equivalent. Here $\Omega_\rho, \Omega_\sigma \in \mathbb{S}\mathcal{H}_\alpha$ are (arbitrary) preimages of $\rho, \sigma \in \mathbb{P}\mathcal{H}_\alpha$ (cf. 2.5 and 2.2).

Note that this implies that the transition probabilities are given by (2.58) if $\mathfrak{A}$ is commutative.

We may assume that $\mathfrak{A}_{\mathbb{R}}$ has a unit. If it hasn't, we use 1.2.1 and (2.2); the special point ω_∞ satisfies $p(\omega_\infty, \rho) = 0$ for all $\rho \neq \omega_\infty$. To see what is happening, we first prove the theorem for finite-dimensional C^*-algebras. By Proposition 2.2.8 these are direct sums of matrix algebras, i.e., $\mathfrak{A} = \oplus_\alpha \mathfrak{M}_{N_\alpha}(\mathbb{C})$. We write $A = \oplus_\alpha A_\alpha$ for $A \in \mathfrak{A}$. The pure state space of $\mathfrak{A}$ is $\mathcal{P} = \cup_\alpha \mathbb{P}\mathbb{C}^{N_\alpha}$. We now take a fixed α; if $\sigma \in \mathbb{P}\mathbb{C}^{N_\alpha} \subset \mathcal{P}$, then $\sigma(A) = \sigma(A_\alpha) = (\Omega_\sigma, A_\alpha \Omega_\sigma)$, where $\Omega_\sigma \in \mathbb{C}^{N_\alpha}$ is defined as in the statement of the theorem. The projection $[\Omega_\sigma]$ onto Ω_σ may be regarded as an element of $\mathfrak{A}$ by adding zero operators. Then

$$[\Omega_\sigma](\rho) = |(\Omega_\sigma, \Omega_\rho)|^2 \tag{2.66}$$

if $\rho \in \mathbb{P}\mathbb{C}^{N_\alpha}$ (i.e., it is equivalent to σ), and $[\Omega_\sigma](\rho) = 0$ otherwise. In particular, $[\Omega_\sigma](\sigma) = 1$, and $[\Omega_\sigma](\rho) < 1$ if $\rho \neq \sigma$. This shows firstly that every pure state is norm-exposed, and secondly that $p(\rho, \sigma)$ vanishes if ρ and σ are inequivalent (note that $0 < [\Omega_\sigma] < \mathbb{I}$ and $\|[\Omega_\sigma]\| = 1$, since $[\Omega_\sigma]$ is a projection).

We now assume that ρ and σ are equivalent, and without loss of generality, put $\mathfrak{A} = \mathfrak{M}_N(\mathbb{C})$. We claim that the infimum in (2.63) is reached for $A = [\Omega_\sigma]$. For suppose there exists an $A \in \mathfrak{A}_{\mathbb{R}} = \mathfrak{M}_N(\mathbb{C})_{\mathbb{R}}$ for which $0 < A < [\Omega_\sigma]$ and $A(\sigma) = (\Omega_\sigma, A\Omega_\sigma) = 1$. Choose a basis $\{\mathbf{e}_1, \ldots, \mathbf{e}_N\}$ in $\mathbb{C}^N$ that projects onto $\{\sigma, \ldots, e_N\}$ in $\mathbb{P}\mathbb{C}^N$. Since $0 < A < [\Omega_\sigma]$ and $[\Omega_\sigma](e_i) = 0$ for $i = 2, \ldots$, it must be that $A(e_i) = 0$ for $i = 2, \ldots, N$. Also, clearly, $A(e_1) = [\Omega_\sigma](e_1)$ (since $e_1 = \sigma$). Then $B = [\Omega_\sigma] - A$ satisfies $B > 0$, and $B(e_i) = (\mathbf{e}_i, B\mathbf{e}_i) = 0$ for all i. The latter is impossible for a positive definite matrix. Hence we can compute p by $p(\rho, \sigma) = [\Omega_\sigma](\rho)$, which, with (2.66), proves (2.65).

The proof of Theorem 2.8.2 for general C^*-algebras follows the same idea; the direct sum of matrix algebras is now replaced by the reduced atomic representation π_{ra} of $\mathfrak{A}$ (see 2.2.7 and the subsequent theory). The projection $[\Omega_\sigma] \in \mathfrak{B}(\mathcal{H}_\sigma)$ is regarded as an element of $\pi_{\mathrm{ra}}(\mathfrak{A})''$ by adding zero operators; hence it lies in $\mathfrak{A}_{\mathbb{R}}^{**}$. As in the finite-dimensional case, this shows that every pure state is norm-exposed, while additionally reducing the proof to the situation where ρ and σ are equivalent.

We then observe that $A(\mathcal{S}(\mathfrak{A}), \mathbb{R}) \subseteq \pi_{\mathrm{ra}}(\mathfrak{A}_{\mathbb{R}})'' \subseteq A_b(\mathcal{S}(\mathfrak{A}), \mathbb{R})$ (with equalities only for finite-dimensional algebras), so that we may take the infimum in (2.63) over all A in $\pi_{ra}(\mathfrak{A}_{\mathbb{R}})''$. The remainder of the proof is then the same as in the finite-dimensional case, since the property of positive definite matrices we used holds for arbitrary positive definite operators on a Hilbert space.

A basis of $\mathcal{P}$ is obtained by using the decomposition $\mathcal{P} = \cup_\alpha \mathbb{P}\mathcal{H}_\alpha$ (see 2.5.4); one chooses an orthonormal basis (in the usual Hilbert space sense) in each $\mathcal{H}_\alpha$, and projects it to $\mathbb{P}\mathcal{H}_\alpha$. This yields a basis (in the sense of 2.7.3) of $\mathbb{P}\mathcal{H}_\alpha$ as a transition probability space. Combining these bases by taking the union over all α then produces a basis of $\mathcal{P}$. The fact that $\mathcal{P}$ is a well-behaved transition probability space then follows from elementary Hilbert space theory. ■

The transition probability between pure states on a C^*-algebra $\mathfrak{A}$ may be related to the norm on $\mathfrak{A}^*$, in that

$$p(\rho, \sigma) = 1 - \tfrac{1}{4}\|\rho - \sigma\|^2. \tag{2.67}$$

If ρ and σ are equivalent, so that they are vector states in the same Hilbert space (cf. the comments following 2.2.6), then (2.67) is equivalent to

$$\|\rho - \sigma\| = \tfrac{1}{2}\|[\Omega_\rho] - [\Omega_\sigma]\|_1; \tag{2.68}$$

either equality follows from a simple calculation with 2×2 matrices. If ρ and σ are inequivalent, one can show that $\|\rho - \sigma\| = 2$.

3 From Pure States to Observables

3.1 Poisson Spaces with a Transition Probability

We have encountered two kinds of structure on the pure state space $\mathcal{P}(\mathfrak{A})$ of a C^*-algebra $\mathfrak{A}$. Firstly, it is a Poisson space (cf. 2.6.2 and 2.6.8), and secondly, as established in 2.8.2, it is a transition probability space. We will now examine how these structures are interrelated. Recall (2.60).

Definition 3.1.1. *The real normed vector space $\mathfrak{A}^{00}_{\mathbb{R}}(\mathcal{P})$, regarded as a subspace of $\ell^\infty(\mathcal{P}, \mathbb{R})$ (with sup-norm), consists of all finite linear combinations of the type $\sum_{i=1}^N c_i p_{\rho_i}$, where $c_i \in \mathbb{R}$ and $\rho_i \in \mathcal{P}$. The closure of $\mathfrak{A}^{00}_{\mathbb{R}}(\mathcal{P})$ is called $\mathfrak{A}^0_{\mathbb{R}}(\mathcal{P})$.*

The double dual of $\mathfrak{A}^0_{\mathbb{R}}(\mathcal{P})$ will play a central role in what follows, so that we use a special symbol:

$$\mathfrak{A}_{\mathbb{R}}(\mathcal{P}) := \mathfrak{A}^0_{\mathbb{R}}(\mathcal{P})^{**}. \tag{3.1}$$

Since $\mathfrak{A}^0_{\mathbb{R}}(\mathcal{P}) \subseteq \ell_0(\mathcal{P}, \mathbb{R})$, one has $\mathfrak{A}_{\mathbb{R}}(\mathcal{P}) \subseteq \ell_0(\mathcal{P}, \mathbb{R})^{**} = \ell^\infty(\mathcal{P}, \mathbb{R})$. The space $\mathfrak{A}_{\mathbb{R}}(\mathcal{P})$ is the function space intrinsically related to a transition probability space $\mathcal{P}$. It is a partially ordered Banach space in the obvious way. We will now identify this space in the case that $\mathcal{P}$ is the pure state space of a C^*-algebra.

According to 1.7.5 there exists a central projection p in $\mathfrak{A}^{**}$ such that $\pi_{ra}(\mathfrak{A})'' \simeq p\mathfrak{A}^{**}$ (cf. 2.2.7). Hence $\pi_{ra}(\mathfrak{A})''$ is contained in $\mathfrak{A}^{**}$ in a natural way. By w^*-continuity, elements of $\mathfrak{A}_{\mathbb{R}} \simeq A(\mathcal{S}(\mathfrak{A}), \mathbb{R})$ are determined by their values on $\mathcal{P}(\mathfrak{A})$.

This is, in general, not the case for arbitrary elements of $\mathfrak{A}^{**}_{\mathbb{R}} \simeq A_b(\mathcal{S}(\mathfrak{A}), \mathbb{R})$. However, $A \in p\mathfrak{A}^{**}_{\mathbb{R}} \subset \mathfrak{A}^{**}_{\mathbb{R}}$ is determined by $\hat{A} \in \ell^{\infty}(\mathcal{P}(\mathfrak{A}), \mathbb{R})$; this follows either from the explicit expression (2.3) or from a more abstract argument. Therefore, the Gelfand transform (1.30) maps $p\mathfrak{A}^{**}_{\mathbb{R}}$, and hence $\pi_{\text{ra}}(\mathfrak{A})''_{\mathbb{R}}$, isometrically into some closed subspace of $\ell^{\infty}(\mathcal{P}(\mathfrak{A}), \mathbb{R})$.

Proposition 3.1.2. *If $\mathcal{P}$ is the pure state space of a C^*-algebra $\mathfrak{A}$, equipped with the transition probabilities (2.65), then the Gelfand transform (1.30) isomorphically maps $\pi_{\text{ra}}(\mathfrak{A})''_{\mathbb{R}} \simeq p\mathfrak{A}^{**}_{\mathbb{R}}$ (as a partially ordered Banach space) to $\mathfrak{A}_{\mathbb{R}}(\mathcal{P})$. In particular, $\mathfrak{A}_{\mathbb{R}}(\mathcal{P}) = \ell^{\infty}(\mathcal{P}, \mathbb{R})$ if $\mathfrak{A}$ is commutative.*

As a visual aid in proving this proposition, we define a (locally nontrivial) fiber bundle $\mathcal{B}(\mathcal{P})$, whose base space B is the space of sectors, equipped with the discrete topology, and whose fiber above a given base point α is $\mathfrak{B}(\mathcal{H}_{\alpha})_{\mathbb{R}}$; here $\mathcal{H}_{\alpha}$ is such that the sector α is $\mathbb{P}\mathcal{H}_{\alpha}$. Moreover, by (2.5.4) the pure state space $\mathcal{P}$ itself may be seen as a fiber bundle over the same base space; now, the fiber above α is $\mathbb{P}\mathcal{H}_{\alpha}$. We will denote the projection of the latter bundle by τ. A cross section s of $\mathcal{B}(\mathcal{P})$ then defines a function $\hat{s}$ on $\mathcal{P}$ by $\hat{s}(\rho) = [s(\tau(\rho))](\rho)$; in this description, we identify a bounded self-adjoint operator H on $\mathcal{H}_{\alpha}$ with the corresponding function $\hat{H}$ on $\mathbb{P}\mathcal{H}_{\alpha}$, cf. (2.43). By (2.44), this identification is isometric if we define the norm of a cross section of $\mathcal{B}(\mathcal{P})$ by $\|s\| = \sup_{\alpha \in B} \|s(\alpha)\|$ (where the right-hand side, of course, contains the operator norm in $\mathfrak{B}(\mathcal{H}_{\alpha})$) and the norm of $\hat{s}$ as the sup-norm in $\ell^{\infty}(\mathcal{P}, \mathbb{R})$.

It follows directly from its definition that the space $\mathfrak{A}^{00}_{\mathbb{R}}(\mathcal{P})$ consists of sections s of $\mathcal{B}(\mathcal{P})$ with finite support for which $s(\alpha)$ has finite rank for each α. Its closure $\mathfrak{A}^{0}_{\mathbb{R}}(\mathcal{P})$ contains all sections for which $\alpha \mapsto \|s(\alpha)\|$ vanishes at infinity, and $s(\alpha)$ is a compact operator. It follows from 1.6.5 that the dual $\mathfrak{A}^{0}_{\mathbb{R}}(\mathcal{P})^*$ may be realized as the space of sections for which $s(\alpha)$ is of trace class and $\alpha \mapsto \|s(\alpha)\|_1$ (cf. (1.61)) is in $\ell^1(B, \mathbb{R})$. The bidual $\mathfrak{A}_{\mathbb{R}}(\mathcal{P})$ then consists of all sections of $\mathcal{B}(\mathcal{P})$ for which $\alpha \mapsto \|s(\alpha)\|$ is in $\ell^{\infty}(B, \mathbb{R})$. It follows from (2.3) that this is precisely the image of the Gelfand transform (1.30) of $\pi_{\text{ra}}(\mathfrak{A})''_{\mathbb{R}}$. ■

If $\mathcal{P}$ is simultaneously a (general, uniform, or smooth) Poisson space (cf. 2.6.2, 2.6.3) and a transition probability space, two function spaces are intrinsically associated with it: $C^{\infty}_{L}(\mathcal{P}, \mathbb{R})$, defined in 2.6.2.3, and $\mathfrak{A}_{\mathbb{R}}(\mathcal{P})$, respectively. The space naturally tied with both structures in concert is therefore

$$\mathfrak{A}_{L}(\mathcal{P}, \mathbb{R}) := \mathfrak{A}_{\mathbb{R}}(\mathcal{P}) \cap C^{\infty}_{L}(\mathcal{P}, \mathbb{R}). \tag{3.2}$$

For example, if $\mathcal{P}$ is a smooth Poisson space (i.e., a Poisson manifold) equipped with the transition probabilities (2.58), then $\mathfrak{A}_{\mathbb{R}}(\mathcal{P}) = \ell^{\infty}(\mathcal{P}, \mathbb{R})$, so that $\mathfrak{A}_{L}(\mathcal{P}, \mathbb{R}) = C^{\infty}_{L,b}(\mathcal{P}, \mathbb{R})$. The corresponding equation for C^*-algebras is (3.6) below.

In general, since elements of $\mathfrak{A}_{L}(\mathcal{P}, \mathbb{R})$ are smooth on each symplectic leaf of $\mathcal{P}$, they generate a well-defined Hamiltonian flow (2.11), which, of course, stays inside a given leaf.

Definition 3.1.3. *A (general, uniform, or smooth) Poisson space that is simultaneously a transition probability space is called* **unitary** *if the Hamiltonian flow on* $\mathcal{P}$ *defined by each element of* $\mathfrak{A}_L(\mathcal{P}, \mathbb{R})$ *preserves the transition probabilities. That is, if* $\rho(t)$ *and* $\sigma(t)$ *are Hamiltonian curves (with respect to a given* $H \in \mathfrak{A}_L(\mathcal{P}, \mathbb{R})$*) through* $\rho(0) = \rho$ *and* $\sigma(0) = \sigma$*, respectively, then*

$$p(\rho(t), \sigma(t)) = p(\rho, \sigma) \tag{3.3}$$

for each t for which both flows are defined.

We infer from (2.46), (2.47), 3.1.2, 2.6.8, and 2.8.2 that the pure state space of a C^*-algebra is unitary. Also, a Poisson manifold with (2.58) is evidently unitary.

Definition 3.1.4. *A (general, uniform, or smooth)* **Poisson space with a transition probability** *is a set* $\mathcal{P}$ *that is a well-behaved transition probability space (Definition 2.7.5) and a unitary (general, uniform, or smooth) Poisson space (Definitions 2.6.2, 2.6.3, and 3.1.3), for which* $\mathfrak{A}_{\mathbb{R}} = \mathfrak{A}_L(\mathcal{P}, \mathbb{R})$ *(defined in (3.2)).*

This definition imposes two closely related compatibility conditions between the Poisson structure and the transition probabilities: Firstly, it makes a definite choice for the space $\mathfrak{A}_{\mathbb{R}}$ appearing in the definition of a Poisson space, and secondly, it imposes the unitarity requirement.

We collect the previous findings in

Theorem 3.1.5.

- *The pure state space of a* C^**-algebra equipped with the* w^**-topology, the transition probabilities (2.63), and the Poisson structure 2.6.8, is a Poisson space with a transition probability.*
- *A Poisson manifold equipped with the transition probabilities (2.58) is a smooth Poisson space with a transition probability.*

3.2 *Identification of the Algebra of Observables*

This section is devoted to the following result, which shows how a unital C^*-algebra $\mathfrak{A}$ can be recovered from its pure state space. We recall that $\mathfrak{A}^0_{\mathbb{R}}(\mathcal{P})$ was defined in 3.1.1, and that $\mathfrak{A}_{\mathbb{R}}(\mathcal{P}) \subseteq \ell^\infty(\mathcal{P}, \mathbb{R})$ (cf. (3.1)). Also, we regard $\mathfrak{A}_{\mathbb{R}}$ as a closed subspace of $C_b(\mathcal{P}(\mathfrak{A}), \mathbb{R}) \subset \ell^\infty(\mathcal{P}(\mathfrak{A}), \mathbb{R})$ through the Gelfand transform (1.30).

Theorem 3.2.1. *Let* $\mathcal{P}(\mathfrak{A})$ *be the pure state space of a unital* C^**-algebra* $\mathfrak{A}$*, equipped with the transition probabilities (2.63) and the* w^**-uniformity inherited from* $\mathfrak{A}^*$*. Then*

$$\mathfrak{A}_{\mathbb{R}} = \mathfrak{A}_{\mathbb{R}}(\mathcal{P}(\mathfrak{A})) \cap C_u(\mathcal{P}(\mathfrak{A}), \mathbb{R}), \tag{3.4}$$

where $C_u(\mathcal{P}(\mathfrak{A}), \mathbb{R})$ *is the space of real-valued uniformly continuous functions on* $\mathcal{P}(\mathfrak{A})$.

Before starting with the proof, we clarify the content of the theorem.

Firstly, the w^*-uniformity on $\mathcal{P}(\mathfrak{A})$ may be defined by its subbase consisting of all subsets of $\mathcal{P} \times \mathcal{P}$ of the type $\{(\rho, \sigma) \in \mathcal{P} \times \mathcal{P} \mid |\rho(A) - \sigma(A)| < \epsilon\}$, where $A \in \mathfrak{A}_{\mathbb{R}}$ and $\epsilon > 0$. It is noteworthy, however, that the subspace of functions in $\ell^\infty(\mathcal{P}(\mathfrak{A}), \mathbb{R})$ that are uniformly continuous with respect to any uniformity on $\mathcal{P}(\mathfrak{A})$ is closed. This generalizes the well-known fact that the subspace of continuous functions relative to any topology on $\mathcal{P}$ is sup-norm closed; the proof of our observation proceeds by the same $\varepsilon/3$ argument.

Secondly, we know from 1.4.5 that $\mathfrak{A}_{\mathbb{R}} \simeq A(\mathcal{S}(\mathfrak{A}), \mathbb{R}) = \mathfrak{A}^{**} \cap C(\mathcal{S}(\mathfrak{A}), \mathbb{R})$. There may, however, exist spurious elements of $\mathfrak{A}^{**}_{\mathbb{R}}$ that happen to be w^*-continuous on $\mathcal{P}(\mathfrak{A})$ but not on $\mathcal{S}(\mathfrak{A})$. Therefore, an arbitrary C^*-algebra $\mathfrak{A}$ does not satisfy $\mathfrak{A}_{\mathbb{R}} = \mathfrak{A}^{**}_{\mathbb{R}} \cap C(\mathcal{P}(\mathfrak{A}), \mathbb{R})$ (although a large class of such algebras does, see below). The theorem shows that these spurious elements fail to be uniformly continuous on $\mathcal{P}(\mathfrak{A})$, and that uniform continuity on $\mathcal{P}(\mathfrak{A})$ can be used to characterize $\mathfrak{A}_{\mathbb{R}}$.

We now pass to the proof of Theorem 3.2.1. According to 3.1.2, we may identify $\mathfrak{A}_{\mathbb{R}}(\mathcal{P})$ with the image of the Gelfand transform of $\pi_{\text{ra}}(\mathfrak{A})''_{\mathbb{R}}$ (or $p\mathfrak{A}^{**}_{\mathbb{R}}$) in $\ell^\infty(\mathcal{P}(\mathfrak{A}), \mathbb{R})$; we denote this image by $\mathfrak{M}_{\mathbb{R}}$. Hence we can write the right-hand side of (3.4) as $\mathfrak{M}_{\mathbb{R}} \cap C_u(\mathcal{P}(\mathfrak{A}), \mathbb{R})$. Since we identify $\mathfrak{A}_{\mathbb{R}}$ with its Gelfand transform, and because $\mathfrak{A}_{\mathbb{R}} \subseteq \pi_{\text{ra}}(\mathfrak{A})''_{\mathbb{R}}$, we can say that $\mathfrak{A}_{\mathbb{R}} \subseteq \mathfrak{M}_{\mathbb{R}}$. The inclusion $\mathfrak{A}_{\mathbb{R}} \subseteq C_u(\mathcal{P}, \mathbb{R})$ is immediate from the definition of the w^*-uniformity, so that $\mathfrak{A}_{\mathbb{R}} \subseteq \mathfrak{M}_{\mathbb{R}} \cap C_u(\mathcal{P}, \mathbb{R})$.

We note that $\mathcal{P}(\mathfrak{A}) = \partial_e \mathcal{N}(\mathfrak{M})$, where $\mathcal{N}(\mathfrak{M})$ is the normal state space of the complexification $\mathfrak{M}$ of $\mathfrak{M}_{\mathbb{R}}$, cf. 1.7 (recall that $\mathfrak{M}$ is a von Neumann algebra). For any von Neumann algebra $\mathfrak{M}$ of the form $\mathfrak{M} = \mathfrak{B}^{**}$ (where $\mathfrak{B}$ is a C^*-algebra) one has

$$\overline{\text{co}}(\partial_e \mathcal{N}(\mathfrak{M})) = \mathcal{S}(\mathfrak{M}), \tag{3.5}$$

which sharpens Theorem 2.1.5, since $\mathcal{N}(\mathfrak{M}) \subseteq \mathcal{S}(\mathfrak{M})$ may be a proper inclusion. We apply this with $\mathfrak{B}_{\mathbb{R}} = \mathfrak{A}^0_{\mathbb{R}}(\mathcal{P})$ (cf. the proof of 3.1.2), for which $\mathfrak{B}^{**}$ indeed equals our $\mathfrak{M}$. As a corollary of Theorem 2.1.5, note that if $L \subset K$ is a closed subset of K for which $\overline{\text{co}}(L) = K$, then $\partial_e K \subseteq L$. It then follows from (3.5) that $\mathcal{P}(\mathfrak{M})^- \subseteq (\partial_e \mathcal{N}(\mathfrak{M}))^- = \mathcal{P}(\mathfrak{A})^-$, where the closures are taken in the w^*-topology on $\mathfrak{M}^*$. Therefore, one can approximate any ρ and σ in $\mathcal{P}(\mathfrak{M})^-$ in the w^*-topology on $\mathfrak{M}^*$ by elements of $\mathcal{P}(\mathfrak{A})$, so that $\rho_\alpha \to \rho$ and $\sigma_\beta \to \sigma$ for nets $\{\rho_\alpha\}$ and $\{\sigma_\beta\}$ in $\mathcal{P}(\mathfrak{A})$. If we choose these such that $\rho = \sigma$ on $\mathfrak{A}_{\mathbb{R}}$, then clearly $\lim_{\alpha,\beta}(\rho_\alpha(A) - \sigma_\beta(A)) = 0$ for all $A \in \mathfrak{A}_{\mathbb{R}}$.

Now choose $B \in \mathfrak{M} \cap C_u(\mathcal{P}(\mathfrak{A}), \mathbb{R})$. By the definition of the w^*-uniformity on $\mathfrak{A}^*_{\mathbb{R}}$, the uniform continuity of B implies that $\lim_{\alpha,\beta}(\rho_\alpha(B) - \sigma_\beta(B)) = 0$. Hence $\rho(B) = \sigma(B)$.

Without proof, we now invoke a deep corollary of the Stone–Weierstrass theorem for C^*-algebras: If $\mathfrak{A}$ and $\mathfrak{B}$ are unital C^*-algebras with $\mathfrak{A} \subseteq \mathfrak{B}$, and $B \in \mathfrak{B}$ is such that $\rho(B) = \sigma(B)$ for any pair ρ, σ coinciding on $\mathfrak{A}$, then $B \in \mathfrak{A}$.

Returning to the previous paragraph, this corollary implies that $B \in \mathfrak{A}_{\mathbb{R}}$. □

Combining Theorem 3.1.5 and (3.4) we infer

Corollary 3.2.2. *The pure state space of a C^*-algebra equipped with the w^*-uniformity, the transition probabilities (2.63), and the Poisson structure 2.6.8 is a uniform Poisson space with a transition probability.*

We briefly return to (3.2). According to Propositions 3.1.2 and 2.6.8, and the comment below (2.44), for pure state spaces $\mathcal{P} = \mathcal{P}(\mathfrak{A})$ of C^*-algebras one has $\mathfrak{A}_{\mathbb{R}}(\mathcal{P}) \cap C_u(\mathcal{P}) \subset C_L^{\infty}(\mathcal{P}, \mathbb{R})$. It then follows from Theorem 3.2.1 that

$$\mathfrak{A}_L(\mathcal{P}(\mathfrak{A}), \mathbb{R}) = \mathfrak{A}_{\mathbb{R}}(\mathcal{P}(\mathfrak{A})) \cap C_u(\mathcal{P}(\mathfrak{A}), \mathbb{R}). \tag{3.6}$$

A unital C^*-algebra $\mathfrak{A}$ is called **perfect** if

$$\mathfrak{A}_{\mathbb{R}} = \mathfrak{A}_{\mathbb{R}}(\mathcal{P}) \cap C(\mathcal{P}(\mathfrak{A}), \mathbb{R}). \tag{3.7}$$

In that case, C_u in (3.6) may be replaced by C.

If $\mathcal{P}(\mathfrak{A})$ is closed (hence compact), then $\mathfrak{A}$ is obviously perfect. Hence commutative C^*-algebras are perfect (cf. 2.1.7, which actually implies 3.2.1 in the commutative case), and so are finite-dimensional C^*-algebras. On the basis of Proposition 2.5.2 one might expect that the unitization $\mathfrak{B}_0(\mathcal{H})_{\mathbb{I}}$ of $\mathfrak{B}_0(\mathcal{H})$ cannot be perfect, but the opposite is true. While 2.5.2 does show that any element of $\mathfrak{B}(\mathcal{H})$ is continuous on all points of $\mathcal{P}(\mathfrak{B}_0(\mathcal{H})_{\mathbb{I}})$ except ω_∞ (cf. (2.2)), only members of $\mathfrak{B}_0(\mathcal{H})_{\mathbb{I}}$ are continuous at ω_∞, too. Finally, deeper analysis shows that $\mathfrak{B}(\mathcal{H})$ is perfect for any Hilbert space $\mathcal{H}$.

3.3 Spectral Theorem and Jordan Product

Given a C^*-algebra $\mathfrak{A}$, one can use Proposition 3.1.2 to endow $\mathfrak{A}_{\mathbb{R}}(\mathcal{P})$, and hence $\mathfrak{A}_{\mathbb{R}}$ (cf. (3.4)), with the structure of a JLB-algebra; cf. 1.1.9. It is enlightening, however, to derive this structure from the pure state space $\mathcal{P} = \mathcal{P}(\mathfrak{A})$. By Theorem 3.1.5 this is a Poisson space with a transition probability; our first goal is to reconstruct the Jordan product on $\mathfrak{A}_{\mathbb{R}}(\mathcal{P})$ from the transition probabilities.

Definition 3.3.1. *Let $\mathcal{P}$ be a well-behaved transition probability space (cf. 2.7.5). A* **spectral resolution** *of an element $A \in \ell^{\infty}(\mathcal{P}, \mathbb{R})$ is an expansion (in the topology of pointwise convergence)*

$$A = \sum_j \lambda_j p_{Q_j}, \tag{3.8}$$

where $\lambda_j \in \mathbb{R}$, and $\{Q_j\}$ is an orthogonal family of orthoclosed subsets of $\mathcal{P}$ (cf. (2.61)) for which $\sum_j p_{Q_j}$ equals the unit function on $\mathcal{P}$.

Proposition 3.3.2. *If $\mathcal{P} = \cup_\alpha \mathbb{P}\mathcal{H}_\alpha$ with transition probabilities (2.65), then any $A \in \mathfrak{A}_{\mathbb{R}}^{00}(\mathcal{P})$ (cf. 3.1.1) has a unique spectral resolution.*

By 2.5.4 and 2.8.2 this applies, in particular, to the pure state space of a C^*-algebra.

Firstly, the case of reducible $\mathcal{P}$ may be reduced to the irreducible one by grouping the ρ_i in $A = \sum_{i=1}^N c_i p_{\rho_i}$ into mutually orthogonal groups, with the property that $(\cup\rho)^{\perp\perp}$ is irreducible if the union is over all ρ_i in a given group. Thus we henceforth assume that $\mathcal{P}$ is irreducible, hence of the form $\mathcal{P} = \mathbb{P}\mathcal{H}$ with the transition probabilities (2.65).

If $\mathcal{P}$ is finite-dimensional, the proposition is simply a restatement of the spectral theorem for Hermitian matrices. In the general case, let A be as above, and $Q := \{\rho_1, \ldots, \rho_N\}^{\perp\perp}$. If $\sigma \in Q$, then $A(\sigma) = \sum_j \lambda_j p_{Q_j}(\sigma)$ for some λ_j and mutually orthogonal $Q_j \subset Q$, since the situation is finite-dimensional. If $\sigma \in Q^{\perp}$, this equation trivially holds, as both sides vanish.

Let us assume, therefore, that σ lies neither in Q nor in $Q^{\perp}$. Define $\varphi_Q(\sigma)$ by the following procedure: Lift σ to a unit vector Σ in $\mathcal{H}$, project Σ onto the subspace defined by Q, normalize the resulting vector to unity, and project back to $\mathbb{P}\mathcal{H}$. In the Hilbert space case relevant to us, the transition probabilities satisfy

$$p(\sigma, \rho) = p(\sigma, \varphi_Q(\sigma))p(\varphi_Q(\sigma), \rho) \tag{3.9}$$

for $\rho \in Q$ and $\sigma \notin Q^{\perp}$. We now compute $A(\sigma)$ by using this equation, followed by the use of the spectral theorem in Q, and subsequently recycle the same equation in the opposite direction. This calculation establishes the proposition for $\sigma \notin Q^{\perp}$. ■

Proposition 3.3.3. *If $\mathcal{P}$ is the pure state space of a C^*-algebra, $A = \sum_j \lambda_j p_{Q_j}$ is the spectral resolution of $A \in \mathfrak{A}^{00}_{\mathbb{R}}(\mathcal{P})$, and A^2 is defined by $A^2 = \sum_j \lambda_j^2 p_{Q_j}$, then the product $\circ$ defined by*

$$A \circ B := \tfrac{1}{4}((A + B)^2 - (A - B)^2) \tag{3.10}$$

turns $\mathfrak{A}^{00}_{\mathbb{R}}(\mathcal{P})$ into a Jordan algebra. Moreover, this Jordan product $\circ$ can be extended to $\mathfrak{A}^{0}_{\mathbb{R}}(\mathcal{P})$ (cf. 3.1.1) by (norm-) continuity, which thereby becomes a JB-algebra. Finally, the bidual $\mathfrak{A}_{\mathbb{R}}(\mathcal{P})$ (with sup-norm inherited from $\ell^{\infty}(\mathcal{P}, \mathbb{R})$) is turned into a JB-algebra by extending $\circ$ by w^*-continuity.

The bilinearity of (3.10) is not obvious, and would not necessarily hold for arbitrary well-behaved transition probability spaces in which a spectral theorem (in the sense of 3.3.2) is valid. In the present case, it follows from the explicit form of the transition probabilities in $\mathbb{P}\mathcal{H}$. The quickest way to establish bilinearity, of course, is to look at a function p_Q (where Q lies in a sector $\mathbb{P}\mathcal{H}$ of $\mathcal{P}$) as the Gelfand transform of a projection operator on $\mathcal{H}$ (cf. (2.43)).

Given bilinearity and the spectral theorem 3.3.2, the proof of (1.2) reduces to showing that $(p_\rho \circ p_\tau) \circ p_\sigma = p_\rho \circ (p_\tau \circ p_\sigma)$ for ρ, σ orthogonal and τ arbitrary. Through the (inverse) Gelfand transform this reduces to a calculation with 3×3 matrices. The first Jordan algebra axiom is trivially satisfied by (3.10).

We now show that the axioms (1.7), (1.8) hold in $\mathfrak{A}^{00}_{\mathbb{R}}(\mathcal{P})$; the norm-closure $\mathfrak{A}^{0}_{\mathbb{R}}(\mathcal{P})$ will then be a JB-algebra. If A is given by (3.8), and $\lambda := \sup_j |\lambda_j|$, then on the one hand $\|A\| \geq \lambda$, since each λ_j is a possible value of A (assumed at any point in Q_j). On the other hand, $|A(\sigma)| \leq \lambda \sum_j p_{Q_j}(\sigma) = \lambda$ by (2.56), so that $\|A\| \leq \lambda$. Hence $\|A\| = \lambda$. With our definition of A^2, this immediately establishes (1.9) and (1.10) (which are equivalent to (1.8)). Axiom (1.7) follows if we assume that $\|A\| \leq 1$ and $\|B\| \leq 1$, use (3.10), the observation that if $f, g \in \ell^{\infty}$ are both positive then $\|f - g\|$ is majorized by $\max\{\|f\|, \|g\|\}$, and the triangle inequality; these steps yield $\|A \circ B\| \leq 1$.

The statement about the bidual is a direct consequence of Lemma 3.3.4 below. ■

Without proof we state a generalization of Proposition 1.7.4.

Lemma 3.3.4. *Let $(\mathfrak{A}_{\mathbb{R}}, \circ)$ be a JB-algebra. Then the Jordan product on $\mathfrak{A}_{\mathbb{R}}$ has a bilinear extension (called $\circ$ as well) to $\mathfrak{A}_{\mathbb{R}}^{**}$ such that the maps $A \mapsto A \circ B$ and $A \mapsto B \circ A$ are w^*-continuous and $\|A \circ B\| \leq \|A\| \, \|B\|$ for all $A, B \in \mathfrak{A}_{\mathbb{R}}^{**}$. An extension with these properties is unique.*

Of course, this discussion includes the situation where $\mathfrak{A}$ is commutative. In that case, the trivial transition probabilities (2.58) and the above construction imply that the Jordan product on $\mathfrak{A}_{\mathbb{R}}(\mathcal{P}) = \ell^{\infty}(\mathcal{P}, \mathbb{R})$ is pointwise multiplication, as it should be.

3.4 Unitarity and Leibniz Rule

The following result shows that the Leibniz rule (1.4) in a JLB-algebra $\mathfrak{A}_{\mathbb{R}}$ is a consequence of the unitarity condition relating the Poisson structure and the transition probabilities on $\mathcal{P}(\mathfrak{A})$.

Proposition 3.4.1. *Let $\mathcal{P}$ be a Poisson space with a transition probability (see 3.1.4) in which every $A \in \mathfrak{A}_{\mathbb{R}}^{00}(\mathcal{P})$ has a unique spectral resolution (in the sense of 3.3.1). Assume that for each $h \in \mathfrak{A}_L(\mathcal{P}, \mathbb{R})$ (cf. (3.2)) the map $A \mapsto \{h, A\}$ is bounded on $\mathfrak{A}_L(\mathcal{P}, \mathbb{R})$ (with sup-norm). If a Jordan product $\circ$ is defined on $\mathfrak{A}_L(\mathcal{P}, \mathbb{R})$ through the transition probabilities, in the manner of Proposition 3.3.3, then $\circ$ and the Poisson bracket satisfy the Leibniz rule.*

The boundedness assumption holds when $\mathcal{P}$ is the pure state space of a C^*-algebra; it is made mainly to simplify the proof. The proposition evidently holds when $\mathfrak{A}_L(\mathcal{P}, \mathbb{R})$ is a Poisson algebra, for which the assumption is violated.

Writing $\delta_h(A)$ for $\{h, A\}$, the boundedness of δ_h implies that the series $\alpha_t(A) = \sum_{n=0}^{\infty} t^n \delta_h^n(A)/n!$ converges uniformly and defines a uniformly continuous one-parameter group of maps on $\mathfrak{A}_L(\mathcal{P}, \mathbb{R})$. On the other hand, if $\sigma(t)$ is the Hamiltonian flow of h on $\mathcal{P}$ (cf. 2.3), then α_t as defined by (2.13) must coincide with the definition above, for they each satisfy the differential equation (2.14) with the same initial condition. In particular, the flow in question must be complete. Moreover, it follows that the Leibniz rule (yet to be established) is equivalent to the property that α_t is a Jordan morphism for each t; this, in turn, can be rephrased by saying that $\alpha_t(A^2) = \alpha_t(A)^2$ for all $A \in \mathfrak{A}_L(\mathcal{P}, \mathbb{R})$.

Let $A \in \mathfrak{A}_{\mathbb{R}}^{00}(\mathcal{P}) \cap \mathfrak{A}_L(\mathcal{P}, \mathbb{R})$. By (3.8) and (2.61), $A = \sum_k \lambda_k p_{e_k}$, where all e_k are orthogonal. Unitarity implies firstly that $\alpha_t(A) = \sum_k \lambda_k p_{e_k(-t)}$, and secondly that the $e_k(-t)$ are orthogonal. Hence $\alpha_t(A)$ is given in its spectral resolution, so that $(\alpha_t(A))^2 = \sum_k \lambda_k^2 p_{e_k(-t)}$. Repeating the first use of unitarity, we find that this equals $\alpha_t(A^2)$. Hence the property holds on $\mathfrak{A}_{\mathbb{R}}^{00}(\mathcal{P})$. Now $\mathfrak{A}_{\mathbb{R}}^{00}(\mathcal{P})$ is dense in $\mathfrak{A}_{\mathbb{R}}(\mathcal{P})$ in the topology of pointwise convergence in $\ell^{\infty}(\mathcal{P}, \mathbb{R})$. But $A_{\lambda} \to A$ pointwise clearly implies $\alpha_t(A_{\lambda}) \to \alpha_t(A)$ pointwise. This, plus the w^*-continuity of the Jordan product (cf. 3.3.4), proves the desired result. ■

We return to the pure state space $\mathcal{P}$ of a unital C^*-algebra. Through the results 3.3.3 and 3.4.1, the fact that $\mathfrak{A}_{\mathbb{R}}(\mathcal{P}) \subset \mathfrak{A}_L(\mathcal{P}, \mathbb{R})$, and the observation that the associator identity (1.6) is a consequence of the special form of the transition probabilities, we have reconstructed $\mathfrak{A}_{\mathbb{R}}(\mathcal{P})$ as a JLB-algebra. The final ingredient on $\mathcal{P}$ that allows one to reconstruct the C^*-algebra $\mathfrak{A}$ whose pure state space it is, is its uniform structure (namely, the w^*-uniformity defined by $\mathfrak{A}_{\mathbb{R}}$). The JLB-algebra $\mathfrak{A}_{\mathbb{R}}$ is given by (3.4), and the C^*-algebra $\mathfrak{A}$ is then constructed as in 1.1.9.

Corollary 3.4.2. *Let $\mathfrak{A}$ be a unital C^*-algebra with pure state space $\mathcal{P}(\mathfrak{A})$, the latter seen as a uniform Poisson space with a transition probability. Then $\alpha : \mathfrak{A}_{\mathbb{R}} \to \mathfrak{A}_{\mathbb{R}}$ is an automorphism (cf. 1.1.3) iff the map $\alpha^* : \mathcal{P} \to \mathcal{P}$, defined by $\alpha^*\rho(A) := \rho(\alpha(A))$,*

1. *is a bijection of $\mathcal{P}$;*
2. *is uniformly continuous, along with its inverse;*
3. *is a Poisson map;*
4. *leaves the transition probabilities invariant.*

This is now obvious, as we have seen that the data preserved by α determine $\mathcal{P}(\mathfrak{A})$, whereas the data preserved by α^* determine $\mathfrak{A}_{\mathbb{R}}$. ■

Corollary 3.4.3. *A bijection of $\mathbb{P}\mathcal{H}$ that preserves transition probabilities is induced by a unitary or an antiunitary operator on $\mathcal{H}$.*

We start with $\mathfrak{A} = \mathfrak{B}_0(\mathcal{H})$, for which $\mathfrak{A}_{\mathbb{R}}(\mathcal{P}) = \mathfrak{B}(\mathcal{H})_{\mathbb{R}}$ (cf. 1.6.5 and 3.1.2). By Proposition 3.3.3, the Jordan structure on $\mathfrak{B}(\mathcal{H})_{\mathbb{R}}$ is therefore determined by the transition probabilities on $\mathcal{P}(\mathfrak{A}) = \mathbb{P}\mathcal{H}$. Hence the given bijection of $\mathcal{P}$ must correspond to a Jordan automorphism of $\mathfrak{B}(\mathcal{H})_{\mathbb{R}}$. The corollary then follows from the following lemma. □

Lemma 3.4.4. *Any Jordan automorphism α of $\mathfrak{B}(\mathcal{H})_{\mathbb{R}}$ is (anti) unitarily implemented. That is, $\alpha(A) = UAU^*$ for some unitary or antiunitary operator U on $\mathcal{H}$.*

To start, extend α to $\mathfrak{B}(\mathcal{H})$ by (complex) linearity. The definition of a Jordan morphism then implies, after some manipulations, that

$$(\alpha(AB) - \alpha(A)\alpha(B))(\alpha(AB) - \alpha(B)\alpha(A)) = 0$$

for all $A, B \in \mathfrak{B}(\mathcal{H})_{\mathbb{R}}$. Since $\mathfrak{B}(\mathcal{H})$ acts irreducibly on $\mathcal{H}$, it follows that α must either be a morphism (i.e., $\alpha(AB) = \alpha(A)\alpha(B)$) or an antimorphism ($\alpha(AB) = \alpha(B)\alpha(A)$). If α is a morphism, one defines the unitary operator U as follows. Take an arbitrary unit vector $\Omega \in \mathcal{H}$; since Ω is cyclic for $\mathfrak{B}(\mathcal{H})$, one may start defining U on vectors of the type $A\Omega$, where $A \in \mathfrak{B}(\mathcal{H})$. Let the range of the projection $\alpha([\Omega])$ be $\mathbb{C}\Omega_{\alpha}$, where Ω_{α} is a unit vector. Then define $UA\Omega := \alpha(A)\Omega_{\alpha}$. The property $\|\alpha(B)\| = \|B\|$ for all $B \in \mathfrak{B}(\mathcal{H})$ (with $B = A[\Omega]$) shows that $\|UA\Omega\| = \|A\Omega\|$, so that U is well-defined and unitary. The property $\alpha(A) = UAU^*$ easily follows from the fact that α is a morphism.

The case where α is an antimorphism can be reduced to the previous paragraph. Define $\tilde{\alpha}$ by $\tilde{\alpha}(A) = \alpha(A^*)$; this is an antilinear morphism of $\mathfrak{B}(\mathcal{H})$. The operator U is then constructed as in the previous paragraph, and evidently turns out to be antilinear. ■

3.5 Orthomodular Lattices

In this section we collect some material from the theory of lattices that will be used in what follows.

Definition 3.5.1. *A* **lattice** *$\mathcal{L}$ is a partially ordered set (poset) in which any two elements x, y have a supremum (or least upper bound) $x \vee y$ (that is, $x \leq x \vee y$ and $y \leq x \vee y$, and if $x \leq z$ and $y \leq z$ for some z, then $x \vee y \leq z$) and an infimum (or greatest lower bound) $x \wedge y$ (i.e., $x \geq x \wedge y$ and $y \geq x \wedge y$, and if $x \geq z$ and $y \geq z$ for some z, then $x \wedge y \geq z$).*

An equivalent definition of a lattice is that it is a set $\mathcal{L}$ equipped with two idempotent, commutative, and associative operations $\vee, \wedge : \mathcal{L} \times \mathcal{L} \to \mathcal{L}$ that satisfy $x \vee (y \wedge x) = x$ and $x \wedge (y \vee x) = x$. The partial ordering is then defined by $x \leq y$ if $x \wedge y = x$. The largest element in the lattice, if it exists, is denoted by I, and the smallest one (if it exists) by o. Hence $o \leq x \leq I$ for all $x \in \mathcal{L}$.

A lattice $\mathcal{L}$ is called **complete** when every subset of $\mathcal{L}$ has a supremum as well as an infimum. An **atom** of a lattice $\mathcal{L}$ with o is an element a for which $o \leq x \leq a$ implies $x = o$ or $x = a$. A lattice with o is called **atomic** if for every $x \neq o$ in L there is an atom $a \neq o$ such that $a \leq x$. All lattices occurring in this section are complete and atomic.

The "classical" example of a lattice is obtained by taking a set S and defining $\mathcal{L}$ as the power set 2^S of S (i.e., the set of all subsets of S). The lattice structure of $\mathcal{L}$ consists of $\vee := \cup$ and $\wedge := \cap$. It follows that $I = S$, whereas $o = \emptyset$ is the empty set. Such a lattice is **distributive**, in that

$$x \vee (y \wedge z) = (x \vee y) \wedge (x \vee z); \tag{3.11}$$

this is equivalent to the same property with $\vee$ and $\wedge$ swapped.

One can weaken the distributivity property by requiring only (3.11) if $x \leq z$; thus a lattice is said to be **modular** if

$$x \leq z \implies x \vee (y \wedge z) = (x \vee y) \wedge z \quad \forall y. \tag{3.12}$$

The canonical example of a nondistributive modular lattice is the collection $L(V)$ of all linear subspaces of a (left) vector space V (over an arbitrary division ring $\mathbb{D}$; the reader may keep $\mathbb{D} = \mathbb{R}$ or $\mathbb{C}$ in mind). The lattice operations are $x \wedge y := x \cap y$, while $x \vee y := x + y$ is the linear span of x and y. Equivalently, the partial order is given by inclusion. Evidently, $I = V$ and $o = 0$.

Definition 3.5.2. *An* **orthocomplementation** *on a lattice $\mathcal{L}$ with o and I is a map $x \mapsto x^{\perp}$, satisfying (for all $x, y \in \mathcal{L}$)*

- $x^{\perp\perp} = x$.
- $x \leq y \iff y^{\perp} \leq x^{\perp}$.
- $x \wedge x^{\perp} = o$.
- $x \vee x^{\perp} = I$.

It follows that $I^{\perp} = o$ and $o^{\perp} = I$, and that

$$(x \vee y)^{\perp} = x^{\perp} \wedge y^{\perp}; \ (x \wedge y)^{\perp} = x^{\perp} \vee y^{\perp} \tag{3.13}$$

(de Morgan's laws). A lattice with an orthocomplementation is called an **orthocomplemented lattice**. For example, in the lattice $\mathcal{L} = 2^S$ an orthocomplementation is given by the set-theoretic complement.

A **lattice homomorphism** between two orthocomplemented lattices is a map preserving $\leq$ and $\perp$ (and hence $\vee$ and $\wedge$). A **lattice isomorphism** is a bijection that with its inverse is a homomorphism; we write $\mathcal{L}_1 \simeq \mathcal{L}_2$ if $\mathcal{L}_1$ and $\mathcal{L}_2$ are isomorphic. Similarly, a **lattice automorphism** is an isomorphism between a lattice and itself.

The following weakening of the modular law (3.12) will soon turn out to be of prime relevance.

Definition 3.5.3. *An orthocomplemented lattice $\mathcal{L}$ is called* **orthomodular** *if (3.12) holds for $y = x^{\perp}$, that is,*

$$x \leq z \implies x \vee (x^{\perp} \wedge z) = z. \tag{3.14}$$

The following reformulation of orthomodularity will be used later on.

Lemma 3.5.4. *An orthocomplemented lattice $\mathcal{L}$ is orthomodular iff $x \leq z$ and $x^{\perp} \wedge z = o$ imply $x = z$.*

If (3.14) holds and $x^{\perp} \wedge z = o$, then $z = x \vee o = x$. Conversely, if $x \leq z$, then $z \vee (x^{\perp} \wedge z) = z$, so that $x \vee (x^{\perp} \wedge z) \leq z$. Assuming that $x^{\perp} \wedge z = o$, one infers $(x \vee (x^{\perp} \wedge z))^{\perp} \wedge z = o$. Now apply the condition stated in the lemma with x replaced by $x \vee (x^{\perp} \wedge z)$. ■

Let $(\, , \,) : V \times V$ be a **Hermitian form** (that is, a nondegenerate sesquilinear form) on V, defined relative to an involution $\lambda \mapsto \overline{\lambda}$ of $\mathbb{D}$ (think of complex conjugation for $\mathbb{D} = \mathbb{C}$, and of the identity map on $\mathbb{R}$). The orthoplement $x^{\perp}$ of $x \in L(V)$ is defined in the obvious way by $x^{\perp} := \{\Psi \in V \mid (\Psi, \Phi) = 0 \, \forall \Phi \in x\}$; this is an element of $L(V)$ as well. One easily verifies that $x^{\perp\perp\perp} = x^{\perp}$ (cf. (2.55) and subsequent text), but in general $x \leq x^{\perp\perp}$, rather than the equality required in Definition 3.5.2.

Therefore, one considers the lattice $\mathcal{L}(V)$ of orthoclosed subspaces of V, that is, $x \in L(V)$ lies in $\mathcal{L}(V)$ iff $x^{\perp\perp} = x$. The lattice operation $\wedge$ is the same as in $L(V)$, but $\vee$ in $\mathcal{L}(V)$ is defined by $x \vee y = (x + y)^{\perp\perp}$ (this is the smallest orthoclosed subspace containing x and y). This lattice is evidently complete. One can show that $\mathcal{L}(V)$ is modular iff V is finite-dimensional. In fact, in general, any finite-dimensional linear subspace of V is orthoclosed, so that $L(V) = \mathcal{L}(V)$ if V is finite-dimensional. Even in the finite-dimensional case, $\perp$ need not be an orthocomplementation on $\mathcal{L}(V)$. It is almost trivial, however, to check the following necessary and sufficient extra condition.

Proposition 3.5.5. *The map $x \mapsto x^\perp$ is an orthocomplementation on $\mathcal{L}(V)$ iff $(x + x^\perp)^\perp = 0$ for all $x \in \mathcal{L}(V)$, which is equivalent to the property $(\Psi, \Psi) = 0 \Leftrightarrow \Psi = 0$ (that is, $(,)$ is* **anisotropic***). If, in addition, $x + x^\perp$ is orthoclosed (implying $x + x^\perp = V$) for all $x \in \mathcal{L}(V)$, then $\mathcal{L}(V)$ is orthomodular.*

To show that the additional assumption implies orthomodularity, note that on this assumption, for any x one has $z = z \wedge V = z \wedge (x + x^\perp)$. If $x \leq z$, this equals $x + z \wedge x^\perp$ by the modular law (3.12) in $L(V)$ (with $y = x^\perp$; recall $+$ is $\vee$ in $L(V)$). Taking the double orthoplement of the equation $z = x + z \wedge x^\perp$ thus found yields $z^{\perp\perp} = z$ for the left-hand side (since $z \in \mathcal{L}(V)$ by assumption) and $(x + z \wedge x^\perp)^{\perp\perp} = x \vee (z \wedge x^\perp)$ by the definition of $\vee$ in $\mathcal{L}(V)$. This proves the orthomodular law (3.14). ■

Corollary 3.5.6. *The lattice $\mathcal{L}(\mathcal{H})$ of all closed subspaces of a Hilbert space is complete, atomic, and orthomodular.*

This follows from Proposition 3.5.5, since a linear subspace of a Hilbert space is closed iff it is orthoclosed. ■

The lattices $\mathcal{L}(V)$ (and in particular $\mathcal{L}(\mathcal{H})$) enjoy the property of irreducibility. Here a lattice is said to be **reducible** if it is (isomorphic to) a nontrivial Cartesian product $\mathcal{L} = \mathcal{L}_1 \times \mathcal{L}_2$ (with lattice operations defined componentwise). If not, it is called **irreducible**. The key tool in analyzing reducibility of orthocomplemented lattices is the **center** $C(\mathcal{L})$ of $\mathcal{L}$. This consists of the elements $c \in \mathcal{L}$ for which $x = (x \wedge c) \vee (x \wedge c^\perp)$ for all $x \in \mathcal{L}$. Clearly, $o, I \in C(\mathcal{L})$.

Proposition 3.5.7. *An orthocomplemented lattice $\mathcal{L}$ is irreducible iff the center is trivial, in that $C(\mathcal{L}) = \{o, I\}$. In general, any $c \in C(\mathcal{L})$ corresponds to a factorization $\mathcal{L} \simeq [o, c] \times [o, c^\perp]$, where the isomorphism is given by $x \leftrightarrow (x \wedge c, x \wedge c^\perp)$. The orthocomplementation in $[o, c] \times [o, c^\perp]$ is defined by $(x, y)^\perp := (x^\perp \wedge c, y^\perp \wedge c^\perp)$.*

Here $[o, c] = \{x \in \mathcal{L} \mid o \leq x \leq c\}$, etc. Note that $I \leftrightarrow (c, c^\perp)$. The proof of this proposition is a straightforward definition-chasing. □

3.6 Lattices Associated with States and Observables

The connection between states and observables is further elucidated by considering various lattices naturally defined in terms of these. Also, one such lattice in particular will play a central role in the axiomatization of pure state spaces.

Proposition 3.6.1. *The collection of projections in a von Neumann algebra $\mathfrak{M}$ forms a complete orthomodular lattice $\mathcal{L}(\mathfrak{M})$, in which the partial ordering is given by the usual order structure inherited from $\mathfrak{M}$ (seen as a partially ordered space, cf. 1.3), and the orthocomplementation is $x^\perp = I - x$, where $I = \mathbb{I}$.*

The lattice $\mathcal{L}(\mathfrak{B}(\mathcal{H}))$ is isomorphic to $\mathcal{L}(\mathcal{H})$.

We first demonstrate the last claim. The isomorphism is obtained by identifying a projection $[\mathcal{K}] \in \mathcal{L}(\mathfrak{B}(\mathcal{H}))$ with the closed subspace $\mathcal{K} \in \mathcal{L}(\mathcal{H})$ onto which it projects (given a projection p, the subspace $\mathcal{K} \subseteq \mathcal{H}$ consists of those $\Psi \in \mathcal{H}$ for which $p\Psi = \Psi$; the fact that $\mathcal{K}$ is closed follows from elementary estimates). The definition of the order $\leq$ and the orthocomplementation $\perp$ in the two lattices then rapidly leads to the conclusion that this identification leads to a lattice isomorphism; note that for projections, $x \leq y$ iff $x = xy$. One then applies Proposition 3.5.5. The proposition itself then follows from Definition 1.7.3 and Theorem 1.7.2, which allow us to regard $\mathcal{L}(\mathfrak{M})$ as a sublattice of $\mathcal{L}(\mathcal{H})$ for some Hilbert space $\mathcal{H}$. The completeness of $\mathcal{L}(\mathfrak{M})$ is equivalent to the property that $\mathfrak{M}$ is strongly closed. Note that $\mathfrak{M}$ is determined by $\mathcal{L}(\mathfrak{M})$ in the sense that $\mathfrak{M} = \mathcal{L}(\mathfrak{M})''$. ■

There are two von Neumann algebras naturally associated with a C^*-algebra $\mathfrak{A}$. Firstly, one can take the bidual $\mathfrak{M} = \mathfrak{A}^{**} = \pi_{\mathrm{u}}(\mathfrak{A})''$ (cf. 1.7.4); through Proposition 3.6.1 this defines the complete orthomodular lattice $\mathcal{L}(\mathfrak{A}^{**})$. The atoms of $\mathcal{L}(\mathfrak{A}^{**})$ are the minimal projections; this lattice is atomic only for a limited class of C^*-algebras. Atomicity holds, for example, if $\mathfrak{A} = \mathfrak{B}_0(\mathcal{H})$, in which case $\mathfrak{A}^{**} = \mathfrak{B}(\mathcal{H})$, so that $\mathcal{L}(\mathfrak{A}^{**}) \simeq \mathcal{L}(\mathcal{H})$ by 3.6.1.

Secondly, one may choose $\mathfrak{M} = \pi_{\mathrm{ra}}(\mathfrak{A})''$. In view of the isomorphism $\pi_{\mathrm{ra}}(\mathfrak{A})''_{\mathbb{R}} \simeq \mathfrak{A}^0_{\mathbb{R}}(\mathcal{P})$ (cf. (3.1) and 3.1.2), we write

$$\mathcal{L}(\mathfrak{A}_{\mathbb{R}}(\mathcal{P})) := \mathcal{L}(\pi_{\mathrm{ra}}(\mathfrak{A})''). \tag{3.15}$$

This is a complete atomic orthomodular lattice.

The lattice $\mathcal{L}(\mathfrak{A}^{**})$ turns out to be isomorphic to a certain lattice defined in terms of the state space $\mathcal{S}(\mathfrak{A})$. This requires the following concept.

Definition 3.6.2. *A* **face** *F of a convex set K is a convex subspace that is closed under "purification". That is, F is a face of K iff given a decomposition $\omega = \lambda\omega_1 + (1-\lambda)\omega_2$ for some $\lambda \in [0,1]$, the condition $\omega_1, \omega_2 \in F$ implies $\omega \in F$, and conversely, $\omega \in F$ implies $\omega_1, \omega_2 \in F$.*

Clearly, a face consisting of a single point is an extreme point of K. The set K is a face, and we regard the empty set $\emptyset$ as a face, too. For example, the faces of an equilateral triangle (interior plus boundary) in $\mathbb{R}^2$ are the empty set, the three corners, the three (closed) sides of the triangle, and the triangle itself.

The set $F(K)$ of all faces of K is partially ordered by inclusion, and has a minimal element $o = \emptyset$ and a maximal element $l = K$. The intersection of an arbitrary family of faces is a face as well. Hence $F(K)$ is a complete lattice with o and l, for which $x \wedge y = x \cap y$, and $x \vee y$ is the intersection of all faces containing $x \cup y$. The atoms of $F(K)$ are the extreme points.

Even if $K = \mathcal{S}(\mathfrak{A})$, the lattice $F(K)$ is not particularly well behaved. As the following result shows, it turns out to be preferable to look at a smaller set of faces.

Proposition 3.6.3. *The collection $\mathcal{F}(\mathcal{S}(\mathfrak{A}))$ of all norm-closed faces in the state space of a unital C^*-algebra is a complete orthomodular lattice under the following operations: $\leq$ is $\subseteq$, and $F^{\perp}$ is the supremum of the set of all norm-closed faces that are orthogonal to F (here we say that F_2 is orthogonal to F_1, or $F_2 \perp F_1$, if there is an element $A \in [0, \mathbb{I}] \subseteq \mathfrak{A}^{**}_{\mathbb{R}} \simeq A_b(\mathcal{S}(\mathfrak{A}), \mathbb{R})$ such that $A(\omega) = 1$ for all $\omega \in F_1$ and $A(\omega) = 0$ for all $\omega \in F_2$). Finally, $\mathcal{F}(\mathcal{S}(\mathfrak{A}))$ is isomorphic to $\mathcal{L}(\mathfrak{A}^{**})$.*

Given a projection $p \in \mathfrak{A}^{**}_{\mathbb{R}}$, the set $F_p := \{\omega \in \mathcal{S}(\mathfrak{A}) \,|\, \omega(p) = 1\}$ is easily shown to be a norm-closed face in $\mathcal{S}(\mathfrak{A})$. Equally easily, given a norm-closed face F, the set $I_F := \{A \in \mathfrak{A}^{**}_{\mathbb{R}} \,|\, \omega(A^*A) = 0 \,\forall \omega \in F\}$ is seen to be the self-adjoint part of a σ-weakly closed left ideal in $\mathfrak{A}^{**}$. A technical argument in the theory of von Neumann algebras shows that any such ideal must be of the form $I_F = \mathfrak{A}^{**}_{\mathbb{R}} p^{\perp}$ for some projection $p^{\perp}$. It is then easily checked that $F_p = F$, and that the correspondence $p \leftrightarrow F_p$ thus established yields an isomorphism of $\mathcal{L}(\mathfrak{A}^{**})$ and $\mathcal{F}(\mathcal{S}(\mathfrak{A}))$ as lattices. Note that under this isomorphism minimal projections correspond to pure states.

We now turn to the orthocomplementation. If $A = 0$ on F_p and $A \in [0, \mathbb{I}]$, then $\sqrt{A} \in I_{F_p}$. Hence $\sqrt{A} = Bp^{\perp}$ for some B, so that $A = Ap^{\perp}$. Since $A^* = A$, this implies $A = p^{\perp}A$, so that $A = p^{\perp}Ap^{\perp}$. Now, $p^{\perp}Ap^{\perp} \leq p^{\perp}\mathbb{I}p^{\perp} = p^{\perp}$, since $A \leq \mathbb{I}$, so that $A \leq p^{\perp}$. A similar argument shows that $A = 1$ on F_p and $A \in [0, \mathbb{I}]$ imply $p \leq A$. Therefore, if $A = 0$ on F_{p_2}, $A = 1$ on F_{p_1}, and $A \in [0, \mathbb{I}]$, then $p_1 \leq A \leq p_2^{\perp}$, from which $p_1 \leq p_2^{\perp}$; we say that $p_1 \perp p_2$. The converse is obvious, so that we have shown that $F_{p_1} \perp F_{p_2}$ is equivalent to $p_1 \perp p_2$. We now notice that $p^{\perp}$ equals the supremum of all q for which $p \perp q$, and conclude that the bijection $p \leftrightarrow F_p$ preserves orthocomplementation. □

We are going to show that the lattice $\mathcal{L}(\mathfrak{A}_{\mathbb{R}}(\mathcal{P}))$ is isomorphic to a certain lattice defined by the transition probabilities on $\mathcal{P}(\mathfrak{A})$. For the moment, however, we return to the general setting of transition probability spaces (cf. 2.7).

Proposition 3.6.4. *The collection of orthoclosed subsets of a well-behaved transition probability space $\mathcal{P}$ forms a complete atomic orthomodular lattice $\mathcal{L}(\mathcal{P})$ under the operations $x \wedge y = x \cap y$, $x \vee y = (x \cup y)^{\perp\perp}$ (equivalently, $\leq$ is $\subseteq$), and $\perp$ is given by (2.55).*

The orthomodularity follows from Lemma 3.5.4: Assume $x \leq z$, and choose a basis $B(z)$ of z containing a basis $B(x)$ of x. It follows from (2.56) and the definition $\wedge = \cap$ that $B(z)\backslash B(x)$ is a basis of $x^{\perp} \wedge z$. If this equals o, then $B(z) = B(x)$, and hence $z = x$.

Equations (2.59) and (2.53) imply that $\rho^{\perp\perp} = \rho$ for all $\rho \in \mathcal{P}$; hence each point of $\mathcal{P}$ lies in $\mathcal{L}(\mathcal{P})$, and the definition of $\leq$ implies that these points are precisely the atoms of $\mathcal{L}(\mathcal{P})$. The completeness of $\mathcal{L}(\mathcal{P})$ is obvious, since arbitrary intersections of orthoclosed subsets are orthoclosed, and the lattice is orthocomplemented. ■

Proposition 3.6.5. *Let $\mathcal{P} = \cup_{\alpha}\mathcal{P}_{\alpha}$ be the decomposition of a well-behaved transition probability space into its sectors (cf. 2.7.2). Then*

$$\mathcal{L}(\mathcal{P}) \simeq \prod_{\alpha} \mathcal{L}(\mathcal{P}_{\alpha}). \tag{3.16}$$

Here each factor $\mathcal{L}(\mathcal{P}_{\alpha})$ is irreducible; in particular, $\mathcal{P}$ is irreducible iff $\mathcal{L}(\mathcal{P})$ is irreducible.

If C is a component of $\mathcal{P}$ (so that $\mathcal{P} = C \cup C^{\perp}$), and $Q, R \in \mathcal{L}(\mathcal{P})$ satisfy $Q \subseteq C$ and $R \subseteq \mathcal{P}\backslash C = C^{\perp}$, then $Q \cup R = Q \vee R$; this follows from repeated application of (3.13), and from $Q^{\perp\perp} = Q$ (etc.). Since sectors are components, this shows that $c := \mathcal{P}_\alpha$ (regarded as an element of $\mathcal{L}(\mathcal{P})$) is in the center $\mathcal{C}(\mathcal{L}(\mathcal{P}))$. Equation (3.16) then follows from Proposition 3.5.7.

Now suppose that $\mathcal{P}$ is irreducible, yet $\mathcal{C}(\mathcal{L}(\mathcal{P}))$ nontrivial. This means that there exists an orthoclosed subset $c \subseteq \mathcal{P}$ (where $c \neq \emptyset$ and $c \neq \mathcal{P}$) such that $Q = (Q \wedge c) \vee (Q \wedge c^\perp)$ for all orthoclosed $Q \subseteq \mathcal{P}$. Since $\mathcal{P}$ is irreducible, one cannot have $\mathcal{P} = c \cup c^\perp$, so that there is an atom ρ lying in neither c nor $c^\perp$. Taking $Q = \rho$ we thus find $Q \wedge c = Q \wedge c^\perp = o$ (recall that $\wedge = \cap$ in $\mathcal{L}(\mathcal{P})$). This shows that such c cannot exist, and therefore $\mathcal{L}(\mathcal{P})$ must be irreducible. ■

Theorem 3.6.6. *If $\mathfrak{A}$ is a C^*-algebra with pure state space $\mathcal{P}(\mathfrak{A})$, one has the lattice isomorphism (cf. (3.15) and 3.6.4)*

$$\mathcal{L}(\mathfrak{A}_{\mathbb{R}}(\mathcal{P})) \simeq \mathcal{L}(\mathcal{P}(\mathfrak{A})). \tag{3.17}$$

By Theorems 2.5.4, 2.8.2, and 3.6.5, one has $\mathcal{L}(\mathcal{P}(\mathfrak{A})) \simeq \prod_\rho \mathcal{L}(\mathbb{P}\mathcal{H}_\rho)$. On the other hand, $\mathcal{L}(\mathfrak{A}_{\mathbb{R}}(\mathcal{P}))$ (which by definition is $\mathcal{L}(\pi_{\mathrm{ra}}(\mathfrak{A})''_{\mathbb{R}})$, cf. Proposition 3.1.2) equals $\mathcal{L}(\oplus_\rho \mathfrak{B}(\mathcal{H}_\rho))$ by (2.3). The center of this lattice is generated by the minimal central projections $[\mathcal{H}_\rho]$, and by 3.5.7 and 3.6.1. One therefore obtains $\mathcal{L}(\mathfrak{A}_{\mathbb{R}}(\mathcal{P})) \simeq \prod_\rho \mathcal{L}(\mathcal{H}_\rho)$. Finally, if $\mathcal{K}$ is a closed subspace of some Hilbert space $\mathcal{H}$, and $\mathcal{K}_1 := \mathcal{K} \cap \mathbb{S}\mathcal{H}$, then $\mathcal{K} \leftrightarrow \tau(\mathcal{K}_1)$ (where $\tau : \mathbb{S}\mathcal{H} \to \mathbb{P}\mathcal{H}$ is the canonical projection, cf. 2.5) establishes an isomorphism between the lattices $\mathcal{L}(\mathcal{H})$ and $\mathcal{L}(\mathbb{P}\mathcal{H})$. ■

The lattice $\mathcal{L}(\mathcal{P}(\mathfrak{A}))$ occurs in an interesting reformulation of the spectral theorem. In preparation, recall from basic measure theory that the σ-algebra $\mathcal{B}(\mathbb{R})$ of Borel subsets of $\mathbb{R}$ is an orthocomplemented lattice in which $\leq$ is $\subseteq$ (hence $B_1 \vee B_2 = B_1 \cup B_2$, $B_1 \wedge B_2 = B_1 \cap B_2$, $o = \emptyset$, and $I = \mathbb{R}$). This lattice is not complete, but merely σ-complete (i.e., $\vee$ and $\wedge$ exist for arbitrary countable families).

Theorem 3.6.7. *For each self-adjoint element A of a C^*-algebra $\mathfrak{A}$ there exists a lattice homomorphism $\varphi_A : \mathcal{B}(\mathbb{R}) \to \mathcal{L}(\mathcal{P}(\mathfrak{A}))$ with the property that*

$$\varphi_A(\vee_{i=1}^\infty B_i) = \vee_{i=1}^\infty \varphi_A(B_i) \tag{3.18}$$

if the B_i are mutually disjoint. For each $\rho \in \mathcal{P}(\mathfrak{A})$ the Gelfand transform $\hat{A}$ then has the spectral resolution

$$\hat{A}(\rho) = \int_{\mathbb{R}} \lambda \, dp^\rho_{\varphi_A}(\lambda), \tag{3.19}$$

where the Borel measure $p^\rho_{\varphi_A}$ on $\mathbb{R}$ is defined by (cf. (2.61))

$$p^\rho_{\varphi_A}(B) := p_{\varphi_A(B)}(\rho). \tag{3.20}$$

This follows from the usual spectral theorem for self-adjoint operators on a Hilbert space, applied to $\pi_{\mathrm{ra}}(A)$ (cf. 2.2.7). It is easily checked that the precise choice of π_{ra} (which, we recall, depends on choosing a pure state in each sector of $\mathcal{P}(\mathfrak{A})$) does not affect any of the statements in the theorem, as different choices lead to equivalent realizations. ■

3.7 The Two-Sphere Property in a Pure State Space

The lattice $\mathcal{L}(\mathcal{P}(\mathfrak{A}))$ plays a central role in the proof of the following characterization of the pure state space of a C^*-algebra as a transition probability space.

Definition 3.7.1. *A well-behaved transition probability space $\mathcal{P}$ (with associated lattice $\mathcal{L}(\mathcal{P})$) is said to have the* **two-sphere property** *if for any two points ρ, σ (with $\rho \neq \sigma$) lying in the same sector of $\mathcal{P}$, the space $\rho \vee \sigma$ is isomorphic as a transition probability space to the two-sphere S^2, with transition probabilities given by $p(z, w) = \frac{1}{2}(1 + \cos\theta(z, w))$, where $\theta(z, w)$ is the angular distance between z and w, measured along a great circle.*

To understand the nature of this property, recall that a two-sphere S^2 with radius 1 may be regarded as the extreme boundary of the unit ball $B^3 \subset \mathbb{R}^3$. The latter is affinely isomorphic to the state space $\mathcal{S}(\mathfrak{M}_2(\mathbb{C}))$ of the C^*-algebra of 2×2 matrices, so that S^2 is the pure state space of this algebra. Concretely, we identify a state on $\mathfrak{M}_2(\mathbb{C})$ with a density matrix ρ on $\mathbb{C}^2$, which may be parametrized as in (1.50). Restricted to the extreme boundary, this parametrization leads to a bijection between $\mathbb{PC}^2$ and S^2. Under this bijection the transition probabilities (2.65) on $\mathbb{PC}^2$ are mapped into the ones stated in 3.7.1. In other words, the two-sphere property states that there exists a fixed two-sphere $S^2_{\text{ref}} \simeq \mathbb{PC}^2$, equipped with the standard Hilbert space transition probabilities $p = p_{\mathbb{C}^2}$ given by (2.65), and a collection of bijections $T_{\rho\vee\sigma} : \rho \vee \sigma \to S^2_{\text{ref}}$, defined for each orthoclosed subspace of the type $\rho \vee \sigma$ (where ρ and $\sigma \neq \rho$ lie in the same sector of $\mathcal{P}$), such that for all $\rho', \sigma' \in \rho \vee \sigma$,

$$p_{\mathbb{C}^2}(T_{\rho\vee\sigma}(\rho'), T_{\rho\vee\sigma}(\sigma')) = p(\rho', \sigma'). \tag{3.21}$$

Theorem 3.7.2. *Let a well-behaved transition probability space $\mathcal{P}$ (with associated lattice $\mathcal{L}(\mathcal{P})$) have the two-sphere property. If $\mathcal{P}$ has no sector of dimension 3, then $\mathcal{P} \simeq \cup_\alpha \mathbb{P}\mathcal{H}_\alpha$ as a transition probability space (for some family $\{\mathcal{H}_\alpha\}$ of Hilbert spaces), where each sector $\mathbb{P}\mathcal{H}_\alpha$ is equipped with the transition probabilities (2.65).*

This statement is not necessarily false when $\mathcal{P}$ does have sectors of dimension 3 (in fact, we believe it to be true in that case as well); unfortunately, the proof below does not work in that special dimension.

If ρ and σ lie in different sectors of $\mathcal{P}$, then $\rho \vee \sigma = \{\rho, \sigma\}$; this follows from repeated application of (3.13) and $\rho^{\perp\perp} = \rho$ (etc.). In any case, it is sufficient to prove the theorem for each sector separately, so we may assume that $\mathcal{P}$ is irreducible. The first step in the proof is then to construct the lattice $\mathcal{L}(\mathcal{P})$ (cf. 3.6.4). The strategy of the proof is to characterize $\mathcal{L}(\mathcal{P})$, and then use the so-called coordinatization theorem in lattice theory to show that $\mathcal{L}(\mathcal{P}) = \mathcal{L}(\mathcal{H})$ for some Hilbert space $\mathcal{H}$.

We already know that $\mathcal{L}(\mathcal{P})$ is orthomodular, atomic, and complete (as established in Proposition 3.6.4); by 3.6.5 it is irreducible (in the sense of 3.5.7) as well. A lattice $\mathcal{L}$ is called **atomistic** if every element is the supremum of the collection of its atoms. If $\mathcal{L}$ is orthomodular, atomic, and complete, then it is atomistic. For let

x be the supremum of the atoms in some $z \in \mathcal{L}$, and assume $x < z$. By (3.14) one then has $x^{\perp} \wedge z \neq o$, so that $x^{\perp} \wedge z$ must contain an atom, which is a contradiction.

To apply the coordinatization theorem, we need to establish a further property of $\mathcal{L}(\mathcal{P})$. An atomistic lattice $\mathcal{L}$ with o is said to have the **covering property** if for an atom $a \in \mathcal{L}$ and an arbitrary element $x \in \mathcal{L}$ with $a \wedge x = o$, the inclusions $x \leq y \leq x \vee a$ for some $y \subset \mathcal{L}$ imply $y = x$ or $y - x \vee a$. For example, the lattices $\mathcal{L}(V)$ in 3.5 have the covering property.

Lemma 3.7.3. *$\mathcal{L}(\mathcal{P})$ has the covering property.*

Consistent with previous notation, we denote atoms of $\mathcal{L}(\mathcal{P})$ (hence points of $\mathcal{P}$) by ρ, σ, and arbitrary elements by Q, Q_i, R, S.

Let $n = \dim(Q)$ (cf. 2.7); for the moment we assume $n < \infty$. We will first use induction to prove that if $\rho \notin Q$, then the element $(\rho \vee Q) \wedge Q^{\perp}$ is an atom.

To start, note that if $Q_1 \leq Q_2$ for orthoclosed Q_1, Q_2 sets of the same finite dimension, then $Q_1 = Q_2$. For an orthoclosed set in $\mathcal{P}$ is determined by a basis of it (cf. (2.62)), which in turn determines its dimension. This implies that $\dim(\rho \vee Q) > \dim(Q)$ if $\rho \notin Q$ (take $Q_1 = Q$ and $Q_2 = \rho \vee Q$). Accordingly, it must be that $(\rho \vee Q) \wedge Q^{\perp} > o$, for equality would imply that $\dim(\rho \vee Q) = \dim(Q)$.

For $n = 1$, Q is an atom. By assumption, $\rho \vee Q$ is S^2; hence $(\rho \vee Q) \wedge Q^{\perp}$ is the antipodal point to Q in $\rho \vee Q$, which is an atom, as desired. Now assume $n > 1$. Choose a basis $\{e_i\}_{i=1,\ldots,\dim(Q)}$ of Q; then $Q = \vee_{i=1}^{n} e_i$. Put $R = \vee_{i=1}^{n-1} e_i$; then $R < Q$, whence $Q^{\perp} < R^{\perp}$, so that $(\rho \vee Q) \wedge Q^{\perp} \leq (\rho \vee Q) \wedge R^{\perp}$. The assumption $(\rho \vee Q) \wedge Q^{\perp} = (\rho \vee Q) \wedge R^{\perp}$ is equivalent, on use of $Q = R \vee e_n$, (3.13), and the associativity of $\wedge$, to $((\rho \vee Q) \wedge R^{\perp}) \wedge e_n^{\perp} = (\rho \vee Q) \wedge R^{\perp}$, which implies that $(\rho \vee Q) \wedge R^{\perp} \leq e_n^{\perp}$. This is not possible, since the left-hand side contains e_n. Hence

$$o < (\rho \vee Q) \wedge Q^{\perp} < (\rho \vee Q) \wedge R^{\perp}. \tag{3.22}$$

It follows from the orthomodularity of $\mathcal{L}(\mathcal{P})$ that if $R \leq S$ and $R \leq Q$, then

$$(S \vee Q) \wedge R^{\perp} = (S \wedge R^{\perp}) \vee (Q \wedge R^{\perp}). \tag{3.23}$$

Since $R < Q$ and $R \leq \rho \vee R$, one has $\rho \vee Q = (\rho \vee R) \vee Q$. Now use (3.23) to find

$$(\rho \vee Q) \wedge R^{\perp} = ((\rho \vee R) \vee Q) \wedge R^{\perp} = ((\rho \vee R) \wedge R^{\perp}) \vee (Q \wedge R^{\perp}).$$

The right-hand side equals $\sigma \vee e_n$, where $\sigma := (\rho \vee R) \wedge R^{\perp}$ is an atom by the induction hypothesis. The equality $\sigma = e_n$ would imply that $\rho \in Q$, hence $\sigma \neq e_n$. But then (3.22) and the two-sphere property imply $0 < \dim((\rho \vee Q) \wedge Q^{\perp}) < 2$, so that $(\rho \vee Q) \wedge Q^{\perp}$ must indeed be an atom.

It follows that $\dim(\rho \vee Q) = \dim(Q)+1$. Hence any $S \subset \mathcal{P}$ satisfying $Q \leq S \leq \rho \vee Q$ must have $\dim(S)$ equal to $\dim(Q)$ or to $\dim(Q) + 1$. In the former case, it must be that $S = Q$ by the dimension argument earlier. Similarly, in the latter case the only possibility is $S = \rho \vee Q$. All in all, we have proved the covering property for finite-dimensional sublattices. A complicated technical argument involving the dimension theory of lattices then shows that the covering property holds for all $x \in \mathcal{L}(\mathcal{P})$. □

Atomistic lattices with the covering property are known as **AC-lattices.** At this stage we can sum up by saying that $\mathcal{L}(\mathcal{P})$ is a complete irreducible orthomodular AC-lattice. This allows us to use the following classical coordinatization theorem; cf. 3.5 for the definition of $\mathcal{L}(V)$. This involves the notion of a **chain**, which is a totally ordered subset of the lattice $\mathcal{L}$. The **length** (also called rank, height, or dimension) of an AC-lattice $\mathcal{L}$ is the cardinality of a maximal chain (which contains o and l) minus 1, which is well-defined because of the covering property. This number coincides with the minimal number of atoms ρ_i for which $l = \vee \rho_i$; for $\mathcal{L} = \mathcal{L}(\mathcal{P})$ it is equal to the dimension of $\mathcal{P}$ (as defined in 2.7).

The coordinatization theorem for AC-lattices is the following.

Theorem 3.7.4. *Let $\mathcal{L}$ be a complete irreducible orthomodular AC-lattice of length ≥ 4. There exists a vector space V over a division ring $\mathbb{D}$ (both unique up to isomorphism), equipped with an anisotropic Hermitian form (defined relative to an involution of $\mathbb{D}$, and unique up to scaling), such that $\mathcal{L} \simeq \mathcal{L}(V)$ as orthocomplemented lattices.*

We omit the lengthy and complicated proof of this theorem. In the context of our lattice $\mathcal{L}(\mathcal{P})$, the essential point is that the division ring $\mathbb{D}$ is constructed by choosing two atoms ρ and $\sigma \neq \rho$, whereupon $\mathbb{D} \simeq (\rho \vee \sigma)\backslash\sigma$. The vector space V is constructed in terms of a basis $\{\mathbf{e}_i\}$, which corresponds to a basis $\{e_i\}$ of $\mathcal{P}$ (or, more generally, of the set of atoms in $\mathcal{L}$); hence the length of $\mathcal{L}$ is equal to the dimension of V. We will need neither the explicit form of the addition and multiplication in $\mathbb{D}$, nor the scalar multiplication in V (which are given in terms of a certain geometric procedure). To proceed, the following information suffices.

Lemma 3.7.5. *Let V be 3-dimensional, and let $\mathcal{L}(V)$ carry a topology for which the lattice operations $\vee$ and $\wedge$ are jointly continuous. Then $\mathbb{D}$ (regarded as a subset of the collection of atoms in $\mathcal{L}(V)$), equipped with the topology inherited from $\mathcal{L}(V)$, is a topological division ring (i.e., addition and multiplication are jointly continuous).*

This is clear from the explicit construction of addition and multiplication in $\mathbb{D}$. ■

Let $F \in \mathcal{L}(\mathcal{P})$ be finite-dimensional. We can define a topology on $[o, F]$ (i.e., the set of all $Q \in \mathcal{L}(\mathcal{P})$ for which $Q \leq F$) through a specification of convergence. Given a net $\{Q_\lambda\}$ in F, we say that $Q_\lambda \to Q$ when eventually $\dim(Q_\lambda) = \dim(Q)$, and if there exists a family of bases $\{e_i^\lambda\}$ for $\{Q_\lambda\}$, and a basis $\{e_j\}$ of Q, such that $\sum_{i,j} p(e_i^\lambda, e_j) \to \dim(Q)$. This notion is actually independent of the choice of all bases involved, since $\sum_j p(\rho, e_j)$ is independent of the choice of the basis in Q for any $\rho \in \mathcal{P}$, and similarly for the bases of Q_λ (to see this, extend $\{e_j\}_{j=1}^{\dim(Q)}$ to a basis $\{e_j\}_{j=1}^{\dim(\mathcal{P})}$, and use (2.56)). An equivalent definition of this convergence is that $Q_\lambda \to Q$ if $p(\rho_\lambda, \sigma) \to 0$ for all $\sigma \in F \wedge Q^\perp$ and all $\{\rho_\lambda\}$ such that $\rho_\lambda \in Q_\lambda$.

Lemma 3.7.6. *The above construction defines a topology on F, which is Hausdorff.*

The first claim is easily verified. For the second, let $Q_\lambda \to Q$ and $Q_\lambda \to R$. Then $p(\rho_\lambda, \sigma) \to 0$ for all $\sigma \in Q^\perp \vee R^\perp = (Q \wedge R)^\perp$, and all $\{\rho_\lambda\}$ for which $\rho_\lambda \in Q_\lambda$. Choose a basis $\{e_j\}$ of Q that extends a basis of $Q \wedge R$. Then $\sum_{j=1}^{\dim(Q\wedge R)} p(\rho_\lambda, e_j) = 1$, but also $\sum_{j=1}^{\dim(Q)} p(\rho_\lambda, e_j) = 1$, since $Q_\lambda \to Q$. Hence $p(\rho_\lambda, \sigma) \to 0$ for all $\sigma \in Q \wedge (Q \wedge R)^\perp$. This leads to a contradiction unless $Q = R$. ■

Lemma 3.7.7. *The restriction of this topology to any two-sphere $\rho \vee \sigma \simeq S^2$ in F induces the usual topology on S^2. Moreover, $\vee$ and $\wedge$ are jointly continuous on any $[o, F]$, where F is a 3-dimensional subspace of $\mathcal{L}(\mathcal{P})$.*

If we restrict the topology to the atoms in F, then $\rho_\lambda \to \rho$ if $p(\rho_\lambda, \rho) \to 1$. On $F = \rho \vee \sigma \simeq S^2$, one can easily show from the explicit form of the transition probabilities p that the convergence $p(\psi_\lambda, \psi) \to 1$ is equivalent to $p(\psi_\lambda, \varphi) \to p(\psi, \varphi)$ for all $\varphi \in \rho \vee \sigma$. Namely, if $p(\psi_\lambda, \psi) \to 1$ in $\mathbb{P}\mathcal{H}$ (for any Hilbert space $\mathcal{H}$; the case of relevance is $\mathcal{H} = \mathbb{C}^2$), then $|(\Psi_\lambda, \Psi)| \to 1$ for arbitrary lifts Ψ_λ, Ψ in $\mathbb{S}\mathcal{H}$. Choose an orthonormal basis $\{\mathbf{e}_i\}$ in $\mathcal{H}$ containing Ψ; the equation $(\Psi, \Phi) = \sum_i (\Psi, \mathbf{e}_i)(\mathbf{e}_i, \Phi)$ then rapidly leads to the conclusion $|(\Psi_\lambda, \Phi)| \to |(\Psi, \Phi)|$ for all $\Phi \in \mathbb{S}\mathcal{H}$. The corresponding topology is the projection of the usual topology on $\mathbb{C}^2$ to $S^2 \simeq \mathbb{P}\mathbb{C}^2$, which demonstrates the first claim.

We turn to the proof of joint continuity of $\vee$ and $\wedge$. Assume that $F \in \mathcal{L}(\mathcal{P})$ is 3-dimensional. We firstly show that $\rho_\lambda \to \rho$ and $\sigma_\lambda \to \sigma$, where ρ and σ are atoms, implies $\rho_\lambda \vee \sigma_\lambda \to \rho \vee \sigma$. Let $\tau_\lambda = (\rho_\lambda \vee \sigma_\lambda)^\perp \wedge F$, and $\tau = (\rho \vee \sigma)^\perp \wedge F$; these are atoms. Let ρ'_λ be the antipodal point to ρ_λ in $\rho_\lambda \vee \sigma_\lambda$ (i.e., $\rho'_\lambda = \rho_\lambda^\perp \wedge (\rho_\lambda \vee \sigma_\lambda)$), and let σ'_λ be antipodal to σ_λ in $\rho_\lambda \vee \sigma_\lambda$. Then $\{\rho_\lambda, \rho'_\lambda, \tau_\lambda\}$ is a basis of F, and so is $\{\sigma_\lambda, \sigma'_\lambda, \tau_\lambda\}$. The definition of a basis and of $\rho_\lambda \to \rho, \sigma_\lambda \to \sigma$ implies that $p(\rho, \tau_\lambda) \to 0$ and $p(\sigma, \tau_\lambda) \to 0$. Hence $p(\tau, \tau_\lambda) \to 1$. Now take an arbitrary atom $\alpha_\lambda \in \tau_\lambda^\perp \wedge F$, and complete to a basis $\{\alpha_\lambda, \alpha'_\lambda, \tau_\lambda\}$, where $\alpha'_\lambda \in \rho_\lambda \vee \sigma_\lambda$. Again, the definition of a basis implies that $p(\alpha_\lambda, \tau) \to 0$. By our second definition of convergence, one therefore has $\rho_\lambda \vee \sigma_\lambda \to \rho \vee \sigma$.

Secondly, we show that $Q_\lambda \to Q$ and $R_\lambda \to R$, where Q and R are two-dimensional subspaces of F, implies $Q_\lambda \wedge R_\lambda \to Q \wedge R$ (we assume $Q \neq R$, so eventually $Q_\lambda \neq R_\lambda$). Let $\alpha = Q^\perp \wedge F$, $\beta = R^\perp \wedge F$, $\gamma = Q \wedge R$, and $\gamma_\lambda = Q_\lambda \wedge R_\lambda$; as a simple dimension count shows, these are all atoms. By assumption, $p(\gamma_\lambda, \alpha) \to 0$ and $p(\gamma_\lambda, \beta) \to 0$. Since $(\alpha \cup \beta)^\perp = (\alpha \vee \beta)^\perp$ by definition of $\vee$, and $(\alpha \vee \beta)$ is two-dimensional, γ is the only point in F that is orthogonal to α and β. Hence $p(\gamma_\lambda, \gamma) \to 1$; if not, the assumption would be contradicted. But this is precisely the definition of $Q_\lambda \wedge R_\lambda \to Q \wedge R$, and the proof is finished. ■

Corollary 3.7.8. *The division ring $\mathbb{D}$ equals $\mathbb{C}$, and the involution relative to which the Hermitian form of Theorem 3.7.4 is defined is complex conjugation.*

It follows from the previous lemma that $\mathbb{D}$ is locally compact and connected. According to the classification of locally compact division rings, there exist only three connected ones: $\mathbb{D} = \mathbb{R}, \mathbb{C}$, and $\mathbb{H}$ (the quaternions). Of these, only $\mathbb{C}$ is homeomorphic to $(\rho \vee \sigma)\backslash\sigma \simeq \mathbb{R}^2$. Note that the algebraic structure is therefore entirely determined by the topology.

Moreover, Lemma 3.7.7 implies that the orthocomplementation $\perp$ is continuous on 3-dimensional subspaces. If one inspects the way the involution $\lambda \mapsto \overline{\lambda}$ of $\mathbb{D}$ is constructed in the proof of Theorem 3.7.4, one immediately infers that this involution (of $\mathbb{C}$ in our case) must be continuous as well. It can be shown that $\mathbb{C}$ possesses only two continuous involutions: complex conjugation and the identity map. The latter cannot define a nondegenerate sesquilinear form (so that in particular, the lattice $\mathcal{L}(V)$ cannot be orthomodular). Hence one is left with complex conjugation. ■

Note that we have used the two-sphere property twice, for different purposes: firstly for deriving the covering property of $\mathcal{L}(\mathcal{P})$, and secondly for identifying $\mathbb{D} = \mathbb{C}$.

With this corollary in hand, the definition of a Hermitian form implies that (Ψ, Ψ) must be real for all $\Psi \in V$, and the anisotropy means that (Ψ, Ψ) must be nonzero and have the same sign for all Ψ. If necessary, one may change the sign of the form so as to make it positive definite. Accordingly, V is equipped with an inner product in the usual sense, that is, it is a pre-Hilbert space. The fact that V is actually a Hilbert space follows from the orthomodularity of $\mathcal{L}(\mathcal{P}) \simeq \mathcal{L}(V)$ by a rather technical result, whose proof we omit.

Proposition 3.7.9. *A pre-Hilbert space V over $\mathbb{C}$ is complete iff the associated orthocomplemented lattice $\mathcal{L}(V)$ is orthomodular.*

We conclude that $\mathcal{L}(\mathcal{P})$ is isomorphic to the projection lattice $\mathcal{L}(\mathcal{H})$ of some complex Hilbert space $\mathcal{H}$. Therefore, their respective collections of atoms $\mathcal{P}$ and $\mathbb{P}\mathcal{H}$ must be isomorphic. Accordingly, we may identify $\mathcal{P}$ and $\mathbb{P}\mathcal{H}$ as sets. Denote the standard transition probabilities (2.65) on $\mathbb{P}\mathcal{H}$ by $p_{\mathcal{H}}$. With p the transition probabilities in $\mathcal{P}$, we will show that $p = p_{\mathcal{H}}$.

Refer to the text following 3.7.1. We may embed S^2_{ref} isometrically in $\mathbb{P}\mathcal{H}$; one then simply has $p = p_{\mathcal{H}}$ on S^2_{ref}. Equation (3.21) then reads

$$p_{\mathcal{H}}(T_{\rho\vee\sigma}(\rho'), T_{\rho\vee\sigma}(\sigma')) = p(\rho', \sigma'); \tag{3.24}$$

in particular, $p_{\mathcal{H}}(T_{\rho\vee\sigma}(\rho'), T_{\rho\vee\sigma}(\sigma')) = 0$ iff $p(\rho', \sigma') = 0$. On the other hand, we know that p and $p_{\mathcal{H}}$ generate isomorphic lattices, which implies that $p_{\mathcal{H}}(\rho', \sigma') = 0$ iff $p(\rho', \sigma') = 0$. Putting this together, we see that

$$p_{\mathcal{H}}(T_{\rho\vee\sigma}(\rho'), T_{\rho\vee\sigma}(\sigma')) = 0$$

iff $p_{\mathcal{H}}(\rho', \sigma') = 0$. A fairly deep generalization of Corollary 3.4.3 states that a bijection $T : \mathbb{P}\mathcal{H}_1 \to \mathbb{P}\mathcal{H}_2$ (where the $\mathcal{H}_i$ are separable) that merely preserves orthogonality (i.e., $p_{\mathcal{H}_2}(T(\rho'), T(\sigma')) = 0$ iff $p_{\mathcal{H}_1}(\rho', \sigma') = 0$) is induced by a unitary or an antiunitary operator $U : \mathcal{H}_1 \to \mathcal{H}_2$. We use this with $\mathcal{H}_1 = \rho \vee \sigma$,

$\mathcal{H}_2 = S^2_{\rm ref}$, and $T = T_{\rho\vee\sigma}$. Since $T_{\rho\vee\sigma}$ is induced by a unitary or an antiunitary map, which preserves $p_{\mathcal{H}}$, we conclude from (3.24) that $p_{\mathcal{H}}(\rho', \sigma') = p(\rho', \sigma')$. Since ρ and σ (and $\rho', \sigma' \in \rho \vee \sigma$) were arbitrary, the proof of Theorem 3.7.2 is finished. ■

3.8 The Poisson Structure on the Pure State Space

We now further investigate how the transition probabilities and the Poisson structure on the pure state of a C^*-algebra are related.

Theorem 3.8.1. *If $\mathcal{P}$ is the pure state space of a C^*-algebra, then the symplectic leaves of $\mathcal{P}$ as a Poisson space coincide with the sectors of $\mathcal{P}$ as a transition probability space.*

This is immediate by combining 2.5.4 and 2.8.2 with 2.6.8. ■

If the C^*-algebra in question is commutative (so that its transition probabilities are (2.58)), it is understood to be equipped with the zero Poisson structure, cf. the comment following 1.1.5. If, however, $\mathfrak{A}_{\mathbb{R}}$ is a Poisson algebra associated to a Poisson space $\mathcal{P}$, then the natural transition probabilities on $\mathcal{P}$ are given by (2.58), and the above result fails: The sectors of $\mathcal{P}$ are its points, whereas its symplectic leaves are nontrivial if the Poisson bracket does not identically vanish.

We now show that the symplectic structure on $\mathbb{P}\mathcal{H}$ is determined by the transition probabilities (2.65) and unitarity; recall Definition 3.1.3.

Theorem 3.8.2. *Let $\mathbb{P}\mathcal{H}$, equipped with the transition probabilities (2.65) and its usual manifold structure, be a unitary Poisson space for which the Poisson structure is symplectic. Then the Poisson structure is determined up to a multiplicative constant, and is given by (2.42) for some $\hbar \neq 0$.*

It follows from (3.2) and Proposition 3.1.2 that $\mathfrak{A}_L(\mathcal{P}, \mathbb{R})$ equals the Gelfand transform of $\mathfrak{B}(\mathcal{H})_{\mathbb{R}}$. According to the definition of unitarity, the Hamiltonian flow generated by any function $\hat{A}$ on $\mathbb{P}\mathcal{H}$ (where A is a bounded self-adjoint operator on $\mathcal{H}$, cf. (2.43)) must preserve the transition probabilities (2.65). Corollary 3.4.3 and Stone's theorem imply that such a flow must be of the form $\psi(t) = e^{-it\hat{C}(A)}\psi$; cf. (2.47), where $C(A)$ is some self-adjoint operator depending on A in an as yet unknown way. Antiunitary flows are excluded, for they cannot satisfy $\hat{\psi}(0) = 0$.

We now compute the Poisson bracket of the functions $\hat{A}$ and $\hat{B}$ (see 2.43). Using (2.8), (2.11), and the preceding paragraph, we obtain $\{\hat{A}, \hat{B}\}(\psi) = \frac{d}{dt}\hat{B}\left(\exp(it\hat{C}(A))\psi\right) |t = 0$. The right-hand side equals $i\widehat{[C(A), B]}(\psi)$. The anti-symmetry of the left-hand side implies that $C(A) = \hbar^{-1}A$ for some $\hbar^{-1} \in \mathbb{R}$. The value $\hbar^{-1} = 0$ is excluded unless $\mathcal{H}$ is one-dimensional, for otherwise the Poisson structure would be degenerate. In other words, the Poisson bracket is given by (2.42). Since the collection of all differentials $d\hat{A}$ spans the cotangent bundle at each point ψ of $\mathbb{P}\mathcal{H}$, the Poisson structure is completely determined. ■

Theorems 3.8.1 and 3.8.2 show that the Poisson structure on the pure state space $\mathcal{P}(\mathfrak{A}) = \cup_\alpha \mathbb{P}\mathcal{H}_\alpha$ of a C^*-algebra (cf. 2.5.4) is to a large extent determined by unitarity. The only freedom resides in a possible sector-dependence of $\hbar$; here $\hbar^{-1} \neq 0$ except in one-dimensional sectors (in which the value of $\hbar$ is irrelevant, as the Poisson bracket identically vanishes at such points). The choice (1.22) for the Poisson bracket on $\mathfrak{A}_\mathbb{R}$ corresponds to taking $\hbar$ to be a sector-independent constant. We may regard $\hbar$ as a function on $\mathcal{P}(\mathfrak{A})$, which is constant on each sector. If $\hat{A}$ denotes an element of $\mathfrak{A}_\mathbb{R}$, the restriction of $\hat{A}$ to a sector $\mathbb{P}\mathcal{H}_\alpha$ corresponds to an operator A_α (cf. (2.43)). The sector in which $\rho \in \mathcal{P}(\mathfrak{A})$ lies is called $\alpha(\rho)$. With this notation, the Poisson bracket is then given by

$$\{\hat{A}, \hat{B}\}(\rho) = \frac{i}{\hbar(\rho)} \widehat{[A_{\alpha(\rho)}, B_{\alpha(\rho)}]}(\rho). \tag{3.25}$$

The following result shows that under a natural topological requirement the sector-dependence of $\hbar$ cannot be arbitrary.

Lemma 3.8.3. *Equip $\mathcal{P}(\mathfrak{A}) = \cup_\alpha \mathbb{P}\mathcal{H}_\alpha$ with the Poisson structure (3.25). Assume that $\mathcal{P}(\mathfrak{A})$ is equipped with a uniformity for which $\mathfrak{A}_\mathbb{R}$ (as defined in (3.4)) is closed under Poisson brackets. Then the function $\hbar$ is uniformly continuous on $\mathcal{P}(\mathfrak{A})$.*

This applies in particular to the w^*-uniformity on $\mathcal{P}(\mathfrak{A})$. Suppose $\hbar$ is not uniformly continuous. We then take $\hat{A}, \hat{B} \in \mathfrak{A}_\mathbb{R}$ in such a way that A_α and B_α are independent of α in a neighborhood of a point σ of discontinuity of $\hbar$, with $[A_{\alpha(\sigma)}, B_{\alpha(\sigma)}] \neq 0$. Then the real-valued function on $\mathcal{P}(\mathfrak{A})$ defined by $\rho \mapsto \hbar(\rho)\{\hat{A}, \hat{B}\}(\rho)$ is certainly uniformly continuous near σ, since its value at ρ is equal to $i\widehat{[A_{\alpha(\rho)}, B_{\alpha(\rho)}]}(\rho)$. But, by assumption, $\{\hat{A}, \hat{B}\}$ is uniformly continuous as well. Because of the factor $\hbar$, the product $\hbar\{\hat{A}, \hat{B}\}$ cannot be uniformly continuous. This leads to a contradiction. ■

We can always rescale the Poisson bracket by multiplying it with $\hbar(\cdot)$; the resulting Poisson structure is then the same in all sectors. In view of Lemma 3.8.3, in the given situation $\mathfrak{A}_\mathbb{R}$ will be closed under the rescaled Poisson bracket as well.

3.9 Axioms for the Pure State Space of a C^*-Algebra

We can sum up part of the preceding discussion as follows.

Theorem 3.9.1. *The pure state space $\mathcal{P}$ of a unital C^*-algebra $\mathfrak{A}$, equipped with the w^*-uniformity, the transition probabilities (2.65), and the Poisson structure (1.22), has the following properties.*

C^*1: $\mathcal{P}$ is a uniform Poisson space with a transition probability (Definition 3.1.4).
C^*2: $\mathcal{P}$ has the two-sphere property (Definition 3.7).
C^*3: The sectors of $\mathcal{P}$ as a transition probability space coincide with the symplectic leaves of $\mathcal{P}$ as a Poisson space.
C^*4: The space $\mathfrak{A}_\mathbb{R}$ (defined through C^*1 by (3.2)) is closed under the Jordan product constructed from the transition probabilities in 3.3.

C^*5: The pure state space $\mathcal{P}(\mathfrak{A})$ of $\mathfrak{A}$ (as defined in 2.1.1) coincides with $\mathcal{P}$.

Recall that $\mathfrak{A}_{\mathbb{R}}$ is the self-adjoint part of $\mathfrak{A}$, and that the norm on $\mathfrak{A}_{\mathbb{R}}$ is equal to the sup-norm that $\mathfrak{A}_{\mathbb{R}}$ inherits from its inclusion in $\ell^{\infty}(\mathcal{P}, \mathbb{R})$. Property C^*1 was established in Corollary 3.2.2; C^*2 is immediate from (2.65); C^*3 holds by Theorem 3.8.1; C^*4 follows from (3.6) and Theorem 3.2.1; and C^*5 holds because the uniformity on $\mathcal{P}(\mathfrak{A})$ used to establish 3.2.1 is precisely the w^*-uniformity. ∎

We now turn things around, and claim that the properties C^*1–C^*5 actually characterize pure state spaces of unital C^*-algebras. Property C^*5 is then regarded as an axiom restricting the possible uniformities on $\mathcal{P}$. As an axiom, the precise meaning of C^*5 is as follows. From Axioms C^*1, C^*2, and C^*4 the space $\mathfrak{A}_{\mathbb{R}}$ emerges as a JB-algebra, which is contained in $C(\mathcal{P}, \mathbb{R})$ as a partially ordered Banach space. Hence each element of $\mathcal{P}$ defines a pure state on $\mathfrak{A}_{\mathbb{R}}$ by evaluation; Axiom C^*5 requires that all pure states of $\mathfrak{A}_{\mathbb{R}}$ be of this form (note that by C^*1, the function space $\mathfrak{A}_{\mathbb{R}} \simeq \hat{\mathfrak{A}}_{\mathbb{R}}$ already separates points).

Theorem 3.9.2. *If a set $\mathcal{P}$ satisfies C^*1–C^*5 (with $\mathcal{P}$ as a transition probability space containing no sector of dimension 3), then there exists a unital C^*-algebra $\mathfrak{A}$, whose self-adjoint part is $\mathfrak{A}_{\mathbb{R}}$ (defined through C^*1). This $\mathfrak{A}$ is unique up to isomorphism, and can be explicitly reconstructed from $\mathcal{P}$, such that*

1. $\mathcal{P} = \mathcal{P}(\mathfrak{A})$ *(i.e., $\mathcal{P}$ is the pure state space of $\mathfrak{A}$).*
2. *The transition probabilities (2.63) coincide with those initially given on $\mathcal{P}$.*
3. *The Poisson structure on each symplectic leaf of $\mathcal{P}$ is proportional to the Poisson structure imposed on the given leaf by (1.22).*
4. *The w^*-uniformity on $\mathcal{P}(\mathfrak{A})$ defined by $\mathfrak{A}$ is contained in the initial uniformity on $\mathcal{P}$.*
5. *The C^*-norm on $\mathfrak{A}_{\mathbb{R}} \subset \mathfrak{A}$ is equal to the sup-norm inherited from the inclusion $\mathfrak{A}_{\mathbb{R}} \subset \ell^{\infty}(\mathcal{P}, \mathbb{R})$.*

As stated after Theorem 3.7.2 (which is an important step in the proof of 3.9.2), we believe that the restriction to dimension $\neq 3$ can be dropped.

The proof of this theorem essentially consists in the description of the construction of $\mathfrak{A}$; practically all the work has already been done. Axioms C^*1 and C^*2 entirely determine $\mathcal{P}$ as a transition probability space by Theorem 3.7.2. Hence $\mathfrak{A}_{\mathbb{R}}(\mathcal{P})$ is determined by Proposition 3.1.2. We now use Axiom C^*3, which implies that each symplectic leaf of $\mathcal{P}$ is a projective Hilbert space $\mathbb{P}\mathcal{H}_{\alpha}$. For the moment let us assume that each leaf $\mathbb{P}\mathcal{H}_{\alpha}$ has a manifold structure (e.g., the usual one) relative to which all functions $\hat{H}$ (cf. (2.43)), where $H \in \mathfrak{B}(\mathcal{H}_{\alpha})_{\mathbb{R}}$, are smooth. Then $\mathfrak{A}_{\mathbb{R}}(\mathcal{P}) \cap C_u(\mathcal{P}, \mathbb{R}) \subset C_L^{\infty}(\mathcal{P}, \mathbb{R})$ by the explicit description of $\mathfrak{A}_{\mathbb{R}}(\mathcal{P})$ in 3.1.2. It then follows from Axiom C^*1, in particular from (3.2), that

$$\mathfrak{A}_{\mathbb{R}} = \mathfrak{A}_{\mathbb{R}}(\mathcal{P}) \cap C_u(\mathcal{P}, \mathbb{R}). \tag{3.26}$$

This space is norm-closed by one of the remarks following 3.2.1. The condition in Proposition 3.3.2 holds, so that we can construct a Jordan product in $\mathfrak{A}_{\mathbb{R}}$ by the procedure in 3.3. By Proposition 3.3.3 and Axiom C^*4, this turns $\mathfrak{A}_{\mathbb{R}}$ into a JB-algebra. At this stage we can already construct the pure state space $\mathcal{P}(\mathfrak{A})$ through

2.1.1 and 1.4.1; the property 3.9.2.1 then holds by Axiom C^*5, whereas 3.9.2.2 follows from Theorem 2.8.2.

We may regard the restriction of $\mathfrak{A}_{\mathbb{R}}$ to a given sector $\mathbb{P}\mathcal{H}_\alpha$ as the Gelfand transform of a Jordan subalgebra of $\mathfrak{B}(\mathcal{H}_\alpha)_{\mathbb{R}}$. This subalgebra must be weakly dense in $\mathfrak{B}(\mathcal{H}_\alpha)_{\mathbb{R}}$, for otherwise Axiom C^*5 cannot hold.

By Axiom C^*3 and a straightforward modification of Theorem 3.8.2 (taking into account that the restriction of $\mathfrak{A}_L(\mathcal{P}, \mathbb{R})$ to $\mathbb{P}\mathcal{H}_\alpha$ is weakly dense in $\mathfrak{B}(\mathcal{H})_{\mathbb{R}}$, rather than coinciding with it), the Poisson structure in each sector of $\mathcal{P}$ is determined up to a constant, which implies 3.9.2.3. By Lemma 3.8.3 and Axiom C^*1 (which, through the definition of a uniform Poisson space, requires that $\mathfrak{A}_{\mathbb{R}}$ be closed under Poisson brackets) we can rescale the Poisson bracket so as to make Planck's "constant" a constant on $\mathcal{P}$. By Proposition 3.4.1 the Leibniz rule (1.4) is then satisfied as a consequence of the unitarity imposed by Axiom C^*1. By the remark after the proof of 3.4.1, the associator identity (1.6) holds for the rescaled Poisson bracket. Hence $\mathfrak{A}_{\mathbb{R}}$ becomes a JLB-algebra by Definition 1.1.5, and the complexification $\mathfrak{A}$ is a C^*-algebra by Proposition 1.1.9.

From 3.9.2.1 and 3.9.2.2 we infer that $\mathfrak{A}_{\mathbb{R}}(\mathcal{P}) = \mathfrak{A}_{\mathbb{R}}(\mathcal{P}(\mathfrak{A}))$. The w^*-uniformity appearing in (3.4) is the weakest uniformity relative to which all elements of $\mathfrak{A}_{\mathbb{R}}$ are uniformly continuous. Property 3.9.2.4 then follows from Theorem 3.2.1 and (3.26). Property 3.9.2.5 is evident from Proposition 3.3.3.

Finally, let us assume that some $\mathbb{P}\mathcal{H}_\alpha$ have an exotic manifold structure such that $\mathfrak{A}_{\mathbb{R}}(\mathcal{P}) \cap C_u(\mathcal{P}, \mathbb{R})$ is not contained in $C^\infty_L(\mathcal{P}, \mathbb{R})$, so that $\mathfrak{A}_{\mathbb{R}} \subset \mathfrak{A}_{\mathbb{R}}(\mathcal{P}) \cap C_u(\mathcal{P}, \mathbb{R})$ is a proper inclusion (rather than the equality (3.26)). It follows from Axiom C^*5 that the weak density mentioned two paragraphs ago must still hold. This weak density suffices for the subsequent arguments to be valid, and we can construct a C^*-algebra $\mathfrak{A}$ with pure state space $\mathcal{P}$. The proper inclusion above would then contradict (3.4). Hence such exotic manifold structures are excluded by the axioms (if they exist at all). ∎

Certain simplifications of this characterization suggest themselves. For example, if one amends C^*1–C^*5 in Theorem 3.9.1 by deleting the word "uniform" from C^*1 and replacing (3.4) in C^*3 by (3.7), then Theorem 3.9.2 is correct if one replaces "(w^*-) uniformity" by "(w^*-) topology" and "C^*-algebra" by "perfect C^*-algebra" (cf. 3.2). Greater simplification is achieved by imposing finite-dimensionality on $\mathcal{P}$ (as a transition probability space, cf. 2.7):

Corollary 3.9.3. *The pure state space $\mathcal{P}$ of a finite-dimensional C^*-algebra is characterized by the following properties:*

QM1: $\mathcal{P}$ is a finite-dimensional Poisson space with a transition probability.
QM2: $\mathcal{P}$ has the two-sphere property (Definition 3.7).
QM3: the sectors of $\mathcal{P}$ as a transition probability space coincide with the symplectic leaves of $\mathcal{P}$ as a Poisson space.

Compare this with the characterization of classical mechanics:

Corollary 3.9.4. *The pure state space P of a classical mechanical system is characterized by the following properties:*

CM1: P is a smooth Poisson space with a transition probability.
CM2: The transition probabilities are $p(\rho, \sigma) = \delta_{\rho\sigma}$.

CHAPTER II

Quantization and the Classical Limit

1 Foundations

1.1 Strict Quantization of Observables

The aim of quantization theory as presented in this book is to relate Poisson algebras or Poisson manifolds to C^*-algebras or their pure state spaces. A slightly awkward feature of the first relationship is that usually Poisson algebras are not Banach spaces; a nonzero Poisson bracket on some Poisson subalgebra $\tilde{\mathfrak{A}}^0_{\mathbb{R}}$ of $C_b^\infty(P, \mathbb{R})$ cannot be extended to the closure $\mathfrak{A}^0_{\mathbb{R}}$ of $\tilde{\mathfrak{A}}^0_{\mathbb{R}}$ in the sup-norm.

Apart from this complication, the following definition is largely motivated by Theorem I.1.1.9; in particular, recall I.(1.22).

Definition 1.1.1. *A* **strict quantization of a Poisson algebra** $\tilde{\mathfrak{A}}^0_{\mathbb{R}}$ *(which is densely contained in the self-adjoint part $\mathfrak{A}^0_{\mathbb{R}}$ of a commutative C^*-algebra $\mathfrak{A}^0$) consists of a collection of points $I_0 \subseteq \mathbb{R}$ that has $0 \notin I_0$ as an accumulation point (we write $I := I_0 \cup \{0\}$), a collection of C^*-algebras $\{\mathfrak{A}^\hbar\}_{\hbar \in I}$, and a collection of linear maps $\{\mathcal{Q}_\hbar : \tilde{\mathfrak{A}}^0_{\mathbb{R}} \to \mathfrak{A}^\hbar_{\mathbb{R}}\}_{\hbar \in I}$ (where $\mathcal{Q}_0$ is the identity map), such that the following conditions hold:*

1. **Rieffel's condition**: *For all $f \in \tilde{\mathfrak{A}}^0_{\mathbb{R}}$, the function $\hbar \mapsto \|\mathcal{Q}_\hbar(f)\|$ is continuous on I. In particular, one has*

$$\lim_{\hbar \to 0} \|\mathcal{Q}_\hbar(f)\| = \|f\|. \tag{1.1}$$

2. **von Neumann's condition**: *For all $f, g \in \tilde{\mathfrak{A}}^0_{\mathbb{R}}$ one has*

$$\lim_{\hbar \to 0} \|\mathcal{Q}_\hbar(f) \circ \mathcal{Q}_\hbar(g) - \mathcal{Q}_\hbar(fg)\| = 0. \tag{1.2}$$

3. Dirac's condition: *For all $f, g \in \tilde{\mathfrak{A}}^0_{\mathbb{R}}$ one has*

$$\lim_{\hbar\to 0} \|\{\mathcal{Q}_\hbar(f), \mathcal{Q}_\hbar(g)\}_\hbar - \mathcal{Q}_\hbar(\{f,g\})\| = 0. \tag{1.3}$$

4. Completeness: *The collection $\{\mathcal{Q}_\hbar(f) \mid f \in \tilde{\mathfrak{A}}^0_{\mathbb{R}}\}$ is dense in $\mathfrak{A}^\hbar_{\mathbb{R}}$ for each $\hbar \in I$.*

A **strict quantization of a Poisson manifold** *P is a strict quantization of some Poisson subalgebra $\tilde{\mathfrak{A}}^0_{\mathbb{R}}$ of $C^\infty_b(P,\mathbb{R})$, equipped with the sup-norm $\|f\| = \|f\|_\infty$, whose closure contains $C_0(P,\mathbb{R})$.*

For a given classical observable $f \in \tilde{\mathfrak{A}}^0$, one construes the operator $\mathcal{Q}_\hbar(f)$ as the quantum observable (at the given value of $\hbar$) whose physical interpretation corresponds to that of its classical counterpart f.

We have suppressed the possible $\hbar$-dependence of the C^*-algebraic operations in $\mathfrak{A}^\hbar$ in our notation. The completeness condition is not crucial: If it fails to be satisfied for a given $\mathfrak{A}^\hbar$, one may simply replace $\mathfrak{A}^\hbar$ in the definition by the C^*-algebra generated by all $\mathcal{Q}_\hbar(f)$.

We may extend $\mathcal{Q}_\hbar$ to a map (denoted by the same symbol) from $\tilde{\mathfrak{A}}^0$ to $\mathfrak{A}^\hbar$ by complex linearity. Conditions 1.1.1.2 and 1.1.1.3 then imply that

$$\lim_{\hbar\to 0} \|\mathcal{Q}_\hbar(f)\mathcal{Q}_\hbar(g) - \mathcal{Q}_\hbar(fg)\| = 0. \tag{1.4}$$

Definition 1.1.2. *A strict quantization $(\mathfrak{A}^\hbar, \{\mathcal{Q}_\hbar\})$ is called a* **strict deformation quantization** *when $\mathcal{Q}_\hbar(\tilde{\mathfrak{A}}^0)$ is closed under multiplication (in $\mathfrak{A}^\hbar$) and $\mathcal{Q}_\hbar$ is* **nondegenerate** *for each $\hbar$ in that $\mathcal{Q}_\hbar(f) = 0$ iff $f = 0$.*

Note that Rieffel's condition implies that a strict quantization is always nondegenerate for small enough $\hbar$, so the last requirement is a modest one. The terminology is justified by the fact that a strict deformation quantization of $\tilde{\mathfrak{A}}^0_{\mathbb{R}}$ allows one to define an associative "deformed" product $\cdot_\hbar$ in $\tilde{\mathfrak{A}}^0$ with the property $\mathcal{Q}_\hbar(f)\mathcal{Q}_\hbar(g) = \mathcal{Q}_\hbar(f \cdot_\hbar g)$ (and, of course, $f \cdot_0 g := fg$). The conditions on a strict quantization may then be rephrased in terms of this product in the obvious way. There are many examples of strict quantization that are not deformation quantizations, in particular those related to pure states (see 1.3).

The maps $\mathcal{Q}_\hbar$ are highly nonunique, depending on what physicists call an operator-ordering prescription. Hence two strict quantizations $(\mathfrak{A}^\hbar_1, \{\mathcal{Q}^1_\hbar\})$, $(\mathfrak{A}^\hbar_2, \{\mathcal{Q}^2_\hbar\})$, where $\mathfrak{A}^\hbar_1 = \mathfrak{A}^\hbar_2$ for all $\hbar$, are called **equivalent** if for each $f \in \tilde{\mathfrak{A}}^0_{\mathbb{R}}$ the function

$$\hbar \mapsto \|\mathcal{Q}^1_\hbar(f) - \mathcal{Q}^2_\hbar(f)\|$$

is continuous on I. It follows that $\lim_{\hbar\to 0} \|\mathcal{Q}^1_\hbar(f) - \mathcal{Q}^2_\hbar(f)\| = 0$. In the next section we will construct an object from a given strict deformation quantization that is invariant under changes to equivalent quantizations.

A strict (deformation) quantization is called **positive** if each $\mathcal{Q}_\hbar$ is positive (that is, $f \geq 0$ in $\mathfrak{A}^0_{\mathbb{R}}$ implies $\mathcal{Q}_\hbar(f) \geq 0$ in $\mathfrak{A}^\hbar_{\mathbb{R}}$). In many physically relevant applications, including the premier example of Weyl quantization, the quantization fails to be positive. However, a nonpositive quantization is sometimes equivalent to a positive one (cf. 2.6.3). If a positive quantization can be extended from $\tilde{\mathfrak{A}}^0_{\mathbb{R}}$ to $\mathfrak{A}^0_{\mathbb{R}}$ (where the property (1.3) is evidently lost) such that it remains positive, the maps $\mathcal{Q}_\hbar : \mathfrak{A}^0 \to \mathfrak{A}^\hbar$ are automatically continuous; see I.1.3.7.

1.2 Continuous Fields of C^*-Algebras

The notion of strict quantization is closely related to an object intrinsic to the theory of C^*-algebras.

Definition 1.2.1. *A* **continuous field of C^*-algebras** $(\mathfrak{C}, \{\mathfrak{A}^x, \varphi_x\}_{x\in X})$ *over a locally compact Hausdorff space X consists of a C^*-algebra $\mathfrak{C}$, a collection of C^*-algebras $\{\mathfrak{A}^x\}_{x\in X}$, and a set $\{\varphi_x : \mathfrak{C} \to \mathfrak{A}^x\}_{x\in X}$ of surjective morphisms, such that:*

1. *The function $x \mapsto \|\varphi_x(A)\|$ is in $C_0(X)$ for all $A \in \mathfrak{C}$.*
2. *The norm of any $A \in \mathfrak{C}$ is $\|A\| = \sup_{x\in X} \|\varphi_x(A)\|$.*
3. *For any $f \in C_0(X)$ and $A \in \mathfrak{C}$ there is an element $fA \in \mathfrak{C}$ for which $\varphi_x(fA) = f(x)\varphi_x(A)$ for all $x \in X$.*

A **section** *of the field is an element $\{A_x\}_{x\in X}$ of $\prod_{x\in X} \mathfrak{A}^x$ for which there is an $A \in \mathfrak{C}$ such that $A_x = \varphi_x(A)$ for all $x \in X$.*

It is clear that $\mathfrak{C}$ may be identified with the space of sections of the field, seen as a C^*-algebra under pointwise scalar multiplication, addition, adjointing, and operator multiplication, by means of $\{\varphi_x(A)\}_{x\in X} \leftrightarrow A$. In particular, $A = B$ iff $\varphi_x(A) = \varphi_x(B)$ for all x.

The simplest example is obtained by taking $\mathfrak{A}^x = \mathfrak{A}$ for all x, and letting $\mathfrak{C} = C_0(X, \mathfrak{A})$ with $\varphi_x(A) := A_x$. Such a field is called **trivial**.

Lemma 1.2.2. *The C^*-algebra $\mathfrak{C}$ of (sections of) a continuous field is* **locally uniformly closed**. *That is, if $A \in \prod_x \mathfrak{A}^x$ is such that for every $y \in X$ and every $\epsilon > 0$ there exists a $B^y \in \mathfrak{C}$ and a neighborhood $\mathcal{N}^y$ of y in which $\|A_x - B^y_x\| < \epsilon$ for all $x \in \mathcal{N}^y$, and also $\lim_{x\to\infty} \|A_x\| = 0$, then $A \in \mathfrak{C}$.*

Alternatively, if the function $x \mapsto \|A_x - C_x\|$ lies in $C_0(X)$ for each $C \in \mathfrak{C}$, then $A \in \mathfrak{C}$.

In the situation of the first part, there is a compact set $K \subseteq X$ for which $\|A_x\| < \epsilon$ outside K, as well as a finite cover $\{\mathcal{N}^{x_1}, \dots, \mathcal{N}^{x_n}\}$ of K. Taking a partition of unity $\{u_i\}$ on K subordinate to this cover, the operator $B := \sum_i u_i B^{x_i}$ lies in $\mathfrak{C}$ because of 1.2.1.3, and satisfies $\sup_{x\in X} \|A_x - B_x\| < \epsilon$. Hence $A \in \mathfrak{C}$ by 1.2.1.2 and the completeness of $\mathfrak{C}$.

Given any $A \in \prod_x \mathfrak{A}^x$ and $y \in X$, because φ_y is surjective there is a $B^y \in \mathfrak{C}$ such that $A_y = B^y_y$. The assumption in the second part then implies that the conditions in the first part are satisfied, such that $A \in \mathfrak{C}$. ■

Proposition 1.2.3. *Suppose one has a family $\{\mathfrak{A}^x\}_{x\in X}$ of C^*-algebras indexed by a locally compact Hausdorff space X, as well as a subset $\tilde{\mathfrak{C}} \subseteq \prod_x \mathfrak{A}^x$ that satisfies the following conditions:*

1. *The set* $\{A_x \mid A \in \tilde{\mathfrak{C}}\}$ *is dense in* $\mathfrak{A}^x$ *for each* $x \in X$.
2. *The function* $x \mapsto \|A_x\|$ *is in* $C_0(X)$ *for each* $A \in \tilde{\mathfrak{C}}$.
3. *The set* $\tilde{\mathfrak{C}}$ *is a* **-algebra (under pointwise operations).*

There exists a unique continuous field of C^**-algebras* $(\mathfrak{C}, \{\mathfrak{A}^x, \varphi_x\}_{x\in X})$ *whose collection of sections contains* $\tilde{\mathfrak{C}}$. *Namely,* $\mathfrak{C}$ *consists of all* $A \in \prod_x \mathfrak{A}^x$ *for which the function* $x \mapsto \|A_x - C_x\|$ *lies in* $C_0(X)$ *for each* $C \in \tilde{\mathfrak{C}}$, *regarded as a* C^**-algebra under pointwise operations, and the norm of 1.2.1.2. Finally,* $\varphi_x(A) := A_x$ *is the evaluation map.*

We first show that $\mathfrak{C}$ as defined above is locally uniformly closed. With the objects A, y, ϵ, B^y, and $\mathcal{N}$ as specified in Lemma 1.2.2, take $C \in \tilde{\mathfrak{C}}$ arbitrary, and define the functions $f_{AC} : x \mapsto \|A_x - C_x\|$ and $f_{BC} : x \mapsto \|B^y_x - C_x\|$. Using the general Banach space inequality

$$|(\|X\| - \|Y\|)| \leq \|X - Y\|, \tag{1.5}$$

one obtains $|f_{AC}(x) - f_{BC}(x)| < \epsilon$ for all $x \in \mathcal{N}$. By assumption, f_{BC} is continuous, so that $|f_{BC}(x) - f_{BC}(y)| < \epsilon$ for all x in some neighborhood $\mathcal{N}'$ of y. Combining the two inequalities yields $|f_{AC}(x) - f_{AC}(y)| < 3\epsilon$ for all $x \in \mathcal{N} \cap \mathcal{N}'$. Hence f_{AC} is continuous at y, which was arbitrary, so that $A \in \mathfrak{C}$ by definition of $\mathfrak{C}$.

Using this property, it is easily shown that $\mathfrak{C}$ is a C^*-algebra, and that condition 3 in Definition 1.2.1 is satisfied. It is clear from 1.2.1.1 and the definition of $\mathfrak{C}$ in 1.2.3 that $\mathfrak{C}$ is maximal. On the other hand, according to the second part of Lemma 1.2.2, $\mathfrak{C}$ is minimal, so that it is unique. ■

We are now in a position to connect Definitions 1.1.1 and 1.2.1.

Theorem 1.2.4. *Suppose one has a strict quantization of a Poisson algebra* $\tilde{\mathfrak{A}}^0_{\mathbb{R}}$, *except perhaps for (1.3). When* I *is not compact, the function* $\hbar \mapsto \|\mathcal{Q}_\hbar(f)\|$ *is assumed to be in* $C_0(I)$ *for all* $f \in \tilde{\mathfrak{A}}^0$. *Furthermore, assume that either* I *is discrete, or that all* $\mathfrak{A}^\hbar$ *are identical for* $\hbar \neq 0$ *and the function* $\hbar \mapsto \mathcal{Q}_\hbar(f)$ *is continuous for all* $f \in \tilde{\mathfrak{A}}^0$.

There exists a unique continuous field of C^**-algebras* $\left(\mathfrak{C}, \{\mathfrak{A}^\hbar, \varphi_\hbar\}_{\hbar\in I}\right)$ *whose collection of sections* $\{\varphi_\hbar(A)\}_{\hbar\in I}$, $A \in \mathfrak{C}$, *contains all* $\{\mathcal{Q}_\hbar(f)\}_{\hbar\in I}$, $f \in \tilde{\mathfrak{A}}^0$. *Moreover, any strict quantization equivalent to the given one leads to the same continuous field.*

One defines $\tilde{\mathfrak{C}} \subset \prod_\hbar \mathfrak{A}^\hbar$ as the complex linear span of all expressions of the form $\hbar \mapsto \mathcal{Q}_\hbar(f_1)\cdots\mathcal{Q}_\hbar(f_n)$, where $f_i \in \tilde{\mathfrak{A}}^0$. We first show that each function of the type $\hbar \mapsto \|\mathcal{Q}_\hbar(f_1)\cdots\mathcal{Q}_\hbar(f_n)\|$ is continuous. It follows from (1.4) that $\lim_{\hbar\to 0}\|\mathcal{Q}_\hbar(f_1)\mathcal{Q}_\hbar(f_2\cdots f_n) - \mathcal{Q}_\hbar(f_1\cdots f_n)\| = 0$, so that by induction one has

$$\lim_{\hbar\to 0}\|\mathcal{Q}_\hbar(f_1)\cdots\mathcal{Q}_\hbar(f_n) - \mathcal{Q}_\hbar(f_1\cdots f_n)\| = 0. \tag{1.6}$$

Equation (1.5) then yields $\lim_{\hbar\to 0}\|\mathcal{Q}_\hbar(f_1)\cdots\mathcal{Q}_\hbar(f_n)\| - \|\mathcal{Q}_\hbar(f_1\cdots f_n)\| = 0$, so that, finally, $\lim_{\hbar\to 0}\|\mathcal{Q}_\hbar(f_1)\cdots\mathcal{Q}_\hbar(f_n)\| = \|f_1\cdots f_n\|$ by (1.1). This proves continuity at $\hbar = 0$. Using (1.6) and the continuity of each function $\hbar \mapsto \mathcal{Q}_\hbar(f)$, the same result follows for polynomials in the $\mathcal{Q}_\hbar(f_i)$.

When I is discrete, continuity away from 0 is trivial. In the alternative case, for monomials this follows from an inductive argument based on the inequalities

$$\begin{aligned} &| \; \|\mathcal{Q}_\hbar(f_1)\cdots\mathcal{Q}_\hbar(f_n)\| - \|\mathcal{Q}_{\hbar'}(f_1)\cdots\mathcal{Q}_{\hbar'}(f_n)\| \; | \\ &\leq \|\mathcal{Q}_\hbar(f_1)\cdots\mathcal{Q}_\hbar(f_n) - \mathcal{Q}_{\hbar'}(f_1)\cdots\mathcal{Q}_{\hbar'}(f_n)\| \\ &\leq \|\mathcal{Q}_\hbar(f_1) - \mathcal{Q}_{\hbar'}(f_1)\| \; \|\mathcal{Q}_\hbar(f_2)\cdots\mathcal{Q}_\hbar(f_n)\| \\ &+ \|\mathcal{Q}_{\hbar'}(f_1)\| \; \|\mathcal{Q}_\hbar(f_2)\cdots\mathcal{Q}_\hbar(f_n) - \mathcal{Q}_{\hbar'}(f_2)\cdots\mathcal{Q}_{\hbar'}(f_n)\|. \end{aligned} \tag{1.7}$$

The extension of this argument to polynomials is a trivial application of the triangle inequality. Since condition 1.2.3.2 is evidently satisfied, one is therefore in the situation of Proposition 1.2.3, and the first claim follows. The second is clear from the proof of 1.2.3 and the definition of equivalent quantizations. ∎

If one wishes to take Definition 1.2.1 as a canonical starting point of the theory of quantization, one might contemplate the following definition of quantization (specialized to the case of Poisson manifolds, for simplicity).

Definition 1.2.5. *Let $I \subseteq \mathbb{R}$ contain 0 as an accumulation point. A* **continuous quantization** *of a Poisson manifold P consists of:*

1. *A continuous field of C^*-algebras $\left(\mathfrak{C}, \{\mathfrak{A}^\hbar, \varphi_\hbar\}_{\hbar\in I}\right)$.*
2. *A Poisson subalgebra $\tilde{\mathfrak{A}}^0$ of $C_b^\infty(P)$ whose closure $\mathfrak{A}^0$ contains $C_0(P)$.*
3. *A linear map $\mathcal{Q} : \tilde{\mathfrak{A}}^0 \to \mathfrak{C}$ that with $\mathcal{Q}_\hbar(f) := \varphi_\hbar(\mathcal{Q}(f))$ for all $f \in \tilde{\mathfrak{A}}^0$ and $\hbar \in I$ satisfies $\mathcal{Q}_0(f) = f$ and $\mathcal{Q}_\hbar(f^*) = \mathcal{Q}_\hbar(f)^*$, and for all $f, g \in \tilde{\mathfrak{A}}^0$ satisfies Dirac's condition (1.3).*

Provided that 1.1.1.4 is satisfied, a continuous quantization is strict. Conversely, Proposition 1.2.4 gives conditions, which will be satisfied in all examples in this book, under which a strict quantization is continuous.

1.3 Coherent States and Berezin Quantization

Having introduced quantization theory from the point of view of observables, we now look at quantization from the dual perspective of pure states. Recall 1.2.1.

Definition 1.3.1. *Relative to a continuous field of C^*-algebras $(\mathfrak{C}, \{\mathfrak{A}^x, \varphi_x\}_{x\in X})$, a* **continuous field of (pure) states** *is a family $\{\omega_x^\lambda\}_{x\in X}^{\lambda\in\Lambda}$, where each ω_x^λ is a (pure) state on $\mathfrak{A}^x$, and Λ is an index set, such that*

1. *For each $\lambda \in \Lambda$ and $A \in \mathfrak{C}$ the function $x \mapsto \omega_x^\lambda(A_x)$ lies in $C_0(X)$.*
2. *For each $x \in X$ the collection $\{\omega_x^\lambda\}^{\lambda\in\Lambda}$ is faithful, in that $\bigcap_{\lambda\in\Lambda} \ker(\pi_{\omega_x^\lambda}) = 0$, where $\pi_{\omega_x^\lambda}(\mathfrak{A}^x)$ is the GNS-representation defined by ω_x^λ (in other words, the representation $\oplus_{\lambda\in\Lambda}\pi_{\omega_x^\lambda}(\mathfrak{A}^x)$ is faithful).*

In the context of quantization theory, the following result allows one to construct continuous fields of pure states by checking a simple condition.

Lemma 1.3.2. *Under the assumptions of Theorem 1.2.4, suppose for each $\hbar \in I$ one has a state $\omega_\hbar$ on $\mathfrak{A}^\hbar$ such that the function $\hbar \mapsto \omega_\hbar(\mathcal{Q}_\hbar(f))$ is continuous on I for each $f \in \tilde{\mathfrak{A}}^0$. Then $\hbar \mapsto \omega_\hbar(A_\hbar)$ is continuous on I for all $A \in \mathfrak{C}$.*

We first assume that $A = \mathcal{Q}_\hbar(f_1)\cdots\mathcal{Q}_\hbar(f_n)$ (the extension to polynomials is trivial). Continuity at $\hbar = 0$ is an immediate consequence of (1.6) and the assumption in 1.3.2. Away from 0, one uses the completeness assumption 1.1.1.4 to approximate $\mathcal{Q}_\hbar(f_1)\cdots\mathcal{Q}_\hbar(f_n)$ by $\mathcal{Q}_\hbar(f)$, and notes that the proof of 1.2.4 establishes the continuity of $\hbar \mapsto \mathcal{Q}_\hbar(f_1)\cdots\mathcal{Q}_\hbar(f_n)$; simply omit the first inequality in (1.7). Combined with the continuity of $\hbar \mapsto \omega_\hbar(\mathcal{Q}_\hbar(f))$, this does the job.

Finally, if A is as specified in the last paragraph of 1.2.3, one uses the last sentence in the proof of 1.2.2, from which the result trivially follows. ■

Given a continuous quantization of a Poisson manifold P (cf. 1.2.5), with $\mathfrak{A}^0 = C_0(P)$, it is natural to take $\Lambda = P$ and $\omega_0^\sigma(f) = f(\sigma)$ for all $\sigma \in P$. Writing $q_\hbar(\sigma)$ for $\omega_\hbar^\sigma$, one may then look at $q_\hbar$ as a map that "quantizes" classical pure states. Such maps may be studied in their own right, even in the absence of a continuous quantization of P. We will do so in the special case that $P = S$ is a symplectic manifold of dimension $2n < \infty$. One may then anticipate that $\mathfrak{A}^\hbar = \mathfrak{B}_0(\mathcal{H}_\hbar)$.

In what follows, the transition probability p is the standard one defined on a projective Hilbert space, given by I.(2.65). The canonical symplectic form on $\mathbb{P}\mathcal{H}$ is denoted by $\omega_\mathcal{H}$ (cf. I.2.5). A measure on a manifold is said to be **locally Lebesgue** if it is equivalent to Lebesgue measure in each local chart.

Definition 1.3.3. *Let $I_0 \subseteq \mathbb{R}$ be as in 1.1.1. In a* **pure state quantization** *of a symplectic manifold (S, ω_S) one specifies, for each $\hbar \in I_0$, a separable Hilbert space $\mathcal{H}_\hbar$, a smooth injection $q_\hbar : S \to \mathbb{P}\mathcal{H}_\hbar$ (cf. I.2.5.1), and a Radon measure $\mu_\hbar$ on S that is locally Lebesgue, such that*

1. *for all $\hbar \in I_0$ and all $\psi \in \mathbb{P}\mathcal{H}_\hbar$ one has*

$$\int_S d\mu_\hbar(\sigma)\, p(q_\hbar(\sigma), \psi) = 1; \tag{1.8}$$

2. *for all fixed $f \in C_c(S)$ and $\rho \in S$, the function*

$$\hbar \mapsto \int_S d\mu_\hbar(\sigma)\, p(q_\hbar(\rho), q_\hbar(\sigma)) f(\sigma)$$

is continuous on I_0 and satisfies

$$\lim_{\hbar \to 0} \int_S d\mu_\hbar(\sigma)\, p(q_\hbar(\rho), q_\hbar(\sigma)) f(\sigma) = f(\rho); \tag{1.9}$$

3. *the map $q_\hbar$ is an approximate symplectomorphism, in that (pointwise)*

$$\lim_{\hbar \to 0} q_\hbar^* \omega_\mathcal{H} = \omega_S. \tag{1.10}$$

Since $f \in C_c(S)$, the requirement 1.3.3.2 is equivalent to the continuity of the function $\hbar \mapsto p(q_\hbar(\rho), q_\hbar(\sigma))$ for fixed ρ and σ. Moreover, we will shortly see that (1.9) and (1.8) imply the conceptually pleasing result

$$\lim_{\hbar\to 0} p(q_\hbar(\rho), q_\hbar(\sigma)) = \delta_{\rho\sigma}. \tag{1.11}$$

The (over) completeness condition (1.8) should be compared with I.(2.56), but note that elements of a basis of a transition probability space are by definition orthogonal, whereas the family $\{q_\hbar(\rho) \mid \rho \in S\}$ becomes approximately orthogonal only in the limit $\lim_{\hbar\to 0}$, as guaranteed by (1.11).

Combining I.(2.56), applied to the transition probability space $\mathbb{P}\mathcal{H}_\hbar$, with (1.8), the volume $\mathrm{vol}_\hbar(S)$ of S with respect to $\mu_\hbar$ is found to be

$$\mathrm{vol}_\hbar(S) = \dim(\mathcal{H}_\hbar). \tag{1.12}$$

In all examples in this book, (1.10) holds without the limit for all $\hbar \in I_0$. In addition, the measure $\mu_\hbar$ will always be of the form

$$\mu_\hbar = c(\hbar)\mu_L, \tag{1.13}$$

where $c : I_0 \to \mathbb{R}\backslash\{0\}$ is some positive continuous function, and the **Liouville measure** μ_L on S is defined by

$$\mu_L(f) := \frac{1}{(2\pi)^n n!}\int_S f\omega_S^n. \tag{1.14}$$

The Liouville measure stands out by its invariance under any Hamiltonian flow, as Proposition I.2.3.3 implies that $\mu_L(f) = \mu_L(\alpha_t^0(f))$ for all t; cf. I.(2.13).

It is clear from (1.13) and (1.12) that $\mathcal{H}_\hbar$ is finite-dimensional iff S is compact, and that only certain discrete values of $\hbar$ are allowed in that case. As $0 \leq p(\cdot,\cdot) \leq 1$, eq. (1.8) then implies $\lim_{\hbar\to 0} c(\hbar) = \infty$, so that $\lim_{\hbar\to 0}\dim(\mathcal{H}_\hbar) = \infty$.

A pure state quantization naturally leads to the quantization of observables.

Definition 1.3.4. *Let $\{\mathcal{H}_\hbar, q_\hbar, \mu_\hbar\}_{\hbar\in I_0}$ be a pure state quantization of a symplectic manifold S. The* **Berezin quantization** *of a function $f \in L^\infty(S)$ is the family of operators $\{\mathcal{Q}_\hbar^B(f)\}_{\hbar\in I_0}$, where $\mathcal{Q}_\hbar^B(f) \in \mathfrak{B}(\mathcal{H}_\hbar)$ is defined by polarizing*

$$\psi(\mathcal{Q}_\hbar^B(f)) := \int_S d\mu_\hbar(\sigma)\, p(q_\hbar(\sigma), \psi) f(\sigma). \tag{1.15}$$

Here $\psi \in \mathbb{P}\mathcal{H}_\hbar$; the integral converges because of (1.8).

Here $L^\infty(S)$ is defined with respect to any locally Lebesgue measure, such as μ_L. If $\mathcal{Q}_\hbar^B$ takes values in $\mathfrak{B}_0(\mathcal{H}_\hbar)$, the left-hand side coincides with the Gelfand transform of $\mathcal{Q}_\hbar^B(f)$ evaluated at ψ, namely $\widehat{\mathcal{Q}_\hbar^B(f)}(\psi)$. If $f \in L^1(S, \mu_\hbar)\cap L^\infty(S)$, the operator $\mathcal{Q}_\hbar^B(f)$ may be written as a Bochner integral

$$\mathcal{Q}_\hbar^B(f) = \int_S d\mu_\hbar(\sigma) f(\sigma)[q_\hbar(\sigma)], \tag{1.16}$$

where $[q_\hbar(\sigma)]$ is the projection onto the one-dimensional subspace in $\mathcal{H}_\hbar$ whose image in $\mathbb{P}\mathcal{H}_\hbar$ is $q_\hbar(\sigma)$. A number of properties of $\mathcal{Q}_\hbar^B$ are immediately evident. Most trivially, (1.9) may be rewritten as

$$\lim_{\hbar\to 0} q_\hbar(\rho)(\mathcal{Q}_\hbar^B(f)) = f(\rho). \tag{1.17}$$

This leads to (1.11), as follows. According to Urysohn's lemma, there is a function $f \in C_0(S, \mathbb{R})$ such that $\|f\|_\infty = 1$, and $f(\rho) = -f(\sigma) = 1$. From I.(2.67) and (1.19) below we infer

$$\left|q_\hbar(\rho)(\mathcal{Q}^B_\hbar(f)) - q_\hbar(\sigma)(\mathcal{Q}^B_\hbar(f))\right| \leq 2\sqrt{1 - p(q_\hbar(\rho), q_\hbar(\sigma))} \leq 2.$$

Letting $\hbar \to 0$, eq. (1.11) then follows from (1.17) and I.(2.53).

Theorem 1.3.5. *Assume that $f \in L^\infty(S, \mathbb{R})$. Then:*

- *$\mathcal{Q}^B_\hbar$ is positive (that is, $f \geq 0$ almost everywhere on S implies $\mathcal{Q}^B_\hbar(f) \geq 0$ in $\mathfrak{B}(\mathcal{H}_\hbar)$).*
- *$\mathcal{Q}^B_\hbar(f)$ is self-adjoint.*
- *If $f \in L^1(S, \mu_\hbar)$, then $\mathcal{Q}^B_\hbar(f) \in \mathfrak{B}_1(\mathcal{H}_\hbar)$ (i.e., $\mathcal{Q}^B_\hbar(f)$ is of trace-class), with*

$$\operatorname{Tr} \mathcal{Q}^B_\hbar(f) = \mu_\hbar(f) = \int_S d\mu_\hbar(\sigma) f(\sigma). \tag{1.18}$$

- *The operator $\mathcal{Q}^B_\hbar(f)$ is bounded by*

$$\|\mathcal{Q}^B_\hbar(f)\| \leq \|f\|_\infty. \tag{1.19}$$

- *If $f \in C_0(S)$, then $\mathcal{Q}^B_\hbar(f) \in \mathfrak{B}_0(\mathcal{H}_\hbar)$ (i.e., $\mathcal{Q}^B_\hbar(f)$ is compact), and $\mathcal{Q}^B_\hbar : C_0(S) \to \mathfrak{B}_0(\mathcal{H}_\hbar)$ is continuous.*

Positivity and self-adjointness are obvious from (1.15). To show that $\mathcal{Q}^B_\hbar(f)$ is trace-class for $f \in L^1(S, \mu_\hbar) \cap L^\infty(S)$, we first assume $f \geq 0$. Then $\mathcal{Q}^B_\hbar(f) \geq 0$, so that the trace norm is $\|\mathcal{Q}^B_\hbar(f)\|_1 = \operatorname{Tr} \mathcal{Q}^B_\hbar(f)$. Choose a basis $\{\mathbf{e}_n\}$ in $\mathcal{H}_\hbar$. Then $\sum_{n=1}^N(\mathbf{e}_n, [q_\hbar(\sigma)]\mathbf{e}_n) = \sum_{n=1}^N p(q_\hbar(\sigma), e_n) \leq 1$ for $N < \infty$. Since $f \in L^1(S, \mu_\hbar)$, the monotone convergence theorem says that $\operatorname{Tr} \mathcal{Q}^B_\hbar(f)$ exists and equals $\int_S d\mu_\hbar(\sigma) f(\sigma)$. Thus $\|\mathcal{Q}^B_\hbar(f)\|_1 = \mu_\hbar(f)$ for $f \geq 0$. For arbitrary f we write $f = f_1 - f_2$, with $f_1, f_2 \geq 0$ a.e. Hence $\|\mathcal{Q}^B_\hbar(f)\|_1 \leq \infty$; linearity of the trace then yields (1.18).

The conclusion from (1.15) that for $f \in L^\infty(S)$ the operator $\mathcal{Q}^B_\hbar(f)$ is bounded, with bound (1.19), uses the following (slightly more general) argument. Let A be a symmetric operator such that $|(\Psi, A\Psi)| \leq c\|\Psi\|^2$ for some $c > 0$ and for all Ψ in its domain. One then replaces Ψ by $\Psi \pm A\Psi/c$, and subtracts the two inequalities thus obtained. This implies the inequality $\|A\Psi\| \leq c\|\Psi\|$, showing that A is bounded with norm $\leq c$. This argument with (1.8) implies (1.19).

Finally, the last claim follows from the second and the third: Start with $f \in C_c(S)$, and use (1.19). ■

There is a clear intuitive connection between the respective conditions 1.1.1.1, 1.1.1.3 on the observable side, and 1.3.3.2, 1.3.3.3 on the pure state side. Moreover, 1.1.1.2 is closely related to (1.11). For the latter equation implies that the projections $[q_\hbar(\sigma)]$ in (1.16) become approximately orthogonal as $h \to 0$, so that the integral should approximate the spectral resolution of $\mathcal{Q}^B_\hbar(f)$. This implies that $\mathcal{Q}^B_\hbar(f)^2$ should approach $\mathcal{Q}^B_\hbar(f^2)$ for small $\hbar$ (cf. I.3.3), which is the essence of von Neumann's condition.

On the other hand, the completeness conditions 1.1.1.4 and 1.3.3.1 are not related. Even if a Berezin quantization satisfies 1.1.1.1–4, it may not define a strict deformation quantization. On the positive side, we have

Proposition 1.3.6. *Let $f \in C_0(S)$, and assume that $\mathcal{H}_\hbar$ is independent of $\hbar$ whenever $\hbar$ varies through a connected subset of I_0. Then Rieffel's condition holds; in particular,*

$$\lim_{\hbar\to 0} \|\mathcal{Q}^B_\hbar(f)\| = \|f\|_\infty. \tag{1.20}$$

We initially assume that $f \in C_c(S)$, and at the end extend the result to $f \in C_0(S, \mathbb{R})$ using the continuity of $\mathcal{Q}^B_\hbar$. The function $\hbar \mapsto \mathcal{Q}^B_\hbar(f)$ from any given connected subset of I_0 to $\mathfrak{B}(\mathcal{H}_\hbar)$ is continuous with respect to the trace norm, hence certainly relative to the operator norm on $\mathfrak{B}(\mathcal{H}_\hbar)$. Therefore, $\hbar \mapsto \|\mathcal{Q}^B_\hbar(f)\|$ is continuous on I_0 by (1.5).

To prove (1.1), note that (1.19) implies

$$\limsup_{\hbar\to 0} \|\mathcal{Q}^B_\hbar(f)\| \leq \|f\|_\infty. \tag{1.21}$$

On the other hand, for $f \in C_0(S)$ we can find $\rho \in S$ for which $\|f\|_\infty = |f(\rho)|$. By (1.9) and the obvious inequality $\|\mathcal{Q}^B_\hbar(f)\| \geq |q_\hbar(\rho)(\mathcal{Q}^B_\hbar(f))|$, we have

$$\liminf_{\hbar\to 0} \|\mathcal{Q}^B_\hbar(f)\| \geq \|f\|_\infty. \tag{1.22}$$

Hence (1.20) follows. ∎

In the examples in this book, the Berezin quantizations constructed from certain pure state quantizations do satisfy all of 1.1.1.1–4. Unfortunately, the proofs of 1.1.1.2–4 seem to involve special features of these examples.

Corollary 1.3.7. *In the situation of Definitions 1.3.3 and 1.3.4, suppose that $\mathcal{H}_\hbar$ is independent of $\hbar$ whenever $\hbar$ varies through a connected subset of I_0, and that the Berezin quantization map $\mathcal{Q}^B_\hbar$, defined on $\mathfrak{A}^0 = C_0(S)$, satisfies 1.1.1.2.*

The collection $\{\omega^\sigma_\hbar\}^{\sigma\in S}_{\hbar\in I}$, where $\omega^\sigma_\hbar := q_\hbar(\sigma)$, is a continuous field of pure states (cf. 1.3.1) relative to the continuous field of C^*-algebras of Theorem 1.2.4.

It is clear from Proposition 1.3.6 and its proof that the assumptions of Theorem 1.2.4 hold. Condition 1.3.3.2 implies that the assumption in 1.3.2 is met. Finally, (1.8) implies that the faithfulness assumption in 1.3.1, where $\mathfrak{A}^x = \mathfrak{A}^\hbar = \mathfrak{B}_0(\mathcal{H}_\hbar)$, is satisfied for $\hbar \neq 0$. Hence the claim follows from Lemma 1.3.2. ∎

This corollary applies to all pure state and Berezin quantizations considered in this book.

1.4 Complete Positivity

Theorem 1.3.5 shows that $\mathcal{Q}^B_\hbar$ is a positive linear map from $C_0(S)$ into $\mathfrak{B}_0(\mathcal{H}_\hbar)$. It has, in fact, a stronger positivity property. The study of this property is further motivated by the idea that a positive map $\mathcal{Q}$ (cf. Definition 1.3.6) generalizes the notion of a state, in that the $\mathbb{C}$ in $\omega : \mathfrak{A} \to \mathbb{C}$ is replaced by a general C^*-algebra $\mathfrak{B}$ in $\mathcal{Q} : \mathfrak{A} \to \mathfrak{B}$. One would like to see whether one can generalize the GNS-construction, and it turns out that for this purpose one needs to impose the stronger positivity property in question.

For a given C^*-algebra $\mathfrak{A}$, and $n \in \mathbb{N}$, we first introduce the C^*-algebra $\mathfrak{M}_n(\mathfrak{A})$. The elements of $\mathfrak{M}_n(\mathfrak{A})$ are $n \times n$ matrices $\mathbb{M}$ with entries in $\mathfrak{A}$; multiplication is done in the usual way, i.e, $(\mathbb{M}\mathbb{N})_{ij} := \mathbb{M}_{ik}\mathbb{N}_{kj}$, with the difference that one now multiplies elements of $\mathfrak{A}$ rather than complex numbers. In particular, the order has to be taken into account. The involution in $\mathfrak{M}_n(\mathfrak{A})$ is, of course, given by $(\mathbb{M}^*)_{ij} = \mathbb{M}^*_{ji}$, in which the involution in $\mathfrak{A}$ replaces the usual complex conjugation in $\mathbb{C}$. One may identify $\mathfrak{M}_n(\mathfrak{A})$ with $\mathfrak{A} \otimes \mathfrak{M}_n(\mathbb{C})$ in the obvious way.

When π is a faithful representation of $\mathfrak{A}$ (which exists by Theorem I.1.1.8), one obtains a faithful realization π_n of $\mathfrak{M}_n(\mathfrak{A})$ on $\mathcal{H}\otimes\mathbb{C}^n$, defined by linear extension of $\pi_n(\mathbb{M})v_i := \pi(\mathbb{M}_{ij})v_j$; we here look at elements of $\mathcal{H}\otimes\mathbb{C}^n$ as n-tuples $(v_1, \ldots, v_n)$, where each $v_i \in \mathcal{H}$. The norm $\|\mathbb{M}\|$ of $\mathbb{M} \in \mathfrak{M}_n(\mathfrak{A})$ is then simply defined to be the norm of $\pi_n(\mathbb{M})$. Since $\pi_n(\mathfrak{M}_n(\mathfrak{A}))$ is a closed *-algebra in $\mathfrak{B}(\mathcal{H}\otimes\mathbb{C}^n)$ (because $n < \infty$), it is obvious that $\mathfrak{M}_n(\mathfrak{A})$ is a C^*-algebra in this norm. The norm is unique by I.1.2.4.4, so that its definition does not depend on the choice of π.

Definition 1.4.1. *Given a linear map $\mathcal{Q} : \mathfrak{A} \to \mathfrak{B}$ between C^*-algebras $\mathfrak{A}$ and $\mathfrak{B}$, and $n \in \mathbb{N}$, define the map $\mathcal{Q}_n : \mathfrak{M}_n(\mathfrak{A}) \to \mathfrak{M}_n(\mathfrak{B})$ by $(\mathcal{Q}_n(\mathbb{M}))_{ij} := \mathcal{Q}(\mathbb{M}_{ij})$. In other words, seen as a map from $\mathfrak{A} \otimes \mathfrak{M}_n(\mathbb{C})$ to $\mathfrak{B} \otimes \mathfrak{M}_n(\mathbb{C})$, one defines $\mathcal{Q}_n$ by linear extension of $\mathcal{Q} \otimes$ id on elementary tensors.*

A linear map $\mathcal{Q} : \mathfrak{A} \to \mathfrak{B}$ between C^*-algebras is called **completely positive** *if $\mathcal{Q}_n$ is positive for all $n \in \mathbb{N}$.*

The point is now that completely positive maps that in addition are normalized (like a state) have a generic structure, which is of central importance for quantization theory. Recall that a **partial isometry** is a linear map $W : \mathcal{H}_1 \to \mathcal{H}_2$ between two Hilbert spaces, with the property that $\mathcal{H}_1$ contains a closed subspace $\mathcal{K}_1$ such that $(W\Psi, W\Phi)_2 = (\Psi, \Phi)_1$ for all $\Psi, \Phi \in \mathcal{K}_1$, and $W = 0$ on $\mathcal{K}_1^\perp$. Hence W is unitary from $\mathcal{K}_1$ to $W\mathcal{K}_1$. It follows that $W^*W = [\mathcal{K}_1]$ and $WW^* = [\mathcal{K}_2]$, where $\mathcal{K}_2$ is the image of W, are projections.

Theorem 1.4.2. *Let $\mathcal{Q} : \mathfrak{A} \to \mathfrak{B}$ be a completely positive map between C^*-algebras with unit, such that $\mathcal{Q}(\mathbb{I}) = \mathbb{I}$. By Theorem I.1.1.8, we may assume that $\mathfrak{B}$ is faithfully represented as a subalgebra $\pi_\chi(\mathfrak{B}) \subseteq \mathfrak{B}(\mathcal{H}_\chi)$, for some Hilbert space $\mathcal{H}_\chi$.*

There exists a Hilbert space $\mathcal{H}^\chi$, a representation π^χ of $\mathfrak{A}$ on $\mathcal{H}^\chi$, and a partial isometry $W : \mathcal{H}_\chi \to \mathcal{H}^\chi$ (with $W^*W = \mathbb{I}$) such that

$$\pi_\chi(\mathcal{Q}(A)) = W^*\pi^\chi(A)W \tag{1.23}$$

for all $A \in \mathfrak{A}$. With $p := WW^*$ (the target projection of W on $\mathcal{H}^\chi$), $\tilde{\mathcal{H}}_\chi := p\mathcal{H}^\chi = W\mathcal{H}_\chi \subset \mathcal{H}^\chi$, and $U : \mathcal{H}_\chi \to \tilde{\mathcal{H}}_\chi$ defined as W, seen as a unitary map from $\mathcal{H}_\chi$ to $\tilde{\mathcal{H}}_\chi$, one has the equivalent relation

$$\tilde{\mathcal{Q}}(A) := U\pi_\chi(\mathcal{Q}(A))U^{-1} = p\pi^\chi(A)p. \tag{1.24}$$

The proof consists in a modification of the GNS-construction (cf. I.1.5). We denote elements of $\mathcal{H}_\chi$ by v, w, with inner product $(v, w)_\chi$.

Construction 1.4.3.

1. *Define the sesquilinear form* $(\, ,\,)_0^\chi$ *on* $\mathfrak{A} \otimes \mathcal{H}_\chi$ *(algebraic tensor product) by (sesqui-)linear extension of*
$$(A \otimes v, B \otimes w)_0^\chi := (v, \pi_\chi(\mathcal{Q}(A^* B))w)_\chi. \tag{1.25}$$
Since $\mathcal{Q}$ *is completely positive, this form is positive semidefinite; denote its null space by* $\mathcal{N}_\chi$.
2. *The form* $(\, ,\,)_0^\chi$ *projects to an inner product* $(\, ,\,)^\chi$ *on* $\mathfrak{A} \otimes \mathcal{H}_\chi/\mathcal{N}_\chi$. *If* $V_\chi : \mathfrak{A} \otimes \mathcal{H}_\chi \to \mathfrak{A} \otimes \mathcal{H}_\chi/\mathcal{N}_\chi$ *is the canonical projection, then by definition*
$$(V_\chi(A \otimes v), V_\chi(B \otimes w))^\chi := (A \otimes v, B \otimes w)_0^\chi. \tag{1.26}$$
The Hilbert space $\mathcal{H}^\chi$ *is the closure of* $\mathfrak{A} \otimes \mathcal{H}_\chi/\mathcal{N}_\chi$ *in this inner product.*
3. *The representation* $\pi^\chi(\mathfrak{A})$ *is initially defined on* $\mathfrak{A} \otimes \mathcal{H}_\chi/\mathcal{N}_\chi$ *by linear extension of*
$$\pi^\chi(A)V_\chi(B \otimes w) := V_\chi(AB \otimes w); \tag{1.27}$$
this is well-defined, because $\mathfrak{A} \otimes \mathbb{I}_\chi \mathcal{N}_\chi \subseteq \mathcal{N}_\chi$. *One has the bound*
$$\|\pi^\chi(A)\| \leq \|A\|, \tag{1.28}$$
so that $\pi^\chi(A)$ *may be defined on all of* $\mathcal{H}^\chi$ *by continuous extension of (1.27). This extension is a representation of* $\mathfrak{A}$ *on* $\mathcal{H}^\chi$.
4. *The map* $W : \mathcal{H}_\chi \to \mathcal{H}^\chi$, *defined by*
$$Wv := V_\chi \mathbb{I} \otimes v, \tag{1.29}$$
is a partial isometry. Its adjoint $W^* : \mathcal{H}^\chi \to \mathcal{H}_\chi$ *is given by (continuous extension of)*
$$W^* V_\chi A \otimes v = \pi_\chi(\mathcal{Q}(A))v, \tag{1.30}$$
from which the properties $W^*W = \mathbb{I}$ *and (1.23) follow.*

We now prove the various claims made in this construction. Firstly, to show that the form defined by (1.25) is positive, we write
$$\sum_{i,j}(A_i \otimes v_i, A_j \otimes v_j)_0^\chi = \sum_{i,j}(v_i, \pi_\chi(\mathcal{Q}(A_i^* A_j))v_j)_\chi. \tag{1.31}$$

Now consider the element $\mathbb{A}$ of $\mathfrak{M}_n(\mathfrak{A})$ with matrix elements $\mathbb{A}_{ij} = A_i^* A_j$. Taking a faithful representation $\pi(\mathfrak{A})$, from which one constructs $\pi_n(\mathfrak{M}_n(\mathfrak{A}))$ as explained above 1.4.1, one sees that
$$(z, \pi_n(\mathbb{A})z) = \sum_{i,j}(z_i, \pi(A_i^* A_j)z_j) = \sum_{i,j}(\pi(A_i)z_i, \pi(A_j)z_j) = \|Az\|^2 \geq 0,$$
where $Az := \sum_i \pi(A_i)z_i$. Hence $\mathbb{A} \geq 0$. Since $\mathcal{Q}$ is completely positive, it must be that $\mathbb{B}$, defined by its matrix elements $\mathbb{B}_{ij} := \mathcal{Q}(A_i^* A_j)$, is positive in $\mathfrak{M}_n(\mathfrak{B})$.

Repeating the above argument with $\mathbb{A}$ and π replaced by $\mathbb{B}$ and π_χ, respectively, one concludes that the right-hand side of (1.31) is positive.

It follows from (1.25) that $(C\otimes v,\, AB\otimes w)_0^\chi = (A^*C\otimes v,\, B\otimes w)_0^\chi$, so that $A\otimes \mathbb{I}_\chi$ leaves $\mathcal{N}_\chi$ stable; compare the corresponding argument for the GNS-construction based on I.(1.58).

To prove (1.28) one uses I.(1.42) in $\mathfrak{M}_n(\mathfrak{A})$. Namely, for an arbitrary collection $A, B_1, \ldots, B_n \in \mathfrak{A}$ we conjugate $0 \leq A^*A\mathbb{I}_n \leq \|A\|^2\mathbb{I}_n$ with the matrix $\mathbb{B}$, whose first row is $(B_1, \ldots, B_n)$, and which has zeros everywhere else; the adjoint $\mathbb{B}^*$ is then the matrix whose first column is $(B_1^*, \ldots, B_n^*)^T$, and all other entries zero. This leads to $0 \leq \mathbb{B}^*A^*A\mathbb{B} \leq \|A\|^2\mathbb{B}^*\mathbb{B}$. Since $\mathcal{Q}$ is completely positive, one has $\mathcal{Q}_n(\mathbb{B}^*A^*A\mathbb{B}) \leq \|A\|^2\mathcal{Q}_n(\mathbb{B}^*\mathbb{B})$. Hence in any representation $\pi_\chi(\mathfrak{B})$ and any vector $(v_1, \ldots v_n) \in \mathcal{H}_\chi \otimes \mathbb{C}^n$ one has

$$\sum_{i,j}(v_i, \pi_\chi(\mathcal{Q}(B_i^*A^*AB_j))v_j)_\chi \leq \|A\|^2 \sum_{i,j}(v_i, \pi_\chi(\mathcal{Q}(B_i^*B_j))v_j)_\chi. \tag{1.32}$$

With $\Psi = \sum_i V_\chi B_i \otimes v_i$, from (1.25), (1.27), and (1.32) one then has

$$\|\pi^\chi(A)\Psi\|^2 \leq \|A\|^2 \sum_{i,j}(v_i, \pi_\chi(\mathcal{Q}(B_i^*B_j))v_j)_\chi = \|A\|^2\ \|\Psi\|^2.$$

To show that W is a partial isometry one merely uses (1.29), (1.25), and $\mathcal{Q}(\mathbb{I}) = \mathbb{I}$. Equation (1.30) is then trivially verified from the defining property $(w,\, W^*\Psi)_\chi = (Ww,\, \Psi)^\chi$ for all $w \in \mathcal{H}_\chi$ and $\Psi \in \mathcal{H}^\chi$.

To verify (1.23), one uses (1.29) and (1.30). Since W is a partial isometry, one has $p = WW^*$ for the projection p onto the image of W, and in this case, $W^*W = \mathbb{I}$ for the projection onto the subspace of $\mathcal{H}_\chi$ on which W is isometric; this subspace is $\mathcal{H}_\chi$ itself. Hence (1.24) follows from (1.23), since

$$U\pi_\chi(\mathcal{Q}(A))U^{-1} = W\pi_\chi(\mathcal{Q}(A))W^* = WW^*\pi^\chi(A)WW^* = p\pi^\chi(A)p. \quad \blacksquare$$

When $\mathcal{Q}$ fails to preserve the unit, the above construction still applies, but W is no longer a partial isometry; one rather has $\|W\|^2 = \|\mathcal{Q}(\mathbb{I})\|$. Thus it is no longer possible to regard $\mathcal{H}_\chi$ as a subspace of $\mathcal{H}^\chi$.

If $\mathfrak{A}$ and perhaps $\mathfrak{B}$ are nonunital, the theorem holds if $\mathcal{Q}$ can be extended (as a positive map) to the unitization of $\mathfrak{A}$ (cf. I.1.2.1), such that the extension preserves the unit $\mathbb{I}$ (perhaps relative to the unitization of $\mathfrak{B}$). When the extension exists but does not preserve the unit, one is in the situation of the previous paragraph.

The relevance of all this to Berezin quantization is as follows.

Proposition 1.4.4. *A positive map between a commutative unital C^*-algebra and a C^*-algebra is completely positive.*

We write $\mathcal{Q} : \mathfrak{A} \to \mathfrak{B}$ for the map in question. By Theorem I.1.2.3 one has $\mathfrak{A} = C(X)$ for some compact Hausdorff space X. We may then identify $\mathfrak{M}_n(C(X))$ with $C(X, \mathfrak{M}_n(\mathbb{C}))$. Take $G \in C(X, \mathfrak{M}_n(\mathbb{C}))$ and pick $\epsilon > 0$. Since X is compact, there is a finite collection of points $x_1, \ldots, x_n$ and a finite cover $\{\mathcal{O}_{x_1}^\epsilon, \ldots, \mathcal{O}_{x_l}^\epsilon\}$ with the property that $\|G(x_i) - G(x)\| < \epsilon$ for all $x \in \mathcal{O}_{x_i}^\epsilon$. Using a partition of unity $\{u_i\}$ subordinate to this cover, one constructs $F_l \in C(X, \mathfrak{M}_n(\mathbb{C}))$ by

$F_l(x) := \sum_{i=1}^{l} u_i(x)G(x_i)$. One then has $\|F_l - G\| < \epsilon$. Hence elements of the form F, where $F(x) = \sum_i f_i(x)M_i$ for $f_i \in C(X)$ and $M_i \in \mathfrak{M}_n(\mathbb{C})$, and the sum is finite, are dense in $C(X, \mathfrak{M}_n(\mathbb{C}))$.

It is easily seen that such F is positive iff all f_i and M_i are positive, so that positive elements G of $C(X, \mathfrak{M}_n(\mathbb{C}))$ can be approximated by positive F's. On such F, one has $\mathcal{Q}_n(F) = \sum_i \mathcal{Q}(f_i) \otimes M_i$. Now, each operator $B_i \otimes M$ is positive in $\mathfrak{M}_n(\mathfrak{B})$ when B_i and M are positive (as can be checked in a faithful representation). Since $\mathcal{Q}$ is positive, it follows that $\mathcal{Q}_n$ maps each positive element of the form $F = \sum_i f_i M_i$ into a positive member of $\mathfrak{M}_n(\mathfrak{B})$.

We know from I.1.3.7 that $\mathcal{Q}$ is continuous; the continuity of $\mathcal{Q}_n$ follows because $n < \infty$. A norm-limit $A = \lim_n A_n$ of positive elements in a C^*-algebra is positive, because by I.(1.39) we have $A_n = B_n^* B_n$, and $\lim B_n = B$ exists because of I.(1.15). Finally, $A = B^*B$ by continuity of multiplication, i.e., by I.(1.14). Hence if $F_k \to G \geq 0$ in $C(X, \mathfrak{M}_n(\mathbb{C}))$, then $\mathcal{Q}_n(G) = \lim_k \mathcal{Q}_n(F_k)$ is a norm-limit of positive elements, which is positive. ∎

The application to Berezin quantization is obvious from I.1.3.5 and 1.4.4: We take $\mathfrak{A} = C_0(S)$, $\mathfrak{B} = \mathfrak{B}_0(\mathcal{H}_\hbar)$, $\mathcal{H}_\chi = \mathcal{H}_\hbar$, and $\mathcal{Q} = \mathcal{Q}_\hbar^B$. Theorem 1.4.2 then applies, for we can extend $\mathcal{Q}_\hbar^B$ to the unitization $C_0(S)_{\mathbb{I}}$ of $C_0(S)$ (which consists of all functions of the form $f + \lambda 1_S$, $f \in C_0(S)$ and $\lambda \in \mathbb{C}$) by linear extension of (1.15). Since $C_0(S) + \mathbb{C}1_S \subset L^\infty(S)$, this extension is still positive by Theorem 1.3.5, and satisfies $\mathcal{Q}_\hbar^B(1_S) = \mathbb{I}$ because of (1.8).

Corollary 1.4.5. *The image $\mathcal{Q}_\hbar^B(C_0(S))$ is closed in $\mathfrak{B}_0(\mathcal{H}_\hbar)$. In particular, if $\mathcal{Q}_\hbar^B(C_c^\infty(S))$ is dense in $\mathfrak{B}_0(\mathcal{H}_\hbar)$, then $\mathcal{Q}_\hbar^B(C_0(S)) = \mathfrak{B}_0(\mathcal{H}_\hbar)$.*

Taking π^χ as in the proof of 1.4.2, the image $\pi^\chi(C_0(S))$ is closed by Theorem I.1.3.10.4, so that $p\pi^\chi(C_0(S))p = \mathcal{Q}_\hbar^B(C_0(S))$ is closed as well. ∎

In the opposite direction, one may ask whether a given positive map $\mathcal{Q}$ can be written in a form similar to (1.16).

Proposition 1.4.6. *Let $\mathcal{Q} : C_0(S) \to \mathfrak{B}(\mathcal{H})$ be positive (where S is a locally compact Hausdorff space), and such that $\mathcal{Q}(f) \in \mathfrak{B}_1(\mathcal{H})$ for all $f \in C_c(S)$. Then there exists a regular Borel measure μ on S and a (weakly) measurable family $\sigma \mapsto \rho(\sigma)$ of density matrices, such that (weakly)*

$$\mathcal{Q}(f) = \int_S d\mu(\sigma) f(\sigma)\rho(\sigma). \tag{1.33}$$

Given the assumptions, the map $f \mapsto \operatorname{Tr}\mathcal{Q}(f)$ defines a positive linear functional, which by the Riesz representation theorem corresponds to a positive regular Borel measure μ on S. Also, for each unit vector $\Psi \in \mathcal{H}$ we obtain a positive linear functional $f \mapsto (\Psi, \mathcal{Q}(f)\Psi)$, hence a positive regular Borel measure μ_Ψ on S. Since $(\Psi, \mathcal{Q}(f)\Psi) \leq \operatorname{Tr}\mathcal{Q}(f)$, we see that μ_Ψ is absolutely continuous with respect to μ. Hence we obtain the Radon–Nikodym derivatives $\rho_\Psi(\sigma) := d\mu_\Psi/d\mu(\sigma)$, and subsequently the operators $\rho(\sigma)$ by polarization. Equation (1.33) follows by construction; the claimed properties of the $\rho(\sigma)$ are then obvious. ∎

Obviously, the given condition is always met if $\mathcal{H}$ is finite-dimensional. It remains to be investigated whether each $\rho(\sigma)$ is a one-dimensional projection, and if so, whether the ensuing map $S \to \mathbb{P}\mathcal{H}$ is smooth. There is, however, a generalization of (1.33) that applies to any positive map on a commutative C^*-algebra.

Definition 1.4.7. *Let X be a set with a σ-algebra Σ of subsets of X. A* **positive-operator-valued measure**, *or* **POVM**, *on X in a Hilbert space $\mathcal{H}$ is a map $\Delta \mapsto A(\Delta)$ from Σ to $\mathfrak{B}(\mathcal{H})^+$ (the set of positive operators on $\mathcal{H}$), satisfying $A(\emptyset) = 0$, $A(X) = \mathbb{I}$, and $A(\cup_i \Delta_i) = \sum_i A(\Delta_i)$ for any countable collection of disjoint $\Delta_i \in \Sigma$ (where the infinite sum is taken in the weak operator topology).*

A **projection-valued measure**, *or* **PVM**, *is a POVM that in addition satisfies $A(\Delta_1 \cap \Delta_2) = A(\Delta_1)A(\Delta_2)$ for all $\Delta_1, \Delta_2 \in \Sigma$.*

Note that the above conditions force $0 \le A(\Delta) \le \mathbb{I}$. A PVM is usually written as $\Delta \mapsto E(\Delta)$; it follows that each $E(\Delta)$ is a projection (take $\Delta_1 = \Delta_2$ in the definition). This notion is familiar from the spectral theorem.

Proposition 1.4.8. *Let X be a locally compact Hausdorff space, with Borel structure Σ. There is a bijective correspondence between positive maps $\mathcal{Q} : C_0(X) \to \mathfrak{B}(\mathcal{H})$ that can be extended to $C_0(X)_{\mathbb{I}}$ by a unit-preserving positive map and POVMs $\Delta \mapsto A(\Delta)$ on X in $\mathcal{H}$, given by*

$$\mathcal{Q}(f) = \int_X dA(x)\, f(x). \tag{1.34}$$

The map $\mathcal{Q}$ is a representation of $C_0(X)$ iff $\Delta \mapsto A(\Delta)$ is a PVM.

The precise meaning of (1.34) will emerge shortly. Given the assumptions, in view of I.1.2.3 we may as well assume that X is compact.

Given $\mathcal{Q}$, for arbitrary $\Psi \in \mathcal{H}$ one constructs a functional $\hat{\mu}_{\Psi,\Psi}$ on $C(X)$ by $\hat{\mu}_{\Psi,\Psi}(f) := (\Psi, \mathcal{Q}(f)\Psi)$. Since $\mathcal{Q}$ is linear and positive, this functional has the same properties. Hence the Riesz representation theorem yields a probability measure $\mu_{\Psi,\Psi}$ on X. For $\Delta \in \Sigma$ one then puts $(\Psi, A(\Delta)\Psi) := \mu_{\Psi,\Psi}(\Delta)$, defining an operator $A(\Delta)$ by polarization. The ensuing map $\Delta \mapsto A(\Delta)$ is easily checked to have the properties required of a POVM.

Conversely, for each pair $\Psi, \Phi \in \mathcal{H}$ a POVM $\Delta \mapsto A(\Delta)$ in $\mathcal{H}$ defines a signed measure $\mu_{\Psi,\Phi}$ on X by means of $\mu_{\Psi,\Phi}(\Delta) := (\Psi, A(\Delta)\Phi)$. This yields a positive map $\mathcal{Q} : C(X) \to \mathfrak{B}(\mathcal{H})$ by $(\Psi, \mathcal{Q}(f)\Phi) := \int_X d\mu_{\Psi,\Phi}(x)\, f(x)$; the meaning of (1.34) is expressed by this equation.

Approximating $f, g \in C(X)$ by step functions, one verifies that the property $E(\Delta)^2 = E(\Delta)$ is equivalent to $\mathcal{Q}(fg) = \mathcal{Q}(f)\mathcal{Q}(g)$; then use I.1.3.7. ∎

Corollary 1.4.9. *Let $\Delta \mapsto A(\Delta)$ be a POVM on a locally compact Hausdorff space X in a Hilbert space $\mathcal{H}_\chi$. There exist a Hilbert space $\mathcal{H}^\chi$, a projection p on $\mathcal{H}^\chi$, a unitary map $U : \mathcal{H}_\chi \to p\mathcal{H}^\chi$, and a PVM $\Delta \mapsto E(\Delta)$ on $\mathcal{H}^\chi$ such that $UA(\Delta)U^{-1} = pE(\Delta)p$ for all $\Delta \in \Sigma$.*

Combine Theorem 1.4.2 with Proposition 1.4.8. ∎

Suppose X is the phase space S of a physical system, and one is in the situation discussed prior to 1.4.5. One then obtains a POVM $\Delta \mapsto A(\Delta)$ on S in $\mathcal{H}_\hbar$ associated to the Berezin quantization map $\mathcal{Q} = \mathcal{Q}_\hbar^B : C_0(S) \to \mathfrak{B}_0(\mathcal{H}_\hbar)$. According to I.1.6.6, one may identify a state on $\mathfrak{B}_0(\mathcal{H}_\hbar)$ with a density matrix ρ on $\mathcal{H}_\hbar$. The physical interpretation of the map $\Delta \mapsto A(\Delta)$ is then contained in the statement that the number

$$p_\rho(\Delta) := \mathrm{Tr}\, \rho A(\Delta) \tag{1.35}$$

is the probability that in a state ρ the system in question is localized in $\Delta \subset S$ (**localization in phase space**). Transferring the situation to $\tilde{\mathcal{H}}_\hbar$ by means of the unitary U in 1.4.2, and writing $\tilde{\rho} := U\rho U^{-1}$, one simply has $p_\rho(\Delta) = \mathrm{Tr}\, \tilde{\rho} E(\Delta)$, where $\Delta \mapsto E(\Delta)$ is the PVM on $\mathcal{H}^\chi$ given by 1.4.9.

When X is a configuration space Q, on the other hand, the Poisson bracket between any two functions on X normally vanishes, so that the conditions (1.2) and (1.3) can be satisfied by taking $\mathcal{Q}$ to be a representation π of $C_0(Q)$ on $\mathcal{H}$. By Proposition 1.4.8, the situation is therefore described by a PVM $\Delta \mapsto E(\Delta)$ on Q in $\mathcal{H}$; the probability that in a state ρ the system is localized in $\Delta \subset Q$ (**localization in configuration space**) is

$$p_\rho(\Delta) := \mathrm{Tr}\, \rho E(\Delta). \tag{1.36}$$

1.5 Coherent States and Reproducing Kernels

One can find an explicit realization of $\mathcal{H}^\hbar := \mathcal{H}^\chi$ and of the projection p in 1.4.2 if a further assumption is made, which is satisfied in many cases of interest.

Definition 1.5.1. *A pure state quantization $\{\mathcal{H}_\hbar, q_\hbar, \mu_\hbar\}_{\hbar\in I_0}$ of S is said to be* **coherent** *if each $q_\hbar(\sigma) \in \mathbb{P}\mathcal{H}_\hbar$ can be lifted to a unit vector $\Psi_\hbar^\sigma \in \mathcal{H}_\hbar$, and the ensuing map $\sigma \mapsto \Psi_\hbar^\sigma$ from S to $\mathcal{H}_\hbar$ is continuous. The unit vectors $\Psi_\hbar^\sigma$ coming from a coherent pure state quantization are called* **coherent states.**

In terms of coherent states, the polarized form of (1.8) is

$$\int_S d\mu_\hbar(\sigma)\, (\Psi_1, \Psi_\hbar^\sigma)(\Psi_\hbar^\sigma, \Psi_2) = (\Psi_1, \Psi_2) \tag{1.37}$$

for all $\Psi_1, \Psi_2 \in \mathcal{H}_\hbar$. We write

$$\mathcal{K}_\hbar(\rho, \sigma) := (\Psi_\hbar^\rho, \Psi_\hbar^\sigma); \tag{1.38}$$

as a consequence of the continuity assumption above, $\mathcal{K}_\hbar$ is jointly continuous. Also, one notices that

$$\mathcal{K}_\hbar(\sigma, \rho) = \overline{\mathcal{K}_\hbar(\rho, \sigma)}. \tag{1.39}$$

Proposition 1.5.2. *Let $\{\mathcal{H}_\hbar, \Psi_\hbar^\sigma, \mu_\hbar\}_{\sigma\in S, \hbar\in I_0}$ be a coherent pure state quantization, with associated Berezin quantization $\mathcal{Q}_\hbar^B$. One may put*

$$\mathcal{H}^\chi = \mathcal{H}^\hbar := L^2(S, d\mu_\hbar); \tag{1.40}$$

$$\pi^\chi(f)\Phi(\sigma) := f(\sigma)\Phi(\sigma) \tag{1.41}$$

in 1.4.2. Furthermore:

- *For each $\hbar \in I_0$, the map $W : \mathcal{H}_\hbar \to \mathcal{H}^\hbar$ defined by*

$$W\Psi(\sigma) := (\Psi_\hbar^\sigma, \Psi) \tag{1.42}$$

is a partial isometry (with $WW^* = p$ a projection, and $W^*W = \mathbb{I}$). We denote its image $W\mathcal{H}_\hbar$ in $\mathcal{H}^\hbar$ by $\tilde{\mathcal{H}}_\hbar$.

- *The projection $p : \mathcal{H}^\hbar \to \tilde{\mathcal{H}}_\hbar$ is given by*

$$p\Phi(\rho) = \int_S d\mu_\hbar(\sigma)\,\mathcal{K}_\hbar(\rho, \sigma)\Phi(\sigma). \tag{1.43}$$

- *The elements of $\tilde{\mathcal{H}}_\hbar \subset \mathcal{H}^\hbar$ may be chosen to be continuous functions.*
- *For each $\rho \in S$, the function $\sigma \mapsto \mathcal{K}_\hbar(\sigma, \rho)$ lies in $\tilde{\mathcal{H}}_\hbar$.*
- *The evaluation map $\Phi \mapsto \Phi(\sigma)$ is continuous for all $\Phi \in \tilde{\mathcal{H}}_\hbar$ and all $\sigma \in S$.*
- *For each $f \in L^\infty(S)$, the operator $\tilde{Q}_\hbar^B(f) := WQ_\hbar^B(f)W^*$ on $\tilde{\mathcal{H}}_\hbar$, which provides an equivalent realization of the Berezin quantization of S, is given by*

$$\tilde{Q}_\hbar^B(f)\Phi(\sigma) = pf(\sigma)\Phi(\sigma). \tag{1.44}$$

The first two claims follow from (1.37). The Cauchy–Schwarz inequality applied to (1.42), and the continuity of $q_\hbar$ prove the third claim. The next claim is immediate from (1.42), since $\mathcal{K}_\hbar(\cdot, \rho) = W\Psi_\hbar^\rho$. To show the continuity of the evaluation map, we write

$$\Phi(\rho) = \int_S d\mu_\hbar(\sigma)\,\mathcal{K}_\hbar(\rho, \sigma)\Phi(\sigma), \tag{1.45}$$

which, as a consequence of (1.43), holds for all $\Phi \in \tilde{\mathcal{H}}_\hbar$ and all $\rho \in S$. The right-hand side is $(\mathcal{K}_\hbar(\cdot, \rho), \Phi)$ (inner product in $\mathcal{H}^\hbar$), which, combined with the previous item, proves the claim. Finally, (1.44) is immediate from the definitions. ■

Comparing, e.g., (1.44) with (1.24), we see how the above construction provides an explicit realization of the objects defined in 1.4.2. As a case in point, we may rewrite (1.35) in an appealing way. Note that because of (1.41), the PVM $\Delta \mapsto E(\Delta)$ in 1.4.9 is given by $E(\Delta) = \chi_\Delta$ (the characteristic function of Δ). Assuming that ρ is a pure state $\rho = [\Psi]$, where $\Psi \in \mathbb{S}\mathcal{H}_\hbar$, the discussion after (1.35) and (1.42) then implies that the probability that the system is localized in Δ is

$$p_{[\Psi]}(\Delta) = \int_\Delta d\mu_\hbar(\sigma)\,|(\Psi_\hbar^\sigma, \Psi)|^2. \tag{1.46}$$

An interesting feature to be abstracted from 1.5.2 is the following.

Definition 1.5.3. *Let S be some set, and let $\mathcal{H}$ be a Hilbert space of functions (of some class) on S. A* **reproducing kernel** *of $\mathcal{H}$ is a function $\mathcal{K} : S \times S \to \mathbb{C}$ such that:*

- *For each $\rho \in S$, the function $\sigma \mapsto \mathcal{K}(\sigma, \rho)$ lies in $\mathcal{H}$.*
- *The reproducing property*

$$\Psi(\rho) = (\mathcal{K}(\cdot, \rho), \Psi) \tag{1.47}$$

holds for all $\Psi \in \mathcal{H}$, $\rho \in S$.

Taking $\Psi = \mathcal{K}(\cdot, \rho)$, we obtain

$$\|\mathcal{K}(\cdot, \rho)\| = \sqrt{\mathcal{K}(\rho, \rho)}; \tag{1.48}$$

in particular, $\mathcal{K}(\rho, \rho) \geq 0$ for all ρ. Putting $\Psi = \mathcal{K}(\cdot, \sigma)$, one observes that $\mathcal{K}(\sigma, \rho) = \overline{\mathcal{K}(\rho, \sigma)}$.

Proposition 1.5.4. *A Hilbert space of functions on S has a (necessarily unique) reproducing kernel iff each evaluation map* $E_\sigma : \Psi \mapsto \Psi(\sigma)$ *is continuous.*

The uniqueness of $\mathcal{K}$ follows by assuming that two reproducing kernels $\mathcal{K}_1$, $\mathcal{K}_2$ exist, and showing that $\|\mathcal{K}_1(\cdot, \rho) - \mathcal{K}_2(\cdot, \rho)\|$ has to vanish because of the reproducing property. The rest is obvious. ■

Lemma 1.5.5. *If* $\mathcal{H}$ *has a reproducing kernel* $\mathcal{K}$*, then strong convergence* $\Psi_n \to \Psi$ *in* $\mathcal{H}$ *implies uniform convergence as functions on all subsets of S where* $\sigma \mapsto \mathcal{K}(\sigma, \sigma)$ *is bounded.*

One has

$$|\Psi_n(\sigma) - \Psi(\sigma)| = |(\mathcal{K}(\cdot, \sigma), \Psi_n - \Psi)| \leq \|\Psi_n - \Psi\| \|\mathcal{K}(\cdot, \sigma)\|.$$

Then use (1.48). ■

This situation becomes particularly interesting when S is a topological space and $\mathcal{K}$ is jointly continuous. In that case, (1.47) and (1.48) imply that $\mathcal{H}$ consists of continuous functions. Moreover, if we equip $C(S)$ with the topology of uniform convergence on compact sets, then Lemma 1.5.5 implies that the canonical injection $\mathcal{H} \hookrightarrow C(S)$ is continuous. This motivates the following abstract considerations, which provide an interesting perspective on the reproducing kernel of $\mathcal{H}$.

Definition 1.5.6. *A* **Hilbert subspace** *of a topological vector space* $\mathcal{V}$ *is a Hilbert space* $\mathcal{H}$ *with continuous linear injection* $\mathcal{H} \hookrightarrow \mathcal{V}$*. In other words,* $\mathcal{H}$ *is a continuously embedded subspace of* $\mathcal{V}$*.*

The Riesz–Fischer theorem then leads to an antilinear map $\theta \mapsto \tilde{\theta}$ from $\mathcal{V}^*$ to $\mathcal{H}$ (and hence to $\mathcal{V}$), defined by the property $\theta(w) = (\tilde{\theta}, w)$ for all $w \in \mathcal{H}$. When $\mathcal{V}^*$ separates points in $\mathcal{V}$, the range $\tilde{\mathcal{V}}^*$ of this map is dense in $\mathcal{H}$. To guarantee this, we assume that $\mathcal{V}$ is locally convex and Hausdorff. In any case, one obtains a positive sesquilinear form Q on $\mathcal{V}^*$ by

$$Q(\theta, \eta) := (\tilde{\eta}, \tilde{\theta}). \tag{1.49}$$

In the situation of the paragraph preceding 1.5.6, the dual of $\mathcal{V} = C(S)$ is the space of complex Radon measures μ on S with compact support. Hence

$$\tilde{\mu}(\rho) = \int_S d\overline{\mu(\sigma)}\, \mathcal{K}(\rho, \sigma) \tag{1.50}$$

from (1.47), so that the quadratic form Q is given by

$$Q(\mu, \nu) = \int_{S\times S} d\nu(\rho)\overline{d\mu(\sigma)}\,\mathcal{K}(\rho, \sigma). \tag{1.51}$$

In particular, the reproducing kernel itself is recovered by $\mathcal{K}(\rho, \sigma) = Q(\delta_\sigma, \delta_\rho)$, where δ_σ is the Dirac measure at σ (i.e., $\delta_\sigma(f) := f(\sigma)$), etc. Hence $\mathcal{K}$ is completely determined by the embedding $\mathcal{H} \hookrightarrow C(S)$. If, in addition, we suppose that $\mathcal{H} \subset L^2(S, d\mu)$ (defined with respect to some Radon measure μ), the projection $p : L^2(S, d\mu) \to \mathcal{H}$ is given by generalizing (1.43) to

$$p\Psi(\rho) = \int_S d\mu(\sigma)\,\mathcal{K}(\rho, \sigma)\Psi(\sigma). \tag{1.52}$$

We know that $\mathcal{K}(\sigma, \sigma) \geq 0$ for all $\sigma \in S$; let us further assume that $\mathcal{K}(\sigma, \sigma) > 0$ for all σ (equivalently, there are no points in S at which all elements of $\mathcal{H}$ vanish). Then one obtains a family of unit vectors Ψ^σ in $\mathcal{H}$, defined by

$$\Psi^\sigma(\rho) := \frac{\mathcal{K}(\rho, \sigma)}{\sqrt{\mathcal{K}(\sigma, \sigma)}}. \tag{1.53}$$

These satisfy the overcompleteness property

$$\int_S d\mu(\sigma)\,\mathcal{K}(\sigma, \sigma)(\Psi_1, \Psi^\sigma)(\Psi^\sigma, \Psi_2) = (\Psi_1, \Psi_2) \tag{1.54}$$

for all $\Psi_1,\ \Psi_2 \in \mathcal{H}$; cf. (1.37), and notice that the inner product $(\ ,\ )$ is the one in $\mathcal{H}$, inherited from $L^2(S, d\mu)$. Hence these unit vectors satisfy the key property (1.8) of coherent states; via the reproducing kernel they are eventually defined through the evaluation map.

The Hilbert space $\mathcal{H}_n$ is defined as the image of $\mathcal{H}$ under the unitary transformation $U : L^2(S, d\mu) \to L^2(S, d\mu_n)$ (where $d\mu_n(\sigma) := d\mu(\sigma)\mathcal{K}(\sigma, \sigma)$) defined by $U\Psi(\sigma) = \Psi(\sigma)/\sqrt{\mathcal{K}(\sigma, \sigma)}$. This space $\mathcal{H}_n$ has a reproducing kernel $\mathcal{K}_n$, namely

$$\mathcal{K}_n(\rho, \sigma) := (\Psi^\rho, \Psi^\sigma) = \frac{\mathcal{K}(\rho, \sigma)}{\sqrt{\mathcal{K}(\rho, \rho)\mathcal{K}(\sigma, \sigma)}}. \tag{1.55}$$

This kernel is normalized, in that $\mathcal{K}_n(\sigma, \sigma) = 1$ for all σ; equivalently, one has $\|\mathcal{K}_n(\cdot, \rho)\| = 1$ in $\mathcal{H}_n$. Its reproducing nature in $\mathcal{H}_n$ may be derived from the corresponding property of $\mathcal{K}$ in $\mathcal{H}$.

A Berezin operator $Q^B(f)$, depending on $f \in L^\infty(S)$, may then be defined on $\mathcal{H}$ (or $\mathcal{H}_n$) as in (1.44), with p given by (1.52) (with $\mathcal{K}$ replaced by $\mathcal{K}_n$). On $\mathcal{H}$ this operator then assumes the form (cf. (1.16))

$$Q^B(f) = \int_S d\mu(\sigma)\,\mathcal{K}(\sigma, \sigma)f(\sigma)[\Psi^\sigma], \tag{1.56}$$

whereas on $\mathcal{H}_n$ one has the same equation with Ψ^σ replaced by $U\Psi^\sigma = \mathcal{K}_n(\cdot, \sigma)$. It remains to be seen, of course, whether one can introduce $\hbar$ in a suitable way, so as to arrive at a pure state quantization or a strict quantization.

2 Quantization on Flat Space

2.1 The Heisenberg Group and its Representations

The manifold $P = T^*\mathbb{R}^n$ is equipped with its canonical cotangent bundle Poisson bracket I.(2.24). Regarding $\nabla f := (\partial f/\partial p, \partial f/\partial q)$ as a vector in $\mathbb{R}^{2n}$, and introducing the $2n \times 2n$ matrix

$$J = \begin{pmatrix} 0 & \mathbb{I} \\ -\mathbb{I} & 0 \end{pmatrix}, \tag{2.1}$$

we can write

$$\{f, g\} = (\nabla f, J \nabla g) \tag{2.2}$$

in terms of the natural inner product in $\mathbb{R}^{2n}$. This Poisson bracket is symplectic; as in I.(2.23), the symplectic form is

$$\omega = dq^i \wedge dp_i. \tag{2.3}$$

A central role in the study of $T^*\mathbb{R}^n$ is played by the so-called **Heisenberg group** H_n. A concrete form of its Lie algebra $\mathfrak{h}_n = \mathbb{R}^{2n+1}$ is obtained by taking the coordinate functions p_i, q^j as well as the unit function on $T^*\mathbb{R}^n$ as basis elements, and equating the Lie bracket with minus the Poisson bracket. This basis is traditionally denoted by $\{P_i, Q^j, Z\}_{i,j=1,\dots,n}$. The Lie brackets are

$$\begin{aligned} [P_i, P_j] &= [Q^i, Q^j] = 0; \\ [P_i, Q^j] &= -\delta_i^j Z; \\ [P_i, Z] &= [Q^j, Z] = 0. \end{aligned} \tag{2.4}$$

Definition 2.1.1. *The Heisenberg group $\tilde{H}_n$ is the unique connected and simply connected Lie group with Lie algebra $\mathfrak{h}_n$.*

Clearly, $\tilde{H}_n = \mathbb{R}^{2n+1}$ is nilpotent, and the exponential map $\mathrm{Exp} : \mathfrak{h}_n \to \tilde{H}_n$ is a diffeomorphism. Following the physics literature, we parametrize $\tilde{H}_n$ by coordinates $u, v \in \mathbb{R}^n$ and $s \in \mathbb{R}$ so that

$$(u, v, s) := \mathrm{Exp}(-uQ + vP + sZ), \tag{2.5}$$

where $uQ := u_i Q^i$, etc. The composition rule in $\tilde{H}_n$ then follows from (2.4) and the CBH-formula $\mathrm{Exp}(A)\mathrm{Exp}(B) = \mathrm{Exp}(A + B + \frac{1}{2}[A, B])$; the higher-order commutators vanish in this case. This yields

$$(u, v, s) \cdot (u', v', s') := (u + u', v + v', s + s' - \tfrac{1}{2}(uv' - vu')), \tag{2.6}$$

where $vu' = v^i u'_i$, etc. Regarding $w := (u, v)$ as a vector in the linear symplectic space $\mathbb{R}^{2n}$, equipped with the (symplectic) form $\omega = dv^i \wedge du_i$ (cf. (2.3)), we may write (2.6) as

$$(w, s) \cdot (w', s') = (w + w', s + s' + \tfrac{1}{2}\omega(w, w')). \tag{2.7}$$

One often works with a version of the Heisenberg group in which the s-coordinate is compactified; the group H_n is the quotient of $\tilde{H}_n$ by the discrete normal subgroup $(0, 0, 2\pi\mathbb{Z})$. Hence the projection $\tau : \tilde{H}_n \to H_n$ is given by $\tau(u, v, s) = (u, v, \exp(-is))$. The composition law in H_n then follows from (2.6) as

$$(u, v, z) \cdot (u', v', z') := \left(u + u', v + v', zz' e^{\frac{1}{2}i(uv' - vu')}\right). \tag{2.8}$$

A Lie algebra anti-isomorphism $\mathsf{P}^{\leq 1} \leftrightarrow \mathfrak{h}_n$ between the Poisson algebra $\mathsf{P}^{\leq 1}$ of polynomials on $T^*\mathbb{R}^n$ of degree ≤ 1 and the Heisenberg Lie algebra is given by

$$\mathcal{P}_{(u,v,s)}(p, q) = vp - uq + s \longleftrightarrow vP - uQ + sZ. \tag{2.9}$$

One may regard $X \in \mathfrak{h}_n$ as a function $\tilde{X}$ on the dual $\mathfrak{h}_n^*$ by putting $\tilde{X}(\theta) := \theta(X)$ for $\theta \in \mathfrak{h}_n^*$; this yields an inclusion $\mathfrak{h}_n \subset C^\infty(\mathfrak{h}_n^*)$. We use coordinates (p, q, c) on $\mathfrak{h}_n^* = \mathbb{R}^{2n+1}$ (where $p, q \in \mathbb{R}^n$ and $c \in \mathbb{R}$), which represent the point $p\hat{P} + q\hat{Q} + c\hat{Z}$. Here $\{\hat{P}_i, \hat{Q}^j, \hat{Z}\}_{i,j=1,\dots,n}$ is the basis of $\mathfrak{h}_n^*$ dual to the given one in $\mathfrak{h}_n$. The functions $\tilde{P}_i, \tilde{Q}^j$ then coincide with the coordinate functions p_i, q^j.

The differentials of all functions $\tilde{X}$ span the cotangent bundle $T^*\mathfrak{h}_n^*$, so that a possible Poisson structure on $\mathfrak{h}_n^*$ is determined by the Poisson brackets of the $\tilde{X}$. Thus one may put

$$\{\tilde{X}, \tilde{Y}\}_- := -\widetilde{[X, Y]}. \tag{2.10}$$

The reason for the minus sign will become clear in III.1.1. This leads to the Poisson bracket (we omit the argument (p, q, c))

$$\{f, g\}_- = c\left(\frac{\partial f}{\partial p_i}\frac{\partial g}{\partial q^i} - \frac{\partial f}{\partial q^i}\frac{\partial g}{\partial p_i}\right), \tag{2.11}$$

cf. I.(2.24). The symplectic leaves of $\mathfrak{h}_n^*$ come in two types. Firstly, one has the manifolds $T^*\mathbb{R}^n_c := \mathbb{R}^{2n} \times \{c\}$ for $c \neq 0$, with symplectic form $\omega_c = c\, dq^i \wedge dp_i$. The "usual" $T^*\mathbb{R}^n$ with Poisson bracket I.(2.24) is the leaf corresponding to $c = 1$. Secondly, each point (p, q) in $T^*\mathbb{R}^n \times \{0\}$ is a leaf.

There is a different way of looking at these leaves. The so-called **coadjoint action** Co of $\tilde{H}_n$ on $\mathfrak{h}_n^*$ is defined by

$$(\mathrm{Co}(u, v, s)\theta)(Y) := \theta(\mathrm{Ad}((u, v, s)^{-1})Y), \tag{2.12}$$

where Ad is the adjoint action of $\tilde{H}_n$ on $\mathfrak{h}_n$. The CBH-formula yields

Proposition 2.1.2. *The coadjoint action of the Heisenberg group is given by*

$$\mathrm{Co}(u, v, s)(p, q, c) = (p + cu, q + cv, c). \tag{2.13}$$

Accordingly, the orbits in $\mathfrak{h}_n^*$ under the coadjoint action coincide with the symplectic leaves of the Poisson structure (2.10).

The result may be recast in the language of Chapter I.

Proposition 2.1.3. *Unless it is defined on a zero-dimensional space, any irreducible representation π^{cl} of the Poisson algebra $C^\infty(\mathfrak{h}_n^*)$ associated to the*

Heisenberg group is equivalent to π_c^{cl}, *for some real* $c \neq 0$, *defined on the symplectic manifold* $(T^*\mathbb{R}^n, c\, dq^i \wedge dp_i)$ *by*

$$\pi_c^{\mathrm{cl}}(f) = f\restriction_c;$$
$$f\restriction_c(p,q) := f(p,q,c). \tag{2.14}$$

If π^{cl} *is zero-dimensional, there is a point* $(p,q) \in T^*\mathbb{R}^n$ *such that*

$$\pi^{\mathrm{cl}}_{(p,q)}(f) = f(p,q,0). \tag{2.15}$$

This is immediate from Theorem I.2.6.7. ■

The corresponding representation of $\mathfrak{h}_n \subset C^\infty(\mathfrak{h}_n^*)$ on $T^*\mathbb{R}^n_c$ is simply

$$P_i \mapsto p_i;$$
$$Q^j \mapsto q^j;$$
$$Z \mapsto c 1_{T^*\mathbb{R}^n}. \tag{2.16}$$

In particular, $C^\infty(T^*\mathbb{R}^n)$ with the canonical Poisson structure I.(2.24) may be seen as the representative $\pi_1^{\mathrm{cl}}(C^\infty(\mathfrak{h}_n^*))$.

Proposition 2.1.3 has an exact parallel in quantum mechanics. Consider the following family of representations of $\tilde{H}_n$. For each real $\lambda \neq 0$, construct the operator $U_\lambda^S(u,v,t)$ on the Hilbert space $L^2(\mathbb{R}^n)$ by

$$U_\lambda^S(u,v,s)\Psi(x) := e^{-i\lambda(s+\frac{1}{2}uv)}e^{iux}\Psi(x-\lambda v). \tag{2.17}$$

It is easily checked that the U_λ^S are unitary, and indeed furnish a representation of $\tilde{H}_n$, called the **Schrödinger representation**. The irreducibility of U_λ^S will be proved in 2.5.5. We see that $U_\lambda^S(0,0,s) = \exp(-i\lambda s)\mathbb{I}$; hence for $\lambda \in \mathbb{Z}$ the representation U_λ^S is defined on H_n as well, satisfying

$$U_n^S(0,0,z) = z^n\mathbb{I}. \tag{2.18}$$

A useful equivalent version of U_λ^S is given by

$$U_\lambda(u,v,s)\Psi(x) := e^{-i\lambda(s+\frac{1}{2}uv-ux)}\Psi(x-v); \tag{2.19}$$

one has $VU_\lambda^S V^* = U_\lambda$ for the unitary $V : L^2(\mathbb{R}^n) \to L^2(\mathbb{R}^n)$ defined by $V\Psi(x) = \lambda^{n/2}\Psi(\lambda x)$. The corresponding representations of the Lie algebra $\mathfrak{h}_n$ are given by (cf. III.(1.69))

$$dU_\lambda^S(Q^i) = -ix^i;$$
$$dU_\lambda^S(P_j) = -\lambda\frac{\partial}{\partial x^i};$$
$$dU_\lambda^S(Z) = -i\lambda\mathbb{I} \tag{2.20}$$

and

$$dU_\lambda(Q^i) = -i\lambda x^i;$$
$$dU_\lambda(P_j) = -\frac{\partial}{\partial x^j};$$
$$dU_\lambda(Z) = -i\lambda\mathbb{I}, \tag{2.21}$$

respectively; here x^i is meant as a multiplication operator, i.e., $(x^i\Psi)(x) = x^i\Psi(x)$. These operators are defined and essentially self-adjoint on $\mathcal{S}(\mathbb{R}^n) \subset L^2(\mathbb{R}^n)$, on which $[dU_\lambda^{(S)}(X), dU_\lambda^{(S)}(Y)] = dU_\lambda^{(S)}([X, Y])$ for all $X, Y \in \mathfrak{h}_n$.

The representation U_λ is of particular use for $\lambda = 1/\hbar$. For later convenience, we introduce the **Weyl operator**

$$U_{\frac{1}{\hbar}}(p, q) := U_{\frac{1}{\hbar}}(p, q, 0) = e^{\frac{i}{\hbar}(pQ_\hbar^S - qP_\hbar^S)}, \tag{2.22}$$

where

$$Q_\hbar^{S,i} := i\hbar dU_{\frac{1}{\hbar}}(Q^i) = idU_\hbar^S(Q^i) = x^i \tag{2.23}$$

and

$$P_{\hbar,i}^S := i\hbar dU_{\frac{1}{\hbar}}(P_i) = idU_\hbar^S(P_i) = -i\hbar\frac{\partial}{\partial x^i} \tag{2.24}$$

are the physicists' **position operator** and **momentum operator**, respectively; cf. (2.20) and (2.37). These operators are both defined and essentially self-adjoint on $\mathcal{S}(\mathbb{R}^n)$, on which domain one has the **canonical commutation relations**

$$[P_{\hbar,i}^S, Q_\hbar^{S,j}] = -i\hbar\delta_i^j \mathbb{I}; \tag{2.25}$$

cf. (2.4). One might add here that

$$i\hbar dU_{\frac{1}{\hbar}}(Z) = \mathbb{I}. \tag{2.26}$$

Theorem 2.1.4. *Unless it is one-dimensional, any irreducible representation U of $\tilde{H}_n$ is equivalent to U_λ^S for some $\lambda \neq 0$. When U is one-dimensional, there is a point $(p, q) \in T^*\mathbb{R}^n$ such that U equals*

$$U_{(p,q)}(u, v, s) = e^{i(uq - vp)}. \tag{2.27}$$

When $U(0, 0, s) = \mathbb{I}$ for all $s \in \mathbb{R}$, the representation must be one-dimensional, so that (2.27) is a restatement of the representation theory of the abelian group $\mathbb{R}^{2n}$. A proof of the remainder of this celebrated theorem will be given at the end of III.3.7. Another appropriate proof is obtained by combining either Corollary 2.6.7 or Proposition III.1.8.4 with Corollary I.2.2.6; the statement in 2.1.4 concerning $\lambda \neq 0$ is equivalent to the uniqueness of the irreducible representation of the C^*-algebra of compact operators. □

2.2 *The Metaplectic Representation*

As we have seen in the previous section, the Heisenberg group is closely related to the Poisson algebra $\mathsf{P}^{\leq 1}$ of polynomials on $T^*\mathbb{R}^n$ of degree ≤ 1. At the next level, the Poisson algebra P^2 of quadratic polynomials on $T^*\mathbb{R}^n$ turns out to be anti-isomorphic to the Lie algebra of the **symplectic group** $Sp(n, \mathbb{R})$. This group consists of the linear Poisson isomorphisms of $T^*\mathbb{R}^n \simeq \mathbb{R}^{2n}$; a matrix $M \in GL(2n, \mathbb{R})$ lies in $Sp(n, \mathbb{R})$ iff $M^T JM = J$ (cf. (2.1)). For the Lie algebra this means that a $2n \times 2n$ matrix X lies in $\mathfrak{sp}(n, \mathbb{R})$ iff $JX + X^T J = 0$

(equivalently, XJ is symmetric). The maximal compact subgroup of $Sp(n, \mathbb{R})$ is $Sp(n, \mathbb{R}) \cap O(2n)$; if $\mathbb{R}^{2n}$ is identified with $\mathbb{C}^n$ through

$$z^j = \frac{(q^j + ip^j)}{\sqrt{2}} \tag{2.28}$$

(where $p^j := p_j$), then $Sp(n, \mathbb{R}) \cap O(2n) = U(n)$. It follows from the theory of noncompact semisimple Lie groups that the homotopic properties of $Sp(n, \mathbb{R})$ are determined by its maximal compact subgroup; hence $Sp(n, \mathbb{R})$ is connected, but not simply connected, since $\pi_1(Sp(n, \mathbb{R})) = \pi_1(U(n)) = \mathbb{Z}$. Note that in terms of complex coordinates the symplectic form (2.3) reads

$$\omega = i dz^i \wedge d\overline{z}^i. \tag{2.29}$$

Hence in terms of the usual inner product on $\mathbb{C}^n$ one has

$$\omega(z, z') = 2\,\mathrm{Im}\,(z, z'). \tag{2.30}$$

This expression renders it self-evident that $U(n) \subset Sp(n, \mathbb{R})$. With $\partial := \partial/\partial z$ and $\overline{\partial} := \partial/\partial\overline{z}$, the Poisson bracket I.(2.24) now reads

$$\{f, g\} = i(\partial f \overline{\partial} g - \overline{\partial} f \partial g). \tag{2.31}$$

Further to the notation $w = (u, v)$, we put $\sigma := (p, q)$; also recall (2.1). For $X \in \mathfrak{sp}(n, \mathbb{R})$ we define the quadratic polynomial

$$\mathcal{P}_X(\sigma) := \tfrac{1}{2}(JX\sigma, \sigma), \tag{2.32}$$

where the inner product is the usual one in $\mathbb{R}^{2n}$. Using (2.2), for $X, X' \in \mathfrak{sp}(n, \mathbb{R})$ one easily verifies that

$$\{\mathcal{P}_X, \mathcal{P}_{X'}\} = -\mathcal{P}_{[X,X']}, \tag{2.33}$$

which proves that (2.32), which is clearly bijective, defines a Lie algebra anti-isomorphism between P^2 and $\mathfrak{sp}(n, \mathbb{R})$.

The group $Sp(n, \mathbb{R})$ acts on $\tilde{H}_n$: the matrix $M \in Sp(n, \mathbb{R})$ maps $(w, s) \in \tilde{H}_n$ to (Mw, s). Writing h for (w, s), we say simply that M maps h into Mh. We may therefore build the semidirect product $Sp(n, \mathbb{R}) \ltimes \tilde{H}_n$, whose elements are pairs (M, h), with $M \in Sp(n, \mathbb{R})$ and $h \in \tilde{H}_n$. The group multiplication is given by $(M, h) \cdot (M', h') := (MM', h \cdot Mh')$, where the product $\cdot$ in $\tilde{H}_n$ is given by (2.6). Note, in particular, that

$$(M, 0) \cdot (e, h) \cdot (M^{-1}, 0) = (e, Mh), \tag{2.34}$$

where e and 0 are the identity elements in $Sp(n, \mathbb{R})$ and $\tilde{H}_n$, respectively. The "mixed" Lie bracket in the Lie algebra $\mathfrak{sp}(n, \mathbb{R}) \ltimes \mathfrak{h}_n$ is

$$[M, (w, s)] = (Mw, s). \tag{2.35}$$

Let $\mathcal{P}_1, \mathcal{P}_2$ be polynomials of degree ≤ 2 in (p_i, q^j). The space $\mathsf{P}^{\leq 2}$ of such polynomials is easily seen to be closed under the Poisson bracket I.(2.24).

Proposition 2.2.1.

- *Under the correspondence (2.9) the Poisson algebra* $\mathsf{P}^{\leq 1}$ *of polynomials of degree* ≤ 1 *is anti-isomorphic to the Lie algebra* $\mathfrak{h}_n$ *of the Heisenberg group* $\tilde{H}_n$.
- *The Poisson algebra* P^2 *of quadratic polynomials is anti-isomorphic to the Lie algebra* $\mathfrak{sp}(n, \mathbb{R})$ *of* $Sp(n, \mathbb{R})$ *under the correspondence* $\mathcal{P}_M(\sigma) \leftrightarrow M$.
- *By linear extension of the preceding two items, the Poisson algebra* $\mathsf{P}^{\leq 2}$ *of polynomials of degree* ≤ 2 *is anti-isomorphic to the Lie algebra* $\mathfrak{sp}(n, \mathbb{R}) \ltimes \mathfrak{h}_n$ *of the semidirect product* $Sp(n, \mathbb{R}) \ltimes \tilde{H}_n$.

The first item was shown in the previous section. The second is proved by (2.33). The third claim follows from (2.35). ∎

One can easily solve the equations of motion for Hamiltonians in $\mathsf{P}^{\leq 2}$. The Hamiltonian flow generated by $\mathcal{P}_{(w,s)}$ (cf. (2.9)) is $\sigma(t) = \sigma + tw$ (cf. (2.13) with $c = 1$), and the flow generated by $\mathcal{P}_X$ is $\sigma(t) = \text{Exp}(tX)\sigma$. These flows are compatible with the natural action ρ^0 of $Sp(n, \mathbb{R}) \ltimes \tilde{H}_n$ on $T^*\mathbb{R}^n$, under which $(M, (w, s))$ maps σ to $\rho^0_{(M,(w,s))}(\sigma) = M\sigma + w$. If $\mathcal{P}_X \leftrightarrow X$ under the isomorphism of 2.2.1 ($X \in \mathfrak{sp}(n, \mathbb{R}) \ltimes \mathfrak{h}_n$), one verifies that $\text{Exp}(tX)$ maps σ to $\text{Exp}(tX)\sigma = \sigma(t)$, where $\sigma \mapsto \sigma(t)$ is the Hamiltonian flow generated by $\mathcal{P}_X$ on $T^*\mathbb{R}^n$. Hence

$$\text{Exp}(X)\sigma = \sigma(1). \tag{2.36}$$

We will now construct an important integrable Hilbert space representation of $\mathfrak{sp}(n, \mathbb{R})$. Let $\mathcal{P}(p_i, q^j, 1)$ be a polynomial on $T^*\mathbb{R}^n$. We define

$$\mathcal{Q}^W_\hbar(\mathcal{P}(p_i, q^j, 1)) := \lambda[\mathcal{P}(P^S_{\hbar,i}, Q^{S,j}_\hbar), \mathbb{I}], \tag{2.37}$$

cf. (2.24) and (2.23). This expression means that one substitutes $P^S_\hbar$, $Q^S_\hbar$, for p, q in $\mathcal{P}$, and symmetrizes; thus $\lambda[\ldots]$ denotes complete symmetrization. For example, $\lambda[A_1, \ldots, A_n] = \sum_{\pi \in S_n} A_{\pi(1)} \cdots A_{\pi(n)}/n!$, where the sum is over all $n!$ elements π of the permutation group S_n.

Given its construction from $U^S_\hbar(\text{Exp}(uQ - vP))$, it follows from standard representation theory that $\mathcal{Q}^W_\hbar(\mathcal{P})$ is well-defined as an unbounded operator on $L^2(\mathbb{R}^n)$ with domain $\mathcal{S}(\mathbb{R}^n)$. If $\mathcal{P}$ is real, then $\mathcal{Q}^W_\hbar(\mathcal{P})$ is symmetric on this domain.

Proposition 2.2.2.

- *Restricted to at most quadratic polynomials,* $\mathcal{Q}^W_\hbar$ *is a Lie algebra homomorphism, in that for all* $\mathcal{P}_1, \mathcal{P}_2 \in \mathsf{P}^{\leq 2}$ *one has*

$$\frac{i}{\hbar}\left[\mathcal{Q}^W_\hbar(\mathcal{P}_1), \mathcal{Q}^W_\hbar(\mathcal{P}_2)\right] = \mathcal{Q}^W_\hbar(\{\mathcal{P}_1, \mathcal{P}_2\}). \tag{2.38}$$

- *Hence* $d\rho^h$, *defined by*

$$d\rho^h(X) := -\frac{i}{\hbar}\mathcal{Q}^W_\hbar(\mathcal{P}_X) \tag{2.39}$$

(where $X \in \mathfrak{sp}(n, \mathbb{R}) \ltimes \mathfrak{h}_n$ *corresponds to* $\mathcal{P}_X$ *under the anti-isomorphism between* $\mathfrak{sp}(n, \mathbb{R}) \ltimes \mathfrak{h}_n$ *and* $\mathsf{P}^{\leq 2}$*, cf. 2.2.1), furnishes a representation of the Lie*

algebra $\mathfrak{sp}(n, \mathbb{R}) \ltimes \mathfrak{h}_n$ *as unbounded operators on the common invariant dense domain* $\mathcal{S}(\mathbb{R}^n) \subset L^2(\mathbb{R}^n)$.

- *For each* $X \in \mathfrak{sp}(n, \mathbb{R}) \ltimes \mathfrak{h}_n$ *the operator* $id\rho^\hbar(X)$ *is essentially self-adjoint on this domain.*
- *There exists a double covering* $Mp(n, \mathbb{R})$ *of* $Sp(n, \mathbb{R})$ *(known as the* **metaplectic group***) and a representation* $\rho^\hbar$ *of* $Mp(n, \mathbb{R}) \ltimes \tilde{H}_n$ *(where the action of* $Mp(n, \mathbb{R})$ *on* $\tilde{H}_n$ *factors through* $Sp(n, \mathbb{R})$ *in the obvious way) on* $L^2(\mathbb{R}^n)$*, whose derived representation of* $\mathfrak{sp}(n, \mathbb{R}) \ltimes \mathfrak{h}_n$ *is* $d\rho^\hbar$.
- *Restricted to* $\tilde{H}_n$*, the representation* $\rho^\hbar$ *coincides with* $U_{1/\hbar}(\tilde{H}_n)$.

The restriction of $\rho^\hbar$ to $Sp(n, \mathbb{R})$ is called the **metaplectic representation**.

A simple calculation shows that the commutation relations (2.38) are satisfied on $\mathcal{S}(\mathbb{R}^n)$; hence the first claim follows from 2.2.1. The equation $d\rho^\hbar(X) = dU_{1/\hbar}(X)$ for $X \in \mathfrak{h}_n$ is immediate from (2.23), (2.24), and (2.26).

A technical result in functional analysis, involving the existence of a dense set of analytic vectors (here given by the linear span of the Hermite polynomials), shows that $d\rho^\hbar(\mathfrak{sp}(n, \mathbb{R}))$ exponentiates to a representation $\rho^\hbar(\tilde{Sp}(n, \mathbb{R}))$, where $\tilde{Sp}(n, \mathbb{R})$ is the unique connected and simply connected covering group of $Sp(n, \mathbb{R})$ (one has $\tilde{Sp}(n, \mathbb{R})/Sp(n, \mathbb{R}) \simeq \mathbb{Z}$). This argument also leads to the essential self-adjointness property mentioned. It can be shown that the metaplectic representation $\rho^\hbar$ is double-valued on $Sp(n, \mathbb{R})$ (that is, $\rho^\hbar(M)\rho^\hbar(M') = \pm\rho^\hbar(MM')$, where the sign depends on M and M'), so that there is a double covering group $Mp(n, \mathbb{R})$ of $Sp(n, \mathbb{R})$ on which $\rho^\hbar$ is single-valued (i.e., is a representation). □

From 2.2.2, (2.22), and (2.34) we have the equivariance property

$$\rho^\hbar(M)U_{\frac{1}{\hbar}}(\sigma)\rho^\hbar(M)^* = U_{\frac{1}{\hbar}}(M\sigma), \tag{2.40}$$

where $M \in Sp(n, \mathbb{R})$. We may reformulate this result in terms of dynamics. We regard a real polynomial h on $T^*\mathbb{R}^n$ as a classical Hamiltonian, denoting its flow by $\sigma \mapsto \sigma(t)$. Its quantization, the quantum Hamiltonian $H_\hbar$, is taken to be the unbounded operator

$$H_\hbar := \mathcal{Q}_\hbar^W(h(P_\hbar^S, Q_\hbar^S)), \tag{2.41}$$

cf. (2.37), (2.24), and (2.23). Let $h = \mathcal{P}_X \in \mathsf{P}^2$. From (2.39) we see that $H_\hbar = i\hbar d\rho^\hbar(X)$, so according to (2.36) we can rewrite (2.40) as

$$e^{itH_\hbar/\hbar}U_{\frac{1}{\hbar}}(\sigma)e^{-itH_\hbar/\hbar} = U_{\frac{1}{\hbar}}(\sigma(-t)). \tag{2.42}$$

We turn to the reducibility of $\rho^\hbar(Sp(n, \mathbb{R}))$. The following result will be of central importance in the construction of the Weyl quantization map in 2.5.

Lemma 2.2.3. *The* **parity operator** P *on* $L^2(\mathbb{R}^n)$*, defined by*

$$P\Psi(x) := \Psi(-x), \tag{2.43}$$

commutes with all $\rho^\hbar(M)$, $M \in Sp(n, \mathbb{R})$*. The eigenspaces* $L^2(\mathbb{R}^n)_\pm \subset L^2(\mathbb{R}^n)$*, characterized by the property* $PL^2(\mathbb{R}^n)_\pm = \pm\mathbb{I}L^2(\mathbb{R}^n)_\pm$*, are irreducible under* $\rho^\hbar(Sp(n, \mathbb{R}))$*. Hence the commutant of* $\rho^\hbar(Sp(n, \mathbb{R}))$ *is spanned by* P *and* $\mathbb{I}$.

Simple computations show that $[P, d\rho^{\hbar}(X)] = 0$ for all $X \in \mathfrak{sp}(2, \mathbb{R})$. Since $Sp(n, \mathbb{R})$ is connected and the exponential map is onto, it follows that $[P, \rho^{\hbar}(M)] = 0$ for all $M \in Sp(n, \mathbb{R})$. Hence $\rho^{\hbar}(Sp(n, \mathbb{R}))$ is reducible, and $\oplus_{\pm} L^2(\mathbb{R}^n)_{\pm}$ obviously decomposes $L^2(\mathbb{R}^n)$. The fact that the $L^2(\mathbb{R}^n)_{\pm}$ are irreducible follows from an uninteresting technical argument. □

This lemma implies that for all $M \in Sp(n, \mathbb{R})$ one has

$$\rho^{\hbar}(M) P \rho^{\hbar}(M)^* = P. \tag{2.44}$$

2.3 Berezin Quantization on Flat Space

After this preparatory material we turn to the quantization of $T^*\mathbb{R}^n$. A suitable choice of the Poisson algebra we wish to quantize turns out to be $\tilde{\mathfrak{A}}^0_{\mathbb{R}} := C^{\infty}_c(T^*\mathbb{R}^n, \mathbb{R})$; this is a dense subspace of $\mathfrak{A}^0_{\mathbb{R}} = C_0(T^*\mathbb{R}^n, \mathbb{R})$ under the sup-norm. We write $\sigma = (p, q)$; the Poisson bracket is given by I.(2.24).

We now construct a Berezin quantization of $\mathfrak{A}^0_{\mathbb{R}}$ from a pure state quantization, as outlined in 1.3. The strategy is generic.

Proposition 2.3.1. *Put $I = \mathbb{R}$ and $\mathcal{H}_{\hbar} := L^2(\mathbb{R}^n)$ for all $\hbar \neq 0$. For each $(p, q) \in T^*\mathbb{R}^n$, define a unit vector $\Psi^{(p,q)}_{\hbar} \in \mathcal{H}_{\hbar}$ by*

$$\Psi^{(p,q)}_{\hbar} := U_{\frac{1}{\hbar}}(p, q)\Psi^0_{\hbar}; \tag{2.45}$$

$$\Psi^0_{\hbar}(x) := (\pi\hbar)^{-n/4} e^{-x^2/(2\hbar)}, \tag{2.46}$$

cf. (2.22). Explicitly, one has

$$\Psi^{(p,q)}_{\hbar}(x) = (\pi\hbar)^{-n/4} e^{-\frac{1}{2}ipq/\hbar} e^{ipx/\hbar} e^{-(x-q)^2/(2\hbar)}. \tag{2.47}$$

Denote the projection of $\Psi^{(p,q)}_{\hbar} \in \mathbb{S}\mathcal{H}_{\hbar}$ to $\mathbb{P}\mathcal{H}_{\hbar}$ by $\psi^{(p,q)}_{\hbar}$. Then the choices

$$q^B_{\hbar}(p, q) := \psi^{(p,q)}_{\hbar}, \tag{2.48}$$

$$d\mu_{\hbar}(p, q) := \frac{d^n p d^n q}{(2\pi\hbar)^n} \tag{2.49}$$

yield a coherent pure state quantization of $T^*\mathbb{R}^n$.

This is established by simple computations. In fact, (1.10), without the limit, and (1.8) are valid for any unit vector $\Psi^0_{\hbar}$; the explicit choice (2.46) is used only to prove (1.9). Here the decisive intermediate result may be expressed in terms of complex variables (see (2.28)) as

$$(\Psi^{(w)}_{\hbar}, \Psi^{(z)}_{\hbar}) = \mathcal{K}_{\hbar}(\overline{z}, w) = e^{\left(-\frac{1}{2}w\overline{w} - \frac{1}{2}z\overline{z} + w\overline{z}\right)/\hbar}; \tag{2.50}$$

cf. (1.38). Hence (1.11) is immediate from the corollary

$$p\left(q^B_{\hbar}(w), q^B_{\hbar}(z)\right) = e^{-|z-w|^2/\hbar}.$$ ■

The Berezin quantization $\mathcal{Q}^B_{\hbar}$ defined by (2.48) (cf. 1.3.4) is given by

$$\mathcal{Q}^B_{\hbar}(f) = \int_{T^*\mathbb{R}^n} \frac{d^n p d^n q}{(2\pi\hbar)^n} f(p, q)[\Psi^{(p,q)}_{\hbar}], \tag{2.51}$$

where $f \in L^\infty(T^*\mathbb{R}^n)$. For $f \in C_0(T^*\mathbb{R}^n, \mathbb{R})$ the Gelfand transform (1.15) is

$$\widehat{\mathcal{Q}_\hbar^B(f)}(\psi) = \int_{T^*\mathbb{R}^n} \frac{d^n p d^n q}{(2\pi\hbar)^n}\, p(\psi_\hbar^{(p,q)}, \psi) f(p,q). \tag{2.52}$$

In terms of the complex variables (2.28), the measure (2.49) reads

$$d\mu_\hbar(z, \overline{z}) = \frac{d^n z d^n \overline{z}}{(2\pi\hbar i)^n}. \tag{2.53}$$

Proposition 2.3.2. *In the context of Proposition 1.5.2 (in which, using complex coordinates, $\mathcal{H}^\hbar = L^2(\mathbb{C}^n, \mu_\hbar)$), the Hilbert space $\tilde{\mathcal{H}}_\hbar$ consists of all functions of the type $\tilde{\Psi}(z, \overline{z}) = \exp(-z\overline{z}/(2\hbar))\Psi(\overline{z})$, where Ψ is an entire function for which $\int_{\mathbb{C}^n} d^n z d^n \overline{z} \exp(-z\overline{z})|\Psi(\overline{z})|^2 < \infty$.*

We call the space of functions of the stated type $\breve{H}^2(\mathbb{C}^n)$; elementary analysis shows that it has $\mathcal{K}_\hbar$ (cf. (2.50)) as a reproducing kernel. By the argument given in the proof of 1.5.2, norm-convergence in $\breve{H}^2(\mathbb{C}^n)$ implies uniform convergence. This shows that $\breve{H}^2(\mathbb{C}^n)$ is complete. Moreover, the fact that entire functions are given by Taylor series (uniformly convergent on compact sets) shows that the functions $\{\tilde{\Phi}_\alpha\}_{|\alpha|=0}^\infty$, where $\alpha := (\alpha_1, \ldots, \alpha_n)$ is a multi-index, with $|\alpha| := \alpha_1 + \cdots + \alpha_n$, and

$$\Phi_\alpha(\overline{z}) := \left(\prod_{i=1}^n \hbar^{\alpha_i} \alpha_i! \right)^{-1/2} \overline{z}^\alpha, \tag{2.54}$$

where $\overline{z}^\alpha := \overline{z}_1^{\alpha_1} \cdots \overline{z}_n^{\alpha_n}$, form an orthonormal basis in $\tilde{\mathcal{H}}_\hbar$. The orthonormality follows from an elementary computation in polar coordinates.

Using (2.47) and (2.28), we write (1.42) as

$$W\Psi(z, \overline{z}) = (\pi\hbar)^{-n/4} e^{-(z\overline{z}+\overline{z}^2)/(2\hbar)} \int d^n x\, \Psi(x) e^{\left(-\frac{1}{2}x^2 + \sqrt{2}x\overline{z}\right)/\hbar}. \tag{2.55}$$

The integral converges uniformly in z on compact sets, so $W\Psi$ is $\exp(-z\overline{z}/(2\hbar))$ times an entire function in $\overline{z}$. The square-integrability of $W\Psi$ follows from the fact that W is a partial isometry. Hence $\tilde{\mathcal{H}}_\hbar \subseteq \breve{H}^2(\mathbb{C}^n)$. For example, $\mathcal{K}_\hbar(\cdot, w) \in \tilde{\mathcal{H}}_\hbar$ for each $w \in \mathbb{C}^n$, as it should be. One computes $(\Phi_\alpha, \mathcal{K}_\hbar(\cdot, w)) = 2^{-n/2}\exp(-\frac{1}{2}w\overline{w})w^\alpha$. It follows that $(\Phi, \mathcal{K}_\hbar(\cdot, w)) = 0$ for all w implies $\Phi = 0$, so that the span of the collection of functions $\mathcal{K}_\hbar(\cdot, w) \in \tilde{\mathcal{H}}_\hbar$, $w \in \mathbb{C}^n$, is dense in $\tilde{\mathcal{H}}_\hbar$. Since these are the images of the coherent states in $\mathcal{H}_\hbar$ under W, the proof is complete. ■

With hindsight, we can now formulate a unitarily equivalent formulation of Berezin quantization on $T^*\mathbb{R}^n$: We start with the Hilbert space $\overline{H}_\hbar^2(\mathbb{C}^n)$ of conjugate-entire functions on $\mathbb{C}^n$, whose inner product with respect to the Gaussian measure on $\mathbb{C}^n$ is finite, namely

$$(\Psi, \Phi) := \hbar^{-n} \int_{\mathbb{C}^n} \frac{d^n z d^n \overline{z}}{(2\pi i)^n} e^{-z\overline{z}/\hbar} \Psi(z)\Phi(\overline{z}) < \infty. \tag{2.56}$$

The functions $\tilde{\Phi}_\alpha$ occurring in the proof of 2.3.2 form an orthonormal basis of $\overline{H}^2_\hbar(\mathbb{C}^n)$. The latter plays the role of $\mathcal{H}$ in Definition 1.5.3. The Berezin quantization in this realization, which we denote by $\breve{Q}^B_\hbar$ to avoid confusion with the equivalent version (2.51), is then given by

$$\breve{Q}^B_\hbar(f) = pfp. \tag{2.57}$$

Here f is regarded as a multiplication operator on $\breve{\mathcal{H}}^\hbar := L^2(\mathbb{C}^n, \hbar^{-n}\mu^\hbar_G)$ (where $\mu^\hbar_G$ is the Gaussian measure occurring in (2.56)), and p is the projection onto the subspace $\overline{H}^2_\hbar(\mathbb{C}^n)$ of entire functions of $\overline{z}$ in $\breve{\mathcal{H}}^\hbar$. Compare with (1.44).

The Hilbert space $\overline{H}^2_\hbar(\mathbb{C}^n)$ has an (unnormalized) reproducing kernel, the so-called **Bergman kernel**, given by

$$\mathcal{K}^u_\hbar(\overline{z}, w) = e^{w\overline{z}/\hbar}. \tag{2.58}$$

Hence by Proposition 1.5.4 each evaluation map $E_z : \Psi \mapsto \Psi(\overline{z})$ is continuous. The coherent states $\Psi^w_\hbar$ are defined as in (1.53). As in the passage from (1.44) to (1.56), we may then rewrite (2.57) as

$$\breve{Q}^B_\hbar(f)\Psi(\overline{z}) = \hbar^{-n}\int_{\mathbb{C}^n}\frac{d^n w d^n\overline{w}}{(2\pi i)^n}e^{-w\overline{w}/\hbar}\mathcal{K}^u_\hbar(\overline{z}, w)f(w, \overline{w})\Psi(\overline{w}). \tag{2.59}$$

As explained in 1.3, this can be transferred to the Hilbert space $\mathcal{H}_n$, which possesses the normalized reproducing kernel (1.55). In the present setting, $\mathcal{H}_n$ coincides with $\tilde{\mathcal{H}}_\hbar$ (cf. 2.3.2), since the rescaled measure μ_n is just the Liouville measure times $\hbar^{-n}$. Hence we indeed have

$$\tilde{Q}^B_\hbar(f) = \int_{\mathbb{C}^n} d\mu_\hbar(w, \overline{w}) f(w, \overline{w})[\Psi^w_\hbar]. \tag{2.60}$$

There is yet another, closely related, way of looking at Berezin quantization, or rather the coherent states behind it. For any Hilbert space $\mathcal{K}$, with inner product $(\,,\,)_\mathcal{K}$, we introduce the **exponential Hilbert space**, or **bosonic Fock space**, $\exp(\mathcal{K})$ as follows. Let the Hilbert space $\otimes^l_s\mathcal{K}$ be the symmetrized tensor product of l copies of $\mathcal{K}$; this is the invariant subspace of $\otimes^l\mathcal{K}$ under the natural action of the permutation group S_l. The closure of the direct sum of all $\otimes^l\mathcal{K}$ is

$$\exp(\mathcal{K}) := \bigoplus_{l=0}^{\infty}\otimes^l_s\mathcal{K}. \tag{2.61}$$

This space is separable iff $\mathcal{K}$ is. The element $1 \in \mathbb{C} = \otimes^0\mathcal{K}$ is denoted by Ω; elements of $\mathcal{K}$ are called w or z. We define a map $\sqrt{\mathrm{Exp}} : \mathcal{K} \to \exp(\mathcal{K})$ by

$$\sqrt{\mathrm{Exp}}(w) := \sum_{l=0}^{\infty}\frac{\otimes^l w}{\sqrt{l!}} = \Omega + w + \frac{w\otimes w}{\sqrt{2!}} + \cdots; \tag{2.62}$$

this is called an **exponential vector**. This map is clearly injective, since the component of $\sqrt{\mathrm{Exp}}(w)$ in $\mathcal{K} \subset \exp(\mathcal{K})$ is w itself. The inner product of two exponential vectors is

$$\left(\sqrt{\mathrm{Exp}}(w), \sqrt{\mathrm{Exp}}(z)\right) = e^{(w,z)_\mathcal{K}}. \tag{2.63}$$

For one thing, this equation easily entails that $\sqrt{\text{Exp}}$ is continuous. It is not difficult to show that the collection of exponential vectors is linearly independent and total; i.e., the linear span $\mathfrak{E}$ of all $\sqrt{\text{Exp}}(w)$, $w \in \mathcal{K}$, is dense in $\exp(\mathcal{K})$.

For $\mathcal{K} = \mathbb{C}^n$ it is clear from the fact that (2.54) provides an orthonormal basis that the map $V_\hbar : \exp(\mathbb{C}^n) \to \overline{H}^2_\hbar(\mathbb{C}^n)$, defined by extension of

$$(V_\hbar w_1 \otimes_s \cdots \otimes_s w_l)(\overline{z}) := \frac{1}{\sqrt{l!\hbar^n}}(z, w_1)_{\mathbb{C}^n} \cdots (z, w_l)_{\mathbb{C}^n} \tag{2.64}$$

is unitary. Hence the subspace $\otimes_s^l \mathbb{C}^n$ of $\exp(\mathbb{C}^n)$ corresponds to the subspace of lth order monomials in $\overline{H}^2_\hbar(\mathbb{C}^n)$. Note that

$$\left(V_\hbar \sqrt{\text{Exp}(w/\sqrt{\hbar})}\right)(\overline{z}) = e^{w\overline{z}/\hbar}; \tag{2.65}$$

cf. (2.58). Hence, with the convention (2.28), the coherent states (2.45) in $L^2(\mathbb{R}^n)$ correspond to the vectors $\sqrt{\text{Exp}}\left(w/\sqrt{\hbar}\right)$, up to normalization. Using (2.63), we may therefore rephrase Proposition 2.3.1 as

Corollary 2.3.3. *For finite-dimensional $\mathcal{K}$ the unit vectors (cf. 1.5.1)*

$$\Psi_\hbar^w := e^{-\frac{1}{2}(w,w)_\mathcal{K}/\hbar} \sqrt{\text{Exp}}\left(w/\sqrt{\hbar}\right) \tag{2.66}$$

define a coherent pure state quantization of $\mathcal{K}$ into $\mathcal{H}_\hbar := \exp(\mathcal{K})$ for all $\hbar \neq 0$.

Conceptually, one should stress that $\mathcal{K}$, although a Hilbert space, is to be seen as a classical phase space. In particular, $q_\hbar(w)$ depends on the phase of w, so that $q_\hbar$ does not quotient to a function on the projective space $\mathbb{P}\mathcal{K}$. In the infinite-dimensional case the conditions (1.8) and (1.9) are not defined in the absence of a Liouville measure on $\mathcal{K}$, but (2.66) makes sense, and comes from a map $q_\hbar$ that satisfies the crucial condition (1.11).

Corollary 2.3.3 and (2.45) suggest that one look for a realization of the representation $U_{\frac{1}{\hbar}}(\tilde{H}_n)$ on $\exp(\mathcal{K})$; what follows holds whatever the dimension of $\mathcal{K}$. For each $z \in \mathcal{K}$ the **annihilation operator** $a(z)$ is an unbounded operator on the dense domain $\mathfrak{E} \subset \exp(\mathcal{K})$ satisfying

$$a(z)\sqrt{\text{Exp}}(w) = (z, w)_\mathcal{K}\sqrt{\text{Exp}}(w). \tag{2.67}$$

The map $z \mapsto a(z)$ is evidently antilinear. It can be shown that $a(z)$ is closable; the domain of its adjoint $a(z)^*$ contains $\mathfrak{E}$. The map $z \mapsto a(z)^*$ is linear; $a(z)^*$ is called a **creation operator**. The domain $\mathfrak{E}$ is evidently invariant under $a(z)$; it can be shown that $a(w)^*\mathfrak{E}$ is contained in the domain of the closure of each $a(z)$. The commutator $[a, a^*]$ is therefore well-defined on $\mathfrak{E}$; it is given by

$$[a(z), a(w)^*] = (z, w)_\mathcal{K}. \tag{2.68}$$

The unbounded operators $\exp(a(z))$ and $\exp(a(z)^*)$ are defined on $\mathfrak{E}$ as well, where their action is given by a strongly convergent power series expansion. From (2.67) one obtains

$$e^{a(z)}\sqrt{\text{Exp}}(w) = e^{(z,w)_\mathcal{K}}\sqrt{\text{Exp}}(w); \tag{2.69}$$

$$e^{a(z)^*}\sqrt{\text{Exp}}(w) = \sqrt{\text{Exp}}(z + w). \tag{2.70}$$

In terms of these, the analogue on $\exp(\mathcal{K})$ of the Weyl operator (2.22) is

$$U_{\frac{1}{\hbar}}(z) := e^{\frac{1}{\sqrt{\hbar}}[a(z)^* - a(z)]}; \tag{2.71}$$

the unitarity of U is obvious from this expression, and $z \mapsto U_{\frac{1}{\hbar}}(z)$ (with (2.28)) yields a representation of the Heisenberg group that is equivalent to the one defined in 2.1 under the same name. We then see from the CBH-formula and (2.68) that we may rewrite (2.66) as

$$\Psi_{\hbar}^{w} = U_{\frac{1}{\hbar}}(w)\Omega. \tag{2.72}$$

The position and momentum operators (2.23), (2.24) may then be expressed in terms of the a and a^* as $Q_{\hbar}^{S,i} = \sqrt{\frac{1}{2}\hbar}(a_i + a_i^*)$ and $P_{\hbar,i}^{S} = \sqrt{\frac{1}{2}\hbar}(a_i - a_i^*)/i$, where $a(z) = a_i\overline{z}^i$, etc.

2.4 Properties of Berezin Quantization on Flat Space

Berezin quantization on flat space has the following pleasant property.

Theorem 2.4.1. *Putting $\tilde{\mathfrak{A}}^0 = C_c^\infty(T^*\mathbb{R}^n)$ and $\mathfrak{A}^\hbar = \mathfrak{B}_0(L^2(\mathbb{R}^n))$ for $\hbar \neq 0$, the Berezin quantization map $\mathcal{Q}_{\hbar}^{B} : \tilde{\mathfrak{A}}_{\mathbb{R}}^0 \to \mathfrak{A}_{\mathbb{R}}^\hbar$ defined by (2.51) is a nondegenerate strict quantization of the Poisson manifold $T^*\mathbb{R}^n$ (with its canonical Poisson bracket I.(2.24)) on $I = \mathbb{R}$. Moreover,*

$$\mathcal{Q}_{\hbar}^{B}(C_0(T^*\mathbb{R}^n)) = \mathfrak{B}_0(L^2(\mathbb{R}^n)). \tag{2.73}$$

Hence $\mathcal{Q}_{\hbar}^{B}$ is a strict deformation quantization, except for (1.3), of $\mathfrak{A}^0 = C_0(T^*\mathbb{R}^n)$.

Before starting with the proof, we note that $\mathcal{Q}_{\hbar}^{B}$ determines a continuous field of C^*-algebras by Proposition 1.2.4; this will be further developed in 2.6.

The nondegeneracy of $\mathcal{Q}_{\hbar}^{B}$ is an easy corollary of 2.3.2. For $\Psi_1, \Psi_2 \in \mathcal{H}_\hbar$ and W defined by (1.42) we have $(\Psi_1, \mathcal{Q}_{\hbar}^{B}(f)\Psi_2) = (W\Psi_1, fW\Psi_2)$, where the inner product is in $\mathcal{H}^\hbar = L^2(S, d\mu_\hbar)$. Since one can construct a basis of the latter Hilbert space consisting of functions of the type $\overline{W\Psi_1}W\Psi_2$, the property $\mathcal{Q}_{\hbar}^{B}(f) = 0$ implies $f = 0$ almost everywhere, which means that $f = 0$ for $f \in C_c^\infty(T^*\mathbb{R}^n)$. The converse is trivial.

The fact that $\mathcal{Q}_{\hbar}^{B}$ maps $C_c^\infty(T^*\mathbb{R}^n, \mathbb{R})$ into $\mathfrak{B}_0(L^2(\mathbb{R}^n))_{\mathbb{R}}$ follows from 1.3.5. To show that $\mathcal{Q}_{\hbar}^{B}(C_c^\infty(T^*\mathbb{R}^n))$ is dense in $\mathfrak{B}_0(L^2(\mathbb{R}^n))$, one observes that $\mathcal{Q}_{\hbar}^{B}(f)$ is Hilbert–Schmidt for $f \in C_c^\infty(T^*\mathbb{R}^n)$. If one assumes that $f(p, q) = f_1(p)f_2(q)$, the kernel $K(x, y)$ of $\mathcal{Q}_{\hbar}^{B}(f)$ factorizes as a function of the variables $x \pm y$. Each factor is then easily seen to be dense in $L^2(\mathbb{R}^n)$ as f_i runs through $C_c^\infty(\mathbb{R}^n)$. Equation (2.73) then follows from 1.4.5.

Rieffel's condition and (1.1) hold by Proposition 1.3.6.

We now turn to the proof of (1.2), using (2.28). For $m \in \mathbb{N}$ we will use

$$\|g\|_{m,\infty} := \sum_{|\alpha|+|\beta|\leq m} \|\partial^\alpha\overline{\partial}^\beta f\|_\infty, \tag{2.74}$$

where α and β are multi-indices, and $\partial^\alpha := \partial_{z^1}^{\alpha_1} \cdots \partial_{z^n}^{\alpha_n}$, etc. For $\Psi \in L^2(\mathbb{R}^n)$ we write $(\Psi_\hbar^z, \Psi) = \exp(-z\overline{z}/(2\hbar))\Psi(\overline{z})$; cf. 2.3.2. From (2.51) and (2.50) we obtain, after a shift of one of the integration variables,

$$(\Psi, Q_\hbar^B(f)Q_\hbar^B(g)\Psi) = \int d\mu_\hbar(z,\overline{z})d\mu_\hbar(\xi,\overline{\xi})\, e^{-(z\overline{z}+\xi\overline{\xi}+z\overline{\xi})/\hbar}$$
$$\times\, \overline{\Psi}(z)\Psi(\overline{z+\xi})f(z,\overline{z})g(z+\xi,\overline{z+\xi}). \qquad (2.75)$$

One now expands $g(z+\xi,\overline{z+\xi})$ in a Taylor series around $(z,\overline{z})$. The zeroth-order term leads to $(\Psi, Q_\hbar^B(fg)\Psi)$. The remainder is $\leq C\|g\|_{1,\infty}|\xi|$, for a constant C of order 1 (further contributions to this constant will be absorbed without change of notation). We take $\|f\|_\infty$ out of the integral, and of the factor $\exp(-\xi\overline{\xi}/\hbar)$ we put $\exp(-\xi\overline{\xi}/(2\hbar))$ into the measure. We then apply Cauchy–Schwarz to the ξ-integral, factorizing the ξ-dependent integrand into $|\xi|$ times the rest. The first of the ensuing two ξ-integrals is a Gaussian integral of $|\xi|^2$, which is proportional to $\hbar$ (which appears under a square root, so it will lead to a factor $\hbar^{1/2}$). There remains an integral over z and ξ. Here we apply Cauchy–Schwarz to the z-integration. The resulting triple integral factorizes after a shift in one of the variables, and can be performed; two of the factors are equal to $\|\Psi\|$. Hence

$$\left|(\Psi, [Q_\hbar^B(f)Q_\hbar^B(g) - Q_\hbar^B(fg)]\Psi)\right| \leq C\|f\|_\infty\|g\|_{1,\infty}\|\Psi\|^2\hbar^{1/2}. \qquad (2.76)$$

By I.2.1.8, I.2.5.3, and I.(1.57), for each $A \in \mathfrak{B}(\mathcal{H})_\mathbb{R}$ there is a unit vector $\Psi \in \mathcal{H}$ such that

$$\|A\| = |(\Psi, A\Psi)|. \qquad (2.77)$$

Hence (1.2) follows from (2.76), which implies

$$\lim_{\hbar\to 0}\|Q_\hbar^B(f)\circ Q_\hbar^B(g) - Q_\hbar^B(fg)\| = 0. \qquad (2.78)$$

The proof of (1.3) is similar. We consider (2.75) with $(\Psi, Q_\hbar^B(f)Q_\hbar^B(g)\Psi)$ replaced by $(\Psi, [Q_\hbar^B(f), Q_\hbar^B(g)]\Psi)$. On the right-hand side one then has the terms $f(z,\overline{z})g(z+\xi,\overline{z+\xi}) - f(z+\xi,\overline{z+\xi})g(z,\overline{z})$, instead of $f(z,\overline{z})g(z+\xi,\overline{z+\xi})$. One now expands $g(z+\xi,\overline{z+\xi})$ as well as $f(z+\xi,\overline{z+\xi})$ in a Taylor series around $(z,\overline{z})$. The zeroth-order term obviously vanishes. The linear term can be evaluated by also expanding $\exp(-z\overline{\xi}/\hbar)\Psi(\overline{z+\xi})$ in powers of $\overline{\xi}$. The ξ-integration can then be performed: The only nonzero contribution comes from factors $\xi\overline{\xi}$. A partial integration in z then shows that the linear term equals $\hbar(\Psi, Q_\hbar^B(-i\{f,g\} + g\partial\overline{\partial}f - f\partial\overline{\partial}g)\Psi)$, where the Poisson bracket is given by (2.31).

The quadratic term contains $\xi\overline{\xi}(f\partial\overline{\partial}g - g\partial\overline{\partial}f)$. In the ξ-integral only the zeroth-order term in $\overline{\xi}$ from $\exp(-z\overline{\xi}/\hbar)\Psi(\overline{z+\xi})$ contributes, and the result may be expressed as $\hbar(\Psi, Q_\hbar^B((f\partial\overline{\partial}g - g\partial\overline{\partial}f)\Psi)$. This cancels the additional term from the linear contribution. Hence the linear term and the expression with $\xi\overline{\xi}$ in the quadratic term together produce $\hbar(\Psi, Q_\hbar^B(-i\{f,g\})\Psi)$. The remainder of the quadratic term has a part proportional to $\overline{\xi}\overline{\xi}$, which vanishes upon integration, and a part proportional to $\xi\xi$. After ξ-integration, and partial integration in z, the latter part is easily seen to be bounded by $\hbar^2\|f\|_{4,\infty}\,\|g\|_{4,\infty}\,\|\Psi\|^2$.

The contribution of the higher-order terms is estimated as follows. Taylor's formula with remainder of third order yields an object bounded by $2\|f\|_{3,\infty}\,\|g\|_{3,\infty}|\xi|^3$. We now proceed as in the proof of (1.2): Practically the only difference is that the Gaussian integral of $|\xi|^2$ in that proof is now replaced by one of $|\xi|^6$, which leads to an overall factor of order $\hbar^{3/2}$. All this leads to the estimate

$$|(\Psi, \frac{i}{\hbar}[\mathcal{Q}_\hbar^B(f), \mathcal{Q}_\hbar^B(g)] - \mathcal{Q}_\hbar^B(\{f,g\})\Psi)| \leq \left(C_1 f\|_{3,\infty}\,\|g\|_{3,\infty}\hbar^{1/2} + C_2\|f\|_{4,\infty}\,\|g\|_{4,\infty}\hbar\right)\|\Psi\|^2. \tag{2.79}$$

Equation (1.3) now follows in the same fashion as (1.2) above. ■

We turn to the equivariance properties of $\mathcal{Q}_\hbar^B$. In preparation:

Definition 2.4.2. *An* **automorphic action** α *of a group* G *on a* C^**-algebra* $\mathfrak{A}$ *is a homomorphism* $x \mapsto \alpha_x$*, such that each* α_x *is an automorphism of* $\mathfrak{A}$*. In other words, apart from the linearity and bijectivity of each* $\alpha_x : \mathfrak{A} \to \mathfrak{A}$ *one has the properties* $\alpha_x \circ \alpha_y = \alpha_{xy}$, $\alpha_x(AB) = \alpha_x(A)\alpha_x(B)$*, and* $\alpha_x(A^*) = \alpha_x(A)^*$.

Consider the natural action ρ^0 of $Sp(n,\mathbb{R}) \ltimes \mathbb{R}^{2n}$ on $T^*\mathbb{R}^n \simeq \mathbb{R}^{2n}$, according to which (M, w) maps $\sigma \in T^*\mathbb{R}^n$ to $\rho^0_{(M,w)}(\sigma) := M\sigma + w$ (cf. 2.2). This leads to an automorphic action α^0 of $Sp(n,\mathbb{R}) \ltimes \mathbb{R}^{2n}$ on $\mathfrak{A}^0 = C_0(T^*\mathbb{R}^n)$, given by

$$\alpha^0_{(M,w)}(f) := f \circ \rho^0_{(M,w)^{-1}}. \tag{2.80}$$

Also, one has an automorphic action $\alpha^\hbar$ of G on $\mathfrak{A}^\hbar = \mathfrak{B}_0(L^2(\mathbb{R}^n))$, given by the representation constructed in 2.2.2. That is,

$$\alpha^\hbar_{(M,w)}(A) := \rho^\hbar(M,w)A\rho^\hbar(M,w)^*. \tag{2.81}$$

Theorem 2.4.3. *For each* $(M, w) \in U(n) \ltimes \mathbb{R}^{2n}$*, where* $U(n) = Sp(n,\mathbb{R}) \cap O(2n)$ *(cf. the text surrounding (2.28)), and all* $f \in C_0(T^*\mathbb{R}^n)$*, one has*

$$\mathcal{Q}_\hbar^B(\alpha^0_{(M,w)}(f)) = \alpha^\hbar_{(M,w)}(\mathcal{Q}_\hbar^B(f)). \tag{2.82}$$

To prove this, we rewrite (2.51) as a weak integral

$$\mathcal{Q}_\hbar^B(f) = \int_{T^*\mathbb{R}^n} \frac{d^npd^nq}{(2\pi\hbar)^n}\, f(p,q)U_{\frac{1}{\hbar}}(p,q)[\Psi_\hbar^0]U_{\frac{1}{\hbar}}(p,q)^*. \tag{2.83}$$

The equivariance under $\mathbb{R}^{2n}$ is obvious from this formula, the last claim in 2.2.2, (2.22), (2.6), and (2.19).

Lemma 2.4.4. *If* $U \in U(n)$ *and* $\rho^\hbar$ *is the metaplectic representation of* $Sp(n,\mathbb{R})$ *on* $\mathcal{H}_\hbar = L^2(\mathbb{R}^n)$*, then*

$$\rho^\hbar(U)[\Psi_\hbar^0]\rho^\hbar(U)^* = [\Psi_\hbar^0]. \tag{2.84}$$

This is most easily proved in the realization on $\tilde{\mathcal{H}}_\hbar$, described in 2.3.2. From (2.55) and (2.50) we have $W\Psi_\hbar^0(z,\bar{z}) = \exp(-z\bar{z}/(2\hbar))$. If $U \in Sp(n,\mathbb{R}) \cap O(2n)$, then $W\rho^{\hbar}(U)W^*$ can be shown to be given by

$$W\rho^{\hbar}(U)W^*\Psi(z,\overline{z}) = \frac{1}{\sqrt{\det(U)}}\Psi(U^{-1}z, \overline{U^{-1}z}), \tag{2.85}$$

from which (2.84) is immediate. □

The equivariance under $U(n)$ follows from this lemma and (2.40). Since each element of a semidirect product factorizes, Theorem 2.4.3 follows. ■

This theorem can be reformulated in terms of dynamics on $T^*\mathbb{R}^n$.

Corollary 2.4.5. *Define a class of classical Hamiltonians on $T^*\mathbb{R}^n$ by*

$$h(p,q) = \tfrac{1}{2}(p, Ap) + \tfrac{1}{2}(q, Aq) + (p, Bq) + (c, p) + (d, q), \tag{2.86}$$

where A and B are real $n \times n$ matrices such that $A^T = A$ and $B^T = -B$, the inner products are in $\mathbb{R}^n$, and $c, d \in \mathbb{R}^n$. Denote the time evolution generated by h on the classical observables by α_t^0 (cf. I.(2.13)). Define the quantum Hamiltonian $H_\hbar$ by

$$H_\hbar := \mathcal{Q}_\hbar^B(h), \tag{2.87}$$

which is an unbounded operator with domain $\mathcal{S}(\mathbb{R}^n)$. Then $H_\hbar$ is essentially self-adjoint on $\mathcal{S}(\mathbb{R}^n)$. The one-parameter automorphism group $\alpha_t^\hbar$ on $\mathfrak{B}_0(L^2(\mathbb{R}^n))$ is defined by

$$\alpha_t^\hbar(A) := e^{itH_\hbar/\hbar}Ae^{-itH_\hbar/\hbar}. \tag{2.88}$$

Then one has

$$\mathcal{Q}_\hbar^B(\alpha_t^0(f)) = \alpha_t^\hbar(\mathcal{Q}_\hbar^B(f)). \tag{2.89}$$

A matrix $X \in \mathfrak{M}_{2n}(\mathbb{C})$ lies in $U(n)$ when it satisfies $JX + X^TJ = 0$ and $X^T + X = 0$. The polynomial (2.32) is then precisely of the form of the quadratic term in (2.86). For h of the form (2.86), one computes

$$\mathcal{Q}_\hbar^B(h) = \tfrac{1}{2}(P_\hbar^S, AP_\hbar^S) + \tfrac{1}{2}(Q_\hbar^S, AQ_\hbar^S) + (P_\hbar^S, BQ_\hbar^S) + (c, P_\hbar^S) + (d, Q_\hbar^S) \tag{2.90}$$

in terms of (2.24) and (2.23). This follows by calculating the matrix elements between coherent states (which indeed lie in the domain of $H_\hbar$). The expression (2.90) coincides with (2.41), and therefore the essential self-adjointness of $H_\hbar$, is a consequence of Proposition 2.2.2. Corollary 2.4.5 now follows from Theorem 2.4.3, exactly as in the derivation of (2.42). ■

2.5 *Weyl Quantization on Flat Space*

Theorem 2.4.3 suggests that one look for a quantization that is equivariant under the full affine symplectic group $Sp(n,\mathbb{R}) \ltimes \mathbb{R}^{2n}$. It is obvious from Lemma 2.4.4, in particular from (2.84) and (2.44), how this may be accomplished: One simply replaces the projection $[\Psi_\hbar^0]$ in (2.83) by (a constant times) the parity operator P.

This leads to the definition of the **Weyl quantization** of a suitable function f on $T^*\mathbb{R}^n$ as the operator on $L^2(\mathbb{R}^n)$ given by

$$\mathcal{Q}_\hbar^W(f) := \int_{T^*\mathbb{R}^n} \frac{d^n p d^n q}{(\pi\hbar)^n} f(p,q) U_{\frac{1}{\hbar}}(p,q) P U_{\frac{1}{\hbar}}(p,q)^*. \tag{2.91}$$

The normalization has been chosen so that $\mathcal{Q}_\hbar^W(1_{T^*\mathbb{R}^n}) = \mathbb{I}$. Note that at least in a heuristic sense, $\mathcal{Q}_\hbar^W(\delta) = (\pi\hbar)^{-n} P$ (where δ is the Dirac delta function on $T^*\mathbb{R}^n \simeq \mathbb{R}^{2n}$), which places the parity operator in a remarkable light.

Since the Fourier transform will play an important role in what follows, we choose the Schwartz space

$$\tilde{\mathfrak{A}}_{\mathbb{R}}^0 = \mathcal{S}(T^*\mathbb{R}^n, \mathbb{R}) \tag{2.92}$$

as the Poisson algebra to be quantized. Clearly, the closure of $\tilde{\mathfrak{A}}^0$ in the sup-norm is $\mathfrak{A}^0 = C_0(T^*\mathbb{R}^n)$. We define (2.91) for $f \in \mathcal{S}(T^*\mathbb{R}^n)$; it is immediate that $\mathcal{Q}_\hbar^W$ maps $\mathcal{S}(T^*\mathbb{R}^n, \mathbb{R})$ into $\mathfrak{B}(L^2(\mathbb{R}^n))_{\mathbb{R}}$.

We will shortly see that $\mathcal{Q}_\hbar^W(f) \in \mathfrak{B}_0(L^2(\mathbb{R}^n))$. Given our motivation for constructing $\mathcal{Q}_\hbar^W$, the following comes as no surprise.

Theorem 2.5.1. *Let α^0 and $\alpha^\hbar$ be as in (2.80) and (2.81), respectively. For each $(M, w) \in Sp(n, \mathbb{R}) \ltimes \mathbb{R}^{2n}$ and all $f \in \mathcal{S}(T^*\mathbb{R}^n)$, one has*

$$\mathcal{Q}_\hbar^W(\alpha^0_{(M,w)}(f)) = \alpha^\hbar_{(M,w)}(\mathcal{Q}_\hbar^W(f)). \tag{2.93}$$

The proof is similar to that of 2.4.3, with (2.44) replacing (2.84). □

Corollary 2.5.2. *Let the classical Hamiltonian h be an arbitrary real polynomial on $T^*\mathbb{R}^n$ of degree ≤ 2 in (p, q). The quantum Hamiltonian $H_\hbar := \mathcal{Q}_\hbar^W(h)$ (see (2.41)) is well-defined as an unbounded operator on the domain $\mathcal{S}(\mathbb{R}^n)$, on which it is essentially self-adjoint. With the one-parameter automorphisms α_t^0 and $\alpha_t^\hbar$ defined as in I.(2.13) and (2.88), respectively, one has*

$$\mathcal{Q}_\hbar^W(\alpha_t^0(f)) = \alpha_t^\hbar(\mathcal{Q}_\hbar^W(f)). \tag{2.94}$$

Equation (2.90) is valid (and proved by the same method) also if $\mathcal{Q}_\hbar^B$ is replaced by $\mathcal{Q}_\hbar^W$, which settles the domain and self-adjointness issues. The corollary then follows from Proposition 2.2.1. ■

The notation $\mathcal{Q}_\hbar^W$ used here and in (2.41) will be justified shortly.

Weyl quantization may be rewritten in various ways. Firstly, one has

$$\mathcal{Q}_\hbar^W(f) = \int_{T^*\mathbb{R}^n} \frac{d^n p d^n q}{(2\pi\hbar)^n} f(p,q) \Omega_\hbar^W(p,q), \tag{2.95}$$

where $\Omega_\hbar^W(p,q) \in \mathfrak{B}(L^2(\mathbb{R}^n))$ is defined by

$$\Omega_\hbar^W(p,q)\Psi(x) := 2^n e^{2ip(x-q)/\hbar}\Psi(2q - x). \tag{2.96}$$

The function f may be recovered from $\mathcal{Q}_\hbar^W(f)$ by the formula

$$f(p,q) = \mathrm{Tr}\, \mathcal{Q}_\hbar^W(f)\Omega_\hbar^W(p,q). \tag{2.97}$$

This equation may be proved by noting that $\mathcal{Q}_\hbar^W(f)$ is trace-class for $f \in \mathcal{S}(T^*\mathbb{R}^n)$ (see below), so that $\mathcal{Q}_\hbar^W(f)\Omega_\hbar^W(p,q)$ is trace-class as well, because $\Omega_\hbar^W(p,q)$ is bounded. If $K(\cdot,\cdot)$ is the kernel of $\mathcal{Q}_\hbar^W(f)\Omega_\hbar^W(p,q)$, its trace is $\int d^n x\, K(x,x)$, which easily leads to (2.97).

More generally, the **Weyl symbol** $\sigma_\hbar^W\lfloor A\rfloor$ of an operator $A \in \mathfrak{B}(L^2(\mathbb{R}^n))$ is a distribution in $\mathcal{S}'(T^*\mathbb{R}^n)$ defined by

$$\sigma_\hbar^W\lfloor A\rfloor(f) := (2\pi\hbar)^n \mathrm{Tr}\, A\mathcal{Q}_\hbar^W(f). \tag{2.98}$$

If $\sigma_\hbar^W\lfloor A\rfloor$ is a locally integrable function, we see from (2.95) that one may write

$$\sigma_\hbar^W\lfloor A\rfloor(p,q) = \mathrm{Tr}\, A\Omega_\hbar^W(p,q). \tag{2.99}$$

Comparing this with (2.97), for $f \in \mathcal{S}(T^*\mathbb{R}^n)$ one infers that

$$\mathcal{Q}_\hbar^W(\sigma_\hbar^W\lfloor A\rfloor) = A. \tag{2.100}$$

(Using distribution theory it is possible to make sense of this equation even when $f \in \mathcal{S}'(T^*\mathbb{R}^n)$.) Hence $\sigma_\hbar^W$ is the inverse of $\mathcal{Q}_\hbar^W$.

Analogously to (2.52), we can write the Gelfand transform of $\mathcal{Q}_\hbar^W(f)$ as

$$\widehat{\mathcal{Q}_\hbar^W(f)}(\psi) = \int_{T^*\mathbb{R}^n} \frac{d^n p\, d^n q}{(2\pi)^n} W_\hbar\lfloor\psi\rfloor(p,q) f(p,q), \tag{2.101}$$

where the (real-valued) **Wigner function** is given by

$$W_\hbar\lfloor\psi\rfloor(p,q) = \hbar^{-n}(\Psi, \Omega_\hbar^W(p,q)\Psi). \tag{2.102}$$

Since $\Omega_\hbar^W(p,q)$ is 2^n times a unitary operator, the Cauchy–Schwarz inequality implies that $\|W_\hbar\lfloor\psi\rfloor\|_\infty \le (2/\hbar)^n$ (if Ψ had not been normalized, the bound would contain an additional factor $\|\Psi\|^2$). It is then easy to show that $W_\hbar\lfloor\psi\rfloor \in L^2(T^*\mathbb{R}^n) \cap C_0(T^*\mathbb{R}^n)$. The expression (2.102) is often written as

$$W_\hbar\lfloor\psi\rfloor(p,q) = \int_{\mathbb{R}^n} d^n v\, e^{ipv}\overline{\Psi(q + \tfrac{1}{2}\hbar v)}\Psi(q - \tfrac{1}{2}\hbar v). \tag{2.103}$$

It may be inferred from (2.102) that $\mathcal{Q}_\hbar^W$ is not positive, since there exist vectors Ψ for which $W_\hbar\lfloor\psi\rfloor$ is not positive definite. For such Ψ, the Wigner function may not even be in $L^1(T^*\mathbb{R}^n)$. Here Berezin quantization is much better behaved. Comparing (2.101) with (2.52), one sees that the Wigner function $W_\hbar\lfloor\psi\rfloor$ in Weyl quantization replaces the positive definite expression $(p,q) \mapsto \hbar^{-n} p(\psi_\hbar^{p,q}, \psi)$ (whose L^1-norm is 1 by (1.8)) in Berezin quantization.

It follows from (2.98) and (2.102) that for a unit vector $\Psi \in L^2(\mathbb{R}^n)$ one has

$$\sigma_\hbar^W\lfloor[\Psi]\rfloor = \hbar^n W_\hbar\lfloor\psi\rfloor, \tag{2.104}$$

or, by (2.100),

$$[\Psi] = \hbar^n \mathcal{Q}_\hbar^W(W_\hbar\lfloor\psi\rfloor). \tag{2.105}$$

Consequently, the transition probabilities I.(2.65) in $\mathcal{P}(\mathfrak{B}_0(L^2(\mathbb{R}^n)))$ may be expressed in terms of the overlap of the pertinent Wigner functions as

$$p(\rho,\sigma) = \hbar^n \int_{T^*\mathbb{R}^n} \frac{d^n p\, d^n q}{(2\pi)^n} W_\hbar\lfloor\rho\rfloor(p,q) W_\hbar\lfloor\sigma\rfloor(p,q); \tag{2.106}$$

note that the integral on the right-hand side is well-defined, since we have just seen that $W_\hbar \in L^2(T^*\mathbb{R}^n)$.

The image of $\mathcal{Q}_\hbar^W$ in $\mathfrak{B}(L^2(\mathbb{R}^n))$ is best studied by rewriting (2.91) as

$$\mathcal{Q}_\hbar^W(f)\Psi(x) = \int_{T^*\mathbb{R}^n} \frac{d^n p d^n y}{(2\pi\hbar)^n} e^{ip(x-y)/\hbar} f\left(p, \tfrac{1}{2}(x+y)\right)\Psi(y). \tag{2.107}$$

In other words, $\mathcal{Q}_\hbar^W(f)$ is an integral operator

$$\mathcal{Q}_\hbar^W(f)\Psi(x) = \int_{\mathbb{R}^n} d^n y\, K_\hbar^W\lfloor f\rfloor(x,y)\Psi(y); \tag{2.108}$$

$$K_\hbar^W\lfloor f\rfloor(x,y) := \hbar^{-n} \grave{f}((x-y)/\hbar, \tfrac{1}{2}(x+y)). \tag{2.109}$$

Here the partial (fiberwise) Fourier transform $\grave{f} \in \mathcal{S}(T\mathbb{R}^n)$ of $f \in \mathcal{S}(T^*\mathbb{R}^n)$ is

$$\grave{f}(v,q) := \int_{T_q^*\mathbb{R}^n} \frac{d^n p}{(2\pi)^n} e^{ipv} f(p,q). \tag{2.110}$$

Proposition 2.5.3. *The map $\mathcal{Q}_\hbar^W$ is an isomorphism between $\mathcal{S}(T^*\mathbb{R}^n)$ and the space $\mathfrak{B}_2(L^2(\mathbb{R}^n))$ of Hilbert–Schmidt operators on $L^2(\mathbb{R}^n)$ with kernel in $\mathcal{S}(\mathbb{R}^{2n})$.*

This is immediate from the above expressions. ■

Corollary 2.5.4.

- *The image $\mathcal{Q}_\hbar^W(\tilde{\mathfrak{A}}^0)$ is a norm-dense subalgebra of $\mathfrak{A} = \mathfrak{B}_0(L^2(\mathbb{R}^n))$, and therefore acts irreducibly on $L^2(\mathbb{R}^n)$.*
- *The quantization $\mathcal{Q}_\hbar^W$ is nondegenerate (cf. 1.1.2).*

Finally, we rewrite (2.91) as

$$\mathcal{Q}_\hbar^W(f) = \int_{\mathbb{R}^{2n}} d^n u d^n v\, \check{f}(u,v) U_\hbar^S(\mathrm{Exp}(-uQ+vP)). \tag{2.111}$$

For $f \in \mathcal{S}(T^*\mathbb{R}^n)$ we have defined the **symplectic Fourier transform** $\check{f} \in \mathcal{S}(\mathbb{R}^{2n})$ by inverting

$$f(p,q) := \int_{\mathbb{R}^{2n}} d^n u d^n v\, \check{f}(u,v) e^{i(uq-vp)}. \tag{2.112}$$

Hence we see (with Weyl) that $\mathcal{Q}_\hbar^W$ corresponds to a particular operator ordering, in which the function $(p,q) \mapsto \exp(iuq - ivp)$ on $T^*\mathbb{R}^n$ (smeared with a test function) is quantized by the operator $U_\hbar^S(\mathrm{Exp}(-uQ+vP))$ on $L^2(\mathbb{R}^n)$. Ignoring the test functions, one may repeatedly differentiate with respect to u and v; the linearity of $\mathcal{Q}_\hbar^W$ then indicates that polynomials $\mathcal{P}$ on $T^*\mathbb{R}^n$ are Weyl-quantized by (2.37). An interesting corollary to (2.111) is

Proposition 2.5.5. *The Schrödinger representation (2.17) is irreducible.*

If $U_\hbar^S$ were reducible, by Schur's lemma there would exist $E \in \mathfrak{B}(L^2(\mathbb{R}^n))$ such that $[E, U_\hbar^S(u,v,z)] = 0$ for all $(u,v,z) \in \mathcal{H}_n$. Equation (2.111) and the definition of a weak integral then imply that $[E, \mathcal{Q}_\hbar^W(f)] = 0$ for all $f \in \mathcal{S}(T^*\mathbb{R}^n)$. But we saw in 2.5.4 that $\mathcal{Q}_\hbar^W(\mathcal{S}(T^*\mathbb{R}^n))$ acts irreducibly on $L^2(\mathbb{R}^n)$; cf. I.2.2.2. ■

2.6 Strict Quantization and Continuous Fields on Flat Space

In this section we show that Weyl quantization is strict, and even continuous, like its Berezin counterpart. The continuous field of C^*-algebras generated by $\mathcal{Q}_\hbar^W$ or $\mathcal{Q}_\hbar^B$ will be described in terms of the Heisenberg group $\tilde{H}_n$.

Theorem 2.6.1. *The Weyl maps $\mathcal{Q}_\hbar^W$ define a strict deformation quantization of $\tilde{\mathfrak{A}}_{\mathbb{R}}^0 = \mathcal{S}(T^*\mathbb{R}^n, \mathbb{R})$ (with Poisson bracket I.(2.24)) over $I = \mathbb{R}$, with $\mathfrak{A}^\hbar = \mathfrak{A} := \mathfrak{B}_0(L^2(\mathbb{R}^n))$ for $\hbar \neq 0$.*

Given that it is strict, the fact that $\mathcal{Q}_\hbar^W$ is a deformation quantization follows from 2.5.4. A key ingredient of the proof of strictness is an estimate we borrow from the theory of pseudo-differential operators.

Lemma 2.6.2. *There exists a constant $C > 0$ such that for all $f \in \mathcal{S}(T^*\mathbb{R}^n)$*

$$\|\mathcal{Q}_1^W(f)\| \leq C\|f\|_{2n+1,\infty}, \tag{2.113}$$

where, for $m \in \mathbb{N}$ (cf. (2.74)),

$$\|f\|_{m,\infty} := \sum_{|\alpha|+|\beta|\leq m} \|\partial_p^\alpha \partial_q^\beta f\|_\infty. \tag{2.114}$$

Here $\partial_p^\alpha := \partial_{p_1}^{\alpha_1} \cdots \partial_{p_n}^{\alpha_n}$; similarly for ∂_q^β.

This lemma is useful also for $\hbar \neq 1$, because $\mathcal{Q}_\hbar^W(f) = \mathcal{Q}_1^W(f_\hbar)$, with $f_\hbar(p, q) := f(\hbar p, q)$. Indeed, it now rapidly follows that $\hbar \mapsto \mathcal{Q}_\hbar^W(f)$ is continuous as a function from $\mathbb{R}\backslash\{0\}$ to $\mathfrak{B}_0(L^2(\mathbb{R}^n))$; this implies the continuity of $\hbar \mapsto \|\mathcal{Q}_\hbar^W(f)\|$ for $\hbar \neq 0$. Also, (1.2) and (1.3) follow straightforwardly from (2.113) by computing $f \cdot_\hbar g$. As in 1.1, this is defined by the property $\mathcal{Q}_\hbar^W(f)\mathcal{Q}_\hbar^W(g) = \mathcal{Q}_\hbar^W(f \cdot_\hbar g)$, and can be computed from (2.109).

To prove continuity at $\hbar = 0$, we use the following facts. Firstly, a simple computation shows that the Wigner function (2.102) of the coherent state (2.46) is a Gaussian:

$$W_\hbar\lfloor\psi_\hbar^0\rfloor(p, q) = \left(\frac{2}{\hbar}\right)^n e^{-(q^2+p^2)/\hbar}. \tag{2.115}$$

Secondly, the connection between Weyl quantization and Berezin quantization is given by

$$\mathcal{Q}_\hbar^B(f) = \mathcal{Q}_\hbar^W(W_\hbar\lfloor\psi_\hbar^0\rfloor * f), \tag{2.116}$$

where $*$ is convolution in $T^*\mathbb{R}^n \simeq \mathbb{R}^{2n}$. This may alternatively be written as

$$\mathcal{Q}_\hbar^B(f) = \mathcal{Q}_\hbar^W(e^{\frac{1}{4}\hbar\Delta_{2n}} f), \tag{2.117}$$

where Δ_{2n} is the Laplacian on $T^*\mathbb{R}^n \simeq \mathbb{R}^{2n}$.

Proposition 2.6.3. *For each $f \in \mathcal{S}(T^*\mathbb{R}^n)$ the function*

$$\hbar \mapsto \|\mathcal{Q}_\hbar^B(f) - \mathcal{Q}_\hbar^W(f)\|$$

is continuous on $\mathbb{R}$. That is, the Weyl and Berezin quantizations of $\mathcal{S}(T^*\mathbb{R}^n)$ are equivalent.

This follows from (2.116) and an application of Lemma 2.6.2. Note that continuity at $\hbar = 0$ simply follows from

$$\lim_{\hbar \to 0} \|\mathcal{Q}_\hbar^W(f) - \mathcal{Q}_\hbar^B(f)\| = 0 \tag{2.118}$$

(from the same lemma), since $\mathcal{Q}_0^W(f) = \mathcal{Q}_0^B(f) = f$. ■

The continuity of $\hbar \mapsto \|\mathcal{Q}_\hbar^W(f)\|$ at $\hbar = 0$ (and, indeed, at any $\hbar$) is now obvious from 2.6.3 and 2.4.1 (or 1.3.6), finishing the proof of 2.6.1. ■

A different and much more general proof of Rieffel's condition for Weyl quantization will be given in Theorem III.3.11.4. For now, we return to the Heisenberg group. One may extend (the inverse of) (2.112) to obtain an isomorphism between $\mathcal{S}(\tilde{H}_n)$ and $\mathcal{S}(\mathfrak{h}_n^*)$. Thus the (symplectic) Fourier transform $\check{f} \in \mathcal{S}(\tilde{H}_n)$ of $f \in \mathcal{S}(\mathfrak{h}_n^*)$ is defined by

$$\check{f}(u, v, s) := \int_{\mathfrak{h}_n^*} \frac{d^n p d^n q dc}{(2\pi)^{2n+1}} e^{i(uq - pv - cs)} f(p, q, c). \tag{2.119}$$

What follows is a special case of a general construction explained in III.1.7. One can define an associative product $\cdot$ on $\mathcal{S}(\tilde{H}_n)$ by convolution, i.e.,

$$\check{f} \cdot \check{g}(u, v, s) := \int_{\tilde{H}_n} d^n u' d^n v' ds' \, \check{f}((u, v, s) \cdot (u', v', s')^{-1}) \check{g}(u', v', s'), \tag{2.120}$$

as well as an involution * by

$$f^*(u, v, s) = \overline{f(u, v, s)^{-1}}. \tag{2.121}$$

A representation U of $\tilde{H}_n$ on a Hilbert space $\mathcal{H}$ defines a linear map $\pi : \mathcal{S}(\tilde{H}_n) \to \mathfrak{B}(\mathcal{H})$ by

$$\pi(\check{f}) := \int_{\tilde{H}_n} d^n u d^n v dt \, \check{f}(u, v, s) U(u, v, s). \tag{2.122}$$

Using (2.120), one easily checks that any representation $U(\tilde{H}_n)$ thus defines a representation π of $\mathcal{S}(\tilde{H}_n)$ as a *-algebra.

We firstly use this construction with $\mathcal{H} = \check{\mathcal{H}} := L^2(\tilde{H}_n, d^n u d^n v ds)$, and $U(\tilde{H}_n)$ defined by

$$\check{U}(u, v, s)\check{\Psi}(u', v', s') = \check{\Psi}((u, v, s)^{-1} \cdot (u', v', s')). \tag{2.123}$$

It is clear that the ensuing representation $\check{\pi}(\mathcal{S}(\tilde{H}_n))$ defined by (2.122) is faithful. One now puts a norm on $\mathcal{S}(\tilde{H}_n)$ by

$$\|\check{f}\| := \|\check{\pi}(\check{f})\|; \tag{2.124}$$

this is evidently a C^*-norm. The completion of $\mathcal{S}(\tilde{H}_n)$ in this norm is denoted by $C^*(\tilde{H}_n)$. All representations of the convolution algebra $\mathcal{S}(\tilde{H}_n)$ extend to $C^*(\tilde{H}_n)$ by continuity. Now recall Definition 1.2.1 and (2.14).

Proposition 2.6.4. *Define $\pi_0^S : C^*(\tilde{H}_n) \to C_0(T^*\mathbb{R}^n)$ by $\pi_0^S(\check{f}) := f_{\restriction 0}$, extended from $\mathcal{S}(\tilde{H}_n)$ to $C^*(\tilde{H}_n)$ by continuity; this yields a representation of $C^*(\tilde{H}_n)$ on*

$L^2(T^*\mathbb{R}^n)$ by multiplication operators. For $\hbar \neq 0$, define a representation $\pi_\hbar^S$ of $C^*(\tilde{H}_n)$ on $L^2(\mathbb{R}^n)$ by putting $U = U_\hbar^S$ in (2.122); see (2.17).

The triples $\left(C^*(\tilde{H}_n), \{\mathfrak{A}^\hbar, \pi_\hbar^S\}_{\hbar\in\mathbb{R}}\right)$ and $\left(C^*(H_n), \{\mathfrak{A}^\hbar, \pi_\hbar^S\}_{\hbar\in 0\cup 1/\mathbb{Z}}\right)$, where $\mathfrak{A}^0 := C_0(T^*\mathbb{R}^n)$ and $\mathfrak{A}^\hbar := \mathfrak{B}_0(L^2(\mathbb{R}^n))$ for $\hbar \neq 0$, are continuous fields of C^*-algebras.

Analogous to (2.111), one derives the remarkable relation

$$\mathcal{Q}_\hbar^W(f_{\restriction\hbar}) = \pi_\hbar^S(\check{f}). \tag{2.125}$$

One may then imitate the method of proof of Rieffel's condition in Theorem 2.6.1, concluding that the function $\hbar \mapsto \|\pi_\hbar^S(\check{f})\|$ lies in $C(\mathbb{R})$ for $\check{f} \in \mathcal{S}(\tilde{H}_n)$. Moreover, one infers from Lemma 2.6.2 and the fact that $f_{\restriction\hbar}$ decreases rapidly in $\hbar$ that this function even lies in $C_0(\mathbb{R})$. Since $\pi_\hbar^S$ is continuous, this property holds for any $\check{f} \in C^*(\tilde{H}_n)$. Hence condition 1.2.1.1 is satisfied.

Consider the Hilbert space $\mathcal{H} := L^2(\mathbb{R}; (2\pi)^{-2n} d\hbar\, |\hbar|^n) \otimes \mathfrak{B}_2(L^2(\mathbb{R}^n))$; elements of $\mathcal{H}$ are functions on $\mathbb{R}$ taking values in $\mathfrak{B}_2(L^2(\mathbb{R}^n))$, with inner product

$$(\Psi, \Phi) = \int_{\mathbb{R}} \frac{d\hbar}{(2\pi)^{2n}} |\hbar|^n \operatorname{Tr} \Psi(\hbar)^* \Phi(\hbar). \tag{2.126}$$

For $\check{\Psi} \in \mathcal{S}(\tilde{H}_n) \subset \check{\mathcal{H}}$ and $\hbar \neq 0$ one then defines the operator $W\check{\Psi}(\hbar)$ on $L^2(\mathbb{R}^n)$ by $W\check{\Psi}(\hbar) := \pi_\hbar^S(\check{\Psi})$. We know from Proposition 2.5.3 that $W\check{\Psi}(\hbar)$ is a Hilbert–Schmidt operator. An explicit calculation, using (2.109), shows that $W : \mathcal{S}(\tilde{H}_n) \to \mathcal{H}$ is unitary, so that W can be extended to a unitary isomorphism from $\check{\mathcal{H}}$ to $\mathcal{H}$. Writing $\pi := W\check{\pi}W^*$ of $C^*(\tilde{H}_n)$, the point is now that

$$\pi(\check{f})\Psi(\hbar) = \pi_\hbar^S(\check{f})\Psi(\hbar) \tag{2.127}$$

for all $\check{f} \in C^*(\tilde{H}_n)$. Using (2.122), (2.123), and (2.17), this is initially proved for $\check{f} \in \mathcal{S}(\tilde{H}_n)$, and extended to $C^*(\tilde{H}_n)$ by continuity. The product of $\pi_\hbar^S(\check{f}) \in \mathfrak{B}(L^2(\mathbb{R}^n))$ and $\Psi(\hbar) \in \mathfrak{B}_2(L^2(\mathbb{R}^n))$ lies in $\mathfrak{B}_2(L^2(\mathbb{R}^n))$, because $\mathfrak{B}_2(\mathcal{H})$ is a (two-sided, nonclosed) ideal in $\mathfrak{B}(\mathcal{H})$.

Condition 1.2.1.2 now follows, since from (2.127), (2.124), the unitarity of W, and 1.2.1.1 just proved, one has

$$\|\check{f}\| = \sup_{\hbar\in\mathbb{R}} \|\pi_\hbar^S(\check{f})\|. \tag{2.128}$$

It follows from 2.5.4 and (2.125) that $\pi_\hbar^S(\mathcal{S}(\tilde{H}_n))$ is dense in $\mathfrak{B}_0(L^2(\mathbb{R}^n))$. For $\hbar \neq 0$ one therefore has

$$\pi_\hbar^S(C^*(\tilde{H}_n)) = \mathfrak{B}_0(L^2(\mathbb{R}^n)), \tag{2.129}$$

since the left-hand side is norm-closed by I.1.3.10.4. This is consistent with (and could alternatively have been derived from) 2.1.4, III.1.7.5, and I.2.2.2.1. Similarly, since $\pi_0^S(\mathcal{S}(\tilde{H}_n))$ is dense in $C_0(T^*\mathbb{R}^n)$, by I.1.3.10.4 one has

$$\pi_0^S(C^*(\tilde{H}_n)) = C_0(T^*\mathbb{R}^n). \tag{2.130}$$

The results just proved imply that $\mathcal{S}(\tilde{H}_n)$, regarded as a subspace of $\prod_{\hbar\in\mathbb{R}} \mathfrak{A}^\hbar$, satisfies the three conditions in Proposition 1.2.3, Moreover, it is obvious from property 1.2.1, the definition of the norm in $C^*(\tilde{H}_n)$, and the continuity of each representation $\pi_\hbar^S$ that the function $\hbar \mapsto \|\pi_\hbar^S(\check{f}) - \pi_\hbar^S(\check{g})\|$ lies in $C_0(\mathbb{R})$ for each $\check{f} \in C^*(\tilde{H}_n)$ and $\check{g} \in \mathcal{S}(\tilde{H}_n)$. In view of the uniqueness part in the statement of Proposition 1.2.3, the continuous field determined by $\mathcal{S}(\tilde{H}_n) \subset \prod_{\hbar\in\mathbb{R}} \mathfrak{A}^\hbar$ through 1.2.3 therefore coincides with the field $\left(C^*(\tilde{H}_n), \{\mathfrak{A}^\hbar, \pi_\hbar^S\}_{\hbar\in\mathbb{R}}\right)$.

The statement about $C^*(H_n)$ is obvious from the comment preceding (2.18). ■

Theorem 2.6.5. *The quantization maps $\mathcal{Q}_\hbar^W$ of Weyl and $\mathcal{Q}_\hbar^B$ of Berezin both satisfy the assumptions of Theorem 1.2.4, and therefore lead to a continuous quantization of $T^*\mathbb{R}^n$ (cf. 1.2.5). The continuous field of C^*-algebras determined by Weyl quantization according to Theorem 1.2.4 coincides with the one determined by Berezin quantization, and is equal to the continuous field $\left(C^*(\tilde{H}_n), \{\mathfrak{A}^\hbar, \pi_\hbar^S\}_{\hbar\in\mathbb{R}}\right)$ of the C^*-algebra of the Heisenberg group.*

First observe that $\lim_{\hbar\to\pm\infty} \|\mathcal{Q}_\hbar^W(f)\| = 0$ for all $f \in \mathcal{S}(T^*\mathbb{R}^n)$. This is most easily proved by (2.108), (2.109), and the inequality $\|A\| \leq \|A\|_2$; see the comment after I.1.6.5. Combining this with Theorem 2.6.1 implies that the first claim in 2.6.5 holds.

Similarly, it follows from (2.51) that $\|\mathcal{Q}_\hbar^B(f)\| \leq \hbar^{-n}\int f$, so that for $f \in C_c^\infty(T^*\mathbb{R}^n)$ one has $\lim_{\hbar\to\pm\infty} \|\mathcal{Q}_\hbar^B(f)\| = 0$. With Theorem 2.4.1, this leads to the second claim in 2.6.5. The continuous fields determined by Weyl and Berezin quantization then coincide by Propositions 2.6.3 and 1.2.3 (used in the context of the proof of 1.2.4).

Lemma 2.6.6. *The continuous field determined by $\mathcal{S}(\tilde{H}_n)$ through Proposition 1.2.3 coincides with the continuous field determined by Weyl quantization according to Theorem 1.2.4.*

It is clear from Proposition 2.6.4 that $\mathcal{S}(\tilde{H}_n)$ satisfies the three conditions in 1.2.3. Similarly, we know from the part of the proof of 2.6.5 that has already been given that the assumptions in 1.2.4 are met. Now note that for any compact set $K \subset \mathbb{R}$ one may choose $\check{f} \in \mathcal{S}(\tilde{H}_n)$ such that $f_{\restriction\hbar}$ does not depend on $\hbar$ for $\hbar \in K$. This shows that the second field defined in 2.6.6 is contained in the first.

Conversely, let $A \in \prod_{\hbar\in K} \mathfrak{A}^\hbar$ lie in the first field. It then satisfies the first ("if") condition in Lemma 1.2.2, where each $B^{\hbar'}$ is of the form $\hbar \mapsto \pi_\hbar^S(\check{f}^{\hbar'})$ for some $\check{f}^{\hbar'} \in \mathcal{S}(\tilde{H}_n)$. Hence for each $\hbar' \in K$ there exists a function $\check{f}^{\hbar'} \in \mathcal{S}(\tilde{H}_n)$ and a neighborhood $\mathcal{N}^{\hbar'}$ such that $\|A_\hbar - \pi_\hbar^S(\check{f}^{\hbar'})\| < \epsilon$ for all $\hbar \in \mathcal{N}^{\hbar'}$. Employing the partition of unity in the proof of 1.2.2, define $C \in \prod_{\hbar\in K} \mathfrak{A}^\hbar$ by $C_\hbar := \sum_j u_j(\hbar)\mathcal{Q}_\hbar^W\left(f_{\restriction\hbar_j}^{\hbar_j}\right)$. Since $f_{\restriction\hbar_j}^{\hbar_j}$ lies in $\mathcal{S}(T^*\mathbb{R}^n)$, the section C lies in the second field because of 1.2.1.3. As $f^{\hbar_j} \in \mathcal{S}(\mathfrak{h}_n^*)$, one can choose the neighborhoods $\mathcal{N}^{\hbar_j}$ small enough so that $\|f^{\hbar_j}(\cdot, \hbar) - f^{\hbar_j}(\cdot, \hbar_j)\|_{2n+1,\infty} < \epsilon/C$ for all $\hbar \in \mathcal{N}^{\hbar_j}$; cf. (2.113). Using (2.125) and (2.113), one finds that $\|A_\hbar - C_\hbar\| < 2\epsilon$ uniformly on K. Since both fields vanish at infinity, this shows that A lies in the second field in 2.6.6. The claim follows. ∎

Theorem 2.6.5 follows from this lemma, since by the proof of Proposition 2.6.4 the first field in 2.6.6 is $\left(C^*(\tilde{H}_n), \{\mathfrak{A}^\hbar, \pi_\hbar^S\}_{\hbar\in\mathbb{R}}\right)$. ∎

Corollary 2.6.7. *The restriction of the continuous field of 2.6.5 to $\mathbb{R}\backslash\{0\}$ is trivial: if $A : \mathbb{R}\backslash\{0\} \to \mathfrak{B}_0(L^2(\mathbb{R}^n))$ is in $C_0(\mathbb{R}\backslash\{0\}, \mathfrak{B}_0(L^2(\mathbb{R}^n)))$, then A is the restriction of some element of $C^*(\tilde{H}_n)$ (seen as an element of $\prod_{\hbar\in\mathbb{R}} \mathfrak{A}^\hbar$) to $\mathbb{R}\backslash\{0\}$.*

As in the paragraph following 2.6.2, for $\hbar \neq 0$ the map $\hbar \mapsto \mathcal{Q}_\hbar^W(f_{\restriction\hbar})$ is continuous as a function from $\mathbb{R}\backslash\{0\}$ to $\mathfrak{B}_0(L^2(\mathbb{R}^n))$. The claim then follows from 2.6.5 and the proof of 1.2.3. ∎

A fascinating perspective on Theorem 2.6.5 will be given in III.3.12.

2.7 The Classical Limit of the Dynamics

We turn our attention to the connection between classical and quantum dynamics on flat space. Equation (2.94) does not hold if $h \notin \mathsf{P}^{\leq 2}$; for general Hamiltonians h one merely has asymptotic results. For the moment we proceed in a more general context, and consider a general strict quantization $\mathcal{Q}_\hbar$, defined with respect to some $\tilde{\mathfrak{A}}^0_{\mathbb{R}} \subseteq C_0(T^*\mathbb{R}^n)$ and $\mathfrak{A}^\hbar \subseteq \mathfrak{B}(L^2(\mathbb{R}^n))$. The sharpest convergence occurs when h itself lies in $\tilde{\mathfrak{A}}^0_{\mathbb{R}}$; then $H_\hbar := \mathcal{Q}_\hbar(h)$ is in $\mathfrak{A}^\hbar$. We use the notation of (2.88) and preceding text.

Proposition 2.7.1. *The flow of $h \in \tilde{\mathfrak{A}}^0_{\mathbb{R}}$ is complete. For $f \in \tilde{\mathfrak{A}}^0_{\mathbb{R}}$, assume that $\alpha_t^0(f) \in \tilde{\mathfrak{A}}^0_{\mathbb{R}}$ for all t. For any strict quantization $\mathcal{Q}_\hbar$ (such as $\mathcal{Q}_\hbar = \mathcal{Q}_\hbar^W$ or $\mathcal{Q}_\hbar = \mathcal{Q}_\hbar^B$), for all fixed t one then has*

$$\lim_{\hbar\to 0} \|\mathcal{Q}_\hbar(\alpha_t^0(f)) - \alpha_t^\hbar(\mathcal{Q}_\hbar(f))\| = 0. \tag{2.131}$$

For $\mathcal{Q}_\hbar = \mathcal{Q}_\hbar^B$ and $\mathcal{Q}_\hbar = \mathcal{Q}_\hbar^W$ we had $\tilde{\mathfrak{A}}^0_{\mathbb{R}} = C_c^\infty(T^*\mathbb{R}^n)$ and $\tilde{\mathfrak{A}}^0_{\mathbb{R}} = \mathcal{S}(T^*\mathbb{R}^n)$, respectively; since the Hamiltonian flow is smooth the assumption that $\alpha_t(f) \in \tilde{\mathfrak{A}}^0_{\mathbb{R}}$ is therefore satisfied in those cases.

The completeness of the flow of h follows, by a standard argument, from the fact that its Hamiltonian vector field ξ_h is bounded on $T^*\mathbb{R}^n$ (the components of ξ_h in canonical coordinates are themselves in $\tilde{\mathfrak{A}}^0_{\mathbb{R}}$). To prove (2.131) we write

$$\mathcal{Q}_\hbar(\alpha_t^0(f)) - \alpha_t^\hbar(\mathcal{Q}_\hbar(f)) = \int_0^t ds\, \frac{d}{ds}\alpha_{t-s}^\hbar(\mathcal{Q}_\hbar(\alpha_s^0(f)))$$
$$= \int_0^t ds\, \alpha_{t-s}^\hbar\left(\mathcal{Q}_\hbar(\{h, \alpha_s^0(f)\}) - \frac{i}{\hbar}[\mathcal{Q}_\hbar(h), \mathcal{Q}_\hbar(\alpha_s^0(f))]\right). \tag{2.132}$$

Using the fact that automorphisms are norm-preserving, we therefore see that the norm $\|\mathcal{Q}_\hbar(\alpha_t^0(f)) - \alpha_t^\hbar(\mathcal{Q}_\hbar(f))\|$ is bounded by

$$\int_0^t ds\, \left\|\mathcal{Q}_\hbar(\{h, \alpha_s^0(f)\}) - \frac{i}{\hbar}[\mathcal{Q}_\hbar(h), \mathcal{Q}_\hbar(\alpha_s^0(f))]\right\|.$$

By (1.3), the integrand vanishes as $\hbar \to 0$. ■

Since most realistic Hamiltonians in physics are not bounded, this instructive result is of limited practical use. Many physically relevant one-particle Hamiltonians are of the form

$$h(p,q) = \frac{(p - eA(q))^2}{2m} + V(q), \tag{2.133}$$

where $m > 0$, $e \in \mathbb{R}$, the function $A : \mathbb{R}^n \to \mathbb{R}^n$ is the **magnetic field potential**, and $V : \mathbb{R}^n \to R$ is a **scalar potential**. It is not necessary to assume that V and A lie in $C^\infty(\mathbb{R}^n)$; for the existence of local solutions $(p(t), q(t))$ to the classical equations of motion with initial value $(p(0), q(0))$ it suffices that ∇V and ∇A be Lipschitz around $q(0)$. A formal application of the Weyl prescription (2.37) indicates that h is quantized by the **Schrödinger operator** (cf. (2.24), (2.23))

$$H_\hbar = h(P_\hbar^S, Q_\hbar^S) = \frac{(P_\hbar^S - eA(Q_\hbar^S))^2}{2m} + V(Q_\hbar^S). \tag{2.134}$$

Theorem 2.7.2. *Given $(p,q) \in T^*\mathbb{R}^n$, assume that*

- *the classical motion $(p(t), q(t))$ with initial conditions $(p(0), q(0)) = (p,q)$ exists for $t_i < t < t_f$;*
- *V and A are $C^3(\mathbb{R}^n)$ in a neighborhood of each point $(p(t), q(t))$;*
- *V and A^2 are $O(\exp(x^2/2))$ for $x \to \infty$.*

If $\hbar < 1$, the expression $H_\hbar$ in (2.133) is symmetric on the domain $\mathcal{D}_0$ consisting of the span of all coherent states (2.47). If $A = 0$, the operator $H_\hbar$ has at least one self-adjoint extension; for arbitrary A, assume this to be the case. By abuse of notation, let the symbol $H_\hbar$ stand for an arbitrary self-adjoint extension of (2.133), generating the unitary one-parameter group $\exp(itH_\hbar/\hbar)$ on $L^2(\mathbb{R}^n)$. Then, with the notation (2.88), I.(2.13), and (2.47), for all $t \in (t_i, t_f)$, for $\mathcal{Q}_\hbar = \mathcal{Q}_\hbar^W$ (and $f \in \mathcal{S}(T^*\mathbb{R}^n)$) or $\mathcal{Q}_\hbar = \mathcal{Q}_\hbar^B$ (and $f \in C_c^\infty(T^*\mathbb{R}^n)$), one has

$$\lim_{\hbar \to 0} \left(\Psi_\hbar^{(p,q)}, [\mathcal{Q}_\hbar(\alpha_t^0(f)) - \alpha_t^\hbar(\mathcal{Q}_\hbar(f))] \Psi_\hbar^{(p,q)} \right) = 0. \tag{2.135}$$

Since $\mathcal{D}_0$ is contained in $\mathcal{S}(\mathbb{R}^n)$, and the growth conditions postulated on V and A imply that the multiplication operators $V(Q_\hbar^S)$, $A_i(Q_\hbar^S)$, and $A(Q_\hbar^S)^2$ map $\mathcal{D}_0$ into itself, it easily follows that $H_\hbar$ is indeed symmetric on $\mathcal{D}_0$. If $A = 0$, then $H_\hbar$ commutes with the conjugation $\Psi \mapsto \overline{\Psi}$ on $L^2(\mathbb{R}^n)$; hence it has equal deficiency indices.

We now write $\sigma = (p,q)$ and $R_\hbar^S = (P_\hbar^S, Q_\hbar^S)$. Given a particular σ, we expand $H_\hbar$ around the solution $\sigma(t)$ of the classical equations of motion $\partial\sigma/dt = \{h, \sigma\}$, with initial value $\sigma(0) = \sigma$. That is,

$$H_\hbar = H_{(2)}(t) + H_3(t), \tag{2.136}$$

with

$$H_{(2)}(t) := H_0 + H_1(t) + H_2(t), \tag{2.137}$$

$$H_0 := h(\sigma(t))\mathbb{I}, \tag{2.138}$$

$$H_1(t) := \frac{\partial h}{\partial \sigma^i}(\sigma(t))(R_\hbar^S - \sigma(t))^i, \tag{2.139}$$

$$H_2(t) := \tfrac{1}{2}\frac{\partial^2 h}{\partial \sigma^i \partial \sigma^j}(\sigma(t))(R_\hbar^S - \sigma(t))^i (R_\hbar^S - \sigma(t))^j, \tag{2.140}$$

while $H_3(t)$ is defined as the remainder.

The operator $H_{(2)}(t)$ has a semiclassical interpretation. Firstly, $h(\sigma(t))$ in (2.138) is just the classical Hamiltonian evaluated at the classical path. This is independent of t. Secondly, writing $S := T^*\mathbb{R}^n$, consider the function $h^{(1)}$ on TS defined by

$$h^{(1)}(v, \sigma) := dh_\sigma(v) = \frac{\partial h}{\partial \sigma^i} v^i. \tag{2.141}$$

One sees that $H_1(t)$ is obtained from $h^{(1)}$ by a "partial" quantization along the trajectory $\sigma(t)$:

$$H_1(t) = h^{(1)}(R_\hbar^S - \sigma(t), \sigma(t)). \tag{2.142}$$

Secondly, for each fixed t and σ define a function $h^{(2)}(t)$ on $T_{\sigma(t)}S$ by

$$h^{(2)}(v, t) := \tfrac{1}{2}(h'')_{ij}(t) v^i v^j, \tag{2.143}$$

where

$$(h'')_{ij}(t) = \frac{\partial^2 h}{\partial \sigma^i \partial \sigma^j}(\sigma(t)). \tag{2.144}$$

Clearly,

$$H_2(t) = h^{(2)}(R_\hbar^S - \sigma(t), t). \tag{2.145}$$

Both $h^{(1)}$ and $h^{(2)}$ generate linearized equations of motion, but they do so in a different sense. The Hamitonian flow $\sigma \mapsto \sigma(t)$ on S generated by h pushes forward to a flow $(v, \sigma) \mapsto (v(t), \sigma(t))$ on the tangent bundle TS. By definition of the pushforward, one may think of the latter as follows: if $\sigma(t, \alpha)$ is a one-parameter family of solutions of the equations of motion on S (where $\alpha \in (-\epsilon, \epsilon)$ for some $\epsilon > 0$) neighboring a given trajectory $\sigma(t) = \sigma(t, 0)$, and $(v, \sigma) \in T_\sigma S$ equals $\partial\sigma(0, \alpha)/\partial\alpha|\alpha = 0$, then $(v(t), \sigma(t)) = \partial\sigma(t, \alpha)/\partial\alpha|\alpha = 0$.

Now, TS is a symplectic manifold in a natural way: The map $B_\sharp : TS \to T^*S$ (cf. I.2.3.6) defines a symplectic form $\omega^* := -B_\sharp^*\omega$ on TS (where ω is the canonical symplectic form on T^*S, cf. Definition I.2.3.8). If (p_i, q^i) are canonical coordinates on S, we denote the coordinates induced on TS by $(\tilde{p}_i, \tilde{q}^i, p_i, q^i)$; these stand for the point $\tilde{p}_i \partial/\partial p_i + \tilde{q}^i \partial/\partial q^i \in T^*_{(p,q)}TS$. In terms of these, the form ω^* is given by

$$\omega^* = d\tilde{q}^i \wedge dp_i + dq^i \wedge d\tilde{p}_i. \tag{2.146}$$

The associated Poisson bracket is

$$\{f, g\}^* = \frac{\partial f}{\partial p_i}\frac{\partial g}{\partial \tilde{q}^i} + \frac{\partial f}{\partial \tilde{p}_i}\frac{\partial g}{\partial q^i} - f \leftrightarrow g. \tag{2.147}$$

Accordingly, the pushforward flow $(v, \sigma) \mapsto (v(t), \sigma(t))$ on TS is Hamiltonian with respect to the Poisson bracket (2.147) and the Hamiltonian (2.141).

Alternatively, if the trajectory $\sigma(t)$ is already known, one can describe the tangent part of the flow $(v, \sigma) \mapsto (v(t), \sigma(t))$ as a Hamiltonian system in the v-variable. This is done as follows. Since in the present case the tangent bundle TS is globally trivial, there is a natural identification of all fibers of TS; in particular, $T_{\sigma(t)}$ is identified with $T_\sigma S$ for all t. The vector space $T_\sigma S$ is a linear symplectic space, whose symplectic form ω_σ is simply the canonical symplectic form ω on S, evaluated at $T_\sigma S$; writing $v = (\tilde{p}, \tilde{q})$, one has $\omega_\sigma = d\tilde{q}^i \wedge d\tilde{p}_i$. The time evolution $v \mapsto v(t)$ (where $v \in T_\sigma S$) then coincides with the Hamiltonian flow on $T_\sigma S$ generated by the time-dependent Hamiltonian $h^{(2)}(t)$ (regarded as a function on $T_\sigma S$ through the above identification). The corresponding Hamiltonian equations of motion are given by

$$\frac{dv}{dt} = \{h^{(2)}(t), v\}. \tag{2.148}$$

Since this system is linear in v, it is solved by

$$v(t) = M(t)v, \tag{2.149}$$

where the $2n \times 2n$ matrix $M(t)$ is the solution of

$$\frac{dM(t)}{dt} = Jh''(t)M(t) \tag{2.150}$$

with initial condition $M(0) = \mathbb{I}_{2n}$; here J is given by (2.1).

We return to the quantum theory. To understand the nature of $H_1(t)$ we define

$$\Psi_\hbar^{(p,q)}(t)_{\text{cl}} := e^{iS(t)/\hbar}\Psi_\hbar^{(p(t),q(t))}, \tag{2.151}$$

with the **classical action**

$$S(t) := \int_0^t ds\, \left[\tfrac{1}{2}(p(s)\dot{q}(s) - \dot{p}(s)q(s)) - h(p(s), q(s))\right]. \tag{2.152}$$

We can evidently write this as

$$\Psi_\hbar^{(p,q)}(t)_{\text{cl}} = U_1^{(p,q)}(t)\Psi_\hbar^{(p,q)}, \tag{2.153}$$

where

$$U_1^{(p,q)}(t) = e^{iS(t)/\hbar}U_{\frac{1}{\hbar}}(p(t), q(t))U_{\frac{1}{\hbar}}(p, q)^* \tag{2.154}$$

is the **classical propagator**. The point is now that the classical equations of motion and the relation

$$U_{\frac{1}{\hbar}}(\sigma)R_\hbar^{S,i}U_{\frac{1}{\hbar}}(\sigma)^* = R_\hbar^{S,i} - \sigma^i\mathbb{I} \tag{2.155}$$

imply that $U_1^{(p,q)}(t)$ is the solution of

$$i\hbar\frac{d}{dt}U_1^{(p,q)}(t) = (H_0 + H_1(t))U_1^{(p,q)}(t) \tag{2.156}$$

with initial condition $U_1^{(p,q)}(0) = \mathbb{I}$.

We now incorporate $H_2(t)$. In terms of the metaplectic representation $\rho^\hbar$ constructed in 2.2.2, we define

$$U_2^{(p,q)}(t) := U_1^{(p,q)}(t)U_{\frac{1}{\hbar}}(p,q)\rho^\hbar(M(t))U_{\frac{1}{\hbar}}(p,q)^*. \tag{2.157}$$

It follows from 2.2.2 and (2.148)–(2.150) that $\rho^\hbar(M(t))$ is the propagator for the Hamiltonian $H^{(2)}(t) := \frac{1}{2}(h'')_{ij}(t)R_\hbar^{S,i}R_\hbar^{S,j}$. Indeed, a short calculation, using $d\rho(M(t))/dt = d\rho^\hbar(\dot{M}(t)M(t)^{-1})\rho^\hbar(M(t))$, and subsequently (2.150), (2.39), and (2.32), shows that

$$i\hbar\frac{d}{dt}\rho^\hbar(M(t)) = H^{(2)}(t)\rho^\hbar(M(t)). \tag{2.158}$$

Consequently, from (2.156), (2.158), and (2.155) one derives

$$i\hbar\frac{d}{dt}U_2^{(p,q)}(t) = H_{(2)}(t)U_2^{(p,q)}(t). \tag{2.159}$$

Hence the object

$$\Psi_\hbar^{(p,q)}(t)_{\mathrm{sc}} := U_2^{(p,q)}(t)\Psi_\hbar^{(p,q)} \tag{2.160}$$

satisfies the **semiclassical Schrödinger equation**

$$i\hbar\frac{d}{dt}\Psi_\hbar^{(p,q)}(t)_{\mathrm{sc}} = H_{(2)}(t)\Psi_\hbar^{(p,q)}(t)_{\mathrm{sc}}. \tag{2.161}$$

We refer to $U_2^{(p,q)}(t)$ as the **semiclassical propagator**. This terminology is motivated by the following result.

Proposition 2.7.3. *With $H_\hbar$, $\Psi_\hbar^{(p,q)}$, and $\Psi_\hbar^{(p,q)}(t)_{\mathrm{sc}}$ given by (2.134), (2.45), and (2.160), respectively, one has*

$$\lim_{\hbar\to 0}\|e^{-itH_\hbar/\hbar}\Psi_\hbar^{(p,q)} - \Psi_\hbar^{(p,q)}(t)_{\mathrm{sc}}\| = 0. \tag{2.162}$$

To prove this, we follow the strategy of the proof of Proposition 2.7.1, and write (with $U(t) := \exp(-itH_\hbar/\hbar)$)

$$\begin{aligned}\left(U(t) - U_2^{(p,q)}(t)\right)\Psi_\hbar^{(p,q)} &= -U(t)\int_0^t ds\,\frac{d}{ds}U(s)^*U_2^{(p,q)}(s)\Psi_\hbar^{(p,q)}\\ &= -\frac{i}{\hbar}U(t)\int_0^t ds\,U(s)(H_\hbar - H_{(2)}(s))U_2^{(p,q)}(s)\Psi_\hbar^{(p,q)},\end{aligned} \tag{2.163}$$

where (2.159) has been used. The existence of the strong derivative d/ds follows from the growth conditions imposed on V and A. We now insert the expansion (2.136), and use the explicit form (2.157) to obtain the estimate

$$\left\|(U(t) - U_2^{(p,q)}(t))\Psi_\hbar^{(p,q)}\right\| \le \frac{1}{\hbar}\int_0^t ds\,\left\|\tilde{H}_3(s)\rho^\hbar(M(s))\Psi_\hbar^{(0,0)}\right\|. \tag{2.164}$$

Here $\tilde{H}_3(s) := U_{1/\hbar}(p(s),q(s))^*H_3(s)U_{1/\hbar}(p(s),q(s))$; this is just $H_3(s)$ with $R_\hbar^S - \sigma(t)$ replaced by $R_\hbar^S$. Using (2.39), one finds $\rho^\hbar(M(s))\Psi_\hbar^{(0,0)}(x)$ to be proportional to $\hbar^{-n/4}\exp(-(Nx,x)/(2\hbar))$, where N is a nonsingular complex matrix (composed from the entries of M) whose real part is positive definite. We then use the explicit action (2.24), (2.23) of the operators in $H_3(s)$, upon which Taylor's formula with remainder and the growth conditions on V and A lead to the conclusion that the integrand in (2.164) is $O(\hbar^{3/2})$. Hence the left-hand side is $O(\hbar^{1/2})$, and (2.162) is proved. ∎

Using (2.157), as well as (2.93) (with (2.80) and (2.81)), we obtain

$$\left(U_2^{(p,q)}(t)\Psi_\hbar^{(p,q)}, \mathcal{Q}_\hbar^W(f)U_2^{(p,q)}(t)\Psi_\hbar^{(p,q)}\right) = \left(\Psi_\hbar^{(0,0)}, \mathcal{Q}_\hbar^W(\alpha^0_{(M(t),\sigma(t))^{-1}}(f))\Psi_\hbar^{(0,0)}\right). \tag{2.165}$$

A short calculation shows that

$$\left(\Psi_\hbar^{(p,q)}, \mathcal{Q}_\hbar^W(f)\Psi_\hbar^{(p,q)}\right) = e^{\hbar\Delta_{2n}/4}f(p,q); \tag{2.166}$$

cf. (2.117). This equation, or a combination of (1.17) and (2.117), implies

$$\lim_{\hbar\to 0}\left(\Psi_\hbar^{(p,q)}, \mathcal{Q}_\hbar^W(f)\Psi_\hbar^{(p,q)}\right) = f(p,q). \tag{2.167}$$

By (2.162) and (2.167) we then obtain

$$\lim_{\hbar\to 0}\left(\Psi_\hbar^{(p,q)}, \alpha_t^\hbar(\mathcal{Q}_\hbar^W(f))\Psi_\hbar^{(p,q)}\right) = \alpha^0_{(M(t),\sigma(t))^{-1}}(f)(0). \tag{2.168}$$

By (2.80), the right-hand side equals $f(\sigma(t)) = f(p(t),q(t))$, as $(M(t),\sigma(t))$ acting on 0 yields just $\sigma(t)$. Theorem 2.7.2 then follows for $\mathcal{Q}_\hbar = \mathcal{Q}_\hbar^W$, since by (2.167) one has

$$\lim_{\hbar\to 0}\left(\Psi_\hbar^{(p,q)}, \alpha_t^0(\mathcal{Q}_\hbar^W(f))\Psi_\hbar^{(p,q)}\right) = f(p(t),q(t)). \tag{2.169}$$

For $\mathcal{Q}_\hbar = \mathcal{Q}_\hbar^B$ we can use 2.4.3 to write

$$U_{\frac{1}{\hbar}}(p,q)^*\mathcal{Q}_\hbar^B(f)U_{\frac{1}{\hbar}}(p,q) = \mathcal{Q}_\hbar^B(\alpha^0_{(-p,-q)}(f)), \tag{2.170}$$

where we have identified $(-p,-q)$ with $(\mathbb{I}_{2n},(p,q))^{-1}$. We apply this with (p,q) replaced by $(p(t),q(t))$. An explicit computation establishes that

$$\lim_{\hbar\to 0}\left(\rho^\hbar(M(t))\Psi_\hbar^{(0,0)}, \mathcal{Q}_\hbar^B(f)\rho^\hbar(M(t))\Psi_\hbar^{(0,0)}\right) = f(0,0); \tag{2.171}$$

cf. (1.17). The desired result then follows as for $\mathcal{Q}_\hbar^W$. ∎

In fact, the above proof for $\mathcal{Q}_\hbar^B$ works for $\mathcal{Q}_\hbar^W$ as well; in either case the essential ingredients of the proof are the equivariance of $\mathcal{Q}_\hbar^B$ and $\mathcal{Q}_\hbar^W$ under translations in $T^*\mathbb{R}^n$ and the fact that the quadratic term $M(t)$ does not contribute to the limit in (2.135).

It is remarkable that while the classical motion generated by h may be incomplete, the quantum evolution generated by $H_\hbar$ is defined for all times. Hence classical incompleteness is generically traded for quantum-mechanical nonuniqueness, for the self-adjoint extension $H_\hbar$ may not be uniquely determined by the formal expression (2.133).

3 Quantization on Riemannian Manifolds

3.1 Some Affine Geometry

We now replace the configuration space $\mathbb{R}^n$ by a general n-dimensional connected manifold Q. In general, whenever it is convenient to employ (local) coordinates q^i on Q, we will use them; recall that $\partial_i := \partial/\partial q^i$.

We start with a geometric structure on the tangent bundle.

Definition 3.1.1. *An* **affine connection** *on the tangent bundle TQ is a collection of linear maps $\nabla_\xi : \Gamma(TQ) \to \Gamma(TQ)$, defined for each vector field $\xi \in \Gamma(TQ)$, such that $\nabla_{f\xi} = f\nabla_\xi$ and $\nabla_\xi f\eta = \xi(f)\eta + f\nabla_\xi\eta$ for all $f \in C^\infty(Q)$ and all $\xi, \eta \in \Gamma(TQ)$.*

It follows from this definition that in local coordinates the covariant derivative can be expressed by

$$\nabla_\xi\eta = \xi^i\nabla_i\eta = \xi^i(\partial_i\eta^j + \Gamma^j_{ik}\eta^k)\partial_j, \tag{3.1}$$

where $\xi = \xi^i\partial_i$ and $\eta = \eta^i\partial_i$ are vector fields, and the **connection coefficients** Γ^j_{ik} are certain functions on Q.

A curve $(v(t), q(t))$ in TQ (covering, as the notation indicates, a curve $q(t)$ in Q) is called **horizontal** if $\nabla_{\dot{q}(t)}v(t) = 0$; although the covariant derivative is defined as acting on vector fields, this condition makes sense because $\nabla_{\dot{q}(t)}$ involves only the behavior of the section it acts on along the curve $q(t)$.

In that case one says that $v(t) \in T_{q(t)}Q$ is the **parallel transport** of $v(0) \in T_{q(0)}Q$, and that $(v(t), q(t))$ is a **horizontal lift** of $q(t)$. Each curve $q(t)$ has a unique lift $\ell_{(v,q)}(q(t))$ through a given point $(v, q) \in TQ$. The collection of all vectors in $T_{(q,v)}TQ$ that are tangent to some horizontal curve through (v, q) forms the **horizontal subspace** $T^{\text{hor}}_{(v,q)}TQ$ of $T_{(v,q)}TQ$. One may equally well speak of the horizontal lift $\ell_{(v,q)}(X)$ of a vector $X \in T_qQ$; this is the unique vector in $T^{\text{hor}}_{(v,q)}(TQ)$ that projects to X under $\tau_{T(TQ)\to TQ}$.

If (v^i, q^i) are canonical coordinates on TQ (standing for the point $v^i\partial_i \in T_qQ$), we denote the coordinates induced on $T(TQ)$ by $(\tilde{v}^i, \tilde{q}^i, v^i, q^i)$; these stand for the point $\tilde{v}^i\partial/\partial v^i + \tilde{q}^i\partial/\partial q^i \in T_{(v,q)}TQ$. In terms of these, it follows from (3.1) that horizontal vectors in $T_{(v,q)}$ are of the generic form

$$\ell_{(v,q)}(w, q) = (-\Gamma^i_{jk}(q)w^jv^k, w^i, v^i, q^i). \tag{3.2}$$

One has a natural isomorphism $T^{\text{hor}}_{(v,q)}TQ \simeq T_qQ$, under which $X \in T^{\text{hor}}_{(v,q)}TQ$ corresponds to $\tau_*X \in T_qQ$; in coordinates, $(-\Gamma^i_{jk}(q)w^jv^k, w^i, v^i, q^i) \simeq (w^i, q^i)$.

In contrast, the **vertical subspace** $T^{\text{ver}}_{(v,q)}TQ \subset T_{(v,q)}TQ$ consists of all tangent vectors to **vertical curves** $(v(t), q)$, which lie in T_qQ. In other words, $T^{\text{ver}}_{(v,q)}TQ := \ker\tau_* \cap T_{(v,q)}TQ$, where $\tau := \tau_{TQ\to Q}$, hence $\tau_* = \tau_{T(TQ)\to TQ}$. Such vertical vectors are of the form $(w^i, 0, v^i, q^i)$. Also here one has a natural isomorphism $T^{\text{ver}}_{(v,q)}TQ \simeq T_qQ$, because $T^{\text{ver}}_{(v,q)}TQ = T(T_qQ) \simeq T_qQ$. In coordinates one has $(w^i, 0, v^i, q^i) \simeq (w^i, q^i)$. Hence the decomposition

$$T_{(v,q)}TQ = T^{\text{hor}}_{(v,q)}TQ \oplus T^{\text{ver}}_{(v,q)}TQ \simeq T_qQ \oplus T_qQ. \tag{3.3}$$

An affine connection on TQ defines a vector field ξ^c on TQ by $\xi^c_X := \ell_X(X)$. The integral curves of ξ^c are the **geodesic flow** on TQ. However, the name ("affinely parametrized") **geodesic** is reserved for a curve in Q that is the projection of such a flow in TQ under $\tau_{TQ\to Q}$. It is customary to denote geodesics by $\gamma(\cdot)$.

Proposition 3.1.2. *With $\dot{\gamma} := d\gamma/dt$, a geodesic satisfies the equation*

$$\nabla_{\dot{\gamma}}\dot{\gamma} = 0. \tag{3.4}$$

This is obvious from the definition of a horizontal lift and of ξ^c. ∎

Putting $w = v$ in (3.2), we see that the coordinate form of the geodesic equation (3.4) is

$$\frac{d}{dt}\dot{\gamma}^i(t) + \Gamma^i_{jk}(\gamma(t))\dot{\gamma}^j(t)\dot{\gamma}^k(t) = 0. \tag{3.5}$$

We write $\gamma(q, v; \cdot)$ for the parametrized geodesic starting at $\gamma(q, v; 0) = q$ with tangent vector $\dot{\gamma}(q, v; 0) = v$. Existence and uniqueness of such a geodesic for small enough t routinely follow from the theory of ordinary differential equations. However, there is no guarantee that a geodesic exists for all t.

An important role in affine geometry is played by the **exponential mapping** exp, which is defined through geodesics. It maps a certain set $\tilde{\mathcal{O}} \subset TQ$ into Q, and is defined by

$$\exp(X) := \gamma(\tau_{TQ\to Q}(X), X; 1). \tag{3.6}$$

The set $\tilde{\mathcal{O}}$ is simply the set of those X for which the geodesic in question is defined at $t = 1$; this is an open subset of TQ, evidently containing the zero section Q. The restriction of exp to $\tilde{\mathcal{O}}_q := T_qQ \cap \tilde{\mathcal{O}}$ is denoted by $\exp_q$. For good global properties of geodesics a special assumption has to be made.

Definition 3.1.3. *A manifold with affine connection is called* **geodesically complete** *when all geodesics exist for arbitrary values of the parameter t.*

Clearly, Q is geodesically complete iff $\tilde{\mathcal{O}} = TQ$; in other words, for all $q \in Q$ the map $\exp_q$ is defined on all of T_qQ. A weaker notion would be completeness at a point q, meaning that $\exp_q$ is defined on T_qQ. The issue of completeness will be taken up further in the next section, where a special form of the affine connection leads to interesting results in this context.

From the tangent bundle we pass to the cotangent bundle. The cotangent bundle $S = T^*Q$ is equipped with the canonical symplectic form I.(2.23) and the associated Poisson bracket I.(2.24). Recall the notation $(p_i, q^j) := p_i dq^i$ for canonical coordinates on T^*Q. The following functions on T^*Q will be of basic importance in what follows. Firstly, a function $\tilde{g} \in C^\infty(Q, \mathbb{R})$ induces the smooth function

$$J_{\tilde{g}} := \tau^*\tilde{g} \tag{3.7}$$

on T^*Q (with $\tau := \tau_{T^*Q\to Q}$). Secondly, a smooth vector field ξ on Q has a **symbol** $J_\xi \in C^\infty(T^*Q)$, defined by

$$J_\xi(\sigma) := \sigma_{\tau(\sigma)}(\xi); \tag{3.8}$$

in coordinates, if $\xi(q) = \xi^i(q)\partial_i$, this reads $J_\xi(p,q) = p_i\xi^i(q)$. The basic Poisson brackets between these functions, which comprise the essence of the canonical Poisson structure on T^*Q, are

$$\{J_{\tilde{g}}, J_{\tilde{h}}\} = 0; \tag{3.9}$$

$$\{J_\xi, J_{\tilde{g}}\} = J_{\xi\tilde{g}}; \tag{3.10}$$

$$\{J_{\xi_1}, J_{\xi_2}\} = J_{[\xi_1,\xi_2]}. \tag{3.11}$$

These functions and Poisson brackets have a group-theoretical interpretation. Firstly, regard $C_c^\infty(Q,\mathbb{R})$ as an abelian group (under addition). The Lie algebra of this group is the same space, equipped with the trivial Lie bracket. Then (3.9) shows that the map $\tilde{g} \mapsto J_{\tilde{g}}$ is a Lie algebra antihomomorphism of $C_c^\infty(Q,\mathbb{R})$ into $C^\infty(T^*Q,\mathbb{R})$.

Secondly, consider the group $\mathrm{Diff}(Q)$ of (smooth) diffeomorphisms of Q with compact support (that is, a diffeomorphism $\varphi \in \mathrm{Diff}(Q)$ is the identity map outside some compact set). It is possible to equip $\mathrm{Diff}(Q)$ with the structure of an infinite-dimensional Lie group (though not one modeled on a Banach manifold). Since one-parameter subgroups of $\mathrm{Diff}(Q)$ by definition generate flows on Q, one infers that, at least formally, the Lie algebra $\mathfrak{diff}(Q)$ of $\mathrm{Diff}(Q)$ is the set $\Gamma_c(TQ)$ of (smooth) vector fields ξ on Q with compact support. In the opposite direction, the exponential map on $\mathfrak{diff}(Q)$ is given by $(\exp\xi)(q) = \varphi_1(q)$, where φ_t is the flow defined by the vector field ξ. Unfortunately, with this identification the Lie bracket in this Lie algebra equals minus the commutator of vector fields; in what follows the notation $[\xi_1,\xi_2]$ stands for the latter (as usual). Evidently, (3.11) shows that the map $\xi \mapsto J_\xi$ is a Lie algebra antihomomorphism of $\mathfrak{diff}(Q)$ into $C^\infty(T^*Q,\mathbb{R})$.

Finally, we can define the semidirect product

$$\mathcal{G}_Q := \mathrm{Diff}(Q) \ltimes C_c^\infty(Q,\mathbb{R}) \tag{3.12}$$

through the natural action of $\mathrm{Diff}(Q)$ on $C_c^\infty(Q,\mathbb{R})$: $\varphi \in \mathrm{Diff}(Q)$ maps $\tilde{g} \in C_c^\infty(Q)$ to $(\varphi^{-1})^*\tilde{g}$. The Lie algebra of $\mathcal{G}_Q$ is denoted by $\mathfrak{g}_Q$. The corresponding "mixed" Lie bracket is $[\xi,\tilde{g}] = -\xi\tilde{g}$; the minus sign reflects the one in φ^{-1} above. Hence (3.10) shows

Proposition 3.1.4. *The map $J : \xi + \tilde{g} \mapsto J_\xi + J_{\tilde{g}}$ is a Lie algebra antihomomorphism of $\mathfrak{g}_Q$ into $C^\infty(T^*Q,\mathbb{R})$.*

We describe the Hamiltonian flow on T^*Q generated by $J_{\tilde{g}}$ and J_ξ.

Proposition 3.1.5. *Define an action ρ_0 of $\mathcal{G}_Q$ on T^*Q by*

$$\rho_0(\tilde{g}) : \sigma \mapsto \sigma - d\tilde{g}(\tau(\sigma));$$

$$\rho_0(\varphi) : \sigma \mapsto (\varphi^{-1})^*\sigma, \tag{3.13}$$

and $\rho_0(\varphi, \tilde{g}) := \rho_0(\tilde{g}) \circ \rho_0(\varphi)$. *If* $X \in \mathfrak{g}_Q$, *then* $\sigma \mapsto \text{Exp}(tX)\sigma$ *is the Hamiltonian flow on* T^*Q *generated by* J_X.

This is shown by a straightforward computation in coordinates. □

This and similar results will be placed in their proper context in III.2.4.

3.2 Some Riemannian Geometry

We now assume that Q is equipped with a Riemannian structure, i.e., with a metric. The following remarks are mainly intended to establish some notation and conventions. The **metric g** provides each tangent space $T_q Q$ with an inner product $\mathbf{g}_q$, that is, a bilinear symmetric positive definite map $\mathbf{g}_q : T_q Q \otimes T_q Q \to Q$. The positive-definiteness means that $\mathbf{g}_q(X, X) \geq 0$ for all $X \in T_q Q$, with $\mathbf{g}_q(X, X) = 0 \Leftrightarrow X = 0$. This, of course, implies that $\mathbf{g}_q$ is nondegenerate. Throughout this chapter $\mathbf{g}$ is assumed to be smooth (C^∞).

The **length** of a parametrized C^1 curve $\{c(t) \mid t \in [t_i, t_f]\}$ is

$$\ell(c) := \int_{t_i}^{t_f} dt \sqrt{\mathbf{g}_{c(t)}(\dot{c}(t), \dot{c}(t))}; \tag{3.14}$$

this is evidently independent of the parametrization. The length of a piecewise C^1 curve is the sum of the lengths of its C^1 pieces. The **distance** d between two points in Q is the infimum over the lengths of all piecewise C^1 curves connecting the points. (If Q were not connected, this definition would apply if the points lie in the same component; if they don't, the distance is ∞.) It is easily shown that this distance defines a metric on Q in the sense of point-set topology, making (Q, d) a metric space in that sense.

Let us use the metric to define a bundle homomorphism $\mathbf{g}^\sharp : T^*Q \to TQ$. This maps T_q^*Q into $T_q Q$, and is defined by the property

$$\mathbf{g}_q(\mathbf{g}^\sharp(\sigma), X) = \sigma(X) \tag{3.15}$$

(where $\sigma \in T_q^*Q$ and $X \in T_q Q$). The nondegeneracy of $\mathbf{g}$ implies firstly that $\mathbf{g}^\sharp$ is well-defined by (3.15), and secondly that it is a bijection; its inverse is denoted by $\mathbf{g}_\sharp : TQ \to T^*Q$. The smoothness of $\mathbf{g}$ then leads to the conclusion that $\mathbf{g}^\sharp$ and $\mathbf{g}_\sharp$ are diffeomorphisms. One application is the definition of the gradient of a function:

$$\nabla f := \mathbf{g}^\sharp(df). \tag{3.16}$$

One writes $\mathbf{g}_{ij}(q) := \mathbf{g}_q(\partial_i, \partial_j)$; the inverse of the matrix $\{\mathbf{g}_{ij}(q)\}$ is denoted by $\{\mathbf{g}^{ij}(q)\}$, so that $\mathbf{g}_{ik}(q)\mathbf{g}^{kj}(q) = \delta_i^j$. If $\sigma = \sigma_i dq^i \in T_q Q$, then $\mathbf{g}^\sharp(\sigma) = \sigma^i \partial_i$, with $\sigma^i = \mathbf{g}^{ij}(q)\sigma_j$; hence $\sigma_i = \mathbf{g}_{ij}(q)\sigma^j$. A similar notation is used for general tensors.

All concepts of the preceding sections apply; recall Definition 3.1.1.

Definition 3.2.1. *The* **Levi-Civita connection**, *or* **covariant derivative**, *is the unique affine connection* ∇ *on* TQ *that is* **torsion-free** *in that*

$$\nabla_\xi \eta - \nabla_\eta \xi = [\xi, \eta], \tag{3.17}$$

and **metric** *in that*

$$\xi \mathbf{g}(\eta_1, \eta_2) = \mathbf{g}(\nabla_\xi \eta_1, \eta_2) + \mathbf{g}(\eta_1, \nabla_\xi \eta_2) \tag{3.18}$$

for all vector fields ξ, η_1, η_2.

For the Levi-Civita connection the object Γ appearing in (3.1) takes the form

$$\Gamma^i_{jk} := \tfrac{1}{2}\mathbf{g}^{il}(\partial_j \mathbf{g}_{kl} + \partial_k \mathbf{g}_{lj} - \partial_l \mathbf{g}_{jk}). \tag{3.19}$$

In this context Γ is known as the **Christoffel symbol.**

The **Riemann curvature tensor R** is defined by

$$\mathbf{R}(\xi, \eta) := [\nabla_\xi, \nabla_\eta] - \nabla_{[\xi,\eta]}, \tag{3.20}$$

where $\mathbf{R}_q(\xi, \eta) : T_q R \to T_q Q$. Remarkably, this is a local expression that indeed defines a tensor. If we write $\mathbf{R}(X, Y)Z^i = \mathbf{R}^i_{jkl} Z^j X^k Y^l$, then from (3.20) and (3.1) one has

$$\mathbf{R}^i_{jkl} = \partial_k \Gamma^i_{jl} - \partial_l \Gamma^i_{jk} + \Gamma^i_{mk}\Gamma^m_{jl} - \Gamma^i_{ml}\Gamma^m_{jk}. \tag{3.21}$$

Lowering the first index, one has the symmetries

$$\mathbf{R}_{ijkl} = -\mathbf{R}_{jikl} = -\mathbf{R}_{ijlk} = \mathbf{R}_{klij}. \tag{3.22}$$

The **Ricci scalar** is defined by

$$\mathbf{R} := \mathbf{g}^{ij}\mathbf{R}^k_{ikj}. \tag{3.23}$$

The Levi-Civita connection leads to geodesics satisfying (3.5) with (3.19). A set $U \subset Q$ is called **geodesically convex** if any two points in U can be joined by a unique geodesic of minimum length, that lies in U. A neighborhood U_q of q is called **normal** if $\exp_q$ is a diffeomorphism between some neighborhood of 0 in $T_q Q$ and U_q. Clearly, a geodesically convex neighborhood is normal.

In local Riemannian geometry one can prove the following

Proposition 3.2.2. *Consider the ball* $B^\epsilon_q := \{X \in T_q Q \mid \mathbf{g}_q(X, X) < \epsilon^2\}$. *For each* q *there exists an* $\epsilon > 0$ *such that* $B^\epsilon_q \subset \tilde{O}_q$, *and* $U^\epsilon_q := \exp_q B^\epsilon_q$ *is geodesically convex.*

We will usually drop the ϵ on U^ϵ_q. On a normal neighborhood U_q of q one can often use **normal coordinates** $q^i_{(n)}$ to simplify computations. These depend on the choice of a fixed orthonormal basis $\{\mathbf{e}_i\}$ of $T_q Q$. By definition, the normal coordinates of a point $\gamma(q, v^i \mathbf{e}_i; 1)$ (assumed to lie in U_q) are $q^i_{(n)} = v^i$. Obviously, the normal coordinates of q itself are $q^i_{(n)} = 0$, and geodesics simply have the form $q^i_{(n)}(t) = tv^i$. One can show that in these coordinates,

$$\mathbf{g}_{ij}(q_{(n)}) = \delta_{ij} - \tfrac{1}{3}\mathbf{R}_{ikjl} q^k_{(n)} q^l_{(n)} + O(q^3_{(n)}). \tag{3.24}$$

Hence $\mathbf{g}_{ij}(q) = \delta^{ij}$ and $\partial_k \mathbf{g}_{ij}(q) = 0$, so that the Christoffel symbols Γ^i_{jk} vanish at q. Furthermore,

$$\partial_k \partial_l \mathbf{g}_{ij}(q) = -\tfrac{1}{3}(\mathbf{R}_{ikjl}(q) + \mathbf{R}_{iljk}(q)). \tag{3.25}$$

A fundamental theorem of global Riemannian geometry gives equivalent forms of completeness. Recall our standing assumption that Q is connected.

Theorem 3.2.3. *The following conditions are equivalent:*

- $(Q, \mathbf{g})$ *is geodesically complete at one point.*
- $(Q, \mathbf{g})$ *is geodesically complete.*
- (Q, d) *is complete (as a metric space).*

If any (hence all) of these conditions is satisfied, then any two points may be joined by a **minimal geodesic***; this is a geodesic whose length equals the distance between the points.*

In view of this theorem, we simply call $(Q, \mathbf{g})$ **complete** iff it is geodesically (hence metrically) complete.

The geodesic of the last claim in this theorem is not necessarily unique.

Definition 3.2.4. *The* **cut locus** $C(q)$ *of a given point q in a complete Riemannian manifold Q is the collection of points q' in Q for which there exists more than one minimal geodesic between q and q'.*

Global Riemannian geometry yields the following decomposition of Q.

Theorem 3.2.5. *In a complete Riemannian manifold Q, let $\mathcal{O}^{\max}_q$ consist of all $X \in T_q Q$ for which $\gamma(q, X; t)$ is minimal for all $t \in [0, 1]$. The cut locus is*

$$C(q) = \exp_q(\partial \mathcal{O}^{\max}_q), \tag{3.26}$$

where $\partial \mathcal{O}^{\max}_q$ is the boundary of $\mathcal{O}^{\max}_q$ in $T_q Q$. The set $U^{\max}_q := \exp_q(\mathcal{O}^{\max}_q)$ is a normal neighborhood of q, which coincides with the set of points in Q that can be connected to q by a unique minimal geodesic. Hence for each q, Q is the disjoint union

$$Q = \exp_q(\mathcal{O}^{\max}_q) \cup C(q). \tag{3.27}$$

Heuristically, $U^{\max}_q$ is the largest neighborhood on which normal coordinates can be defined.

Corollary 3.2.6. *Let $(Q, \mathbf{g})$ be complete. For each given $q \in Q$, the set of all points $q' \in Q$ for which there is a unique minimal geodesic between q and q' is open and dense in Q.*

3.3 Hamiltonian Riemannian Geometry

We move on to perturbations of geodesics. Some of this material is interesting in its own right; other parts will be used in the study of Weyl quantization.

Definition 3.3.1. *A* **Jacobi field** $\mathcal{J}$ *along a geodesic* γ *is a vector field satisfying the* **equation of geodesic deviation** *(or* **Jacobi equation***)*

$$(\nabla_{\dot{\gamma}})^2\mathcal{J} + \mathbf{R}(\mathcal{J}, \dot{\gamma})\dot{\gamma} = 0. \tag{3.28}$$

This is a second-order differential equation, whose solution $\mathcal{J}(t)$ is determined by the initial data $\mathcal{J}(0)$ and $\nabla_{\dot{\gamma}}\mathcal{J}(0)$.

To derive and interpret (3.28), one looks at a family $\{\gamma(\cdot, \alpha)\}_{\alpha\in[0,\epsilon]}$ of geodesics, smoothly depending on a parameter α, such that $\gamma(t, 0) = \gamma(t)$. The value of $\mathcal{J}$ at $\gamma(t)$ is defined by $\mathcal{J}(\gamma(t)) := d\gamma(t, \alpha)/d\alpha|\alpha = 0$. The geodesic equation (3.4) satisfied by $\gamma(\cdot, \alpha)$ for each α leads to $\nabla_{\mathcal{J}}\nabla_{\dot{\gamma}}\dot{\gamma} = 0$. Since $[\partial_t, \partial_\alpha] = 0$, one has $[\dot{\gamma}, \mathcal{J}] = 0$, which, in view of the fact that the Levi-Civita connection ∇ is torsion-free, implies $\nabla_{\mathcal{J}}\dot{\gamma} = \nabla_{\dot{\gamma}}\mathcal{J}$. Combined with (3.20) this results in (3.28).

A Jacobi field along $\gamma(q, v; \cdot)$ is generically denoted by $\mathcal{J}(q, v; \cdot)$. In normal coordinates based at q the equation of geodesic deviation (3.28) at q reads

$$\frac{d^2}{dt^2}\mathcal{J}^i(q) + \tfrac{2}{3}\mathbf{R}^i_{jkl}(q)\dot{\gamma}^j(q)\dot{\gamma}^l(q)\mathcal{J}^k(q) = 0. \tag{3.29}$$

Our aim is to show that the evolution equations (3.4) and (3.28) may be brought into Hamiltonian form. Consider the classical Hamiltonian h_* on T^*Q, defined by

$$h_*(\sigma) := \tfrac{1}{2}\mathbf{g}^{-1}(\sigma, \sigma). \tag{3.30}$$

In coordinates this reads

$$h_*(p, q) = \tfrac{1}{2}\mathbf{g}^{ij}(q)p_i p_j. \tag{3.31}$$

For simplicity we have put a possible mass parameter m equal to 1; cf. (2.133). Also, we have omitted a possible potential energy from (3.31); the metric tensor already represents a (static) gravitational field in which the particle moves. The Hamiltonian flow $\sigma \mapsto \sigma(t)$ on T^*Q generated by h_* is known as **cogeodesic flow**. This terminology is explained by the following

Proposition 3.3.2. *Suppose that* $\sigma(t)$ *satisfies the Hamiltonian equation of motion* $d\sigma(t)/dt = \xi_{h_*}(\sigma(t))$. *Then* $\gamma_\sigma(t) := \tau_{T^*Q\to Q}(\sigma(t))$ *is a geodesic on* Q *with tangent vector field* $\dot{\gamma}_\sigma(t) = \mathbf{g}^\sharp(\sigma(t))$. *Accordingly,* $\sigma(t)$ *is equal to the parallel transport of* σ *along* γ_σ.

This is most easily proved by a coordinate calculation; in a local chart, one needs to establish that the motion $(p(t), q(t))$ is such that $q(t)$ is a geodesic with $\dot{q}(t) = \mathbf{g}^\sharp(p(t))$. This follows from I.(2.24), (3.5), and (3.19). ∎

Proposition 3.3.2 suggests that it is more natural to transfer the situation from T^*Q to TQ by the isomorphism $\mathbf{g}^\sharp$. Thus the Hamiltonian on TQ, which we denote by h, has the two equivalent expressions

$$h(X) = \tfrac{1}{2}\mathbf{g}_{\tau(X)}(X, X); \tag{3.32}$$

$$h(v, q) = \tfrac{1}{2}\mathbf{g}_{ij}v^i v^j. \tag{3.33}$$

Proposition 3.3.3. *The Hamiltonian flow $(v, q) \mapsto (v(t), q(t))$ on TQ generated by h coincides with the geodesic flow, and is given by parallel transporting the tangent vector $v = v(0)$ along the geodesic $q(t) = \gamma(q, v; t)$ on Q (with $q = q(0)$).*

In the coordinates (v^i, q^i), the Poisson bracket on TQ reads

$$\{f, g\} = \mathbf{g}^{ij}\left(\frac{\partial f}{\partial v^i}\frac{\partial g}{\partial q^j} - \frac{\partial f}{\partial q^i}\frac{\partial g}{\partial v^j} + \mathbf{g}^{mn} v^l(\partial_j \mathbf{g}_{ml} - \partial_m \mathbf{g}_{jl})\frac{\partial f}{\partial v^n}\frac{\partial g}{\partial v^i}\right). \tag{3.34}$$

The claim is then easily derived from (3.5). ■

Alternatively, one may regard (p_i, q^i) as canonical coordinates on TQ, which are related to the noncanonical ones (v^i, q^i) by

$$(p_i, q^i) = (\mathbf{g}_{ij}(q)v^j, q^i). \tag{3.35}$$

In either case, the Hamiltonian equations of motion derived from the canonical Poisson bracket I.(2.24) come out as

$$\begin{aligned} \dot{q} &= v; \\ \nabla_v v &= 0. \end{aligned} \tag{3.36}$$

One sees that Q is complete iff the Hamiltonian h is complete in the sense of Definition 3.1.3.

To find the Hamiltonian form of (3.28) we recall the discussion surrounding (2.147), which equally well applies to the present case $S = TQ$. Hence $T(TQ)$ is a symplectic manifold, and the pushforward of the geodesic flow on TQ to $T(TQ)$ is generated by $h^{(1)}$, which is constructed from h in (3.33) by (2.141).

Theorem 3.3.4. *The Hamiltonian equations of motion on $T(TQ)$ generated by $h^{(1)}$ take the form (3.36), supplemented by*

$$\begin{aligned} \nabla_v X^{\text{hor}} &= X^{\text{ver}}; \\ \nabla_v X^{\text{ver}} + \mathbf{R}(X^{\text{hor}}, v)v &= 0, \end{aligned} \tag{3.37}$$

where we have decomposed $X \in T_{(v,q)}$ as $X = X^{\text{hor}} + X^{\text{ver}}$, and have identified X^{hor} and X^{ver} with elements of $T_q Q$ in accordance with (3.3) and preceding text. Hence the Jacobi equation (3.28) along a given geodesic $q(\cdot)$ is Hamiltonian on $T(TQ)$, if we use (3.3) in the opposite direction to identify $\mathcal{J}(t)$ and $\nabla_v \mathcal{J}(t)$ in $T_{q(t)}Q$ with a horizontal and a vertical vector in $T_{(\dot{q}(t), q(t))}$, respectively.

We give a computational proof. The coordinates $(\tilde{v}^i, \tilde{q}^i, v^i, q^i)$ on $T(TQ)$ (cf. the paragraph after (3.1)) are not canonical with respect to the symplectic structure on $T(TQ)$; they are related to canonical coordinates $(\tilde{p}_i, \tilde{q}^i, p_i, q^i)$ by

$$(\tilde{p}_i, \tilde{q}^i, p_i, q^i) = (\mathbf{g}_{ij}(q)\tilde{v}^j + \partial_l \mathbf{g}_{ij} v^j \tilde{q}^l, \tilde{q}^i, \mathbf{g}_{ij}(q)v^j, q^i); \tag{3.38}$$

cf. (3.35). The Hamiltonian (2.141) on TS derived from (3.33), expressed in canonical coordinates, then reads

$$h^{(1)}(\tilde{p}^i, \tilde{q}^i, p^i, q^i) = \mathbf{g}^{ij}(q)p_i\tilde{p}_j - \tfrac{1}{2}\mathbf{g}^{ij}\mathbf{g}^{kl}\partial_j \mathbf{g}_{jk} p_i p_l \tilde{q}^l. \tag{3.39}$$

From (2.147), (3.39), and (3.38) one finds that $d\tilde{q}^i/dt = \tilde{v}^i$. Using the coordinate expressions for horizontal and vertical vectors, and making the identification with vectors in T_qQ mentioned in the theorem, the time-derivative d/dt may be converted into a directional derivative along the curve $q(t)$. Using $\dot{q}(t) = v(t)$, this leads to the first member of (3.37). Note here that the coordinate expression of $X + \nabla_v X$, where $X \in T_qQ$ and $\nabla_v X \in T_qQ$ are embedded in $T_{(q,v)}TQ$ as horizontal and vertical vectors, respectively, is simply $(\tilde{v}^i = v(X^i), \tilde{q}^i = X^i)$.

The proof of the second member of (3.37) is a straightforward but lengthy computation, which may be simplified by working in normal coordinates on Q based at q. Using the same simplification, the Hamiltonian equation of motion for $\tilde{v}^i$ at q is calculated to be $d\tilde{v}^i/dt + \frac{2}{3}\mathbf{R}^i_{jkl}v^j v^l \tilde{q}^k$. Converting the time-derivative into a directional derivative as in the previous paragraph and comparing with (3.29) then leads to the second member of (3.37). $\square$

3.4 Weyl Quantization on Riemannian Manifolds

Our goal is the quantization of a suitable subspace of $\mathfrak{A}^0 := C_0(T^*Q)$. The most natural way of doing this is based on a Riemannian generalization of the kernel (2.109) characterizing Weyl quantization. Hence we start by generalizing the partial Fourier transform (2.110) to the Riemannian setting. The invariant measure on Q is called μ, the one on the fiber T_qQ is μ_q, and the measure on T_q^*Q is denoted by μ_q^*. In coordinates one has

$$\begin{aligned} d\mu(q) &= d^n q\sqrt{\det \mathbf{g}(q)}; \\ d\mu_q(v) &= d^n v\sqrt{\det \mathbf{g}(q)}; \\ d\mu_q^*(p) &= \frac{d^n p}{(2\pi)^n\sqrt{\det \mathbf{g}(q)}}. \end{aligned} \tag{3.40}$$

Here $\det \mathbf{g}(q)$ denotes the determinant of the matrix $\mathbf{g}_{ij}(q)$ in given coordinates. The natural measure on T^*Q constructed from (3.40) coincides with the Liouville measure μ_L, since the factors $\sqrt{\det \mathbf{g}(q)}$ cancel. That is, for $f \in L^1(T^*Q)$ one has

$$\int_{T^*Q} d\mu_L(p,q)\, f(p,q) = \int_Q d\mu(q) \int_{T_q^*Q} d\mu_q^*(p) f(p,q). \tag{3.41}$$

The fiberwise Fourier transform of a suitable function f on T^*Q is the function $\grave{f}$ on TQ defined by

$$\grave{f}(X) := \int_{T_q^*Q} d\mu_q^*(\theta)\, e^{i\theta(X)} f(\theta), \tag{3.42}$$

where $X \in T_qX$. In coordinates, this simply amounts to (cf. (2.110))

$$\grave{f}(v,q) = \int d\mu_q^*(p)\, e^{ipv} f(p,q). \tag{3.43}$$

The fiberwise convolution $\grave{f} * \grave{g}$ is the Fourier transform of the pointwise product fg, which gives

$$\grave{f} * \grave{g}(v, q) = \int d\mu_q(v')\, \grave{f}(\tfrac{1}{2}v - v', q)\grave{g}(\tfrac{1}{2}v + v', q). \tag{3.44}$$

Similarly, the Fourier transform of the Poisson bracket I.(2.24) is

$$\{\grave{f}, \grave{g}\}(v, q) = i \int_{T_qQ} d\mu_q(v')\, \grave{f}(\tfrac{1}{2}v - v', q) \\ \times \left[\frac{\overleftarrow{\partial}}{\partial q^i}(\tfrac{1}{2}v + v')^i + (-\tfrac{1}{2}v + v')^i \frac{\overrightarrow{\partial}}{\partial q^i} \right] \grave{g}(\tfrac{1}{2}v + v', q). \tag{3.45}$$

Definition 3.4.1. *Let M_1 and M_2 be manifolds, and $\iota : M_1 \hookrightarrow M_2$ an embedding (i.e., an injective immersion). The* **pullback** *ι^*TM_2 is the manifold*

$$\iota^*TM_2 := \{(X, m) \in TM_2 \times M_1 \mid \tau_{TM_2 \to M_2}(X) = \iota(M_1)\}, \tag{3.46}$$

containing M_1 as a distinguished submanifold (the so-called **zero section***) through the identification $(0, m) \equiv m$; compare III.(2.2).*

The **normal bundle** *of the embedding ι is the manifold*

$$N^\iota M_1 := \iota^*TM_2/\iota_*TM_1, \tag{3.47}$$

containing M_1 as the zero section by the identification inherited from ι^*TM_2.

These definitions may be rephrased as follows. Firstly, the pullback $\iota^*(TM_2)$ is just the restriction $TM_2 \upharpoonright \iota(M_1)$ of TM_2 to $\iota(M_1) \subset M_2$; this is the union $\cup_{m \in M_1} V_m$ of the vector spaces $V_m := \tau_{TM_2 \to M_2}^{-1}(\iota(m))$, with topology inherited from TM_2. Secondly, the pushforward $\iota_*(TM_1) \subset TM_2$ is a subspace of $\iota^*(TM_2)$; it is the union $\cup_{m \in M_1} V_m^t$, where the vector space $V_m^t \subset V_m$ consists of all vectors that are tangent to $\iota(M_1)$. Finally, the quotient ι^*TM_2/ι_*TM_1 is $\cup_{m \in M_1} V_m/V_m^t$, equipped with the quotient topology.

The normal bundle is isomorphic to a subbundle of $TM_2 \upharpoonright M_1$, but not naturally so. The following is a fundamental theorem of differential geometry.

Theorem 3.4.2. *Let $\iota(M_1)$ be a closed submanifold of a manifold M_2.*

- *There exist a* **tubular neighborhood** *$\mathcal{N}^\iota(M_1)$ of $M_1 \subset N^\iota M_1$ (where M_1 is identified with the zero section), a neighborhood $\mathcal{N}_\iota(M_1) \subset M_2$ of $\iota(M_1)$, and a diffeomorphism $\varphi : \mathcal{N}^\iota(M_1) \to \mathcal{N}_\iota(M_1)$ satisfying $\varphi(m) = \iota(m)$ for all $m \in M_1$.*

Let, in addition, M_2 have a Riemannian metric, and define $(TM_2 \upharpoonright M_1)^\perp$ as the union $\cup_{m \in M_1} V_m^\perp$, where V_m is the orthogonal complement $T_{\iota(m)}\iota(M_1)^\perp$ of $T_{\iota(m)}\iota(M_1)$ in $T_{\iota(m)}M_2$ (with topology inherited from TM_2).

- *There is a diffeomorphism $\eta_\iota : N^\iota M_1 \to (TM_2 \upharpoonright M_1)^\perp$ such that η_ι is linear on each vector space V_m/V_m^t and $\eta_\iota(V_m/V_m^t) = V_m^\perp$ for all m.*

- *The tubular neighborhood $\mathcal{N}^\iota(M_1)$ may be chosen in such a way that the diffeomorphism φ of the first part of this theorem is given by the restriction of* $\exp \circ \eta_\iota : N^\iota M_1 \to N$ *to* $\mathcal{N}^\iota(M_1)$.

A special case is that of $M_1 \subset M_2$ a submanifold of M_2, and ι the inclusion map. We further specialize to the case $M_1 = Q$, $M_2 = Q \times Q$, and $\iota = \delta$, the diagonal embedding defined by $\delta(q) := (q, q)$.

Lemma 3.4.3.

- *The normal bundle $N^\delta Q$ of the diagonal embedding is isomorphic to TQ.*
- *Equip $Q \times Q$ with the Riemannian metric $\mathbf{g} \oplus \mathbf{g}$; then $\nu_\delta = \exp \circ \eta_\delta : TQ \to Q \times Q$ is given on $X_q \in T_q Q$ by*

$$\nu_\delta(X_q) := \exp \circ \eta_\delta(X_q) = (\exp_q(\tfrac{1}{2}X_q), \exp_q(-\tfrac{1}{2}X_q)). \tag{3.48}$$

One identifies $T_{(q,q)}(Q \times Q)$ with $T_q Q \oplus T_q Q$. The fiber of $\delta^* T(Q \times Q)$ at a point (q, q) is $T_q Q \oplus T_q Q$. The fiber of the pushforward bundle $\delta_* TQ$ at (q, q), on the other hand, consists of all vectors of the type $X \dot{+} X$, $X \in T_q Q$. Hence $N^\delta Q \simeq TQ$ by the definition (3.47). With the metric $\mathbf{g} \oplus \mathbf{g}$, the orthogonal decomposition of $X \dot{+} Y$ has the component $\frac{1}{2}(X+Y) \dot{+} \frac{1}{2}(X+Y)$ in $\delta_* TQ$ and $\frac{1}{2}(X-Y) \dot{+} \frac{1}{2}(Y-X)$ in $(T(Q \times Q) \restriction \delta(Q))^\perp$. Hence η_δ maps $X_q \in T_q Q$ to $\frac{1}{2}X_q \dot{+} -\frac{1}{2}X_q \in T_{(q,q)} Q \times Q$, and (3.48) follows. ■

To appreciate the following quantization prescription it is helpful to understand the geometric meaning of the diffeomorphism (3.48): Namely, $\nu_\delta^{-1}(q, q')$ in TQ is the tangent vector to the geodesic from q' to q at its midpoint.

We are now in a position to define the (generalized) Weyl quantization map $\mathcal{Q}_\hbar^W$. We take the dense subalgebra of $C_0(T^* Q)$ of quantizable functions to be

$$\tilde{\mathfrak{A}}^0 := C^\infty_{\rm PW}(T^* Q). \tag{3.49}$$

These are by definition the functions f on $T^* Q$ whose Fourier transform $\check{f}$ is in $C^\infty_c(TQ)$; the motivation for this choice will become clear shortly. The space $C^\infty_{\rm PW}(T^* Q, \mathbb{R})$ is a Poisson subalgebra of $C^\infty(T^* Q, \mathbb{R})$; this follows from an inspection of (3.45), using the fact that C^∞_c is closed under convolution and pointwise multiplication.

The map $\mathcal{Q}_\hbar^W$ takes values in $\mathfrak{A}^\hbar := \mathfrak{B}_0(L^2(Q))$, where the Hilbert space $L^2(Q)$ is defined with respect to the Riemannian measure μ on Q; cf. (3.40).

Definition 3.4.4. *The* **Weyl quantization** *of $f \in C^\infty_{\rm PW}(T^* Q)$ is given, for $\hbar \neq 0$, by the integral operator*

$$\mathcal{Q}_\hbar^W(f)\Psi(x) := \int_Q d\mu(y)\, K_\hbar^W \lfloor f \rfloor(x, y)\Psi(y). \tag{3.50}$$

For $(x, y) \notin \mathcal{N}_\delta(Q)$ we define $K_\hbar^W \lfloor f \rfloor(x, y) := 0$, whereas for $(x, y) \in \mathcal{N}_\delta(Q)$ we put

$$K_\hbar^W \lfloor f \rfloor(x, y) := \hbar^{-n} \kappa(\nu_\delta^{-1}(x, y)) \check{f}(\nu_\delta^{-1}(x, y)/\hbar), \tag{3.51}$$

with ν_δ given by (3.48). Here κ is a smooth function on TQ with the following properties:

- *$\kappa = 1$ in a neighborhood $\tilde{\mathcal{N}}^\delta(Q) \subset \mathcal{N}^\delta(Q)$ of Q (regarded as the zero section in TQ).*
- *κ has support in $\mathcal{N}^\delta(Q)$.*
- *$\kappa(-v, q) = \kappa(v, q)$.*

For later convenience, we shall in fact assume that for each q the support of $\kappa(\cdot, q)$ is contained in a geodesically convex (hence normal) neighborhood of $0 \in T_q Q$. As in 2.5, we write the argument of Ψ as x rather than q to avoid confusion with the argument (p, q) of f.

Proposition 3.4.5. *For $f \in C^\infty_{\mathrm{PW}}(T^*Q, \mathbb{R})$ one has $\mathcal{Q}_\hbar^W(f) \subset \mathfrak{B}_0(L^2(Q))_\mathbb{R}$.*

By definition of $C^\infty_{\mathrm{PW}}(T^*Q)$, the kernel of $\mathcal{Q}_\hbar^W(f)$ is in $C_c^\infty(Q \times Q)$, so that $\mathcal{Q}_\hbar^W(f)$ is a Hilbert–Schmidt operator, hence compact. Also, (3.42), (3.48), and the symmetry of κ in v guarantee that $K_\hbar^W$ is Hermitian, so that for real f the operator $\mathcal{Q}_\hbar^W(f)$ is self-adjoint. ∎

The presence of the cutoff function κ implies that the kernel $K_\hbar^W$ is smooth; unfortunately, $\mathcal{Q}_\hbar^W(f)$ seems to depend on the choice of this function (as well as of $\tilde{\mathcal{N}}^\delta(Q)$). However, since $\grave{f}$ has compact support by our choice of $\tilde{\mathfrak{A}}^0$, there is a value $\hbar_0 > 0$ (that depends on f) such that $\mathcal{Q}_\hbar^W(f)$ does not depend on these choices for $\hbar \in (0, \hbar_0)$. Namely, $\hbar_0$ is the smallest value of $\hbar$ for which $\hbar_0 \mathrm{supp}\,(\grave{f})$ lies in $\tilde{\mathcal{N}}^\delta(Q)$. If the tubular neighborhoods may be chosen as $\mathcal{N}^\delta(Q) = TQ$ and $\mathcal{N}_\delta(Q) = Q \times Q$, then one may obviously put $\kappa = 1$. This is possible for $Q = \mathbb{R}^n$ (with flat metric), in which case (3.50) and (3.51) reduce to (2.108), and (2.109), respectively.

The same conclusion of κ-independence formally holds true if $\grave{f}$ is a distribution with compact support; if f is polynomial in the momenta p^i, the support of $\grave{f}$ is localized at the zero section of TQ, and $\mathcal{Q}_\hbar^W(f)$, now defined as an unbounded operator on the domain $C_c^\infty(Q)$, is independent of κ for any $\hbar$. This will be further explored in 3.6.

As in the flat case (cf. (2.101)) there is a Wigner function.

Proposition 3.4.6. *The Gelfand transform of $\mathcal{Q}_\hbar^W(f)$ is given by*

$$\widehat{\mathcal{Q}_\hbar^W(f)}(\psi) = \int_{T^*Q} d\mu_L(p, q) W_\hbar \lfloor \psi \rfloor (p, q) f(p, q), \tag{3.52}$$

with Wigner function (cf. (2.103))

$$W_\hbar \lfloor \psi \rfloor (p, q) = \int_{T_q Q} d\mu_q(v) \kappa(\hbar v, q) J(q, v; \tfrac{1}{2}\hbar) \times e^{ipv} \overline{\Psi(\gamma(q, v; \tfrac{1}{2}\hbar))} \Psi(\gamma(q, v; -\tfrac{1}{2}\hbar)), \tag{3.53}$$

where J is a Jacobian defined in (3.55) below.

To prove this, we initially assume that $\tau_{T^*Q\to Q}(\text{supp}\,(f))$ is contained in a suitably small geodesically convex set $U \subset P$, on which we use coordinates q^i. The linearity of $\mathcal{Q}_\hbar^W$, the fact that $\tau_{T^*Q\to Q}(\text{supp}\,(f))$ is compact, and the existence of (smooth) partitions of unity on Q then imply the result for general $f \in \tilde{\mathfrak{A}}^0_{\mathbb{R}}$. We change integration variables in

$$\widehat{\mathcal{Q}_\hbar^W(f)(\psi)} = \int_{Q\times Q} d\mu(q_1)d\mu(q_2)\, K_\hbar^W \lfloor f \rfloor(q_1, q_2)\overline{\Psi(q_1)}\Psi(q_2): \tag{3.54}$$

If an arbitrary function $F \in C(Q \times Q)$ has support inside $U \times U$, we put

$$\int_{U\times U} d\mu(q_1)d\mu(q_2)\, F(q_1, q_2) =$$

$$\int_{TU} d\mu(q)d\mu_q(v)\, J(q, v; \tfrac{1}{2})F(\gamma(q, v; \tfrac{1}{2}), \gamma(q, v; -\tfrac{1}{2})), \tag{3.55}$$

which, with the property $J(q, tv; \frac{1}{2}) = J(q, v; \frac{1}{2}t)$, defines J in (3.53). The proposition then follows from (3.54), (3.55), (3.50), (3.51), and (3.43). ■

The Jacobian J will be studied in detail in the next section; we will find that $J(q, v, \frac{1}{2}\hbar) = 1 + O(\hbar^2)$. Also, J will be seen to have the symmetry property $J(-v, q; t) = J(v, q; t)$, which, with (3.53), confirms that $W_\hbar\lfloor\psi\rfloor$ is real for real-valued f. Given that $\mathcal{Q}_\hbar^W(f)$ is bounded for $f \in \tilde{\mathfrak{A}}^0_{\mathbb{R}}$, this property is equivalent to the self-adjointness of $\mathcal{Q}_\hbar^W(f)$.

3.5 Proof of Strictness

Recalling Definitions 1.1.1 and 1.2.5, the aim of this section is to prove

Theorem 3.5.1. *The map $\mathcal{Q}_\hbar^W$ defined in 3.4.4 is a nondegenerate strict and continuous quantization of $\tilde{\mathfrak{A}}^0_{\mathbb{R}} = C^\infty_{\text{PW}}(T^*Q, \mathbb{R})$, so that $\mathfrak{A}^0 = C_0(T^*Q)$, and $\mathfrak{A}^\hbar = \mathfrak{B}_0(L^2(Q))$ for $\hbar \neq 0$ (with the possible exception of the completeness condition 1.1.1.4).*

It is clear that 1.1.1.4 is not satisfied if the cutoff function κ in (3.51) is not equal to unity. If $Q \times Q$ is diffeomorphic to TQ by the map ν_δ (cf. (3.48)), the quantization $\mathcal{Q}_\hbar^W$ does satisfy 1.1.1.4, since the collection $\{\mathcal{Q}_\hbar^W(f)\}$ is dense in the set of Hilbert–Schmidt operators on $L^2(Q)$.

The nondegeneracy is obvious: $\mathcal{Q}_\hbar^W(f) = 0$ implies $\check{f} = 0$ by (3.51), which implies $f = 0$. Continuity follows from strictness by Theorem 1.2.4.

In the following discussion we will assume that $\hbar > 0$; the arguments for $\hbar < 0$ are a trivial modification. The necessary computations are greatly simplified by the possibility of localization.

Lemma 3.5.2. *Let $f, g \in \tilde{\mathfrak{A}}^0_{\mathbb{R}}$. If the projection $\tau_{T^*Q\to Q}(\text{supp}\,(f))$ of the support of f is disjoint from that of g, then there is $\hbar_{f,g} > 0$ such that $\mathcal{Q}_\hbar^W(f)\mathcal{Q}_\hbar^W(g) = 0$ for $\hbar \in (0, \hbar_{f,g})$.*

It follows from (3.51) and the fact that $\grave{f}$ has compact support (particularly in the fiber direction) that $K_\hbar^W\lfloor f\rfloor(x,z)$ is nonzero only if the (Riemannian) distance from both x and z to $\tau_{T^*Q\to Q}(\,\mathrm{supp}\,(f))$ is $O(\hbar)$; similarly for $K_\hbar^W\lfloor g\rfloor(z,y)$. Hence for fixed x, y the kernel $\int d\mu(z)\, K_\hbar^W\lfloor f\rfloor(x,z)K_\hbar^W\lfloor g\rfloor(z,y)$ of the product $\mathcal{Q}_\hbar^W(f)\mathcal{Q}_\hbar^W(g)$ vanishes for sufficiently small $\hbar < \hbar(x,y)$, for a certain $\hbar(x,y) > 0$. Since $\grave{f}$ and $\grave{g}$ have compact support also on in the base direction, this vanishing can be achieved uniformly in (x,y). ■

For the reasons stated in the proof of 3.4.6, Lemma 3.5.2 allows us to assume that $\tau_{T^*Q\to Q}(\,\mathrm{supp}\,(f))$ and $\tau_{T^*Q\to Q}(\,\mathrm{supp}\,(g))$ are contained in an arbitrarily small open set $U\subset Q$. For U we choose a geodesically convex set U_q (cf. the paragraph following 3.4.6).

Since (3.28) is a second-order differential equation, for given $X, Y\in T_{\gamma(0)}Q$ there exists a unique Jacobi field $\mathcal{J}$ for which $\mathcal{J}(0)=X$ and $\nabla_{\dot\gamma}\mathcal{J}(0)=Y$. If we write $\mathcal{J}=\mathcal{J}^i\partial_i$ in given coordinates, one may equally well pose the initial conditions $\mathcal{J}^i(0)=X^i$, $\dot{\mathcal{J}}^i(0)=Y^i$, with unique solution $\mathcal{J}^i(t)$. We write $\mathcal{J}_{(j)}(q,v;\cdot)$ for the Jacobi field with initial conditions

$$\begin{aligned}\mathcal{J}^i_{(j)}(q,v;0)&=\delta^i_j;\\ \dot{\mathcal{J}}^i_{(j)}(q,v;0)&=0,\end{aligned}\tag{3.56}$$

and $\tilde{\mathcal{J}}_{(j)}(q,v;\cdot)$ for the Jacobi field with initial conditions

$$\begin{aligned}\tilde{\mathcal{J}}^i_{(j)}(q,v;0)&=0;\\ \left(\frac{d}{dt}\tilde{\mathcal{J}}^i_{(j)}\right)(q,v;0)&=\delta^i_j.\end{aligned}\tag{3.57}$$

The $n\times n$ matrices $M(q,v;t)$ and $\tilde M(q,v;t)$ are then defined by their matrix elements $M(q,v;t)^i_j:=\mathcal{J}^i_{(j)}(q,v;t)$ and $\tilde M(q,v;t)^i_j:=\tilde{\mathcal{J}}^i_{(j)}(q,v;t)$, respectively. These are combined in the $2n\times 2n$ matrix

$$M_2:=\begin{pmatrix} M(q,v;t) & \tilde M(q,v;t)\\ M(q,v;-t) & \tilde M(q,v;-t)\end{pmatrix}.\tag{3.58}$$

Lemma 3.5.3.

- *The Jacobian in (3.55) is given (for arbitrary t, as long as the geodesics in question exist) by*

$$\begin{aligned}J(q,v;t)=|t^{-n}|[\det\mathbf{g}(\gamma(q,v;t))\det\mathbf{g}(\gamma(q,v;-t))]^{1/2}\\ \times\det\mathbf{g}(q)^{-1}|\det M_2(q,v;t)|.\end{aligned}\tag{3.59}$$

- *If $G\in C(Q)$ has support in U, and $q\in U$ is such that U is contained in the image of the exponential map on T_qQ, then*

$$\int_U d\mu(q')\,G(q')=\int_{T_qQ}d\mu_q(v)\,\tilde J(q,v;1)G(\gamma(q,v;1)),\tag{3.60}$$

with Jacobian

$$\tilde{J}(q, v; t) = |t^{-n}| [\det \mathbf{g}(\gamma(q, v; t)) / \det \mathbf{g}(q)]^{1/2} | \det \tilde{M}(q, v; t)|. \tag{3.61}$$

To derive (3.55), one passes from the coordinates (q_1^i, q_2^i) to q^i, v^i via the expression $q_1 = \gamma(q, v; t)$, $q_2 = \gamma(q, v; -t)$ (where $t = \frac{1}{2}$ in the special case above). The definition of a Jacobi field implies that

$$\partial q_1^i(q, v; t)/\partial q^j = \mathcal{J}_{(j)}^i(q, v; t); \qquad \partial q_1^i(q, v; t)/\partial v^j = \tilde{\mathcal{J}}_{(j)}^i(q, v; t);$$
$$\partial q_2^i(q, v; t)/\partial q^j = \mathcal{J}_{(j)}^i(q, v; -t); \quad \partial q_2^i(q, v; t)/\partial v^j = \tilde{\mathcal{J}}_{(j)}^i(q, v; -t), \tag{3.62}$$

which leads to (3.55). The derivation of (3.60) is analogous. ■

We assume the support conditions on f and g stated after 3.5.2, and take an arbitrary $\Psi \in L^2(Q)$. From (3.50), (3.51), Lemma 3.5.3, and the property $\gamma(q, v, \hbar t) = \gamma(q, \hbar v, t)$, we obtain

$$\left(\Psi, [\mathcal{Q}_\hbar^W(f)\mathcal{Q}_\hbar^W(g) - \mathcal{Q}_\hbar^W(fg)]\Psi\right) = \int_U d\mu(q) \int_{T_qQ} d\mu_q(v) \int_{T_qQ} d\mu_q(v')$$
$$\overline{\Psi(\gamma(q, v; \hbar/2))}\Psi(\gamma(q, v; -\hbar/2)) I(\hbar, q, v, v'), \tag{3.63}$$

with

$$I(\hbar, q, v, v') = J(q, v; \hbar/2) \Big[\hbar^{2n} \tilde{J}(q, v'; \hbar)$$
$$\times K_\hbar^W \lfloor f \rfloor (\gamma(q, v; \hbar/2), \gamma(q, v'; \hbar)) K_\hbar^W \lfloor g \rfloor (\gamma(q, v'; \hbar), \gamma(q, v; -\hbar/2))$$
$$- \check{f}(q, \tfrac{1}{2}v - v')\grave{g}(q, \tfrac{1}{2}v + v') \Big]. \tag{3.64}$$

For a fixed value (q, v, v'), we now make a Taylor expansion of $I(\hbar, q, v, v')$ in $\hbar$. Here I is a function on $TU \otimes TU$, so we may proceed in any coordinate system.

By evaluating (3.28) in normal coordinates at $t = 0$ (cf. (3.29)) it follows immediately that $\tilde{\mathcal{J}}_j^i(q, v, \hbar) = \hbar\delta_j^i(1 + O(\hbar^2))$ and $\mathcal{J}_j^i(q, v, \hbar) = \delta_j^i(1 + O(\hbar^2))$. Combined with the explicit form (3.24) of the metric in normal coordinates, we thus infer from (3.59) and (3.61) that (in any coordinates)

$$J(q, v; \hbar/2) = 1 + O(\hbar^2);$$
$$\tilde{J}(q, v'; \hbar) = 1 + O(\hbar^2). \tag{3.65}$$

To deal with the terms involving $K_\hbar^W$ in (3.64) we use (3.25) and perform a Taylor expansion of $K_\hbar^W \lfloor f \rfloor$ around the point $\left(\gamma(q, \frac{1}{2}v - v'; \hbar/2), \gamma(q, \frac{1}{2}v - v'; -\hbar/2)\right)$, and of $\mathcal{Q}_\hbar^W(g)$ around $\left(\gamma(q, \frac{1}{2}v + v'; \hbar/2), \gamma(q, \frac{1}{2}v + v'; -\hbar/2)\right)$. Using (3.51), the result is then rewritten in terms of the $\check{f}$ and $\grave{g}$. The $O(1)$ term vanishes. In computing the $O(\hbar)$ term, one encounters expressions of the type

$$\left(\frac{\partial}{\partial q_1^i} + \frac{\partial}{\partial q_2^i}\right) K_\hbar^W \lfloor f \rfloor (q_1 = \gamma(q, \tfrac{1}{2}v - v'; \hbar/2), q_2 = \gamma(q, \tfrac{1}{2}v - v'; -\hbar/2)), \tag{3.66}$$

to be expressed in normal coordinates as indicated above. This is done by inverting (3.62) and expanding in powers of $\hbar$. The result is that $\hbar^n$ times (3.66) equals $(\partial \grave{f}/\partial q^i)(q, \frac{1}{2}v - v') + O(\hbar)$. Using (3.45), one then finds that the integrand of the analogue of (3.63), with $\mathcal{Q}_\hbar^W(f)\mathcal{Q}_\hbar^W(g) - \mathcal{Q}_\hbar^W(fg)$ replaced by $i[\mathcal{Q}_\hbar^W(f), \mathcal{Q}_\hbar^W(g)]/\hbar - \mathcal{Q}_\hbar^W(\{f, g\})$, is $O(\hbar)$.

In either case one is left with an expression of the type (3.63), with I replaced by a remainder $I^{(1)}$ of $O(\hbar)$. One then replaces the integration variables (q, v, v') in (3.63) by (q', v, v'), with $q' = \gamma(q, v; \hbar/2)$; this introduces a Jacobian, which is $1 + O(\hbar^2)$, as in the argument leading to (3.65). This Jacobian may be absorbed into $I^{(1)}$, which then remains $O(\hbar)$. Then apply the Cauchy–Schwarz inequality to the q'-integration, splitting the integrand into $\overline{\Psi(q')}$ and the rest. This takes out a term $(\int_U d\mu(q')\, |\Psi(q')|^2)^{1/2} \leq \|\Psi\|$. The second factor produced by the Cauchy–Schwarz inequality is majorized by taking out another factor $\|\Psi\|$, and bounding the rest of the q'-integrand by taking its supremum over q'. This leads to

$$\left|\left(\Psi, [\mathcal{Q}_\hbar^W(f)\mathcal{Q}_\hbar^W(g) - \mathcal{Q}_\hbar^W(fg)]\Psi\right)\right| \leq C\hbar\|\Psi\|^2\|\grave{f}\|_{1,\infty}\|\grave{g}\|_{1,\infty} \\ \times \sup_{q\in U}[\mu_q(\mathrm{supp}\,(\grave{f}) \cap T_qU)\mu_q(\mathrm{supp}\,(\grave{g}) \cap T_qU)], \quad (3.67)$$

and a similar inequality for $|(\Psi, (i[\mathcal{Q}_\hbar^W(f), \mathcal{Q}_\hbar^W(g)]/\hbar - \mathcal{Q}_\hbar^W(\{f, g\})\Psi)|$, in which the norms $\|\cdot\|_{1,\infty}$ in (3.67) are replaced by $\|\cdot\|_{2,\infty}$. Hence (1.2) and (1.3) follow as in the proof of 2.4.1.

It remains to prove Rieffel's condition. Since a very general proof of this property will be given in Theorem III.3.11.4, we will merely sketch how the proof in flat space may be generalized.

Firstly, continuity at $\hbar \neq 0$ can be proved in several ways, e.g., by proving continuity with respect to the Hilbert–Schmidt norm of $\mathcal{Q}_\hbar^W(f)$. To prove continuity for $\hbar \to 0$, we shall construct a positive map $\mathcal{Q}_\hbar^{\mathrm{pos}}$, which is equivalent to $\mathcal{Q}_\hbar^W$ in the sense that the function $\hbar \mapsto \|\mathcal{Q}_\hbar^W(f) - \mathcal{Q}_\hbar^{\mathrm{pos}}(f)\|$ is continuous on $\mathbb{R}\backslash\{0\}$ and $\lim_{\hbar\to 0}\|\mathcal{Q}_\hbar^W(f) - \mathcal{Q}_\hbar^{\mathrm{pos}}(f)\| = 0$. This map may be shown to satisfy (1.1), which then implies the same for $\mathcal{Q}_\hbar^W$. In the proof of 2.6.1 we had $\mathcal{Q}_\hbar^{\mathrm{pos}} = \mathcal{Q}_\hbar^B$; in the present case the construction of $\mathcal{Q}_\hbar^{\mathrm{pos}}$ is motivated by $\mathcal{Q}_\hbar^B$ on flat space, but unlike $\mathcal{Q}_\hbar^W$ it holds no intrinsic significance on curved spaces.

We define $\mathcal{Q}_\hbar^{\mathrm{pos}}(f)$ (where $f \in \tilde{\mathfrak{A}}_{\mathbb{R}}^0$) through its Gelfand transform $\widehat{\mathcal{Q}_\hbar^{\mathrm{pos}}(f)}$, defined as a function on $\mathbb{P}L^2(Q)$, by

$$\widehat{\mathcal{Q}_\hbar^{\mathrm{pos}}(f)}(\psi) = \int_{T^*Q} d\mu_L(p, q)\, W_\hbar^{\mathrm{pos}}\lfloor\psi\rfloor(p, q)f(p, q), \quad (3.68)$$

where

$$W_\hbar^{\mathrm{pos}}\lfloor\psi\rfloor(p, q) := \hbar^{-n}\left|\int_{T_qQ}\frac{d\mu_q(v)}{(\pi\hbar)^{n/4}}\right. \\ \left.\kappa(v, q)\sqrt{\tilde{J}(q, v; 1)}e^{-ipv/\hbar}e^{-v^2/(2\hbar)}\Psi(\gamma(q, v; 1))\right|^2, \quad (3.69)$$

where $v^2 := \mathbf{g}_q(v, v)$, and $\tilde{J}$ is given by (3.61). It is easy to see that (3.69) defines a bounded operator $\mathcal{Q}_\hbar^{\text{pos}}(f)$ on $L^2(Q)$ (see 3.5.4 below). In flat space one has $\mathcal{Q}_\hbar^{\text{pos}} = \mathcal{Q}_\hbar^B$; in general, (3.69) depends on the cutoff function κ, cf. (3.53).

Lemma 3.5.4. *The map $\mathcal{Q}_\hbar^{\text{pos}} : \tilde{\mathfrak{A}}_\mathbb{R}^0 \to \mathfrak{B}(L^2(Q))$ is positive and takes values in $\mathfrak{B}_0(L^2(Q))_\mathbb{R}$. It satisfies, for all $f \in \tilde{\mathfrak{A}}_\mathbb{R}^0$,*

$$\lim_{\hbar \to 0} \|\mathcal{Q}_\hbar^{\text{pos}}(f) - \mathcal{Q}_\hbar^W(f)\| = 0 \tag{3.70}$$

and

$$\lim_{\hbar \to 0} \|\mathcal{Q}_\hbar^{\text{pos}}(f)\| = \|f\|_\infty. \tag{3.71}$$

The positivity of $\mathcal{Q}_\hbar^{\text{pos}}$ is obvious from (3.69). Since f and κ are compactly supported in q and v, respectively, $\mathcal{Q}_\hbar^{\text{pos}}(f)$ is an integral operator with smooth compactly supported kernel; hence it is Hilbert–Schmidt and therefore compact. Self-adjointness is immediate from the reality of $W_\hbar^{\text{pos}}$.

The proof of (3.70) and (3.71) is very tedious, and will be omitted.

Given this lemma, the corresponding argument in the proof of 2.6.1 leads to

$$\lim_{\hbar \to 0} \|\mathcal{Q}_\hbar^W(f)\| = \|f\|_\infty, \tag{3.72}$$

and the proof of Theorem 3.5.1 is finished. □

The continuous field of C^*-algebras defined by $\mathcal{Q}_\hbar^W$ through Theorems 3.5.1 and 1.2.4 will be identified in III.3.12.

3.6 Commutation Relations on Riemannian Manifolds

We would like to quantize certain unbounded smooth functions f on the phase space T^*Q. This can be done by the prescription (3.50), (3.51) if f is polynomial in p. The domain on which the ensuing unbounded operator $\mathcal{Q}_\hbar^W(f)$ is defined is initially taken to be $C_c^\infty(Q)$, since on this domain the formal manipulations used in computing $\mathcal{Q}_\hbar^W(f)$ are well-defined. In this section we examine certain intrinsically defined functions on T^*Q of order zero and one in the canonical momenta (the Hamiltonian, which is of order two, will be dealt with in the next section).

Proposition 3.6.1. *The Weyl quantizations (in the sense of 3.4.4) of $f = J_{\tilde{g}}$ (cf. (3.7)) and of $f = J_\xi$ (cf. (3.8)), defined on $C_c^\infty(Q) \subset L^2(Q)$, are given by*

$$\mathcal{Q}_\hbar^W(J_{\tilde{g}}) = \tilde{g}; \tag{3.73}$$

$$\mathcal{Q}_\hbar^W(J_\xi) = -i\hbar(\xi + \tfrac{1}{2}\nabla \cdot \xi). \tag{3.74}$$

Here $\tilde{g}$ and $\frac{1}{2}\nabla \cdot \xi$ ($:= \frac{1}{2}\nabla_i \xi^i$) are multiplication operators.

The computation of $\mathcal{Q}_\hbar^W(\cdots)\Psi(x)$ is best done in normal coordinates q_n^i based at x (cf. 3.2). In these coordinates the point q and the vector $X_q \in T_qQ$ for which $(\exp_q(\frac{1}{2}X_q), \exp_q(-\frac{1}{2}X_q)) = (0, v)$ (see (3.48)) are given simply by $q_n^i = \frac{1}{2}v^i$ and

$X_q^i = -v^i$. Hence from (3.50), (3.51), and (3.43) one obtains

$$\mathcal{Q}_\hbar^W(f)\Psi(x) = \int_Q d^n v \int_{\mathbb{R}^n} \frac{d^n p}{(2\pi)^n} \kappa(-v, \tfrac{1}{2}v) e^{-ipv} \sqrt{\frac{\det \mathbf{g}(v)}{\det \mathbf{g}(\frac{1}{2}v)}} f(\hbar p, \tfrac{1}{2}v)\Psi(v); \tag{3.75}$$

for the functions f under study this expression will turn out to be independent of the cutoff function κ. For f polynomial in p one has the (oscillatory) integral

$$\int \frac{d^n p}{(2\pi)^n} e^{-ipv} p^{i_1} \cdots p^{i_n} = \left(i \frac{\partial}{\partial v^{i_1}}\right) \cdots \left(i \frac{\partial}{\partial v^{i_n}}\right) \delta^{(n)}(v), \tag{3.76}$$

where $\delta^{(n)}$ is the n-dimensional Dirac delta distribution. This leads to (3.73) and (3.74). ■

Straightforward computation leads to the following "canonical" commutation relations (valid on the domain $C_c^\infty(Q)$):

$$\frac{i}{\hbar}\left[\mathcal{Q}_\hbar^W(J_{\tilde{g}}), \mathcal{Q}_\hbar^W(J_{\tilde{h}})\right] = 0; \tag{3.77}$$

$$\frac{i}{\hbar}\left[\mathcal{Q}_\hbar^W(J_\xi), \mathcal{Q}_\hbar^W(J_{\tilde{g}})\right] = \mathcal{Q}_\hbar^W(J_{\xi\tilde{g}}); \tag{3.78}$$

$$\frac{i}{\hbar}\left[\mathcal{Q}_\hbar^W(J_{\xi_1}), \mathcal{Q}_\hbar^W(J_{\xi_2})\right] = \mathcal{Q}_\hbar^W(J_{[\xi_1,\xi_2]}). \tag{3.79}$$

These reflect the classical Poisson brackets (3.9)-(3.11), in that Dirac's relation $i\left[\mathcal{Q}_\hbar^W(f), \mathcal{Q}_\hbar^W(g)\right]/\hbar = \mathcal{Q}_\hbar^W(\{f,g\})$ is satisfied for the functions in question. The "canonical" commutation relations may be interpreted in terms of a certain representation $\rho_\hbar$ of the group $\mathcal{G}_Q$ (see (3.12)) on $L^2(Q)$.

Proposition 3.6.2. *The linear action of $\mathcal{G}_Q$ on $L^2(Q)$ defined by*

$$\rho_\hbar(\tilde{g})\Psi(q) := e^{-i\tilde{g}(q)/\hbar}\Psi(q), \tag{3.80}$$

$$\rho_\hbar(\varphi)\Psi(q) := \sqrt{\frac{d\mu(\varphi^{-1}(q))}{d\mu(q)}}\,\Psi(\varphi^{-1}(q)), \tag{3.81}$$

and $\rho_\hbar(\varphi, \tilde{g}) := \rho_\hbar(\tilde{g}) \circ \rho_\hbar(\varphi)$, is unitary, hence a representation. The derived representation of the Lie algebra $\mathfrak{g}_Q$ is given in terms of the map J (cf. 3.1.4) by

$$i\hbar d\rho_\hbar(X) = \mathcal{Q}_\hbar^W(J_X). \tag{3.82}$$

The Radon–Nikodym derivative under the square root exists because φ^{-1} is a diffeomorphism, under which the locally Lebesgue measure class is invariant. Given the square root, unitarity is immediate from the definitions. For the remaining calculation one combines the identity

$$\partial_i \log \sqrt{\det \mathbf{g}} = \Gamma_{ij}^j \tag{3.83}$$

with (3.40) and (3.1). □

The representation $\rho_\hbar$ induces a *-automorphic action $\alpha^\hbar$ of $\mathcal{G}_Q$ on $\mathfrak{A}^\hbar = \mathfrak{B}_0(L^2(Q))$ by

$$\alpha^\hbar_{(\varphi,\tilde{g})}(A) := \rho_\hbar(\varphi,\tilde{g})A\rho_\hbar(\varphi,\tilde{g})^*. \tag{3.84}$$

Writing $\alpha^\hbar_\varphi := \alpha^\hbar_{(\varphi,0)}$ and $\alpha^\hbar_{\tilde{g}} := \alpha^\hbar_{(\mathrm{id},\tilde{g})}$, the kernel of $A = \mathcal{Q}^W_\hbar(f)$ (cf. (3.51)) transforms as

$$\begin{aligned}(\alpha^\hbar_\varphi(K^W_\hbar\lfloor f\rfloor))(x,y) &= \sqrt{\frac{d\mu(\varphi^{-1}(x))}{d\mu(x)}\frac{d\mu(\varphi^{-1}(y))}{d\mu(y)}}\\ &\quad\times K^W_\hbar\lfloor f\rfloor(\varphi^{-1}(x),\varphi^{-1}(y));\end{aligned} \tag{3.85}$$

$$\left(\alpha^\hbar_{\tilde{g}}(K^W_\hbar\lfloor f\rfloor)\right)(x,y) = e^{i[\tilde{g}(y)-\tilde{g}(x)]/\hbar}K^W_\hbar\lfloor f\rfloor(x,y). \tag{3.86}$$

The classical analogue of this automorphism is evidently given by

$$\alpha^0_{(\varphi,\tilde{g})}(f) := f\circ\rho_0((\varphi,\tilde{g})^{-1}), \tag{3.87}$$

where ρ_0 is defined in (3.13).

An **isometry** of $(Q,\mathbf{g})$ is a diffeomorphism φ for which $\varphi^*\mathbf{g} = \mathbf{g}$; an **infinitesimal isometry** is a vector field ξ on Q for which $L_\xi\mathbf{g} = 0$ (where L is the Lie derivative).

Theorem 3.6.3. *If φ is an isometry of $(Q,\mathbf{g})$, then*

$$\alpha^\hbar_\varphi(\mathcal{Q}^W_\hbar(f)) = \mathcal{Q}^W_\hbar(\alpha^0_\varphi(f)) \tag{3.88}$$

for all $f\in C^\infty_{\mathrm{PW}}(T^*Q)$ (cf. (3.49)), and $\hbar$ small enough so that $\mathcal{Q}^W_\hbar(f)$ is independent of the cutoff κ (cf. the comments following 3.4.4). If κ is invariant under the (pushforward) action of φ to TQ, eq. (3.88) holds for all $\hbar\neq 0$.

If ξ is an infinitesimal isometry whose flow is complete, then, under the same conditions on f, on the domain $C^\infty_c(Q)$ one has

$$\frac{i}{\hbar}\left[\mathcal{Q}^W_\hbar(J_\xi),\mathcal{Q}^W_\hbar(f)\right] = \mathcal{Q}^W_\hbar(\{J_\xi,f\}). \tag{3.89}$$

If φ is an isometry, the Radon–Nikodym derivatives in (3.85) equal unity. Equation (3.88) then follows from (3.85), (3.51), (3.42), (3.48), and the property

$$\exp_{\varphi(q)}(\pm\tfrac{1}{2}\varphi_*(X_q)) = \varphi(\exp_q(\pm\tfrac{1}{2}X_q)), \tag{3.90}$$

which, because φ is an isometry, holds by the definition of the exponential map. Equation (3.89) follows from (3.88), (3.82), Proposition 3.1.5, and the following interesting result. ∎

Proposition 3.6.4. *If a vector field ξ on Q is complete, then $\mathcal{Q}^W_\hbar(J_\xi)$ is essentially self-adjoint on the domain $C^\infty_c(Q)$.*

By (3.82) we might as well consider $d\rho_\hbar(\xi)$. Let φ_t be the flow generated by ξ; since ξ is complete, the flow exists for all t. Then (3.81) defines a one-parameter unitary group $t\mapsto\rho_\hbar(\varphi_t)$ on $L^2(Q)$. A routine calculation shows that $d\rho_\hbar(\varphi_t)\Psi/dt|t = 0$ exists for all $\Psi\in C^\infty_c(Q)$, and equals $\xi\Psi$. Hence $C^\infty_c(Q)$ is contained in the domain of the generator of the unitary group, and this generator equals $-i\xi$ on $C_c^\infty(Q)$. Furthermore, by (3.81), $\rho_\hbar(\varphi_t)$ leaves the dense domain $C_c^\infty(Q)$ invariant for all t, since φ_t is a diffeomorphism. A lemma in functional analysis states that if $t \mapsto \exp(itH)$ is a unitary group in a Hilbert space that leaves a dense linear subspace $\mathcal{D}_0 \subset D(H)$ invariant, then H is essentially self-adjoint on $\mathcal{D}_0$. This implies the proposition. ■

3.7 *The Quantum Hamiltonian and its Classical Limit*

We extend Proposition 3.6.1 to the most important function on T^*Q that is quadratic in the momenta.

Definition 3.7.1. *Given a Riemannian metric* **g** *on Q, the* **Laplace–Beltrami operator** Δ *is an elliptic second-order differential operator on Q, defined by*

$$(\Psi, \Delta\Phi) = -\int_Q d\mu(q)\mathbf{g}_q(\nabla\Psi(q), \nabla\Phi(q)), \tag{3.91}$$

where $\Psi, \Phi \in C_c^\infty(Q)$; the gradient ∇ is defined in (3.16).

In coordinates, one has

$$\Delta = \mathbf{g}^{ij}\nabla_i\partial_j = \frac{1}{\sqrt{\det \mathbf{g}}}\partial_i(\sqrt{\det \mathbf{g}}\,\mathbf{g}^{ij}\partial_j); \tag{3.92}$$

here ∂_i acts on everything to its right, including the (omitted) Ψ. In flat space Δ clearly reduces to the Laplacian.

Proposition 3.7.2. *The Weyl quantization of the Hamiltonian h_* (cf. (3.31)) is given (on the domain $C_c^\infty(Q) \subset L^2(Q)$) by*

$$H_\hbar := \mathcal{Q}_\hbar^W(h) = -\tfrac{1}{2}\hbar^2(\Delta - \tfrac{1}{3}\mathbf{R}). \tag{3.93}$$

Here the Ricci scalar **R** *(cf. (3.23)) is seen as a multiplication operator.*

The proof of (3.93) follows the same steps as in 3.6.1; here one additionally uses (3.25). □

The functional analysis of the first term of (3.93) is given by a result of the same type as 3.6.4, but somewhat deeper.

Theorem 3.7.3. *When $(Q, \mathbf{g})$ is complete (cf. 3.1.3), the Laplace–Beltrami operator is essentially self-adjoint on $C_c^\infty(Q)$.*

The symbol Δ stands for the differential operator (3.92) defined on the domain $C_c^\infty(Q)$; its closure is denoted by $\overline{\Delta}$. We can look at (3.91) as the definition of Δ as a quadratic form with initial domain $C_c^\infty(Q)$.

It is easily verified that Δ is symmetric. It is evident from the definition (3.91) that Δ, and therefore $\overline{\Delta}$, is negative. It follows that $\overline{\Delta}$ has equal deficiency indices, so that self-adjoint extensions exist. (This conclusion also follows from the fact that Δ commutes with the conjugation $\Psi \mapsto \overline{\Psi}$ on $L^2(Q)$.) The domain $D(\overline{\Delta})$ of

$\overline{\Delta}$ is the set of vectors $\Psi \in L^2(Q)$ for which there exists a sequence $\Psi_j \in C_c^\infty(Q)$ such that $\Psi_j \to \Psi$, and $\Delta\Psi_j$ converges to an element in $L^2(Q)$; the latter is then by definition $\overline{\Delta}\Psi$. The domain $D(\Delta^*)$ of the adjoint is the collection of vectors in $L^2(Q) \subset \mathcal{D}(Q)'$ (the distributional dual of $C_c^\infty(Q) = \mathcal{D}(Q)$ with the Schwartz topology) for which $\Delta^*\Psi$ lies in $L^2(Q)$; here Δ^* is given by the expression (3.92), understood in the sense of weak (distributional) derivatives. The theorem states that $D(\overline{\Delta}) = D(\Delta^*)$ if Q is complete.

The following fact will be used: If A is a positive closed operator, then the dimension of $\ker(A - \lambda)$ is constant for $\lambda \in \mathbb{C}\backslash[0, \infty)$. Suppose $\overline{\Delta}$ is not self-adjoint. Then the deficiency indices are nonzero (and equal), so that the equation $\Delta^*\Psi = i\Psi$ has a nonzero solution in $D(\Delta^*)$. By the above fact, there is a nonzero solution $\Psi = \Psi_1$ of $\Delta^*\Psi = \Psi$. The theory of elliptic PDE's shows that $\Psi_1 \in C^\infty(Q)$ ("elliptic regularity"), so that the weak derivatives in Δ^* are actually strong ones. Abbreviating the right-hand side of (3.91) as $(\nabla\Psi, \nabla\Phi)$, the idea of the proof is to write

$$(\nabla\Psi_1, \nabla\Psi_1) = -(\Delta^*\Psi_1, \Psi_1) = -(\Psi_1, \Psi_1) \leq 0,$$

forcing $\nabla\Psi_1 = 0$ (in $L^2(Q)$) and therefore $\Delta^*\Psi_1 = \Psi_1 = 0$. However, the partial integration leading to the first equality is not a priori justified unless Ψ_1 has compact support (hence if Q were compact the proof would be finished here). Hence one uses the following device. Pick a fixed $q_0 \in Q$, and define a family of functions $j_k : Q \to [0, 1]$ by $j_k(q) = j(d(q, q_0)/k)$ $(k \in \mathbb{N})$, where d is the distance function on $Q \times Q$ (see 3.2), and $j : [0, \infty) \to [0, 1]$ is a smooth cutoff function that is 1 in $[0, 1]$ and 0 on $[2, \infty)$. At this point the completeness of (Q, g) is used: It follows from Theorem 3.2.3, guaranteeing metric completeness, that each j_k has compact support (since a closed and bounded set in a finite-dimensional complete metric space is compact). One clearly has $j_k \to 1$ pointwise.

The distance $d(q, q_0)$ is a differentiable function of q except at q_0 and at the cut locus $C(q_0)$ (cf. 3.2.4). By Corollary 3.2.6 the set $C(q_0)$ is of μ-measure zero, so that each component of $\nabla d(\cdot, q_0)$ is well-defined as an element of $L^2_{\text{loc}}(Q)$. The triangle inequality leads to $|d(q_1, q_0) - d(q_2, q_0)| \leq d(q_1, q_2)$. This Lipschitz condition implies that in normal coordinates centered at q_0 the metric is absolutely continuous with respect to each variable, with $|\partial_i d(q, q_0)| \leq 1$. Hence, by the chain rule, $|\partial_i j_k(q)| \leq \|j'\|_\infty/k$, so that

$$\lim_{k\to\infty} \|\partial_i j_k\|_\infty = 0. \tag{3.94}$$

Trivially, $(j_k\Psi_1, j_k\Psi_1) \geq 0$. Moving the second j_k to the left, replacing the second Ψ_1 by $\Delta^*\Psi_1$, and performing a partial integration (now allowed, since $j_k^2\Psi_1$ has compact support), one rewrites this inequality as

$$\|j_k \nabla \Psi_1\|_2^2 \leq 2|(\Psi_1 \nabla j_k, j_k \nabla \Psi_1)|,$$

where $\|\cdot\|_2$ is the norm derived from $(\ ,\ )$. The Cauchy–Schwarz inequality then yields $\|j_k \nabla \Psi_1\|_2^2 \leq 2\|\Psi_1 \nabla j_k\|_2 \, \|j_k \nabla \Psi_1\|_2$, which in turn leads to the bound $\|j_k \nabla \Psi_1\|_2 \leq 2\|\Psi_1 \nabla j_k\|_2$. Accordingly, by (3.94) one has $\lim_k \|j_k \nabla \Psi_1\|_2 = 0$.

Fatou's lemma and $\lim_k j_k = 1$ (pointwise) then imply that $\| \nabla \Psi_1 \|_2 = 0$; as we have seen before, this implies $\Psi_1 = 0$. ■

Self-adjointness of $\mathcal{Q}_\hbar^W(h)$ in (3.93) then follows from a mild assumption.

Proposition 3.7.4. *Let $(Q, \mathbf{g})$ be complete with Ricci scalar $\mathbf{R}$ bounded. Then $\mathcal{Q}_\hbar^W(h)$ is essentially self-adjoint on $C_c^\infty(Q)$ and self-adjoint on $D(\overline{\Delta})$.*

This is immediate from the Kato–Rellich theorem on perturbations of self-adjoint operators. □

In order to generalize the proof of Theorem 2.7.2 to Riemannian spaces $(Q, \mathbf{g})$, we have to make a simplifying assumption, namely that $Q \simeq \mathbb{R}^n$ as a manifold. If $(Q, \mathbf{g})$ is complete, by Theorem 3.2.5 this would follow from the assumption that the cut locus $C(q_0)$ is empty for some point $q_0 \in Q$. The globally defined coordinates x^i on $\mathbb{R}^n$ may then taken to be normal coordinates based at q_0. However, given the assumption that $Q \simeq \mathbb{R}^n$, we need not assume that the cut loci defined by the metric are empty, and neither is it necessary that the classical motion defined by h be complete. (The case that $(Q \simeq \mathbb{R}^n, \mathbf{g})$ is complete and $\mathbf{R}$ is bounded is, of course, covered; cf. 3.7.4.)

In the present case the notation $L^2(Q)$ stands for $L^2(\mathbb{R}^n, d^n x \sqrt{\det \mathbf{g}(x)})$. We can define normalized coherent states in $L^2(Q)$ by

$$\Psi_\hbar^{(p,q)}(x) := (\pi\hbar)^{-n/4}(\det \mathbf{g}(x))^{-1/4} e^{ip(x-\frac{1}{2}q)/\hbar} e^{-(x-q)^2/(2\hbar)}; \tag{3.95}$$

this slightly generalizes (2.47).

We shall merely assume that $\mathbf{g}$ is a metric on $Q = \mathbb{R}^n$ for which $\mathbf{g}^{ij}$ and its derivatives are $O(\exp(x^2/2))$ for $x \to \infty$. If $\hbar < 1$, the operator $H_\hbar$ in (3.93) is then symmetric on the domain $\mathcal{D}_0$ consisting of the span of all coherent states (3.95), and has one or more self-adjoint extensions (since it commutes with complex conjugation). As in 2.7.2, with slight abuse of notation the unitary one-parameter group $\exp(itH_\hbar/\hbar)$ on $L^2(Q)$ is understood to be generated by an arbitrary self-adjoint extension of (3.93). We use the notation (2.88) and I.(2.13).

Theorem 3.7.5. *Let $(Q, \mathbf{g})$ be as detailed in the preceding paragraph. Fix $(p, q) \in T^*Q$, assuming that the cogeodesic motion $(p(t), q(t))$ with initial conditions $(p(0), q(0)) = (p, q)$ exists for $t_i < t < t_f$. Then with $\mathcal{Q}_\hbar^W$ defined by 3.4.4 and $\Psi_\hbar^{(p,q)}$ given by (3.95), for all $t \in (t_i, t_f)$ one has*

$$\lim_{\hbar \to 0} \left(\Psi_\hbar^{(p,q)}, \left[\mathcal{Q}_\hbar^W(\alpha_t^0(f)) - \alpha_t^\hbar(\mathcal{Q}_\hbar^W(f)) \right] \Psi_\hbar^{(p,q)} \right) = 0. \tag{3.96}$$

The coordinates (p_i, q^i) are globally defined on T^*Q, and from (3.74), (3.1), and (3.83) we obtain

$$\begin{aligned} Q_\hbar^i &:= \mathcal{Q}_\hbar^W(q^i) = x^i; \\ P_{\hbar,i} &:= \mathcal{Q}_\hbar^W(p_i) = -i\hbar \det \mathbf{g}(x)^{-1/4} \frac{\partial}{\partial x^i} \det \mathbf{g}(x)^{1/4}, \end{aligned} \tag{3.97}$$

defined as operators on $\mathcal{D}_0$ or on $C_c^\infty(Q)$ (cf. the comment after (3.92)). We may then write

$$H_\hbar = \tfrac{1}{2}\lambda\left[\mathcal{Q}_\hbar^W(\mathbf{g}^{ij}), \mathcal{Q}_\hbar^W(p_i)\mathcal{Q}_\hbar^W(p_j)\right] = \tfrac{1}{2}\mathbf{g}^{ij}(Q_\hbar)\circ(P_{\hbar,i}P_{\hbar,j}) \tag{3.98}$$

(rather than the complete symmetrization $\frac{1}{2}\lambda\left[\mathcal{Q}_\hbar^W(\mathbf{g}^{ij}), \mathcal{Q}_\hbar^W(p_i), \mathcal{Q}_\hbar^W(p_j)\right]$, as might have been expected on the basis of the flat-space case (2.37)). The transformation $V: L^2(Q)\to L^2(\mathbb{R}^n)$ defined by $V\Psi(x) = \det\mathbf{g}(x)^{1/4}\Psi(x)$ is clearly unitary, and satisfies

$$\begin{aligned} VQ_\hbar^i V^* &= Q_\hbar^{S,i};\\ VP_{\hbar,i}V^* &= P_{\hbar,i}^S, \end{aligned} \tag{3.99}$$

cf. (2.23), (2.24). In particular, one infers that $Q_\hbar^i$ and $P_{\hbar,i}$ are essentially self-adjoint on $\mathcal{D}_0$ or on $C_c^\infty(Q)$. Moreover, the canonical commutation relations are the same as in the flat-space case; see (2.25). From (3.98) and (3.99) we obtain

$$VH_\hbar V^* = \tfrac{1}{2}\left[\mathbf{g}^{ij}(Q_\hbar^S)P_{\hbar,i}^S P_{\hbar,j}^S - i\hbar\partial_j\mathbf{g}^{ij}(Q_\hbar^S)P_{\hbar,i}^S - \tfrac{1}{2}\hbar^2\partial_i\partial_j\mathbf{g}^{ij}(Q_\hbar^S)\right]. \tag{3.100}$$

The final virtue of V is that up to a phase, it maps the coherent states (3.95) into their flat-space analogues (2.47). Hence one can transfer the entire situation to $L^2(\mathbb{R}^n)$.

Lemma 3.7.6. *With $\Psi_\hbar^{(p,q)}$ the coherent state (3.95), one has for all $(p,q)\in T^*Q$, all $(\varphi,\tilde{g})\in\mathcal{G}_Q$, and all $f\in C_{\rm PW}^\infty(T^*Q)$*

$$\lim_{\hbar\to 0}\left(\Psi_\hbar^{(p,q)}, [\alpha_{(\varphi,\tilde{g})}^\hbar(\mathcal{Q}_\hbar^W(f)) - \mathcal{Q}_\hbar^W(\alpha_{(\varphi,\tilde{g})}^0(f))]\Psi_\hbar^{(p,q)}\right) = 0. \tag{3.101}$$

Denote the Fourier transform (cf. (3.42)) of $\alpha_{(\varphi,\tilde{g})}^0(f)$ by $\alpha_{(\varphi,\tilde{g})}^0(\check{f})$. From (3.87), for $X\in T_qQ$ we obtain the expression

$$\alpha_\varphi^0(\check{f})(X) = \frac{d\mu(\varphi^{-1}(q))}{d\mu(q)}\check{f}(\varphi_*^{-1}X); \tag{3.102}$$

$$\alpha_{\tilde{g}}^0(\check{f})(X) = e^{-i(X\tilde{g})(q)}\check{f}(X). \tag{3.103}$$

As in the proof of Theorem 3.5.1, Lemma 3.5.2 allows the assumption that the set $\tau_{T^*Q\to Q}(\operatorname{supp}(f))$ is contained in an arbitrarily small open set $U\subset Q$, which we choose to be some geodesically convex set U. It is clear from (3.85), (3.102), and (3.95) that both terms in (3.101) vanish in the limit $\hbar\to 0$ if $\varphi(q)\notin\tau_{T^*Q\to Q}(\operatorname{supp}(f))$; hence we assume the converse.

We treat $\varphi=(\varphi,0)$ and $\tilde{g}=(\mathrm{id},\tilde{g})$ separately; the result for the two combined follows in an obvious way. We start with α_φ. We write both terms in (3.101) in the form (3.54), change variables firstly by $q_i\mapsto\varphi(q_i)$, and secondly by (3.55), with subsequent rescaling $v\mapsto\hbar v$. We now write q' for the object q in (3.55) to avoid confusion with the label q on $\Psi_\hbar^{(p,q)}$. In the first term we then use (3.85) and (3.51). In the second term we have the expression

$$\hbar^n K_\hbar^W\lfloor\alpha^0(f)\rfloor(\varphi(\gamma(q',v;\tfrac{1}{2}\hbar)),\varphi(\gamma(q',v;-\tfrac{1}{2}\hbar))).$$

By definition, $\varphi(q') = \varphi(\gamma(q', v; 0))$ and $\varphi_*(v) = \frac{1}{2}d\gamma(q', v; \frac{1}{2}\hbar)/d\hbar|\hbar = 0$. Hence

$$\varphi(\gamma(q', v; \pm\tfrac{1}{2}\hbar)) = \gamma(\varphi(q'), \varphi_*(v); \pm\tfrac{1}{2}\hbar) + O(\hbar^2),$$

where the order symbol is meant in the sense of smooth functions evaluated on both sides. From (3.51) and (3.102) we then see that the expression above equals $[d\mu(q')/d\mu(\varphi(q'))]\check{f}(v, q') + O(\hbar)$. In the first term in (3.101) we expand the terms involving $\sqrt{\ldots}$ (see (3.85)) and $\check{f}$ around $\hbar = 0$. To $O(1)$ the first and the second term are then seen to cancel out. The remainder of $O(\hbar)$ is easily shown to vanish for $\hbar \to 0$, since $\check{f}$ has compact support.

The argument for $\alpha_{\tilde{g}}$ is analogous. This time the cancellation of the $O(1)$ term is effected by (3.86), (3.103), and the fact that

$$[\tilde{g}(\gamma(q', v; -\tfrac{1}{2}\hbar)) - \tilde{g}(\gamma(q', v; -\tfrac{1}{2}\hbar))]/\hbar = -v\tilde{g}(q') + O(\hbar),$$

which follows because by definition $v = \dot{\gamma}(q, v; 0)$. ∎

One then proceeds in precisely the same fashion as in the proof of 2.7.2, and obtains (2.162). The final stage is analogous to the procedure to prove 2.7.2 for $\mathcal{Q}_\hbar = \mathcal{Q}_\hbar^B$, except that the use of 2.4.3 is replaced by Lemma 3.7.6. This is possible because (up to a phase) the operators $U_{\frac{1}{\hbar}}(p(t), q(t))$ used in that step are of the form $V\rho_\hbar(\varphi, \tilde{g})V^*$; cf. 3.6.2. ∎

CHAPTER III

Groups, Bundles, and Groupoids

1 Lie Groups and Lie Algebras

1.1 Lie Algebra Actions and the Momentum Map

This section describes the main class of examples of Poisson manifolds that are not symplectic. Here G is a Lie group, $\mathfrak{g}$ its Lie algebra, and $\mathfrak{g}^*$ is the dual of $\mathfrak{g}$.

Definition 1.1.1. *The* $(\pm)$ **Lie–Poisson structure** *on* $\mathfrak{g}^*$ *is given by the Poisson bracket*

$$\{f, g\}_{\pm}(\theta) := \pm\theta([df_\theta, dg_\theta]); \tag{1.1}$$

here the differential df_θ *of* $f \in C^\infty(\mathfrak{g}^*, \mathbb{R})$ *at* $\theta \in \mathfrak{g}^*$*, which is a linear map from* $T_\theta\mathfrak{g}^* \simeq \mathfrak{g}^*$ *to* $\mathbb{R}$*, is identified with an element of* $\mathfrak{g} \simeq \mathfrak{g}^{**}$*, so that the right-hand side of (1.1) is the Lie bracket in* $\mathfrak{g}$*. The space* $\mathfrak{g}^*$ *equipped with the Poisson bracket (1.1) is denoted by* $\mathfrak{g}^*_\pm$*; hence* $C^\infty(\mathfrak{g}^*_\pm, \mathbb{R})$ *stands for the associated Poisson algebra.*

For $X \in \mathfrak{g}$ let $\tilde{X}$ be the linear function on $\mathfrak{g}^*$ defined by $\tilde{X}(\theta) := \theta(X)$; clearly $\tilde{X} \in C^\infty(\mathfrak{g}^*, \mathbb{R})$. From (1.1) one then obtains

$$\{\tilde{X}, \tilde{Y}\}_{\pm} = \pm\widetilde{[X, Y]}; \tag{1.2}$$

cf. II.(2.10) and surrounding text. In fact, the Poisson structure is determined by the special case (1.2). For let $\{T_a\}$ be a basis of $\mathfrak{g}$, with $[T_a, T_b] = C^c_{ab}T_c$, and dual basis $\{\omega^a\}$ of $\mathfrak{g}^*$, defined by $\omega^a(T_b) = \delta^a_b$. We then have global coordinates θ_a on $\mathfrak{g}^*$ (so that $\theta = \theta_a\omega^a$), and $\tilde{T}_a$ is simply the coordinate function θ_a. We know that $\{f, g\}$ depends linearly on df and dg; cf. I.(2.4). Since $df = (\partial f/\partial\theta_a)d\tilde{T}_a$, the claim follows. Evidently, $\{\tilde{T}_a, \tilde{T}_b\}_{\pm} = \pm C^c_{ab}\tilde{T}_c$, or $\{\theta_a, \theta_b\}_{\pm} = \pm C^c_{ab}\theta_c$. Thus, omitting the argument θ, (1.1) may be written as

$$\{f, g\}_\pm = \pm C^c_{ab}\theta_c \frac{\partial f}{\partial \theta_a}\frac{\partial g}{\partial \theta_b}. \tag{1.3}$$

We now look at the representation theory of $C^\infty(\mathfrak{g}^*_-, \mathbb{R})$ (in the sense of I.2.6.1). By Corollary I.2.6.5 a representation of $C^\infty(\mathfrak{g}^*_-, \mathbb{R})$ corresponds to a symplectic manifold S and a smooth map $J : S \to \mathfrak{g}^*$, such that $\{J^*f, J^*g\}_S = J^*\{f, g\}_-$ for all $f, g \in C^\infty(\mathfrak{g}^*, \mathbb{R})$. That is, $J : S \to \mathfrak{g}^*_-$ is a Poisson map. Let

$$J_X := J^*\tilde{X}, \tag{1.4}$$

which is in $C^\infty(S, \mathbb{R})$; in other words, $J_X(\sigma) := (J(\sigma))(X)$.

Proposition 1.1.2. *A smooth map $J : S \to \mathfrak{g}^*_-$ is Poisson iff $\{J_X, J_Y\}_S = -J_{[X,Y]}$ for all $X, Y \in \mathfrak{g}$.*

If B^S is the Poisson tensor on S, then $\{J^*f, J^*g\}_S = B^S_\sigma(J^*df, J^*dg\}$. As in the previous paragraph, this implies that

$$\{J^*f, J^*g\}_S(\sigma) = \frac{\partial f}{\partial \theta_a}(J(\sigma))\frac{\partial g}{\partial \theta_b}(J(\sigma))\{J^*\tilde{T}_a, J^*\tilde{T}_b\}(\sigma). \tag{1.5}$$

By assumption $\{J^*\tilde{T}_a, J^*\tilde{T}_b\}_S(\sigma) = -C^c_{ab}\tilde{T}_c(J(\sigma))$, so that the right-hand side of (1.5) is $\{f, g\}_-(J(\sigma))$. ∎

Define $\xi_X := \xi_{J_X}$, which is the Hamiltonian vector field of J_X. Assuming that J is indeed a Poisson morphism, the Jacobi identity on the Poisson bracket of S (or (I.2.9)) implies that

$$[\xi_X, \xi_Y] = -\xi_{[X,Y]}; \tag{1.6}$$

here the left-hand side contains the commutator of vector fields, whereas on the right-hand side [,] stands for the Lie bracket in $\mathfrak{g}$.

Let us refer to a linear map $X \mapsto \xi_X$ of $\mathfrak{g}$ into the space of vector fields $\Gamma(TS)$ on S satisfying (1.6) as a **$\mathfrak{g}$-action** on S. Here $\Gamma(TS)$ may be regarded as the Lie algebra of the diffeomorphism group of S, whose Lie bracket is minus the commutator (cf. I.3.3), so (1.6) corresponds to a Lie algebra homomorphism as appropriate. When various $\mathfrak{g}$-actions play a role we sometimes write ξ^S_X for ξ_X. If $\{T_a\}$ is a basis of $\mathfrak{g}$, we abbreviate $\xi_a := \xi_{T_a}$. One speaks of a **Poisson $\mathfrak{g}$-action** when S is a Poisson manifold and $L_{\xi_X}B = 0$ for all $X \in \mathfrak{g}$, where B is the Poisson tensor. When (S, ω) is symplectic, which is the only case we shall consider in the context of $\mathfrak{g}$-actions, this condition is equivalent to $L_{\xi_X}\omega = 0$ for all X.

We infer from I.(2.10) that a representation of $C^\infty(\mathfrak{g}^*_-, \mathbb{R})$ on S leads to a Poisson $\mathfrak{g}$-action on S. Conversely, one may ask whether a given $\mathfrak{g}$-action on a symplectic manifold S is related to a representation of $C^\infty(\mathfrak{g}^*_-, \mathbb{R})$ on S.

Definition 1.1.3. *A* **momentum map** *for a $\mathfrak{g}$-action $X \mapsto \xi_X$ on S is a map $J : S \to \mathfrak{g}^*$ for which*

$$\xi_{J_X} = \xi_X \tag{1.7}$$

for all $X \in \mathfrak{g}$; here J_X is defined by (1.4).

This definition applies to general Poisson manifolds, but we will use it only when S is symplectic. We will occasionally write J_a for J_{T_a}. A **Hamiltonian $\mathfrak{g}$-action** on a symplectic manifold is a $\mathfrak{g}$-action given by a momentum map as in (1.7). It is clear from (1.7) and I.(2.10) that a Hamiltonian $\mathfrak{g}$-action is Poisson. When a momentum map J exists, (1.7) is equivalent to

$$i_{\xi_X}\omega = dJ_X \tag{1.8}$$

for all $X \in \mathfrak{g}$; to prove this, contract both sides with ξ_f (which is the most general type of local vector field, since S is symplectic) and use I.(2.19).

Conversely, when a $\mathfrak{g}$-action is Poisson, the properties $d\omega = 0$ and $L_{\xi_X}\omega = 0$ and the identity $L_\xi = i_\xi d + di_\xi$ imply $di_{\xi_X}\omega = 0$, so that by Poincaré's lemma a function J_X satisfying (1.8) must exist at least locally.

Proposition 1.1.4. *Sufficient conditions for the existence of a momentum map for a Poisson $\mathfrak{g}$-action on a symplectic manifold (S, ω) are $H^1_{\mathrm{dR}}(S, \mathbb{R}) = 0$ or $\mathfrak{g} = [\mathfrak{g}, \mathfrak{g}]$ (equivalently, $H^1(\mathfrak{g}, \mathbb{R}) = 0$).*

Here $H^1_{\mathrm{dR}}(S, \mathbb{R}) := Z^1_{\mathrm{dR}}(S, \mathbb{R})/B^1_{\mathrm{dR}}(S, \mathbb{R})$ is the first de Rham cohomology group of S; recall that $Z^1_{\mathrm{dR}}(S, \mathbb{R})$ and $B^1_{\mathrm{dR}}(S, \mathbb{R})$ are the spaces of all closed and all exact 1-forms on S, respectively. The sufficiency of the condition $H^1_{\mathrm{dR}}(S, \mathbb{R}) = 0$ is evident from the paragraph preceding this proposition.

The vector space $H^1(\mathfrak{g}, \mathbb{R})$ is the **first cohomology group of** $\mathfrak{g}$; since $B^1(\mathfrak{g}, \mathbb{R})$ is identically zero, $H^1(\mathfrak{g}, \mathbb{R})$ is defined as the subspace $Z^1(\mathfrak{g}, \mathbb{R}) \subset \mathfrak{g}^*$ of all θ for which $\theta([X, Y]) = 0$ for all $X, Y \in \mathfrak{g}$. The equivalence between the conditions $\mathfrak{g} = [\mathfrak{g}, \mathfrak{g}]$ and $H^1(\mathfrak{g}, \mathbb{R}) = 0$ is obvious.

If $\mathfrak{g} = [\mathfrak{g}, \mathfrak{g}]$, then an arbitrary $X \in \mathfrak{g}$ can be written as $X = \sum_i X_i$ with $X_i = [Y_i, Z_i]$ for appropriate $Y_i, Z_i \in \mathfrak{g}$. If $X = [Y, Z]$, we choose $J_X = \omega(\xi_Y, \xi_Z)$, which, by an elementary calculation, using (1.6) and $d\omega = 0$, satisfies (1.8) and hence (1.7). For arbitrary X we define J_X by linear extension of this expression. ■

The existence of a momentum map J in itself does not imply that J preserves the Poisson bracket. To detect the extent to which it does we define a function Γ on $\mathfrak{g} \times \mathfrak{g} \times S$ by

$$\{J_X, J_Y\}_S = -J_{[X,Y]} - \Gamma(X, Y). \tag{1.9}$$

It is clear that Γ is bilinear and antisymmetric in X, Y. Taking the Poisson bracket of both sides of (1.9) with an arbitrary $f \in C^\infty(S)$, and using (1.7), (1.6), and the Jacobi identity, we infer that $\{\Gamma(X, Y), f\} = 0$ for all X, Y. Since S is symplectic, this shows that Γ does not depend on its argument in S. A bilinear function $\Gamma : \mathfrak{g} \otimes \mathfrak{g} \to \mathbb{R}$ satisfying

$$\Gamma(X, Y) = -\Gamma(Y, X), \tag{1.10}$$

$$\Gamma(X, [Y, Z]) + \Gamma(Z, [X, Y]) + \Gamma(Y, [Z, X]) = 0 \tag{1.11}$$

is called a **2-cocycle** on $\mathfrak{g}$ (with values in $\mathbb{R}$). The space of all 2-cocycles on $\mathfrak{g}$ is denoted by $Z^2(\mathfrak{g}, \mathbb{R})$. It follows from the Jacobi identity on both $\{\, ,\, \}_S$ and $[\, ,\,]$ that Γ as defined in (1.9) is indeed an element of $Z^2(\mathfrak{g}, \mathbb{R})$.

We are now motivated to define a modified Lie–Poisson bracket on $\mathfrak{g}^*$ by

$$\{f, g\}_{\pm}^{\Gamma} := \{f, g\}_{\pm} \pm \Gamma(df, dg); \tag{1.12}$$

this is indeed a Poisson bracket on account of (1.11). Generalizing (1.3), in coordinates one has

$$\{f, g\}_{\pm}^{\Gamma} = \pm\left(C_{ab}^{c}\theta_c + \Gamma_{ab}\right)\frac{\partial f}{\partial\theta_a}\frac{\partial g}{\partial\theta_b}, \tag{1.13}$$

$$\Gamma_{ab} := \Gamma(T_a, T_b). \tag{1.14}$$

As for $\Gamma = 0$, one shows that this modified Poisson bracket is determined by the special case

$$\{\tilde{X}, \tilde{Y}\}_{\pm}^{\Gamma} = \pm\left(\widetilde{[X, Y]} + \Gamma(X, Y)\right). \tag{1.15}$$

Definition 1.1.5. *The space $\mathfrak{g}^*$ equipped with the Poisson bracket (1.13) is denoted by $\mathfrak{g}^*_{(\Gamma)\pm}$; we sometimes write $C^{\infty}_{\Gamma}(\mathfrak{g}^*_{\pm})$ for the associated Poisson algebra.*

Generalizing Proposition 1.1.2, one easily proves

Proposition 1.1.6. *A smooth map $J : S \to \mathfrak{g}^*_{(\Gamma)-}$ is Poisson iff (1.9) holds.*

The essence of the preceding discussion may now be summarized as follows.

Theorem 1.1.7. *There is a bijective correspondence between representations π of $C^{\infty}_{\Gamma}(\mathfrak{g}^*_{-})$ (in the sense of I.2.6.1) and Hamiltonian $\mathfrak{g}$-actions with given complete momentum map and associated 2-cocycle Γ. Given $\pi : C^{\infty}_{\Gamma}(\mathfrak{g}^*_{-}) \to C^{\infty}(S)$ one constructs a Poisson map $J : S \to \mathfrak{g}^*_{(\Gamma)-}$ by I.2.6.5, and subsequently defines the $\mathfrak{g}$-action $X \mapsto \xi_X$ by (1.7). Conversely, given a $\mathfrak{g}$-action with associated complete momentum map J (yielding a 2-cocycle Γ), one puts $\pi = J^*$.*

The fact that $X \mapsto \xi_X$ is indeed a $\mathfrak{g}$-action follows from the argument leading to (1.6); even when $\Gamma \neq 0$ the additional term in (1.12) is a constant function, so that the Jacobi identity on the Poisson bracket still implies (1.6). It is Poisson by I.(2.10), and Hamiltonian with 2-cocycle Γ by construction. In the converse the fact that J^* is a representation is immediate from 1.1.6. ∎

A **strongly Hamiltonian** $\mathfrak{g}$**-action** is a Hamiltonian $\mathfrak{g}$-action possessing a momentum map $J : S \to \mathfrak{g}^*_{-}$ that is Poisson; in other words, there exists a J for which $\Gamma = 0$ in (1.9). A Hamiltonian $\mathfrak{g}$-action with 2-cocycle Γ may alternatively be described as a strongly Hamiltonian action of a certain Γ-dependent Lie algebra containing $\mathfrak{g}$.

Definition 1.1.8. *The* **central extension** $\mathfrak{g}_{\Gamma}$ *of a Lie algebra $\mathfrak{g}$ by $\mathbb{R}$ relative to some $\Gamma \in Z^2(\mathfrak{g}, \mathbb{R})$ is $\mathfrak{g}_{\Gamma} := \mathfrak{g} \oplus \mathbb{R}$ as a vector space, equipped with the Lie bracket*

$$[X, Y]_{\Gamma} = [X, Y] + \Gamma(X, Y)T_0; \tag{1.16}$$

$$[X, T_0]_{\Gamma} = 0 \tag{1.17}$$

for $X, Y \in \mathfrak{g}$, and T_0 a basis vector of the extension $\mathbb{R}$.

The Jacobi identity for $[\,,\,]_\Gamma$ is a consequence of (1.11). We have an embedding $\iota : \mathfrak{g} \hookrightarrow \mathfrak{g}_\Gamma$ by $\iota(X) = X \dot{+} 0$ (which is not a Lie algebra homomorphism unless $\Gamma = 0$), as well as a quotient $\mathfrak{g} \simeq \mathfrak{g}_\Gamma/\mathbb{R}$ as Lie algebras.

Proposition 1.1.9. *There is a bijective correspondence between Poisson maps $J : S \to \mathfrak{g}^*_{(\Gamma)-}$ (or, equivalently, Hamiltonian $\mathfrak{g}$-actions with 2-cocycle Γ) and Poisson maps $J_\Gamma : S \to \mathfrak{g}_{\Gamma-}$ (or strongly Hamiltonian $\mathfrak{g}_\Gamma$-actions) in which $(J_\Gamma(\sigma))(T_0) = 1$ for all $\sigma \in S$ (equivalently, $\pi(\tilde{T}_0) = 1_S$, so that $\xi_{T_0} = 0$). This correspondence preserves irreducibility.*

Let ω^0 be the basis element in $\mathfrak{g}_\Gamma$ dual to T_0. Then $J_1 : \mathfrak{g}^*_{(\Gamma)-} \to \mathfrak{g}_{\Gamma-}$ given by $J_1(\theta) := \theta \dot{+} \omega^0$ is a Poisson map (where $\mathfrak{g}^*$ is embedded in $\mathfrak{g}^*_\Gamma$ as the annihilator of the extension $\mathbb{R}$); this follows from (1.15), (1.16), (1.2), and $J_1^*\tilde{T}_0 = 1_{\mathfrak{g}^*}$.

Given $J : S \to \mathfrak{g}^*_{(\Gamma)-}$, one constructs $J_\Gamma : S \to \mathfrak{g}_{\Gamma-}$ by $J_\Gamma := J_1 \circ J$. Conversely, when a given J_Γ is as stated, the equality $J_\Gamma^*(\tilde{T}_0) = 1_S$ and Proposition 1.1.6 imply that $J_X(\sigma) := (J_\Gamma)_{\iota(X)}(\sigma)$ with (1.4) is a Poisson map $J : S \to \mathfrak{g}^*_{(\Gamma)-}$. Finally, Definition I.2.6.6 and the fact that $\xi_{T_0} = 0$ lead to the last part of the proposition. ■

As a by-product of the proof we have

Proposition 1.1.10. *The canonical identification of $C^\infty(\mathfrak{g}^*_{\Gamma-})/\ker(J_1^*)$ with $C^\infty_\Gamma(\mathfrak{g}^*_-)$ is a Poisson isomorphism.*

This is immediate from the definitions and (1.13). ■

The theory so far has been concerned with a Hamiltonian $\mathfrak{g}$-action with given momentum map J. However, when some J exists, then any map $J' = J + \theta_0$, where $\theta_0 \in \mathfrak{g}^*$, is obviously a momentum map for the given $\mathfrak{g}$-action as well. Having found a particular J for which $\Gamma \neq 0$, when can we redefine $J \mapsto J'$ so as to make J' a Poisson morphism?

A 2-cocycle $\Gamma \in Z^2(\mathfrak{g}, \mathbb{R})$ is said to be **trivial** when

$$\Gamma(X, Y) = \theta_0([X, Y]) \tag{1.18}$$

for some $\theta_0 \in \mathfrak{g}^*$. The subspace of trivial 2-cocycles is called $B^2(\mathfrak{g}, \mathbb{R})$. The quotient $H^2(\mathfrak{g}, \mathbb{R}) := Z^2(\mathfrak{g}, \mathbb{R})/B^2(\mathfrak{g}, \mathbb{R})$ is the **second cohomology group of** $\mathfrak{g}$.

Proposition 1.1.11. *If a momentum map of a Hamiltonian $\mathfrak{g}$-action defines a 2-cocycle satisfying (1.18), then the $\mathfrak{g}$-action is strongly Hamiltonian. In particular, when $H^2(\mathfrak{g}, \mathbb{R}) = 0$ any Hamiltonian $\mathfrak{g}$-action is strongly Hamiltonian.*

Given (1.18), the redefined momentum map $J' = J + \theta_0 : S \to \mathfrak{g}^*_-$ is a Poisson map by (1.9). The condition $H^2(\mathfrak{g}, \mathbb{R}) = 0$ implies that any 2-cocycle Γ is given by (1.18). ■

Combining 1.1.11 and 1.1.4 we obtain

Corollary 1.1.12. *Let $H^1(\mathfrak{g}, \mathbb{R}) = H^2(\mathfrak{g}, \mathbb{R}) = 0$. Any Poisson $\mathfrak{g}$-action is strongly Hamiltonian, and is associated with a unique Poisson momentum map $J : S \to \mathfrak{g}^*_-$.*

If one has a Poisson momentum map J, one may still shift $J \mapsto J + \theta_0$, preserving the Poisson property, iff θ_0 annihilates $[\mathfrak{g}, \mathfrak{g}]$; when $\mathfrak{g} = [\mathfrak{g}, \mathfrak{g}]$ this forces $\theta_0 = 0$. ■

The conditions $H^i(\mathfrak{g}, \mathbb{R}) = 0$, $i = 1, 2$, are satisfied, for example, when G is semisimple. The consistency between 1.1.9 and 1.1.11 is guaranteed by

Proposition 1.1.13. *If $\Gamma(X, Y) = \theta_0([X, Y])$ for some $\theta_0 \in \mathfrak{g}^*$, then $\mathfrak{g}_\Gamma$ is isomorphic to the trivial extension $\mathfrak{g}_0 = \mathfrak{g} \oplus \mathbb{R}$ as a direct sum of Lie algebras. In particular, when $H^2(\mathfrak{g}, \mathbb{R}) = 0$, any central extension of $\mathfrak{g}$ by $\mathbb{R}$ is trivial.*

If $\Gamma(X, Y) = \theta_0([X, Y])$, then $X \mapsto X + \theta_0(X)T_0$ for $X \in \mathfrak{g}$ and $T_0 \mapsto T_0$ is the desired isomorphism between $\mathfrak{g}_\Gamma$ and $\mathfrak{g}_0$. ■

1.2 Hamiltonian Group Actions

We will now relate $\mathfrak{g}$-actions to G-actions, where G is a Lie group with Lie algebra $\mathfrak{g}$. Recall that a (left) **action** L of a group G on a manifold S is a map $L : G \times S \to S$, satisfying $L(e, \sigma) = \sigma$ and $L(x, L(y, \sigma)) = L(xy, \sigma)$ for all $\sigma \in S$ and $x, y \in G$. If G is a Lie group, we assume that L is smooth, unless the contrary is explicitly stated. We write $L_x(\sigma) = x\sigma := L(x, \sigma)$.

Given a Lie group action, one defines a linear map $X \mapsto \xi_X$ by

$$\xi_X f(\sigma) := \frac{d}{dt} f(\text{Exp}(tX)\sigma)_{|t=0}, \tag{1.19}$$

where $\text{Exp} : \mathfrak{g} \to G$ is the usual exponential map. One sees that (1.6) holds, so that $X \mapsto \xi_X$ is a Lie algebra homomorphism from $\mathfrak{g}$ into the Lie algebra $\Gamma(TS)$ of $\text{Diff}(S)$. We say that the ξ_X generate the G-action, and call ξ_X a **generator** defined by X.

Conversely, one may ask whether a representation of $C^\infty(\mathfrak{g}^*_-, \mathbb{R})$ on S is derived from a G-action, in which case the representation is called **integrable**. This question is partly answered by the following statement.

Theorem 1.2.1. *Let $X \mapsto \xi_X$ be a homomorphism as above, and suppose the flow of each ξ_X, $X \in \mathfrak{g}$, is complete (this is the case iff there is a basis $\{T_a\}$ of $\mathfrak{g}$ such that the flow of each ξ_a is complete). Then the ξ_X generate an action of the simply connected Lie group $\tilde{G}$ whose Lie algebra is $\mathfrak{g}$.*

The construction of the $\tilde{G}$-action is, of course, done with the flow of the generators; that is, if $\sigma \mapsto \sigma(t)$ is the flow generated by ξ_X, one puts $\text{Exp}(tX) : \sigma \mapsto \sigma(t)$. Such one-parameter groups generate $\tilde{G}$ (which is connected), but it remains to check that one indeed obtains a smooth group action. □

Note that the statement about the basis is nontrivial, since in principle the sum of two complete vector fields may be incomplete. Clearly, the hypothesis is automatically satisfied when S is compact, or more generally when all ξ_X have compact support. In any case, if the theorem leads to a $\tilde{G}$-action and if no ξ_X is identically zero, there is a discrete normal subgroup $D_S \subset \tilde{G}$ such that D_S is the maximal subgroup of $\tilde{G}$ that acts trivially on S. If $G = \tilde{G}/D$ for some discrete central subgroup $D \subset \tilde{G}$ (recall that any Lie group with Lie algebra $\mathfrak{g}$ is of this form), then the ξ_X generate a G-action if $D \subseteq D_S$.

When the $\mathfrak{g}$-action associated to a G-action on a symplectic manifold S is Hamiltonian, one speaks of a **Hamiltonian group action.** Similarly, a **strongly Hamiltonian group action** is an action for which a momentum map $J : S \to \mathfrak{g}^*$ exists that is a Poisson map; cf. (1.19) and 1.1.3. It is immediate from the comment preceding I.2.3.5 that a Hamiltonian G-action automatically consists of Poisson maps. Further to this, the conditions for a G-action on a symplectic manifold to be (strongly) Hamiltonian are entirely determined by the properties of the associated $\mathfrak{g}$-action, and are therefore given by Propositions 1.1.4 and 1.1.11 and Corollary 1.1.12.

The Hamiltonian version of **Noether's theorem** is as follows.

Proposition 1.2.2. *Given a Hamiltonian G-action on a symplectic manifold S, when $h \in C^\infty(S, \mathbb{R})$ is G-invariant (i.e., $h(x\sigma) = h(\sigma)$ for all $x \in G$), each J_X is constant along the Hamiltonian flow lines of h.*

Putting $x = \text{Exp}(tX)$ in $h(x\sigma) = h(\sigma)$, evaluating d/dt at $t = 0$, and using (1.7) leads to $\{J_X, h\} = 0$, which by I.(2.8) and I.(2.11) implies the claim. ∎

In view of this, in physics the components J_X of the momentum map usually play the role of conserved charges.

A Hamiltonian G-action enjoys a certain equivariance property. Recall the **adjoint action** Ad of G on $\mathfrak{g}$, defined by $\text{Ad}(x)Y := xYx^{-1}$ (more precisely, if $Y = d\gamma(t)/dt|t = 0$, then $\text{Ad}(x)Y = dx\gamma(t)x^{-1}|t = 0$). The derived representation of $\mathfrak{g}$ is then given by $\text{ad}(X)Y = [X, Y]$ (where we simply write ad for the awkward $d\text{Ad}$). The **coadjoint action** Co of G on $\mathfrak{g}^*$ is defined by $(\text{Co}(x)\theta)(Y) := \theta(\text{Ad}(x^{-1})Y)$, with derived action of $\mathfrak{g}$ on $\mathfrak{g}^*$ written as $\text{co} := d\text{Co}$. One has $\text{ad}(T_a)T_b = C_{ab}^c T_c$, whence

$$\text{co}(T_a)\theta_b = -C_{ab}^c\theta_c. \tag{1.20}$$

As a first application of these definitions we note:

Proposition 1.2.3. *The Lie–Poisson structure is invariant under the coadjoint action (in other words, the map* $\text{Co}(x)$ *is a Poisson map for each* $x \in G$*).*

By the comment following (1.2) it suffices to show that $\{\tilde{X} \circ \text{Co}(x), \tilde{Y} \circ \text{Co}(x)\}_\pm = \{\tilde{X}, \tilde{Y}\}_\pm \circ \text{Co}(x)$ for all $X, Y \in \mathfrak{g}$ and $x \in G$. Since $\tilde{X} \circ \text{Co}(x) = \text{Ad}(x^{-1})X$ etc., this is evident from the fact that the adjoint action is an automorphism of the Lie algebra. ∎

Given a choice J of a momentum map associated to a Hamiltonian G-action on S via the derived $\mathfrak{g}$-action, we define $\gamma : G \times S \to \mathfrak{g}^*$ by

$$\gamma(x, \sigma) = J(x\sigma) - \mathrm{Co}(x)J(\sigma). \tag{1.21}$$

Lemma 1.2.4. *The function γ is independent of σ, and satisfies*

$$\gamma(xy) = \gamma(x) + \mathrm{Co}(x)\gamma(y). \tag{1.22}$$

A smooth map $\gamma : G \to \mathfrak{g}^*$ with property (1.22) is called a **1-cocycle on G with values in $\mathfrak{g}^*$**; the space of such 1-cocycles is denoted by $Z^1(G, \mathrm{Co}, \mathfrak{g}^*)$. The proof that (1.21) is independent of σ is similar to the argument after (1.9). For arbitrary $f \in C^\infty(S)$ we compute $\{J_Y \circ L_x, f\}_S$ for fixed $x \in G$ and $Y \in \mathfrak{g}$, and use (1.7) and the invariance of the Poisson bracket under L_x. This shows that $\{J_Y \circ L_x, f\}_S = \{J_{\mathrm{Ad}(x^{-1})Y}, f\}_S$. Since S is symplectic, γ_Y (in obvious notation) is therefore constant in σ for all Y. The property (1.22) is then immediate from the definition and the σ-independence of γ and the equality $\mathrm{Co}(xy) = \mathrm{Co}(x)\mathrm{Co}(y)$. ■

Corollary 1.2.5. *A momentum map J for a Hamiltonian G-action is equivariant with respect to the modified coadjoint G-action on $\mathfrak{g}^*$ defined by*

$$\mathrm{Co}^\gamma(x)\theta := \mathrm{Co}(x)\theta + \gamma(x), \tag{1.23}$$

where γ is given by (1.21). That is, $J \circ L_x = \mathrm{Co}^\gamma(x) \circ J$ for all $x \in G$ (recall that $\gamma(x, \sigma) = \gamma(x)$).

In particular, for a strongly Hamiltonian G-action the momentum map J is Co-**equivariant**, *or simply* **equivariant**, *in that $J \circ L_x = \mathrm{Co}(x) \circ J$ for all x. Moreover, infinitesimal* Co-*equivariance (in the sense that $J : S \to \mathfrak{g}^*_-$ is a Poisson map) is equivalent to global* Co-*equivariance.*

Note that (1.22) guarantees that Co^γ is an (affine) action.

Only the final claim is not immediately obvious. It is clear from (1.9) and (1.21) that the 2-cocycle Γ defined by the momentum map of the $\mathfrak{g}$-action corresponding to the group action is given in terms of γ by

$$\Gamma(X, Y) = -\frac{d}{dt}\gamma(\mathrm{Exp}(tX))(Y)_{|t=0}. \tag{1.24}$$

If we put $y = \mathrm{Exp}(tY)$ in (1.22) and differentiate with respect to t, the right-hand side vanishes when the $\mathfrak{g}$-action is strongly Hamiltonian (as $\Gamma = 0$ in that case). Hence the vanishing of the left-hand side says that $\gamma(x)$ is constant in x; since $\gamma(e) = 0$, γ identically vanishes. The equivariance is then stated by (1.21). ■

It should be remarked here that the antisymmetry of Γ as defined by (1.24) is not automatic; it is guaranteed, however, when γ is of the form (1.21).

Theorem 1.2.6. *Assume that G is simply connected, and let $\Gamma \in Z^2(\mathfrak{g}, \mathbb{R})$ and $\gamma \in Z^1(G, \mathrm{Co}, \mathfrak{g}^*)$ be related by (1.24). There is a bijective correspondence between representations of $C^\infty_\Gamma(\mathfrak{g}^*_-)$ and Hamiltonian G-actions with 1-cocycle γ whose momentum map J is complete:*

- *Given a representation $\pi = J^* : C^\infty_\Gamma(\mathfrak{g}^*_-) \to C^\infty(S)$, define the vector fields $\xi_X := \xi_{J_X}$ (with (1.4)) on S; then the corresponding G-action exists and satisfies (1.21).*
- *Given a Hamiltonian G-action on S with momentum map J and 1-cocycle γ (defined by (1.21)), the corresponding representation π is given by $\pi = J^*$.*

This follows from Theorems 1.1.7 and 1.2.1, and the fact that γ is uniquely determined by Γ and (1.24). To see this, define $\tilde{\Gamma} : \mathfrak{g} \to \mathfrak{g}^*$ by $(\tilde{\Gamma}(X))(Y) := \Gamma(Y, X)$. This leads to an affine map $\mathrm{co}^\Gamma : \mathfrak{g}^* \to \mathfrak{g}^*$, given by $\mathrm{co}^\Gamma(X)\theta = \mathrm{co}(X)\theta + \tilde{\Gamma}(X)$. It follows from (1.11) that $\mathrm{co}^\Gamma(X)$ is a Lie algebra homomorphism; note that the Lie bracket of two affine maps $A_i = L_i + v_i$, where L is linear, is defined by $[A_1, A_2]\theta := (A_1A_2 - A_2A_1)\theta + A_1v_2 - A_2v_1$. As in the linear case, when G is simply connected there is a unique affine action $\mathrm{Co}^\gamma(G)$ on $\mathfrak{g}^*$ whose derivative is $\mathrm{co}^\Gamma(\mathfrak{g})$; it is given by (1.23), where γ satisfies (1.22), which is equivalent to (1.11). This γ then has to coincide with the same symbol defined by (1.21), stripped of its vacuous σ-dependence. ∎

When $G = \tilde{G}/D$ (in the notation used after 1.2.1) is not simply connected, one has to assume integrability of the G-action. In turn, this guarantees integrability of Γ to γ; given its existence, γ is uniquely determined by the property $\tilde{\gamma} = \gamma \circ \tau_{\tilde{G}\to G}$.

Recall the definition of the space of 1-cocycles $Z^1(G, \mathrm{Co}, \mathfrak{g}^*)$ below (1.22); define $B^1(G, \mathrm{Co}, \mathfrak{g}^*) \subset Z^1(G, \mathrm{Co}, \mathfrak{g}^*)$ as the subspace of maps of the form

$$\gamma(x) = \mathrm{Co}(x)\theta_0 - \theta_0 \tag{1.25}$$

for some $\theta_0 \in \mathfrak{g}^*$. The 2-cocycle Γ derived from $\gamma \in B^1(G, \mathrm{Co}, \mathfrak{g}^*)$ by (1.24) lies in $B^2(\mathfrak{g}, \mathbb{R})$; it is remarkable that such a Γ is automatically antisymmetric. In general, elements of $Z^2(\mathfrak{g}, \mathbb{R})$ derived from $\gamma \in Z^1(G, \mathrm{Co}, \mathfrak{g}^*)$ by (1.24) may fail to be antisymmetric. Elements of $Z^1(G, \mathrm{Co}, \mathfrak{g}^*)$ that do give rise to an antisymmetric Γ are called **symplectic cocycles,** forming the space $Z^1_s(G, \mathrm{Co}, \mathfrak{g}^*)$. The **first cohomology group of G relative to the coadjoint representation** is $H^1_s(G, \mathrm{Co}, \mathfrak{g}^*) := Z^1_s(G, \mathrm{Co}, \mathfrak{g}^*)/B^1(G, \mathrm{Co}, \mathfrak{g}^*)$.

Proposition 1.2.7. *When G is simply connected, $H^2(\mathfrak{g}, \mathbb{R})$ and $H^1_s(G, \mathrm{Co}, \mathfrak{g}^*)$ are isomorphic.*

The proof of Theorem 1.2.6 shows that any $\Gamma \in Z^2(\mathfrak{g}, \mathbb{R})$ corresponds to a unique $\gamma \in Z^1(G, \mathrm{Co}, \mathfrak{g}^*)$. The claim then easily follows from the paragraph preceding the proposition. ∎

More generally, further to 1.1.11 we have

Proposition 1.2.8. *When a Hamiltonian G-action with momentum map J satisfies (1.21) with (1.25), the action is strongly Hamiltonian. When $H^1_s(G, \mathrm{Co}, \mathfrak{g}^*) = 0$, any Hamiltonian action of G is strongly Hamiltonian.*

When γ is of the form (1.25), the redefined momentum map $J' = J + \theta_0$ is equivariant, as is clear from (1.21). If $H^1_s(G, \mathrm{Co}, \mathfrak{g}^*) = 0$, then any γ is of this form. ∎

Note that the affine map $A : \theta \mapsto \theta + \theta_0$ satisfies $A\mathrm{Co}^\gamma(x)A^{-1} = \mathrm{Co}(x)$ for all x, as well as $\{A^* f, A^* g\}_\pm^\Gamma = A^*\{f, g\}_\pm$.

Corollary 1.2.9. *A Hamiltonian action of a compact Lie group is always strongly Hamiltonian.*

The cohomology group $H^1(G, \mathrm{Co}, \mathfrak{g}^*) := Z^1(G, \mathrm{Co}, \mathfrak{g}^*)/B^1(G, \mathrm{Co}, \mathfrak{g}^*)$ is zero when G is compact. Hence $H_s^1(G, \mathrm{Co}, \mathfrak{g}^*) \subset H^1(G, \mathrm{Co}, \mathfrak{g}^*)$ must be trivial as well. □

The group analogue of 1.1.12 is

Corollary 1.2.10. *Let $H^1(\mathfrak{g}, \mathbb{R}) = H^2(\mathfrak{g}, \mathbb{R}) = 0$, and let G act on a symplectic manifold by Poisson maps. There exists a unique equivariant momentum map J associated with this action.*

Example 1.2.11. *Let $S = T^*\mathbb{R}^n$ with its canonical symplectic structure.*

1. *The action of $G = \mathbb{R}^n$ (whose Lie algebra $\mathfrak{g} = \mathbb{R}^n$ has a basis $\{T_i\}_{i=1,\ldots,n}$) on S, given by $a : (p, q) \mapsto (p, q + a)$, is Hamiltonian, with equivariant momentum map $J_i(p, q) = p_i$.*
2. *The action of $G = SO(n)$ (whose Lie algebra $\mathfrak{g} = \wedge^2(\mathbb{R}^n)$ has a natural basis $\{T_{ij}\}_{i<j=1,\ldots,n}$) on S, given by $R : (p, q) \mapsto (Rp, Rq)$, is Hamiltonian, with equivariant momentum map $J_{ij}(p, q) = p_i q_j - p_j q_i$, $q_i := q^i$.*
3. *Let the abelian Lie algebra $\mathfrak{g} = \mathbb{R}^{2n}$ have a basis $\{P_i, Q^j\}_{i,j=1,\ldots,n}$. The corresponding Lie group $G = \mathbb{R}^{2n}$ is parametrized by $(u, v) := \mathrm{Exp}(-uQ + vP)$; cf. II.(2.5). It acts on S by*

$$(u, v) : (p, q) \mapsto (p + u, q + v);$$

see II.(2.13), in which we have put $c = 1$. This action is Hamiltonian, with momentum map $J_{P_i}(p, q) = p_i$ and $J_{Q^i}(p, q) = q^i$. The 2-cocycle Γ is

$$\Gamma(P_i, P_j) = \Gamma(Q^i, Q^j) = 0;$$
$$\Gamma(P_i, Q^j) = -\delta_i^j. \tag{1.26}$$

The central extension $\mathfrak{g}_\Gamma$ is the Heisenberg Lie algebra $\mathfrak{h}_n$; see II.(2.4).
4. *Finally, the map J of Proposition II.3.1.4 is an equivariant momentum map for the group action II.(3.13).*

The first two examples are a special case of Lemma 2.3.1 below. As to the third, it should be remarked that since G is abelian, it must be that $B^2(\mathbb{R}^{2n}, \mathbb{R}) = 0$; cf. (1.18). Also, (1.11) is identically satisfied, so that a 2-cocycle on $\mathbb{R}^{2n}$ is simply an antisymmetric bilinear map on $\mathbb{R}^{2n}$. The dimension of the space of antisymmetric $m \times m$ matrices is $\frac{1}{2}m(m-1)$; hence $H^2(\mathbb{R}^{2n}, \mathbb{R}) = \mathbb{R}^{n(2n-1)}$.

1.3 Multipliers and Central Extensions

Definition 1.1.8 and the ensuing discussion have an analogue at the group level. The best way to approach this matter is via the following concept.

Definition 1.3.1. *A* **central extension** *of a Lie group G by $U(1)$ is a short exact sequence*

$$e \to U(1) \xrightarrow{\varphi} \mathsf{G} \xrightarrow{\tau} G \to e, \tag{1.27}$$

in which G *is a Lie group, φ and τ are smooth homomorphisms, and $\varphi(U(1))$ is contained in the center of* G *(by definition of an exact sequence, the image of each map is the kernel of the next).*

It is most natural to analyze this structure in terms of principal fiber bundles; the reader unfamiliar with this notion may either skip the following geometric discussion and resume at Theorem 1.3.3, or jump ahead and read 2.1 before proceeding.

It is quite easy to see that the group G in (1.27) is a principal $U(1)$-bundle $\mathsf{G}(G, U(1), \tau)$ over the base G; in particular, the action of $U(1)$ on G is given by $R_z(\chi) := \chi\varphi(z)$. (Identifying $U(1)$ with $\mathbb{T}$, we write its elements as z.) We now choose a section $s : G \to \mathsf{G}$ of this bundle (that is, $\tau(s(x)) = x$ for all $x \in G$). Since τ is a homomorphism, it must be that $\tau(s(x)s(y)) = \tau(s(xy))$. Hence there exists a function $c : G \times G \to U(1)$ such that

$$s(x)s(y) = c(x, y)s(xy) \tag{1.28}$$

for all $x, y \in G$. Since $s((xy)z) = s(x(yx))$ by associativity of the multiplication in G, one must have the identity

$$c(x, y)c(xy, z) = c(x, yz)c(y, z) \tag{1.29}$$

for all $x, y, z \in G$. We may restrict ourselves to sections satisfying $s(e) = e$ (where on the left-hand side $e \in G$ and on the right-hand side $e \in \mathsf{G}$). Then

$$c(e, x) = c(x, e) = 1 \tag{1.30}$$

for all x. Moreover, while s may not be globally smooth, it may always be chosen so as to be (Borel) measurable, and smooth in a neighborhood of $e \in G$. In that case c is smooth near (e, e). This motivates the following

Definition 1.3.2. *A* **multiplier** *on a Lie group G is a measurable function $c : G \times G \to U(1)$ that is smooth near (e, e) and satisfies (1.29) and (1.30).*

The set of all multipliers on G is called $Z^2(G, U(1))$; this is a vector space when the group operation in the abelian group $U(1)$ is written additively.

As explained in 2.1, a section s leads to a trivialization $\psi_s : \mathsf{G} \to G \times U(1)$ of G, which is generally discontinuous with respect to the product manifold structure. According to (2.3) one has $\psi_s^{-1}(x, z) = s(x)z$. Transferring the group operations from G to $G \times U(1)$, using (1.28) one obtains

$$\begin{aligned} (x, z) \cdot (y, w) &= (xy, zwc(x, y)); \\ (x, z)^{-1} &= (x^{-1}, \overline{zc(x, x^{-1})}). \end{aligned} \tag{1.31}$$

The c-**extension** G_c of G is $G \times U(1)$ with the above group operations, and with the manifold structure inherited from G via ψ_s. Thus G_c, which is a Lie group, is simply a trivialized version of G, depending on c via the choice of the section s. Conversely, one may start from a multiplier.

Theorem 1.3.3. *Let c be a multiplier on G. Equip the set $G_c := G \times U(1)$ with the group law (1.31). There exists a manifold structure on G_c that turns it into a Lie group with the given multiplication. If c is smooth on $G \times G$, the manifold structure is that of the direct product.*

Associativity of the group multiplication in G_c is a consequence of (1.29). The remainder of the proof is a technical exercise in the definition of a Lie group; the idea is that the product manifold structure may be used in a neighborhood of $(e, 1)$; the group operations are then smooth because c is smooth near (e, e). This local manifold structure is subsequently transferred to all of G_c using the group law. □

The embedding $\iota : G \hookrightarrow G_c$ by $\iota(x) = (x, 1)$ is not a homomorphism unless $c = 1$. When necessary for unambiguity we will denote the subgroup $U(1) \subset G_c$ defining the central extension by $U_c(1)$.

If one starts from the diagram (1.27) and then passes to G_c via a section s, one may examine the effect of a change in s on c and hence on G_c. Given some measurable function $b : G \to U(1)$ that is smooth near e, one may pass from s to s', defined by $s'(x) := s(x)R_{b(x)} = s(x)\varphi(b(x))$. This leads to the replacement of c by

$$c'(x, y) = \frac{b(x)b(y)}{b(xy)} c(x, y). \tag{1.32}$$

Two multipliers c, c' related by (1.32) for some b are called **equivalent**. Thus an appropriate cohomology theory is defined through the subspace $B^2(G, U(1))$ of $Z^2(G, U(1))$, consisting of multipliers of the form

$$c(x, y) = \frac{b(xy)}{b(x)b(y)} \tag{1.33}$$

for some measurable function $b : G \to U(1)$ that is smooth near e. Hence one forms the cohomology group $H^2(G, U(1)) := Z^2(G, U(1))/B^2(G, U(1))$. Equivalent multipliers then define the same element of $H^2(G, U(1))$. In particular, the multiplier c in (1.33) is equivalent to 1.

The connection between equivalent multipliers and isomorphic group extensions is now as follows.

Proposition 1.3.4.

1. *Two multipliers c and c' are related by (1.32) iff the extensions G_c and $G_{c'}$ are isomorphic as Lie groups.*
2. *In particular, when c is of the form (1.33), the corresponding c-extension is isomorphic to the direct product of G and $U(1)$.*
3. *Thus when $H^2(G, U(1)) = 0$, any c-extension of G is trivial.*
4. *When G is simply connected, every multiplier is equivalent to one that is smooth on $G \times G$, so that as a manifold G_c is a trivial $U(1)$-bundle over G (cf. 2.1).*

The map $(x, z) \mapsto (x, zb(x)^{-1})$ provides the desired isomorphism from G_c to $G_{c'}$. Conversely, when G_c and $G_{c'}$ are isomorphic to G, they must be related by an isomorphism $\varphi : G_c \to G_{c'}$ of the form $\varphi(x, z) = (x, \tilde{\phi}(x)z)$, where $\tilde{\phi} : G \to U_c(1)$. Since φ is in particular a group homomorphism, the choice $b(x) := \tilde{\varphi}(x)^{-1}$ satisfies (1.32). The second point follows from the first by choosing $c' = 1$.

When $H^2(G, U(1)) = 0$, any c is given by (1.33), which implies 1.3.4.3.

The last statement is a consequence of Theorem 1.3.3 and the fact that a $U(1)$ bundle over a contractible space is necessarily (isomorphic to) a trivial bundle. ∎

Under certain conditions there is a correspondence between extensions of Lie algebras and of Lie groups. We identify $\mathbb{R}$ in 1.1.8 with the Lie algebra $\mathfrak{u}_c(1)$ of $U_c(1)$, and write $\text{Exp} : \mathfrak{u}(1) \to U(1)$ for the exponential map, conventionally realized as $\text{Exp}(X) = \exp(-iX)$. In a neighborhood $\mathcal{N}_e \times \mathcal{N}_e$ of (e, e) we can write $c = \text{Exp}(\chi)$, where $\chi : \mathcal{N}_e \times \mathcal{N}_e \to \mathfrak{u}_c(1)$ (for simply connected G this can be done on all of $G \times G$). Then define $\Gamma : \mathfrak{g} \otimes \mathfrak{g} \to \mathbb{R}$ by

$$\Gamma(X, Y) := \frac{d}{ds}\frac{d}{dt}\left[\chi(\text{Exp}(tX), \text{Exp}(sY)) - \chi(\text{Exp}(sY), \text{Exp}(tX))\right]_{|s=t=0}. \tag{1.34}$$

For example, in the setting of Example 1.2.11.3 the multiplier $c : \mathbb{R}^{2n} \times \mathbb{R}^{2n} \to U(1)$ is given by

$$c((u, v), (u', v')) = e^{i(uv' - vu')/2}. \tag{1.35}$$

Through (1.34) (with $c = \exp(-i\chi)$) and II.(2.5) this indeed reproduces the 2-cocycle Γ in (1.26). The Heisenberg group H_n is nothing but the central extension $\mathbb{R}^{2n}_c$ defined by Γ; cf. II.(2.8) and 1.3.6 below. The multiplier $c'((u, v), (u', v')) := \exp(iuv')$ leads to the same Γ; it is related to c by (1.32), with $b(u, v) = \exp(-\frac{1}{2}iuv)$.

Lemma 1.3.5. *The map Γ defined by (1.34) is an element of $Z^2(\mathfrak{g}, \mathbb{R})$. If $c \in B^2(G, U(1))$, then $\Gamma \in B^2(\mathfrak{g}, \mathbb{R})$.*

We write $\Gamma(X, Y) = \chi_*(X, Y) - \chi_*(Y, X)$ for $\chi_* : \mathfrak{g} \times \mathfrak{g} \to \mathbb{R}$. With this notation, (1.29) combined with its cyclic permutations in x, y, z implies $\chi_*(X, [Y, Z]) +$ cycl. $= \chi_*([Y, Z], X) +$cycl., which leads to (1.11). The second claim is immediate from (1.33) and (1.18). ∎

We now discuss the inverse process of passing from Γ to c.

Proposition 1.3.6.

1. *When $\tilde{G}$ is simply connected there exists a multiplier $c \in Z^2(\tilde{G}, U(1))$ that is related to a given 2-cocycle $\Gamma \in Z^2(\mathfrak{g}, \mathbb{R})$ by (1.34).*
2. *When $G = \tilde{G}/D$ (where $\tilde{G}$ is simply connected and $D = \pi_1(G)$ is a central subgroup of $\tilde{G}$), such a $c \in Z^2(G, U(1))$ exists iff $D \subseteq Z(\tilde{G}_c)/U_c(1)$ (here $\tilde{G}_c$ is the central extension given by the previous item and 1.3.3, and we have identified $\tilde{G}$ with $\tilde{G}_c/U_c(1)$).*

3. *When c exists, its equivalence class in $H^2(\tilde{G}, U(1))$ or $H^2(G, U(1))$ is uniquely determined by* Γ.

By Lie's third theorem there is a simply connected Lie group $\tilde{G}_\Gamma$ with Lie algebra $\mathfrak{g}_\Gamma$ (unique up to isomorphism); as a manifold $\tilde{G}_\Gamma = \tilde{G} \times \mathbb{R}$. Define $\tilde{c} : \tilde{G} \times \tilde{G} \to \mathbb{R}$ by $(x, 0) \cdot (y, 0) = (xy, \tilde{c}(x, y))$, where $\cdot$ is the multiplication in $\tilde{G}_\Gamma$ that comes with its construction. The associativity of $\cdot$ implies that $\tilde{c}$ satisfies (1.29) (if the group law in $\mathbb{R}$ is written multiplicatively). If $\tau : \mathbb{R} \to U(1)$ is the covering projection, we put $c := \tau \circ \tilde{c}$ and verify that c satisfies (1.29), since τ is a homomorphism. This is the desired multiplier c, and $\tilde{G}_c$, defined as in 1.3.3, is a quotient of $\tilde{G}_\Gamma$ by the central subgroup $\mathbb{Z} \subset \mathbb{R}$. In particular, the Lie algebra of $\tilde{G}_c$ is $\mathfrak{g}_\Gamma$. This proves the first claim.

As to the not simply connected case, the necessity of the stated condition is obvious, for D must lie in the center $Z(\tilde{G})$ of $\tilde{G}$. To prove sufficiency, consider $D_\Gamma \subset \tilde{G}_\Gamma$; as a set $D_\Gamma := D \times \mathbb{R}$, which is a subgroup of $\tilde{G}_\Gamma$. The assumption implies that D_Γ is abelian, so that there must be an isomorphism $\phi : D_\Gamma \to D \times \mathbb{R}$, where this time the symbol $\times$ stands for the direct product of groups. Hence $\hat{D}_\Gamma := \phi^{-1}(D \times \mathbb{Z})$ is a discrete central subgroup of $\tilde{G}_\Gamma$. Then one easily infers that the Lie group $G_c := \tilde{G}_\Gamma / \hat{D}_\Gamma$ is a central extension of G. Its multiplier c is defined by the property $(x, 0) \cdot (y, 0) = (xy, c(x, y))$, proving its existence.

Finally, uniqueness in cohomology follows from Lie's third theorem in combination with 1.3.4.1. ■

Given G and Γ, this proposition gives conditions for the existence of a central extension G_c with Lie algebra $\mathfrak{g}_\Gamma$.

Corollary 1.3.7. *When G is simply connected one has*

$$H^2(G, U(1)) \simeq H^2(\mathfrak{g}, \mathbb{R}) \simeq H^1_s(G, \mathrm{Co}, \mathfrak{g}^*). \tag{1.36}$$

The first isomorphism is clear from 1.3.5 and 1.3.6; the second one follows from Proposition 1.2.7. ■

We return to symplectic geometry. The group analogue of 1.1.9 is

Corollary 1.3.8. *Let a Hamiltonian G-action on a connected symplectic manifold with* Co^γ*-equivariant momentum map be given, with* Γ *defined by (1.24). Assume that G and* Γ *are such that a central extension* G_c *(defined through 1.3.6 and 1.3.3) with Lie algebra* $\mathfrak{g}_\Gamma$ *exists. Then the* G_c*-action obtained from the G-action through projection on* $G = G_c/U_c(1)$ *is strongly Hamiltonian.*

This is immediate from (1.12), 1.1.8, and 1.1.6. ■

According to Proposition 1.2.8, the special case $\gamma \in B^1(G, \mathrm{Co}, \mathfrak{g}^*)$ implies that the G-action has an equivariant momentum map; in other words, it is strongly Hamiltonian. This is consistent with 1.3.8, for $\gamma \in B^1(G, \mathrm{Co}, \mathfrak{g}^*)$ implies $\Gamma \in B^2(\mathfrak{g}, \mathbb{R})$ by the proof of 1.2.7; this, in turn, leads to $c \in B^2(G, U(1))$ by 1.3.6.3, which means that $G_c \simeq G \times U(1)$ as a Lie group by 1.3.4.

1.4 The (Twisted) Lie–Poisson Structure

We now turn to an analysis of the Poisson algebra $C^{\infty}_{\Gamma}(\mathfrak{g}^*_{\pm})$, starting with an interesting realization of it. This involves the geometry of T^*G, which we briefly review first. We denote the right and left-invariant vector fields on G by ξ^R_X and ξ^L_X, respectively; i.e.,

$$\xi^L_X f(y) := \frac{d}{dt} f(y\mathrm{Exp}(tX))_{|t=0}. \tag{1.37}$$

$$\xi^R_X f(y) := \frac{d}{dt} f(\mathrm{Exp}(tX)y)_{|t=0}. \tag{1.38}$$

For the commutator one has

$$[\xi^{L,R}_X, \xi^{L,R}_Y] = \pm\xi^{L,R}_{[X,Y]}; \tag{1.39}$$

here and in what follows the upper sign enters for "L", and the lower one for "R". We write $\xi^{L,R}_a := \xi^{L,R}_{T_a}$. One sees that

$$\xi^L_X(y) = \xi^R_{\mathrm{Ad}(y)X}(y). \tag{1.40}$$

The left or right **Maurer–Cartan form** $\theta^{MC}_{L,R}$ is an element of $\Lambda^1(G)\otimes\mathfrak{g}$ (i.e., a $\mathfrak{g}$-valued 1-form on G), defined by

$$\theta^{MC}_{L,R}(\xi^{L,R}_X) := X. \tag{1.41}$$

The connection between the two follows from (1.40) as

$$\mathrm{Ad}(x)\theta^{MC}_L(x) = \theta^{MC}_R(x). \tag{1.42}$$

In terms of a basis $\{T_a\}$ of $\mathfrak{g}$ we expand $\theta^{MC}_{L,R}(x) = \theta^a_{L,R}(x)T_a$, defining a collection of left- or right-invariant 1-forms $\theta^a_{L,R}(x)$. Define the G-actions L and R on G by

$$L_x(y) := xy; \tag{1.43}$$

$$R_x(y) := yx^{-1}. \tag{1.44}$$

One may then equivalently define $\theta^a_L(x) := L^*_{x^{-1}}\omega^a$ and $\theta^a_R(x) := R^*_{x^{-1}}\omega^a$, where the ω^a form a basis of $\mathfrak{g}^* = T^*_eG$ dual to the basis $\{T_a\}$. The **Maurer–Cartan equations**

$$d\theta^a_{L,R}(x) = \mp\tfrac{1}{2}C^a_{bc}\theta^b_{L,R}(x)\wedge\theta^c_{L,R}(x) \tag{1.45}$$

are an immediate consequence of (1.39).

One defines two (globally valid) trivializations

$$TG \simeq \mathfrak{g}\times G; \tag{1.46}$$

in the **left trivialization** one maps $Y\in T_xG$ to $(L_{x^{-1}})_*Y \in \mathfrak{g}\times G$, whereas in the **right trivialization** one maps the same Y to $(R_x)_*Y$. Conversely, $(Y,x)_L$ stands for $(L_x)_*Y \in T_xG$, and $(Y,x)_R$ corresponds to $(R_{x^{-1}})_*Y \in T_xG$. For example, the left trivialization of $\xi^L_Y(x)$ is $(Y,x)_L$, and the left trivialization of $\xi^R_Y(x)$ is $(\mathrm{Ad}(x^{-1})Y, x)_L$; cf. (1.40).

Similar to (1.46), one has

$$T^*G \simeq \mathfrak{g}^* \times G; \tag{1.47}$$

in the left trivialization $\sigma \in T^*_x G$ is mapped to $(L^*_x\sigma, x)$ in $\mathfrak{g}^* \times G$, and in the right trivialization σ is mapped to $(R^*_{x^{-1}}\sigma, x)$. We write $(\theta, x)_L$ for $L^*_{x^{-1}}\theta \in T^*_x G$, and $(\theta, x)_R$ for $R^*_x\theta \in T^*_x G$. The connection between these trivializations is

$$(\theta, x)_L = (\mathrm{Co}(x)\theta, x)_R. \tag{1.48}$$

Given a cocycle $\Gamma \in Z^2(\mathfrak{g}, \mathbb{R})$ (regarded as an element of $\wedge^2_e(G)$), one can define a 2-form $\hat{\Gamma}_L$ on G by $\hat{\Gamma}_L(x) := L^*_{x^{-1}}\Gamma$; similarly, $\hat{\Gamma}_R(x) := R^*_x\Gamma$. In other words, $\hat{\Gamma}_{L,R}(\xi^{L,R}_X, \xi^{L,R}_Y) = \Gamma(X, Y)$. Hence $\hat{\Gamma}_{L,R}(x) = \Gamma_{ab}\theta^a_{L,R}(x) \wedge \theta^b_{L,R}(x)$; cf. (1.14). Thus we obtain a 2-form $\hat{\Gamma}^*_{L,R} := \tau^*\hat{\Gamma}_{L,R}$ on T^*G, where $\tau := \tau_{T^*G\to G}$ (this notation will be used throughout this section).

We define G-actions $\lambda_x := L^*_{x^{-1}}$ and $\rho_x := R^*_{x^{-1}}$ on T^*G; in the trivializations defined above their expressions are

$$\rho_x(\theta, y)_L := (\mathrm{Co}^\gamma(x)\theta, yx^{-1})_L; \tag{1.49}$$
$$\lambda_x(\theta, y)_L := (\theta, xy)_L; \tag{1.50}$$
$$\rho_x(\theta, y)_R := (\theta, yx^{-1})_R; \tag{1.51}$$
$$\lambda_x(\theta, y)_R := (\mathrm{Co}^\gamma(x)\theta, xy)_R; \tag{1.52}$$

here we assume that Γ is related to γ by (1.24). To derive these expressions, one uses relations of the type (cf. (1.42))

$$R^*_x\theta^{MC}_L = \mathrm{Ad}(x)\theta^{MC}_L. \tag{1.53}$$

Recall the coordinates θ_a on $\mathfrak{g}^*$ introduced after (1.2).

Proposition 1.4.1. *Let ω be the canonical symplectic form on T^*G (cf. I.2.3.8), and equip T^*G with the 2-form $\omega^{L,R}_\Gamma := \omega + \hat{\Gamma}^*_{L,R}$.*

- *The form $\omega^{L,R}_\Gamma$ is symplectic.*
- *In the above trivializations the corresponding Poisson bracket on T^*G is given by*

$$\{f, g\}^{T^*G}_{\Gamma^*} = \frac{\partial f}{\partial\theta_a}\xi^{L,R}_a g - \xi^{L,R}_a f\frac{\partial g}{\partial\theta_a} \pm (C^c_{ab}\theta_c + \Gamma_{ab})\frac{\partial f}{\partial\theta_a}\frac{\partial g}{\partial\theta_b}. \tag{1.54}$$

- *The actions ρ and λ commute and are Hamiltonian, with Co^γ-equivariant momentum mappings J^R and J^L, respectively, given by*

$$J^R(\theta, x)_L = -\theta; \tag{1.55}$$
$$J^L(\theta, x)_L = \mathrm{Co}^\gamma(x)\theta; \tag{1.56}$$
$$J^R(\theta, x)_R = -\mathrm{Co}^\gamma(x^{-1})\theta; \tag{1.57}$$
$$J^L(\theta, x)_R = \theta. \tag{1.58}$$

The 2-form $\hat{\Gamma}^*_{L,R}$ is closed as a consequence of (1.11). For an arbitrary manifold Q, it is easily verified that the form $\omega + \tau^*\alpha$ on T^*Q is symplectic for any closed

2-form α on Q. To derive (1.54), one first observes that Θ in I.(2.22) is given by $\Theta(\theta, x) = \theta_a \wedge \tau^*\theta^a_{L,R}(x)$, and then uses the Maurer–Cartan equation to show that

$$\omega = -d\theta_a \wedge \tau^*\theta^a_{L,R} \pm \tfrac{1}{2}C^a_{ab}\theta_c\tau^*\theta^a_{L,R} \wedge \tau^*\theta^b_{L,R}. \tag{1.59}$$

Equation (1.54) then follows from I.(2.19). Given this Poisson bracket, one verifies that the momentum maps (1.58) generate the actions (1.52); this, in turn, guarantees that the group actions in question are Hamiltonian. The equivariance of J^R and J^L is trivially verified. ∎

Let $C^\infty_\Gamma(T^*G)^R$ stand for the set of ρ-invariant functions in $C^\infty(T^*G)$, with Poisson bracket (1.54). This is a Poisson subalgebra, since each ρ_x is a Poisson map.

Corollary 1.4.2. *The map $(J^L)^* : C^\infty_\Gamma(\mathfrak{g}^*_-) \to C^\infty_\Gamma(T^*G)^R$ is an isomorphism of Poisson algebras.*

We see from (1.52) that in the right trivialization, $C^\infty_\Gamma(T^*G)^R$ consists of those $f(p, x)_R$ that are independent of x. One then immediately infers from (1.54), where the lower sign applies, that such functions satisfy the Poisson bracket (1.12); cf. (1.3). Then use (1.58). ∎

The obvious generalization of Proposition 1.2.3 is

Proposition 1.4.3. *The Poisson structure (1.12) is invariant under the G-action (1.23); in other words, the map $\mathrm{Co}^\gamma(x)$ is a Poisson map for each $x \in G$.*

Proceeding as in the proof of 1.2.3, the claim follows if

$$\gamma(x)([X, Y]) = \Gamma(\mathrm{Ad}(x^{-1})X, \mathrm{Ad}(x^{-1})Y) - \Gamma(X, Y). \tag{1.60}$$

To prove this, we write the left-hand side as $d[\gamma(x)(\mathrm{Ad}(\mathrm{Exp}(tX))Y)]/dt$ at $t = 0$. The expression in square brackets equals $[\mathrm{Co}(\mathrm{Exp}(-tX))\gamma(x)](Y)$. Using (1.22), this equals $[\gamma(\mathrm{Exp}(-tX)x) - \gamma((\mathrm{Exp}(-tX))](Y)$. Writing $xx^{-1}\mathrm{Exp}(-tX)x$ for $\mathrm{Exp}(-tX)x$ and using (1.22) once again, as well as (1.24), we eventually obtain (1.60). ∎

A **coadjoint orbit** $\mathcal{O}$ in $\mathfrak{g}^*$ is an orbit under the coadjoint action. Similarly, a Co^γ-orbit in $\mathfrak{g}^*$ is defined with respect to the action (1.23). The Co^γ-orbit $\mathcal{O}^\gamma_\theta$ through $\theta \in \mathfrak{g}^*$ is of the form $\mathcal{O}^\gamma_\theta = G/G^\gamma_\theta$, where G^γ_θ is the stability group of θ under the Co^γ-action; we see from (1.23) and (1.24) that its Lie algebra is

$$\mathfrak{g}^\gamma_\theta = \{X \in \mathfrak{g} \mid \theta([X, Y]) - \Gamma(X, Y) = 0 \,\forall\, Y \in \mathfrak{g}\}. \tag{1.61}$$

Theorem 1.4.4. *Let γ be a symplectic cocycle, and define Γ by (1.24). The symplectic leaves of $\mathfrak{g}^*$ with respect to the Poisson structure (1.12) are the Co^γ-orbits of G. In particular, the symplectic leaves of the Lie–Poisson structure coincide with the coadjoint orbits.*

We know from I.2.4.7 and I.2.3.7 that the tangent space at some point θ of a given leaf L_θ is spanned by the Hamiltonian vector fields $\xi_f(\theta)$. These depend linearly on $df = (\partial f/\partial\theta_a)d\theta_a = (\partial f/\partial\theta_a)d\tilde{T}_a$. We now use an ancillary result.

Lemma 1.4.5. *The Hamiltonian flow on $\mathfrak{g}^*_\pm$ (computed with respect to the Poisson structure (1.12)) generated by $\tilde{X}$ is $\theta(t) = \mathrm{Co}^\gamma(\mathrm{Exp}(\mp tX))\theta$. In other words, $\xi^\pm_{\tilde{X}} = \mp\xi_X$, where ξ_X is defined by (1.19) and the Co^γ-action (1.23) on $\mathfrak{g}^*$, and $\xi^\pm_{\tilde{X}}$ is the Hamiltonian vector field defined by the Poisson structure (1.12).*

For clarity we start with the proof for $\gamma = \Gamma = 0$, choosing the minus sign for concreteness. Since the linear functions separate points in $\mathfrak{g}^*$, it suffices to compute

$$\frac{d}{dt}\tilde{Y}(\mathrm{Co}(\mathrm{Exp}(tX))\theta) = -\theta(\mathrm{Ad}(\mathrm{Exp}(-tX))[X, Y]) = \{\tilde{X}, \tilde{Y}\}_-(\theta(t)),$$

from which the claim follows. For general γ one in addition uses the identity

$$\frac{d}{dt}\gamma(\mathrm{Exp}(tX))(Y) = -\Gamma(X, Y) - \gamma(\mathrm{Exp}(tX))([X, Y]), \tag{1.62}$$

which follows by putting $x = \mathrm{Exp}(tX)$ and $y = \mathrm{Exp}(sX)$ in (1.22), dividing by s and letting $s \to 0$, and subsequently applying (1.60).

The calculation for the plus sign is analogous. ■

By this lemma, the Hamiltonian vector fields $\xi_{\tilde{T}_a}$ span the tangent space at θ to the Co^γ-orbit through θ; this implies the claim locally. Globally, since G is connected, it is generated by the image of $\mathrm{Exp}(\mathfrak{g})$ in G. This ends the proof of Theorem 1.4.4. ■

It follows that any Co^γ-orbit (and in particular any coadjoint orbit) $\mathcal{O}^\gamma$ is an even-dimensional symplectic manifold; for $\gamma = 0$ the immersion of $\mathcal{O} := \mathcal{O}^0$ in $\mathfrak{g}^*_\pm$ defines the ($\pm$) **Lie symplectic form** $\omega^{\mathcal{O}}_\pm$ on $\mathcal{O}$. Equipped with this form, we denote $\mathcal{O}$ by $\mathcal{O}_\pm$. (As in the general case of symplectic leaves, $\mathcal{O}$ is not necessarily a submanifold of $\mathfrak{g}^*$; cf. the text following I.2.4.4.) For general γ we see from I.(2.19) and (1.2) that $\omega^{\mathcal{O}^\gamma}_\pm$ is given by

$$(\omega^{\mathcal{O}^\gamma}_\pm)_\theta(\xi^\pm_X, \xi^\pm_Y) = \mp(\theta([X, Y]) + \Gamma(X, Y)); \tag{1.63}$$

since this action is trivially transitive, (1.63) suffices to define $\omega^{\mathcal{O}^\gamma}_\pm$. It is clear from (1.63) or 1.2.3 that $\omega^{\mathcal{O}^\gamma}_\pm$ is invariant under the Co^γ-action.

Lemma 1.4.5 has the following

Corollary 1.4.6. *The momentum map for the coadjoint action of G on a coadjoint orbit $\mathcal{O}_\pm$ is given by $J_\pm(\theta) = \mp\theta$.*

Recall Definition 1.1.5. Theorem 1.4.4 leads to

Corollary 1.4.7. *Let $\pi : C^\infty_\Gamma(\mathfrak{g}^*_-) \to C^\infty(S)$ be an irreducible representation (in the sense of I.2.6.6). Then S must be (symplectomorphic to) a Co^γ-orbit in $\mathfrak{g}^*$ (equipped with the symplectic structure (1.63)), or a covering space thereof.*

This is immediate from Theorems I.2.6.7 and 1.4.4. The symplectomorphism in question is given by the momentum map. ■

Note that Proposition II.2.1.2 is a special case of 1.4.7.

Corollary 1.4.8. *Let S be a connected symplectic space with a transitive Hamiltonian action of a Lie group G. There is a $\gamma \in Z^1(G, \text{Co}, \mathfrak{g}^*)$ such that S is (symplectomorphic to) a Co^γ-orbit in $\mathfrak{g}^*$ (equipped with the symplectic structure (1.63)), or a covering space thereof.*

If, in addition, $H^1(\mathfrak{g}, \mathbb{R}) = H^2(\mathfrak{g}, \mathbb{R}) = 0$, then S is (symplectomorphic to) a coadjoint orbit in $\mathfrak{g}^*$, or a covering space thereof.

If the G-action is Hamiltonian, there is a momentum map J. The transitivity of the G-action on S implies that J^* is irreducible in the sense of Definition I.2.6.6. Then apply 1.4.7. The second claim is then immediate from Corollary 1.2.10. ■

Finally, the central extension G_c introduced in 1.3.3 may be used to shed light on the Poisson structure (1.54) on T^*G, which in this context we write as $(T^*G)_c$.

Proposition 1.4.9. *Let $U_c(1)$ act on T^*G_c (equipped with the canonical cotangent bundle Poisson structure) by lifting the action $h : x \mapsto xh^{-1}$ on G_c. An equivariant momentum map $J_c : T^*G_c \to \mathfrak{u}_c(1) = \mathbb{R}$ for this action is given in the right trivialization by $J_c(\theta_0, \theta_1, \ldots, \theta_n) = -\theta_0$. Then $(T^*G)_c \simeq J_c^{-1}(1)/U_c(1)$ as Poisson manifolds.*

Realizing that the additional structure constants of G_c (compared with G) are given by $C^0_{ab} = \Gamma_{ab}$ and $C^i_{0j} = 0$ for all $i, j = 0, \ldots, n$, this follows from (1.54), first applied to G_c and then to G. ■

This statement will be properly understood in the setting of 2.3 and IV.1.5.

1.5 Projective Representations

We specialize the discussion to the setting relevant to quantum mechanics: The symplectic manifold S is a projective Hilbert space $\mathbb{P}\mathcal{H}$ (see I.2.5), and the G-action $L_x : \psi \mapsto x\psi$ on $\mathbb{P}\mathcal{H}$ should preserve the transition probabilities I.(2.65) for all $x \in G$. In all cases of interest it turns out that requiring smoothness of a Lie group action on $\mathbb{P}\mathcal{H}$ would force $\mathcal{H}$ to be finite-dimensional, so we here assume the action to be merely continuous.

This continuity may be restated as follows. We equip the group $\mathcal{U}(\mathcal{H})$ of all unitary operators on $\mathcal{H}$ with the strong operator topology (or the weak one, which coincides with the strong topology on the unitaries). Denote the central subgroup of all multiples of $\mathbb{I}$ by $\mathbb{T}\mathbb{I}$, and form the quotient $\mathcal{U}(\mathcal{H})/\mathbb{T}\mathbb{I}$, endowed with the quotient topology. Continuity of the G-action is then equivalent to continuity of L, seen as a homomorphism from G into $\mathcal{U}(\mathcal{H})/\mathbb{T}\mathbb{I}$.

By I.3.4.3, for each $x \in G$ there exists a unitary or an antiunitary operator $U(x)$ on $\mathcal{H}$ such that $\tau \circ U(x) = L_x \circ \tau$ (where $\tau := \tau_{\mathbb{S}\mathcal{H} \to \mathbb{P}\mathcal{H}}$). For G connected, all $U(x)$ must be unitary, since in a connected Lie group each x is a square, and a square of either a unitary or an antiunitary operator is unitary. Two different U's projecting to the same map on $\mathbb{P}\mathcal{H}$ must differ by a phase. Now consider

$$\mathsf{G} := \{(x, U) \in G \times \mathcal{U}(\mathcal{H}) \mid L_x = \tau(U)\}, \tag{1.64}$$

where $\tau(U)$ is the map of $\mathbb{P}\mathcal{H}$ defined by U through the canonical projection of $\mathbb{S}\mathcal{H}$ to $\mathbb{P}\mathcal{H}$. Inheriting the group operations and topology from $G \times \mathcal{U}(\mathcal{H})$, one verifies that G is a Lie group, which by the previous paragraph is a central extension of G by $U(1)$; cf. (1.27). Following the discussion after Definition 1.3.2, we choose a measurable section $s : G \to \mathsf{G}$, smooth near e, associated with a multiplier c on G. The trivialization $\mathsf{G} \simeq G \times U(1)$ (as a set) defined by s leads to the choice of a representative $U(x)$ for each map L_x. This choice obviously satisfies

$$U(x)U(y) = c(x, y)U(xy). \tag{1.65}$$

Hence U is a **projective representation** of G on $\mathcal{H}$ with multiplier c; we sometimes say that U is a c**-representation.**

For example, the abelian group $\mathbb{R}^{2n}$ has a projective representation on $L^2(\mathbb{R}^n)$ given by

$$U_1^S(u, v)\Psi(x) := e^{iu(x-\frac{1}{2}v)}\Psi(x - v); \tag{1.66}$$

cf. II.(2.17). The multiplier is given by (1.35), which should not be surprising in view of the definition of the Heisenberg groups $\tilde{H}_n$ and $H_n = \mathbb{R}^{2n}_c$.

A redefinition $U'(x) = b(x)U(x)$, where $b : G \to U(1)$, leads to the modification (1.32); we say that U and U' are **equivalent**. Clearly, U is equivalent to a representation iff c is of the form (1.33).

Proposition 1.5.1. *There is a bijective correspondence between c-representations U of G and representations U_c of G_c in which $U_c(1)$ is represented by the defining representation (times the identity operator). This correspondence preserves irreducibility.*

Given a c-representation $U(G)$, define $U_c(G_c)$ by $U_c(x, z) := zU(x)$. Conversely, if a representation $U_c(G_c)$ satisfies $U_c(e, z) = z\mathbb{I}$, then $U(x) := U_c(x, 1)$ satisfies (1.65). The last claim is obvious from Schur's lemma and the fact that $U_c(1)$ is a central subgroup of G_c. ■

Indeed, the projective representation $U_1^S(\mathbb{R}^{2n})$ defined in (1.66) is the restriction of $U_1^S(H_n)$ (see II.(2.17)) to $\mathbb{R}^{2n}$, identified with $(\mathbb{R}^{2n}, 1) \subset H_n$.

The classical analogue of 1.5.1 is Proposition 1.1.9.

Proposition 1.5.2. *If $H^2(G, U(1)) = 0$, then any projective representation of G is equivalent to a representation.*

As already pointed out, we see from (1.32) that U is equivalent to a representation iff c is of the form (1.33). ■

We now look at the corresponding concepts for Lie algebras. A **projective representation** of a Lie algebra $\mathfrak{g}$ on a complex vector space V is a linear map $R : \mathfrak{g} \to L(V)$ (the space of linear maps on V) such that

$$[R(X), R(Y)] = R([X, Y]) - i\Gamma(X, Y)\mathbb{I} \tag{1.67}$$

for some 2-cocycle $\Gamma \in Z^2(\mathfrak{g}, \mathbb{R})$; the Jacobi identity on the commutator on the left-hand side enforces (1.11). One may speak of a Γ**-representation** of $\mathfrak{g}$. If R

is modified to $R'(X) := R(X) + i\theta_0(X)\mathbb{I}$ for some (fixed) $\theta_0 \in \mathfrak{g}$, then (1.67) is satisfied for R', with

$$\Gamma'(X, Y) = \Gamma(X, Y) + \theta_0([X, Y]); \tag{1.68}$$

cf. (1.18). Such representations R and R' are called **equivalent.**

Proposition 1.5.3. *There is a bijective correspondence between Γ-representations of $\mathfrak{g}$ (on complex vector spaces) and representations of $\mathfrak{g}_\Gamma$ in which the generator T_0 is represented by $-i\mathbb{I}$; cf. Definition 1.1.8.*

A Γ-representation $R(\mathfrak{g})$ defines a representation $R_\Gamma(\mathfrak{g}_\Gamma)$ by $R_\Gamma(X) := R(X)$ and $R_\Gamma(T_0) := -i\mathbb{I}$, and vice versa. ■

Proposition 1.5.4. *If $\Gamma(X, Y) = \theta_0([X, Y])$ for some $\theta_0 \in \mathfrak{g}^*$, then a Γ-representation is equivalent to a representation. Hence when $H^2(\mathfrak{g}, \mathbb{R}) = 0$ any Γ-representation is equivalent to a representation.*

Compare with 1.1.13. This is obvious from (1.68): A projective representation whose Γ is of the above form is equivalent to a representation by the shift $R'(X) := R(X) - i\theta_0(X)\mathbb{I}$. ■

In relating (projective) representations of a Lie group G to (projective) representations of its Lie algebra $\mathfrak{g}$, we need to discuss a technical point. This discussion is necessary, because when $\mathcal{H}$ is infinite-dimensional, the G-action on $\mathbb{P}\mathcal{H}$ provided by a representation is not necessarily smooth or even C^1. In other words, for a given $\Psi \in \mathcal{H}$ the map from G to $\mathcal{H}$ defined by $x \mapsto U(x)\Psi$ may not be differentiable, so that the curve $U(\text{Exp}(tX))\Psi$ is not necessarily C^1; this would make it difficult to define the generating vector field ξ_X at Ψ.

To simplify the discussion somewhat we assume that U, when it is projective, defines a multiplier c that is smooth on $G \times G$ (the case where $c = 1$ is therefore included). As we have seen in 1.3.4, when G is simply connected this can always be achieved.

Definition 1.5.5. *A* **smooth vector** *for a (projective) representation U is an element $\Psi \in \mathcal{H}$ for which the map $x \mapsto U(x)\Psi$ is smooth.*

It can be shown that the set $\mathcal{H}_U^\infty$ of smooth vectors for U is a dense linear subspace of $\mathcal{H}$.

Proposition 1.5.6. *Under the above conditions on U and $\mathcal{H}$, the subspace $\mathcal{H}_U^\infty$ is stable under $U(G)$. For each $X \in \mathfrak{g}$ the operator $dU(X)$, defined by*

$$dU(X)\Psi := \frac{d}{dt} U(\text{Exp}(tX))\Psi_{|t=0} \tag{1.69}$$

is essentially self-adjoint on $\mathcal{H}_U^\infty$. Finally, $\mathcal{H}_U^\infty$ is stable under $dU(\mathfrak{g})$.

The stability of $\mathcal{H}_U^\infty$ follows from (1.65) and the smoothness of c and of group multiplication. The second claim is then shown as in the proof of II.3.6.4. The last point is evident. ■

It follows from (1.65) and (1.34) that on $\mathcal{H}_U^\infty$ one has

$$[dU(X), dU(Y)] = dU([X, Y]) - i\Gamma(X, Y)\mathbb{I}, \tag{1.70}$$

where Γ is defined by (1.34) with $c = \exp(-i\chi)$; cf. (1.67).

Conversely, given a (projective) representation R of $\mathfrak{g}$ by symmetric operators on some common domain D, one may ask whether there is a (projective) representation $U(\tilde{G})$ such that $R = dU$. As in the classical case, such a representation of $\mathfrak{g}$ is then called **integrable**.

When $\mathcal{H}$ is finite-dimensional and G is simply connected, every representation of $\mathfrak{g}$ by skew-Hermitian matrices is integrable; this is already a difference with the case of group actions on general finite-dimensional manifolds, caused by the fact that the flow of a skew-Hermitian matrix is always complete. In the infinite-dimensional case further conditions are required.

1.6 The Twisted Enveloping Algebra

Recall the definition I.(2.43) of the function $\hat{A} \in C^\infty(\mathbb{P}\mathcal{H})$, where $A \in \mathfrak{B}(\mathcal{H})_\mathbb{R}$. Since $dU(X)$ tends to be unbounded, the functions $\widehat{idU(X)}$ are defined only on $\mathbb{P}\mathcal{H}_U^\infty$; cf. the preceding section. One can topologize $\mathcal{H}_U^\infty$ so that $\mathbb{P}\mathcal{H}_U^\infty$ is a Fréchet submanifold of $\mathbb{P}\mathcal{H}$, and the G-action restricted to $\mathbb{P}\mathcal{H}_U^\infty$ is smooth, with smooth momentum mapping. Since this is technically involved, we will merely state a key result.

Theorem 1.6.1. *Let $\mathcal{H}$ and U be as stated above 1.5.5. Then*

$$J_X := i\hbar\widehat{dU(X)}, \tag{1.71}$$

defined on $\mathbb{P}\mathcal{H}_U^\infty$, is a momentum map for the G-action on $\mathbb{P}\mathcal{H}$ derived from the representation U on $\mathcal{H}$. It satisfies

$$\{J_X, J_Y\}_\hbar = -J_{[X,Y]} - \hbar\Gamma(X, Y)1_{\mathbb{P}\mathcal{H}}, \tag{1.72}$$

where Γ is defined by (1.34) or (1.70).

With $\psi = \tau_{\mathbb{S}\mathcal{H}\to\mathbb{P}\mathcal{H}}(\Psi)$ one obtains $\xi_X(\psi) = v(X\Psi)$ (where v is defined after I.(2.30)), so that (1.71) follows from I.(2.45). Equation (1.72) is then derived from (1.69), I.(2.42), and (1.70). ∎

Hamiltonian G-actions on $\mathbb{P}\mathcal{H}$ will, in general, fail to preserve the transition probabilities I.(2.65), and will therefore not be given by a (projective) representation $U(G)$ on $\mathcal{H}$. For this reason the Poisson algebra $C_\Gamma^\infty(\mathfrak{g}_-^*)$ is not useful in quantum mechanics.

There are three algebras that do play a role in quantum mechanics analogous to the job performed by $C_\Gamma^\infty(\mathfrak{g}_-^*)$ in classical mechanics. One will be constructed in this section, the other two in the next.

Definition 1.6.2. *The* **enveloping algebra** $\mathcal{U}(\mathfrak{g}_\mathbb{C})$ *of $\mathfrak{g}$ is the quotient of the complexified tensor algebra $\mathcal{T}(\mathfrak{g}_\mathbb{C}) = \bigoplus_{n=0}^\infty \otimes^n \mathfrak{g}_\mathbb{C}$ (where $\otimes^0\mathfrak{g}_\mathbb{C} := \mathbb{C}$) by the* *two-sided ideal $\mathcal{I}$ generated by all elements of the form $X \otimes Y - Y \otimes X - [X, Y]$, where $X, Y \in \mathfrak{g}_{\mathbb{C}}$.*

In other words, $\mathcal{T}(\mathfrak{g}_{\mathbb{C}})$ is the complex vector space consisting of linear combinations of complex elements of the form $X_1 \otimes \cdots \otimes X_k$, $X_i \in \mathfrak{g}_{\mathbb{C}}$, with algebra product given by concatenation. The product in $\mathcal{U}(\mathfrak{g}_{\mathbb{C}})$ is the one inherited from $\mathcal{T}(\mathfrak{g}_{\mathbb{C}})$. The quotienting procedure imposes the relation $X \otimes Y - Y \otimes X - [X, Y] = 0$ for all $X, Y \in \mathfrak{g}_{\mathbb{C}}$ in $\mathcal{U}(\mathfrak{g}_{\mathbb{C}})$. We denote the image of $X_1 \otimes \cdots \otimes X_k \in \mathcal{T}(\mathfrak{g}_{\mathbb{C}})$ in $\mathcal{U}(\mathfrak{g}_{\mathbb{C}})$ simply by $X_1 \cdots X_k$, so that $XY - YX = [X, Y]$ in $\mathcal{U}(\mathfrak{g}_{\mathbb{C}})$.

Definition 1.6.3. *The* **twisted enveloping algebra** $\mathcal{U}_{\Gamma}(\mathfrak{g}_{\mathbb{C}})$ *of* $\mathfrak{g}$ *relative to* $\Gamma \in Z^2(\mathfrak{g}, \mathbb{R})$ *is the quotient of* $\mathcal{T}(\mathfrak{g}_{\mathbb{C}})$ *by the two-sided ideal* $\mathcal{I}_{\Gamma}$ *generated by all elements of the form* $X \otimes Y - Y \otimes X - [X, Y] + i\Gamma(X, Y)$.

Here $\Gamma(X, Y)$ is seen as an element of $\otimes^0 \mathfrak{g}_{\mathbb{C}}$. This time one has the relation $X \otimes Y - Y \otimes X - [X, Y] + i\Gamma(X, Y) = 0$ in $\mathcal{U}_{\Gamma}(\mathfrak{g}_{\mathbb{C}})$.

Proposition 1.6.4. *The twisted enveloping algebra* $\mathcal{U}_{\Gamma}(\mathfrak{g}_{\mathbb{C}})$ *is isomorphic to the quotient of* $\mathcal{U}(\mathfrak{g}_{\Gamma})$ *by the two-sided ideal generated by* $T_0 + i$.

This is obvious. ■

The algebra $\mathcal{U}_{\Gamma}(\mathfrak{g}_{\mathbb{C}})$ (and hence its special case $\mathcal{U}(\mathfrak{g}_{\mathbb{C}})$) has a natural involution (cf. I.1.1), given by linear extension of $(X_1 \cdots X_k)^* = (-1)^k X_k \cdots X_1$ (as well as $\lambda^* = \overline{\lambda}$ for $\lambda \in \mathbb{C}$); this is well-defined, and descends from a similarly defined evolution on $\mathcal{T}(\mathfrak{g}_{\mathbb{C}})$ because the relation mentioned in 1.6.3 is stable under it.

Definition 1.6.5. *As a real vector space, the Jordan–Lie algebra* $\mathfrak{A}_{\Gamma}^{\hbar}(\mathfrak{g})$*, defined for each* $\hbar \neq 0$*, is the subspace of elements of* $\mathcal{U}_{\Gamma/\hbar}(\mathfrak{g}_{\mathbb{C}})$ *that are invariant under the involution. Its Jordan product and Poisson bracket are given by the projection of the operations* $A \circ B := \frac{1}{2}(A \otimes B + B \otimes A)$ *and* $\{A, B\}_{\hbar} = i(A \otimes B - B \otimes A)/\hbar$*, defined on* $\mathcal{T}(\mathfrak{g}_{\mathbb{C}})$*, to* $\mathcal{U}_{\Gamma/\hbar}(\mathfrak{g}_{\mathbb{C}})$*, respectively.*

Note that the projection of these operations to $\mathfrak{A}_{\Gamma}^{\hbar}(\mathfrak{g})$ is well-defined, since $A \circ \mathcal{I}_{\Gamma/\hbar} \subset \mathcal{I}_{\Gamma/\hbar}$ and $\{B, \mathcal{I}_{\Gamma}\}_{\hbar} \subset \mathcal{I}_{\Gamma/\hbar}$ for all $A, B \in \mathcal{T}(\mathfrak{g}_{\mathbb{C}})$. One verifies I.(1.6). The analogue of Proposition 1.4.3 is

Proposition 1.6.6. *One obtains an automorphic group action* $\beta^{\gamma/h}$ *of* G *on* $\mathcal{U}_{\Gamma/\hbar}(\mathfrak{g}_{\mathbb{C}})$ *and thence on* $\mathfrak{A}_{\Gamma}^{\hbar}(\mathfrak{g})$ *by defining*

$$\beta_x^{\gamma}(X) := \mathrm{Ad}(x)X - i\gamma(x^{-1})(X) \tag{1.73}$$

on $\mathfrak{g} \subset \mathcal{T}(\mathfrak{g}_{\mathbb{C}})$*, extending this action to* $\mathcal{T}(\mathfrak{g}_{\mathbb{C}})$ *by* $\beta_x^{\gamma}(1) = 1$ *and* $\beta_x^{\gamma}(X_1 \otimes \cdots \otimes X_k) := \beta_x^{\gamma} X_1 \otimes \cdots \otimes \beta_x^{\gamma} X_k$*, projecting the action to* $\mathcal{U}_{\Gamma/\hbar}(\mathfrak{g}_{\mathbb{C}})$*, and finally restricting it to* $\mathfrak{A}_{\Gamma}^{\hbar}(\mathfrak{g})$.

The fact that one indeed has a group action follows from (1.22). On $\mathcal{T}(\mathfrak{g}_{\mathbb{C}})$ the action is automorphic by construction. The fact that it quotients well to $\mathcal{U}_{\Gamma}(\mathfrak{g}_{\mathbb{C}})$ follows from the property

$$\beta_x^{\gamma}(X \otimes Y - Y \otimes X - [X, Y] + i\Gamma(X, Y)) = \mathrm{Ad}(x)X \otimes \mathrm{Ad}(x)Y -$$

$$\mathrm{Ad}(x)Y \otimes \mathrm{Ad}(x)X - [\mathrm{Ad}(x)X, \mathrm{Ad}(x)Y] + i\Gamma(\mathrm{Ad}(x)X, \mathrm{Ad}(x)Y), \tag{1.74}$$

which is a consequence of (1.60) and the fact that $\mathrm{Ad}(x)$ is an automorphism of $\mathfrak{g}$. Finally, the restriction to $\mathfrak{A}_\Gamma^\hbar(\mathfrak{g})$ is well-defined because $\beta_x^\gamma(X^*) = (\beta_x^\gamma(X))^*$ etc. ∎

The Jordan–Lie algebra $\mathfrak{A}_\Gamma^\hbar(\mathfrak{g})$ is a quantum analogue of the Poisson algebra $\mathsf{P}_\Gamma(\mathfrak{g}_-^*)$ of (real) polynomials on $\mathfrak{g}^*$, equipped with the Poisson bracket (1.12) inherited from $C_\Gamma^\infty(\mathfrak{g}_-^*)$. A quantization map is constructed as follows.

Theorem 1.6.7. *The map $\mathcal{Q}_\hbar^\lambda : \mathsf{P}_\Gamma(\mathfrak{g}_-^*) \to \mathfrak{A}_\Gamma^\hbar(\mathfrak{g})$ defined by $\mathcal{Q}_\hbar(1_{\mathfrak{g}^*}) := 1$ and linear extension of*

$$\mathcal{Q}_\hbar^\lambda(\tilde{X}_1 \cdots \tilde{X}_l) := (i\hbar)^l \lambda[X_1, \ldots, X_l] \tag{1.75}$$

satisfies

$$\{\mathcal{Q}_\hbar^\lambda(A), \mathcal{Q}_\hbar^\lambda(B)\}_\hbar = \mathcal{Q}_\hbar^\lambda(\{A, B\}_-^\Gamma) + O(\hbar); \tag{1.76}$$

$$\mathcal{Q}_\hbar^\lambda(A) \circ \mathcal{Q}_\hbar^\lambda(B) = \mathcal{Q}_\hbar^\lambda(AB) + O(\hbar); \tag{1.77}$$

$$\mathcal{Q}_\hbar^\lambda(\mathsf{P}_\Gamma(\mathfrak{g}_-^*)) = \mathfrak{A}_\Gamma^\hbar(\mathfrak{g}). \tag{1.78}$$

Here the symmetrization operation λ is defined after II.(2.37). The first two equations are a matter of checking the definitions. The third one follows from the fact that symmetrization establishes a vector space isomorphism between $\mathcal{U}_\Gamma(\mathfrak{g}_\mathbb{C})$ and the symmetric tensor algebra $\mathcal{S}(\mathfrak{g})$; this follows from the Poincaré–Birkhoff–Witt theorem, which is well known for $\mathcal{U}(\mathfrak{g}_\mathbb{C})$, hence valid for $\mathcal{U}(\mathfrak{g}_{\Gamma\mathbb{C}})$, and holds for $\mathcal{U}_\Gamma(\mathfrak{g}_\mathbb{C})$ in view of Proposition 1.6.4. ∎

Proposition 1.6.8. *With β^γ and Co^γ given by (1.73) and (1.23), respectively, for all $x \in G$ and $A \in \mathsf{P}(\mathfrak{g}^*)$ one has the equivariance property*

$$\beta_x^{\gamma/\hbar}(\mathcal{Q}_\hbar^\lambda(A)) = \mathcal{Q}_\hbar^\lambda(\mathrm{Co}^\gamma(x^{-1})^* A). \tag{1.79}$$

The proof is a simple calculation. □

Due to the integrability problem, not all representations of $\mathfrak{A}_\Gamma^\hbar(\mathfrak{g})$ are related to unitary G-actions. In addition, $\mathfrak{A}_\Gamma^\hbar(\mathfrak{g})$ has the drawback of not being a JLB-algebra, so that much of the functional-analytic apparatus developed in Chapter I is not available.

1.7 Group C^*-Algebras

We will now construct an object free from these drawbacks. To simplify the notation, we assume that the multiplier c is globally smooth; when it isn't, one should replace $C_c^\infty(G)$ in the discussion below by the space $B_c^\infty(G)$ of bounded measurable functions with compact support that are smooth near e. We also assume that G is **unimodular**; that is, each left Haar measure is also right-invariant. This assumption is not necessary, but simplifies most of the formulae. We denote Haar measure by dx; it is unique up to normalization. When G is compact we choose the normalization so that $\int_G dx = 1$. The Banach space $L^1(G)$ and the Hilbert space $L^2(G)$ are defined with respect to Haar measure.

Given a multiplier c, we define the **(twisted) convolution**

$$f * g(x) := \int_G dy\, c(xy^{-1}, y) f(xy^{-1}) g(y). \tag{1.80}$$

This certainly makes sense for $f, g \in C_c^\infty(G)$. The associativity of $*$ is a consequence of the associativity of group multiplication and the invariance properties of the Haar measure (and, when $c \neq 1$, of (1.29)). Moreover, we can define an involution on $C_c^\infty(G)$ by

$$f^*(x) := \overline{c(x, x^{-1}) f(x^{-1})}. \tag{1.81}$$

The property $(f * g)^* = g^* * f^*$ reflects the law $(xy)^{-1} = y^{-1}x^{-1}$ in a group (for $c \neq 1$ one in addition needs (1.29)). Hence $C_c^\infty(G)$ has been turned into a $*$-algebra.

A representation π of $C_c^\infty(G)$ on a Hilbert space $\mathcal{H}$ is defined as a morphism $\pi : C_c^\infty(G) \to \mathfrak{B}(\mathcal{H})$. An example of a representation is $\pi_L : C_c^\infty(G) \to \mathfrak{B}(L^2(G))$, defined by

$$\pi_L(f)\Psi := f * \Psi. \tag{1.82}$$

In Lemma 1.7.2 we will see that this operator is bounded. Moreover, one easily verifies that $\pi_L(f * g) = \pi_L(f)\pi_L(g)$ and $\pi_L(f^*) = \pi_L(f)^*$. Introducing the **left-regular representation** U_L of G on $L^2(G)$ by

$$U_L(y)\Psi(x) := c(y, y^{-1}x)\Psi(y^{-1}x), \tag{1.83}$$

it follows that

$$\pi_L(f) = \int_G dx\, f(x) U_L(x). \tag{1.84}$$

Definition 1.7.1. *The* **(twisted) reduced group C^*-algebra** *$C_r^*(G, c)$ is the smallest C^*-algebra in $\mathfrak{B}(L^2(G))$ containing $\pi_L(C_c^\infty(G))$. In other words, $C_r^*(G, c)$ is the closure of the latter in the norm $\|f\|_r := \|\pi_L(f)\|$. We write $C_r^*(G)$ for $C_r^*(G, 1)$.*

Perhaps the simplest example of a reduced group algebra is obtained by taking $G = \mathbb{R}^n$. Since the Fourier transform $f \mapsto \hat{f}$ turns convolution into pointwise multiplication, the algebra $C_r^*(\mathbb{R}^n)$ is commutative. Indeed, the left-regular representation π_L on $L^2(\mathbb{R}^n)$ is Fourier-transformed into the action on $L^2(\mathbb{R}^n)$ by multiplication operators. Hence

$$\|f\|_r = \|\hat{f}\|_\infty, \tag{1.85}$$

so that by the Riemann–Lebesgue lemma and the Stone–Weierstrass theorem,

$$C_r^*(\mathbb{R}^n) \simeq C_0(\mathbb{R}^n). \tag{1.86}$$

This generalizes to arbitrary abelian Lie groups G (and, more generally, to locally compact abelian groups). Let $\hat{G}^c$ be the set of all irreducible c-representations of

G; for $c = 1$ this is the set of characters, and we write $\hat{G} := \hat{G}^1$. It is well known that $\hat{G}$ is itself a locally compact abelian group, in terms of which the Fourier transform $\hat{f}$ (which is a function on $\hat{G}$) of $f \in L^1(G)$ may be defined as

$$\hat{f}(\gamma) := \int_G dx\, f(x) U_\gamma(x). \tag{1.87}$$

By the same arguments as for $\mathbb{R}^n$, one obtains

$$C_r^*(G) \simeq C_0(\hat{G}). \tag{1.88}$$

We return to the general, possibly twisted case.

Lemma 1.7.2. *Let U be an arbitrary continuous c-representation of G on a Hilbert space $\mathcal{H}$. Then $\pi(f)$, defined by*

$$\pi(f) := \int_G dx\, f(x) U(x), \tag{1.89}$$

is bounded, with

$$\|\pi(f)\| \leq \|f\|_1. \tag{1.90}$$

Since U is unitary, we have $|(\Psi, \pi(f)\Psi)| \leq (F, F)_{L^2(G)}$ for all $\Psi \in \mathcal{H}$, where $F(x) := \|\Psi\|\sqrt{|f(x)|}$. The Cauchy–Schwarz inequality then leads to $|(\Psi, \pi(f)\Psi)| \leq \|f\|_1 \|\Psi\|^2$. The argument in the proof of II.1.3.5 then leads to (1.90). (A more sophisticated proof uses properties of Bochner integrals to argue that $\|\pi(f)\| \leq \int_G dx\, |f(x)|\, \|U(x)\| = \|f\|_1$.) ■

The following result generalizes the correspondence between U_L in (1.83) and π_L in (1.84) to arbitrary representations.

Theorem 1.7.3. *There is a bijective correspondence between nondegenerate representations π of the *-algebra $C_c^\infty(G)$ that satisfy (1.90) and continuous c-representations U of G. This correspondence is given in one direction by (1.89), and in the other by*

$$U(x)\pi(f)\Omega := \pi(f^x)\Omega, \tag{1.91}$$

where $f^x(y) := c(x, x^{-1}y) f(x^{-1}y)$. This bijection preserves direct sums, and therefore irreducibility.

It is technically convenient to extend the *-algebra $C_c^\infty(G)$ to a Banach algebra $L^1(G, c)$; this is $L^1(G)$ as a Banach space. The operations (1.80) and (1.81) are easily seen to be continuous on the L^1-norm, so that they may be extended from $C_c^\infty(G)$ to $L^1(G)$. Recall from I.1.5.2 that any nondegenerate representation of a C^*-algebra is a direct sum of cyclic representations; the same can be shown to be true of $L^1(G, c)$. Thus Ω in (1.91) stands for a cyclic vector of a certain cyclic summand of $\mathcal{H}$, and (1.89) defines U on a dense subspace of this summand; it will be shown that U is unitary, so that it can be extended to all of $\mathcal{H}$ by continuity.

Given U, it follows from easy calculations, using (1.65), that $\pi(f)$ in (1.89) indeed defines a representation. It is bounded by Lemma 1.7.2. The proof of nondegeneracy makes use of the existence of an approximate unit in $L^1(G, c)$, which heuristically converges to the Dirac delta function δ_e. This is constructed as follows. Consider a basis of neighborhoods $\mathcal{N}_\lambda$ of e, partially ordered by inclusion. Choose $\mathbb{I}_\lambda = N_\lambda \chi_{\mathcal{N}_\lambda}$, which is the characteristic function of $\mathcal{N}_\lambda$ times a normalization factor ensuring that $\|\mathbb{I}_\lambda\|_1 = 1$. One can then show that $\lim_\lambda \mathbb{I}_\lambda * f = f$ (in $\|\cdot\|_1$) for all $f \in L^1(G, c)$ (and similarly for $f * \mathbb{I}_\lambda$); for $c \neq 1$ it is at this point that the continuity of c near e is used. Since π is continuous, one has $\lim_\lambda \pi(\mathbb{I}_\lambda) = \mathbb{I}$ strongly, proving that π must be nondegenerate.

To go in the opposite direction we use the approximate unit once more; it follows from (1.91) (from which the continuity of U is obvious) that $U(x)\pi(f)\Omega = \lim_\lambda \pi(\mathbb{I}_\lambda^x)\pi(f)\Omega$. Hence $U(x) = \lim_\lambda \pi(\mathbb{I}_\lambda^x)$ strongly on a dense domain. The property (1.65) then follows from (1.91) and (1.80). Since $\|\pi(\mathbb{I}_\lambda^x)\| \leq \|\mathbb{I}_\lambda^x\|_1 = 1$, we infer that $\|U(x)\| \leq 1$ for all x. Hence also $\|U(x^{-1})\| \leq 1$. From (1.65) we derive

$$U(x)^{-1} = \overline{c(x, x^{-1})}U(x^{-1}), \tag{1.92}$$

so that $\|U(x)^{-1}\| \leq 1$. We see that $U(x)$ and $U(x)^{-1}$ are both contractions; this is possible only when $U(x)$ is unitary.

Finally, if U is reducible, there is a projection E such that $[E, U(x)] = 0$ for all $x \in G$ (see I.2.2.2). It follows from (1.89) that $[\pi(f), E] = 0$ for all f; hence π is reducible. Conversely, if π is reducible, then $[E, \pi(\mathbb{I}_\lambda^x)] = 0$ for all $x \in G$; by the previous paragraph this implies $[E, U(x)] = 0$ for all x. □

This theorem suggests looking at a slightly different object from $C_r^*(G, c)$. Inspired by I.1.5.7 one makes the following

Definition 1.7.4. *The* **(twisted) group C^*-algebra** *$C^*(G, c)$ is the closure of the convolution algebra $C_c^\infty(G)$ in the norm*

$$\|f\| := \|\pi_u(f)\| = \sup_\pi \|\pi(f)\|, \tag{1.93}$$

where π_u is the direct sum of all nondegenerate representations π of $C_c^\infty(G)$ that are bounded as in (1.90). We write $C^*(G)$ for $C^*(G, 1)$.

By Theorem 1.7.3 the representations π occurring in the sum are those associated with representations $U(G)$ via (1.89).

Corollary 1.7.5. *There is a bijective correspondence between nondegenerate representations π of the C^*-algebra $C^*(G, c)$ and continuous c-representations U of G, given by (continuous extension of) (1.89) and (1.91). This correspondence preserves irreducibility.*

Hence one may alternatively define $C^*(G, c)$ as the closure of $C_c^\infty(G)$ in the norm (1.93), where now the sum is over all representations π of $C_c^\infty(G)$ that correspond to an irreducible representation $U(G)$ via (1.89).

The second part follows from the last statement of the first part and the faithfulness of the reduced atomic representation; cf. I.2.2.7 etc. Hence one obtains the same norm in (1.93) by restricting the π's to be irreducible. ■

Since a Lie group is separable as a topological space (separability being part of the definition of a manifold used in this book), the algebras $C_c^\infty(G)$ (inheriting the norm of $L^1(G)$) and hence $C^*(G)$ are (norm) separable. Therefore, all irreducible representations are on separable Hilbert spaces, and one would obtain the same C^*-algebra by restricting the π's in Definition 1.7.4 to be on separable Hilbert spaces.

In conjunction with (1.85), the second definition of $C^*(G)$ stated in 1.7.5 implies that for abelian groups, $C^*(G)$ always coincides with $C_r^*(G)$. The reason is that for $\gamma \in \hat{G}$ one has $\pi_\gamma(f) = \hat{f}(\gamma) \in \mathbb{C}$, so that the norms (1.93) and (1.85) coincide. For future reference we single out (cf. (1.86))

$$C^*(\mathbb{R}^n) \simeq C_0(\mathbb{R}^n). \tag{1.94}$$

In any case, looking at 1.7.1, we see that

$$C_r^*(G,c) = \pi_L(C^*(G,c)) \simeq C^*(G,c)/\ker(\pi_L). \tag{1.95}$$

A Lie group is said to be **amenable** when the equality $C_r^*(G) = C^*(G)$ holds; in other words, $\pi_L(C^*(G))$ is faithful iff G is amenable. This turns out to imply that also $C_r^*(G,c) = C^*(G,c)$ for arbitrary multipliers c; we shall not prove this remarkable result. We have just seen that all locally compact abelian groups are amenable, so that the previous comment implies that the Heisenberg group $\tilde{H}_n$ is amenable. Hence the object $C^*(\tilde{H}_n)$ constructed in II.2.6 is indeed the group C^*-algebra of $\tilde{H}_n$. It follows from the Peter–Weyl theorems in the next section that all compact groups are amenable. It may be shown that also all solvable Lie groups are amenable, as are direct products of the amenable groups mentioned.

To provide an alternative characterization of amenability we first describe the connection between the representation theories of $C^*(G,c)$ and of $C_r^*(G,c)$.

Definition 1.7.6. *The* c-**unitary dual** $\hat{G}^c$ *of a group* G *is the collection of equivalence classes of irreducible* c*-representations of* G*. In other words (cf. Corollary 1.7.5),* $\hat{G}^c$ *is the set of equivalence classes of irreducible representations of* $C^*(G,c)$.

The **reduced unitary dual** $\hat{G}_r^c$ *is the set of equivalence classes of irreducible representations of* $C_r^*(G,c)$*. For* $c = 1$ *we write* $\hat{G}_{(r)} := \hat{G}_{(r)}^1$*, and speak of the* **(reduced) unitary dual**.

The earlier definition of the unitary dual of an abelian group is evidently a special case of 1.7.6. The following notion provides the key to describing $\hat{G}_r^c$. We say that a representation $U_1(G)$ is **weakly contained** in a representation $U_2(G)$ when $\ker \pi_2(C^*(G,c)) \subseteq \ker \pi_1(C^*(G,c))$; here U_i is related to π_i by (1.89).

For example, every subrepresentation properly contained in a representation is weakly contained in it. However, the notion of weak containment is more general than proper containment. Consider the regular representation U_L of $\mathbb{R}^n$ on $L^2(\mathbb{R}^n)$; since the associated representation $\pi_L(C^*(\mathbb{R}))$ is faithful (see (1.94) and the preceding discussion), its kernel is $\{0\}$. Hence every irreducible representation of $\mathbb{R}^n$ is weakly contained in U_L, although none is properly contained in it.

Proposition 1.7.7. *The reduced c-unitary dual $\hat{G}^c_r$ consists of those irreducible representations that are weakly contained in the left-regular representation $U_L(G)$. Consequently, G is amenable iff all its irreducible representations are weakly contained in U_L.*

This directly follows from the above definitions. ■

Remarkably, one may show that the above condition for amenability is equivalent to the weak containment of merely the trivial representation in U_L. Either way, when G is a noncompact semisimple Lie group it can be shown that the trivial representation is neither properly nor weakly contained in the (left- or right-) regular representation. Hence such groups are not amenable.

A comparison between Theorem 1.2.6 and Corollary 1.7.5 indicates that $C^*(G)$ is a quantum analogue of $C^\infty(\mathfrak{g}^*_-)$. More generally, Corollaries 1.4.7 and 1.7.5 suggest that $C^*(G,c)$ is a quantum analogue of $C^\infty_\Gamma(\mathfrak{g}^*_-)$, and that Co^γ-orbits are a classical version of projective irreducible representations. In particular, coadjoint orbits are analogous to irreducible representations.

In addition, we can formulate an "integrated" version of Proposition 1.6.6 (and thereby a quantum version of 1.4.3):

Proposition 1.7.8. *One obtains an automorphic group action $\alpha^{(c)}$ of G on $C^*(G,c)$ by putting*

$$\alpha^{(c)}_x(f) : y \mapsto c(x, \text{Ad}(x^{-1})y)\overline{c(y,x)}f(\text{Ad}(x^{-1})y) \tag{1.96}$$

for $f \in C^\infty_c(G)$, and extending to $C^*(G,c)$ by continuity.

In the universal representation one has

$$\pi_u(\alpha^{(c)}_x(f)) = U_u(x)\pi_u(f)U_u(x)^{-1}, \tag{1.97}$$

which firstly shows that $\alpha^{(c)}_x$ is an automorphism, secondly that it can be extended to $C^*(G,c)$, and thirdly that (1.96) defines a group action. ■

1.8 A Generalized Peter–Weyl Theorem

Further to the left-regular representation $U_L(G)$ in (1.83), which is a c-representation, consider the **right-regular representation** $U_R(G)$ on the Hilbert space $L^2(G)$, defined by

$$U_R(y)\Psi(x) := \overline{c(x,y)}\Psi(xy); \tag{1.98}$$

this is a $\overline{c}$-representation. Note that

$$\pi_R^-(f)\Psi := \int_G dx\, f(x)U_R(x)^{-1}\Psi = \Psi * f, \tag{1.99}$$

where convolution (1.80) is defined with respect to c. It immediately follows from (1.99) and (1.82) that U_L and U_R commute; this may also be verified directly, using (1.29).

Recall Definition 1.7.6. Each $\gamma \in \hat{G}^c$ has a **conjugate** $\overline{\gamma} \in \hat{G}^{\overline{c}}$; a representative $U_{\overline{\gamma}}$ of the class $\overline{\gamma}$ is obtained by defining a representative U_γ of γ on the conjugate Hilbert space $\overline{\mathcal{H}}_\gamma$ rather than on $\mathcal{H}_\gamma$ (hence $\mathcal{H}_{\overline{\gamma}} = \overline{\mathcal{H}}_\gamma$).

After these definitions we recall (a version of) the **Peter–Weyl theorem**. This theorem states that for a compact Lie group G one has

$$L^2(G) \simeq \bigoplus_{\gamma\in\hat{G}} \mathcal{H}_\gamma \otimes \mathcal{H}_{\overline{\gamma}}, \tag{1.100}$$

under which decomposition

$$U_L(G) \otimes U_R(G) \simeq \bigoplus_{\gamma\in\hat{G}} U_\gamma(G) \otimes U_{\overline{\gamma}}(G). \tag{1.101}$$

The direct sum is, of course, meant in the Hilbert space sense. This is usually stated and proved for $c = 1$, but is, in fact, valid for any multiplier; see below. One may identify $\mathcal{H}_\gamma \otimes \mathcal{H}_{\overline{\gamma}}$ with $\mathfrak{M}_{d_\gamma}(\mathbb{C})$ (where d_γ is the dimension of the representations in the class γ) as Hilbert spaces by letting $v \otimes w \in \mathcal{H}_\gamma \otimes \mathcal{H}_{\overline{\gamma}}$ correspond to the operator mapping $u \in \mathcal{H}_\gamma$ to $(w, u)v$, and extending by linearity. The inner product on $\mathfrak{M}_{d_\gamma}(\mathbb{C})$ is then given by $(M, N) = \mathrm{Tr}\, M^*N$. We accordingly rewrite (1.100) as

$$L^2(G) \simeq \hat{L}^2(G) := \bigoplus_{\gamma\in\hat{G}} \mathfrak{M}_{d_\gamma}(\mathbb{C}). \tag{1.102}$$

Writing $\hat{U}_{L,R}(x)$ for the operator on $\hat{L}^2(G)$ that is equivalent to $U_{L,R}(x)$ on $L^2(G)$ under the isomorphism (1.102), we may rephrase (1.101) as

$$\hat{U}_L(x)\Psi(\gamma) = U_\gamma(x)\Psi(\gamma); \tag{1.103}$$

$$\hat{U}_R(x)\Psi(\gamma) = \Psi(\gamma)U_\gamma(x)^*. \tag{1.104}$$

The essential step in the proof of the Peter–Weyl theorem consists in showing that the **Plancherel transform** $V : L^2(G) \to \hat{L}^2(G)$, defined by

$$\hat{\Psi}(\gamma) := V\Psi(\gamma) = \sqrt{d_\gamma}\int_G dx\, \Psi(x)U_\gamma(x), \tag{1.105}$$

is unitary. The inverse transform can then be computed from unitarity as

$$V^{-1}\hat{\Psi}(x) = \sum_{\gamma\in\hat{G}} \sqrt{d_\gamma}\,\mathrm{Tr}\,[\hat{\Psi}(\gamma)U_\gamma(x)^*]. \tag{1.106}$$

The Peter–Weyl theorem (with multiplier) has an interesting reformulation, also valid for $c \neq 1$, which in a certain sense is a quantization of Theorem 1.4.4.

Theorem 1.8.1. *For a compact Lie group G one has*

$$C^*_r(G, c) \simeq C^*(G, c) \simeq \bigoplus_{\gamma\in\hat{G}^c} \mathfrak{M}_{d_\gamma}(\mathbb{C}). \tag{1.107}$$

Here the direct sum of matrix algebras includes those sums $\oplus_\gamma M_\gamma$ of matrices for which the function $\gamma \mapsto \|M_\gamma\|$ is in $\ell_0(\hat{G}^c)$.

Note that the definition of the direct sum is different from the one in (1.102). The proof below uses some elementary aspects of the theory of induced group representations. This subject will be studied in great generality in 2.9; for the moment we need just a very special and simple case. Let H be a compact subgroup of a unimodular locally compact group G_c, and let U_χ be a 1-dimensional representation of H. (In the application below, G_c will indeed be a central extension, but for the moment it is arbitrary as stated above. We will, accordingly, denote its elements simply by x.) The Hilbert space $\mathcal{H}^\chi \subset L^2(G_c)$ is defined as the set of functions $\Psi \in L^2(G_c)$ that satisfy the equivariance condition

$$\Psi(xh) = U_\chi(h^{-1})\Psi(x) \tag{1.108}$$

for (almost) all $x \in G_c, h \in H$; the inner product on $\mathcal{H}^\chi$ is the one inherited from $L^2(G_c)$. In other words, $\mathcal{H}^\chi$ is the subspace of $L^2(G_c)$ that transforms trivially under $U_R \otimes U_\chi(H)$, where $U_R(h)\Psi(x) := \Psi(xh)$. The left-regular representation (1.83) (with $c = 1$) restricts to a representation $U^\chi(G_c)$ on $\mathcal{H}^\chi$, which is said to be **induced** by U_χ. In other words, for $\Psi \in \mathcal{H}^\chi$ one has

$$U^\chi(y)\Psi(x) := \Psi(y^{-1}x). \tag{1.109}$$

The projection $V^\chi : L^2(G_c) \to \mathcal{H}^\chi$ defined by

$$V_\chi\Psi(x) := \int_H dh\, \Psi(xh)U_\chi(h) \tag{1.110}$$

obviously intertwines U_L and U^χ, i.e., $V_\chi \circ U_L = U^\chi \circ V_\chi$. Moreover, when H is abelian, Fourier analysis on H shows that $\oplus_{\chi\in\hat{H}} V_\chi = \mathbb{I}$, so that for such H

$$\begin{aligned} L^2(G_c) &\simeq \oplus_{\chi\in\hat{H}}\mathcal{H}^\chi; \\ U_L(G_c) &\simeq \oplus_{\chi\in\hat{H}}U^\chi(G_c). \end{aligned} \tag{1.111}$$

An equivalent realization of U^χ may be defined on $L^2(G_c/H)$ (defined with respect to a suitably normalized G-invariant measure, which exists because G_c and H are unimodular), as follows. Choose a cross section $s : G_c/H \to G_c$ (i.e., $\tau \circ s = \mathrm{id}$, with $\tau := \tau_{G_c \to G_c/H}$), and define $V^\chi : L^2(G_c) \to L^2(G_c/H)$ by

$$V^\chi\Psi(q) := \Psi(s(q)), \tag{1.112}$$

with adjoint

$$(V^\chi)^*\Phi(x) = U_\chi(x^{-1}s(\tau(x)))\Phi(\tau(x)). \tag{1.113}$$

It follows that V^χ is a partial isometry, which is unitary on the image of $(V^\chi)^*$, which is $\mathcal{H}^\chi$. Putting $U_s^\chi := V^\chi U^\chi (V^\chi)^*$, one obtains

$$U_s^\chi(x)\Phi(q) = U_\chi(s(q)^{-1}xs(x^{-1}q))\Phi(x^{-1}q). \tag{1.114}$$

We will apply this to the case where G_c is as defined in 1.3.3 and $H = U_c(1)$, so that $G_c/H = G$. In the following result G is not necessarily compact.

Lemma 1.8.2. *The representation $U_c(G_c)$ associated (by 1.5.1) with the c-representation $U_L(G)$ on $\mathcal{H} = L^2(G)$ defined by (1.83) is equivalent to the representation $U^1(G_c)$ induced by the defining representation U_1 of $U_c(1)$.*

This is verified using the cross section $s : G \to G_c$ given by $s(x) = (x, 1)$. The property

$$U^1(e, z) = z\mathbb{I} \tag{1.115}$$

follows from (1.108), (1.109), and the fact that $U_c(1)$ is central. ■

Let us now assume that G (and hence G_c) is compact. The well-known **Frobenius reciprocity theorem** states that the number of times a given irreducible representation $U(G_c)$ occurs in $\mathcal{H}^\chi$ is equal to the number of times $U_\chi(H)$ occurs in $U(G_c \upharpoonright H)$ (i.e., the restriction of U to H).

Hence a given irreducible representation $U_\gamma(G_c)$ occurs in the decomposition of $U_c(G_c)$ with multiplicity equal to the number of times the defining representation of $U_c(1)$ occurs in U_γ. By 1.5.1 this multiplicity must equal d_γ, since $U_c(1)$ is always in the defining representation times the unit matrix. Hence by 1.5.1 all c-irreducible representations γ of G occur in $L^2(G)$ with multiplicity d_γ, as in the case $c = 1$.

Clearly, the Hilbert space $\mathfrak{M}_{d_\gamma}(\mathbb{C})$ carries a c-representation $U_\gamma^{(2)}(G)$ given by $U_\gamma^{(2)}(x)M := U_\gamma(x)M$, which is the irreducible c-representation $U_\gamma(G)$ with multiplicity d_γ. Here U_γ is some representative of γ; everything that follows depends on the choice of this representative, but other choices lead to equivalent statements.

We recall the orthogonality relations for a compact group K: Given an irreducible representation $U_\kappa(K)$ of dimension d_κ one has

$$d_\kappa \int_K dx\, (\Psi_1, U_\kappa(x)\Psi_2)(U_\kappa(x)\Psi_3, \Psi_4) = (\Psi_1, \Psi_4)(\Psi_3, \Psi_2). \tag{1.116}$$

It follows from these relations for G_c that $P_\gamma : L^2(G) \to \mathfrak{M}_{d_\gamma}(\mathbb{C})$, defined by $P_\gamma\Psi := \hat{\Psi}(\gamma)$ (see (1.105)), is a partial isometry (note that $L^2 \subseteq L^1$ on compact spaces, so that P_γ is well-defined). Trivially, $P_\gamma U_L = U_\gamma^{(2)} P_\gamma$. From the preceding two paragraphs we conclude that the map $V : L^2(G) \to \hat{L}^2(G)$ given by $V := \oplus_{\gamma \in \hat{G}^c} P_\gamma$ is unitary, and satisfies $VU_L = \hat{U}_L V$. Of course, V is the Plancherel transform (1.105).

This shows that, as in the case $c = 1$, the left-regular representation U_L on $L^2(G)$ contains all irreducible c-representations. It then follows from 1.7.1 and the comment after I.2.2.7 that $C_r^*(G, c) = C^*(G, c)$.

For $f \in C_c^\infty(G)$ we have from (1.84) and (1.105) that

$$\hat{\pi}_L(f) := V\pi_L(f)V^{-1} = \oplus_{\gamma \in \hat{G}^c} \pi_\gamma(f) \otimes \mathbb{I}_{d_\gamma}; \tag{1.117}$$

cf. (1.89). The map $f \mapsto V\pi_L(f)V^{-1}$ is a *-isomorphism from $C_c^\infty(G)$ into $\bigoplus_{\gamma \in \hat{G}^c} \mathfrak{M}_{d_\gamma}(\mathbb{C})$, seen as direct sum of matrix algebras, since π_L is a faithful representation and V is unitary. It can therefore be extended by continuity. The irreducibility statement in 1.7.5 implies that $\pi_\gamma(C_r^*(G, c)) = \mathfrak{M}_{d_\gamma}(\mathbb{C})$.

Finally, to prove that the direct sum in (1.107) should be defined as stated, first note that $\pi_L(f) \in \mathfrak{B}_1(L^2(G))$ for $f \in C_c^\infty(G)$. Hence $\pi_L(f) \in \mathfrak{B}_0(L^2(G))$ by I.(1.62), so that $\pi_L(C^*(G)) \subset \mathfrak{B}_0(L^2(G))$ by the continuity of π_L. Since V is unitary, it follows that $V\pi_L(C^*(G))V^{-1} \in \mathfrak{B}_0(\hat{L}^2(G))$. It is then easy to adapt the standard proof that the eigenvalues of a compact self-adjoint operator (ordered from large to small) go to zero to conclude that $\lim_{\gamma\to\infty} \|\pi_\gamma(f)\| = 0$.

Theorem 1.8.1 follows. ■

We write π^k for the representation of $C^*(G_c)$ corresponding to the representation $U^k(G_c)$ induced by $U_k(U_c(1))$ (see (1.89)), where $k \in \mathbb{Z}$ and $U_k(z) := z^k$ for $z \in \mathbb{T} = U(1)$. For G possibly noncompact, the first stage of the above proof leads to

Corollary 1.8.3. *For each $k \in \mathbb{Z}$ there are isomorphisms*

$$C^*_r(G, c^k) \simeq \pi^k(C^*(G_c)) \simeq C^*(G_c)/\ker(\pi^k). \tag{1.118}$$

Explicitly, under the first isomorphism the function $\pi^k(f) \in C^*_r(G, c^k)$ (where $f \in C^\infty_c(G_c) \subset C^*(G_c)$) is

$$\pi^k(f) : x \mapsto \int_{\mathbb{T}} dz\, z^k f(x, z). \tag{1.119}$$

Here dz is the normalized Haar measure on $\mathbb{T}$. This corollary is proved by a straightforward generalization of Lemma 1.8.2: Given a c^k-representation U of G, one defines an associated representation U_{c^k} of G_c by $U_{c^k}(x, z) := z^k U(x)$, and verifies that $U_{c^k} \simeq U^k$. ■

For $k = 1$ one should compare 1.8.3 with 1.1.10. As we have seen in 1.3, the multiplier c is a derived object, the intrinsic object being the central extension (1.27). Hence $C^*_r(G, c)$ is not quite intrinsic either, but Corollary 1.8.3 shows how to define the intrinsic analogue of $C^*_r(G, c)$: It is $C^*(\mathrm{G})/\ker(\pi^1)$. This C^*-algebra is, of course, isomorphic to $C^*_r(G, c)$, and also to any $C^*_r(G, c')$, where c' is equivalent to c. The case of general k will be used in the next section.

Corollary 1.8.3 is closely related to the decomposition

$$C^*_r(G_c) \simeq \oplus_{k\in\mathbb{Z}}\, \pi^k(C^*(G_c)), \tag{1.120}$$

which follows from $C^*_r(G_c) = \pi_L(C^*(G_c))$ and (1.111). Equation (1.118) shows that $C^*_r(G, c)$ is isomorphic to a (closed 2-sided) ideal in $C^*_r(G_c)$, namely the one that is isomorphic to $\pi^1(C^*(G_c))$ by (1.120).

As an application of (1.118) we prove

Proposition 1.8.4. *Let c be the multiplier on $\mathbb{R}^{2n}$ given in (1.35). Then for all $k \in \mathbb{Z}\backslash\{0\}$ there are isomorphisms*

$$C^*(\mathbb{R}^{2n}, c^k) \simeq C^*_r(\mathbb{R}^{2n}, c^k) \simeq \mathfrak{B}_0(L^2(\mathbb{R}^n)). \tag{1.121}$$

We will not prove the first isomorphism here; the proof is identical to that of Theorem 3.7.1 below. As to the second, we saw (after (1.35)) that $\mathbb{R}^{2n}_c = H_n$. We use the notation of Lemma 1.8.2. Using (1.115), II.(2.18), and Theorem II.2.1.4, or direct calculation, one shows that $U^k(H_n)$ is a multiple of the irreducible Schrödinger representation $U^S_k(H_n)$ defined in II.(2.17). The second isomorphism in (1.121) then follows from 1.8.3 and II.(2.129) (with $\tilde{H}_n$ replaced by H_n).

Here is a direct proof as well. Define a map $K^k : \mathcal{S}(\mathbb{R}^{2n}) \to \mathcal{S}(\mathbb{R}^{2n})$ by

$$K_f^k(x, y) := \int_{\mathbb{R}^n} d^n v \, e^{-\frac{1}{2}ikv(x+y)} f(x - y, v), \tag{1.122}$$

with inverse

$$f(u, v) = k^n \int_{\mathbb{R}^n} \frac{d^n q}{(2\pi)^n} \, e^{ikqv} K_f^k(q + \tfrac{1}{2}u, q - \tfrac{1}{2}u). \tag{1.123}$$

The well-known properties of the Fourier transform show that K^k is an isomorphism (of vector spaces to begin with). We identify the image $\mathcal{S}(\mathbb{R}^{2n})$ of K^k with $\mathcal{S}(\mathbb{R}^n \times \mathbb{R}^n)$, regarded as a space of kernels on $\mathbb{R}^n$, and as such as a subspace of $\mathfrak{B}_0(L^2(\mathbb{R}^n))$; cf. II.2.5.3. One calculates from (1.80) and (1.81) with (1.35) that K^k is a *-isomorphism between dense subspaces of $C_r^*(\mathbb{R}^{2n}, c^k)$ and $\mathfrak{B}_0(L^2(\mathbb{R}^n))$. Then extend K^k by continuity. ■

Note that the multiplier has a crucial effect: Without it, one has the isomorphism $C^*_{(r)}(\mathbb{R}^{2n}) \simeq C_0(\mathbb{R}^{2n})$ by Fourier transformation; cf. (1.86).

In the light of Example 1.2.11.3, the classical analogue of Proposition 1.8.4 (for $k = 1$, say) states that the Poisson algebra $C^\infty_\Gamma(\mathbb{R}^{2n}, \mathbb{R})$ (where $\mathbb{R}^{2n}$ is seen as the dual Lie algebra $\mathfrak{g}^*$ for $G = \mathbb{R}^{2n}$) is $C^\infty(T^*\mathbb{R}^n, \mathbb{R})$ equipped with the canonical Poisson bracket I.(2.24); cf. 1.9.6.

To close this section, we "quantize" the realization of $C^\infty_\Gamma(\mathfrak{g}^*_-)$ stated in 1.4.2. Denote the operators in a C^*-algebra $\mathfrak{A} \subset \mathfrak{B}(L^2(G))$ that commute with $U_R(G)$ (cf. (1.98)) by $\mathfrak{A}^R$. The following result is nontrivial even for $c = 1$.

Proposition 1.8.5. *When G is compact, $\pi_L(C^*(G, c)) = \mathfrak{B}_0(L^2(G))^R$.*

The Plancherel transform (1.105) maps U_R into $\hat{U}_R := VU_RV^{-1}$, given by $\hat{U}_R(x)\hat{\Psi}(\gamma) = \hat{\Psi}(\gamma)U_\gamma(x)^{-1}$. The result then follows from (1.117). ■

1.9 The Group C^*-Algebra as a Strict Quantization

When G is compact the C^*-algebra $C^*(G)$ turns out to be related to the (complexified) Poisson algebra $C^\infty(\mathfrak{g}^*_-)$ by a strict quantization.

Analogously to II.(3.49), we define $C^\infty_{\text{PW}}(\mathfrak{g}^*)$ as the class of functions on $\mathfrak{g}^*$ whose Fourier transform $\check{f}$ is in $C^\infty_c(\mathfrak{g})$. Here the Fourier transform of $f \in L^1(\mathfrak{g}^*)$ is defined by (cf. II.(3.42))

$$\check{f}(X) := \int_{\mathfrak{g}^*} \frac{d^n\theta}{(2\pi)^n} \, e^{i\theta(X)} f(\theta), \tag{1.124}$$

where $d^n\theta$ is Lebesgue measure on $\mathfrak{g}^* \simeq \mathbb{R}^n$, whose normalization is fixed by that of the Haar measure dx on G, as follows. When f has support near e, we can write

$$\int_G dx \, f(x) = \int_{\mathfrak{g}} d^n X \, J(X) f(\text{Exp}(X)), \tag{1.125}$$

where $d^n X$ is a Lebesgue measure on $\mathfrak{g}$, and J is some Jacobian. The normalization is now fixed by the condition $J(0) = 1$. In turn, the normalization of the Lebesgue measure $d^n\theta$ on $\mathfrak{g}^*$ is fixed by requiring the inversion formula

$$f(\theta) = \int_{\mathfrak{g}} d^n X\, e^{-i\theta(X)} \check{f}(X). \tag{1.126}$$

As in the argument after II.(3.49), one infers that $C^\infty_{\mathrm{PW}}(\mathfrak{g}^*_\pm, \mathbb{R})$ is a Poisson subalgebra of $C^\infty(\mathfrak{g}^*_\pm, \mathbb{R})$.

We choose a smooth cutoff function κ on $\mathfrak{g}$ that equals 1 in a neighborhood $\tilde{\mathcal{N}}$ of 0, is invariant under inversion $X \mapsto -X$, and has support in the neighborhood $\mathcal{N}$ of 0 on which Exp is a diffeomorphism; cf. II.3.4.4. When G is compact one may assume that κ is Ad-invariant; i.e., satisfies $\kappa(\mathrm{Ad}(y)X) = \kappa(X)$ for all $y \in G$. This may always be achieved by averaging.

Definition 1.9.1. *For an n-dimensional Lie group G, the* **Rieffel quantization** $\mathcal{Q}^R_\hbar : C^\infty_{\mathrm{PW}}(\mathfrak{g}^*_-) \to C^*(G)$ *is defined as follows: For* $x \notin \mathrm{Exp}(\mathcal{N})$ *we put* $\mathcal{Q}^R_\hbar(f)(x) = 0$, *whereas for* $x \in \mathrm{Exp}(\mathcal{N})$ *we put*

$$\mathcal{Q}^R_\hbar(f)(x) := \hbar^{-n}\kappa(\mathrm{Exp}^{-1}(x))\check{f}(\mathrm{Exp}^{-1}(x)/\hbar). \tag{1.127}$$

Analogously to $\mathcal{Q}^W_\hbar$ in II.3.4, the restriction $f \in C^\infty_{\mathrm{PW}}(\mathfrak{g}^*)$ implies that for small enough $\hbar$ the operator $\mathcal{Q}^R_\hbar(f)$ is independent of κ. When G is exponential (in that $\mathrm{Exp} : \mathfrak{g} \to G$ is a diffeomorphism) the cutoff κ can be omitted altogether. For general $f \in C^\infty(\mathfrak{g}^*)$ the object $\mathcal{Q}^R_\hbar(f)$ is a distribution on G. In particular, when f is a polynomial, one obtains a distribution with support at e. One may identify the set of such distributions with the enveloping algebra $\mathcal{U}(\mathfrak{g}_\mathbb{C})$, but even on this space $\mathcal{Q}^R_\hbar$ does not coincide with $\mathcal{Q}^\lambda_\hbar$ in (1.78) unless the Jacobian J appearing in (1.125) equals unity. However, for unimodular groups, and therefore in particular for compact groups, one has $J(X) = 1 + O(X^2)$, and this property suffices to guarantee that, at least formally, $\mathcal{Q}^R_\hbar(\tilde{X}) = \mathcal{Q}^\lambda_\hbar(\tilde{X})$ for all $X \in \mathfrak{g}$. Since $\mathcal{Q}^R_\hbar(\tilde{X})$ is not defined as an element of $C^*(G)$, one may pass to a representation $\pi(C^*(G))$, related to $U(G)$ by (1.89). Formally, one then has

$$\pi(\mathcal{Q}^R_\hbar(\tilde{X})) = i\hbar dU(X). \tag{1.128}$$

Theorem 1.9.2. *Let* $\mathfrak{A}^0 = C_0(\mathfrak{g}^*)$ *and* $\mathfrak{A}^\hbar = C^*(G)$ *for* $\hbar \notin 0$. *When G is compact, the map* $\mathcal{Q}^R_\hbar$ *in 1.9.1 yields a strict and continuous quantization of* $\tilde{\mathfrak{A}}^0_\mathbb{R} := C^\infty_{\mathrm{PW}}(\mathfrak{g}^*_-, \mathbb{R})$ *on* $I = \mathbb{R}$, *up to condition II.1.1.1.4.*

Strictness, which implies continuity (cf. II.1.2.5) by Theorem II.1.2.4, will follow from the fact that $\mathcal{Q}^R_\hbar$ is a special case of the generalized Weyl quantization prescription on Riemannian manifolds (cf. II.3.4).

Lemma 1.9.3. *A compact Lie group G admits a right-invariant Riemannian metric* **g** *such that the exponential map* $\exp_e$ *obtained from* **g** *coincides with the map* Exp *defined by the Lie group structure.*

Choose an inner product $(\, , \,)$ on $T_eG = \mathfrak{g}$ that is invariant under the adjoint action of G, and define **g** by the property $\mathbf{g}_x(\xi^R_X, \xi^R_Y) := (X, Y)$; cf. (1.38). this is evidently right-invariant, but due to the Ad-invariance of $(\, , \,)$ it is left-invariant as well. Such metrics are called **bi-invariant**.

For any right-invariant metric $\mathbf{g}$ and any point of G one has the identity

$$\mathbf{g}\left(\nabla_{\xi_X^R}\xi_Y^R, \xi_Z^R\right) = \tfrac{1}{2}\left\{\mathbf{g}\left(\left[\xi_X^R, \xi_Y^R\right], \xi_Z^R\right) - \mathbf{g}\left(\left[\xi_Y^R, \xi_Z^R\right], \xi_X^R\right) + \mathbf{g}\left(\left[\xi_Z^R, \xi_X^R\right], \xi_Y^R\right)\right\}, \tag{1.129}$$

obtained from II.(3.17) and II.(3.18) with various permutations of the entries, using the x-independence of $\mathbf{g}_x(\xi_X^R, \xi_Y^R)$. When $\mathbf{g}$ is Ad-invariant, the last two terms cancel, upon which the nondegeneracy of $\mathbf{g}$ implies that $\nabla_{\xi_X^R}\xi_Y^R = \frac{1}{2}\left[\xi_X^R, \xi_Y^R\right]$ due to the nondegeneracy of $\mathbf{g}$. In particular, $\nabla_{\xi_X^R}\xi_X^R = 0$, so that by II.(3.4) we infer that the curves $x(t) = \mathrm{Exp}(tY)x$ are geodesic for all $Y \in \mathfrak{g}$. The claim follows. ∎

We identify $C^*(G)$ with $\pi_L(C^*(G))$ (see (1.95) and 1.8.1)), which in turn is expressed as in 1.8.5. Also, we identify $C^\infty_{\mathrm{PW}}(\mathfrak{g}^*) \subset C^\infty(\mathfrak{g}^*)$ with $C^\infty_{\mathrm{PW}}(T^*G)^R \subset C^\infty(T^*G)^R$, as in 1.4.2 (with $\Gamma = 0$). Choosing a metric on G as in the lemma, it follows from II.(3.51), II.(3.48), (1.82), (1.80) (with $c = 1$), and (1.127) that under the above identifications one has $\mathcal{Q}_\hbar^R = \mathcal{Q}_\hbar^W$ on $C^\infty_{\mathrm{PW}}(T^*G)^R$ (confirming II.3.6.3). The theorem then follows from Theorem II.3.5.1. ∎

Proposition 1.9.4. *Suppose that G admits a metric $\mathbf{g}$ as specified in Lemma 1.9.3, and that the cutoff κ is* Ad*-invariant (these assumptions are satisfied when G is compact). With $\alpha := \alpha^{(1)}$ given by (1.96), for all $x \in G$ and $f \in \tilde{\mathfrak{A}}^0$ one has*

$$\alpha_x(\mathcal{Q}_\hbar^R(f)) = \mathcal{Q}_\hbar^R(\mathrm{Co}(x^{-1})^* f). \tag{1.130}$$

The metric $\mathbf{g}$ of Lemma 1.9.3 is bi-invariant, and $\mathbf{g}_e$ is invariant under the adjoint representation $\mathrm{Ad}(G)$. Identifying $\mathfrak{g} \simeq \mathbb{R}^n$ this implies that $\mathrm{Ad}(G) \subseteq SO(n)$; in particular, $\mathrm{Ad}(G)$ and $\mathrm{Co}(G)$ leave the Lebesgue measures invariant on $\mathfrak{g}$ and $\mathfrak{g}^*$, respectively. The claim then follows from (1.127), (1.124), and the Ad-invariance of the cutoff κ.

Alternatively, the claim follows from Proposition 1.2.3, equation (1.52), and Theorem II.3.6.3. ∎

This proposition is a "bounded" version of 1.6.8.

Theorem 1.9.2 can be generalized to the twisted case, at the cost of $\hbar$ being defined only at a discrete set. Let $\Gamma \in Z^2(\mathfrak{g}, \mathbb{R})$ be related to $c \in Z^2(G, U(1))$ by (1.34). This leads to a Poisson algebra $C^\infty_\Gamma(\mathfrak{g}^*_-, \mathbb{R})$ (see 1.1.5), a central extension G_c with Lie algebra $\mathfrak{g}_\Gamma$ (see 1.3.3 and 1.3.6), a group C^*-algebra $C^*(G_c)$, and a twisted group C^*-algebra $C^*(G, c)$ (cf. 1.7).

Theorem 1.9.5. *Let $\mathfrak{A}^0 := C_0(\mathfrak{g}^*)$ and $\mathfrak{A}^\hbar := C^*(G, c^{1/\hbar})$ for $\hbar = 1/k$, where $k \in \mathbb{Z}$, and let $\tilde{\mathfrak{A}}^0_{\mathbb{R}} := C^\infty_{\mathrm{PW}}(\mathfrak{g}^*_{(\Gamma)-}, \mathbb{R})$ be equipped with the Poisson bracket (1.12), (1.13), taking the minus sign. When G is compact, the map $\mathcal{Q}_\hbar^R$, defined in 1.9.1 with $C^*(G)$ replaced by $C^*(G, c^{1/\hbar})$, is a strict quantization of $\tilde{\mathfrak{A}}^0_{\mathbb{R}}$ on $I_0 = 1/\mathbb{Z}$ (except possibly for the completeness condition II.1.1.1.4).*

The signs may be checked from (1.128), (1.67), and (1.15). The proof is based on the analogy between 1.1.10 and 1.8.3. Extend $f \in C^\infty_{\mathrm{PW}}(\mathfrak{g}^*, \mathbb{R})$ to a function $\tilde{f} \in C^\infty_{\mathrm{PW}}(\mathfrak{g}^*_\Gamma, \mathbb{R})$, such that $f(\theta) = \tilde{f}(1, \theta)$ and

$$\tilde{f}(\theta_0 \neq 1, \theta) < \tilde{f}(\theta_0 = 1, \theta) = f(\theta); \tag{1.131}$$

in particular, one has

$$\|f\|_\infty = \|\tilde{f}\|_\infty. \tag{1.132}$$

In view of (1.13) this automatically means that

$$\{\tilde{f}, \tilde{g}\}_-(1, \theta) = \{f, g\}_-^\Gamma(\theta), \tag{1.133}$$

since the left-hand side does not involve derivatives with respect to θ_0.

We denote the map of Definition 1.9.1 as defined on $C^\infty_{\rm PW}(\mathfrak{g}^*, \mathbb{R})$, taking values in $C^*(G, c^{1/\hbar})$, by $\mathcal{Q}^R_\hbar$, whereas the map defined in the same way, but now on $C^\infty_{\rm PW}(\mathfrak{g}^*_\Gamma, \mathbb{R})$, taking values in $C^*(G_c)$, is written as $\tilde{\mathcal{Q}}^R_\hbar$. A short computation using (1.119) and an elementary oscillatory integral shows that

$$\pi^{1/\hbar}(\tilde{\mathcal{Q}}^R_\hbar(\tilde{f})) = \mathcal{Q}^R_\hbar(f) \tag{1.134}$$

for $\hbar \in 1/\mathbb{Z}$ small enough so that the right-hand side is independent of κ. In particular, the left-hand side depends only on the value of $\tilde{f}$ at $\theta_0 = 1$; this is a special case of the fact that, for $\hbar$ small enough, $\pi^k(\tilde{\mathcal{Q}}^R_\hbar(\tilde{f}))$ depends only on $\tilde{f}(\theta_0 = k\hbar)$. This follows by a calculation similar to the one leading to (1.134).

Theorem 1.9.2 applied to G_c implies that $\lim_{\hbar\to 0}\|\tilde{\mathcal{Q}}^R_\hbar(\tilde{f})\| = \|\tilde{f}\|_\infty$. On the other hand, according to (1.120) one has $\|A\| = \sup_{k\in\mathbb{Z}}\|\pi^k(A)\|$ for all $A \in C^*(G_c)$. Combining the two of these equations with the last remark of the preceding paragraph and the property (1.131), we conclude that

$$\lim_{\hbar\to 0}\|\tilde{\mathcal{Q}}^R_\hbar(\tilde{f})\| = \lim_{\hbar\to 0}\|\pi^{1/\hbar}(\tilde{\mathcal{Q}}^R_\hbar(\tilde{f}))\| = \|\tilde{f}\|_\infty. \tag{1.135}$$

Together with (1.132) and (1.134) this proves II.(1.1) for $\mathcal{Q}^R_\hbar$.

Conditions II.(1.2) and II.(1.3) in Theorem 1.9.5 now follow from (1.134), Corollary 1.8.3, (1.133), Theorem 1.9.2 (once again applied to G_c), and the inequality $\|\pi^k(A)\| \leq \|A\|$ in $C^*(G_c)$; cf. I.1.5. ∎

The obvious generalization of 1.9.4 (in which Co is replaced by Co^γ, where γ and Γ are related by (1.24)) is not valid except in special cases (see below).

While proved for compact G, Theorem 1.9.5 may hold in other situations.

Proposition 1.9.6. *Let $G \simeq \mathfrak{g}^* = \mathbb{R}^{2n}$, with Γ given by (1.26) and c defined in (1.35). Then the statement of Theorem 1.9.5 holds (without the final qualification).*

Using Proposition 1.8.4 one obtains that $\mathcal{Q}^R_\hbar = \mathcal{Q}^W_\hbar$ (cf. II.2.5), so the proposition follows from Theorem II.2.6.1. ∎

In this case one does have the "twisted equivariance property"

$$\alpha_x(\mathcal{Q}^R_\hbar(f)) = \mathcal{Q}^R_\hbar(\text{Co}^\gamma(x^{-1})^* f) \tag{1.136}$$

for all $x \in \mathbb{R}^{2n}$ and $f \in C^\infty_{\rm PW}(T^*\mathbb{R}^n)$ (or $\mathcal{S}(T^*\mathbb{R}^n)$), where γ is related to Γ by (1.24). This follows by direct computation; in (1.23) only the term $\gamma(x)$ contributes, yielding $\text{Co}^\gamma(u, v) : (p, q) \mapsto (p + u, q + v)$. Alternatively, one uses the corresponding property II.(2.93) of $\mathcal{Q}^W_\hbar$.

1.10 Representation Theory of Compact Lie Groups

Following the discussion of the Weyl quantization of $\mathfrak{g}^*$, we turn our attention to the possible quantization of coadjoint orbits $\mathcal{O}$ in $\mathfrak{g}^*$. In view of later applications to gauge theories we restrict ourselves to the case that G is compact. We start with a brief review of the relevant representation theory, assuming familiarity with the standard Cartan-Weyl approach. Throughout this section G is a compact connected Lie group unless stated otherwise, and all representations are finite-dimensional.

Firstly, let G be abelian; it then has to be a **torus** $G = T = \mathbb{T}^r = U(1)^r$. Each irreducible representation of T is one-dimensional, and is a **character** $U_\lambda : \mathbb{T}^r \to \mathbb{C}$ (a character of an arbitrary group is a one-dimensional representation). The label λ of the character is an element of $\mathfrak{t}^* \simeq \mathbb{R}^r$ (the dual of the Lie algebra $\mathfrak{t} \simeq \mathbb{R}^r$ of T), related to U_λ by

$$dU_\lambda(X) = -i\lambda(X). \tag{1.137}$$

It follows that $\lambda \in \Lambda := \mathbb{Z}^r \subset \mathfrak{t}^*$. Conversely, each $\lambda \in \Lambda$ defines an irreducible representation of T by exponentiation, so that we have found a bijective correspondence between the unitary dual $\hat{T}$ and the lattice $\Lambda \subset \mathfrak{t}^*$.

For a general Lie group, we note that (1.61) (with $\Gamma = 0$) implies that $\theta \in \mathfrak{g}^*$ satisfies $\theta([X, Y]) = 0$ for all $X, Y \in \mathfrak{g}_\theta$ (where $\mathfrak{g}_\theta$ is the Lie algebra of the stabilizer G_θ of θ under the coadjoint action). In other words, $\theta : \mathfrak{g}_\theta \to \mathbb{R}$ is a Lie algebra homomorphism.

Definition 1.10.1. *A coadjoint orbit $\mathcal{O} \in \mathfrak{g}^*$ is called* **integral** *if for some (hence all) $\theta \in \mathcal{O}$ the functional $\theta \upharpoonright \mathfrak{g}_\theta$ exponentiates to a character of G_θ.*

In other words, θ is integral iff there is a character U_θ of G_θ such that $\theta = idU_\theta$ on $\mathfrak{g}_\theta$. If this holds for one $\theta \in \mathcal{O}$, it holds for all, since one has $U_{\mathrm{Co}(x)\theta} = U_\theta \circ \mathrm{Ad}(x^{-1})$.

Obviously, if G is a torus T, its coadjoint action is trivial, so that its coadjoint orbits are the points of $\mathfrak{t}^*$; the integral orbits are precisely the elements of the lattice Λ. Consequently, one has a bijective correspondence between $\hat{T}$ and the set of integral coadjoint orbits of T. The following theorem generalizes this idea.

Theorem 1.10.2. *There exists a bijective correspondence between the unitary dual $\hat{G}$ and the set of integral coadjoint orbits in $\mathfrak{g}^*$.*

We will merely sketch the proof in explaining how this parametrization of $\hat{G}$ is related to the Cartan–Weyl theory. This theory starts by choosing a **maximal torus** T (i.e., a maximal connected abelian subgroup) of G, with associated **Weyl group** $W := N(T)/T$ (where $N(T)$ is the normalizer of T). The integer $r := \dim(T)$ is called the **rank** of G; it does not depend on the choice of T, since all maximal tori are conjugate. The Weyl group acts on T by conjugation, and hence it acts on $\mathfrak{t}$ and $\mathfrak{t}^*$. The latter action is the projection of the coadjoint action of $N(T)$. It maps $\Lambda \subset \mathfrak{t}^*$ (called the **weight lattice** in the present context; elements of Λ are traditionally called **weights**) into itself; the W-action on Λ coincides with the natural W-action on $\hat{T}$ under the identification of $\hat{T}$ with Λ explained above. The Cartan–Weyl theory states

Theorem 1.10.3. *Let T be some maximal torus in G. There exists a bijective correspondence between the unitary dual $\hat{G}$ and the set of W-orbits in $\hat{T}$; that is, $\hat{G} \simeq \hat{T}/W = \Lambda/W$.*

The relationship between this parametrization of $\hat{G}$ and the one in 1.10.2 follows from an important lemma in the structure theory of compact Lie groups, which we state without proof.

Lemma 1.10.4. *In the notation of 1.10.3 there is a bijection*

$$G/\mathrm{Ad}(G) \simeq T/W. \tag{1.138}$$

Denoting the set of coadjoint orbits in $\mathfrak{g}^*$ by $\mathfrak{g}^*/G$, one therefore has

$$\mathfrak{g}^*/G \simeq \mathfrak{t}^*/W. \tag{1.139}$$

The concrete association of a coadjoint orbit in $\mathfrak{g}^*$ with a W-orbit in $\mathfrak{t}^*$ is as follows. Restrict the adjoint representation of G (extended to the complexification $\mathfrak{g}_{\mathbb{C}}$) to T, and decompose $\mathfrak{g}_{\mathbb{C}}$ under $\mathrm{Ad}(T)$ as $\mathfrak{g}_{\mathbb{C}} = \mathfrak{t}_{\mathbb{C}} \oplus \mathfrak{t}_{\mathbb{C}}^{\perp}$, where $\mathfrak{t}_{\mathbb{C}}^{\perp}$ is the sum of all eigenspaces with nonzero eigenvalues; this leads to a decomposition $\mathfrak{g} = \mathfrak{t} \oplus \mathfrak{t}^{\perp}$, where $\mathfrak{t}^{\perp} := \mathfrak{t}_{\mathbb{C}}^{\perp} \cap \mathfrak{g}$. This coincides with the orthogonal decomposition of $\mathfrak{g}$ under an arbitrary $\mathrm{Ad}(G)$-invariant inner product $(\, , \,)$. One may, for example, take a faithful representation U of G, and define the invariant inner product by

$$(X, Y) := -\mathrm{Tr}\, dU(X) dU(Y). \tag{1.140}$$

An arbitrary compact Lie group is of the form $G = S \times \mathbb{T}^k$, where S is semisimple (i.e., a product of simple factors with finite discrete center) and $\mathbb{T}^k$ is a torus. If G is semisimple, the adjoint representation is faithful, and may be used in (1.140); this defines the **Killing form** on $\mathfrak{g}$. More generally, all invariant metrics have the property that the direct summands in $\mathfrak{g}$ are mutually orthogonal, and that the metric restricted to a given simple summand is proportional to the Killing form. For concreteness' sake, in what follows we assume that $(\, , \,)$ restricted to the semisimple part of $\mathfrak{g}$ coincides with the Killing form.

The extension $\theta(\lambda) \in \mathfrak{g}^*$ of $\lambda \in \mathfrak{t}^*$ obtained by putting $\theta(\lambda) = 0$ on $\mathfrak{t}^{\perp}$ and $\theta(\lambda) = \lambda$ on $\mathfrak{t}$ is therefore independent of the choice of the metric on $\mathfrak{g}$. Thus the coadjoint orbit $\mathcal{O}_{\lambda} := \mathcal{O}_{\theta(\lambda)}$ associated to λ is the coadjoint orbit through $\theta(\lambda)$; it is obvious from the definition of the W-action on $\mathfrak{t}^*$ that all points of the W-orbit of λ are mapped into $\mathcal{O}_{\lambda}$.

To go in the opposite direction one needs (1.138) to show that the stabilizer of any point in $\mathcal{O}$ is connected, and that it contains a maximal torus. As any two maximal tori are conjugate to each other, and $G_{\mathrm{Co}(x)\theta} = xG_{\theta}x^{-1}$, there accordingly exists a $\theta \in \mathcal{O}$ for which $T \subseteq G_{\theta}$. Hence we can define $\lambda(\theta) = \theta \restriction \mathfrak{t}$. Note that (1.61) and $T \subseteq G_{\theta}$ imply that $\theta \restriction \mathfrak{t}^{\perp} = 0$.

If $G_{\theta} = T$ the coadjoint orbit $\mathcal{O}_{\theta}$ through θ is said to be **regular**; it is of maximal dimension among all coadjoint orbits. Otherwise, it is called **singular**. For regular orbits one immediately sees that $G_{\mathrm{Co}(x)\theta} \doteq T$ implies that $x \in N(T)$, so that different choices of θ for which $G_{\theta} = T$ map into the same W-orbit of $\lambda(\theta)$. It follows from (1.138) that the same is true for the singular orbits.

Theorem 1.10.2 now follows from Theorem 1.10.3 and Lemma 1.10.4 by restricting the isomorphism to weights and integral orbits. ∎

A functional $\lambda \in \mathfrak{t}^*$ is called **regular** when $w\lambda = \lambda$ for $w \in W$ implies $w = e$ (and **singular** otherwise); this defines the sets $\mathfrak{t}^*_{\mathrm{r}}$ and $\Lambda_{\mathrm{r}} := \mathfrak{t}^*_{\mathrm{r}} \cap \Lambda$ of regular elements and **regular weights** in $\mathfrak{t}^*$, respectively. In the context of 1.10.2, elements of $\mathfrak{t}^*_{\mathrm{r}}$ evidently correspond to regular coadjoint orbits, and similarly for the singular case.

Each connected component C of $\mathfrak{t}^*_{\mathrm{r}}$ is called a **Weyl chamber**; this is an open convex cone in $\mathfrak{t}^*$. Singular weights clearly lie on the boundary of some Weyl chamber. One singles out an arbitrary Weyl chamber C_{d}, and declares a weight **dominant** if it lies in the closure $\overline{C}_{\mathrm{d}}$. The point is now that each W-orbit intersects a given closed Weyl chamber $\overline{C}$ in exactly one point. Hence Theorem 1.10.3 may now be restated:

Corollary 1.10.5. *In the notation of 1.10.3 there is a bijection between $\hat{G}$ and the set $\Lambda_{\mathrm{d}} := \Lambda \cap \overline{C}_{\mathrm{d}}$ of dominant weights.*

Any Hilbert space $\mathcal{H}$ carrying a representation $U(G)$ decomposes under $U(T)$ as $\mathcal{H} \simeq \oplus_{\lambda\in\Delta_0(U)}\mathcal{H}_\lambda$, where each $\mathcal{H}_\lambda$ carries the representation $U_\lambda(T)$ (perhaps with multiplicity). The set $\Delta_0(U) \subset \Lambda$ contains the **weights of** U. This applies, in particular, to the adjoint representation Ad. The nonzero weights of Ad are called **roots**; one writes Δ for $\Delta_0(\mathrm{Ad})\backslash\{0\}$, with elements generically denoted by α.

The decomposition of $\mathfrak{g}_{\mathbb{C}}$ under Ad takes the form $\mathfrak{g}_{\mathbb{C}} = \oplus_{\alpha\in\Delta}\mathfrak{g}_\alpha \oplus \mathfrak{t}_{\mathbb{C}}$, where each $\mathfrak{g}_\alpha$ is one-dimensional. Writing $\mathfrak{g}_\alpha = \mathbb{C}E_\alpha$ for some nonzero vector E_α, we have

$$[X, E_\alpha] = -i\alpha(X)E_\alpha \tag{1.141}$$

for $X \in \mathfrak{t}$. It follows that if $\alpha \in \Delta$, then $-\alpha \in \Delta$, since $\overline{\mathfrak{g}}_\alpha = \mathfrak{g}_{-\alpha}$ (where the complex conjugation is the usual one on $\mathfrak{g}_{\mathbb{C}} = \mathfrak{g} \oplus i\mathfrak{g}$).

Given a choice of C_{d}, a root is called **positive** if $(\alpha, \lambda) > 0$ for all $\lambda \in C_{\mathrm{d}}$ (here the inner product on $\mathfrak{g}$ has been transferred to $\mathfrak{g}^*$ in the usual way). The collection of positive roots is called Δ^+. A root lies either in Δ^+ or in $-\Delta^+$. Singular dominant weights λ have the property that $(\alpha, \lambda) = 0$ for some $\alpha \in \Delta^+$; a weight is regular iff $(\alpha, \lambda) \neq 0$ for all roots α.

It is not difficult to show from (1.141) and the Jacobi identity that $\mathfrak{g}_{\mathbb{C}}$ has a basis $\{H_j, E_\alpha, E_{-\alpha}\}_{j=1,\dots,r;\alpha\in\Delta^+}$, normalized such that $(E_\alpha, E_{-\alpha}) = 1$, satisfying

$$\begin{aligned}
&[H_j, H_k] = 0;\\
&[H_j, E_{\pm\alpha}] = \mp i\alpha_j E_{\pm\alpha};\\
&[E_\alpha, E_{-\alpha}] = -i\alpha_j H_j;\\
&[E_\alpha, E_\beta] = N_{\alpha,\beta}E_{\alpha+\beta} \quad (\beta \neq -\alpha),
\end{aligned} \tag{1.142}$$

where $\alpha \in \Delta^+$, $\beta \in \Delta$, $\alpha_j := \alpha(H_j)$, and the $N_{\alpha,\beta}$ are constants that vanish iff $\alpha + \beta$ is not a root (in which case $E_{\alpha+\beta}$ is, of course, not defined).

The bijection in 1.10.5 is now implemented by the following fact:

Corollary 1.10.6. *In the notation of 1.10.3, a Hilbert space $\mathcal{H}_\gamma^{\text{hw}}$ carrying an irreducible representation U_γ corresponding to a dominant weight γ has a unit vector Ψ_γ, unique up to a phase, on which*

$$dU_\gamma(X)\Psi_\gamma = -i\gamma(X)\Psi_\gamma \tag{1.143}$$

for all $X \in \mathfrak{t}$, whereas for all $\alpha \in \Delta^+$ one has

$$dU_\gamma(E_\alpha)\Psi_\gamma = 0. \tag{1.144}$$

The unit vector Ψ_γ is called a **highest weight vector**; it is unique up to a phase. It is easily inferred that

$$(\Psi_\gamma, dU_\gamma(E_{\pm\alpha})\Psi_\gamma) = 0 \tag{1.145}$$

for all $\alpha \in \Delta^+$, since the Lie brackets (1.142) imply that $dU(E_{-\alpha})\Psi_\gamma$ must either be zero or a vector with weight $\gamma - \alpha \neq \gamma$.

One may now see the correspondence in 1.10.2 in a clearer light. Let $J : \mathbb{P}\mathcal{H}_\gamma \to \mathfrak{g}^*$ be the momentum map for the G-action on $\mathbb{P}\mathcal{H}_\gamma$ defined by U_γ, given by (1.71); this may be rewritten as

$$J_X(\psi) = i(\Psi, dU_\gamma(X)\Psi), \tag{1.146}$$

where the unit vector Ψ is a lift of $\psi \in \mathbb{P}\mathcal{H}_\gamma$ to $\mathbb{S}\mathcal{H}_\gamma^{\text{hw}}$.

Proposition 1.10.7. *The coadjoint orbit $\mathcal{O}_\gamma$ corresponding to an irreducible representation U_γ with highest weight vector Ψ_γ contains $J\psi_\gamma$.*

In fact, $J : \mathbb{P}U_\gamma(G)\Psi_\gamma \to \mathcal{O}_\gamma$ is a symplectomorphism when $\mathbb{P}U_\gamma(G)\Psi_\gamma$ inherits the usual symplectic structure of $\mathbb{P}\mathcal{H}_\gamma^{\text{hw}}$ (with $\hbar = 1$), and $\mathcal{O}_\gamma$ is endowed with the (minus) Lie symplectic form (1.63).

Equations (1.146), (1.143), (1.144), and (1.145) imply that $(J(\psi_\gamma))(X)$ equals $\gamma(X)$ for $X \in \mathfrak{t}$ and equals 0 for $X \in \mathfrak{t}^\perp$. Hence $J(\psi_\gamma)$ is precisely the element $\theta(\gamma) \in \mathfrak{g}^*$ discussed after the proof of 1.10.4, proving the first claim.

By (1.146), the stability group $G_{J(\psi_\gamma)}$ of $J(\psi_\gamma)$ consists of those $x \in G$ for which $(U_\gamma(x)\Psi_\gamma, dU_\gamma(Y)U_\gamma(x)\Psi_\gamma) = (\Psi_\gamma, dU_\gamma(Y)\Psi_\gamma)$ for all $Y \in \mathfrak{g}$. Since U_γ is irreducible, this implies that Ψ_γ and $U_\gamma(x)\Psi_\gamma$ define the same element of $\mathbb{P}\mathcal{H}_\gamma$, proving that $G_{J(\psi_\gamma)} \subseteq G_{\psi_\gamma}$. The opposite inclusion is trivial from the equivariance of J, which can either be checked directly from (1.146), or may more abstractly be inferred from 1.2.5, it having been realized from (1.72) (with $\Gamma = 0$) and 1.1.2 that J is a Poisson map on $\mathbb{P}\mathcal{H}_\gamma$. ■

For general $\hbar$ one would have a factor $\hbar$ on the right-hand side of (1.146).

It is actually quite easy to give an explicit description of the Lie algebra $\mathfrak{g}_\gamma$ of $G_\gamma := G_{\theta(\gamma)}$. From (1.61) (or the above proof), (1.142), and the previously discussed fact that $\gamma(E_\alpha) = 0$ for all $\alpha \in \Delta$ we infer that

$$\mathfrak{g}_\gamma = \mathfrak{t} \bigoplus \oplus_{\alpha \in \Delta^+ | (\gamma,\alpha)=0} \mathfrak{g}_\alpha^{\mathbb{R}}, \tag{1.147}$$

where $\mathfrak{g}_\alpha^{\mathbb{R}} := \mathfrak{g} \cap (\mathfrak{g}_\alpha \oplus \mathfrak{g}_{-\alpha})$. It follows that the dimension $\dim(\mathcal{O}_\gamma)$ of the orbit through a dominant weight γ is given by

$$\dim(\mathcal{O}_\gamma) = \dim(\mathfrak{g}) - \dim(\mathfrak{t}) - 2\,\mathrm{Card}\,\{\alpha \in \Delta^+ | (\gamma, \alpha) = 0\}. \qquad (1.148)$$

It follows from Proposition 1.10.7 that $\mathbb{P}U_\gamma(G)\Psi_\gamma$ is a symplectic submanifold of $\mathbb{P}\mathcal{H}_\gamma^{\text{hw}}$; this is not necessarily true if $\mathbb{C}\Psi_\gamma$ is replaced by an arbitrary one-dimensional subspace of $\mathcal{H}_\gamma^{\text{hw}}$. However, the same statement as in 1.10.7 evidently applies to any vector of the type $U_\gamma(\tilde{w})\Psi_\gamma$, where $\tilde{w}$ is a lift of $w \in W = N(T)/T$ to $N(T) \subset G$. For J maps all vectors of this type into the same coadjoint orbit. In fact, $U_\gamma(\tilde{w})\Psi_\gamma$ has weight $w\gamma$, showing that all weights $w\gamma$, $w \in W$, occur in $\mathcal{H}_\gamma^{\text{hw}}$.

1.11 Berezin Quantization of Coadjoint Orbits

Coadjoint orbits of compact Lie groups are interesting partly because they lead to coherent pure state quantizations indexed by a discrete set $I \ni \hbar$; cf. II.1.5.1 and II.1.3.3. We will use the label γ to denote a dominant weight in $\Lambda_{\text{d}} \subset \mathfrak{t}^*$, as well as the corresponding element $\theta(\gamma)$ of the coadjoint orbit $\mathcal{O}_\gamma \subset \mathfrak{g}^*$.

Theorem 1.11.1. *Let G be a compact connected Lie group, and $\mathcal{O}_\gamma$ an integral coadjoint orbit (cf. 1.10.1), corresponding to a highest weight $\gamma \in \Lambda_{\text{d}}$. For $\hbar = 1/k$, $k \in \mathbb{N}$, define $\mathcal{H}_\hbar := \mathcal{H}_{\gamma/\hbar}^{\text{hw}}$, i.e., the carrier space of the irreducible representation $U_{\gamma/\hbar}(G)$ with highest weight $\gamma/\hbar = k\gamma$. The map $q_\hbar : \mathcal{O}_\gamma \to \mathbb{P}\mathcal{H}_\hbar$, given by*

$$q_\hbar(\mathrm{Co}(x)\gamma) := \tau_{\mathcal{H}_\hbar \to \mathbb{P}\mathcal{H}_\hbar}(U_{\gamma/\hbar}(x)\Psi_{\gamma/\hbar}), \qquad (1.149)$$

is well-defined and injective. Together with

$$\mu_\hbar = d_{\gamma/\hbar}\mu_L, \qquad (1.150)$$

where $d_\lambda := \dim(\mathcal{H}_\lambda^{\text{hw}})$, this provides a pure state quantization of $\mathcal{O}_\gamma$ (equipped with minus the Lie symplectic structure) on $I_0 := 1/\mathbb{N}$.

One should note here that $k\gamma \in \overline{C}_{\text{d}}$ when $\gamma \in \overline{C}_{\text{d}}$, since Weyl chambers are convex cones. In what follows, $\mathcal{O}_\gamma$ stands for $(\mathcal{O}_\gamma)_-$; see the notation introduced before (1.63).

The map $q_\hbar$ is well-defined and injective by the equation $G_{k\gamma} = G_\gamma$ plus the argument on stability groups used in the proof of 1.10.7. In fact, if we define $J_\hbar : \mathbb{P}\mathcal{H}_\hbar \to \mathfrak{g}^*$ by (1.71), with $\hbar = 1/k$ (equivalently, by (1.146) with the right-hand side divided by k), it follows from 1.10.7 that $J_\hbar$ takes values in $\mathcal{O}_\gamma$ and is a left inverse of $q_\hbar$.

We start from the fact that the Haar measure on G (with total mass 1) pushes forward to the Liouville measure derived from the Lie symplectic structure under the canonical projection $G \to \mathcal{O}_\gamma \simeq G/G_\gamma$. Using the invariance of the Haar measure and the unitarity of $U_{k\gamma}$, we then have

$$\int_{\mathcal{O}_\gamma} d\mu_L(\sigma)\, p(\rho_\hbar, \sigma_\hbar) f(\sigma) = \int_G dx\, |(\Psi_{k\gamma}, U_{k\gamma}(x)\Psi_{k\gamma})|^2 f_\gamma(yx) \qquad (1.151)$$

for all $f \in C(\mathcal{O}_\gamma)$, where $f_\gamma = \tau^*_{G\to G/G_\gamma} f \in C(G)$ is a right-G_γ-invariant function, and y is such that $\tau_{G\to G/G_\gamma}(y) = \rho \in \mathcal{O}_\gamma$. Choosing $f = 1$, the orthogonality relations (1.116) for compact groups and (1.150) then imply II.(1.8).

Equation II.(1.10), even without the limit, follows from 1.10.7. To prove II.(1.9) we need a lemma.

Lemma 1.11.2. *Let γ_i be dominant weights with highest weight representations and vectors U_{γ_i} and Ψ_{γ_i}, respectively ($i = 1, 2$). Then for each $x \in G$ one has*

$$(\Psi_{\gamma_1}, U_{\gamma_1}(x)\Psi_{\gamma_1}) \cdot (\Psi_{\gamma_2}, U_{\gamma_2}(x)\Psi_{\gamma_2}) = (\Psi_{\gamma_1+\gamma_2}, U_{\gamma_1+\gamma_2}(x)\Psi_{\gamma_1+\gamma_2}). \tag{1.152}$$

This is immediate from 1.10.6 and the connectedness of G. ■

This lemma implies that

$$(\Psi_{k\gamma}, U_{k\gamma}(x)\Psi_{k\gamma}) = (\Psi_\gamma, U_\gamma(x)\Psi_\gamma)^k. \tag{1.153}$$

Using (1.151), we can write the left-hand side of II.(1.9) as

$$\lim_{\hbar\to 0} \int_{\mathcal{O}_\gamma} d\mu_L(\sigma)\, p(\rho_\hbar, \sigma_\hbar) f(\sigma) = \lim_{k\to\infty} \int_G d\mu_k(x) f_\gamma(yx),$$

where μ_k is a probability measure on G defined by

$$d\mu_k(x) := d_{k\gamma}\, dx |(\Psi_\gamma, U_\gamma(x)\Psi_\gamma)|^{2k}. \tag{1.154}$$

It is obvious that each μ_k is right-G_γ-invariant. It follows from (1.144), (1.145), and the fact that the exponential map is surjective for compact Lie groups, that $|(\Psi_\gamma, U_\gamma(x)\Psi_\gamma)|$, which is evidently ≤ 1, equals 1 iff $x \in G_\gamma$. Hence for large k the support of μ_k is increasingly concentrated on G_γ. This suggests that

$$\lim_{k\to\infty} \mu_k(f) = \int_{G_\gamma} dh\, f(h) \tag{1.155}$$

for all $f \in C(G)$, where dh is the normalized Haar measure on G_γ. This is confirmed by more detailed analysis (cf. the proof of 1.11.4 below). For the right-G_γ-invariant function $f_\gamma \in C(G)^{G_\gamma}$ one therefore obtains

$$\lim_{k\to\infty} \mu_k(f_\gamma) = f_\gamma(e).$$

This proves II.(1.9), which finishes the proof of 1.11.1. □

The Berezin quantization $\mathcal{Q}^B_\hbar$ associated with the pure state quantization in 1.11.1 (cf. II.1.3.4) is defined on $\tilde{\mathfrak{A}}^0 := C^\infty(\mathcal{O}_\gamma)$. By II.(1.16), one has

$$\mathcal{Q}^B_{1/k}(f) = d_{k\gamma} \int_G dx\, f_\gamma(x)[U_{k\gamma}(x)\Psi_{k\gamma}]; \tag{1.156}$$

this is an element of $\mathfrak{A}^{1/k} = \mathfrak{M}_{d_{k\gamma}}(\mathbb{C})$. The most important property of $\mathcal{Q}^B_\hbar$ is its G-equivariance. For $x \in G$ we write

$$\alpha^{1/k}_x(A) := U_{k\gamma}(x) A U_{k\gamma}(x)^*, \tag{1.157}$$

where $A \in \mathfrak{A}^{1/k}$, and for $f \in \mathfrak{A}^0$ we put

$$\alpha_x^0(f) := \mathrm{Co}(x^{-1})^* f. \tag{1.158}$$

Proposition 1.11.3. *For all $k \in \mathbb{N}$, $x \in G$, and $f \in L^\infty(\mathcal{O}_\gamma)$ one has*

$$\mathcal{Q}^B_{1/k}(\alpha_x^0(f)) = \alpha_x^{1/k}(\mathcal{Q}^B_{1/k}(f)). \tag{1.159}$$

This is immediate from (1.156), (1.158), the fact that $(\mathrm{Co}(y)^* f)_\gamma = L_y^* f_\gamma$, the right-invariance of the Haar measure, and (1.157). ■

Theorem 1.11.4. *The Berezin quantization (1.156), defined on the space $C^\infty(\mathcal{O}_\gamma, \mathbb{R})$, is strict.*

Recall that $\mathcal{O}_\gamma := (\mathcal{O}_\gamma)_-$. Rieffel's condition II.1.1.1.1 follows from Theorem 1.11.1 and Proposition II.1.3.6. The completeness condition II.1.1.1.4 is an easy consequence of Schur's lemma and the irreducibility of $U_{k\gamma}$.

We will now prove von Neumann's condition II.1.1.1.2 and Dirac's condition II.1.1.1.3. We pick a unit vector Φ_k in each $\mathcal{H}_{k\gamma}^{\mathrm{hw}}$ and use the invariance of the Haar measure and (1.116) to write (using the notation of the proof of 1.11.1)

$$\begin{aligned}&\left(\Phi_k, (\mathcal{Q}^B_{1/k}(f)\mathcal{Q}^B_{1/k}(g) - \mathcal{Q}^B_{1/k}(fg))\Phi_k\right)\\ &= d_{k\gamma}\int_G dx\, f_\gamma(x)(\Phi_k, U_{k\gamma}(x)\Psi_{k\gamma})I_k(x),\end{aligned} \tag{1.160}$$

where

$$I_k(x) := d_{k\gamma}\int_G dy\,(\Psi_{k\gamma}, U_{k\gamma}(y)\Psi_{k\gamma})F_\gamma^x(y), \tag{1.161}$$

$$F_\gamma^x(y) := (U_{k\gamma}(xy)\Psi_{k\gamma}, \Phi_k)[g_\gamma(xy) - g_\gamma(x)]. \tag{1.162}$$

In the notation used after (1.151), the function F_γ^x on G corresponds to a function F^x on $\mathcal{O}_\gamma$.

Using (1.153), we can write $(\Psi_{k\gamma}, U_{k\gamma}(y)\Psi_{k\gamma}) = \exp(-nS_\gamma(y))$, where $S_\gamma(y) := -\log(\Psi_\gamma, U_\gamma(y)\Psi_\gamma)$ (in view of the exponentiation, the choice of the branch cut of the logarithm is irrelevant). The function S_γ is right-G_γ-invariant; we denote the corresponding function on G/G_γ by S. We identify G/G_γ with $\mathcal{O}_\gamma$, so that the coset $[G_\gamma] \in G/G_\gamma$ is identified with $\gamma \in \mathcal{O}_\gamma$.

Putting $S_\gamma^+(y) := -\log|(\Psi_\gamma, U_\gamma(y)\Psi_\gamma)|$, the absolute value of $\exp(-nS)$ is $\exp(-nS^+)$. As in the argument preceding (1.155), we see that S^+ takes values in $[0, \infty]$ and assumes its unique absolute minimum 0 at γ. Since F_γ^x in (1.161) is bounded, a standard argument implies that to $O(\exp(-n))$ we may replace the integration over G/G_γ by one over any neighborhood of γ.

We identify $T_\gamma\mathcal{O}_\gamma$ with $\mathfrak{g}/\mathfrak{g}_\gamma$, and use complex coordinates $\{z_\alpha, \overline{z}_\alpha\}_{\alpha\in\Delta_\gamma^+}$, where Δ_γ^+ consists of those positive roots for which $(\gamma, \alpha) \neq 0$. By the definition of a highest weight, this implies that $(\gamma, \alpha) > 0$ for all $\alpha \in \Delta_\gamma^+$. The coordinates $(z_\alpha, \overline{z}_\alpha)$ correspond to the point in $\mathcal{O}_\gamma$ given by

$$\mathrm{Co}\left[\mathrm{Exp}\left(i\sum_{\alpha\in\Delta_\gamma^+}(z_\alpha E_\alpha - \overline{z}_\alpha E_{-\alpha})\right)\right]\gamma.$$

A simple computation, using (1.142) and 1.10.6, leads to

$$S(z_\alpha, \bar{z}_\alpha) = \sum_{\alpha \in \Delta_\gamma^+} (\gamma, \alpha) z_\alpha \bar{z}_\alpha + O(|z|^3). \tag{1.163}$$

Hence to $O(\exp(-k))$ we may approximate $I_k(x)$ by

$$d_{k\gamma} \int_{\mathfrak{g}/\mathfrak{g}_\gamma} \left(\prod_{\alpha \in \Delta_\gamma^+} \frac{dz_\alpha d\bar{z}_\alpha}{2\pi} \right) J(z_\alpha, \bar{z}_\alpha) e^{-k \sum_{\alpha \in \Delta_\gamma^+} (\gamma, \alpha) z_\alpha \bar{z}_\alpha} F^x(z_\alpha, \bar{z}_\alpha),$$

where J is a Jacobian, and F_γ^x has been extended to $\mathfrak{g}/\mathfrak{g}_\gamma$ by, say, the exponential map. If we omit the factor $[\ldots]$ in (1.162), the integral (1.161) can be evaluated, using the orthogonality relations (1.116). On the other hand, we can compute the above integral to lowest order in the steepest descent approximation; this avoids the need to compute $J(0)$. Comparing the results computes the prefactor in the steepest descent approximation as unity. As a by-product we obtain the asymptotic expression for $k \to \infty$

$$d_{k\gamma} \sim \frac{\prod_{\alpha \in \Delta_\gamma^+} (\gamma, \alpha)}{J(0)} k^{\frac{1}{2} \dim(\mathcal{O}_\gamma)}, \tag{1.164}$$

where $\dim(\mathcal{O}_\gamma)$ is given by (1.148). (Comparison with the Weyl dimension formula then yields $J(0) = \prod_{\alpha \in \Delta^+} (\alpha, \delta)$, where $\delta := \frac{1}{2} \sum_{\alpha \in \Delta^+} \alpha$.)

Thus the steepest descent approximation to the above integral, and therefore to (1.161), reads

$$I_k(x) = \sum_{l=0}^{N} \frac{1}{l!} D^l (J F^x)(0) + O(k^{-N-1}), \tag{1.165}$$

where, abbreviating $\partial_\alpha := \partial / \partial z_\alpha$ and $\bar{\partial}_\alpha := \bar{\partial} / \bar{\partial} z_\alpha$, we have put

$$D := \sum_{\alpha \in \Delta_\gamma^+} \frac{1}{(\gamma, \alpha)} \partial_\alpha \bar{\partial}_\alpha . \tag{1.166}$$

Substituting this expansion in (1.160) we see that

$$\left(\Phi_k, (Q_{1/k}^B(f) \circ Q_{1/k}^B(g) - Q_{1/k}^B(fg)) \Phi_k \right) = O(1/k). \tag{1.167}$$

To analyze the remainder of $O(1/k)$ we note that the lth term in the expansion leads to an x-integrand in (1.160) of the form

$$(U_{k\gamma}(x^{-1})\Phi_k, \Psi_{k\gamma}) \overline{(U_{k\gamma}(x^{-1})\Phi_k, \Psi^{(l_1)})} f \partial^{l_2} \bar{\partial}^{l_3} g_\gamma(x),$$

where $l_i \leq l$ and the $\Psi^{(l_1)}$ are given by the action of products of $dU_{k\gamma}(E_\alpha)$ and $dU_{k\gamma}(E_{-\alpha})$ on $\Psi_{k\gamma}$. The important point is now that the orthogonality relations (1.116) (applied to the x-integration) then imply that the $O(k^{-N-1})$ term is bounded by $C \|\Phi_k\|^2 / k^{N+1}$ for some constant C. Hence II.(1.2) follows by II.(2.77).

To prove II.(1.3) we need the $l = 1$ term in (1.165). We substitute (1.166), and perform some partial integrations in the remaining x-integral (using the invariance of the Haar measure). We abbreviate $A := (\Phi_k, U_{k\gamma}(x)\Psi_{k\gamma})$; then (1.144) implies that $\partial_\alpha A$ and $\overline{\partial}_\alpha \overline{A}$ vanish at $z_\alpha = \overline{z}_\alpha = 0$. Terms of the form $\partial_\alpha \overline{\partial}_\alpha A$ (or $\overline{A}$) drop out in the commutator, as do contributions from J (whose first derivatives at 0 already vanish identically). What remains is

$$\left(\Phi_k, ik[\mathcal{Q}^B_{1/k}(f), \mathcal{Q}^B_{1/k}(g)] - \mathcal{Q}^B_{1/k}(\{f,g\}_-)\Phi_k\right) = O(1/k), \tag{1.168}$$

where, in the realization of f, g as G_γ-invariant functions f_γ, g_γ on G,

$$\{f_\gamma, g_\gamma\}_\pm := \pm i \sum_{\alpha \in \Delta_\gamma} \frac{1}{(\gamma, \alpha)} \xi^L_\alpha f_\gamma \, \xi^L_{-\alpha} g_\gamma. \tag{1.169}$$

Here the left-invariant vector fields $\xi^L_{\pm\alpha}$ on G are defined as in (1.37), the element $E_{\pm\alpha}$ of $\mathfrak{g}_\mathbb{C}$ having been expressed in terms of elements of $\mathfrak{g}$. Also, Δ_γ is $\Delta^+_\gamma \cup \Delta^-_\gamma$, i.e., the set of all roots α for which $(\gamma, \alpha) \neq 0$.

To finish the proof, we remark that (1.169) is precisely the Lie–Poisson bracket on $\mathcal{O}_\gamma$; this may be verified at the point $\gamma \in \mathcal{O}_\gamma$ (or $e \in G$) by direct computation from (1.3), from which the general statement follows by the G-invariance of the Poisson structure.

It is manifest that the right-hand side of (1.169) is left-G-invariant if f_γ and g_γ are; its right-G_γ-invariance is not so obvious. The latter may be verified at the infinitesimal level from (1.147), (1.142), and the fact that for $\beta \neq \pm\alpha$ one has

$$N_{-\alpha-\beta,\beta} = -N_{\alpha,\beta}. \tag{1.170}$$

This follows from the Ad($\mathfrak{g}$)-invariance of the inner product on $\mathfrak{g}_\mathbb{C}$, combined with the normalization of the E_α. Invariance of (1.169) under $\mathfrak{g}_\gamma$ implies invariance under G_γ, which is connected.

The higher-order terms in (1.168) are dealt with as in the above proof of II.(1.2). This proves II.(1.3), finishing the proof of Theorem 1.11.4. ∎

It is possible to regard $\sigma_\hbar := q_\hbar(\sigma)$, defined in (1.149) for $\hbar = 1/k$, as a state $\tilde{\sigma}_\hbar$ on the group algebra $C^*(G)$ by

$$\tilde{\sigma}_\hbar(A) := \sigma_\hbar(\pi_{\gamma/\hbar}(A)). \tag{1.171}$$

The following result is analogous to II.(2.167).

Proposition 1.11.5. *With $\mathcal{Q}^R_\hbar$ defined in 1.9.1, for all $\sigma \in \mathcal{O}_\gamma$ and $f \in C^\infty_{\text{PW}}(\mathfrak{g}^*)$ one has*

$$\lim_{\hbar \to 0} \tilde{\sigma}_\hbar(\mathcal{Q}^R_\hbar(f)) = f(\sigma) \tag{1.172}$$

along the sequence $\hbar = 1/k$, $k \in \mathbb{N}$.

This follows from a straightforward calculation. One starts by using (1.124), (1.125), and rescaling $X \mapsto X/k$. The k-dependence is firstly in $J(X/k)\kappa(X/k)$, which goes to 1 for $k \to \infty$. Secondly, one uses (1.153) and subsequently

$$\lim_{k \to \infty} (\Psi_\gamma, U_\gamma(\text{Exp}(X/k))\Psi_\gamma)^k = e^{(\Psi_\gamma, dU_\gamma(X)\Psi_\gamma)}.$$

This can be computed by (1.143) and (1.144). The result then follows from the well-known representation of the delta function as an oscillatory integral. ∎

Finally, we remark that the results in this section have an obvious yet somewhat cumbersome generalization: If the orbit $\mathcal{O}_\gamma$ is not integral, but such that $\mathcal{O}_{\gamma/c}$ is integral for some $c \in \mathbb{R}\backslash\{0\}$, we can construct a strict quantization for the values $\hbar = c/k, k \in \mathbb{N}$.

2 Internal Symmetries and External Gauge Fields

2.1 Bundles

Many constructions where some form of symmetry plays a role, and in particular the mathematical description of gauge field theories, involve the notion of a (smooth locally trivial fiber) bundle. We have already encountered the tangent bundle and the cotangent bundle of a manifold; here is a general definition.

Definition 2.1.1. *A* **bundle** $\mathsf{B}(Q, F, \tau)$ *consists of manifolds* B *(the* **total space***), Q (the* **base***), F (the* **typical fiber***), and a smooth surjection $\tau : \mathsf{P} \to Q$ with the following property: Each $q \in Q$ has a neighborhood $\mathcal{N}_\alpha$ such that there is a diffeomorphism $\psi_\alpha : \tau^{-1}(\mathcal{N}_\alpha) \to \mathcal{N}_\alpha \times F \subset Q \times F$ for which $\tau = \tau_Q \circ \psi_\alpha$ (where $\tau_Q : Q \times F \to Q$ is the projection onto the first factor).*

The maps ψ_α are called **local trivializations.** To avoid cumbersome expressions we shall often say "$\mathsf{B} \simeq Q \times F$ (locally)", omitting reference to $\mathcal{N}$. Similarly, we then loosely write "$\psi : \mathsf{B} \to Q \times F$ (locally)". We factorize $\psi_\alpha = (\tau, \psi_\alpha^F)$ so that ψ_α^F restricted to $\tau^{-1}(q)$ provides a diffeomorphism between the latter and the typical fiber F. Each subset $\tau^{-1}(q)$ is called a **fiber** of B. One may think of B as Q with a copy of F attached at each point.

Throughout this chapter Q will be physically interpreted as the space on which a particle moves, or perhaps as some more general configuration space.

Two bundles $\mathsf{B}_i(Q_i, F_i, \tau_i)$ $(i = 1, 2)$ are said to be **isomorphic** if there is a diffeomorphism $\psi : \mathsf{B}_1 \to \mathsf{B}_2$ that preserves fibers. Such a bundle isomorphism defines a diffeomorphism of the base spaces and typical fibers in question. The bundle is said to be **trivial** if there is a bundle isomorphism $\psi : \mathsf{P} \to Q \times F$. Any bundle over a contractible base is trivial.

By definition, a **section** of B is a map $s : Q \to \mathsf{B}$ satisfying $\tau \circ s = \text{id}$. It can be shown that (Borel) measurable sections always exist, whereas the existence of smooth sections is not guaranteed (they certainly exist if B is trivial). However, one can always choose smooth **local sections** $s_\alpha : \mathcal{N}_\alpha \to \mathsf{B}$. In the spirit of the paragraph before the last, we may say "$s : Q \to \mathsf{B}$ (locally)" when s is actually defined on some $\mathcal{N} \subset Q$.

Definition 2.1.2. *Given two bundles B_1 and B_2 over the same base Q, with projections τ_1, τ_2 and typical fibers F_1 and F_2, respectively, the* **fiber product** *of B_1 and B_2 is*

$$\mathsf{B}_1 *_Q \mathsf{B}_2 := \{(x, y) \in \mathsf{B}_1 \times \mathsf{B}_2 \mid \tau_1(x) = \tau_2(y)\}, \tag{2.1}$$

with manifold structure inherited from the Cartesian product. This may be regarded as a bundle over Q with projection $\tau(x, y) = \tau_1(x) = \tau_2(y)$ and typical fiber $F_1 \times F_2$.

Let $\mathsf{B}(Q, F, \tau)$ be a bundle over Q and let $f : M \to Q$ be a smooth map from some manifold M to Q. Then the **pullback bundle**

$$f^*\mathsf{B} := \mathsf{B} *_Q M = \{(x, y) \in \mathsf{B} \times M \mid \tau(x) = f(y)\} \tag{2.2}$$

is a bundle over M under projection $\tau_{(2)}$ onto the second variable and typical fiber F.

Hence $\mathsf{B}_1 *_Q \mathsf{B}_2$ can be equipped with a bundle structure in three different ways: It is a bundle over Q as explained above, it is a bundle $\tau_1^*\mathsf{B}_2$ over B_1 with typical fiber F_2 under the projection $\tau_{(1)}$ onto the first variable, and finally it is a bundle $\tau_2^*\mathsf{B}_1$ over B_2 with typical fiber F_1 under the projection $\tau_{(2)}$ onto the second variable.

One can specialize the bundle structure. For example, in a **vector bundle** each fiber is a (topological) vector space (where the linear operations are smooth with respect to the ambient manifold structure), and the local trivializations respect the linear structure in the obvious sense. Clearly, T^*Q and TQ are vector bundles. Even when it is nontrivial, a vector bundle always admits a smooth global section, namely the **zero section** $s_0(q) := 0$. We will generically denote vector bundles by the letter V, unless the typical fiber is a Hilbert space, in which case we write H, and speak of a **Hilbert bundle.**

When the B_i in 2.1.2 are both vector bundles V_i (with finite-dimensional or Hilbert fibers), one may form two different vector bundles over Q from $\mathsf{V}_1 *_Q \mathsf{V}_2$ by declaring the typical fiber to be either the tensor product $V_1 \otimes V_2$ or the direct sum $V_1 \oplus V_2$. One accordingly writes $\mathsf{V}_1 \otimes \mathsf{V}_2$ or $\mathsf{V}_1 \oplus \mathsf{V}_2$. One can also form the **dual bundle** V^* of a vector bundle V, whose typical fiber is the dual V^* of V, and whose local trivializations are dual to those of V.

Here is the "mother" of all bundles in which group actions play a role.

Definition 2.1.3. *A* **principal bundle** *$\mathsf{P}(Q, H, \tau)$ is a bundle for which the typical fiber is a Lie group H (the* **structure group***) with smooth (left) action R on P such that $Q = \mathsf{P}/H$, and $\psi_\alpha \circ R_h = R_h^Q \circ \psi_\alpha$, where the action $R_h^Q : Q \times H \to Q \times H$ on the right-hand side stands for $R_h^Q(q, k) = (q, kh^{-1})$.*

To stress the role of H and Q, one may speak of a **principal H-bundle over Q** for clarity. It follows that the H-action R_h must be free, and that $\tau(R_h(x)) = \tau(x)$. In agreement with the above, we will write R_h as $x \mapsto xh^{-1}$ for $x \in \mathsf{P}$. In contrast with a vector bundle, it can be shown that a principal bundle admits smooth global sections iff it is trivial. In a trivial bundle $\mathsf{P} = Q \times H$ one obviously has $R_h(q, k) = (q, kh^{-1})$.

In a principal bundle a given local trivialization ψ_s is equivalent to a smooth local section s: Given s one can put $\psi_s(s(q)) = (q, e)$ and subsequently extend ψ_s by H-equivariance; that is,

$$\psi_s(s(q)h) = (q, h). \tag{2.3}$$

Conversely, given $\psi = \psi_s$ one defines $s(q) := \psi^{-1}(q, e)$. If various local sections s_α are involved, we will write ψ_α for ψ_{s_α}.

If Q is covered by open sets of the type $\mathcal{N}_\alpha$ and $q \in \mathcal{N}_\alpha \cap \mathcal{N}_\beta$, it must be that

$$s_\beta(q) = s_\alpha(q) g_{\alpha\beta}(q) \tag{2.4}$$

(no sum over α), where the smooth maps $g_{\alpha\beta} : \mathcal{N}_\alpha \cap \mathcal{N}_\beta \to H$ are called **transition functions**. More generally, two different systems of local trivializations are related in this way.

In an interesting special case one takes P to be the universal covering space $\tilde{Q}$ of Q, so that $H = \pi_1(Q)$ is the first homotopy group of Q (regarding discrete groups as zero-dimensional Lie groups). For another example the reader could now skip ahead to 2.7.

Definition 2.1.4. *Given a principal H-bundle P over Q and a smooth H-action L on some manifold M, the* **associated bundle** $\mathsf{M} = \mathsf{P} \times_H M$ *is* $(\mathsf{P} \times M)/H$, *where the H-action on $\mathsf{P} \times M$ is given by* $h : (x, m) \mapsto (xh^{-1}, L_h(m))$. *This is a bundle over Q with typical fiber M and projection* $\tau_{\mathsf{M} \to M}([x, m]_H) = \tau(x)$, *which is well-defined in being independent of the representative* $(x, m) \in \mathsf{P} \times M$ *in the equivalence class* $[x, m]_H \in (\mathsf{P} \times M)/H$.

The following result will be used on many occasions.

Proposition 2.1.5. *A section* $\Psi^{(L)} : Q \to \mathsf{M}$ *of a bundle* M *associated to a principal bundle* $\mathsf{P}(Q, H, \tau)$ *may alternatively be represented:*

- *As a map* $\Psi^L : \mathsf{P} \to M$ *that is H-equivariant in that*

$$\Psi^L(xh^{-1}) = L(h)\Psi^L(x). \tag{2.5}$$

  This is related to $\Psi^{(L)}$ *by* $\Psi^{(L)}(\tau(x)) = [x, \Psi^L(x)]_H$, *which is independent of the choice of* $x \in \tau^{-1} \circ \tau(x)$ *because of (2.5).*
- *Given a section* $s : Q \to \mathsf{P}$, *as a map* $\Psi_s^L : Q \to M$, *in terms of which*

$$\begin{aligned} \Psi^{(L)}(q) &= [s(q), \Psi_s^L(q)]_H; \\ \Psi_s^L(q) &= \Psi^L(s(q)); \\ \Psi^L(x) &= L(h_s(x))\Psi_s^L(\tau(x)), \end{aligned} \tag{2.6}$$

  where $h_s(x)$ *is determined by* $xh_s(x) = s(\tau(x))$.

This follows directly from the definitions involved. Note that $h_s(xk) = k^{-1}h_s(x)$, ensuring the consistency of the relation between Ψ_s^L and Ψ^L. ■

The space of smooth compactly supported sections of a vector bundle V (where compact support is defined with reference to the zero vector in each fiber) is denoted by $\Gamma(\mathsf{V})$; when M is a vector space V, the symbol $\Gamma(\mathsf{P} \times_H V)$ will specifically refer to the first realization discussed above. The second realization will be called $\Gamma_s(\mathsf{P} \times_H V)$.

We will see that interesting classical phase spaces arise by taking M to be a coadjoint orbit of G in this construction. Alternatively, in classical as well as quantum mechanics one encounters the case where M is a linear space V, carrying a linear H-action L; in that case one speaks of an **associated vector bundle**. The case relevant to quantum mechanics is that in which V is a Hilbert space $\mathcal{H}_\chi$ carrying a representation $U_\chi(H)$; cf. 2.9.

A (local) trivialization $\psi_\alpha = (\tau, \psi_\alpha^H)$ of P leads to a (local) trivialization ψ_α^M : $\mathsf{M} \to Q \times M$ of an associated bundle M by putting

$$\psi_\alpha^M([x, m]_H) := \big(\tau(x), L(\psi_\alpha^H(x))m\big). \tag{2.7}$$

This is well-defined, since $\psi_\alpha^H(xh) = \psi_\alpha^H(x)h$. Conversely, $(\psi_\alpha^M)^{-1}(q, m) = [s_\alpha(q), m]_H$, in terms of the (local) section associated with ψ_α; cf. 2.3.

If two vector bundles V_1, V_2 are both associated to a principal H-bundle P over Q by H-actions L_i, then $\mathsf{V}_1 \otimes \mathsf{V}_2$ and $\mathsf{V}_1 \oplus \mathsf{V}_2$ are associated to P by the actions $L_1 \otimes L_2$ and $L_1 \oplus L_2$, respectively. The dual bundle V^* of an associated vector bundle V defined by an H-action L^* is defined by the dual H-action $L_h^* := (L_{h^{-1}})^*$ on V^*.

2.2 Connections

In preparation for the definition of a connection on a principal bundle, note that the tangent bundle $T\mathsf{P}$ of a principal bundle $T\mathsf{P}$ has a natural subbundle

$$V\mathsf{P} := \{v \in T\mathsf{P} \mid \tau_* v = 0\}. \tag{2.8}$$

Elements of $V\mathsf{P}$ are called **vertical vectors**, and the linear space $V_x\mathsf{P} := T_x\mathsf{P} \cap V\mathsf{P}$ is called the **vertical tangent space** at x. It is easily seen that $V\mathsf{P}$ is stable under the lifted H-action R_*. Indeed, $V\mathsf{P}$ is spanned by vector fields of the type ξ_X given by the H-action on P; cf. (1.19). It is customary to define

$$\xi_X^f := -\xi_X^{\mathsf{P}}, \tag{2.9}$$

referred to as a **fundamental vector field**. The vector $\xi_X^f(x)$ is evidently tangent to the curve $x(t) = x\mathrm{Exp}(tX)$. One easily shows that

$$(R_h)_* \xi_X^f = \xi_{\mathrm{Ad}(h)X}^f. \tag{2.10}$$

A **lift** of $X \in T_{\tau(x)}Q$ to $x \in \mathsf{P}$ is an element $l_x(X) \in T_xX$ for which $\tau_* l_x(X) = X$; a lift is evidently unique up to the addition of vertical vectors. Each lift H_x of $T_{\tau(x)}Q$ to $x \in \mathsf{P}$ satisfies $H_x\mathsf{P} \oplus V_x\mathsf{P} = T_x\mathsf{P}$, but there is no canonical choice of such a complement to $V_x\mathsf{P}$.

Definition 2.2.1. *A* **connection** *on a principal H-bundle is a smooth assignment $x \mapsto H_x\mathsf{P} \subset T_x\mathsf{P}$ such that $H_x\mathsf{P} \oplus V_x\mathsf{P} = T_x\mathsf{P}$ and*

$$H_{R_h(x)}\mathsf{P} = (R_h)_* H_x\mathsf{P}. \tag{2.11}$$

Elements of H_x are called **horizontal vectors**, and each H_x is called the **horizontal subspace** of $T_x\mathsf{P}$. The collection of all H_x is the **horizontal subbundle** $H\mathsf{P}$ of $T\mathsf{P}$. The **horizontal lift** $\ell_x(X)$ of $X \in T_{\tau(x)}Q$ to $T_x\mathsf{P}$ is the unique vector in $H_x\mathsf{P}$ satisfying $\tau_*(\ell_x(X)) = X$. From (2.11) we infer

$$\ell_{R_h(x)}(X) = (R_h)_*\ell_x(X). \tag{2.12}$$

Similarly, a horizontal lift $\ell(q(\cdot))$ of a curve $q(\cdot)$ in Q is a curve $x(\cdot)$ in P for which $\tau(x(t)) = q(t)$ and $\dot{x}(t) \in H_{x(t)}\mathsf{P}$ for all t. Such a lift is unique if one specifies $x = x(0)$ (at which $\tau(x) = q(0)$). The **parallel transport** of $x \in \mathsf{P}$ to $\tau^{-1}(q(t))$ along a C^1 curve $q(\cdot)$ in Q (with $\tau(x) = q(0)$) is by definition $x(t)$, where $x(\cdot)$ is the horizontal lift of $q(\cdot)$ through x.

These notions may be transferred to any bundle M associated to P. The horizontal lift of a C^1 curve $q(\cdot)$ in Q through $[x, m]_H \in \mathsf{M}$ (with $\tau(x) = q(0)$) is the curve $[x(\cdot), m]_H$, where $x(\cdot)$ is the horizontal lift of $q(\cdot)$ through x. Similarly, the parallel transport of $[x, m]_H$ to $\tau_{\mathsf{M}\to Q}^{-1}(q(t))$ is the point

$$\mathsf{p}_{q\to q(t)}[x, m]_H := [x(t), m]_H, \tag{2.13}$$

where $x(t)$ is as defined above.

Proposition 2.2.2. *There is a bijective correspondence between connections on* P *and smooth sections* $\mathbf{A}$ *of* $\Lambda^1(\mathsf{P}) \otimes \mathfrak{h}$ *(i.e., fields of* $\mathfrak{h}$*-valued 1-forms on* P*), called* **connection 1-forms,** *satisfying*

$$\mathbf{A}(\xi_X^f) = X, \tag{2.14}$$

$$R_h^*\mathbf{A} = \mathrm{Ad}(h)\mathbf{A} \tag{2.15}$$

for all $X \in \mathfrak{h}$ *and* $h \in H$.

Given the H_x, one defines $\mathbf{A}$ by (2.14) and $\mathbf{A}_x(X) = 0$ for all $X \in H_x\mathsf{P}$; equation (2.15) follows from (2.14), (2.10), and (2.11). Given $\mathbf{A}$, one defines $H_x\mathsf{P}$ as the subspace of $T_x\mathsf{P}$ annihilated by $\mathbf{A}_x$; (2.11) follows from (2.15). ∎

In a local trivialization $\psi : \mathsf{P} \to Q \times H$ (locally) associated to a section $s : Q \to \mathsf{P}$ (locally; cf. (2.3)) we can write

$$(\psi^{-1})^*\mathbf{A}(q, h) = \mathrm{Ad}(h^{-1})s^*\mathbf{A}(q) + \theta_L^{MC}(h), \tag{2.16}$$

where $s^*\mathbf{A} \in \Lambda^1(Q) \otimes \mathfrak{h}$ and θ_L^{MC} is defined in (1.41). This expression is enforced by 2.1.3, (2.14), (2.15), and (1.53). Connections on a trivial principal bundle $Q \times H$ are, then, necessarily of the above form.

A connection 1-form $\mathbf{A}$ determines a projection $\tau_{\mathrm{v}} : T\mathsf{P} \to V\mathsf{P}$ (mapping $T_x\mathsf{P}$ onto $V_x\mathsf{P}$) by

$$\tau_{\mathrm{v}}(X) = \xi_{\mathbf{A}(X)}^f. \tag{2.17}$$

The complementary projection $\tau_{\mathrm{h}} : T\mathsf{P} \to H\mathsf{P}$ is given by

$$\tau_{\mathrm{h}}(X) = X - \xi_{\mathbf{A}(X)}^f. \tag{2.18}$$

The **curvature** of $\mathbf{A}$ is an $\mathfrak{h}$-valued 2-form $\mathbf{F}$ on P, defined by

$$\mathbf{F}(X, Y) := d\mathbf{A}(\tau_{\mathrm{h}}(X), \tau_{\mathrm{h}}(Y)). \tag{2.19}$$

Some rearrangements lead to the expression

$$\mathbf{F} = d\mathbf{A} + [\mathbf{A}, \mathbf{A}]. \tag{2.20}$$

Writing $s^*\mathbf{A} = T_i A^i_\mu dq^\mu$ and $s^*\mathbf{F} = T_i F^i_{\mu\nu} dq^\mu dq^\nu$ (in terms of coordinates q^μ on Q and a basis $\{T_i\}$ of $\mathfrak{h}$, omitting explicit reference to the section s) one therefore has the physicists' formula

$$F^i_{\mu\nu} = \partial_\mu A^i_\nu - \partial_\nu A^i_\mu + C^i_{jk} A^j_\mu A^k_\nu. \tag{2.21}$$

A most important aspect of a connection on P is that it defines a certain first-order differential operator on all vector bundles associated to P. We write $\mathsf{V} := \mathsf{P} \times_H V$ (defined relative to some H-action L on V). Recall that the space of all compactly supported smooth sections of V is denoted by $\Gamma(\mathsf{V})$.

As in II.3.1.1, a **covariant derivative** on a vector bundle $\mathsf{V}(Q, V, \tau)$ is a linear association $\xi \mapsto \nabla_\xi$, where $\xi \in \Gamma(TQ)$ and $\nabla_\xi : \Gamma(\mathsf{V}) \to \Gamma(\mathsf{V})$ satisfies $\nabla_{f\xi} = f\nabla_\xi$ and $\nabla_\xi(f\Psi) = \xi(f)\Psi + f\nabla_\xi\Psi$ for all $f \in C^\infty(Q)$ and $\Psi \in \Gamma(\mathsf{V})$. One may look at a covariant derivative as a map $\nabla : \Gamma(\mathsf{V}) \to \Gamma(\wedge^1(Q) \otimes \mathsf{V})$, so that $\nabla_\xi\Psi = \nabla\Psi(\xi)$.

We can (locally) choose a **moving frame**, that is, a collection of sections $s_i : Q \to \mathsf{V}$, such that $\{s_i(q)\}_{i=1,\ldots,\dim(V)}$ is a basis of $\tau^{-1}(q)$ for all q (in some open subset of Q). A function $\psi : Q \to \mathbb{R}^{\dim(V)}$ then defines a section $\Psi : q \mapsto \psi^i s_i(q)$ of V. With obvious abuse of notation we may then write

$$\nabla\psi^i = d\psi^i + A^i_j\psi^j, \tag{2.22}$$

where A is a matrix-valued 1-form on Q, defined by the property $\nabla s_i = A^j_i s_j$; it evidently depends on the moving frame. This dependence is controlled by the transformation property

$$\tilde{A} = MAM^{-1} + MdM^{-1}, \tag{2.23}$$

where $\tilde{A}$ is the 1-form determined by a moving frame $\tilde{s}_i(q) := s_j(q)M^{-1}(q)^j_i$, where $M : Q \to GL(\mathbb{R}^{\dim(V)})$ (locally). This property, which is immediate from the definition of A, guarantees that $\nabla\tilde{\psi} = M\nabla\psi$, where $\tilde{\psi} = M\psi$ defines the same section Ψ as ψ does, but in terms of the frame $\tilde{s}$.

By slight abuse of terminology (cf. the case of a principal bundle) one often refers to A as a connection, and instead of saying that there is a covariant derivative on V one says that V has a connection.

Proposition 2.2.3. *A connection (with associated 1-form* $\mathbf{A}$*) on* $\mathsf{P}(Q, H, \tau)$ *defines a covariant derivative* $\nabla^{\mathbf{A}}$ *on any vector bundle* $\mathsf{V}(Q, H, \tau)$ *associated to* P *by*

$$\nabla^{\mathbf{A}}_\xi \Psi^{(L)}(q) := \lim_{t\to 0}\left[\mathsf{p}_{q(t)\to q(0)}\Psi^{(L)}(q(t)) - \Psi^{(L)}(q)\right]/t, \tag{2.24}$$

when the limit exists; here $q(\cdot)$ *is a curve through* q *with tangent vector* ξ_q*, and the limit is independent of the specific choice of the curve.*

In terms of the realization Ψ^L *(cf. 2.1.5) one then has (with abuse of notation)*

$$\nabla^{\mathbf{A}}_\xi \Psi^L(x) := \ell_x(\xi_{\tau(x)})\Psi^L(x). \tag{2.25}$$

In terms of the realization Ψ_s^L *the covariant derivative is*

$$\nabla_\xi^A \Psi_s^L = [\xi + dL(A(\xi))]\Psi_s^L, \tag{2.26}$$

where $A := s^*\mathbf{A}$.

Here $dL(X) := dL(\mathrm{Exp}(tX))/dt|t = 0$ for $X \in \mathfrak{h}$; hence $dL(s^*\mathbf{A}) \in \Lambda^1(Q) \otimes dL(\mathfrak{h})$ (locally). Note that the right-hand side of (2.24) is well-defined as the difference between two elements of a vector space, in that both terms in the numerator lie in the same fiber $\tau_{\mathsf{V}\to Q}^{-1}(q)$ because of the parallel transport operation (2.13) involved. The properties of a covariant derivative then easily follow from the ordinary Leibniz rule.

We take the horizontal lift $x(\cdot)$ of $q(\cdot)$ in V through x, and note that (2.13) implies that

$$\mathsf{p}_{q(t)\to q(0)}\left[x(t), \Psi^L(x(t))\right]_H = \left[x, \Psi^L(x(t))\right]_H, \tag{2.27}$$

from which the equivalence of (2.24) and (2.25) is immediate. Note that (2.25) is well-defined in that $\nabla_\xi^{\mathbf{A}}\Psi^L(ph^{-1}) = L(h)\nabla_\xi^{\mathbf{A}}\Psi^L(p)$ because of (2.12).

To derive (2.26), which is obviously a special case of (2.22), we notice that $\nabla_\xi^A\Psi_s^L(q) = \ell_{s(q)}(X)\Psi^L(s(q))$, use (2.18) to write $\ell_{s(q)}(X) = s_*(X) - \xi^f_{A(s_*(X))}$, and then use the definition of ξ^f and the equivariance of Ψ^L. ■

It follows most easily from (2.26) and a well-known identity for the exterior derivative d that in a given trivialization the curvature $s^*\mathbf{F}$ of $\mathbf{A}$ is related to the covariant derivative by (cf. II.(3.20))

$$dL(s^*\mathbf{F})(\xi, \eta) = \left[\nabla_\xi^A, \nabla_\eta^A\right] - \nabla_{[\xi,\eta]}^A. \tag{2.28}$$

In complete analogy to the special case of an affine connection (see II.3.1), given a covariant derivative ∇ on some vector bundle $\mathsf{V}(Q, V, \tau)$ one may define the horizontal lift ℓ of a vector or of a curve. In terms of the matrix-valued 1-form A appearing in (2.22) and the identification $\tau^{-1}(q) \simeq \mathbb{R}^{\dim(V)}$ given by a moving frame (that is, the components v^i are defined by $v =: v^i s_i(q)$), one has $\ell_v(X) = (-v^j A^i_j(X), X)$.

The following construction will not be used until 3.10, but logically fits in at this point. We follow the notation of Definition 2.1.2, except that the general bundle B is now a vector bundle V.

Proposition 2.2.4. *Let* $f^*\mathsf{V}$ *be the vector bundle defined by (2.2). Then a covariant derivative on* V *pulls back to a covariant derivative on* $f^*\mathsf{V}$.

Sections of $f^*\mathsf{V}$ have the form $\Psi(x) = (\Psi_1(x), x)$, where $\Psi_1(x) \in \tau^{-1}(f(x))$. Choose a (local) moving frame $\{s_i\}$ on V, with associated connection A, and define $\psi_1 : M \to \mathbb{R}^{\dim(V)}$ (locally) by $\Psi_1(x) = \psi_1^i(x)s_i(f(x))$. The desired covariant derivative is then given by $\nabla\Psi(x) = (\nabla\Psi_1(x), x)$, where

$$\nabla\Psi_1(x) = \left[d\psi_1^i(x) + f^*A^i_j(x)\psi_1^j(x)\right] s_i(f(x)). \tag{2.29}$$

This is well-defined (cf. (2.22) and subsequent text): A change of moving frame and corresponding change in A and ψ_1 does not affect the total expression, because of properties like $f^*M = M \circ f$. ■

The corresponding horizontal lift ℓ in $f^*\mathsf{V}$ is given by

$$\ell_{(v,x)}(X) = \ell_v(f_* X)\dot{+}X, \tag{2.30}$$

seen as an element of $T_v\mathsf{V} \oplus T_x M \supset T_{(v,x)}(\mathsf{V} *_Q M)$; one immediately sees that (2.30) is indeed tangent to $\mathsf{V} *_Q M$.

2.3 Cotangent Bundle Reduction

In preparation for the main theorem of this section we will first look at connections more specifically from the point of view of the cotangent bundle $T^*\mathsf{P}$ and the momentum map. We start with a more general statement.

Lemma 2.3.1. *The pullback of an H-action on a manifold M to the cotangent bundle T^*M is strongly Hamiltonian, with momentum map $J : T^*M \to \mathfrak{h}^*_-$ given by*

$$J_X(\sigma) = \sigma(\xi_X). \tag{2.31}$$

Note that in order to obtain a (left) action on T^*M one has to put $h \mapsto R^*_{h^{-1}}$, if the H-action on M is denoted by R. The claim most easily follows from I.(2.24), (1.19), and (1.7) in canonical coordinates, in which

$$J_X(p,q) = p_i \xi^i_X(q). \tag{2.32}$$

Here ξ^i_X is defined by $\xi_X(q) := \xi^i_X(q)\partial/\partial q^i$. ■

The annihilator of $V\mathsf{P}$ (cf. (2.8)) in $T^*\mathsf{P}$ is

$$V^0\mathsf{P} := \{\sigma \in T^*\mathsf{P} \,|\, \sigma(X) = 0 \;\forall X \in V\mathsf{P}\}. \tag{2.33}$$

Elements of $V^0\mathsf{P}$ are called **horizontal 1-forms**, and $V^0_x\mathsf{P}$ is the **horizontal cotangent space** at x. Let R^* be the pullback of the H-action on P to $T^*\mathsf{P}$ (if we speak of "the H-action R^* on $T^*\mathsf{P}$" we mean the left action in question, i.e., we silently incorporate the shift $h \mapsto h^{-1}$, as in the comment preceding (2.32)). By the above lemma this action is strongly Hamiltonian, with momentum map $J : T^*\mathsf{P} \to \mathfrak{h}^*_-$. From (2.9) and (2.31) we conclude that

$$V^0\mathsf{P} = J^{-1}(0). \tag{2.34}$$

The projection dual to $\tau_{\mathrm{h}} : T\mathsf{P} \to H\mathsf{P}$ is $\tau^*_{\mathrm{h}} : T^*\mathsf{P} \to V^0\mathsf{P}$ (that is, $\tau^*_{\mathrm{h}}(\omega) = \omega \circ \tau_{\mathrm{h}}$). From (2.18), (2.31), (2.14), and (2.9) we see that

$$\tau^*_{\mathrm{h}}(\omega) = \omega + J(\omega) \circ \mathbf{A}; \tag{2.35}$$

here $J(\omega) \in \mathfrak{h}^*$ hits the $\mathfrak{h}$-part of $\mathbf{A} \in \Lambda^1(\mathsf{P}) \otimes \mathfrak{h}$.

In a (local) trivialization we have $T^*\mathsf{P} \simeq T^*Q \times T^*H$ (locally). We then put $T^*H \simeq \mathfrak{h}^* \times H$ in either the left or the right trivialization (see 1.4), and choose coordinates θ_i ($i = 1, \ldots, \dim(H)$) on $\mathfrak{h}^*$ such that $\theta = \theta_i \omega^i$ in terms of a basis $\{\omega^i\}$ of $\mathfrak{h}^*$ dual to a basis $\{T_i\}$ of $\mathfrak{h}$. Choosing also canonical coordinates (p_μ, q^μ) on T^*Q (locally) we have the quadruple $(p, q, h, \theta)_{L,R}$ representing $p_\mu dq^\mu + \theta_i \theta^i_{L,R}(h)$. From 2.1.3, (1.49), (1.51), (1.55), and (1.57) we obtain

$$R^*_{k^{-1}}(p, q, \theta, h)_L = (p, q, \mathrm{Co}(k)\theta, hk^{-1})_L, \tag{2.36}$$

$$J(p, q, \theta, h)_L = -\theta, \tag{2.37}$$

$$R^*_{k^{-1}}(p, q, \theta, h)_R = (p, q, \theta, hk^{-1})_R, \tag{2.38}$$

$$J(p, q, \theta, h)_R = -\mathrm{Co}(h^{-1})\theta. \tag{2.39}$$

Hence from (2.35), (2.16), (2.36) etc., and (1.42) one derives the coordinate expression of τ^*_{h}:

$$\tau^*_{\mathrm{h}}(p_\mu, q^\mu, \theta_i, h)_L = (p_\mu - \mathrm{Ad}(h^{-1})^k_j A^j_\mu(q)\theta_k, q^\mu, 0, h)_L; \tag{2.40}$$

$$\tau^*_{\mathrm{h}}(p_\mu, q^\mu, \theta_i, h)_R = (p_\mu - A^j_\mu(q)\theta_j, q^\mu, 0, h)_R. \tag{2.41}$$

With (2.37) and (2.39) this confirms (2.34). In this trivialization the canonical Poisson bracket on $T^*\mathsf{P}$ (cf. I.(2.24)) reads simply

$$\{f, g\} = \{f, g\}^{T^*Q} + \{f, g\}^{T^*H}, \tag{2.42}$$

where the first term is given by I.(2.24) and the second by (1.54) (with $\Gamma = 0$ and G replaced by H).

The aim of the following considerations is to factorize $\mathfrak{h}^*$ from $T^*\mathsf{P}$ in an intrinsic way, so as to facilitate the study of the momentum map. Recall Definition 2.1.2.

Definition 2.3.2. *Let P be a principal H-bundle over Q. Then H acts on $\mathsf{P}_H := \mathsf{P} *_Q TQ$ and $\mathsf{P}^0_V := \mathsf{P} *_Q T^*Q$ through its action on P (combined with the trivial action on $T^{(*)}Q$); e.g., on P^0_V one has $h : (x, \sigma) \mapsto (xh^{-1}, \sigma)$. With this action, and projection onto the second factor, $\mathsf{P}_H = \tau^*_{TQ\to Q}\mathsf{P}$ and $\mathsf{P}^0_V = \tau^*_{T^*Q\to Q}\mathsf{P}$ are principal H-bundles over TQ and T^*Q, respectively.*

Lemma 2.3.3. *Regarding all spaces in question as bundles over P with the obvious projections, there are natural H-equivariant isomorphisms between*

- *$\mathsf{P} \times \mathfrak{h}$ and $V\mathsf{P}$, where H acts on the former by $\rho^A := R \times \mathrm{Ad}$ and on the latter by R_*;*
- *P^0_V and $V^0\mathsf{P}$, where H acts on the former as specified in 2.3.2 and on the latter by $(R^*)^{-1}$.*

The first isomorphism is given by $(x, X) \leftrightarrow \xi^f_X(x)$. The desired equivariance follows from (2.10). Given $(x, \sigma) \in \mathsf{P}^0_V$, the map $\tau^*_x : T^*_{\tau(x)}Q \to T^*_x\mathsf{P}$ produces an element $\tau^*_x\sigma \in V^0_x\mathsf{P}$. Given $\omega \in V^0_x\mathsf{P}$ one defines $\sigma_\omega \in T^*_{\tau(x)}Q$ by

$$\sigma_\omega(X) := \omega(l_x(X)); \tag{2.43}$$

the nonuniqueness of the lift does not matter, since ω annihilates vertical vectors. One easily checks that these maps are diffeomorphisms, and are each other's inverse. The H-equivariance follows from the property $\tau \circ R_h = \tau$. ∎

Further to 2.3.3, one has

Lemma 2.3.4. *For any given connection, $H\mathsf{P}$ is H-equivariantly isomorphic to P_H as a bundle over P.*

The isomorphism is obtained by letting $(x, v) \in H\mathsf{P}$ correspond to $\ell_x(v)$. The equivariance follows from $(R_h)_* \ell_x(v) = \ell_{R_h(x)}(v)$ (see (2.11)). ∎

Let H act on $\mathsf{P} \times \mathfrak{h}^*$ by $\rho^C := R \times \mathrm{Co}$; that is,

$$\rho_h^C(x, \sigma, \theta) = (xh^{-1}, \sigma, \mathrm{Co}(h)\theta). \tag{2.44}$$

Proposition 2.3.5. *A connection on P leads to H-equivariant isomorphisms*

$$T\mathsf{P} \simeq \mathsf{P}_H \oplus (\mathsf{P} \times \mathfrak{h}); \tag{2.45}$$

$$T^*\mathsf{P} \simeq \mathsf{P}_V^0 \oplus (\mathsf{P} \times \mathfrak{h}^*). \tag{2.46}$$

Here all objects are seen as vector bundles over P, on which the associated diffeomorphism is trivial.

All claims are immediate from 2.3.2, 2.3.3, and 2.3.4. ∎

Corollary 2.3.6. *Regarding the objects involved as manifolds (rather than bundles), (2.46) is an H-equivariant diffeomorphism*

$$T^*\mathsf{P} \simeq \mathsf{P}_V^0 \times \mathfrak{h}^*, \tag{2.47}$$

and similarly for (2.45). The momentum map of the H-action (2.44) on $\mathsf{P}_V^0 \times \mathfrak{h}^*$ is

$$J(x, \sigma, \theta) = -\theta. \tag{2.48}$$

To prove (2.48) one adopts a local trivialization, compares (2.44) with (2.36), and uses (2.37). ∎

We shall need the explicit form of the isomorphism $\psi_{\mathbf{A}} : T^*\mathsf{P} \to \mathsf{P}_V^0 \times \mathfrak{h}^*$, which follows from (2.35); one has

$$\psi_{\mathbf{A}}(\omega) = (\tau_{T^*\mathsf{P}\to\mathsf{P}}(\omega), \sigma_{\omega+J(\omega)\circ\mathbf{A}}, -J(\omega)), \tag{2.49}$$

with $\sigma_{\ldots}$ given by (2.43). The inverse is

$$\psi_{\mathbf{A}}^{-1}(x, \sigma, \theta) = \tau^*_{T_x\mathsf{P}\to TQ}(\sigma) + \theta \circ \mathbf{A}_x. \tag{2.50}$$

The canonical Poisson structure I.(2.24) is invariant under the H-action on $T^*\mathsf{P}$ (any diffeomorphism of a manifold M pulled back to T^*M is a Poisson map). Consequently, one obtains a Poisson structure on $(T^*\mathsf{P})/H$. The associated Poisson algebra is the classical algebra of observables of a particle moving on $Q = \mathsf{P}/H$ with all possible "classical" charges $\mathcal{O} \subset \mathfrak{h}^*$. There is a natural isomorphism between $C^\infty((T^*\mathsf{P})/H)$ and $C^\infty(T^*\mathsf{P})^H$, the space of smooth functions on $T^*\mathsf{P}$ that are invariant under the H-action R^*.

Theorem 2.3.7. *Each symplectic leaf of the Poisson manifold $(T^*\mathsf{P})/H$ (inheriting the canonical Poisson structure of $T^*\mathsf{P}$) is of the form*

$$(\widetilde{T^*\mathsf{P}})^{\mathcal{O}} := J^{-1}(\mathcal{O})/H, \tag{2.51}$$

where $\mathcal{O} \subset \mathfrak{h}^*$ is a coadjoint orbit of H.

Any connection on P *leads to a diffeomorphism*

$$(T^*\mathsf{P})/H \simeq \mathsf{P}^0_V \times_H \mathfrak{h}^*, \tag{2.52}$$

seen as the vector bundle over T^*Q associated to the principal H-bundle P^0_V (cf. 2.3.2) by the coadjoint representation on $\mathfrak{h}^*$. Consequently, for each $\mathcal{O}$ and a given connection $\mathbf{A}$ one obtains a diffeomorphism

$$\psi^{\mathcal{O}}_{\mathbf{A}} : (\widetilde{T^*\mathsf{P}})^{\mathcal{O}} \to \mathsf{P}^0_V \times_H \mathcal{O}, \tag{2.53}$$

where $(\widetilde{T^*\mathsf{P}})^{\mathcal{O}}$ is a bundle over T^*Q (with typical fiber $\mathcal{O}$) associated to P^0_V.

The first claim is clear from (2.38), (2.39), 1.4.4, and (2.42). The second and third follow from the first and 2.3.6. The map $\psi^{\mathcal{O}}_{\mathbf{A}}$ is simply the reduction of $\psi_{\mathbf{A}}$ in (2.49); that is, the latter quotients to a map $[\psi_{\mathbf{A}}]_H$ from $(T^*\mathsf{P})/H$ to $\mathsf{P}^0_V \times_H$, which provides the diffeomorphism in (2.52), and $\psi^{\mathcal{O}}_{\mathbf{A}}$ is then the restriction of $[\psi_{\mathbf{A}}]_H$ to the symplectic leaf $T^*\mathsf{P}^{\mathcal{O}}$ in $(T^*\mathsf{P})/H$. ■

The tilde on $(\widetilde{T^*\mathsf{P}})^{\mathcal{O}}$ signifies that the quotient in (2.51) is taken with the entire group H, rather than with its identity component H^0; cf. IV.1.6.

From Corollary I.2.6.10 we now have

Corollary 2.3.8. *Up to equivalence, each irreducible representation of the Poisson algebra $C^\infty((T^*\mathsf{P})/H, \mathbb{R})$ (in the sense of I.2.6.6) is realized on a symplectic manifold of the type $(\widetilde{T^*\mathsf{P}})^{\mathcal{O}}$, or, equivalently, of the type $\mathsf{P}^0_V \times_H \mathcal{O}$ (or a covering space thereof).*

The space $\mathsf{P}^{\mathcal{O}}_V \times_H \mathfrak{h}^*$ becomes a Poisson manifold by declaring the diffeomorphism (2.52) to be a Poisson map. Similarly:

Definition 2.3.9. *The symplectic space $\mathsf{P}^0_V \times_H \mathcal{O}$, equipped with the Poisson structure that it inherits as a symplectic leaf of $(\widetilde{T^*\mathsf{P}})^{\mathcal{O}}$, is denoted by $(\widetilde{T^*\mathsf{P}})^{\mathcal{O}}_{\mathbf{A}}$.*

Equivalently, the Poisson structure on $(\widetilde{T^*\mathsf{P}})^{\mathcal{O}}_{\mathbf{A}}$ could be defined by stipulating that the diffeomorphism $\psi^{\mathcal{O}}_{\mathbf{A}}$ in (2.53) be a Poisson map.

The corresponding Poisson bracket is easily computed in a local trivialization of the associated bundles in question corresponding to a local trivialization of P (see (2.7)). Both $(T^*\mathsf{P})/H$ and $\mathsf{P}^0_V \times_H \mathfrak{h}^*$ trivialize to $T^*Q \times \mathfrak{h}^*$ (locally), and from (2.49) with either (2.37) and (2.40), or (2.39) and (2.41) (the difference between these expressions fades when passing to the quotient by H and taking local trivializations), we infer that locally the diffeomorphism $\tilde{\psi}_A : (T^*\mathsf{P})/H \to \mathsf{P}^0_V \times_H \mathfrak{h}^*$ is given by

$$\tilde{\psi}_A(p, q, \theta) = (p_\mu - A^j_\mu(q)\theta_j, q^\mu, \theta_i). \tag{2.54}$$

It follows from (2.42) and 1.4.2 that the Poisson bracket on $\mathsf{P}^0_V \times_H \mathfrak{h}^*$ is

$$\{f, g\}^{\mathcal{O}}_* = \frac{\partial f}{\partial p_\mu}\frac{\partial g}{\partial q^\mu} - \frac{\partial f}{\partial q^\mu}\frac{\partial g}{\partial p_\mu} - \theta_i F^i_{\mu\nu}(q)\frac{\partial f}{\partial p_\mu}\frac{\partial g}{\partial p_\nu}$$

$$+ C^k_{ij}\theta_k \left(-\frac{\partial f}{\partial \theta_i}\frac{\partial g}{\partial \theta_j} + A^i_\mu(q)\frac{\partial f}{\partial p_\mu}\frac{\partial g}{\partial \theta_j} + A^j_\mu \frac{\partial f}{\partial \theta_i}\frac{\partial g}{\partial p_\mu} \right); \tag{2.55}$$

for brevity we have omitted the argument (p, q, θ) on both sides. One notices that this is the sum of the canonical Poisson brackets on T^*Q and $\mathfrak{h}^*_-$ and terms in which the connection $\mathbf{A}$ and its curvature $\mathbf{F}$ (see (2.21)) enter.

As an immediate corollary one infers that

$$J^{-1}(0)/H \simeq T^*Q \tag{2.56}$$

as symplectic manifolds, where T^*Q is equipped with the canonical symplectic structure I.2.23.

2.4 Bundle Automorphisms and the Gauge Group

We have seen in II.3.3 that when H is trivial the group $\mathcal{G}_Q$ defined by II.(3.12) plays an important role in the construction of observables on T^*Q. In particular, the map J in II.3.1.4 is the momentum map for the action of $\mathcal{G}_Q$ on T^*Q defined in II.3.1.5 (this observation is a special case of the results below).

If H is nontrivial and P is a principal H-bundle over Q, neither $\mathrm{Diff}(Q) \ltimes C^\infty_c(Q, \mathbb{R})$ nor $\mathrm{Diff}(\mathsf{P}) \ltimes C^\infty_c(\mathsf{P}, \mathbb{R})$ is sensitive to the bundle structure. We now prepare ourselves for the specification of the correct group.

Definition 2.4.1. *The group* $\mathrm{Aut}(\mathsf{P})$ *is the group of smooth bundle isomorphisms of* P *with compact support on* Q. *In other words,* $\mathrm{Aut}(\mathsf{P})$ *is the restriction of* $\mathrm{Diff}(\mathsf{P})$ *to those* C^∞ *diffeomorphisms* φ *that satisfy*

$$R_h \circ \varphi = \varphi \circ R_h \tag{2.57}$$

for all $h \in H$, *and for which the projection (to* Q *under* τ*) of the set where* φ *differs from the identity is compact.*

The **gauge group** $\mathrm{Gau}(\mathsf{P}) \subset \mathrm{Aut}(\mathsf{P})$ *consists of those* $\varphi \in \mathrm{Aut}(\mathsf{P})$ *satisfying* $\tau \circ \varphi = \tau$.

Proposition 2.4.2. *The space* $\mathrm{Aut}(\mathsf{P})$ *is isomorphic to* $\Gamma(\mathsf{P} \times_H \mathsf{P})$, *that is, to the collection of sections of the bundle* $\mathsf{P} \times_H \mathsf{P}$ *associated to* P *by the given* H*-action on* P. *Here the group operation in* $\Gamma(\mathsf{P} \times_H \mathsf{P})$ *(realized as* H*-equivariant maps* $\varphi^{\mathrm{Ad}} : \mathsf{P} \to \mathsf{P}$*) is* $\varphi_1^{\mathrm{Ad}}\varphi_2^{\mathrm{Ad}} = \varphi_1^{\mathrm{Ad}} \circ \varphi_2^{\mathrm{Ad}}$.

The gauge group $\mathrm{Gau}(\mathsf{P})$ *is isomorphic to* $\Gamma(\mathsf{P} \times_H H)$, *where the associated* **adjoint bundle** $\mathsf{P} \times_H H$ *is defined with respect to the adjoint action of* H *on itself. The group operation in* $\Gamma(\mathsf{P} \times_H H)$ *(realized as* H*-equivariant maps* $g := g^{\mathrm{Ad}} : \mathsf{P} \to H$*) is pointwise multiplication.*

Hence for trivial $\mathsf{P} = Q \times H$ *the gauge group is isomorphic to* $C^\infty_c(Q, H)$ *(with pointwise multiplication as the group operation).*

The first claim is immediate from Proposition 2.1.5. Define a map $g : \mathsf{P} \to H$ by the property

$$\varphi_g(x) = R_{g(x)^{-1}}(x) = xg(x). \tag{2.58}$$

Since by definition one has

$$g(xh^{-1}) = \mathrm{Ad}(h)g(x) = hg(x)h^{-1}, \tag{2.59}$$

the second claim follows. ■

The relation between Aut(P) and Diff(Q) is described by the exact sequence of groups

$$1 \to \mathrm{Gau}(\mathsf{P}) \to \mathrm{Aut}(\mathsf{P}) \to \mathrm{Diff}(Q) \to 1; \tag{2.60}$$

that is, the image of each homomorphism is the kernel of the next one. The second arrow is given by inclusion, and the third is the map $\varphi \mapsto \varphi_Q$, where φ_Q is the element of Diff(Q) defined by φ through the bundle projection τ; that is,

$$\tau \circ \varphi = \varphi_Q \circ \tau. \tag{2.61}$$

As in II.3.3 we say, in a somewhat loose sense, that the Lie algebra $\mathfrak{diff}(\mathsf{P})$ of Diff(P) is the collection $\Gamma(T\mathsf{P})$ of smooth compactly supported vector fields on P. Since Aut(P) and Gau(P) are, equally loosely, Lie subgroups of Diff(P), we can discuss their respective Lie algebras. In preparation, note that the space $(T\mathsf{P})/H$ is a bundle over Q, with projection inherited from $\tau_{T\mathsf{P}\to\mathsf{P}\to Q}$. The space $\Gamma(T\mathsf{P})^H$ consists of all H-invariant vector fields ξ on P for which $\tau(\mathrm{supp}\,(\xi))$ is compact. It is a Lie algebra under (minus) the commutator borrowed from $\Gamma(T\mathsf{P})$.

Proposition 2.4.3. *One may identify $\Gamma((T\mathsf{P})/H)$ and $\Gamma(T\mathsf{P})^H$, upon which the Lie algebra $\mathfrak{aut}(\mathsf{P})$ of* Aut(P) *is isomorphic to $\Gamma((T\mathsf{P})/H)$.*

The Lie algebra $\mathfrak{gau}(\mathsf{P})$ of Gau(P) *of the gauge group is isomorphic to $\Gamma(\mathsf{P}\times_H \mathfrak{h})$, where $\mathsf{P}\times_H \mathfrak{h}$ is the vector bundle (over Q) associated to* P *by the adjoint representation of H on $\mathfrak{h}$, and the Lie bracket on $\Gamma(\mathsf{P}\times_H \mathfrak{h})$ is the pointwise bracket in $\mathfrak{h}$.*

Hence for trivial $\mathsf{P} = Q \times H$ the Lie algebra of the gauge group is isomorphic to $C_c^\infty(Q, \mathfrak{h})$ (with pointwise Lie bracket).

Since an element $\xi^{\mathsf{P}} \in \mathfrak{aut}(\mathsf{P})$ by definition satisfies

$$(R_h)_*\xi^{\mathsf{P}} = \xi^{\mathsf{P}}, \tag{2.62}$$

it is clear that $\mathfrak{aut}(\mathsf{P}) = \Gamma(T\mathsf{P})^H$. With $[X]_H$ denoting the equivalence class of $X \in T\mathsf{P}$ in $(T\mathsf{P})/H$, the map $\Gamma(T\mathsf{P})^H \ni \xi \mapsto \tilde{\xi} \in \Gamma((T\mathsf{P})/H)$ defined by $\tilde{\xi}(q) = [\xi(s(q))]_H$ is therefore independent of the section $s : Q \to \mathsf{P}$. Conversely, one puts $\xi(x) = \tau^{-1}_{T\mathsf{P}\to\mathsf{P}}(x) \cap \tau^{-1}_{T\mathsf{P}\to(T\mathsf{P})/H}(\tilde{\xi}(\tau_{\mathsf{P}\to Q}(x)))$; this intersection consists of one point, since the H-action is free. These two maps provide a bijection between $\Gamma((T\mathsf{P})/H)$ and $\mathfrak{aut}(\mathsf{P})$.

The Lie algebra $\mathfrak{gau}(\mathsf{P})$ of Gau(P) comprises all vertical H-invariant vector fields on P; the second claim therefore follows from Lemma 2.3.3. Alternatively, it is obvious from Proposition 2.4.2. ■

The relationship between $\mathsf{P}\times_H \mathsf{P}$ and $(T\mathsf{P})/H$ will be elucidated in 3.8.8.

For later use, we record that elements $\lambda := \lambda^{\mathrm{Ad}} \in \Gamma(\mathsf{P} \times_H \mathfrak{h})$ of $\mathfrak{gau}(\mathsf{P})$ satisfy (cf. (2.59))

$$\lambda(xh^{-1}) = \mathrm{Ad}(h)\lambda(x). \tag{2.63}$$

The diagram (2.60) infinitesimalizes to an exact sequence of Lie algebras

$$0 \to \mathfrak{gau}(\mathsf{P}) \to \mathfrak{aut}(\mathsf{P}) \to \mathfrak{diff}(Q) \to 0, \tag{2.64}$$

where we recall that $\mathfrak{diff}(Q) = \Gamma(TQ)$ (with minus the commutator as its Lie bracket). The corresponding exact sequence of vector bundles (all over Q) is

$$0 \to \mathsf{P} \times_H \mathfrak{h} \to (T\mathsf{P})/H \to TQ \to 0; \tag{2.65}$$

taking sections, we recover (2.64). A connection $\mathbf{A}$ on P is then equivalent to a splitting of the sequence (2.64) (in the sense of a map from $\mathfrak{diff}(Q)$ to $\mathfrak{aut}(\mathsf{P})$ that is a left inverse to the arrow in the opposite direction), since $\xi \in \mathfrak{diff}(Q)$ has a horizontal lift $\ell(\xi)$, which lies in $\mathfrak{aut}(\mathsf{P})$ because of (2.11).

As shown in (2.3), a (local) section $s : Q \to \mathsf{P}$ is equivalent to a (local) trivialization $\psi_s : \mathsf{P} \to Q \times H$. In a fixed such trivialization ψ_s, a gauge transformation $\varphi_g : \mathsf{P} \to \mathsf{P}$ then induces a (local) diffeomorphism $g^s : Q \times H \to Q \times H$ by requiring that $g^s \circ \psi_s = \psi_s \circ g$ ("active picture"). This yields

$$g^s(q, h) = (q, g_s(q)h), \tag{2.66}$$

where g_s is related to g as in 2.1.5; that is,

$$g_s(q) = g(s(q)). \tag{2.67}$$

In the "passive picture" the gauge transformation φ_g defines a new section s_g by the property $\psi_{s_g} = \psi_s \circ \varphi_g^{-1}$. Using (2.59) this gives (cf. (2.4))

$$s_g(q) = s(q)g_s(q). \tag{2.68}$$

It easily follows from (2.14), (2.15), and (2.57) that $\varphi \in \mathrm{Aut}(\mathsf{P})$ has the property that the pullback $(\varphi^{-1})^*\mathbf{A}$ is a connection 1-form if $\mathbf{A}$ is. This particularly applies to $\varphi_g \in \mathrm{Gau}(\mathsf{P})$. It is interesting to compute the action of $\mathrm{Gau}(\mathsf{P})$ in a local trivialization given by a section s.

Proposition 2.4.4. *For $\varphi_g \in \mathrm{Gau}(\mathsf{P})$ (cf. 2.4.2) one has*

$$s^*\mathbf{A}^g(q) = \mathrm{Ad}(g_s(q))s^*\mathbf{A}(q) + g_s(q)dg_s^{-1}(q), \tag{2.69}$$

where $\mathbf{A}^g := (\varphi_g^{-1})^*\mathbf{A}$. The right-hand side of (2.69) describes both the value of $(\varphi_g^{-1})^*\mathbf{A}$ in the fixed trivialization ψ_s defined by s (active picture), and the value of $\mathbf{A}$ in the trivialization defined by the transformed section $s_{g^{-1}}$ (passive picture).

Alternatively, if two sections s_α and s_β are related by (2.4), and $A_\alpha := s_\alpha^*\mathbf{A}$ etc., then, writing $g_{\beta\alpha} = g_{\alpha\beta}^{-1}$, one has

$$A_\beta(q) = \mathrm{Ad}(g_{\beta\alpha}(q))A_\alpha(q) + g_{\beta\alpha}(q)dg_{\alpha\beta}(q). \tag{2.70}$$

The last claim follows from $s^* \circ (\varphi_g^{-1})^* = s_{g^{-1}}^*$, as is immediate from (2.68).

To derive (2.69), which should be compared with (2.23), one evaluates the left-hand side on $X = dq(t)/dt|t = 0$; this yields $\mathbf{A}(d/dt\, s(q(t))g_s^{-1}(s(q(t))|t = 0)$, where we used (2.67). The differentiation d/dt firstly hits $s(q(t))$; one uses (2.15) to obtain the first term on the right-hand side of (2.69). It secondly hits $g_s^{-1}(\ldots)$; this time one inserts $g_s(q)^{-1}g_s(q)$ after $s(q)$, and uses (2.14) to find the second term. ∎

The second term in (2.69) equals $(g_s^{-1})^*\theta_L^{MC}(q)$, and is, of course, also equal to $-dg_s(q)g_s^{-1}(q)$. Suppressing the dependence of $g(q)$ on the section s, physicists write (2.69) as

$$A_\mu^g(q) = g(q)A_\mu(q)g(q)^{-1} + g(q)\partial_\mu g(q)^{-1}. \tag{2.71}$$

In any case, the second term drops out in the transformation of the curvature: From (2.20) and (2.69), or (2.21) and (2.71), one infers that

$$s^*\mathbf{F}^g(q) := s^* \circ (\varphi_g^{-1})^*\mathbf{F}(q) = \mathrm{Ad}(g_s(q))s^*\mathbf{F}(q). \tag{2.72}$$

2.5 *Construction of Classical Observables*

In this section we construct a complete set of classical observables on a symplectic leaf $(\widetilde{T^*\mathsf{P}})^{\mathcal{O}}$ of $(T^*\mathsf{P})/H$ in terms of the momentum map of a certain group action. In preparation, consider the following general construction (which includes (2.78) below as a special case).

Definition 2.5.1. *Let L_l be an H-action on M, denoting the bundle associated to* P *through this action by* M^l *(here l is some label). Then* $\mathrm{Aut}(\mathsf{P})$ *acts on* M^l *by*

$$\mathsf{L}^l(\varphi) : [x, m]_H \mapsto [\varphi(x), m]_H. \tag{2.73}$$

This action is well-defined on account of (2.57).

Returning to the main theme, our basic group is the semidirect product

$$\mathcal{G}_{\mathsf{P}}^{\tau} := \mathrm{Aut}(\mathsf{P}) \ltimes C_c^\infty(Q, \mathbb{R}), \tag{2.74}$$

which is defined through the action $\varphi : \tilde{g} \mapsto (\varphi_Q^{-1})^*\tilde{g}$, where $\varphi \in \mathrm{Aut}(\mathsf{P})$ and $\tilde{g} \in C_c^\infty(Q, \mathbb{R})$, and φ_Q is the diffeomorphism of Q defined by φ; see (2.61).

Theorem 2.5.2. *Consider the $\mathcal{G}_{\mathsf{P}}^{\tau}$-action on $T^*\mathsf{P}$ defined by*

$$\rho_0(\tilde{g}) : \omega \mapsto \omega - d\tau^*\tilde{g}(\omega); \tag{2.75}$$

$$\rho_0(\varphi) : \omega \mapsto (\varphi^{-1})^*\omega, \tag{2.76}$$

and $\rho_0(\varphi, \tilde{g}) := \rho_0(\tilde{g}) \circ \rho_0(\varphi)$, where $\tilde{g} \in C_c^\infty(Q, \mathbb{R})$ and $\varphi \in$ $\mathrm{Aut}(\mathsf{P})$. *This action is strongly Hamiltonian and commutes with the H-action R^*, so that there is a reduced strongly Hamiltonian action $\rho_0^{\mathcal{O}}$ on $(\widetilde{T^*\mathsf{P}})^{\mathcal{O}}$.*

Recall (2.53). The actions $\tilde{\rho}_0^{\mathcal{O}}(\tilde{g}) := \psi_{\mathbf{A}}^{\mathcal{O}} \circ \rho_0(\tilde{g})$ and $\tilde{\rho}_0^{\mathcal{O}}(\varphi) := \psi_{\mathbf{A}^\varphi} \circ \rho_0(\varphi)$, where $\mathbf{A}^\varphi := (\varphi^{-1})^*\mathbf{A}$, on $\mathsf{P}_{\mathsf{V}}^0 \times_H \mathcal{O}$ are then given by

$$\tilde{\rho}_0^{\mathcal{O}}(\tilde{g}) : [x, \sigma, \theta]_H \mapsto [x, \sigma - d\tilde{g}, \theta]_H; \tag{2.77}$$

$$\tilde{\rho}_0^{\mathcal{O}}(\varphi) : [x, \sigma, \theta]_H \mapsto [\varphi(x), (\varphi_Q^{-1})^*\sigma, \theta]_H. \tag{2.78}$$

In particular, for a gauge transformation φ_g *one simply has*

$$\tilde{\rho}_0^{\mathcal{O}}(\varphi_g) : [x, \sigma, \theta]_H \mapsto [x, \sigma, \mathrm{Co}(g(x))\theta]_H, \tag{2.79}$$

where $g \in \Gamma(\mathsf{P} \times_H H)$ *is related to* φ_g *by (2.58).*

Finally, each $\tilde{\rho}_0^{\mathcal{O}}(\varphi) : (\widetilde{T^*\mathsf{P}})^{\mathcal{O}}_{\mathbf{A}} \mapsto (\widetilde{T^*\mathsf{P}})^{\mathcal{O}}_{\mathbf{A}^\varphi}$ *is a Poisson map; cf. 2.3.9.*

Here $[x, \sigma, \theta]_H \in \mathsf{P}^0_V \times_H \mathcal{O}$ is the H-equivalence class of $(x, \sigma, \theta) \in \mathsf{P}^0_V \times \mathcal{O}$. Note that the reduced actions $\rho_0^{\mathcal{O}}(\varphi)$ and $\rho_0^{\mathcal{O}}(g)$ are well-defined because of (2.57) and (2.59), respectively. Equation (2.77) is obvious. To derive (2.78) one uses the property $\varphi_* \xi_X^f = \xi_X^f$, which follows from (2.57). Specializing (2.78) to a gauge transformation, the associated diffeomorphism φ_Q of Q is the identity, so that φ_g maps (x, σ, θ) into $(xg(x), \sigma, \theta)$; cf. (2.58). Since $[xh, \sigma, \theta]_H = [x, \sigma, \mathrm{Co}(h)\theta]_H$ by definition of the H-equivalence class in question, (2.79) follows.

The fact that the group action on $T^*\mathsf{P}$ is strongly Hamiltonian follows from 2.3.1, and from the existence of an equivariant momentum map: For $\mathrm{Aut}(\mathsf{P})$ one has (2.31), whereas for the $C_c^\infty(Q, \mathbb{R})$-action on $T^*\mathsf{P}$ one has

$$J_{\tilde{g}}(\omega) = \tau^*_{T^*\mathsf{P} \to Q} \tilde{g}(\omega). \tag{2.80}$$

Compare with II.(3.7).

Since the action ρ_0 on $T^*\mathsf{P}$ commutes with the H-action on this space, the momentum map J for ρ_0 reduces to a well-defined map $J^{\mathcal{O}}$ on the reduced space $(\widetilde{T^*\mathsf{P}})^{\mathcal{O}}$. Because of the definition of the Poisson structure and the reduced group action $\rho_0^{\mathcal{O}}$ on $(\widetilde{T^*\mathsf{P}})^{\mathcal{O}}$, the map $J^{\mathcal{O}}$ is an equivariant momentum map for $\rho_0^{\mathcal{O}}$. This shows that $\rho_0^{\mathcal{O}}$ is strongly Hamiltonian.

The last claim is obvious from 2.3.9 and the rest of the theorem. ∎

Let $\psi : \mathsf{P} \to Q \times H$ be a (local) trivialization of P; see the text preceding (2.36). A local expression for the action $\rho_0^{\mathcal{O}}(\varphi_g) : (\widetilde{T^*\mathsf{P}})^{\mathcal{O}} \to (\widetilde{T^*\mathsf{P}})^{\mathcal{O}}$ of the gauge group may be derived from (2.66). Pulling this action back to $T^*\mathsf{P}$ and subsequently to $(\widetilde{T^*\mathsf{P}})^{\mathcal{O}} \simeq T^*Q \times \mathcal{O}$ (locally), one obtains

$$\rho_0^{\mathcal{O}}(\varphi_g) : (\sigma_q, \theta)_s \mapsto (\sigma_q + \theta(dg_s^{-1}(q)g_s(q)), \mathrm{Co}(g_s(q))\theta)_s. \tag{2.81}$$

Here $\sigma_q \in T_q^*Q$, and $dg_s^{-1}(q)g_s(q) \in \mathfrak{h}$; cf. the proof of 2.4.4. Also, we have explicitly indicated the dependence on the section s. In other words, the point $(p_\mu, q^\mu, \theta_i)_s$ is mapped to $(p_\mu + (\partial_\mu g(q)^{-1} g_s(q))^i \theta_i, q, \mathrm{Co}(g_s(q))_i^j \theta_j)_s$.

Relative to a fixed section s, a p-form f^A on $T^*Q \times \mathcal{O}$ (perhaps defined only locally), also depending on the connection $A = s^*\mathbf{A}$, is said to be **gauge-covariant** when

$$\rho_0^{\mathcal{O}}(\varphi_g)^* f^{A^g} = f^A. \tag{2.82}$$

This property, sacred in physics, states the fact that f is an expression of a function on $(\widetilde{T^*\mathsf{P}})^{\mathcal{O}}$, depending on the connection $\mathbf{A}$, in a local trivialization. A case in point is the **covariant momentum**, which is a 1-form defined by

$$p^A(\sigma_q, \theta)_s := \sigma_q - \theta(A(q)). \tag{2.83}$$

In coordinates this reads $p^A_\mu(p,q,\theta) = p_\mu - \theta_i A^i_\mu(q)$. It then follows from (2.81) and (2.69) (or (2.71)) that this function is indeed gauge-covariant. More generally, an expression of the type $f \circ p^A$ is gauge-covariant.

Similarly, we could compare (local) expressions for a function on $(\widetilde{T^*\mathsf{P}})^{\mathcal{O}}$ in two (local) trivializations whose (local) sections are related by (2.4). With slight abuse of notation we define

$$\rho_0^{\mathcal{O}}(g_{\beta\alpha}) : \big(\sigma_q, \theta\big) \mapsto (\sigma_q + \theta(dg_{\alpha\beta}(q)g_{\beta\alpha}(q)), \mathrm{Co}(g_{\beta\alpha}(q))\theta\big). \tag{2.84}$$

With the relation (2.70), the condition for gauge-covariance is

$$f^{A_\beta} \circ \rho_0^{\mathcal{O}}(g_{\beta\alpha}) = f^{A_\alpha}. \tag{2.85}$$

It follows from (2.54) and (2.82), or from direct calculation, that the local expression for $\tilde{\rho}_0^{\mathcal{O}}(\varphi_g) : (\widetilde{T^*\mathsf{P}})^{\mathcal{O}}_{\mathbf{A}} \to (\widetilde{T^*\mathsf{P}})^{\mathcal{O}}_{\mathbf{A}^g}$, cf. (2.79) and (2.69), is simply

$$(\tilde{\rho}_0^{\mathcal{O}}(\varphi_g))(\sigma_q, \theta)_s^{\mathbf{A}} = (\sigma, \mathrm{Co}(g_s(q))\theta)_s^{\mathbf{A}^g}, \tag{2.86}$$

where the explicit dependence on the connection has been displayed. In other words, the momentum on $\mathsf{P}_V^0 \times_H \mathcal{O}$ is gauge-covariant. The fact that $\tilde{\rho}_0^{\mathcal{O}}(\varphi_g)$ is a Poisson map may then be verified from I.(2.15), (2.86), and (2.55).

A (local) trivialization ψ_s, corresponding to a (local) section s of P (cf. (2.3)), induces a (local) trivialization $(\widetilde{T^*\mathsf{P}})^{\mathcal{O}} \simeq T^*Q \times \mathcal{O}$, whose inherited Poisson structure is simply the sum of the canonical bracket I.(2.24) on T^*Q and minus the Lie–Poisson bracket (1.3) on $\mathcal{O}$; that is,

$$\{f,g\}^{\mathcal{O}} = \frac{\partial f}{\partial p_\mu}\frac{\partial g}{\partial q^\mu} - \frac{\partial f}{\partial q^\mu}\frac{\partial g}{\partial p_\mu} - C^k_{ij}\theta_k \frac{\partial f}{\partial \theta_i}\frac{\partial g}{\partial \theta_j}. \tag{2.87}$$

Theorem 2.5.3. *Relative to an arbitrary connection* $\mathbf{A}$ *on* P *and a (local) section* s*, the equivariant momentum map* $J^{\mathcal{O}}$ *of the reduced* $\mathcal{G}_{\mathsf{P}}^{\tau}$*-action* $\rho_0^{\mathcal{O}}$ *on* $(\widetilde{T^*\mathsf{P}})^{\mathcal{O}}$ *(cf. 2.5.2) is as follows. With slight abuse of notation we simply write* $J^{\mathcal{O}}$ *for* $(\psi_s^{-1})^* J^{\mathcal{O}}$*, putting the* s*-dependence into the argument of* $J^{\mathcal{O}}$*.*

For the $C_c^\infty(Q,\mathbb{R})$*-action (2.75) on* $(\widetilde{T^*\mathsf{P}})^{\mathcal{O}}$ *one has*

$$J^{\mathcal{O}}_{\tilde{g}}(\sigma_q, \theta)_s = \tilde{g}(q). \tag{2.88}$$

For general $\xi \in \mathfrak{aut}(\mathsf{P})$*, identified with a vector field* ξ^{P} *on* P*, with (2.83),* $\xi^Q := \tau_* \xi^{\mathsf{P}}$*, and* $A := s^*\mathbf{A}$*, one has*

$$J^{\mathcal{O}}_{\xi^{\mathsf{P}}}(\sigma_q, \theta)_s = p^A_{(\sigma_q,\theta)}(\xi^Q) + \theta \circ \mathbf{A}_{s(q)}(\xi^{\mathsf{P}}). \tag{2.89}$$

For pure gauge transformations this specializes to

$$J^{\mathcal{O}}_{\lambda}(\sigma_q, \theta)_s = \theta(\lambda_s(q)), \tag{2.90}$$

where $\lambda \in \Gamma(\mathsf{P} \times_H \mathfrak{h})$ *and* $\lambda_s := \lambda \circ s$*; cf. 2.4.3 and 2.1.5.*

Finally, for the horizontal lift of some vector field ξ^Q *on* Q *one has*

$$J^{\mathcal{O}}_{\ell(\xi^Q)}(\sigma_q, \theta)_s = p^A_{(\sigma_q,\theta)}(\xi^Q). \tag{2.91}$$

The equivariance of $J^{\mathcal{O}}$ is immediate from 2.5.2, and may be verified by explicit computation using the above formulae. The expression (2.80) for the momentum map of the $C_c^\infty(Q, \mathbb{R})$-action on $T^*\mathsf{P}$ entails (2.88). To derive the remaining expressions we decompose

$$\xi^{\mathsf{P}} = \ell(\xi^Q) + \xi^f_{\mathbf{A}(\xi^{\mathsf{P}})}, \tag{2.92}$$

where we recall that $\xi^Q = \tau_* \xi^{\mathsf{P}}$, and $\ell(\xi^Q) = s_* \xi^Q - \xi^f_{A(\xi^Q)}$. This follows from the identity $\xi^{\mathsf{P}} = s_* \xi^Q + (\xi^{\mathsf{P}} - s_* \xi^Q)$, in which the second term is vertical, and (2.17). It is obvious from (1.19) and (2.9) that $(\psi_s)_* \xi^f_X = \xi^L_X$; with (2.31), this leads to (2.90). Also, the property $(\psi_s \circ s)_* \xi^Q = \xi^Q$, combined with the previous equation and (2.31), leads to (2.91). Equation (2.89) then follows from (2.92), (2.91), and (2.90). ■

One verifies that $J^{\mathcal{O}}$ is gauge-covariant in the sense of (2.82); in case the section s explicitly occurs, according to Proposition 2.4.4 one should interpret the symbol A^g in (2.82) by substituting $s_{g^{-1}}$ for s; cf. (2.68). For example, the invariance of the last term in (2.89) may be checked using (2.15) and (2.62). Similarly, the gauge-invariance of (2.90) follows from (2.68) and (2.63).

Since the map $\tilde{\rho}_0^{\mathcal{O}}(\varphi)$ in (2.78) maps $(\widetilde{T^*\mathsf{P}})^{\mathcal{O}}_{\mathbf{A}}$ to $(\widetilde{T^*\mathsf{P}})^{\mathcal{O}}_{\mathbf{A}^\varphi}$, which has a different Poisson structure, there is no concept of a momentum map unless $\mathbf{A}^\varphi = \mathbf{A}$ (i.e., the connection is invariant under $(\varphi^{-1})^*$). In that case the momentum map $\tilde{J}^{\mathcal{O}}$ is given by

$$\tilde{J}^{\mathcal{O}}_{\xi^{\mathsf{P}}}([x, \sigma_{\tau(x)}, \theta]_H) = \sigma_{\tau(x)}(\xi^Q) + \theta \circ \mathbf{A}_x(\xi^{\mathsf{P}}), \tag{2.93}$$

which is well-defined, and easily follows from (2.89) and (2.50).

The momentum map for $\tilde{\rho}_0^{\mathcal{O}}(\tilde{g})$ always exists, and, analogously to (2.88), is given by

$$\tilde{J}^{\mathcal{O}}_{\tilde{g}}([x, \sigma_{\tau(x)}, \theta]_H) = \tilde{g}(\tau(x)). \tag{2.94}$$

2.6 *The Classical Wong Equations*

We turn to the Hamiltonian. In the spirit of II.3.3 we assume that there is a Riemannian metric $\mathbf{g}_Q$ on Q. On each symplectic space $(\widetilde{T^*\mathsf{P}})^{\mathcal{O}}_{\mathbf{A}} = \mathsf{P}^0_V \times_H \mathcal{O}$ one then has a natural Hamiltonian

$$h^{\mathcal{O}}([x, \sigma, \theta]_H) := \tfrac{1}{2}\mathbf{g}_Q^{-1}(\sigma, \sigma); \tag{2.95}$$

cf. II.(3.30). In coordinates this is simply

$$h^{\mathcal{O}}(p, q, \theta) = \tfrac{1}{2}\mathbf{g}_Q^{\mu\nu}(q) p_\mu p_\nu, \tag{2.96}$$

where $\mathbf{g}_Q^{\mu\nu} = (\mathbf{g}^Q_{\mu\nu})^{-1}$. Since (2.95) explicitly depends neither on $\theta \in \mathcal{O}$ nor on the connection $\mathbf{A}$, the Hamiltonian $h^{\mathcal{O}}$ is evidently invariant under the gauge transformation (2.79) or (2.86). Since, by Theorem 2.5.2, the diffeomorphism $\tilde{\rho}_0^{\mathcal{O}}(\varphi_g)$ is a Poisson map from $(\widetilde{T^*\mathsf{P}})^{\mathcal{O}}_{\mathbf{A}}$ to $(\widetilde{T^*\mathsf{P}})^{\mathcal{O}}_{\mathbf{A}^g}$, it follows from I.2.3.5 and the gauge-invariance of $h^{\mathcal{O}}$ that the Hamiltonian flow on $(\widetilde{T^*\mathsf{P}})^{\mathcal{O}}_{\mathbf{A}^g}$ is the image under $\tilde{\rho}_0^{\mathcal{O}}(\varphi_g)$ of the Hamiltonian flow on $(\widetilde{T^*\mathsf{P}})^{\mathcal{O}}_{\mathbf{A}}$.

From the Poisson bracket (2.55) one then obtains the equation of motion

$$\dot{q}^\mu = \mathbf{g}_Q^{\mu\nu}(q)p_\nu; \tag{2.97}$$

$$\dot{p}_\mu = -\tfrac{1}{2}\partial_\mu \mathbf{g}_Q^{\rho\sigma}(q)p_\rho p_\sigma + F^i_{\mu\nu}(q)p^\nu\theta_i; \tag{2.98}$$

$$\dot{\theta}_i = -C^k_{ij}\theta_k p^\mu A^j_\mu(q). \tag{2.99}$$

These are known as the **Wong equations**, describing the motion of a "colored" particle in the external gauge field $\mathbf{A}$. The word "color" refers to the classical charge, represented by the coadjoint orbit $\mathcal{O}$. The equations have a simple geometric interpretation. Firstly, the projection of the motion to T^*Q (cf. the remark following (2.53)) is cogeodesic motion distorted by the **Lorentz force**, that is, by the last term in (2.98). Secondly, the motion $\theta(t)$ in the fiber $\mathcal{O}$ is given by parallel transport in the bundle $\mathsf{P} \times_H \mathcal{O}$ associated to P by the coadjoint action of H on $\mathcal{O}$. Indeed, looked at as an equation in the vector bundle $\mathsf{P}^0_V \times_H \mathfrak{h}^*$, with covariant derivative $\nabla^{\mathbf{A}}$ defined by the connection $\mathbf{A}$ (see 2.2.3), equation (2.99) combined with (2.97) reads $\nabla^{\mathbf{A}}_{\dot{q}}\theta = 0$.

We will show that the Hamiltonian flow on each leaf is the reduction of a single Hamiltonian flow on $T^*\mathsf{P}$.

Proposition 2.6.1. *There is a bijective correspondence between H-invariant Riemannian metrics $\mathbf{g}$ on P satisfying*

$$\mathbf{g}_x(\xi^f_{\mathrm{Ad}(h)X}, \xi^f_{\mathrm{Ad}(h)Y}) = \mathbf{g}_x(\xi^f_X, \xi^f_Y) \tag{2.100}$$

for all $x \in \mathsf{P}$ and $X, Y \in \mathfrak{h}$, and triples $(\mathbf{g}^Q, \{\mathbf{g}^H_q\}_{q\in Q}, \mathbf{A})$, where $\mathbf{g}^Q$ is a Riemannian metric on Q, each $\mathbf{g}^H_q$ is a bi-invariant Riemannian metric on H (the dependence on q being smooth in that $\mathbf{g}^H_q(X, Y) \in C^\infty(Q)$ for all $X, Y \in T_eH = \mathfrak{h}$), and $\mathbf{A}$ is a connection on $\mathsf{P}(Q, H, \tau)$.

Given $\mathbf{g}$, for each $q \in Q$ one defines a bilinear form on $T_eH = \mathfrak{h}$ by

$$\mathbf{g}^H_q(X, Y) := \mathbf{g}_{s(q)}(\xi^f_X, \xi^f_Y); \tag{2.101}$$

by (2.10) and the right invariance of $\mathbf{g}$ this is independent of the section s. Since $\mathbf{g}^H_q$ is Ad-invariant by (2.100), one subsequently obtains a bi-invariant metric (with the same name) on H by left or right translation. The smooth dependence on q is immediate from the smoothness of $\mathbf{g}$.

A connection on P is constructed by defining $H_x\mathsf{P} \subset T_x\mathsf{P}$ as the orthogonal complement of $V_x\mathsf{P}$; condition (2.11) is satisfied because $\mathbf{g}$ is H-invariant. Equivalently, the connection 1-form $\mathbf{A}$ may be directly constructed as

$$\mathbf{A}_x = (\mathbf{g}^H_x)^\sharp \circ J \circ \mathbf{g}_\sharp, \tag{2.102}$$

where $\mathbf{g}_\sharp : T\mathsf{P} \to T^*\mathsf{P}$ is defined below II.(3.15), J is the momentum map for the H-action on $T^*\mathsf{P}$, and $(\mathbf{g}^H_x)^\sharp : \mathfrak{h}^* \to \mathfrak{h}$ is obtained from $\mathbf{g}^H_{\tau(x)}$ in the usual way. Equation (2.14) is then satisfied because of (2.31), whereas (2.15) is a consequence of the equivariance of J.

Finally, the metric $\mathbf{g}_Q$ is constructed from $\mathbf{g}$ by $\mathbf{g}_Q(X,Y) := \mathbf{g}(\ell(X), \ell(Y))$, where the point to which one lifts X, Y horizontally is immaterial in view of the H-invariance of $\mathbf{g}$. In other words, $\mathbf{g}_Q$ is the unique metric on Q for which horizontal lifting is an isometry.

In the opposite direction one defines $\mathbf{g}$ on $V\mathsf{P}$ by reading (2.101) from right to left, declares $H\mathsf{P}$ to be orthogonal to $V\mathsf{P}$, and manufactures the metric on $H\mathsf{P}$ as in the previous paragraph. ■

In a trivialization $\mathsf{P} \simeq Q \times H$ (locally) the correspondence is

$$\begin{aligned} \mathbf{g}^H_{ij} &= \mathbf{g}_{ij}; \\ A^i_\mu &= \mathbf{g}^{ij}_H \mathbf{g}_{\mu j}; \\ \mathbf{g}^Q_{\mu\nu} &= \mathbf{g}_{\mu\nu} - \mathbf{g}^{ij}_H \mathbf{g}_{\mu j} \mathbf{g}_{\nu i}, \end{aligned} \tag{2.103}$$

where $\mathbf{g}^{ij}_H$ is the inverse of $\mathbf{g}^H_{ij}$ (which may differ from $\mathbf{g}^{ij}$). All expressions depend on $q \in Q$ but not on $h \in H$. These comments equally well apply to the expressions below. In the opposite direction we obtain

$$\begin{aligned} \mathbf{g}_{ij} &= \mathbf{g}^H_{ij}; \\ \mathbf{g}_{i\mu} &= \mathbf{g}_{\mu i} = \mathbf{g}^H_{ij} A^j_\mu; \\ \mathbf{g}_{\mu\nu} &= \mathbf{g}^Q_{\mu\nu} + \mathbf{g}^H_{ij} A^i_\mu A^j_\nu. \end{aligned} \tag{2.104}$$

In preparation for the following theorem, we mention the obvious fact that any H-invariant function f on $T^*\mathsf{P}$ is is well-defined on $(T^*\mathsf{P})/H$, and therefore defines a reduced function $f^{\mathcal{O}}$ on each leaf $(\widetilde{T^*\mathsf{P}})^{\mathcal{O}}$ by restriction.

Theorem 2.6.2. *Let the equivalent data in 2.6.1 satisfy the condition that $\mathbf{g}^H_q$ be independent of q. Then the Hamiltonian h_* on $T^*\mathsf{P}$, defined by II.(3.30) through the metric $\mathbf{g}$ on P, reduces to a function $h^{\mathcal{O}}_*$ on each leaf $\mathsf{P}^0_V \times_H \mathcal{O} \simeq (\widetilde{T^*\mathsf{P}})^{\mathcal{O}}$, which differs from $h^{\mathcal{O}}$ in (2.95) by a constant. In other words, the equations of motion of $h^{\mathcal{O}}_*$ are the Wong equations (2.97)–(2.99).*

This is most quickly established in local coordinates; inverting $\mathbf{g}$ using (2.104), one obtains

$$h_*(p,q,\theta,h)_R = \tfrac{1}{2}\mathbf{g}^{\mu\nu}_Q(q)(p_\mu - \theta_i A^i_\mu(q))(p_\nu - \theta_j A^j_\nu(q)) + \mathbf{g}^{ij}_H(q)\theta_i\theta_j. \tag{2.105}$$

Using the inverse of (2.54), the reduced Hamiltonian on $\mathsf{P}^0_V \times_H \mathcal{O}$ is

$$h^{\mathcal{O}}_*(p,q,\theta) = \tfrac{1}{2}\mathbf{g}^{\mu\nu}_Q(q)p_\mu p_\nu + \mathbf{g}^{ij}_H(q)\theta_i\theta_j. \tag{2.106}$$

When $\mathbf{g}^{ij}_H(q)$ is independent of q, one computes from (2.55) that the last term in (2.105) Poisson-commutes with every function on $\mathsf{P}^0_V \times_H \mathcal{O}$. This computation exploits the fact that $C^i_{jk}\mathbf{g}^H_{il}$ is totally antisymmetric, which is a restatement of the Ad-invariance of $\mathbf{g}^H$. Since $\mathsf{P}^0_V \times_H \mathcal{O}$ is a symplectic space, this means that the term in question must be a constant. ■

Rather than on $(\widetilde{T^*\mathsf{P}})^{\mathcal{O}}_{\mathbf{A}}$ one can work on $(\widetilde{T^*\mathsf{P}})^{\mathcal{O}}$, equipped with the Poisson bracket (2.87); cf. (2.53). This is more natural when the bundle P is trivial, i.e.,

$\mathsf{P} \simeq Q \times H$, for in that case one has $(\widetilde{T^*\mathsf{P}})^{\mathcal{O}} \simeq T^*Q \times \mathcal{O}$. Also, formulating the dynamics on $(\widetilde{T^*\mathsf{P}})^{\mathcal{O}}$ is better suited for studying the relation between the classical and the quantum theory; see 2.11 below. The following considerations equally well apply to a local trivialization of a possibly nontrivial bundle P.

Instead of (2.106), which generates the Hamiltonian flow on $(\widetilde{T^*\mathsf{P}})^{\mathcal{O}}$ as transformed by the diffeomorphism (2.54), and relative to the Poisson bracket (2.55), one now uses the Hamiltonian $h_A^{\mathcal{O}} := h^{\mathcal{O}} \circ \psi_{\mathbf{A}}^{\mathcal{O}}$. By (2.96), the coordinate expression of $h_A^{\mathcal{O}}$ is

$$h_A^{\mathcal{O}}(p, q, \theta) = \tfrac{1}{2}\mathbf{g}_Q^{\mu\nu}(q)(p_\mu - \theta_i A_\mu^i(q))(p_\nu - \theta_j A_\nu^j(q)). \tag{2.107}$$

This may equally well be obtained from (2.105), omitting the last term (which does not contribute to the equations of motion).

The comment following (2.83) evidently applies; one has

$$h_{A^g}^{\mathcal{O}} \circ \rho_0^{\mathcal{O}}(\varphi_g) = h_A^{\mathcal{O}}, \tag{2.108}$$

so that the gauge transformation $\rho_0^{\mathcal{O}}(\varphi_g)$ maps the Hamiltonian flow on $(\widetilde{T^*\mathsf{P}})^{\mathcal{O}}$ generated by $h_A^{\mathcal{O}}$ into the Hamiltonian flow generated by $h_{A^g}^{\mathcal{O}}$.

From (2.107) and (2.87) one deduces the Hamiltonian equations

$$\dot{q}^\mu = \mathbf{g}_Q^{\mu\nu}(q)p_\nu^A; \tag{2.109}$$

$$\dot{p}_\mu = -\tfrac{1}{2}\partial_\mu \mathbf{g}_Q^{\rho\sigma}(q)p_\rho^A p_\sigma^A + \mathbf{g}_Q^{\rho\sigma}(q)p_\rho^A \partial_\mu A_\sigma^i(q)\theta_i; \tag{2.110}$$

$$\dot{\theta}_i = -C_{ij}^k \theta_k \mathbf{g}_Q^{\rho\sigma}(q)p_\rho^A A_\sigma^j(q), \tag{2.111}$$

where $p_\mu^A := p_\mu - \theta_i A_\mu^i(q)$, as before. A different form of (2.111) is obtained by transferring the motion on $\mathcal{O}$ to a certain flow $h(\cdot)$ on H; this correspondence will be used in the proof of Theorem 2.11.1. We assume that $\mathcal{O} = \mathcal{O}_\theta$ is the coadjoint orbit through θ, and note that $\dot{h}h^{-1}$ and the connection form A are elements of the Lie algebra $\mathfrak{h}$.

Proposition 2.6.3. *Let the flow $h(\cdot)$ in H be the solution of*

$$\dot{h}h^{-1} = -\mathbf{g}_Q^{\rho\sigma}(q)(p_\rho - \theta(\mathrm{Ad}(h^{-1})A_\rho(q))A_\sigma(q), \tag{2.112}$$

with initial condition $h(0) = h$. If $\theta(t)$ and $h(t)$ are related by $\theta(t) = \mathrm{Co}(h(t))\theta$, so that

$$\theta_i(t) = \theta(\mathrm{Ad}(h(t)^{-1}T_i), \tag{2.113}$$

then $\theta_i(t)$ solves (2.111) if $h(t)$ solves (2.112).

One evaluates both sides of (2.112) in the coadjoint representation, acts on θ, and uses (1.20). ■

2.7 The H-Connection

Let us illustrate the preceding concepts in the case that the principal bundle P is $G(G/H, H, \tau)$, where G is a Lie group having H as a closed subgroup (hence H

is automatically a Lie group), τ is the canonical projection from G to G/H, and the H-action on G is given by

$$R_h(x) := xh^{-1}. \tag{2.114}$$

Many nontrivial bundles are obtained in this way. The simplest and most famous one will be studied in 2.12.

The restriction to H of the adjoint representation $\mathrm{Ad}(G)$ on $\mathfrak{g}$ quotients to a representation $\mathrm{Is}(H)$ on $\mathfrak{g}/\mathfrak{h}$, called the **isotropic representation**. This leads to the vector bundle $G \times_H (\mathfrak{g}/\mathfrak{h})$ over G/H associated with $G(G/H, H, \tau)$. Let $\mathfrak{h}^0 \subset \mathfrak{g}^*$ be the annihilator of $\mathfrak{h} \subset \mathfrak{g}$; clearly, $\mathfrak{h}^0 \simeq (\mathfrak{g}/\mathfrak{h})^*$. The **coisotropic representation** $\mathrm{Ci}(H)$ on $\mathfrak{h}^0$ is the restriction of the representation $\mathrm{Co}(G)$ on $\mathfrak{g}^*$ to H. This defines the associated vector bundle $G \times_H \mathfrak{h}^0$.

Lemma 2.7.1. *There are diffeomorphisms (which are bundle isomorphisms whose associated diffeomorphism of G/H is the identity)*

$$T(G/H) \simeq G \times_H (\mathfrak{g}/\mathfrak{h}); \tag{2.115}$$
$$T^*(G/H) \simeq G \times_H \mathfrak{h}^0. \tag{2.116}$$

Furthermore, there is an isomorphism

$$G_V^0 \simeq G \times \mathfrak{h}^0 \tag{2.117}$$

as principal H-bundles over $T^*(G/H)$. Here the H-action on $G \times \mathfrak{h}^0$ is given by $R_h(x, \theta) = (xh^{-1}, \mathrm{Ci}(h)\theta)$, which defines $G \times \mathfrak{h}^0$ as a principal H-bundle over $T^*(G/H)$ through the isomorphism (2.116).

Recall that G_V^0 is defined by 2.3.2 (with P replaced by G). The canonical left action L of G on G/H pushes forward to a G-action L_* on $T(G/H)$. One identifies $T_{[e]_H}(G/H)$ with $\mathfrak{g}/\mathfrak{h}$ in the obvious way. Under the isomorphism (2.115) the point $[x, Y]_H \in G \times_H (\mathfrak{g}/\mathfrak{h})$ then corresponds to $(L_x)_* Y \in T(G/H)$. One verifies the independence of the representative (x, Y).

Similarly, $T^*_{[e]_H}(G/H) \simeq \mathfrak{h}^0$, and $[x, \theta]_H \in G \times_H \mathfrak{h}^0$ corresponds to $L^*_{x^{-1}}\theta \in T^*(G/H)$.

Finally, (2.117) is obtained by letting $(x, [y, \theta]_H) \in G_V^0$ correspond to the point $(x, \mathrm{Is}(x^{-1}y)\theta) \in G \times \mathfrak{h}^0$; note that $x^{-1}y \in H$ by the definition of G_V^0, and that this is evidently independent of the choice of $(y, \theta) \sim (yh^{-1}, \mathrm{Is}(h)\theta)$. ■

Recall (2.51). Using the right trivialization of T^*G (see 1.4), and noticing that the momentum map for the H-action on T^*G (from the right) is simply the restriction of the momentum map of the G-action to $\mathfrak{h} \subset \mathfrak{g}$, we see from (1.51) and (1.57) that

$$(\widetilde{T^*G})^{\mathcal{O}} = \{(\theta, [x]_H) \in \mathfrak{g}^* \times G/H \mid (\mathrm{Co}(x^{-1})\theta) \restriction \mathfrak{h} \in -\mathcal{O}\}. \tag{2.118}$$

Here $[x]_H := xH$; since $\mathcal{O}$ is stable under $\mathrm{Co}(H)$, the right-hand side is independent of the choice of x.

On the other hand, in the left trivialization of T^*G (1.55) and (1.49) yield

$$(\widetilde{T^*G})^{\mathcal{O}} = \{[\theta, x]_H\, \theta \restriction \mathfrak{h} \in -\mathcal{O}\}, \tag{2.119}$$

where the H-equivalence classes are the orbits of the H-action given by

$$h : (\theta, x) \mapsto \big(\mathrm{Co}(h)\theta, xh^{-1}\big). \tag{2.120}$$

We note that G acts on itself by left multiplication; this action evidently commutes with the H-action (2.114), allowing us to regard G as a subgroup of the group $\mathrm{Aut}(G)$ of all bundle automorphisms of $G(G/H, H, \tau)$. According to (1.52), in the right trivialization the reduced G-action $\lambda^{\mathcal{O}}$ on $(\widetilde{T^*G})^{\mathcal{O}}$ is

$$\lambda_x^{\mathcal{O}}(\theta, [y]_H) = (\mathrm{Co}(x)\theta, [xy]_H), \tag{2.121}$$

whereas in the left trivialization, (1.50) yields

$$\lambda_x^{\mathcal{O}}([\theta, y]_H) = [\theta, xy]_H. \tag{2.122}$$

By (1.58) and (1.56), the momentum map $J^{\mathcal{O}} : (\widetilde{T^*G})^{\mathcal{O}} \to \mathfrak{g}^*_-$ for this action in the right and the left trivialization, respectively, is

$$J^{\mathcal{O}}(\theta, [x]_H) = \theta; \tag{2.123}$$
$$J^{\mathcal{O}}([\theta, x]_H) = \mathrm{Co}(x)\theta. \tag{2.124}$$

Choosing an orthonormal basis $\{T_a\}$ of $\mathfrak{g}$ (with ensuing coordinates θ_a on $\mathfrak{g}^*$, cf. 1.1) allows one to write (2.123) as

$$J_a^{\mathcal{O}}(\theta, [x]_H) = \theta_a^R. \tag{2.125}$$

Here $J_a^{\mathcal{O}}(\cdot) := J^{\mathcal{O}}(\cdot)(T_a)$ and θ_a^R is the coordinate function θ_a relative to the right trivialization, regarded as a function on $(\widetilde{T^*G})^{\mathcal{O}}$ (it is, of course, equally well a function on T^*G and on $(T^*G)/H$).

We now further specialize the discussion to the following situation.

Definition 2.7.2. *A closed subgroup $H \subset G$ is called* **reductive** *if there exists a linear space $\mathfrak{m} \subset \mathfrak{g}$ such that $\mathrm{Ad}(H)\mathfrak{m} = \mathfrak{m}$ and*

$$\mathfrak{g} = \mathfrak{h} \oplus \mathfrak{m}. \tag{2.126}$$

One calls (2.126) a **reductive decomposition** of $\mathfrak{g}$. Any compact subgroup is reductive, for one can equip $\mathfrak{g}$ with an H-invariant inner product, and define $\mathfrak{m}$ to be the orthogonal complement of $\mathfrak{h}$.

Choose a basis $\{T_i\}$ of $\mathfrak{h}$. Given a reductive decomposition (2.126) of $\mathfrak{g}$, let $\{\theta^i\}$ be elements of $\mathfrak{g}^*$ with the properties $\theta^i(T_j) = \delta^i_j$ for all $i, j = 1, \ldots, \dim(H)$ and $\theta^i(X) = 0$ for all $X \in \mathfrak{m}$.

Proposition 2.7.3. *Let H be a reductive subgroup of G with associated reductive decomposition (2.126). The H-***connection**

$$\mathbf{A}^H(x) := \sum_{i=1}^{\dim(H)} \theta_L^i(x) \otimes T_i \tag{2.127}$$

is independent of the choice of basis (within the class of bases considered), and defines a connection 1-form on $G(G/H, H, \tau)$.

Thus $\mathbf{A}^H$ is the restriction of the left Maurer–Cartan form θ_L^{MC} to $\mathfrak{h}$ (see 1.4) in an adapted basis. The defining properties of a connection are easily verified: (2.15) follows from (1.53), whereas (2.14) is equivalent to $[\mathfrak{h}, \mathfrak{m}] \subseteq \mathfrak{m}$ (which is immediate from the reductivity of H). The basis independence of $\mathbf{A}^H$ will be clear from the proof of the following corollary. ■

Let us note that $\mathbf{A}^H$ is a G-**invariant connection**, because θ_L^{MC} is G-invariant under the canonical left action of G on T^*G.

Corollary 2.7.4. *There is a bijective correspondence between reductive decompositions (2.126) and G-invariant connections on $G(G/H, H, \tau)$.*

We use the left trivializations $TG \simeq \mathfrak{g} \times G$ and $T^*G \simeq \mathfrak{g}^* \times G$; cf. (1.46) and (1.47). The vertical tangent vectors in T_xG are those of the form $(X, x)_L$, where $X \in \mathfrak{h}$, whereas the horizontal ones are defined by $\mathbf{A}^H$ to be the vectors $(Y, x)_L$, where $Y \in \mathfrak{m}$. Hence the choice of $\mathfrak{m}$ as a complement to $\mathfrak{h}$ in $\mathfrak{g}$ defines the connection, and vice versa. In view of (1.50), the G-invariance of a connection $\mathbf{A}$ forces it to be of the form $\mathbf{A}^H$. ■

Using (2.117), the diffeomorphism (2.47) implemented by $\mathbf{A}^H$ assumes the simple form

$$T^*G \simeq G \times \mathfrak{h}^0 \times \mathfrak{h}^*. \tag{2.128}$$

Comparing with (1.47) one sees that this amounts to a factorization $\mathfrak{g}^* \simeq \mathfrak{h}^0 \times \mathfrak{h}^*$ as a manifold; as a vector space this actually sharpens to $\mathfrak{g}^* \simeq \mathfrak{h}^0 \oplus \mathfrak{h}^*$. It is evident from (2.7.4) that an H-connection provides such a decomposition, for apart from the canonical embedding $\mathfrak{h}^0 \hookrightarrow \mathfrak{g}^*$ it defines an embedding $\mathfrak{h}^* \hookrightarrow \mathfrak{g}^*$ through the identification of $\mathfrak{h}^*$ with $\mathfrak{m}^0 \subset \mathfrak{g}^*$.

Using (2.128), we denote elements of $G_V^0 \times \mathfrak{h}^*$ by (x, μ, θ), where $x \in G$, $\mu \in \mathfrak{h}^0$, and $\theta \in \mathfrak{h}^* = \mathfrak{m}^0$. The H-action on $G_V^0 \times \mathfrak{h}^*$ is

$$R_h(x, \mu, \theta) = (xh^{-1}, \mathrm{Ci}(h)\mu, \mathrm{Co}(h)\theta), \tag{2.129}$$

which is consistent with (1.49). The momentum map for this action, given in general form by (2.48), then simply reads $J(x, \mu, \theta) = -\theta$, which complies with (1.55). Obviously, the references to (1.49) and (1.55) are on the understanding that the pertinent G-action on T^*G is restricted to H. Hence by (2.53) one has the symplectomorphism

$$(\widetilde{T^*G})^{\mathcal{O}} \simeq (G \times \mathfrak{h}^0) \times_H \mathcal{O} \tag{2.130}$$

for the symplectic leaf $(\widetilde{T^*G})^{\mathcal{O}}$ in $(T^*G)/H$. For $\mathcal{O} = \{0\}$ this reproduces (2.116). The bundle projection of $(\widetilde{T^*G})^{\mathcal{O}}$ onto $T^*(G/H)$ is given simply by

$$\tau_{(\widetilde{T^*G})^{\mathcal{O}} \to T^*(G/H)}([x, \mu, \theta]_H) = [x, \mu]_H. \tag{2.131}$$

In the present formulation, the reduced G-action (2.122) on $(\widetilde{T^*G})^{\mathcal{O}}$ is

$$\lambda_x^{\mathcal{O}}([y, \mu, \theta]_H) = [xy, \mu, \theta]_H. \tag{2.132}$$

This is a special case of (2.78); the 1-form $\sigma \in T^*Q$ appearing in the latter is not to be confused with $\mu \in \mathfrak{h}^0$ in (2.132). In the present context and notation the momentum map (2.124) of this action reads

$$J^{\mathcal{O}}([x, \mu, \theta]_H) = \mathrm{Co}(x)(\mu + \theta). \tag{2.133}$$

This depends on the connection, which enabled us to identify $\theta \in \mathfrak{h}^*$ with an element of $\mathfrak{g}^*$.

It is often more informative to give the momentum map relative to a (local) trivialization of P, and hence of $(\widetilde{T^*G})^{\mathcal{O}} \simeq T^*(G/H) \times \mathcal{O}$ (locally). Specializing (2.89) to the case at hand we obtain, using (2.127), (1.42), and (2.83),

$$J_a^{\mathcal{O}}(\sigma_q, \theta)_s = p^A_{(\sigma_q,\theta)}(\xi_a^Q) + \mathrm{Co}(s(q))^i_a \theta_i, \tag{2.134}$$

where the index a on $J^{\mathcal{O}}$ and ξ^Q stands for T_a. Specializing 2.6.1 to the situation at hand, we have

Proposition 2.7.5. *An* Ad(H)*-invariant inner product* (,) *on* $\mathfrak{g}$ *defines*

- *a left-G-invariant and right-H-invariant Riemannian metric* $\mathbf{g}$ *on* G;
- *a G-invariant Riemannian metric* $\mathbf{g}^{G/H}$ *on* G/H;
- *a bi-invariant Riemannian metric* $\mathbf{g}^H$ *on* H;
- *an H-connection* $\mathbf{A}^H$ *on* $G(G/H, H, \tau)$.

The metric $\mathbf{g}$ *satisfies (2.100), and the four objects listed are related as in Proposition 2.6.1, where* $\mathbf{g}^H_q$ *is independent of* q.

The metric $\mathbf{g}$ is obtained from (,) by left translation and the identification $\mathfrak{g} = T_eG$. The restriction of (,) to $\mathfrak{m} \simeq \mathfrak{g}/\mathfrak{h}$ yields an inner product on $T_{[H]}G/H$, and subsequently a G-invariant Riemannian metric $\mathbf{g}^{G/H}$ by left translation. The bi-invariant metric on H is defined by left or right- translating the restriction of (,) to $\mathfrak{h}$. The H-connection is obtained by declaring $\mathfrak{m}$ to be the orthogonal complement of $\mathfrak{h}$; cf. 2.7.4. Equation (2.100) is a restatement of the Ad(H)-invariance of (,). It is easily verified that $\mathbf{g}^{G/H}$, $\mathbf{g}^H$, and $\mathbf{A}$ defined in 2.6.1 coincide with the ones above. ■

We are therefore in the situation of Theorem 2.6.2. With respect to the basis $\{T_a\}$ of $\mathfrak{g}$ introduced earlier, the free Hamiltonian II.(3.30) on T^*G in the left trivialization is simply

$$h_*(\theta, x)_L = \tfrac{1}{2} \sum_{a=1}^{\dim(G)} \theta_a^2. \tag{2.135}$$

If (,) happens to be Ad(G)-invariant, the expression for $h_*(\theta, x)_R$ in the right trivialization is the same. In that case, equations (1.58) and (2.135) imply

$$h_* = \tfrac{1}{2} \sum_{a=1}^{\dim(G)} (J_a^L)^2, \tag{2.136}$$

where J^L is the momentum map for the left action of G on T^*G. Being right-H-invariant, this momentum map reduces to the momentum map $J^{\mathcal{O}}$ of the reduced

G-action on $(\widetilde{T^*G})^{\mathcal{O}}$; cf. (2.125). Hence the reduced Hamiltonian $h_*^{\mathcal{O}}$ on $(\widetilde{T^*G})^{\mathcal{O}}$, defined as in (2.106), is given by essentially the same expression:

$$h_*^{\mathcal{O}} = \tfrac{1}{2} \sum_{a=1}^{\dim(G)} (J_a^{\mathcal{O}})^2. \tag{2.137}$$

2.8 The Quantum Algebra of Observables

In 2.3 we defined the Poisson algebra $C^\infty((T^*\mathsf{P})/H)$, and noticed the isomorphism $C^\infty((T^*\mathsf{P})/H) \simeq C^\infty(T^*\mathsf{P})^H$. We now turn to the quantization of this algebra. For simplicity we assume that we have an H-invariant measure μ on P that is locally Lebesgue. This may be the measure obtained from an H-invariant Riemannian metric $\mathbf{g}$ on P, cf. 2.6.1 and II.(3.40). In any case, one may define the Hilbert space $L^2(\mathsf{P}) := L^2(\mathsf{P}, \mu)$.

On $L^2(\mathsf{P})$ we have a representation U_R of H, given by

$$U_R(h)\Psi(x) = \Psi(xh). \tag{2.138}$$

Hence we can define the C^*-algebras $\mathfrak{B}_0(L^2(\mathsf{P}))^H$ and $\mathfrak{B}(L^2(\mathsf{P}))^H$ of compact and bounded operators on $L^2(\mathsf{P})$ that commute with each $U_R(h)$, $h \in H$. The latter is not particularly useful in the present context, whereas the former is empty unless H is compact. We will therefore proceed on the assumption that H is compact. This assumption leads to a particularly clean analogy with the classical case. The noncompact situation will be treated, with new techniques, in 3.7.1.

According to II.3.4 one may think of $\mathfrak{B}_0(L^2(\mathsf{P}))_{\mathbb{R}}$ as the quantization of the Poisson algebra $C^\infty_{\mathrm{PW}}(T^*\mathsf{P}, \mathbb{R})$; a quantization map is provided by the generalized Weyl quantization $\mathcal{Q}_\hbar^W$ in II.3.4.4. In analogy with Theorem 1.9.2 (corresponding to the special case $\mathsf{P} = H = G$) we are led to

Theorem 2.8.1. *Let $\mathcal{Q}_\hbar^W : C^\infty_{\mathrm{PW}}(T^*\mathsf{P}) \to \mathfrak{B}_0(L^2(\mathsf{P}))$ be as defined in II.3.4.4, and assume that H is compact and κ is H-invariant.*

With $\tilde{\mathfrak{A}}^0 = C^\infty_{\mathrm{PW}}((T^*\mathsf{P})/H)$, so that $\mathfrak{A}^0 = C_0((T^*\mathsf{P})/H)$, and $\mathfrak{A}^\hbar = \mathfrak{B}_0(L^2(\mathsf{P}))^H$ for $\hbar \in \mathbb{R}\backslash\{0\}$, the map $\mathcal{Q}_\hbar^W : \tilde{\mathfrak{A}}^0 \to \mathfrak{A}^\hbar$ defines a nondegenerate strict quantization of $(T^*\mathsf{P})/H$, with the possible exception of the completeness condition II.1.1.1.4.

The H-invariance of κ may be achieved by averaging over the compact group H, but this is not essential: Since for $\hbar$ small enough $\mathcal{Q}_\hbar^W(f)$ is independent of κ, it would suffice to use any quantization that maps $C^\infty_{\mathrm{PW}}(T^*\mathsf{P}, \mathbb{R})^H$ into $\mathfrak{B}_0(L^2(\mathsf{P}))^H_{\mathbb{R}}$ and coincides with $\mathcal{Q}_\hbar^W$ for small $\hbar$.

The theorem is immediate from Theorems II.3.5.1 and II.3.6.3. ∎

It is not difficult at all to determine the structure of the (complexified) quantum algebra of observables $\mathfrak{B}_0(L^2(\mathsf{P}))^H$. In preparation, we note that the H-invariant measure μ on P defines a unique measure ν on Q, satisfying

$$\int_{\mathsf{P}} d\mu(x)\, f(x) = \int_Q d\nu(q) \int_H dh\, f(s(q)h) \tag{2.139}$$

for any $f \in L^1(\mathsf{P})$ and any measurable section $s : Q \to \mathsf{P}$. This measure is locally Lebesgue; if μ comes from a Riemannian metric $\mathbf{g}$, then ν is just the Riemannian measure determined by $\mathbf{g}^Q$ (cf. 2.6.1). This measure is used in the construction of $L^2(Q) := L^2(Q, \nu)$.

Proposition 2.8.2. *Each measurable section* $s : Q \to \mathsf{P}$ *determines an isomorphism* $\mathfrak{B}_0(L^2(\mathsf{P}))^H \simeq \mathfrak{B}_0(L^2(Q)) \otimes C^*(H)$.

Here the tensor product is defined as the norm-closure of the algebraic tensor product in the natural representation on $L^2(Q) \otimes L^2(H)$.

The section s determines a trivialization of P by (2.3), which leads to a unitary transformation $U_s : L^2(\mathsf{P}) \to L^2(Q \times H) \simeq L^2(Q) \otimes L^2(H)$ defined by $U_s\Psi(q, h) := \Psi(s(q)h)$. Consider the space $\mathfrak{B}_2(L^2(\mathsf{P}))^H$ of H-invariant Hilbert–Schmidt operators on $L^2(\mathsf{P})$, whose elements K are characterized by a kernel $K \in L^2(\mathsf{P} \times \mathsf{P})^H$ satisfying

$$K(xh, yh) = K(x, y) \tag{2.140}$$

almost everywhere. We construct a map $\rho : L^2(\mathsf{P} \times \mathsf{P})^H \to L^2(Q \times Q \times H)$ by $(\rho(K))(q, q', h) = K(s(q)h, s(q'))$. We then identify $L^2(Q \times Q \times H)$ with $\mathfrak{B}_2(L^2(Q)) \otimes L^2(H)$, where $\mathfrak{B}_2(L^2(Q))$ and $L^2(H)$ are seen as (dense) subspaces of $\mathfrak{B}_0(L^2(Q))$ and $C^*(H)$ (in turn identified with $C^*_r(H)$, see 1.7). It is then verified that $U_s K U_s^* = \rho(K)$. Since the norm-closures of $\mathfrak{B}_2(L^2(\mathsf{P}))^H$ and $\mathfrak{B}_2(L^2(Q)) \otimes L^2(H)$ are $\mathfrak{B}_0(L^2(\mathsf{P}))^H$ and $\mathfrak{B}_0(L^2(Q)) \otimes C^*(H)$, respectively, the claim follows. ■

Corollary 2.8.3. *Up to equivalence there is a bijective correspondence between the irreducible representations* π^χ *of* $\mathfrak{B}_0(L^2(\mathsf{P}))^H$ *and the irreducible representations* U_χ *of* H, $\chi \in \hat{H}$.

This follows from 2.8.2, 1.7.5, and I.2.2.6. ■

An analogous statement holds for arbitrary representations, but the stated form is helpful in understanding the analogy with the classical result 2.3.8.

The representations $\pi^\chi(\mathfrak{B}_0(L^2(\mathsf{P}))^H)$ may be explicitly realized in various forms. The first one is π_s^χ on

$$\mathcal{H}_s^\chi = L^2(Q) \otimes \mathcal{H}_\chi. \tag{2.141}$$

The proof of Proposition 2.8.2 leads to

$$\pi_s^\chi(K)\Psi_s^\chi(q) = \int_Q d\nu(q') \int_H dh\, K(s(q)h, s(q'))U_\chi(h)\Psi_s^\chi(q'). \tag{2.142}$$

One may realize $\mathcal{H}^\chi$ in a fashion that is directly analogous to the realization of the symplectic manifold $(\widetilde{T^*\mathsf{P}})^{\mathcal{O}}$ as the associated bundle $\mathsf{P}_V^0 \times_H \mathcal{O}$. The following construction is valid as it stands whether or not H is compact, as long as H is unimodular.

Definition 2.8.4. *Consider the Hilbert bundle*

$$\mathsf{H}^\chi = \mathsf{P} \times_H \mathcal{H}_\chi \tag{2.143}$$

associated to P *by the representation* U_χ *of* H *on* $\mathcal{H}_\chi$ *(here* U_χ *may be reducible, but for convenience we use notation pertinent to the irreducible case). There is a natural inner product on the space of sections* $\Gamma(\mathsf{H}^\chi)$*, given by*

$$(\Psi^{(\chi)}, \Phi^{(\chi)}) := \int_Q d\nu(q)\, (\Psi^{(\chi)}(q), \Phi^{(\chi)}(q))_\chi, \tag{2.144}$$

where $(\,,\,)_\chi$ *is the inner product in the fiber* $\tau^{-1}_{\mathsf{H}^\chi \to Q}(q) \simeq \mathcal{H}_\chi$ *(different identifications of the fiber with* $\mathcal{H}_\chi$ *lead to the same inner product). The Hilbert space* $L^2(\mathsf{H}^\chi)$ *is the completion of* $\Gamma(\mathsf{H}^\chi)$ *under this inner product.*

As explained in 2.1.5, we may realize $\Gamma(\mathsf{H}^\chi)$ *as the set of smooth functions* $\Psi^\chi : \mathsf{P} \to \mathcal{H}_\chi$ *satisfying the equivariance condition (2.5); that is,*

$$\Psi^\chi(xh^{-1}) = U_\chi(h)\Psi^\chi(x). \tag{2.145}$$

Moreover, the projection of the support of Ψ^χ *to* Q *must be compact. In this realization the inner product is given by*

$$(\Psi^\chi, \Phi^\chi) := \int_Q d\nu(\tau(x))\, (\Psi^\chi(x), \Phi^\chi(x))_\chi; \tag{2.146}$$

the integrand indeed depends only on x *through* $\tau(x)$ *because of (2.145). The Hilbert space* $\mathcal{H}^\chi$ *is the completion of* $\Gamma(\mathsf{H}^\chi)$ *in this inner product.*

We return to the case that H is compact. Then $\mathcal{H}^\chi$ is a subspace of $L^2(\mathsf{P}) \otimes \mathcal{H}_\chi$: the latter carries a representation $U_R \otimes U_\chi$ of H (cf. (2.138)), and it follows by definition that $\mathcal{H}^\chi$ is the subspace of $L^2(\mathsf{P}) \otimes \mathcal{H}_\chi$ transforming trivially under $U_R \otimes U_\chi$. We already encountered a special case of this in 1.8.

Proposition 2.8.5. *The representation* $\pi^\chi(\mathfrak{B}_0(L^2(\mathsf{P}))^H)$ *on* $\mathcal{H}^\chi$*, defined by*

$$\pi^\chi(K)\Psi^\chi(x) = \int_{\mathsf{P}} d\mu(y)\, K(x, y)\Psi^\chi(y) \tag{2.147}$$

(initially defined on $\mathfrak{B}_2(L^2(\mathsf{P}))^H$ *and extended to* $\mathfrak{B}_0(L^2(\mathsf{P}))^H$ *by continuity), corresponds to* $U_\chi(H)$ *as in 2.8.3.*

Note that the left-hand side satisfies (2.145), and that the integrand on the right-hand side is a function of $\tau(y)$ because of (2.140). The expression may therefore be rewritten as

$$\pi^\chi(K)\Psi^\chi(x) = \int_Q d\nu(\tau(y)) \int_H dh\, K(xh, y)U_\chi(h)\Psi^\chi(y). \tag{2.148}$$

To relate π^χ to the realization (2.142) we realize $\Gamma(\mathsf{H}^\chi)$ in the second manner mentioned in 2.1.5. In this realization the inner product on $\Gamma_s(\mathsf{H}^\chi)$ is simply the one in (2.141), which therefore is the closure of $\Gamma_s(\mathsf{H}^\chi)$. The relation (2.6) between Ψ_s^χ and Ψ^χ defines a unitary map $T_s^\chi : \mathcal{H}^\chi \to \mathcal{H}_s^\chi$, given by

$$T_s^\chi \Psi^\chi(q) = \Psi^\chi(s(q)); \tag{2.149}$$

$$(T_s^\chi)^{-1}\Psi_s^\chi(x) = U_\chi(h_s(x))\Psi_s^\chi(\tau(x)). \tag{2.150}$$

For all $A \in \mathfrak{B}_0(L^2(\mathsf{P}))^H$ one then has $T_s^\chi \pi^\chi(A)(T_s^\chi)^{-1} = \pi_s^\chi(A)$. ■

In the spirit of the idea of bundles, it is more elegant to take not a single (possibly discontinuous) section s, but rather a collection of smooth local sections $s_\alpha : \mathcal{N}_\alpha \to \mathsf{P}$, relative to a cover $\{\mathcal{N}_\alpha\}$ of Q; cf. 2.1. On regions of overlap $\mathcal{N}_\alpha \cap \mathcal{N}_\beta$, the appropriate sections are related by (2.4). This leads to yet another realization of the carrier space $\mathcal{H}^\chi$, which we denote by $\mathcal{H}^\chi_{\{s\}}$. This is defined as the closure (under the inner product (2.152) below) of the space of all objects $\Psi^\chi_{\{s\}}$, defined as follows. A vector $\Psi^\chi_{\{s\}}$ consists of a collection $\{\Psi^\chi_\alpha\}$ of smooth functions $\Psi^\chi_\alpha : \mathcal{N}_\alpha \to \mathcal{H}_\chi$, which are related on overlap regions by

$$\Psi^\chi_\alpha(q) = U_\chi(g_{\alpha\beta}(q))\Psi^\chi_\beta(q) \tag{2.151}$$

(no sum over β). Furthermore, one requires that $(\Psi^\chi_{\{s\}}, \Psi^\chi_{\{s\}}) < \infty$ in terms of the inner product on $\mathcal{H}^\chi_{\{s\}}$, defined by

$$(\Psi^\chi_{\{s\}}, \Phi^\chi_{\{s\}}) := \sum_\alpha \int_Q d\nu(q)\, \phi_\alpha(q)(\Psi^\chi_\alpha(q), \Phi^\chi_\alpha(q))_\chi. \tag{2.152}$$

Here $\{\phi_\alpha\}$ is a partition of unity subordinate to the cover $\{\mathcal{N}_\alpha\}$; the inner product is independent of its precise choice.

Proposition 2.8.6. *The realization $\pi^\chi_{\{s\}}$ is given by ($q \in U_\beta$)*

$$\pi^\chi_{\{s\}}(K)\Psi^\chi_\beta(q) = \sum_\alpha \int_Q d\nu(q')\, \phi_\alpha(q') \int_H dh\, K(s_\beta(q)h, s_\alpha(q'))U_\chi(h)\Psi^\chi_\alpha(q'). \tag{2.153}$$

If $g_\alpha : \mathcal{N}_\alpha \to H$ is such that $s(q) = s_\alpha(q)g_\alpha(q)$, the realizations $\pi^\chi_{\{s\}}$ and π^χ_s are intertwined by the unitary $V_{s,\{s\}} : \mathcal{H}^\chi_s \to \mathcal{H}^\chi_{\{s\}}$, given by $V_{s,\{s\}}\Psi^\chi_\alpha(q) := U_\chi(g_\alpha(q))\Psi^\chi_s(q)$. With (2.142) this leads to (2.153). ■

The classical inclusion $(\widetilde{T^*\mathsf{P}})^{\mathcal{O}} \subset (T^*\mathsf{P})/H$ has a quantum analogue; the Hilbert space $\mathcal{H}^\chi$ carrying an irreducible representation $\pi^\chi(\mathfrak{B}_0(L^2(\mathsf{P}))^H)$ may be naturally realized as a subspace of $L^2(\mathsf{P})$. This is done through the following result (of which (1.111) is a special case).

Proposition 2.8.7. *For compact H one has the decomposition*

$$L^2(\mathsf{P}) \simeq \bigoplus_{\chi\in\hat{H}} \mathcal{H}^\chi \otimes \mathcal{H}_{\overline{\chi}}, \tag{2.154}$$

under which

$$\mathfrak{B}_0(L^2(\mathsf{P}))^H \otimes \pi_R(C^*(H)) \simeq \bigoplus_{\chi\in\hat{H}} \pi^\chi(\mathfrak{B}_0(L^2(\mathsf{P}))^H) \otimes \pi_{\overline{\chi}}(C^*(H)). \tag{2.155}$$

Here one could replace $\pi_R(C^*(H))$ and $\pi_{\overline{\chi}}(C^*(H))$ by $U_R(H)$ and $U_{\overline{\chi}}(H)$, respectively. This proposition is proved by mapping $\mathcal{H}^\chi \otimes \mathcal{H}_{\overline{\chi}}$ into a subspace of $L^2(\mathsf{P})$ so as to intertwine $\pi^\chi \otimes \pi_{\overline{\chi}}$ with the defining representation of $\mathfrak{B}_0(L^2(\mathsf{P}))^H$ tensored with $U_R(C^*(H))$. Define $V^\chi : \mathcal{H}^\chi \otimes \mathcal{H}_{\overline{\chi}} \to L^2(\mathsf{P})$ by linear extension of

$$V^\chi(\Psi^\chi \otimes v)(x) := \sqrt{d_\chi}(\Psi^\chi(x), v)_\chi, \tag{2.156}$$

where $v \in \mathcal{H}_{\overline{\chi}}$ and $d_\chi = \dim(\mathcal{H}_\chi)$. Note that V^χ is indeed linear, since $\mathcal{H}_{\overline{\chi}} = \overline{\mathcal{H}}_\chi$. Equation (2.139) and the orthogonality relations (1.116) then imply that V^χ is a partial isometry. Using (2.140) and (2.145) one verifies that

$$V^\chi \circ \pi^\chi(A) \otimes U_{\overline{\chi}}(h) = A \otimes U_R(h) \circ V^\chi \tag{2.157}$$

for all $A \in \mathfrak{B}_0(L^2(\mathsf{P}))^H$ and $h \in H$.

The simplest way to prove that $\oplus_{\chi \in \hat{H}} V^\chi = \mathbb{I}$ is to use the isomorphism 2.8.2; the operator U_s featured in the proof of 2.8.2 accomplishes

$$U_s \mathfrak{B}_0(L^2(\mathsf{P}))^H \otimes \pi_R(C^*(H)) U_s^* = \mathfrak{B}_0(L^2(Q)) \otimes \pi_L(C^*(H)) \otimes \pi_R(C^*(H)), \tag{2.158}$$

where π_R and π_L are defined via (1.89) by the right- and the left-regular representations (1.98) and (1.83) of H on $L^2(H)$ (with $c = 1$), respectively. The desired result then follows from the Peter–Weyl decomposition (1.100) of $L^2(H)$. ■

One may select a copy of $\mathcal{H}^\chi$ by picking a fixed unit vector $v \in \mathcal{H}_\chi$. The operator P_v^χ on $L^2(\mathsf{P})$ defined by

$$P_v^\chi \Psi(x) = d_\chi \int_H dh\, (v, U_\chi(h) v)_\chi \Psi(xh) \tag{2.159}$$

lies in the commutant of $\mathfrak{B}_0(L^2(\mathsf{P}))^H$, and is a projection for which $P_v^\chi L^2(\mathsf{P}) \simeq \mathcal{H}^\chi$ and $P_v^\chi A \simeq \pi^\chi(A)$ for all $A \in \mathfrak{B}_0(L^2(\mathsf{P}))^H$.

2.9 *Induced Group Representations*

We turn to an important application of principal bundles and their associated vector bundles. In what follows we do not assume that H is compact, unless stated otherwise.

Recall (2.73), which in the present case defines a H-action U^χ on the Hilbert bundle H^χ associated to P by a representation $U_\chi(H)$ on a Hilbert space $\mathcal{H}_\chi$.

Definition 2.9.1. *Let a principal bundle* $\mathsf{P}(Q, H, \tau)$ *and a representation* $U_\chi(H)$ *on* $\mathcal{H}_\chi$ *be given. The* **induced representation** $U^{(\chi)}$ *of the group* Aut(P) *of automorphisms of* P *on the Hilbert space* $L^2(\mathsf{H}^\chi)$ *is given by*

$$U^{(\chi)}(\varphi)\Psi^{(\chi)}(q) := \sqrt{\frac{d\nu(\varphi_Q^{-1}(q))}{d\nu(q)}}\, \mathsf{U}^\chi(\varphi)\Psi^{(\chi)}(\varphi_Q^{-1}(q)). \tag{2.160}$$

It is essential here that the measure ν on Q is locally Lebesgue, for this guarantees that $\nu \circ \varphi_Q^{-1}$ and ν are equivalent. This means that these measures have the same null sets; one says that ν is **quasi-invariant** under Diff(Q). In view of the square root it is easily checked that $U^{(\chi)}(\varphi)$ is unitary, hence defines a representation.

In the realization $\mathcal{H}^\chi$ of H-equivariant functions $\Psi^\chi : \mathsf{P} \to \mathcal{H}_\chi$ satisfying (2.145) this reads

$$U^\chi(\varphi)\Psi^\chi(x) = \sqrt{\frac{d\nu(\varphi_Q^{-1}(\tau(x)))}{d\nu(\tau(x))}}\, \Psi^\chi(\varphi^{-1}(x)). \tag{2.161}$$

In the realization $\mathcal{H}^{\chi}_{\{s\}}$ one has

$$U^{\chi}_{\{s\}}(\varphi)\Psi^{\chi}_{\alpha}(q) = \sqrt{\frac{d\nu(\varphi_Q^{-1}(q))}{d\nu(q)}}\, U_{\chi}(h_{\beta}(\varphi^{-1}(s_{\alpha}(q))))\Psi^{\chi}_{\beta}(\varphi_Q^{-1}(q)), \tag{2.162}$$

where we assume that $q \in \mathcal{N}_{\alpha}$ and $\varphi_Q^{-1}(q)) \in \mathcal{N}_{\beta}$, and $h_{\beta}(x) \in H$ is defined by the property $xh_{\beta}(x) = s_{\beta}(\tau(x))$; cf. 2.1.5.

In the realization $\mathcal{H}^{\chi}_{s}$ one of course has the same expression, with the indices α and β omitted.

These representations of Aut(P) may be extended to representations of $\mathcal{G}^{\tau}_{\mathsf{P}}$ (see (2.74)). For any $\hbar \neq 0$ the (additive) group $C^{\infty}_{c}(Q, \mathbb{R})$ is represented on $L^2(\mathsf{H}^{\chi})$ by the appropriate analogue of II.(3.81), namely

$$\rho^{(\chi)}_{\hbar}(\tilde{g})\Psi^{(\chi)}(q) := e^{-i\tilde{g}(q)/\hbar}\Psi^{(\chi)}(q). \tag{2.163}$$

Essentially the same expression is valid on $\mathcal{H}^{\chi}_{s}$ or on $\mathcal{H}^{\chi}_{\{s\}}$; on $\mathcal{H}^{\chi}$ one has

$$\rho^{\chi}_{\hbar}(\tilde{g})\Psi^{\chi}(x) = e^{-i\tilde{g}(\tau(x))/\hbar}\Psi^{\chi}(x). \tag{2.164}$$

We then obtain a representation $\rho^{\chi}_{\hbar}(\mathcal{G}^{\tau}_{\mathsf{P}})$ by letting $\rho^{\chi}_{\hbar}(\mathrm{Aut}(\mathsf{P}))$ coincide with $U^{\chi}(\mathrm{Aut}(\mathsf{P}))$, and putting $\rho^{\chi}_{\hbar}(\varphi, \tilde{g}) = \rho^{\chi}_{\hbar}(\tilde{g})\rho^{\chi}_{\hbar}(\varphi)$. These induced representations are the quantum analogues of the reduced actions $\rho^{\mathcal{O}}_{0}(\mathcal{G}^{\tau}_{\mathsf{P}})$ on $(\widetilde{T^*\mathsf{P}})^{\mathcal{O}}$ in 2.5.2; in the quantum case the relevant action is on the space of sections of the appropriate associated bundle, rather than on the associated bundle itself (as in the classical case).

When H is compact this construction of induced representations can be reformulated as follows. Firstly, $\mathcal{G}^{\tau}_{\mathsf{P}}$ is contained in $\mathcal{G}_{\mathsf{P}}$ (see II.(3.12)) by the inclusions $\mathrm{Aut}(\mathsf{P}) \subset \mathrm{Diff}(\mathsf{P})$ and $C^{\infty}_{c}(Q, \mathbb{R}) \simeq C^{\infty}_{c}(\mathsf{P}, \mathbb{R})^{H} \subset C^{\infty}_{c}(\mathsf{P}, \mathbb{R})$. Now define a representation $\rho_{\hbar}$ of $\mathcal{G}^{\tau}_{\mathsf{P}} \subset \mathcal{G}_{\mathsf{P}} = \mathrm{Diff}(\mathsf{P}) \ltimes C^{\infty}_{c}(\mathsf{P}, \mathbb{R})$ on $L^2(\mathsf{P})$ by restriction of the representation $\rho_{\hbar}(\mathcal{G}_{\mathsf{P}})$ given by II.3.6.2 (with Q replaced by P). Subsequently, extend $\rho_{\hbar}(\mathcal{G}^{\tau}_{\mathsf{P}})$ to a representation $\rho_{\hbar} \otimes I_{\chi}$ on $L^2(\mathsf{P}) \otimes \mathcal{H}_{\chi}$, where $\mathbb{I}_{\chi}$ is the unit operator on $\mathcal{H}_{\chi}$. The restriction of $\rho_{\hbar}(\mathcal{G}^{\tau}_{\mathsf{P}}) \otimes \mathbb{I}_{\chi}$ to the subspace $\mathcal{H}^{\chi} \subset L^2(\mathsf{P}) \otimes \mathcal{H}_{\chi}$ is then given by $\rho^{\chi}_{\hbar}$. Note that this restriction is well-defined, because $\rho_{\hbar}(\mathcal{G}^{\tau}_{\mathsf{P}})$ lies in the commutant of $U_R(H)$.

We give yet another description of $\rho^{\chi}_{\hbar}$.

Lemma 2.9.2. *Let H be compact. Under the decomposition (2.154) the restriction of $\rho_{\hbar}(\mathcal{G}^{\tau}_{\mathsf{P}})$ to $\mathcal{H}^{\chi} \otimes \mathcal{H}_{\overline{\chi}}$ is $\rho^{\chi}_{\hbar} \otimes \mathbb{I}_{\overline{\chi}}$.*

This follows from (2.159) and the orthogonality relations (1.116). ■

As explained at the beginning of II.3.6, one can extend the Weyl quantization prescription $\mathcal{Q}^{W}_{\hbar}$ (initially defined on $C^{\infty}_{PW}(T^*\mathsf{P}, \mathbb{R})$) to certain unbounded functions. This equally well applies to the restriction of $\mathcal{Q}^{W}_{\hbar}$ to $C^{\infty}_{PW}(T^*\mathsf{P}, \mathbb{R})^{H}$, with H compact. A representation $\pi^{\chi}(\mathfrak{B}_0(L^2(\mathsf{P}))^{H})$ may be extended to the H-invariant unbounded operators on $L^2(\mathsf{P})$ thus encountered in an obvious way. The following result is the "quantization" of Theorem 2.5.3, whose notation we use.

Theorem 2.9.3. *Let J be the momentum map for the $\mathcal{G}^{\tau}_{\mathsf{P}}$-action (2.75), (2.76) on $T^*\mathsf{P}$. Relative to an arbitrary connection $\mathbf{A}$ on P one has*

$$\pi^{\chi}\left(\mathcal{Q}^W_{\hbar}(J_{\tilde{g}})\right) = \tilde{g}; \tag{2.165}$$

$$\pi^{\chi}\left(\mathcal{Q}^W_{\hbar}(J_{\xi^{\mathsf{P}}})\right) = -i\hbar\left[\nabla^{\mathbf{A}}_{\xi^Q} + \tfrac{1}{2}\nabla\cdot\xi^Q - dU_{\chi}(\mathbf{A}(\xi^{\mathsf{P}}))\right]; \tag{2.166}$$

$$\pi^{\chi}\left(\mathcal{Q}^W_{\hbar}(J_{\lambda})\right) = i\hbar dU_{\chi}(\lambda); \tag{2.167}$$

$$\pi^{\chi}\left(\mathcal{Q}^W_{\hbar}(J_{\ell(\xi^Q)})\right) = -i\hbar\left[\nabla^{\mathbf{A}}_{\xi^Q} + \tfrac{1}{2}\nabla\cdot\xi^Q\right], \tag{2.168}$$

defined as unbounded operators on the domain $\Gamma(\mathsf{H}^{\chi}) \subset L^2(\mathsf{H}^{\chi})$.

It should be clear what these expressions stand for in the various realizations $\mathcal{H}^{\chi}$, $\mathcal{H}^{\chi}_s$, and $\mathcal{H}^{\chi}_{\{s\}}$. For example, $\tilde{g}$ and $\nabla\cdot\xi^Q$ are functions on Q, which on $\mathcal{H}^{\chi}_s$ and $\mathcal{H}^{\chi}_{\{s\}}$ are realized as multiplication operators; on $\mathcal{H}^{\chi}$, though, one should pull these functions back to P with $\tau^*_{\mathsf{P}\to Q}$. In all cases $\tilde{g}$ should more properly be written as $\tilde{g}\otimes\mathbb{I}_{\chi}$. On $\mathcal{H}^{\chi}$ the object $dU_{\chi}(F)$, where F is $\mathbf{A}(\xi^{\mathsf{P}})$ or λ, acts like $dU_{\chi}(F(x))\Psi^{\chi}(x)$, whereas on $\mathcal{H}^{\chi}_s$ and $\mathcal{H}^{\chi}_{\{s\}}$ one should replace F by $F\circ s$ (recall that $\lambda\circ s = \lambda_s$) and tensor with the unit operator on $L^2(Q)$. The covariant derivative ∇_{ξ^Q} is defined in 2.2.3; on $\mathcal{H}^{\chi}_s$ one has, from (2.26),

$$\nabla^{\mathbf{A}}_{\mu}\Psi^{\chi}_s(q) = (\partial_{\mu} + A^i_{\mu}(q)dU_{\chi}(T_i))\Psi^{\chi}_s(q), \tag{2.169}$$

where $A = s^*\mathbf{A}$. The corresponding expression on $\mathcal{H}^{\chi}_{\{s\}}$ is obtained by replacing Ψ^{χ}_s by Ψ^{χ}_{α} and A by $s^*_{\alpha}\mathbf{A}$.

The proof of 2.9.3 starts with an equation of independent interest, namely

$$i\hbar dU^{\chi}(X) = \pi^{\chi}(\mathcal{Q}^W_{\hbar}(J_X)), \tag{2.170}$$

for all X in the Lie algebra $\mathfrak{g}^{\tau}_{\mathsf{P}}$ of $\mathcal{G}^{\tau}_{\mathsf{P}}$. This equation is defined on $\Gamma(\mathsf{H}^{\chi})$, and follows directly from II.(3.82) and Lemma 2.9.2. Now note that on $\mathcal{H}^{\chi}$ one simply has II.(3.73) and II.(3.74). This easily leads to (2.165)–(2.168); for example, (2.166) follows from (2.25) and (2.92). ■

Equations (2.165), (2.166), (2.167), and (2.168) are to be compared with their classical counterparts (2.88), (2.89), (2.90), and (2.91), respectively.

Proposition 2.9.4. *The operators in (2.165)–(2.168) are essentially self-adjoint on $\Gamma(\mathsf{H}^{\chi}) \subset L^2(\mathsf{H}^{\chi})$.*

This follows as in the proof of Proposition II.3.6.4. □

Since J is equivariant (cf. 2.5.3) and U^{χ} is a representation, on account of (2.170) the following equation holds on $\Gamma(\mathsf{H}^{\chi})$ for all $X, Y \in \mathfrak{g}^{\tau}_{\mathsf{P}}$:

$$\frac{i}{\hbar}\left[\pi^{\chi}\left(\mathcal{Q}^W_{\hbar}(J_X)\right), \pi^{\chi}(\mathcal{Q}^W_{\hbar}(J_Y))\right] = \pi^{\chi}\left(\mathcal{Q}^W_{\hbar}(\{J_X, J_Y\}\right); \tag{2.171}$$

this is a version of Dirac's condition II.(1.3).

This is the right place to mention the quantum-mechanical counterpart of the discussion on gauge covariance in 2.5. Firstly, the analogue of the classical condition (2.82) in the "active picture" is as follows. Let $s_{g^{-1}}(q) = s(q)g_s^{-1}(q)$, as in

2.4.4, and define the unitary operator $U_g : L^2(Q)\otimes\mathcal{H}_\chi \to: L^2(Q)\otimes\mathcal{H}_\chi$ (regarded as a map from $\mathcal{H}_s^\chi$ to $\mathcal{H}_{s_{g^{-1}}}^\chi$) by $U_g\Psi_s^\chi(q) := U_\chi(g_s(q))\Psi_s^\chi(q)$. An operator O^A that involves $dU_\chi(A)$, where $A = s^*\mathbf{A}$, is said to be **gauge-covariant** if

$$U_g O^A U_g^* = O^{A^g}, \tag{2.172}$$

where A^g is related to A by (2.69). The covariant derivative (2.169) is a case in point. The meaning of this condition is similar to that of (2.82); it is satisfied when O^A is of the form $T_s^\chi O^{\mathbf{A}}(T_s^\chi)^{-1}$; cf. (2.149) and (2.150).

Secondly, the quantum counterpart of (2.85) in the "passive picture" is that an operator O^A on $\mathcal{H}_{\{s\}}^\chi$ satisfy

$$U_{g_{\alpha\beta}} O^{A_\beta} U_{g_{\alpha\beta}}^* = O^{A_\alpha}, \tag{2.173}$$

where $U_{g_{\alpha\beta}}\Psi_\beta^\chi := \Psi_\alpha^\chi$, as defined as in (2.151).

We now specialize the construction of induced representations to the case that the principal bundle P is $G(G/H, H, \tau)$; see 2.7. Given a representation U_χ of H on a Hilbert space $\mathcal{H}_\chi$, we construct the associated vector bundle $\mathsf{H}^\chi := G \times_H \mathcal{H}_\chi$ as in the general case. A central ingredient in the definition of the induced Hilbert space $\mathcal{H}^\chi$ is the measure ν, which is constructed from an H-invariant measure on G. We use a right-invariant Haar measure $dx := d\mu(x)$ on G for this purpose. In the present context, the space $\mathcal{H}^\chi$ is the closure (in the inner product (2.146)) of the space of smooth functions $\Psi^\chi : G \to \mathcal{H}_\chi$ whose projected support on G/H is compact and that satisfy (2.145).

Recall from 2.7 that $G \subset \text{Aut}(G)$. The induced action of Aut(P) on H^χ given by (2.73) specializes to $G \ni y$ by

$$\mathsf{U}^\chi(y)[x, v]_H = [yx, v]_H. \tag{2.174}$$

Equation (2.161) then specializes to

$$U^\chi(y)\Psi^\chi(x) = \sqrt{\frac{d\nu(\tau(y^{-1}x))}{d\nu(\tau(x))}}\,\Psi^\chi(y^{-1}x). \tag{2.175}$$

We continue to denote points in the base space $Q = G/H$ by q, and denote the canonical left action of G on G/H by $y : q \mapsto yq$, where $y \in G$. On $\mathcal{H}_{\{s\}}^\chi$ we then have, in the notation of (2.162),

$$U_{\{s\}}^\chi(y)\Psi_\alpha^\chi(q) = \sqrt{\frac{d\nu(y^{-1}q)}{d\nu(q)}}\,U_\chi(s_\alpha(q)^{-1}ys_\beta(y^{-1}q))\Psi_\beta^\chi(y^{-1}q). \tag{2.176}$$

As in the general case, the realization of $U_s^\chi(y)$ on $\mathcal{H}_s^\chi = L^2(G/H)\otimes\mathcal{H}_\chi$ (defined with respect to a single measurable section $s : G/H \to G$) is obtained from (2.176) by simply omitting the indices α and β.

This special case of induced group representations is called **Mackey induction**; compare with the corresponding classical theory described in 2.7. The formulae for Mackey induction simplify in the case that G and H are unimodular, which implies that ν is not merely quasi-invariant but actually invariant under the canonical left action of G on G/H. Hence in that case the square roots in (2.175) and (2.176) are identically 1 and can be omitted.

Combining (2.125) and (2.170), we infer that

$$\pi^{\chi}(\mathcal{Q}_{\hbar}^{W}(\theta_a^R)) = i\hbar dU^{\chi}(T_a), \tag{2.177}$$

where θ_a^R is regarded as a function on $(T^*G)/H$; cf. the comment after (2.125). To put this in perspective, let us return to Weyl quantization on T^*G (where G is seen as a Riemannian manifold). Using II.(3.74) (in which the divergence term vanishes in view of the invariance of the measure) and (1.58), we obtain

$$\mathcal{Q}_{\hbar}^{W}(\theta_a^R) = i\hbar dU_L(T_a) \tag{2.178}$$

on $C_c^{\infty}(G) \subset L^2(G)$, in terms of the left-regular representation U_L defined in (1.83) (with $c = 1$). In view of 2.8.7, equation (2.177) therefore follows from (2.178), at least when H is compact. For later reference (cf. (2.192)) we give the corresponding formula in the left trivialization of T^*G. Defining θ_a^L as θ_a relative to the left trivialization, we infer from (2.178) and the relation $\theta_a^R(\sigma_x) = \text{Ad}(x^{-1})_a^b\theta_b^L(\sigma_x)$, where $\sigma_x \in T_x^*G$ (cf. (1.42)), that

$$\mathcal{Q}_{\hbar}^{W}(\theta_a^L) = -i\hbar dU_R(T_a). \tag{2.179}$$

Combining (2.177) and (2.166), and assuming that G and H are unimodular, we obtain the geometric expression

$$dU^{\chi}(X)\Psi^{\chi}(y) = -\nabla^{\mathbf{A}}_{\xi_X^{G/H}} + dU_{\chi}(\mathbf{A}_y(\xi_X^R)) \tag{2.180}$$

on $\mathcal{H}^{\chi}$; on $\mathcal{H}_s^{\chi}$ (and analogously on $\mathcal{H}_{\{s\}}^{\chi}$) one replaces $\Psi^{\chi}(y)$ by $\Psi_s^{\chi}(q)$ and $\mathbf{A}_y(\xi_X^R)$ by $\mathbf{A}_{s(q)}(\xi_X^R)$. The right-hand side is, of course, independent of $\mathbf{A}$.

We now assume a reductive decomposition (2.126); recall 2.7.3. Specializing to the associated H-connection $\mathbf{A}^H$ on the bundle $G(G/H, H, \tau)$, equation (2.166) becomes (cf. the corresponding classical expression (2.134))

$$\pi^{\chi}\left(\mathcal{Q}_{\hbar}^{W}(J_a)\right) = -i\hbar\left[\nabla^{\mathbf{A}^H}_{\xi_a^Q} - \text{Co}(s(q))_a^i dU_{\chi}(T_i)\right]. \tag{2.181}$$

See IV.2.8 for applications of Mackey induction in physics.

2.10 The Quantum Wong Hamiltonian

We now look at the quantization of the Hamiltonian, assuming that H is compact.

Definition 2.10.1. *Given a Riemannian metric* $\mathbf{g}^Q$ *on* Q *and a connection* $\mathbf{A}$ *on* $\mathsf{P}(Q, H, \tau)$, *the* **Laplace–Bochner operator** $\Delta_{\chi}^{\mathbf{A}}$ *is a second-order differential operator on the space of sections* $\Gamma(\mathsf{H}^{\chi})$ *of the Hilbert bundle* H^{χ} *associated to* P. *For* $\Psi^{\chi}, \Phi^{\chi} \in \Gamma(\mathsf{H}^{\chi})$ *it is defined by the property*

$$(\Psi^{\chi}, \Delta_{\chi}^{\mathbf{A}}\Phi^{\chi}) := -\int_Q d\nu(\tau(x))\mathbf{g}_Q^{-1}(\nabla^{\mathbf{A}}\Psi^{\chi}(x), \nabla^{\mathbf{A}}\Phi^{\chi}(x))_{\chi}. \tag{2.182}$$

Cf. II.3.7.1; recall that elements of $\Gamma(\mathsf{H}^\chi)$ by definition have compact support. An analogous definition can be given on $\Gamma_s(\mathsf{H}^\chi)$ or $\Gamma_{\{s\}}(\mathsf{H}^\chi)$, where $\tau(x)$ and x are replaced by q. In any case, the expression $(\ldots)_\chi$ in (2.182) is an element of $T_q^*Q \otimes T_q^*Q$, so that the integrand is a scalar.

In coordinates, on $\Gamma_s(\mathsf{H}^\chi)$ or $\Gamma_{\{s\}}(\mathsf{H}^\chi)$ one has (cf. (2.169))

$$\Delta_\chi^{\mathbf{A}} = \mathbf{g}_Q^{\mu\nu}(\nabla_\mu + dU_\chi(A_\mu))(\partial_\nu + dU_\chi(A_\nu)), \tag{2.183}$$

where ∇ is the Levi-Civita connection defined by $\mathbf{g}^Q$. Here $dU_\chi(A_\mu)$ may be rewritten as $A^i_\mu dU_\chi(T_i)$, where A^i_μ is a multiplication operator on $L^2(Q)$. Note that $\Delta_\chi^{\mathbf{A}}$ is gauge-covariant; cf. (2.172).

Proposition 2.10.2. *In the notation and circumstances of Theorem 2.6.2 the quantum Hamiltonian $H_\hbar^\chi := \pi^\chi(\mathcal{Q}_\hbar^W(h_*))$ in the sector χ (for the moment defined as an unbounded operator on the domain $\Gamma(\mathsf{H}^\chi)$) is*

$$H_\hbar^\chi = -\tfrac{1}{2}\hbar^2\left(\Delta_\chi^{\mathbf{A}} - \tfrac{1}{3}\mathbf{R}_Q + \tfrac{1}{12}\mathbf{F}^2 - C^\chi\right), \tag{2.184}$$

where $\mathbf{R}_Q$ is the Ricci scalar on Q, the **Yang–Mills Hamiltonian** *is given in an arbitrary section and coordinate system by*

$$\mathbf{F}^2 := \mathbf{g}_{ij}^H \mathbf{g}_Q^{\mu\rho}\mathbf{g}_Q^{\nu\sigma} F^i_{\mu\nu} F^j_{\rho\sigma}, \tag{2.185}$$

and the constant C^χ is

$$C^\chi := C_2^\chi(H) + \tfrac{1}{12}\mathbf{g}_H^{kl} C^i_{jk} C^j_{il}. \tag{2.186}$$

Here the **Casimir element** $C_2^\chi(H)$ in the representation $U_\chi(H)$ is defined through $\sum_i (dU_\chi(T_i))^2 = -C_2^\chi(H)\mathbb{I}_\chi$; it is a (positive) constant, as U_χ is irreducible. It is possible to give an intrinsic definition of $\mathbf{F}^2$, but note that (2.185) is independent of the section in which $F = s^*\mathbf{F}$ is computed because of (2.72) and the Ad-invariance of $\mathbf{g}_H$. The terms $\mathbf{R}_Q$ and $\mathbf{F}^2$ in (2.184) depend on q, and are to be seen as multiplication operators on $\mathcal{H}^\chi$.

Equation (2.184) follows from II.(3.93) and two identities. Firstly, if Δ is the Laplace–Beltrami operator on P defined by $\mathbf{g}$, one has

$$\Delta = \Delta_\chi^{\mathbf{A}} - C_2^\chi(H) \tag{2.187}$$

on $\mathcal{H}^\chi$ (seen as a subspace of $L^2(\mathsf{P}) \otimes \mathcal{H}_\chi$). This easily follows from 2.6.1 if one decomposes $\nabla\Psi$ and $\nabla\Phi$ in II.(3.91) in a horizontal and a vertical part; cf. the text surrounding (2.35). Secondly, the Ricci scalars on P and Q are related by the famous identity

$$\mathbf{R} = \mathbf{R}_Q - \tfrac{1}{4}\mathbf{F}^2 + \tfrac{1}{4}\mathbf{g}_H^{kl} C^i_{jk} C^j_{il}. \tag{2.188}$$

This may be verified from II.(3.23), II.(3.21), (2.21), and (2.104). One obtains some additional terms, whose sum vanishes on account of the Ad-invariance and q-independence of $\mathbf{g}_H$. ■

One tool in the analysis of the possible self-adjointness of $H_\hbar^\chi$ is the following. If $\Psi \in \oplus_\lambda \mathcal{H}_\lambda$, then Ψ_λ denotes the component of Ψ in $\mathcal{H}_\lambda$.

Lemma 2.10.3. *Let $\mathcal{H} = \oplus_\lambda \mathcal{H}_\lambda$, and let a closed operator A_λ be given for each λ. The operator $A := \oplus_\lambda A_\lambda$ is defined on the domain $\mathcal{D}(A)$ consisting of all $\Psi \in \mathcal{H}$ for which $\Psi_\lambda \in \mathcal{D}(A_\lambda)$ and $\sum_\lambda A_\lambda \Psi_\lambda \in \mathcal{H}$ (this operator is easily seen to be closed). Then A is self-adjoint iff each A_λ is self-adjoint.*

The adjoint A^* of A is easily seen to be $A^* = \oplus_\lambda A_\lambda^*$. If thc equation $A^*\Psi = \pm i\Psi$ has no solution in $\mathcal{D}(A^*)$, then none of the equations $A_\lambda^*\Psi = \pm i\Psi_\lambda$ can have a solution, and vice versa. ∎

Corollary 2.10.4. *If $(\mathsf{P}, \mathbf{g})$ is complete with Ricci scalar $\mathbf{R}$ bounded, then $H_\hbar^\chi$ is essentially self-adjoint on $\Gamma(\mathsf{H}^\chi)$ for all $\chi \in \hat{H}$.*

In particular, $H_\hbar^\chi$ is essentially self-adjoint on $\Gamma(\mathsf{H}^\chi)$ when Q is compact.

This follows from II.3.7.4, 2.8.7, 2.10.3, and the inclusion $C_c^\infty(\mathsf{P}) \subset D(\oplus_\chi \overline{H}_\hbar^\chi)$, where $\overline{H}_\hbar^\chi$ is the closure of $H_\hbar^\chi$ as defined in 2.10.2. ∎

When Q is compact the hypothesis is obviously satisfied.

This result, and similarly Proposition 2.9.4, gives a hint as to why vector bundles are relevant in quantum mechanics. For one might question this relevance on the grounds that $\mathcal{H}^\chi$ is the same for all vector bundles over Q with fiber $\mathcal{H}_\chi$; it is the dense subspace $\Gamma(\mathsf{H}^\chi)$ that is sensitive to the topology of the bundle. We therefore conclude that this topology is relevant for the specification of the domain of the key observables of the quantum theory, such as the Hamiltonian.

We may look at the quantum Hamiltonian $H_\hbar^\chi$ from a different point of view. If h_* is the usual classical Hamiltonian on $T^*\mathsf{P}$ (defined with respect to an H-invariant metric $\mathbf{g}$ on P), under the appropriate hypotheses the operator $H_\hbar = \mathcal{Q}_\hbar^W(h_*)$ defines a self-adjoint operator on $L^2(\mathsf{P})$, interpreted as the quantum Hamiltonian of a particle moving on P. This leads to an action of $\mathbb{R}$ as a one-parameter automorphism group $\alpha_t^\hbar$ on $\mathfrak{B}_0(L^2(\mathsf{P}))$, defined as in II.(2.88). By the H-invariance of $H_\hbar$ this restricts to $\mathfrak{B}_0(L^2(\mathsf{P}))^H$. One may then ask whether α_t is **implemented** in a representation $\pi^\chi(\mathfrak{B}_0(L^2(\mathsf{P}))^H)$, in that there exists a unitary group $t \mapsto U_{\text{dyn}}^\chi(t)$ on $\mathcal{H}^\chi$ for which

$$\pi^\chi(\alpha_t(A)) = U_{\text{dyn}}^\chi(t)^* \pi^\chi(A) U_{\text{dyn}}^\chi(t) \tag{2.189}$$

for all $A \in \mathfrak{B}_0(L^2(\mathsf{P}))^H$ and all $t \in \mathbb{R}$. This is evidently the case in the present circumstances, and the Hamiltonian remerges as the generator of U_{dyn}^χ, that is, $U_{\text{dyn}}^\chi(t) = \exp(-itH_\hbar^\chi/\hbar)$.

Specializing the theory to the case where the principal bundle P is chosen to be $G(G/H, H, \tau)$, as in 2.7, allows one to give a purely group-theoretic formulation of the various geometric objects encountered. We first look at the situation on G itself.

Lemma 2.10.5. *Let a unimodular Lie group G be equipped with a left-invariant metric $\mathbf{g}$, with corresponding orthonormal basis $\{T_a\}$ of $\mathfrak{g} = T_eG$. Then the Laplace–Beltrami operator Δ on $C_c^\infty(G) \subset L^2(G)$ is given in terms of the Casimir* *element*

$$C_2(G) := -\sum_a T_a^2 \tag{2.190}$$

and the right-regular representation (1.98) (with $c = 1$) by

$$\Delta = -dU_R(C_2(G)). \tag{2.191}$$

We start from the definition II.(3.91), and write $df = \theta^a_L \xi^L_a f$ (cf. 1.4). Since $\mathbf{g}^{-1}(\theta^a, \theta^b) = \delta^{ab}$, the Riemannian volume element defined by the metric must be proportional to a left-invariant Haar measure (unique up to scale), which is also right-invariant by unimodularity. Hence we can partially integrate in the right-hand side of II.(3.91), and obtain (2.191) on account of $dU_R(X) = \xi^L_X$, $X \in \mathfrak{g}$. ■

Comparing (2.191) with II.(3.93) and (2.135), we infer that

$$\mathcal{Q}^W_\hbar(h_*) = \tfrac{1}{2}\sum_a \mathcal{Q}^W_\hbar((\theta^L_a)^2) = -\tfrac{1}{2}\hbar^2\left(\sum_a dU_R(T_a^2) - \tfrac{1}{3}\mathbf{R}_G\right); \tag{2.192}$$

note that the Ricci scalar $\mathbf{R}_G$ is a constant. To put this in perspective, one should recall (2.179), inferring that $\mathcal{Q}^W_\hbar((\theta^L_a)^2) \neq (\mathcal{Q}^W_\hbar(\theta^L_a))^2$.

We now assume that H is a reductive subgroup of G, so that we are in the situation of Proposition 2.7.5. Comparing (2.191) and (2.187), and using the H-equivariance of elements of $\mathcal{H}^\chi$, we conclude that

$$\Delta^{\mathbf{A}^H}_\chi = \sum_{a=1}^{\dim(\mathfrak{m})} dU_R(T_a^2), \tag{2.193}$$

where the T_a occurring in the sum form a basis of $\mathfrak{m}$. This expression may be substituted into the Hamiltonian (2.184); in the present case the last three terms on the right-hand side of (2.184) are constants. If, in addition, the inner product $(\, ,\,)$ on $\mathfrak{g}$ is $\mathrm{Ad}(G)$-invariant, one has

$$\Delta^{\mathbf{A}^H}_\chi = \sum_{a=1}^{\dim(\mathfrak{g})} dU^\chi(T_a^2) + C_2^\chi(H), \tag{2.194}$$

which should be compared both with its classical counterpart (2.137) and with the more general quantum formula (2.187).

2.11 From the Quantum to the Classical Wong Equations

Let us now investigate the possible classical limit of the dynamics generated by the quantum Hamiltonian $H^\chi_\hbar$ on $\mathcal{H}^\chi$. As in Theorem II.3.7.5 we assume that $Q = \mathbb{R}^n$, so that the bundle $\mathsf{P} = \mathbb{R}^n \times H$ is necessarily trivial. Accordingly, given a coadjoint orbit $\mathcal{O} \subset \mathfrak{h}^*$, one has $(\widetilde{T^*\mathsf{P}})^{\mathcal{O}} = T^*Q \times \mathcal{O}$. Also, $\mathcal{H}^\chi_s = L^2(Q) \otimes \mathcal{H}_\chi$ with respect to the natural section $s : Q \to \mathsf{P}$ given by $s(q) = (q, e)$; cf. (2.141).

Recall that H is a compact Lie group. As in 1.10, we assume that the object χ labeling the irreducible representation $U_\chi(H)$ is a highest weight, corresponding to an integral coadjoint orbit $\mathcal{O}_\chi \subset \mathfrak{h}^*$. We extend $\chi \in \mathfrak{t}^*$ to an element $\theta(\chi)$ of $\mathfrak{h}^*$, as explained in 1.10, and denote the stabilizer of $\theta(\chi)$ under the coadjoint action by H_χ. Then $\mathcal{O}_\chi \simeq H/H_\chi$.

Inspired by 1.11 we put $\hbar = 1/k$, $k \in \mathbb{N}$, and study the Hamiltonians $H_{1/k}^{k\chi}$ on $\mathcal{H}^{k\chi} = L^2(Q) \otimes \mathcal{H}_{k\chi}$ as $k \to \infty$. The corresponding time evolution is then expected to converge to the flow on $(\widetilde{T^*\mathsf{P}})^{\mathcal{O}}$ generated by the classical Hamiltonian $h_A^{\mathcal{O}_\chi}$ given by (2.107).

The precise formulation of this convergence will be in terms of pure state quantizations, as in Theorem II.2.7.2. Combining the coherent states II.(3.95) in $L^2(\mathbb{R}^n)$ with their counterparts $U_{k\chi}(h)\Psi_{k\chi} \in \mathcal{H}_{k\chi}$, where $h \in H$ (see (1.149)), we define unit vectors in $L^2(\mathbb{R}^n) \otimes \mathcal{H}_{k\chi}$ by

$$\Psi_{1/k}^{(p,q,h)} := \Psi_{1/k}^{(p,q)} \otimes U_{k\chi}(h)\Psi_{k\chi}. \tag{2.195}$$

Recall that $\Psi_{k\chi}$ is a normalized highest weight vector in $\mathcal{H}_{k\chi}$.

Analogously to f_γ in 1.11, for $f \in C(T^*\mathbb{R}^n \times \mathcal{O}_\chi)$ we define $f_\chi \in C(T^*\mathbb{R}^n \times H)$ as the pullback of f under the canonical projection from $T^*\mathbb{R}^n \times H$ to $T^*\mathbb{R}^n \times \mathcal{O}_\chi$; the function f_χ is right-H_χ-invariant. The Berezin quantization corresponding to the pure state quantization (2.195) is then given by

$$Q_\hbar^B(f) := d_{\chi/\hbar} \int_{T^*\mathbb{R}^n} \frac{d^n p d^n q}{(2\pi\hbar)^n} \int_H dh\, f_\chi(p,q,h) \left[\Psi_\hbar^{(p,q,h)}\right], \tag{2.196}$$

defined for $f \in C_c^\infty(T^*\mathbb{R}^n \times \mathcal{O}_\chi, \mathbb{R})$, taking values in $\mathfrak{A}^\hbar = \mathfrak{B}_0(L^2(\mathbb{R}^n)) \otimes \mathfrak{M}_{d_{\chi/\hbar}}(\mathbb{C})_\mathbb{R}$. In (2.196) and all subsequent expressions in this section, $\hbar = 1/k$.

Theorem 2.11.1 below generalizes Theorems II.2.7.2 and II.3.7.5. As before, we write $\alpha_t^0(f)$ for the function $(p,q,\theta) \mapsto f(p(t), q(t), \theta(t))$, where the time evolution is defined for $t \in (t_i, t_f)$ as the Hamiltonian flow on $T^*\mathbb{R}^n \times \mathcal{O}_\chi$ generated by $h_A^{\mathcal{O}_\chi}$; in other words, $(p(t), q(t), \theta(t))$ is the solution of the system (2.109)–(2.111) with initial conditions (p,q,θ). Since the terms $\frac{1}{2}\hbar^2(\frac{1}{3}\mathbf{R}_Q - \frac{1}{12}\mathbf{F}^2 + C^\chi)$ in (2.184) do not contribute for $\hbar \to 0$ (as will be clear from the proof below), in what follows one may replace $H_\hbar^{\chi/\hbar}$ by

$$\tilde{H}_\hbar^{\chi/\hbar} := \tfrac{1}{2}\mathbf{g}^{\mu\nu}(P_{\hbar,\mu}^{\mathrm{cov}} - A_\mu^i T_i^\hbar)(P_{\hbar,\nu}^S - A_\nu^j T_j^\hbar). \tag{2.197}$$

Here $P_{\hbar,\mu}^{\mathrm{cov}} := -i\hbar\nabla_\mu$, $P_{\hbar,\nu}^S := -i\hbar\partial_\mu$ (as in (II.2.24)), and

$$T_j^\hbar := i\hbar dU_{\chi/\hbar}(T_j); \tag{2.198}$$

this is, of course, consistent with (2.107), II.(2.24), and (2.167) with (2.90). We assume that $\mathbf{g}$ and A are C^3 near the classical trajectory $(p(t), q(t), \theta(t))$, and such that each multiplication operator occurring in $H_\hbar^{\chi/\hbar}$ is $O(\exp(\frac{1}{2}x^2))$ for $x \to \infty$. Then for $\hbar < 1$ the operator $H_\hbar^{\chi/\hbar}$ is symmetric on the domain $\mathcal{D}_c$ consisting of the linear span of all states (2.195). Subsequently, we assume that each operator $H_\hbar^{\chi/\hbar}$ thus defined has at least one self-adjoint extension, which we denote by the same symbol. This abuse of notation is justified by the fact that as in II.2.7.2, for times that the classical flow exists different self-adjoint extensions will have the same classical limit. We then put

$$\alpha_t^\hbar(A) := e^{\frac{i}{\hbar}tH_\hbar^{\chi/\hbar}} A e^{-\frac{i}{\hbar}tH_\hbar^{\chi/\hbar}}. \tag{2.199}$$

Theorem 2.11.1. *Let* $\mathbf{g}$ *and* A *be as specified above. For given* $(p, q, h) \in T^*\mathbb{R}^n \times H$ *(corresponding to* $(p, q, \mathrm{Co}(h)\theta(\chi)) \in T^*\mathbb{R}^n \times \mathcal{O}_\chi)$*), for all* $t \in (t_i, t_f)$ *and all* $f \in C_c^\infty(T^*\mathbb{R}^n \times \mathcal{O}_\chi, \mathbb{R})$*, one has*

$$\lim_{\hbar\to 0} \left(\Psi_\hbar^{(p,q,h)}, [\mathcal{Q}_\hbar^B(\alpha_t^0(f)) - \alpha_t^\hbar(\mathcal{Q}_\hbar^B(f))]\Psi_\hbar^{(p,q,h)} \right) = 0 \tag{2.200}$$

along the sequence $\{\hbar = 1/k\}_{k\in\mathbb{N}}$.

This may, of course, be restated as

$$\lim_{\hbar\to 0} \psi_\hbar^{(p,q,\theta)} \left(\mathcal{Q}_\hbar^B(\alpha_t^0(f)) - \alpha_t^\hbar(\mathcal{Q}_\hbar^B(f)) \right) = 0. \tag{2.201}$$

Equation (2.200) is proved along the lines of the proof of II.(2.135), of which it is obviously a generalization. For simplicity we restrict the argument to the case where the metric $\mathbf{g}$ on Q is the flat Euclidean one; nontrivial metrics are easily incorporated by the method of proof of Theorem II.3.7.5.

As in II.(2.136), II.(2.137) we expand $H_\hbar^{\chi/\hbar} = H_{(2)}(t) + H_3(t)$. Here

$$H_0 := h_A^{\mathcal{O}_\chi}(p(t), q(t), \theta(t))\mathbb{I}, \tag{2.202}$$

where the classical Hamiltonian $h_A^{\mathcal{O}_\chi}(p, q, \theta)$ is given by (2.107), and subsequently the argument (p, q, θ) is replaced by the solution $(p(t), q(t), \theta(t))$ of the Wong equations (2.109)–(2.111) with initial conditions (p, q, θ). Also,

$$H_1(t) := p_\mu^A(t)(\delta P_\mu - \partial_\nu A_\mu^j(t)\theta_j(t)\delta Q^\nu - A_\mu^j(t)\delta T_j), \tag{2.203}$$

in which $p_\mu^A(t) := p_\mu(t) - \theta_i(t)A_\mu^i(t)$, where $A_\mu^i(t) := A_\mu^i(q(t))$; furthermore, $\delta P_\mu := P_{\hbar,\mu}^S - p_\mu(t)$, along with $\delta Q^\nu := Q_\hbar^{S,\nu} - q^\nu(t)$ (cf. II.2.23)) and $\delta T_j := T_j^\hbar - \theta_j(t)$; cf. (2.198). Finally,

$$\begin{aligned} H_2(t) := {} & \tfrac{1}{2}\left(\delta P_\mu - \partial_\nu A_\mu^i(t)\theta_i(t)\delta Q^\nu - A_\mu^i(t)\delta T_i\right)^2 \\ & - \tfrac{1}{2} p_\mu^A(t)\partial_\rho\partial_\sigma A_\mu^j(t)\theta_j(t)\theta_i(t)A_\mu^i(t)\delta Q^\rho\delta Q^\sigma. \end{aligned} \tag{2.204}$$

Generalizing II.(2.154), we introduce the classical propagator

$$U_1^{(p,q,h)}(t) = e^{\frac{i}{\hbar}(S(t)-\int A)} U_{\frac{1}{\hbar}}(p(t), q(t)) U_{\chi/\hbar}(h(t)) U_{\chi/\hbar}(h)^* U_{\frac{1}{\hbar}}(p, q)^*, \tag{2.205}$$

in which $\int A$ is shorthand for $\int_0^t ds\, \dot{q}^\mu(s)A_\mu^i(s)\theta_i(s)$; the classical action $S(t)$ is given by II.(2.152), with h replaced by $h_A^{\mathcal{O}_\chi}$; and $h(t)$ is the solution of (2.112) with initial condition h (the parameter appearing in (2.200)). Using the Wong equations (2.109), (2.110), and (2.112), as well as II.(2.155), the relation

$$\frac{d}{dt}U_{\chi/\hbar}(h(t)) = dU_{\chi/\hbar}(\dot{h}(t)h(t)^{-1})U_{\chi/\hbar}(h(t)),$$

and (2.113), with θ replaced by $\theta(\chi)$, one verifies that $U_1^{(p,q,h)}(t)$ satisfies II.(2.156) with initial condition $U_1^{(p,q,h)}(0) = \mathbb{I}$.

The next step in the proof of II.2.7.2 is not as easily generalized, since there no longer exists a simple expression for the propagator of a Hamiltonian that is quadratic in $P^S_{\hbar,\mu}$, $Q^{S,\nu}_\hbar$, and $T^\hbar_j$. In any case, we introduce

$$H^{(2)}(t) := \tfrac{1}{2}\left(P^S_{\hbar,\mu} - \partial_\nu A^i_\mu(t)\theta_i(t)Q^{S,\nu}_\hbar - A^i_\mu(t)\tilde{T}^\hbar_i(t)\right)^2 \\ - \tfrac{1}{2}p^A_\mu(t)\partial_\rho\partial_\sigma A^j_\mu(t)\theta_j(t)A^i_\mu(t)\theta_i(t)Q^{S,\rho}_\hbar Q^{S,\sigma}_\hbar, \tag{2.206}$$

in which

$$\tilde{T}^\hbar_i(t) := i\hbar dU_{\chi/\hbar}(\mathrm{Ad}(h(t)^{-1})T_i) - \theta_i(t). \tag{2.207}$$

In view of (2.113), we could write $(\theta(\chi))(\mathrm{Ad}(h(t)^{-1})T_i)$ instead of $\theta_i(t)$ on the right-hand side. It is not difficult to prove that $H^{(2)}(t)$ is essentially adjoint on $\mathcal{D}_c$, since each vector (2.195) is analytic for $H^{(2)}(t)$, and the linear span of such vectors is dense.

The unitary operator $U_D^{(2)}(t)$, which plays the role of $\rho^\hbar(M(t))$ in the present context (cf. II.(2.158)), is defined as the solution of

$$i\hbar\frac{d}{dt}U_D^{(2)}(t) = H^{(2)}(t)U_D^{(2)}(t) \tag{2.208}$$

with initial condition $U_D^{(2)}(0) = \mathbb{I}$. An explicit form of this operator is

$$U_D^{(2)}(t) = \sum_{k=0}^{\infty}\left(\frac{-i}{\hbar}\right)^k \int_0^t ds_1 \int_0^{s_1} ds_2 \cdots \int_0^{s_{k-1}} ds_k\, H^{(2)}(s_1)\cdots H^{(2)}(s_k), \tag{2.209}$$

the sum converging strongly on $\mathcal{D}_c$. As in II.(2.157) we define

$$U_2^{(p,q,h)}(t) := U_1^{(p,q,h)}(t)U_{\frac{1}{\hbar}}(p,q)U_{\chi/\hbar}(h)U_D^{(2)}(t)U_{\chi/\hbar}(h)^*U_{\frac{1}{\hbar}}(p,q)^*. \tag{2.210}$$

Using II.(2.155) and the fact that $U_{\chi/\hbar}$ is a representation, one obtains the relation

$$H_2(t) = U_{\frac{1}{\hbar}}(p(t),q(t))U_{\chi/\hbar}(h(t))H^{(2)}(t)U_{\chi/\hbar}(h(t))^*U_{\frac{1}{\hbar}}(p(t),q(t))^*. \tag{2.211}$$

This suffices to prove II.(2.159) and II.(2.161), with the label (p,q) replaced by (p,q,h).

The analogue of Proposition II.2.7.3 may now be proved by the same method, substituting the series (2.209) for $\rho^\hbar(M(s))$ in II.(2.164) and replacing $\Psi^{(0,0)}_\hbar$ by $\Psi^{(0,0)}_\hbar \otimes \Psi_{\chi/\hbar}$. The essential point is that $P^S_{\hbar,\mu}$, $Q^{S,\nu}_\hbar$, and $\tilde{T}^\hbar_i(t)$ each contribute a "fluctuation" factor $\sqrt{\hbar}$ to the norm on the right-hand side of II.(2.164). For $\tilde{T}^\hbar_i(t)$ this is a consequence of (1.142)–(1.145) and the fact that

$$(\Psi_{\chi/\hbar}, \tilde{T}^\hbar_i(t)\Psi_{\chi/\hbar}) = 0. \tag{2.212}$$

The appropriate generalization of II.(2.170) is

$$U_{\frac{1}{\hbar}}(p,q)^*U_{\chi/\hbar}(h)^*\mathcal{Q}^B_\hbar(f)U_{\chi/\hbar}(h)U_{\frac{1}{\hbar}}(p,q) = \mathcal{Q}^B_\hbar(f\circ(p,q,h)), \tag{2.213}$$

where $f\circ(p,q,h):(p',q',\theta)\mapsto f(p'+p,q'+q,\mathrm{Co}(h)\theta)$. This is a consequence of II.2.4.3, (1.159), and the definition (2.196). As explained after the proof of II.2.7.2, this equivariance property lies at the heart of (2.200). To finish the proof of 2.11.1 we therefore need the analogue of II.(2.171), namely

$$\lim_{\hbar\to 0}\left(U_D^{(2)}(t)\Psi_\hbar^{(0,0,e)},\,\mathcal{Q}_\hbar^B(f)U_D^{(2)}(t)\Psi_\hbar^{(0,0,e)}\right)=f_\chi(0,0,e). \tag{2.214}$$

This can indeed be proved from (2.209) and a lengthy combinatorial argument. Theorem 2.11.1 then follows from II.2.7.3, (2.210), (2.213), and (2.214). □

2.12 *The Dirac Monopole*

It is now time to illustrate some of the abstract concepts introduced in this chapter in an example that is simple yet instructive. We will work in the setting of 2.7, specializing to $G=SO(3)$ and $H=SO(2)$, seen as the subgroup of rotations around the z-axis. The coset G/H is a two-sphere S^2, and the principal bundle $SO(3)(S^2,SO(2),\tau)$ is a quotient by the discrete group $\mathbb{Z}_2$ of the famous **Hopf fibration** $S^3(S^2,SO(2),\tau)$.

We use the standard generators T_a, $a=1,2,3$, of the Lie algebra $\mathfrak{g}=\mathfrak{so}(3)\simeq\mathbb{R}^3$. The commutation relations in $\mathfrak{so}(3)$ are $[T_a,T_b]=\epsilon_{abc}T_c$, where ϵ_{abc} is the fully antisymmetric symbol with $\epsilon_{123}=1$. We label elements x of $SO(3)$ by the Euler angles (ϕ,θ,ψ), so that

$$x=R(\phi,\theta,\psi):=\mathrm{Exp}(\phi T_3)\mathrm{Exp}(\theta T_2)\mathrm{Exp}(\psi T_3), \tag{2.215}$$

where $0\le\alpha<2\pi$, $0\le\beta<\pi$, $0\le\gamma<2\pi$. Since $H=\{\mathrm{Exp}(\psi)\}$ is the stability group of the point $\mathbf{e}_z=(0,0,1)$ in $\mathbb{R}^3$ with respect to the defining action of $SO(3)$ on $\mathbb{R}^3$, we may realize G/H as $G\mathbf{e}_z=S^2$. The bundle projection $\tau:SO(3)\to SO(3)/SO(2)$ is then given by $\tau(x)=x\mathbf{e}_z$. This yields

$$\tau(R(\phi,\theta,\psi))=(\sin\phi\sin\theta,-\cos\phi\sin\theta,\cos\theta). \tag{2.216}$$

We denote this point in S^2 by (ϕ,θ). (These coordinates are related to the usual spherical coordinates (ϕ_s,θ_s) by $\phi_s=\phi-\frac{1}{2}\pi$, $\theta_s=\theta$.)

The standard bi-invariant metric $\mathbf{g}$ on G is defined by declaring that $\{T_a\}$ be an orthonormal basis. By Proposition 2.7.5 this defines an $SO(3)$-invariant metric $\mathbf{g}^{S^2}$ on S^2, as well as an $SO(3)$-invariant connection on $SO(3)(S^2,SO(2),\tau)$. The explicit form of the connection will be determined shortly. The metric $\mathbf{g}^{S^2}$ is diagonal in (ϕ,θ), and is easily seen to be given by

$$\mathbf{g}^{S^2}=d\theta^2+(\sin\theta)^2d\phi^2; \tag{2.217}$$

this coincides with the pullback of the Euclidean metric on $\mathbb{R}^3$ to the unit sphere. (One could introduce an arbitrary radius r of the two-sphere by multiplying $\mathbf{g}$ with r^2, which leads to an overall factor r^2 in (2.217) as well.)

Choosing $\mathfrak{m}$ to be the linear span of T_1 and T_2, the decomposition (2.126) is reductive, as is easily verified from the commutation relations. This is consistent with 2.7.5, since the decomposition in question is indeed orthogonal with respect to $\mathbf{g}$. Hence by 2.7.3 and 2.7.4, or alternatively by 2.7.5, there is an associated H-connection $\mathbf{A}^H$ on the bundle $SO(3)(S^2, SO(2), \tau)$. To describe $\mathbf{A}^H$ explicitly we introduce two sections $s_\pm : \mathcal{N}_\pm \to SO(3)$ of the bundle. Here $\mathcal{N}_+$ and $\mathcal{N}_-$ consist of S^2 minus the northpole ($\theta = 0$) and minus the southpole ($\theta = \pi$), respectively. The sections in question are defined by

$$s_\pm(\phi, \theta) = R(\phi, \theta, \pm\phi); \tag{2.218}$$

evidently, s_- is discontinuous at the southpole, whereas s_+ is so at the northpole. The transition function g_{-+} in (2.4) relating s_+ and s_- is

$$g_{-+}(\phi, \theta) = \mathrm{Exp}(2\phi T_3) = R(0, 0, 2\phi). \tag{2.219}$$

Proposition 2.12.1. *In the gauges $s_\pm$ the H-connection $A_\pm := s_\pm^* \mathbf{A}^H$ is given by*

$$A_\pm(\phi, \theta) = \frac{\cos\theta \pm 1}{\sin\theta} \omega^\phi \otimes T_3, \tag{2.220}$$

where $\omega^\phi := \sin\theta d\phi$.

The result is stated in the given form because ω^ϕ rather than $d\phi$ has unit norm with respect to $\mathbf{g}$. We see that A_- would be singular at the southpole ($\theta = \pi$), whereas A_+ would be singular at the northpole ($\theta = 0$); happily, these points do not lie in the relevant domain of definition. Combining (2.127) with (2.14), one sees that

$$\mathbf{A}^H(x) = \theta_L^3(x) \otimes T_3. \tag{2.221}$$

If $x_{(n)}^a$ are the coordinates of $\mathrm{Exp}(x_{(n)}^a T_a)$ in an arbitrary Lie group G (whenever these coordinates, which are normal in the sense of II.3.1, are defined), one has the relation

$$\theta_L^a(x) = M(x)_b^a dx_{(n)}^b, \tag{2.222}$$

where

$$M(x) := \left(1 - e^{-\mathrm{Ad}(x)}\right) \mathrm{Ad}(x)^{-1}, \tag{2.223}$$

regarded as a matrix acting on $\mathfrak{g}$ relative to the basis $\{T_a\}$. For $G = SO(3)$ the matrix M can be calculated explicitly, yielding

$$M(x)_b^a = \delta_b^a \frac{\sin\|x\|}{\|x\|} + \frac{x_{(n)}^b x_{(n)}^a}{\|x\|^2}\left(1 - \frac{\sin\|x\|}{\|x\|}\right) + \frac{\epsilon_{abc} x_{(n)}^c}{\|x\|^2}(1 - \cos\|x\|), \tag{2.224}$$

where $\|x\|^2 := x_{(n)}^a x_{(n)}^a$. We now use the identity

$$\mathrm{Exp}(\phi T_3)\mathrm{Exp}(\theta T_2)\mathrm{Exp}(-\phi T_3) = \mathrm{Exp}(-\theta \sin\phi\, T_1 + \theta \cos\phi\, T_2). \tag{2.225}$$

Hence the normal coordinates of $s_-(\phi, \theta)$ are $(-\theta \sin\phi, \theta\cos\phi, 0)$; note that therefore $\|x\|^2 = \theta^2$. The object A_- can now be computed from (2.221), (2.222), and (2.224). To find A_+ we apply the gauge transformation (2.70), using (2.219); cf. the comment following the proof of 2.4.4. This yields $A_+ = A_- + 2d\phi \otimes T_3$, and one obtains (2.220). ∎

Being gauge-invariant (for an abelian structure group), the curvature $F = s_{\pm}^* \mathbf{F}^H = dA_{\pm}$ is the same for A_+ and A_-, and is given by

$$F(\phi, \theta) = -\sin\theta \, d\theta \wedge d\phi \otimes T_3; \tag{2.226}$$

this is $-T_3$ tensored with the volume 2-form on S^2 with respect to the $SO(3)$-invariant metric $\mathbf{g}^{S^2}$ defined by $\mathbf{g}$ (cf. 2.7.5). The field $A_{\pm}$ describes a **magnetic monopole** (of unit strength) located at the origin of $\mathbb{R}^3$.

By the theory of the H-connection, in particular Corollary 2.7.4, combined with the uniqueness of the reductive decomposition (2.126) of $\mathfrak{so}(3)$, the magnetic monopole field is the unique $SO(3)$-invariant connection on the bundle $SO(3)(S^2, SO(2), \tau)$. (Note that the local forms $A_{\pm}$ are only $SO(3)$-invariant up to a gauge transformation.)

Since $H = SO(2)$, whose coadjoint action on $\mathfrak{h}^* = \mathbb{R}$ is trivial, the coadjoint orbits of H are simply points in $\mathbb{R}$. Physically being the electric charge, these orbits are traditionally denoted by the symbol e. From (2.53) or (2.130), combined with (2.116), we conclude that the reduced space $(T^*G)^e$ is simply T^*S^2 as a manifold. Identifying $(T^*G)^e$ with T^*S^2 automatically implements the diffeomorphism (2.53), so that the Poisson bracket on $(T^*G)^e$ is given by (2.55). Since H is abelian, the C_{ij}^k in that expression vanish, so that from (2.226) we obtain (with $q^1 = \phi$ and $q^2 = \theta$)

$$\{f, g\}_*^e = \frac{\partial f}{\partial p_\mu}\frac{\partial g}{\partial q^\mu} - \frac{\partial f}{\partial q^\mu}\frac{\partial g}{\partial p_\mu} + e\sin\theta\left(\frac{\partial f}{\partial p_\theta}\frac{\partial g}{\partial p_\phi} - \frac{\partial f}{\partial p_\phi}\frac{\partial g}{\partial p_\theta}\right). \tag{2.227}$$

We now compute the momentum map J^e for the reduced $SO(3)$-action on $(T^*G)^e$ with respect to the sections $s_{\pm}$. This is done by specializing (2.134) to the case at hand. Firstly, for $e = 0$ one has

$$J^0(p_\phi, p_\theta, \phi, \theta)_{\pm} = \begin{pmatrix} -\sin\phi\cot\theta \, p_\phi + \cos\phi \, p_\theta \\ \cos\phi\cot\theta \, p_\phi + \sin\phi \, p_\theta \\ p_\phi \end{pmatrix}. \tag{2.228}$$

This follows from the well-known expression for the vector fields $\xi_a := \xi_{T_a}^{S^2}$ generating the $SO(3)$-action on S^2; these are nothing but J^0, with $p_{\dots}$ replaced by $\partial/\partial_{\dots}$. Because $\theta = 0$ in (2.81), the "free" momentum map (2.228) is independent of the section.

Furthermore, since the coadjoint representation of $SO(3)$ is the same as its defining representation (as is its adjoint one), the term $\mathrm{Co}(s_{\pm}(\phi, \theta))_a^3\theta_3$ occurring in (2.134) is simply $R(\phi, \theta, \pm\phi)_{a3}$. Seen as a vector in $\mathbb{R}^3$ this is given by (2.216), and evidently coincides with the unit vector pointing at (ϕ, θ). Using (2.220) one therefore obtains (omitting the argument $(p_\phi, p_\theta, \phi, \theta)$)

$$J_{\pm}^e = J^0 + e\begin{pmatrix} \dfrac{\sin\phi}{\sin\theta}(1 \pm \cos\theta) \\ \dfrac{-\cos\phi}{\sin\theta}(1 \pm \cos\theta) \\ \pm 1 \end{pmatrix}. \tag{2.229}$$

Since by (2.81) the map $\rho_0^e(\varphi_{g_{+-}})$ sends $(p_\phi, p_\theta, \phi, \theta)$ to $(p_\phi + 2e, p_\theta, \phi, \theta)$, one has the relation $\rho_0^e(\varphi_{g_{+-}})^* J_+^e = J_-^e$. Hence J^e is gauge-covariant, as it should be; cf. the comment following the proof of Theorem 2.5.3.

The Hamiltonian h^e on $G^e = T^*S^2$ is equivalently given by (2.107) or by (2.137); using (2.217) and (2.220) one obtains

$$h_\pm^e = \tfrac{1}{2}\left[p_\theta^2 + (\sin\theta)^{-2}(p_\phi - e(cos\theta \pm 1))^2\right]. \qquad (2.230)$$

Using (2.229) it may be verified that (2.137) equals $h_\pm^e + (eg)^2$; cf. the comment following (2.107). The gauge-covariance of $h_\pm^e$ is verified as in the case of $J_\pm^e$.

By the third Wong equation (2.99), whose right-hand side evidently vanishes, the charge e is conserved in time; this is obvious anyway, because the motion cannot leave the coadjoint orbit $\mathcal{O}$. The extra term in (2.227) then leads to a perturbation of the cogeodesic motion on T^*S^2.

The fact that H is abelian allows the introduction of a free parameter $g \in \mathbb{R}\backslash\{0\}$ in the definition of the principal bundle we started from. The group H_g, isomorphic to $SO(2)$, but parametrized by $\beta \in [0, 2\pi g)$, acts on $\mathsf{P} = SO(3)$ by $R_\beta R(\phi, \theta, \psi) := R(\phi, \theta, \psi - \beta/g)$. This leads to a modified bundle $\mathsf{P}(S^2, H_g, \tau)$, and has the effect that the right-hand side of (2.220) (and hence of (2.226)) should be multiplied by g. Consequently, the parameter e in (2.227), (2.229), and (2.230) becomes eg.

In the general context of mechanics on the bundle $SO(3)(S^2, SO(2), \tau)$, by Corollary 2.3.8 the significance of the classical parameter $e \in \mathfrak{so}(2)^* = \mathbb{R}$ is that it classifies the irreducible representations of the Poisson algebra $C^\infty((T^*SO(3))/SO(2), \mathbb{R})$. As we have seen (cf. Theorem 2.8.1), one should think of the JLB-algebra $\mathfrak{A}_\mathbb{R} = \mathfrak{B}_0(L^2(SO(3)))_\mathbb{R}^{SO(2)}$ as the quantization of this Poisson algebra. By Corollary 2.8.3, the irreducible representations of $\mathfrak{A}$ are classified by the unitary dual $\widehat{SO(2)} = \mathbb{Z}$ of $SO(2)$. Hence each integer $n \in \mathbb{Z}$ corresponds to an irreducible representation π^n of $\mathfrak{A}$. For $g = 1$ this integer is the quantum counterpart of the classical charge $e \in \mathbb{R}$; the rescaling $T_3 \mapsto T_3/g$ means that in this consideration e should be replaced by eg.

Labeling elements of $H = SO(2)$ by $\beta \in [0, 2\pi)$, the representation U_n is

$$U_n(\beta) = e^{-in\beta}. \qquad (2.231)$$

The Hilbert bundle H^n that the representation U_n associates to the principal bundle $SO(3)(S^2, SO(2), \tau)$ (cf. 2.8) is a line bundle over S^2 (that is, the typical fiber is $\mathbb{C}$). The Hilbert space $L^2(\mathsf{H}^n)$ of square-integrable sections of this line bundle carries both the irreducible representation π^n of the algebra of observables $\mathfrak{A}_\mathbb{R}$ and the induced representation U^n of $SO(3)$; cf. 2.8 and 2.9. It is therefore a central object in the quantum mechanics of a charged particle moving on S^2.

The first realization of $L^2(\mathsf{H}^n)$ is the space $\mathcal{H}^n$ of L^2-functions on $SO(3)$ satisfying the equivariance condition (2.145); in Euler angles this condition reads $\Psi^n(\phi, \theta, \psi + \beta) = \exp(in\beta)\Psi^n(\phi, \theta, \psi)$. Hence Ψ^n is $\exp(in\psi)$ times a function of (ϕ, θ). Therefore, this realization has no particular advantage over the other two, which directly work with functions on S^2.

The Hilbert space $\mathcal{H}^n_\pm := \mathcal{H}^n_{s_\pm}$, where one has to choose a sign, is

$$\mathcal{H}^n_\pm = L^2(S^2) = L^2\left([0, 2\pi] \times [0, \pi], \sin\theta d\theta d\phi\right); \tag{2.232}$$

cf. (2.141). The most important operators on $\mathcal{H}^n_\pm$ are the angular momenta

$$i\hbar dU^n(T_a) = \pi^n(\mathcal{Q}^W_\hbar(J_a)); \tag{2.233}$$

cf. (2.170). For $n = 0$ one obtains the well-known angular momentum operators (here expressed in shifted spherical coordinates; cf. (2.216))

$$\pi^0_\pm(\mathcal{Q}^W_\hbar(J)) = -i\hbar \begin{pmatrix} -\sin\phi\cot\theta\,\dfrac{\partial}{\partial\phi} + \cos\phi\,\dfrac{\partial}{\partial\theta} \\ \cos\phi\cot\theta\,\dfrac{\partial}{\partial\phi} + \sin\phi\,\dfrac{\partial}{\partial\theta} \\ \dfrac{\partial}{\partial_\phi} \end{pmatrix}. \tag{2.234}$$

As with its classical counterpart (2.228), this is independent of the section.

For general n, one may compute the left-hand side of (2.233) directly from the definition (1.69), or one evaluates the right-hand side using (2.181), (2.169), and (2.220). Either way, one obtains

$$\pi^n_\pm(\mathcal{Q}^W_\hbar(J)) = \pi^0(\mathcal{Q}^W_\hbar(J)) + n\hbar \begin{pmatrix} \dfrac{\sin\phi}{\sin\theta}(1 \pm \cos\theta) \\ \dfrac{-\cos\phi}{\sin\theta}(1 \pm \cos\theta) \\ \pm 1 \end{pmatrix}. \tag{2.235}$$

This expression, then, is the quantization of (2.229). It follows from Proposition 2.9.4 that these operators are essentially self-adjoint on the domain $\Gamma(\mathsf{H}^n)$ of smooth sections of the line bundle H^n. For the specification of this domain in the realization of $L^2(\mathsf{H}^n)$ as $\mathcal{H}^n_-$ one needs to take into account that the section s_- is discontinuous at the southpole. This complication may be resolved by observing that for $\theta \neq 0, \pi$ the section $\Psi^n_- \in C^\infty(\mathcal{N}_-)$ is related to $\Psi_+ \in C^\infty(\mathcal{N}_+)$ by

$$\Psi^n_-(\phi, \theta) = e^{-2in\phi}\Psi^n_+(\phi, \theta), \tag{2.236}$$

compare (2.151) combined with (2.219) and (2.231). Accordingly, Ψ^n_- is in $\Gamma(\mathsf{H}^n)$ iff it is in $C^\infty((0, 2\pi) \times (0, \pi))$ and in addition satisfies the boundary conditions

$$\lim_{\theta\to 0} \frac{\partial}{\partial\phi}\Psi^n_-(\phi, \theta) = 0; \tag{2.237}$$

$$\lim_{\theta\to\pi} \left(\frac{\partial}{\partial\phi} + 2in\right)\Psi^n_-(\phi, \theta) = 0; \tag{2.238}$$

$$\lim_{\phi\to 0} \frac{\partial^k}{\partial\phi^k}\Psi^n_-(\phi, \theta) = \lim_{\phi\to 2\pi} \frac{\partial^k}{\partial\phi^k}\Psi^n_-(\phi, \theta) \tag{2.239}$$

for all $k \in \{0 \cup \mathbb{N}\}$. It is interesting to verify that condition (2.238) guarantees that the operators $\pi^n_-(\mathcal{Q}^W_\hbar(J))$ are well-defined on $\Gamma(\mathsf{H}^n)$, in that the differential operators conspire with the boundary condition so as to effectively replace $(1 - \cos\theta)/\sin\theta$ in the second term by $(1 + \cos\theta)/\sin\theta$ near $\theta = \pi$.

On $\mathcal{H}^n_+$ the conditions for Ψ^n_+ to be in $\Gamma(\mathsf{H}^n)$ are analogous: One requires (2.237) for $\theta \to \pi$ rather than $\theta \to 0$, and in (2.238) one replaces $\theta \to \pi$ by $\theta \to 0$ and $+2in$ by $-2in$. Condition (2.239) is the same.

The story on $\mathcal{H}^n_{\{s\}}$ is entirely analogous: The pair (Ψ^n_-, Ψ^n_+) comprising $\Psi^n_{\{s\}}$ is related by (2.236), and to be in $\Gamma(\mathsf{H}^n)$ one imposes the above conditions for $\mathcal{H}^n_\pm$.

The quantum Hamiltonian $H^n_\hbar$ is given by (2.184), which may be evaluated using either (2.194) or (2.183) with (2.217) and (2.220), as well as (2.226) and $\mathbf{R}_{S^2} = 1$, $C^n_2(SO(2)) = n^2$. There is not much point in writing down the resulting expression. The important facts are that $H^n_\hbar$ is essentially self-adjoint on $\Gamma(\mathsf{H}^n)$ (cf. Corollary 2.10.4), and that the parameter e, or, more generally, eg, in the classical Hamiltonian h^e is replaced by $n\hbar$ in the quantum Hamiltonian. The same substitution applies to the angular momentum (2.235). This phenomenon is the **Dirac quantization condition** $eg = n\hbar$, expressing the fact that in quantum theory electric charge is quantized in the presence of a magnetic monopole field.

3 Lie Groupoids and Lie Algebroids

3.1 Groupoids

A groupoid is a certain generalization of a group, in which multiplication is only partially defined. When it is defined, it is associative, and each element has an inverse in a suitable sense. Lie groupoids, which are groupoids with an appropriate smooth structure, and their infinitesimal objects, Lie algebroids, enable one to give a unified description of a large class of examples in quantization theory. Although the definition of a groupoid appears complicated, it will become clear through the examples that one is studying a most natural object.

Definition 3.1.1. *A* **groupoid** $\mathsf{G}(Q, \tau_s, \tau_t, \iota, \cdot, I)$, *sometimes written as* $\mathsf{G} \overset{\leftarrow}{\Rightarrow} Q$, *consists of a set* G *(the* **total space***), a set* Q *(the* **base***), a map* $\tau_s : \mathsf{G} \to Q$ *(the* **source projection***), a map* $\tau_t : \mathsf{G} \to Q$ *(the* **target projection***), a map* $\iota : Q \hookrightarrow \mathsf{G}$ *(the* **inclusion***), and a multiplication* $\cdot : \mathsf{G}_2 \to \mathsf{G}$, *where*

$$\mathsf{G}_2 := \mathsf{G} *_Q \mathsf{G} = \{(\gamma_1, \gamma_2) \in \mathsf{G} \times \mathsf{G} \mid \tau_s(\gamma_1) = \tau_t(\gamma_2)\}, \tag{3.1}$$

and a map $I : \mathsf{G} \to \mathsf{G}$ *(the* **inversion***), subject to the following conditions. We write* $\gamma_1\gamma_2$ *for* $\gamma_1 \cdot \gamma_2$.

1. *If* $(\gamma_1, \gamma_2) \in \mathsf{G}_2$ *(so that* $\gamma_1\gamma_2$ *is defined), then* $\tau_s(\gamma_1\gamma_2) = \tau_s(\gamma_2)$ *and* $\tau_t(\gamma_1\gamma_2) = \tau_t(\gamma_1)$.
2. *If* $\tau_s(\gamma_1) = \tau_t(\gamma_2)$ *and* $\tau_s(\gamma_2) = \tau_t(\gamma_3)$ *(so that* $(\gamma_1\gamma_2)\gamma_3$ *and* $\gamma_1(\gamma_2\gamma_3)$ *are defined), then* $(\gamma_1\gamma_2)\gamma_3 = \gamma_1(\gamma_2\gamma_3)$.
3. *One has* $\tau_s(\iota(q)) = \tau_t(\iota(q)) = q$ *for all* $q \in Q$, *and* $\gamma\iota(\tau_s(\gamma)) = \iota(\tau_t(\gamma))\gamma = \gamma$ *for all* $\gamma \in \mathsf{G}$.

4. *Writing* $\gamma^{-1} := I(\gamma)$, *the inversion satisfies* $\tau_s(\gamma^{-1}) = \tau_t(\gamma)$, $\tau_t(\gamma^{-1}) = \tau_s(\gamma)$, $\gamma^{-1}\gamma = \iota(\tau_s(\gamma))$, *and* $\gamma\gamma^{-1} = \iota(\tau_t(\gamma))$.

Note that the second equation in 3.1.1.3 is well-defined on account of its predecessor; similarly, the third and fourth equations in 3.1.1.4 are meaningful because of the first and the second ones, respectively. It follows from 3.1.1.3 that τ_s and τ_t are onto (i.e., surjective).

It can be shown from these axioms that the inverse is unique, so that one could omit I from the data specifying a groupoid, and require its existence subject to the last condition above. It also follows that I is involutive with respect to multiplication; that is,

$$(\gamma^{-1})^{-1} = \gamma; \tag{3.2}$$

$$(\gamma_1\gamma_2)^{-1} = \gamma_2^{-1}\gamma_1^{-1}. \tag{3.3}$$

In addition, 3.1.1.3 and 3.1.1.4 imply

$$\gamma\gamma^{-1}\gamma = \gamma. \tag{3.4}$$

In fact, when $\gamma_0 \in \iota(Q)$ and $(\gamma, \gamma_0) \in \mathsf{G}_2$, so that $\gamma\gamma_0$ is defined, then $\gamma\gamma_0 = \gamma$. This follows since $\gamma_0 = \iota(\tau_s(\gamma_1))$ for some γ_1; then $\gamma_0 = \gamma_1^{-1}\gamma_1$ by 3.1.1.4, but since $\tau_s(\gamma) = \tau_t(\gamma_0) = \tau_t(\gamma_1^{-1}\gamma_1) = \tau_s(\gamma_1)$, we see that $\gamma_0 = \iota(\tau_s(\gamma_1)) = \gamma^{-1}\gamma$, so that finally $\gamma\gamma_0 = \gamma\gamma^{-1}\gamma = \gamma$ by 3.1.1.4, as claimed. Similarly, when $(\gamma_0, \gamma) \in \mathsf{G}_2$, then $\gamma_0\gamma = \gamma$.

Hence elements of $\iota(Q)$ act like units for the partially defined multiplication, and one sometimes calls an element of the form $\iota(q)$ a **unit**, referring to $\mathsf{G}_0 := \iota(Q)$ as the **unit space** in G. For $\gamma_0 \in \mathsf{G}$ to be a unit it suffices to find one γ for which $\gamma\gamma_0 = \gamma$, for in that case the above argument implies that $\gamma_0 = \iota(\tau_s(\gamma))$.

One thinks of elements of Q as "objects", and of elements of G as "arrows". The arrow γ then points from $\tau_s(\gamma)$ to $\tau_t(\gamma)$, and has an inverse γ^{-1} pointing in the opposite direction. Arrows are composed from right to left; two arrows γ_1, γ_2 can be composed to $\gamma_1\gamma_2$ iff the endpoint of γ_2 matches the starting point of γ_1. The arrow $\iota(q)$ connects q with itself, but it may not be the only arrow to do so. The collection of all arrows connecting q with itself is the **isotropy group**

$$\mathsf{G}_q := \tau_s^{-1}(q) \cap \tau_t^{-1}(q) \tag{3.5}$$

of q; this is clearly a group under the multiplication inherited from G. We conclude from the preceding paragraph that $\iota(q)$ is the unit element of the group G_q in the usual sense.

An equivalent definition of a groupoid is obtained by saying that a groupoid consists of a set G, a subset $\mathsf{G}_2 \subset \mathsf{G} \times \mathsf{G}$, a map $\gamma \mapsto \gamma^{-1}$ from G to G, and a map $\cdot : \mathsf{G}_2 \to \mathsf{G}$, such that: (i) G_2 contains $\{(\gamma, \gamma^{-1}) \mid \gamma \in \mathsf{G}\}$; (ii) if $(\gamma_1, \gamma_2) \in \mathsf{G}_2$ and $(\gamma_2, \gamma_3) \in \mathsf{G}_2$, then $(\gamma_1\gamma_2, \gamma_3)$ and $(\gamma_1, \gamma_2\gamma_3)$ are in G_2, and $(\gamma_1\gamma_2)\gamma_3 = \gamma_1(\gamma_2\gamma_3)$; (iii) $(\gamma^{-1})^{-1} = \gamma$; (iv) if $(\gamma_1, \gamma_2) \in \mathsf{G}_2$, then $(\gamma_1\gamma_2)\gamma_2^{-1} = \gamma_1$ and $\gamma_1^{-1}(\gamma_1\gamma_2) = \gamma_2$.

The connection with Definition 3.1.1 is then established by identifying Q with G_0 through ι, which in turn is identified with the set $\{\gamma\gamma^{-1} \mid \gamma \in \mathsf{G}\}$. The projections

τ_t and τ_s are given by $\tau_t(\gamma) = \gamma\gamma^{-1}$ and $\tau_s(\gamma) = \gamma^{-1}\gamma$. The equivalence of the two definitions is then easily checked.

We see that a group $\mathsf{G} = G$ is a special case of a groupoid, in which Q consists of one point e, identified with its image $\iota(e) \in G$, which is the unit in the group. Since $\tau_t(\gamma) = \tau_s(\gamma) = e$ for all $\gamma \in G$, all elements can be multiplied, and the group axioms follow from 3.1.1. At the opposite extreme we have

Definition 3.1.2. *The* **pair groupoid** $Q \times Q \overset{\leftarrow}{\Rightarrow} Q$ *is defined by the operations* $\tau_s(q_1, q_2) := q_2$, $\tau_t(q_1, q_2) := q_1$, $\iota(q) := (q, q)$, $(q_1, q_2) \cdot (q_2, q_3) := (q_1, q_3)$, *and* $(q_1, q_2)^{-1} := (q_2, q_1)$.

This time $\mathsf{G}_q = (q, q)$ consists of one point only.

When $\mathsf{H} \subset \mathsf{G}$ is closed under multiplication and inverses, we can form the **subgroupoid** $\mathsf{H}(\tau_s(\mathsf{H}), \tau_s, \tau_t, \iota, \cdot, I)$; note that $\tau_t(\mathsf{H}) = \tau_s(\mathsf{H})$. A subgroupoid of the pair groupoid $Q \times Q$ that contains all identities is evidently the same as an equivalence relation on Q; this follows from straightforward definition-chasing.

More generally, a groupoid $\mathsf{G} \overset{\leftarrow}{\Rightarrow} Q$ gives rise to an equivalence relation on Q: the map $\tau_t \times \tau_s : \mathsf{G} \to Q \times Q$ is a morphism (in the obvious sense) of G into the pair groupoid $Q \times Q \overset{\leftarrow}{\Rightarrow} Q$, whose image is a subgroupoid of the latter. By 3.1.1.3 one has $\tau_t \times \tau_s(\iota(q)) = (q, q)$, so that the image contains all identities. From the previous paragraph one therefore obtains an equivalence relation $\sim$ on Q. This relation is simply that $q \sim q'$ iff there exists a $\gamma \in \mathsf{G}$ for which $\tau_s(\gamma) = q$ and $\tau_t(\gamma) = q'$. Hence the equivalence class of q is the same as the **orbit** $\mathsf{G}q$ of q under G, if we define the latter as the set of all $q' \in Q$ for which there is a $\gamma \in \mathsf{G}$ satisfying $\tau_s(\gamma) = q$ and $\tau_t(\gamma) = q'$. If $\mathsf{G}q = Q$, then G is said to be **transitive**.

Proposition 3.1.3.

- *A groupoid is the disjoint union of transitive subgroupoids.*
- *A transitive subgroupoid* $\mathsf{G} \overset{\leftarrow}{\Rightarrow} Q$ *is isomorphic to a groupoid of the form* $Q \times H \times Q \overset{\leftarrow}{\Rightarrow} Q$, *where the group* H *is isomorphic to the isotropy group* G_{q_0} *of an arbitrary point* $q_0 \in Q$.

Here the groupoid operations in $Q \times H \times Q \overset{\leftarrow}{\Rightarrow} Q$ *are* $\tau_s(q_1, h, q_2) := q_2$, $\tau_t(q_1, h, q_2) := q_1$, $\iota(q) := (q, e, q)$, $(q_1, h, q_2) \cdot (q_2, k, q_3) := (q_1, hk, q_3)$, *and* $(q_1, h, q_2)^{-1} := (q_2, h^{-1}, q_1)$.

When G is not transitive, each orbit $\mathsf{G}q$ in Q gives rise to a transitive subgroupoid $\mathsf{G}^q := \tau_s^{-1}(\mathsf{G}q) = \tau_t^{-1}(\mathsf{G}q)$. When $\mathsf{G} \overset{\leftarrow}{\Rightarrow} Q$ is transitive, one chooses an arrow $\gamma_0(q) : q_0 \to q$ for each q, in terms of which $(q_1, h, q_2) \mapsto \gamma_0(q_1)h\gamma_0(q_2)^{-1}$ is an isomorphism between $Q \times \mathsf{G}_{q_0} \times Q$ and G. ■

Definition 3.1.4. *Let a group* G *act on a set* Q. *The* **action groupoid** $G \times Q \overset{\leftarrow}{\Rightarrow} Q$ *is defined by the operations* $\tau_s(x, q) := x^{-1}q$ *and* $\tau_t(x, q) := q$, *so that the product* $(x, q) \cdot (y, q')$ *is defined when* $q' = x^{-1}q$. *Then* $(x, q) \cdot (y, x^{-1}q) := (xy, q)$. *The inclusion is* $\iota(q) := (e, q)$, *and for the inverse one has* $(x, q)^{-1} := (x^{-1}, x^{-1}q)$.

Hence (x, q) is an arrow from $x^{-1}q$ to q. The isotropy group G_q coincides with the usual isotropy group G_q of the G-action.

Definition 3.1.5. *A* **Lie groupoid** *is a groupoid* $\mathsf{G}(Q, \tau_s, \tau_t, \iota, \cdot, I)$, *where* G *and* Q *are manifolds, the maps* τ_s *and* τ_t *are surjective submersions, and multiplication and inclusion are smooth.*

Proposition 3.1.6. *In a Lie groupoid:*

1. *The inclusion* ι *is an immersion.*
2. *The inverse* I *is a diffeomorphism.*
3. G_2 *is a closed submanifold of* $\mathsf{G} \times \mathsf{G}$.
4. *For each* $q \in Q$ *the fibers* $\tau_s^{-1}(q)$ *and* $\tau_t^{-1}(q)$ *are submanifolds of* G.
5. *The isotropy group of any point* q *is a Lie group.*

We omit the proof, which is a nontrivial exercise in differential geometry.

In the second definition of a groupoid given above one obtains a Lie groupoid by requiring that G be a manifold and that inversion and multiplication be smooth.

A Lie group is clearly a Lie groupoid. When Q is a manifold, the pair groupoid $Q \times Q \overset{\leftarrow}{\Rightarrow} Q$ is a Lie groupoid. Thirdly, when the G-action on Q is smooth, the action groupoid $G \times Q \overset{\leftarrow}{\Rightarrow} Q$ is a Lie groupoid.

A new example of a Lie groupoid may be constructed if one is given a principal H-bundle. Recall that the equivalence class $[x, y]_H$ is defined by the equivalence relation $(x, y) \sim (xh, yh)$ for all $h \in H$.

Definition 3.1.7. *The* **gauge groupoid** $\mathsf{P} \times_H \mathsf{P} \overset{\leftarrow}{\Rightarrow} Q$ *of a principal bundle* $\mathsf{P}(Q, H, \tau)$ *is defined by the projections* $\tau_s([x, y]_H) := \tau(y)$ *and* $\tau_t([x, y]_H) := \tau(x)$, *and the inclusion* $\iota(\tau(x)) := [x, x]_H$. *Accordingly, the multiplication* $[x, y]_H \cdot [x', y']_H$ *is defined when* y *and* x' *lie in the same fiber of* P, *in which case* $[x', y']_H = [y, z]_H$ *for some* $z = y'h$, $h \in H$. *Then* $[x, y]_H \cdot [y, z]_H := [x, z]_H$. *Finally, the inverse is* $[x, y]_H^{-1} := [y, x]_H$.

The isotropy group G_q consists of all $[x, y]_H$ for which $\tau(x) = \tau(y) = q$. Each $\gamma \in \mathsf{G}_q$ is of the form $[s(q)h, s(q)]_H$, where s is an arbitrary section of P. Hence $[s(q)h_1, s(q)]_H \cdot [s(q)h_2, s(q)]_H = [s(q)h_1h_2, s(q)]_H$, so that for all $q \in Q$ one concludes that $\mathsf{G}_q \simeq H$ as a group.

It is plain that a gauge groupoid is transitive. If the bundle P is trivial and brought into the form $\mathsf{P} = Q \times H$, there is a smooth isomorphism $\mathsf{P} \times_H \mathsf{P} \overset{\leftarrow}{\Rightarrow} Q \simeq Q \times H \times Q \overset{\leftarrow}{\Rightarrow} Q$ (see 3.1.3). This is given by $[(q_1, h), (q_2, e)]_H \mapsto (q_1, h, q_2)$. More generally, when P is nontrivial, a section $s : Q \to \mathsf{P}$ leads to a nonsmooth isomorphism of the above type through Proposition 3.1.3. For one may choose $\gamma_0(q) := [s(q), s(q_0)]_H$ to obtain the isomorphism $[s(q_1)h, s(q_2)]_H \mapsto (q_1, h, q_2)$.

We learn from Proposition 2.4.2 that the space of sections of the total space $\mathsf{P} \times_H \mathsf{P}$ of a gauge groupoid may be identified with the group of automorphisms of the bundle $\mathsf{P}(Q, H, \tau)$.

Proposition 3.1.8. *Let* $G(G/H, H, \tau)$ *be the principal* H*-bundle defined in 2.7. Then the gauge groupoid* $G \times_H G \overset{\leftarrow}{\Rightarrow} G/H$ *and the action groupoid* $G \times (G/H) \overset{\leftarrow}{\Rightarrow} G/H$ *are isomorphic (in the obvious sense).*

The correspondence $[x, y]_H \leftrightarrow (xy^{-1}, \tau(x))$, combined with the identity map on the base Q, is easily seen to provide the desired isomorphism. ■

3.2 Half-Densities on Lie Groupoids

As we have seen in 1.7, one can associate a C^*-algebra $C^*(G)$ with a Lie group G, and establish a correspondence between representations of G and representations of $C^*(G)$. One can generalize this to Lie groupoids. The material in this section is preparatory for this purpose.

For a vector bundle V over a manifold M with n-dimensional typical fiber V, the bundle $\mathsf{A}(\mathsf{V})$ is defined as $\wedge^n\mathsf{V}$ minus the zero section. This is a principal $\mathbb{C}^*$-bundle over M, whose fiber at x is the n-fold antisymmetric tensor product of V_x (the fiber of V at x), with 0 omitted (here $\mathbb{C}^*$ is $\mathbb{C}\backslash\{0\}$, seen as a multiplicative group). We write $\mathsf{A}M := \mathsf{A}(TM)$.

Definition 3.2.1. *For $\alpha > 0$, the* **bundle of α-densities** $|\Lambda|^\alpha(\mathsf{V})$ *is the line bundle over M associated to $\mathsf{A}(\mathsf{V})(M, \mathbb{C}^*, \tau)$ by the representation $z \mapsto |z|^{-\alpha}$ of $\mathbb{C}^*$ on $\mathbb{C}$. An* **α-density** *on V is a section of the bundle of α-densities. We put $|\Lambda|^\alpha M := |\Lambda|^\alpha(TM)$.*

A 1-density is called simply a density; of interest to us are the cases $\alpha = 1$ and $\alpha = \frac{1}{2}$. According to Proposition 2.1.5, we describe a section Ψ^α of $|\Lambda|^\alpha M$ as a (smooth and compactly supported) map $\Psi^\alpha : \mathsf{A}M \to \mathbb{C}$ satisfying $\Psi^\alpha(\lambda X) = |\lambda|^\alpha \Psi^\alpha(X)$ for all $\lambda \in \mathbb{R}$ and all $X \in \mathsf{A}M$. Such a section may be represented by an equivalence class $\Psi^\alpha = [f, \nu]_\alpha$, where $f \in C_c^\infty(M)$ and ν is a positive measure on M that is locally Lebesgue. The equivalence relation defining the class $[f, \nu]_\alpha$ is $(f, \nu) \sim (g, \mu)$ when $g = f(d\nu/d\mu)^\alpha$ (the Radon–Nikodym derivative $d\nu/d\mu$ is well-defined, since μ and ν are equivalent). It follows from the multiplicative property of the Radon–Nikodym derivative that this is indeed an equivalence relation. The section defined by the class $[f, \nu]_\alpha$ takes the form

$$[f, \nu]_\alpha(\partial_1 \wedge \cdots \wedge \partial_n) := f(x)\left(\frac{d\nu}{d\mu_L}(x)\right)^\alpha, \tag{3.6}$$

where $\partial_1 \wedge \cdots \wedge \partial_n \in \mathsf{A}_x M$, and $d\mu_L := dx^1 \cdots dx^n$ is the Lebesgue measure in the particular chart used. The point of introducing densities is that they can be integrated over M (even if it is not orientable) without specifying a measure, i.e., when Ψ^1 is a density, the object

$$\int_M \Psi^1 := \int_M d^n x\, \Psi^1\left(\wedge_{i=1}^n \frac{\partial}{\partial x^i}\right) \tag{3.7}$$

is independent of the choice of (local) coordinates; cf. the usual definition of the integral of an n-form over M. In the realization of Ψ^α as an equivalence class $[f, \nu]_\alpha$ this simply reads $\int_M [f, \nu]_1 = \int_M d\nu f$. Similarly, since $\Psi^{\alpha_1}\Psi^{\alpha_2}$ is an element of $\Gamma(|\Lambda|^{\alpha_1+\alpha_2} M)$, the product of two half-densities may be integrated on M; here

$$\int_M [f, \nu]_{\frac{1}{2}} [g, \mu]_{\frac{1}{2}} = \int_M d\nu \sqrt{\frac{d\mu}{d\nu}} fg = \int_M d\mu \sqrt{\frac{d\nu}{d\mu}} fg. \tag{3.8}$$

Accordingly, one can form the **Hilbert space of half-densities** $L^2(M)$.

Proposition 3.2.2. *Each bundle $|\Lambda|^\alpha M$ is trivial, and any measure μ on M that is locally Lebesgue defines a trivialization ψ_μ of $|\Lambda|^\alpha M$. Accordingly, μ establishes a natural bijection between $\Gamma(|\Lambda|^\alpha M)$ and $C_c^\infty(M)$.*

Since the notation of 2.1.4 would contain no information about α, we write elements of $|\Lambda|^\alpha M$ as equivalence classes $[X, \lambda]_\alpha$ (rather than $[X, \lambda]_{\mathbb{C}^*}$). In a given local chart we define a local trivialization of $|\Lambda|^\alpha M$ by

$$\psi_\mu([\partial_1 \wedge \cdots \wedge \partial_n, \lambda]_\alpha) := \left(\frac{d\mu_L}{d\mu}\right)^\alpha \lambda. \tag{3.9}$$

By the $\mathbb{C}^*$-equivariance property $[\rho X, \lambda]_\alpha = [X, |\rho|^{-\alpha}\lambda]_\alpha$ this is actually independent of the chart, and defines a global trivialization.

A section $[f, \nu]_\alpha$ then trivializes to $f_\mu^\nu : M \to \mathbb{C}$, given by

$$f_\mu^\nu(x) = f(x)\left(\frac{d\nu}{d\mu}(x)\right)^\alpha. \tag{3.10}$$

Conversely, a function $f : M \to \mathbb{C}$ corresponds to a section $[f, \mu]_\alpha$. ∎

When applicable, integration of such trivialized sections is then done with respect to μ, on which the numerical value of the integral does not depend.

Let now $\mathsf{G} \overset{\leftarrow}{\rightrightarrows} Q$ be a Lie groupoid. Recall from 3.1 that $q \sim q'$ on Q when there exists $\gamma \in \mathsf{G}$ for which $\tau_s(\gamma) = q$ and $\tau_t(\gamma) = q'$.

Lemma 3.2.3. *The fibers $\mathsf{G}_\gamma^s := \tau_s^{-1}(\tau_s(\gamma))$ and $\mathsf{G}_\gamma^t := \tau_t^{-1}(\tau_t(\gamma))$ are diffeomorphic; their common dimension is denoted by d^γ.*

The fibers G_γ^s and $\mathsf{G}_{\gamma'}^s$ (and similarly G_γ^t and $\mathsf{G}_{\gamma'}^t$) are diffeomorphic when $\tau_s(\gamma)$ and $\tau_s(\gamma')$ are equivalent.

If $q \sim q'$ by such a γ, then R_γ, defind by

$$R_\gamma(\gamma') := \gamma'\gamma \tag{3.11}$$

when $(\gamma', \gamma) \in \mathsf{G}_2$, maps $\tau_s^{-1}(\tau_t(\gamma))$ into $\tau_s^{-1}(\tau_s(\gamma))$; since γ has an inverse, this map is a diffeomorphism. Similarly, L_γ, defined by

$$L_\gamma(\gamma') := \gamma\gamma' \tag{3.12}$$

whenever $(\gamma, \gamma') \in \mathsf{G}_2$, maps $\tau_t^{-1}(\tau_s(\gamma))$ diffeomorphically to $\tau_t^{-1}(\tau_t(\gamma))$. The inversion $I : \gamma \mapsto \gamma^{-1}$ is a diffeomorphism between $\tau_s^{-1}(\tau_s(\gamma))$ and $\tau_t^{-1}(\tau_s(\gamma))$, which is equal to $\tau_t^{-1}(\tau_t(\gamma^{-1})$. Thus we have

$$\begin{aligned} R_{\gamma^{-1}} &: \mathsf{G}_\gamma^s \to \mathsf{G}_{\gamma^{-1}}^s; \\ L_{\gamma^{-1}} &: \mathsf{G}_\gamma^t \to \mathsf{G}_{\gamma^{-1}}^t; \\ I &: \mathsf{G}_\gamma^s \to \mathsf{G}_{\gamma^{-1}}^s. \end{aligned} \tag{3.13}$$

Hence $R_\gamma \circ I : \mathsf{G}_\gamma^t \to \mathsf{G}_\gamma^s$ is a diffeomorphism; its inverse is $L_\gamma \circ I : \mathsf{G}_\gamma^s \to \mathsf{G}_\gamma^t$.

Since $\tau_t(\gamma^{-1}) = \tau_s(\gamma)$ and $\tau_t(\gamma)$ are equivalent points in Q (namely by γ), the last claim follows as well. ∎

Lemma 3.2.4. *For $\gamma_1 \in \tau_t^{-1}(\tau_s(\gamma))$ there are canonical isomorphisms*

$$\begin{aligned} T_{\gamma_1^{-1}}\mathsf{G}^s_{\gamma_1^{-1}} &\simeq T_{\gamma_1}\mathsf{G}^t_{\gamma_1};\\ T_{\gamma\gamma_1}\mathsf{G}^t_{\gamma\gamma_1} &\simeq T_{\gamma_1}\mathsf{G}^t_{\gamma_1};\\ T_{\gamma\gamma_1}\mathsf{G}^s_{\gamma\gamma_1} &\simeq T_{\gamma}\mathsf{G}^s_{\gamma};\\ T_{\gamma_1^{-1}}\mathsf{G}^t_{\gamma_1^{-1}} &\simeq T_{\gamma}\mathsf{G}^t_{\gamma}. \end{aligned} \tag{3.14}$$

These isomorphisms are given by the pushforwards of the inversion map, of $L_{\gamma^{-1}}$, of $R_{\gamma_1^{-1}}$, and of $L_{\gamma\gamma_1}$, respectively. ■

The space $\mathsf{A}^s\mathsf{G}$ (or $\mathsf{A}^t\mathsf{G}$) is defined as the line bundle over G whose fiber at γ is the complexified d^γ-fold antisymmetric tensor product of $T_\gamma\mathsf{G}^s_\gamma$ (or $T_\gamma\mathsf{G}^t_\gamma$); this fiber is evidently one-dimensional. Note that $T_\gamma\mathsf{G}^s_\gamma$ is the kernel of $(\tau_s)_*$, etc. We can then form the tensor product $\mathsf{A}^s\mathsf{G}\otimes\mathsf{A}^t\mathsf{G}$.

Definition 3.2.5. *The symbol $\mathsf{A}^{s\otimes t}\mathsf{G}$ stands for the line bundle $\mathsf{A}^s\mathsf{G}\otimes\mathsf{A}^t\mathsf{G}$ with the zero section omitted; this is a principal $\mathbb{C}^*$-bundle over G.*

An $s\otimes t$-density on G is a smooth compactly supported section of the line bundle $\sqrt{|\Lambda|}^{s\otimes t}\mathsf{G}$ associated to $\mathsf{A}^{s\otimes t}\mathsf{G}$ by the representation $z\mapsto |z|^{-\frac{1}{2}}$ of $\mathbb{C}^*$.

One may equivalently define $\sqrt{|\Lambda|}^{s\otimes t}\mathsf{G}$ as $\sqrt{|\Lambda|}^{s}\otimes\sqrt{|\Lambda|}^{t}$, where $\sqrt{|\Lambda|}^{s,t}$ is the line bundle associated to $\mathsf{A}^{s,t}\mathsf{G}$, minus the zero section, seen as a principal $\mathbb{C}^*$-bundle over G, by the representation $z\mapsto|z|^{-\frac{1}{2}}$.

Pulling back, Lemma 3.2.4 leads to an isomorphism

$$\begin{aligned} \lambda_{\gamma_1}: \sqrt{|\Lambda|}^s_{\gamma\gamma_1}\mathsf{G}\otimes\sqrt{|\Lambda|}^t_{\gamma\gamma_1}\mathsf{G}\otimes\sqrt{|\Lambda|}^s_{\gamma_1^{-1}}\mathsf{G}\otimes\sqrt{|\Lambda|}^t_{\gamma_1^{-1}}\mathsf{G}\\ \to\sqrt{|\Lambda|}^s_{\gamma}\mathsf{G}\otimes\sqrt{|\Lambda|}^t_{\gamma}\mathsf{G}\otimes|\Lambda|^t_{\gamma_1}\mathsf{G}_\gamma. \end{aligned} \tag{3.15}$$

Similarly, the pullback of the inversion map leads to an isomorphism

$$\lambda^-_\gamma: \sqrt{|\Lambda|}^s_{\gamma^{-1}}\mathsf{G}\otimes\sqrt{|\Lambda|}^t_{\gamma^{-1}}\mathsf{G}\to\sqrt{|\Lambda|}^s_{\gamma}\mathsf{G}\otimes\sqrt{|\Lambda|}^t_{\gamma}\mathsf{G}. \tag{3.16}$$

3.3 The Convolution Algebra of a Lie Groupoid

After this preparation we come to the definition of convolution.

Proposition 3.3.1. *Let $\Psi^{s\otimes t}$, $\Phi^{s\otimes t}$ be $s\otimes t$-densities on G. Convolution on the groupoid is defined by*

$$\Psi^{s\otimes t}*\Phi^{s\otimes t}(\gamma)=\int_{\tau_t^{-1}(\tau_s(\gamma))}\lambda_1(\Psi^{s\otimes t}\Phi^{s\otimes t}), \tag{3.17}$$

where $\lambda_1(\Psi^{s\otimes t}\Phi^{s\otimes t})$ is defined for $\gamma_1\in\tau_t^{-1}(\tau_s(\gamma))$ by $\lambda_1(\Psi^{s\otimes t}\Phi^{s\otimes t}): \gamma_1\mapsto \lambda_{\gamma_1}(\Psi^{s\otimes t}(\gamma\gamma_1)\otimes\Phi^{s\otimes t}(\gamma_1^{-1}))$. Involution is defined by

$$(\Psi^{s\otimes t})^*(\gamma):=\lambda^-_\gamma(\overline{\Psi^{s\otimes t}(\gamma^{-1})}). \tag{3.18}$$

With these operations, the vector space $\Gamma(\sqrt{|\Lambda|}^{s\otimes t}\mathsf{G})$ is a *-algebra.

The fact that $\Gamma(\sqrt{|\Lambda|}^{s\otimes t}\mathsf{G})$ is closed under convolution follows from 3.1.6.3, the smoothness of groupoid multiplication (cf. 3.1.5), and the smoothness of the isomorphism λ_1. Similarly, closedness under involution is a consequence of the smoothness of inversion and of the isomorphism λ^-.

To verify associativity of the convolution product, one uses the associativity of multiplication in the groupoid and a self-evident property of λ_1.

The equalities $(\Psi^{s\otimes t})^{**} = \Psi^{s\otimes t}$ and $(\Psi^{s\otimes t} \otimes \Phi^{s\otimes t})^* = (\Phi^{s\otimes t})^*(\Psi^{s\otimes t})^*$ follow from (3.2) and (3.3), respectively. ■

Up to the maps λ_1 and λ^-, there is therefore a direct correspondence between the key properties of the groupoid operations (i.e., the associativity of multiplication and the involutive nature of the inversion) and those in the corresponding convolution *-algebra (where the role of the inversion is played by the adjoint).

Although it shows that one can always associate a convolution *-algebra with a Lie groupoid, the definition above is not easy to use in practice. Fortunately, one may trivialize the bundle $\sqrt{|\Lambda|}^{s\otimes t}\mathsf{G}$. What follows adapts the general considerations in 3.2 on measures and α-densities to groupoids.

Definition 3.3.2. *A* **t-system** *on a Lie groupoid $\mathsf{G} \rightrightarrows Q$ is a family $\{\mu_q^t\}_{q\in Q}$ of positive measures such that:*

1. *The measure μ_q^t is defined on $\tau_t^{-1}(q) \subset \mathsf{G}$ (or, equivalently, on G with support in $\tau_t^{-1}(q) \subset \mathsf{G}$).*
2. *Each μ_q^t is locally Lebesgue (recall from 3.1.6 that each fiber $\tau_t^{-1}(q)$ is a manifold).*
3. *For each $f \in C_c^\infty(\mathsf{G})$ the map $q \mapsto \int_{\tau_t^{-1}(q)} d\mu_q^t(\gamma) f(\gamma)$ from Q to $\mathbb{C}$ is smooth.*

Similarly, an **s-system** *is defined as above, with t replaced by s.*

If, in addition, the family of measures defining a t-system is invariant under all maps L_γ, the t-system is called a **left Haar system**. *Similarly, a* **right Haar system** *is an s-system invariant under all maps R_γ.*

It is clear from (3.13) that the inversion I maps a given t-system $\{\mu_q^t\}$ into an associated s-system $\{\mu_q^s\} = \{I_*\mu_q^t\}$, by which a left Haar system is mapped into a right Haar system (and vice versa). This is possible because of the diffeomorphism $\mathsf{G}_\gamma^s \simeq \mathsf{G}_\gamma^t$ discussed in Lemma 3.2.3.

Proposition 3.3.3. *Every Lie groupoid $\mathsf{G} \rightrightarrows Q$ possesses a left Haar system. Consequently, the bundle $\sqrt{|\Lambda|}^{s\otimes t}\mathsf{G}$ is trivial.*

The proof of the first claim will be given at the end of 3.8. For the second, it is enough to have a t-system. The construction of the global trivialization generalizes (3.9). For $\gamma \in \mathsf{G}$, choose local coordinates $\{x^i\}$ and $\{y^i\}$ on G_γ^t and G_γ^s, respectively, with associated Lebesgue measures $d\mu_L^t := dx^1\cdots dx^n$ and $d\mu_L^s := dy^1\cdots dy^n$. (These coordinates may be different even if $\mathsf{G}_\gamma^t = \mathsf{G}_\gamma^s$.) The map

$$\psi_\mu([\partial_1^x \wedge \cdots \wedge \partial_{d_\gamma}^x \otimes \partial_1^y \wedge \cdots \wedge \partial_{d_\gamma}^y, \lambda]) := \lambda \sqrt{\frac{d\mu_L^s}{d\mu^s} \frac{d\mu_L^t}{d\mu^t}} \tag{3.19}$$

is well-defined (in not depending on the representative in the equivalence class), smooth, independent of the chart, and defines a global trivialization. ■

One may represent a section $\Psi^{s\otimes t}$ of $\sqrt{|\Lambda|}^{s\otimes t}\mathsf{G}$ by an equivalence class that we denote by $[f, \{\nu_q^t\}]_{\frac{1}{2}\otimes\frac{1}{2}}$, where $f \in C_c^\infty(\mathsf{G})$ and $\{\nu_q^t\}$ is a t-system on G. With $\{\nu_q^s\}$ the s-system associated to the given t-system, the equivalence relation is $(f, \{\nu_q^t\}) \sim (g, \{\mu_q^t\})$ when for all $\gamma \in \mathsf{G}$ one has

$$g(\gamma) = f(\gamma) \left(\frac{d\nu_{\tau_t(\gamma)}^s}{d\mu_{\tau_t(\gamma)}^s} \frac{d\nu_{\tau_t(\gamma)}^t}{d\mu_{\tau_t(\gamma)}^t}(\gamma) \right)^{\frac{1}{2}}.$$

For $\partial_1^x \wedge \cdots \partial_{d_\gamma}^y \in \mathsf{A}_\gamma^{s\otimes t}\mathsf{G}$, the section in question is then duly given as a $\mathbb{C}^*$-equivariant map from $\mathsf{A}^{s\otimes t}\mathsf{G}$ to $\mathbb{C}$ by (cf. (3.6))

$$[f, \{\nu_q^t\}]_{\frac{1}{2}\otimes\frac{1}{2}}(\partial_1^x \wedge \cdots \wedge \partial_{d_\gamma}^x \otimes \partial_1^y \wedge \cdots \wedge \partial_{d_\gamma}^y) := f(\gamma) \left(\frac{d\nu_{\tau_s(\gamma)}^s}{d\mu_L^s} \frac{d\nu_{\tau_t(\gamma)}^t}{d\mu_L^t}(\gamma) \right)^{\frac{1}{2}}. \tag{3.20}$$

As in (3.10), a fixed t-system $\{\nu_q^t\}$ on G leads to a trivialization of $s\otimes t$-densities as complex-valued functions on G, tied to the trivialization of $\sqrt{|\Lambda|}^{s\otimes t}\mathsf{G}$ defined by $\{\nu_q^t\}$. Generalizing (3.10), a section $[f, \{\nu_q^t\}]_{\frac{1}{2}\otimes\frac{1}{2}}$ as above is trivialized by $f_\mu^\nu : \mathsf{G} \to \mathbb{C}$ given by

$$f_\mu^\nu(\gamma) := f(\gamma) \left(\frac{d\nu_{\tau_s(\gamma)}^s}{d\mu^s} \frac{d\nu_{\tau_t(\gamma)}^t}{d\mu^t}(\gamma) \right)^{\frac{1}{2}}. \tag{3.21}$$

Proposition 3.3.4. *A left Haar system $\{\mu_q^t\}$ on a Lie groupoid $\mathsf{G} \rightrightarrows Q$ defines an isomorphism between the convolution *-algebra $\Gamma(\sqrt{|\Lambda|}^{s\otimes t}\mathsf{G})$ and $C_c^\infty(\mathsf{G})$, turned into a *-algebra as follows. Convolution and involution are defined as*

$$f * g(\gamma) := \int_{\tau_t^{-1}(\tau_s(\gamma))} d\mu_{\tau_s(\gamma)}^t(\gamma_1)\, f(\gamma\gamma_1)g(\gamma_1^{-1}); \tag{3.22}$$

$$f^*(\gamma) := \overline{f(\gamma^{-1})}. \tag{3.23}$$

The isomorphism is given by letting $f \in C_c^\infty(\mathsf{G})$ correspond to the equivalence class $[f, \{\mu_q^t\}]_{\frac{1}{2}\otimes\frac{1}{2}}$ in $\Gamma(\sqrt{|\Lambda|}^{s\otimes t}\mathsf{G})$. The fact that (3.22) is thus mapped into (3.17) follows from the definition of λ_{γ_1} in (3.15) by Lemma 3.2.4, and the comment following 3.3.2. Similarly, the correspondence between (3.23) and (3.18) follows from the definition of λ^- in (3.16) and the fact that the right Haar system occurring in, e.g., (3.21) is defined from the given left Haar system by the inversion map. ■

Let us look at some examples of the prceeding abstract constructions.

Proposition 3.3.5. *If* $\mathsf{G} = G$ *is a Lie group, then each locally Lebesgue measure defines a t-system, which in this case is the same as an s-system. A left-invariant Haar measure on G provides a left Haar system. The ensuing convolution algebra is the group algebra, restricted to* $C_c^\infty(G)$.

This is obvious, as $G = \tau_t^{-1}(e) = \tau_s^{-1}(e)$. A measure on G defines a left Haar system iff it is left-invariant in the usual sense. ■

Note that the right Haar system defined by a left Haar system coincides with it when G is unimodular.

Combined with 3.3.4, this proposition shows that one may define the group algebra of a Lie group without specifying a Haar measure, but since one still needs the isomorphism (3.15) in (3.17), there is not much advantage in this.

Proposition 3.3.6. *On a pair Lie groupoid* $Q \times Q \rightrightarrows Q$ *any measure* ν *on* Q *that is locally Lebesgue defines a left Haar system.*

*The corresponding pair groupoid *-algebra is* $C_c^\infty(Q \times Q)$, *with operations*

$$f * g(q_1, q_2) = \int_Q d\nu(q)\, f(q_1, q)g(q, q_2); \tag{3.24}$$

$$f^*(q_1, q_2) = \overline{f(q_2, q_1)}. \tag{3.25}$$

To construct the left Haar system one identifies $\tau_t^{-1}(q) = \{q\} \times Q$ with Q for each q. The above formulae then follow from (3.22), (3.23), and 3.1.2. ■

In particular, when Q is a finite set with cardinality n, the convolution algebra is simply $\mathfrak{M}_n(\mathbb{C})$.

Proposition 3.3.7. *On a gauge groupoid* $\mathsf{P} \times_H \mathsf{P} \rightrightarrows Q$ *an H-invariant measure* μ *on* P *which is locally Lebesgue produces a left Haar system.*

*The corresponding gauge groupoid *-algebra* $C_c^\infty(\mathsf{P} \times_H \mathsf{P})$ *is given by*

$$f * g([x, y]_H) = \int_{\mathsf{P}} d\mu(z)\, f([x, z]_H)g([z, y]_H); \tag{3.26}$$

$$f^*([x, y]_H) = \overline{f([y, x]_H)}. \tag{3.27}$$

When H is compact one may identify $C_c^\infty(\mathsf{P} \times_H \mathsf{P})$ *with* $C_c^\infty(\mathsf{P} \times \mathsf{P})^H$, *seen as a subalgebra of the convolution algebra* $C_c^\infty(\mathsf{P} \times \mathsf{P})$ *of the pair Lie groupoid* $\mathsf{P} \times \mathsf{P} \rightrightarrows \mathsf{P}$; *cf. 3.3.6.*

We identify $\tau_t^{-1}(q)$ with P through the choice of a measurable section $s : Q \to \mathsf{P}$; that is, we let $[s(q), x]_H \in \tau_t^{-1}(q)$ correspond to $x \in \mathsf{P}$. Since $[x, s(q)]_H \cdot [s(q), y]_H = [s(\tau(x)), yh_s(x)]_H$, where h_s is defined below (2.6), the first claim follows. The second follows from the above consideration and the H-invariance of the measure. ■

3.4 Action *-Algebras

We now look at the convolution algebra of an action Lie groupoid.

Proposition 3.4.1. *In an action Lie groupoid $G \times Q \rightrightarrows Q$ each left-invariant Haar measure dx on G leads to a left Haar system. The operations in the action groupoid *-algebra $C_c^\infty(G \times Q)$ are*

$$f * g(x, q) = \int_G dy\, f(xy, q) g(y^{-1}, y^{-1}x^{-1}q); \tag{3.28}$$

$$f^*(x, q) = \overline{f(x^{-1}, x^{-1}q)}. \tag{3.29}$$

To construct the left Haar system one identifies $\tau_t^{-1}(q)$ with G, letting $(x, q) \in \tau_t^{-1}(q)$ correspond to $x \in G$. Equations (3.28) and (3.29) then follow from (3.22), (3.23), and Definition 3.1.4. ■

We will now construct a seemingly different *-algebraic structure on the function space $C_c^\infty(G \times Q)$, which turns out to be isomorphic to the one above. The data for the definition still constitute a smooth action of a Lie group on a manifold Q; the associated action of G on $C_c^\infty(Q)$ is denoted by

$$\alpha_x(\tilde{f}) : q \mapsto \tilde{f}(x^{-1}q). \tag{3.30}$$

We look at $C_c^\infty(Q)$ as a commutative *-algebra in the obvious way, i.e., as a dense subalgebra of the C^*-algebra $C_0(Q)$.

To exhibit the natural structure of the construction, we first consider the following generalization. Recall Definition II.2.4.2.

Definition 3.4.2. *A* **smooth C^*-dynamical system** *consists of a Lie group G, a dense subalgebra $\tilde{\mathfrak{A}}$ of some C^*-algebra $\mathfrak{A}$, and an automorphic action α of G on $\mathfrak{A}$, such that for each fixed $A \in \tilde{\mathfrak{A}}$ the function $x \mapsto \alpha_x(A)$ from G to $\mathfrak{A}$ is smooth.*

It follows that $x \mapsto \alpha_x(A)$ is continuous for all $A \in \mathfrak{A}$. (More generally, when G is merely locally compact and the latter continuity property is satisfied for $\tilde{\mathfrak{A}} = \mathfrak{A}$, one speaks of a **$C^*$-dynamical system.**) The term "dynamical system" comes from the example $G = \mathbb{R}$ and $\mathfrak{A} = C_0(S)$, where $\mathbb{R}$ acts on S and $\alpha_t(f) : \sigma \mapsto \sigma(t)$; cf. I.(2.13). Another example is, of course, provided by $\mathfrak{A} = C_0(Q)$ with $\tilde{\mathfrak{A}} = C_c^\infty(Q)$, where the G-action on $\mathfrak{A}$ is defined as in (3.30). The smoothness of the G-action on Q then implies that one indeed has a smooth C^*-dynamical system. In any case, given a smooth C^*-dynamical system, one considers the space $C_c^\infty(G, \tilde{\mathfrak{A}})$, made into a *-algebra by the operations

$$F * G(x) := \int_G dy\, F(y)\alpha_y(G(y^{-1}x)); \tag{3.31}$$

$$F^*(x) := \alpha_x(F(x^{-1})^*). \tag{3.32}$$

In the nonunimodular case one here needs to assume that the Haar measure is left-invariant. The *-algebra $C_c^\infty(G, \tilde{\mathfrak{A}})$ thus defined is called the (smooth) **crossed product *-algebra**, or simply the **crossed product**, of G and $\tilde{\mathfrak{A}}$, and is denoted by $C^*(G, \tilde{\mathfrak{A}})$. If $\tilde{\mathfrak{A}} = C_c^\infty(Q)$, we call $C^*(G, C_c^\infty(Q))$ an **action *-algebra**.

Lemma 3.4.3. *The action groupoid *-algebra $C_c^\infty(G \times Q)$ is isomorphic to the action *-algebra $C^*(G, C_c^\infty(Q))$.*

A function $F \in C^*(G, C_c^\infty(Q))$ defines $f \in C_c^\infty(G \times Q)$ by $f(x, q) = F(x)(q)$. The definition of the function spaces in question easily implies that this correspondence is bijective. Under this correspondence, a shift in the integration variable in (3.28) and the left invariance of the Haar measure reproduces (3.31). The equality of (3.29) and (3.32) is immediate. ■

Generalizing 1.7.3, we have

Theorem 3.4.4. *There is a bijective correspondence between*

- *nondegenerate representations π of the crossed product $C^*(G, \tilde{\mathfrak{A}})$ that are bounded as in*

$$\|\pi(F)\| \leq \|F\|_I := \int_G dx\, \|F(x)\|; \tag{3.33}$$

- *pairs $(U, \tilde{\pi})$, where U is a representation of G, and $\tilde{\pi}$ is a nondegenerate representation of $\mathfrak{A}$ that for all $x \in G$ and $A \in \mathfrak{A}$ satisfies the* **covariance condition**

$$U(x)\tilde{\pi}(A)U(x)^* = \tilde{\pi}(\alpha_x(A)). \tag{3.34}$$

This correspondence is given in one direction by

$$\pi(F) = \int_G dx\, \tilde{\pi}(F(x))U(x); \tag{3.35}$$

in the other direction one defines $AF : x \mapsto AF(x)$ and $\tilde{\alpha}_x(f) : y \mapsto \alpha_x(f(x^{-1}y))$, and puts

$$U(x)\pi(F)\Omega = \pi(\tilde{\alpha}_x(F))\Omega; \tag{3.36}$$

$$\tilde{\pi}(A)\pi(F)\Omega = \pi(AF)\Omega, \tag{3.37}$$

where Ω is a cyclic vector for a cyclic summand of $\pi(C^*(G, \tilde{\mathfrak{A}}))$.

The proof of this theorem is analogous to that of 1.7.3. The analogue of the Banach algebra $L^1(G, c)$ used in that proof is $L^1(G, \mathfrak{A})$, the closure of $C_c^\infty(G, \tilde{\mathfrak{A}})$ in the norm (3.33). The rest of the proof may be read off from 1.7.3. □

We return to G-actions on a manifold Q.

Definition 3.4.5. *Given a G-action on Q, a* **smooth system of imprimitivity** *of G on Q is a pair $(U(G), \tilde{\pi}(C_c^\infty(Q)))$, where U is a continuous representation of G, and $\tilde{\pi}$ is a nondegenerate representation of $C_c^\infty(Q)$ (seen as a commutative *-algebra in the obvious way), satisfying the covariance condition*

$$U(x)\tilde{\pi}(\tilde{f})U(x)^* = \tilde{\pi}(\alpha_x(\tilde{f})). \tag{3.38}$$

The meaning of the conditions on the pair $(U, \tilde{\pi})$ may be clarified by expressing them in infinitesimal form. For $X \in \mathfrak{g}$, $\tilde{f} \in C_c^\infty(Q)$ we put

$$Q_\hbar^\pi(\tilde{X}) := i\hbar dU(X); \tag{3.39}$$

$$\mathcal{Q}_\hbar^\pi(\tilde{f}) := \tilde{\pi}(\tilde{f}), \tag{3.40}$$

and obtain (on the domain $\mathcal{H}_U^\infty$; see III.1.5.5)

$$\frac{i}{\hbar}[\mathcal{Q}_\hbar^\pi(\tilde{f}), \mathcal{Q}_\hbar^\pi(\tilde{g})] = 0; \tag{3.41}$$

$$\frac{i}{\hbar}[\mathcal{Q}_\hbar^\pi(\tilde{X}), \mathcal{Q}_\hbar^\pi(\tilde{Y})] = \mathcal{Q}_\hbar^\pi(-\widetilde{[X, Y]}); \tag{3.42}$$

$$\frac{i}{\hbar}[\mathcal{Q}_\hbar^\pi(\tilde{X}), \mathcal{Q}_\hbar^\pi(\tilde{f})] = \mathcal{Q}_\hbar^\pi(\xi_X \tilde{f}). \tag{3.43}$$

As in 1.1 we have $\tilde{X} \in C^\infty(\mathfrak{g}^*, \mathbb{R})$; also recall the definition (1.19) of ξ_X. Equation (3.41) is evident from the fact that $\tilde{\pi}(C_c^\infty(Q))$ is a representation (but is a weaker property); equation (3.42) is equivalent to (1.70) (with $\Gamma = 0$); and finally, (3.43) is an infinitesimal restatement of the covariance condition (3.38). These commutation relations may be seen as a version of Dirac's condition II.(1.3); cf. 3.11.

Corollary 3.4.6. *There is a bijective correspondence between*

- *nondegenerate representations π of the action *-algebra $C^*(G, C_c^\infty(Q))$ (or, equivalently, of the action groupoid *-algebra $C_c^\infty(G \times Q)$) that are bounded as in*

$$\|\pi(F)\| \le \|F\|_I = \sup_{q \in Q} \int_G dx\, |f(x, q)|; \tag{3.44}$$

- *smooth systems of imprimitivity of G on Q.*

Further to 3.1.8 we naturally have

Proposition 3.4.7. *The action groupoid *-algebra $C_c^\infty(G \times (G/H))$ and the gauge groupoid *-algebra $C_c^\infty(G \times_H G)$ are isomorphic.*

For simplicity we prove this only for unimodular G. We identify $C^\infty(G \times_H G)$ with $C^\infty(G \times G)^H$ in the obvious way, so that $C_c^\infty(G \times_H G)$ is identified with a certain subspace of $C^\infty(G \times G)^H$. One then establishes the desired bijection $f \leftrightarrow \hat{f}$ between $C_c^\infty(G \times (G/H))$ and $C_c^\infty(G \times_H G)$ by $\hat{f}([x, y]_H) = f(xy^{-1}, \tau(x))$, with inverse $f(x, q) = \hat{f}([s(q), x^{-1}s(q)]_H)$. Here $s : G/H \to G$ is an arbitrary section, on which the right-hand side clearly does not depend. Using the invariance of the Haar measure under inversion and right translation, one verifies that this bijection duly intertwines the *-algebraic operations stated in 3.4.1 and 3.3.7. ∎

Corollary 3.4.8. *There is a bijective correspondence between nondegenerate representations π of the gauge groupoid *-algebra $C_c^\infty(G \times_H G)$ that (for unimodular G) are bounded as in*

$$\|\pi(f)\| \le \sup_{y \in G} \int_G dx\, |f([y, x]_H)|, \tag{3.45}$$

and smooth systems of imprimitivity $(U(G), \tilde{\pi}(C_c^\infty(G/H)))$.

Using the invariance of the Haar measure under inversion, it is easy to see that the bound (3.45) is equivalent to (3.44). ∎

Using (3.27) and the fact that π by definition satisfies $\pi(f^*) = \pi(f)^*$, one infers that the bound (3.45) is equivalent to the same expression in which $f([y, x]_H)$ is replaced by $f([x, y]_H)$. The significance of this comment will become clear in the light of (3.57). See also Corollary 3.7.6.

3.5 *Representations of Groupoids*

In 1.7 the (reduced or full) C^*-algebra of a Lie group G was defined in direct relationship with the representations of G. This motivates the following definition of a representation of a general groupoid.

Definition 3.5.1. *A* **representation** *U of a groupoid $\mathsf{G} \overset{\leftarrow}{\Rightarrow} Q$ consists of a family $\{\mathcal{H}\}_{q \in Q}$ of Hilbert spaces indexed by the base Q and a collection $\{U(\gamma)\}_{\gamma \in \mathsf{G}}$ of maps such that*

1. *$U(\gamma) : \mathcal{H}_{\tau_s(\gamma)} \to \mathcal{H}_{\tau_t(\gamma)}$ is unitary.*
2. *$U(\gamma_1\gamma_2) = U(\gamma_1)U(\gamma_2)$ whenever $(\gamma_1, \gamma_2) \in \mathsf{G}_2$.*
3. *$U(\gamma^{-1}) = U(\gamma)^*$ for all $\gamma \in \mathsf{G}$.*

Hence, with reference to the third paragraph after (3.4), one thinks of a representation of $\mathsf{G} \overset{\leftarrow}{\Rightarrow} Q$ as a functor converting the points of Q into Hilbert spaces and the arrows in G into unitary maps connecting these Hilbert spaces.

The simplest example of a representation is obtained by choosing $\mathcal{H}_q = \mathbb{C}$ for all $q \in Q$ and $U(\gamma) := 1$ for all $\gamma \in \mathsf{G}$. This representation is evidently not faithful if there are nontrivial isotropy groups G_q.

For all our applications, and also for the purpose of defining the groupoid C^*-algebra, it will be sufficient to assume that all $\mathcal{H}_q$ are separable and of the same dimension. Let us give an example. Since each isotropy group G_q is a Lie group, we can form $\mathcal{H}_q := L^2(\mathsf{G}_q, \mu_q^H)$, where μ_q^H is a left or right Haar measure on G_q. For $\Psi \in \mathcal{H}_{\tau_s(\gamma)}$ we then define $U_{lr}(\gamma)\Psi \in \mathcal{H}_{\tau_t(\gamma)}$ by

$$U_{lr}(\gamma)\Psi(h) := \sqrt{\frac{d\mu^H_{\tau_s(\gamma)}(\gamma^{-1}h\gamma)}{d\mu^H_{\tau_t(\gamma)}(h)}}\,\Psi(\gamma^{-1}h\gamma). \tag{3.46}$$

The Radon–Nikodym derivative occurring here makes sense, since $\mathsf{G}_{\tau_t(\gamma)}$ and $\mathsf{G}_{\tau_s(\gamma)}$ are diffeomorphic by $h \mapsto \gamma^{-1}h\gamma$, so that the measure class of the Haar measure is preserved (recall that a Haar measure on a Lie group is locally Lebesgue). It is easily verified that this indeed furnishes a representation of G in the sense of 3.5.1.

In a variation on this example, in the base of each transitive subgroupoid one picks a point q_0 of G (cf. the proof of 3.1.3), and a function $\gamma_0 : \mathsf{G}q_0 :\to \tau_s^{-1}(q_0)$ satisfying $\tau_s(\gamma_0(q)) = q_0$ and $\tau_t(\gamma_0(q)) = q$. One then replaces (3.46) by

$$U_{q_0}(\gamma)\Psi(h) := \sqrt{\frac{d\mu^H_{\tau_s(\gamma)}(\gamma^{-1}h\gamma_0(\tau_t(\gamma))\gamma_0(\tau_s(\gamma))^{-1})}{d\mu^H_{\tau_t(\gamma)}(h)}}$$

$$x\Psi(\gamma^{-1}h\gamma_0(\tau_t(\gamma))\gamma_0(\tau_s(\gamma))^{-1}). \tag{3.47}$$

Since it depends on the function γ_0, this representation is not really intrinsic (unlike its predecessor), but different choices of γ_0 lead to equivalent versions. In any case, it will be reconsidered shortly.

In the **regular representation** one takes $\mathcal{H}_q := L^2(\tau_t^{-1}(q), \mu_q^t)$. Defined for a general Lie groupoid, this representation is given by

$$U_L^t(\gamma)\Psi(\gamma') = \Psi(\gamma^{-1}\gamma'). \tag{3.48}$$

This makes sense, since $\gamma^{-1}\gamma' \in \tau_t^{-1}(\tau_s(\gamma))$ when $\tau_t(\gamma) = \tau_t(\gamma')$, and is unitary because of the left invariance of the left Haar measure.

To relate representations of $\mathsf{G} \overset{\leftarrow}{\Rightarrow} Q$ in the above sense to representations of the convolution algebra $C_c^\infty(\mathsf{G})$ by a *-algebra of bounded operators on a Hilbert space in the usual sense, one has to choose a measure ν on Q. As we shall see, it is not sufficient to limit one's attention to measures that are locally Lebesgue.

Definition 3.5.2. *Let a Borel measure ν on Q and a left Haar system $\{\mu_q^t\}_{q\in Q}$ on G, with associated right Haar system $\{\mu_q^s\}$, be given. One obtains measures $\nu\times\mu^t$ and $\nu\times\mu^s$ on G, defined by*

$$\nu\times\mu^t(f) := \int_Q d\nu(q)\int_{\tau_t^{-1}(q)} d\mu_q^t(\gamma)\, f(\gamma); \tag{3.49}$$

$$\nu\times\mu^s(f) := \int_Q d\nu(q)\int_{\tau_s^{-1}(q)} d\mu_q^s(\gamma)\, f(\gamma). \tag{3.50}$$

The measure ν is said to be **quasi-invariant** *when $\nu\times\mu^t$ and $\nu\times\mu^s$ are equivalent, and* **invariant** *when they are equal.*

If ν is quasi-invariant, the Radon–Nikodym derivative

$$\rho := \frac{d\nu\times\mu^s}{d\nu\times\mu^t} \tag{3.51}$$

is well-defined on G, equaling unity in the invariant case. An example of a quasi-invariant measure on the base of an arbitrary Lie groupoid is a measure that is supported and locally Lebesgue on an arbitrary orbit G_q. A measure that is locally Lebesgue on Q (so that it is supported on all of Q) is quasi-invariant as well, since $\nu\times\mu^t$ and $\nu\times\mu^s$ are both locally Lebesgue on G. More examples are given in the following proposition, whose main goal it is to examine when a quasi-invariant measure is invariant. Here the measure ν on Q is as specified above.

Proposition 3.5.3.

- *A left Haar measure on a Lie group G is invariant iff G is unimodular.*
- *For a pair Lie groupoid $Q\times Q \overset{\leftarrow}{\Rightarrow} Q$, a measure is invariant iff it is a multiple of the fixed measure on Q defining the left Haar system (cf. 3.3.6).*
- *For a gauge groupoid $\mathsf{P}\times_H \mathsf{P} \overset{\leftarrow}{\Rightarrow} Q$, a measure ν is invariant when H is unimodular, and when ν is related to μ (the measure defining the left Haar system, as in 3.3.7) by (2.139).*

- *For an action Lie groupoid $G \times Q \rightrightarrows Q$, a measure ν is invariant iff it is G-invariant.*

This follows directly from the definitions. ∎

If in a given representation U of $\mathsf{G} \rightrightarrows Q$ one has $\mathcal{H}_q = \mathcal{K}$ for all q, one can form $L^2(Q, \nu) \otimes \mathcal{K}$, which will be the Hilbert space carrying the representation π of the convolution algebra $C_c^\infty(\mathsf{G})$ associated to U.

In general, assuming that each $\mathcal{H}_q$ is separable, the subset $Q_n \subseteq Q$ for which $\dim(\mathcal{H}_q) = n$ for all $q \in Q_n$ should be measurable for each $n \in \mathbb{N} \cup \infty$. One then considers the space Γ of all functions $\Psi : Q \to \{\mathcal{H}_q\}$ satisfying $\Psi_q \in \mathcal{H}_q$; such a function is called a **section of the field** $\{\mathcal{H}_q\}$. The idea is to turn a suitable subset of Γ into a Hilbert space with inner product

$$(\Psi, \Phi) := \int_Q d\nu(q)\, (\Psi_q, \Phi_q)_q, \tag{3.52}$$

where $(\ ,\ )_q$ is the inner product in $\mathcal{H}_q$.

Such a suitable subset is obtained by specifying a sequence of sections Ψ_n satisfying the two conditions that firstly the function $q \mapsto (\Psi_n(q), \Psi_m(q))_q$ be measurable for all n, m, and secondly that for each fixed q the Ψ_n span $\mathcal{H}_q$. There then exists a unique maximal linear subspace Γ_0 of Γ that contains $\{\Psi_n\}$, and for which all functions $q \mapsto (\Psi_q, \Phi_q)_q$ are measurable. The **direct integral**

$$\mathcal{H} = \int_Q^\oplus d\nu(q)\, \mathcal{H}_q \tag{3.53}$$

is then by definition the subset of Γ_0 of functions Ψ for which $(\Psi, \Psi) < \infty$, as defined by (3.52). It depends on the choice of the sequence $\{\Psi_n\}$, but in all practical applications it is clear that all reasonable such choices lead to the same result, so that this dependence will be suppressed.

The simplest example of a direct integral is $\mathcal{H} = \int_{\mathbb{R}} dx \mathcal{H}_x$, where $\mathcal{H}_x = \mathbb{C}$ for all x. Choosing the sequence $\{\Psi_n\}$ to consist of a single strictly positive measurable function then leads to $\mathcal{H} = L^2(\mathbb{R})$. When $\mathcal{H}_q = \mathcal{K}$ for all q, one takes $\{\Psi_n\}$ to be a strictly positive measurable function on Q tensored with a basis in $\mathcal{K}$. The corresponding direct integral is nothing but $\mathcal{H} = L^2(Q, \nu) \otimes \mathcal{K}$. Since Q is a finite-dimensional manifold, $\mathcal{H}$ is separable.

As an example relevant to Lie groupoids, we construct the direct integral

$$\mathcal{H}_{(q_0)} = \int_{\mathsf{G}q_0}^\oplus d\nu(q)\, L^2(\mathsf{G}_q, \mu_q^H) \tag{3.54}$$

over the Hilbert spaces $\mathcal{H}_q = L^2(\mathsf{G}_q, \mu_q^H)$ considered in the context of (3.46), with ν supported on a given orbit $\mathsf{G}q_0$. One picks the function γ_0 mentioned before (3.47) so that it is measurable, and identifies each stability group G_q with G_{q_0} through $\mathsf{G}_q \ni h \mapsto \gamma_0(q)^{-1} h \gamma_0(q) \in \mathsf{G}_{q_0}$. Inserting the appropriate Radon–Nikodym derivative, this leads to a unitary map $V_q : L^2(\mathsf{G}_{q_0}, \mu_{q_0}^H) \to \mathcal{H}_q$. One then picks a basis $\{\mathbf{e}_n\}$ in $L^2(\mathsf{G}_{q_0})$ and a strictly positive measurable function f on Q, and defines the sequence $\{\Psi_n\}$ as $\Psi_n(q, h) := f(q) V_q \mathbf{e}_\mu(h)$. The resulting direct integral clearly consists of those sections Ψ of the given field for which $q \mapsto V_q^{-1}\Psi_q$ lies in $L^2(\mathsf{G}q_0 \times \mathsf{G}_{q_0})$.

In a third example, we consider the field $\mathcal{H}_q = L^2(\tau_t^{-1}(q), \mu_q^t)$ featured in (3.48), and a measure ν on Q that is locally Lebesgue. One then tacitly chooses the sequence $\{\Psi_n\}$ in such a way that (cf. (3.49))

$$\int_Q^{\oplus} d\nu(q)\, L^2(\tau_t^{-1}(q), \mu_q^t) = L^2(\mathsf{G}, \nu \times \mu^t). \tag{3.55}$$

3.6 The C^*-Algebra of a Lie Groupoid

In preparation for the definition of the C^*-algebra of a Lie groupoid we show how a representation of a Lie groupoid determines a representation of its groupoid *-algebra.

Proposition 3.6.1. *Assume that one has a Lie groupoid* $\mathsf{G} \rightrightarrows Q$ *with*

- *a left Haar system* $\{\mu_q^t\}$;
- *a representation* U *of* G *on a collection* $\{\mathcal{H}_q\}$ *of separable Hilbert spaces* $\mathcal{H}$ *for which the function*

$$\gamma \mapsto (\Psi_{\tau_s(\gamma)}, U(\gamma)\Psi_{\tau_t(\gamma)})_{\mathcal{H}_{\tau_t(\gamma)}}$$

  is measurable for all $\Psi, \Phi \in \mathcal{H}$;
- *a quasi-invariant measure* ν *on* Q;
- *an associated direct integral* $\mathcal{H} = \int_Q^{\oplus} d\nu(q)\, \mathcal{H}_q$.

For each $f \in C_c^\infty(\mathsf{G})$ *the operator* $\pi(f)$ *on* $\mathcal{H}$*, defined by*

$$\pi(f)\Psi_q := \int_{\tau_t^{-1}(q)} d\mu_q^t(\gamma)\, \sqrt{\rho}(\gamma) f(\gamma) U(\gamma) \Psi_{\tau_s(\gamma)}, \tag{3.56}$$

where ρ *is given by (3.51), is bounded, with*

$$\|\pi(f)\| \le \|f\|_I := \max\{\|f\|_s, \|f\|_t\}; \tag{3.57}$$

$$\|f\|_{s,t} := \sup_{q \in Q} \int_{\tau_{s,t}^{-1}(q)} d\mu_q^{s,t}(\gamma) |f(\gamma)|. \tag{3.58}$$

Then π *is a nondegenerate representation of the groupoid* **-algebra* $C_c^\infty(\mathsf{G})$ *on* $\mathcal{H}$.

To derive the bound, we generalize the proof of Lemma 1.7.2. Writing $F(\gamma) := \sqrt{\rho(\gamma)|f(\gamma)|}\|\Psi_{\tau_s(\gamma)}\|$ and $G(\gamma) := \sqrt{|f(\gamma)|}\|\Psi_{\tau_s(\gamma)}\|$, we use 3.5.1.1 to majorize $|(\Psi, \pi(f)\Psi)|$ by $\int_{\mathsf{G}} d\nu \times \mu^t FG$. Applying the Cauchy–Schwarz inequality and using the argument in the proof of II.1.3.5 then leads to (3.57).

To verify that π preserves multiplication, one writes out $\pi(f * g)$ using (3.56) and (3.22), changes variables $\gamma_1 \mapsto \gamma_2 = \gamma_1^{-1}$ and subsequently $\gamma \mapsto \gamma' = \gamma\gamma_2^{-1}$. The range of (γ', γ_2) is then $\gamma' \in \tau_t^{-1}(q)$ and $\gamma_2 \in \tau_t^{-1}(\tau_s(\gamma'))$. Finally, one needs 3.5.1.2 and the properties

$$\rho(\gamma\gamma') = \rho(\gamma)\rho(\gamma');$$

$$\rho(\gamma^{-1}) = \rho(\gamma)^{-1}. \tag{3.59}$$

The first is a nontrivial consequence of the left invariance of $\{\mu_q^t\}$ and the right invariance of $\{\mu_q^s\}$, and the second is immediate from the definition of $\{\mu_q^s\}$.

The proof that π preserves the involution follows from 3.5.1.3 and the fact that the measure $\nu \times \mu^t \sqrt{\rho}$ on G is invariant under inversion.

The nondegeneracy of π is an easy consequence of the surjectivity of $\tau_{s,t}$ and the unitarity of U. ■

An important example of such a representation is the regular one.

Theorem 3.6.2.

1. *The regular representation π_L^t of $C_c^\infty(\mathsf{G})$ on $\mathcal{H}_L^t = L^2(\mathsf{G}, \nu \times \mu^t)$, given by (3.48), (3.56), and a locally Lebesgue measure ν on Q, is equivalent to $\pi_L^s(C_c^\infty(\mathsf{G}))$ on $\mathcal{H}_L^s := L^2(\mathsf{G}, \nu \times \mu^s)$, given by*
$$\pi_L^s(f)\Psi = f * \Psi. \tag{3.60}$$
Here the convolution is given by (3.22).
2. *The regular representation is faithful on $C_c^\infty(\mathsf{G})$. It may be decomposed as a direct integral over the* **reduced regular representations** *π_q^s, $q \in Q$, defined on $L^2(\tau_s^{-1}(q), \mu_q^s)$ by $\pi_q^s(f)\Psi = f * \Psi$, that is, by the restriction of (3.60).*
3. *For given $q_0 \in Q$, the representation $\pi_{q_0}^s(C_c^\infty(\mathsf{G}))$ is equivalent to the representation π_{q_0} on $\mathcal{H}_{(q_0)}$ (see (3.54)) given by (3.56) with (3.47).*
4. *Finally, $\pi_{q_0}^s$ may be realized on the Hilbert space $L^2(\tau_s^{-1}(q_0))$ of half-densities on $\tau_s^{-1}(q_0)$, so that it can be defined without the choice of a (left or right) Haar system.*

The representation π_L^t is given on $\mathcal{H}_L^t = L^2(\mathsf{G}, \nu \times \mu^t)$ (cf. (3.55)) by

$$\pi_L^t(f)\Psi(\gamma) = \int_{\mathsf{G}_\gamma^t} d\mu_{\tau_t(\gamma)}^t(\gamma_1)\sqrt{\rho}(\gamma_1) f(\gamma_1)\Psi(\gamma_1^{-1}\gamma). \tag{3.61}$$

Now perform the unitary transformation $V : \mathcal{H}_L^t \to \mathcal{H}_L^s := L^2(\mathsf{G}, \nu \times \mu^s)$, defined by $V\Psi(\gamma) := \rho(\gamma)^{-\frac{1}{2}}\Psi(\gamma)$. Using (3.59), changing integration variables, and using the left invariance of μ^t one obtains $V\pi_L^t V^* = \pi_L^s$; cf. (3.60).

(Incidentally, the price for the simplicity of (3.60) is that compared with (3.48) the corresponding representation $U_L^s(\mathsf{G})$ on $L^2(\tau_s^{-1}(q), \mu_q^s)$ now contains an additional Radon–Nikodym derivative.) In analogy to (3.55), one decomposes $\mathcal{H}_L^s$ as a direct integral over Q by

$$L^2(\mathsf{G}, \nu \times \mu^s) = \int_Q^\oplus d\nu(q)\, L^2(\tau_s^{-1}(q), \mu_q^s); \tag{3.62}$$

$$\pi_L^s = \int_Q^\oplus d\nu(q)\, \pi_q^s. \tag{3.63}$$

The latter means that $\pi_L^s(f)\Psi_q = \pi_q^s(f)\Psi_q$ for (almost) all $q \in Q$, where π_q^s is the operator on $\mathcal{H}_q = L^2(\tau_s^{-1}(q), \mu_q^s)$ specified in the theorem.

To prove the equivalence between $\pi^s_{q_0}$ and π_{q_0}, one uses 3.1.3 to realize both representations in the model $\mathsf{G}q_0 \times \mathsf{G}_{q_0} \times \mathsf{G}q_0$, where they coincide.

Finally, analogously to (3.15), we can use (3.14) to obtain an isomorphism

$$\tilde{\lambda}_{\gamma_1} : \sqrt{|\Lambda|}^s_{\gamma\gamma_1}\mathsf{G} \otimes \sqrt{|\Lambda|}^t_{\gamma\gamma_1}\mathsf{G} \otimes \sqrt{|\Lambda|}^s_{\gamma_1^{-1}} \to \sqrt{|\Lambda|}^s_{\gamma}\mathsf{G} \otimes |\Lambda|_{\gamma_1}\mathsf{G}_\gamma. \tag{3.64}$$

In a slight modification of (3.17), if Ψ is a half-density on $\tau_s^{-1}(q_0)$ and $\Phi^{s\otimes t}$ is an $s \otimes t$-density on G, we can write convolution as

$$\Phi^{s\otimes t} * \Psi(\gamma) = \int_{\tau_t^{-1}(\tau_s(\gamma))} \tilde{\lambda}_1(\Phi^{s\otimes t}\Phi), \tag{3.65}$$

where the definition of $\tilde{\lambda}_1$ in terms of $\tilde{\lambda}_{\gamma_1}$ is analogous to that of λ_1 in terms of λ_{γ_1}. The left hand side is then by definition equal to $\pi^s_q(\Phi^{s\otimes t})\Psi$. ■

One should mention that π^s_q may or may not be reducible, and that these representations may or may not be equivalent when q varies.

For example, for a group there only is one q, and π^s_e is the left-regular representation, which is reducible.

For a pair groupoid $Q \times Q \overset{\leftarrow}{\Rightarrow} Q$ the reduced regular representation $\pi^s_{q_0}$ may be realized on $L^2(Q)$ (defined with respect to a locally Lebesgue measure ν). For $f \in C^\infty_c(Q \times Q)$ it takes the form

$$\pi^s(f)\Psi(q) = \int_Q d\nu(q')f(q,q')\Psi(q'). \tag{3.66}$$

We have written π^s for $\pi^s_{q_0}$, since this representation does not depend on q_0. We see from Proposition 3.3.6 that π^s is faithful and irreducible; the representations π^s_q are trivially equivalent for all $q \in Q$.

For an action groupoid $G \times Q \overset{\leftarrow}{\Rightarrow} Q$ all possibilities may occur, depending on the group action. For unimodular G the regular representation is realized on $L^2(G\times Q)$. If, for simplicity, we assume that Q has a G-invariant locally Lebesgue measure, and define $L^2(Q)$ accordingly, we have

$$\pi^s_L(f)\Psi(x,q) = \int_G dy\, f(xy,q)\Psi(y^{-1}, y^{-1}x^{-1}q). \tag{3.67}$$

By Corollary 3.4.6 there are representations $U^s_L(G)$ and $\tilde{\pi}^s_L(C^\infty_c(Q))$ associated to $\pi^s_L(C^\infty_c(G \times Q))$. We infer from (3.36) that

$$U^s_L = U_L \otimes U^Q_L, \tag{3.68}$$

where U_L is defined on $L^2(G)$ by (1.83) (with $c = 1$), and U^Q_L is (analogously) defined on $L^2(Q)$ by $U^Q_L(x)\Psi(q) := \Psi(x^{-1}q)$. From (3.37) we see that

$$\tilde{\pi}^s_L(\tilde{f})\Psi(x,q) = \tilde{f}(q)\Psi(x,q). \tag{3.69}$$

To reduce this we perform a unitary transformation U on $L^2(G \times Q)$, defined by $V\Psi(x,q) := \Psi(x,xq)$; this step is necessary because x and q are mixed up in

$\tau_s^{-1}(q)$, and hence in (3.68). Then $\pi_L^{\tilde{s}} := V\pi_L^s V^{-1}$ is given by

$$\pi_L^{\tilde{s}}(f)\Psi(x,q) = \int_G dy\, f(xy^{-1}, xq)\Psi(y,q), \tag{3.70}$$

with associated representations

$$U_L^{\tilde{s}} = U_L \otimes \mathbb{I}; \tag{3.71}$$

$$\tilde{\pi}_L^{\tilde{s}}(\tilde{f})\Psi(x,q) = \tilde{f}(xq)\Psi(x,q). \tag{3.72}$$

This evidently reduces as a direct integral over q; the representation $\pi_{q_0}^s$ is simply obtained by fixing $q = q_0$ in (3.70). One sees that $\pi_{q_0}^s$ and $\pi_{q_1}^s$ are inequivalent when q_0 and q_1 lie in orbits of different type.

We are now in a position to paraphrase Definitions 1.7.1 and 1.7.4.

Definition 3.6.3. *The* **reduced groupoid C^*-algebra** $C_r^*(\mathsf{G})$ *of a Lie groupoid* $\mathsf{G} \rightrightarrows Q$ *is the completion of* $C_c^\infty(\mathsf{G})$ *in the norm*

$$\|f\|_r := \|\pi_L(f)\| = \sup_{q\in Q} \|\pi_q(f)\|. \tag{3.73}$$

Here π_L stands for any of the realizations of the regular representation discussed in 3.6.2; likewise for π_q.

The **groupoid C^*-algebra** $C^*(\mathsf{G})$ *of a groupoid* G *is the closure of* $C_c^\infty(\mathsf{G})$ *in the norm*

$$\|f\| := \|\pi_u(f)\| = \sup_{\pi} \|\pi(f)\|, \tag{3.74}$$

where π_u is the direct sum of all nondegenerate bounded representations π of $C_c^\infty(\mathsf{G})$ satisfying (3.57).

In the definition of $C^*(\mathsf{G})$ the bound (3.57), which depends on the choice of a left Haar system, may be replaced by an appropriate intrinsic continuity condition. In the presence of a given Haar system one obviously has the inequalities (cf. (3.57))

$$\|f\|_r \le \|f\| \le \|f\|_I. \tag{3.75}$$

3.7 Examples of Lie Groupoid C^*-Algebras

We will now determine the structure of the C^*-algebras of some of the Lie groupoids we have been looking at so far.

Theorem 3.7.1.

- *When $\mathsf{G} = G$ is a unimodular Lie group, the (reduced) groupoid C^*-algebra $C_{(r)}^*(G)$ coincides with the (reduced) group C^*-algebra defined in 1.7.1 and 1.7.4.*
- *For a pair Lie groupoid $Q \times Q \rightrightarrows Q$ one has*

$$C^*(Q\times Q) \simeq C_r^*(Q\times Q) \simeq \mathfrak{B}_0(L^2(Q)). \tag{3.76}$$

- *For a gauge groupoid* $\mathsf{P} \times_H \mathsf{P} \rightrightarrows Q$ *each measurable section* $s : Q \to \mathsf{P}$ *determines isomorphisms*

$$C_r^*(\mathsf{P} \times_H \mathsf{P}) \simeq \mathfrak{B}_0(L^2(Q)) \otimes C_r^*(H); \tag{3.77}$$
$$C^*(\mathsf{P} \times_H \mathsf{P}) \simeq \mathfrak{B}_0(L^2(Q)) \otimes C^*(H). \tag{3.78}$$

In particular, for compact H *one has* $C^*(\mathsf{P} \times_H \mathsf{P}) \simeq \mathfrak{B}_0(L^2(\mathsf{P}))^H$.

Note that there is no ambiguity in the definition of the tensor product, as one of the factors is $\mathfrak{B}_0(\mathcal{H})$.

For a Lie group $\|f\|_l = \|f\|_1$, so the first claim is obvious from 1.7.4.

The reduced regular representation π^s of $C_c^\infty(Q \times Q)$ is faithful; cf. (3.66). The norm-closure of $\pi^s(C_c^\infty(Q \times Q))$ is $\mathfrak{B}_0(L^2(Q))$, so that one infers (3.76) for C_r^*.

To prove the same result for the full groupoid C^*-algebra, one needs to show that every nondegenerate bounded representation π of $C_c^\infty(Q \times Q)$ on some Hilbert space $\mathcal{H}$ is a multiple of π^s on $L^2(Q)$. This can be done by a method whose significance will emerge at the end of IV.2.4.

Given $(\pi, \mathcal{H})$, we equip the algebraic tensor product $\overline{C_c^\infty(Q)} \otimes \mathcal{H}$ with a sesquilinear form $(\, ,\,)_0$, defined by linear extension of

$$(g \otimes \Psi, f \otimes \Phi)_0 := (\Psi, \pi(f \times \overline{g})\Phi)_{\mathcal{H}}, \tag{3.79}$$

where $f \times \overline{g}(q, q') := f(q)\overline{g}(q')$. This form is easily seen to be positive semidefinite; if $\mathcal{N}_0$ is its null space, the completion of $(\overline{C_c^\infty(Q)} \otimes \mathcal{H})/\mathcal{N}_0$ in the inherited inner product is a Hilbert space, denoted by $\mathcal{H}^0$. Subsequently, define a linear map $U : C_c^\infty(Q) \otimes \mathcal{H}^0 \to \mathcal{H}$ by linear extension of

$$Uf \otimes [g \otimes \Psi] := \pi(f \times \overline{g})\Psi. \tag{3.80}$$

Here $[g \otimes \Psi] \in \mathcal{H}^0$ is the image of $g \otimes \Psi \in \overline{C_c^\infty(Q)} \otimes \mathcal{H}$ under the canonical projection. The map U is well-defined: Firstly, if $[g \otimes \Psi] = 0$ then $\pi(f \times \overline{g})\Psi = 0$, as can be checked using (3.79). Secondly, π is bounded, so that the right-hand side exists for all $\Psi \in \mathcal{H}$, and accordingly for all $[g \otimes \Psi] \in \mathcal{H}^0$. Moreover, using (3.79) and the property $\pi(f \times \overline{g})^* = \pi(g \times \overline{f})$, as well as (3.24) and (3.25), one verifies that U satisfies

$$(U\Omega_1, U\Omega_2)_{\mathcal{H}} = (\Omega_1, \Omega_2)_{L^2(Q)\otimes\mathcal{H}^0} \tag{3.81}$$

for all $\Omega_1, \Omega_2 \in C_c^\infty(Q) \otimes \mathcal{H}^0$. Since in addition, the image of U is dense in $\mathcal{H}$ as a consequence of the nondegeneracy of π, it follows that U can be extended to a unitary map from $L^2(Q) \otimes \mathcal{H}^0$ to $\mathcal{H}$, which we call U as well. The point is that U intertwines $\pi^s \otimes \mathbb{I}$ and π in that $U\pi^s(f) \otimes \mathbb{I} = \pi(f)U$, as is trivially verified from (3.80). One concludes that π is equivalent to the direct sum of $\dim(\mathcal{H}^0)$ copies of π^s. This implies (3.76).

We now come to the last claim of the theorem, which obviously generalizes Proposition 2.8.2. The section s establishes an isomorphism $\tau_s^{-1}(q_0) \simeq \mathsf{P}$ by identifying $[x, s(q_0)]_H \in \tau_s^{-1}(q_0)$ with $x \in \mathsf{P}$. Hence, using Proposition 3.5.3, we have $L^2(\tau_s^{-1}(q_0)) = L^2(\mathsf{P}, \mu)$ for some fixed H-invariant locally Lebesgue measure μ on P. The reduced regular representation $\pi_r := \pi_{q_0}^s$ on $L^2(\mathsf{P}, \mu)$ depends neither on q_0 nor on s. It is faithful, and reads

$$\pi_r(f)\Psi(x) = \int_{\mathsf{P}} d\mu(y)\, f([x,y]_H)\Psi(y). \tag{3.82}$$

As in the proof of 2.8.2, we in addition use s to map $L^2(\mathsf{P})$ into $L^2(Q)\otimes L^2(H)$ (in case H is not unimodular, one should note that the measure dh occurring in (2.139) is a right Haar measure, whereas the one used to define $L^2(H)$ is a left Haar measure μ_L). One then obtains

$$U_s\pi_r(f)U_s^*\Psi(q,h) = \int_Q d\nu(q') \int_H d\mu_L(k) f_s(q,k,q')\Psi(q',k^{-1}h), \tag{3.83}$$

where $f_s(q,k,q') := f([s(q)k, s(q')]_H)$. Hence $U_s\pi_r U_s^*$ factorizes into the product of the defining representation (3.66) in the argument $(q,q') \in Q\times Q$ and the left-regular representation π_L in the argument $h\in H$; cf. (1.84). With Definition 1.7.1 this immediately implies the isomorphism (3.77).

Let $C_c^s(Q\times H\times Q)$ be the image of $C_c^\infty(\mathsf{P}\times_H\mathsf{P})$ under the map $f\mapsto f_s$, and define $C_c^s(Q)$ as the space of functions on Q of the type $f(q) = f^{\mathsf{P}}(s(q))$, where $f^{\mathsf{P}}\in C_c^\infty(\mathsf{P})$; the space $C_c^s(Q\times Q)$ is defined similarly.

Rather than with $C_c^s(Q\times H\times Q)$, we may work with its subspace $C_c^s(Q\times Q)\otimes C_c^\infty(H)$. This is justified by the fact that the two spaces in question have the same closure in the norm $\|\cdot\|_I$ (cf. (3.57)). In particular, $C_c^s(Q\times H\times Q)$ and $C_c^s(Q\times Q)\otimes C_c^\infty(H)$ have the same closure in the C^*-norm (3.74).

Transferring the *-algebraic operations from $\mathsf{P}\times_H\mathsf{P}$, one sees that as a *-algebra $C_c^s(Q\times Q)\otimes C_c^\infty(H)$ is the direct product of $C_c^s(Q\times Q)$ (with operations (3.24) and (3.25)) and $C_c^\infty(H)$ (with operations (1.80) and (1.81), in which $c=1$). This reflects the fact that s leads to an isomorphism $\mathsf{P}\times_H\mathsf{P}\simeq Q\times H\times Q$ as groupoids over Q, as explained after Definition 3.1.7.

The argument used to prove (3.76) works equally well when $C_c^\infty(Q)$ is replaced by $C_c^s(Q)$. Combining this with Theorem 1.7.3 and the above factorization, one is led to (3.78). The final claim then follows from Proposition 2.8.2 or 3.3.7. ■

Recall 2.8.4. Generalizing 2.8.3 and 2.8.5, we have

Corollary 3.7.2. *Up to equivalence there is a bijective correspondence between the nondegenerate representations π^χ of $C^*(\mathsf{P}\times_H\mathsf{P})$ and the representations U_χ of H. Here π^χ is realized on $\mathcal{H}^\chi$ by*

$$\pi^\chi(f)\Psi^\chi(x) = \int_{\mathsf{P}} d\mu(y)\, f([x,y]_H)\Psi^\chi(y). \tag{3.84}$$

The representation $\pi^\chi(C^*(\mathsf{P}\times_H\mathsf{P}))$ is irreducible iff $U_\chi(H)$ is irreducible.

Choosing a section $s: Q\rightarrow\mathsf{P}$ and following the proof of (3.78) above, one sees, by first restricting to $C_c^s(Q\times Q)\otimes C_c^\infty(H)$, that

$$\pi^\chi(C^*(\mathsf{P}\times_H\mathsf{P}))\simeq\mathfrak{B}_0(L^2(Q))\otimes\pi_\chi(C^*(H)).$$

Then use Corollaries 1.7.5 and I.2.2.6. ■

The C^*-algebra of an action groupoid $G \times Q \rightrightarrows Q$ is called an **action C^*-algebra**, denoted by $C^*(G, Q)$ (which is short for $C^*(G, C_0(Q))$). As a slight variation on Definition 3.4.5 we put

Definition 3.7.3. *A **system of imprimitivity** of G on Q in a Hilbert space $\mathcal{H}$ is a pair $(U, \tilde{\pi})$ where U is a continuous representation of G on $\mathcal{H}$, and π is a nondegenerate representation of $C_0(Q)$, satisfying the covariance condition (3.38).*

There is, in fact, no real difference between 3.4.5 and 3.7.3.

Corollary 3.7.4. *There is a bijective correspondence between nondegenerate representations π of $C^*(G, Q)$ and systems of imprimitivity of G on Q. This correspondence is given by continuous extension of (3.35)–(3.37).*

Combine Corollary 3.4.6 and Definition 3.6.3. ■

The structure of action C^*-algebras is in general fairly complicated, except for the following corollary to Proposition 3.4.7.

Corollary 3.7.5. *One has the isomorphisms*

$$C^*(G, G/H) \simeq C^*(G \times_H G) \simeq \mathfrak{B}_0(L^2(G/H)) \otimes C^*(H), \tag{3.85}$$

and similarly for the reduced C^*-algebras (i.e., C^* is replaced by C^*_r).

Since the *-algebras $C_c^\infty(G \times (G/H))$ and $C_c^\infty(G \times_H G)$ are isomorphic by 3.4.7, they in particular have the same representation theory. Hence all claims follow from Definition 3.6.3 and Theorem 3.7.1. ■

In particular, we may look at the C^*-algebra $\mathfrak{B}_0(L^2(\mathbb{R}^n))$ in two different ways: It is the C^*-algebra of the pair groupoid $\mathbb{R}^n \times \mathbb{R}^n \rightrightarrows \mathbb{R}^n$, as well as the action C^*-algebra defined by the canonical action of $\mathbb{R}^n$ on itself.

What follows is the **quantum transitive imprimitivity theorem**.

Corollary 3.7.6. *There is a bijective correspondence between systems of imprimitivity $(U^\chi(G), \tilde{\pi}(C_0(G/H)))$ and representations U_χ of H. The system $(U^\chi, \tilde{\pi}^\chi)$ is irreducible (in the sense that the only bounded operators commuting with all $U^\chi(x)$ and $\tilde{\pi}^\chi(\tilde{f})$ are $\lambda\mathbb{I}$, $\lambda \in \mathbb{C}$) iff $U_\chi(H)$ is irreducible.*

Up to equivalence, the pair $(U^\chi, \tilde{\pi}^\chi)$ is realized on $\mathcal{H}^\chi$ (cf. 2.8.4), where $U^\chi(G)$ is the induced representation defined in (2.175), and $\tilde{\pi}^\chi$ is given by

$$\tilde{\pi}^\chi(\tilde{f})\Psi^\chi(x) = \tilde{f}([x]_H)\Psi^\chi(x). \tag{3.86}$$

In the realization of the carrier space as $\mathcal{H}^\chi_{\{s\}}$ (cf. 2.8) one has $(U^\chi_{\{s\}}, \tilde{\pi}^\chi_{\{s\}})$, given by (2.176) and

$$\tilde{\pi}^\chi_{\{s\}}(\tilde{f})\Psi^\chi_\alpha(q) = \tilde{f}(q)\Psi^\chi_\alpha(q), \tag{3.87}$$

and analogously on $\mathcal{H}^\chi_s$; see (2.141).

Combine Corollaries 3.7.4, 3.7.5, 3.4.8, and 3.7.2. The irreducibility of $(U^\chi, \tilde{\pi}^\chi)$ is equivalent to the irreducibility of $\pi^\chi(C^*(G \times_H G))$. ■

A deeper understanding of this result will be achieved in IV.2.7.

We close this section with a remarkable consequence of Corollary 3.7.6, namely Theorem II.2.1.4. We are going to apply 3.7.6 with $H = \{e\}$ and $G = Q = \mathbb{R}^n$, in which case there is only one irreducible system of imprimitivity, as $\hat{H}$ consists of a single element.

A representation $U_{(\lambda)}(\tilde{H}_n)$ on a Hilbert space $\mathcal{H}$ for which

$$dU_{(\lambda)}(Z) = -i\lambda\mathbb{I}, \tag{3.88}$$

where $\lambda \in \mathbb{R}\backslash\{0\}$, defines a system of imprimitivity $(U(\mathbb{R}^n), \tilde{\pi}(C_0(\mathbb{R}^n)))$ on $\mathcal{H}$ by

$$U(v) := U_{(\lambda)}(\mathrm{Exp}(vP));$$
$$\tilde{\pi}(\tilde{f}) := \int_{\mathbb{R}^n} d^n u\, U_{(\lambda)}\left(\mathrm{Exp}(-\frac{u}{\lambda}Q)\right) \hat{f}(u), \tag{3.89}$$

where $\hat{f}$ is the Fourier transform of $\tilde{f}$. (Here $\hat{f} \in L^1(\mathbb{R}^n)$, so that $\tilde{\pi}$ thus defined is to be extended from the image of the Fourier transform of $L^1(\mathbb{R}^n)$ in $C_0(\mathbb{R}^n)$ to $C_0(\mathbb{R}^n)$ by continuity.) This may be verified using II.(2.6).

Conversely, given a system of imprimitivity $(U(\mathbb{R}^n), \tilde{\pi}(C_0(\mathbb{R}^n)))$ on $\mathcal{H}$ and a real number $\lambda \neq 0$, one obtains a representation $U_{(\lambda)}(\tilde{H}_n)$ on $\mathcal{H}$ for which (3.88) holds by

$$U_{(\lambda)}(\mathrm{Exp}(-uQ)) := \tilde{\pi}(\tilde{e}_{\lambda u});$$
$$U_{(\lambda)}(\mathrm{Exp}(vP)) := U(v), \tag{3.90}$$

where $\tilde{e}_p(x) := \exp(ipx)$. Here the representation $\tilde{\pi}(C_0(\mathbb{R}^n))$ has been extended to $C_b(\mathbb{R}^n)$ by the functional calculus obtained from the spectral theorem.

These two constructions are each other's inverse; the uniqueness of the irreducible representation satisfying (3.88) follows from the uniqueness of the irreducible system of imprimitivity $(U(\mathbb{R}^n), \tilde{\pi}(C_0(\mathbb{R}^n)))$. In summary, the classification of the irreducible representations of the Heisenberg group with nonzero central element follows from the uniqueness of the irreducible representation of the group with one element. ■

3.8 Lie Algebroids

The construction of the Lie algebra of a Lie group as an "infinitesimal" object can be generalized to the setting of Lie groupoids. Like a Lie algebra, the object in question may be defined in its own right.

Definition 3.8.1. *A* **Lie algebroid** $\mathsf{V} \xrightarrow{TQ}_{Q}$ *on a manifold Q is a vector bundle $\mathsf{V}(Q, V, \tau)$ over Q, which apart from the bundle projection $\tau : \mathsf{V} \to Q$ is equipped with a vector bundle map $\tau_a : \mathsf{V} \to TQ$ (called the* **anchor***), as well as with a Lie bracket $[\, ,\,]_{\mathsf{V}}$ on the space $\Gamma(\mathsf{V})$ of (smooth compactly supported) sections of V, satisfying*

$$\tau_a \circ [s_1, s_2]_{\mathsf{V}} = [\tau_a \circ s_1, \tau_a \circ s_2], \tag{3.91}$$

where the right-hand side is the usual commutator of vector fields on $\Gamma(TQ)$, *and*

$$[s_1, f s_2]_{\mathsf{V}} = f[s_1, s_2]_{\mathsf{V}} + (\tau_a \circ s_1 f)s_2 \tag{3.92}$$

for all $s_1, s_2 \in \Gamma(\mathsf{V})$ *and* $f \in C^\infty(Q)$.

It is part of the definition of a bundle map that the anchor is fiber-preserving and linear on each fiber.

When Q is a point, one has $\Gamma(\mathsf{V}) = \mathsf{V} := \mathfrak{g}$, and the only nontrivial requirement is that $[\,,\,]_{\mathfrak{g}}$ be a Lie bracket. Hence in that case the Lie algebroid $\mathfrak{g}$ is simply a real Lie algebra. The next simplest example is

Definition 3.8.2. *The* **pair algebroid** $TQ \overset{\to}{\to}{}^{TQ}_{Q}$ *consists of the tangent bundle with its usual projection and commutator, and anchor* $\tau_a = \mathrm{id}$.

The property (3.92) then reads $[\xi_1, f\xi_2] = f[\xi_1, \xi_2] + (\xi_1 f)\xi_2$, which is, indeed, identically satisfied.

Definition 3.8.3. *Let* $\mathsf{P}(Q, H, \tau)$ *be a principal* H*-bundle over* Q. *The* **gauge algebroid** $(T\mathsf{P})/H \overset{\to}{\to}{}^{TQ}_{Q}$ *is defined by the obvious projections (both inherited from* τ*), and the Lie bracket on* $\Gamma((T\mathsf{P})/H)$ *obtained by identifying this space with* $\Gamma(T\mathsf{P})^H$ *(as in the proof of Proposition 2.4.3), and borrowing the commutator from* $\Gamma(T\mathsf{P})$.

Thus the bundle V is the central term in the exact sequence (2.65).

Definition 3.8.4. *Suppose one has a* $\mathfrak{g}$*-action on a manifold* Q*; see 1.1. The* **action algebroid** $\mathfrak{g} \times Q \overset{\to}{\to}{}^{TQ}_{Q}$ *has* $\mathsf{V} = \mathfrak{g} \times Q$ *(as a trivial bundle over* Q*), with anchor* $\tau_a(X, q) := -\xi_X(q)$. *Identifying sections of* $\mathfrak{g} \times Q$ *with* $\mathfrak{g}$*-valued functions* $X(\cdot)$ *on* Q, *the Lie bracket on* $\Gamma(\mathfrak{g} \times Q)$ *is*

$$[X, Y]_{\mathfrak{g}\times Q}(q) := [X(q), Y(q)]_{\mathfrak{g}} + \xi_Y X(q) - \xi_X Y(q). \tag{3.93}$$

Similar to Proposition 3.1.8, we have

Proposition 3.8.5. *Let* $G(G/H, H, \tau)$ *be the principal* H*-bundle defined in 2.7. The gauge algebroid* $(TG)/H \overset{\to}{\to}{}^{T(G/H)}_{G/H}$ *and the action algebroid* $\mathfrak{g} \times (G/H) \overset{\to}{\to}{}^{T(G/H)}_{G/H}$ *are isomorphic (in the obvious sense).*

The quotient of the right trivialization $TG \simeq \mathfrak{g} \times G$ by H provides a diffeomorphism $(TG)/H \simeq \mathfrak{g} \times (G/H)$. ∎

We now explain how one may associate a Lie algebroid $\mathfrak{G} \overset{\to}{\to}{}^{TQ}_{Q}$ with a given Lie groupoid $\mathsf{G} \overset{\leftarrows}{\rightrightarrows} Q$.

A **left-invariant vector field** ξ^L on G is a vector field satisfying $(\tau_t)_*\xi^L = 0$ and $(L_\gamma)_*\xi^L(\gamma') = \xi^L(\gamma\gamma')$ for all $(\gamma, \gamma') \in \mathsf{G}_2$. Note that the second condition is well-defined because of the first one. The space of all left-invariant vector fields on G is denoted by $\Gamma(T\mathsf{G})^L$.

This definition may be restated in terms of the corresponding flow: If $\xi(\gamma) = \dot{\gamma}$ for some flow $\gamma(\lambda)$, then ξ is a left-invariant vector field iff $\tau_t(\gamma(\lambda))$ is independent of λ and $\gamma'(\gamma(\lambda)) = (\gamma'\gamma)(\lambda)$ whenever $\gamma'\gamma$ is defined; once again, the second condition is well-defined because of the first one. We call such a $\gamma(\cdot)$ a **left-invariant flow** on G.

Lemma 3.8.6.

1. *The vector space* $\Gamma(T\mathsf{G})^L$ *is a Lie algebra under the usual commutator borrowed from* $\Gamma(T\mathsf{G})$.
2. *A left-invariant vector field is determined by its values on the unit space* $\mathsf{G}_0 = \iota(Q)$.
3. *The tangent bundle of* G *at the unit space has a decomposition*

$$T_{\iota(q)}\mathsf{G} = T_{\iota(q)}\mathsf{G}_0 \oplus T^t_{\iota(q)}\mathsf{G}, \tag{3.94}$$

$$T^t\mathsf{G} := \ker(\tau_t)_* \subset T\mathsf{G}. \tag{3.95}$$

To prove 3.8.6.1 it suffices to remark that since for any smooth map φ the commutator satisfies $\varphi_*[\xi_1, \xi_2] = [\varphi_*\xi_1, \varphi_*\xi_2]$, the space $\Gamma(T\mathsf{G})^L$ is closed under the commutator.

Since $\gamma = \gamma(\gamma^{-1}\gamma)$, left invariance implies that $\xi^L(\gamma) = (L_\gamma)_*\xi^L(\gamma^{-1}\gamma)$, which proves 3.8.6.2.

Because nonzero elements of $T_{\iota(q)}(\mathsf{G}_0)$ are tangent to curves $\iota(q(s))$, for which $\tau_t(\iota(q(s))) = q(s) \neq q$ for small enough s, it follows that $T_{\iota(q)}\mathsf{G}_0 \cap T^t_{\iota(q)}\mathsf{G} = 0$. Now note that the image of $T_{\iota(q)}\mathsf{G}$ under $\ker(\tau_t)_*$ on the one hand equals T_qQ, and on the other hand is isomorphic to the quotient $T_{\iota(q)}\mathsf{G}/T^t_{\iota(q)}\mathsf{G}$ as a vector space. A dimension count then establishes (3.94). ■

Definition 3.8.7. *The Lie algebroid* $\mathfrak{G} \overset{\to TQ}{\to_Q}$ *of a Lie groupoid* $\mathsf{G} \overset{\leftarrow}{\rightrightarrows} Q$ *is given by the following (cf. Definition II.3.4.1).*

- *The vector bundle* $\mathsf{V} = \mathfrak{G}$ *over* Q *is the normal bundle* $N^\iota Q$ *defined by the embedding* $\iota : Q \hookrightarrow \mathsf{G}$*; accordingly, the projection* $\tau : N^\iota Q \to Q$ *is given by* τ_s *or* τ_t *(these projections coincide on* G_0*).*
- *Identifying* $N^\iota_{\iota(q)}Q$ *with* $T^t_{\iota(q)}\mathsf{G}$ *by (3.94), the anchor is given by* $\tau_a := (\tau_s)_* : T\mathsf{G} \to TQ$ *(restricted to* $\ker(\tau_t)_*$*).*
- *Identifying a section of* $N^\iota_{\iota(q)}Q$ *with a left-invariant vector field on* G *through the previous item and 3.8.6.2, so that* $\Gamma(\mathfrak{G}) = \Gamma(N^\iota Q) \simeq \Gamma(T\mathsf{G})^L \subset \Gamma(T\mathsf{G})$*, the Lie bracket* $[\,,\,]_{\mathfrak{G}}$ *is given by the commutator on* $\Gamma(T\mathsf{G})$ *(this is consistent because of 3.8.6.1).*

The required equality (3.91) is automatically satisfied (as it holds for all vector fields on G). To verify (3.92), note that from 3.1.1.4, for $f \in C^\infty(Q)$ and $\xi \in \Gamma(T\mathsf{G})$ one has $(L_\gamma)_*(f\xi(\gamma^{-1}\gamma)) = f(\tau_s(\gamma))(L_\gamma)_*\xi(\gamma^{-1}\gamma)$. Hence the action of $C^\infty(Q)$ on $\Gamma(T\mathsf{G})^L$ is given by $(f\xi^L)(\gamma) = f(\tau_s(\gamma))\xi^L(\gamma)$. Equation (3.92) then follows as in the case of the pair algebroid.

Rather than defining $\mathfrak{G}$ in terms of the normal bundle N^ι, one may put

$$\mathfrak{G}' := \iota^* T^t\mathsf{G} = T^t\mathsf{G} *_{\mathsf{G}} Q = \{(X, q) \in T^t\mathsf{G} \times Q \mid \tau_{T\mathsf{G}\to\mathsf{G}}(X) = \iota(q)\}; \tag{3.96}$$

cf. 2.1.2. This is simply the restriction of $T^t\mathsf{G}$ to $\iota(Q)$, seen as a bundle over Q through projection onto the second variable. The anchor is defined as $\tau_a := (\tau_s)_*$, as above, and the Lie bracket is obtained by extending sections of $\mathfrak{G}'$ to left-invariant vector fields on G. The isomorphism between $\mathfrak{G}$ and $\mathfrak{G}'$ is then obvious from Definition 3.8.7 and (3.94).

Proposition 3.8.8.

- *A Lie algebra* $\mathfrak{g}$ *is the Lie algebroid of a Lie group* G.
- *The pair algebroid* $TQ \overset{\to TQ}{\to_Q}$ *is the Lie algebroid of the pair groupoid* $Q \times Q \overset{\leftarrow}{\Rightarrow} Q$.
- *The gauge algebroid* $(T\mathsf{P})/H \overset{\to TQ}{\to_Q}$ *is the Lie algebroid of the gauge groupoid* $\mathsf{P} \times_H \mathsf{P} \overset{\leftarrow}{\Rightarrow} Q$.
- *The action algebroid* $\mathfrak{g} \times Q \overset{\to TQ}{\to_Q}$ *is the Lie algebroid of the action groupoid* $G \times Q \overset{\leftarrow}{\Rightarrow} Q$.

For a Lie group G the base Q consists of a point, which ι maps to $e \in G$; the normal bundle is T_eG. The construction of the Lie algebroid then amounts to the usual identification of $\mathfrak{g} = T_eG$ with the space of left-invariant vector fields on G.

The Lie algebroid of a pair groupoid is identified by Lemma II.3.4.3, since ι is the diagonal embedding. The isomorphism $N^{\iota}_{\iota(q)}Q \simeq T^t_{\iota(q)}\mathsf{G}$ identifies $X \in T_qQ$ with $0\dot{+}X \in T_{(q,q)}(Q \times Q) \simeq T_qQ \oplus T_qQ$. Hence the anchor is the identity, the left-invariant vector fields are of the form $\xi^L(q,q') = 0\dot{+}\xi(q')$, and the Lie bracket is simply the usual one on $\Gamma(TQ)$.

In discussing the gauge groupoid, one first notes that when $\mathsf{P}(Q,H,\tau)$ is a principal fiber bundle, the tangent bundle TQ has the following description. One defines $(T\mathsf{P})/I$ as the bundle over P whose fiber at x is $T_x\mathsf{P}/V_x\mathsf{P}$; see (2.8). In view of (2.10), the H-action on $(T\mathsf{P})/I$ (pushed forward from the H-action on P) is well-defined, and one has $TQ \simeq ((T\mathsf{P})/I)/H$. (The dual of this isomorphism is (2.56).)

We apply the same procedure to the principal H-bundle $\mathsf{P} \times \mathsf{P}$ over $\mathsf{P} \times_H \mathsf{P}$. Identifying $T_{(x,y)}\mathsf{P} \times \mathsf{P}$ with $T_x\mathsf{P} \oplus T_y\mathsf{P}$, the role of $V_x\mathsf{P}$ in the definition of I is now played by the space of all vectors of the form $\xi^f_X(x)\dot{+}\xi^f_X(y) \in T_{(x,y)}\mathsf{P} \times \mathsf{P}$, $X \in \mathfrak{h}$. The vector bundle $T^t(\mathsf{P} \times_H \mathsf{P})$ over Q is then a double quotient $((V\mathsf{P} \times \mathsf{P})/I)/H$.

The restriction of $T^t(\mathsf{P} \times_H \mathsf{P})$ to G_0 equals the H-quotient of the restriction of $(V\mathsf{P} \times \mathsf{P})/I$ to the diagonal. The fiber $T_{(x,x)}(V\mathsf{P} \times \mathsf{P})/I$ is isomorphic to $T_x\mathsf{P}$ through the identification of the equivalence class $[X\dot{+}Y] \in (V_x\mathsf{P} \oplus T_x\mathsf{P})/I$ with $Y - X \in T_x\mathsf{P}$. Taking the H-quotient, and using the isomorphism $N^{\iota}_{\iota(q)}Q \simeq T^t_{\iota(q)}(\mathsf{P} \times_H \mathsf{P})$ given by (3.94), we arrive at $N^{\iota}Q \simeq (T\mathsf{P})/H$.

Following the steps in the above derivation, one immediately infers from its definition that the anchor is the canonical projection from $(T\mathsf{P}/H)$ to TQ, and that the Lie bracket is as stated.

In the action groupoid the identification $N^{\iota}Q \simeq T_eG = \mathfrak{g}$ is immediate from Definition 3.1.4; the normal bundle is automatically identified with $T_eG \subset T_{(e,q)}G \times Q$. The anchor then follows from (1.19). To compute the Lie bracket on $\Gamma(\mathfrak{g} \times Q)$ one notes that since $(y,q) = L_{(y,q)}(e, y^{-1}q)$, a section $\xi : q \mapsto X(q) \in \mathfrak{g}$ defines a left-invariant vector field on $G \times Q$ by $\xi^L(y,q) := L_{(y,q)*}\xi(e, y^{-1}q) = \xi^L_{X(y^{-1}q)}(y)$. Here ξ^L is defined in (1.37), and

$\xi^L_{\cdot}(y) \in T_yG$, regarded as a subspace of $T_{(y,q)}G \times Q$. The expression (3.93) then easily follows. ■

To close this section, we provide the missing proof of the first claim in Proposition 3.3.3. Indeed, a given strictly positive smooth density ρ on the vector bundle $\mathfrak{G}$ associated to $\mathsf{G} \overset{\leftarrow}{\Rightarrow} Q$ by 3.8.7 can be (uniquely) extended to a left-invariant density $\tilde{\rho}$ on the vector bundle $T^t\mathsf{G}$, which in turn yields a left Haar system by $\mu^t_q(f) := \int_{\tau_t^{-1}(q)} \tilde{\rho} f$. ■

3.9 The Poisson Algebra of a Lie Algebroid

We saw in 3.6 that one can associate a C^*-algebra to a Lie groupoid. The classical analogue is the construction of a Poisson algebra of a Lie algebroid. This generalizes the Lie–Poisson structure on $C^\infty(\mathfrak{g}^*, \mathbb{R})$ introduced in 1.1.

Proposition 3.9.1. *Given a Lie algebroid* $\mathsf{V} \overset{\to}{\to}{}^{TQ}_{Q}$ *with anchor* τ_a, *the dual vector bundle* V^* *is a Poisson manifold* $\mathsf{V}^*_\pm$ *under the Poisson bracket on* $C^\infty(\mathsf{V}^*, \mathbb{R})$ *defined by the following special cases:*

$$\{f, g\}_\pm = 0; \tag{3.97}$$

$$\{\tilde{s}, f\}_\pm = \pm\tau_a \circ sf; \tag{3.98}$$

$$\{\tilde{s}_1, \tilde{s}_2\}_\pm = \pm\widetilde{[s_1, s_2]_{\mathsf{V}}}. \tag{3.99}$$

Here $f := \tau^*_{\mathsf{V}^* \to Q}\tilde{f} \in C^\infty(\mathsf{V}^*, \mathbb{R})$ *is defined by* $\tilde{f} \in C^\infty(Q, \mathbb{R})$, *and similarly for* g. *Also,* $\tilde{s} \in C^\infty(\mathsf{V}^*, \mathbb{R})$ *is defined by a section* $s \in \Gamma(\mathsf{V})$ *through* $\tilde{s}(\theta) := \theta(s(\tau_{\mathsf{V}^* \to Q}(\theta)))$.

Note that the function $\tilde{s}$ is linear (in the sense of being linear on each fiber of V^*), and that any such (smooth) function is of this form. Hence the collection of differentials $df, d\tilde{s}$ spans the cotangent space at every point of V^*, so that the Poisson bracket is indeed completely defined by (3.97)–(3.99).

In a local trivialization of V one has $s(q) = s^a(q)\mathbf{e}_a$ (where $\{\mathbf{e}_a\}$ is a basis of the typical fiber V of V), hence $\tilde{s}(q, \theta) = \theta_a s^a(q)$ in terms of the coordinates θ_a on V^* defined by the dual basis. We write $[\mathbf{e}_a, \mathbf{e}_b](q) = C^c_{ab}(q)\mathbf{e}_c$ and $\tau_a(\mathbf{e}_a, q) = A^\mu_a(q)\partial/\partial q^\mu$, in terms of which the Poisson tensor is given by $B_{(\theta,q)}(dq^\mu, dq^\nu) = 0$, $B_{(\theta,q)}(d\theta_a, dq^\mu) = \pm A^\mu_a(q)$, and $B_{(\theta,q)}(d\theta_a, d\theta_b) = \pm C^c_{ab}(q)\theta_c$. The conditions (3.91) and (3.92) then lead to identities on C and A that are used to prove the Jacobi identity I.(2.6). ■

Proposition 3.9.1 has the following converse.

Proposition 3.9.2. *If a Poisson manifold* V^* *is a vector bundle over* Q, *such that the Poisson bracket of two linear functions is linear, then* V^* *is the dual of a Lie algebroid* V, *and the Poisson bracket on* $C^\infty(\mathsf{V}^*, \mathbb{R})$ *is the one in 3.9.1.*

The Lie bracket on $\Gamma(\mathsf{V})$ is defined by reading (3.99) from right to left. To define the anchor, we note that the Leibniz rule yields

$$\{\tilde{s}_1, f\tilde{s}_2\} = f\{\tilde{s}_1, \tilde{s}_2\} + \{\tilde{s}_1, f\}\tilde{s}_2. \tag{3.100}$$

We take f as described in 3.9.1. Since the left-hand side and the first term on the right are linear, as is $\tilde{s}_2$, it follows that $\{\tilde{s}_1, f\}$ is constant on the fibers, defining a function on Q. Applying the Leibniz rule to $\{\tilde{s}, fg\}$, the map $\tilde{f} \mapsto \{\tilde{s}, f\}$ is seen to be a derivation on $C^\infty(Q, \mathbb{R})$, so it must be that $\{\tilde{s}_1, f\} = \xi_s \tilde{f}$ for some vector field ξ_s on Q.

Hence $s \mapsto \xi_s$ is a map from $\Gamma(\mathsf{V})$ to $\Gamma(TQ)$. To prove that it is given by a bundle map $\tau_a : \mathsf{V} \to TQ$, we must show that $\xi_{\tilde{f}s} = \tilde{f}\xi_s$ for all $\tilde{f} \in C^\infty(Q)$. This follows from the Leibniz rule $\{f\tilde{s}, g\} = f\{\tilde{s}, g\} + \{f, g\}\tilde{s}$, which may be rewritten as $\xi_{\tilde{f}s} g - \tilde{f}\xi_s \tilde{g} = \{f, g\}\tilde{s}$. The left-hand side is a function on Q, whereas the right-hand side is linear; this is possible only when (3.97) holds. Therefore, $\xi_s = \tau_a \circ s$ for some bundle map τ_a.

The Jacobi identity on the Poisson bracket and the definition of τ_a imply (3.91). Finally, (3.100) is equivalent to (3.92). ∎

Combining Propositions 3.9.1 and 3.9.2, we conclude that there is a complete equivalence between Lie algebroids and linear Poisson structures on vector bundles. We now apply this to our usual list of examples; cf. 3.8.8.

Proposition 3.9.3.

- *A Lie algebra $\mathfrak{g}$ yields the $\pm$ Lie–Poisson structure (1.1) on $\mathfrak{g}^*$.*
- *The pair algebroid $TQ \overset{\to TQ}{\to_Q}$ leads to $\pm$ the canonical Poisson bracket I.(2.24) on T^*Q.*
- *The gauge algebroid $(T\mathsf{P})/H \overset{\to TQ}{\to_Q}$ is associated with $\pm$ the Poisson structure on $(T^*\mathsf{P})/H$ specified prior to 2.3.7.*
- *The $\pm$ Poisson bracket on $\mathfrak{g}^* \times Q$ associated to the action algebroid $\mathfrak{g} \times Q \overset{\to TQ}{\to_Q}$ is given by*

$$\{f, g\}_\pm = \pm C_{ab}^c \theta_c \frac{\partial f}{\partial \theta_a}\frac{\partial g}{\partial \theta_b} \pm \xi_a f \frac{\partial g}{\partial \theta_a} \mp \frac{\partial f}{\partial \theta_a}\xi_a g. \tag{3.101}$$

Compare 1.1 for the notation used in (3.101).

The first claim is obvious. The second is most easily proved in a local trivialization, using canonical coordinates. The section $s(q) = (\partial/\partial q^\mu, q)$ then leads to $\tilde{s}(p, q) = p_\mu$. The brackets (3.97)–(3.99) thus specialize to $\{q^\mu, q^\nu\}_+ = 0$, $\{p_\mu, q^\nu\}_+ = (\partial/\partial q^\mu)q^\nu = \delta_\mu^\nu$, and $\{p_\mu, p_\nu\}_+ = \widetilde{[\partial/\partial q^\mu, \partial/\partial q^\nu]} = 0$, respectively. This proves the claim. Note that the linear function $\tilde{\xi} \in C^\infty(T^*Q, \mathbb{R})$ corresponding to $\xi \in \Gamma(TQ)$ is simply the usual **symbol** of the vector field.

The third point follows from the second, quotienting by H.

To prove (3.101) one simply verifies that it includes (3.97)–(3.99) with (3.93) as special cases; do not forget the minus sign in the anchor. ∎

It is worth giving the main special cases of (3.101). For $X \in \mathfrak{g}$ we regard $\tilde{X}$ as a function on $\mathfrak{g}^* \times Q$ by $\tilde{X}(\theta, q) = \theta(X)$, and as before, $\tilde{f} \in C_c^\infty(Q, \mathbb{R})$ defines a function on $\mathfrak{g}^* \times Q$ by $\tilde{f}(\theta, q) = \tilde{f}(q)$. We then have

$$\{\tilde{f}, \tilde{g}\}_- = 0; \tag{3.102}$$

$$\{\tilde{X}, \tilde{Y}\}_- = -\widetilde{[X, Y]}; \tag{3.103}$$

$$\{\tilde{X}, \tilde{f}\}_{-} = \xi_X \tilde{f}. \tag{3.104}$$

As will become clear in 3.11 below, these brackets are the classical counterpart of the commutation relations (3.41), (3.42), and (3.43). We denote the **action Poisson algebra** on $\mathfrak{g}^* \times Q$ with bracket (3.101) by $C^\infty(\mathfrak{g}^*_\pm \times Q, \mathbb{R})$; the corresponding Poisson manifold is, of course, written as $\mathfrak{g}^*_\pm \times Q$.

Proposition 3.9.4. *The symplectic leaves of $\mathfrak{g}^*_- \times Q$ are classified by pairs $(\mathcal{O}^G, \mathcal{O}^H)$, where $\mathcal{O}^G$ is a G-orbit in Q, the group $H \subseteq G$ is the stabilizer of an arbitrary point in $\mathcal{O}^G$, and $\mathcal{O}^H$ is a coadjoint orbit in $\mathfrak{h}^*$. The leaf $L_{(\mathcal{O}^G,\mathcal{O}^H)}$ corresponding to $(\mathcal{O}^G, \mathcal{O}^H)$ is given by*

$$L_{(\mathcal{O}^G,\mathcal{O}^H)} = \{(\theta, q) \in \mathfrak{g}^* \times Q \mid q \in \mathcal{O}^G,\ (-\mathrm{Co}(s(q)^{-1})\theta \restriction \mathfrak{h}^*) \in \mathcal{O}^H\}, \tag{3.105}$$

where $s : \mathcal{O}^G \simeq G/H \to G$ is an arbitrary section of the bundle $G(G/H, H, \tau)$.

One infers from (3.101) that any Hamiltonian flow starting in $\mathfrak{g}^*_- \times \mathcal{O}^G$ stays in this subspace. Using the right trivialization of T^*G one sees that $\mathfrak{g}^* \times \mathcal{O}^G$ is diffeomorphic to $(T^*G)/H$. Equipping the latter, and therefore $\mathfrak{g}^* \times \mathcal{O}^G$, with the Poisson structure inherited from T^*G, and comparing (1.54) with (3.101), one infers that the injection of $\mathfrak{g}^* \times \mathcal{O}^G$ into $\mathfrak{g}^*_- \times \mathcal{O}^G$ is a Poisson map. The proposition then follows from Theorem 2.3.7, applied to the bundle $\mathsf{P}(Q, H, \tau) = G(G/H, H, \tau)$, and (2.118). ∎

We saw in 3.4 that an action *-algebra is a special case of a crossed product *-algebra. Similarly, the Poisson algebra of an action Lie algebroid is a special case of the following classical analogue of a crossed product *-algebra.

Definition 3.9.5. *Let a Lie group G act on a Poisson manifold P by Poisson maps L. The* **semidirect product** *of P and T^*G (equipped with the canonical Poisson structure I.(2.24)) is the quotient $T^*G \times_G P$ under the product action $\rho \times L$ of G (cf. (1.51)), equipped with the unique Poisson structure making the canonical projection $\tau : T^*G \times P \to T^*G \times_G P$ a Poisson map.*

The diffeomorphism $[(\theta, x)_R, \sigma]_G \mapsto (\theta, x\sigma)$ between $T^*G \times_G P$ and $\mathfrak{g}^* \times P$ equips the latter with a Poisson structure; the associated Poisson algebra $C^\infty(\mathfrak{g}^*_- \times P, \mathbb{R})$ is called a **crossed product Poisson algebra**.

In self-evident notation, the Poisson bracket is explicitly given by

$$\{f, g\}_\pm = -C^c_{ab}\theta_c \frac{\partial f}{\partial \theta_a}\frac{\partial g}{\partial \theta_b} - \xi_a f \frac{\partial g}{\partial \theta_a} + \frac{\partial f}{\partial \theta_a}\xi_a g + \{f, g\}^P. \tag{3.106}$$

Hence putting $P = Q$ with the zero Poisson structure shows that (3.101) is indeed a special case of (3.106); this is the classical version of Lemma 3.4.3.

In the following classical analogue of Theorem 3.4.4 the boundedness condition (3.33) is replaced by an integrability condition.

Theorem 3.9.6. *There is a bijective correspondence between*

- *Poisson maps $J : S \to \mathfrak{g}^*_- \times P$ for which the associated $\mathfrak{g}$-action is integrable (here S is a symplectic manifold);*

- *Pairs consisting of a strongly Hamiltonian G-action on S and a Poisson map $\rho : S \to P$ such that the* **classical covariance condition**

$$\xi^S_X \rho^*(\tilde{f}) = \rho^*(\xi^P_X \tilde{f}) \tag{3.107}$$

holds for all $X \in \mathfrak{g}$ and $\tilde{f} \in C^\infty(P, \mathbb{R})$; cf. (1.19).

It is not necessary that S be connected; the proof works for each component separately. Given J, one defines a $\mathfrak{g}$-action on S by $X \mapsto \xi_{J^*\tilde{X}}$, where the definition of $\tilde{X}$ is similar to the one given prior to (3.102). When this action is integrable, the corresponding G-action is strongly Hamiltonian by definition. Also, the restriction of J^* to $C^\infty(P)$ evidently defines a representation on S. Condition (3.107) is then satisfied because of (3.106) and the fact that J is a Poisson map.

Conversely, a strongly Hamiltonian G-action is associated with a Poisson map $J_{(1)} : S \to \mathfrak{g}^*_-$ (see 1.2). Writing $J_{(2)} := \rho$, one obtains a map $J = (J_{(1)}, J_{(2)}) : S \to \mathfrak{g}^*_- \times P$. Using (3.106) and an argument similar to the proof of 1.1.2 one shows that J is a Poisson map. □

One would like to sharpen this result by saying that there is a bijective correspondence between representations J^* of the Poisson algebra $C^\infty(\mathfrak{g}^*_- \times P, \mathbb{R})$ (which by definition implies that the Poisson map J is complete) and pairs as stated in 3.9.6, for which in addition the map ρ is complete (so that ρ^* is a representation of $C^\infty(P, \mathbb{R})$ on S). This works in one direction when G is simply connected, for in that case the completeness of $\tau_{(1)} \circ J : S \to \mathfrak{g}^*$ (where $\tau_{(1)} : \mathfrak{g}^* \times P \to \mathfrak{g}^*$ is the projection onto the first variable) implies that the $\mathfrak{g}$-action on S is integrable by Theorem 1.2.1. However, in the opposite direction it is in general not clear that the completeness of $J_{(1)}$ and $J_{(2)} = \rho$ implies that of J. Paraphrasing the first ingredient of 3.9.6, we obtain

Definition 3.9.7. *Given a G-action on Q, a* **classical system of imprimitivity** *of G on Q consists of a symplectic manifold S, along with a strongly Hamiltonian G-action on S and a nonzero representation $\tilde{\pi} : C^\infty(Q, \mathbb{R}) \to C^\infty(S, \mathbb{R})$ (where Q has the zero Poisson structure), such that the integrated classical covariance condition*

$$\alpha^S_x(\tilde{\pi}(\tilde{f})) = \tilde{\pi}(\alpha^Q_x(\tilde{f})) \tag{3.108}$$

holds for all $x \in G$ and $\tilde{f} \in C^\infty(Q, \mathbb{R})$. Here $\alpha^S_x(f)(\sigma) := f(x^{-1}\sigma)$ and $\alpha^Q_x(\tilde{f})(q) = \tilde{f}(x^{-1}q)$.

By I.2.6.5 there exists a complete Poisson map $J_{(2)} : S \to Q$ for which $\tilde{\pi} = J^*_{(2)}$. Condition (3.108) is evidently equivalent to the G-equivariance of $J_{(2)}$, that is, one has $J_{(2)}(x\sigma) = xJ_{(2)}(\sigma)$ for all $x \in G$ and $\sigma \in S$. Moreover, equation (3.108) implies that $\xi^S_X \tilde{\pi}(\tilde{f}) = \tilde{\pi}(\xi^Q_X \tilde{f})$, cf. (3.107), and is equivalent to this condition when G is connected.

This time we have spoken of a representation of $C^\infty(Q)$, rather than merely a Poisson map $\rho : S \to Q$. This is justified by the classical version of Corollary 3.4.6, which specializes and sharpens Theorem 3.9.6:

Corollary 3.9.8. *When G is connected there is a bijective correspondence between classical systems of imprimitivity of G on Q and representations π of the Poisson algebra $C^\infty(\mathfrak{g}^*_- \times Q, \mathbb{R})$ whose associated $\mathfrak{g}$-action is integrable.*

As remarked above, the integrability condition is automatically satisfied when G is simply connected. The new issue relative to the proof of 3.9.6 and the subsequent comment is that in this special case the completeness of $J_{(1)}$ and $J_{(2)} = \rho$ does imply the completeness of J. The proof of this requires advanced techniques in symplectic geometry that we have not developed. □

In analogy to Proposition 3.7.5 we have a result suggested by 3.8.5.

Proposition 3.9.9. *The Poisson manifolds $(T^*G)/H$ and $\mathfrak{g}^*_- \times (G/H)$, and the Poisson algebras $C^\infty((T^*G)/H, \mathbb{R})$ and $C^\infty(\mathfrak{g}^*_- \times (G/H), \mathbb{R})$, are isomorphic.*

The isomorphism mentioned in the proof of 3.8.5 can be "dualized", defining a Poisson map. ■

Paraphrasing the comment after 3.7.5, it therefore follows that we may look at the Poisson algebra $C^\infty(T^*\mathbb{R}^n, \mathbb{R})$ in two different ways: It is the Poisson algebra of the pair algebroid $T\mathbb{R}^n$, as well as the Poisson algebra of the action algebroid defined by the canonical action of $\mathbb{R}^n$ on itself.

The classical analogue of 3.7.6 is

Corollary 3.9.10. *When G is connected, each coadjoint orbit $\mathcal{O}$ of H leads to an irreducible classical system of imprimitivity of G on G/H in $(\widetilde{T^*G})^{\mathcal{O}}$.*

Recall the definition (2.51), with $\mathsf{P} = G$, of $(\widetilde{T^*G})^{\mathcal{O}}$, as well as (2.118) and (2.130). The claim follows from Corollaries 2.3.8 and 3.9.8. ■

In view of the possibility of covering spaces in 2.3.8, this analogy is not quite perfect. The connectedness assumption and the restriction to irreducible representations will, however, be removed in IV.1.6.4.

In order to generalize Theorem 1.1.7, we need an appropriate concept of the action of a Lie algebroid on a manifold.

Definition 3.9.11. *A (left)* **groupoid action** *of $\mathsf{G} \rightrightarrows Q$ on a space S consists of maps $J_\rho : S \to Q$ and $L : \mathsf{G} *_Q S \to S$, where*

$$\mathsf{G} *_Q S := \{(\gamma, \sigma) \in \mathsf{G} \times S \mid \tau_s(\gamma) = J_\rho(\sigma)\}. \tag{3.109}$$

Writing $\gamma\sigma := L(\gamma, \sigma)$, these maps must satisfy

$$J_\rho(\gamma\sigma) = \tau_t(\gamma); \tag{3.110}$$

$$\iota(J_\rho(\sigma))\sigma = \sigma; \tag{3.111}$$

$$\gamma(\gamma'\sigma) = (\gamma\gamma')\sigma, \tag{3.112}$$

whenever $(\gamma', \sigma) \in \mathsf{G} *_Q S$ and $(\gamma, \gamma') \in \mathsf{G}_2$.

When G is a Lie groupoid, one speaks of a **smooth groupoid action** *if S is a manifold and J_ρ and L are smooth.*

As to the last definition, note that the surjectivity of τ_s implies that $\mathsf{G} *_Q S$ is a submanifold of $\mathsf{G} \times Q$.

For example, taking $S = \mathsf{G}$ and $J_\rho = \tau_t$ leads to $\mathsf{G} *_Q S = \mathsf{G}_2$, and L is simply multiplication in the groupoid. Alternatively, choosing $S = Q$ and $J_\rho = \mathrm{id}$ reproduces the action of G on its base.

In the smooth case one obtains a linear map $s \mapsto \xi_s$ from $\Gamma(\mathfrak{G})$ to $\Gamma(TS)$ (where $\mathfrak{G}$ is the Lie algebroid of G), defined as follows. Identifying the section $s \in \Gamma(\mathfrak{G})$ with a left-invariant vector field ξ^L on G, which in turn corresponds to a left-invariant flow $\gamma(\cdot)$ on G (i.e., $\xi^L(\gamma) = \dot{\gamma}$), we put

$$\xi_s(\sigma) := \frac{d}{d\lambda}(\iota(J_\rho(\sigma)))(\lambda)^{-1}\sigma_{|\lambda=0}. \tag{3.113}$$

It follows from 3.1.1.4, the left invariance of $\gamma(\cdot)$, and (3.111) that $(\iota(J_\rho(\sigma)))(\lambda)^{-1}\sigma$ lies in $\mathsf{G} *_Q S$, so that ξ_s is well-defined. The definition of the Lie bracket on $\Gamma(\mathfrak{G})$ implies that $s \mapsto \xi_s$ is a Lie algebra homomorphism. Finally, (3.110) and the definition of the anchor τ_a in 3.8.7 entail

$$(J_\rho)_*\xi_s = \tau_a \circ s \tag{3.114}$$

for all $s \in \Gamma(\mathfrak{G})$. Thus we are led to

Definition 3.9.12. *An* **action** *of a Lie algebroid $\mathsf{V} \overset{\to TQ}{\to} Q$ on a manifold S consists of a smooth map $J_\rho : S \to Q$ and a Lie algebra homomorphism $s \mapsto \xi_s$ from $\Gamma(\mathsf{V})$ into $\Gamma(TS)$ such that (3.114) holds for all $s \in \Gamma(\mathsf{V})$.*

This definition is further motivated by the thought that the pair algebroid TS is the most natural Lie algebroid; to express the idea that an action should "preserve" the anchor τ_a one in addition needs the map J_ρ. One could, equivalently, use an antihomomorphism $s \mapsto \xi_s$, in which case the condition on the anchor reads $(J_\rho)_*\xi_s = -\tau_a \circ s$. Thus one has generalized the definition of an action of a Lie algebra $\mathfrak{g}$ on a manifold; cf. 1.1.

Proposition 3.9.13. *Let $\mathsf{V} \overset{\to TQ}{\to} Q$ be a Lie algebroid with associated Poisson algebra $C^\infty(\mathsf{V}^*, \mathbb{R})$. A representation $\pi : C^\infty(\mathsf{V}^*, \mathbb{R}) \to C^\infty(S, \mathbb{R})$ (in the sense of I.2.6.1), where S is a symplectic manifold, leads to a V-action on S.*

Given π, one obtains a Poisson map $J : S \to \mathsf{V}^*$ by I.2.6.5, and subsequently defines the V-action by $J_\rho := \tau \circ J$ and $\xi_s := \xi_{J^*\tilde{s}}$ (i.e., the Hamiltonian vector field of $J^*\tilde{s}$). Equations I.(2.15) and (3.99) then imply that $[\xi_{s_1}, \xi_{s_2}] = [s_1, s_2]_{\mathsf{V}}$, whereas I.(2.8), I.(2.15), and (3.98) imply the condition on the anchor. ∎

One may then define a strongly Hamiltonian V-action on a symplectic manifold essentially as an action given by a "momentum map" J as above. This, then, leads to the obvious generalization of Theorem 1.1.7 from Lie algebras to Lie algebroids, which is a classical counterpart of the correspondence between representations of a Lie groupoid and representations of the associated C^*-algebra; cf. 3.6.

3.10 A Generalized Exponential Map

The theory of Lie groupoids and algebroids suggests a unifying principle behind the various strict quantizations we have discussed so far. It turns out that these may be formulated in terms of a generalized exponential map $\mathrm{Exp}^W : \mathfrak{G} \to \mathsf{G}$ from a Lie algebroid into a corresponding Lie groupoid.

Lemma 3.10.1. *The vector bundles $T^t\mathsf{G}$ and $\tau_s^*\mathfrak{G}$ (over G) are isomorphic.*

Recall (3.95) and 2.1.2. The pullback bundle

$$\tau_s^*\mathfrak{G} := \{(X, \gamma) \in \mathfrak{G} \times \mathsf{G} \mid \tau(X) = \tau_s(\gamma)\} \tag{3.115}$$

is a vector bundle over G with projection onto the second variable. The isomorphism of 3.10.1 is proved via the vector bundle isomorphism $\mathfrak{G} \simeq \mathfrak{G}'$; see (3.96) and subsequent text. Thus replacing $\mathfrak{G}$ in (3.115) by $\mathfrak{G}'$, one checks that $(L_{\gamma^{-1}})_* : T^t_\gamma\mathsf{G} \to T^t_{\gamma^{-1}\gamma}\mathsf{G}$ is the desired bundle isomorphism (note that the inverse is $(L_\gamma)_*$, and cf. Lemma 3.2.3). ■

For a Lie group G we have $T^tG = TG$, and Lemma 3.10.1 simply reproduces the left trivialization $TG \simeq \mathfrak{g} \times G$. For a pair Lie groupoid $Q \times Q \overset{\leftarrow}{\Rightarrow} Q$ we identify $T^t(Q \times Q)$ with $Q \times TQ$, where the first Q is seen as the zero section in TQ; the projection is $\tau_{T^t(Q\times Q)\to Q\times Q}(q, Y) = (q, \tau_{TQ\to Q}(Y))$. On the other hand, the lemma says that $Q \times TQ$ should be isomorphic to $\{(Y, q', q) \in TQ \times Q \times Q \mid \tau_{TQ\to Q}(Y) = q\}$, with projection $\tau_{\tau_s^*TQ\to Q\times Q}(Y, q', q) = (q', q)$. This isomorphism is immediately obvious. Similarly for a gauge groupoid.

For an action Lie groupoid $G \times Q \overset{\leftarrow}{\Rightarrow} Q$ we identify $T^t(G \times Q)$ with $TG \times Q$ (where Q is the zero section of TQ), with projection $\tau_{T^t(G\times Q)\to G\times Q}(Y, q) = (\tau_{TG\to G}(Y), q)$. The lemma identifies $TG \times Q$ with $\{(Y, x^{-1}q, x, q) \in \mathfrak{g} \times Q \times G \times Q\}$ with (Y, x, q) through the left trivialization of TG.

Let us now assume that $\mathfrak{G}$ has a covariant derivative (or, equivalently, a connection), with associated horizontal lift $\ell^{\mathfrak{G}}$; cf. the paragraph following (2.28). By Proposition 2.2.4 and Lemma 3.10.1 one then obtains a connection on $T^t\mathsf{G}$ (seen as a vector bundle over G, whose projection is borrowed from $T\mathsf{G}$). Going through the definitions, one obtains that the associated horizontal lift ℓ of a tangent vector $X = \dot{\gamma} := d\gamma(t)/dt|t = 0$ in $T_\gamma\mathsf{G}$ to $Y \in T^t_\gamma\mathsf{G}$ is

$$\ell_Y(\dot{\gamma}) = \frac{d}{dt}\left[L_{\gamma(t)*}\ell^{\mathfrak{G}}_{(L_{\gamma^{-1}})_*Y}(\tau_s(\gamma(t)))\right]_{t=0}, \tag{3.116}$$

which is an element of $T_Y(T^t\mathsf{G})$ (here $\ell^{\mathfrak{G}}(\ldots)$ lifts a curve).

Example 3.10.2.

- *For a Lie group $\mathsf{G} = G$ the base space is a point, so that no connection needs to be chosen, and a horizontal lift is always zero. In the left trivialization (where $(X, x) := d/dt(x\mathrm{Exp}(tX)/dt)|t = 0$; cf. (1.37)), the expression (3.116) then reads*

$$\ell_{(Y,x)}(X, x) = \frac{d}{dt}(Y, x\mathrm{Exp}(tX))_{t=0} = (0, Y, X, x). \tag{3.117}$$

Here $T(TG) \simeq T(\mathfrak{g} \times G) \simeq \mathfrak{g} \times \mathfrak{g} \times TG \simeq \mathfrak{g} \times \mathfrak{g} \times \mathfrak{g} \times G$.

- *A connection in a pair Lie algebroid* $TQ \overset{\to TQ}{\to}_Q$ *is the same as an affine connection, with horizontal lift* $\ell^{TQ} : TQ \to T(TQ)$*; cf. II.3.2.1. Using the identification* $T^t(Q \times Q) \simeq Q \times TQ$ *as above, the left-hand side of (3.116) is of the form* $\ell_{(q',Y)}(\dot{q}', \dot{q})$*, where* $Y \in T_q Q$*. The right-hand side then assumes the form* $d/dt(q'(t), \ell_Y^{TQ}(q(t))|t = 0$*, so that*

$$\ell_{(q',Y)}(\dot{q}', \dot{q}) = (\dot{q}, \ell_Y^{TQ}(\dot{q})). \tag{3.118}$$

This is a vector in $T(T^t(Q \times Q)) \simeq TQ \times T(TQ)$*, as it should be. An analogous computation may be done for gauge groupoids.*

- *For an action Lie groupoid* $G \times Q \overset{\leftarrow}{\Rightarrow} Q$ *we just saw that* $T^t(G \times Q) \simeq TG \times Q \simeq \mathfrak{g} \times G \times Q$*. Using the notation of (3.117), the right-hand side of (3.116) is a vector in* $T_{(Y,x)}(TG) \times T_q Q$*, namely*

$$\ell_{(Y,x,q)}(X, x, \dot{q}) = (0, Y, X, x, \dot{q}). \tag{3.119}$$

Since the bundle $T^t\mathsf{G} \to \mathsf{G}$ has a connection, one can define the geodesic flow $X \mapsto X(t)$ on $T^t\mathsf{G}$ in precisely the same way as on a tangent bundle with affine connection; see II.3.1. To recapitulate, the flow $X(t)$ is the solution of

$$\dot{X}(t) = \ell_{X(t)}(X(t)), \tag{3.120}$$

with initial condition $X(0) = X$.

Definition 3.10.3. *Let the Lie algebroid* $\mathfrak{G} \overset{\to TQ}{\to}_Q$ *of a Lie groupoid* $\mathsf{G} \overset{\leftarrow}{\Rightarrow} Q$ *be equipped with a connection. Relative to the latter, the* **left exponential map** $\mathrm{Exp}^L : \mathfrak{G} \to \mathsf{G}$ *is defined by*

$$\mathrm{Exp}^L(X) := \gamma_{X'}(1) = \tau_{T^t\mathsf{G}\to\mathsf{G}}(X'(1)), \tag{3.121}$$

whenever the geodesic flow $X'(t)$ *on* $T^t\mathsf{G}$ *(defined by the connection on* $T^t\mathsf{G}$ *pulled back from the one on* $\mathfrak{G}$*) is defined at* $t = 1$*. Here* $X' \in \mathfrak{G}' = T^t\mathsf{G} \restriction \mathsf{G}_0$ *is the image of* X *under the isomorphism* $\mathfrak{G}' \simeq \mathfrak{G}$*; cf. (3.96) etc.*

Our goal, however, is to define a "symmetrized" version of Exp^L.

Lemma 3.10.4. *For all* $X \in \mathfrak{G}$ *for which* $\mathrm{Exp}^L(X)$ *is defined one has*

$$\tau_t(\mathrm{Exp}^L(X)) = \tau(X). \tag{3.122}$$

We write X for X' in (3.121). One has $\tau_t(\gamma_X(0)) = \tau(X)$ and

$$\frac{d}{dt}\tau_t(\gamma_X(t)) = (\tau_t \circ \tau_{T^t\mathsf{G}\to\mathsf{G}})_* \ell_{X(t)}(X(t)) = (\tau_t)_* X(t) = 0,$$

since $\ell_X(Y)$ covers Y, and $X(t) \in T^t\mathsf{G} = \ker(\tau_t)_* \cap T\mathsf{G}$. ∎

We combine this with the obvious $\tau(\frac{1}{2}X) = \tau(-\frac{1}{2}X)$ to infer that

$$\tau_t(\mathrm{Exp}^L(\tfrac{1}{2}X)) = \tau_t(\mathrm{Exp}^L(-\tfrac{1}{2}X)) = \tau_s(\mathrm{Exp}^L(-\tfrac{1}{2}X)^{-1}).$$

Thus the (groupoid) multiplication in (3.123) below is well-defined.

Definition 3.10.5. *The* **Weyl exponential map** $\mathrm{Exp}^W : \mathfrak{G} \to \mathsf{G}$ *is defined by*

$$\mathrm{Exp}^W(X) := \mathrm{Exp}^L(-\tfrac{1}{2}X)^{-1}\mathrm{Exp}^L(\tfrac{1}{2}X). \tag{3.123}$$

The following result is closely related to the tubular neighborhood theorem II.3.4.2, and includes Lemma II.3.4.3 as a special case.

Theorem 3.10.6. *The maps* Exp^L *and* Exp^W *are diffeomorphisms from a neighborhood* $\mathcal{N}^\iota$ *of* $Q \subset \mathfrak{G}$ *(as the zero section) to a neighborhood* $\mathcal{N}_\iota$ *of* $\iota(Q)$ *in* G*, such that* $\mathrm{Exp}^L(q) = \mathrm{Exp}^W(q) = \iota(q)$ *for all* $q \in Q$.

The property $\mathrm{Exp}^L(q) = \iota(q)$ is immediate from Definition 3.10.3. The pushforward of Exp^L at q is $\mathrm{Exp}^L_* : T_q\mathfrak{G} \to T_{\iota(q)}\mathsf{G}$. Now recall the decomposition (3.94). For X tangent to $Q \subset \mathfrak{G}$ one immediately sees that $\mathrm{Exp}^L_*(X) = \iota_* X$. For X tangent to the fiber $\tau^{-1}(q)$, which we identify with $T^t_{\iota(q)}\mathsf{G}$ (cf. (3.96) etc.), one has $\mathrm{Exp}^L_*(X) = X'$, as follows by the standard argument used to prove that $\exp_q$ in the theory of affine geodesics is a local diffeomorphism: For a curve $X(s) = sX$ in $T^t_{\iota(q)}\mathsf{G}$ one has $\mathrm{Exp}^L(X(s)) = \gamma_{X'(s)}(1) = \gamma_{X'}(s)$, so that $d/ds[\mathrm{Exp}^L(X(s))]|s = 0 = X'$. Since Exp^L_* is a bijection at q, the inverse function theorem implies that Exp^L is a local diffeomorphism. Since it maps Q pointwise to $\iota(Q)$, the local diffeomorphisms can be patched together to yield a diffeomorphism of the neighborhoods stated in 3.10.6; we omit the details of this last step, since it is identical to the proof of the tubular neighborhood theorem.

As for Exp^W, we have $\mathrm{Exp}^W_*(X) = \iota_* X$ for $X \in T_q Q \subset T_q\mathfrak{G}$. Also,

$$\frac{d}{ds}[\mathrm{Exp}^L(-\tfrac{1}{2}sX)^{-1}\mathrm{Exp}^L(\tfrac{1}{2}sX)]_{s=0} = -\tfrac{1}{2}I_*X' + \tfrac{1}{2}X',$$

where I_* is the pushforward of the inversion in G. The right-hand side lies in $\ker((\tau_s)_* + (\tau_t)_*) \subset T\mathsf{G}$, and every element in this kernel is of the stated form. Similarly to (3.94), one may prove the decomposition

$$T_{\iota(q)}\mathsf{G} = T_{\iota(q)}\mathsf{G}_0 \oplus \ker((\tau_s)_* + (\tau_t)_*)(\iota(q)). \tag{3.124}$$

It follows that Exp^W_* is a bijection at q, and the second part of the theorem is derived as for Exp^L. ■

Our standard list of examples illustrates Definition 3.10.5.

Proposition 3.10.7.

- *For a Lie group G no connection is needed, and one has*

$$\mathrm{Exp}^L(X) = \mathrm{Exp}^W(X) = \mathrm{Exp}(X), \tag{3.125}$$

  where $X \in \mathfrak{g}$ *and* $\mathrm{Exp} : \mathfrak{g} \to G$ *is the usual exponential map.*
- *For a pair Lie groupoid* $Q \times Q \overset{\leftarrow}{\Rightarrow} Q$ *one chooses an affine connection* ∇ *on* TQ*, with associated exponential map* $\exp : TQ \to Q$*. Then*

$$\mathrm{Exp}^L(X) = (\tau(X), \exp_{\tau(X)}(X)); \tag{3.126}$$

$$\mathrm{Exp}^W(X) = (\exp_{\tau(X)}(-\tfrac{1}{2}X), \exp_{\tau(X)}(\tfrac{1}{2}X)), \tag{3.127}$$

where $X \in TQ$ and $\tau := \tau_{TQ\to Q}$.

- *For a gauge groupoid $\mathsf{P}\times_H \mathsf{P} \overset{\leftarrow}{\rightrightarrows} Q$ one chooses an H-invariant affine connection on $T\mathsf{P}$, with exponential map $\exp : T\mathsf{P} \to \mathsf{P}$. This induces a connection on $(T\mathsf{P})/H$, in terms of which*

$$\mathrm{Exp}^L([X]_H) = [\tau(X), \exp_{\tau(X)}(X)]_H; \tag{3.128}$$

$$\mathrm{Exp}^W([X]_H) = [\exp_{\tau(X)}(-\tfrac{1}{2}X), \exp_{\tau(X)}(\tfrac{1}{2}X)]_H, \tag{3.129}$$

where $\tau := \tau_{T\mathsf{P}\to\mathsf{P}}$, and $[X]_H \in (T\mathsf{P})/H$ is the equivalence class of $X \in T\mathsf{P}$ under the H-action on $T\mathsf{P}$.

- *For an action groupoid $G \times Q \overset{\leftarrow}{\rightrightarrows} Q$ the trivial connection on $\mathfrak{g}\times Q \to Q$ yields*

$$\mathrm{Exp}^L(X, q) = (\mathrm{Exp}(X), q); \tag{3.130}$$

$$\mathrm{Exp}^W(X, q) = (\mathrm{Exp}(X), \mathrm{Exp}(\tfrac{1}{2}X)q). \tag{3.131}$$

We infer from (3.117) and (3.120) that the geodesic flow in $T^tG = TG \simeq \mathfrak{g}\times G$ is determined by the differential equation $(\dot{Y}, \dot{x}) = (0, Y)$; this suggestive notation should actually read $(0, Y, Y, x)$. Recalling that we work in the left trivialization, this equation is solved by $(Y(t), x(t)) = (Y, x\mathrm{Exp}(tY))$. Now $X \in \mathfrak{g}$ is identified with $X' \in T_eG$, so that (3.125) follows.

The geodesic flow on $T^t(Q\times Q)$ is $(X(t), Y(t)) = (X(0), Y(t))$, where $Y(t)$ is the flow on TQ determined by the affine connection. This immediately leads to (3.126), and hence to (3.127), since $(x, y_-)^{-1}(x, y_+) = (y_-, y_+)$.

An H-invariant connection on $T\mathsf{P}$ by definition satisfies $(R_h)_*\nabla_\xi\eta = \nabla_\xi\eta$ for all $h \in H$ and all $\xi, \eta \in \Gamma(T\mathsf{P})^H$. This implies H-invariance of the geodesic flow on $T\mathsf{P}$ in that $(R_h)_*(X(t)) = ((R_h)_*X)(t)$ for all h and X. The exponential map is then H-invariant in the sense of II.(3.90), which, with (3.126), leads to (3.128).

Equation (3.130) follows from (3.119), in analogy with the derivation of (3.126). Subsequently, (3.131) is derived from (3.123) and the definitions of multiplication and inversion in an action groupoid (see 3.1.4). ■

3.11 The Groupoid C^*-Algebra as a Strict Quantization

Theorem II.2.6.1 shows that $C^\infty_{\mathrm{PW}}(T^*\mathbb{R}^n)$, regarded as a subalgebra of the complexified Poisson algebra $C^\infty(T^*\mathbb{R}^n)$, is quantized by $\mathfrak{B}_0(L^2(\mathbb{R}^n))$. This is generalized to arbitrary Riemannian manifolds Q in Theorem II.3.5.1, in which $C^\infty_{\mathrm{PW}}(T^*Q)$ is quantized by $\mathfrak{B}_0(L^2(Q))$. For compact G, in Theorem 1.9.2 the complexified Lie–Poisson algebra $C^\infty_{\mathrm{PW}}(\mathfrak{g}^*_-)$ is quantized into the group C^*-algebra $C^*(G)$. When $\mathsf{P}(Q, H, \tau)$ is a principal fiber bundle with compact structure group H, we saw in Theorem 2.8.1 that the complexified Poisson algebra $C^\infty_{\mathrm{PW}}((T^*\mathsf{P})H)$ is quantized by $\mathfrak{B}_0(L^2(\mathsf{P}))^H$.

If we look at this list, as well as at Propositions 3.8.8 and 3.9.3 and Theorem 3.7.1, we discern that in all cases an appropriate subspace of the Poisson algebra $C^\infty(\mathfrak{G}^*)$ canonically associated to a given Lie algebroid $\mathfrak{G} \overset{\to TQ}{\to Q}$ (see 3.9.1) is quantized by the C^*-algebra $C^*(\mathsf{G})$ (defined in 3.6.3) of a Lie groupoid G whose Lie algebroid is $\mathfrak{G}$. In the cases at hand one has $C^*(\mathsf{G}) = C^*_r(\mathsf{G})$.

In all cases the quantization map $\mathcal{Q}_\hbar$ is a special case of the following construction. We start by defining a fiberwise Fourier transform $\check{f} \in C^\infty(\mathfrak{G})$ of suitable $f \in C^\infty(\mathfrak{G}^*)$. This transform depends on the choice of a family $\{\mu_q^L\}_{q\in Q}$ of Lebesgue measures, where μ_q^L is defined on the fiber $\tau^{-1}(q)$. We will discuss the normalization of each μ_q^L in due course; for the moment we merely assume that the q-dependence is smooth in the obvious (weak) sense.

For a function $\check{f}$ on $\mathfrak{G}$ that is L^1 on each fiber we put

$$f(\theta) = \int_{\tau^{-1}(q)} d\mu_q^L(X)\, e^{-i\theta(X)}\, \check{f}(X), \tag{3.132}$$

where $\tau := \tau_{\mathfrak{G}\to Q}$ and $X \in \tau^{-1}(q)$. Each μ_q^L determines a Lebesgue measure μ_q^{L*} on the fiber $\tau^{-1}_{\mathfrak{G}^*\to Q}(q)$ of $\mathfrak{G}^*$, whose normalization is fixed by requiring that the inverse to (3.132) be given by

$$\check{f}(X) = \int_{\tau^{-1}_{\mathfrak{G}^*\to Q}(q)} d\mu_q^{L*}(\theta)\, e^{i\theta(X)}\, f(\theta). \tag{3.133}$$

The fiberwise Fourier transforms II.(3.42) and (1.124) are clearly special cases of (3.133). On the action Lie algebroid $\mathfrak{g} \times Q$, equipped with the trivial connection, we simply have

$$\check{f}(X, q) = \int_{\mathfrak{g}^*} \frac{d^n\theta}{(2\pi)^n}\, e^{i\theta(X)}\, f(\theta, q). \tag{3.134}$$

As in (1.124), the normalization of $d^n\theta$ is determined by the normalization of the Haar measure on G.

Having constructed a Fourier transform, we define the class $C^\infty_{\mathrm{PW}}(\mathfrak{G}^*)$ as consisting of those smooth functions on $\mathfrak{G}^*$ whose Fourier transform is in $C^\infty_c(\mathfrak{G})$; cf. II.(3.49). Generalizing the procedure in Definition II.3.4.4, we pick a function $\kappa \in C^\infty(\mathfrak{G}, \mathbb{R})$ with support in $\mathcal{N}^\iota$ (cf. 3.10.6), equaling unity in some smaller tubular neighborhood of Q, as well as satisfying $\kappa(-X) = \kappa(X)$ for all $X \in \mathfrak{G}$.

Definition 3.11.1. *Let* G *be a Lie groupoid with Lie algebroid* $\mathfrak{G}$. *For* $\hbar \neq 0$ *the* $\pm$ **Weyl quantization** *of* $f \in C^\infty_{\mathrm{PW}}(\mathfrak{G}^*)$ *is the element* $\mathcal{Q}_\hbar^W(f)_\pm \in C^\infty_c(\mathsf{G})$ *(regarded as a dense subalgebra of* $C^*(\mathsf{G})$ *or* $C^*_r(\mathsf{G})$*), defined by* $\mathcal{Q}_\hbar^W(f)_\pm(\gamma) := 0$ *when* $\gamma \notin \mathcal{N}_\iota$*, and by*

$$\mathcal{Q}_\hbar^W(f)_\pm(\mathrm{Exp}^W(X)) := \hbar^{-n}\kappa(X)\check{f}(\pm X/\hbar). \tag{3.135}$$

Here the Weyl exponential $\mathrm{Exp}^W : \mathfrak{G} \to \mathsf{G}$ *is defined in (3.123), and the cutoff function* κ *is as specified above.*

This definition is possible by virtue of Theorem 3.10.6. By our choice of $C^\infty_{\mathrm{PW}}(\mathfrak{G}^*)$, the operator $\mathcal{Q}_\hbar^W(f)_\pm$ is independent of κ for small enough $\hbar$.

Proposition 3.11.2. *For real* f *the operator* $\mathcal{Q}_\hbar^W(f)_\pm$ *is self-adjoint in* $C^*_{(r)}(\mathsf{G})$.

This is immediate from (3.23) and (3.123). ∎

It is evident from Proposition 3.10.7 that the previously constructed Weyl quantization maps II.(2.108) with II.(2.109) on $T^*\mathbb{R}^n$, II.(3.50) with II.(3.51) on T^*Q, and the map defined in Theorem 2.8.1 on $(T^*\mathsf{P})/H$ correspond to $\mathcal{Q}_\hbar^W(f)_-$, whereas the quantization defined in (1.127) on $\mathfrak{g}^*$ corresponds to $\mathcal{Q}_\hbar^W(f)_+$. In all cases, $\mathcal{Q}_\hbar^W(\cdot)_\pm$ defines a strict quantization of the appropriate subalgebra of $C^\infty(\mathfrak{G}^*_\mp, \mathbb{R})$ into $C^*(\mathsf{G})$. A new case is the action Lie groupoid $G \times Q \overset{\leftarrow}{\Rightarrow} Q$. Here the cutoff κ is independent of q, and coincides with the function appearing in (1.127).

Theorem 3.11.3. *For small enough $\hbar$, a function $f \in C^\infty_{\text{PW}}(\mathfrak{g}^* \times Q)$ is quantized according to (3.135) by*

$$\mathcal{Q}_\hbar^W(f)_\pm(\text{Exp}(X), q) = \int_{\mathfrak{g}^*} \frac{d^n\theta}{(2\pi\hbar)^n}\, e^{\frac{i}{\hbar}\theta(X)} f(\pm\theta, \text{Exp}(-\tfrac{1}{2}X)q). \tag{3.136}$$

When $G = \mathbb{R}^n$ and Q has a G-invariant measure, $\mathcal{Q}_\hbar^W(\cdot)_\pm$ defines a strict quantization of the Poisson algebra $C^\infty_{\text{PW}}(\mathfrak{g}^*_\mp \times Q, \mathbb{R})$ of the action Lie algebroid $\mathbb{R}^n \times Q$ into the action C^*-algebra $C^*(\mathbb{R}^n, Q) = C^*_r(\mathbb{R}^n, Q)$ on $I = \mathbb{R}$.

Equation (3.136) follows from (3.135), (3.131), and (3.134). Conditions II.1.1.1.1 and 2 hold by Theorem 3.11.4 below. We prove II.1.1.1.3. For an action C^*-algebra $C^*(G, Q)$ the bound (3.75) reads

$$\|f\| \leq \|f\|_I = \sup_{q\in Q} \int_G dx\, |f(x, q)|, \tag{3.137}$$

where $f \in C^\infty_c(G \times Q)$; cf. (3.44) and (3.57). We put $\mathcal{Q}_\hbar^W(f) := \mathcal{Q}_\hbar^W(f)_+$, and substitute (3.136) and (3.137) in $\|\mathcal{Q}_\hbar^W(f)\mathcal{Q}_\hbar^W(g) - \mathcal{Q}_\hbar^W(fg)\|_I$ (do not confuse $f \in C^*(G, Q)$ in (3.137) with $f \in C^\infty_{\text{PW}}(\mathfrak{g}^*_- \times Q)$ used in this step). One rescales some integration variables so that $\hbar$ occurs only in expressions of the generic form $f(\theta, \text{Exp}(\hbar X)q)$, where $X \in \mathfrak{g} = \mathbb{R}^n$. One then Taylor-expands f and g in $\hbar$, e.g.,

$$f(\theta, \text{Exp}(\hbar X)q) = f(\theta, q) + \hbar X^a \xi_a^Q f(\theta, q) + O(\hbar^2); \tag{3.138}$$

cf. (1.19). Expressions of the form $X^a \exp(i\theta_b X^b)$ in the $O(\hbar)$ term in (3.138) are rewritten as $-i\partial/\partial\theta_a \exp(i\theta_b X^b)$, upon which one partially integrates in θ. Two integrations in the $O(\hbar)$ term can then be done explicitly, and using (3.101) (in which the structure constants C of course vanish) and $\check{f}, \check{g} \in C^\infty_c(\mathbb{R}^n \times Q)$ one proves II.(1.3) via (3.137). ∎

When $Q = G/H$, one may use Proposition 3.9.9 and Corollary 3.7.5 to show that the prescription (3.135) applied to $C^\infty_{\text{PW}}(\mathfrak{g}^*_- \times Q)$ (mapping it into $C^*(G, G/H)$), that is, (3.136), coincides with its application to $C^\infty_{\text{PW}}((T^*G)/H)$ (thereby mapped into $C^*(G \times_H G)$). Specifically, one should use $\mathcal{Q}_\hbar^W(\cdot)_+$ on $\mathfrak{g}^*_- \times (G/H)$ and $\mathcal{Q}_\hbar^W(\cdot)_-$ on $(T^*G)/H$; this is because the two relevant Poisson brackets stated in 3.9.3 differ by a sign. Taking $G = \mathbb{R}^n$ and H trivial, we see that Theorem 3.11.3 is essentially Theorem II.2.6.1.

To further understand the prescription (3.136), we pass to some representation π of $C^*(G, Q)$, for example, to the regular representation π_L^s, cf. (3.67). This has the advantage that $\mathcal{Q}_\hbar^W$ may be extended to certain unbounded functions. With $\tilde{X}$ and $\tilde{f}$ as defined prior to (3.102), and $U(G)$ and $\tilde{\pi}(C_0(Q))$ associated to $\pi(C^*(G, Q))$ as in 3.4.6 or 3.7.4, easy formal manipulations (which in the case of π_L^s are valid, on, say, $C_c^\infty(G \times Q) \subset L^2(G \times Q)$) yield

$$\pi(\mathcal{Q}_\hbar^W(\tilde{X})_+) = i\hbar dU(X); \tag{3.139}$$

$$\pi(\mathcal{Q}_\hbar^W(\tilde{f})_+) = \tilde{\pi}(\tilde{f}). \tag{3.140}$$

Comparing this with (3.39) and (3.40), we see that $\mathcal{Q}_\hbar^\pi(g) = \pi(\mathcal{Q}_\hbar^W(g))$ when g is either $\tilde{X}$ or $\tilde{f}$. In particular, if we omit the arbitrary representation π, eqs. (3.41)–(3.43) and (3.102)–(3.104) lead to a strong version of Dirac's condition II.(1.3),

$$\frac{i}{\hbar}\left[\mathcal{Q}_\hbar^W(f)_+, \mathcal{Q}_\hbar^W(g)_+\right] = \mathcal{Q}_\hbar^W(\{f, g\}_-)_+. \tag{3.141}$$

To reiterate, this is valid when f and g are of the form $\tilde{f}$ or $\tilde{X}$, and strictly speaking holds in any representation π of $C^*(G, Q)$ on a suitable domain (e.g., $\mathcal{H}_U^\infty$) of the carrier space of π.

Motivated by these examples, one would like (3.135) to provide a strict quantization for any Lie groupoid G. However, Dirac's condition II.(1.3) has been proved only in cases featuring a good correspondence between the symplectic leaves of $\mathfrak{G}^*$ and the irreducible representations of $C_r^*(\mathsf{G})$; cf. the proofs of Theorems II.2.6.1, II.3.5.1, 1.9.2, and 2.8.1. The other conditions, though, always hold.

Theorem 3.11.4. *The map* $(\mathcal{Q}_\hbar^W)_\pm : C_{\mathrm{PW}}^\infty(\mathfrak{G}_\mp^*, \mathbb{R}) \to C_r^*(\mathsf{G})_\mathbb{R}$ *defined by (3.135) satisfies conditions II.1.1.1.1 and 2 of Rieffel and von Neumann.*

3.12 The Normal Groupoid of a Lie Groupoid

The essence of the proof of Theorem 3.11.4 is to regard $\mathfrak{G}$ as a Lie groupoid, and glue it to G so as to obtain a new Lie groupoid containing both G and $\mathfrak{G}$.

Definition 3.12.1. *Let* $\mathsf{G} \rightrightarrows Q$ *be a Lie groupoid with associated Lie algebroid* $\mathfrak{G} \xrightarrow{\tau} {}^{TQ}_{Q}$ *as defined in 3.8.7. The* **normal groupoid** G_N *is a Lie groupoid with base* $\mathbb{R} \times Q$, *defined by the following structures.*

- *As a set,* $\mathsf{G}_N = \mathfrak{G} \cup \{\mathbb{R}\backslash\{0\} \times \mathsf{G}\}$. *We write elements of* G_N *as pairs* $(\hbar, u)$, *where* $u \in \mathfrak{G}$ *for* $\hbar = 0$ *and* $u \in \mathsf{G}$ *for* $\hbar \neq 0$. *Thus* $\mathfrak{G}$ *is identified with* $\{0\} \times \mathfrak{G}$.
- *As a groupoid,* $\mathsf{G}_N = \{0 \times \mathfrak{G}\} \cup \{\mathbb{R}\backslash\{0\} \times \mathsf{G}\}$. *Here* $\mathfrak{G}$ *is regarded as a Lie groupoid over* Q, *with* $\tau_s = \tau_t = \tau$ *and addition in the fibers as the groupoid multiplication. The groupoid operations in* $\mathbb{R}\backslash\{0\} \times \mathsf{G}$ *are those in* G. *In other words,*

$$\tau_s(0, X) := \tau_t(0, X) = (0, \tau(X)); \tag{3.142}$$

$$\tau_s(\hbar, \gamma) := (\hbar, \tau_s(\gamma)); \tag{3.143}$$

$$\tau_t(\hbar, \gamma) := (\hbar, \tau_t(\gamma)); \tag{3.144}$$

$$\iota(0, q) := (0, q); \tag{3.145}$$
$$\iota(\hbar, q) := (\hbar, \iota(q)); \tag{3.146}$$
$$(0, X) \cdot (0, Y) := (0, X + Y); \tag{3.147}$$
$$(\hbar, \gamma_1) \cdot (\hbar, \gamma_2) := (\hbar, \gamma_1\gamma_2); \tag{3.148}$$
$$(0, X)^{-1} := (0, -X); \tag{3.149}$$
$$(\hbar, \gamma)^{-1} := (\hbar, \gamma^{-1}). \tag{3.150}$$

Here $\hbar \neq 0$; in (3.145) it is understood that $Q \subset \mathfrak{G}$ as the zero section.

- *The smooth structure on G_N is as follows. To start, the open subset $\mathcal{O}_1 := \mathbb{R}\backslash\{0\} \times \mathsf{G} \subset \mathsf{G}_N$ inherits the product manifold structure. Let $Q \subset \mathcal{N}^\iota \subset \mathfrak{G}$ and $\iota(Q) \subset \mathcal{N}_\iota \subset \mathsf{G}$, as in Theorem 3.10.6.*
  Let $\mathcal{O} := \{(\hbar, X) \mid \hbar X \in \mathcal{N}^\iota\}$; this is an open subset of $\mathbb{R} \times \mathfrak{G}$, containing $\{0\} \times \mathfrak{G}$. Define $\phi : \mathcal{O} \to \mathsf{G}_N$ by

$$\phi(0, X) := (0, X);$$
$$\phi(\hbar, X) := (\hbar, \mathrm{Exp}^W(\hbar X)). \tag{3.151}$$

  Since $\mathrm{Exp}^W : \mathcal{N}^\iota \to \mathcal{N}_\iota$ is a diffeomorphism (cf. 3.10.6), we see that $\mathcal{O}$ is a bijection from $\mathcal{O}$ to $\mathcal{O}_2 := \{0 \times \mathfrak{G}\} \cup \{\mathbb{R}\backslash\{0\} \times \mathcal{N}_\iota\}$. This defines the smooth structure on $\mathcal{O}_2$ in terms of the smooth structure on $\mathcal{O}$. Since $\mathcal{O}_1$ and $\mathcal{O}_2$ cover G_N, this specifies the smooth structure on G_N.

The fact that G_N is a Lie groupoid follows from the corresponding property of G. The given chart is defined in terms of the Weyl exponential, which depends on the choice of a connection in $\mathfrak{G}$. However, one may verify that any (smooth) connection, or, indeed, any (Q-preserving) diffeomorphisms between $\mathcal{N}^\iota$ and $\mathcal{N}_\iota$, leads to an equivalent smooth structure on G_N. For example, we could have used Exp^L instead of Exp^W. Also, the smoothness of Exp^W makes the above manifold structure on G_N well-defined, in that open subsets of $\mathcal{O}_1 \cap \mathcal{O}_2$ are assigned the same smooth structure.

The normal groupoid of a pair Lie groupoid $Q \times Q \rightrightarrows Q$ is known as the **tangent groupoid** of Q, and is sometimes described by saying that one "blows up" the diagonal $\delta(Q)$ in $Q \times Q$. Convergence in $(Q \times Q)_N$ in the manifold topology is as follows: If $\hbar_n \to 0$ then $(\hbar_n, q_n, q_n') \to (0, X)$ iff $q_n \to \tau(X)$, $q_n' \to \tau(X)$, and $\dot{\gamma}_n(\frac{1}{2})/\hbar_n \to X$, where γ_n is an affinely parametrized geodesic with $\gamma_n(0) = q_n$ and $\gamma_n(1) = q_n'$. This convergence is independent of the affine connection defining the geodesic in question. In local coordinates, where $X = (v^\mu, q^\mu)$, the convergence condition is simply that $q_n^\mu \to q^\mu$, $(q_n')^\mu \to q^\mu$, and $((q_n')^\mu - q_n^\mu)/\hbar_n \to v^\mu$.

We now pick a left Haar system $\{\mu_q^t\}_{q \in Q}$ on $\mathsf{G} \rightrightarrows Q$; cf. 3.3.3. The vector bundle $\mathfrak{G}$, regarded as a Lie groupoid as in 3.12.1, has a left Haar system consisting of the family $\{\mu_q^L\}_{q \in Q}$ of Lebesgue measures on each fiber, already used in the construction of the Fourier transform. Since we have a Lie groupoid, the Radon–Nikodym derivative $J_q(X) := d\mu_q^t(\mathrm{Exp}^W(X))/d\mu_q^L(X)$ is well-defined and strictly positive on $\mathcal{N}^\iota$ (since both measures are locally Lebesgue on spaces with the same dimension). We now fix the normalization of the μ_q^L by requiring that $\lim_{X \to 0} J_q(X) = 1$

for all q. This leads to a left Haar system for G_N, given by

$$\mu^t_{(0,q)} := \mu^L_q;$$
$$\mu^t_{(\hbar,q)} := \hbar^{-n}\mu^t_q, \tag{3.152}$$

where n is the dimension of the typical fiber of $\mathfrak{G}$. The factor $\hbar^{-n}$ is necessary in order to satisfy condition 3.3.2.3 at $\hbar = 0$, as is easily verified using the manifold structure on G_N defined in 3.12.1.

To avoid confusion between functions on G_N and on $\mathfrak{G}^*$, we denote the former by f, g. Thus the *-algebraic structure on $C_c^\infty(\mathsf{G}_N)$ defined by (3.22) and (3.23) with 3.12.1 and (3.152) becomes

$$\mathsf{f} * \mathsf{g}(0, X) = \int_{\tau^{-1}\circ\tau(X)} d\mu^L_{\tau(X)}(Y)\,\mathsf{f}(0, X - Y)\mathsf{g}(0, Y); \tag{3.153}$$

$$\mathsf{f} * \mathsf{g}(\hbar, \gamma) = \hbar^{-n} \int_{\tau_t^{-1}(\tau_s(\gamma))} d\mu^t_{\tau_s(\gamma)}(\gamma_1)\,\mathsf{f}(\hbar, \gamma\gamma_1)\mathsf{g}(\hbar, \gamma_1^{-1}); \tag{3.154}$$

$$\mathsf{f}^*(0, X) = \overline{\mathsf{f}(0, -X)}; \tag{3.155}$$

$$\mathsf{f}^*(\hbar, \gamma) = \overline{\mathsf{f}(\hbar, \gamma^{-1})}. \tag{3.156}$$

The reduced normal groupoid C^*-algebra $C_r^*(\mathsf{G}_N)$ is the closure of the *-algebra $C_c^\infty(\mathsf{G}_N)$ in the norm (3.73); cf. Definition 3.6.3.

Let $\mathfrak{J}_\hbar$ be the ideal in $C_r^*(\mathsf{G}_N)$ generated by those functions in $C_c^\infty(\mathsf{G}_N)$ that vanish at $\hbar$. The canonical map $\mathsf{f} \mapsto [\mathsf{f}]_\hbar$ from $C_r^*(\mathsf{G}_N)$ to $C_r^*(\mathsf{G}_N)/\mathfrak{J}_\hbar$ is given by $[\mathsf{f}]_\hbar(\cdot) = \mathsf{f}(\hbar, \cdot)$. However, in view of the factor $\hbar^{-n}$ in (3.154), for $\hbar \neq 0$ this map is only a *-homomorphism from $C_r^*(\mathsf{G}_N)$ to $C_r^*(\mathsf{G})$ if we add a factor $\hbar^{-n}$ to the definition (3.22) of convolution on G. Since we would like to identify $C_r^*(\mathsf{G}_N)/\mathfrak{J}_\hbar$ with $C_r^*(\mathsf{G})$, in which convolution is defined in the usual, $\hbar$-independent, way, we should therefore renormalize the canonical projection.

For $\hbar = 0$ one has $C^*(\mathsf{G}_N)/\mathfrak{J}_0 \simeq C_r^*(\mathfrak{G})$, which in turn is isomorphic to $C_0(\mathfrak{G}^*)$ by the fiberwise Fourier transform (3.132). This motivates the definition of $\varphi_0 : C_r^*(\mathsf{G}_N) \to C_0(\mathfrak{G}^*)$ for $\hbar = 0$, and $\varphi_\hbar : C_r^*(\mathsf{G}_N) \to C_r^*(\mathsf{G})$ when $\hbar \neq 0$, by continuous extension from $\mathsf{f} \in C_c^\infty(\mathsf{G}_N)$ of

$$\varphi_0(\mathsf{f}) : \theta \mapsto \acute{\mathsf{f}}(0, \theta),$$
$$\varphi_\hbar(\mathsf{f}) : \gamma \mapsto \hbar^{-n}\mathsf{f}(\hbar, \gamma). \tag{3.157}$$

Here $\acute{\mathsf{f}}(0, \theta)$ and $\mathsf{f}(0, X)$ are related as $f(\theta)$ and $\grave{f}(X)$ are in (3.132).

Theorem 3.12.2. *Let $\mathsf{G} \rightrightarrows Q$ be a Lie groupoid, with associated Lie algebroid $\mathfrak{G} \xrightarrow{\to TQ}_{Q}$. The triple $\left(C_r^*(\mathsf{G}_N), \{\mathfrak{A}^\hbar, \varphi_\hbar\}_{\hbar\in\mathbb{R}}\right)$, where $\mathfrak{A}^0 = C_0(\mathfrak{G}^*)$, and $\mathfrak{A}^\hbar = C_r^*(\mathsf{G})$ for $\hbar \neq 0$, is a continuous field of C^*-algebras.*

To prove this, we need some standard concepts in the theory of C^*-algebras. A **primitive ideal** in a C^*-algebra $\mathfrak{A}$ is an ideal that occurs as the kernel of an irreducible representation of $\mathfrak{A}$.

Definition 3.12.3. *The* **primitive spectrum** Prim($\mathfrak{A}$) *of a C^*-algebra $\mathfrak{A}$ is a topological space whose elements are the primitive ideals in $\mathfrak{A}$, and whose topology* *is defined by the following closure operation: The closure of $S \subset \mathrm{Prim}(\mathfrak{A})$ is the set of all primitive ideals of $\mathfrak{A}$ containing the intersection of the elements of S (which is an ideal in $\mathfrak{A}$).*

The topology defined here is known as the **Jacobson topology**; it can be defined in a much wider context. We let $\pi_{\mathfrak{J}}$ be the irreducible representation annihilating $\mathfrak{J}$; i.e., $\mathfrak{J} = \ker(\pi_{\mathfrak{J}})$. One sees that $\mathfrak{J}_n \to \mathfrak{J}$ in the Jacobson topology when $\pi_{\mathfrak{J}_n}(A) = 0$ for all sufficiently large n implies $\pi_{\mathfrak{J}}(A) = 0$. For example, when X is a locally compact Hausdorff space, one may identify $\mathrm{Prim}(C_0(X))$ with X by identifying $\mathfrak{J}_x := \{f \mid f(x) = 0\}$ with x. The Jacobson topology then coincides with the original topology on X.

Lemma 3.12.4. *Let $\mathfrak{C}$ be a C^*-algebra, and let $\psi : \mathrm{Prim}(\mathfrak{C}) \to X$ be a continuous and open map from the primitive spectrum $\mathrm{Prim}(\mathfrak{C})$ (equipped with the Jacobson topology) to a locally compact Hausdorff space X. Define $\mathfrak{J}_x := \cap\psi^{-1}(x)$; i.e., $A \in \mathfrak{J}_x$ iff $\pi_{\mathfrak{J}}(A) = 0$ for all $\mathfrak{J} \in \psi^{-1}(x)$. Note that $\mathfrak{J}_x$ is an ideal in $\mathfrak{C}$.*
Taking $\mathfrak{A}^x = \mathfrak{C}/\mathfrak{J}_x$ and $\tau_x : \mathfrak{C} \to \mathfrak{A}^x$ to be the canonical projection, the triple $(\mathfrak{C}, \{\mathfrak{A}^x, \tau_x\}_{x\in X})$ is a continuous field of C^*-algebras.

We omit the long and difficult proof of this lemma, and instead apply it, with $\mathfrak{C} = C^*_r(\mathsf{G}_N)$ and $X = I = \mathbb{R}$. In order to verify the assumption in 3.12.4, we first note that $\mathfrak{J}_0 \simeq C_0(\mathbb{R}\backslash\{0\}) \otimes C^*_r(\mathsf{G})$, as follows from a glance at the topology of G_N. Hence $\mathrm{Prim}(\mathfrak{J}_0) = \mathbb{R}\backslash\{0\} \times \mathrm{Prim}(C^*_r(\mathsf{G}))$, with the product topology. Since $C^*(\mathsf{G}_N)/\mathfrak{J}_0 \simeq C_0(\mathfrak{G}^*)$, one has $\mathrm{Prim}(C^*(\mathsf{G}_N)/\mathfrak{J}_0) \simeq \mathfrak{G}^*$.
We need a second lemma, proved by straightforward definition-chasing.

Lemma 3.12.5. *Let $\mathfrak{J}$ be an ideal in a C^*-algebra $\mathfrak{A}$, and decompose*

$$\mathrm{Prim}(\mathfrak{A}) = \mathrm{Prim}_{\mathfrak{J}}(\mathfrak{A}) \cup \mathrm{Prim}^{\mathfrak{J}}(\mathfrak{A}), \tag{3.158}$$

where $\mathrm{Prim}_{\mathfrak{J}}(\mathfrak{A})$ consists of those primitive ideals containing $\mathfrak{J}$, and $\mathrm{Prim}^{\mathfrak{J}}(\mathfrak{A})$ is its complement. Then one has the homeomorphisms $\mathrm{Prim}_{\mathfrak{J}}(\mathfrak{A}) \simeq \mathrm{Prim}(\mathfrak{A}/\mathfrak{J})$ and $\mathrm{Prim}^{\mathfrak{J}}(\mathfrak{A}) \simeq \mathrm{Prim}(\mathfrak{J})$. Moreover, $\mathrm{Prim}_{\mathfrak{J}}(\mathfrak{A})$ is closed and $\mathrm{Prim}^{\mathfrak{J}}(\mathfrak{A})$ is open in $\mathrm{Prim}(\mathfrak{A})$.

We apply this lemma with $\mathfrak{A} = C^*_r(\mathsf{G}_N)$ and $\mathfrak{J} = \mathfrak{J}_0$. Then $C^*_r(\mathsf{G}_N)/\mathfrak{J}_0 \simeq C_0(\mathfrak{G}^*)$, and a glance at the topology of G_N shows that $\mathfrak{J}_0 \simeq C_0(\mathbb{R}\backslash\{0\}) \otimes C^*_r(\mathsf{G})$. Thus the decomposition (3.158) reads

$$\mathrm{Prim}(C^*_r(\mathsf{G}_N)) \simeq \mathfrak{G}^* \cup \{\mathbb{R}\backslash\{0\} \times \mathrm{Prim}(C^*_r(\mathsf{G}))\}. \tag{3.159}$$

For example, when $\mathsf{G} = Q \times Q \overset{\leftarrow}{\Rightarrow} Q$ is a pair Lie groupoid, the right-hand side of the decomposition (3.159) is $T^*Q \cup \mathbb{R}\backslash\{0\}$. In that case the topology is easily computed: The closure of a set $(0, \hbar_0)$ is $T^*Q \cup (0, \hbar_0]$. This illustrates the fact that in general, the primitive spectrum is not Hausdorff.
Equation (3.159) with 3.12.5 does not provide the full topology on $\mathrm{Prim}(C^*_r(\mathsf{G}_N))$, but it is sufficient to know that $\mathfrak{G}^*$ is not open. If it were, $\mathbb{R}\backslash\{0\} \times \mathrm{Prim}(\mathsf{G})$ would be closed, and this possibility can be excluded using the convergence criterion mentioned after 3.12.3. Using (3.159), we can define a map $\psi : \mathrm{Prim}(C^*_r(\mathsf{G}_N)) \to \mathbb{R}$

by $\psi(\mathfrak{J}) = 0$ for all $\mathfrak{J} \in \mathfrak{G}^*$ and $\psi(\hbar, \mathfrak{J}) = \hbar$ for $\hbar \neq 0$ and $\mathfrak{J} \in \text{Prim}(C^*_r(\mathsf{G}))$. It is clear from the preceding considerations that ψ is continuous and open. Using this in Lemma 3.12.4, one sees that $\mathfrak{J}_\hbar$ is indeed the ideal in $C^*_r(\mathsf{G}_N)$ generated by those $f \in C^\infty_c(\mathsf{G}_N)$ that vanish at $\hbar$. Hence $\mathfrak{A}^0 \simeq C_0(\mathfrak{G}^*)$ as above, and $\mathfrak{A}^\hbar \simeq C^*_r(\mathsf{G})$ for $\hbar \neq 0$. Theorem 3.12.2 then follows from Lemma 3.12.4 and the argument leading to (3.157). ■

We now prove Theorem 3.11.4. Recalling the cutoff κ in 3.11.3, for each $f \in C^\infty_{\text{PW}}(\mathfrak{G}^*)$ we define a function $\mathcal{Q}(f)$ on G_N by

$$\begin{aligned} \mathcal{Q}(f)(0, X) &:= \check{f}(X); \\ \mathcal{Q}(f)(\hbar, \text{Exp}^W(X)) &:= \kappa(X)\check{f}(X/\hbar); \\ \mathcal{Q}(f)(\hbar, \gamma) &:= 0 \ \forall \gamma \notin \mathcal{N}_\iota. \end{aligned} \tag{3.160}$$

It is clear from Definition 3.12.1 that $\mathcal{Q}(f)$ is smooth on G_N. Although $\mathcal{Q}(f)$ does not have compact support in $\hbar$, using (3.75) and (3.152) one may argue that it lies in $C^*_r(\mathsf{G}_N)$. Comparing (3.160), (3.157), and (3.135), we have $\mathcal{Q}^W_\hbar(f)_+ = \varphi_\hbar(\mathcal{Q}(f))$. Hence $\{\mathcal{Q}^W_\hbar(f)_+\}_{\hbar\in\mathbb{R}}$ is a section of the continuous field of Theorem 3.12.2. Definition II.1.2.1 then implies Rieffel's condition II.1.1.1.1, as well as II.(1.4), which in turn implies von Neumann's condition II.1.1.1.2.

To also cover functions of the type $\mathcal{Q}^W_\hbar(f)_-$, one equips G_N with a different smooth structure, obtained by replacing $\text{Exp}^W(X)$ in 3.12.1 by $\text{Exp}^W(-X)$. The original "+" smooth structure is equivalent to the modified "−" one by the diffeomorphism $(0, X) \mapsto (0, -X)$ and $(\hbar, \gamma) \mapsto (\hbar, \gamma)$. ■

Corollary 3.12.6. *When II.(1.3) holds, the field $(C^*_r(\mathsf{G}_N), \{\mathfrak{A}^\hbar, \varphi_\hbar\}_{\hbar\in\mathbb{R}})$ of Theorem 3.12.2, the space $\tilde{\mathfrak{A}}^0 = C^\infty_{\text{PW}}(\mathfrak{G}^*)$, and the map $\mathcal{Q}$ in (3.160) define a continuous quantization of $\mathfrak{G}^*$ (cf. II.1.2.5).*

We now see that the continuous field of Theorem II.2.6.5 is a special case of the one in Theorem 3.12.2, with G given by the pair groupoid $\mathbb{R}^n \times \mathbb{R}^n$, so that G_N is the tangent groupoid of $\mathbb{R}^n$. More generally, the continuous field generated by Weyl quantization on a Riemannian manifold Q (cf. Theorem II.3.5.1) is given by putting $\mathsf{G} = Q \times Q$ in 3.12.2.

CHAPTER IV

Reduction and Induction

1 Reduction

1.1 Basics of Constraints and Reduction

We start with a geometric description of symplectic reduction in a rather general form, and subsequently relate this to the notion of a constraint.

Recall Definition I.2.4.1 and subsequent paragraph.

Definition 1.1.1. *Let* (S, ω) *be a symplectic manifold, and let* C *be a closed submanifold of* S*. The* **null distribution** $\mathcal{N}_C$ *on* C *is the kernel of the restriction* $\omega_C = \iota^*\omega$ *of* ω *to* C *(here* $\iota : C \hookrightarrow \mathsf{S}$ *is the canonical embedding).*

Note that although ω is by definition closed and nondegenerate, its restriction to C, while closed, may be degenerate. Namely, given a vector $X \in T_\sigma C$, no vector $Y \in T_\sigma \mathsf{S}$ for which $\omega(X, Y) \neq 0$ may be tangent to C.

We denote the annihilator in $T^*\mathsf{S}$ of a subbundle $V \subset T\mathsf{S}$ by V^0. For example, $\mathcal{N}_C^0$ consists of all 1-forms α on S that satisfy $\alpha(X) = 0$ for all $X \in \mathcal{N}_C$. The **symplectic orthogonal complement** in $T\mathsf{S}$ of V, on the other hand, is called $V^\perp$; it consists of all $Y \in T\mathsf{S}$ such that $\omega(X, Y) = 0$ for all $X \in V$ (assuming, of course, that X and Y lie in the same fiber of $T\mathsf{S}$). In this notation we obviously have

$$\mathcal{N}_C = TC \cap TC^\perp. \tag{1.1}$$

Theorem 1.1.2. *When the rank of* ω_C *is constant on* C*, the null distribution* $\mathcal{N}_C$ *is smooth and completely integrable; hence* $\mathcal{N}_C$ *defines the* **null foliation** Φ_C *of* C*. When the space*

$$\mathsf{S}^C := C/\Phi_C \tag{1.2}$$

of leaves of this foliation is a manifold in its natural topology, there is a unique symplectic form ω^C on S^C satisfying

$$\tau^*\omega^C = \omega_C, \tag{1.3}$$

where $\tau := \tau_{C\to\mathsf{S}^C}$ maps σ to the leaf of the null foliation in which it lies.

We omit the technical proof that $\mathcal{N}_C$ is smooth (when the rank of ω_C varies on C the null distribution is not necessarily smooth). Use of the well-known identity $d\omega_C(X_0, X_1, X_2) = X_0\,\omega_C(X_1, X_2) - \ldots - \omega([X_1, X_2], X_0)$ for any 2-form ω_C, combined with $d\omega_C = 0$, shows that $\mathcal{N}_C$ is involutive. Consequently, it is completely integrable by Frobenius's theorem.

Using the identity $L_X = d\iota_X + \iota_X d$, one shows that ω is invariant under any flow along the leaves of Φ_C. Thus we can define ω^C by

$$\omega^C(X, Y) := \omega_C(\tilde{X}, \tilde{Y}), \tag{1.4}$$

where $\tilde{X}$ is a preimage of X in TC under τ_*. By construction, ω^C does not depend on the particular choice of the preimage, since any two possible choices differ by an element of $\mathcal{N}_C$, which is annihilated by ω_C. The 2-form ω^C satisfies (1.3) by construction, and is then seen to be closed because ω_C is closed and τ_* is surjective (alternatively, one uses the identity above). Finally, ω^C is nondegenerate by construction; or note that its rank is equal to the dimension of S^C. The uniqueness of ω^C is obvious, for any ω^C satisfying (1.3) must satisfy (1.4). □

We now pass to the elementary theory of constraints. In an expression of the type "$\xi \in TC$" one refers, of course, to the restriction of a vector field ξ to C.

Lemma 1.1.3. *For $f \in C^\infty(\mathsf{S}, \mathbb{R})$ the property $df = 0$ on C is equivalent to $\xi_f \in TC^\perp$.*

Here ξ_f is the Hamiltonian vector field of f; recall I.(2.8). The claim is immediate from I.(2.21). ■

Lemma 1.1.4. *The property $df \in (TC^\perp)^0$ is equivalent to $\xi_f \in TC$.*

The condition $df \in (TC^\perp)^0$ implies that $Xf = 0$ for all $X \in TC^\perp$. Now take $X = \xi_g$ (at some point), and use Lemma 1.1.3 and the equality $\xi_f g = -\xi_g f$. This proves that $\xi_f \in TC$, because the map $g \mapsto \xi_g$ is surjective, since S is symplectic. This argument works in the opposite direction as well. ■

Definition 1.1.5. *A function $\varphi \in C^\infty(\mathsf{S}, \mathbb{R})$ satisfying $\varphi = 0$ on C is called a* **constraint**. *A* **first-class constraint** *is a constraint satisfying $d\varphi \in (TC^\perp)^0$. A constraint that is not first class is called* **second class**.

This definition is of great significance, as there is a fundamental difference between situations with first- and second-class constraints.

Proposition 1.1.6. *Each of the following conditions is necessary and sufficient for a function $\varphi \in C^\infty(\mathsf{S}, \mathbb{R})$ to be a first-class constraint (up to a possible constant):*

1. *Its Hamiltonian vector field ξ_φ lies in $TC \cap TC^\perp$.*

2. *The Poisson bracket $\{\varphi, \varphi'\}$ with any other constraint φ' is itself a constraint.*

Moreover, at each point of C the space $TC \cap TC^{\perp}$ is spanned by the Hamiltonian vector fields of first-class constraints.

The first characterization is immediate from 1.1.5, 1.1.3, and 1.1.4. The second follows from I.(2.19), I.(2.8), and the first one (using 1.1.3 once again). The final claim holds because S is symplectic, so that $f \mapsto \xi_f$ is surjective onto TC; then use 1.1.3 and 1.1.4 in the opposite direction. ∎

Hence the Hamiltonian flow generated by the first-class constraints sweeps out the leaves of the null foliation of C. In physics this flow is regarded as unphysical, corresponding to the fact that the Hamiltonian equations of motion are underdetermined on C. Passing from C to S^C is then a means of eliminating redundancy and indeterminism. Note that using the comment after I.(2.16), one may reconfirm the invariance of ω under flows tangent to the null distribution.

There are four special cases of interest (which are neither exhaustive nor mutually exclusive).

Definition 1.1.7. *A submanifold C of a symplectic manifold (S, ω) is called*

- **isotropic** *when $TC \subseteq TC^{\perp}$;*
- **coisotropic** *when $TC^{\perp} \subseteq TC$;*
- **Lagrangian** *when it is at the same time isotropic and coisotropic, in other words, when $TC^{\perp} = TC$, so that $\omega_C = 0$;*
- **symplectic** *when $TC \cap TC^{\perp} = 0$, so that ω_C on C is symplectic.*

Some authors ascribe theomorphic status to Lagrangian submanifolds, but in this book they hardly play a role. Locally a symplectic submanifold C may be described as the set on which a collection of second-class constraints vanishes, whereas a coisotropic submanifold is locally described as the null set of a set of first-class constraints.

Definition 1.1.8. *A* **weak observable** *on* S *is a function $f \in C^{\infty}(\mathsf{S}, \mathbb{R})$ for which $df \restriction C$ lies in $\mathcal{N}_C^0$.*

The restriction of a weak observable to C is evidently constant on the leaves of the null foliation of C.

Proposition 1.1.9. *A smooth function f is a weak observable*

1. *iff ξ_f lies in $TC + TC^{\perp}$;*
2. *iff its Poisson bracket with any first-class constraint vanishes on C.*

The proof of the first characterization is like that of 1.1.4, adding the fact that $(TC \cap TC^{\perp})^{\perp}$ equals $TC + TC^{\perp}$. The second follows from the equation $\{\varphi, f\} = \xi_{\varphi} f$ in combination with (1.1) and 1.1.6.1. ∎

Proposition 1.1.10. *When C is coisotropically embedded in* S*, the collection $\mathfrak{A}_w^C$ of weak observables is a Poisson algebra.*

Since $\mathfrak{A}_w^C$ is characterized by a differential condition, it is obvious that it is closed under the pointwise product as well as under linear operations. Closure under the Poisson bracket is proved as follows. In the coisotropic case, Proposition 1.1.9 states that f is a weak observable iff $\xi_f \in TC$. To check whether $\xi_{\{f,g\}} \in TC$ when ξ_f and ξ_g are, we compute $\xi_{\{f,g\}}\varphi$ for an arbitrary constraint that is first class by assumption. Using I.(2.8), the Jacobi identity, and then applying 1.1.9.2 twice, one shows that $\xi_{\{f,g\}}\varphi = 0$ on C. ■

Now assume that the reduced space S^C constructed in 1.1.2 is a manifold.

Definition 1.1.11. *The* **reduced representation** $\pi^C(f)$ *of a weak observable* f *is the unique element of* $C^\infty(\mathsf{S}^C, \mathbb{R})$ *that satisfies*

$$\tau^*_{C\to\mathsf{S}^C}\pi^C(f) = f \restriction C. \tag{1.5}$$

The Poisson algebra of observables of the system whose constraint hypersurface is S is then defined as

$$\mathfrak{A}^C := \pi^C(\mathfrak{A}_w^C). \tag{1.6}$$

Using the tubular neighborhood theorem, one easily sees that any smooth function on a submanifold C may be extended to a smooth function defined on a neighborhood of C. When C is closed, a further smooth extension to S is always possible. Hence $\mathfrak{A}^C \simeq C^\infty(\mathsf{S}, \mathbb{R})$ as Poisson algebras.

It follows from 1.1.9 that when C is coisotropic, so that $TC + TC^\perp = TC$, the Hamiltonian flow of f does not leave C when it starts there, and that this flow projects onto the flow of $\pi^C(f)$ in S^C. For general C, it can be shown that one can always decompose $f = f_1 + f_2$ (at least in a neighborhood of C) such that f_1 satisfies $\xi_{f_1} \in TC$ (so that its Hamiltonian flow stays in C) and f_2 vanishes on C. In physics one is given a Hamiltonian h on C that should be a weak observable by construction (of C), and one subsequently tries to extend h to S such that ξ_h is tangent to C.

1.2 Special Symplectic Reduction

We are now going to describe a special case of the construction in the previous section that includes many physically relevant examples.

Definition 1.2.1. *In* **special symplectic reduction** *one starts from:*

- *A pair of symplectic manifolds* (S, ω_S) *and* (S_ρ, ω_ρ).
- *A Poisson manifold* P *(we denote the same manifold equipped with minus the Poisson bracket by* P^-*).*
- *A pair of Poisson morphisms* $J : S \to P^-$ *and* $J_\rho : S_\rho \to P$.

The total space is then $\mathsf{S} = S \times S_\rho$, *equipped with the symplectic form* $\omega := \omega_S + \omega_\rho$, *and the constraint manifold* C *in* S *is*

$$C = S *_P S_\rho := \{(\sigma, \alpha) \in S \times S_\rho | J(\sigma) = J_\rho(\alpha)\}. \tag{1.7}$$

This situation is denoted by

$$S \xrightarrow{J} P \xleftarrow{J_\rho} S_\rho. \tag{1.8}$$

The reduction of S by C is described by the following

Theorem 1.2.2. *Let either J_* or $(J_\rho)_*$ (or both) be surjective at all points relevant to $S *_P S_\rho$. Then $S *_P S_\rho$ is a coisotropic submanifold of $S \times S_\rho$, and ω_C is (locally) of constant rank.*

The null distribution $\mathcal{N}_C$ of $S *_P S_\rho$ is spanned by the collection of vector fields $\hat{\xi}_f$, where $f \in C^\infty(P, \mathbb{R})$ and

$$\hat{\xi}_f\, g := \{J^* f - J_\rho^* f, g\}. \tag{1.9}$$

Here the Poisson bracket is the one corresponding to the symplectic form ω.

We first show that C is a submanifold. Define $f : S \times S_\rho \to P \times P$ by $f(\sigma, \alpha) := (J(\sigma), J_\rho(\alpha))$, and let D be the diagonal in $P \times P$. Then $C = f^{-1}(D)$ and $f_*^{-1} T_{(p,p)} D = T_{(\sigma,\alpha)} C$ whenever $J(\sigma) = J_\rho(\alpha) = p$. The surjectivity condition implies that f intersects D transversally, which in turn guarantees that C is a submanifold of S.

We next prove that C is coisotropic. Let $X \in T_\sigma S$ and $Y \in T_\alpha S_\rho$; then $X + Y \in T_{(\sigma,\alpha)} C$ iff $J_* X = (J_\rho)_* Y$. The dimension of $T_{(\sigma,\alpha)} C$ at any point $(\sigma, \alpha) \in C$ equals $\dim S + \dim S_\rho - (\text{rank } J_*)(\sigma)$, so that

$$\dim(T_{(\sigma,\alpha)} C^\perp) = (\text{rank } J_*)(\sigma). \tag{1.10}$$

Let $\mathcal{N}_{(\sigma,\alpha)}$ denote the linear span of the collection of vector fields $\hat{\xi}_f$ taken at (σ, α), where f runs through $C^\infty(P, \mathbb{R})$. Then

$$\dim(\mathcal{N}_{(\sigma,\alpha)}) = (\text{rank } J_*)(\sigma). \tag{1.11}$$

We now argue that $\mathcal{N}_{(\sigma,\alpha)} \subseteq T_{(\sigma,\alpha)} C^\perp$, so that (1.11) and (1.10) imply that

$$\mathcal{N}_{(\sigma,\alpha)} = T_{(\sigma,\alpha)} C^\perp. \tag{1.12}$$

Namely, let $X + Y \in T_{(\sigma,\alpha)} C$, as above; then, since $J_* X = (J_\rho)_* Y$, one has

$$\omega_{(\sigma,\alpha)}(X + Y, \hat{\xi}_f) = d(J^* f - J_\rho^* f)_{(\sigma,\alpha)}(X + Y) = 0.$$

Moreover, $\mathcal{N}_{(\sigma,\alpha)} \subset T_{(\sigma,\alpha)} C$ by a similar calculation, which uses Proposition I.2.3.5. Therefore, according to (1.12) the submanifold C is coisotropically immersed in $S \times S_\rho$, and one has $\mathcal{N} = \mathcal{N}_C$. It then follows from (1.11) and the condition stated in the theorem that ω_C has constant rank on each connected component of $S \times S_\rho$. The above argument is symmetric in J and J_ρ. ■

We are therefore in a position to apply Theorem 1.1.2, obtaining a reduced symplectic space (S^C, ω^C), which we assume to be a manifold. To indicate the dependence on the given data, we will denote the reduced space by $(S_J^\rho, \omega_J^\rho)$, where

$$S_J^\rho := (S *_P S_\rho)/\Phi. \tag{1.13}$$

Here $\Phi = \Phi_C$ is the null foliation generated by $\mathcal{N}_C$. When S and J are fixed, we sometimes simply write (S^ρ, ω^ρ) for $(S^\rho_J, \omega^\rho_J)$. The collection of weak observables is then called $\mathfrak{A}^\rho_w$.

It follows from 1.1.5 or 1.1.6 that the constraints defined by C (cf. (1.7)) are precisely the functions of the form $J^*f - J^*_\rho f$, where $f \in C^\infty(P, \mathbb{R})$.

Inspired by the theory of von Neumann algebras, we define the **Poisson commutant** of some subspace $\mathfrak{A}_s \subset C^\infty(S, \mathbb{R})$ by

$$\mathfrak{A}'_s := \{g \in C^\infty(S, \mathbb{R}) \mid \{f, g\} = 0 \, \forall f \in \mathfrak{A}_s)\}. \tag{1.14}$$

It follows from the Jacobi identity and the Leibniz rule that $\mathfrak{A}'_s$ is a Poisson algebra (even when $\mathfrak{A}_s$ isn't). The operation $\mathfrak{A}_s \mapsto \mathfrak{A}''_s$ plays the role of the "weak closure" of $\mathfrak{A}_s$; the previous remark implies that $\mathfrak{A}''_s$ is always a Poisson algebra. In general, $\mathfrak{A}_s$ may be strictly contained in $\mathfrak{A}''_s$ even when the former is a Poisson algebra. Similary, one defines the **Poisson center** of $\mathfrak{A}_s$ as $\mathfrak{A}_s \cap \mathfrak{A}'_s$. This may not be a Poisson algebra, but its "weak closure" $\mathfrak{A}''_s \cap \mathfrak{A}'_s$ is.

An important subspace of $C^\infty(S, \mathbb{R})$ is $J^*C^\infty(P, \mathbb{R})'$. This may be regarded as a Poisson subalgebra of $C^\infty(S \times S_\rho, \mathbb{R})$ under the obvious embedding of $C^\infty(S, \mathbb{R})$ in the latter. Combining 1.1.8 and 1.1.5 (or 1.1.9 and 1.1.6), one infers that

$$J^*C^\infty(P, \mathbb{R})' \subseteq \mathfrak{A}^\rho_w. \tag{1.15}$$

Hence Definition 1.1.11 applies. We write

$$\pi^\rho_J : J^*C^\infty(P, \mathbb{R})' \to C^\infty(S^\rho, \mathbb{R}),$$

or simply π^ρ, for π^C. Denoting a point in S^ρ by an equivalence class $[\sigma, \alpha]_\Phi$ under the null foliation, one has simply

$$\pi^\rho(f)([\sigma, \alpha]_\Phi) = f(\sigma). \tag{1.16}$$

Because $f \in J^*C^\infty(P, \mathbb{R})'$, this is independent of the choice of σ in the given equivalence class. The same construction applies, of course, to Poisson subalgebras $\mathfrak{A}_s$ of $J^*C^\infty(P, \mathbb{R})'$.

Corollary 1.2.3. *In the context of Definition 1.2.1, suppose one has a second Poisson manifold P_2 and a Poisson map $J_2 : S \to P_2$ such that $J_2^*C^\infty(P_2, \mathbb{R}) \subseteq$*

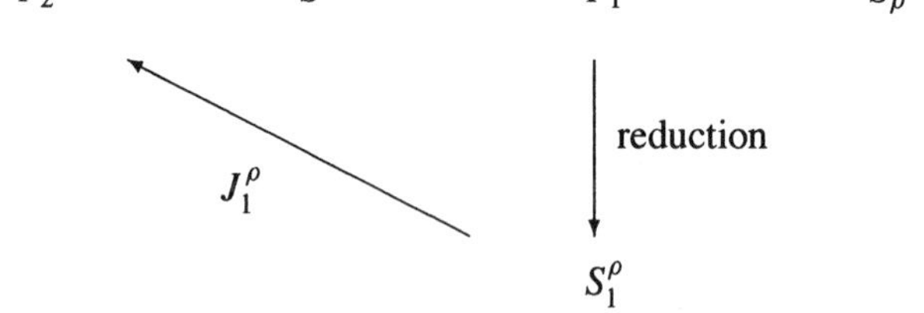

FIGURE 1. Special symplectic reduction; $S_1^\rho := S^\rho_{J_1}$ etc.

$J^*C^\infty(P,\mathbb{R})'$. The map $J^\rho : S^\rho \to P_2$ defined by

$$J^\rho([\sigma,\alpha]_\Phi) = J_2(\sigma) \tag{1.17}$$

is well-defined, and is a Poisson map.

Relabeling (J, P) as (J_1, P_1) for clarity, and writing S_1^ρ and J_1^ρ for $S^\rho = S_{J_1}^\rho$ and J^ρ, respectively, we can summarize this situation pictorially as in Figure 1. Equivalently, there is a reduced representation $\pi_2^\rho := (J_1^\rho)^*$ of $C^\infty(P_2,\mathbb{R})$ on S^ρ that is given in terms of π^ρ by

$$\pi_2^\rho = \pi^\rho \circ J_2^*. \tag{1.18}$$

1.3 Classical Dual Pairs

In this section we define various duality relationships of increasing strength between Poisson manifolds. The strongest of these will relate the respective representation theories of these manifolds to each other.

A foliation Φ of S is always understood to be smooth in the sense that its associated distribution $T\Phi$ (consisting of the vectors in TS which are tangent to the pertinent leaf of Φ) is smooth as defined in I.2.4.1.

Definition 1.3.1. *A* **symplectically complete foliation** *Φ of a symplectic manifold S is a foliation with the property that the distribution $T\Phi^\perp$ is completely integrable.*

Here is an alternative characterization.

Proposition 1.3.2. *A foliation Φ is symplectically complete iff the space $C^\infty(S,\mathbb{R})^\Phi$ of all smooth functions on S that are constant on each leaf of Φ is a Poisson algebra.*

A function f is constant along Φ iff $df \in T\Phi^0$. As in 1.1.4, this is equivalent to $\xi_f \in T\Phi^\perp$. As in I.2.4, when $C^\infty(S,\mathbb{R})^\Phi$ is a Poisson algebra one uses I.(2.9) and Lemma I.2.4.2 to show that the distribution $T\Phi^\perp$ is completely integrable.

Conversely, complete integrability implies that $T\Phi^\perp$ is involutive, so that the first step of the proof establishes the claim in the opposite direction. □

The foliation generated by $T\Phi^\perp$ is called $\Phi^\perp$. Recall (1.14).

Proposition 1.3.3. *Let Φ be a symplectically complete foliation, and assume that the functions in $C^\infty(S,\mathbb{R})^\Phi$ separate the leaves of Φ.*

Then $(C^\infty(S,\mathbb{R})^\Phi)' = C^\infty(S,\mathbb{R})^{\Phi^\perp}$.

As before, one shows that $f \in C^\infty(S,\mathbb{R})^{\Phi^\perp}$ implies $\xi_f \in T\Phi$. Hence, without the additional assumption, one has $C^\infty(S,\mathbb{R})^{\Phi^\perp} \subseteq (C^\infty(S,\mathbb{R})^\Phi)'$ by I.(2.8).

Let $g \in (C^\infty(S,\mathbb{R})^\Phi)'$; by I.(2.8) this is equivalent to $\xi_g f = 0$ for all $f \in C^\infty(S,\mathbb{R})^\Phi$. The assumption now implies that $\xi_g \in T\Phi$. Since Φ is smooth, so is $\Phi^\perp$. Hence any two points in a connected leaf of $\Phi^\perp$ can be joined by a (piecewise) smooth curve $c(\cdot)$. Using I.(2.21) one computes $dg(c(t))/dt = \omega(\xi_g, \dot{c}(t))$. This vanishes when $\xi_g \in T\Phi$, so that g is constant along $c(\cdot)$, so $g \in C^\infty(S,\mathbb{R})^{\Phi^\perp}$. ■

The assumption in 1.3.3 is satisfied when the leaf space S/Φ is a manifold. We now look at the case where P is a Poisson manifold, and $J : S \to P^-$ is a Poisson map for which the level sets of J define a foliation Φ of S; that is, the leaf of Φ through $\sigma \in S$ is $J^{-1}(J(\sigma))$. This is, for example, the case when J is a submersion, which guarantees that each subspace $J^{-1}(\alpha)$ is a submanifold of S. In addition, we assume that S/Φ is a manifold. When all this holds, one has the equality

$$\tau^*_{S\to S/\Phi}C^\infty(S/\Phi, \mathbb{R}) = C^\infty(S, \mathbb{R})^\Phi = J^*C^\infty(P, \mathbb{R}). \tag{1.19}$$

Since $J^*C^\infty(P, \mathbb{R})$ is evidently a Poisson algebra, the foliation Φ is symplectically complete by Proposition 1.3.2.

The situation is particularly neat when the associated foliation $\Phi^\perp$ is itself given by the level sets of a Poisson map $J_2 : S \to P_2$. In view of the symmetry between (J, P) and J_2, P_2 we relabel the former as (J_1, P_1) in what follows.

Definition 1.3.4. *A* **classical dual pair** (S, P_1, P_2, J_1, J_2) *consists of a connected symplectic manifold S and a pair of Poisson manifolds* P_1, P_2, *together with Poisson maps* $J_1 : S \to P_1^-$ *and* $J_2 : S \to P_2$, *such that:*

1. *The level sets of* J_1 *and* J_2 *define foliations* Φ_1 *and* Φ_2 *of S, respectively, with the property that* $\Phi_2 = \Phi_1^\perp$ *(and hence* $\Phi_1 = \Phi_2^\perp$*).*
2. *The leaf spaces* S/Φ_1 *and* S/Φ_2 *are manifolds.*
3. *The maps* J_1 *and* J_2 *are surjective submersions.*
4. *The level sets* $J_1^{-1}(\sigma_1)$ *and* $J_2^{-1}(\sigma_2)$ *are connected for all* $\sigma_i \in P_i$ $(i = 1, 2)$.
5. *The level sets* $J_1^{-1}(\sigma_1)$ *and* $J_2^{-1}(\sigma_2)$ *are simply connected for all* σ_i.
6. *The maps* J_1 *and* J_2 *are complete.*

We denote classical dual pairs by

$$P_2 \xleftarrow{J_2} S \xrightarrow{J_1} P_1, \tag{1.20}$$

using this notation also when not all of the above conditions are satisfied.

It follows from 1.3.3 and (1.19) that when 1.3.4.1–3 are obeyed one has

$$\begin{aligned} J_1^*C^\infty(P_1, \mathbb{R})' &= J_2^*C^\infty(P_2, \mathbb{R}), \\ J_2^*C^\infty(P_2, \mathbb{R})' &= J_1^*C^\infty(P_1, \mathbb{R}). \end{aligned} \tag{1.21}$$

As in the comment after (1.19), we infer that Φ_1 and Φ_2 are symplectically complete.

Lemma 1.3.5. *When conditions 1, 3, and 4 in 1.3.4 are satisfied, the foliation* Φ_2 *(or* Φ_1*) coincides with the foliation defined by all Hamiltonian vector fields of the form* $\xi_{J_1^*f}$, *(or* $\xi_{J_2^*f}$*), where* $f \in C^\infty(P_1, \mathbb{R})$ *(or* $f \in C^\infty(P_2, \mathbb{R})$*).*

This is immediate from the proof of 1.3.2. ∎

Corollary 1.3.6. *In a classical dual pair with connected leaves there is a bijective correspondence* $L_1 \leftrightarrow L_2 := J_2(J^{-1}(L))$ *between the symplectic leaves* L_1 *and* L_2 *in* P_1 *and* P_2, *respectively.*

Given $\alpha \in P_1$, Lemma 1.3.5 for Φ_1 shows that the leaf $J_1^{-1}(\alpha)$ of Φ_1 is generated by the Hamiltonian flow of the vector fields $\xi_{J_1^* f}$, where $f \in C^\infty(P_2, \mathbb{R})$. By I.2.3.5 and I.2.4.3 the set $J_2(J_1^{-1}(\alpha))$ is a (connected) symplectic leaf in P_2. When α' lies on the same leaf as α, there is a (piecewise) smooth Hamiltonian curve c in P_1 that connects α' and α. Using I.2.4.3 once again, and subsequently 1.3.5 for Φ_2, we infer that $J_1^{-1}(c)$ lies in a single leaf of Φ_2. Hence $J_2(J_1^{-1}(\alpha)) = J_2(J_1^{-1}(\alpha'))$, and $J_1^{-1}(L_1)/\Phi_2 \simeq J^2(J_1^{-1}(L_1))$. Thus we obtain a bijection between the symplectic leaves reached in this way. Because J_1 and J_2 are surjections, all symplectic leaves are included. ■

We saw in I.2.6 that a symplectic leaf in P may be regarded as an irreducible representation of the Poisson algebra $C^\infty(P, \mathbb{R})$, so that (up to possible covering spaces of the symplectic leaves in question) Corollary 1.3.6 expresses a bijective correspondence between the irreducible representations of two Poisson algebras connected by a classical dual pair. This result can be generalized to all representations. The following concept is central to this generalization.

Definition 1.3.7. *Two Poisson manifolds P_1, P_2 are called* **Morita equivalent** *when they form part of a classical dual pair (S, P_1, P_2, J_1, J_2).*

Despite the terminology, this definition fails to define an equivalence relation in the class of all (finite-dimensional) Poisson manifolds, because not all Poisson manifolds are Morita equivalent to themselves.

Proposition 1.3.8. *Morita equivalence defines an equivalence relation in the subclass of all Poisson manifolds that are Morita equivalent to themselves.*

Reflexivity (i.e., $P \overset{M}{\sim} P$) being satisfied by definition, symmetry (that is, $P_1 \overset{M}{\sim} P_2$ implies $P_2 \overset{M}{\sim} P_1$) holds because from the diagram $P_2 \overset{J_2}{\leftarrow} S \overset{J_1}{\rightarrow} P_1$ one obtains $P_1 \overset{J_1}{\leftarrow} S^- \overset{J_2}{\rightarrow} P_2$. Finally, transitivity is true by the following argument. When $P_1 \overset{M}{\sim} P_2$ by S_1 and $P_2 \overset{M}{\sim} P_3$ by S_2, so that one has

$$P_1 \overset{J_{11}}{\longleftarrow} S_1 \overset{J_{12}}{\longrightarrow} P_2 \overset{J_{22}}{\longleftarrow} S_2 \overset{J_{23}}{\longrightarrow} P_3,$$

one obtains a symplectic manifold S_{12}^{22} by special symplectic reduction from the middle three spaces in the diagram. Using Corollary 1.2.3 with respect to J_{11} and J_{33}, this leads to the classical dual pair $P_1 \leftarrow S_{12}^{22} \rightarrow P_3$. ■

Here are some simple examples of Morita equivalence.

Proposition 1.3.9.

- *If S is a connected and simply connected symplectic manifold, and P is a connected manifold with the zero Poisson structure, then $S \times P$ is Morita equivalent to P. In particular, S is Morita equivalent to a point.*
- *Two connected symplectic manifolds S_1 and S_2 are Morita equivalent iff their fundamental groups are isomorphic.*

In the first case the pertinent classical dual pair is

$$S \times P \xleftarrow{J_2} S \times T^*P \xrightarrow{J_1} P,$$

with $J_2 = (\tau_{(1)}, \tau_{T^*P\to P} \circ \tau_{(2)})$ and $J_1 = \tau_{T^*P\to P} \circ \tau_{(2)}$ (here $\tau_{(i)}$ is the projection onto the ith variable).

We pass to the second example. Let (S_1, ω_1) and (S_2, ω_2) have isomorphic fundamental groups $\pi_1(S_1) \simeq \pi_2(S)$, with isomorphism $\phi : \pi_1(S_1) \to \pi_2(S)$, and denote the universal covering spaces by $\tilde{S}_i$. Then $\pi_1(S)$ acts on $\tilde{S}_1 \times \tilde{S}_2$ by $x : (\sigma_1, \sigma_2) \mapsto (x\sigma_1, \phi(x)\sigma_2)$. Since $\pi_i(S_i)$ is discrete, the form $\tilde{\omega}_i := \tau^*_{\tilde{S}_i\to S_i}\omega_i$ on $\tilde{S}_i$ is symplectic ($i = 1, 2$). Equip $\tilde{S}_1 \times \tilde{S}_2$ with the symplectic form $\tilde{\omega}_{12} := \tilde{\omega}_2 - \tilde{\omega}_1$. The quotient $S_{12} := (\tilde{S}_1 \times \tilde{S}_2)/\pi_1(S_1)$ has a unique symplectic form ω_{12} whose pullback to $\tilde{S}_1 \times \tilde{S}_2$ under the canonical projection is $\tilde{\omega}_{12}$.

The classical dual pair is now given by

$$S_2 \xleftarrow{J_2} S_{12} \xrightarrow{J_1} S_1,$$

with the obvious projections $J_i : S_{12} \to S_i$. One easily verifies all pertinent properties. For example, the completeness of the J_i follows from the path lifting lemma of homotopy theory. Also, for each $\sigma_1 \in S_1$ the leaf $J_1^{-1}(\sigma_1)$ is homeomorphic to $\tilde{S}_2$, which is indeed connected and simply connected; analogously, $J_2^{-1}(\sigma_2) \simeq \tilde{S}_1$.

Conversely, assume that one has a classical dual pair $S_2 \xleftarrow{J_2} S \xrightarrow{J_1} S_1$. Some algebraic topology then shows that $\pi_1(S_i) \simeq \pi_1(S)$ for $i = 1, 2$. □

1.4 The Classical Imprimitivity Theorem

We now state and prove the **classical imprimitivity theorem.**

Recall the definition of completeness after I.2.6.1. In the situation of 1.3.7 the pullbacks J_1^* and J_2^* are representations of $C^\infty(P_1, \mathbb{R})$ and $C^\infty(P_2, \mathbb{R})$ on S, respectively (cf. I.2.6.1).

Theorem 1.4.1. *Let P_1 and P_2 be Morita-equivalent Poisson manifolds. There is a bijective correspondence between the representations of $C^\infty(P_1, \mathbb{R})$ and $C^\infty(P_2, \mathbb{R})$ preserving irreducibility. Equivalently (by Corollary I.2.6.5), there is a bijection between complete Poisson maps $J_\rho : S_\rho \to P_1$ and $J_\sigma : S_\sigma \to P_2$ preserving symplectic leaves (or their covering spaces). This correspondence arises as follows.*

Let $P_2 \xleftarrow{J_2} S \xrightarrow{J_1} P_1$ be a classical dual pair implementing the Morita equivalence between P_1 and P_2. When $J_\sigma : S_\sigma \to P_2$ is a complete Poisson map (where S_σ is symplectic), there exist a symplectic manifold S_ρ and a complete Poisson map $J_\rho : S_1 \to P_1$ such that S_σ is symplectomorphic to the reduced space S_1^ρ obtained by special symplectic reduction (and $J_\sigma \simeq J_1^\rho$), where S_1^ρ and J_1^ρ are defined in Figure 1.

In the opposite direction, given a complete Poisson map $J_\rho : S_\rho \to P_1$ (where S_ρ is symplectic) there exist a symplectic manifold S_σ and a complete Poisson map

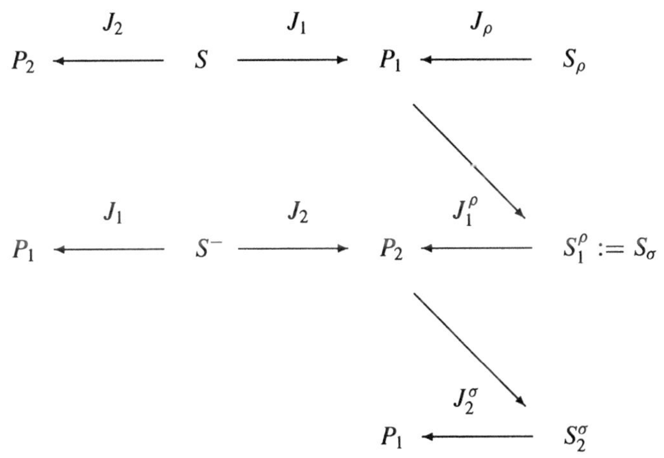

FIGURE 2. Classical imprimitivity theorem: $S_2^\sigma \simeq S_\rho$ and $J_2^\sigma \simeq J_\rho$

$J_\sigma : S_2 \to P_2$ such that S_ρ is symplectomorphic to the reduced space $(S^-)_2^\sigma$ (and $J_\rho \simeq J_2^\sigma$), defined as in Figure 1, but now with respect to $P_1 \overset{J_1}{\leftarrow} S^- \overset{J_2}{\rightarrow} P_2$.

Taking $S_\sigma = S_1^\rho$ and $J_\sigma = J_1^\rho$ as just defined, one has $(S^-)_2^\sigma \simeq S_\rho$ and $J_2^\sigma \simeq J_\rho$. Conversely, taking $S_\rho = (S^-)_2^\sigma$ and $J_\rho = J_2^\sigma$, one has $S_1^\rho \simeq S_\sigma$ and $J_1^\rho \simeq J_\sigma$.

See Figure 2. We write S_2^σ for $(S^-)_2^\sigma$.

The existence of the Poisson maps $J_1^\rho : S_1^\rho \to P_2$ and $J_2^\sigma : S_2^\sigma \to P_1$ follows from Corollary 1.2.3. The reinterpretation of the data (1.20) as $P_1 \overset{J_1}{\leftarrow} S^- \overset{J_2}{\rightarrow} P_2$ will be used again in the proof of Theorem 1.8.1.

Starting from $J_\rho : S_\rho \to P_1$, constructing $J_1^\rho : S_1^\rho \to P_2$ as indicated, which we relabel as $J_\sigma : S_\sigma \to P_2$, and subsequently defining $J_2^\sigma : S_2^\sigma \to P_1$, the essence of the proof of the theorem consists in the construction of a symplectomorphism $\varphi : S_2^\sigma \to S_\rho$.

Consider the space $S *_{P_1} S_\rho *_{P_2} S$, defined as the subset of $S \times S_\rho \times S^-$ consisting of triples $(\sigma_1, \alpha, \sigma_2)$ satisfying $J_1(\sigma_1) = J_\rho(\alpha)$ and $J_2(\sigma_1) = J_2(\sigma_2)$. By construction, the space S^σ is obtained from this by a double foliation (cf. 1.2.2). The first one, Φ_I on $S \times S_\rho$, is generated by the Hamiltonian vector fields defined by the functions $J_1^* f - J_\rho^* f$, where $f \in C^\infty(P_1, \mathbb{R})$; we denote points of the leaf space S^ρ by $[\sigma_1, \alpha]_{\Phi_I}$. The second foliation, Φ_{II} on $S \times S^-$, is generated by the Hamiltonian vector fields defined by the functions $\tau_{(1)}^* J_2^* g - \tau_{(2)}^* J_2^* g$, with $g \in C^\infty(P_2, \mathbb{R})$ and $\tau_{(1)}, \tau_{(2)}$ the projections onto the first and second variables in $S \times S^-$, respectively. Its leaf space has elements $[\sigma_1, \sigma_2]_{\Phi_{II}}$. By (1.17) the equivalence classes $[[\sigma_1, \alpha]_{\Phi_I}, \sigma_2]_{\Phi_{II}}$ correspond to elements of $S_2^{\overset{\rho}{1}}$.

Take a triple $(\sigma_1, \alpha, \sigma_2) \in S *_{P_1} S_\rho *_{P_2} S$, projecting to $[[\sigma_1, \alpha]_{\Phi_I}, \sigma_2]_{\Phi_{II}}$. By Lemma 1.3.5 the Hamiltonian vector fields defined by the functions $J_1^* f$, with $f \in C^\infty(P_1, \mathbb{R})$, generate a foliation that coincides with the foliation by the connected level sets of J_2. Recall that $J_2(\sigma_1) = J_2(\sigma_2)$; for simplicity we initially assume that there is a smooth Hamiltonian curve $c_{\sigma_1\to\sigma_2}(\cdot)$ in S connecting σ_1 and σ_2, whose tangent vector is $\xi_{J_1^*f}$, for some $f \in C^\infty(P_1, \mathbb{R})$. Using a cutoff function if necessary, we may pick an f with compact support, so that ξ_f is complete in P_1. Let $\sigma_2 = c_{\sigma_1\to\sigma_2}(t_0)$ (and, of course, $\sigma_1 = c_{\sigma_1\to\sigma_2}(0)$). We can move α in S_ρ along the Hamiltonian curve $c_\alpha(\cdot)$ that is generated by $-J_\rho^*f$ and starts at $c_\alpha(0) = \alpha$. Let $\tilde{\alpha} := c_\alpha(t_0)$; this makes sense, since by assumption the map J_ρ is complete. The definition of $[\ldots]_{\Phi_I}$ then implies that

$$[[\sigma_1, \alpha]_{\Phi_I}, \sigma_2]_{\Phi_{II}} = [[\sigma_2, \tilde{\alpha}]_{\Phi_I}, \sigma_2]_{\Phi_{II}}. \tag{1.22}$$

Let now σ_2 move around a closed Hamiltonian curve $c_{\sigma_2\to\sigma_2}(\cdot)$, generated by some J_1^*g, where $g \in C^\infty(P_1, \mathbb{R})$, and, say, $c_{\sigma_2\to\sigma_2}(0) = c_{\sigma_2\to\sigma_2}(1) = \sigma_2$. We put $c_{\sigma_2\to\sigma_2} := \{c_{\sigma_2\to\sigma_2}(t) | \, t \in [0, 1]\}$. According to I.2.3.5 with $J = J_1$, the curve $c_1(\cdot) := J_1(c_{\sigma_2\to\sigma_2}(\cdot))$ in P_1 is Hamiltonian, being generated by g. The curve $c_1 := \{c_1(t) | \, t \in [0, 1]\}$ is closed, since $c_1(0) = c_1(1) = J_1(y)$. Our assumption that the level sets of J_2 are simply connected implies that $c_{\sigma_2\to\sigma_2}$, and hence c_1, is contractible. Using I.2.3.5 with $J = J_\rho$, one infers that the Hamiltonian curve c_ρ in S_ρ that is generated by J_ρ^*g and starts at $c_\rho(0) = \tilde{\alpha}$, covers c_1. The latter being contractible, the monodromy lemma of homotopy theory (or a direct argument) implies that $c_\rho(1)$ must be closed; i.e., $c_\rho(1) = \tilde{\alpha}$. This, in turn, guarantees that $\tilde{\alpha}$ is independent of the Hamiltonian path from σ_1 to y, which in addition implies that $\tilde{\alpha}$ is independent of the choice of (σ_1, α) in the class $[\sigma_1, \alpha]_{\Phi_I}$. Finally, it is clear from the construction that $\tilde{\alpha}$ does not change when a different representative of the given class $[\ldots]_{\Phi_{II}}$ is chosen.

When the Hamiltonian curve $c_{\sigma_1\to\sigma_2}(\cdot)$ is merely piecewise smooth, one simply uses the above argument for each smooth piece, with the same conclusion. Thus we may define

$$\varphi([[\sigma_1, \alpha]_{\Phi_I}, \sigma_2]_{\Phi_{II}}) := \tilde{\alpha}, \tag{1.23}$$

where $\tilde{\alpha}$ is determined by (1.22). We have just seen that this map is well-defined. We now use Lemma 1.3.5 once more, this time saying that the foliation determined by the Hamiltonian vector fields $\xi_{J_2^*f}$, where $f \in C^\infty(P_2, \mathbb{R})$, coincides with the foliation by the level sets of J_1. Since $J_1(\sigma_2) = J_\rho(\tilde{\alpha})$, we infer that the equivalence class $[\ldots]_{\Phi_{II}}$ is uniquely determined by $\tilde{\alpha}$, and conclude that φ is a bijection (this could alternatively be established by a dimension count; cf. the proof of 1.2.2). Moreover, φ is a Poisson map, as is obvious from Theorem 1.2.2. Hence φ is a symplectomorphism. By (1.17) we have

$$J_2^\sigma([[\sigma_2, \tilde{\alpha}]_{\Phi_I}, \sigma_2]_{\Phi_{II}}) = J_1(\sigma_2).$$

But $J_1(\sigma_2) = J_\rho(\tilde{\alpha})$, so that under the above symplectomorphism the map J_2^σ is transformed into J_ρ.

The construction can, of course, be carried out in the opposite direction as well. All relevant constructions preserve completeness, and the proof of bijectivity is finished.

When $S_\rho = L$ is a symplectic leaf of P_1 and $J_\rho = \iota$ is the inclusion map, the fiber product $S *_{P_1} L$ coincides with $J_1^{-1}(L)$, and the foliation Φ_I is nothing but the foliation Φ_2 by the level sets of J_2. By (1.17) we have $J_1^\rho([\sigma]_{\Phi_2}) = J_2(\sigma)$ for $\sigma \in J_1^{-1}(L)$; since $J_2(J_1^{-1}(L)) \simeq J_1^{-1}(L)/\Phi_2 = S_1^\rho$, we conclude that J_1^ρ injects S_1^ρ into P_2. The proof of Corollary 1.3.6 then shows that $J_2(J_1^{-1}(L))$ is a symplectic leaf in P_2, so that S_1^ρ is symplectomorphic to such a leaf by J_1^ρ.

When $S_\rho = \tilde{L}$ is a covering space of L, one analogously obtains that S_1^ρ covers $J_2(J_1^{-1}(L))$. Theorem I.2.6.7 then shows that the bijection preserves irreducibility, as claimed. ■

1.5 Marsden–Weinstein Reduction

We will now look at a further specialization of the reduction procedure in 1.2; under suitable assumptions this will produce examples of classical dual pairs as well. Throughout this section H is a Lie group, and until the last paragraph we consider a strongly Hamiltonian H-action on a symplectic manifold (S, ω), with Co-equivariant momentum map $J : S \to \mathfrak{h}^*_-$.

Lemma 1.5.1. *The map $J_* : T_\sigma S \to T_{J(\sigma)}\mathfrak{h}^* \simeq \mathfrak{h}^*$ is surjective iff the stabilizer H_σ is discrete.*

By III.(1.7) an element $X \in \mathfrak{h}$ is annihilated by the image of $J_*(\sigma)$ iff $\xi_X(\sigma) = 0$. Hence the dimension of H_σ equals the dimension of the annihilator of this image. This common dimension is 0 when $J_*(\sigma)$ is surjective. ■

We now take a coadjoint orbit $\mathcal{O}$ in $\mathfrak{h}^*$, and specialize Definition 1.2.1 to the case $P = \mathfrak{h}^*_+$ and $S_\rho = \mathcal{O}_+$, with $J_\rho = \iota_\mathcal{O}$ the inclusion map (we will omit the "+" whenever it is convenient). In other words, the situation is

$$S \xrightarrow{J} \mathfrak{h}^*_+ \xleftarrow{\iota_\mathcal{O}} \mathcal{O}_+. \tag{1.24}$$

Recall that a map between two manifolds is proper when the inverse image of every compact set is compact.

Definition 1.5.2. *A Lie group action $L : H \times S \to S$ on a manifold is called* **proper** *when the map $(x, \sigma) \mapsto (x\sigma, \sigma)$ from $H \times S$ to $S \times S$ is proper. In other words, for sequences $\{\sigma_n\}$ in S and $\{x_n\}$ in H the convergence of $\{\sigma_n\}$ and of $\{x_n\sigma_n\}$ must imply that $\{x_n\}$ has a convergent subsequence.*

Roughly speaking, this means that nearby points in S can be mapped into each other by $x \in G$ only if x is near e. We collect some relevant properties of proper group actions.

Proposition 1.5.3.

- *The action of a compact group is always proper.*
- *Under a proper action the stabilizer H_σ of every point $\sigma \in S$ is compact.*
- *The quotient of a manifold by a proper and free action is a manifold.*

The first claim is immediate from the definition. The second follows because $(\sigma, \sigma) \in S \times S$ is compact, so that its inverse image $\{(H_\sigma\sigma, \sigma)\}$ must be compact, too. The third statement is a nontrivial theorem of differential geometry, whose proof we omit. □

Theorem 1.5.4. *Given a strongly Hamiltonian action of a Lie group H on a symplectic manifold (S, ω), with* Co*-equivariant momentum map $J : S \to \mathfrak{h}^*_-$, assume that there is a $\theta \in \mathcal{O}$ such that H acts freely and properly on $J^{-1}(\theta)$. Then the hypothesis in Theorem 1.2.2 is met, and the reduced space $S^{\mathcal{O}}_J := S^{\iota\mathcal{O}}_J$ is a symplectic manifold. There are symplectomorphisms*

$$S^{\mathcal{O}}_J \simeq J^{-1}(\mathcal{O})/H^0 \simeq J^{-1}(\theta)/(H_\theta \cap H^0), \tag{1.25}$$

where H^0 is the identity component of H, and H_θ is the stabilizer of θ under the coadjoint action.

Since the H-action is free, Lemma 1.5.1 and the equivariance of J (cf. III.1.2.5) imply that the hypothesis of Theorem 1.2.2 is satisfied.

We first take $\mathcal{O} = \{0\}$, in which case (1.9) and preceding text implies the first isomorphism in (1.25); recall that the functions $J^*\tilde{X}$ generate the H^0-action. For general orbits we perform a **shifting trick** that reduces the situation to the zero orbit. Namely, we consider $\tilde{S} = S \times \mathcal{O}$, on which H acts by the product of the original action and the coadjoint action. By III.1.4.6 the momentum map $\tilde{J} : \tilde{S} \to \mathfrak{h}^*_-$ is given by $\tilde{J}(\sigma, \theta) = J(\sigma) - \theta$, so that $S *_{\mathfrak{h}^*} \mathcal{O} = \tilde{J}^{-1}(0)$. This trick establishes the first isomorphism in (1.25) in the general case. The second isomorphism follows from the transitivity of the H-action on $\mathcal{O}$ and the equivariance of J. ■

It should be mentioned that although under the stated assumptions $J^{-1}(\mathcal{O})$ can be shown to be coisotropically embedded in S, the quotient $J^{-1}(\mathcal{O})/H^0$ does not coincide with the quotient $J^{-1}(\mathcal{O})/\Phi$ by the null foliation Φ; in fact, when H and H_θ are connected, it can be shown that $J^{-1}(\mathcal{O})/\Phi \simeq (J^{-1}(\mathcal{O})/H^0) \times \mathcal{O}$. Also, unless H_θ is connected, the space $J^{-1}(\theta)/(H_\theta \cap H^0)$ is not the same as the quotient of $J^{-1}(\theta)$ (which is equally well coisotropically embedded in S) by its null foliation, for the latter quotient is $J^{-1}(\theta)/(H_\theta \cap H^0)^0$.

The reduced space $J^{-1}(\mathcal{O})/H$ is called the **Marsden–Weinstein quotient** of S with respect to $\mathcal{O}$ (and the given group action). As expressed by (1.25), when H is connected, this quotient is essentially the same as the reduced space $S^{\mathcal{O}}_J$. To show how they are related when H fails to be connected, we form the discrete group $\pi_0(H) := H/H^0$, and see that

$$\tilde{S}^{\mathcal{O}}_J := J^{-1}(\mathcal{O})/H \simeq S^{\mathcal{O}}_J/\pi_0(H). \tag{1.26}$$

Hence $\tilde{S}^{\mathcal{O}}_J$ is symplectic, as is $S^{\mathcal{O}}_J$ (this follows only because $\pi_0(H)$ is discrete).

One may look at Marsden–Weinstein quotients from a different angle.

Theorem 1.5.5. *Let the H-action on S be free, proper, and strongly Hamiltonian. Then the space S/H is a manifold with a unique Poisson structure for which the canonical projection $\tau : S \to S/H$ is a Poisson map.*

The symplectic leaves of S/H are the connected components of the Marsden–Weinstein quotients $J^{-1}(\mathcal{O})/H$, where $\mathcal{O} \subset \mathfrak{h}^*$ is a coadjoint orbit.

The Poisson structure on S/H is obtained by identifying $C^\infty(S/H, \mathbb{R})$ with the space $C^\infty(S, \mathbb{R})^H$ of H-invariant smooth functions on S; this space is a Poisson algebra, since the symplectic and hence the Poisson structure on S is H-invariant (see III.1.2). The proof of the second claim uses the following

Lemma 1.5.6. *Let Φ_1 and Φ_2 be the foliations of S by the level sets of J and by the H-orbits, respectively. Then $\Phi_1^\perp = \Phi_2$ (hence $\Phi_2^\perp = \Phi_1$).*

By the definition III.(1.7) of the momentum map, the distribution $T\Phi_2$ is generated by the vector fields ξ_{J_X}, where X runs through $\mathfrak{h}$. The claim is then obvious from III.(1.8). ■

Corollary 1.5.7. *Under the assumptions of 1.5.4, specializing (1.20) to the case $S/H \overset{\tau}{\leftarrow} S \overset{J}{\rightarrow} \mathfrak{h}_+^*$ satisfies conditions 1.3.4.1–3.*

Continuing the proof of 1.5.5, Lemmas 1.3.5 and 1.5.6 show that $T\Phi_1$ is spanned by Hamiltonian vector fields of the form $\xi_{\tau^* f}$, where $f \in C^\infty(S/H, \mathbb{R})$. Using the equivariance of J, one sees that $\tau(J^{-1}(\theta)) = J^{-1}(\mathcal{O})/H$; cf. (1.25). Since τ is a Poisson map, Proposition I.2.3.5 applies, showing that all Hamiltonian vector fields in S/H are tangent to the subspaces $J^{-1}(\mathcal{O})/H$. It follows that each connected component of the latter must be a symplectic leaf, and it is plain that all leaves are then of this form. ■

For later use we collect more precise information about this situation.

Proposition 1.5.8. *In the situation of 1.5.4 and 1.5.7 the maps τ and J are complete.*

Pick a function $g \in C^\infty(S/H, \mathbb{R})$ with complete Hamiltonian flow $[\sigma]_H(\cdot)$, and take any smooth curve $c(\cdot)$ covering $[\sigma]_H(\cdot)$. For the Hamiltonian flow $\sigma(\cdot)$ of $\tau^* g \in C^\infty(S, \mathbb{R})^H$ through $c(0)$ we make the ansatz $\sigma(t) = x(t)c(t)$, where $x(\cdot)$ is some curve in H starting at the identity. Using the equation of motion I.(2.11) (with h replaced by $\tau^* g$) and the H-invariance of $\xi_{\tau^* h}$, one obtains the equation

$$\xi^f_{x^{-1}\dot{x}} = \dot{c} - \xi_{\tau^* h} \tag{1.27}$$

along $c(\cdot)$; here $\xi^f_X(\sigma) := -d\,\mathrm{Exp}(\lambda X)\sigma/d\lambda|\lambda = 0$ for $X \in \mathfrak{h}$ (cf. III.(2.9)), and we have written $x^{-1}\dot{x}$ for the cumbersome $(L_{x^{-1}})_*\dot{x}$; cf. III.(1.43).

Applying I.2.3.5, one sees that $\tau_*(\dot{c} - \xi_{\tau^* h}) = 0$, so that there exists a curve $X(\cdot)$ in $\mathfrak{h}$ for which $\dot{c} - \xi_{\tau^* h} = \xi^f_X$ along $c(\cdot)$. Comparing with (1.27), we see that $x(t)^{-1}\dot{x}(t) = X(t)$. This can be solved for all t, so that $\sigma(t)$ exists for all t as well. Hence τ is complete.

We turn to the question of the completeness of the momentum map J. Choose a function $f \in C^\infty(\mathfrak{h}^*, \mathbb{R})$ with complete Hamiltonian flow in $\mathfrak{h}^*_-$. Denote the flow of $J^* f$ in S by $\sigma(\cdot)$. According to Lemma 1.3.5, the $J^* \tilde{X}$ generate the group action on S, so we can make the ansatz $\sigma(t) = x(t)\sigma$ for some flow $x(\cdot)$ in H. Applying J to both sides and using its equivariance as well as Proposition I.2.3.5, we obtain $x(t)\theta = \theta(t)$, where $\theta(\cdot)$ is the Hamiltonian flow of f in $\mathfrak{h}^*_-$. Using III.(1.54) (with $\Gamma = 0$) and III.(1.58) we infer that $x(t) = \tau_{T^*H \to H}(\alpha(t))$, where $\alpha(\cdot)$ is the

Hamiltonian flow on T^*H (with its canonical symplectic structure) generated by $(J^L)^*f$.

The momentum map $J^L : T^*H \to \mathfrak{h}^*_-$ for the left action of H on T^*H coincides with the projection $\tau^\rho : T^*H \to (T^*H)/H$, when we identify the quotient $(T^*H)/H$ under the right action with $\mathfrak{h}^*$ (using the right trivialization of T^*H). Hence J^L is complete by the first part of the proposition.

It follows that $x(\cdot)$ and therefore $\sigma(\cdot)$ is defined for all t. ■

We now recognize that the projective Hilbert space $\mathbb{P}\mathcal{H}$, looked at in I.2.5 as a symplectic leaf in the Poisson manifold $\mathcal{H}^*/U(1)$, may alternatively be described as a Marsden–Weinstein quotient. The group $U(1)$ acts on $\mathcal{H}$ by $z : \Psi \mapsto z\Psi$ (where $z \in \mathbb{C}$ with $|z| = 1$). Using I.(2.35), I.(2.37), and III.(1.8), as well as the standing convention $T = -i$ for the single generator T of $\mathfrak{u}(1)$, one easily derives that the momentum map $J = J_T$ for this action is

$$J(\Psi) = (\Psi, \Psi). \tag{1.28}$$

Thus $\mathbb{P}\mathcal{H} \simeq J^{-1}(1)/U(1)$; cf. the closing comment of I.2.5.

To close this section, we remark that the Marsden–Weinstein reduction procedure is easily generalized to the case where the momentum map J is Co^γ-equivariant; cf. III.1.2.5. Instead of a coadjoint orbit in $\mathfrak{h}^*$ we simply take a Co^γ-orbit (cf. III.1.4.4), and proceed as in the Co-equivariant case.

1.6 Kazhdan–Kostant–Sternberg Reduction

We can now see the theory in III.2.3 in the light of Marsden–Weinstein reduction. Indeed, given a principal bundle $\mathsf{P}(Q, H, \tau)$, the hypotheses in 1.5.4 evidently apply to the case $S = T^*\mathsf{P}$, and we infer that the spaces $(\widetilde{T^*\mathsf{P}})^{\mathcal{O}}$ defined in III.(2.51) are nothing but Marsden–Weinstein quotients.

We may generalize the construction of the reduced space $(\widetilde{T^*\mathsf{P}})^{\mathcal{O}}$, in replacing the inclusion $\iota_{\mathcal{O}} : \mathcal{O} \to \mathfrak{h}^*_+$ by a general Poisson map $J_\rho : S_\rho \to \mathfrak{h}^*_+$, where S_ρ is symplectic. Such maps usually come from a strongly Hamiltonian H-action on S_ρ whose associated momentum map is minus J_ρ; we assume that we are indeed in this slightly more special situation. Hence we fill out Figure 1 as in Figure 3, where J is the momentum map for the H-action on $T^*\mathsf{P}$ pulled back from the given action on P, and $\tau^R := \tau_{T^*\mathsf{P}\to(T^*\mathsf{P})/H}$ is the canonical projection. We refer to this situation as (generalized) **Kazhdan–Kostant–Sternberg reduction.**

We may look at this instance of special symplectic reduction as a special case of Marsden–Weinstein reduction: As in the proof of 1.5.4 we take $\tilde{S} := T^*\mathsf{P} \times S_\rho$, equipped with the product H-action. The associated momentum map $\tilde{J} : \tilde{S} \to \mathfrak{h}^*_-$ is simply $\tilde{J} = J - J_\rho$, so that the submanifold $T^*\mathsf{P} *_{\mathfrak{h}^*} S_\rho$ of $T^*\mathsf{P} \times S_\rho$ (defined as in (1.7)) coincides with $\tilde{J}^{-1}(0)$. Hence for connected H the reduced space is

$$(T^*\mathsf{P})^\rho = (T^*\mathsf{P} *_{\mathfrak{h}^*} S_\rho)/H. \tag{1.29}$$

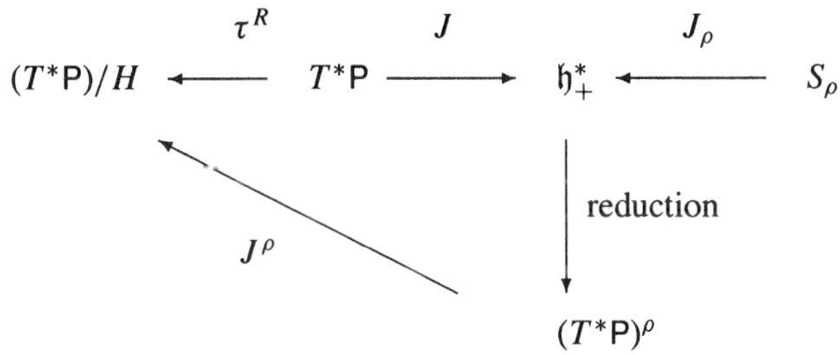

FIGURE 3. Kazhdan–Kostant–Sternberg reduction

We will use the right-hand side as the definition of the reduced space also when H is not connected; as in (1.26) we write the reduced space as

$$(\widetilde{T^*\mathsf{P}})^\rho := (T^*\mathsf{P})^\rho/\pi_0(H). \tag{1.30}$$

Theorem III.2.5.2 then generalizes: The group $\mathcal{G}_\mathsf{P}^\tau$ defined in III.(2.74) acts on $(\widetilde{T^*\mathsf{P}})^\rho$ in strongly Hamiltonian fashion by the obvious generalization of III.(2.75)–III.(2.76). Theorem III.2.3.7 generalizes to

Theorem 1.6.1. *Each connection* $\mathbf{A}$ *on a principal bundle* $\mathsf{P}(Q, H, \tau)$ *defines a projection* $\tau_\mathbf{A} : (\widetilde{T^*\mathsf{P}})^\rho \to T^*Q$ *that makes* $(\widetilde{T^*\mathsf{P}})^\rho$ *a bundle over* T^*Q *with typical fiber* S_ρ.

Using the ($\mathbf{A}$-dependent) factorization III.(2.47), we denote points of $T^*\mathsf{P}$ by triples (x, σ, θ); recall that $x \in \mathsf{P}$, $\sigma \in T^*Q$, and $\theta \in \mathfrak{h}^*$, with the constraint $\tau_{T^*Q\to Q}(\sigma) = \tau_{\mathsf{P}\to Q}(x)$. From III.(2.48) and Definition III.2.3.2 we have

$$(\widetilde{T^*\mathsf{P}})^\rho \simeq (\mathsf{P} *_Q T^*Q \times S_\rho)/H, \tag{1.31}$$

where the H-action is $h : (x, \sigma, \alpha) \mapsto (xh^{-1}, \sigma, h\alpha)$. Hence $\tau_\mathbf{A}(x, \sigma, \alpha) := \sigma$ is the desired projection. ■

Theorem 1.4.1 leads to the following classical imprimitivity-type theorem.

Theorem 1.6.2. *Let* P *and* H *be connected and simply connected. Given a complete Poisson map* $J_\rho : S_\rho \to \mathfrak{h}^*_+$, *one obtains a strongly Hamiltonian* H*-action on* S_ρ *and a complete Poisson map* J^ρ *from the corresponding reduced space* $(T^*\mathsf{P})^\rho$ *to* $(T^*\mathsf{P})/H$.

Conversely, for any complete Poisson map $J : S \to (T^*\mathsf{P})/H$ *(where* S *is symplectic) there exists a symplectic manifold* S_ρ *and a strongly Hamiltonian* H*-action on* S_ρ, *with momentum map* $-J_\rho$, *such that* S *is symplectomorphic to the reduced space* $(T^*\mathsf{P})^\rho$.

This correspondence is bijective.

By Corollary 1.5.7, the diagram $(T^*\mathsf{P})/H \xleftarrow{\tau^R} T^*\mathsf{P} \xrightarrow{J} \mathfrak{h}^*_+$ satisfies conditions 1, 2, 3, and 6 in 1.3.4; the maps τ^R and J are complete by Proposition 1.5.8. When P and H are connected and simply connected, we see from III.(2.48) and from the fact that the level sets of τ^R are the H-orbits (as the H-action is free) that the level sets of J and τ^R are connected and simply connected. Hence we have a classical dual pair, and Theorem 1.4.1 applies.

This almost leads to the theorem: Rather than H-actions one obtains $\mathfrak{h}$-actions. However, the $\mathfrak{h}$-action on $(T^*\mathsf{P})^\rho$ is integrable, because it is derived from an integrable $\mathfrak{h}$-action on $T^*\mathsf{P}$ (namely the H-action pulled back from the given H-action on P). At the other side, the $\mathfrak{h}$-action on S_ρ is integrable by Theorem III.1.2.1, for H is simply connected and J_ρ is complete. ■

We now take $\mathsf{P} = G = H$ in the bundle $\mathsf{P}(Q, H, \tau)$, where G is a Lie group, so that the base is a point. Recall the momentum maps J^L and J^R for the left and the right action of G on T^*G, respectively; cf. III.1.4, and in particular III.(1.55)–III.(1.58). The top line

$$(T^*\mathsf{P})/H \xleftarrow{\tau^R} T^*\mathsf{P} \xrightarrow{J} \mathfrak{h}^*_+ \tag{1.32}$$

of Figure 3 then specializes to

$$\mathfrak{g}^*_- \xleftarrow{J^L} T^*G \xrightarrow{J^R} \mathfrak{g}^*_+, \tag{1.33}$$

since the momentum map $J^L : T^*G \to \mathfrak{g}^*_-$ for the left action of G on T^*G coincides with the projection $\tau^R : T^*G \to (T^*G)/G$, when we identify the quotient $(T^*G)/G$ under the right action with $\mathfrak{g}^*$; cf. the proof of 1.5.8.

The structure of (1.33) is illuminated by an easy calculation.

Proposition 1.6.3. *Pick a coadjoint orbit $\mathcal{O}_+$ in $\mathfrak{g}^*_+$, and specialize Figure 3 to $\mathsf{P} = H = G$ and $S_\rho = \mathcal{O}$, with $J_\rho = \iota_\mathcal{O}$ the inclusion map.*

The reduced space $(\widetilde{T^*G})^{\mathcal{O}_+}$ is symplectomorphic to $\mathcal{O}_+$, and $J^\mathcal{O}$ is equivalent to $-\iota_\mathcal{O} : \mathcal{O}_+ \to \mathfrak{g}^*_-$.

Using the left trivialization, we identify T^*G with $\mathfrak{g}^*_+ \times G$, with Poisson bracket III.(3.101); cf. Proposition III.3.9.9. From III.(1.55) we then obtain

$$T^*G *_{\mathfrak{g}^*_+} \mathcal{O} = \{(-\theta, y, \theta) | \, y \in G, \, \theta \in \mathcal{O}\}. \tag{1.34}$$

The reduced space $(\widetilde{T^*G})^\mathcal{O}$ is the orbit space of the G-action $x : (-\theta, y, \theta) \mapsto (-\mathrm{Co}(x)\theta, yx^{-1}, \mathrm{Co}(x)\theta)$; cf. III.(1.49). One easily sees that the desired symplectomorphism from $(\widetilde{T^*G})^{\mathcal{O}_+}$ to $\mathcal{O}_+$ is given by $[-\theta, y, \theta]_G \mapsto y\theta$, so that $[-y^{-1}\theta, y, y^{-1}\theta]_G \mapsto \theta$.

From (1.17) one has $J^\mathcal{O}([-y^{-1}\theta, y, y^{-1}\theta]_G) = J^L(-y^{-1}\theta, y)_L$; on account of III.(1.56), this equals $-\theta$. ■

We now assume that the bundle $\mathsf{P}(Q, H, \tau)$ is of the form $G(G/H, H, \tau)$, where G is a Lie group with closed subgroup H acting on G from the right; see III.2.7. This time we use the right trivialization $T^*G \simeq \mathfrak{g}^*_- \times G$, so that (1.32) becomes

$$\mathfrak{g}^*_- \times G/H \xleftarrow{\tau^R} \mathfrak{g}^*_- \times G \xrightarrow{J^R_{|\mathfrak{h}}} \mathfrak{h}^*_+; \tag{1.35}$$

compare with Proposition III.3.9.9. Here

$$\tau^R(\theta, x) = (\theta, [x]_H); \tag{1.36}$$

$$J^R_{\restriction\mathfrak{h}}(\theta, x) = -(\mathrm{Co}(x^{-1})\theta) \restriction \mathfrak{h}, \tag{1.37}$$

where $[x]_H$ is the coset $xH \in G/H$, and J^R is the momentum map for the H-action on T^*G from the right; see III.(1.57). As always, the restriction $J^R_{\restriction\mathfrak{h}}$ of J^R to $\mathfrak{h}$ is just the momentum map for the right action of H on T^*G. Writing $\tau^R = (\tau^R_{(1)}, \tau^R_{(2)})$, we see from III.(1.58) that $\tau^R_{(1)} : T^*G \to \mathfrak{g}^*_-$ equals the momentum map of the left action of G on T^*G, that is,

$$\tau^R_{(1)} = J^L. \tag{1.38}$$

Given a strongly Hamiltonian H-space S_ρ, with momentum map $-J_\rho : S_\rho \to \mathfrak{h}^*_-$, the reduced space $(\widetilde{T^*G})^\rho$ (cf. (1.26) or (1.30)) consists of equivalence classes $[\theta, x, \alpha]_H$, where

$$J_\rho(\alpha) = -(\mathrm{Co}(x^{-1})\theta) \restriction \mathfrak{h}, \tag{1.39}$$

and the H-action is given by (cf. III.(1.51))

$$h : (\theta, x, \alpha) \mapsto (\theta, xh^{-1}, h\alpha). \tag{1.40}$$

Since $G \subset \mathrm{Aut}(G) \subset \mathcal{G}^\tau_G$ (cf. the paragraphs following III.(2.118) and (1.30)), one obtains a "reduced" action λ^ρ of G on the reduced space $(\widetilde{T^*G})^\rho$. By III.(1.52) the explicit form of this action is

$$\lambda^\rho_x([\theta, y, \alpha]_H) = [\mathrm{Co}(x)\theta, xy, \alpha]_H; \tag{1.41}$$

this generalizes III.(2.121). By III.(1.58) the momentum map for λ^ρ is

$$J^\rho_{(1)}([\theta, y, \alpha]_H) = \theta. \tag{1.42}$$

Given (1.40), this is clearly well-defined, generalizing III.(2.123). See Figure 4, in which $J^\rho_R = -J^\rho_{(1)}$. The minus sign in front of J^L is explained by Proposition 1.6.3: It changes $-\iota_\mathcal{O}$ to $\iota_\mathcal{O}$.

Specializing to the case where S_ρ is a coadjoint orbit $\mathcal{O}_+$ in $\mathfrak{h}^*_+$ and $J_\rho = \iota_\mathcal{O}$ is the inclusion map, we recover the reduced space $(\widetilde{T^*G})^{\mathcal{O}_+}$ already encountered in

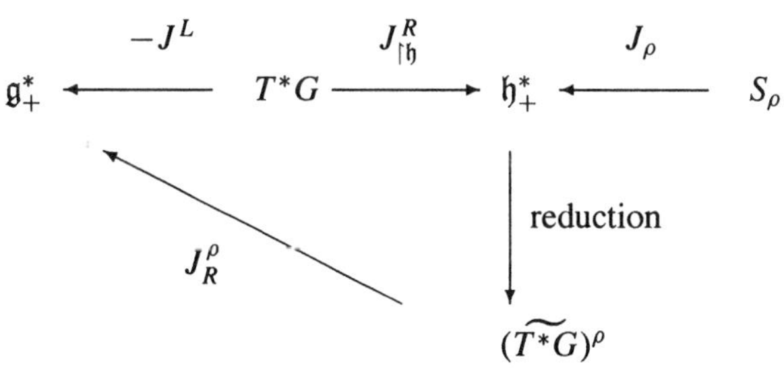

FIGURE 4. The reduced space $(\widetilde{T^*G})^\rho$

III.(2.118). From our present perspective one has

$$(\widetilde{T^*G})^{\mathcal{O}_+} := (J^R_{\restriction\mathfrak{h}})^{-1}(\mathcal{O})/H \simeq (T^*G)^{\mathcal{O}_+}/\pi_0(H). \tag{1.43}$$

Recall Definition III.3.9.7. A classical system of imprimitivity for a given Lie group G is called **transitive** when the G-action on Q is transitive; hence $Q = G/H$ for some (closed) subgroup $H \subset G$. Theorem 1.6.2 now has a refinement, in which the special assumptions on $\mathsf{P} = G$ and H are dropped. In particular, we do not assume that G or H is connected. The following **classical transitive imprimitivity theorem** is the classical analogue of Corollary III.3.7.6.

Theorem 1.6.4. *Let S_ρ carry a strongly Hamiltonian H-action. Then the Kazhdan–Kostant–Sternberg reduced space $(\widetilde{T^*G})^\rho$ carries a (transitive) classical system of imprimitivity for G on G/H.*

Conversely, each symplectic manifold carrying a transitive classical system of imprimitivity for G on G/H is a reduced space of the type $(\widetilde{T^*G})^\rho$ (up to a G-equivariant symplectomorphism), and the correspondence $S_\rho \leftrightarrow (\widetilde{T^*G})^\rho$ thus achieved is bijective.

1.7 Proof of the Classical Transitive Imprimitivity Theorem

This section contains the proof of Theorem 1.6.4.

Since the proof of 1.4.1 as it stands breaks down without the connectedness assumptions, we seek a modification of the construction of the respective reduced spaces. We have already replaced the reduced space $(T^*G)^\rho$ as originally defined by the Marsden–Weinstein quotient $(\widetilde{T^*G})^\rho \simeq (T^*G)^\rho/\pi_0(H)$.

When H is connected, Corollary 1.2.3 yields a complete Poisson map J^ρ : $(T^*G)^\rho \to (T^*G)/H \simeq \mathfrak{g}^*_- \times G/H$. Using the explicit description of $(T^*G)^\rho$ in the previous section, as well as (1.17) and (1.36), this map is explicitly given by

$$J^\rho([\theta, x, \alpha]_H) = \tau^R(\theta, x, \alpha) = (\theta, [x]_H). \tag{1.44}$$

One sees from (1.40) that this map is indeed well-defined, and remains so when H is no longer connected. Hence we obtain (using the same notation) a complete Poisson map $J^\rho : (\widetilde{T^*G})^\rho \to \mathfrak{g}^*_- \times G/H$.

Splitting J^ρ into $J^\rho_{(1)} : (\widetilde{T^*G})^\rho \to \mathfrak{g}^*_-$ and $J^\rho_{(2)} : (\widetilde{T^*G})^\rho \to G/H$, we see that $J^\rho_{(1)}$ is the momentum map for the G-action λ^ρ on $(\widetilde{T^*G})^\rho$ specified in (1.41). In particular, it follows that the strongly Hamiltonian $\mathfrak{g}$-action on $(\widetilde{T^*G})^\rho$ defined by $J^\rho_{(1)}$ is integrable. Thus one obtains a classical system of imprimitivity on $(\widetilde{T^*G})^\rho$ defined by the G-action λ^ρ and the representation $\tilde{\pi} = (J^\rho_{(2)})^*$ of $C^\infty(G/H, \mathbb{R})$; the covariance property is easily verified from (1.44) and (1.41).

Reducing in the opposite direction, we will exploit the following insight.

Lemma 1.7.1. *Consider the situation in Figure 5, where*

$$J = (J_{(1)}, J_{(2)}) : S \to \mathfrak{g}^*_- \times G/H$$

is a complete Poisson map corresponding to a given classical system of imprimitivity of G on G/H. The space $\mathfrak{g}^*_+ \times G$ is T^*G^- in the right trivialization.

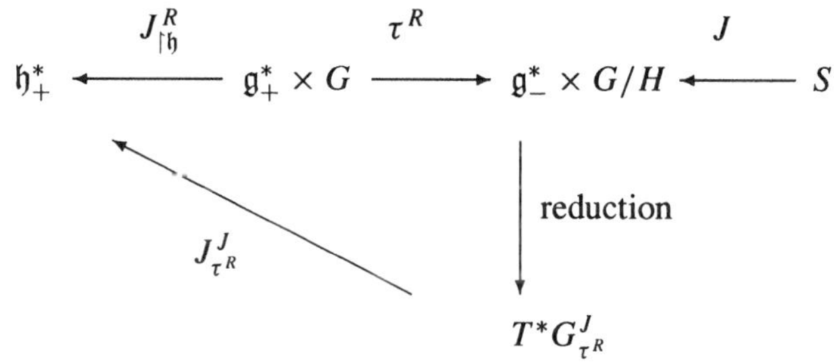

FIGURE 5. $T^*G^J_{\tau^R} \simeq J^{-1}_{(2)}([e]_H)/\mathcal{N}_{G/H}$ and $J^J_{\tau^R} \simeq [-J_{\restriction\mathfrak{h}}]_H$

When G is connected, the reduced space $T^*G^J_{\tau^R}$ is a manifold that is symplectomorphic to

$$\widehat{T^*G^J_{\tau^R}} := J^{-1}_{(2)}([e]_H)/\Phi_{G/H}, \tag{1.45}$$

where $\Phi_{G/H}$ is the foliation of S generated by the distribution $\mathcal{N}_{G/H}$ spanned by all Hamiltonian vector fields of the form $\xi_{J^*_{(2)}\tilde{f}}$, where $\tilde{f} \in C^\infty(G/H, \mathbb{R})$.

Moreover, minus the reduced Poisson map $-J^J_{\tau^R} : T^*G^J_{\tau^R} \to \mathfrak{h}^*_-$ is equivalent to the well-defined quotient

$$J_{(12)} := [J_{(1)\restriction\mathfrak{h}} \restriction J^{-1}_{(2)}([e]_H)]_{\Phi_{G/H}} \tag{1.46}$$

of the momentum map $J_{(1)} \restriction \mathfrak{h}$ of the restriction of the G-action on S to H.

The reduced space $T^*G^J_{\tau^R}$ by definition consists of equivalence classes $[\theta, x, \sigma]_\Phi$, where θ is determined by $J_{(1)}(\sigma) = \theta$, and x is constrained by $J_{(2)}(\sigma) = [x]_H$.

Recall that Φ is the foliation defined by all vector fields $\xi_{(\tau^R)^*f} - \xi_{J^*f}$, where $f \in C^\infty(\mathfrak{g}^* \times G/H, \mathbb{R})$. Since G is connected, the distribution spanned by these vector fields is simply the sum of $\mathcal{N}_{G/H}$ and the vectors tangent to the G-orbits of the action $x : (\theta, y, \sigma) \mapsto (\mathrm{Co}(x)\theta, xy, x\sigma)$. This action is derived from (1.38) and III.(1.52). By III.(1.54) in the right trivialization, the flow of $\xi_{(\tau^R)^*f}$ leaves x in $(\theta, x) \in T^*G$ inert.

The constraints on (θ, x, σ) and the G-covariance of $J_{(2)}$ (which holds by definition of a classical system of imprimitivity) imply that $J_{(2)}(x^{-1}\sigma) = [e]_H$. It is then easily verified that the map $[\theta, x, \sigma]_\Phi \mapsto [x^{-1}\sigma]_{\Phi_{G/H}}$ defines a symplectomorphism from $T^*G^J_{\tau^R}$ to $J^{-1}_{(2)}([e]_H)/\Phi_{G/H}$ (given that these spaces are manifolds, see below). Note that $J^{-1}_{(2)}([e]_H)$ is indeed stable under the Hamiltonian flow of any $J^*_{(2)}\tilde{f}$, $\tilde{f} \in C^\infty(G/H, \mathbb{R})$. Denoting this flow by $\sigma(\cdot)$, we use Proposition I.2.3.5 to compute

$$\frac{d}{dt} J_{(2)}\sigma(t) = \frac{d}{dt} q(t) = 0,$$

where $q(t)$ is the Hamiltonian flow of $\tilde{f}$ in G/H. Since $C^\infty(G/H, \mathbb{R})$ is commutative as a Poisson algebra, one has $q(t) = q(0)$ for any initial condition.

By (1.17), III.(1.57), and the equivariance of $J_{(1)}$ we have

$$J^J_{\tau^R}([\theta, x, \sigma]_\Phi) = J^R(\theta, x) \restriction \mathfrak{h} = -J_{(1)}(x^{-1}\sigma) \restriction \mathfrak{h}. \tag{1.47}$$

Hence under the above symplectomorphism, under which $J^J_{\tau^R}$ is transformed to, say, $\tilde{J}^J_{\tau^R}$, we find that $\tilde{J}^J_{\tau^R}([\sigma]_H) = -J_{(1)}(\sigma) \restriction \mathfrak{h}$, where $\sigma \in J^{-1}_{(2)}([e]_H)$.

It is instructive to verify that $J_{(1)} \restriction \mathfrak{h}$, restricted to $J^{-1}_{(2)}([e]_H)$, is constant on the leaves of the foliation $\Phi_{G/H}$. Picking $X \in \mathfrak{h}$ and $\tilde{f} \in C^\infty(G/H, \mathbb{R})$ we compute, using the notation in the paragraph before the last,

$$\frac{d}{dt} J_{(1)X}(\sigma(t)) = \{J^*_{(2)}\tilde{f}, J_{(1)X}\}(\sigma(t)) = -\xi^S_X J^*_{(2)}\tilde{f}(\sigma(t)),$$

where we have used the antisymmetry of the Poisson bracket and III.(1.7). By I.2.3.5 (as above) and the equivariance of $J_{(2)}$ under the action of G, and hence of H, this equals $-\xi^{G/H}_X f(q(0))$, where, for general $X \in \mathfrak{g}$, the vector field $\xi^{G/H}_X$ is defined by the left action of G on G/H. This vanishes, since $q(0) = [e]_H$ and $X \in \mathfrak{h}$.

Finally, $J^{-1}_{(2)}([e]_H)/\Phi_{G/H}$ is a manifold. It is immediate from its equivariance that $J_{(2)}$ is a surjective submersion, so that $J^{-1}_{(2)}([x]_H)$ is a submanifold of S for any $x \in G$ (this is, of course, consistent with the fact that $T^*G *_{(T^*G)/H} S$ is a submanifold of $T^*G \times S$, since τ^R is a surjective submersion; cf. the proof of Theorem 1.2.2).

It follows from I.(2.9), the fact that $J_{(2)}$ is a Poisson map, and the commutativity of the Poisson algebra $C^\infty(G/H, \mathbb{R})$ that $\xi_{J^*_{(2)}\tilde{f}} \mapsto d\tilde{f}([e]_H)$ is an isomorphism between the abelian Lie algebra spanned by the $\xi_{J^*_{(2)}\tilde{f}}$, restricted to $J^{-1}_{(2)}([e]_H)$, and the vector space $T^*_{[e]_H}(G/H)$. Under this isomorphism the foliation $\Phi_{G/H}$ of $J^{-1}_{(2)}([e]_H)$ is given by the orbits of a Lie group that acts freely and properly; hence the quotient space is a manifold. ■

In view of this lemma, we declare that even when G is not connected, the reduced space defined by the triple

$$T^*G^- \xrightarrow{\tau^R} (T^*G)/H \xleftarrow{J} S \tag{1.48}$$

is $\widehat{T^*G}^J_{\tau^R}$ rather than $T^*G^J_{\tau^R}$. As in the proof of Lemma 1.7.1, one obtains a Poisson map $J_{(1)} \restriction \mathfrak{h} : \widehat{T^*G}^J_{\tau^R} \to \mathfrak{h}^*_-$, which is complete when J is.

Following the proof of Theorem 1.4.1 (with J_1 and J_2 interchanged) we now construct the reduced space depicted in Figure 6. Here $\mathfrak{g}^*_- \times G/H$ is $(T^*G)/H$ in the right trivialization, whereas $\mathfrak{g}^*_+ \times G$ is T^*G in the left trivialization. The tilde in the name $\widetilde{\mathfrak{g}^*_+ \times G}^{(12)}_{J^R_{\restriction\mathfrak{h}}}$ of the reduced space has the same significance as in (1.30).

Elements of this reduced space have the form $[\theta, x, [\sigma]_{\Phi_{G/H}}]_H$, where $\theta \in \mathfrak{g}^*$, $x \in G$, and $[\sigma]_{\Phi_{G/H}} \in \widehat{T^*G}^J_{\tau^R}$ (where $\sigma \in S$) are constrained by

$$J_{(1)}(\sigma) \restriction \mathfrak{h} = \theta \restriction \mathfrak{h}, \tag{1.49}$$

$$J_{(2)}(\sigma) = [e]_H. \tag{1.50}$$

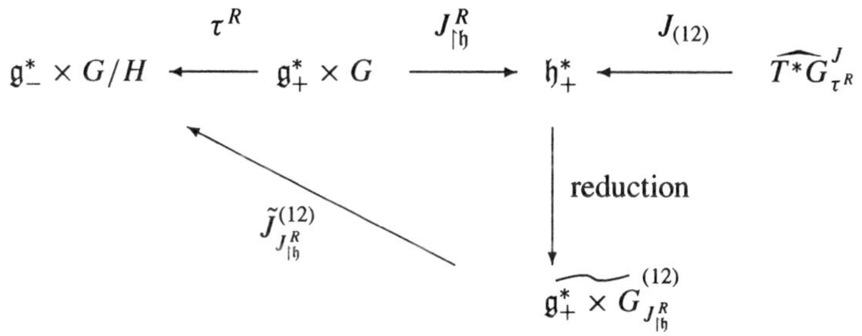

FIGURE 6. Classical imprimitivity: $\widetilde{\mathfrak{g}^*_+ \times G}^{(12)}_{J^R_{\restriction\mathfrak{h}}} \simeq S$ and $\tilde{J}^{(12)}_{J^R_{\restriction\mathfrak{h}}} \simeq J$

The H-action defining the orbit space is given by

$$h : (\theta, x, [\sigma]_{\Phi_{G/H}}) \mapsto (\mathrm{Co}(h)\theta, xh^{-1}, [h\sigma]_{\Phi_{G/H}}); \tag{1.51}$$

see III.(1.49). Equation (1.49) has the implication that there exists a unique lift $\sigma_\theta \in S$ of $[\sigma]_{\Phi_{G/H}}$ satisfying $J_{(1)}(\sigma_\theta) = \theta$. This follows by regarding $\tilde{f} \in C^\infty(G/H, \mathbb{R})$ as a function on $\mathfrak{g}^*_- \times G/H$, so that $J^*\tilde{f} = J^*_{(2)}\tilde{f}$; according to Proposition I.2.3.5, the flow $\sigma(t)$ of $J^*\tilde{f}$ on S projects to the flow of $\tilde{f}$ on $\mathfrak{g}^*_- \times G/H$ under J. Using III.(1.54) with "L" and III.(1.37), one sees that the variable in G/H as well as $\theta \restriction \mathfrak{h}$ remain fixed, whereas any θ with given restriction to $\mathfrak{h}$ can be reached with such a flow. This implies the existence of σ_θ; uniqueness follows, for example, from the argument at the end of the proof of Lemma 1.7.1.

Consider the map $\varphi : \widetilde{\mathfrak{g}^*_+ \times G}^{(12)}_{J^R_{\restriction\mathfrak{h}}} \to S$, defined by

$$\varphi([\theta, x, [\sigma]_{\Phi_{G/H}}]_H) = x\sigma_\theta. \tag{1.52}$$

This is clearly independent of the particular point in the H-orbit, and a dimension count shows that φ is locally a diffeomorphism. To prove that φ is injective, assume that $x\sigma_\theta = x'\sigma'_{\theta'}$. Equation (1.50) implies that $x' = xh^{-1}$ for some $h \in H$; applying $J_{(1)}$ to both sides and using the G-equivariance of this map leads to $\theta' = h\theta$, and applying $J_{(2)}$ finally shows that the primed variables are related to the unprimed ones by the H-action (1.51). Hence φ is injective. The G-covariance of $J_{(2)}$ implies that φ is surjective; hence φ is a diffeomorphism. The definition of the Poisson bracket on $\widetilde{\mathfrak{g}^*_+ \times G}^{(12)}_{J^R_{\restriction\mathfrak{h}}}$ and the fact that G acts on S by Poisson maps easily imply that φ is a Poisson map. In conclusion, φ is a symplectomorphism.

By (1.17) and III.(1.48) the map $\tilde{J}^{(12)}_{J^R_{\restriction\mathfrak{h}}}$ in Figure 6 is given by

$$\tilde{J}^{(12)}_{J^R_{\restriction\mathfrak{h}}}([\theta, x, [\sigma]_{\Phi_{G/H}}]_H) = \tau^R(\theta, x)_L = (\mathrm{Co}(x)\theta, [x]_H). \tag{1.53}$$

Using (1.49), (1.50), and the G-equivariance of both $J_{(1)}$ and $J_{(1)}$, one sees that the right-hand side of (1.53) equals $J(x\sigma_\theta)$. Hence by (1.52) the map φ intertwines $\tilde{J}^{(12)}_{J^R_{\restriction\mathfrak{h}}}$ and J.

This proves Theorem 1.6.4, except for the bijectivity claim. To prove this claim we have to show that when we start from a Poisson map $J_\rho : S_\rho \to \mathfrak{h}^*_+$, construct $J^\rho : (\widetilde{T^*G})^\rho \to \mathfrak{g}^*_- \times G/H$, and subsequently put $S = (\widetilde{T^*G})^\rho$ and $J = J^\rho$ in (1.48), the ensuing reduced space is symplectomorphic to S_ρ, with $J_{(1)} \upharpoonright \mathfrak{h}$ being equivalent to $-J_\rho$.

We use the description of S given around (1.40). One has

$$J^\rho_{(2)}([\theta, x, \alpha]_H) = [x]_H, \tag{1.54}$$

from which we see that the constraint $J^\rho_{(2)} = [e]_H$ forces $x \in H$. Hence we can label points in $(J^\rho_{(2)})^{-1}([e]_H)$ by (θ, e, α); by (1.39) one has

$$J_\rho(\alpha) = -\theta \upharpoonright \mathfrak{h}. \tag{1.55}$$

Once again using the argument at the end of the proof of Lemma 1.7.1, we conclude that the map $[\theta, e, \alpha]_{\Phi_{G/H}} \mapsto \alpha$ from $(J^\rho_{(2)})^{-1}([e]_H)/\Phi_{G/H}$ to S_ρ is a symplectomorphism. Using the final claim in Lemma 1.7.1 (and the comment ending the paragraph following the lemma) and (1.42), we infer from (1.55) that this symplectomorphism intertwines $J_{(1)} \upharpoonright \mathfrak{h}$ and $-J_\rho$, so that it intertwines the corresponding integrated H-actions as well.

This finishes the proof of 1.6.4. ∎

1.8 Reduction in Stages

For the time being we return to the setting of special symplectic reduction in order to prove a theorem on (special) **symplectic reduction in stages.** The statement appears to be rather complicated, but the thrust of the result will become obvious when the application to Marsden–Weinstein reduction is considered.

Let $S \xrightarrow{J} P \xleftarrow{J_\rho} S_\rho$ be as specified in 1.2.1; since there will be quite a few spaces and maps in what follows, we relabel these objects as $S_1 \xrightarrow{J_{11}} P_1 \xleftarrow{J_\rho} S_\rho$. Now assume that S_ρ is itself a reduced space; this means that there are data $S_2 \xrightarrow{J_{23}} P_3 \xleftarrow{J_{33}} S_3$ as in 1.2.1, so that $S_\rho = S^{33}_{23} := S^{J_{33}}_{J_{23}}$. The right-hand side is, of course, defined as in (1.13), and in what follows one could replace equality by symplectomorphism. In addition, we require that there be a map $J_{21} : S_2 \to P_1$ for which the reduced Poisson map $J^{33}_{23} : S^{33}_{23} \to P_1$ (defined as in (1.17)) coincides with $J_\rho : S_\rho \to P_1$. The reduced space S^ρ is consequently called $S^{^{33}_{23}}_{11}$.

Since in particular we are given Poisson maps $J_{11} : S_1 \to P_1^-$ and $J_{21} : S_2 \to P_1$, we are in the setting of Definition 1.2.1 for the third time, now with data $S_2 \xrightarrow{J_{21}} P_1^- \xleftarrow{J_{11}} S_1$. Theorem 1.2.2 then leads to a reduced symplectic space that we denote by S^{11}_{21}.

Theorem 1.8.1. *With the above notation:*

1. *There is a Poisson map $J^{11}_{21} : S^{11}_{21} \to P_3^-$. We are in the setting of 1.2.1 and 1.2.2 for the fourth time, with data $S^{21}_{11} \xrightarrow{J_{11}} P_3 \xleftarrow{J_{33}} S_3$.*

2. *The resulting reduced space* $S^{33}_{\substack{11\\21}}$ *is symplectomorphic to* $S^{\substack{33\\23}}_{11}$.
3. *Suppose one has a Poisson manifold* P_2 *and a Poisson map* $J_{12} : S_1 \to P_2$, *such that* $J_{12}^* C^\infty(P_2, \mathbb{R}) \subseteq J^* C^\infty(P_1, \mathbb{R})'$. *By 1.2.3 this leads to a Poisson map* $J^{\substack{33\\23}}_{11} : S^{\substack{33\\23}}_{11} \to P_2$.
   There exists a Poisson map $J^{33}_{\substack{11\\21}} : S^{33}_{\substack{11\\21}} \to P_2$ *that is equivalent to* $J^{\substack{33\\23}}_{11}$ *(under the symplectomorphism mentioned above).*

See Figure 7.

The Poisson map J^{11}_{21} is constructed as in Figure 1, with (clockwise) P_2, J_2, S, J_2, P_1, J_ρ, S_ρ, S^ρ_1, and J^ρ replaced by P_3, J_{23}, S_2^-, J_{21}, P_1, J_{11}, S_1, S^{11}_{21}, and J^{11}_{21}, respectively.

The symplectomorphism $S^{33}_{\substack{11\\21}} \simeq S^{\substack{33\\23}}_{11}$ is a consequence of the fact that both spaces are symplectomorphic to $(S_1 *_{P_1} S_2 *_{P_3} S_3)/\Phi$. Here $S_1 *_{P_1} S_2 *_{P_3} S_3$ consists of those triples (x, y, z) in $S_1 \times S_2 \times S_3$ for which $J_{11}(x) = J_{21}(y)$ and $J_{23}(y) = J_{33}(z)$, and Φ is the foliation generated by the distribution spanned by all vector fields of the form $J_{11}^* f - J_{21}^* f + J_{23}^* g - J_{33}^* g$, where $f \in C^\infty(P_1, \mathbb{R})$ and $g \in C^\infty(P_3, \mathbb{R})$. This follows from the definition of the reduced space, as well as from the equations (in self-explanatory notation) $J^{33}_{23}([y, z]_3) = J_{21}(y)$ and $J^{11}_{21}([x, y]_1) = J_{23}(y)$; cf. (1.17).

To define $J^{33}_{\substack{11\\21}}$ we notice that there is a Poisson map $J^{21}_{11} : S^{11}_{21} \to P_2$ that is obtained by reinterpreting the data $S_2 \xrightarrow{J_{21}} P_1^- \xleftarrow{J_{11}} S_1$ as $S_1 \xrightarrow{J_{11}} P_1 \xleftarrow{J_{21}} S_2$, and subsequently looking at S^{11}_{21} as S^{21}_{11}. One is then back at Figure 1, with the clockwise data listed a paragraph ago replaced by P_2, J_{12}, S_1, J_{11}, P_1, J_{21}, S_2, S^{21}_{11}, and J^{21}_{11}. The Poisson map $J^{33}_{\substack{11\\21}}$ is then constructed once again as in Figure 1, this time with entries P_2, J^{21}_{11}, S^{11}_{21}, J^{11}_{21}, P_3, J_{33}, S_3, $S^{33}_{\substack{11\\21}}$, and $J^{33}_{\substack{11\\21}}$. ∎

Thus the theorem is in essence a consequence of the associativity of the fiber product, namely,

$$(S_1 *_{P_1} S_2) *_{P_3} S_3 = S_1 *_{P_1} (S_2 *_{P_3} S_3). \tag{1.56}$$

We apply Theorem 1.8.1 to the Marsden–Weinstein case. Given a strongly Hamiltonian H-space S_ρ, with equivariant momentum map $-J_\rho : S_\rho \to \mathfrak{h}^*_-$, the reduced space $(\widetilde{T^*G})^\rho$ has been defined in 1.6; cf. Figure 4.

Given a strongly Hamiltonian G-space S, with equivariant momentum map $J : S \to \mathfrak{g}^*_-$, we specialize (1.8) to $S \xrightarrow{J} \mathfrak{g}^*_+ \xleftarrow{J^\rho_R} (\widetilde{T^*G})^\rho$, denoting the reduced space defined by these data by

$$\tilde{S}^{\rho}_{J}{}^{\!R} := (S *_{\mathfrak{g}^*} (\widetilde{T^*G})^\rho)/G \simeq S^{\rho}_{J}{}^{\!R}/\pi_0(G). \tag{1.57}$$

Here the G-action on $S \times (\widetilde{T^*G})^\rho$ is the product of the given action and the action λ^ρ given in (1.41), and $S^{\rho}_{J}{}^{\!R}$ is defined as in (1.13), where the foliation Φ coincides with the foliation by the orbits of the G^0-action.

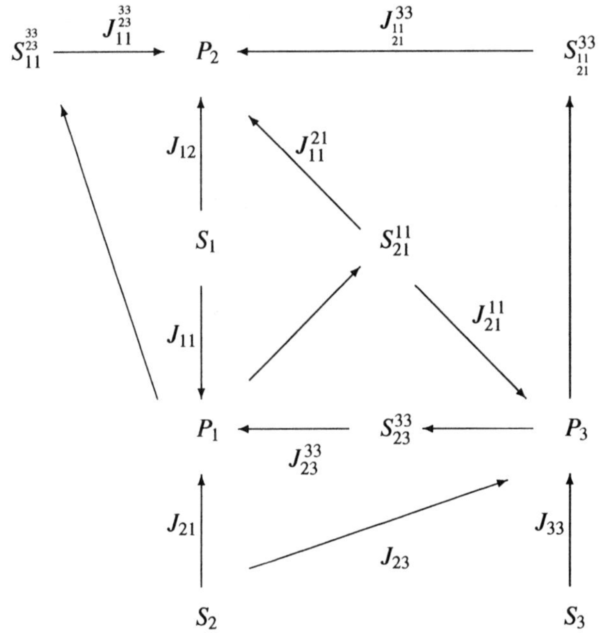

FIGURE 7. Reduction in stages: $S^{33}_{\underset{21}{11}} \simeq S^{33}_{\overset{23}{11}}$ and $J^{33}_{\underset{21}{11}} \simeq J^{33}_{\overset{23}{11}}$

Instead of T^*G one may consider an arbitrary strongly Hamiltonian H-space S with equivariant momentum map J_H, and specialize the diagram in (1.8) to $S \xrightarrow{J_H} \mathfrak{h}^*_+ \xleftarrow{J_\rho} S_\rho$. This leads to a reduced space

$$\tilde{S}^\rho_{J_H} := (S *_{\mathfrak{h}^*} S_\rho)/H \simeq S^\rho_{J_H}/\pi_0(H), \tag{1.58}$$

whose elements are equivalence classes $[\sigma, \alpha]_H$, constrained by $J_H(\sigma) = J_\rho(\alpha)$, where the equivalence relation on $S \times S_\rho$ is given by the orbits of the H-action $h : (\sigma, \alpha) \mapsto (h\sigma, h\alpha)$. Compare with (1.13).

Theorem 1.8.2. *Suppose one has a strongly Hamiltonian G-action on a symplectic manifold S, with equivariant momentum map J, as well as a strongly Hamiltonian H-space S_ρ, with equivariant momentum map $-J_\rho$.*

The reduced space $\tilde{S}^{\tilde{\rho}}_J$ defined in (1.57) is symplectomorphic to the space $\tilde{S}^\rho_{J\restriction\mathfrak{h}}$ defined in (1.58), where $J_H = J_{\restriction\mathfrak{h}}$ is the momentum map for the H-action on S obtained by restricting the given G-action.

Consider the specialization of Figure 1 shown in Figure 8.

Lemma 1.8.3. *In Figure 8 one has the symplectomorphism $\widetilde{T^*G}^J_{-J^L} \simeq S$ that leads to the equivalence $J^J_{-J^L} \simeq J_{\restriction\mathfrak{h}}$.*

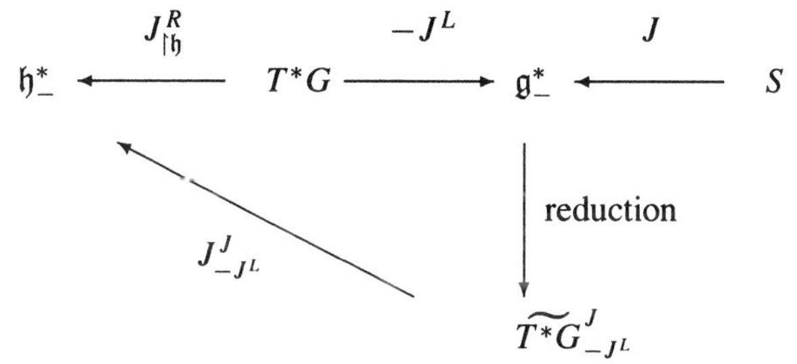

FIGURE 8. $\widetilde{T^*G}^{J}_{-J^L} \simeq S$ and $J^{J}_{-J^L} \simeq J_{\restriction\mathfrak{h}}$

Using (1.7) and III.(1.58), one sees that in the right trivialization of T^*G one has

$$T^*G *_{\mathfrak{g}^*} S = \{(-J(\sigma), y, \sigma) | \sigma \in S, y \in G\}.$$

By 1.2.2, specifically (1.9), and III.(1.52), for connected G the null foliation Φ of this space coincides with the foliation by the orbits of the G-action $x : (\theta, y, \sigma) \mapsto (\text{Co}(x)\theta, xy, x\sigma)$. This also defines the equivalence classes for general G. The equivariance of J and the fact that the G-action preserves the symplectic structure of S then imply that the map $[(-J(\sigma), y, \sigma)]_{\Phi} \mapsto y^{-1}\sigma$ is a symplectomorphism.

Choose $H = G$. Using (1.17), and subsequently III.(1.57) and the equivariance of J, one obtains

$$J^{J}_{-J^L}([(-J(\sigma), y, \sigma)]_{\Phi}) = J^R(-J(\sigma), y) = \text{Co}(y^{-1})J(\sigma) = J(y^{-1}\sigma).$$

For $H = G$ the last claim of the lemma is then immediate from the previous paragraph. For general H one simply restricts J to $\mathfrak{h}$. ■

Using this lemma, we fill out Figure 7 as in Figure 9; this immediately leads to Theorem 1.8.4 for connected H and G (where Marsden–Weinstein reduction coincides with special symplectic reduction).

To proceed for general H and G we take a closer look at the proof of Theorem 1.8.1. In the case at hand the space $S_1 *_{P_1} S_2 *_{P_3} S_3$ consists of those triples $(\sigma, \omega, \alpha) \in S \times T^*G \times S_\rho$ for which $J(\sigma) = -J^L(\omega)$ and $J^R_{\restriction\mathfrak{h}}(\omega) = J_\rho(\alpha)$. In the right trivialization of T^*G, where $\omega = (\theta, y)$, these conditions read $J(\sigma) = -\theta$ and $-(\text{Co}(y^{-1})\theta) \restriction \mathfrak{h} = J_\rho(\alpha)$; see III.(1.57) and III.(1.58). The foliation Φ coincides with the foliation by the orbits of the $G^0 \times H^0$-action

$$(x, h) : (\sigma, \theta, y, \alpha) \mapsto (x\sigma, \text{Co}(x)\theta, xyh^{-1}, h\alpha);$$

cf. III.(1.51) and III.(1.52). The symplectomorphism $S_J^{\overset{\rho}{R}} \to S^{\rho}_{J_{\restriction\mathfrak{h}}}$ is then given by 1.8.1 as

$$[\sigma, -J(\sigma), y, \alpha]_{\Phi} \mapsto [y^{-1}\sigma, \alpha]_{H^0}.$$

From this we see that replacing $(T^*G)^{\mathcal{O}_+}$ by $(\widetilde{T^*G})^{\mathcal{O}_+}$ amounts to replacing H^0 by H in the last-mentioned equivalence class, too. This is precisely what is needed

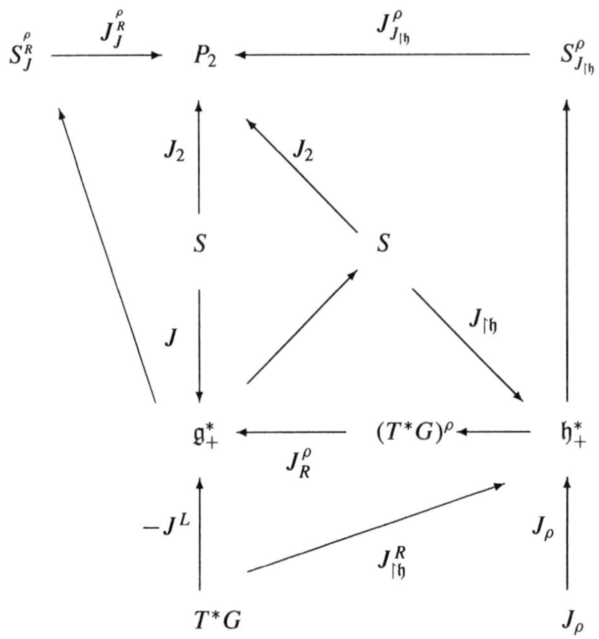

FIGURE 9. Marsden–Weinstein reduction in stages: $S_J^{\overset{\mathcal{O}}{\rho}} \simeq S^{\mathcal{O}}_{J\restriction\mathfrak{h}}$

to prove the theorem, which now follows, because Lemma 1.8.3 holds whether or not G is connected. ■

Applying Theorem 1.8.2 to the special case (1.43) we obtain

Corollary 1.8.4. *Suppose one has a strongly Hamiltonian G-action on a symplectic manifold S with equivariant momentum map J, leading to the reduced space $\tilde{S}_J^{\overset{\mathcal{O}}{R}}$ defined in (1.57).*

This reduced space is symplectomorphic to the Marsden–Weinstein quotient $\tilde{S}^{\mathcal{O}}_{J\restriction\mathfrak{h}} = J_{\restriction\mathfrak{h}}^{-1}(\mathcal{O})/H$.

A different application of Theorem 1.8.1 leads to

Proposition 1.8.5. *Let G_1 and G_2 be Lie groups, and let S carry a strongly Hamiltonian action of $G_1 \times G_2$, with equivariant momentum map $J = J_1 \oplus J_2 : S \to \mathfrak{g}_1^* \oplus \mathfrak{g}_2^*$. (When G_1 and G_2 are compact this is equivalent to having commuting actions of G_1 and G_2, both of which are strongly Hamiltonian.)*

Then the Marsden–Weinstein quotient $\tilde{S}_1^0 := J_1^{-1}(0)/G_1$ carries a strongly Hamiltonian G_2-action, with equivariant momentum map J_2^0, and similarly for $1 \leftrightarrow 2$. One has the symplectomorphisms

$$(J_1^0)^{-1}(0)/G_1 \simeq (J_2^0)^{-1}(0)/G_2 \simeq J^{-1}(0)/G_1 \times G_2. \tag{1.59}$$

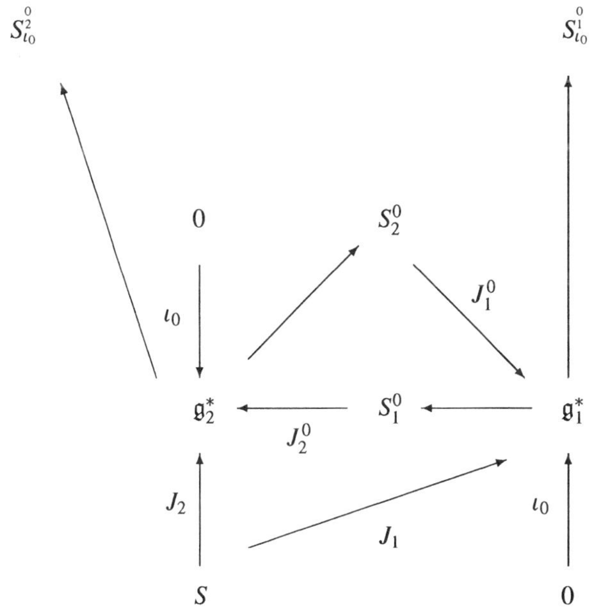

FIGURE 10. Reduction in stages: $\overset{0}{S^1_{\iota_0}} \simeq \overset{0}{S^2_{\iota_0}}$

The $G_1 \times G_2$-equivariance of J implies that J_1 is G_1-equivariant and G_2-invariant, and similarly for $1 \leftrightarrow 2$. (When two commuting and strongly Hamiltonian actions of compact Lie groups G_1 and G_2 are given, one may conversely achieve that $J = J_1 \oplus J_2$ is $G_1 \times G_2$-equivariant by averaging J_1 over G_2 and J_2 over G_1.) With III.(1.19), III.(1.7), and I.(2.8), the invariance property guarantees that $J_1^* C^\infty(\mathfrak{g}_1^*, \mathbb{R})$ and $J_2^* C^\infty(\mathfrak{g}_2^*, \mathbb{R})$ Poisson-commute. The first symplectomorphism in (1.59) now follows by specializing Figure 7 to Figure 10; the second follows from the proof of Theorem 1.8.1. ■

1.9 Coadjoint Orbits of Nilpotent Groups

We will now look at a situation where $(\widetilde{T^*G})^{\mathcal{O}_+}$ (see (1.43)) is (symplectomorphic to) a coadjoint orbit of G, and where every coadjoint orbit is of this form. In such a situation Corollary 1.8.4 is particularly useful, stating that any Marsden–Weinstein quotient with respect to G may more simply be constructed by reducing with respect to a suitable subgroup $H \subset G$.

As we shall see in this section, such a scenario applies when G is **nilpotent**; recall that this means that $[X_1, [X_2, \ldots [X_{k-1}, X_k] \ldots]]$ vanishes for all $X_i \in \mathfrak{g}$

and all integers $k \geq k_0$, for some $k_0 \geq 1$. For example, the Heisenberg group $\tilde{H}_n$ is nilpotent.

Theorem 1.9.1. *Let a Lie group G be nilpotent, connected, and simply connected, and pick a coadjoint orbit $\mathcal{O}^G_{\tilde{\theta}}$ through $\tilde{\theta} \in \mathfrak{g}^*$.*

There exists a subgroup $P_{\tilde{\theta}} \subseteq G$ of dimension $\dim(G) - \frac{1}{2}\dim(\mathcal{O}^G_{\tilde{\theta}})$, and a point $\alpha \in \mathfrak{p}^*_{\tilde{\theta}}$ (which is stable under $\text{Co}(P_{\tilde{\theta}})$, and therefore is a coadjoint orbit), for which the reduced space $(\widetilde{T^*G})^{\mathcal{O}_+}$, defined as in (1.43) with $H = P_{\tilde{\theta}}$ and $\mathcal{O} = \alpha$, is symplectomorphic to $(\mathcal{O}^G_{\tilde{\theta}})_+$. Moreover, the reduced G-action on $(\widetilde{T^*G})^{\alpha}$ (with associated momentum map $-J^{\alpha}_R$, where $J^{\alpha}_R := J^{\alpha}_{J_R\restriction \mathfrak{p}_{\tilde{\theta}}}$) is equivalent to the coadjoint action on $\mathcal{O}^G_{\tilde{\theta}}$ (with momentum map $-\iota_{\mathcal{O}^G_{\tilde{\theta}}}$).

Note that $P_{\tilde{\theta}}$ is a proper subgroup unless $\mathcal{O}^G_{\tilde{\theta}}$ is zero-dimensional.

The idea of the proof is that the map J^{α}_R in Figure 4 (with $S_\rho \to \alpha, \mathfrak{h} \to \mathfrak{p}_{\tilde{\theta}}$, and $J^{\rho}_R \to J^{\alpha}_R$) is a symplectomorphism. To show this, one relies on a fundamental fact in the theory of nilpotent Lie groups, which we state without proof.

Lemma 1.9.2. *For every $\tilde{\theta} \in \mathfrak{g}^*$ there exists a subalgebra $\mathfrak{p}_{\tilde{\theta}}$ of $\mathfrak{g}$ of dimension $\frac{1}{2}(\dim(\mathfrak{g}) + \dim(\mathfrak{g}_{\tilde{\theta}}))$ (where $\mathfrak{g}_{\tilde{\theta}}$ is the Lie algebra of the stabilizer $G_{\tilde{\theta}}$ of $\tilde{\theta}$ under the coadjoint action) that contains $\mathfrak{g}_{\tilde{\theta}}$, and has the property that $\tilde{\theta}([X, Y]) = 0$ for all $X, Y \in \mathfrak{p}_{\tilde{\theta}}$.*

The Lie algebra $\mathfrak{p}_{\tilde{\theta}}$ is called a **polarizing subalgebra** of $\mathfrak{g}$. Its defining property is equivalent to the stability of $\alpha := \tilde{\theta} \restriction \mathfrak{p}_{\tilde{\theta}}$ under $\text{Co}(P_{\tilde{\theta}})$, where $P_{\tilde{\theta}}$ is the connected and simply connected Lie group with Lie algebra $\mathfrak{p}_{\tilde{\theta}}$. Indeed, in the theorem we take $P_{\tilde{\theta}}$ and α as indicated above.

According to III.(2.119), in the left trivialization of T^*G the reduced space $(\widetilde{T^*G})^{\alpha}$ consists of H-equivalence classes $[-\theta, x]_{P_{\tilde{\theta}}}$, with $\theta \restriction \mathfrak{p}_{\tilde{\theta}} = \alpha$ and $x \in G$. The $P_{\tilde{\theta}}$-action defining the equivalence classes is given by III.(2.120).

The crucial technical point is the equality

$$\{\theta \in \mathfrak{g}^* \mid (\theta \restriction \mathfrak{p}_{\tilde{\theta}}) = \tilde{\theta} \restriction \mathfrak{p}_{\tilde{\theta}}\} = \text{Co}(P_{\tilde{\theta}})\tilde{\theta}. \tag{1.60}$$

It is obvious that the right-hand side is contained in the left-hand side, for

$$(\text{Co}^G(h)\tilde{\theta}) \restriction \mathfrak{p}_{\tilde{\theta}} = \text{Co}^{P_{\tilde{\theta}}}(h)(\tilde{\theta}) \restriction \mathfrak{p}_{\tilde{\theta}}) \tag{1.61}$$

for all $h \in P_{\tilde{\theta}}$; for clarity we have denoted the coadjoint action on $\mathfrak{g}^*$ and on $\mathfrak{p}^*_{\tilde{\theta}}$ by Co^G and $\text{Co}^{P_{\tilde{\theta}}}$, respectively. The opposite inclusion is proved by observing that the left-hand side of (1.60) is a copy of $\mathbb{R}^{\dim(\mathfrak{g})-\dim(\mathfrak{p}_{\tilde{\theta}})}$. It is a nontrivial fact about connected, simply connected, and nilpotent groups that the orbit $\text{Co}(P_{\tilde{\theta}})\tilde{\theta}$ is homeomorphic to $\mathbb{R}^{\dim(\mathfrak{p}_{\tilde{\theta}})-\dim(\mathfrak{g}_{\tilde{\theta}})}$. By Lemma 1.9.2 the dimensions match, so that (1.61) has been proved.

By (1.17) and III.(2.124) one has

$$J^{\alpha}_R([-\theta, x]_{P_{\tilde{\theta}}}) = \text{Co}(x)\theta. \tag{1.62}$$

Combining the information collected so far, one infers from Corollary 1.2.3 that J_R^α is a symplectomorphism between $(\widetilde{T^*G})^\alpha$ and $\mathcal{O}_{\tilde{\theta}}^G$.

The reduced G-action on $(\widetilde{T^*G})^\alpha$ is given by III.(2.122); the final claim in Theorem 1.9.1 is then immediate from (1.62). □

Consider, for example, the Heisenberg group $\tilde{H}_n$ introduced in II.2.1; its coadjoint orbits are listed after II.(2.11). For the zero-dimensional orbit $\tilde{\theta} = (p, q, 0)$ one has $\mathfrak{g}_{\tilde{\theta}} = \mathfrak{p}_{\tilde{\theta}} = \mathfrak{h}_n$, so that the claims of the theorem are self-evident. According to II.(2.13), a $2n$-dimensional orbit through $\tilde{\theta} = (0, 0, c \neq 0)$ has stabilizer $\mathfrak{g}_{\tilde{\theta}} = \mathbb{R}Z$, and one may choose $\mathfrak{p}_{\tilde{\theta}} = \mathbb{R}Q \oplus \mathbb{R}Z$, which is abelian. (Alternatively, one could pick, e.g., $\mathbb{R}P \oplus \mathbb{R}Z$, which illustrates the fact that polarizing subalgebras are not necessarily unique.) The left-hand side of (1.60) is $\{(0, \mathbb{R}, c)\}$, which by II.(2.13) indeed coincides with the right-hand side. Thus one verifies without any difficulty that the coadjoint orbits of $\tilde{H}_n$ are indeed as described by Theorem 1.9.1.

We return to the general case. The quantum counterpart of Theorem 1.9.1 is the following. Recall Mackey induction from III.2.9.

Theorem 1.9.3. *Let U be an irreducible representation of a connected and simply connected nilpotent Lie group G. There exists a point $\tilde{\theta} \in \mathfrak{g}^*$ and a polarizing subalgebra $\mathfrak{p}_{\tilde{\theta}}$ such that U is equivalent to the representation $U^{\tilde{\theta}}$ induced from the one-dimensional representation*

$$U_{\tilde{\theta}}(\mathrm{Exp}(X)) := e^{-i\tilde{\theta}(X)} \tag{1.63}$$

of the connected and simply connected group $P_{\tilde{\theta}}$ with Lie algebra $\mathfrak{p}_{\tilde{\theta}}$.

Two induced representations $U^{\tilde{\theta}_i}$ $(i = 1, 2)$ of this type are equivalent iff $\tilde{\theta}_1$ and $\tilde{\theta}_2$ lie in the same coadjoint orbit. In particular, different choices of the polarizing subalgebra $\mathfrak{p}_{\tilde{\theta}}$ lead to equivalent representations.

Note that $U_{\tilde{\theta}}$ is indeed a representation of $P_{\tilde{\theta}}$ as a consequence of the property mentioned at the end of Lemma 1.9.2.

We will not prove this theorem, merely illustrating it for the Heisenberg group. Choosing $\tilde{\theta} = (0, 0, c \neq 0)$ and $\mathfrak{p}_{\tilde{\theta}} = \mathbb{R}Q \oplus \mathbb{R}Z$ as above, one computes from II.(2.6) and III.(2.176), in which one takes the section $s(q) = (0, q, 0)$, that $U^{(0,0,c)}$ coincides with U_c as defined in II.(2.19), with $\lambda = c$. The one-dimensional representation II.(2.27) corresponds, of course, to $\tilde{\theta} = (p, q, 0)$. Theorem II.2.1.4 is now seen to be a corollary of Theorem 1.9.3.

1.10 Coadjoint Orbits of Semidirect Products

Another illustrative and physically relevant situation where Corollary 1.8.4 applies with force is given by semidirect product Lie groups $G = L \ltimes_\rho V$, where V is a vector space carrying a linear L-action ρ. Following the habit of physicists, we denote elements x of G by pairs $x = (\Lambda, v) \in L \times V$; elements of the dual V^* are generically called p. The multiplication rule in G is

$$(\Lambda_1, v_1)(\Lambda_2, v_2) := (\Lambda_1\Lambda_2, v_1 + \rho(\Lambda_1)v_2). \tag{1.64}$$

The Lie algebra $\mathfrak{g} = \mathfrak{l} \oplus V$ then has the bracket

$$[(X, v), (Y, w)] = ([X, Y], d\rho(X)w - d\rho(Y)v). \tag{1.65}$$

The coadjoint action of G on $\mathfrak{g}^* = \mathfrak{l}^* \oplus V^*$ is

$$\text{Co}(\Lambda_1, v)(\theta, p) = (\text{Co}(\Lambda)\theta + (\rho^*(\Lambda)p) \wedge v, \rho^*(\Lambda)p), \tag{1.66}$$

where ρ^* is the dual action of G on V^*; that is, $(\rho^*(\Lambda)p)(v) := p(\rho(\Lambda^{-1})v)$. For $p \in V^*$ and $v \in V$, the element $p \wedge v \in \mathfrak{l}^*$ is defined by

$$p \wedge v(X) := p(d\rho(X)v). \tag{1.67}$$

In terms of the dual ρ_p^* of the map $\rho_p : \mathfrak{g} \to V^*$, defined for fixed $p \in V^*$ by $\rho_p(X) := -d\rho^*(X)p$, one has the equivalent definition $p \wedge v = \rho_p^* v$.

The stabilizer of $p \in V^*$ under ρ^* is denoted by L_p (as usual), with Lie algebra $\mathfrak{l}_p$. Now recall Definition III.2.1.4 and the first paragraph of III.2.7. Given a coadjoint orbit $\mathcal{O}^{L_p}$ in $\mathfrak{l}_p^*$, we can form the bundle $L \times_{L_p} \mathcal{O}^{L_p}$ associated to the principal bundle $L(L/L_p, L_p, \tau)$ by the coadjoint action of L_p on $\mathcal{O}^{L_p}$. This associated bundle has base L/L_p, which we identify with the L-orbit $\mathcal{O}_p^L = \rho^*(L)p$ in V^*, and typical fiber $\mathcal{O}^{L_p}$.

We now fix a pair $(\tilde{\theta}, \tilde{p}) \in \mathfrak{g}^*$, for simplicity denoting the coadjoint orbit $\mathcal{O}^{L_{\tilde{p}}}_{\tilde{\theta}\restriction\mathfrak{l}_{\tilde{p}}}$ in $\mathfrak{l}_{\tilde{p}}^*$ by $\mathcal{O}_{\tilde{\theta}}^{L_{\tilde{p}}}$. The first characterization of the coadjoint orbits of $L \ltimes_\rho V$ is as follows. Note the similarity with Proposition III.3.9.4.

Proposition 1.10.1. *The coadjoint orbit $\mathcal{O}^G_{(\tilde{\theta},\tilde{p})}$ of $G = L \ltimes_\rho V$ through $(\tilde{\theta}, \tilde{p}) \in \mathfrak{g}^*$ is a fiber bundle over $L \times_{L_{\tilde{p}}} \mathcal{O}_{\tilde{\theta}}^{L_{\tilde{p}}}$ with typical fiber $\mathfrak{l}_{\tilde{p}}^0$ (the annihilator of $\mathfrak{l}_{\tilde{p}}$ in $\mathfrak{l}^*$). This provides a bijection between the set of coadjoint orbits in $\mathfrak{g}^*$ and the pairs $(\mathcal{O}^L, \mathcal{O}_{\tilde{\theta}}^{L_{\tilde{p}}})$ consisting of an L-orbit $\mathcal{O}^L$ in V^* and a coadjoint orbit $\mathcal{O}_{\tilde{\theta}}^{L_{\tilde{p}}}$ in $\mathfrak{l}_{\tilde{p}}^*$ (where $L_{\tilde{p}}$ is the stabilizer of an arbitrary point $\tilde{p}$ in $\mathcal{O}^L$).*

The bundle projection $\tau_{\mathcal{O}^G} : \mathcal{O}^G_{(\tilde{\theta},\tilde{p})} \to L \times_{L_{\tilde{p}}} \mathcal{O}_{\tilde{\theta}}^{L_{\tilde{p}}}$ is given by

$$\tau_{\mathcal{O}^G}(\text{Co}(\Lambda, v)(\tilde{\theta}, \tilde{p})) := [\Lambda, \tilde{\theta} \restriction \mathfrak{l}_{\tilde{p}}]_{L_{\tilde{p}}}. \tag{1.68}$$

This projection is well-defined because of a property analogous to (1.61). The coadjoint orbit $\mathcal{O}^G_{(\tilde{\theta},\tilde{p})}$ in $\mathfrak{g}^*$ determines a pair $(\mathcal{O}^L_{\tilde{p}}, \mathcal{O}_{\tilde{\theta}}^{L_{\tilde{p}}})$; conversely, a pair $\left(\mathcal{O}^L, \mathcal{O}_{\tilde{\theta}}^{L_{\tilde{p}}}\right)$ as stated corresponds to an orbit $\text{Co}^G_{(\tilde{\theta},\tilde{p})}$ with the property that $\tilde{\theta} \restriction \mathfrak{l}_{\tilde{p}}$ lies in $\mathcal{O}_{\tilde{\theta}}^{L_{\tilde{p}}}$, and $\tilde{p}$ has $L_{\tilde{p}}$ as its stabilizer. The fact that this correspondence is bijective and independent of all choices follows from elementary verifications, using the equality

$$\{\tilde{p} \wedge v \mid v \in V\} = \mathfrak{l}_{\tilde{p}}^0, \tag{1.69}$$

which is easily checked. ∎

We now turn to an alternative description of the coadjoint orbits of $L \ltimes_\rho V$ that places the bijective correspondence stated in Proposition 1.10.1 in a new light. To formulate this result, we first infer from (1.66) that for each $p \in V^*$ and each coadjoint orbit $\mathcal{O}^{L_p}$ in $\mathfrak{l}_p^*$ the set $\mathcal{O}^{L_p} \dot{+} p \subset \mathfrak{l}_p^* \oplus V^*$ is a coadjoint orbit of $L_p \ltimes_\rho V^*$.

Theorem 1.10.2. *The coadjoint orbit $\mathcal{O}^G_{(\tilde{\theta},\tilde{p})}$ (with the "+" Lie symplectic structure) is symplectomorphic to the reduced space $(\widetilde{T^*G})^{\mathcal{O}_+}$, defined as in (1.43) with $G = L \ltimes_\rho V$,*

$$H := L_{\tilde{p}} \ltimes_\rho V, \tag{1.70}$$

and $\mathcal{O} = \mathcal{O}^{L_{\tilde{p}}}_{\tilde{\theta}} \dot{+} \tilde{p}$. Moreover, the reduced G-action on $(\widetilde{T^*G})^{\mathcal{O}_+}$ (with associated momentum map $-J^{\mathcal{O}}_R := -J^{\mathcal{O}}_{J^R_{|\mathfrak{h}}}$) is equivalent to the coadjoint action on $\mathcal{O}^G_{(\tilde{\theta},\tilde{p})}$ (with momentum map $-\iota_{\mathcal{O}^G_{\tilde{\theta}}}$).

What follows is a more complicated version of the proof of Theorem 1.9.1.

We see from III.(2.119) that in the left trivialization of T^*G the reduced space $(\widetilde{T^*G})^{\mathcal{O}_+}$ consists of H-equivalence classes $[-\theta, -\tilde{p}, \Lambda, v]_H$, with $\theta \restriction \mathfrak{l}_{\tilde{p}} \in \mathcal{O}^{L_{\tilde{p}}}_{\tilde{\theta}}$, and according to III.(2.120), (1.64), and (1.66) the H-action is given by

$$(\Lambda_1, w) : (-\theta, -\tilde{p}, \Lambda_2, v) \mapsto (-\mathrm{Co}(\Lambda_1)\theta - \tilde{p}\wedge w, -\tilde{p}, \Lambda_2\Lambda_1^{-1}, v - \rho(\Lambda_2\Lambda_1^{-1})w). \tag{1.71}$$

According to (1.17) and III.(2.124) one has

$$J^{\mathcal{O}}_R([-\theta, -p, \Lambda, v]_H) = \mathrm{Co}(\Lambda, v)(\theta, p), \tag{1.72}$$

which is given by (1.66). Using (1.69), it is now easily verified that $J^{\mathcal{O}}_R$ is injective. Since (Λ, v) varies freely in G, it is obvious from (1.72) that $J^{\mathcal{O}}_R((\widetilde{T^*G})^{\mathcal{O}_+})$ contains $\mathcal{O}^G_{(\tilde{\theta},\tilde{p})}$. Combined with Proposition 1.10.1, the restriction $\tilde{\theta} \restriction \mathfrak{l}_{\tilde{p}} \in \mathcal{O}^{L_{\tilde{p}}}_{\tilde{\theta}}$ implies that the image of $J^{\mathcal{O}}_R$ is precisely $\mathcal{O}^G_{(\tilde{\theta},\tilde{p})}$.

Hence $J^{\mathcal{O}}_R : (\widetilde{T^*G})^{\mathcal{O}_+} \to \mathcal{O}^G_{(\tilde{\theta},\tilde{p})}$ is a diffeomorphism that is even a symplectomorphism by Corollary 1.2.3.

The reduced G-action on $(\widetilde{T^*G})^{\mathcal{O}_+}$ being given by III.(2.122), the final claim is immediate from (1.66) and (1.72). ■

To state the quantum versions of Proposition 1.10.1 and Theorem 1.10.2, we take $\tilde{p} \in V^*$ and $\sigma \in \hat{L}_{\tilde{p}}$, so that $U_\sigma(L_{\tilde{p}})$ is an irreducible representation of $L_{\tilde{p}}$. Define H as in (1.70), and note that $U_\sigma(L_{\tilde{p}})$ extends to an irreducible representation $U_{\tilde{p},\sigma}(H)$ by

$$U_{\tilde{p},\sigma}(\Lambda, v) := e^{i\tilde{p}v} U_\sigma(\Lambda). \tag{1.73}$$

Theorem 1.10.3. *Suppose that the semidirect product G is* **regular** *in that each L-orbit in V^* is (relatively) open in its closure. Then the representation $U^{\tilde{p},\sigma}(G)$ induced by an irreducible representation $U_{\tilde{p},\sigma}(H)$ of the above type is irreducible for any choice of $\tilde{p}$ and σ, and for every irreducible representation $U(G)$ there exists a pair $(\tilde{p}, \sigma)$ such that U is equivalent to $U^{\tilde{p},\sigma}$.*

Two representations $U^{\tilde{p}_1,\sigma_1}(G)$ and $U^{\tilde{p}_2,\sigma_2}(G)$ are equivalent iff the $\tilde{p}_i$ lie in the same orbit $\mathcal{O}^L$ (so that $\tilde{p}_2 = \rho^*(\Lambda)\tilde{p}_1$ for some $\Lambda \in L$), and $U_{\sigma_2} \circ \mathrm{Ad}_\Lambda$ is equivalent to U_{σ_1}. In other words, the unitary dual $\hat{G}$ is parametrized by pairs $(\mathcal{O}^L, \sigma)$, where $\mathcal{O}^L$ is an L-orbit in V^* and σ is a member of the unitary dual of the stabilizer of an arbitrary point in $\mathcal{O}^L$.

Let U be a representation of G on a Hilbert space $\mathcal{H}$. One easily sees from (1.64) that

$$U(\Lambda)U(v)U(\Lambda)^* = U(\rho(\Lambda)v), \tag{1.74}$$

where $\Lambda := (\Lambda, 0)$ and $v := (e, v)$ etc. By III.(1.89) the restriction of U to V defines a representation π of the group C^*-algebra $C^*(V)$. Using III.(1.88) with G replaced by V, this yields a representation $\tilde{\pi}(C_0(V^*))$. We see from (1.74) that the pair $(U(L), \tilde{\pi}(C_0(V^*)))$ is a system of imprimitivity of L on V^* in $\mathcal{H}$; cf. Definition III.3.7.3. Conversely, such a system determines a representation $U(G)$ on $\mathcal{H}$, and the correspondence thus obtained is bijective. Theorem 1.10.3 then follows from Theorem 2.7.3 below and Corollary III.3.7.4. ∎

The general description of the explicit form of induced representations in III.2.9 simplifies somewhat, because G/H is now equal to $L/L_{\tilde{p}} \simeq \mathcal{O}^L_{\tilde{p}}$. To obtain a (measurable global, or smooth local) section $s : G/H \to G$ we merely need to choose a (...) section $b : L/L_{\tilde{p}} \to L$ (where the name "b", standing for "boost", comes from physics), in terms of which s is given by $s(p) := (b(p), 0)$. The carrier space of the realization $U_s^{\tilde{p},\sigma}$ (cf. the text below III.(2.176)) is then

$$\mathcal{H}_b^{\tilde{p},\sigma} = L^2(\mathcal{O}^L_{\tilde{p}}) \otimes \mathcal{H}_\sigma, \tag{1.75}$$

where $\mathcal{H}_\sigma$ is the carrier space of $U_\sigma(L_{\tilde{p}})$, and we have replaced the suffix s by b. Assuming that $\mathcal{O}^L_{\tilde{p}}$ possesses an L-invariant measure (which will be the case in our applications), one is able to simplify III.(2.176) to

$$U_b^{\tilde{p},\sigma}(\Lambda, v)\Psi_b^{\tilde{p},\sigma}(p) = e^{ipv}U_\sigma\left(b(p)^{-1}\Lambda b(\rho^*(\Lambda^{-1})p)\right)\Psi_b^{\tilde{p},\sigma}(\rho^*(\Lambda^{-1})p). \tag{1.76}$$

Note that the argument of U_σ indeed lies in $L_{\tilde{p}}$.

We return to the classical setting. From 1.10.2 we are led to a third description of the coadjoint orbits in question. Recall that we identify the L-orbit $\mathcal{O}^L_{\tilde{p}}$ in V^* with $L/L_{\tilde{p}}$; this leads to an embedding (i.e., an injective homomorphism) $\chi : V \to C^\infty(L/L_{\tilde{p}}, \mathbb{R})$ (as additive groups), given by $\chi(v) : p \mapsto -p(v)$, where $p \in \mathcal{O}^L_{\tilde{p}}$. Using the natural embedding $L \subset \mathrm{Aut}(L)$, where L is seen as the total space of the principal bundle $L(L/L_{\tilde{p}}, L_{\tilde{p}}, \tau)$ (cf. III.2.7), we observe that χ extends to an embedding $\tilde{\chi} : L \ltimes_\rho V \to \mathrm{Aut}(L) \ltimes C^\infty(L/L_{\tilde{p}}, \mathbb{R})$. This enables us to regard G as a subgroup of $\mathrm{Aut}(L) \ltimes C^\infty(L/L_{\tilde{p}}, \mathbb{R})$. In particular, G acts on the reduced space $(\widetilde{T^*L})^{\mathcal{O}_{\tilde{\theta}}^{L_{\tilde{p}}}}$ by restriction of the action ρ_0 of $\mathrm{Aut}(L) \ltimes C^\infty(L/L_{\tilde{p}}, \mathbb{R})$ defined in Theorem III.2.5.2.

Theorem 1.10.4. *The coadjoint orbit $\left(\mathcal{O}^G_{(\tilde{\theta},\tilde{p})}\right)_+$ is symplectomorphic to the reduced space $(\widetilde{T^*L})^{\mathcal{O}_{\tilde{\theta}}^{L_{\tilde{p}}}}$, defined as in (1.43) with $G = L$, $H = L_{\tilde{p}}$, and $\mathcal{O} = \mathcal{O}^{L_{\tilde{p}}}_{\tilde{\theta}}$.*

The G-action on $(\widetilde{T^*L})^{\mathcal{O}_{\tilde{\theta}}^{L_{\tilde{p}}}}$ explained above is equivalent to the coadjoint action on $\mathcal{O}^G_{(\tilde{\theta},\tilde{p})}$.

Given $\tilde{p} \in V^*$, we regard T^*L as a submanifold of T^*G by the embedding $(\theta, l) \mapsto (\theta, -\tilde{p}, l, 0)$; this is a Poisson map. The momentum map $J^R_{\restriction \mathfrak{l}_{\tilde{p}}}$ is simply the restriction of $J^R_{\restriction \mathfrak{h}}$ (where $\mathfrak{h}$ is the Lie algebra of the group H specified in 1.10.2) to $\mathfrak{l}_{\tilde{p}}$, regarded as a subalgebra of $\mathfrak{h}$ by the embedding $\theta \mapsto (\theta, 0)$. One then easily infers from (1.71) that $(J^R_{\restriction \mathfrak{l}_{\tilde{p}}})^{-1}(\mathcal{O}^{L_{\tilde{p}}}_{\tilde{\theta}})/L_{\tilde{p}}$ is diffeomorphic to $(J^R_{\restriction \mathfrak{h}})^{-1}(\mathcal{O}^{L_{\tilde{p}}}_{\tilde{\theta}} \dot{+} \tilde{p})/H$ under the bijection

$$\varphi : [-\theta, \Lambda]_{L_{\tilde{p}}} \mapsto [-\theta, -\tilde{p}, \Lambda, 0]_{L_{\tilde{p}} \ltimes_\rho V}, \tag{1.77}$$

where $\theta \restriction \mathfrak{l}_{\tilde{p}} \in \mathcal{O}^{L_{\tilde{p}}}_{\theta}$. Since the embedding $T^*L \hookrightarrow T^*G$ above is a Poisson map, this diffeomorphism is a Poisson map, and therefore a symplectomorphism, by definition of the quotient Poisson structure. The first claim then follows from Theorem 1.10.2.

Comparing (1.72) with (1.77), and using (1.66) with $v = 0$, one sees that the pertinent symplectomorphism $\tilde{\varphi} : \mathcal{O}^G_{(\tilde{\theta},\tilde{p})} \to (\widetilde{T^*L})^{\mathcal{O}_{\tilde{\theta}}^{L_{\tilde{p}}}}$ is given by

$$\tilde{\varphi}(\theta, \rho^*(\Lambda)\tilde{p}) = [-\text{Co}(\Lambda^{-1})\theta, \Lambda]_{L_{\tilde{p}}}. \tag{1.78}$$

We now prove that φ intertwines the G-actions in question. Firstly, it is obvious from (1.77), (1.64) and III.(2.122) that φ intertwines the L-actions. Secondly, we note that in the left trivialization the one-form $df(x) \in T^*_x G$ (where $f \in C^\infty(G, \mathbb{R})$) is represented by $(\theta_f, x)_L$, where $\theta_f(X) := \xi^L_X f$. Regarding $\chi(v)$ as a right-$L_{\tilde{p}}$-invariant function on L, it then follows from III.(1.37) and (1.69) that in the left trivialization, $d\chi(v)$ at Λ is represented by $(\tilde{p} \wedge \rho(\Lambda^{-1})v, \Lambda)_L$. Hence the action III.(2.77) reads $v : (-\theta, \Lambda)_L \mapsto (-\theta - \tilde{p} \wedge \rho(\Lambda^{-1})v, \Lambda)_L$. This quotients to the action

$$v : [-\theta, \Lambda]_{L_{\tilde{p}}} \mapsto [-\theta - \tilde{p} \wedge \rho(\Lambda^{-1})v, \Lambda]_{L_{\tilde{p}}}$$

on $(\widetilde{T^*L})^{\mathcal{O}_{\tilde{\theta}}^{L_{\tilde{p}}}}$. The second claim in 1.10.4 then follows from (1.77), (1.71), III.(2.122), (1.66), and (1.72); cf. the end of the proof of 1.10.2.

As a check on this computation we note that by definition of χ, the momentum map III.(2.80) reads $J_{\chi(a)}([-\theta, \Lambda]_{L_{\tilde{p}}}) = -(\rho^*(\Lambda)\tilde{p})(a)$. By (1.66) and Corollary III.1.4.6 this coincides with $J_{(0,a)}(\text{Co}(\Lambda, 0)(\theta, \tilde{p}))$, where J is the momentum map for the coadjoint action of G on its coadjoint orbit $\left(\mathcal{O}^G_{(\tilde{\theta},\tilde{p})}\right)_+$. We conclude from (1.72) and (1.77) that $J^{\mathcal{O}}_R \circ \varphi$ intertwines $J_{\chi(a)}$ and $J_{(0,a)}$. ∎

We infer from this proof that the G-action on $(\widetilde{T^*L})^{\mathcal{O}_{\tilde{\theta}}^{L_{\tilde{p}}}}$ is given by

$$(\Lambda_1, v) : [-\theta, \Lambda_2]_{L_{\tilde{p}}} \mapsto [-\theta - \tilde{p} \wedge \rho(\Lambda_1\Lambda_2)^{-1}v, \Lambda_1\Lambda_2]_{L_{\tilde{p}}}. \tag{1.79}$$

This formula may alternatively be derived from (1.66) and (1.78).

Applying Theorem III.2.3.7 (or its generalization 1.6.1), we conclude that $\mathcal{O}^G_{(\tilde{\theta},\tilde{p})}$ is a bundle over $T^*\mathcal{O}^L_{\tilde{p}}$ with typical fiber $\mathcal{O}^{L_{\tilde{p}}}_{\tilde{\theta}}$; the bundle projection depends on the choice of a connection on the principal bundle $L(L/L_{\tilde{p}}, L_{\tilde{p}}, \tau)$. When L_p is reductive one may choose the H-connection; see III.2.7.

Alternatively, since $(\widetilde{T^*L})^{\mathcal{O}_{\tilde{\theta}}^{L_{\tilde{p}}}}$ is by definition equal to $(J^R_{\restriction \mathfrak{l}_{\tilde{p}}})^{-1}(\mathcal{O}_{\tilde{\theta}}^{L_{\tilde{p}}})/L_{\tilde{p}}$, it may be regarded as a bundle $L \times_{L_{\tilde{p}}} \tau_{\restriction}^{-1}(-\mathcal{O}_{\tilde{\theta}}^{L_{\tilde{p}}})$, where $\tau_{\restriction} : \mathfrak{l}^* \to \mathfrak{l}^*_{\tilde{p}}$ is the restriction map $\tau_{\restriction}(\theta) := \theta \restriction \mathfrak{l}_{\tilde{p}}$. As always, we identify $L/L_{\tilde{p}}$ with the L-orbit $\mathcal{O}^L_{\tilde{p}}$ through $\tilde{p}$ in V^*. This is the bundle over $L/L_{\tilde{p}}$ that is associated to the principal bundle $L(L/L_{\tilde{p}}, L_{\tilde{p}}, \tau)$ by the coadjoint action of $L_{\tilde{p}}$ on $\tau_{\restriction}^{-1}(-\mathcal{O}_{\tilde{\theta}}^{L_{\tilde{p}}}) \subset \mathfrak{l}^*$; note that this subset is indeed stable under the restriction of $\mathrm{Co}(L)$ to $L_{\tilde{p}}$. This bundle structure is evidently independent of the choice of a connection, and may also be derived from Proposition 1.10.1. For example, when $\tilde{\theta}_0 \in \mathfrak{l}^0_{\tilde{p}}$, so that $\mathcal{O}_{\tilde{\theta}_0}^{L_{\tilde{p}}} = \{0\}$, one infers from III.(2.116) that

$$\mathcal{O}^G_{(\tilde{\theta}_0, \tilde{p})} \simeq (\widetilde{T^*L})^0 \simeq T^*\mathcal{O}^L_{\tilde{p}}. \tag{1.80}$$

By III.(2.56) this is even a symplectomorphism.

As a simple example, consider $L = SO(3)$ and $V = \mathbb{R}^3$, with ρ the defining representation of $SO(3)$. The latter coincides with the coadjoint action of $SO(3)$ on $\mathfrak{l}^* = \mathbb{R}^3$, as well as with the action ρ^* on $V^* \simeq \mathbb{R}^3$ (where $\mathbb{R}^3$ and its dual have been identified through the Euclidean inner product). Hence both the coadjoint orbits in $\mathfrak{l}^*$ and the $\rho^*(SO(3))$-orbits in V^* are either spheres S^2_r with radius $r > 0$ or the origin ($r = 0$). The expression $p \wedge v$ in (1.67) is easily seen to coincide with the usual exterior product of p and v.

We take $G = E(3) := SO(3) \ltimes_\rho \mathbb{R}^3$, which is the **Euclidean group**. When $\tilde{p} = 0$ one has $L_0 = SO(3)$, so that the G-coadjoint orbit $\mathcal{O}^G_{(\tilde{\theta},0)}$ is simply the $SO(3)$-orbit through $\tilde{\theta}$, whose Lie symplectic structure is r times the volume form on S^2. For $\tilde{p} \neq 0$ we may choose $\tilde{p} = r_3 := (0, 0, r)$, so that $L_{\tilde{p}} = SO(2)$. By Theorem 1.10.4 the orbit $\mathcal{O}^G_{(\tilde{\theta},r_3)}$ is a bundle over T^*S^2 with typical fiber $\tilde{\theta}_3 =: e \in \mathbb{R}$, which is just a point. Thus $\mathcal{O}^G_{(\tilde{\theta},r_3)}$ is T^*S^2 as a manifold, which as a symplectic space is the space T^*G^e discussed in III.2.12. In particular, the Poisson bracket on $\mathcal{O}^G_{(\tilde{\theta},r_3)}$ is III.(2.227).

For a more complicated illustration of the formalism see 3.1.

We close this section with a result on reduction in stages for semidirect products. Theorem 1.10.2 and Corollary 1.8.4 have the

Corollary 1.10.5. *Let $G = L \ltimes_\rho V$ act on a symplectic manifold S in strongly Hamiltonian fashion, with equivariant momentum map $J : S \to \mathfrak{g}^*_-$, and choose a coadjoint orbit $\mathcal{O} := \mathcal{O}^G_{(\tilde{\theta},\tilde{p})}$ in $\mathfrak{g}^*_+$.*

The Marsden–Weinstein quotient $\tilde{S}^{\mathcal{O}_+}_J = J^{-1}(\mathcal{O})/G$ is symplectomorphic to $(\widetilde{J^{-1}_{\restriction V}(\tilde{p})/V})^{\mathcal{O}_{\tilde{\theta}}^{L_{\tilde{p}}}}_{J \restriction L_{\tilde{p}}}$, that is, to the symplectic space obtained by first reducing S by V with respect to the coadjoint orbit $\tilde{p} \in V^*$, and then reducing by $L_{\tilde{p}}$ with respect to the coadjoint orbit $\mathcal{O}^{L_{\tilde{p}}} \subset \mathfrak{l}^*_{\tilde{p}}$.

Applying 1.10.2 and 1.8.1 in succession shows that we need to prove only that $\widetilde{(J_{\restriction V}^{-1}(\tilde{p})/V)}^{\mathcal{O}_{\tilde{\theta}}^{L_{\tilde{p}}}}_{J \restriction L_{\tilde{p}}}$ coincides with the reduction of S by $H = L_{\tilde{p}} \ltimes_{\rho} V$ with respect to the orbit $\mathcal{O} = \mathcal{O}_{\tilde{\theta}}^{L_{\tilde{p}}} \dot{+} \tilde{p}$ in $\mathfrak{h}_{+}^{*}$.

The G-equivariance of J implies that $J_{\restriction V}(l\sigma) = \rho^{*}(l) J_{\restriction V}(\sigma)$. It follows that $L_{\tilde{p}}$ maps $J_{\restriction V}^{-1}(\tilde{p})$ into itself. Moreover, the rule (1.64) easily implies that the action of $L_{\tilde{p}}$ quotients to an action on the reduced space $J^{-1}(\tilde{p})/V$. It is quite straightforward to verify that the reduced $L_{\tilde{p}}$-action is strongly Hamiltonian, with equivariant momentum map given by the quotient of $J_{\restriction \mathfrak{l}_{\tilde{p}}}$.

Hence the second reduction is well-defined. By the same argument, the map $[[\sigma]_V]_{L_{\tilde{p}}} \mapsto [\sigma]_{L_{\tilde{p}} \ltimes_{\rho} V}$, where $\sigma \in J_{\restriction V}^{-1}(\tilde{p})$, is well-defined, and is almost trivially seen to be a diffeomorphism. Using Theorem 1.5.4 one infers that it is even a symplectomorphism. ∎

Note that Theorem 1.10.4 follows from 1.10.5 by taking $S = T^{*}G$ and observing that $J_{\restriction V}^{-1}(\tilde{p})/V$ is symplectomorphic to $T^{*}L$; in our derivation of 1.10.4 from 1.10.2 this was used in the opposite direction.

1.11 *Singular Marsden–Weinstein Reduction*

In this section we look at what happens to Theorem 1.5.4 when H does not act freely. We do assume that the strongly Hamiltonian H-action on S is proper, as reasonable results are available only in that case. We look only at reduction from $0 \in \mathfrak{h}^{*}$; by the shifting trick in the proof of 1.5.4 this entails no loss of generality. The reduced space is therefore $\tilde{S}^{0} = J^{-1}(0)/H$; for simplicity we omit the subscript J in the notation $\tilde{S}_{J}^{0}$ (cf. (1.26)).

When the H-action fails to be free, but each stabilizer H_{σ} is discrete for $\sigma \in J^{-1}(0)$, the space $J^{-1}(0)$ is still a submanifold of S by Lemma 1.5.1, and the possible singularities in the reduced space $J^{-1}(0)/H$ come from taking the quotient by H. This case is, of course, included in what follows.

When the dimension of some stabilizer is greater than zero, there is no guarantee that $J^{-1}(0)$ is a submanifold of S. For a given subgroup $K \subseteq H$, define

$$S_K := \{\sigma \in S \mid H_{\sigma} = K\}; \tag{1.81}$$

$$S_{[K]} := H S_K = \{\sigma \in S \mid H_{\sigma} \text{ is conjugate to } K\}. \tag{1.82}$$

These spaces are empty when K is not compact; cf. Proposition 1.5.3. Using the compactness of each stabilizer H_{σ}, it can be shown that S_K and $S_{[K]}$ are submanifolds of S. We can say more:

Lemma 1.11.1. *The space S_K is a symplectic manifold, with symplectic form inherited from S.*

The subspace $T_{\sigma} S_K$ of $T_{\sigma} S$, where $\sigma \in S_K$, consists of the K-invariant vectors (since σ is K-invariant, the pushforward of the K-action maps $T_{\sigma} S$ into itself). Since K is compact, we can choose a K-invariant inner product $(\ ,\ )$ on $T_{\sigma} S$, and by linear symplectic geometry there exists an invertible linear map J on $T_{\sigma} S$ such that $(X, JY) = \omega_\sigma(X, Y)$ for all X, Y. By the K-invariance of the symplectic form ω and of the inner product, the map J commutes with the K-action, so that the restriction of J to $T_\sigma S_K$ is well-defined. Then the assumption that for some $X \in T_\sigma S_K$ the number $\omega_\sigma(X, Y)$ vanishes for all $Y \in T_\sigma S_K$ implies that $(X, JY) = 0$ for all $Y \in T_\sigma S_K$; hence $(X, X) = 0$, so that $X = 0$. The claim follows. ∎

This lemma will be used in the proof of Proposition 1.11.3.

Proposition 1.11.2. *Let $\sigma(\cdot)$ be the Hamiltonian flow of $f \in C^\infty(S, \mathbb{R})^H$. For any t for which the flow exists, the point $\sigma(t)$ lies in $J^{-1}(0) \cap S_{[H_\sigma]}$, where $\sigma = \sigma(0)$. Moreover, any two points in a connected component of $J^{-1}(0) \cap S_K$ may be connected by a piecewise smooth Hamiltonian curve, where the pieces are generated by H-invariant Hamiltonians.*

By Proposition III.1.2.2 the set $J^{-1}(0)$ is invariant under the flow. The H-invariance of f implies that $h(\sigma(t)) = (h\sigma)(t)$ for all $h \in H$, so that $H_{\sigma(t)} \subseteq H_\sigma$. Inverting the flow leads to the opposite inclusion, so that $H_{\sigma(t)} = H_\sigma$. It follows that the flow preserves $J^{-1}(0) \cap S_{H_\sigma}$. Using the stability of $J^{-1}(0)$ under H resulting from the equivariance of the momentum map and the second equality in (1.82), one obtains

$$J^{-1}(0) \cap S_{[H_\sigma]} = H(J^{-1}(0) \cap S_{H_\sigma}). \tag{1.83}$$

Since we have just seen that the right-hand side is preserved by Hamiltonian flows, the first claim follows. The second claim results from the equality

$$T_\sigma(J^{-1}(0) \cap S_{H_\sigma}) = \{\xi_f(\sigma) | \, f \in C^\infty(S, \mathbb{R})^H\}. \tag{1.84}$$

Here the inclusion of the right-hand side in $T_\sigma J^{-1}(0)$ follows from Noether's Theorem (Proposition III.1.2.2). Its inclusion in $T_\sigma S_{H_\sigma}$ follows from the first line in the proof of 1.11.1 combined with I.(2.8) and the H-invariance of the Poisson bracket. The proof of Corollary 1.5.7 then yields the equality in (1.84). ∎

The reduced space $\tilde{S}^0$ is trivially a disjoint union

$$\tilde{S}^0 = \cup_{[K]} \tilde{S}^0_{[K]}, \tag{1.85}$$

where $[K]$ varies over all conjugacy classes in H, and

$$\tilde{S}^0_{[K]} := (J^{-1}(0) \cap S_{[K]})/H. \tag{1.86}$$

Proposition 1.11.2 suggests that from a Hamiltonian point of view this is an interesting decomposition, since the flow of an H-invariant Hamiltonian on S projects to a flow that stays inside a given subspace $\tilde{S}^0_{[K]}$.

Proposition 1.11.3. *Let S'_K be the union of those components of S_K whose intersection with $J^{-1}(0)$ is not empty.*

The natural action of the group $N_H(K)/K$ on S'_K is free and strongly Hamiltonian; denote its equivariant momentum map by J_K. The (regular) Marsden–Weinstein quotient $J_K^{-1}(0)/(N_H(K)/K)$ is homeomorphic to $\tilde{S}^0_{[K]}$ (with *the quotient topology). Since the former is a symplectic manifold, the space $\tilde{S}^0_{[K]}$ thereby becomes a symplectic manifold as well.*

Here $N_H(K)$ is the normalizer of K in H (that is, the collection of elements of H that commute with all members of K). It is clear from the definitions that $N_H(K)/K$ acts on S_K by restricting and quotienting the H-action, and that this action is free. From the second equality in (1.82) we then infer that $S_{[K]}/H \simeq S_K/(N_H(K)/K)$. Taking intersections with $J^{-1}(0)$ leads to the desired homeomorphism.

To interpret $(J^{-1}(0) \cap S_K)/(N_H(K)/K)$ as a Marsden–Weinstein quotient, we first note that Lemma 1.11.1 implies that S'_K is symplectic. Secondly, since J is equivariant and S'_K consists of K-stable points, the momentum map for the H-action restricted to S'_K takes values in the space $(\mathfrak{h}^*)^K$ of Co(K)-invariant points in $\mathfrak{h}^*$. Since J_X for $X \in \mathfrak{k}$ generates the K-action on S'_K, which is trivial, each J_X must be constant on each component. The constants are all zero, as S'_K intersects $J^{-1}(0)$. Hence $J \upharpoonright S'_K$ takes values in $(\mathfrak{h}^*)^K \cap \mathfrak{k}^0$, where $\mathfrak{k}^0$ is the annihilator of $\mathfrak{k}$ in $\mathfrak{h}$.

Now observe that $(\mathfrak{h}^*)^K \cap \mathfrak{k}^0$ is naturally isomorphic to the dual of the Lie algebra of $N_H(K)/K$: This is immediate from the definition of the normalizer and of the (co) adjoint action, combined with the isomorphism $T^*_{[e]_K}(H/K) \simeq \mathfrak{k}^0$ (cf. the proof of Lemma III.2.7.1).

We conclude that $J \upharpoonright S'_K$ may be interpreted as the momentum map J_K for the $N_H(K)/K$-action on S'_K. Since this action is free (and evidently proper), Theorem 1.5.4 applies. ∎

Since the reduced space $\tilde{S}^0$ is not (necessarily) a manifold, there is no self-evident definition of the space of "smooth" functions $C^\infty(\tilde{S}^0, \mathbb{R})$. In the present context the following approach is appropriate.

Definition 1.11.4. *A continuous function f on $\tilde{S}^0$ is said to be smooth when there exists an H-invariant smooth function on S whose restriction to $J^{-1}(0)$ quotients to f. In other words,*

$$C^\infty(\tilde{S}^0, \mathbb{R}) := C^\infty(S, \mathbb{R})^H / \mathcal{J}_0^H, \tag{1.87}$$

where $\mathcal{J}_0^H := \mathcal{J}_0 \cap C^\infty(S, \mathbb{R})^H$, and $\mathcal{J}_0$ is the ideal of smooth functions on S that vanish on $J^{-1}(0)$.

When $\tilde{S}^0$ is a manifold, one recovers the usual definition of $C^\infty(\tilde{S}^0, \mathbb{R})$. Otherwise, the main advantage of Definition 1.11.4 is that one obtains a Poisson algebra.

Proposition 1.11.5. *The space $C^\infty(\tilde{S}^0, \mathbb{R})$ is a Poisson algebra under the Poisson bracket inherited from $C^\infty(S, \mathbb{R})$. This bracket coincides with the one corresponding to the symplectic structure on each subspace $\tilde{S}^0_{[K]}$. In other words, the inclusion of each symplectic manifold $\tilde{S}^0_{[K]}$ in $\tilde{S}^0$ is a Poisson map.*

We already know that $C^\infty(S, \mathbb{R})^H$ is a Poisson algebra under the bracket inherited from $C^\infty(S, \mathbb{R})$; cf. the proof of Theorem 1.5.5. We need to prove that $\mathcal{J}_0^H$ is a Poisson ideal in $C^\infty(S, \mathbb{R})^H$. It is trivially an ideal with respect to pointwise multiplication. To show that in addition it is a Lie algebra ideal under the Poisson bracket, we pick $f \in C^\infty(S, \mathbb{R})^H$ and $g \in \mathcal{J}_0^H$, and study $\{f, g\}$. This function is in $C^\infty(S, \mathbb{R})^H$ because of the H-invariance of the Poisson bracket, and vanishes on $J^{-1}(0)$ by I.(2.8), 1.5.7, and 1.3.5.

This proves the first claim. The remainder is obvious from Proposition 1.11.3. ∎

In the case that $\tilde{S}^0$ is not a manifold, one cannot define the Hamiltonian flow of $h \in C^\infty(\tilde{S}^0, \mathbb{R})$ as the solution of I.(2.11), since the notion of a tangent vector is problematic at singular points. However, one can simply say that $\sigma(\cdot)$ is the Hamiltonian flow of h iff I.(2.14) is satisfied for all $f \in C^\infty(\tilde{S}^0, \mathbb{R})$. Existence of the flow may be proved by lifting the flow to S, and uniqueness is eventually a consequence of the fact that due to the properness of the H-action on S, the function space $C^\infty(\tilde{S}^0, \mathbb{R})$ separates points in $\tilde{S}^0$.

We sum up.

Theorem 1.11.6. *Assume that one has a proper and strongly Hamiltonian H-action on a symplectic manifold S, with equivariant momentum map J.*

- *The Marsden–Weinstein quotient $\tilde{S}^0 = J^{-1}(0)/H$ can be decomposed as a (disjoint) union of symplectic manifolds $\tilde{S}^0_{[K]}$, defined by (1.86) with (1.82), referred to as the* **symplectic pieces** *of $\tilde{S}^0$.*
- *The function space $C^\infty(\tilde{S}^0, \mathbb{R})$, defined by (1.87), is a Poisson algebra.*
- *The inclusion of each $\tilde{S}^0_{[K]}$ in $\tilde{S}^0$ is a Poisson map.*
- *Any Hamiltonian flow on $\tilde{S}^0$ preserves the decomposition in question; in fact, any two points in a connected component of a given subspace $\tilde{S}^0_{[K]}$ can be connected by a piecewise smooth Hamiltonian curve.*

Combine (1.85) with Propositions 1.11.3 and 1.11.5. The second claim in the final item follows from Proposition 1.11.2. ∎

It follows that the pair $(\tilde{S}^0, \mathfrak{A}_{\mathbb{R}} = C^\infty(\tilde{S}^0, \mathbb{R}))$ is a Poisson space in the sense of Definition I.2.6.2.

The abstract theory may be illustrated by what is probably the simplest nontrivial example. Consider the standard action of $H = SO(2)$ on $Q = \mathbb{R}^2$, given by

$$\theta : (q^1, q^2) \mapsto (q^1 \cos\theta - q^2 \sin\theta, q^1 \sin\theta + q^2 \cos\theta).$$

According to Lemma III.2.3.1 this lifts to a strongly Hamiltonian action on $S = T^*\mathbb{R}^2$, in which (p_1, p_2) transforms in exactly the same way as (q^1, q^2). This lifted action is not free at the point $(0, 0, 0, 0)$, whose stabilizer is $SO(2)$; the stabilizer of all other points is trivial. The momentum map is

$$J(p, q) = q^1 p_2 - q^2 p_1. \tag{1.88}$$

Hence the level set $J^{-1}(0)$ consists of those (p, q) for which $p = tq$ or $q = 0$ for some $t \in \mathbb{R}$; this set is $\{(\mathbb{R}^2\backslash(0, 0)) \times \mathbb{R}\} \cup \mathbb{R}^2$, where (q^1, q^2, t) in $(\mathbb{R}^2\backslash(0, 0)) \times \mathbb{R}$ stands for (q^1, q^2, tp_1, tp_2), and the second $\mathbb{R}^2$ represents the points $(p_1, p_2, 0, 0)$. Hence the quotient $J^{-1}(0)/SO(2)$ may be identified with $\mathbb{R}^+ \times \mathbb{R}$, which may be thought of as the cotangent bundle $T^*\mathbb{R}^+$. The singularity in the reduced space takes the form of a boundary.

The topology on the reduced space may be computed by noting that the copy of $\mathbb{R}^2$ in $T^*\mathbb{R}^2$ defined by the equations $p_2 = q^2 = 0$ is contained in $J^{-1}(0)$, and has the property that every $SO(2)$-orbit in $J^{-1}(0)$ cuts it in two points. These are related by the action of $\mathbb{Z}^2$ that maps (p_1, q^1) to $(\pm p_1, \pm q^1)$. Since the $\mathbb{R}^2$ in question evidently has the canonical symplectic structure, and $\mathbb{Z}^2$ acts on it by Poisson maps, one infers that

$$S^0 := J^{-1}(0)/SO(2) \simeq \mathbb{R}^2/\mathbb{Z}^2 \tag{1.89}$$

as symplectic spaces. Here $\mathbb{R}^2/\mathbb{Z}^2$ is seen as a topological space with the quotient topology, at the same time being the union of the symplectic manifolds $(0, 0)$ and $(\mathbb{R}^2\backslash(0, 0))/\mathbb{Z}^2$. Thus the isomorphism (1.89) means that one has a homeomorphism in the usual sense, under which the appropriate symplectic subspaces are mapped into each other symplectomorphically.

In summary, the decomposition (1.85) consists of $(0, 0)$ and $(\mathbb{R}^2\backslash(0, 0))/\mathbb{Z}^2$, each of which is a symplectic manifold in its own right. Using invariant theory, it may be shown that smooth functions on S^0 (in the sense of 1.11.4) must correspond to smooth functions h on S that depend only on the $SO(2)$-invariants $(p, p) := p_1^2 + p_2^2$, (q, q), and (q, p). Since $dh = 0$ at $(0, 0, 0, 0)$, one verifies that the two symplectic pieces of S^0 are indeed stable under Hamiltonian flows; cf. Proposition 1.11.2.

This example illustrates a deeper property of singular Marsden–Weinstein reduction, which we will not prove.

Proposition 1.11.7. *Under the assumptions of 1.11.6, in each connected component of $\tilde{S}^0$ one of the symplectic pieces $\tilde{S}^0_{[K^o]}$ is open and dense.*

Since the Poisson bracket on $C^\infty(\tilde{S}^0, \mathbb{R})$ is evidently determined by the symplectic form on $\tilde{S}^0_{[K^o]}$, Proposition 1.11.5 implies that the symplectic structure on all other symplectic subspaces is determined by $\tilde{S}^0_{[K^o]}$.

Finally, we give an example in which the group action is not proper, but the claims of Theorem 1.11.6 nonetheless hold. We continue with the symplectic manifold $S = T^*\mathbb{R}^2$, but now consider an action of $H = \mathbb{R}$, namely

$$t : (p_1, p_2, q^1, q^2) \mapsto (p_1, p_2, q^1 + p_1 t, q^2 - p_2 t). \tag{1.90}$$

An equivariant momentum map for this action is

$$J(p_1, p_2, q^1, q^2) = \tfrac{1}{2}(p_1^2 - p_2^2). \tag{1.91}$$

It follows that J_* fails to be surjective at all "singular" points of the form $(0, 0, q^1, q^2)$, at which it is identically zero; the $\mathbb{R}$-action is not proper precisely at these singular points. The singular points have stabilizer $\mathbb{R}$, whereas the stability group of all other points is trivial. This opens the possibility that $J^{-1}(0)$ might not be a submanifold of S, and this is indeed the case.

The Marsden–Weinstein quotient $S^0 = J^{-1}(0)/\mathbb{R}$ is connected as a topological space, but it does not have a constant dimension as a "manifold": if we look at S^0

as fibered over the subspace $p_1 = \pm p_2$ of $\mathbb{R}^2$, then the fiber above $(0, 0)$ is two-dimensional, whereas at all other points it is one-dimensional. By Definition 1.11.4, the space $C^\infty(S^0, \mathbb{R})$ consists of smooth functions $f \in C^\infty(S, \mathbb{R})$, restricted to $J^{-1}(0)$, that satisfy $\{J, f\} = 0$ on $J^{-1}(0)$. It follows that such an f arbitrarily depends on the p_i, but depends on the q^i through the combination $q^1 p_2 + q^2 p_1$. A study of the Hamiltonian flow on S defined by such functions, and therefore of the corresponding flow on S^0 obtained by projection, shows that S^0 may be decomposed into five symplectic pieces. These are given by the equations $p_1 = p_2 > 0$, $p_1 = p_2 < 0$, $p_1 = -p_2 > 0$, $p_1 = -p_2 < 0$, and $p_1 = p_2 = 0$. Any point in a given piece cannot leave the piece under a Hamiltonian flow. Hence we have the same situation as for proper group actions.

2 Induction

2.1 *Hilbert C^*-Modules*

What follows is the most important mathematical concept in the quantization theory of classical systems obtained by special symplectic reduction (see 1.2).

Definition 2.1.1. *A* **Hilbert C^*-module** *over a C^*-algebra $\mathfrak{B}$ consists of*

- *A complex linear space $\mathcal{E}$.*
- *A right action π_R of $\mathfrak{B}$ on $\mathcal{E}$ (i.e., π_R maps $\mathfrak{B}$ linearly into the space of all linear operators on $\mathcal{E}$, and satisfies $\pi_R(AB) = \pi_R(B)\pi_R(A)$), for which we shall write $\Psi B := \pi_R(B)\Psi$, where $\Psi \in \mathcal{E}$ and $B \in \mathfrak{B}$.*
- *A sesquilinear map $(\, ,\,)_{\mathfrak{B}} : \mathcal{E} \times \mathcal{E} \to \mathfrak{B}$, linear in the second and antilinear in the first entry, satisfying*

$$\langle \Psi, \Phi \rangle_{\mathfrak{B}}^* = \langle \Phi, \Psi \rangle_{\mathfrak{B}}; \tag{2.1}$$

$$\langle \Psi, \Phi B \rangle_{\mathfrak{B}} = \langle \Psi, \Phi \rangle_{\mathfrak{B}} B; \tag{2.2}$$

$$\langle \Psi, \Psi \rangle_{\mathfrak{B}} \geq 0; \tag{2.3}$$

$$\langle \Psi, \Psi \rangle_{\mathfrak{B}} = 0 \;\Leftrightarrow\; \Psi = 0, \tag{2.4}$$

for all $\Psi, \Phi \in \mathcal{E}$ and $B \in \mathfrak{B}$.

The space $\mathcal{E}$ is complete in the norm

$$\|\Psi\| := \|\langle \Psi, \Psi \rangle_{\mathfrak{B}}\|^{\frac{1}{2}}. \tag{2.5}$$

We say that $\mathcal{E}$ is a **Hilbert $\mathfrak{B}$-module**, *and write $\mathcal{E} \rightleftharpoons \mathfrak{B}$.*

One checks that (2.5) is indeed a norm: $\|\Psi\|^2$ equals $\sup\{\omega(\langle \Psi, \Psi \rangle_{\mathfrak{B}})\}$, where the supremum is taken over all states ω on $\mathfrak{B}$. Since each map $\Psi \mapsto \sqrt{\omega(\langle \Psi, \Psi \rangle_{\mathfrak{B}})}$ is a seminorm (i.e., a norm except for positive definiteness) by (2.3), the supremum is a seminorm, which is actually positive definite because of Corollary I.1.4.4 (which applies because $\langle \Psi, \Psi \rangle_{\mathfrak{B}}$ is self-adjoint by (2.1) or (2.3)) and (2.4).

The $\mathfrak{B}$-action on $\mathcal{E}$ is automatically nondegenerate: The property $\Psi B = 0$ for all $B \in \mathfrak{B}$ implies that $\langle \Psi, \Psi \rangle_{\mathfrak{B}} B = 0$ for all B, hence $\langle \Psi, \Psi \rangle_{\mathfrak{B}} = 0$ (when $\mathfrak{B}$ is unital this follows by taking $B = \mathbb{I}$; otherwise, one uses an approximate unit in $\mathfrak{B}$), so that $\Psi = 0$ by (2.4).

When all conditions in 2.1.1 are met except (2.4), so that $\| \cdot \|$ defined by (2.5) is only a seminorm, one simply takes the quotient of $\mathcal{E}$ by its subspace of all null vectors and completes, obtaining a Hilbert C^*-module in that way.

It is useful to note that (2.1) and (2.2) imply that

$$\langle \Psi B, \Phi \rangle_{\mathfrak{B}} = B^* \langle \Psi, \Phi \rangle_{\mathfrak{B}}. \tag{2.6}$$

Example 2.1.2.

1. *Any C^*-algebra $\mathfrak{A}$ is an $\mathfrak{A}$-module $\mathfrak{A} \rightleftharpoons \mathfrak{A}$ over itself, with $\langle A, B \rangle_{\mathfrak{A}} := A^* B$. Note that the norm (2.5) coincides with the C^*-norm by I.(1.15).*
2. *Any Hilbert space $\mathcal{H}$ is a Hilbert $\mathbb{C}$-module $\mathcal{H} \rightleftharpoons \mathbb{C}$ in its inner product.*
3. *Let H be a Hilbert bundle over a compact Hausdorff space Q. The space of continuous sections $\mathcal{E} = \Gamma_0(\mathsf{H})$ of H is a Hilbert C^*-module $\Gamma_0(\mathsf{H}) \rightleftharpoons C(Q)$ over $\mathfrak{B} = C(Q)$; for $\Psi, \Phi \in \Gamma_0(\mathsf{H})$ the function $\langle \Psi, \Phi \rangle_{C(Q)}$ is defined by*

$$\langle \Psi, \Phi \rangle_{C(Q)} : q \mapsto (\Psi(q), \Phi(q)), \tag{2.7}$$

where the inner product is the one in the fiber $\tau^{-1}(q)$. The right action of $C(Q)$ on $\Gamma_0(\mathsf{H})$ is defined by stipulating that Ψf is the section that maps q to $f(q)\Psi(q)$.

In the third example the norm in $\Gamma_0(\mathsf{H})$ is $\|\Psi\| = \sup_q (\Psi(q), \Psi(q))^{\frac{1}{2}}$, so that it is easily seen that $\mathcal{E}$ is complete.

Many Hilbert C^*-modules of interest will be constructed in the following way. A **pre-C^*-algebra** is a *-algebra satisfying all properties of a C^*-algebra except perhaps completeness. Given a pre-C^*-algebra $\tilde{\mathfrak{B}}$, define a **pre-Hilbert $\tilde{\mathfrak{B}}$-module** $\tilde{\mathcal{E}} \rightleftharpoons \tilde{\mathfrak{B}}$ as in Definition 2.1.1, except that the final completeness condition is omitted.

Proposition 2.1.3. *In a pre-Hilbert $\tilde{\mathfrak{B}}$-module (and hence in a Hilbert $\mathfrak{B}$-module) one has the inequalities*

$$\|\Psi B\| \leq \|\Psi\| \, \|B\|; \tag{2.8}$$

$$\langle \Psi, \Phi \rangle_{\mathfrak{B}} \langle \Phi, \Psi \rangle_{\mathfrak{B}} \leq \|\Phi\|^2 \, \langle \Psi, \Psi \rangle_{\mathfrak{B}}; \tag{2.9}$$

$$\|\langle \Psi, \Phi \rangle_{\mathfrak{B}}\| \leq \|\Psi\| \, \|\Phi\|. \tag{2.10}$$

To prove (2.8) one uses (2.6), I.(1.42), I.(1.41), and I.(1.15). For (2.9) we substitute $\Phi\langle \Phi, \Psi \rangle_{\mathfrak{B}} - \Psi$ for Ψ in the inequality $\langle \Psi, \Psi \rangle_{\mathfrak{B}} \geq 0$. Expanding, the first term equals $\langle \Psi, \Phi \rangle_{\mathfrak{B}} \langle \Phi, \Phi \rangle_{\mathfrak{B}} \langle \Phi, \Psi \rangle_{\mathfrak{B}}$. Then use I.(1.42), and replace Φ by $\Phi / \|\Phi\|$. Equations I.(1.15), (2.1), and (2.9) then imply (2.10). ■

Corollary 2.1.4. *A pre-Hilbert $\tilde{\mathfrak{B}}$-module $\tilde{\mathcal{E}} \rightleftharpoons \tilde{\mathfrak{B}}$ can be completed to a Hilbert $\mathfrak{B}$-module.*

One first completes $\tilde{\mathcal{E}}$ in the norm (2.5), obtaining $\mathcal{E}$. Using (2.8), the $\tilde{\mathfrak{B}}$-action on $\tilde{\mathcal{E}}$ extends to a $\mathfrak{B}$-action on $\mathcal{E}$. The completeness of $\mathfrak{B}$ and (2.10) then allow one to extend the $\tilde{\mathfrak{B}}$-valued sesquilinear form on $\tilde{\mathcal{E}}$ to a $\mathfrak{B}$-valued one on $\mathcal{E}$. It is easily checked that the required properties hold by continuity. ■

In Example 2.1.2, it is almost trivial to see that $\mathfrak{A}$, $\mathcal{H}$, and $\Gamma_0(\mathsf{H})$ are the closures of $\tilde{\mathfrak{A}}$ (defined over $\tilde{\mathfrak{A}}$), of a dense subspace $\tilde{\mathcal{E}}$, and of $\Gamma(\mathsf{H})$ (defined over $C_c^\infty(X)$), respectively.

A Hilbert C^*-module $\mathcal{E} \rightleftharpoons \mathfrak{B}$ defines a certain C^*-algebra $C^*(\mathcal{E}, \mathfrak{B})$ that plays an important role in the induction theory in 2.2. A map $A : \mathcal{E} \to \mathcal{E}$ for which there exists a map $A^* : \mathcal{E} \to \mathcal{E}$ such that

$$\langle \Psi, A\Phi \rangle_{\mathfrak{B}} = \langle A^*\Psi, \Phi \rangle_{\mathfrak{B}} \tag{2.11}$$

for all $\Psi, \Phi \in \mathcal{E}$ is called **adjointable.**

Theorem 2.1.5.

1. *An adjointable map is automatically $\mathbb{C}$-linear, $\mathfrak{B}$-linear (that is, $(A\Psi)B = A(\Psi B)$ for all $\Psi \in \mathcal{E}$ and $B \in \mathfrak{B}$), and bounded.*
2. *The adjoint of an adjointable map is unique, and the map $A \mapsto A^*$ defines an involution on the space $C^*(\mathcal{E}, \mathfrak{B})$ of all adjointable maps on $\mathcal{E}$.*
3. *Equipped with this involution, and with the norm I.(1.17), defined with respect to the norm (2.5) on $\mathcal{E}$, the space $C^*(\mathcal{E}, \mathfrak{B})$ is a C^*-algebra.*
4. *Each element $A \in C^*(\mathcal{E}, \mathfrak{B})$ satisfies the bound*
$$\langle A\Psi, A\Psi \rangle_{\mathfrak{B}} \leq \|A\|^2 \langle \Psi, \Psi \rangle_{\mathfrak{B}} \tag{2.12}$$
for all $\Psi \in \mathcal{E}$.
5. *The (defining) action of $C^*(\mathcal{E}, \mathfrak{B})$ on $\mathcal{E}$ is nondegenerate.*

We write $C^*(\mathcal{E}, \mathfrak{B}) \to \mathcal{E} \rightleftharpoons \mathfrak{B}$.

The property of $\mathbb{C}$-linearity is immediate. To establish $\mathfrak{B}$-linearity one uses (2.6); this also shows that $A^* \in C^*(\mathcal{E}, \mathfrak{B})$ when $A \in C^*(\mathcal{E}, \mathfrak{B})$.

To prove boundedness, fix $\Psi \in \mathcal{E}$ and define $T_\Psi : \mathcal{E} \to \mathfrak{B}$ by $T_\Psi \Phi := \langle A^*A\Psi, \Phi \rangle_{\mathfrak{B}}$. It is clear from (2.10) that $\|T_\Psi\| \leq \|A^*A\Psi\|$, so that T_Ψ is bounded. On the other hand, since A is adjointable, one has $T_\Psi \Phi = \langle \Psi, A^*A\Phi \rangle_{\mathfrak{B}}$, so that, using (2.10) once again, one has $\|T_\Psi \Phi\| \leq \|A^*A\Phi\| \, \|\Psi\|$. Hence $\sup\{\|T_\Psi\| \mid \|\Psi\| = 1\} < \infty$ by the principle of uniform boundedness (here it is essential that $\mathcal{E}$ is complete). It then follows from (2.5) that $\|A\| < \infty$.

Uniqueness and involutivity of the adjoint are proved as for Hilbert spaces; the former follows from (2.4), the latter in addition requires (2.1).

The space $C^*(\mathcal{E}, \mathfrak{B})$ is norm-closed, since one easily verifies from (2.11) and (2.5) that if $A_n \to A$ then A_n^* converges to some element, which is precisely A^*. As a norm-closed space of linear maps on a Banach space, $C^*(\mathcal{E}, \mathfrak{B})$ is a Banach algebra, so that its satisfies I.(1.14). To check the remaining axiom, one infers from (2.5) and the definition (2.11) of the adjoint that $\|A\|^2 \leq \|A^*A\|$; using I.(1.14) and the argument leading to I.(1.16), one first obtains $\|A^*\| = \|A\|$, and subsequently I.(1.15).

Finally, it follows from (2.3), I.(1.39), and (2.11) that for fixed $\Psi \in \mathcal{E}$ the map $A \mapsto \langle \Psi, A\Psi \rangle_{\mathfrak{B}}$ from $C^*(\mathcal{E}, \mathfrak{B})$ to $\mathfrak{B}$ is positive. Replacing A by A^*A in I.(1.40) and using I.(1.15) and (2.11) then leads to (2.12).

To prove the final claim, we note that for fixed $\Psi, \Phi \in \mathcal{E}$, the map $Z \mapsto \Psi\langle \Phi, Z \rangle_{\mathfrak{B}}$ is in $C^*(\mathcal{E}, \mathfrak{B})$. When the right-hand side vanishes for all Ψ, Φ, it follows from (2.2) that $A\langle \Phi, Z \rangle_{\mathfrak{B}} = 0$ for all A in the C^*-algebra in $\mathfrak{B}$ generated by $\langle \mathcal{E}, \mathcal{E} \rangle_{\mathfrak{B}}$. In any C^*-algebra, the property $AB = 0$ for all A implies $B = 0$; use an approximate unit if necessary. Hence $\langle \Phi, Z \rangle_{\mathfrak{B}} = 0$ for all $\Phi \in \mathcal{E}$. Taking $\Phi = Z$, we conclude that $Z = 0$ by (2.4). ■

Under a further assumption (which is by no means always met in our examples) one can completely characterize $C^*(\mathcal{E}, \mathfrak{B})$. A Hilbert C^*-module over $\mathfrak{B}$ is called **self-dual** when every bounded $\mathfrak{B}$-linear map $\varphi : \mathcal{E} \to \mathfrak{B}$ is of the form $\varphi(\Psi) = \langle \Phi, \Psi \rangle_{\mathfrak{B}}$ for some $\Phi \in \mathcal{E}$.

Proposition 2.1.6. *In a self-dual Hilbert C^*-module $\mathcal{E} \rightleftharpoons \mathfrak{B}$ the C^*-algebra $C^*(\mathcal{E}, \mathfrak{B})$ coincides with the space $\mathcal{L}(\mathcal{E})^{\mathfrak{B}}$ of all bounded $\mathbb{C}$-linear and $\mathfrak{B}$-linear maps on $\mathcal{E}$.*

In view of Theorem 2.1.5 we need to show only that a given map $A \in \mathcal{L}(\mathcal{E})^{\mathfrak{B}}$ is adjointable. Indeed, for fixed $\Psi \in \mathcal{E}$ define $\varphi_{A,\Psi} : \mathcal{E} \to \mathfrak{B}$ by $\varphi_{A,\Psi}(Z) := \langle \Psi, AZ \rangle_{\mathfrak{B}}$. By self-duality this must equal $\langle \Phi, Z \rangle_{\mathfrak{B}}$ for some Φ, which by definition is $A^*\Psi$. ■

In the context of Example 2.1.2.1, one may wonder what $C^*(\mathfrak{A}, \mathfrak{A})$ is. The map $\rho : \mathfrak{A} \to \mathfrak{B}(\mathfrak{A})$ given by I.(1.27) is easily seen to map $\mathfrak{A}$ into $C^*(\mathfrak{A}, \mathfrak{A})$. This map is isometric (hence injective). Using (2.11), one infers that $A\rho(B) = \rho(AB)$ for all $A, B \in \mathfrak{A}$. Hence $\rho(\mathfrak{A})$ is an ideal in $C^*(\mathfrak{A}, \mathfrak{A})$. When $\mathfrak{A}$ has a unit, one therefore has $C^*(\mathfrak{A}, \mathfrak{A}) = \rho(\mathfrak{A}) \simeq \mathfrak{A}$; cf. the proof of I.1.2.1.

When $\mathfrak{A}$ has no unit, $C^*(\mathfrak{A}, \mathfrak{A})$ is the so-called **multiplier algebra** of $\mathfrak{A}$. One may compute this object by taking a faithful nondegenerate representation $\pi : \mathfrak{A} \to \mathfrak{B}(\mathcal{H})$; it can be shown that $C^*(\mathfrak{A}, \mathfrak{A})$ is isomorphic to the idealizer of $\pi(\mathfrak{A})$ in $\mathfrak{B}(\mathcal{H})$ (this is the set of all $B \in \mathfrak{B}(\mathcal{H})$ for which $B\pi(A) \in \pi(\mathfrak{A})$ for all $A \in \mathfrak{A}$). One thus obtains

$$C^*(C_0(X), C_0(X)) = C_b(X); \tag{2.13}$$

$$C^*(\mathfrak{B}_0(\mathcal{H}), \mathfrak{B}_0(\mathcal{H})) = \mathfrak{B}(\mathcal{H}). \tag{2.14}$$

Equation (2.13) follows by taking $\pi(C_0(X))$ to be the representation on $L^2(X)$ by multiplication operators (where L^2 is defined by a measure with support X), and (2.14) is obtained by taking $\pi(\mathfrak{B}_0(\mathcal{H}))$ to be the defining representation; see the paragraph following I.1.6.3.

In Example 2.1.2.2 the C^*-algebra $C^*(\mathcal{H}, \mathbb{C})$ coincides with $\mathfrak{B}(\mathcal{H})$, because every bounded operator has an adjoint. Its subalgebra $\mathfrak{B}_0(\mathcal{H})$ of compact operators has an analogue in the general setting of Hilbert C^*-modules as well; see 2.4.

2.2 *Rieffel Induction*

Given a Hilbert $\mathfrak{B}$-module $\mathcal{E}$, the goal of the **Rieffel induction** procedure described in this section is to construct a representation π^{χ} of $C^*(\mathcal{E}, \mathfrak{B})$ from a representation π_{χ} of $\mathfrak{B}$. In order to explicate that the induction procedure is a generalization of the GNS-construction I.1.5.4, we first induce from a state ω_{χ} on $\mathfrak{B}$, rather than from a representation π_{χ}.

Construction 2.2.1. *Suppose one has a Hilbert C^*-module $\mathcal{E} \rightleftharpoons \mathfrak{B}$.*

1. *Given a state ω_{χ} on $\mathfrak{B}$, define the sesquilinear form $\widetilde{(\, ,\,)}_0^{\chi}$ on $\mathcal{E}$ by*

$$\widetilde{(\Psi, \Phi)}_0^{\chi} := \omega_{\chi}(\langle \Psi, \Phi \rangle_{\mathfrak{B}}). \tag{2.15}$$

Since ω_{χ} and $\langle\, ,\, \rangle_{\mathfrak{B}}$ are positive (cf. (2.3)), this form is positive semidefinite. Its null space is

$$\tilde{\mathcal{N}}_{\chi} = \{\Psi \in \mathcal{E} \mid \widetilde{(\Psi, \Psi)}_0^{\chi} = 0\}. \tag{2.16}$$

2. *The form $\widetilde{(\, ,\,)}_0^{\chi}$ projects to an inner product $\widetilde{(\, ,\,)}^{\chi}$ on the quotient $\mathcal{E}/\tilde{\mathcal{N}}_{\chi}$. If $\tilde{V}_{\chi} : \mathcal{E} \to \mathcal{E}/\tilde{\mathcal{N}}_{\chi}$ is the canonical projection, then by definition*

$$\widetilde{(\tilde{V}_{\chi}\Psi, \tilde{V}_{\chi}\Phi)}^{\chi} := \widetilde{(\Psi, \Phi)}_0^{\chi}. \tag{2.17}$$

The Hilbert space $\tilde{\mathcal{H}}^{\chi}$ is the closure of $\mathcal{E}/\tilde{\mathcal{N}}_{\chi}$ in this inner product.

3. *The representation $\tilde{\pi}^{\chi}(C^*(\mathcal{E}, \mathfrak{B}))$ is firstly defined on $\mathcal{E}/\tilde{\mathcal{N}}_{\chi} \subset \tilde{\mathcal{H}}^{\chi}$ by*

$$\pi^{\chi}(A)\tilde{V}_{\chi}\Psi := \tilde{V}_{\chi}A\Psi; \tag{2.18}$$

it follows that $\tilde{\pi}^{\chi}$ is well-defined and continuous. Since $\mathcal{E}/\tilde{\mathcal{N}}_{\chi}$ is dense in $\tilde{\mathcal{H}}^{\chi}$, the operator $\tilde{\pi}^{\chi}(A)$ may be defined on all of $\tilde{\mathcal{H}}^{\chi}$ by continuous extension of (2.18), where it satisfies I.(1.19) and I.(1.20).

The GNS-construction I.1.5.4 is a special case of 2.2.1, obtained by choosing $\mathcal{E} = \mathfrak{B} = \mathfrak{A}$, as explained in Example 2.1.2.1.

The analogue of I.(1.58) and the property (2.11) imply that $A\tilde{\mathcal{N}}_{\chi} \subseteq \tilde{\mathcal{N}}_{\chi}$, so that (2.18) is well-defined. The continuity of $\tilde{\pi}^{\chi}$ follows from (2.18) and (2.17), which imply that $\|\tilde{\pi}^{\chi}(A)\tilde{V}_{\chi}\Psi\|^2 = \widetilde{(A\Psi, A\Psi)}_0^{\chi}$. Using (2.15), (2.12), and (2.10) in succession, one obtains

$$\|\tilde{\pi}^{\chi}(A)\| \leq \|A\|. \tag{2.19}$$

On the other hand, I.(1.51) applied to $\mathfrak{B}$, used with the definition of $\|A\|$ for $A \in C^*(\mathcal{E}, \mathfrak{B})$, implies that

$$\|A\| = \sup\{\|\tilde{\pi}^{\chi}(A)\|, \omega_{\chi} \in \mathcal{S}(\mathfrak{B})\}. \tag{2.20}$$

As a corollary, one infers a useful property that will be used, e.g., in the proof of Theorem 2.3.3.

Lemma 2.2.2. *Let $A \in C^*(\mathcal{E}, \mathfrak{B})$ satisfy $\langle \Psi, A\Psi \rangle_{\mathfrak{B}} \geq 0$ for all $\Psi \in \mathcal{E}$. Then $A \geq 0$.*

It follows from (2.20) that $\oplus_{\omega_\chi \in S(\mathfrak{B})} \tilde{\pi}^\chi$ is faithful; the condition on A implies that $\tilde{\pi}^\chi(A) \geq 0$ for all ω_χ. ∎

An illustration of the construction of the induced space is obtained by specializing Example 2.1.2.3 to the case $\mathsf{H} = \mathsf{H}^\chi$, which is a vector bundle over Q with typical fiber $\mathcal{H}_\chi$; cf. III.(2.143). Hence we take $\mathcal{E} = \Gamma_0(\mathsf{H}^\chi)$, the space of continuous sections $\Gamma_0(\mathsf{H}^\chi)$ of H^χ, and $\mathfrak{B} = C(Q)$, made into a Hilbert C^*-module by (2.7) and the canonical right action of $C(Q)$ on $\Gamma_0(\mathsf{H}^\chi)$. For any pure state $\omega_\chi = q$ on $C(Q)$, the induced space is just $\mathcal{H}_\chi$.

When one starts from a representation $\pi_\chi(\mathfrak{B})$ rather than from a state, the general construction proceeds as follows.

Construction 2.2.3. *Start from a Hilbert C^*-module $\mathcal{E} \rightleftharpoons \mathfrak{B}$.*

1. *Given a representation $\pi_\chi(\mathfrak{B})$ on a Hilbert space $\mathcal{H}_\chi$, with inner product $(\, ,\,)_\chi$, the sesquilinear form $(\, ,\,)_0^\chi$ is defined on $\mathcal{E} \otimes \mathcal{H}_\chi$ (algebraic tensor product) by sesquilinear extension of*

$$(\Psi \otimes v, \Phi \otimes w)_0^\chi := (v, \pi_\chi(\langle \Psi, \Phi \rangle_{\mathfrak{B}})w)_\chi, \tag{2.21}$$

where $v, w \in \mathcal{H}_\chi$. This form is positive semidefinite, because $(\, ,\,)_\chi$ and $\langle\, ,\,\rangle_{\mathfrak{B}}$ are. The null space is

$$\mathcal{N}_\chi = \{\tilde{\Psi} \in \mathcal{E} \otimes \mathcal{H}_\chi \mid (\tilde{\Psi}, \tilde{\Psi})_0^\chi = 0\}. \tag{2.22}$$

As in I.(1.58), we may equally well write

$$\mathcal{N}_\chi = \{\tilde{\Psi} \in \mathcal{E} \otimes \mathcal{H}_\chi \mid (\tilde{\Psi}, \tilde{\Phi})_0^\chi = 0 \,\forall \tilde{\Phi} \in \mathcal{E} \otimes \mathcal{H}_\chi\}. \tag{2.23}$$

2. *The form $(\, ,\,)_0^\chi$ projects to an inner product $(\, ,\,)^\chi$ on the quotient $\mathcal{E} \otimes \mathcal{H}_\chi/\mathcal{N}_\chi$, defined by*

$$(V_\chi \tilde{\Psi}, V_\chi \tilde{\Phi})^\chi := (\tilde{\Psi}, \tilde{\Phi})_0^\chi, \tag{2.24}$$

where $V_\chi : \mathcal{E} \otimes \mathcal{H}_\chi \to \mathcal{E} \otimes \mathcal{H}_\chi/\mathcal{N}_\chi$ is the canonical projection. The Hilbert space $\mathcal{H}^\chi$ is the closure of $\mathcal{E} \otimes \mathcal{H}_\chi/\mathcal{N}_\chi$ in this inner product.

3. *The representation $\pi^\chi(C^*(\mathcal{E}, \mathfrak{B}))$ is then defined on $\mathcal{H}^\chi$ by continuous extension of*

$$\pi^\chi(A)V_\chi \tilde{\Psi} := V_\chi(A \otimes \mathbb{I}_\chi \tilde{\Psi}), \tag{2.25}$$

where $\mathbb{I}_\chi$ is the unit operator on $\mathcal{H}_\chi$; this is well-defined, and the extension in question is possible, since

$$\|\pi^\chi(A)\| \leq \|A\|. \tag{2.26}$$

To prove that the form defined in (2.21) is positive semidefinite, we assume that $\pi_\chi(\mathfrak{B})$ is cyclic (if not, the argument below is repeated for each cyclic summand; see I.1.5.2). With $\tilde{\Psi} = \sum_i \Psi_i v_i$ and $v_i = \pi_\chi(B_i)\Omega$ (where Ω is a cyclic vector for $\pi_\chi(\mathfrak{B})$), one then uses (2.21), (2.6), and (2.2) to obtain $(\tilde{\Psi}, \tilde{\Psi})_0^\chi = (v, \pi_\chi(\langle \Phi, \Phi \rangle_{\mathfrak{B}})v)_\chi$ with $\Phi := \sum_i \Psi_i B_i$. Hence $(\tilde{\Psi}, \tilde{\Psi})_0^\chi \geq 0$ by

(2.3) and the positivity of π_χ. By (2.11) and (2.23), the operator $A \otimes \mathbb{I}_\chi$ maps $\mathcal{N}_\chi$ to itself, so that (2.25) is well-defined.

To prove continuity, one computes $\|\pi^\chi(A)V_\chi\tilde{\Psi}\|^2 = (v, \pi_\chi(\langle A\Phi, A\Phi\rangle_{\mathfrak{B}})v)_\chi$ from (2.24) and (2.25); according to (2.12) and the property $\|\pi_\chi(A)\| \leq \|A\|$ (cf. the text after I.1.5.1), this is bounded by $\|A\|^2(v, \pi_\chi(\langle\Phi, \Phi\rangle_{\mathfrak{B}})v)_\chi$. Since the second factor equals $\|V_\chi\tilde{\Psi}\|^2$, this proves (2.26). ■

Similarly, π^χ is faithful and nondegenerate when π_χ is.

To interrelate the above two formulations, one assumes that π_χ is cyclic, with cyclic vector Ω_χ. Then define a linear map $\tilde{U} : \mathcal{E} \to \mathcal{E} \otimes \mathcal{H}_\chi$ by

$$\tilde{U}\Psi := \Psi \otimes \Omega_\chi. \tag{2.27}$$

According to (2.15), (2.21), and I.(1.57), this map has the property

$$(\tilde{U}\Psi, \tilde{U}\Phi)_0^\chi = \widetilde{(\Psi, \Phi)_0}^\chi. \tag{2.28}$$

By (2.17) and (2.24) the map $\tilde{U}$ therefore quotients to a unitary isomorphism $U : \tilde{\mathcal{H}}^\chi \to \mathcal{H}^\chi$, which by (2.18) and (2.25) duly intertwines $\tilde{\pi}^\chi$ and π^χ.

Of course, any subspace of $C^*(\mathcal{E}, \mathfrak{B})$ may be subjected to the induced representation π^χ. This particularly applies when one has a given (pre-) C^*-algebra $\mathfrak{A}$ and a morphism $\pi : \mathfrak{A} \to C^*(\mathcal{E}, \mathfrak{B})$, leading to an induced representation $\pi^\chi(\mathfrak{A})$ on $\mathcal{H}^\chi$. Further to an earlier comment, one verifies that π^χ is nondegenerate when π and π_χ are. With slight abuse of notation we will write $\pi^\chi(A)$ for $\pi^\chi(\pi(A))$. The situation is depicted in Figure 11.

We turn to a practical method of obtaining alternative and more explicit realizations of $\pi^\chi(C^*(\mathcal{E}, \mathfrak{B}))$.

Proposition 2.2.4. *Suppose one has a Hilbert space $\mathcal{H}_*^\chi$ (with inner product denoted by $(\, , \,)_*^\chi$) and a linear map $\tilde{U} : \mathcal{E} \otimes \mathcal{H}_\chi \to \mathcal{H}_*^\chi$ satisfying*

$$(\tilde{U}\tilde{\Psi}, \tilde{U}\tilde{\Phi})_*^\chi = (\tilde{\Psi}, \tilde{\Phi})_0^\chi \tag{2.29}$$

for all $\tilde{\Psi}, \tilde{\Phi} \in \mathcal{E} \otimes \mathcal{H}_\chi$. Then $\tilde{U}$ quotients to an isometric map between $\mathcal{E} \otimes \mathcal{H}_\chi / \mathcal{N}^\chi$ and the image of $\tilde{U}$ in $\mathcal{H}_*^\chi$. When the image is dense this map extends to a unitary

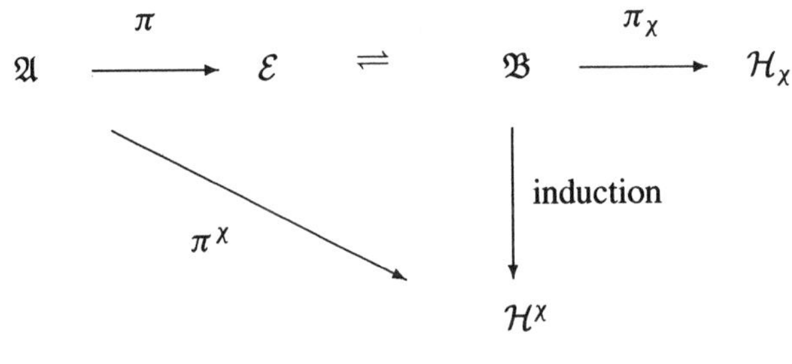

FIGURE 11. Rieffel induction

isomorphism $U : \mathcal{H}^\chi \to \mathcal{H}^\chi_*$. Otherwise, U is unitary between $\mathcal{H}^\chi$ and the closure of the image of $\tilde{U}$.

In any case, the representation $\pi^\chi(C^*(\mathcal{E}, \mathfrak{B}))$ is equivalent to the representation $\pi^\chi_*(C^*(\mathcal{E}, \mathfrak{B}))$, defined by continuous extension of

$$\pi^\chi_*(A)\tilde{U}\tilde{\Psi} := \tilde{U}(A \otimes I_\chi \tilde{\Psi}). \tag{2.30}$$

It is obvious that $\mathcal{N}^\chi = \ker(\tilde{U})$, so that, comparing with (2.25), one indeed has $U \circ \pi^\chi = \pi^\chi_* \circ U$. ■

As an abstract illustration of this technique, consider the space $\mathcal{L}(\overline{\mathcal{E}}, \mathcal{H}_\chi)^{\mathfrak{B}}$ of all antilinear maps $f : \mathcal{E} \to \mathcal{H}_\chi$ satisfying

$$f(\Psi B) = \pi_\chi(B^*)f(\Psi) \tag{2.31}$$

for all $\Psi \in \mathcal{E}$ and $B \in \mathfrak{B}$. Define a map $\tilde{U} : \mathcal{E} \otimes \mathcal{H}_\chi \to \mathcal{L}(\overline{\mathcal{E}}, \mathcal{H}_\chi)^{\mathfrak{B}}$ by linear extension of

$$\tilde{U}(\Psi \otimes v) : \Phi \mapsto \pi_\chi(\langle \Phi, \Psi \rangle_{\mathfrak{B}})v. \tag{2.32}$$

Taking the inner product of $(\tilde{U}(\Psi \otimes v))(\Phi)$ with an arbitrary vector $w \in \mathcal{H}_\chi$, one sees from (2.21) that $\tilde{U}(\tilde{\Psi})$ is the zero map iff $\tilde{\Psi} \in \mathcal{N}_\chi$. The image of $\tilde{U}$ may be equipped with an inner product designed to satisfy (2.29), i.e., we put

$$(\tilde{U}(\Psi \otimes v), \tilde{U}(\Phi \otimes w))^\chi_* := (v, \pi_\chi(\langle \Psi, \Phi \rangle_{\mathfrak{B}})w)_\chi. \tag{2.33}$$

Comparing with (2.21), one sees that (2.29) is indeed satisfied, so that the completion of $\tilde{U}(\mathcal{E} \otimes \mathcal{H}_\chi)$ in this inner product may be identified with the Hilbert space $\mathcal{H}^\chi_*$ of the preceding paragraphs. As we shall see, in practical applications one can sometimes obtain a direct characterization of the space $\mathcal{H}^\chi_*$ thus defined.

So far, we have presented the simplest version of Rieffel induction, in which $\mathcal{E}$ is a Hilbert C^*-module. One may consider the following generalizations.

Firstly, it is not necessary that $\mathcal{E} \rightleftharpoons \mathfrak{B}$ be complete. When $\tilde{\mathcal{E}} \rightleftharpoons \tilde{\mathfrak{B}}$ isn't, an operator A satisfying (2.11) need neither be bounded on $\tilde{\mathcal{E}}$, nor automatically satisfy (2.12). Let an adjointable operator A on $\tilde{\mathcal{E}}$ satisfy

$$\langle A\Psi, A\Psi \rangle_{\mathfrak{B}} \leq C_A^2 \langle \Psi, \Psi \rangle_{\mathfrak{B}} \tag{2.34}$$

for some positive number C_A. Using the reasoning leading to the bound (2.19), one sees that this bound is still satisfied, with $\|A\|$ replaced by C_A. Moreover, defining $\|A\|$ as the smallest number C_A for which (2.34) holds, one can still derive the equality (2.20) in the same way. This equality, then, implies that $\|\cdot\|$ thus defined is a norm on the space $\tilde{C}^*(\tilde{\mathcal{E}}, \tilde{\mathfrak{B}})$ of all maps on $\mathcal{E}$ satisfying (2.11) and (2.34). The proof that $C^*(\mathcal{E}, \mathfrak{B})$ is a C^*-algebra in the complete case may then by copied, showing that in the above norm $\tilde{C}^*(\tilde{\mathcal{E}}, \tilde{\mathfrak{B}})$ is a pre-C^*-algebra.

Using (2.24), (2.15), and (2.5), one shows that $\|\tilde{V}\Psi\| \leq \|\Psi\|$, where the norm on the left-hand side is in $\tilde{\mathcal{H}}^\chi$, and the norm on the right-hand side is the one defined in (2.5). It follows that the induced space $\mathcal{H}^\chi$ (or $\tilde{\mathcal{H}}^\chi$) obtained by Rieffel-inducing from a pre-Hilbert C^*-module is the same as the induced space constructed from its completion.

Secondly, whether or not $\tilde{\mathcal{E}} \rightleftharpoons \tilde{\mathfrak{B}}$ is complete, one may drop the positivity condition (2.3), as long as $\pi_\chi(\langle\Psi, \Psi\rangle_{\mathfrak{B}})$ is a positive operator on $\mathcal{H}_\chi$ for all $\Psi \in \tilde{\mathcal{E}}$. For the latter condition is sufficient to guarantee that the form (2.21) is positive semidefinite. In that case $\tilde{\mathcal{E}} \rightleftharpoons \tilde{\mathfrak{B}}$ is called **π_χ-positive.**

This suggests that one may generalize the Rieffel induction procedure by altogether omitting the C^*-algebra $\mathfrak{B}$ and its representation π^χ. The price one pays for the absence of a $\mathfrak{B}$-action on $\mathcal{E}$ is the stringent positivity condition (2.36) below (which in Rieffel induction is automatically satisfied).

Construction 2.2.5. *Suppose one has a vector space $\tilde{\mathcal{E}}$ and a Hilbert space $\mathcal{H}_\chi$, where $\tilde{\mathcal{E}}$ is equipped with a sesquilinear form $\langle\,,\,\rangle_{\mathfrak{B}(\mathcal{H}_\chi)}$ that takes values in $\mathfrak{B}(\mathcal{H}_\chi)$, and for all $\Psi, \Phi \in \tilde{\mathcal{E}}$ satisfies*

$$\langle\Psi, \Phi\rangle^*_{\mathfrak{B}(\mathcal{H}_\chi)} = \langle\Phi, \Psi\rangle_{\mathfrak{B}(\mathcal{H}_\chi)}; \tag{2.35}$$

$$\sum_{i,j=1}^{n} (v_i, \langle\Psi_i, \Psi_j\rangle v_j)_\chi \geq 0 \tag{2.36}$$

for each $n \in \mathbb{N}$ and all $v_1, \ldots, v_n \in \mathcal{H}_\chi$ and $\Psi_1, \ldots, \Psi_n \in \tilde{\mathcal{E}}$. In other words, the matrix $\mathbb{M} \in \mathfrak{M}_n(\mathfrak{B}(\mathcal{H}_\chi))$ with entries $\mathbb{M}_{ij} = \langle\Psi_i, \Psi_j\rangle_{\mathfrak{B}(\mathcal{H}_\chi)}$ is positive (cf. II.1.4). Then

1. *The form $(\,,\,)_0^\chi$ on $\tilde{\mathcal{E}} \otimes \mathcal{H}_\chi$ is defined by*
$$(\Psi \otimes v, \Phi \otimes w)_0^\chi := (v, \langle\Psi, \Phi\rangle_{\mathfrak{B}(\mathcal{H}_\chi)} w)_\chi. \tag{2.37}$$
2. *The induced Hilbert space $\mathcal{H}^\chi$ is the closure of $\tilde{\mathcal{E}} \otimes \mathcal{H}_\chi / \mathcal{N}_\chi$ (where the null space $\mathcal{N}_\chi$ is defined as in (2.22)) in the inner product $(\,,\,)^\chi$ inherited from $(\,,\,)_0^\chi$, defined as in (2.24).*
3. *The induced action $\pi^\chi(A)$ on $\tilde{\mathcal{H}}^\chi$ of an adjointable operator A on $\tilde{\mathcal{E}}$ satisfying (2.11) and (2.12) with $\mathfrak{B} \to \mathfrak{B}(\mathcal{H}_\chi)$ is defined as in (2.25).*
4. *The induced action $\pi^\chi(B)$ on $\tilde{\mathcal{H}}^\chi$ of an operator $B \in \mathfrak{B}(\mathcal{H}_\chi)$ satisfying*
$$B\langle\Psi, \Phi\rangle_{\mathfrak{B}(\mathcal{H}_\chi)} = \langle\Psi, \Phi\rangle_{\mathfrak{B}(\mathcal{H}_\chi)} B \tag{2.38}$$
for all $\Psi, \Phi \in \tilde{\mathcal{E}}$ is defined by continuous extension of
$$\pi^\chi(B) V_\chi \tilde{\Psi} := V_\chi \mathbb{I} \otimes B\tilde{\Psi}, \tag{2.39}$$
where $\mathbb{I}$ is the unit operator on $\tilde{\mathcal{E}}$.

The form $(\,,\,)_0^\chi$ is positive because of the assumption (2.36). Equation (2.38) implies that B maps $\mathcal{N}_\chi$ into itself, so that $\pi^\chi(B)$ is well-defined. Using I.(1.40), one easily proves the inequality

$$(\mathbb{I} \otimes B\tilde{\Psi}, \mathbb{I} \otimes B\tilde{\Psi})_0^\chi \leq \|B\|^2 (\tilde{\Psi}, \tilde{\Psi})_0^\chi; \tag{2.40}$$

as in the proof of (2.26) this leads to the bound

$$\|\pi^\chi(B)\| \leq \|B\|. \tag{2.41}$$

This equally well holds for adjointable operators A on $\tilde{\mathcal{E}}$. ∎

This generalized induction procedure is known as **Fell induction**. Construction II.1.4.3 is a special case of 2.2.5: Given a completely positive map $\mathcal{Q} : \mathfrak{A} \to \mathfrak{B}$ and a representation $\pi_\chi(\mathfrak{B})$ on $\mathcal{H}_\chi$, we take $\tilde{\mathcal{E}} = \mathfrak{A}$, and define

$$\langle A, B \rangle_{\mathfrak{B}(\mathcal{H}_\chi)} := \pi_\chi(\mathcal{Q}(A^* B)). \tag{2.42}$$

The Hilbert space $\mathcal{H}^\chi$ and the representation π^χ constructed in II.1.4.3 are then exactly the same as the objects defined in 2.2.5.

Interestingly, the map W in II.(1.29) is a special case of a map intrinsically defined in Fell induction in general. Namely, in the setting of Construction 2.2.5, pick a $\Phi \in \tilde{\mathcal{E}}$, and define $W_\Phi : \mathcal{H}_\chi \to \mathcal{H}^\chi$ by

$$W_\Phi v := V_\chi \Phi \otimes v. \tag{2.43}$$

Using (2.35) and (2.37), one computes its adjoint $W_\Phi^* : \mathcal{H}^\chi \to \mathcal{H}_\chi$ as

$$W_\Phi^* V_\chi \Psi \otimes v = \langle \Phi, \Psi \rangle_{\mathfrak{B}(\mathcal{H}_\chi)} v. \tag{2.44}$$

Comparing (2.43) and (2.44) with II.(1.29) and II.(1.30), respectively, one sees that in the special case II.1.4.3 one has to put $\Phi = \mathbb{I}$.

One may pass from Rieffel induction to Fell induction by defining

$$\langle \Psi, \Phi \rangle_{\mathfrak{B}(\mathcal{H}_\chi)} := \pi_\chi(\langle \Psi, \Phi \rangle_{\mathfrak{B}}). \tag{2.45}$$

Condition (2.36) then holds by the argument in the proof of 2.2.3, and may alternatively be derived from Lemma 2.2.2. This is especially useful when one has found a candidate for $\langle\, ,\, \rangle_{\mathfrak{B}}$ that fails to be positive, but that is π_χ-positive. Alternatively, one may have a family of C^*-algebras $\mathfrak{B}^{(n)}$ and sesquilinear forms $\langle\, ,\, \rangle_{\mathfrak{B}^{(n)}}$ for which one would like to take a limit $n \to \infty$, which makes no sense at the level of C^*-algebras. It may then nonetheless be the case that the expressions (2.37), defined via (2.45), do converge. We will see Fell induction in action in 2.10, 3.3, 3.5, and 3.8. In these applications one has $\mathcal{H}_\chi = \mathbb{C}$, for which condition (2.36) reads simply

$$\langle \Psi, \Psi \rangle_{\mathbb{C}} \geq 0. \tag{2.46}$$

A detailed comparison between classical reduction and quantum induction will be given in 2.9.

2.3 *The C^*-Algebra of a Hilbert C^*-Module*

In preparation for the quantum imprimitivity theorem in the next section, and also as a matter of independent interest, we introduce the analogue for Hilbert C^*-modules of the C^*-algebra $\mathfrak{B}_0(\mathcal{H})$ of compact operators on a Hilbert space. This is the C^*-algebra most canonically associated to a Hilbert C^*-module.

Definition 2.3.1. *The collection $C_0^*(\mathcal{E}, \mathfrak{B})$ of "compact" operators on a Hilbert C^*-module $\mathcal{E} \rightleftharpoons \mathfrak{B}$ is the C^*-algebra generated by the adjointable maps of the type $T^{\mathfrak{B}}_{\Psi,\Phi}$, where $\Psi, \Phi \in \mathcal{E}$, and*

$$T^{\mathfrak{B}}_{\Psi,\Phi} Z := \Psi \langle \Phi, Z \rangle_{\mathfrak{B}}. \tag{2.47}$$

We write $C_0^*(\mathcal{E}, \mathfrak{B}) \rightleftharpoons \mathcal{E} \rightleftharpoons \mathfrak{B}$, *and call this a* **quantum dual pair**.

The word "compact" appears between quotation marks because in general, elements of $C_0^*(\mathcal{E}, \mathfrak{B})$ need not be compact operators. The significance of the notation introduced at the end of the definition will emerge from Theorem 2.3.3 below. Using the (trivially proved) properties

$$(T^{\mathfrak{B}}_{\Psi,\Phi})^* = T^{\mathfrak{B}}_{\Phi,\Psi}; \tag{2.48}$$

$$AT^{\mathfrak{B}}_{\Psi,\Phi} = T^{\mathfrak{B}}_{A\Psi,\Phi}; \tag{2.49}$$

$$T^{\mathfrak{B}}_{\Psi,\Phi}A = T^{\mathfrak{B}}_{\Psi,A^*\Phi}, \tag{2.50}$$

where $A \in C^*(\mathcal{E}, \mathfrak{B})$, one verifies without difficulty that $C_0^*(\mathcal{E}, \mathfrak{B})$ is a (closed 2-sided) ideal in $C^*(\mathcal{E}, \mathfrak{B})$, so that it is a C^*-algebra by Theorem 2.1.5. From (2.8) and (2.10) one obtains the bound

$$\|T^{\mathfrak{B}}_{\Psi,\Phi}\| \leq \|\Psi\| \, \|\Phi\|. \tag{2.51}$$

One sees from the final part of the proof of Theorem 2.1.5 that $C_0^*(\mathcal{E}, \mathfrak{B})$ acts nondegenerately on $\mathcal{E}$. When $C_0^*(\mathcal{E}, \mathfrak{B})$ has a unit, it must coincide with $C^*(\mathcal{E}, \mathfrak{B})$.

Proposition 2.3.2.

1. *When* $\mathcal{E} = \mathfrak{B} = \mathfrak{A}$ *(see Example 2.1.2.1) one has*

$$C_0^*(\mathfrak{A}, \mathfrak{A}) \simeq \mathfrak{A}. \tag{2.52}$$

   This leads to the quantum dual pair $\mathfrak{A} \rightleftharpoons \mathfrak{A} \rightleftharpoons \mathfrak{A}$.
2. *For* $\mathcal{E} = \mathcal{H}$ *and* $\mathfrak{B} = \mathbb{C}$ *(see Example 2.1.2.2) one obtains*

$$C_0^*(\mathcal{H}, \mathbb{C}) = \mathfrak{B}_0(\mathcal{H}), \tag{2.53}$$

   whence the quantum dual pair $\mathfrak{B}_0(\mathcal{H}) \rightleftharpoons \mathcal{H} \rightleftharpoons \mathbb{C}$.

One has $T^{\mathfrak{A}}_{\Psi,\Phi} = \rho(\Psi\Phi^*)$; see I.(1.27). Since $\rho : \mathfrak{A} \to \mathfrak{B}(\mathfrak{A})$ is an isometric morphism, the map φ from the linear span of all $T^{\mathfrak{A}}_{\Psi,\Phi}$ to $\mathfrak{A}$, defined by linear extension of $\varphi(T^{\mathfrak{A}}_{\Psi,\Phi}) = \Psi\Phi^*$, is an isometric morphism as well. It is, in particular, injective. When $\mathfrak{A}$ has a unit it is obvious that φ is surjective; in the nonunital case the existence of an approximate unit implies that the linear span of all $\Psi\Phi^*$ is dense in $\mathfrak{A}$. Extending φ to $C_0^*(\mathfrak{A}, \mathfrak{A})$ by continuity, one sees from I.1.3.10.4 that $\varphi(C_0^*(\mathfrak{A}, \mathfrak{A})) = \mathfrak{A}$.

Equation (2.53) follows from Definition I.1.6.3 and the fact that the linear span of all $T^{\mathbb{C}}_{\Psi,\Phi}$ is $\mathfrak{B}_f(\mathcal{H})$. ■

In Example 2.1.2.3 one derives that $C_0^*(\Gamma_0(\mathsf{H}), C(Q))$ is the C^*-algebra of the continuous field of C^*-algebras over Q determined by H (in which $\mathfrak{A}^q = \mathfrak{M}_n(\mathbb{C})$ for all $q \in Q$).

A Hilbert C^*-module $\mathcal{E}$ over $\mathfrak{B}$ is called **full** when the collection $\{\langle\Psi, \Phi\rangle_{\mathfrak{B}}\}$, where Ψ, Φ run over $\mathcal{E}$, is dense in $\mathfrak{B}$. A similar definition applies to pre-Hilbert C^*-modules.

Given a complex linear space $\mathcal{E}$, the **conjugate space** $\overline{\mathcal{E}}$ is equal to $\mathcal{E}$ as a real vector space, but has the conjugate action of complex scalars.

Theorem 2.3.3. *Let $\mathcal{E}$ be a full Hilbert $\mathfrak{B}$-module. The expression*

$$\langle \Psi, \Phi \rangle_{C_0^*(\mathcal{E},\mathfrak{B})} := T^{\mathfrak{B}}_{\Psi,\Phi} \tag{2.54}$$

in combination with the right action $\pi_{\mathrm{R}}(A)\Psi := A^*\Psi$, where $A \in C_0^*(\mathcal{E}, \mathfrak{B})$, defines $\overline{\mathcal{E}}$ as a full Hilbert C^*-module over $C_0^*(\mathcal{E}, \mathfrak{B})$. In other words, from $\mathcal{E} \rightleftharpoons \mathfrak{B}$ one obtains $\overline{\mathcal{E}} \rightleftharpoons C_0^*(\mathcal{E}, \mathfrak{B})$. The left action $\pi_{\mathrm{L}}(B)\Psi := \Psi B^*$ of $\mathfrak{B}$ on $\overline{\mathcal{E}}$ implements the isomorphism

$$C_0^*(\overline{\mathcal{E}}, C_0^*(\mathcal{E}, \mathfrak{B})) \simeq \mathfrak{B}. \tag{2.55}$$

We define $\mathfrak{A}$ to be $C_0^*(\mathcal{E}, \mathfrak{B})$; in the references to (2.1) etc. below one should substitute $\mathfrak{A}$ for $\mathfrak{B}$ when appropriate. The properties (2.1), (2.2), and (2.3) follow from (2.48), (2.50), and Lemma 2.2.2, respectively.

To prove (2.4), we use (2.54) with $\Phi = \Psi$, (2.47) with $Z = \Psi$, (2.2), (2.6), and (2.5) to show that $\langle \Psi, \Psi \rangle_{\mathfrak{A}} = 0$ implies $\| \langle \Psi, \Psi \rangle^3_{\mathfrak{B}} \| = 0$. Since $\langle \Psi, \Psi \rangle_{\mathfrak{B}}$ is positive by (2.3), this implies $\langle \Psi, \Psi \rangle_{\mathfrak{B}} = 0$; hence $\Psi = 0$ by (2.4).

It follows from (2.6) and (2.50) that each $\pi_{\mathrm{L}}(B)$ is adjointable with respect to $\langle\, , \,\rangle_{\mathfrak{A}}$. Moreover, applying (2.5), (2.54), (2.51), and (2.8) one verifies that $\pi_{\mathrm{L}}(B)$ is a bounded operator on $\overline{\mathcal{E}}$ with respect to $\| \cdot \|_{\mathfrak{A}}$, whose norm is majorized by the norm of B in $\mathfrak{B}$. The map π_{L} is injective because $\mathcal{E}$ is nondegenerate as a right $\mathfrak{B}$-module.

Let $\overline{\mathcal{E}}_c$ be the completion of $\overline{\mathcal{E}}$ in $\| \cdot \|_{\mathfrak{A}}$; we will shortly prove that $\overline{\mathcal{E}}_c = \overline{\mathcal{E}}$. It follows from the previous paragraph that $\pi_{\mathrm{L}}(B)$ extends to an operator on $\overline{\mathcal{E}}_c$ (denoted by the same symbol), and that π_{L} maps $\mathfrak{B}$ into $C^*(\overline{\mathcal{E}}_c, \mathfrak{A})$. It is trivial from its definition that π_{L} is a morphism. Now observe that

$$\pi_{\mathrm{L}}(\langle \Psi, \Phi \rangle_{\mathfrak{B}}) = T^{\mathfrak{A}}_{\Psi,\Phi}, \tag{2.56}$$

for the definitions in question imply that

$$T^{\mathfrak{A}}_{\Psi,\Phi} Z = \Psi \langle \Phi, Z \rangle_{\mathfrak{A}} = T^{\mathfrak{B}}_{Z,\Phi} \Psi = Z \langle \Phi, \Psi \rangle_{\mathfrak{B}}. \tag{2.57}$$

The fullness of $\mathcal{E} \rightleftharpoons \mathfrak{B}$ and the definition of $C_0^*(\overline{\mathcal{E}}_c, \mathfrak{A})$ imply that $\pi_{\mathrm{L}} : \mathfrak{B} \to C_0^*(\overline{\mathcal{E}}_c, \mathfrak{A})$ is an isomorphism. In particular, it is norm-preserving by I.1.3.10.5.

The space $\mathcal{E}$ is equipped with two norms by applying (2.5) with $\mathfrak{B}$ or with $\mathfrak{A}$; we write $\| \cdot \|_{\mathfrak{B}}$ and $\| \cdot \|_{\mathfrak{A}}$. From (2.54) and (2.51) one derives

$$\| \Psi \|_{\mathfrak{A}} \leq \| \Psi \|_{\mathfrak{B}}. \tag{2.58}$$

For $\Psi \in \mathcal{E}$ we now use (2.5), the isometric nature of π_{L}, and (2.56) to obtain $\| \Psi \|_{\mathfrak{B}} = \| T^{\mathfrak{A}}_{\Psi,\Psi} \|^{\frac{1}{2}}$. From (2.51) with $\mathfrak{B}$ replaced by $\mathfrak{A}$ one then derives the converse inequality to (2.58), so that $\| \Psi \|_{\mathfrak{A}} = \| \Psi \|_{\mathfrak{B}}$. Hence $\overline{\mathcal{E}}_c = \overline{\mathcal{E}}$, as $\mathcal{E}$ is complete in $\| \cdot \|_{\mathfrak{B}}$ by assumption. In other words, the completeness of $\mathcal{E}$ as a Hilbert $\mathfrak{B}$-module is equivalent to the completeness of $\overline{\mathcal{E}}$ as a Hilbert $\mathfrak{A}$-module.

We have now proved (2.55). Finally, noticing that as a Hilbert C^*-module over $\mathfrak{A}$ the space $\overline{\mathcal{E}}$ is full by definition of $C_0^*(\mathcal{E}, \mathfrak{B})$, the proof of Theorem 2.3.3 is complete. ■

For later reference we record the remarkable identity

$$\langle Z, \Phi \rangle_{C_0^*(\mathcal{E},\mathfrak{B})} \Psi = Z \langle \Phi, \Psi \rangle_{\mathfrak{B}}, \tag{2.59}$$

which is a restatement of (2.57).

2.4 The Quantum Imprimitivity Theorem

Our aim in this section is to prove an operator-algebraic version of Theorem 1.4.1, in which special symplectic reduction is replaced by Rieffel induction. This theorem will be based on the following concept.

Definition 2.4.1. *Two C^*-algebras $\mathfrak{A}$ and $\mathfrak{B}$ are* **Morita equivalent** *when there exists a full Hilbert C^*-module $\mathcal{E}$ over $\mathfrak{B}$ under which $\mathfrak{A} \simeq C_0^*(\mathcal{E}, \mathfrak{B})$. We write $\mathfrak{A} \overset{M}{\sim} \mathfrak{B}$ and $\mathfrak{A} \rightleftharpoons \mathcal{E} \rightleftharpoons \mathfrak{B}$.*

This definition is better behaved than its classical counterpart 1.3.7, for we have

Proposition 2.4.2. *Morita equivalence is an equivalence relation in the class of all C^*-algebras.*

The reflexivity property $\mathfrak{B} \overset{M}{\sim} \mathfrak{B}$ follows from (2.52), which establishes the quantum dual pair $\mathfrak{B} \rightleftharpoons \mathfrak{B} \rightleftharpoons \mathfrak{B}$. Symmetry is implied by (2.55), proving that $\mathfrak{A} \rightleftharpoons \mathcal{E} \rightleftharpoons \mathfrak{B}$ implies $\mathfrak{B} \rightleftharpoons \overline{\mathcal{E}} \rightleftharpoons \mathfrak{A}$.

The proof of transitivity is more involved. When $\mathfrak{A} \overset{M}{\sim} \mathfrak{B}$ and $\mathfrak{B} \overset{M}{\sim} \mathfrak{C}$ we have the chain of quantum dual pairs

$$\mathfrak{A} \rightleftharpoons \mathcal{E}_1 \rightleftharpoons \mathfrak{B} \rightleftharpoons \mathcal{E}_2 \rightleftharpoons \mathfrak{C}.$$

We then form the linear space $\mathcal{E}_1 \otimes_{\mathfrak{B}} \mathcal{E}_2$ (which is the quotient of $\mathcal{E}_1 \otimes \mathcal{E}_2$ by the ideal $\mathcal{I}_{\mathfrak{B}}$ generated by all vectors of the form $\Psi_1 B \otimes \Psi_2 - \Psi_1 \otimes B\Psi_2$), which carries a right action $\pi_{\mathrm{R}}^{\otimes}(\mathfrak{C})$ given by

$$\pi_{\mathrm{R}}^{\otimes}(C)(\Psi_1 \otimes_{\mathfrak{B}} \Psi_2) := \Psi_1 \otimes_{\mathfrak{B}} (\Psi_2 C). \tag{2.60}$$

Moreover, we can define a sesquilinear map $\langle\, ,\, \rangle_{\mathfrak{C}}^{\otimes}$ on $\mathcal{E}_1 \otimes_{\mathfrak{B}} \mathcal{E}_2$ by

$$\langle \Psi_1 \otimes_{\mathfrak{B}} \Psi_2, \Phi_1 \otimes_{\mathfrak{B}} \Phi_2 \rangle_{\mathfrak{C}}^{\otimes} := \langle \Psi_2, \langle \Psi_1, \Phi_1 \rangle_{\mathfrak{B}} \Phi_2 \rangle_{\mathfrak{C}}. \tag{2.61}$$

With (2.60) this satisfies (2.1) and (2.2); as explained prior to (2.6), one may therefore construct a Hilbert C^*-module, denoted by $\mathcal{E}_{\otimes} \rightleftharpoons \mathfrak{C}$. (Remarkably, if one looks at (2.61) as defined on $\mathcal{E}_1 \otimes \mathcal{E}_2$, the null space of (2.5) is easily seen to contain $\mathcal{I}_{\mathfrak{B}}$, but in fact coincides with it, so that in constructing $\mathcal{E}_{\otimes}$ one only needs to complete $\mathcal{E}_1 \otimes_{\mathfrak{B}} \mathcal{E}_2$.)

Apart from the right action $\pi_{\mathrm{R}}^{\otimes}(\mathfrak{C})$, the space $\mathcal{E}_{\otimes}$ carries a left action $\pi_L^{\otimes}(\mathfrak{A})$: The operator

$$\pi_{\mathrm{L}}^{\otimes}(A)(\Psi_1 \otimes_{\mathfrak{B}} \Psi_2) := (A\Psi_1) \otimes_{\mathfrak{B}} \Psi_2 \tag{2.62}$$

is bounded on $\mathcal{E}_1 \otimes_{\mathfrak{B}} \mathcal{E}_2$ and extends to $\mathcal{E}_{\otimes}$. We now claim that

$$C_0^*(\mathcal{E}_{\otimes}, \mathfrak{C}) = \pi_L^{\otimes}(\mathfrak{A}). \tag{2.63}$$

Using (2.47), the definition of $\otimes_{\mathfrak{B}}$, and (2.2), it is easily shown that

$$\pi_{\mathrm{L}}^{\otimes}(T^{\mathfrak{B}}_{\Psi_1\langle\Psi_2,\Phi_2\rangle_{\mathfrak{B}},\Phi_1})\Omega_1 \otimes_{\mathfrak{B}} \Omega_2 = \Psi_1 \otimes_{\mathfrak{B}} \langle\Psi_2, \Phi_2\langle\Phi_1, \Omega_1\rangle_{\mathfrak{B}}\rangle_{\mathfrak{B}}\Omega_2. \tag{2.64}$$

Now use the assumption $C_0^*(\mathcal{E}_2, \mathfrak{C}) = \mathfrak{B}$; as in (2.54), with $\mathfrak{B}$ and $\mathcal{E}$ replaced by $\mathfrak{C}$ and $\mathcal{E}_2$, this yields $\langle\Psi, \Phi\rangle_{\mathfrak{B}} = T^{\mathfrak{C}}_{\Psi,\Phi}$. Substituting this in the right-hand side of (2.64), and using (2.47) with $\mathfrak{B}$ replaced by $\mathfrak{C}$, the right-hand side of (2.64) becomes $\Psi_1 \otimes_{\mathfrak{B}} \Psi_2\langle\Phi_2\langle\Phi_1, \Omega_1\rangle_{\mathfrak{B}}, \Omega_2\rangle_{\mathfrak{C}}$. Using $\Psi B^* = \pi_L(B)\Psi$ (see 2.3.3), (2.11) with $\mathfrak{C}$ instead of $\mathfrak{B}$, (2.61), and (2.47) with $\mathfrak{B}$ replaced by $\mathfrak{C}$, we eventually obtain

$$T^{\mathfrak{C}}_{\Psi_1\otimes_{\mathfrak{B}}\Psi_2,\Phi_1\otimes_{\mathfrak{B}}\Phi_2} = \pi_{\mathrm{L}}^{\otimes}(T^{\mathfrak{B}}_{\Psi_1\langle\Psi_2,\Phi_2\rangle_{\mathfrak{B}},\Phi_1}). \tag{2.65}$$

This leads to the inclusion $C_0^*(\mathcal{E}_{\otimes}, \mathfrak{C}) \subseteq \pi_L^{\otimes}(\mathfrak{A})$. To prove the opposite inclusion, one picks a double sequence $\{\Psi_2^i, \Phi_2^i\}$ such that $\sum_i^N T^{\mathfrak{C}}_{\Psi_2^i,\Phi_2^i}$ is an approximate unit in $\mathfrak{B} = C_0^*(\mathcal{E}_2, \mathfrak{C})$. One has $\lim_N \sum_i^N \Psi_2^i\langle\Phi_2^i, Z\rangle_{\mathfrak{C}} = Z$ from (2.47), and a short computation using (2.47) with (2.61) then yields

$$\lim_N \sum_i^N T^{\mathfrak{C}}_{\Psi_1\otimes_{\mathfrak{B}}\Psi_2^i,\Phi_1\otimes_{\mathfrak{B}}\Phi_2^i} = \pi_{\mathrm{L}}^{\otimes}(T^{\mathfrak{B}}_{\Psi_1,\Phi_1}).$$

Hence $\pi_L^{\otimes}(\mathfrak{A}) \subseteq C_0^*(\mathcal{E}_{\otimes}, \mathfrak{C})$, and combining both inclusions one obtains (2.65).

Therefore, one has the quantum dual pair $\mathfrak{A} \rightleftharpoons \mathcal{E}_{\otimes} \rightleftharpoons \mathfrak{C}$, implying that $\mathfrak{A} \overset{M}{\sim} \mathfrak{C}$. This proves transitivity. ∎

The following simple example of this concept will have nontrivial consequences.

Proposition 2.4.3. *For any Hilbert space $\mathcal{H}$, the C^*-algebra $\mathfrak{B}_0(\mathcal{H})$ of compact operators on $\mathcal{H}$ is Morita equivalent to $\mathbb{C}$, with quantum dual pair $\mathfrak{B}_0(\mathcal{H}) \rightleftharpoons \mathcal{H} \rightleftharpoons \mathbb{C}$. In particular, the matrix algebra $\mathfrak{M}_n(\mathbb{C})$ is Morita equivalent to $\mathbb{C}$.*

This is immediate from (2.53). In the finite-dimensional case one has $\mathfrak{M}_n(\mathbb{C}) \rightleftharpoons \mathbb{C}^n \rightleftharpoons \mathbb{C}$, where $\mathfrak{M}_n(\mathbb{C})$ and $\mathbb{C}$ act on $\mathbb{C}^n$ in the usual way. The double Hilbert C^*-module structure is completed by specifying

$$\langle z, w\rangle_{\mathbb{C}} = \overline{z}^i w^i;$$
$$(\langle z, w\rangle_{\mathfrak{M}_n(\mathbb{C})})_{ij} = z^i\overline{w}^j, \tag{2.66}$$

from which one easily verifies (2.59). ∎

In practice, the following way to construct quantum dual pairs, and therefore Morita equivalences, is useful.

Proposition 2.4.4. *Suppose one has*

- *two pre-C^*-algebras $\tilde{\mathfrak{A}}$ and $\tilde{\mathfrak{B}}$;*
- *a full pre-Hilbert $\tilde{\mathfrak{B}}$-module $\tilde{\mathcal{E}}$;*
- *a left action of $\tilde{\mathfrak{A}}$ on $\tilde{\mathcal{E}}$, such that $\overline{\tilde{\mathcal{E}}}$ can be made into a full pre-Hilbert $\tilde{\mathfrak{A}}$-module with respect to the right action $\pi_{\mathrm{R}}(A)\Psi := A^*\Psi$;*
- *the identity*

$$\langle\Psi, \Phi\rangle_{\tilde{\mathfrak{A}}} Z = \Psi\langle\Phi, Z\rangle_{\tilde{\mathfrak{B}}} \tag{2.67}$$

(for all $\Psi, \Phi, Z \in \tilde{\mathcal{E}}$) relating the two Hilbert C^*-module structures;

- *the bounds (for all $A \in \tilde{\mathfrak{A}}$ and $B \in \tilde{\mathfrak{B}}$)*

$$\langle \Psi B, \Psi B \rangle_{\tilde{\mathfrak{A}}} \leq \|B\|^2 \langle \Psi, \Psi \rangle_{\tilde{\mathfrak{A}}}; \tag{2.68}$$

$$\langle A\Psi, A\Psi \rangle_{\tilde{\mathfrak{B}}} \leq \|A\|^2 \langle \Psi, \Psi \rangle_{\tilde{\mathfrak{B}}}. \tag{2.69}$$

Then $\mathfrak{A} \overset{M}{\sim} \mathfrak{B}$, with quantum dual pair $\mathfrak{A} \rightleftharpoons \mathcal{E} \rightleftharpoons \mathfrak{B}$, where $\mathcal{E}$ is the completion of $\tilde{\mathcal{E}}$ as a Hilbert $\mathfrak{B}$-module.

Using Corollary 2.1.4 we first complete $\tilde{\mathcal{E}}$ to a Hilbert $\mathfrak{B}$-module $\mathcal{E}$. By (2.69), which implies $\|A\Psi\| \leq \|A\| \, \|\Psi\|$ for all $A \in \tilde{\mathfrak{A}}$ and $\Psi \in \tilde{\mathcal{E}}$, the action of $\tilde{\mathfrak{A}}$ on $\tilde{\mathcal{E}}$ extends to an action of $\mathfrak{A}$ on $\mathcal{E}$. Similarly, we complete $\overline{\tilde{\mathcal{E}}}$ to a Hilbert $\mathfrak{A}$-module $\overline{\mathcal{E}}_c$; by (2.68) the left action $\pi_{\mathrm{L}}(B)\Psi := \Psi B^*$ extends to an action of $\mathfrak{B}$ on $\overline{\mathcal{E}}_c$. As in the proof of Theorem 2.3.3, one derives (2.58) and its converse for $\Psi \in \tilde{\mathcal{E}}$, so that the $\mathfrak{B}$-completion $\mathcal{E}$ of $\tilde{\mathcal{E}}$ coincides with the $\mathfrak{A}$-completion $\overline{\mathcal{E}}_c$ of $\overline{\tilde{\mathcal{E}}}$; that is, $\overline{\mathcal{E}}_c = \overline{\mathcal{E}}$.

Since $\overline{\mathcal{E}}$ is a full pre-Hilbert $\tilde{\mathfrak{A}}$-module, the $\mathfrak{A}$-action on $\mathcal{E}$ is injective, hence faithful. It follows from (2.67), Theorem 2.3.3, and (once again) the fullness of $\overline{\mathcal{E}}$ that $\mathfrak{A} \simeq C_0^*(\mathcal{E}, \mathfrak{B})$. In particular, each $A \in \mathfrak{A}$ automatically satisfies (2.11). ∎

Clearly, (2.67) is inspired by (2.59), into which it is turned after use of 2.4.4.

For example, one may take $\mathfrak{A} = \mathfrak{B}_0(L^2(Q))$ (where Q is a manifold), whose dense subalgebra $\tilde{\mathfrak{A}}$ consists of the Hilbert–Schmidt operators with kernel in $C_c^\infty(Q \times Q)$. This subalgebra acts on $\tilde{\mathcal{E}} = C_c^\infty(Q)$ in the obvious way. Further taking $\tilde{\mathfrak{B}} = \mathfrak{B} = \mathbb{C}$, with self-evident action on $\tilde{\mathcal{E}}$, one generalizes (2.66) to

$$\langle \Psi, \Phi \rangle_{\mathbb{C}} = (\Psi, \Phi);$$

$$\langle \Psi, \Phi \rangle_{C_c^\infty(Q \times Q)}(q, q') = \Psi(q)\overline{\Phi(q')}. \tag{2.70}$$

The bounds (2.68) and (2.69) are trivially satisfied, so that in this case Proposition 2.4.4 reconfirms 2.4.3.

After this preparation, we pass to the **quantum imprimitivity theorem**; cf. its classical analogue Theorem 1.4.1.

Theorem 2.4.5. *There is a bijective correspondence between the nondegenerate representations of Morita-equivalent C^*-algebras $\mathfrak{A}$ and $\mathfrak{B}$, preserving direct sums and irreducibility. This correspondence is as follows.*

Let the pertinent quantum dual pair be $\mathfrak{A} \rightleftharpoons \mathcal{E} \rightleftharpoons \mathfrak{B}$. When $\pi_\sigma(\mathfrak{A})$ is a representation on a Hilbert space $\mathcal{H}_\sigma$, there exists a representation $\pi_\chi(\mathfrak{B})$ on a Hilbert space $\mathcal{H}_\chi$ such that π_σ is equivalent to the Rieffel-induced representation π^χ defined by (2.25) and the above quantum dual pair.

In the opposite direction, a given representation $\pi_\chi(\mathfrak{B})$ is equivalent to the Rieffel-induced representation π^σ, defined with respect to some representation $\pi_\sigma(\mathfrak{A})$ and the quantum dual pair $\mathfrak{B} \rightleftharpoons \overline{\mathcal{E}} \rightleftharpoons \mathfrak{A}$.

Taking $\pi_\sigma(\mathfrak{A}) = \pi^\chi(\mathfrak{A})$ as just defined, one has $\pi^\sigma(\mathfrak{B}) \simeq \pi_\chi(\mathfrak{B})$. Conversely, taking $\pi_\chi(\mathfrak{B}) = \pi^\sigma(\mathfrak{B})$, one has $\pi^\chi(\mathfrak{A}) \simeq \pi_\sigma(\mathfrak{A})$.

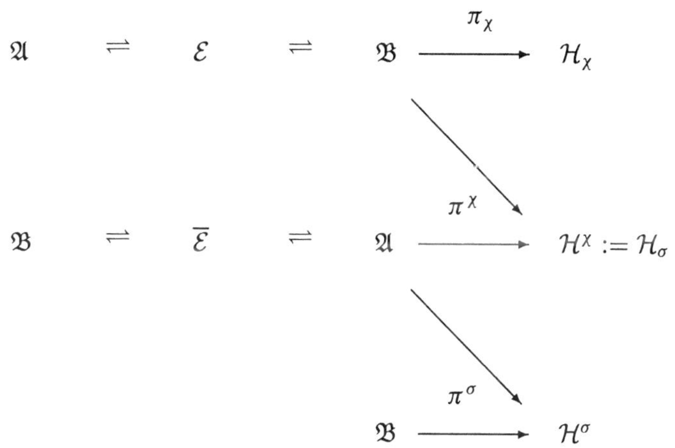

FIGURE 12. Quantum imprimitivity theorem: $\mathcal{H}^\sigma \simeq \mathcal{H}_\chi$ and $\pi^\sigma \simeq \pi_\chi$

See Figure 12. The idea of the proof is the same as in 1.4.1; its execution is, in fact, simpler. Starting with $\pi_\chi(\mathfrak{B})$, we construct $\pi^\chi(\mathfrak{A})$ with Rieffel induction from the quantum dual pair $\mathfrak{A} \rightleftharpoons \mathcal{E} \rightleftharpoons \mathfrak{B}$, relabel this representation as $\pi_\sigma(\mathfrak{A})$, and move on to construct $\pi^\sigma(\mathfrak{B})$ from Rieffel induction with respect to the quantum dual pair $\mathfrak{B} \rightleftharpoons \overline{\mathcal{E}} \rightleftharpoons \mathfrak{A}$. We then construct a unitary map $U : \mathcal{H}^\sigma \to \mathcal{H}_\chi$ that intertwines π^σ and π_χ.

We first define $\tilde{U} : \overline{\mathcal{E}} \otimes \mathcal{E} \otimes \mathcal{H}_\chi \to \mathcal{H}_\chi$ by linear extension of

$$\tilde{U}\Psi \otimes \Phi \otimes v := \pi_\chi(\langle \Psi, \Phi \rangle_{\mathfrak{B}})v. \tag{2.71}$$

Note that $\tilde{U}$ is indeed $\mathbb{C}$-linear. Using (2.71), the properties I.(1.20) and I.(1.19) with φ replaced by π_χ, (2.21), and (2.47), one obtains

$$(\tilde{U}\Psi_1 \otimes \Phi_1 \otimes v_1, \tilde{U}\Psi_2 \otimes \Phi_2 \otimes v_2)_\chi = (\Phi_1 \otimes v_1, T^{\mathfrak{B}}_{\Psi_1,\Psi_2}\Phi_2 \otimes v_2)^\chi_0. \tag{2.72}$$

Now use the assumption $\mathfrak{A} = C^*_0(\mathcal{E}, \mathfrak{B})$ to use (2.59), and subsequently (2.24) and (2.25), all read from right to left. The right-hand side of (2.72) is then seen to be equal to $(V_\chi\Phi_1 \otimes v_1, \pi^\chi(\langle \Psi_1, \Psi_2 \rangle_{\mathfrak{A}})V_\chi\Phi_2 \otimes v_2)^\chi$. Now put $\pi^\chi = \pi_\sigma$ and $\mathcal{H}^\chi = \mathcal{H}_\sigma$, and use (2.21) and (2.24) from right to left, with χ replaced by σ. This shows that

$$(\tilde{U}\Psi_1 \otimes \Phi_1 \otimes v_1, \tilde{U}\Psi_2 \otimes \Phi_2 \otimes v_2)_\chi = (V_\sigma(\Psi_1 \otimes V_\chi\Phi_1 \otimes v_1), V_\sigma(\Psi_2 \otimes V_\chi\Phi_2 \otimes v_2))^\sigma. \tag{2.73}$$

In particular, $\tilde{U}$ annihilates $\Psi \otimes \tilde{\Phi}$, where $\tilde{\Phi} \in \mathcal{E} \otimes \mathcal{H}_\chi$, whenever $\tilde{\Phi} \in \mathcal{N}_\chi$ or $\Psi \otimes V_\chi\tilde{\Phi} \in \mathcal{N}_\sigma$. Hence we see from the construction firstly of $\mathcal{H}^\chi = \mathcal{H}_\sigma$ from $\mathcal{E} \otimes \mathcal{H}_\chi$, and secondly of $\mathcal{H}^\sigma$ from $\overline{\mathcal{E}} \otimes \mathcal{H}_\sigma$ (cf. 2.2.3), that $\tilde{U}$ descends to an isometry $U : \mathcal{H}^\sigma \to \mathcal{H}_\chi$, defined by linear extension of

$$UV_\sigma(\Psi \otimes V_\chi\Phi \otimes v) := \tilde{U}\Psi \otimes \Phi \otimes v = \pi_\chi(\langle \Psi, \Phi \rangle_{\mathfrak{B}})v. \tag{2.74}$$

Using the assumptions that the Hilbert C^*-module $\mathcal{E} \rightleftharpoons \mathfrak{B}$ is full and that the representation $\pi_\chi(\mathfrak{B})$ is nondegenerate, we see that the range of $\tilde{U}$ and hence of U is dense in $\mathcal{H}_\chi$, so that U is unitary.

To verify that U intertwines π^σ and π_χ, we use (2.74) and (2.25), with χ replaced by σ, to compute

$$U\pi^\sigma(B)V_\sigma(\Psi \otimes V_\chi\Phi \otimes v) = \pi_\chi(\langle \pi_L(B)\Psi, \Phi\rangle_{\mathfrak{B}})v, \tag{2.75}$$

where the left action of $B \in \mathfrak{B}$ on $\Psi \in \overline{\mathcal{E}}$ is as defined in 2.3.3. Thus writing $\pi_L(B)\Psi = \Psi B^*$, using (2.6), I.(1.19) with φ replaced by π_χ, and (2.74) from right to left, the right-hand side of (2.75) is seen to be $\pi_\chi(B)UV_\sigma(\Psi \otimes V_\chi\Phi \otimes v)$. Hence $U\pi^\sigma(B) = \pi_\chi(B)U$ for all $B \in \mathfrak{B}$.

Using the proof that the Morita equivalence relation is symmetric (see 2.4.2), one immediately sees that the construction works in the opposite direction as well.

It is easy to verify that $\pi_\chi = \pi_{\chi^1} \oplus \pi_{\chi^2}$ leads to $\pi^\chi = \pi^{\chi^1} \oplus \pi^{\chi^2}$. This also proves that the bijective correspondence $\pi_\chi(\mathfrak{B}) \leftrightarrow \pi^\chi(\mathfrak{A})$ preserves irreducibility: When π_χ is irreducible and π^χ isn't, one puts $\pi^\chi = \pi_\sigma$ as above, decomposes $\pi_\sigma = \pi_{\sigma^1} \oplus \pi_{\sigma^2}$, then decomposes the induced representation $\pi^\sigma(\mathfrak{B})$ as $\pi^\sigma = \pi^{\sigma^1} \oplus \pi^{\sigma^2}$, and thus arrives at a contradiction, since $\pi^\sigma \simeq \pi_\chi$. ∎

Combined with Proposition 2.4.3, this theorem leads to a new proof of Corollary I.2.2.6. Furthermore, in the light of the example given after the proof of Proposition 2.4.4, the first part of the proof of Theorem III.3.7.1 is now seen to be an application of Theorem 2.4.5.

2.5 *Quantum Marsden–Weinstein Reduction*

We come to a class of examples of Hilbert C^*-modules and Rieffel induction which is of central importance to applications in physics. What follows may be seen as the quantum counterpart of the Marsden–Weinstein symplectic reduction procedure in 1.5. For simplicity, proofs are given only for the unimodular case (recall that every compact group is unimodular).

For clarity of presentation, we do not start with the most general assumptions; the following result will be generalized in due course.

Theorem 2.5.1. *Let U be a representation of a compact Lie group H on a Hilbert space $\mathcal{H}$, with corresponding representation π of the group C^*-algebra $C^*(H)$; cf. III.(1.89). The formula*

$$\pi_{\mathrm{R}}(f) = \int_H dh\, f(h)U(h)^{-1} \tag{2.76}$$

defines a right action π_{R} of $C^*(H)$ by continuous extension from $f \in C_c^\infty(H)$. In conjunction with the map $\langle\, ,\, \rangle_{C^*(H)} : \mathcal{H} \times \mathcal{H} \to C^*(H)$, defined by

$$\langle \Psi, \Phi\rangle_{C^*(H)} : h \mapsto (\Psi, U(h)\Phi), \tag{2.77}$$

one obtains a pre-Hilbert C^*-module $\mathcal{H} \rightleftharpoons C^*(H)$. Completion therefore produces a Hilbert $C^*(H)$-module.

Using III.(1.80) with $c = 1$, it is trivial to check that (2.76) defines a right action. The verification of (2.1) and (2.2) is equally straightforward. Because U is continuous, the function defined by (2.77) lies in $C(H)$, and therefore in $C^*(H)$. For $\Psi \neq 0$ this function is nonzero, as it is nonzero at least at the identity e. Hence (2.4) follows.

To prove (2.3), consider the isomorphism $C^*(H) \simeq \pi_{\mathrm{L}}(C^*(H))$; this follows from III.(1.95) and the fact that compact groups are amenable, so that $C^*(H) \simeq C^*_r(H)$. Picking $\Omega \in L^2(H)$, a shift of variables shows that

$$(\Omega, \pi_{\mathrm{L}}(\langle\Psi, \Psi\rangle_{C^*(H)})\Omega)_{L^2(H)} = \|\pi_{\mathrm{R}}(\Omega)\Psi\|^2; \tag{2.78}$$

on the right-hand side Ω is regarded as an element of $C^*(H)$, which is justified, since for compact H one has the inclusion $L^2(H) \subset C^*(H)$. This proves the desired positivity.

A different proof of positivity, which does not use the isomorphism between $C^*(H)$ and $C^*_r(H)$, is as follows. Consider the function 1_H. Assuming that the Haar measure dh is normalized to unit volume, one sees from III.(1.80) that $1_H * 1_H = 1_H$, whereas III.(1.81) shows that 1_H is self-adjoint in $C^*(H)$. Hence 1_H lies in the positive cone of $C^*(H)$ by I.(1.39). Accordingly, for any representation $\pi_\rho(C^*(H))$ on a Hilbert space $\mathcal{H}_\rho$, with inner product $(\,,\,)_\rho$, one has

$$\int_H dh\, (\Psi_\rho, U(h)\Psi_\rho)_\rho = (\Psi_\rho, \pi_\rho(1_H)\Psi_\rho)_\rho \geq 0.$$

Applying this with $\pi_\rho = \pi \otimes \pi_\chi$, where π_χ is arbitrary, and choosing $\Psi_\rho = \Psi \otimes \Psi_\chi$ for some $\Psi_\chi \in \mathcal{H}_\chi$, one obtains

$$(\Psi_\chi, \pi_\chi(\langle\Psi, \Psi\rangle_{C^*(H)})\Psi_\chi)_\chi \geq 0.$$

Hence $\langle\Psi, \Psi\rangle_{C^*(H)} \geq 0$ by the proof of Theorem I.1.1.8. We have now verified all conditions for a pre-Hilbert C^*-module; completion is possible by Corollary 2.1.4. ∎

To see what the completion may look like, consider the example $H = U(1)$ in the regular representation $\pi = \pi_{\mathrm{L}}$ on $\mathcal{H} = L^2(U(1))$. Fourier-transforming $L^2(U(1))$ to ℓ^2, one infers that $C^*(U(1)) \simeq \ell_0$; cf. I.1.6.1 and III.(1.86). From (2.77) one derives that on ℓ^2 one has

$$\langle\hat{\Psi}, \hat{\Phi}\rangle_{\ell_0} : n \mapsto \overline{\hat{\Psi}_n}\hat{\Phi}_n. \tag{2.79}$$

Since the norm in ℓ_0 is the sup-norm, it follows that the Hilbert C^*-norm (2.5) on ℓ^2 is the sup-norm as well. We conclude that the completion of $\ell^2 \rightleftharpoons \ell_0$ is $\ell_0 \rightleftharpoons \ell_0$ (sine the commutative C^*-algebra ℓ_0 is already complete).

In Theorem 2.5.6 we will generalize this example to arbitrary Lie groups with multiplier. Partly in preparation for this generalization, and partly as a matter of interest for physics, we first consider a generalization of 2.5.1 to representations with multiplier (see III.1.3 and III.1.5).

Proposition 2.5.2. *Let c be a multiplier on a compact Lie group H, and let $U(H)$ be a $\overline{c}$-representation on a Hilbert space $\mathcal{H}$. Replacing $C^*(H)$ by the twisted group algebra $C^*(H,c)$, all statements and formulae of Theorem 2.5.1 hold.*

The argument is essentially the same as for 2.5.1. In the first proof of positivity one should, of course, include the factor c in III.(1.83). Using III.(1.65), III.(1.92), and III.(1.29), which leads to the cancellation of all factors c, (2.78) still follows. In the second proof of positivity one should take π_χ to be a representation of $C^*(H,c)$, so that π_ρ is a representation of $C^*(H)$ (since U and U_χ have multiplier $\overline{c}$ and c, respectively, and $\overline{c}c = 1$). ■

Here it is crucial that (2.76) contains $U(h)^{-1}$ rather than $U(h^{-1})$. In view of Theorems III.1.4.4 and III.1.9.5, and the last paragraph in 1.5, we may regard the above "twisted" version of the construction in 2.5.1 as a quantum analogue of Marsden–Weinstein reduction for momentum maps that are not Co-equivariant.

We now turn to Rieffel induction in the case at hand. While an almost trivial matter, this is nonetheless fairly instructive, especially for the purpose of comparison with the noncompact case treated below. For simplicity we look only at ordinary representations ($c = 1$).

Proposition 2.5.3. *In the situation of Theorem 2.5.1, apply Construction 2.2.3 with $\mathfrak{B} = C^*(H)$ and a nondegenerate representation $\pi_\chi(C^*(H))$, or, equivalently, a representation $U_\chi(H)$.*

The induced space $\mathcal{H}^\chi$ is isomorphic to $p_{\mathrm{id}}(\mathcal{H}\otimes\mathcal{H}_\chi)$, where p_{id} is the projection onto the subspace of $\mathcal{H}\otimes\mathcal{H}_\chi$ transforming trivially under the representation $U\otimes U_\chi(H)$.

Any bounded operator A on $\mathcal{H}$ commuting with $U(H)$ satisfies (2.11) and (2.12), so that its induced representative $\pi^\chi(A)$ on $\mathcal{H}^\chi$ may be defined. Under the isomorphism of the previous paragraph, $\pi^\chi(A)$ is the restriction of $A\otimes\mathbb{I}_\chi$ to $P_{\mathrm{id}}(\mathcal{H}\otimes\mathcal{H}_\chi)$.

By the discussion surrounding (2.34), we may start from the pre-Hilbert C^*-module $\mathcal{H} \rightleftharpoons C^*(H)$. Using III.(1.89) with π replaced by π_χ and (2.77), one computes the inner product (2.21) as

$$(\Psi\otimes v, \Phi\otimes w)_0^\chi = \int_H dh\, (\Psi, U(h)\Phi)(v, U_\chi(h)w)_\chi. \tag{2.80}$$

For arbitrary $\tilde{\Psi},\tilde{\Phi}\in\mathcal{H}\otimes\mathcal{H}_\chi$ one therefore has

$$(\tilde{\Psi},\tilde{\Phi})_0^\chi = \int_H dh\,(\tilde{\Psi}, U\otimes U_\chi(h)\tilde{\Phi})_{\mathcal{H}\otimes\mathcal{H}_\chi}. \tag{2.81}$$

Since the integrand is bounded and $\int_H dh = 1$, one may bring the integral over H inside the inner product. The well-known expression

$$p_{\mathrm{id}} = \int_H dh\, U(h) \tag{2.82}$$

for the projection on the trivial representation of a compact group gives

$$(\tilde{\Psi}, \tilde{\Phi})_0^\chi = (\tilde{\Psi}, p_{\mathrm{id}}\tilde{\Phi})_{\mathcal{H}\otimes\mathcal{H}_\chi}. \tag{2.83}$$

Hence the null space $\mathcal{N}_\chi$ is the orthogonal complement $(p_{\mathrm{id}}(\mathcal{H}\otimes\mathcal{H}_\chi))^\perp$, and since for a closed subspace $\mathcal{K} \subset \mathcal{H}$ one has $\mathcal{H}/\mathcal{K}^\perp \simeq \mathcal{K}$, the first claim follows.

To prove (2.12) for $A \in U(H)'$ we take a vector state ω_χ on $C^*(H)$ such that $\omega_\chi(B) = (\Omega_\chi, \pi_\chi(B)\Omega_\chi)_\chi$, and use (2.83) to obtain

$$\omega_\chi(\langle A\Psi, A\Psi\rangle_{C^*(H)}) = (A \otimes \mathbb{I}_\chi \Psi \otimes \Psi_\chi, p_{\mathrm{id}}(A \otimes \mathbb{I}_\chi \Psi \otimes \Psi_\chi))_{\mathcal{H}\otimes\mathcal{H}_\chi}. \tag{2.84}$$

The assumption $A \in U(H)'$ implies that $A \otimes \mathbb{I}_\chi$ commutes with p_{id}, so that the right-hand side is bounded by $\|A\|^2(\Psi\otimes\Psi_\chi, p_{\mathrm{id}}(\Psi\otimes\Psi_\chi))_{\mathcal{H}\otimes\mathcal{H}_\chi}$. In this expression we rewrite the second factor by using (2.84) with $A = \mathbb{I}$ from right to left. This yields

$$\omega_\chi(\langle A\Psi, A\Psi\rangle_{C^*(H)}) \leq \|A\|^2\omega_\chi(\langle \Psi, \Psi\rangle_{C^*(H)}).$$

Since this is true for all vector states, (2.12) follows.

Finally, the identification of $\pi^\chi(A)$ with $A \restriction p_{\mathrm{id}}(\mathcal{H} \otimes \mathcal{H}_\chi)$ is obvious. ∎

Let us now examine a possible generalization of the construction in Theorem 2.5.1 to the case where the Lie group H is merely locally compact. When H is noncompact, the function defined by (2.77) will not, in general, lie in $C^*(H)$ for all $\Psi, \Phi \in \mathcal{H}$. For example, for $\mathcal{H} = L^2(\mathbb{R})$ and $U = U_R$, equation III.(1.94) and its derivation implies that $\langle\Psi, \Psi\rangle_{C^*(\mathbb{R})}$ as defined by (2.77) lies in $C^*(\mathbb{R})$ iff Ψ is such that its Fourier transform lies in $C_0(\mathbb{R})$. There certainly exist functions $\Psi \notin L^1(\mathbb{R})$, $\Psi \in L^2(\mathbb{R})$, for which this is not the case.

The way out is simply to try to find a dense subspace of $\mathcal{H}$ that does have the required property. As this subspace will not be stable under the action of $C^*(H)$, one in addition needs to identify a suitable dense subalgebra of $C^*(H)$. It simplifies the discussion to do this once and for all, taking $C_c^\infty(H)$.

Theorem 2.5.4. *Let U be a representation of a Lie group H on a Hilbert space $\mathcal{H}$ containing a dense subspace $\tilde{\mathcal{E}}$ that for each $f \in C_c^\infty(H)$ is stable under $\pi_{\mathrm{R}}(f)$ (as defined by (2.76)) and for which the function $h \mapsto (\Psi, U(h)\Phi)$ lies in $C_c^\infty(H)$ for all $\Psi, \Phi \in \tilde{\mathcal{E}}$.*

For all $\Psi \in \tilde{\mathcal{E}}$ the function $h \mapsto (\Psi, U(h)\Psi)$ is a positive element of the reduced group algebra $C_r^*(H)$.

Hence when H is amenable, the operator $\langle\Psi, \Psi\rangle_{C^*(H)}$, defined as in (2.77), is positive for all $\Psi \in \tilde{\mathcal{E}}$, so that the right action (2.76) makes $\tilde{\mathcal{E}} \rightleftharpoons C_c^\infty(H)$ a pre-Hilbert $C_c^\infty(H)$-module, which may be completed to a Hilbert C^*-module $\mathcal{E} \rightleftharpoons C^*(H)$. Under the assumption of amenability, a bounded operator A on $\mathcal{H}$ that, along with its adjoint A^*, leaves $\tilde{\mathcal{E}}$ stable and commutes with $U(H)$ satisfies (2.11) and (2.12).

More generally, the diagram $\tilde{\mathcal{E}} \rightleftharpoons C_c^\infty(H)$ defined by (2.76) and (2.77) is π_χ-positive when $U_\chi(H)$ is weakly contained in the (left or right) regular representation.

The first proof of positivity of Theorem 2.5.1 generalizes to the noncompact case; the only change is that one should take $\Phi \in C_c^\infty(H) \subset L^2(H)$. (For a bounded operator, positivity on a dense subspace of a Hilbert space implies positivity.) The second proof of positivity has to be modified by the introduction of an approximation method, as follows.

It can be shown that an amenable group H has a family of subsets $\{U_j\}_{j\in J}$, where J is a directed index set, with the following properties:

- Each U_j is measurable, with finite Haar measure $\mu(U_j)$.
- The U_j eventually fill up H in the precise sense that the family of functions $g_j \in L^1(H) \subset C^*(H)$ defined by $g_j = (\mu(U_j))^{-1/2}\chi_{U_j}$ (with χ_E the characteristic function of a Borel set E) satisfies $\lim_j g_j * g_j^* = 1_H$ pointwise on H.

Clearly, each $g_j * g_j^*$ is a positive element of $C^*(H)$.

In the second proof of positivity one now replaces 1_H by $g_j * g_j^*$. Using the bound $g_j * g_j^* \leq 1_H$ and the Lebesgue dominated convergence theorem, one can interchange $\lim_j$ and the H-integration, and the result follows.

The first part of the second claim follows from our definition "$C^*(H) = C_r^*(H)$" of amenability. The fact that A as specified satisfies (2.11) is trivial. The proof of (2.12) uses the above approximation technique as well. The operator $p_{\text{id}}^j := \int_H dh\, U \otimes U_\chi(h) g_j * g_j^*(h)$ is well-defined, as g_j has compact support. One then proceeds as in the proof of Proposition 2.5.3, replacing p_{id} by p_{id}^j. Taking the limit in j yields the claim.

The final point is obvious from Proposition III.1.7.7, according to which one may work with $C_r^*(H)$ even when H is not amenable. ■

As in the noncompact case, the above construction may be generalized.

Proposition 2.5.5. *Let c be a multiplier on a Lie group H, and let $U(H)$ be a $\overline{c}$-representation on a Hilbert space $\mathcal{H}$. Replacing $C_c^\infty(H)$ by the space $B_c^\infty(H)$ of bounded measurable functions with compact support that are smooth near e (cf. III.1.7), regarded as a dense subalgebra of the twisted group algebra $C^*(H, c)$, all statements and formulae of Theorem 2.5.4 hold.*

The verification of this claim is similar to that of 2.5.5. □

The amenability of H is sufficient, but by no means necessary, for the positivity of $\langle\Psi, \Psi\rangle_{C^*(H)}$.

Theorem 2.5.6. *Let c be a multiplier on a Lie group H, and apply the construction in 2.5.4 to the case $\mathcal{H} = L^2(H)$ and the $\overline{c}$-representation $U = U_R$, defined by III.(1.98), and $\tilde{\mathcal{E}} = B_c^\infty(G)$ (or, when $c = 1$, by $C_c^\infty(H)$).*

The ensuing Hilbert C^*-module is $C^*(H, c) \rightleftharpoons C^*(H, c)$; cf. Example 2.1.2.1. In particular, the completion of $\tilde{\mathcal{E}}$ in the norm (2.5) is $C^*(H, c)$.

Let ω_χ be a vector state on $C^*(H, c)$, as in the proof of 2.5.3. Analogously to (2.78), one derives

$$\omega_\chi(\langle\Psi, \Psi\rangle_{C^*(H,c)}) = \|\pi_\chi(\Psi)\Omega_\chi\|^2. \qquad (2.85)$$

This firstly proves positivity, and secondly shows that the norm (2.5) of Ψ coincides with its norm as an element of $C^*(H, c)$. The rest is obvious from III.(1.99) and

$$\langle\Psi, \Phi\rangle_{C^*(H,c)} = \Psi^* * \Phi, \tag{2.86}$$

which follows from III.(1.80), III.(1.81), and III.(1.29). ∎

The C^*-algebra of "compact" operators is therefore given by (2.52).

Rieffel induction with noncompact groups differs essentially from the compact case. Proposition 2.5.3 breaks down, because the H-integration may no longer be brought inside the inner product, and the would-be projection p_{id} does not exist. However, (2.80) and (2.81) are still valid, and are often computable. A simple example is given around (2.123).

2.6 Induction in Stages

This section parallels its classical counterpart 1.8. We return to general induction in order to prove a theorem on **Rieffel induction in stages**. The comments on its classical version Theorem 1.8.1 apply here as well.

Theorem 2.6.1. *Suppose that $\mathfrak{B}$ and $\mathfrak{C}$ are C^*-algebras, and that one has a Hilbert C^*-module $\mathcal{E}_2 \rightleftharpoons \mathfrak{C}$, a homomorphism $\pi_2 : \mathfrak{B} \to C^*(\mathcal{E}_2, \mathfrak{C})$, and a representation $\pi_\gamma(\mathfrak{C})$ on a Hilbert space $\mathcal{H}_\gamma$. On Rieffel induction, these data lead to an induced space $\mathcal{H}^\gamma$ and an induced representation $\pi^\gamma(\mathfrak{B})$ on $\mathcal{H}^\gamma$; cf. Figure 11.*

Now assume that one in addition has a Hilbert C^*-module $\mathcal{E}_1 \rightleftharpoons \mathfrak{B}$ and a morphism $\pi_1 : \mathfrak{A} \to C^*(\mathcal{E}_1, \mathfrak{B})$ (where $\mathfrak{A}$ is a C^*-algebra), and take $\mathcal{H}_\chi$ and π^χ in Figure 11 (with $\mathcal{E}$ and π replaced by $\mathcal{E}_1$ and π_1) to be equal to $\mathcal{H}^\gamma$ and π^γ, respectively. We denote the corresponding induced Hilbert space $\mathcal{H}^\chi$ and representation π^χ by $\mathcal{H}^{\pi^\gamma}$ and π^{π^γ}, respectively.

Define a $\mathfrak{C}$-valued sesquilinear form $\langle\, ,\, \rangle_{\mathfrak{C}}^{\otimes}$ on $\mathcal{E}_1 \otimes \mathcal{E}_2$ by sesquilinear extension of (cf. (2.61))

$$\langle\Psi_1 \otimes \Psi_2, \Phi_1 \otimes \Phi_2\rangle_{\mathfrak{C}}^{\otimes} := \langle\Psi_2, \pi_2(\langle\Psi_1, \Phi_1\rangle_{\mathfrak{B}})\Phi_2\rangle_{\mathfrak{C}}^{\otimes}, \tag{2.87}$$

and let $\mathfrak{C}$ act on $\mathcal{E}_1 \otimes \mathcal{E}_2$ from the right by $\mathbb{I} \otimes \pi_{\mathrm{R}}$ (where π_{R} is the given right action of $\mathfrak{C}$ on $\mathcal{E}_2$). Then the conditions (2.1) and (2.2) (with $\mathfrak{B}$ replaced by $\mathfrak{C}$) are met, formula (2.5) defines a seminorm on $\mathcal{E}_1 \otimes \mathcal{E}_2$, and one obtains a Hilbert C^*-module $\mathcal{E}_1 \tilde{\otimes} \mathcal{E}_2$ over $\mathfrak{C}$ by removing the null space and completing; cf. the text preceding (2.6). Moreover, the given action π_1 of $\mathfrak{A}$ on $\mathcal{E}_1$ leads to a morphism $\pi_1^{\otimes} : \mathfrak{A} \to C^*(\mathcal{E}_1 \tilde{\otimes} \mathcal{E}_2, \mathfrak{C})$, obtained by quotienting and extending $\pi_1 \otimes \mathbb{I}$.

The induced space $\mathcal{H}_{\otimes}^{\gamma}$ defined by Rieffel induction from the Hilbert $\mathfrak{C}$-module $\mathcal{E}_1 \tilde{\otimes} \mathcal{E}_2$ and the representation $\pi_\gamma(\mathfrak{C})$ is equivalent to $\mathcal{H}^{\pi^\gamma}$, and the corresponding induced representation $\pi_{\otimes}^{\gamma}(\mathfrak{A})$ is equivalent to $\pi^{\pi^\gamma}(\mathfrak{A})$.

All statements except those in the final paragraph are easily verified from the pertinent definitions. The situation is summarized in Figure 13, which should be compared with its classical counterpart Figure 7.

Let $\Psi_i, \Phi_i \in \mathcal{E}_i$ $(i = 1, 2)$, and $v, w \in \mathcal{H}_\gamma$. We denote the element $V_\gamma \Psi_2 \otimes v \in \mathcal{H}^\gamma$ by $[\Psi_2 \otimes v]_{II}$, and, similarly, for $\Omega \in \mathcal{H}^\gamma$ we denote the projection of $\Psi_1 \otimes \Omega$

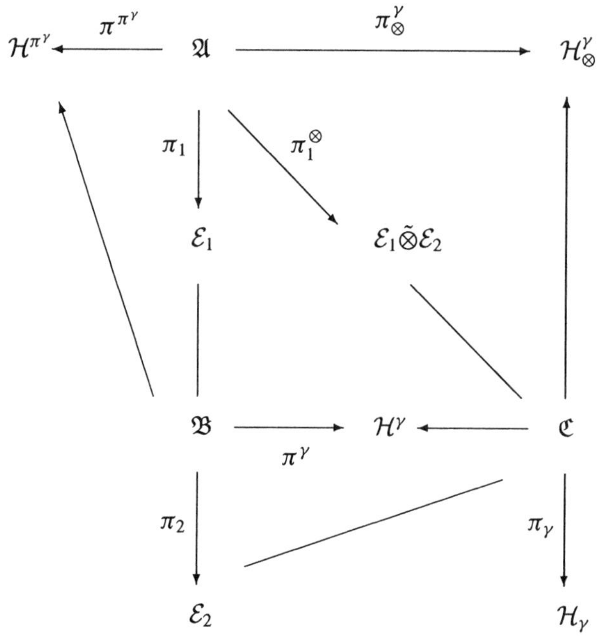

FIGURE 13. Rieffel induction in stages: $\mathcal{H}^{\pi^\gamma} \simeq \mathcal{H}^\gamma_\otimes$ and $\pi^{\pi^\gamma} \simeq \pi^\gamma_\otimes$

to $\mathcal{H}^{\pi^\gamma}$ by $[\Psi \otimes \Omega]_I$. Also, the projection of $\Psi_1 \otimes \Psi_2$ to $\mathcal{E}_1\tilde{\otimes}\mathcal{E}_2$ is called $\Psi_1\tilde{\otimes}\Psi_2$, and the projection of $\alpha \otimes v$ (where $\alpha \in \mathcal{E}_1\tilde{\otimes}\mathcal{E}_2$) to $\mathcal{H}^\gamma_\otimes$ is written as $[\alpha \otimes v]_{III}$. The essential point is the equality

$$([\Psi_1\otimes[\Psi_2\otimes v]_{II}]_I, [\Phi_1\otimes[\Phi_2\otimes w]_{II}]_I)^{\pi^\gamma} = (v, \pi_\gamma(\langle\Psi_2, \pi_2(\langle\Psi_1, \Phi_1\rangle_{\mathfrak{B}})\Phi_2\rangle_{\mathfrak{C}})w)_\gamma, \tag{2.88}$$

where the left-hand side is an inner product in $\mathcal{H}^{\pi^\gamma}$, the right-hand side being in $\mathcal{H}_\gamma$. This follows by applying (2.21) twice. Using (2.21), (2.87), and (2.1) (for $\langle\, ,\, \rangle_{\mathfrak{C}}$), one computes that the right-hand side coincides with the inner product of $[\Psi_1\tilde{\otimes}\Psi_2 \otimes v]_{III}$ and $[\Phi_1\tilde{\otimes}\Phi_2 \otimes w]_{III}$ in $\mathcal{H}^\gamma_\otimes$.

From this, one easily infers that the map from the pertinent dense subspace of $\mathcal{H}^{\pi^\gamma}$ to $\mathcal{H}^\gamma_\otimes$ that sends $[\Psi_1 \otimes [\Psi_2 \otimes v]_{II}]_I$ to $[\Psi_1\tilde{\otimes}\Psi_2 \otimes v]_{III}$ is well-defined, and extends to a unitary isomorphism U between $\mathcal{H}^{\pi^\gamma}$ and $\mathcal{H}^\gamma_\otimes$. One then trivially sees from their definitions that π^{π^γ} and $\pi^\gamma_\otimes$ are intertwined by U. ■

With trivial modifications, the theorem might equally well have been formulated and proved in terms of pre-Hilbert C^*-modules and pre-C^*-algebras.

In complete analogy with the classical case (cf. 1.8), we may specialize Theorem 2.6.1 to the case of quantum Marsden–Weinstein reduction in stages. We repeat the setting of Theorem 2.5.4, with H replaced by G.

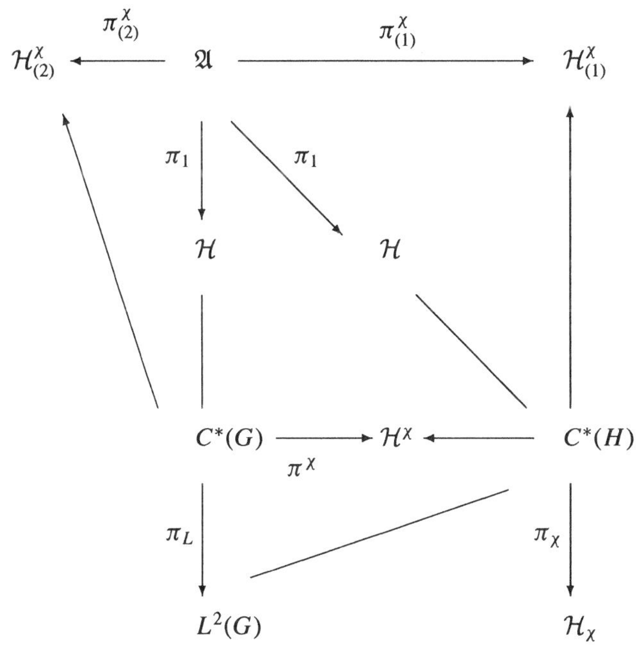

FIGURE 14. Quantum Marsden–Weinstein reduction in stages: $\mathcal{H}^{\chi}_{(1)} \simeq \mathcal{H}^{\chi}_{(2)}$ and $\pi^{\chi}_{(1)} \simeq \pi^{\chi}_{(2)}$

Theorem 2.6.2. *Suppose G is a Lie group, with closed subgroup H. Let U be a representation of G on a Hilbert space $\mathcal{H}$ containing a dense subspace $\tilde{\mathcal{E}}$ that for each $f \in C_c^{\infty}(G)$ is stable under $\pi_{\mathrm{R}}(f)$ (as defined by (2.76) with H replaced by G), and for which the function $\langle \Psi, \Phi \rangle_{C^*(G)} : x \mapsto (\Psi, U(x)\Phi)$ lies in $C_c^{\infty}(G)$ for all $\Psi, \Phi \in \tilde{\mathcal{E}}$, such that $\langle \Psi, \Psi \rangle_{C^*(G)}$ is positive for all Ψ. This defines a pre-Hilbert C^*-module $\tilde{\mathcal{E}} \rightleftharpoons C_c^{\infty}(G)$, which may be completed to a Hilbert C^*-module $\mathcal{E} \rightleftharpoons C^*(G)$.*

Now choose a representation $U_{\chi}(H)$, and apply Rieffel induction to $\mathcal{E} \rightleftharpoons C^*(G)$ from the representation $\pi^{\chi}(C^*(G))$ on $\mathcal{H}^{\chi}$ (cf. Definition III.2.8.4 with P replaced by G), corresponding to the representation $U^{\chi}(G)$ induced from $U_{\chi}(H)$; see III.(2.175). Call the doubly induced Hilbert space $\mathcal{H}^{\chi}_{(2)}$, with $\pi^{\chi}_{(2)}$ the corresponding doubly induced representation of some (pre) C^*-algebra $\mathfrak{A} \subseteq C^*(\mathcal{E}, C^*(G))$.

Restrict $U(G)$ to H, and Rieffel-induce on $\mathcal{E} \rightleftharpoons C^*(H)$ from $U_{\chi}(H)$, obtaining a Hilbert space $\mathcal{H}^{\chi}_{(1)}$ and an induced representation $\pi^{\chi}_{(1)}(\mathfrak{A})$.

There exists a unitary $U : \mathcal{H}^{\chi}_{(2)} \to \mathcal{H}^{\chi}_{(1)}$ intertwining $\pi^{\chi}_{(1)}(\mathfrak{A})$ and $\pi^{\chi}_{(2)}(\mathfrak{A})$.

We specialize Figure 13 to Figure 14. In this specialization we have replaced $\mathcal{E}_1 \tilde{\otimes} \mathcal{E}_2$ by $\mathcal{H}$, and $\pi_1^{\otimes}$ by π_1. This is justified by the following quantum counterpart of Lemma 1.8.3.

Lemma 2.6.3. *Define the pre-Hilbert C^*-module $\tilde{\mathcal{E}}\tilde{\otimes}C_c^\infty(G) \rightleftharpoons C_c^\infty(H)$ as in the proof of 2.6.1, with $\mathcal{E}_1$ replaced by $\tilde{\mathcal{E}}$ as specified in 2.6.2, and $\mathcal{E}_2 = C_c^\infty(G) \subset L^2(G)$ carrying the left-regular representation π_L of $C_c^\infty(G) \subset C^*(G)$.*

This module is equivalent (in the obvious sense) to $\tilde{\mathcal{E}} \rightleftharpoons C_c^\infty(H)$, defined by restricting $\tilde{\mathcal{E}} \rightleftharpoons C_c^\infty(G)$ from G to H.

Using (2.87) with $\mathfrak{B} = C^*(G)$ and $\pi_2 = \pi_L$, one easily shows that

$$\langle \Psi_1 \otimes \Psi_2, \Phi_1 \otimes \Phi_2\rangle_{C_c^\infty(H)} = \langle \pi_{\mathrm{R}}(\Psi_2)\Psi_1, \pi_{\mathrm{R}}(\Phi_2)\Phi_1\rangle_{C_c^\infty(H)}, \tag{2.89}$$

where π_{R} is defined as in (2.76) with H replaced by G. Hence we define $\tilde{V} : \tilde{\mathcal{E}} \otimes C_c^\infty(G) \to \tilde{\mathcal{E}}$ by linear extension of

$$\tilde{V}\Psi_1 \otimes \Psi_2 := \pi_{\mathrm{R}}(\Psi_2)\Psi_1. \tag{2.90}$$

By (2.89) this map quotients well to $V : \tilde{\mathcal{E}}\tilde{\otimes}C_c^\infty(G) \to \tilde{\mathcal{E}}$, and carries the pre-Hilbert C^*-module $\tilde{\mathcal{E}}\tilde{\otimes}C_c^\infty(G) \rightleftharpoons C_c^\infty(H)$ into $\tilde{\mathcal{E}} \rightleftharpoons C_c^\infty(H)$. ∎

Theorem 2.6.2 now follows from 2.6.1 and 2.6.3. ∎

This particular case of Theorem 2.6.1 has a further specialization that historically was the first example of a theorem on induction in stages.

Corollary 2.6.4. *Let K be a Lie group, and let $H \subset G \subset K$ be closed subgroups. The representation $U^\gamma(K)$ induced from a representation $U_\gamma(G)$ that is itself induced from a representation $U_\chi(H)$ (so that $U_\gamma(G) \simeq U^\chi(G)$) is equivalent to the representation $U^\chi(K)$ directly induced from $U_\chi(H)$.*

Apply Theorem 2.6.2 with $\mathfrak{A} = C^*(K)$, $\mathcal{H} = L^2(K)$, $\tilde{\mathcal{E}} = C_c^\infty(K)$, and $\pi_1 = \pi_L$. We will show in the next section that the positivity condition is met; see the argument following (2.92), with H replaced by G and P by K. ∎

2.7 The Imprimitivity Theorem for Gauge Groupoids

Recall the gauge groupoid $\mathsf{P} \times_H \mathsf{P} \rightrightarrows Q$ defined by a principal bundle $\mathsf{P}(Q, H, \tau)$ (cf. III.3.1.7) and its C^*-algebra $C^*(\mathsf{P} \times_H \mathsf{P})$ (see III.3.6.3). Using Proposition 2.4.3 and Theorem III.3.7.1, particularly III.(3.78), one may infer that $C^*(\mathsf{P} \times_H \mathsf{P})$ and $C^*(H)$ are Morita equivalent. The aim of this section is to restate this result with the machinery developed in this chapter, by directly constructing a quantum dual pair. This provides considerable insight into the situation, using the material of 2.4 in an instructive way.

Theorem 2.7.1. *For any principal bundle $\mathsf{P}(Q, H, \tau)$ with associated gauge groupoid $\mathsf{P} \times_H \mathsf{P} \rightrightarrows Q$, the groupoid C^*-algebra $C^*(\mathsf{P} \times_H \mathsf{P})$ is Morita equivalent to $C^*(H)$.*

We apply the construction of Theorem 2.5.4 with $\mathcal{H} = L^2(\mathsf{P})$, $U = U_R$, and $\tilde{\mathcal{E}} = C_c^\infty(\mathsf{P})$; see III.2.8 and III.(2.138). The presence of an H-invariant measure μ on P is not essential, but simplifies some of the formulae. For the same reason, we retain our standing assumption that H is unimodular. Hence the right action $\pi_{\mathrm{R}}(C_c^\infty(H))$ is given by specializing (2.76), yielding

$$\pi_{\mathrm{R}}(f)\Psi(x) = \int_H dh\, \Psi(xh^{-1})f(h), \tag{2.91}$$

and the map $\langle\Psi, \Phi\rangle_{C^*(H)}$ is defined by (2.77), which specializes to

$$\langle\Psi, \Phi\rangle_{C^*(H)} : h \mapsto \int_{\mathsf{P}} d\mu(x)\, \overline{\Psi(x)}\Phi(xh). \tag{2.92}$$

It is easily shown that $\langle\Psi, \Psi\rangle_{C^*(H)}$ is positive in $C^*(H)$ (whether or not H is amenable). Indeed, we may proceed as in the proof of positivity in Theorem 2.5.6. Generalizing (2.85), we derive

$$\omega_\chi(\langle\Psi, \Psi\rangle_{C^*(H)}) = \int_Q d\nu(q)\, \| \int_H dh\, \Psi(s(q)h)U_\chi(h)\Omega_\chi \|^2, \tag{2.93}$$

where $s : Q \rightarrow \mathsf{P}$ is an arbitrary measurable section, and we have used III.(2.139).

Thus one obtains a pre-Hilbert C^*-module $C_c^\infty(\mathsf{P}) \rightleftharpoons C_c^\infty(H)$, which by Corollary 2.1.4 may be completed to a Hilbert C^*-module that in analogy with 2.5.6 is denoted by $\mathcal{C}(\mathsf{P}) \rightleftharpoons C^*(H)$. (When q is a point, the above construction reduces to 2.5.6, so that $\mathcal{C}(\mathsf{P}) = C^*(H)$. However, when $\mathsf{P}(Q, H, \tau) = G(G/H, H, \tau)$, the space $\mathcal{C}(G)$ is different from $C^*(G)$.)

Applying (2.47) to the case at hand, one derives that for $\Psi, \Phi, Z \in C_c^\infty(\mathsf{P})$ the operator $T_{\Psi,\Phi}^{C^*(H)}$ is given by

$$T_{\Psi,\Phi}^{C^*(H)} Z(x) = \int_{\mathsf{P}} d\mu(y)\, K_{\Psi,\Phi}^{C^*(H)}(x, y)Z(y), \tag{2.94}$$

with

$$K_{\Psi,\Phi}^{C^*(H)}(x, y) := \int_H dh\, \Psi(xh)\overline{\Phi(yh)}. \tag{2.95}$$

Since the H-action on P is proper, the integrand in (2.95) has compact support in h. Noting that $K_{\Psi,\Phi}^{C^*(H)}$ is invariant under the H-action $h : (x, y) \mapsto (xh^{-1}, yh^{-1})$, one infers that $K_{\Psi,\Phi}^{C^*(H)}$ lies in $C_c^\infty(\mathsf{P} \times_H \mathsf{P})$.

In view of the bound (2.51) (from whose derivation one sees that it holds in a pre-Hilbert C^*-module as well), the operator $T_{\Psi,\Phi}^{C^*(H)}$ may be extended to $\mathcal{C}(\mathsf{P})$ by continuity, and may be defined for all $\Psi, \Phi \in \mathcal{C}(\mathsf{P})$. In particular, one obtains the bound (2.12).

Since $K_{\Psi,\Phi}^{C^*(H)} \in C_c^\infty(\mathsf{P} \times_H \mathsf{P})$, we may regard $T_{\Psi,\Phi}^{C^*(H)}$ as an element of $C^*(\mathsf{P} \times_H \mathsf{P})$. However, in order to identify $C_0^*(\mathcal{C}(\mathsf{P}), C^*(H))$ with $C^*(\mathsf{P} \times_H \mathsf{P})$, we need to show that the norm of $T_{\Psi,\Phi}^{C^*(H)}$ in $C_0^*(\mathcal{C}(\mathsf{P}), C^*(H))$ coincides with its norm III.(3.74) in $C^*(\mathsf{P} \times_H \mathsf{P})$. This indeed follows from the isomorphism III.(3.78). Choosing a section $s : Q \rightarrow \mathsf{P}$, one passes from $K_{\Psi,\Phi}^{C^*(H)}(x, y)$ to $K_{\Psi,\Phi,s}^{C^*(H)}(q, h, q')$ as indicated in III.(3.83), and computes its operator norm I.(1.17) using (2.5) with $\mathfrak{B} = C^*(H)$. This computation yields a norm that coincides with the norm in $\mathfrak{B}_0(L^2(Q)) \otimes C^*(H)$, proving the claim.

Passing to the C^*-algebra generated by all $T_{\Psi,\Phi}^{C^*(H)}$, Definitions 2.3.1 and III.3.6.3 show that $C_0^*(\mathcal{C}(\mathsf{P}), C^*(H))$ is isomorphic to $C^*(\mathsf{P} \times_H \mathsf{P})$. Hence the theorem follows from Definition 2.4.1. ■

It follows from this proof that one has a quantum dual pair

$$C^*(\mathsf{P} \times_H \mathsf{P}) \rightleftharpoons \mathcal{C}(\mathsf{P}) \rightleftharpoons C^*(H)$$

in which $\pi_{\mathrm{R}}(C^*(H))$ and $\langle\, ,\, \rangle_{C^*(H)}$ are given by (2.91) and (2.92), respectively, and the left action $\pi_{\mathrm{L}}(C^*(\mathsf{P} \times_H \mathsf{P}))$ coincides with the representation π_r given in III.(3.82). Finally, according to (2.95) and (2.59) one has

$$\langle \Psi, \Phi \rangle_{C^*(\mathsf{P} \times_H \mathsf{P})} : (x, y) \mapsto \int_H dh\, \Psi(xh)\overline{\Phi(yh)}. \tag{2.96}$$

These formulae are all understood to be defined on the pre-quantum dual pair used above, namely $C_c^\infty(\mathsf{P} \times_H \mathsf{P}) \rightleftharpoons C_c^\infty(G) \rightleftharpoons C_c^\infty(H)$, and extended to the completion displayed above by continuity.

In particular, Corollary III.3.7.2 follows from Theorem 2.4.5. Moreover, the theory of induced group representations of III.2.9 may be reinterpreted in the light of Theorems 2.7.1 and 2.4.5. To do this, we explain how the construction in Definitions III.2.8.4 and III.2.9.1 is a special case of Rieffel induction.

We start from Construction 2.2.3, in which we take the Hilbert C^*-module $\mathcal{E} \rightleftharpoons \mathfrak{B}$ to be $\mathcal{C}(\mathsf{P}) \rightleftharpoons C^*(H)$, defined by (2.76) and (2.77). Furthermore, we take $\mathcal{H}_\chi$ to be the carrier space of a representation $U_\chi(H)$ (or, equivalently, of $\pi_\chi(C^*(H))$. As explained in 2.2, one obtains the same induced space $\mathcal{H}^\chi$ if one starts from suitable dense subspaces $\tilde{\mathcal{C}}(\mathsf{P})$ and $C_c^\infty(H)$ of $\mathcal{C}(\mathsf{P})$ and $C^*(H)$, respectively.

When H is compact, we are in the situation of Proposition 2.5.3. Hence we start from the pre-Hilbert C^*-module $L^2(\mathsf{P}) \rightleftharpoons C^*(H)$, and obtain the induced space $\mathcal{H}^\chi$ as the subspace of all $\tilde{\Psi} \in L^2(\mathsf{P}) \otimes \mathcal{H}_\chi$ satisfying $p_{\mathrm{id}}\tilde{\Psi} = \tilde{\Psi}$. Since $U = U_R$, this condition is nothing but the equivariance condition III.(2.145). For $\varphi \in \mathrm{Aut}(\mathsf{P})$ and $\Psi \in L^2(\mathsf{P})$ we define $U_L(\varphi)\Psi$ by the right-hand side of III.(2.161), with Ψ^χ replaced by Ψ. This defines an operator $U_L(\varphi)$ on $L^2(\mathsf{P})$, which is easily seen to be adjointable as a consequence of the fact that elements of Aut(P) by definition commute with the H-action on P. We identify $L^2(\mathsf{P}) \otimes \mathcal{H}_\chi$ with $L^2(\mathsf{P}, \mathcal{H}_\chi)$, and use the description of $\pi^\chi(A)$ in 2.5.3. It is then obvious that $\pi^\chi(U_L(\varphi))$ coincides with $U^\chi(\varphi)$ as defined in III.(2.161).

The noncompact case is slightly more involved, since the induced space is no longer a subspace of $L^2(\mathsf{P}) \otimes \mathcal{H}_\chi$. We are now in the situation of Theorem 2.5.4, with $\mathcal{H} = L^2(\mathsf{P})$, $\tilde{\mathcal{E}} = C_c^\infty(\mathsf{P})$, and $U = U_R$. To identify the reduced space, we use the method of Proposition 2.2.4. We take $\mathcal{H}_*^\chi$ to be what is, prophetically, called $\mathcal{H}^\chi$ in Definition III.2.8.4. Consider the map $\tilde{U} : C_c^\infty(\mathsf{P}) \otimes \mathcal{H}_\chi \to \mathcal{H}_*^\chi$ defined by linear extension of

$$\tilde{U}\Psi \otimes v(x) := \int_H dh\, \Psi(xh) U_\chi(h) v. \tag{2.97}$$

Note that the equivariance condition III.(2.145) is indeed satisfied by the left-hand side, as follows from the invariance of the Haar measure. Using (2.80) or (2.81), one verifies the fundamental equation (2.29). It is clear that $\tilde{U}(C_c^\infty(\mathsf{P}) \otimes \mathcal{H}_\chi)$ is dense in $\mathcal{H}^\chi$, so by Proposition 2.2.4 one obtains an isomorphism $U : \mathcal{H}^\chi \to \mathcal{H}_*^\chi$. Morover, using (2.30), one may repeat the last sentence of the discussion of the compact case.

Had we not known Definition III.2.8.4 beforehand, it would have been possible to obtain the form of $\mathcal{H}^\chi$ on the basis of the method explained after (2.31). For the image of $\tilde{U}$ as defined in (2.32) coincides with the image of $\tilde{U}$ as defined in (2.97), under the identification of $\mathcal{H}_*^\chi$ with a subspace of $\mathcal{L}(\bar{\tilde{\mathcal{E}}}, \mathcal{H}_\chi)^{\mathfrak{B}}$ via the inner product in $L^2(\mathsf{P})$.

As in the second half of III.2.9, we now specialize to Mackey induction, where the principal bundle $\mathsf{P}(Q, H, \tau)$ is $G(G/H, H, \tau)$; cf. III.2.7. As always, the explicit expressions below apply to the unimodular case.

Corollary 2.7.2. *The action C^*-algebra $C^*(G, G/H)$ is Morita equivalent to the group C^*-algebra $C^*(H)$, with quantum dual pair $C^*(G, G/H) \rightleftharpoons \mathcal{C}(G) \rightleftharpoons C^*(H)$. This pair is obtained by completing $C_c^\infty(G \times G/H) \rightleftharpoons C_c^\infty(G) \rightleftharpoons C_c^\infty(H)$, defined by the maps (2.91), (2.92), and*

$$\pi_{\mathrm{L}}(f)\Psi(x) = \int_G dy\, f(xy^{-1}, [x]_H)\Psi(y); \tag{2.98}$$

$$\langle \Psi, \Phi \rangle_{C^*(G,G/H)} : (x, [y]_H) \mapsto \int_H dh\, \Psi(yh)\overline{\Phi(x^{-1}yh)}. \tag{2.99}$$

Every nondegenerate representation π of $C^*(G, G/H)$ is equivalent to an induced representation π^χ, realized on $\mathcal{H}^\chi$ (see Definition III.2.8.4 with P replaced by G), where U_χ is a representation of H. Explicitly, one has

$$\pi^\chi(f)\Psi^\chi(x) = \int_G dy\, f(xy^{-1}, [x]_H)\Psi^\chi(y); \tag{2.100}$$

cf. (2.98). The representation $\pi^\chi(C^*(G, G/H))$ is irreducible iff $U_\chi(H)$ is irreducible.

This follows upon combining Theorems 2.7.1 and 2.4.5, including the explicit construction of induced representations above, with Corollary III.3.7.5. Equations (2.98) and (2.99) are derived from the formulae stated after Theorem 2.7.1, specialized to the case at hand using Proposition III.3.4.7. Similarly, the stated form of π^χ follows from III.(3.84). ∎

Corollary III.3.7.6 is then seen to be a consequence of Corollaries 2.7.2 and III.3.7.4, combined with Definition III.3.7.3.

Under a regularity assumption on the group action (which is always satisfied when G is compact), it is possible to classify the irreducible representations of an arbitrary action C^*-algebra. This classification is the quantum version of Proposition III.3.9.4.

Theorem 2.7.3. *Let a smooth action of a Lie group G on a manifold Q be* **regular**, *in that each orbit is (relatively) open in its closure. Then the irreducible representations of $C^*(G, Q)$ are classified by pairs $(\mathcal{O}, U_\chi(H))$, where $\mathcal{O}$ is a*

G-orbit in Q, the group H is the stabilizer G_{q_0} of an arbitrary point q_0 in $\mathcal{O}$, and U_χ is an irreducible representation of H.

The representation $\pi_{\mathcal{O}}^{\chi}$ corresponding to such a pair may be realized on the Hilbert space $\mathcal{H}^\chi$ defined in III.2.8.4, and is given by (2.100), with $[x]_H$ replaced by xq_0. Equivalently, the system of imprimitivity $(U(G), \tilde{\pi}^\chi(C_0(Q)))$ corresponding to the representation $\pi^\chi(C^*(G, Q))$ by Corollary III.3.7.4 is given by $U = U^\chi$ (cf. III.(2.175)) and

$$\tilde{\pi}^\chi(\tilde{f})\Psi^\chi(x) = \tilde{f}(xq_0)\Psi^\chi(x). \tag{2.101}$$

We start from a given irreducible system of imprimitivity $(U(G), \tilde{\pi}(C_0(Q)))$ on some Hilbert space $\mathcal{H}$. Since G is a Lie group and Q is a manifold, it can be shown that $C^*(G, Q)$, and therefore $\mathcal{H}$ are separable. Using the spectral theorem applied to $\tilde{\pi}(C_0(Q))$, one can bring $\mathcal{H}$ into the form

$$\mathcal{H} \simeq \oplus_i L^2(Q, \nu_i) \otimes \mathcal{H}_i, \tag{2.102}$$

where the ν_i are mutually singular Borel measures, and the $\mathcal{H}_i$ are multiplicity spaces. The representation $\tilde{\pi}$ is then given in terms of multiplication operators in the obvious way. The fact that $\tilde{\pi}$ can be extended to a system of imprimitivity easily implies that each measure ν_i is quasi-invariant under G (that is, its measure class is G-invariant).

A measure ν on a G-space Q is said to be **ergodic** when a G-invariant Borel function is constant almost everywhere. Equivalently, for a G-invariant Borel set B either its complement $Q\backslash B$ or B itself must have measure zero. It is quite obvious that $(U, \tilde{\pi})$ can be irreducible only when there is a single term in (2.102) in which the measure ν is ergodic; for otherwise one could decompose the carrier space as $L^2(Q, \nu) = L^2(Q\backslash B, \nu) \oplus L^2(B, \nu)$.

Let S be the smallest closed set in Q for which $\nu(Q\backslash S) = 0$; this set is automatically G-invariant, and we can write $L^2(S, \nu)$ instead of $L^2(Q, \nu)$. Suppose S contains two orbits $Gq_1 \neq Gq_2$. First assume that $[q_1]_G^-$ (i.e., the closure of Gq_1 regarded as a point $[q_1]_G$ in Q/G) does not contain $[q_2]_G$. Then $\tau_{Q\to Q/G}^{-1}([q_1]_G^-)$, which is a closed G-invariant subset of S, does not contain Gq_2. This contradicts either ergodicity or the definition of S. Hence $[q_2]_G \in [q_1]_G^-$. We combine this inclusion with the regularity assumption. Accordingly, there is an open set $\mathcal{N}$ in S that contains Gq_1^- (the closure of Gq_1 in Q, hence in S) but is disjoint from Gq_2. The G-translate $G\mathcal{N}$ of $\mathcal{N}$ has the same properties, so that its complement $S\backslash G\mathcal{N}$ is a closed G-invariant subset of S, which does not contain Gq_1. This again leads to a contradiction with either ergodicity or the definition of S. Hence ν must be concentrated on a single orbit in Q.

Furthermore, the regularity assumption turns out to be equivalent to the statement that each orbit (equipped with the topology inherited from Q) is homeomorphic to G/G_{q_0}. Hence $\tilde{\pi}(C_0(Q)) \simeq C_0(Gq_0) \simeq C_0(G/G_{q_0})$, so that the situation reduces to the representation theory of the action C^*-algebra $C^*(G, G/H)$, with $H = G_{q_0}$.

The theorem now follows from the second half of Corollary 2.7.2. □

2.8 Covariant Quantization

Let us return to quantization theory, and ask in some generality what happens to Berezin quantization in the presence of a classical symmetry group. The following notion, which generalizes Definition III.3.7.3, is natural in this context.

Definition 2.8.1. *Given a smooth G-action on a manifold X, a* **generalized system of imprimitivity** *of G on X in a Hilbert space $\mathcal{H}$ is a pair $(U, \mathcal{Q})$ in which U is a representation of G on $\mathcal{H}$, and $\mathcal{Q} : C_0(X) \to \mathfrak{B}(\mathcal{H})$ is a positive map, satisfying the covariance condition*

$$U(x)\mathcal{Q}(\tilde{f})U(x)^* = \mathcal{Q}(\alpha_x(\tilde{f})), \tag{2.103}$$

where $\alpha_x(\tilde{f}) : y \mapsto \tilde{f}(x^{-1}y)$. The system is called **transitive** *when $X = G/H$ with the natural G-action.*

When X is compact, or when $\mathcal{Q}$ may be extended to $C_0(X)_{\mathbb{I}}$, this condition may be equivalently stated in terms of the POVM $\Delta \mapsto A(\Delta)$ associated to $\mathcal{Q}$ (cf. II.1.4.8) as

$$U(x)A(\Delta)U(x)^{-1} = A(x\Delta). \tag{2.104}$$

Every (ordinary) system of imprimitivity is evidently a generalized one as well, since a representation is a particular example of a positive map. A class of examples of truly generalized transitive systems of imprimitivity arises as follows. Let $(U(G), \tilde{\pi}(C_0(G/H)))$ be a system of imprimitivity on a Hilbert space $\mathcal{K}$, and suppose that $U(G)$ is reducible. Pick a projection p in the commutant of $U(G)$; then $(pU(G), p\tilde{\pi}p)$ is a generalized system of imprimitivity on $\mathcal{H} = p\mathcal{K}$. Of course, $(U, \tilde{\pi})$ is described by Corollary III.3.7.6, and must be of the form $(U^{\chi}, \tilde{\pi}^{\chi})$. Under an innocent technical assumption, this class turns out to exhaust all possibilities. What follows generalizes Corollary III.3.7.6 (or the second part of Corollary 2.7.2) to the case where the representation $\tilde{\pi}$ is replaced by a positive map $\mathcal{Q}$.

Theorem 2.8.2. *Let $(U(G), \mathcal{Q}(C_0(G/H)))$ be a transitive generalized system of imprimitivity on $\mathcal{H}$, where $\mathcal{Q}$ may be extended to the unitization of $C_0(G/H)$ such that the extension preserves the unit.*

There exists a representation $U_{\chi}(H)$, with corresponding induced representation $U^{\chi}(G)$ on $\mathcal{H}^{\chi}$ and system of imprimitivity $(U^{\chi}, \tilde{\pi}^{\chi})$ as described in Corollary III.3.7.6, and a projection p on $\mathcal{H}^{\chi}$ in the commutant of $U^{\chi}(G)$, such that $(pU^{\chi}(G), p\tilde{\pi}^{\chi}p)$ and $(U(G), \mathcal{Q}(C_0(G/H)))$ are equivalent.

We apply Theorem II.1.4.2. To avoid confusion, we denote the Hilbert space $\mathcal{H}^{\chi}$ and the representation π^{χ} in Construction II.1.4.3 by $\tilde{\mathcal{H}}^{\chi}$ and $\tilde{\pi}^{\chi}$, respectively; the space defined in III.2.8.4 and the induced representation of III.(2.175) will still be called $\mathcal{H}^{\chi}$ and π^{χ}, as in the formulation of the theorem above. Indeed, our goal is to show that $(\tilde{\pi}^{\chi}, \tilde{\mathcal{H}}^{\chi})$ may be identified with $(\pi^{\chi}, \mathcal{H}^{\chi})$. We identify $\mathfrak{B}$ in II.1.4.2 and II.1.4.3 with $\mathfrak{B}(\mathcal{H})$, where $\mathcal{H}$ is specified in 2.8.2; we therefore omit the representation π_{χ} occurring in II.1.4.2 etc., putting $\mathcal{H}_{\chi} = \mathcal{H}$. For $x \in G$ we define a linear map $\tilde{U}(x)$ on $C_0(G/H) \otimes \mathcal{H}$ by linear extension of

$$\tilde{U}(x) f \otimes \Psi := \alpha_x(f) \otimes U(x)\Psi. \tag{2.105}$$

Since $\alpha_x \circ \alpha_y = \alpha_{xy}$, and U is a representation, $\tilde{U}$ is clearly a G-action. Using the covariance condition (2.103) and the unitarity of $U(x)$, one verifies that

$$(\tilde{U}(x) f \otimes \Psi, \tilde{U}(x) g \otimes \Phi)_0^{\chi} = (f \otimes \Psi, g \otimes \Phi)_0^{\chi}, \tag{2.106}$$

where $(\, ,\,)_0^{\chi}$ is defined in II.(1.25). Hence $\tilde{U}(G)$ quotients to a representation $\tilde{U}^{\chi}(G)$ on $\tilde{\mathcal{H}}^{\chi}$. One checks that $(\tilde{U}^{\chi}, \tilde{\pi}^{\chi})$ is a system of imprimitivity on $\tilde{\mathcal{H}}^{\chi}$ (compute on $C_0(G/H) \otimes \mathcal{H}$ and then pass to the quotient). By Corollary III.3.7.6, this system must be of the form $(U^{\chi}, \tilde{\pi}^{\chi})$ (up to equivalence).

Finally, the projection p defined in II.1.4.2 commutes with all $\tilde{U}^{\chi}(x)$. This is verified from II.(1.29), II.(1.30), and (2.103). The claim follows. ∎

The power of this result is clear in the following application. Let a phase space S be a coadjoint orbit in $\mathfrak{g}^*$, so that $S \simeq G/H$, equipped with the Lie symplectic structure (see 3.1 for examples). A pure state quantization $\{q_\hbar, \mathcal{H}_\hbar, \mu_\hbar\}$ (cf. II.1.3.3) of S leads to an associated Berezin quantization $\mathcal{Q}_\hbar^B$. This quantization is called **covariant** when each $\mathcal{H}_\hbar$ carries an irreducible representation $U_\hbar(G)$, such that

$$\mathcal{Q}_\hbar^B(\alpha_x^0(f)) = U_\hbar(x)\mathcal{Q}_\hbar^B(f)U_\hbar(x)^*, \tag{2.107}$$

for all $f \in C_0(G/H)$ and $x \in G$, where α^0 is defined by III.(1.158). We have already seen examples of this in III.1.11. Theorem 2.8.2 then implies

Corollary 2.8.3. *The representation $U_\hbar(G)$ occurring in a covariant Berezin quantization of $C_0(G/H)$ must be an irreducible subrepresentation of an induced representation $U^{\chi}(G)$. When p is the corresponding projection on $\mathcal{H}_\hbar \subset \mathcal{H}^{\chi}$, the Berezin quantization $\mathcal{Q}_\hbar^B$ must have the form $\mathcal{Q}_\hbar^B(f) = p\tilde{\pi}^{\chi}(f)p$, where*

$$\tilde{\pi}^{\chi}(f)\Psi^{\chi}(x) = f([x]_H)\Psi^{\chi}(x). \tag{2.108}$$

Cf. III.(3.86). In particular, if $\mathcal{H}^{\chi} = L^2(G/H)$ (i.e., one has induced from the trivial representation of H), then $\tilde{\pi}^{\chi}(f)$ is simply f, seen as a multiplication operator, and its Berezin quantization consists in squeezing this operator into a subspace.

In view of (2.104) and the discussion following II.1.4.9, Theorem 2.8.2 and its Corollary 2.8.3 describe **covariant localization in phase space**. There is a corresponding theory of **covariant localization in configuration space**, which we now briefly discuss. To set the stage, we start on Euclidean space $Q = \mathbb{R}^3$. In elementary quantum mechanics, a particle moving on $\mathbb{R}^3$ with spin $j \in \mathbb{N}$ is described by the Hilbert space

$$\mathcal{H}^j = L^2(\mathbb{R}^3) \otimes \mathcal{H}_j, \tag{2.109}$$

where $\mathcal{H}_j = \mathbb{C}^{2j+1}$ carries the irreducible representation $U_j(SO(3))$. The basic physical observables are represented by the unbounded operators Q_k^S (position), P_k^S (momentum), and J_k^S (angular momentum), where $k = 1, 2, 3$. These operators satisfy the commutation relations (say, on the domain $\mathcal{S}(\mathbb{R}^3) \otimes \mathcal{H}_j$)

$$[Q_k^S, Q_l^S] = 0; \tag{2.110}$$
$$[P_k^S, Q_l^S] = -i\hbar\delta_{kl}; \tag{2.111}$$
$$[J_k^S, Q_l^S] = i\hbar\epsilon_{klm} Q_m^S; \tag{2.112}$$
$$[P_k^S, P_l^S] = 0; \tag{2.113}$$
$$[J_k^S, J_l^S] = i\hbar\epsilon_{klm} J_m^S; \tag{2.114}$$
$$[J_k^S, P_l^S] = i\hbar\epsilon_{klm} P_m^S, \tag{2.115}$$

justifying their physical interpretation.

The momentum and angular momentum operators are best defined in terms of a unitary representation U^j of the Euclidean group $E(3) = SO(3) \ltimes \mathbb{R}^3$ on $\mathcal{H}^j$, given by

$$U^j(R, a)\Psi(q) = U_j(R)\Psi(R^{-1}(q - a)). \tag{2.116}$$

In terms of the standard generators P_k and T_k of $\mathbb{R}^3$ and $SO(3)$, respectively, one then has $P_k^S = i\hbar dU^j(P_k)$ and $J_k^S = i\hbar dU^j(T_k)$; the commutation relations (2.113)–(2.115). follow from III.(1.70) and (1.65).

Moreover, we define a representation $\tilde{\pi}^j$ of $C_0(\mathbb{R}^3)$ on $\mathcal{H}^j$ by

$$\tilde{\pi}^j(\tilde{f}) = \tilde{f} \otimes \mathbb{I}_j, \tag{2.117}$$

where $\tilde{f}$ is seen as a multiplication operator on $L^2(\mathbb{R}^3)$. The associated PVM $\Delta \mapsto E(\Delta)$ on $\mathbb{R}^3$ in $\mathcal{H}^j$ (see II.1.4.8) is $E(\Delta) = \chi_\Delta \otimes \mathbb{I}_j$, in terms of which the position operators are given by $Q_k^S = \int_{\mathbb{R}^3} dE(x)x_k$; cf. the spectral theorem for unbounded operators. Equation (2.110) then reflects the commutativity of $C_0(\mathbb{R}^3)$, as well as the fact that $\tilde{\pi}^j$ is a representation.

Identifying $Q = \mathbb{R}^3$ with $G/H = E(3)/SO(3)$ in the obvious way, one checks that the canonical left action of $E(3)$ on $E(3)/SO(3)$ is identified with its defining action on $\mathbb{R}^3$. It is not hard to then verify from (2.116) that $(U^j(E(3)), \tilde{\pi}^j(C_0(\mathbb{R}^3)))$ is a system of imprimitivity as defined in III.3.7.3. The commutation relations (2.111), (2.112) are a consequence of the covariance relation III.(3.38).

Rather than using the unbounded operators Q_k^S, P_k^S, and J_k^S and their commutation relations, we therefore state the situation in terms of $(U^j(E(3)), \tilde{\pi}^j(C_0(\mathbb{R}^3)))$. Such a pair, or, equivalently, a nondegenerate representation π^j of the action C^*-algebra $C^*(E(3), \mathbb{R}^3)$ (cf. III.3.7.4), then by definition describes a quantum system that is localizable in $\mathbb{R}^3$ and covariant under the defining action of $E(3)$. It is natural to require that π^j be irreducible, in which case the quantum system itself is said to be irreducible.

Proposition 2.8.4. *An irreducible quantum system that is localizable in $\mathbb{R}^3$ and covariant under $E(3)$ is completely characterized by its spin $j \in \mathbb{N}$. The corresponding system of imprimitivity $(U^j(E(3)), \tilde{\pi}^j(C_0(\mathbb{R}^3)))$ is equivalent to the one described by (2.109), (2.116), and (2.117).*

This follows from Corollary III.3.7.6. The induced representation $U_s^j(E(3))$ defined by the section $s : \mathbb{R}^3 \to E(3)$ given by $s(q) := (e, q)$ (see III.2.9) is precisely (2.116); cf. Lemma 3.3.2 below. The corresponding representation III.(3.86) of $C_0(\mathbb{R}^3)$ is (2.117). ■

This is a neat explanation of spin in quantum mechanics, though it would be a mistake to think that spin has no classical counterpart.

Generalizing this approach to an arbitrary homogeneous configuration space $Q = G/H$, a nondegenerate representation π of $C^*(G, G/H)$ on a Hilbert space $\mathcal{H}$ describes a quantum system that is localizable in G/H and covariant under the canonical action of G on G/H. By Corollary III.3.7.4 this is equivalent to a system of imprimitivity $(U(G), \tilde{\pi}(C_0(G/H)))$ on $\mathcal{H}$, and by Proposition II.1.4.8 one may instead assume that one has a PVM $\Delta \mapsto E(\Delta)$ on G/H in $\mathcal{H}$ and a unitary representation $U(G)$ that satisfy

$$U(x)E(\Delta)U(x)^{-1} = E(x\Delta) \tag{2.118}$$

for all $x \in G$ and $\Delta \in \Sigma$; cf. (2.104). The physical interpretation of the PVM is given by II.(1.36); the operators defined in III.(3.39) play the role of quantized momentum observables. Generalizing Proposition 2.8.4, we have

Theorem 2.8.5. *An irreducible quantum system that is localizable in $Q = G/H$ and covariant under the canonical action of G is completely characterized by an element $\chi \in \hat{H}$ of the unitary dual of H. The corresponding system of imprimitivity $(U^{\chi}(G), \tilde{\pi}^{j}(C_0(G/H)))$ is equivalent to the one described by III.2.8.4, III.(2.175), and III.(3.86).*

This is immediate from Corollary III.3.7.6. ■

With an analogous notion of classical localization, the corresponding situation in classical mechanics is described by Theorem 1.6.4. Thus the different possibilities allowed by Theorem 2.8.5 do not correspond to inequivalent quantizations of the classical phase space $T^*(G/H)$ (for which notion cf. 3.9), but rather to quantizations of various symplectic leaves in $(T^*G)/H$. For by Theorem 1.5.5 the latter coincide with the spaces defined in (1.43), which in turn carry all irreducible transitive classical systems of imprimitivity for G on G/H (up to possible covering spaces).

It is suggested by III.1.11 and III.2.10 that for $n \in \mathbb{N}$ the Hilbert space $\mathcal{H}^{n\chi}$ carrying a system of imprimitivity $(U^{\chi}(G), \tilde{\pi}^{\chi}(C_0(G/H)))$ should be seen as the quantization of the symplectic leaf $(\widetilde{T^*G})^{\mathcal{O}_\chi}$ of $(T^*G)/H$ for the value $\hbar = 1/n$ of Planck's constant; here χ is an integral weight. In particular, for fixed $\hbar = 1$, different values of χ correspond to the quantization of different symplectic manifolds. The phase space $T^*(G/H)$ is quantized by $L^2(G/H)$; none of the nontrivially induced representations should be seen as the quantization of $T^*(G/H)$.

2.9 *The Quantization of Constrained Systems*

We will now look at Rieffel induction as a quantum analogue of the classical special symplectic reduction procedure of 1.2. For the purpose of comparing the two, the second approach in 2.2 to Rieffel induction is more appropriate; see Construction 2.2.3.

Firstly, $\mathfrak{B}_{\mathbb{R}}$ should be regarded as the quantization of the Poisson algebra $C^\infty(P, \mathbb{R})$, whereas the Hilbert $\mathfrak{B}$-module $\mathcal{E}$ is the simultaneous quantum analogue of the Poisson map $J : S \to P^-$ and its pullback $J^* : C^\infty(P, \mathbb{R}) \to C^\infty(S, \mathbb{R})$. The representation $\pi_\chi(\mathfrak{B})$ on $\mathcal{H}_\chi$ is the quantum counterpart of the Poisson map $J_\rho : S_\rho \to P$; under favorable conditions, the Hilbert space $\mathcal{H}_\chi$ may even be thought of as the quantization of the symplectic manifold S_ρ. Quotienting $\mathcal{E} \otimes \mathcal{H}_\chi$ by the null space $\mathcal{N}_\chi$, which is the decisive step in the construction of the induced Hilbert space $\mathcal{H}^\chi$, is clearly the analogue of quotienting by the null foliation Φ in (1.13).

Recall Definition 1.1.8. A suitable quantum analogue is given by

Definition 2.9.1. *A* **weak quantum observable** *is a linear map A on $\mathcal{E} \otimes \mathcal{H}_\chi$ that for all $\tilde{\Psi}, \tilde{\Phi} \in \mathcal{E} \otimes \mathcal{H}_\chi$ satisfies*

$$(A\tilde{\Psi}, \tilde{\Phi})_0^\chi = (\tilde{\Psi}, A\tilde{\Phi})_0^\chi; \tag{2.119}$$

$$(A\tilde{\Psi}, A\tilde{\Psi})_0^\chi \leq \|A\|^2 (\tilde{\Psi}, \tilde{\Psi})_0^\chi. \tag{2.120}$$

The collection of all weak quantum observables is called $\mathfrak{A}_w^\chi$.

Using (2.23), we see that (2.119) implies that $A\mathcal{N}_\chi \subseteq \mathcal{N}_\chi$. Hence we may define $\pi^\chi(A)V\tilde{\Psi} := VA\tilde{\Psi}$, generalizing (2.25). Because of (2.120), the analogue of (2.19) still holds.

If we weaken 2.9.1 to allow for operators defined on a dense domain $\tilde{\mathcal{E}} \otimes \mathcal{H}_\chi$, thereby giving up (2.120), the operator $\pi^\chi(A)$ may be unbounded, with (initial) domain $V(\tilde{\mathcal{E}} \otimes \mathcal{H}_\chi)$. In view of (2.24), condition (2.119) guarantees that $\pi^\chi(A)$ is then symmetric on this domain. Moreover, the vector space $\mathfrak{A}_w^\chi$ (defined with or without (2.120)) is easily seen to be a Jordan–Lie algebra under the operations I.(1.22).

The quantum version of (1.15) is the inclusion

$$C^*(\mathcal{E}, \mathfrak{B})_{\mathbb{R}} \subset \mathfrak{A}_w^\chi, \tag{2.121}$$

where $C^*(\mathcal{E}, \mathfrak{B})$ is identified with $C^*(\mathcal{E}, \mathfrak{B}) \otimes \mathbb{I}_\chi$. This follows, because (2.11) and (2.21) guarantee that a self-adjoint element A of $C^*(\mathcal{E}, \mathfrak{B})$ satisfies (2.119). Finally, (2.25) is evidently the quantum analogue of (1.16).

It is instructive to reconsider the example $\mathcal{E} = \Gamma_0(\mathsf{H}^\chi)$ (given after 2.2.2) in the light of the analogy with special symplectic reduction. In Definition 1.2.1 we take $S = (\widetilde{T^*\mathsf{P}})^\chi$, which is defined as in (1.30), in which we replace S_ρ by S_χ to avoid notational ambiguity, and $P = Q$ (with zero Poisson structure). Regarding S as a bundle over T^*Q (cf. Theorem 1.6.1), and therefore over Q, the Poisson map $J : S \to Q$ is the bundle projection. Finally, we reduce from $S_\rho = q$, seen as a symplectic leaf in Q, and ρ the inclusion map. To compute the reduced space we look locally: By Theorem 1.6.1 we locally have $S \simeq T^*Q \times S_\chi$. The constraint hypersurface $S *_Q S_\rho$ is the fiber $J^{-1}(q)$ of S above q. The null distribution $\mathcal{N}$ is tangent to the fibers of T^*Q, so that finally the reduced space is $S^q \simeq S_\chi$.

However, mesmerized by the analogy between special symplectic reduction and Rieffel induction, one should not overlook a cardinal difference between the classical and the quantum reduction procedure. In the classical case one had to restrict the full space $S \times S_\rho$ to the constraint hypersurface $S *_P S_\rho$ before taking the quotient by the null foliation; the latter would, indeed, not even be defined without this restriction. In the quantum case, on the other hand, one passes straight from $\mathcal{E} \otimes \mathcal{H}_\chi$ to the quotient $\mathcal{E} \otimes \mathcal{H}_\chi / \mathcal{N}_\chi$. There is no analogue of the constraint hypersurface; in other words, in quantum mechanics there is no need to impose the constraints. (This comment applies only to the case where all constraints are first class, which is appropriate here in view of Theorem 1.2.2 and the comment preceding 1.1.8.)

In an effort to better mimic the classical procedure, one may introduce an additional step in Rieffel induction, in which one at the very beginning passes from $\mathcal{E} \otimes \mathcal{H}_\chi$ to $\mathcal{E} \otimes_{\mathfrak{B}} \mathcal{H}_\chi$. Here the tensor product over $\mathfrak{B}$ means that $\mathcal{E} \otimes_{\mathfrak{B}} \mathcal{H}_\chi$ consists of equivalence classes $[\Psi \otimes v]_{\mathfrak{B}}$ under the equivalence relation $\Psi B \otimes v \sim \Psi \otimes \pi_\chi(B) v$ for all $B \in \mathfrak{B}$. In other words, this new first step would consist in dividing $\mathcal{E} \otimes \mathcal{H}_\chi$ by the vector space generated by all expressions of the form $\Psi B \otimes v - \Psi \otimes \pi_\chi(B) v$. Although there is a notational similarity between $S *_P S_\rho$ and $\mathcal{E} \otimes_{\mathfrak{B}} \mathcal{H}_\chi$, this step still amounts to taking a quotient rather than imposing the constraints in some way or another. More importantly, this first step is entirely unnecessary, because all vectors $\Psi B \otimes v - \Psi \otimes \pi_\chi(B) v$ lie in the null space $\mathcal{N}_\chi$, as a simple computation shows. Hence they automatically disappear in the construction of $\mathcal{H}^\chi$.

In physics one would interpret the above discussion in the light of the quantization of constrained systems. Here "quantization" may refer to any procedure deserving that name (such as the methods of strict or pure state quantization discussed in this book). The problem is to quantize the symplectic manifold S^C defined by (1.2), given that it has been obtained by symplectic reduction from S. One could, of course, try to quantize S^C without this knowledge, but in case it is simpler and less problematic to quantize S, it makes sense to try to quantize S first, and then use some method that mimics symplectic reduction in order to construct what should be the quantization of S^C. See Figure 15.

For the purpose of this discussion we assume that C is coisotropically embedded in S; in other words, that all constraints are first class (we also put $\hbar = 1$). The traditional approach is to quantize S as if there were no constraints. This yields

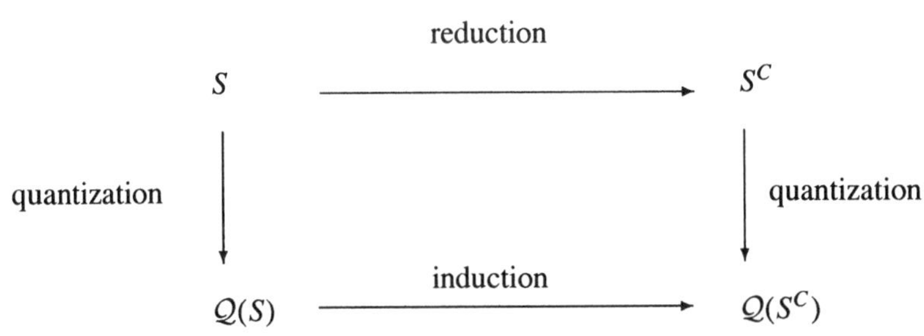

FIGURE 15. Constrained quantization: $\mathcal{Q}$ is some quantization procedure

a Hilbert space $\mathcal{Q}(S) = \mathcal{H}$ and some quantization $\mathcal{Q}(C^\infty(S, \mathbb{R}))$ of the classical observables as operators on $\mathcal{H}$. In particular, the constraints φ_i are quantized by operators $\mathcal{Q}(\varphi_i)$, which are usually unbounded and symmetric on some common dense domain. For example, when C is defined by Marsden–Weinstein reduction at zero, that is, when $C = J^{-1}(0)$ for some strongly Hamiltonian H-action on S with equivariant momentum map $J : S \to \mathfrak{h}^*_-$, the constraints φ_i are the components $J_i, i = 1, \ldots, \dim(H)$. In that case one hopes that $\mathcal{Q}(J_i) = idU(T_i)$, where U is a representation of H on $\mathcal{H}$; cf. III.(1.128) and III.(2.167).

The **Dirac method** of quantizing systems with first-class constraints now consists in defining the "physical state space" $\mathcal{Q}(S^C) = \mathcal{H}_D$ as

$$\mathcal{H}_D := \{\Psi \in \mathcal{H} \mid \mathcal{Q}(\varphi_j)\Psi = 0 \, \forall i\}. \tag{2.122}$$

When $f \in C^\infty(S, \mathbb{R})$ is a weak observable, it is hoped that $[\mathcal{Q}(f), \mathcal{Q}(\varphi_i)]$ is proportional to a linear combination of the $\mathcal{Q}(\varphi_j)$; cf. 1.1.9.2. If so, $\mathcal{Q}(f)$ leaves $\mathcal{H}_D$ stable, so that one may define the "physical observable" $\mathcal{Q}^C(f)$ as the restriction of $\mathcal{Q}(f)$ to $\mathcal{H}_D$.

The Dirac method attempts to quantize the first step of classical reduction, namely imposing the constraints. The second step of quotienting by the null foliation, i.e., passing from C to C/Φ_C, has no counterpart. Compare this with the opposite state of affairs in the method based on Rieffel induction.

In the case $C = J^{-1}(0)$ with H compact, the Dirac method is successful. The condition $dU(T_i)\Psi = 0$ for all i (where the T_i form a basis of $\mathfrak{h}$) is equivalent to $U(h)\Psi = \Psi$ for all $h \in H^0$ (the component of e in H); in other words, the physical state space $\mathcal{H}_D$ is the subspace $\mathcal{H}_0$ of $\mathcal{H}$ that transforms trivially under H^0. When $\mathcal{Q}(f)$ commutes with all $U(h)$, the physical observable $\mathcal{Q}^C(f)$ is the restriction of $\mathcal{Q}(f)$ to $\mathcal{H}_0$. Indeed, we see from Proposition 2.5.3 that $\mathcal{H}_0$ coincides with the Hilbert space $\mathcal{H}^0$ obtained by Rieffel induction from the trivial representation U_{id} of H on $\mathcal{H}_{\mathrm{id}} = \mathbb{C}$, and that the physical observable $\mathcal{Q}^C(f)$ is nothing but the induced representative $\pi^{\mathrm{id}}(\mathcal{Q}(f))$.

However, when the quantum constraints $\mathcal{Q}(\varphi_j)$ fail to have zero in their discrete spectrum or fail to have a joint eigenvector for this value, the space $\mathcal{H}_D$ is empty. For Marsden–Weinstein reduction at zero this habitually happens when H is not compact. Examples are provided by Theorem 2.5.4. However, probably the simplest example is $S = T^*\mathbb{R}^n$, with its standard symplectic structure, equipped with the constraint $p_1 = 0$. Analogously to Example III.1.2.11, this constraint is the momentum map for the $\mathbb{R}$-action given by

$$a : (p_1, p_2, \ldots, q^1, q^2, \ldots) \mapsto (p_1, p_2, \ldots, q^1 + a, q^2, \ldots).$$

It is easy to show that Marsden–Weinstein reduction at zero leads to the reduced phase space $S^C \simeq T^*\mathbb{R}^{n-1}$. All functions of $(p_2, \ldots, q^2, \ldots)$ are weak observables, their reduced action on S^C being the obvious one.

Quantizing S by the Hilbert space $\mathcal{H} = L^2(\mathbb{R}^n)$, with $\mathcal{Q}(p_1) = -i\partial/\partial x^1$, it is clear that the spectrum of the quantized constraint is absolutely continuous, and equal to $\mathbb{R}$. Hence $\mathcal{H}_D = \{0\}$, which is not the desired quantization $L^2(\mathbb{R}^{n-1})$ of S^C.

The method based on quantum induction handles this problem quite effortlessly. In the context of Theorem 2.5.4 we take $\mathcal{H} = L^2(\mathbb{R}^n)$, as in the Dirac method. The representation $U(\mathbb{R})$ is taken to be

$$U(a)\Psi(x^1, x^2, \ldots) := \Psi(x^1 - a, x^2, \ldots); \tag{2.123}$$

this representation satisfies $\mathcal{Q}(J_T) = idU(T) = \mathcal{Q}(p_1)$, where T is the standard generator of $\mathbb{R}$. Choosing $\tilde{\mathcal{E}} = C_c^\infty(\mathbb{R}^n)$, one verifies the properties required in 2.5.4. We now induce from the trivial representation U_{id} of $\mathbb{R}$, so that $\mathcal{H}_\chi = \mathcal{H}_{\text{id}} = \mathbb{C}$. Hence

$$\begin{aligned}(\Psi, \Phi)_0^{\text{id}} &= \int_{\mathbb{R}} da \int_{\mathbb{R}^n} d^n x\, \overline{\Psi(x^1 - a, x^2, \ldots)}\Phi(x^1, x^2, \ldots) \\ &= \overline{\hat{\Psi}(0, x^2, \ldots)}\hat{\Phi}(0, x^2, \ldots),\end{aligned} \tag{2.124}$$

where the Fourier transform (indicated by the hat) is taken only in the first variable. To identify the induced space $\mathcal{H}^{\text{id}}$ we use Proposition 2.2.4. Our guess is $\mathcal{H}_*^{\text{id}} = L^2(\mathbb{R}^{n-1})$, consisting of the square-integrable functions of $(x^2, \ldots)$. This is proved by constructing $\tilde{U} : \tilde{\mathcal{E}} \to L^2(\mathbb{R}^{n-1})$ as

$$\tilde{U}\Psi(x^2, \ldots) := \hat{\Psi}(0, x^2, \ldots). \tag{2.125}$$

It is evident from (2.124) that (2.29) is satisfied.

We identify $L^2(\mathbb{R}^n)$ with $L^2(\mathbb{R}) \otimes L^2(\mathbb{R}^{n-1})$ in the well-known way. Bounded operators A on $L^2(\mathbb{R}^n)$ that do not act on the first variable in Ψ have the form $A = \mathbb{I}_1 \otimes A_2$, where $\mathbb{I}_1$ is the unit operator on $L^2(\mathbb{R})$, and $A_2 \in \mathfrak{B}(L^2(\mathbb{R}^{n-1}))$. Such operators lie in $\tilde{C}^*(\tilde{\mathcal{E}}, \tilde{\mathfrak{B}})$ (cf. the text following (2.34)), and may be Rieffel-induced to bounded operators $\pi^{\text{id}}(A)$ on $\mathcal{H}^{\text{id}}$. One easily sees from (2.30) that $\pi^{\text{id}}(A)$ is simply A_2.

Thus we have a satisfactory quantization of the classical situation, which, despite its simplicity, is not so easily amenable to treatment by other constrained quantization techniques.

2.10 *Quantization of Singular Reduction*

Basing constrained quantization on Rieffel induction allows one to quantize singular Marsden–Weinstein quotients. It is interesting to see how the singularities in the reduced space, and particularly its decomposition into symplectic pieces, is reflected in the quantum theory. In the absence of a general theory of this reflection, we approach this matter through specific examples.

Consider the case $S = T^*\mathbb{R}^2$ and $H = SO(2)$ discussed in 1.11. A suitable quantization of S is given by the Hilbert space $\mathcal{H} = L^2(\mathbb{R}^2)$, carrying the representation of $SO(2)$ given by $U(h)\Psi(q) := \Psi(h^{-1}q)$, in terms of the defining action of $SO(2)$ on $\mathbb{R}^2$. Mapping $L^2(\mathbb{R}^2)$ to $L^2(\mathbb{R}^+, r\,dr) \otimes L^2(SO(2))$ in the usual way, the representation U is the tensor product of the unit $\mathbb{I}$ on $L^2(\mathbb{R}^+, r\,dr)$ and the left-regular representation U_L on the second factor. We then induce from the trivial representation U_{id} of $SO(2)$, reflecting the zero in $S^0 = J^{-1}(0)/H$. According to Proposition 2.5.3, the induced space $\mathcal{H}^{\text{id}}$ is $L^2(\mathbb{R}^+, r\,dr)$. Adjointable operators on $\mathcal{H}$ are necessarily $SO(2)$-invariant, so that they are linear combinations of operators of the form $A = A_1 \otimes A_2$, where A_2 commutes with U_L. The induced representation is then given by linear extension of

$$\pi^{\mathrm{id}}(A_1 \otimes A_2) = (1_{SO(2)}, A_2 1_{SO(2)})_{L^2(SO(2))} A_1. \tag{2.126}$$

The identification of $\mathcal{H}^{\mathrm{id}}$ with $L^2(\mathbb{R}^+, r\,dr)$ is a pleasant result, since we have seen that the reduced space S^0 was $T^*\mathbb{R}^+$. On the other hand, the origin 0 is a null set with respect to the measure $r\,dr$, so that at first sight the quantum theory contains no analogue of the lower-dimensional symplectic piece of S^0.

However, the singular structure of the classical reduced phase space is reflected in the domains of unbounded observables. It is remarkable that unbounded operators, notably differential operators, of the form $\pi^{\mathrm{id}}(A)$ on $\mathcal{H}^{\mathrm{id}}$ that would not be essentially self-adjoint on the natural domain $C_c^\infty(\mathbb{R}^+\backslash\{0\})$ are often essentially self-adjoint on the domain $V_\chi C_c^\infty(\mathbb{R}^2)$ inherited from $\mathcal{H}$ (where V_χ is defined below (2.24)). The domain of self-adjointness will then generically be such that wave functions in this domain vanish near the origin, providing a quantum analogue of the fact that motion in the highest-dimensional (and, indeed, in any) symplectic piece of S^0 cannot cross the barrier to the lower-dimension piece(s).

For example, the Hamiltonian $H = -\Delta + V(r)$ on $L^2(\mathbb{R}^2)$ is $SO(2)$-invariant and essentially self-adjoint on $C_c^\infty(\mathbb{R}^2)$. The unitary map $U : L^2(\mathbb{R}^+, r\,dr) \to L^2(\mathbb{R}^+, dr)$, defined by $U\Psi(r) := \sqrt{r}\Psi(r)$, transforms $\pi^{\mathrm{id}}(H)$ into

$$U\pi^{\mathrm{id}}(H)U^* = -\frac{d^2}{dr^2} - \frac{1}{4r^2} + V(r). \tag{2.127}$$

While the analysis of this expression is quite straightforward for any reasonable potential V, the free case $V = 0$ already suffices to illustrate the main point. Defined on $C_c^\infty(\mathbb{R}^+\backslash\{0\})$, the operator (2.127) then has deficiency indices (1, 1), so that it is not essentially self-adjoint. However, defined on $V_\chi C_c^\infty(\mathbb{R}^2)$, which consists of functions of the type $\Psi(r) = \sqrt{r} f(r^2)$ with $f \in C_c^\infty(\mathbb{R}_0^+)$, the operator in question is indeed essentially self-adjoint. The closure of the latter operator is an extension of the closure of the former, to whose domain one adds functions of the indicated type in order to achieve essential self-adjointness. The boundary condition $\Psi(0) = 0$ corresponds to a hard wall potential at the origin.

We now turn to the quantization of the example at the end of 1.11. As above, the quantization of $S = T^*\mathbb{R}^2$ is taken to be $\mathcal{H} = L^2(\mathbb{R}^2)$, of which we take the Fourier transform $\Psi \mapsto \hat{\Psi}$, so that we work in p-space. The $\mathbb{R}$-action (1.90) on S is quantized by the representation $\hat{U}(\mathbb{R})$, given by

$$\hat{U}(t)\hat{\Psi}(p) := e^{\frac{1}{2}it(p_1^2 - p_2^2)}\hat{\Psi}(p). \tag{2.128}$$

This is motivated by the fact that the generator $idU(T)$ (where T is the standard generator of $\mathbb{R}$) on $\mathcal{H}$ is

$$Q(\varphi) := i\,dU(T) = \frac{1}{2}\left(-\frac{\partial^2}{\partial(x^1)^2} + \frac{\partial^2}{\partial(x^2)^2}\right), \tag{2.129}$$

which is the quantization of the single classical constraint $\varphi = J$ in the Schrödinger representation; cf. (1.91).

It is possible now to follow the procedure of Theorem 2.5.4 with $H = \mathbb{R}$, with a suitable replacement of $C_c^\infty(H)$. However, it is easier to approach this situation using Fell induction rather than Rieffel induction. Hence we apply Construction 2.2.5, in which $\tilde{\mathcal{E}} \subset \mathcal{L}^2(\mathbb{R})$ is taken to consist of those $\hat{\Psi} \in C_c^\infty(\mathbb{R}^2)$ for which there exists an $\epsilon > 0$ (depending on Ψ) such that $\hat{\Psi} = O(|p|^\epsilon)$ for $|p| := \sqrt{p_1^2 + p_2^2} \rightarrow 0$. The motivation for this choice of $\tilde{\mathcal{E}}$ will emerge shortly. Since the classical reduced phase space is $J^{-1}(0)/\mathbb{R}$, we put $\mathcal{H}_\chi = \mathcal{H}_{\text{id}} = \mathbb{C}$ in 2.2.5, so that also $\mathfrak{B}(\mathcal{H}_\chi) = \mathbb{C}$. With $v = w = 1$ in (2.37), the form $(\Psi, \Phi)_0^\chi$ then coincides with $\langle \Psi, \Phi \rangle_{\mathbb{C}}$. Proceeding as if we were performing Rieffel induction from the trivial representation of $\mathbb{R}$, we define $(\Psi, \Phi)_0^{\text{id}}$ by (2.80) with $U_{\text{id}}(h) = 1$. Using (2.128), this yields

$$(\Psi, \Phi)_0^{\text{id}} = \int_{\mathbb{R}} dt \int_{\mathbb{R}^2} \frac{d^2 p}{(2\pi)^2} \, e^{\frac{1}{2} it(p_1^2 - p_2^2)} \overline{\hat{\Psi}(p)} \hat{\Phi}(p), \tag{2.130}$$

defined as an oscillatory integral. This can be computed as such, resulting in

$$(\Psi, \Phi)_0^{\text{id}} = \int_{\mathbb{R}} \frac{dk}{2\pi |k|} \left[\overline{\hat{\Psi}(k, k)} \hat{\Phi}(k, k) + \overline{\hat{\Psi}(k, -k)} \hat{\Phi}(k, -k) \right]. \tag{2.131}$$

This expression is well-defined by virtue of our choice of $\tilde{\mathcal{E}}$.

Using the method of Proposition 2.2.4, the induced space $\mathcal{H}^{\text{id}}$ may be identified with

$$\mathcal{H}_*^{\text{id}} := L^2(\mathbb{R}, dk/2\pi |k|) \otimes \mathbb{C}^2. \tag{2.132}$$

For we may define $\tilde{U} : \tilde{\mathcal{E}} \rightarrow \mathcal{H}_*^{\text{id}}$ by

$$\tilde{U}\Psi(k) := (f_k^+, \Psi) \dot{+} (f_k^-, \Psi), \tag{2.133}$$

where the functions $f_k^\pm \in C^\infty(\mathbb{R}^2)$ are defined for $k \in \mathbb{R}$ by

$$f_k^\pm(x^1, x^2) := e^{ik(x^1 \mp x^2)}, \tag{2.134}$$

and the pairing on the right-hand side of (2.133) is defined as if Ψ and $f_k^\pm$ both were in $L^2(\mathbb{R}^2)$. The right-hand side of (2.133) equals $\hat{\Psi}(k, k) \dot{+} \hat{\Psi}(k, -k)$, so that one easily verifies (2.29) from (2.131). (Equation (2.125) may be rewritten in a similar way, since the right-hand side equals $(f_{x^2,\ldots}, \Psi)$, where $f_{x^2,\ldots}(\tilde{x}^1, \tilde{x}^2, \ldots)$ is $\delta(x^2 - \tilde{x}^2, \ldots)$, but the gain in doing so is small.)

This way of writing the map $\tilde{U}$ is of fundamental importance, as the $f_k^\pm$ form the complete set of linearly independent solutions of the quantum constraint equation $\mathcal{Q}(\varphi) f = 0$; cf. (2.129). These solutions do not lie in L^2, so that the Dirac method fails in this example, but we see that the solutions of the quantized constraint equations still play a formal role in the quantization procedure.

An operator A that commutes with all $U(t)$ is adjointable. Now suppose that $A : \tilde{\mathcal{E}} \rightarrow \tilde{\mathcal{E}}$ satisfies the stronger condition of commuting with $\mathcal{Q}(\varphi)$. When the action of A may be extended to an action on each $f_k^\pm$, it follows that $A f_k^\pm$ is again a solution of the quantized constraint equations, so that it may be expanded in terms of the solutions $f_k^{\pm}$. The induced representation $\pi^{\mathrm{id}}(A)$ may then be expressed in terms of the expansion coefficients. To make this argument precise, suppose that A as above is such that $(f_k, A\Psi) = (\mathsf{A}^* f_k, \Psi)$, where $\mathsf{A}^* f_k := \overline{(\mathsf{A}(k,\cdot), f_\cdot)_*}$, for some kernel A. Here f_k is regarded as a 2-component vector, so that each $\mathsf{A}(k, k')$ is a 2×2 matrix, the inner product $(\, ,\,)_*$ being in $L^2(\mathbb{R}, dk/2\pi|k|) \otimes \mathbb{C}^2$. It then follows from (2.30) and (2.133) that

$$\pi_*^{\mathrm{id}}(A)\Psi^{\mathrm{id}}(k) = (\mathsf{A}(k,\cdot), \Psi^{\mathrm{id}})_*. \tag{2.135}$$

We now examine the question to what extent the quantum theory reflects the decomposition of the classical reduced phase space S^0 into its symplectic pieces; see the final paragraph of 1.11. Firstly, as in the previous example, the lowest-dimensional piece $p_1 = p_2 = 0$ does not occur in the quantum theory. Secondly, we have seen that smooth functions on S^0 have to descend from smooth functions on S that depend only on p_1, p_2, and $q^1 p_2 + q^2 p_1$. In the Schrödinger representation on $L^2(\mathbb{R}^2)$ (position space) these are quantized by

$$\begin{aligned} \mathcal{Q}(p_1) &= -i\partial/\partial x^1; \\ \mathcal{Q}(p_2) &= -i\partial/\partial x^2; \\ \mathcal{Q}(q^1 p_2 + q^2 p_1) &= -i(x^1\partial/\partial x^2 + x^2\partial/\partial x^1), \end{aligned} \tag{2.136}$$

respectively. The induced action on $\mathcal{H}_*^{\mathrm{id}}$ is then given by

$$\begin{aligned} \pi_*^{\mathrm{id}}(\mathcal{Q}(p_1)) &= \mathrm{diag}\,(k, k); \\ \pi_*^{\mathrm{id}}(\mathcal{Q}(p_2)) &= \mathrm{diag}\,(k, -k); \\ \pi_*^{\mathrm{id}}(\mathcal{Q}(q^1 p_2 + q^2 p_1)) &= -ik\,\mathrm{diag}\left(\frac{d}{dk}, -\frac{d}{dk}\right). \end{aligned} \tag{2.137}$$

Here k is seen as a multiplication operator on $L^2(\mathbb{R}, dk/2\pi|k|)$, and the diagonal is meant to be of a 2×2 matrix. The operators in (2.137) are defined and essentially self-adjoint on the space of C_c^∞-functions in $\mathcal{H}_*^{\mathrm{id}}$. It is obvious from these expressions that the four subspaces $L^2(\mathbb{R}^+, dk/2\pi|k|) \otimes \mathbf{e}_i$, $L^2(\mathbb{R}^-, dk/2\pi|k|) \otimes \mathbf{e}_i$, where $i = 1, 2$, of $L^2(\mathbb{R}, dk/2\pi|k|) \otimes \mathbb{C}^2$ do not mix under the action of these operators. More precisely, each of these spaces is stable under the group generated by the Lie algebra spanned by the operators in (2.137). Each of these four sectors is plainly the quantum counterpart of the appropriate symplectic piece in S^0.

3 Applications in Relativistic Quantum Theory

3.1 Coadjoint Orbits of the Poincaré Group

We will now apply the formalism of 1.10 to the case that G is the Poincaré group. Assuming that the reader is familiar with the theory of special relativity, we recall some notation. **Minkowski space** $\mathcal{M}$ is $\mathbb{R}^4$ equipped with the metric tensor $\mathbf{g}_{\mathcal{M}} := \mathrm{diag}\,(1, -1, -1, -1)$; this tensor is used to raise and lower indices as explained in II.3.2. Greek indices run from 0 to 3, whereas Latin ones go from 1 to 3. The pairing between $\mathcal{M}$ and its dual $\mathcal{M}^*$ is $pv := p(v) := p_\mu v^\mu = p_0 v^0 + p_i v^i$, but $p^2 := \mathbf{g}_M(p, p) = p_0^2 - \mathbf{p}^2$, where $\mathbf{p}^2 := p_i p_i$, etc. In general, $\mathbf{p}$ stands for (p_1, p_2, p_3).

Furthermore, $O(1, 3)$ is the subgroup of $GL(4, \mathbb{R})$ consisting of elements that leave $\mathbf{g}_{\mathcal{M}}$ invariant. For convenience of notation, the **Lorentz group** L is defined as the connected component of $O(3, 1)$ containing the identity (this subgroup of $O(3, 1)$ is often called $L_+^\uparrow$). The Lie algebra $\mathfrak{l}$ of the Lorentz group consists of those real 4×4 matrices $M^\mu_{\ \nu}$ for which the components $M_{\mu\nu}$ are antisymmetric.

The dual $\mathfrak{l}^*$ is identified with $\mathfrak{l}$ under the pairing $M(N) := M^{\mu\nu} N_{\mu\nu}$. This is useful, because $p \wedge v$ defined in (1.67) is now the matrix with entries

$$(p \wedge v)_{\mu\nu} = \tfrac{1}{2}(p_\mu v_\nu - p_\nu v_\mu). \tag{3.1}$$

Moreover, the coadjoint action of L is given by $\mathrm{Co}(M) : N \mapsto MNM^T$.

A convenient basis of $\mathfrak{l}$ is $\{J_i, B_i\}$, where $J_i := \epsilon_{ijk} E_{jk}$ and $B_i := E_{0i} - E_{i0}$; here $E_{\mu\nu}$ is the matrix with entry 1 at position $\mu\nu$, and 0 elsewhere. The commutation relations are $[J_i, J_k] = \epsilon_{ijk} J_k$, $[B_i, B_j] = -\epsilon_{ijk} J_k$, and $[J_i, B_j] = \epsilon_{ijk} B_k$. Physically, the J_i generate rotations, whereas the B_i generate boosts. Hence $\mathfrak{l}$ has the reductive decomposition

$$\mathfrak{l} = \mathfrak{so}(3) \oplus \mathfrak{b}, \tag{3.2}$$

in which $\mathfrak{so}(3)$ and $\mathfrak{b}$ are the linear spans of the J_i and of the B_i, respectively.

The **Poincaré group** is $P = L \ltimes_\rho \mathcal{M}$, in which ρ is the defining action $\rho(\Lambda) q^\mu = \Lambda q^\mu := \Lambda^\mu_{\ \nu} q^\nu$ of $L \subset O(3, 1)$ on $\mathcal{M}$. The dual action is $\rho^*(\Lambda) p_\mu = \Lambda p_\mu = \Lambda_\mu^{\ \nu} p_\nu$. We will often omit the symbols ρ and ρ^*.

In principle, each coadjoint orbit of P is the "covariant" phase space of some relativistic particle. As we have seen in Proposition 1.10.1, the first step in the classification of these orbits is the study of the $\rho^*(L)$-orbits in $\mathcal{M}^*$. The classification of the latter is well known: The orbit types are

$$\begin{aligned}
\mathcal{O}_0^L &:= \mathcal{O}_{(0,0,0,0)}^L = \{0\};\\
\mathcal{O}_{m^2,\pm}^L &:= \mathcal{O}_{(\pm m,0,0,0)}^L = \{p \in \mathcal{M}^* \mid p^2 = m^2,\ \pm p_0 > 0\} \simeq L/SO(3);\\
\mathcal{O}_{0,\pm}^L &:= \mathcal{O}_{(\pm 1,0,0,-1)}^L = \{p \in \mathcal{M}^* \mid p^2 = 0,\ \pm p_0 > 0\} \simeq L/E(2);\\
\mathcal{O}_{-m^2}^L &:= \mathcal{O}_{(0,0,0,m)}^L = \{p \in \mathcal{M}^* \mid p^2 = -m^2\} \simeq L/SO(1,2),
\end{aligned} \tag{3.3}$$

where $m > 0$. Here $SO(3)$ is a subgroup of L in the obvious way (which corresponds to the decomposition (3.2)). The embedding in L of the Euclidean group $E(2) := SO(2) \ltimes_\rho \mathbb{R}^2$ in dimension 2 is specified by looking at its generators: A basis $\{T_i\}$ of $E(2)$ is obtained by putting

$$\begin{aligned}
T_1 &:= B_1 - J_2;\\
T_2 &:= B_2 + J_1;\\
T_3 &:= J_3.
\end{aligned} \tag{3.4}$$

This time $\mathfrak{l}$ has the decomposition

$$\mathfrak{l} = \mathfrak{e}(2) \oplus \mathfrak{m}, \tag{3.5}$$

where $\mathfrak{m}$ is the linear span of $T_4 := B_1 + J_2$, $T_5 := B_2 - J_1$, and $T_6 := B_3$. This decomposition fails to be reductive, but on the other hand $\mathfrak{m}$ is a Lie subalgebra of $\mathfrak{l}$, unlike the subspace $\mathfrak{b}$ in the massive case (3.2).

To further describe the data called for in 1.10.1, one should classify the coadjoint orbits of $SO(3)$, $E(2)$, and $SO(1,2)$. The P-orbits related to the L-orbit $\mathcal{O}^L_{-m^2}$ describe tachyonic particles, which are believed not to exist in nature. In what follows, we will therefore look only at the coadjoint orbits that are related to the L-orbits $\mathcal{O}^L_{m^2,\pm}$ and $\mathcal{O}^L_{0,\pm}$; the vacuum $\mathcal{O}^L_0$ is relevant to (quantum) field theory, but not to particle dynamics.

Proposition 3.1.1. *The physically relevant coadjoint orbits of the Poincaré group are as follows.*

- *One family, $\mathcal{O}^P_{m,\pm,s}$, related to the $\rho^*(L)$-orbits $\mathcal{O}^L_{m^2,\pm}$ with $m > 0$, is further labeled by a parameter $s \geq 0$. Each value of the* **spin** *s labels a coadjoint orbit S^2_s of $SO(3)$. One has diffeomorphisms*

$$\mathcal{O}^P_{m,\pm,s} \simeq T^*\mathbb{R}^3 \times S^2_s. \tag{3.6}$$

- *A second family $\mathcal{O}^P_{0,\pm,s}$, related to the $\rho^*(L)$-orbits $\mathcal{O}^L_{0,\pm}$, is further labeled by a parameter $h \in \mathbb{R}$. Each value of the* **helicity** *h labels a coadjoint orbit $(0,0,h)$ of $E(2)$. There are diffeomorphisms*

$$\mathcal{O}^P_{0,\pm,s} \simeq T^*(L/E(2)). \tag{3.7}$$

We start with the massive case; the parameter m is the mass of the particle living on any of the P-orbits to be described. The stabilizer of $\tilde{p} = (\pm m, 0, 0, 0)$ is $L_{\tilde{p}} = SO(3)$. As we have seen in 1.10, the coadjoint orbits of $SO(3)$ are the two-spheres S^2_r with radius r; the case $r = 0$ is included here as the origin. By Proposition 1.10.1 the orbit type of the P-orbit $\mathcal{O}^P_{\tilde{M},\tilde{p}}$ depends on the restriction of $\tilde{M}$ to $\mathfrak{so}(3) \simeq \mathbb{R}^3$. This restriction passes through a unique $SO(3)$ coadjoint orbit S^2_s, and we correspondingly label the P-orbit as $\mathcal{O}^P_{m,\pm,s}$.

The simplest case is $s = 0$; the orbit $\mathcal{O}^P_{m,\pm,0}$ is the one through $(0, \tilde{p}) \in \mathfrak{p}^*$. We infer from (1.66) and (3.1) that the stability group of the above point in this orbit is $SO(3) \times \mathbb{R}$, where $\mathbb{R} \subset \mathcal{M}$ is the zeroth copy $\mathbb{R}e_0$. By (1.80) we then simply have $\mathcal{O}^P_{m,\pm,0} \simeq T^*(L/SO(3))$. The space $L/SO(3)$, and therefore the orbit $\mathcal{O}^L_{m^2,\pm}$, is diffeomorphic to $\mathbb{R}^3$, because $SO(3)$ is the maximal compact subgroup of L. Hence $\mathcal{O}^P_{m,\pm,0} \simeq T^*\mathbb{R}^3$. This is the usual phase space of a spinless particle, with the difference that in the present description the base $\mathbb{R}^3$ is momentum space, whereas the fiber of the cotangent bundle is position space.

For $s \neq 0$ the stability group is $SO(2) \times \mathbb{R}$, where $SO(2) \subset SO(3)$ is the stabilizer of some point in S^2_s, and $\mathbb{R}$ is as above. We are now in a position to use Theorem 1.10.4 (and especially the subsequent comments) in its full glory. Since (3.2) is a reductive decomposition, there exists an L-invariant connection on the principal bundle $L(L/SO(3), SO(3), \tau)$, namely the H-connection; see III.2.7. Hence Theorem III.2.3.7 asserts that there is an L-equivariant identification of $\mathcal{O}^P_{m,\pm,s}$ with a bundle over $T^*\mathbb{R}^3$ with typical fiber S^2_s, providing a Lorentz-invariant splitting of spin and orbital degrees of freedom. Since the base of this bundle is contractible, we therefore obtain (3.6). Note that the symplectic structure on the right-hand side of (3.6) does not factorize; this is what physicists call spin–orbit coupling.

We pass to the massless case. As we have seen, the stabilizer of $\tilde{p} = (\pm 1, 0, 0, -1)$ is $E(2)$. Its coadjoint orbits may be described either in analogy to those of $E(3)$ (cf. 1.10), or by using Proposition 1.10.1, or by direct calculation from (1.66), in which $p \wedge v$ equals the number $p_1 v^2 - p_2 v^1$. Either way, identifying $\mathfrak{e}(2)^*$ with $\mathbb{R}^3$ using the basis $\{T_i\}$ (cf. (3.4)), one derives that each point $(0, 0, h)$ is an orbit, and that the remaining orbits are cylinders $C_r = S^1_r \times \mathbb{R}$, where the circle S^1_r of radius $r > 0$ lies in the p-plane, and $\mathbb{R}$ is the z-axis. The stabilizer of $(0, 0, h)$ is $E(2)$ itself, whereas the stabilizer of a point in C_r is $\mathbb{R}$ (embedded in the $\mathbb{R}^2$ of $SO(2) \ltimes_\rho \mathbb{R}^2$ in a way depending on the choice of the point).

The only orbits that are believed to be of physical relevance are the points $(0, 0, h)$. The P-orbit corresponding to such a point is denoted by $\mathcal{O}^P_{0,\pm,h}$. The stabilizer of a point in $\mathcal{O}^P_{0,\pm,h}$ is conjugate to $E(2) \ltimes_\rho \mathbb{R}$, where $\mathbb{R}$ lies in $\mathcal{M}$ as $\mathbb{R}(1, 0, 0, -1)$. This group is equal to $SO(2) \ltimes_{\tilde{\rho}} \mathbb{R}^3$, where the $SO(2)$-action $\tilde{\rho}$ on $\mathbb{R}^3$ is given by rotations in the (x, y)-plane. By Theorems 1.10.4 and III.2.3.7 we infer that $\mathcal{O}^P_{0,\pm,h}$ is diffeomorphic to $T^*(L/E(2))$. ∎

However, since $E(2)$ is not a reductive subgroup of L, one cannot choose an L-invariant connection $\mathbf{A}$ on the bundle $L(L/E(2), E(2), \tau)$. Hence the pertinent diffeomorphism cannot be chosen in a Lorentz-invariant way. Moreover, the Poisson structure on $T^*(L/E(2))$ is not the canonical one; it depends on h. Indeed, using coordinates p, q on the cotangent bundle that are canonical with respect to the standard symplectic form, but interchanging p and q in the light of the comment concluding the treatment of $\mathcal{O}^P_{m,\pm,0}$ above, the Poisson bracket on $T^*(L/E(2))$ is

$$\{f, g\}^h = \frac{\partial f}{\partial q_i}\frac{\partial g}{\partial p^i} - \frac{\partial f}{\partial p^i}\frac{\partial g}{\partial q_i} - hF_{ij}(p)\frac{\partial f}{\partial q_i}\frac{\partial g}{\partial q_j}, \tag{3.8}$$

where F is the curvature of the connection $\mathbf{A}$; cf. III.(2.55). In particular, the canonical position coordinates do not Poisson-commute.

3.2 *Orbits from Covariant Reduction*

According to Theorem 1.10.2, the coadjoint orbits $\mathcal{O}^P$ of P may be obtained by Kazhdan–Kostant–Sternberg reduction from a particular coadjoint orbit of a subgroup $H \subset P$ of the form (1.70). It is interesting to see what happens when one instead reduces from a coadjoint orbit of the Lorentz group. In particular, we shall investigate how one may recover the orbits $\mathcal{O}^P$ in that way. Apart from being an illustration of symplectic reduction and the theory of constraints, this turns out to be the classical analogue of a standard procedure in relativistic quantum theory, to be discussed in the next section.

Lemma 3.2.1. *Let $\mathcal{O}^L$ be a coadjoint orbit of the Lorentz group L. The reduced space $(T^*P)^{\mathcal{O}^L}$, defined by (1.43) with G replaced by P and H by L, is symplectomorphic to $T^*\mathcal{M}^* \times \mathcal{O}^L_+$, in such a way that the reduced P-action III.(2.122) on $(T^*P)^{\mathcal{O}^L}$ becomes the natural action on $T^*\mathcal{M}^* \times \mathcal{O}^L_+$.*

Here the "natural P-action" is the product of the pullback to $(T^*\mathcal{M})^- \simeq T^*\mathcal{M}^*$ of the defining action on $\mathcal{M}$ and the coadjoint L-action on $\mathcal{O}^L$ (on which $\mathcal{M}$ acts trivially). The symplectic structure on $T^*\mathcal{M}^* \times \mathcal{O}^L_+$ is not the direct product one, although its restriction to each factor is the one indicated by the notation; see below.

The proof is a simple calculation; as in the proof of 1.10.2 one uses the left trivialization of T^*P. The space $J_{\mathfrak{l}}^{-1}(\mathcal{O}^L)$ consists of points (M, p, Λ, v) in $-\mathcal{O}^L \times \mathcal{M}^* \times L \times \mathcal{M}$; elements of the reduced space are equivalence classes of such points under the L-action derived from III.(1.49) and (1.64). The map $[M, p, \Lambda, v)]_L \mapsto (-\Lambda p, v, -\mathcal{O}^L(\Lambda)M)$ is a diffeomorphism from $J_{\mathfrak{l}}^{-1}(\mathcal{O}^L)/L$ to $\mathcal{M}^* \times \mathcal{M} \times \mathcal{O}^L$. (This map is a combination of minus the momentum map J^L and the identity on $\mathcal{M}$; cf. III.(1.56), (1.66), and Figure 4.)

The symplectic structure on this space may be computed from III.(1.54) and (1.65). The claim about the reduced P-action may be verified from III.(1.50) and (1.64). ∎

Our aim is now to construct the physically relevant coadjoint orbits of P by symplectic reduction of $T^*\mathcal{M}^* \times \mathcal{O}^L_+$. In the simplest case $\mathcal{O}^L_+ = \{0\}$ this is done by imposing the constraints

$$\varphi := \tfrac{1}{2}(p^2 - m^2) = 0 \tag{3.9}$$

and $\pm p_0 > 0$. The constraint φ is a momentum map for the $\mathbb{R}$-action on $T^*\mathcal{M}^*$ given by

$$t : (p_\mu, q^\mu) \mapsto (p_\mu, q^\mu - p^\mu t).$$

The symplectic reduction $(T^*\mathcal{M}^*)^\varphi$ of $T^*\mathcal{M}^*$ with respect to this constraint is symplectomorphic to $T^*\mathbb{R}^3$ (with p-space as the base); for one may map $(p_\mu, q^\mu) \in C$ (the subspace of $T^*\mathcal{M}^*$ where $\varphi = 0$) to $(p_i, q^i - q^0 p^i/p_0)$. Similarly, one has $T^*\mathcal{O}^L_{m^2,\pm} \simeq T^*\mathbb{R}^3$, so that finally $(T^*\mathcal{M}^*)^\varphi \simeq \mathcal{O}^P_{m,\pm,0}$ by (3.6). One may put $m = 0$ here.

This construction may be generalized to arbitrary values of the spin. We use the matrices M as parameters on $\mathcal{O}$, in addition employing $K_i := M_{0i}$ and $R_i := \frac{1}{2}\epsilon_{ijk}M_{jk}$, as well as $K^2 := K_iK_i$ and $R^2 := R_iR_i$. It may be shown that the conditions

$$\begin{aligned} K_iR_i &= 0; \\ R^2 - K^2 &= s^2 \end{aligned} \tag{3.10}$$

select a coadjoint orbit $\mathcal{O}^L_s$ in $\mathfrak{l}^*$; this is the only type of orbit we shall need in order to reach the physical orbits in $\mathfrak{p}^*$. These orbits are plainly four-dimensional.

The massive case is qualitatively different from the massless situation, but we present the results in united fashion.

Theorem 3.2.2. *Let $m \geq 0$, and consider the constraints φ in (3.9) and $\varphi_\nu := p^\mu M_{\mu\nu}$. In addition, impose $\pm p_0 > 0$.*

- *The symplectic reduction of $T^*\mathcal{M}^* \times \mathcal{O}_s^L$ by φ is $T^*\mathcal{O}^L_{m^2,\pm} \times \mathcal{O}_s^L$. The functions φ_ν are well-defined on this reduced space (and in what follows are regarded as functions on $T^*\mathcal{O}^L_{m^2,\pm} \times \mathcal{O}_s^L$).*
- *For $m > 0$ the constraints $\varphi_\nu = 0$ are second class, and the subspace of $T^*\mathcal{O}^L_{m^2,\pm} \times \mathcal{O}_s^L$ on which they hold is symplectomorphic to the coadjoint orbit $(\mathcal{O}^P_{m,\pm,s})_+$ of P.*
- *For $m = 0$ the constraints $\varphi_\nu = 0$ are first class, and the symplectic reduction of $T^*\mathcal{O}^L_{0,\pm} \times \mathcal{O}_s^L$ by these constraints is symplectomorphic to the union of the coadjoint orbits $(\mathcal{O}^P_{0,\pm,s})_+$ and $(\mathcal{O}^P_{0,\pm,-s})_+$.*
- *The reduction of the natural P-action on $T^*\mathcal{M}^* \times \mathcal{O}_s^L$ (cf. 3.2.1) to the reduced spaces above is equivalent to the coadjoint action.*

The first claim of the proposition is immediate from the calculation preceding 3.2.2. In what follows $\tilde{p}$ is either $(\pm m, 0, 0, 0)$ or $(\pm 1, 0, 0, -1)$, and $L_{\tilde{p}}$ denotes either $SO(3)$ (for $m > 0$) or $E(2)$ (when $m = 0$). In the former case $\mathfrak{l}^0_{\tilde{p}}$ may be identified with $\mathfrak{b}$ (cf. (3.2)).

We identify $T^*\mathcal{O}^L_{m^2,\pm}$ with $L \times_{L_{\tilde{p}}} \mathfrak{l}^0_{\tilde{p}}$; cf. (3.2) and III.(2.116). The constraints $p^\mu M_{\mu\nu} = 0$ are then the components of $\Phi : (L \times_{L_{\tilde{p}}} \mathfrak{l}^0_{\tilde{p}}) \times \mathcal{O}_s^L \rightarrow \mathcal{M}^*$, defined by $\Phi([\Lambda, N]_{L_{\tilde{p}}}, M) := M\Lambda\tilde{p}$.

Contracting Φ with $\tilde{p}$, or φ_ν with p^ν, one immediately sees that at most three components of φ_ν are independent. Since M itself satisfies the condition of lying in $\mathcal{O}_s^L$, only two components of Φ are actually independent. The solution set C_s^m of these constraints consists of those $([\Lambda, N]_{L_{\tilde{p}}}, M)$ for which $N \in \mathfrak{l}^0_{\tilde{p}}$ and $M \in \Lambda(\mathfrak{l}_{\tilde{p}} \cap \mathcal{O}_s^L)$. This is well-defined, for $L_{\tilde{p}}$ maps $\mathfrak{l}_{\tilde{p}} \cap \mathcal{O}_s^L$ into itself, so that changing Λ by $\Lambda\tilde{\Lambda}$ makes no difference as long as $\tilde{\Lambda} \in L_{\tilde{p}}$.

For $m > 0$ the set $\mathfrak{l}_{\tilde{p}} \cap \mathcal{O}_s^L$ contains all matrices M in $\mathfrak{l}$ for which $K_i = 0$ and $R^2 = s^2$; with $\mathfrak{l}_{\tilde{p}} = \mathfrak{so}(3)$, identified with its dual, it is clear that this set is the coadjoint orbit S_s^2 of $SO(3)$. We now define $\psi : C_s^m \rightarrow L \times_{SO(3)} \tau_\upharpoonright^{-1}(S_s^2)$ (where $\tau_\upharpoonright : \mathfrak{l}^* \rightarrow \mathfrak{l}^*_\upharpoonright$ is the restriction map) by

$$\psi([\Lambda, N]_{SO(3)}, M) := [\Lambda, N - \Lambda^{-1}M]_{SO(3)}. \tag{3.11}$$

It is obvious from the above description of C_s^m that ψ is a diffeomorphism. Now recall from 1.10 that $L \times_{L_{\tilde{p}}} \tau_\upharpoonright^{-1}(-\mathcal{O}^{L_{\tilde{p}}}_{\tilde{\theta}})$ is symplectomorphic to $\mathcal{O}^G_{(\tilde{\theta},\tilde{p})}$; see the paragraph preceding (1.80). Applied to the case at hand, this means that $L \times_{SO(3)} \tau_\upharpoonright^{-1}(S_s^2)$ is symplectomorphic to $\mathcal{O}^P_{m,\pm,s}$. Chasing the definitions of the relevant symplectic structures, one may (tediously) verify that C_s^m is in fact symplectomorphic to $\mathcal{O}^P_{m,\pm,s}$. This proves, in particular, that C_s^m is symplectic.

For $m = 0$ the set $\mathfrak{l}_{\tilde{p}} \cap \mathcal{O}_s^L$ consists of all matrices M in $\mathfrak{l}$ for which $K_1 = -R_2$, $K_2 = R_1$, $R_3 = \pm 1$, and $K_3 = 0$. Hence C_s^0 is the union of two components

$C_s^{0,\pm}$. This time the analogue of (3.11) fails to be a diffeomorphism. The selection of two independent constraints Φ_1, Φ_2 is made via the choice of a (local) section $b : L/E(2) \simeq \mathcal{O}^L_{0,\pm} \to L$. The Φ_i in question are then simply the first and second components of Φ^b, defined by $\Phi^b([\Lambda, N]_{E(2)}, M) := b(\Lambda\tilde{p})^{-1}\Phi([\Lambda, N]_{E(2)}, M)$.

The following claims may be verified by a straightforward but tedious local analysis. The two independent constraints generate a free and proper action of $\mathbb{R}^2$ on $T^*\mathcal{O}^L_{0,\pm} \times \mathcal{O}^L_s$; the momentum map J of this action is, of course, given by $J_i = \Phi^b_i$. We are therefore in the situation of Marsden–Weinstein reduction at zero, concluding from Theorems 1.5.4 and 1.2.2 that $J^{-1}(0) = C^0_s$ is coisotropically embedded in $T^*\mathcal{O}^L_{0,\pm} \times \mathcal{O}^L_s$. While the explicit form of the $\mathbb{R}^2$-action depends on b, the Marsden–Weinstein quotient $J^{-1}(0)/\mathbb{R}^2$ does not. The analogue of the map (3.11) quotients well to $J^{-1}(0)/\mathbb{R}^2$, and is a diffeomorphism. The remainder of the argument is then as in the massive case. It is elementary to verify that the pertinent P-actions are intertwined by the symplectomorphism between C^m_s and $\mathcal{O}^P_{m,\pm,s}$ constructed above. □

3.3 Representations of the Poincaré Group

In this section we relate the irreducible representations of the Poincaré group P to its **covariant representations**; these are by definition the ones that are induced from some finite-dimensional representation of the Lorentz group L. We will first quickly go through the classification and realization of the irreducible representations of P, assuming that the reader has seen this material before.

In order to apply Theorem 1.10.3 we need to verify that P is a regular semidirect product. All L-orbits in $\mathcal{M}^*$ are closed, except $\mathcal{O}^L_{0,\pm}$. The closure of the latter is obtained by adding the origin, so that it is obvious that the regularity condition is met. Hence the irreducible representations of P are classified by 1.10.3. As in the classical case, we will concentrate on the representations related to the L-orbits $\mathcal{O}^L_{m^2,\pm}$, for $m^2 \geq 0$.

The orbits $\mathcal{O}^L_{m^2,\pm}$ possess an L-invariant measure ν, which, upon identification of $\mathcal{O}^L_{m^2,\pm}$ with $\mathbb{R}^3$ by simply omitting the variable

$$\pm\, p_0 = \omega_{\mathbf{p}} := \sqrt{\mathbf{p}^2 + m^2}, \tag{3.12}$$

is $d\nu(p) = dp_1 dp_2 dp_3/(16\pi^3\omega_{\mathbf{p}})$. Hence we can form the representation space (1.75) and the representation (1.76). For $m > 0$ and $L_{\tilde{p}} = SO(3)$ the label σ is the spin s, taking values in $\mathbb{N} \cup 0$, so that $\mathcal{H}_s = \mathbb{C}^{2s+1}$ is the space carrying the well-known irreducible representation $U_s(SO(3))$. For $m = 0$ we are interested only in irreducible representations of $L_{\tilde{p}} = E(2)$ that correspond to the $SO(2)$-orbit $(0, 0)$ in $\mathbb{R}^2$. Such representations U_h are labeled by the helicity $h \in \hat{SO}(2) = \mathbb{Z}$, and are realized on $\mathcal{H}_h = \mathbb{C}$.

Comparing these representations with the coadjoint orbits of the classical theory in 3.1, one sees that the only difference in the parametrization lies in the fact that in quantum theory spin and helicity assume integral values. (It can be shown that the projective representations of P are given by representations of its covering group

$\tilde{P} = SL(2, \mathbb{C}) \ltimes_{\tilde{\rho}} \mathcal{M}$, where the action $\tilde{\rho}$ projects to an action of L via the well-known covering projection from $SL(2, \mathbb{C})$ onto L. For the above classification the replacement of L by $SL(2, \mathbb{C})$ has the consequence that s and h may now be half-integers as well.) The following summary should be compared with Proposition 3.1.1.

Proposition 3.3.1.

- *The Poincaré group P is a regular semidirect product.*
- *For $m > 0$, P has a family of irreducible representations $U^{m,\pm,s}$, where $s \in \mathbb{N} \cup 0$, which is realized on $\mathcal{H}_b^{m,\pm,s} = L^2(\mathcal{O}^L_{m^2,\pm}) \otimes \mathbb{C}^{2s+1}$ by (1.76).*
- *For $m = 0$, P has, among others, the series of irreducible representations $U^{0,\pm,h}$, where $h \in \mathbb{Z}$, realized on $\mathcal{H}_b^{0,\pm,h} = L^2(\mathcal{O}^L_{0,\pm})$.*

Here $b : \mathcal{O}^L_{m^2,\pm} \simeq L/L_{\tilde{p}} \to L$ is a measurable section, as explained above (1.75). The explicit form of these representations is given by (1.76).

Our goal is now to relate these representations to the covariant representations of P. These are defined by taking $G = P$ and $H = L$ in III.2.9 and taking the representation $U_\lambda(L)$ from which one induces to be finite-dimensional. This is motivated by relativistic field theory, in which the basic fields of a theory generically transform according to such a covariant representation. However, since L is semisimple and noncompact, it can be shown that its finite-dimensional representations are all nonunitary, except the trivial one. Thus we discard the symbols $\mathcal{H}_\lambda$ and U_λ, suggesting unitarity, in favor of S_λ and $\mathcal{R}_\lambda$, respectively. The corresponding induced representations $\mathcal{R}^\lambda(P)$ are nonunitary as well, except the trivially induced representation $\mathcal{R}^{\mathrm{id}}$.

This is not a problem, because the induced space $\mathcal{H}^\lambda$ will be regarded primarily as a symplectic manifold, henceforth called S^λ. The point, then, is that $\mathcal{R}^\lambda$ defines a strongly Hamiltonian action of P. The relationship between the covariant and irreducible representations will, accordingly, be achieved by symplectic reduction. The notion of irreducibility, however, is the one from unitary representation theory (rather than the transitivity of the P-action).

In any case, one may construct the induced space $S^\lambda = \mathcal{H}^\lambda$ as a Hilbert space in the same way as for unitary induction; see, in particular, the text below III.(2.176). The next lemma should be compared with 3.2.1. The analogue of (3.14) for $E(3)$ instead of P has already been encountered in (2.116).

Lemma 3.3.2. *The Hilbert space S^λ and the induced representation $\mathcal{R}^\lambda(P)$ may be realized as*

$$S^\lambda = L^2(\mathcal{M}) \otimes S_\lambda; \tag{3.13}$$
$$\mathcal{R}^\lambda(\Lambda, v)\Psi^\lambda(q) = U_\lambda(\Lambda)\Psi^\lambda(\Lambda^{-1}(q - v)), \tag{3.14}$$

or, after a Fourier transform on Ψ^λ, as

$$\hat{S}^\lambda = L^2(\mathcal{M}^*) \otimes S_\lambda; \tag{3.15}$$
$$\hat{\mathcal{R}}^\lambda(\Lambda, v)\hat{\Psi}^\lambda(p) = e^{ipv} U_\lambda(\Lambda)\hat{\Psi}^\lambda(\Lambda^{-1}p). \tag{3.16}$$

Here we have simply written S^λ for S_s^λ, etc.; the inner product on S^λ (and similarly on $\hat{S}^\lambda$) is given by

$$(\Psi, \Phi) := \int_{\mathcal{M}} d^4x \ (\Psi_\lambda(x), \Phi_\lambda(x))_\lambda \,. \tag{3.17}$$

One notes that $P/L \simeq \mathcal{M}$, so that $L^2(P/L) = L^2(\mathcal{M}) := L^2(\mathbb{R}^4)$, and chooses the section $s : P/L \to P$ to be $s(q) = (1, q)$. Since the Lebesgue measure on $\mathcal{M}$ is Lorentz-invariant, III.(2.176) then simplifies as stated. ■

The definition of the symplectic form with respect to which the action $\hat{\mathcal{R}}^\lambda(P)$ is strongly Hamiltonian, and the corresponding identification of the canonical variables, depend on the form of $\mathcal{R}_\lambda$. We shall discuss only two cases, of relevance to physics. Firstly, we take $\mathcal{R}_\lambda = \mathcal{R}_{\text{id}}$, where $\mathcal{R}_{\text{id}}(\Lambda) = 1$ for all $\Lambda \in L$. The inner product on $S_{\text{id}} = \mathbb{C}$ is the usual one, so that I.(2.35) with (3.17) defines a symplectic form relative to which the action $\mathcal{R}^{\text{id}}(P)$ is strongly Hamiltonian; cf. Theorem III.1.6.1. As we have already remarked, this case is rather atypical, in that $\mathcal{R}^{\text{id}}(P)$ is unitary.

Secondly, we pick the **vector representation** $\mathcal{R}_\lambda = \mathcal{R}_V$, defined on $S_V := \mathbb{C}^4$ by

$$\mathcal{R}_V(\Lambda)A^\mu := \Lambda^\mu{}_\nu A^\nu. \tag{3.18}$$

Elements of S_V are traditionally denoted by A, since in physics gauge fields are examples of fields transforming according to $\mathcal{R}^V$. The complex structure and inner product on S_V are not the usual ones on $\mathbb{C}^4$, however; they are defined by

$$i(A_0, A_1, A_2, A_3) := (-iA_0, iA_1, iA_2, iA_3); \tag{3.19}$$

$$(A, B)_V := A_0\overline{B}_0 + \overline{A}_1 B_1 + \overline{A}_2 B_2 + \overline{A}_3 B_3. \tag{3.20}$$

The point of these definitions is that I.(2.35) now defines a Lorentz-invariant symplectic form on S_V, for (with slight abuse of notation) one has

$$\omega(A, B) := 2\,\text{Im}\,(A, B)_V = -2\,\text{Im}\,\overline{A}_\mu B^\mu. \tag{3.21}$$

The same definition, but now applied to the inner product (3.17), then defines a symplectic form on S^V that is invariant under the P-action $\mathcal{R}^V$. Consequently, this action is strongly Hamiltonian.

In both cases the identification of S^λ as a phase space is equivalent to regarding $L^2(\mathcal{M}) \otimes \mathbb{C}^N$, where $N = 1$ or 4, as the cotangent bundle of $L^2(\mathcal{M}, \mathbb{R}^N)$; cf. III.(2.28)–(2.31). This comment, however, would not apply to arbitrary representations $\mathcal{R}_\lambda$.

In all cases, the first step of the reduction of S^λ is performed by imposing the infinite number of constraints

$$(\Box + m^2)\Psi^\lambda(q) = 0 \tag{3.22}$$

on S^λ, where $\Box := \partial_\mu \partial^\mu$ is the d'Alembertian. On $\hat{S}^\lambda$ these constraints read

$$(p^2 - m^2)\hat{\Psi}^\lambda(p) = 0, \tag{3.23}$$

cf. (3.9). Unfortunately, this equation has no solutions in the given phase space. Although conceptually this situation is purely classical, one is reminded of the discussion about Dirac's constrained quantization method in 2.9. In the present context the problem arises because of the infinite number of degrees of freedom.

Regarding $\hat{\Psi}^\lambda$ as an infinite-component coordinate q^i, $i \in \mathcal{M}^*$, the form (3.23) shows that the constraints in question amount to putting $q^i = 0$ for a certain subset $i \in \mathcal{M}^*_C \subset \mathcal{M}^*$. If the index set $\mathcal{M}^*$ were finite, the reduced phase space would simply be $T^*\mathbb{R}^{\mathcal{M}^*\backslash\mathcal{M}^*_C}$. Formally regarding $\mathcal{M}^*_C$ as the set where (3.12) holds, the space $\mathcal{M}^*\backslash\mathcal{M}^*_C$ may be identified with $\mathbb{R}^3$ through the elimination of p_0 as an independent variable. To make this argument precise, one first expands solutions of (3.23) with given sign $\pm$ of the energy p_0 by

$$\Psi^\lambda_\pm(q) = \int_{\mathbb{R}^3} \frac{d^3p}{16\pi^3\omega_{\mathbf{p}}}\, e^{-ipq}\,\tilde{\Psi}^\lambda_\pm(\mathbf{p}), \tag{3.24}$$

where $p_0 = \pm\omega_{\mathbf{p}}$. One subsequently declares the reduced space to be

$$\begin{aligned} S^{m,\pm,\lambda} &:= \{\Psi^\lambda_\pm \in \mathcal{S}'(\mathcal{M}) \otimes S_\lambda \mid \\ &(\Box + m^2)\Psi^\lambda = 0,\ \tilde{\Psi}^\lambda_\pm \in L^2(\mathbb{R}^3, d^3p/(16\pi^3\omega_{\mathbf{p}})) \otimes S_\lambda\}; \end{aligned} \tag{3.25}$$

the inner product $(\Psi_\pm, \Phi_\pm)$ in $S^{m,\pm,\lambda}$ is by definition equal to the one $(\tilde{\Psi}_\pm, \tilde{\Phi}_\pm)$ in $L^2(\mathbb{R}^3, d^3p/(16\pi^3\omega_{\mathbf{p}}))\otimes S_\lambda$. Here the fact that all solutions of (3.23) in $\mathcal{S}'(\mathcal{M})\otimes S_\lambda$ admit an expansion of the type (3.24) is a consequence of (3.23) and the fact that the Fourier transform maps $\mathcal{S}'(\mathcal{M})$ into $\mathcal{S}'(\mathcal{M}^*)$. The extension of $L^2(\mathcal{M})$ to $\mathcal{S}'(\mathcal{M})$ is possible, since in classical physics one is not tied to the choice of $L^2(\mathcal{M}^*)$ as the unconstrained phase space, and is free to enlarge it.

It follows that one may identify $S^{m,\pm,\lambda}$ with $L^2(\mathcal{O}^L_{m^2,\pm}) \otimes S_\lambda$. The reduced P-action $\mathcal{R}^{m,\pm,\lambda}$ is given as in (3.16), with the understanding that p_0 is given by (3.12). We look at $S^{m,\pm,\lambda}$ as a symplectic manifold in a manner analogous to the interpretation of S^λ explained above, and we regard $\mathcal{R}^{m,\pm,\lambda}$ as a strongly Hamiltonian action, rather than as a representation.

Notice, however, that $S^{m,\pm,\mathrm{id}} = \mathcal{H}^{m,\pm,0}$ and $\mathcal{R}^{m,\pm,\mathrm{id}} = U^{m,\pm,0}$, which is, exceptionally, a unitary representation. This brings us in a position to justify the choice of (3.25) as the reduced phase space by a completely different argument. Here S_λ and U_λ "go along for the ride", so we omit them.

Proposition 3.3.3. *Regarding $L^2(\mathcal{M})$ as a Hilbert space, Fell induction on the basis of the quantum constraint (3.22), supplemented by $\pm p_0 > 0$, yields the induced Hilbert space $L^2(\mathcal{O}^L_{m^2,\pm})$.*

The representation (3.14) on $L^2(\mathcal{M})$ is thereby induced to a representation on $L^2(\mathcal{O}^L_{m^2,\pm})$, which coincides with the irreducible representation $U^{m,\pm,0}(P)$.

The following construction may be seen as the quantum counterpart of the reduction of $T^*\mathcal{M}^*$ by the constraint (3.9). The construction is possible, and provides an unexpected quantum twist to an otherwise classical situation, because the action $\mathcal{R}^{\mathrm{id}}(P)$ is not merely strongly Hamiltonian, but unitary.

The quantum reduction procedure is essentially the same as the one discussed in 2.10 for the constraint $\frac{1}{2}(p_1^2 - p_2^2)$ on $T^*\mathbb{R}^2$; see (2.128) etc. Thus we consider the representation $\hat{U}(\mathbb{R})$ on $L^2(\mathcal{M}^*)$ defined by

$$\hat{U}(t)\hat{\Psi}(p) := e^{\frac{1}{2}it(p^2-m^2)}\hat{\Psi}(p). \tag{3.26}$$

We then perform Fell induction, applying Construction 2.2.5. This time we may simply take $\tilde{\mathcal{E}} \subset L^2(\mathcal{M}^*)$ to be $C_c^\infty(\mathcal{M}^*)$. Adding the condition $\pm p_0 > 0$, and rescaling the inner product by a factor of 4 (which could have been avoided by rescaling the constraint if desired), the induced space then emerges as $S_F^{\mathrm{id}} = L^2(\mathcal{O}^L_{m^2,\pm})$, on which the representation $U_F^{\mathrm{id}}(P)$ Fell-induced from $U^{\mathrm{id}}(P)$ is equal to the irreducible representation $U^{m,\pm,0}(P)$. ■

We have added the suffix F here in order to distinguish between induction from representations of L and Fell induction. In the current analogue of the functions (2.134), the label k is replaced by $\mathbf{p}$, and one now has $f_{\mathbf{p}}^{\pm}(q^\mu) = e^{ipq}$, where $p_0 = \pm\omega_{\mathbf{p}}$. These functions do not lie in $\mathcal{H}$, yet they nonetheless form a complete set of linearly independent solutions of the constraints (3.22).

3.4 The Origin of Gauge Invariance

For scalar fields we have achieved our goal of relating the covariant action $\mathcal{R}^{\mathrm{id}}(P)$ to the irreducible representation $\mathcal{R}^{m,\pm,\mathrm{id}}(P)$. For nontrivially induced actions $\mathcal{R}^\lambda$ it is necessary to impose further constraints in order to achieve irreducibility. This will eventually bring us to the main theme of this section, namely the relationship between masslessness and gauge invariance.

The first step towards irreducibility consists in defining new variables by means of the (bounded and invertible) map U_b, defined on $S^{m,\pm,\lambda}$ by

$$U_b\tilde{\Psi}_\pm^\lambda(p) := \mathcal{R}_\lambda(b(p)^{-1})\tilde{\Psi}_\pm^\lambda(p), \tag{3.27}$$

where $p = (\pm\omega_{\mathbf{p}}, \mathbf{p})$, and $b : \mathcal{O}^L_{m^2,\pm} \to L$ is a section, as before. This map is obviously not unitary, but it is a symplectomorphism when the symplectic form on $S^{m,\pm,\lambda}$ has been defined appropriately. It is customary in physics to denote the left hand side of (3.27) by $a_\pm^\lambda(p)$, suppressing the b-dependence. The point of the transformation (3.27) is that the P-action $\mathcal{R}_b^{m,\pm,\lambda} := U_b\mathcal{R}^{m,\pm,\lambda}U_b^*$ is given by

$$\mathcal{R}_b^{m,\pm,\lambda}(\Lambda, v)a_\pm^\lambda(p) = e^{ipv}\mathcal{R}_\lambda\left(b(p)^{-1}\Lambda b(\Lambda^{-1}p)\right)a_\pm^\lambda(\Lambda^{-1}p). \tag{3.28}$$

Compare with (1.76), recalling that the argument of U_σ in that formula, and therefore the argument of $\mathcal{R}_\lambda$ in (3.28), lies in $L_{\tilde{p}}$ (which, we recall, is $SO(3)$ for $m > 0$ and $E(2)$ for $m = 0$). The only difference is that U_σ in (1.76) is a unitary irreducible representation of $L_{\tilde{p}}$, whereas the restriction $\mathcal{R}_\lambda(L \upharpoonright L_p)$ of $\mathcal{R}_\lambda(L)$ to $L_{\tilde{p}}$ is possibly nonunitary and reducible.

Let us briefly discuss the massive case, which is well understood.

Proposition 3.4.1. *Suppose that $\mathcal{R}_\lambda(L \upharpoonright SO(3))$ contains $U_s(SO(3))$, and let $P_{\lambda\to s}$ be the projector on S_λ whose image is the subspace carrying $U_s(L \upharpoonright SO(3))$.*

The constraints

$$P^{\perp}_{\lambda\to s}\mathcal{R}_{\lambda}(b(p)^{-1})\tilde{\Psi}^{\lambda}_{\pm}(p)=0 \tag{3.29}$$

are Poincaré covariant, in that $\tilde{\Psi}^{\lambda}_{\pm}$ satisfies (3.29) iff $\mathcal{R}^{m,\pm,\lambda}(\Lambda,v)\tilde{\Psi}^{\lambda}_{\pm}$ satisfies (3.29). In particular, the left-hand side of (3.29) does not explicitly depend on the section b.

The subspace of $S^{m,\pm,\lambda}$ satisfying the constraints (3.29) is precisely the Hilbert space $\mathcal{H}^{m,\pm,s}$ carrying the irreducible representation $U^{m,\pm,s}(P)$. In particular, this subspace is symplectic, so that the constraints (3.29) are second class.

The first point is obvious from (3.28) and the fact that $P^{\perp}_{\lambda\to s}$ commutes with all $\mathcal{R}_{\lambda}(\cdot)$. The remainder follows from the discussion after (3.27). ■

For example, the representation $\mathcal{R}_V(L\restriction SO(3))$ reduces as $U_0\oplus U_1$, under which the carrier space $\mathbb{C}^4$ decomposes as $\mathbb{C}\mathbf{e}_0\oplus\mathbb{C}^3$. The projection $P^{\perp}_{V\to 1}$ equals $[\mathbf{e}_0]$, so that with $\tilde{p}=(m,0,0,0)$, we may write $\tilde{p}^{\mu}A_{\mu}=0$ for $P^{\perp}_{V\to 1}A=0$. Since $(\Lambda^{-1})_{\nu}{}^{\mu}=\Lambda^{\mu}{}_{\nu}$, and by definition of b one has $b(p)^{\mu}{}_{\nu}\tilde{p}^{\nu}=p_{\mu}$, (3.29) may be written as the covariant equation $p_{\mu}\tilde{A}^{\mu}(p)=0$.

Returning to our original starting point, one may impose (3.29) and (3.23) or (3.22) on S^{λ}, and, choosing a sign of the energy p_0, conclude that the reduced phase space is $\mathcal{H}^{m,\pm,s}$. All values of $s\in\mathbb{N}$ can be reached in this way, and any covariant wave equation for massive fields is of this form. (To reach $s\in\mathbb{N}/2$ one needs to start with a representation of $SL(2,\mathbb{C})$ rather than of L.) Wave equations are usually posed in $\mathcal{M}$ rather than $\mathcal{M}^*$; for example, in the situation discussed above, the complete set of constraints consists of $(\Box+m^2)A^{\mu}=0$ and $\partial_{\mu}A^{\mu}=0$.

For massless fields a similar procedure may be used: For any given helicity $h\in\mathbb{Z}$ there exist representations of L whose restriction to $E(2)$ contains U_h. For example, to reach $U_{\pm1}$ one may use the representation on antisymmetric tensor fields $F_{\mu\nu}$. However, in physics a key role in the description of massless fields with helicity ±1 is played by vector fields A_{μ}. Let us therefore investigate the restriction of $\mathcal{R}_V(L)$ to $E(2)$.

Proposition 3.4.2. *Equip $S_V=\mathbb{C}^4$ with the symplectic form (3.21); the representation $\mathcal{R}_V(L)$ defined in (3.18) defines a strongly Hamiltonian L-action on S_V. Seen as a representation, the restriction of $\mathcal{R}_V$ to $E(2)$ is indecomposable.*

The symplectic reduction S^0_V of S_V by the constraint $\tilde{p}^{\mu}A_{\mu}=0$, where $\tilde{p}=(1,0,0,-1)$, is $\mathbb{C}^2$ with its usual symplectic form I.(2.35). The reduced $E(2)$-action on S^0_V is the representation $U_1\oplus U_{-1}$.

Choosing a basis $\{\mathbf{u}_1:=\mathbf{e}_1,\mathbf{u}_2:=\mathbf{e}_2,\mathbf{u}_{\pm}:=\frac{1}{2}(\mathbf{e}_0\pm\mathbf{e}_3)\}$, and taking the generators (3.4) in the defining representation, one calculates that $\mathcal{P}:=\mathbb{C}\mathbf{u}_-=\mathbb{C}\tilde{p}$ is invariant under $E(2)$, as is the span $\mathcal{J}$ of $\{\mathbf{u}_1,\mathbf{u}_2,\mathbf{u}_-\}$. The representation is indecomposable, because $T_a\mathbf{u}_a=\mathbf{u}_-$ for $a=1,2$.

Since $\tilde{p}^{\mu}=(1,0,0,1)$, the solution of the constraints is $\mathcal{J}$. One computes that the null space of the symplectic form on $\mathcal{J}$ is $\mathcal{P}$, so that $S^0_V=\mathcal{J}/\mathcal{P}\simeq\mathbb{C}^2$. Explicit computation of the action of the generators yields the final claim. □

This reduction of S_V^0 is of the Marsden–Weinstein type: With $H = \mathbb{C}$ acting on $\mathbb{C}^4$ by

$$\lambda : A_\mu \mapsto A_\mu - i\lambda \tilde{p}_\mu, \tag{3.30}$$

one computes from III.(1.8) and I.(2.32) and the subsequent text line that

$$J(A) := -2\tilde{p}^\mu A_\mu \tag{3.31}$$

is an equivariant momentum map for this action. Hence $\mathcal{J} = J^{-1}(0)$ and $S_V^0 = J^{-1}(0)/H$. The space $\mathcal{P}$ consists precisely of those vectors in S_V that are "pure gauge", that is, of the form $A_\mu = \lambda \tilde{p}_\mu$ for some $\lambda \in H$.

The symplectic orthogonal complement of a subspace $\mathcal{V}$ (which is not necessarily linear) of a linear symplectic space $\mathcal{K}$ (such as a Hilbert space with symplectic form I.(2.35)) is (cf. 1.1)

$$\mathcal{V}^\perp := \{z \in \mathcal{K} \mid \omega(z, w) = 0 \,\forall\, w \in \mathcal{V}\}. \tag{3.32}$$

Applying this to $\mathcal{K} = S_V$ and $\mathcal{V} = \mathcal{P}$, one sees that $\mathcal{P}^\perp = \mathcal{J}$.

Before proceeding on the basis of these insights, we should first pay attention to a subtle point concerning real fields. The representation $\mathcal{R}_V(L)$ is real, in that $\mathbb{R}^4 \subset \mathbb{C}^4$ is invariant under L. In physics one defines a **real vector field** with mass $m \geq 0$ as an element of

$$S^{m,\mathbb{R},V} := \left(S^{m,+,V} \oplus S^{m,-,V}\right)_{\mathbb{R}}. \tag{3.33}$$

The Fourier coefficients $\tilde{A}_\pm$ are now dependent variables, related by $\tilde{A}_-(\mathbf{p}) = \overline{\tilde{A}_+(\mathbf{p})}$. This leads to an identification of the real Hilbert space $S^{m,V,\mathbb{R}}$ with the complex Hilbert space $S^{m,+,V}$, endowing $S^{m,\mathbb{R},V}$ with the structure of a complex Hilbert space, as well as with the action $\mathcal{R}^{m,+,V}(P)$ (rather than with the restriction of $\mathcal{R}^{m,+,V} \oplus \mathcal{R}^{m,-,V}$ to the real subspace in question, as might have been expected). This juggling is justified by the physical requirement of positive energy. In any case, the symplectic form on $S^{m,\mathbb{R},V}$ may be expressed as

$$\omega(A, B) = -\int_\Sigma d^3\sigma(q)\, A_\mu(q)\, \overset{\leftrightarrow}{\partial_0}\, B^\mu(q), \tag{3.34}$$

where $\Sigma \subset \mathcal{M}$ is an arbitrary Cauchy surface for (3.22), and we have written A for $V(A)$, etc. The norm in $S^{m,\mathbb{R},V}$ (seen as a Hilbert manifold) is

$$\|A\|^2 := \int_{\mathbb{R}^3} \frac{d^3p}{16\pi^3\omega_{\mathbf{p}}} \sum_{\mu=0}^{3} |\tilde{A}_\mu(\mathbf{p})|^2. \tag{3.35}$$

This norm is not Poincaré invariant, but the topology it induces is. A similar procedure applies to scalar fields.

After these preparations we come to our main point of relating masslessness to gauge symmetry.

Definition 3.4.3. *Let $\mathcal{G}_c \subset S'(\mathbb{R}^4)/\mathbb{R}$ consist of all real solutions λ of the wave equation $\Box\lambda = 0$ on $\mathcal{M}$, modulo the constants, and let the* **gauge group** *$\mathcal{G}$ consist of those $\lambda \in \mathcal{G}_c$ whose (weak) derivative $\partial\lambda$ (seen as a four-vector with components*

$\partial_\mu\lambda$) lies in $S^{0,\mathbb{R},V}$. The Lie algebra $\mathfrak{g}$ is identified with $\mathcal{G}$, which becomes a Hilbert space in the norm

$$\|\lambda\|^2 = (\lambda, \lambda)_{\mathcal{G}} := (\partial\lambda, \partial\lambda)_{S^{0,\mathbb{R},V}} = \int_{\mathbb{R}^3} \frac{d^3p}{(2\pi)^3}\, \omega_{\mathbf{p}} |\tilde{\lambda}(\mathbf{p})|^2. \tag{3.36}$$

Here $\omega_{\mathbf{p}} = \sqrt{\mathbf{p}^2}$, and $\tilde{\lambda}$ equals $\tilde{\lambda}_+$, defined as in (3.24).

The following result transfers the setting of Proposition 3.4.2 from S_V to $S := S^{0,\mathbb{R},V}$, at the same time being an infinite-dimensional analogue of the part of Theorem 3.2.2 concerning $m = 0$.

Theorem 3.4.4. *The action of the gauge group $\mathcal{G}$ on $S^{0,\mathbb{R},V}$, given by*

$$\lambda : A_\mu \mapsto A_\mu + \partial_\mu\lambda, \tag{3.37}$$

is strongly Hamiltonian. The Marsden–Weinstein quotient $S^0 := J^{-1}(0)/\mathcal{G}$ with respect to this action is symplectomorphic to the Hilbert space $\mathcal{H}^{0,+,1} \oplus \mathcal{H}^{0,+,-1}$ (cf. 3.3.1), with symplectic form I.(2.35). The action $\mathcal{R}^{0,+,V}$ of P on S reduces to an action on S^0 that is equivalent to the representation $U^{0,+,1} \oplus U^{0,+,-1}$.

The last-mentioned representation of P is reducible, but becomes irreducible when spatial reflections are included. Particles transforming under $U^{0,+,1} \oplus U^{0,+,-1}$ are called **photons**.

Our choice of the (Hilbert) manifold structure on S and $\mathcal{G}$ implies that the action (3.37) is smooth, since addition in a Hilbert space is smooth. Following the steps leading to (3.31), one then verifies that a momentum map for (3.37) is given by

$$\widetilde{J(A)} : \mathbf{p} \mapsto -p^\mu \tilde{A}_\mu(\mathbf{p})/\mathbf{p}^2, \tag{3.38}$$

where $p_0 = \sqrt{\mathbf{p}^2}$. When A lies in $S^{0,\mathbb{R},V}$, the function $\widetilde{J(A)}$ is indeed an element of $\mathfrak{g}^* \simeq \mathcal{G}$. Since the coadjoint action is trivial, one verifies that J is equivariant.

In the setting of (real) Hilbert manifolds, the theory of Marsden–Weinstein reduction is essentially the same as for finite-dimensional manifolds. Definition 3.4.3 and (3.37) easily imply that the $\mathcal{G}$-action is free. However, since Lemma 1.5.1 was proved using a dimension-counting argument, which does not generalize to the infinite-dimensional case, we now prove directly that 0 is a regular value of J. The derivative $J_*(B)$ at A is independent of A, and equal to $J(B)$. Hence J_* is surjective for all A, so that 0 is regular by definition.

Using the second formulation of properness in Definition 1.5.2, one immediately shows that the $\mathcal{G}$-action on S is proper, for $\lambda_n \to \lambda$ in $\mathcal{G}$ precisely when $\partial_\mu\lambda_n \to \partial_\mu\lambda$ in S.

The Marsden–Weinstein quotient S^0 therefore exists as a symplectic manifold by Theorem 1.5.4. The set $J^{-1}(0)$ consists of all $A \in S$ satisfying $\partial^\mu A_\mu = 0$, or, equivalently, $p^\mu \tilde{A}(\mathbf{p}) = 0$. (In the present context this equation should not be thought of as a gauge-fixing condition, but as Gauss's law.) This is a closed subspace of S, which easily implies that S^0 is a Hilbert space. The identification of the reduced space and the reduced P-action proceeds in complete analogy with the proof of Proposition 3.4.2. ∎

The comments following (3.31) may be repeated verbatim: defining $\mathcal{P} \subset S^{0,\mathbb{R},V}$ as the fields that are pure gauge, i.e., of the form $A_\mu = \partial_\mu \lambda$ for some $\lambda \in \mathcal{G}$, and putting $\mathcal{J} := J^{-1}(0)$, one has $\mathcal{J} = \mathcal{P}^\perp$ and $S^0 \simeq \mathcal{J}/\mathcal{P}$.

3.5 *Quantum Field Theory of Photons*

On the basis of the general idea that symplectic reduction is to be quantized by induction (in the sense of Rieffel or Fell), we shall formulate an analogue of Theorem 3.4.4 in quantum field theory. In preparation, we define an important class of C^*-algebras.

Definition 3.5.1. *The* **CCR algebra** $\mathfrak{W}(\mathcal{K})$ *over a Hilbert space* $\mathcal{K}$ *is the twisted group* C^**-algebra* $C^*(\mathcal{K}_d, c)$*, where* $\mathcal{K}_d$ *is* $\mathcal{K}$ *as an additive group, equipped with the discrete topology, and the multiplier is*

$$c(z, w) := e^{-\frac{1}{2} i\omega(z,w)}. \tag{3.39}$$

Here the symplectic form ω *on* $\mathcal{K}$ *is defined by I.(2.35).*

It is customary to write $W(z)$ for the function f satisfying $f(z) = 1$ and $f(w) = 0$ for all $w \neq z$. We assume that the Haar measure on $\mathcal{K}_d$ is normalized so that each point has measure 1. From III.(1.80) and III.(1.81) we then have the relations

$$W(z)W(w) = e^{-\frac{1}{2} i\omega(z,w)} W(z + w); \tag{3.40}$$
$$W(z)^* = W(-z). \tag{3.41}$$

Equation (3.40) describes the **canonical commutation relations in Weyl form**, which explains the names CCR and $\mathfrak{W}$. Note that substitution of (3.41) into (3.40) shows that $W(z)$ is unitary. We will denote the linear span of all $W(z)$ by $\widetilde{\mathfrak{W}(\mathcal{K})}$; this is plainly a dense subalgebra of $\mathfrak{W}(\mathcal{K})$, playing an important role in what follows.

It can be shown that the CCR-algebra is simple, so that all nondegenerate representations are faithful. A most important representation of $\mathfrak{W}(\mathcal{K})$ is the **Fock representation** π_F on the bosonic Fock space $\exp(\mathcal{K})$ defined in II.(2.61). Recalling II.(2.67) etc., this representation is defined by continuous extension of

$$\pi_F(W(z)) := e^{a(z)^* - a(z)} \tag{3.42}$$

from the span $\mathfrak{E}$ of the exponential vectors; cf. II.(2.62). On use of the CBH-formula, II.(2.69), and II.(2.70), this is equivalent to

$$\pi_F(W(z))\sqrt{\mathrm{Exp}}(w) = e^{-\frac{1}{2}(z,z)-(z,w)}\sqrt{\mathrm{Exp}}(w + z). \tag{3.43}$$

For example, when $\mathcal{K} = \mathbb{C}^n$, the operator $\pi_F(W(z))$ coincides with $U_1(z)$ as defined in II.(2.71).

While the preceding paragraph in conjunction with II.2.3 relates the CCR-algebra to Berezin quantization (at least when $\mathcal{K}$ is finite-dimensional), there are equally close links between the CCR-algebra and Weyl quantization; cf. the comments following II.(2.112).

Proposition 3.5.2. *Let U be an operator on $\mathcal{K}$ that preserves the symplectic form I.(2.35); this is the case, for example, when U is unitary. Then linear and continuous extension of $\alpha_U(W(z)) := W(Uz)$ defines an automorphism of $\mathfrak{W}(\mathcal{K})$.*

Linear extension of $U^F\sqrt{\text{Exp}}(z) := \sqrt{\text{Exp}}(Uz)$ defines an operator U^F on $\mathfrak{E} \subset \exp(\mathcal{K})$. When U is unitary, the map U^F is unitary, so that it may be extended to all of $\exp(\mathcal{K})$. It then implements α_U, in that $U^F\pi_F(A)(U^F)^* = \pi_F(\alpha_U(A))$ for all $A \in \mathfrak{W}(\mathcal{K})$. In particular, a group representation $U_\chi(G)$ on $\mathcal{K}$ leads to a representation $U_\chi^F(G)$ on $\exp(\mathcal{K})$.

It is clear that (3.40) is preserved by α_U, and so is (3.41) by the linearity of U. The unitarity of U^F for unitary U is immediate from II.(2.63).

It is not obvious that U^F as given is well-defined, since the exponential vectors form an overcomplete set. However, putting $U^F(w_1 \otimes_s \cdots \otimes_s w_n) := Uw_1 \otimes_s \cdots \otimes_s Uw_n$, and using II.(2.62), defines the same operator in an unambiguous way. The implementing property follows from a simple calculation, using (3.43). ■

The following result will be used in relationship to the weak algebra of observables of the quantum field theory of photons. Restricting ω to $\mathcal{V}$, one obtains a C^*-algebra $\mathfrak{W}(\mathcal{V})$ by Definition 3.5.1. It is clear that the linear span of all $W(z)$, $z \in \mathcal{V}$, may be regarded as a subspace of $\mathfrak{W}(\mathcal{K})$ by extending the functions in question to $\mathcal{K}$ with the value 0 outside $\mathcal{V}$. Looking in the Fock representation, one sees also that the completion $\mathfrak{W}(\mathcal{V})$ is a subalgebra of $\mathfrak{W}(\mathcal{K})$. Similarly, $\mathfrak{W}(\mathcal{V}^\perp)$ may be defined as in 3.5.1; cf. (3.32).

Proposition 3.5.3. *The commutant $\mathfrak{W}(\mathcal{V})'$ of $\mathfrak{W}(\mathcal{V})$ in $\mathfrak{W}(\mathcal{K})$ is $\mathfrak{W}(\mathcal{V}^\perp)$.*

The inclusion $\mathfrak{W}(\mathcal{V}^\perp) \subseteq \mathfrak{W}(\mathcal{V})'$ is immediate from (3.40); the hard part of the proof is the opposite inclusion.

For $f \in \ell^\infty(\mathcal{K}_d)$ one has the inequalities

$$\|f\|_\infty \le \|f\|_2 \le \|f\|, \tag{3.44}$$

where the first norm is the sup-norm, the second norm is in $\ell^2(\mathcal{K}_d)$ (with respect to the Haar measure on $\mathcal{K}_d$), and the third is in $\mathfrak{W}(\mathcal{K})$. The first inequality is obvious (given the discreteness of the underlying measure space), and the second follows from the existence of the state ω_0, defined by continuous extension of $\omega_0(f) = f(0)$; indeed, $\|f\|_2^2 = \omega_0(f^*f)$. It follows that $\mathfrak{W}(\mathcal{K})$ as a Banach space (with its C^*-norm) is continuously embedded in $\ell_0(\mathcal{K}_d)$ (with sup-norm), for any element of the former is the limit of a Cauchy sequence in $\ell_c(\mathcal{K}_d)$; by (3.44) this sequence must also converge in the sup-norm, so that its limit must lie in $\ell_0(\mathcal{K}_d)$.

Now take an arbitrary $f \in \mathfrak{W}(\mathcal{K})$, and a Cauchy sequence f_n in $\ell_c(\mathcal{K}_d)$ converging to f in $\mathfrak{W}(\mathcal{K})$. It then follows from (3.40) that the commutator $[f_n, W(z)]$ is the function $f_n^{(z)} : w \mapsto 2if_n(w-z)\sin(-\frac{1}{2}\omega(w,z))$. Now, $\lim_n f_n^{(z)}$ exists in $\mathfrak{W}(\mathcal{K})$, hence in $\ell_0(\mathcal{K}_d)$. The function $w \mapsto \sin(-\frac{1}{2}\omega(w,z))$ lies in $\ell_b(\mathcal{K}_d)$, which is the multiplier algebra of $\ell_0(\mathcal{K}_d)$; cf. (2.13). Hence $f_n^{(z)} \to f^{(z)}$ (defined like $f_n^{(z)}$, with f_n replaced by f) is in $\ell_0(\mathcal{K}_d)$. By uniqueness of the limit, we infer $f_n^{(z)} \to f^{(z)}$ in $\mathfrak{W}(\mathcal{K})$. We conclude that $[f, W(z)] = f^{(z)}$.

Now, f is in $\mathfrak{W}(\mathcal{V})'$ iff $[f, W(z)]$ vanishes for all $z \in \mathcal{V}$. The preceding paragraph then yields $\| f^{(z)} \| = 0$, whereupon (3.44) implies that $f^{(z)}$ identically vanishes for such z. Therefore (evaluating $f^{(z)}$ at $w = w' + z$, and using $\omega(z, z) = 0$), f must vanish whenever its argument does not lie in $\mathcal{V}^{\perp}$, and the proposition follows. ∎

Applied to the cases $\mathcal{K} = S_V$ and $\mathcal{K} = S^{0,\mathbb{R},V}$ of the preceding section, we see that in either case Proposition 3.5.3 implies the equality

$$\mathfrak{W}(\mathcal{P})' = \mathfrak{W}(\mathcal{J}). \tag{3.45}$$

Note that $\mathfrak{W}(\mathcal{P})$ is abelian, because ω vanishes on $\mathcal{P}$. For $\mathcal{P} \subset S_V$ this is because $\tilde{p}^{\mu} \tilde{p}_{\mu} = 0$, and for $\mathcal{P} \subset S^{0,\mathbb{R},V}$ the reason is that $\partial_{\mu} \partial^{\mu} \lambda = 0$.

To interpret this result, we look in the Fock representation; in what follows $\mathcal{K}$ is either S_V or $S^{0,\mathbb{R},V}$, and H is $\mathbb{C}$ or $\mathcal{G}$, respectively. The Hilbert space $\exp(\mathcal{K})$ carries a representation $U_F(H)$, defined by

$$U_F(\lambda) := \pi_F(W(\lambda)), \tag{3.46}$$

where $\lambda \in \mathcal{P}$; we have identified H with the corresponding subspace $\mathcal{P} \subset \mathcal{K}$. It is obvious from (3.40) and the fact that ω vanishes on $\mathcal{P}$ that U_F defines a linear action of H, and the unitarity of each $W(\lambda)$ implies that this action is unitary. (Note that U_F is not an example of a representation of the type U_{χ}^F mentioned in 3.5.2.) The explicit form (3.43) shows that $U_F(\lambda)$ performs the gauge transformation (3.37) on the argument of $\sqrt{\mathrm{Exp}}$; the remaining term may be thought of as a factor included to make $U_F(\lambda)$ unitary. Identifying $\mathfrak{W}(\mathcal{K})$ with its faithful Fock representative, (3.45) then means that the gauge-invariant subalgebra of $\mathfrak{W}(\mathcal{K})$ is $\mathfrak{W}(\mathcal{J})$.

We are now going to "quantize" Proposition 3.4.2 and Theorem 3.4.4. We would like to use the specialization of Rieffel induction to quantum Marsden–Weinstein reduction explained in 2.5. In particular, we take our cue from Theorem 2.5.4, in which we put $\mathcal{H} = \exp(\mathcal{K})$, $U = U_F$, and $\tilde{\mathcal{E}} = \mathfrak{E}$; recall from II.2.3 that $\mathfrak{E}$ is the linear span of all exponential vectors. However, for $\mathcal{K} = S_V$ and $H = \mathbb{C}$ it is not clear how to construct a pre-Hilbert $C_c^{\infty}(H)$-module that is stable under the action of a suitable subalgebra of $\mathfrak{W}(S_V)$. For $\mathcal{K} = S^{0,\mathbb{R},V}$ and $H = \mathcal{G}$ the group C^*-algebra does not even exist; infinite-dimensional topological vector spaces do not support nontrivial translation-invariant Borel measures, so that $\mathcal{G}$ does not have a Haar measure. We therefore use Fell induction, in which we fill out the data as if we were performing Rieffel induction from the trivial representation $U_{\mathrm{id}}(H)$; cf. (2.81).

We first specialize to $\mathcal{K} = S_V$ and $H = \mathbb{C}$; the following theorem is the quantum counterpart of Proposition 3.4.2.

Proposition 3.5.4. *In Construction 2.2.5, put $\tilde{\mathcal{E}} = \mathfrak{E}$, $\mathcal{H}_{\chi} = \mathfrak{B}(\mathcal{H}_{\chi}) = \mathcal{H}_{\mathrm{id}} = \mathbb{C}$, and*

$$(\Psi, \Phi)_0^{\mathrm{id}} := \int_{\mathbb{C}} \frac{d\lambda\, d\overline{\lambda}}{2\pi i} (\Psi, U_F(\lambda)\Phi). \tag{3.47}$$

Here we have written $(\Psi, \Phi)_0^{\mathrm{id}}$ for $\langle \Psi, \Phi \rangle_{\mathbb{C}}$, which is the same by (2.37), in order to stress the analogy with (2.80) or (2.81).

1. *This form is finite for $\Psi, \Phi \in \mathfrak{E}$, and satisfies (2.35) and (2.46).*
2. *The induced space $\mathcal{H}_F^{\mathrm{id}} := \mathcal{H}^\chi$ may therefore be constructed as in 2.2.5.2. This space is naturally isomorphic to* $\exp(S_V^0)$*; recall that $\mathcal{J}/\mathcal{P} \simeq S_V^0$ is the classical reduced space.*
3. *Each gauge transformation $U_F(\lambda)$ is adjointable, and acts trivially on the induced space, in that for all $\lambda \in \mathbb{C}$ one has*
$$U_F^{\mathrm{id}}(\lambda) := \pi^{\mathrm{id}}(U_F(\lambda)) = \mathbb{I}. \tag{3.48}$$
4. *The pre-C^*-algebra $\pi_F(\widetilde{\mathfrak{W}(\mathcal{J})})$ leaves $\mathfrak{E}$ stable, and consists of adjointable operators. The induced representative $\pi_F^{\mathrm{id}}(\widetilde{\mathfrak{W}(\mathcal{J})})$ defined by 2.2.5.3, where $\pi_F^{\mathrm{id}}(A) := \pi^{\mathrm{id}}(\pi_F(A))$, is isomorphic to $\mathfrak{W}(S_V^0)$.*
5. *Defining $\mathcal{R}_V^F(E(2))$ on $\mathfrak{E}$ by taking $U = \mathcal{R}_V(\Lambda)$ in 3.5.2, where $\Lambda \in E(2)$, each $\mathcal{R}_V^F(\Lambda)$ leaves $\mathfrak{E}$ stable and is adjointable. The induced action U_F^{id} on $\mathcal{H}_F^{\mathrm{id}}$, given by $U_F^{\mathrm{id}}(\Lambda) := \pi^{\mathrm{id}}(\mathcal{R}_V^F(\Lambda))$, is linear and unitary, and equivalent to the representation $(U_1 \oplus U_{-1})^F$ (cf. 3.5.2).*
6. *Since the $E(2)$-action $\mathcal{R}_V$ on S_V is symplectic, it defines an automorphic action α of $E(2)$ on $\mathfrak{W}(S_V)$, as explained in 3.5.2. Because $\mathcal{J} \subset S_V$ is stable under $\mathcal{R}_V(E(2))$, this action restricts to $\mathfrak{W}(\mathcal{J})$. In the representation π_F^{id} the latter automorphism group is implemented by U_F^{id}.*

We will show that (3.47) is finite by explicit calculation. The property (2.35) holds because $H = \mathbb{C}$ is unimodular. Equation (2.46) follows as in the proof of Theorem 2.5.4 (as $\mathbb{C}$ is amenable); a different proof is given below.

Claim 3 follows from the property
$$(\Psi, U_F(\lambda)\Phi)_0^{\mathrm{id}} = (\Psi, \Phi)_0^{\mathrm{id}} \tag{3.49}$$
for all $\lambda \in \mathcal{G}$ and $\Psi, \Phi \in \mathfrak{E}$. This is a simple consequence of (3.47) and the translation invariance of the Haar measure on $\mathbb{C}$.

To express (3.47) in a convenient form, we decompose $A = A^L + A^T$, where $A^T = (0, A_1, A_2, 0)$ lies in the orthogonal complement (in the Hilbert space sense) in $\mathcal{J}$ of $\mathcal{P}$, and $A^L = (A_0, 0, 0, A_3)$ is the orthogonal complement of A^T in S_V. Recall that Ω is defined below II.(2.61). A Gaussian integration results in
$$\left(\sqrt{\mathrm{Exp}}(A), \sqrt{\mathrm{Exp}}(B)\right)_0^{\mathrm{id}} = \left(\sqrt{\mathrm{Exp}}(A^L), \Omega\right)_0^{\mathrm{id}} \left(\Omega, \sqrt{\mathrm{Exp}}(B^L)\right)_0^{\mathrm{id}} e^{(A^T, B^T)}. \tag{3.50}$$

The identification of the reduced space uses the method of Proposition 2.2.4. Our guess is $\mathcal{H}_*^{\mathrm{id}} = \exp(S_V^0)$, and this is proved by defining $\tilde{U} : \mathfrak{E} \to \exp(S_V^0)$ by linear extension of
$$\tilde{U}\sqrt{\mathrm{Exp}}(A) := \left(\Omega, \sqrt{\mathrm{Exp}}(A^L)\right)_0^{\mathrm{id}} \sqrt{\mathrm{Exp}}([A^T]_{\mathcal{P}}), \tag{3.51}$$
where $[A^T]_{\mathcal{P}}$ is the equivalence class of $A^T \in \mathcal{J}$ in $\mathcal{J}/\mathcal{P} \simeq \mathbb{C}^2$; cf. 3.4.2. Since (A^T, B^T) equals $([A^T]_{\mathcal{P}}, [B^T]_{\mathcal{P}})_{\mathbb{C}^2}$, it is obvious from (3.50) that (2.29) holds. Moreover, this construction establishes (2.46), for $\mathcal{H}_*^{\mathrm{id}}$ is a Hilbert space. It is clear

that $\tilde{U}\mathfrak{E}$ is dense in $\exp(S_V^0)$, so that the guess of the latter as the induced space has been vindicated, proving 3.5.4.2

It is easily seen that an operator on $\mathfrak{E}$ is adjointable iff it commutes with all $U_F(\lambda)$. By (3.45) the intersection of $\pi_F(\widetilde{\mathfrak{W}(S_V)})$ with the space of all adjointable operators on $\mathfrak{E}$ is $\pi_F(\mathfrak{M}(\mathcal{J}))$; this implies the first part of 3.5.4.4.

All remaining claims in 3.5.4 easily follow from (2.30); the use of exponential vectors has reduced these verifications to Proposition 3.4.2. ∎

After this warm-up we turn to the quantization of Theorem 3.4.4. Thus we specialize the discussion preceding 3.5.4 to the case $\mathcal{K} = S^{0,\mathbb{R},V}$ and $H = \mathcal{G}$. In order to define an integral of the type (3.47), we first use (3.46), (3.43), and (3.36) to compute the integrand in the attempted generalization of (3.47) as

$$\left(\sqrt{\mathrm{Exp}}(A), U_F(\lambda)\sqrt{\mathrm{Exp}}(B)\right) = e^{(A,B)}e^{-\frac{1}{2}\|\lambda\|^2}e^{(A,\partial\lambda)-(\partial\lambda,B)}. \tag{3.52}$$

Knowing that topological vector spaces support certain Gaussian measures, this suggests combining the Gaussian factor in (3.52) with the nonexistent flat measure on $\mathcal{G}$. Unfortunately, the ensuing combination defines a set function on $\mathcal{G}$ that is merely finitely additive, and therefore fails to be a measure (which by definition is countably additive). To have a measure, it is necessary to enlarge $\mathcal{G}$. Partly for later use, we present the general setting.

Recall Definition II.1.5.6 and the subsequent theory.

Theorem 3.5.5. *Let $\mathcal{H}$ be a real Hilbert subspace of a quasi-complete locally convex Hausdorff vector space $\mathcal{V}$, and take $\gamma > 0$. Suppose that $\mathcal{V}$ carries a Radon measure μ_γ whose Fourier transform is given by*

$$\int_{\mathcal{V}} d\mu_\gamma(v)\, e^{i\theta(v)} = e^{-\frac{1}{2}\gamma Q(\theta,\theta)}. \tag{3.53}$$

Here $\theta \in \mathcal{V}^*$, and the quadratic form Q on $\mathcal{V}^*$ is defined by II.(1.49). A measure with this property is called **Gaussian**, *with* **covariance** *Q, and is uniquely determined by its Fourier transform (3.53).*

1. *The map $\tilde{\theta} \mapsto \gamma^{-1/2}\hat{\theta}$ from $\tilde{\mathcal{V}}^*$ to $L^2(\mathcal{V}, \mu_\gamma)$, defined by $\hat{\theta}(v) := \theta(v)$, is well-defined and isometric, so that it extends to an isometry $w \mapsto \gamma^{-1/2}\hat{w}$ from $\mathcal{H}$ to $L^2(\mathcal{V}, \mu_\gamma)$. For each $w \in \mathcal{H}$ this defines $\hat{w}$ as an element of $L^2(\mathcal{V}, \mu_\gamma)$; we write $(w, v) := \hat{w}(v)$, which makes sense for almost all $v \in \mathcal{V}$ with respect to μ_γ.*
2. *In (3.53) one may replace θ by $\hat{w}$, yielding*

$$\int_{\mathcal{V}} d\mu_\gamma(v)\, e^{i(w,v)} = e^{-\frac{1}{2}\gamma(w,w)}. \tag{3.54}$$

3. *The translate of μ_γ by $w \in \mathcal{V}$ is disjoint from μ_γ when $w \notin \mathcal{H}$, and equivalent to μ_γ when $w \in \mathcal{H}$, with Radon–Nikodym derivative given by the* **general Cameron–Martin formula**

$$d\mu_\gamma(v+w) = e^{-\left[\frac{1}{2}(w,w)+(w,v)\right]/\gamma}\, d\mu_\gamma(v). \tag{3.55}$$

An elementary computation yields $(\hat{\theta}, \hat{\theta})_{L^2(\mathcal{V},\mu_\gamma)} = \gamma Q(\theta, \theta)$, so that the first claim follows by II.(1.49). In particular, the property $\theta(v) = 0$ for all $v \in \mathcal{H}$ implies $\theta(v) = 0$ for μ_γ-almost all $v \in \mathcal{V}$, so that the map $\tilde{\theta} \mapsto \hat{\theta}$ is well-defined.

To derive (3.54), take a sequence $\{\theta_n\}$ in $\mathcal{V}^*$ for which $\tilde{\theta}_n \to w$ in $\mathcal{H}$, so that $\hat{\theta}_n \to \hat{w}$ in $L^2(\mathcal{V}, \mu_\gamma)$ by 3.5.5.1. Hence $\exp(i\hat{\theta}_n) \to \exp(i\hat{w})$. Since $f_n \to f$ in L^2 implies $(f_n, 1) \to (f, 1)$ when $1 \in L^2$, one obtains (3.54) from (3.53).

The last claim is the **general Cameron–Martin theorem.** The proof of the disjointness property is beyond the scope of this book. In view of its importance for what follows, we do outline the proof of (3.55). For simplicity we put $\gamma = 1$ and $\mu := \mu_1$. First take $w = \tilde{\eta}$, where $\eta \in \mathcal{V}^*$, and take the Fourier transform of each side of (3.55). Using (3.53), the left hand side is immediately found to be

$$\int_{\mathcal{V}} d\mu(v + \tilde{\eta})\, e^{i\theta(v)} = e^{-\frac{1}{2}Q(\theta,\theta) - iQ(\eta,\theta)}. \tag{3.56}$$

The right-hand side of (3.55) may be heuristically computed by formally applying (3.53) with θ replaced by $\theta + i\eta$. The result indeed equals the left-hand side.

To proceed rigorously, we remark that a continuous linear (hence measurable) map $\varphi : \mathcal{V}_1 \to \mathcal{V}_2$ between two quasi-complete locally convex Hausdorff vector spaces pushes a Radon measure μ on $\mathcal{V}_1$ forward to a Radon measure $\varphi_*\mu$ on $\mathcal{V}_2$; the definition $\varphi_*\mu(B) := \mu(\varphi^{-1}(B))$ (where B is a Borel set in $\mathcal{V}_2$) implies that for each $f \in L^1(\mathcal{V}_2, \varphi_*\mu)$ one has

$$\int_{\mathcal{V}_1} d\mu(v)\, f(\varphi(v)) = \int_{\mathcal{V}_2} d\varphi_*\mu(u)\, f(u). \tag{3.57}$$

In particular, the pushforward of a Gaussian measure on $\mathcal{V}_1$ is again Gaussian, with covariance Q_φ given by $Q_\varphi(\alpha, \beta) = Q(\varphi^*\alpha, \varphi^*\beta)$. Here $\varphi^* : \mathcal{V}_2^* \to \mathcal{V}_1^*$ is the transpose of $\varphi : \mathcal{V}_1 \to \mathcal{V}_2$.

We apply this with $\mathcal{V}_1 = \mathcal{V}$ and $\mathcal{V}_2 = \mathbb{R}^2$, with $\varphi : \mathcal{V} \to \mathbb{R}^2$ given by $\varphi(v) := (\theta(v), \eta(v))$. Since θ and η lie in $\mathcal{V}^*$, this map is continuous. Applying (3.57) and II.(1.49), the Fourier transform of the right-hand side of (3.55) becomes

$$e^{-\frac{1}{2}(\tilde{\eta},\tilde{\eta})} \int_{\mathcal{V}} d\mu(v)\, e^{i\theta(v) - \eta(v)} = e^{-\frac{1}{2}Q(\eta,\eta)} \int_{\mathbb{R}^2} d\varphi_*\mu(x)\, e^{ix_1 - x_2}. \tag{3.58}$$

Inspecting all possibilities, one verifies that

$$\int_{\mathbb{R}^2} d\nu(x)\, e^{i\alpha(x) - \beta(x)} = e^{-\frac{1}{2}Q_2(\alpha + i\beta, \alpha + i\beta)} \tag{3.59}$$

for any Gaussian measure ν on $\mathbb{R}^2$ with covariance Q_2; in other words, for $\mathcal{V} = \mathbb{R}^2$ equation (3.53) is correct also for complex $\theta \in \mathcal{V}^*_{\mathbb{C}}$. Note that Q_2 is seen as a bilinear (rather than a sesquilinear) form, as we have assumed that $\mathcal{H}$ is real. To compute (3.58), we now take $\alpha = (1, 0)$ and $\beta = (0, 1)$, so that $\varphi^*\alpha = \theta$ and $\varphi^*\beta = \eta$. Combining (3.58) and, (3.59), one recovers the right-hand side of (3.56).

Hence (3.55) follows for $w \in \tilde{\mathcal{V}}^*$. For general $w \in \mathcal{H}$ the result then follows by a continuity argument similar to the one used in proving (3.54). □

Key differences with the finite-dimensional situation are that generically one has $\mu_\gamma(\mathcal{H}) = 0$, and that the measures μ_γ tend to be disjoint for different γ. In applying induction methods to the quantization of gauge theories, one is given the gauge group as a Hilbert Lie group $\mathcal{H}$, but the space $\mathcal{V}$ in 3.5.5 has to be guessed, and is generally not unique. However, any successful choice leads to (3.55) for $w \in \mathcal{H}$, and this turns out to guarantee that the gauge group acts trivially in the induced space.

Corollary 3.5.6. *The inclusion $\mathcal{G} \hookrightarrow \mathcal{G}_c$ (cf. 3.4.3) is continuous, and $\mathcal{V} = \mathcal{G}_c$ carries a Radon measure satisfying (3.53). Hence for $\lambda \in \mathcal{G}_c$ and $\eta \in \mathcal{G}$ one has*

$$\int_{\mathcal{G}_c} d\mu_1(\lambda)\, e^{i(\eta,\lambda)} = e^{-\frac{1}{2}\|\eta\|^2}; \tag{3.60}$$

$$d\mu_1(\lambda+\eta) = e^{-\frac{1}{2}\|\eta\|^2-(\eta,\lambda)} d\mu_1(\lambda). \tag{3.61}$$

The dual of $\mathcal{G}_c$ is $\mathcal{G}_c^* = \mathcal{S}_0(\mathbb{R}^4)/\mathcal{I}_\Box$, where $\mathcal{S}_0(\mathbb{R}^4)$ consists of those $f \in \mathcal{S}(\mathbb{R}^4)$ whose Fourier transform $\acute{f}$ vanishes at 0, and $\mathcal{I}_\Box$ is the annihilator in $\mathcal{S}(\mathbb{R}^4)$ of the space of solutions of $\Box\lambda = 0$ in $\mathcal{S}'(\mathbb{R}^4)$. The first claim then follows from the estimate $|(\lambda, f)| \leq \|\lambda\|\,\|\tilde{f}\|$, where $\tilde{f}(\mathbf{p}) := \acute{f}(\omega_\mathbf{p}, \mathbf{p})/\omega_\mathbf{p}^2$, and $\|\tilde{f}\|$ is defined by the right-hand side of (3.36). Note that $\|\tilde{f}\| < \infty$ because $\acute{f}(0) = 0$, so that the constant term in the Hermite expansion of $\acute{f}$ vanishes.

The existence of μ_1 eventually follows from the fact that $\mathcal{G}_c^*$ is a nuclear space (a property it inherits from $\mathcal{S}(\mathbb{R}^4)$), so that its dual $\mathcal{G}_c = \mathcal{G}_c^{**}$ is conuclear. □

By 3.5.5.1, the right-hand side of (3.52) is meaningful for $\lambda \in \mathcal{G}_c$. For example, $(A, \partial\lambda)$ stands for $-i\tilde{\alpha}(\lambda)$, where $\alpha \in \mathcal{G}$ is defined by (cf. 3.4.3) $\tilde{\alpha}(\mathbf{p}) := (\tilde{A}_0(\mathbf{p})/\omega_\mathbf{p}) + p_i\tilde{A}_i(\mathbf{p})/\omega_\mathbf{p}^2$. We then postulate that the counterpart of (3.47) is

$$\left(\sqrt{\operatorname{Exp}}(A), \sqrt{\operatorname{Exp}}(B)\right)_0^{\mathrm{id}} := e^{(A,B)} \int_{\mathcal{G}_c} d\mu_1(\lambda) e^{(A,\partial\lambda)-(\partial\lambda,B)}. \tag{3.62}$$

The following theorem closely parallels Proposition 3.5.4, and quantizes Theorem 3.4.4. This time $\mathfrak{E}$ stands for the pertinent subspace of $\exp(S^{0,\mathbb{R},V})$; the spaces $\mathcal{J}$ and $\mathcal{P}$ are defined at the end of 3.4.

Theorem 3.5.7. *Apply Construction 2.2.5 with $\tilde{\mathcal{E}} = \mathfrak{E}$, $\mathcal{H}_\chi = \mathfrak{B}(\mathcal{H}_\chi) = \mathcal{H}_{\mathrm{id}} = \mathbb{C}$, and define $(\Psi, \Phi)_0^{\mathrm{id}}$ by sesquilinear extension of (3.62).*

1. *This form is finite for $\Psi, \Phi \in \mathfrak{E}$, and satisfies (2.35) and (2.46).*
2. *The induced space $\mathcal{H}_F^{\mathrm{id}}$ is naturally isomorphic to* $\exp(S^0)$, *where the classical reduced space $S^0 \simeq \mathcal{H}^{0,+,1} \oplus \mathcal{H}^{0,+,-1}$ is defined in 3.4.4.*
3. *Each gauge transformation $U_F(\lambda)$ is adjointable, and, satisfying (3.48), acts trivially on the induced space.*
4. *The pre-C*-algebra $\pi_F(\widetilde{\mathfrak{W}}(\mathcal{J}))$ leaves $\mathfrak{E}$ stable, and consists of adjointable operators. The induced representative $\pi_F^{\mathrm{id}}(\widetilde{\mathfrak{W}}(\mathcal{J}))$ is isomorphic to $\widetilde{\mathfrak{W}}(S^0)$.*

5. *Each* $(\mathcal{R}^{0,+,V}(\Lambda, v))^F$, *where* $(\Lambda, v) \in P$, *leaves* $\mathfrak{E}$ *stable and is adjointable. The induced P-action* $(\mathcal{R}^{0,+,V})^{\mathrm{id}}_F$ *is unitary, and equivalent to the representation* $(U_{0,+,1} \oplus U_{0,+,-1})^F$.
6. *The automorphic P-action* α *on* $\mathfrak{W}(S)$ *obtained from the symplectic P-action* $\mathcal{R}^{0,+,V}$ *on* S *restricts to* $\mathfrak{W}(\mathcal{J})$. *In the representation* π^{id}_F *this restriction is implemented by* U^{id}_F.

Compared with the proof of Proposition 3.5.4, the components $A^{L,T}$ of A are now given in momentum space by

$$\tilde{A}^T(\mathbf{p}) = (0, A_i(\mathbf{p}) - p_i p_j A_j(\mathbf{p})/\mathbf{p}^2);$$
$$\tilde{A}^L(\mathbf{p}) = (A_0, p_i p_j A_j(\mathbf{p})/\mathbf{p}^2). \tag{3.63}$$

The integral in (3.62) is computed by transforming the Gaussian integration over $\mathcal{G}_c$ to one over $\mathbb{R}^2$ with the aid of (3.57) and the subsequent expression for Q_φ; cf. the proof of 3.5.5. Moreover, (3.61) guarantees that (3.49) still holds. Otherwise, similar steps as in the proof of 3.5.4 lead to a tedious verification of all claims. □

In view of this theorem we conclude that the induced Hilbert space and the various representations it carries describes a quantum field theory of photons.

3.6 Classical Yang–Mills Theory on a Circle

In order to explain certain topological features of quantum field theory in the simplest possible model, we shift our attention from Minkowski space in dimension 3+1 to the cylinder $\mathbb{R} \times S^1$. Here $\mathbb{R}$ stands for the time axis, whereas the circle S^1 represents space. The field theory to be studied comes from a relativistically invariant model, but this time we are not interested in aspects of special relativity and the Poincaré group. The model will therefore be presented in a partial gauge fixing ("$A_0 = 0$") that breaks relativistic invariance. We look at the circle S^1 as the interval $[0, 1]$ with boundary points identified; it is parametrized by $\alpha \in [0, 1)$.

Any principal H-bundle over the circle is isomorphic to the trivial bundle $P = S^1 \times H$. Recall from the comment after III.(2.16) that a connection on a trivial bundle $P = Q \times H$ is an element of $\Lambda^1(Q) \otimes \mathfrak{h}$. When $Q = S^1$, the space $\Lambda^1(S^1)$ of smooth 1-forms is simply $C^\infty(S^1, \mathbb{R})$, so that a (smooth) connection on $S^1 \times H$ is an element of $C^\infty(S^1, \mathfrak{h})$. Similarly, according to Proposition III.2.4.2, the space of smooth gauge transformations may be identified with $C^\infty(S^1, H)$.

Although the classical theory can be defined on the basis of smooth connections and gauge transformations, the corresponding quantum theory requires a more general class. It clarifies matters to include certain nonsmooth connections and gauge transformations already at the classical level. For the following definition we equip $\mathfrak{h}$ with an $\mathrm{Ad}(H)$-invariant inner product; by the compactness of H, this is always possible. The real Hilbert space $L^2(S^1, \mathfrak{h}) = L^2([0, 1]) \otimes \mathfrak{h}$ is then defined with respect to the Lebesgue measure and the above inner product on $\mathfrak{h}$.

Definition 3.6.1. *Let H be a compact connected Lie group.*

- *The configuration space $\mathcal{A}_{\mathbb{R}}$ of classical Yang–Mills theory on a circle with structure group H is $L^2(S^1, \mathfrak{h})$. The phase space is $S = T^*\mathcal{A}_{\mathbb{R}}$.*
- *The gauge group $\mathcal{G}$ of this theory is the Hilbert manifold $\mathcal{H}_1(S^1, H)$, consisting of all $g \in C(S^1, H)$ whose (weak) derivative $\dot{g} := g^{-1}dg/d\alpha$ lies in $L^2(S^1, \mathfrak{h})$.*

Elements g of $\mathcal{G}$ are loops in H, so that $\mathcal{G}$ is a (Sobolev) **loop group** of H. It can be shown that $g \in \mathcal{G}$ is absolutely continuous, and that $\dot{g}$ exists almost everywhere. Physically, one could say that $\mathcal{G}$ consists of all continuous loops with finite kinetic energy. To obtain an alternative characterization of $\mathcal{G}$, take a faithful representation $U(H)$ on some $\mathbb{C}^n$; then $\mathfrak{M}_n(\mathbb{C})$ is a normed space in the usual way, so that one can define the Hilbert space $\mathcal{H}_1(S^1, \mathfrak{M}_n(\mathbb{C}))$ as the completion of $C^\infty(S^1, \mathfrak{M}_n(\mathbb{C}))$ in the $p = 1$ Sobolev norm. The gauge group $\mathcal{H}_1(S^1, G)$ is the subset of $\mathcal{H}_1(S^1, \mathfrak{M}_n(\mathbb{C}))$ consisting of those functions that take values in $U(H)$. This endows $\mathcal{H}_1(S^1, H)$ with the structure of a Hilbert manifold. The continuous inclusion $\mathcal{H}_1(S^1, H) \subset C(S^1, H)$ is then a consequence of the Sobolev embedding theorem, from which it also follows that $\mathcal{H}_1(S^1, H)$ is not contained in any $C^p(S^1, H)$ for $p > 0$.

The Lie algebra $\mathfrak{g}$ of $\mathcal{G}$ is $\mathcal{H}_1(S^1, \mathfrak{h})$, with dual $\mathfrak{g}^* = \mathcal{H}_1(S^1, \mathfrak{h}^*)$. We will often use the notation

$$\|g\|^2 := (\dot{g}, \dot{g})_{L^2(S^1, \mathfrak{h})}. \tag{3.64}$$

We write elements of S as pairs (E, A), where $E \in L^2(S^1, \mathfrak{h}^*)$ and $A \in \mathcal{A}_{\mathbb{R}}$. Using the inner product on $\mathfrak{h}$, we identify $\mathfrak{h}^*$ with $\mathfrak{h}$, and subsequently identify S with $\mathcal{A} = L^2(S^1, \mathfrak{h}_{\mathbb{C}})$. Rather than II.(2.28), we use the convention

$$Z := A + \tfrac{1}{2} iE. \tag{3.65}$$

This has the advantage that the **complex connection** Z behaves in the same way as A under gauge transformations; cf. 3.6.2 below. (Whereas in the previous sections no confusion was possible, we will now denote complex connections by Z or W, and real ones by A.)

Lemma 3.6.2. *The action III.(2.69) of $\mathcal{G}$ on $\mathcal{A}_{\mathbb{R}}$ pulls back to a $\mathcal{G}$-action on S. Writing $Ad(g)A$ for the function $\alpha \mapsto \mathrm{Ad}(g(\alpha))A(\alpha)$, and similarly for $\mathrm{Co}(g)E$, this action is given by*

$$g : (E, A) \mapsto (\mathrm{Co}(g)E, A^g), \tag{3.66}$$
$$A^g := \mathrm{Ad}(g)A + gdg^{-1} = g(A - \dot{g})g^{-1}. \tag{3.67}$$

On complex connections (3.65) this reads $Z \mapsto Z^g$, where Z^g is defined as in (3.67). This action is smooth, proper, and strongly Hamiltonian.

The smoothness of this action is a technical exercise in Hilbert manifold theory that we omit. The main point is that $\mathcal{A}_{\mathbb{R}}$ and $\mathcal{G}$ have been defined precisely so that $\dot{g}$ lies in $\mathcal{A}_{\mathbb{R}}$ when $g \in \mathcal{G}$; cf. 3.4.4. Properness follows as in 3.4.4. The last point follows from Lemma III.2.3.1, as usual. □

The $\mathcal{G}$-action is not free, unless H is abelian. This may be handled by realizing that $\mathcal{G}$ is isomorphic to the semidirect product $H \ltimes_\rho \mathcal{G}_e$, where $\mathcal{G}_e$ is the subgroup of **based loops** or **based gauge transformations**, i.e., loops starting and ending at $e \in H$. The structure group H acts on $\mathcal{G}_e$ by

$$\tau(h)g : \alpha \mapsto hg(\alpha)h^{-1}. \tag{3.68}$$

A homomorphism from $\mathcal{G}$ to $H \ltimes_\rho \mathcal{G}_e$ is given by $g \mapsto (g(0), gg(0)^{-1})$, with inverse $(h, g) \mapsto gh$. Now the action of $\mathcal{G}_e$ on $\mathcal{A}_\mathbb{R}$ is free, and Marsden–Weinstein reduction (at 0) of S by the above $\mathcal{G}$-action, which is our goal, may be carried out in two steps; see Corollary 1.10.5. One firstly reduces by the $\mathcal{G}_e$-action, yielding a Marsden–Weinstein quotient that is duly a manifold, and secondly performs singular reduction by $\mathcal{G}/\mathcal{G}_e \simeq H$. As we shall see, the first step yields a finite-dimensional reduced space, and the second step is easy.

Another interesting feature of Marsden–Weinstein reduction in the present situation is that $\mathcal{G}$ may well be disconnected.

Proposition 3.6.3. *The group $\pi_0(\mathcal{G}) := \mathcal{G}/\mathcal{G}^0$ (where $\mathcal{G}^0$ is the identity component of $\mathcal{G}$) is isomorphic to the first homotopy group $\pi_1(H)$ of H.*

To put this in perspective, consider the loop group $LH = C(S^1, H)$, equipped with the topology of uniform convergence (with respect to the metric topology of H inherited from the Riemannian structure, or from $H \simeq U(H)$ as above). This topology coincides with the compact-open topology, so that one has $\pi_0(LH) = \pi_1(H)$ by definition of π_1.

The group $\pi_1(H)$ is isomorphic to a discrete subgroup D of the center of the universal covering group $\tilde{H}$ of H (i.e., $H = \tilde{H}/D$). Under this isomorphism an element $[\delta] \in \pi_1(H)$ is the equivalence class of loops in H that are homotopic to the projection (from $\tilde{H}$ to H) of a path from e to δ in $\tilde{H}$. We thus label the components $LH^{[\delta]}$ of LH by $\delta \in D$. Since the inclusion $\mathcal{G} \subset LH$ is continuous with respect to the manifold topology on $\mathcal{G}$, the proposition will follow if each intersection $\mathcal{G}^{[\delta]} := \mathcal{G} \cap LH^{[\delta]}$ is connected in the topology of $\mathcal{G}$. By the reasoning in the previous paragraph, this follows from the obvious fact that any two paths with finite kinetic energy in $\tilde{H}$ between e and δ are homotopy-equivalent in the topology of $\mathcal{H}_1(S^1, \tilde{H})$. ■

For later use, we infer from this proof that

$$LH_e^{[e]} := LH_e \cap LH^{[e]} \simeq L\tilde{H}_e, \tag{3.69}$$

where the based loop group LH_e is defined similarly to $\mathcal{G}_e$.

Elements of $\mathcal{G}^{[e]}$ are called **small gauge transformations**, whereas members of the other $\mathcal{G}^{[\delta]}$ are **large gauge transformations**. For example, if $H = U(1)$, it follows that

$$\pi_0(\mathcal{G}) = \pi_0(LU(1)) = \mathbb{Z}. \tag{3.70}$$

The members of a given component $\mathcal{G}^{[n]}$, $n \in \mathbb{Z}$, are labeled by the winding number of the loop. An example of an element of $\mathcal{G}^{[n]}$ is, of course,

$$g_n : \alpha \mapsto e^{-2\pi i n \alpha}. \tag{3.71}$$

The fact that $\mathcal{G}$ is disconnected when $\pi_1(H)$ is nontrivial is important mainly in the quantum theory of this model, but even in the classical theory it is enlightening to calculate both S_J^0 and $\tilde{S}_J^0$; cf. (1.26). We put $\mathcal{G}_e^{[e]} := \mathcal{G}_e \cap \mathcal{G}^{[e]}$.

Lemma 3.6.4. *Under the $\mathcal{G}$-action III.(2.69) on $\mathcal{A}_{\mathbb{R}}$, there are the diffeomorphisms $\mathcal{A}_{\mathbb{R}}/\mathcal{G}_e^{[e]} \simeq \tilde{H}$, $\mathcal{A}_{\mathbb{R}}/\mathcal{G}_e \simeq H$, and $\mathcal{A}_{\mathbb{R}}/\mathcal{G} \simeq H/\mathrm{Ad}(H)$.*

We shall first sketch the proof of the second diffeomorphism; the first is then obvious from (3.69). We define a map $\hat{\mathcal{W}} : \mathcal{A}_{\mathbb{R}} \to C([0,1], H)$ as the solution of the differential equation (valid for almost every α)

$$\left(\frac{\partial}{\partial \alpha} + A\right) \hat{\mathcal{W}}_A(\alpha) = 0, \tag{3.72}$$

with initial condition $\hat{\mathcal{W}}_A(0) = e$; we have written $\hat{\mathcal{W}}_A$ for $\hat{\mathcal{W}}(A)$. According to the theory of product integration, the solution is absolutely continuous in α, and may be written as a uniform limit

$$\hat{\mathcal{W}}_A(\alpha) = \lim_{N\to\infty} \prod_{n=0}^{N-1} \mathrm{Exp}\left[\mathcal{P}A\left((1-\frac{n+1}{N})\alpha\right) - \mathcal{P}A\left((1-\frac{n}{N})\alpha\right)\right]. \tag{3.73}$$

Here $\mathcal{P}A(\alpha) := \int_0^\alpha d\beta\, A(\beta)$ is the primitive of A.

Simple manipulations show that the function $g\hat{\mathcal{W}}_A : \alpha \mapsto g(\alpha)\hat{\mathcal{W}}_A(\alpha)$ on S^1 satisfies $d(g\hat{\mathcal{W}}_A)/d\alpha = A^g g\hat{\mathcal{W}}_A$, where A^g is given in III.(2.69) or 3.6.2. When $g \in \mathcal{G}_e$ one has the initial condition $g\hat{\mathcal{W}}_A(\alpha) = e$, so that $g\hat{\mathcal{W}}_A = \hat{\mathcal{W}}_{A^g}$. Since $g(1) = g(0) = e$, this implies $\hat{\mathcal{W}}_A(1) = \hat{\mathcal{W}}_{A^g}(1)$. Accordingly, the **Wilson loop**

$$\mathcal{W}(A) := \hat{\mathcal{W}}_A(1) \tag{3.74}$$

is invariant under based gauge transformations. In other words, for all $g \in \mathcal{G}_e$ one has $\mathcal{W}(A^g) = \mathcal{W}(A)$.

Suppose that $\mathcal{W}(A) = \mathcal{W}(B)$. Then $g := \hat{\mathcal{W}}_A\hat{\mathcal{W}}_B^{-1}$ lies in $\mathcal{G}_e$. Using (3.72), one sees that $dg/d\alpha = -Ag + gB$; in other words, $A = B^g$. Noting that for compact connected groups Exp is surjective, we conclude that $A \mapsto \mathcal{W}(A)$ induces a bijection from $\mathcal{A}_{\mathbb{R}}/\mathcal{G}_e$ to H. It is a nontrivial technical task to prove that this bijection is a diffeomorphism; we omit this part of the proof.

The third diffeomorphism follows from the isomorphism $H \simeq \mathcal{G}/\mathcal{G}_e$ (by the discussion following the proof of 3.6.2) and the intertwining property

$$\mathcal{W}(\mathrm{Ad}(h)A) = h\mathcal{W}(A)h^{-1}. \tag{3.75}$$

Here $h \in H$, seen as a subgroup of $\mathcal{G}$. This is the case $\alpha = 1$ of the property $\hat{\mathcal{W}}_{\mathrm{Ad}(h)A}(\alpha) = h\hat{\mathcal{W}}_A(\alpha)h^{-1}$, which is immediate from (3.72). □

Hence a gauge-invariant function of $A \in \mathcal{A}_{\mathbb{R}}$ is a function of $\mathcal{W}(A)$.

Theorem 3.6.5. *An equivariant momentum map for the $\mathcal{G}$-action (3.66) is given by*

$$J_\lambda(E, A) = (D_A E, \lambda)_{\mathcal{A}_{\mathbb{R}}}, \tag{3.76}$$

where $\lambda \in \mathfrak{g}$ *and* $D_A E := dE + [A, E]$.

There are symplectomorphisms

$$J^{-1}(0)/\mathcal{G}_e^{[e]} \simeq T^*\tilde{H}; \tag{3.77}$$

$$J^{-1}(0)/\mathcal{G}_e \simeq T^*H; \tag{3.78}$$

$$J^{-1}(0)/\mathcal{G} \simeq T^*(H/\mathrm{Ad}(H)). \tag{3.79}$$

The d in dE above is a weak derivative; the integral $(dE, \lambda)_{\mathcal{A}_\mathbb{R}}$ is well-defined, since $d\lambda$ lies in $L^2(S^1, \mathfrak{h})$. Similarly, $([A, E], \lambda)_{\mathcal{A}_\mathbb{R}}$ is well-defined because A and E are in $L^2(S^1, \mathfrak{h})$ and $\lambda \in C(S^1, \mathfrak{h})$.

In somewhat symbolic notation, the Poisson bracket on $C^\infty(T^*\mathcal{A}_\mathbb{R}, \mathbb{R})$ is

$$\{f, g\} = \left(\frac{\delta f}{\delta E}, \frac{\delta g}{\delta A}\right)_{\mathcal{A}_\mathbb{R}} - \left(\frac{\delta f}{\delta A}, \frac{\delta g}{\delta E}\right)_{\mathcal{A}_\mathbb{R}}. \tag{3.80}$$

One derives from (3.66) that the generator of λ is given by

$$\xi_\lambda f = \left(\frac{\delta f}{\delta E}, \mathrm{co}(\lambda)E\right)_{\mathcal{A}_\mathbb{R}} - \left(\frac{\delta f}{\delta A}, D_A\lambda\right)_{\mathcal{A}_\mathbb{R}}. \tag{3.81}$$

Combining (3.80), (3.81), III.(1.7), and I.(2.8), one verifies (3.76).

The symplectomorphisms all follow from III.(2.56), which is valid also for Hilbert manifolds, and Lemma 3.6.4; see below for comments on the potentially singular third case. ∎

An explicit expression for the symplectomorphism (3.78) may be derived as follows. By (3.76), the condition $(E, A) \in J^{-1}(0)$ forces $D_A E = 0$; in physics this is seen as the **Gauss law** constraint of Yang–Mills theory; for abelian H one simply has $dE = 0$. Gauss's law implies that $E(\alpha) = \mathrm{Co}(\hat{\mathcal{W}}_A(\alpha))\tilde{E}$, where $\tilde{E} \in \mathfrak{h}^*$ is independent of α.

This suggests that it is convenient to use the variables $(\tilde{E}, A)$, where

$$\tilde{E}(\alpha) := \mathrm{Co}\left(\hat{\mathcal{W}}_A^{-1}(\alpha)\right) E(\alpha). \tag{3.82}$$

Define a map $\varphi_S : T^*\mathcal{A}_\mathbb{R} \to T^*H$ by $\varphi_S(E, A) := (\tilde{E}_1, \mathcal{W}(A))$, where $\tilde{E}_1 := \int_0^1 d\alpha\, \tilde{E}(\alpha)$, and T^*H is identified with $\mathfrak{h}^* \times H$ through the left trivialization. As in the proof of (3.6.4), one verifies that on $J^{-1}(0)$ this map quotients to a diffeomorphism from $J^{-1}(0)/\mathcal{G}_e$ to T^*H, and checks from (3.80), and III.(1.54) that this diffeomorphism is a Poisson map.

We may implement (3.78) in a more elegant way (which will be essential in quantum theory) on the basis of the following notion.

Definition 3.6.6. *Let H be a compact connected Lie group. The* **complexification** *$H_\mathbb{C}$ of H is the unique Lie group that contains K as a closed subgroup, and has the property that for any complex-analytic Lie group K and real Lie group homomorphism $\psi : H \to K$ there exists a unique complex-analytic homomorphism $\psi_\mathbb{C} : H_\mathbb{C} \to K$ that on H (regarded as a subgroup of $H_\mathbb{C}$) coincides with ψ.*

The Lie algebra of $H_{\mathbb{C}}$ is the complexification $\mathfrak{h}_{\mathbb{C}}$ of $\mathfrak{h}$ (avoiding a potential ambiguity of notation); when H is simply connected, $H_{\mathbb{C}}$ is the connected and simply connected Lie group whose Lie algebra is $\mathfrak{h}_{\mathbb{C}}$. For $H = U(1)$ one has $H_{\mathbb{C}} = \mathbb{C}^*$. It can be shown that H is always a maximal compact subgroup of $H_{\mathbb{C}}$, and that every finite-dimensional representation of H extends to a holomorphic representation of $H_{\mathbb{C}}$. In the present context the following property of $H_{\mathbb{C}}$ is crucial.

Proposition 3.6.7. *The cotangent bundle T^*H is diffeomorphic to $H_{\mathbb{C}}$.*

Identifying T^*H with $\mathfrak{h}^* \times H$ (in the left trivialization), and subsequently equating $\mathfrak{h}^*$ with $\mathfrak{h}$ through the Ad-invariant inner product on $\mathfrak{h}$, it turns out that $\varphi_H : (X, h) \mapsto h\text{Exp}(-\frac{1}{2}iX)$ is a diffeomorphism from T^*H to $H_{\mathbb{C}}$. □

This diffeomorphism equips $H_{\mathbb{C}}$ with a symplectic structure.

The differential equation (3.72) still makes sense when $A \in \mathcal{A}_{\mathbb{R}}$ is replaced by a complex connection $Z \in \mathcal{A}$. Thus the Wilson loop map $\mathcal{W} : \mathcal{A}_{\mathbb{R}} \to H$ may be analytically continued to a map $\mathcal{W}_{\mathbb{C}} : \mathcal{A} \to H_{\mathbb{C}}$.

Proposition 3.6.8. *The map $\mathcal{W}_{\mathbb{C}}$, restricted to $J^{-1}(0)$ (seen as a subspace of $\mathcal{A}$ through the identification of $\mathcal{A}$ with $T^*\mathcal{A}_{\mathbb{R}}$ explained after (3.64)), quotients to a symplectomorphism from $J^{-1}(0)/\mathcal{G}_e$ to $H_{\mathbb{C}}$.*

We regard $E(\alpha)$ as an element of $\mathfrak{h}$ rather than of $\mathfrak{h}^*$ (cf. 3.6.7); replacing Co in (3.82) by Ad, this applies to $\tilde{E}$ as well. Given $(E, A) \in T^*\mathcal{A}_{\mathbb{R}}$ and $Z \in \mathcal{A}$, related to (E, A) by (3.65), consider the functions $f, g : [0, 1] \to H_{\mathbb{C}}$ defined by $f(\alpha) := \hat{\mathcal{W}}_A(\alpha)\hat{\mathcal{W}}_{\frac{1}{2}i\tilde{E}}(\alpha)$ and $g(\alpha) := \hat{\mathcal{W}}_Z(\alpha)$. Using (3.82), with Co replaced by Ad, and (3.72), one verifies that f and g satisfy the same first-order differential equation. Since $f(0) = g(0) = e$, it follows that $f(1) = g(1)$. On $J^{-1}(0)$ the function $\tilde{E}$ is constant, so that $\hat{\mathcal{W}}_{\frac{1}{2}i\tilde{E}}(1) = \text{Exp}(-\frac{1}{2}i\tilde{E}_1)$. Hence

$$f(1) = \mathcal{W}(A)\text{Exp}(-\tfrac{1}{2}i\tilde{E}_1) = g(1) = \mathcal{W}_{\mathbb{C}}(Z).$$

The claim then follows from the proof of 3.6.7 and the paragraph below (3.82). ■

Note that (3.75) with A replaced by Z reconfirms (3.79). When H is nonabelian, the space $H/\text{Ad}(H)$ is not a manifold; the cotangent bundle is then defined as follows. Let Ad* be the pullback to T^*H of the adjoint action of H on itself; in a trivialization $T^*H \simeq \mathfrak{h}^* \times H$ this is given by $\text{Ad}^*(h) : (\theta, k) \mapsto (\text{Co}(h)\theta, hkh^{-1})$. The momentum map J^{Ad} for Ad* is given by $J^{\text{Ad}}_X(\sigma) = \sigma(\xi^R_X - \xi^L_X)$. One then puts $T^*(H/\text{Ad}(H)) := (J^{\text{Ad}})^{-1}(0)/\text{Ad}^*(H)$ as a topological space; away from the singularities of $H/\text{Ad}(H)$, this is the usual cotangent bundle.

The structure of $T^*(H/\text{Ad}(H))$ may be described with the theory in 1.11. We will not do so here, other than saying that by Lemma III.1.10.4, the space $H/\text{Ad}(H)$ is homeomorphic to the **Stieffel chamber** T/W. For example, for $H = SU(2)$ the latter is the closed interval [0, 1], so that the singularities are merely boundary points. This is true in other cases as well. In the abelian case this phenomenon does not arise; for example, for $H = U(1)$ we simply have $\tilde{S}^0_J \simeq T^*S^1$.

In any casc, Theorem 3.6.5 is spectacular, showing that the reduced phase space of a particular field theory is finite-dimensional. In the present model this feature is peculiar to the one-dimensionality of space.

3.7 Quantum Yang–Mills Theory on a Circle

We now turn to the quantum theory of the model in the previous section, expecting to see that quantum induction naturally leads to $L^2(H)$ or $L^2(H/\mathrm{Ad}(H))$.

The CCR-algebra $\mathfrak{W}(\mathcal{A})$ is defined as in 3.5.1. The gauge group $\mathcal{G}$ acts on $\mathfrak{W}(\mathcal{A})$ by automorphisms, defined by (extension of)

$$\alpha_g(W(Z)) := e^{i\omega(\dot{g},Z)}W(\mathrm{Ad}(g)Z). \tag{3.83}$$

The analogue of the gauge-invariant algebra $\mathfrak{W}(\mathcal{J})$ in the quantum field theory of photons is now the subalgebra $\mathfrak{W}(\mathcal{A})^{\mathcal{G}}$ of $\mathfrak{W}(\mathcal{A})$ consisting of those $B \in \mathfrak{W}(\mathcal{A})$ for which $\alpha_g(B) = B$ for all $g \in \mathcal{G}$.

Proposition 3.7.1. *There exists a representation $U_F(\mathcal{G})$ of the gauge group on the Fock space* $\exp(\mathcal{A})$ *that on exponential vectors takes the form*

$$U_F(g)\sqrt{\mathrm{Exp}}(Z) = e^{-\frac{1}{2}\|\dot{g}\|^2+(\dot{g},Z)}\sqrt{\mathrm{Exp}}(Z^g), \tag{3.84}$$

and implements the automorphism (3.83) in the Fock representation $\pi_F(\mathfrak{W}(\mathcal{A}))$.

Here and in what follows, inner products of the type $(\dot{g}, Z)$ are in $\mathcal{A}$. After the proof of 3.7.3 below we show that (3.84) is indeed the restriction of an operator on $\exp(\mathcal{A})$. Granted this, unitarity follows from II.(2.63), the representation property comes from (3.67), and finally (3.83) with (3.43) leads to the implementing property $\pi_F(\alpha_g(W(Z))) = U_F(g)\pi_F(W(Z))U_F(g)^*$. □

As in (3.46) (combined with (3.43)), we see that U_F performs a gauge transformation on the argument of $\sqrt{\mathrm{Exp}}$, with additional factors guaranteeing unitarity. It is not of the form (3.46), because the automorphisms (3.83) are not inner.

Apart from the fact that it has led us to (3.84), which plays a central role in what follows, the CCR-algebra is hardly of any use in quantum Yang–Mills theory. This is because it fails to contain such crucial observables as the quantized functions of the Wilson loop. In order to construct the latter, as well as to verify that (3.84) is well-defined, we return to the setting of Definition II.1.5.6 and Theorem 3.5.5.

Proposition 3.7.2.

1. *The obvious inclusion of $\mathcal{H} = \mathcal{A}_{\mathbb{R}} = L^2(0,1)\otimes\mathfrak{h}$ into the Schwartz space $\mathcal{V} = \mathcal{D}'(0,1)\otimes\mathfrak{h}$ is continuous.*
2. *The ensuing map from $(\mathcal{D}'(0,1)\otimes\mathfrak{h})^* = \mathcal{D}(0,1)\otimes\mathfrak{h}$ to $\mathcal{A}_{\mathbb{R}}$, defined after II.1.5.6, is the natural inclusion.*
3. *The assumption of Theorem 3.5.5 is satisfied.*
4. *The support of μ_γ is contained in the set of distributions that are the (weak) derivative of a continuous function.*

The first claim is immediate from the Cauchy-Schwarz inequality, and the second is obvious. The third claim follows because $\mathcal{D}'(0,1) \otimes \mathfrak{h}$ is conuclear. The fourth statement is a well-known property of white noise, whose proof we omit. □

We may, accordingly, construct the Hilbert space $L^2(\mathcal{D}'(0,1) \otimes \mathfrak{h}, \mu_\gamma)$. The map $A \mapsto \mathrm{Ad}(g)A$ may be extended from $\mathcal{A}_{\mathbb{R}}$ to $\mathcal{D}'(0,1) \otimes \mathfrak{h}$ by dualizing its restriction to $\mathcal{D}(0,1) \otimes \mathfrak{h} \subset \mathcal{A}_{\mathbb{R}}$, so that the gauge transformation (3.67) applies to $L^2(\mathcal{D}'(0,1) \otimes \mathfrak{h}, \mu_\gamma)$ as well. For each $g \in \mathcal{G}$ we then define an operator $U_\gamma(g)$ on $L^2(\mathcal{D}'(0,1) \otimes \mathfrak{h}, \mu_\gamma)$ by

$$U_\gamma(g)\Psi(A) := e^{\left[-\frac{1}{4}\|\dot{g}\|^2 - \frac{1}{2}(A, dgg^{-1})\right]/\gamma}\Psi(A^{g^{-1}}). \tag{3.85}$$

It follows from the uniqueness of a measure satisfying (3.53), and the invariance of the inner product in $\mathcal{A}_{\mathbb{R}}$, that μ_γ is invariant under $A \mapsto \mathrm{Ad}(g)A$. Using this in conjunction with (3.55), which applies in view of Definition 3.6.1, one shows that $U_\gamma(g)$ is unitary. It is then easily checked that (3.85) defines a representation of the gauge group $\mathcal{G}$ on $L^2(\mathcal{D}'(0,1) \otimes \mathfrak{h}, \mu_\gamma)$.

Lemma 3.7.3. *There exists a unique unitary operator $V_\gamma : \exp(\mathcal{A}) \to L^2(\mathcal{D}'(0,1) \otimes \mathfrak{h}, \mu_\gamma)$ that maps $\sqrt{\mathrm{Exp}}(Z)$ to*

$$V_\gamma\sqrt{\mathrm{Exp}}(Z) = \Psi_Z : A \mapsto e^{-\frac{1}{2}(\overline{Z},Z)+\gamma^{-1/2}(A,Z)}. \tag{3.86}$$

Here (A, Z) is defined as in 3.5.5.1. The uniqueness of V_γ is clear, as the exponential vectors are total in $\exp(\mathcal{A})$. The fact that V_γ is well-defined follows from an alternative expression. Note that the linear span of all vectors of the form $\otimes^n Z$, where $Z \in \mathcal{A}$ with $\|Z\| = 1$, and $n \in \mathbb{N}$, is dense in $\exp(\mathcal{A})$. We put

$$\tilde{V}_\gamma \otimes^n Z : A \mapsto \frac{1}{\sqrt{n!}} H_n\left(\gamma^{-1/2}(A, Z)\right), \tag{3.87}$$

where H_n is a Hermite polynomial. This operator is clearly well-defined, and easily shown to be bounded, hence extendible to $\exp(\mathcal{A})$. A somewhat lengthy but elementary computation then shows that $\tilde{V}_\gamma = V_\gamma$. □

Since one computes from (3.84)–(3.86) that $V_{1/4}U_F(g)V_{1/4}^{-1} = U_{1/4}(g)$, it is clear that (3.84) is well-defined.

We now turn to the construction of the quantized Wilson loop. Using Proposition 3.7.2.4, it can be shown that the limit in (3.73) exists for almost every $A \in \mathcal{D}'(0,1) \otimes \mathfrak{h}$ with respect to μ_γ. In particular, the Wilson loop $\mathcal{W}(A)$ is well-defined for such A.

Definition 3.7.4. *Let $f \in C^\infty(H, \mathbb{R})$, and define $\mathcal{W}_f \in C^\infty(\mathcal{A}_{\mathbb{R}})$ by*

$$\mathcal{W}_f(A) := f(\mathcal{W}(A)). \tag{3.88}$$

The observable $\mathcal{Q}_{1/4}(\mathcal{W}_f)$ on $L^2(\mathcal{D}'(0,1) \otimes \mathfrak{h}, \mu_{1/4})$ is the operator defined by

$$\mathcal{Q}_{1/4}(\mathcal{W}_f)\Psi(A) := f(\mathcal{W}(A))\Psi(A). \tag{3.89}$$

On $\exp(\mathcal{A})$ *one has the operator*

$$\mathcal{Q}_F(\mathcal{W}_f) := V_{1/4}^{-1}\mathcal{Q}_{1/4}(\mathcal{W}_f)V_{1/4}. \tag{3.90}$$

To understand the choice $\gamma = 1/4$, consider the abelian case. When $H = U(1)$, identified with $\mathbb{T} \subset \mathbb{C}$, the (classical) Wilson loop is a numerical function on $\mathcal{A}_{\mathbb{R}}$ (or on S), which coincides with the function $A \mapsto \exp(-A_1)$. Here $A_1 := \int_0^1 A$; more generally, for $f \in \mathcal{A}_{\mathbb{R}}$ one has the linear functions

$$A_f(E, A) := (A, f); \tag{3.91}$$

$$E_f(E, A) := (E, f). \tag{3.92}$$

Note that $\{E_f, A_g\} = (f, g)$ by (3.80). These linear functions are quantized on Fock space $\exp(\mathcal{A})$ by

$$\mathcal{Q}_F(A_f) := \tfrac{1}{2}\left(a(f) + a(f)^*\right); \tag{3.93}$$

$$\mathcal{Q}_F(E_f) := -i\left(a(f) - a(f)^*\right). \tag{3.94}$$

It follows from II.(2.68) and the above Poisson bracket that Dirac's original condition $i[\mathcal{Q}_F(E_f), \mathcal{Q}_F(A_f)] = \mathcal{Q}(\{E_f, A_g\})$ is satisfied. For $H = U(1)$ we define $\mathcal{Q}_{1/4}(\mathcal{W})$ as in (3.89), with the function f simply omitted (or taken to be the identity map), and subsequently define $\mathcal{Q}_F(\mathcal{W})$ by (3.90). A partial justification of Definition 3.7.4 then lies in the equality

$$\mathcal{Q}_F(\mathcal{W}) = e^{-\mathcal{Q}_F(A_1)}. \tag{3.95}$$

This may be derived on a vector $\sqrt{\mathrm{Exp}}(Z)$ by use of (3.93), the CBH-formula, II.(2.68), II.(2.69), II.(2.70), 3.7.4, and (3.86).

Our aim is to quantize Theorem 3.6.5. Since the motivation for the steps to follow is similar to that for the steps leading from Theorem 3.4.4 to 3.5.7, we will not repeat the arguments given there. The counterpart of (3.52) for Yang–Mills on the circle is found, from (3.84), II.(2.63), and (3.67), to be

$$\left(\sqrt{\mathrm{Exp}}(W), U_F(g)\sqrt{\mathrm{Exp}}(Z)\right) = e^{-\frac{1}{2}\|\dot{g}\|^2}e^{(W, Z^g)+(\dot{g}, Z)}. \tag{3.96}$$

In the situation of Definition II.1.5.6, we take $\mathcal{H} = \mathcal{A}_{\mathbb{R}}$ and $\mathcal{V} = C([0, 1], \mathfrak{h})_0$, seen as a Banach space in the sup-norm; the suffix $_0$ indicates that $\mathcal{V}$ consists of the continuous paths in $\mathfrak{h}$ satisfying $X(0) = 0$. Here $\mathcal{A}_{\mathbb{R}}$ is seen as a Hilbert subspace of $C([0, 1], \mathfrak{h})_0$ by identifying it with the set of all $X \in C([0, 1], \mathfrak{h})_0$ whose norm with respect to the inner product

$$(X, Y)_1 := \int_0^1 d\alpha\, X'(\alpha)Y'(\alpha) \tag{3.97}$$

is finite. Equivalently, $\mathcal{A}_{\mathbb{R}}$ is injected into $C([0, 1], \mathfrak{h})_0$ by the primitive mapping $A \mapsto \mathcal{P}A$ defined below (3.73). The pertinent analogue of Corollary 3.5.6 is then as follows.

Proposition 3.7.5. *The inclusion* $\mathcal{A}_{\mathbb{R}} \hookrightarrow C([0, 1], \mathfrak{h})_0$ *by* $\mathcal{P}$ *is continuous, and for each* $\gamma > 0$ *there is a Radon measure* $\mu_\gamma^{W_0}$ *on* $C([0, 1], \mathfrak{h})_0$ *satisfying (3.53).*

Since $\mathcal{P}A(\alpha) = (A, \chi_{[0,\alpha]})$, one has $\|\mathcal{P}A\|_\infty \leq \|A\|$ by the Cauchy-Schwarz inequality. The second claim, whose proof we omit, defines the **Wiener measure on** $C([0, 1], \mathfrak{h})_0$ with variance γ. □

Our interest, of course, lies in a suitable measure on a completion of the gauge group $\mathcal{G}$ in some topology, rather than of its Lie algebra $\mathcal{A}_{\mathbb{R}}$. For this completion we take the loop group $LH = C(S^1, H)$. Assuming that $H \subset \mathfrak{M}_n(\mathbb{C})$ for some n, this loop group is seen as a submanifold of the Banach space $C([0, 1], \mathfrak{M}_n(\mathbb{C}))$. The same comment applies to its subspace $C([0, 1], H)_e$ of continuous paths that start at e. We define **Ito's map** $\hat{\mathcal{I}} : C([0, 1], \mathfrak{h})_0 \to C([0, 1], H)_e$ by

$$\hat{\mathcal{I}}_X(\alpha) := \lim_{N\to\infty} \prod_{n=0}^{N-1} \mathrm{Exp}\left[X\left((1 - \frac{n+1}{N})\alpha\right) - X\left((1 - \frac{n}{N})\alpha\right)\right]. \tag{3.98}$$

It can be shown that the limit exists for almost every X with respect to $\mu_\gamma^{W_0}$. The image of $\mu_\gamma^{W_0}$ under Ito's map is the **Wiener measure** $\mu_\gamma^{W_e}$ **on** $C([0, 1], H)_e$ with variance γ. Ito's map is a bijection up to null sets of $\mu_\gamma^{W_0}$ and $\mu_\gamma^{W_e}$. Comparing (3.98) with (3.73), it is clear that $\hat{\mathcal{I}} \circ \mathcal{P} = \hat{\mathcal{W}}$; cf. the comment preceding 3.7.4. Hence the Wiener measure $\mu_\gamma^{W_e}$ may equivalently be defined as the image of the measure μ_γ on $\mathcal{D}'(0, 1) \otimes \mathfrak{h}$ under $\hat{\mathcal{W}}$; see Proposition 3.7.2.

Let $C([0, 1], H)_e^h$ be the space of continuous paths g in H for which $g(0) = e$ and $g(1) = h$. Each such space carries a Radon measure $\mu_\gamma^{W_e^h}$, characterized by the disintegration property

$$\int_{C([0,1],H)_e} d\mu_\gamma^{W_e}\, f(\gamma) = \int_H dh \int_{C([0,1],H)_e^h} d\mu_\gamma^{W_e^h}\, f(\gamma) \tag{3.99}$$

for all $f \in L^1(C([0, 1], H)_e, \mu_\gamma^{W_e})$. In particular, this assigns a measure to the based loop group $LH_e = C([0, 1], H)_e^e$. As to the loop group LH itself, as in the case of $\mathcal{G}$ (cf. 3.6), we can write $LH = H \ltimes_\rho LH_e$ as groups.

Definition 3.7.6. *Writing $LH \simeq H \times LH_e$ as Borel spaces, the* **Wiener measure on** *LH with variance γ is the direct product $\mu_\gamma^W = \mu_H \times \mu_\gamma^{W_e^h}$ of the Haar measure μ_H on H and the measure $\mu_\gamma^{W_e^h}$ on LH_e.*

Note that neither $\mu_\gamma^{W_e^h}$ nor μ_γ^W is a probability measure, unlike $\mu_\gamma^{W_0}$ and $\mu_\gamma^{W_e}$.

The behavior of all these Wiener measures under translations follows from the general Cameron–Martin formula (3.55) for $\mu_\gamma^{W_0}$ (which applies because of 3.7.5) and the (almost sure) bijectivity of Ito's map. For example, for $\chi \in \mathcal{H}_1(S^1, H)$ one obtains

$$d\mu_\gamma^{W_e^{h\chi(1)}}(g\chi) = e^{-\left[\frac{1}{2}\|\dot{\chi}\|^2 + (\dot{g}, \mathrm{Ad}(\chi)\dot{\chi})\right]/\gamma} d\mu_\gamma^{W_e^h}(g), \tag{3.100}$$

where the second term in the exponential is defined by Theorem 3.5.5.1 and Proposition 3.7.2. The Radon-Nikodym derivative $d\mu_\gamma^W(g\chi)/d\mu_\gamma^W(g)$, where once again $\chi \in \mathcal{H}_1(S^1, H)$, will be equally important; it is given by the same expression. The translate of μ_γ^W (etc.) by $\chi \notin \mathcal{H}_1(S^1, H)$ is singular with respect to μ_γ^W.

After this intermezzo we return to Yang–Mills theory. Motivated by (3.96) and the surrounding discussion, we define $(\,,\,)_0^{\mathrm{id}}$ on $\mathfrak{E} \subset \exp(\mathcal{A})$ by sesquilinear extension of

$$\left(\sqrt{\mathrm{Exp}}(W), \sqrt{\mathrm{Exp}}(Z)\right)_0^{\mathrm{id}} := \int_{LH} d\mu_1^W(g)\, e^{(W,Z^g)+(\dot{g},Z)}. \tag{3.101}$$

The expressions of the type $(\dot{g}, Z)$, which in (3.96) were well-defined for $g \in \mathcal{H}_1(S^1, H)$ as inner products, make sense for general $g \in LH$ by 3.5.5.1 with 3.7.2; cf. (3.100) and subsequent comment. Indeed, the justification of (3.101) lies in the property (3.49), with λ replaced by g; this follows from the counterpart of (3.100) for μ_1^W. As in 3.5.7.3, it will follow that the gauge group $\mathcal{G}$ is trivially represented in the induced space (to be defined in Theorem 3.8.1 below). Since the Cameron–Martin formula, and therefore (3.49), is not valid for all $h \in LH$, the choice of the gauge group in 3.6.1 (instead of LH) has hereby to some extent been justified. In this connection it is worthwhile to remark further that the representation U_F in (3.84) cannot be extended from $\mathcal{H}_1(S^1, H)$ to $C(S^1, H)$.

3.8 Induction in Quantum Yang–Mills Theory on a Circle

We now come to our main result on quantum Yang–Mills theory on a cylinder. The context of the following theorem is the same as in 3.5.7, and the notation is similar. We write $\mathcal{H}_F^{\mathrm{id}}$ for the induced space and π^{id} for the corresponding induced representation; this convention is used in order to be able to write $\pi_F^{\mathrm{id}}(A) := \pi^{\mathrm{id}}(\mathcal{Q}_F(A))$ for the quantization of some observable given as an operator $\mathcal{Q}_F(A)$ on $\exp(\mathcal{A})$.

Theorem 3.8.1. *Apply Construction 2.2.5 with* $\tilde{\mathcal{E}} = \mathfrak{E} \subset \exp(\mathcal{A})$, $\mathcal{H}_\chi = \mathfrak{B}(\mathcal{H}_\chi) = \mathcal{H}_{\mathrm{id}} = \mathbb{C}$, *and define* $(\Psi, \Phi)_0^{\mathrm{id}}$ *by sesquilinear extension of (3.101).*

1. *This form is finite on* $\mathfrak{E}$, *and satisfies (2.35) and (2.46).*
2. *The induced space* $\mathcal{H}_F^{\mathrm{id}}$ *is naturally isomorphic to* $L^2(H)^{\mathrm{Ad}(H)}$; *this is the subspace of* $L^2(H)$ *that is invariant under* $U_{\mathrm{Ad}}(H)$, *defined by*
$$U_{\mathrm{Ad}}(k)\Psi(h) := \Psi(k^{-1}hk). \tag{3.102}$$
3. *For each* $g \in \mathcal{G}$ *the gauge transformation* $U_F(g)$ *is adjointable, and acts trivially on the induced space.*
4. *Let* $f \in C^\infty(H)$ *be a class function, and recall (3.90). The operator* $\pi_F^{\mathrm{id}}(\mathcal{W}_f)$ *is the restriction of the multiplication operator* f *on* $L^2(H)$ *to* $L^2(H)^{\mathrm{Ad}(H)}$.
5. *Replacing* LH *by* LH_e *in (3.101) yields an induced space* $\mathcal{H}_{F,e}^{\mathrm{id}}$ *that is naturally isomorphic to* $L^2(H)$. *For all* $f \in C^\infty(H)$ *one has* $\pi_{F,e}^{\mathrm{id}}(\mathcal{W}_f) = f$ *as multiplication operators on* $L^2(H)$.
6. *Replacing* LH *by* $LH_e^{[e]}$ *leads to the induced space* $L^2(\tilde{H})$.

The first item is proved as in 3.5.7, granted that a map $\tilde{U}$ may be constructed; this will be done below. We compute the integral in (3.101) in two steps. According to Definition 3.7.6, the integration over LH factorizes into an integral over of LH_e, and an additional integration over H. We start with the former.

The computation will be based on the connection between the Wiener measure and the **heat equation** on H. For fixed $\gamma > 0$, this is the PDE $df/dt - \frac{1}{2}\gamma\Delta_H f = 0$, where $\Delta_H := \sum_i T_i^2$ is the **Laplacian** on H. Here the orthonormal basis $\{T_i\}$ of $\mathfrak{h}$, seen as a set of right-invariant vector fields on H, is defined with respect to the $\mathrm{Ad}(H)$-invariant inner product on $\mathfrak{h}$ already used in the definition of $L^2([0,1],\mathfrak{h})$; the Laplacian depends on this inner product, but not on the basis. It can be shown that the fundamental solution ρ_γ is in $C^\infty(H)$ for $t > 0$.

Given k Borel sets $B_1, \ldots, B_k$ in H, and k elements $\alpha_1, \ldots, \alpha_k$ in $[0, 1]$, define a class of subsets of $C([0, 1], H)_e$ by

$$C^{B_1,\ldots,B_k}_{\alpha_1,\ldots,\alpha_k} := \{g \in C([0,1],H)_e \mid g(\alpha_1) \in B_1, \ldots, g(\alpha_k) \in B_k\}. \tag{3.103}$$

Using the theory of stochastic processes, it can be shown that

$$\mu^{W_e}_\gamma\left(C^{B_1,\ldots,B_k}_{\alpha_1,\ldots,\alpha_k}\right) = \prod_{i=1}^{k}\int_{B_i} dh_i \prod_{j=1}^{k}\rho_\gamma(h_j h_{j-1}^{-1}, \alpha_j - \alpha_{j-1}), \tag{3.104}$$

where $h_0 := e$ and $\alpha_0 := 0$. Similarly, one has

$$\mu^{W_e^h}_\gamma\left(C^{B_1,\ldots,B_{k-1}}_{\alpha_1,\ldots,\alpha_{k-1}}\right) = \prod_{i=1}^{k-1}\int_{B_i} dh_i \prod_{j=1}^{k}\rho_\gamma(h_j h_{j-1}^{-1}, \alpha_j - \alpha_{j-1}), \tag{3.105}$$

where, of course, $\alpha_k := 1$ and $h_k := h$. Therefore, one has

$$\mu^{W_e^h}_\gamma\left(C([0,1],H)_e^h\right) = \rho_\gamma(h, 1) = \rho_1(h, \gamma). \tag{3.106}$$

Lemma 3.8.2. *The integral (3.101) over LH_e yields*

$$\int_{LH_e} d\mu_1^{W_e}(g)\, e^{(W,Z^g)+(\dot{g},Z)} = e^{\frac{1}{2}[(W,\overline{W})+(\overline{Z},Z)]}\rho_1(\mathcal{W}_{\mathbb{C}}(Z)^{-1}\mathcal{W}_{\mathbb{C}}(\overline{W}), 1). \tag{3.107}$$

To derive this result, one starts with $Z, W \in \mathcal{A}_{\mathbb{R}}$. Then substitute the identity $Z = -(d\hat{\mathcal{W}}_Z/d\alpha)\hat{\mathcal{W}}_Z^{-1}$ (which is immediate from (3.72)), and use (3.100) with $\chi = \hat{\mathcal{W}}_Z$. In terms of the new variable $\tilde{g} = g\chi$ the integral is over $C([0,1],H)_e^{\mathcal{W}(Z)}$, since $\chi(0) = e$ and $\chi(1) = \hat{\mathcal{W}}_Z(1) = \mathcal{W}(Z)$. Perform the transformation $\tilde{g} \mapsto \tilde{g}^{-1}$, under which the Wiener measure is invariant. The integral is now over the space $C([0,1],H)_e^{\mathcal{W}(Z)^{-1}}$. Repeating the above trick for W, and using (3.106), leads to (3.107). Of course, at this stage one has $\overline{W} = W$ and $\overline{Z} = Z$.

We now use the nontrivial fact that the fundamental solution $\rho_\gamma(\cdot, t)$ of the heat equation on H has a unique analytic continuation to $H_{\mathbb{C}}$ (containing H as a subgroup; see 3.6.6). Since we know from II.(2.62) and (3.101) that the left hand side of (3.107) is analytic in Z and antianalytic in W, and that the complexified Wilson loop $\mathcal{W}_{\mathbb{C}} : \mathcal{A} \to H_{\mathbb{C}}$ is analytic, the result follows. (Alternatively, one may extend the Cameron–Martin formula by analytic continuation in χ). □

We refer to the induced space defined by (3.101), with LH replaced by LH_e, as $\mathcal{H}^{\mathrm{id}}_{F,e}$, with corresponding induced representation $\pi^{\mathrm{id}}_{F,e}$ of the set of adjointable operators on $\mathfrak{E}$. We now wish to apply Proposition 2.2.4. In order to put the result in a neat form, we define an appropriate analogue of the coherent states II.(2.47) for compact connected Lie groups. We formulate this in terms of $H_{\mathbb{C}}$ rather than T^*H; cf. Proposition 3.6.7.

Definition 3.8.3. *Let H be a compact connected Lie group. For $z \in H_{\mathbb{C}}$, the* **Hall coherent state** $\tilde{\Psi}_{\hbar}^{z}$ *in $L^2(H)$ is the function $h \mapsto \rho_{\hbar}(h^{-1}z, 1)$, where ρ_{γ} is the analytic continuation of the fundamental solution of the heat equation on H.*

The linear span of all $\tilde{\Psi}_{\hbar}^{z}$ (as z varies in $H_{\mathbb{C}}$) is dense in $L^2(H)$. The Hall coherent states are not normalized, and may or may not provide a pure state quantization of T^*H in the sense of II.(1.3.3); in the present context this is entirely irrelevant. Note that $\rho_{\hbar}(h^{-1}z, 1) = \rho_1(h^{-1}z, \hbar)$.

Our guess for $\mathcal{H}_{F,e}^{\text{id}}$ is $L^2(H)$; this plays the role of $\mathcal{H}_*^{\chi}$ in 2.2.4. We define $\tilde{U}_e : \mathfrak{E} \to L^2(H)$ by linear extension of

$$\tilde{U}_e \sqrt{\text{Exp}}(Z) := e^{\frac{1}{2}(\overline{Z}, Z)} \tilde{\Psi}_{1/2}^{W_{\mathbb{C}}(Z)}. \tag{3.108}$$

Equation (2.29) follows from (3.107) and the identity

$$\rho_{\gamma}(h, 1) = \int_H dk\, \rho_{\gamma}(hk^{-1}, \tfrac{1}{2}) \rho_{\gamma}(k, \tfrac{1}{2}). \tag{3.109}$$

This identity follows from the definition of the heat kernel, as well as from the properties $\rho_{\gamma}(hk, t) = \rho_{\gamma}(kh, t)$ (from the Ad-invariance of the Laplacian Δ_H) and $\rho_{\gamma}(h^{-1}, t) = \rho_{\gamma}(h, t)$ (from the invariance of Δ_H under $T_i \mapsto -T_i$). One should compare (3.108) with its classical analogue 3.6.8.

The definition of the Wiener measure $\mu_{\gamma}^{W_e^e}$ on LH_e and the invariance of the inner product on $\mathcal{A}_{\mathbb{R}}$ under $\text{Ad}(H)$ imply that $\mu_{\gamma}^{W_e^e}$ is invariant under the (outer) automorphisms $\tau(h)$ defined in (3.68). It follows that for $h \in H$, identified with a constant function on S^1, the operator $U_F(h)$ on $\exp(\mathcal{A})$ is adjointable with respect to the inner product (3.101), with LH replaced by LH_e. By (3.84) and (3.64) one simply has $U_F(h)\sqrt{\text{Exp}}(Z) = \sqrt{\text{Exp}}(\text{Ad}(h)Z)$, as $\dot{h} = 0$. Using (2.30), (3.108), and the Ad(H)-invariance of the complexified heat kernel, the representation $U_F(H)$, induced to $L^2(H)$, is equal to U_{Ad}; see (3.102).

To complete the induction with respect to (3.101), the integration over H mentioned in the second paragraph of the proof must still be performed. This is done at the present stage, and is a special instance of quantum Marsden–Weinstein reduction (itself a special case of Rieffel induction; see 2.5). We are in the setting of Theorem 2.5.1, with $\mathcal{H} = L^2(H)$ and $U = U_{\text{Ad}}$. Inducing from the trivial representation of H, Proposition 2.5.3 leads to the induced space $L^2(H)^{\text{Ad}(H)}$ announced in 3.8.1. This proves 3.8.1.2.

The claim 3.8.1.3 follows from the Cameron–Martin formula (3.100) for μ_1^W, as we explained at the end of 3.7.

To prove 3.8.1.5, which immediately implies 3.8.1.4, we first note that 3.8.3, (3.106), and (3.100) imply that $\tilde{U}_e$ in (3.108) may be rewritten as

$$\tilde{U}_e \sqrt{\text{Exp}}(Z) : h \mapsto e^{-\frac{1}{2}(\overline{Z}, Z)} \int_{C([0,1],H)_e^{h^{-1}}} d\mu_{1/2}^{W_e^{h^{-1}}}(g) e^{2(g, Z)}. \tag{3.110}$$

Now recall Lemma 3.7.3. By (3.86), with $\gamma = 1/4$, and (3.110), the action of the operator $\tilde{U}_{e,1/4} := V_{1/4}\tilde{U}_e V_{1/4}^{-1}$ on $\Psi \in V_{1/4}\mathfrak{E}$ is given by

$$\tilde{U}_{e,1/4}\Psi(h) = \int_{C([0,1],H)_e^{h^{-1}}} d\mu_{1/2}^{W_e^{h^{-1}}}(g)\Psi(\dot{g}). \tag{3.111}$$

Hence by (2.30), (3.90), and (3.88) one has

$$\pi_e^{\mathrm{id}}(\mathcal{Q}_F(\mathcal{W}_f))\tilde{U}_{e,1/4}\Psi(h) = \int_{C([0,1],H)_e^{h^{-1}}} d\mu_{1/2}^{W_e^{h^{-1}}}(g)\, f(\mathcal{W}(\dot{g}))\Psi(\dot{g}). \tag{3.112}$$

However, one rapidly derives from (3.74) and (3.72) that $\mathcal{W}(\dot{g}) = g^{-1}(1)$. Since $g(1) = h^{-1}$ by definition of the space over which one integrates in (3.112), we conclude that $f(\mathcal{W}(\dot{g})) = f(h)$. It follows that

$$\pi_e^{\mathrm{id}}(\mathcal{Q}_F(\mathcal{W}_f))\tilde{U}_{e,1/4}\Psi(h) = f(h)\tilde{U}_{e,1/4}\Psi(h).$$

Since the vectors $\tilde{U}_{e,1/4}\Psi$ are dense in $L^2(H)$ and f is bounded as a function, hence as a multiplication operator, this proves 3.8.1.5.

If in (3.101) one integrates over $LH_e^{[e]}$ rather than LH_e, the isomorphism (3.69) shows that the above argument still goes through, with H replaced by its universal covering group $\tilde{H}$. It follows from the structure theory of compact Lie groups that $\tilde{H}$ is the direct product of $\mathbb{R}^m$ and a compact group; in the former, one uses the coherent states II.(2.47), and in the latter, one employs those of Hall.

This proves 3.8.1.5, concluding the proof of Theorem 3.8.1. ■

3.9 Vacuum Angles in Constrained Quantization

In this final section we will explain why the discussion so far of the quantization of Yang–Mills theory on a cylinder has been incomplete. More generally, whenever the classical reduced space is given as a Marsden–Weinstein quotient with respect to a disconnected group H, there turns out to be a certain freedom in the induction process quantizing the reduced space.

For simplicity, we initially assume that H is a finite-dimensional unimodular Lie group, acting on a symplectic manifold S in strongly Hamiltonian fashion, with equivariant momentum map $J : S \to \mathfrak{h}^*$. Recall that H^0 is the component of H containing the identity e, and that $\pi_0(H) := H/H^0$. One then considers the reduced spaces $\tilde{S}_J^0 = J^{-1}(0)/H$ and $S_J^0 = J^{-1}(0)/H^0$. By (1.26) the former may be obtained by a two-step reduction process: one firstly reduces S by H^0, which results in S_J^0, and secondly reduces S_J^0 by $\pi_0(H)$, which yields $\tilde{S}_J^0 \simeq S_J^0/\pi_0(H)$. Note that the incompletely reduced space S_J^0 is already symplectic.

We now abstract the second step, and consider the reduction of a general symplectic manifold S_0 by a discrete group D whose given action on S_0 consists of Poisson maps. We regard D as a zero-dimensional Lie group, with Lie algebra $\{0\}$. The momentum map J of the D-action may then be thought of as being identically zero, so that $J^{-1}(0) = S_0$. The reduced space is then simply S_0/D. In particular, there are no constraints.

Following our general strategy, we wish to quantize S_0/D by Rieffel (or Fell) induction. Hence we assume that we have a unitary representation $U_0(D)$ on some Hilbert space $\mathcal{H}_0$ that "quantizes" the D-action on S_0. Although on the classical side D possesses only the trivial coadjoint orbit $\{0\}$, on the quantum side it will have nontrivial irreducible representations. At first sight, one should Rieffel-induce from the trivial representation of D, but in the absence of classical constraints there is actually no good reason not to induce from an arbitrary (irreducible) representation $U_\theta(D)$ defined on a Hilbert space $\mathcal{H}_\theta$.

The induced space $\mathcal{H}_0^\theta$ is constructed as in Proposition 2.5.3, amended if necessary when D is not compact (cf. the end of 2.5). One chooses a suitable dense subspace $\tilde{\mathcal{E}} \subset \mathcal{H}_0$ (when D is compact one may take $\tilde{\mathcal{E}} = \mathcal{H}_0$), and considers the sesquilinear form on $\tilde{\mathcal{E}} \otimes \mathcal{H}_\theta$ defined by

$$(\tilde{\Psi}, \tilde{\Phi})_0^\theta = \sum_{\delta \in D} (\tilde{\Psi}, U_0 \otimes U_\theta(\delta)\tilde{\Phi})_{\mathcal{H}_0 \otimes \mathcal{H}_\theta}; \tag{3.113}$$

cf. (2.81). Quotienting by the null space of this form and completing then leads to the induced space $\mathcal{H}_0^\theta$ in the standard way. This space carries an induced representation π_0^θ of the algebra of weak observables $\mathfrak{A}_w$ of the model in question; see 2.9.1. The algebra of observables

$$\mathfrak{A}_{\mathbb{R}}^\theta := \pi_0^\theta(\mathfrak{A}_w) \tag{3.114}$$

in general explicitly depends on $\theta \in \hat{D}$, even when U_θ is one-dimensional.

For example, take a not simply connected Lie group G with universal covering group $\tilde{G}$, so that $G \simeq \tilde{G}/\pi_1(G)$. Here $D = \pi_1(G)$ is a discrete subgroup of the center of $\tilde{G}$. The cotangent bundle T^*G is then symplectomorphic to $(T^*\tilde{G})/\pi_1(G)$, where $\pi_1(G)$ acts on $T^*\tilde{G}$ by pullback of its action on $\tilde{G}$. Physically, this describes a particle moving on $Q = \tilde{G}$, with phase space $S = T^*G$.

To quantize, we take $\mathcal{H}_0 = L^2(\tilde{G})$, on which $U_0(\pi_1(G))$ acts as the right-regular representation; that is, $U_0(\delta)\Psi(x) := \Psi(x\delta)$. We now choose a $\theta \in \widehat{\pi_1(G)}$, and realize that we are in the situation discussed in the paragraph containing (2.97), with $\mathsf{P} = \tilde{G}$, $\tilde{\mathcal{E}} = C_c^\infty(\tilde{G})$, $H = \pi_1(G)$, and $U_\chi = U_\theta$. It follows from that discussion that the induced space $\mathcal{H}_0^\theta$ obtained from the induction process on $L^2(\tilde{G})$ is isomorphic to the Hilbert space $\mathcal{H}^\theta$ carrying the representation U^θ of $\tilde{G}$ that is Mackey-induced by $U_\theta(\pi_1(G))$.

In the realization $\mathcal{H}_s^\theta$ given by a section $s : G \to \tilde{G}$ (cf. the text after III.(2.162)), the induced space is simply $L^2(G)\otimes\mathcal{H}_\theta$. The corresponding induced representation $\pi_s^\theta(\tilde{G})$ is then given by III.(2.176). Moreover, according to Corollary 2.7.2 the space $\mathcal{H}^\theta$ carries an irreducible representation π^θ of the action C^*-algebra $C^*(\tilde{G}, G)$. Conversely, every irreducible representation of $C^*(\tilde{G}, G)$ is equivalent to one of this form.

Noting that $C^*(\tilde{G}, G)_{\mathbb{R}}$ may be thought of as the quantum algebra of observables of a particle moving on G, we see that a quantum particle moving on a not simply connected Lie group has a family of superselection sectors labeled by $\widehat{\pi_1(G)}$.

For a simple illustration of this scheme, we take $\tilde{G} = \mathbb{R}$ and $D = 2\pi\mathbb{Z}$, so that $G = U(1)$. This is the setting for a particle moving on a circle. Note that the group $U(1)$ plays a double role: It is the configuration space Q of the particle, as well as the unitary dual $\hat{\mathbb{Z}}$. A configuration space variable will be called β, whereas an element of $\hat{\mathbb{Z}}$ is denoted by θ. Both variables take values in $[0, 2\pi)$. The representation $U_\theta(\mathbb{Z})$ corresponding to $\theta \in \hat{\mathbb{Z}}$ is

$$U_\theta(n) := e^{in\theta}, \tag{3.115}$$

defined on $\mathcal{H}_\theta = \mathbb{C}$. Hence we obtain a family of induced representations $U^\theta(\mathbb{R})$ and $\pi^\theta(C^*(\mathbb{R}, U(1)))$, the latter irreducible. As recalled above, these representations may be realized on the Hilbert space $\mathcal{H}_s^\theta$, which is the same for all θ and equal to $L^2(U(1))$. We choose the section $s : U(1) \to \mathbb{R}$ to be $s(\beta) = \beta$. From III.(2.176) the explicit form of $U_s^\theta(\mathbb{R})$ is

$$U_s^\theta(2\pi n + \beta')\Psi_s^\theta(\beta) = e^{in\theta}\Psi_s^\theta(\beta - \beta') \tag{3.116}$$

when $\beta - \beta' \in [0, 2\pi)$; in the case that $\beta - \beta' \in (-2\pi, 0)$ one has

$$U_s^\theta(2\pi n + \beta')\Psi_s^\theta(\beta) = e^{i(n+1)\theta}\Psi_s^\theta(\beta - \beta' + 2\pi). \tag{3.117}$$

The corresponding representation $\pi_s^\theta(C^*(\mathbb{R}, U(1)))$ is most easily described through Corollary III.3.7.4 and Theorem III.3.4.4. With $U_s^\theta(\mathbb{R})$ given above, it remains to state the representation $\tilde{\pi}_s^\theta(C_0(U(1)))$. By III.(3.87) and subsequent comment, this is given by

$$\tilde{\pi}_s^\theta(\tilde{f})\Psi_s^\theta(\beta) = \tilde{f}(\beta)\Psi_s^\theta(\beta). \tag{3.118}$$

It follows from (3.116), (3.117), and III.(1.69) that the associated representation dU_s^θ of the generator T of $\mathbb{R}$ is

$$p_\theta := idU_s^\theta(T) = -i\frac{d}{d\beta}. \tag{3.119}$$

The θ-dependence of this operator lies in its domain; by Proposition III.1.5.6 one should initially define p_θ on the space of smooth vectors for $U_s^\theta(\mathbb{R})$, on which it is essentially self-adjoint. It is a simple technical matter to show that the domain of the self-adjoint closure $\overline{p_\theta}$ of p_θ thus defined is

$$\mathcal{D}(\overline{p_\theta}) = \mathcal{D}_\theta := \{\Psi \in AC([0, 2\pi]) \mid \Psi(2\pi) = e^{-i\theta}\Psi(0)\}, \tag{3.120}$$

where AC stands for the space of absolutely continuous functions.

Physicists like to see the θ-dependence of p_θ in the explicit form of the operator. This is achieved by the unitary transformation $V_\theta : L^2(U(1)) \to L^2(U(1))$, defined by $V_\theta\Psi(\beta) := \exp(i\beta\theta/(2\pi))\Psi(\beta)$. One obtains

$$V_\theta\overline{p_\theta}V_\theta^{-1} = -i\frac{d}{d\beta} - \frac{\theta}{2\pi}, \tag{3.121}$$

which is self-adjoint on the domain $\mathcal{D}_0$; cf. (3.120).

Comparing (3.121) with the classical covariant momentum III.(2.83), one is tempted to interpret the term $\theta/(2\pi)$ as an external electromagnetic potential A. This interpretation is correct, and provides a physical realization of the superselection sector θ. The fact that the particle undergoes scattering despite the fact that the field strength F of the potential in question vanishes is called the **Aharonov–Bohm effect**. At a time when it was thought that only the field strength (rather than more general gauge-invariant functions of the potential A) was an element of "physical reality", this effect was considered very surprising. (The Wilson loop defined by this A is nontrivial.) The truly surprising feature of the Aharonov–Bohm effect is its periodicity in θ with period 2π. In the above description this is an immediate consequence of the periodicity of (3.115).

After this intermezzo we return to the main theme. The discussion at the beginning of this section suggests that the space $\tilde{S}^0_J$ admits a family of **inequivalent quantizations** $\mathcal{H}^\theta_0$, where $\theta \in \widehat{\pi_0(H)}$. These are defined as follows. As always, one starts with a representation $U(H)$ on a Hilbert space $\mathcal{H}$ that quantizes the given H-action on S. Then take a representation U_θ of $\pi_0(H)$ on $\mathcal{H}_\theta$, leading to a representation $\tilde{U}_\theta(H)$ by

$$\tilde{U}_\theta(h) := U_\theta(\tau_{H \to H/H^0}(h)). \tag{3.122}$$

Rather than inducing from the trivial representation of H, one now induces from $\tilde{U}_\theta(H)$. As we have seen, classical Marsden–Weinstein reduction from a disconnected group may be split into two steps. The quantum induction procedure may be split up in a similar way. In the first step one induces from the trivial representation of H^0. This is done by putting the form

$$(\Psi, \Phi)^0_0 := \int_{H^0} dh\, (\Psi, U(h)\Phi) \tag{3.123}$$

on some domain $\tilde{\mathcal{E}} \subseteq \mathcal{H}$ and constructing the induced space, now called $\mathcal{H}_0$, and the induced representation π_0 as usual. The operator $U(h)$ is adjointable for all $h \in H$ (and not merely for all $h \in H^0$). Since H^0 is trivially represented on $\mathcal{H}_0$, it follows that

$$U_0([h]_{H^0}) := \pi_0(U(h)) \tag{3.124}$$

defines a representation of $\pi_0(H)$ on $\mathcal{H}_0$. The second step of the induction procedure then consists in induction on $\mathcal{H}_0$ with respect to the representations U_0 and $\tilde{U}_\theta$ of $D = \pi_0(H)$, in the way explained prior to (3.113).

It is instructive to illustrate this two-step procedure in the example of Yang–Mills theory on a cylinder, with structure group $H = U(1)$. Recall (3.70) and (3.115); it follows that the inequivalent quantizations of $S = T^*\mathcal{A}_{\mathbb{R}}$ are labeled by $\theta \in U(1)$.

In the first step of the induction procedure we integrate over $LU(1)^0 = LU(1)^{[0]}$. We denote the left-hand side of (3.101), with $LU(1)$ replaced by $LU(1)^0$, by $(\,,\,)^0_0$. With $\mathcal{A} = L^2(S^1)$, this leads to the expression

$$\left(\sqrt{\mathrm{Exp}}(W), \sqrt{\mathrm{Exp}}(Z)\right)^0_0 = e^{(W,Z)} \int_{LU(1)^0} d\mu_1^W(g)\, e^{(\dot{g}, Z-\overline{W})}. \tag{3.125}$$

We see from (3.69) that for the present purpose we may put $H = \mathbb{R}$. In that case the fundamental solution of the heat equation is $\rho_\gamma(x,t) = (2\pi\gamma t)^{-1/2}\exp(-x^2/(2\gamma t))$. Using this in Lemma 3.8.2, or calculating directly, one obtains

$$\left(\sqrt{\mathrm{Exp}}(W), \sqrt{\mathrm{Exp}}(Z)\right)_0^0 = (2\pi)^{-1/2} e^{\frac{1}{2}\left[(W^L, \overline{W}^L) + (\overline{Z}^L, Z^L)\right] + \overline{W}_1 Z_1}. \quad (3.126)$$

We put $\pi_F^0(A) := \pi_0(\mathcal{Q}_F(A))$, where $\mathcal{Q}_F(A)$ is some adjointable operator on $\exp(\mathcal{A})$. Recall (3.93) and (3.94).

Proposition 3.9.1. *The induced space $\mathcal{H}_0$ is naturally isomorphic to $L^2(\mathbb{R})$. On this space one has*

$$\pi_F^0(A_1) = x; \quad (3.127)$$

$$\pi_F^0(E_1) = -i\frac{d}{dx}; \quad (3.128)$$

$$\pi_F^0(\mathcal{W}) = e^{ix}. \quad (3.129)$$

Here we have identified the generator of $U(1)$ with $-i$, as usual. The first two operators are defined and essentially self-adjoint on the linear span $\tilde{\mathcal{E}}_H$ of all Hermite polynomials.

Compare the first two expressions with II.(2.23) and II.(2.24).

As always, we employ Proposition 2.2.4, this time omitting the suffix $*$. The guess $\mathcal{H}_0 = L^2(\mathbb{R})$ is substantiated by defining $\tilde{U}_0 : \mathfrak{E} \to L^2(\mathbb{R})$ by linear extension of

$$\tilde{U}_0\sqrt{\mathrm{Exp}}(Z) : x \mapsto \pi^{-1/2} e^{\frac{1}{2}(\overline{Z}, Z)} e^{(x - Z_1)^2}. \quad (3.130)$$

One checks (2.29) from (3.126) and a standard Gaussian integration. Since $\tilde{U}_0\mathfrak{E}$ coincides with $\tilde{\mathcal{E}}_H$ as defined above, and the latter is dense in $L^2(\mathbb{R})$, we conclude that the guess of $L^2(\mathbb{R})$ for the induced space was a good one.

In the classical abelian theory both A_1 and E_1 are gauge-invariant under small gauge transformations. In the quantum theory this is reflected by the fact that on the domain $\mathfrak{E}$ the operators $\mathcal{Q}_F(A_1)$ and $\mathcal{Q}_F(E_1)$ (see (3.93) and (3.94)) commute with $U_F(g)$, where $g \in LU(1)^0$, and are therefore adjointable. The induced action of $a(1)$ is found from II.(2.67), (2.30), and (3.130) to be

$$\pi_0(a(1)) = x + \tfrac{1}{2}\frac{d}{dx}; \quad (3.131)$$

by construction, this operator is defined on the domain $\tilde{U}_0\mathfrak{E}$. Since the induction procedure preserves the adjoint of adjointable operators, it follows from (3.131) that $\pi_0(a(1)^*) = x - \frac{1}{2}d/dx$. From (3.93) and (3.94) we then obtain (3.127) and (3.128), respectively. Recalling the definition of $\mathcal{Q}_F(\mathcal{W})$ in the abelian case, given prior to (3.95), Equation (3.129) is then obvious. ■

We now compute the representation $U_0(\mathbb{Z})$ defined in (3.124). Any $g \in LU(1)^{[n]}$ is of the form $g = g^0 g_n$, where $g^0 \in LU(1)^0$, and g_n is defined in (3.71). By (3.84), (2.30), and (3.130) one obtains

$$U_0(n)\Psi(x) = \Psi(x + 2\pi n). \quad (3.132)$$

Theorem 3.9.2. *The induced space $\mathcal{H}_0^\theta$ defined by induction from the representation $\tilde{U}_\theta(LU(1))$, given by (3.122) and (3.115), is naturally isomorphic to $L^2(U(1))$. Writing $\pi_F^\theta(A) := \pi_0^\theta(Q_F(A))$, one has*

$$\pi_F^\theta(\mathcal{W}) = e^{i\beta}, \tag{3.133}$$

seen as a multiplication operator. Moreover, the quantized "electric field" is

$$\pi_F^\theta(E_1) = -i\frac{d}{d\beta}; \tag{3.134}$$

cf. (3.119). This operator is defined and essentially self-adjoint on the domain provided by Rieffel induction, and its closure is $\overline{p_\theta}$.

Since (3.132) is the right-regular representation of $\pi_1(U(1)) = 2\pi\mathbb{Z}$ on $\tilde{G} = \mathbb{R}$, we are exactly in the situation leading to (3.115). In the present case, (3.113), with $\mathcal{H}_\theta = \mathbb{C}$, reads

$$(\Psi, \Phi)_0^\theta = 2\pi \sum_{n\in\mathbb{Z}} e^{in\theta} \int_{\mathbb{R}} dx\, \overline{\Psi(x)}\Phi(x + 2\pi n), \tag{3.135}$$

where we have normalized the Haar measure on $\mathbb{Z}$ as 2π times the counting measure. Both $\pi_F^0(E_1)$ and $\pi_F^0(\mathcal{W})$ are adjointable with respect to (3.135). (The operator $\pi_F^0(A_1)$ isn't, because among the multiplication operators on $L^2(\mathbb{R})$ only functions with period 2π are adjointable.) The map (2.97), transferred from $\mathcal{H}^\chi$ to $\mathcal{H}_s^\chi$, then reads

$$\tilde{U}_s^\theta \Psi(\beta) = 2\pi \sum_{n\in\mathbb{Z}} e^{in\theta}\Psi(\beta + 2\pi n), \tag{3.136}$$

where $\Psi \in \tilde{\mathcal{E}}_H \subset L^2(\mathbb{R})$. Equation (3.133) then follows from (3.129), (2.30), and (3.136). Similarly, from (3.128), (2.30), and (3.136) one obtains (3.134). It is crucial that this unbounded operator is defined on the natural domain $\tilde{U}_s^\theta \tilde{\mathcal{E}}_H$ provided by Rieffel induction. It follows from (3.136) that functions in this domain are smooth, and satisfy the boundary condition $\Psi(2\pi) = \exp(-i\theta)\Psi(0)$. The final claim then follows from standard functional analysis. ∎

On the basis of these considerations one expects that any quantum gauge theory on a compact space whose gauge group $\mathcal{G}$ is disconnected possesses inequivalent quantizations labeled by the unitary dual of $\pi_0(\mathcal{G})$. In physics one refers to elements of $\widehat{\pi_0(\mathcal{G})}$ as **vacuum angles** or **θ-angles**. Such angles do not label superselection sectors (defined as inequivalent representations of the algebra of observables). Rather, each vacuum angle defines its own algebra of observables (3.114). Whatever their physical interpretation, in the description suggested here, vacuum angles emerge if one constructs the algebra of observables by induction from a nontrivial representation of the gauge group.

Notes

Chapter I

I.1.1 Jordan algebras were introduced by Jordan [1932] in connection with quantum mechanics (of which Jordan had been one of the founders), and were further studied by Jordan et al. [1934]. There is a substantial literature on such algebras; an interesting modern textbook is Faraut and Korányi [1994].

Jordan–Lie algebras appeared in Grgin and Petersen [1974], who claimed that in certain cases the associator identity (1.6) follows from the other axioms. It was added as an extra postulate by Emch [1984], who also noted that for $\hbar^2 \neq 0$ the Jacobi identity (1.5) follows from the other axioms. Also cf. Ayupov et al. [1997]. For Poisson algebras see the notes to 2.3.

The study of infinite-dimensional Jordan algebras was initiated by von Neumann [1936]. JB-algebras were introduced by Alfsen et al. [1978]. See Emch [1972] and especially Emch [1984] for a nice overview with historical perspective. Axiom (1.7) can actually be derived from (1.9) and (1.10); see the comment on p. 111 of Alfsen and Shultz [1976], Alvermann [1985], and Rodriguez Palacios [1988]. Hanche-Olsen and Størmer [1984] is a textbook on Jordan algebra and JB-algebras, and Upmeier [1987] presents an overview with many applications. These works also describe a structure theory (mostly already present in Alfsen et al. [1978]), whose main conclusion is that any JB-algebra $\mathfrak{A}$ contains an ideal $\mathfrak{J}$ such $\mathfrak{A}/\mathfrak{J}$ is isomorphic (as a JB-algebra) to a norm-closed Jordan subalgebra of $\mathfrak{B}(\mathcal{H})_{\mathbb{R}}$ for some Hilbert space $\mathcal{H}$.

The history of C^*-algebras is told by Kadison [1982, 1994]; also see the Introduction in Bratteli and Robinson [1987]. Standard references are Dixmier [1977], Pedersen [1979], Takesaki [1979], and Kadison and Ringrose [1983, 1986]. The most extensive analysis of the Gelfand–Neumark Theorem 1.1.8 is in Doran and Belfi [1986]. See Connes [1994], Doran [1994], Fillmore [1996], and Davidson [1996] for modern surveys of C^*-algebras. Introductions that relate C^*-algebras to quantum mechanicsand that are more oriented towards physicists are Bratteli and Robinson [1987, 1981], Thirring [1981, 1983], Emch [1972, 1984], Haag [1996], and Landi [1997].

As shown by Araki and Elliott [1973], axiom (1.14) can actually be derived from (1.15); see Doran and Belfi [1986] for an exhaustive study of the axioms and their consequences. The relevance of C^*-algebras to quantum mechanics was recognized by Segal [1947], and received considerable impetus from the work of Haag and his collaborators on algebraic quantum field theory; see Haag and Kastler [1964] and Haag [1996].

Theorem 1.1.9 is taken from Landsman [1997], who provides an alternative proof of the second half. Namely, as shown by Wright [1977], the complexification $\mathfrak{A}$ of a JB-algebra $\mathfrak{A}_{\mathbb{R}}$ can be normed and made into a so-called Jordan C^*-algebra (alternatively called JB^*-algebra). One then adds the Lie structure and uses the result of Rodriguez Palacios [1988] that if the Jordan product in a JB^*-algebra $\mathfrak{A}$ is the anticommutator of an associative product, then $\mathfrak{A}$, equipped with this associative product and the original norm, is a C^*-algebra.

If $\hbar^2 < 0$ one can turn $\mathfrak{A}$ itself into an associative algebra through (1.23) with the i omitted. This leads to a so-called **real C^*-algebra**, or **R^*-algebra**, which is isomorphic to an algebra of bounded operators on some real or quaternionic Hilbert space; see Goodearl [1982]. Attempts to model quantum mechanics on such spaces have been unsuccessful; cf. Beltrametti and Cassinelli [1984] and references therein.

I.1.2 This material may be found in all textbooks on C^*-algebras. Takesaki [1979], Kadison and Ringrose [1983], and Davidson [1996] are particularly efficient. Palmer [1994] is an encyclopedic treatise on Banach algebras.

Most of the theory holds for general JB-algebras; see Alfsen et al. [1978] and Hanche-Olsen and Størmer [1984].

I.1.3 Recall that a **partial ordering** $\leq$ on a set is a binary relation von satisfying: (i) $x \leq x$ for all x; (ii) if $x \leq y$ and $y \leq x$ then $x = y$; (iii) if $x \leq y$ and $y \leq z$ then $x \leq z$. One writes $x < y$ if $x \leq y$ and $x \neq y$; also, $y \geq x$ (or $y > x$) is the same as $x \leq y$ (or $x < y$). The general theory of partially ordered topological vector spaces is given by Wong and Ng [1973] and by Asimow and Ellis [1980].

All C^*-algebraic results in this section may be found in the standard textbooks.

I.1.4 Definition 1.4.1 is due to von Neumann [1932] (for $\mathfrak{A} = \mathfrak{B}(\mathcal{H})$) and Segal [1947] (for general C^*-algebras); both were motivated by quantum mechanics.

All books on C^*-algebras discuss the basic properties of states and state spaces. The general theory of compact convex sets may be found in in Alfsen [1971] and Asimow and Ellis [1980]. Kadison and Ringrose [1983] discuss unital C^*-algebras and their state spaces in the light of this general theory; Alfsen et al. [1978], Asimow and Ellis [1980], and Hanche-Olsen and Størmer [1984] do so for general unital JB-algebras.

The use of more general compact convex sets and partially ordered Banach spaces than those provided by C^*-algebras is central to the so-called operational approach to quantum mechanics, for which we refer to Haag and Kastler [1964], Schwinger [1970], Davies and Lewis [1971], Hartkämper and Neumann [1974], Mielnik [1974], Davies [1976], Gudder [1979], Beltrametti and Cassinelli [1981], Holevo [1982], Ludwig [1985], Lahti and Bugajski [1980, 1985], and Busch et al. [1995]. The starting point is the duality between the state space K, assumed to be a compact convex set, and the partially ordered Banach space of observables $A(K, \mathbb{R})$ or $A_b(K, \mathbb{R})$. Theorem 1.4.5 is a special case of this theory, and should be seen in its light. The connection between this approach and the theory of JB-algebras has been studied by Kummer [1991].

The decomposition of φ used in the proof of 1.4.5 is Thm. 4.3.6 in Kadison and Ringrose [1983]. The final step of the proof of Theorem 1.1.9 was inspired by the proof of Lemma 8.5 in Alfsen et al. [1978].

I.1.5 Representations and the GNS- (Gelfand–Neumark–Segal) construction are discussed in all books on C^*-algebras. For more on the universal representation and the closely related proof of Theorem 1.1.8, see Dixmier [1977], Pedersen [1979], or Kadison and Ringrose [1983, 1985].

I.1.6 For the Riesz representation theorem see, for example, Pedersen [1989]. A detailed proof of Theorem 1.6.5 may be found in Reed and Simon [1972], Takesaki [1979], or Pedersen [1989].

I.1.7 For the history of von Neumann algebras cf. Kadison [1958, 1982]. Von Neumann developed his theory of "rings of operators" partly in order to generalize his own Hilbert space formalism of quantum mechanics (von Neumann [1932]). He eventually came to believe that quantum mechanics should be described by so-called type II_1 factors; see von Neumann [1981]. Bub [1981] provides historical comments.

A clear discussion of the various topologies on $\mathfrak{B}(\mathcal{H})$ is in Takesaki [1979] or Pedersen [1989]. The theory of von Neumann algebras, which started with Theorem 1.7.2 due to von Neumann himself, is covered by Pedersen [1979], Takesaki [1979], Kadison and Ringrose [1983, 1985], and Connes [1994]. In addition to the first three of these books, cf. Dixmier [1977] for the description of the bidual of a C^*-algebra $\mathfrak{A}$ as a von Neumann algebra, i.e., Proposition 1.7.4. Proposition 1.7.5 is Theorem 10.1.12 in Kadison and Ringrose [1985]; equivalent statements are in Pedersen [1979] and Takesaki [1979]. It states the "universal" property of the universal representation. Some authors use the term W^*-algebra for an abstract C^*-algebra that is the dual of a Banach space, a von Neumann algebra then meaning a W^*-algebra realized on a Hilbert space.

I.2.1 The insight that an extreme point of a convex set (as defined by Minkowski) is precisely a pure state in the sense of quantum mechanics is due to von Neumann. Extreme points of general compact convex sets are studied in Alfsen [1971], Asimow and Ellis [1980], and Pedersen [1989]; in connection with C^*-algebras see Pedersen [1979], Takesaki [1979], and Bratteli and Robinson [1987]. Kadison and Ringrose [1983] is particularly efficient, and contains a proof of the **Krein–Milman Theorem** 2.1.5. For (2.2) see Dixmier [1977], §3.2.4.

I.2.2 Most of this section is standard; for the reduced atomic representation see Pedersen [1979], Kadison and Ringrose [1983], and Akemann and Shultz [1985]. Proposition 2.2.8 is given, for example, in Kadison and Ringrose [1986], Prop. 6.6.6, and in Davidson [1996], Thm. III.1.1.

I.2.3 Poisson algebras and Poisson manifolds in the setting of function spaces go back to Lie [1890]; the modern era started with Kirillov [1976] and Lichnerowicz [1977]. See Marsden and Ratiu [1994] (p. 293) for historical comments, and Huebschmann [1990] for extensive references. The abstract concept of a Poisson algebra appeared implicitly in Falk [1951]; Dirac [1950, 1964] deserves major credit as well. The theory of Poisson manifolds can be found in Weinstein [1983], Libermann and Marle [1987], Marsden and Ratiu [1994], and Vaisman [1994]; also see the survey by Weinstein [1998].

Equation (2.5) is equivalent to the statement that the so-called Schouten (or Nijenhuis) bracket of B with itself vanishes; see the books quoted in the previous paragraph.

For technical properties of flows in connection with classical mechanics, see Abraham and Marsden [1985]. For the theory in infinite dimension, cf. Marsden [1974] and Chernoff and Marsden [1974]. A symplectic structure of the type we consider is called a strong symplectic structure in the literature (see the two books just quoted). A closed 2-form ω is called **strongly symplectic** if the map $B_\sharp$, defined by reading (2.18) from right to left, is an isomorphism. In contrast, for a **weak symplectic structure** this map is merely injective.

There is a huge literature on symplectic manifolds; Abraham and Marsden [1985], Arnold [1989], Arnold and Givental [1990], and Sławianowski [1991] are particularly useful in the context of classical mechanics.

I.2.4 For Definition 2.4.1 see, for example, Libermann and Marle [1987], App. 3. Here one also finds the singular Frobenius theorem 2.4.2 (due to Sussmann [1973]) as Thm. 3.10. The ordinary Frobenius theorem is included as Thm. 4.2; for the latter also cf. Choquet-Bruhat et al. [1982]. A readable review of the theory of (singular) distributions and foliations is Dazord [1985].

Theorem 2.4.7 is due to Kirillov [1976]; our discussion follows Marsden and Ratiu [1994], §10.6, where further details concerning the connection with singular foliation theory may be found (note that these authors use the terminology "symplectic stratification theorem", although the singular foliation obtained is not a stratification in the sense of Goreski and McPherson [1988] or Sjamaar and Lerman [1991]). Other detailed treatments are in Libermann and Marle [1987], §III.12, and Vaisman [1994], Ch. 2, who includes examples.

I.2.5 The insight that the (normal) pure states of an irreducible quantum system correspond to points in $\mathbb{P}\mathcal{H}$ goes back (at least) to Weyl [1931], p. 75. For $\mathbb{P}\mathcal{H}$ as a symplectic manifold see, e.g., Cirelli et al. [1983], Abbati et al. [1984], Cirelli et al. [1994], or Marsden and Ratiu [1994]. For infinite-dimensional manifolds in general see Marsden [1974], Abraham and Marsden [1985], Choquet-Bruhat et al. [1982], or Lang [1995].

Proposition 2.5.2 is due to Cirelli et al. [1983]. It follows from (2.68) and Prop. 2.6.15 of Bratteli and Robinson [1987] that the norm-topology relative to both $\mathbb{P}\mathcal{H} \subset \mathfrak{B}_0(\mathcal{H})^*$ and $\mathbb{P}\mathcal{H} \subset \mathfrak{B}(\mathcal{H})^*$ coincides with the manifold topology on $\mathbb{P}\mathcal{H}$ as well. Theorem 2.5.4 is similar to Prop. 4.2 in Roberts and Roepstorff [1969]; we have added the appearance of $\mathbb{P}\mathcal{H}$ with its manifold topology.

Our construction of $\mathbb{P}\mathcal{H}$ as a symplectic leaf in $\mathcal{H}^*/U(1)$ is not standard, but forms an instance of the general procedure of reduction; see IV.1.5, in particular Theorem IV.1.5.5. If a compact Lie group G acts smoothly on a manifold M, then M_H (the collection of points in M with stability group H) and M_H/G are manifolds (M/G may not be). See Bredon [1972]. We could have worked with $\mathcal{H}/U(1)$; this is not a Poisson manifold, but a Poisson space in the sense of Definition 2.6.2. Also see IV.1.11.

The projected Schrödinger equation (2.46) in the given (symplectic) context goes back at least to Hermann [1973]. It is physically not very interesting for bounded H. If the Hamiltonian H is unbounded, the function $\hat{H}$ and its Hamiltonian vector field $\xi_{\hat{H}}$ are defined only on a dense submanifold of $\mathbb{P}\mathcal{H}$, namely the projection of the domain of H. This situation can be handled by the theory of densely defined vector fields on infinite-dimensional manifolds, see Marsden [1974] and Chernoff and Marsden [1974]. Even in that case the flow $\psi(t)$ is defined on all of $\mathbb{P}\mathcal{H}$.

Using the standard complex structure J on $\mathcal{H}$ (defined by $JV(\Phi) = V(i\Phi)$) or on $\mathbb{P}\mathcal{H}$ (where similarly, $J = i$ in each local chart (2.28)), one can define a Kähler metric $\mathbf{g}$ by

$$\mathbf{g}(X, Y) = \tfrac{1}{2}\omega(X, JY).$$

See, e.g., Griffiths and Harris [1978] ($N < \infty$) and Marsden [1974] ($N = \infty$) for the mathematics, and Strocchi [1966], Hermann [1973], Marsden [1974], Cirelli et al. [1983], Cirelli and Lanzavecchia [1984], Abbati et al. [1984], Marsden and Ratiu [1994], Cirelli et al. [1994], Hughston [1995], and Ashtekar and Schilling [1997] for applications in quantum mechanics. On $\mathcal{H}$ the Kähler metric is given by

$$\mathbf{g}(V(\Phi), V(\Omega)) = \hbar \mathrm{Re}\,(\Phi, \Omega).$$

On $\mathbb{P}\mathcal{H}$ one has the remarkable equation, found by Cirelli et al. [1990],

$$\hbar \mathbf{g}(\xi_{\hat{A}}, \xi_{\hat{B}})(\psi) = \widehat{A \circ B}(\psi) - \hat{A}(\psi)\hat{B}(\psi).$$

This relates the Fubini–Study metric on $\mathbb{P}\mathcal{H}$ to the Jordan product $\circ$ on $\mathfrak{B}(\mathcal{H})$, much as the Fubini–Study symplectic form on $\mathbb{P}\mathcal{H}$ is related to the Poisson bracket on $\mathfrak{B}(\mathcal{H})$, cf. (1.22). Since this relation is not easily generalized to other C^*-algebras than $\mathfrak{B}(\mathcal{H})$ or $\mathfrak{B}_0(\mathcal{H})$, it plays no role in our approach.

I.2.6 The point of view in this section, as well as Proposition 2.6.4, originate in Landsman [1996a]. The definition of a Poisson space was partly inspired by that of a stratified symplectic space in Sjamaar and Lerman [1991].

Here is the technical argument alluded to at the end of the proof of Proposition 2.6.4. If necessary, one adds the unit function 1_P to $\mathfrak{A}$, and extends π by linearity and the rule $\pi(1_P) = 1_S$; this still defines a representation, and $\tilde{J}$ remains multiplicative on the extended algebra $\tilde{\mathfrak{A}}$ (cf. Lemma 2.3.26 in Bratteli and Robinson [1987]). Evidently, $\tilde{J}_\sigma(1_P) = 1$. The multiplicativity of $\tilde{J}$ immediately implies that it is positive on $\tilde{\mathfrak{A}} \cap C(P)$, and by the previous equation $\tilde{J}$ must therefore be continuous with norm 1. Hence it can be extended to all of $C(P)$, where it remains multiplicative. By continuity, $|(\pi(f))(\sigma)| \leq \|f\|_\infty$ for all $\sigma \in S$, so that $\|\pi(f)\|_\infty \leq \|f\|_\infty$. Hence $\pi : \mathfrak{A} \to C_b^\infty(S) \subset C_b(S)$ is continuous as a map between Banach spaces. We extend π to all of $C(P)$ by putting $(\pi(f))(\sigma) = J(\sigma)(f)$; this is precisely its extension by continuity. Since multiplication is continuous in the sup-norm, $\pi = J^*$ is a Jordan morphism of $C(P)$ into $C_b(S)$ (where the Jordan product $\circ$ is pointwise multiplication). The continuity of J now follows (cf. Thm. 3.4.3 in Kadison and Ringrose [1983]): Since P, being compact and Hausdorff, is completely regular, a subbase for the topology of P is given by $\{f^{-1}(\mathcal{O})\}$, where f and $\mathcal{O}$ range over $C(P)$ and the open sets in $\mathbb{R}$, respectively (cf. Kelley [1955], p. 117). Now, $J^{-1}(f^{-1}(\mathcal{O})) = \pi(f)^{-1}(\mathcal{O})$, which is open since $\pi(f)$ is continuous. Hence J is continuous.

If P is not compact, the proof undergoes only minor changes. The Stone–Weierstrass theorem (see, e.g., Pedersen [1989]) now says that $\mathfrak{A} \cap C_c(P)$ is dense in $C_0(P)$. It follows that $\tilde{J}_\sigma$ is normalized, and the remainder of the argument is the same. Note that general positive functionals defined on a dense subalgebra of $C_0(P)$ may not be extendible to positive functionals on the unitization of this subalgebra; the argument needs the preservation of multiplicativity on this extension to conclude positivity (and hence boundedness).

Definition 2.6.6 is taken from Landsman [1996a] (written in 1992); it was rediscovered in Gotay et al. [1996]. Theorem 2.6.7, in the special case that P is a Poisson manifold, is in Landsman [1996a].

I.2.7 Transition probabilities were introduced by Born [1926] in the context of quantum-mechanical collision theory. Curiously, he initially thought that (in modern notation) the transition probability between two unit vectors Ψ and Φ was given by the inner product (Ψ, Φ), and stated the correct expression $|(\Psi, \Phi)|^2$ only as a note added in proof. The abstract notion of a transition probability space is due to von Neumann [1981], who thereby went beyond the general situation in quantum mechanics laid out in von Neumann [1932]. The condition of symmetry (which has nothing to do with the invariance of the laws of physics under time-inversion, cf. Haag [1996]) was not included in his definition of a transition probability space. The concept was revived by Mielnik [1968], who introduced the notion of a basis and proved Proposition 2.7.4. Further work is in Zabey [1975], Belinfante [1976], and Pulmannová [1986]; see Beltrametti and Cassinelli [1984] for a concise review.

Here is an example of a transition probability space that is not well-behaved (mentioned, with an error, in Zabey [1975], who attributes it to Mielnik). The elements of $\mathcal{P}$ are equivalence classes of subsets of $\mathbb{R}$ of the type $\rho \subseteq [0, n]$ (where the integer $n \geq 3$) of Lebesgue measure $\mu(\rho) = 1$; two subsets are equivalent when they differ by a null set. The transition probabilities are defined by $p(\rho, \sigma) = \mu(\rho \cap \sigma)$. Note that $\dim(\mathcal{P}) = n$. For $\mathcal{Q} \subseteq \mathcal{P}$, let $\tilde{\mathcal{Q}} \subseteq [0, n]$ denote the union of all members of $\mathcal{Q}$. Then $\mathcal{Q}^{\perp}$ consists of all elements of $\mathcal{P}$ that are disjoint (up to null sets) from $\tilde{\mathcal{Q}}$. If $\mu(\tilde{\mathcal{Q}}) > n-1$ there are no such elements, so that $\mathcal{Q}^{\perp\perp} = \mathcal{P}$ in that case. If $\mu(\tilde{\mathcal{Q}}) \leq n-1$, on the other hand, $\mathcal{Q}^{\perp\perp}$ consists of all elements of $\mathcal{P}$ that are contained in $\tilde{\mathcal{Q}}$. Therefore, if $\tilde{\mathcal{Q}} \subset [0, n]$ is a given subset and $\mathcal{Q}$ is defined as the collection of all elements of $\mathcal{P}$ that lie in $\tilde{\mathcal{Q}}$, then $\mathcal{Q}$ is orthoclosed iff $\mu(\tilde{\mathcal{Q}}) \leq n-1$. However, $\mathcal{Q}$ has a basis only if $\mu(\tilde{\mathcal{Q}})$ is integral. For another example see Beltrametti and Cassinelli [1984].

Any orthogonal subset of $\mathcal{P}$ is a sample space in the sense of classical probability theory. The second requirement of 2.7.5 may then be rephrased by saying that any maximal "classical" subspace of $\mathcal{P}$ is complete, in the sense that an arbitrary point in $\mathcal{P}$ must make a transition to some member of the given subspace. Moreover, this kind of completeness continues to hold if one restricts to orthoclosed subspaces.

The existence of a multitude of maximal classical subspaces in a nontrivial transition probability space is very hard to interpret. In the case of quantum mechanics, the so-called Copenhagen interpretation (see Jammer [1974]), which is based on the somewhat obscure philosophy of complementarity of Niels Bohr, says that the choice of some such subspace is determined by the experimental arrangement set up by a physicist. The author's opinion is laid out in Landsman [1991, 1995b].

I.2.8 The expression (2.63) was proposed by Mielnik [1969] for arbitrary convex sets K, but this formula does not actually define a transition probability without the extra condition we have added. The expression is motivated by operational considerations about filters and preparation procedures in quantum mechanics (see the references in the notes to 1.4). We will not give this motivation here, since in our approach transition probabilities are fundamental and irreducible properties of pure state spaces, from which a possible operational interpretation of the theory is to be derived, rather than the converse.

Theorem 2.8.2 may be generalized, stating that the pure state space of a JB-algebra $\mathfrak{A}_{\mathbb{R}}$ is a symmetric transition probability space under (2.63). Firstly, it follows from equation (4.3) and Cor. 7.3 in Alfsen and Shultz [1978] that every pure state in a JB-algebra is norm-exposed, so that by 2.8.1, equation (2.63) indeed defines a transition probability. Secondly, we need to show that the transition probabilities thus defined are symmetric. As explained on p. 159 of Alfsen and Shultz [1978], and also in Prop. 1.13 of Alfsen and Shultz [1979], in JB-algebras there is a bijective correspondence between pure states ρ of $\mathfrak{A}_{\mathbb{R}}$ and minimal idempotents (projections) $\hat{\rho}$ in $\mathfrak{A}_{\mathbb{R}}^{**}$; here an idempotent p in a JB-algebra is an element satisfying $p^2 = p$ (hence $0 \leq p \leq \mathbb{I}$), and the minimality of p means that there is no nonzero projection q such that $q \leq p$ (our notation is different from the reference cited). This correspondence is given by the equation $\hat{\rho}(\rho) = 1$, which uniquely determines one entry given the other. It then follows from Thm. 2.17 in Alfsen and Shultz [1976] that the transition probability $p(\rho, \sigma)$ as defined by (2.63) is given by $\hat{\sigma}(\rho)$. The symmetry of the transition probabilities then follows from equation (4.5) and Cor. 7.3 in Alfsen and Shultz [1978].

Equations (4.3) and (4.5) in Alfsen and Shultz [1978], which are central to the above proof, are two of the three "pure state properties", which they show to be satisfied by the pure state space of a JB-algebra. In a slightly more general context, these or closely related properties were first postulated by Gunson [1967] and Pool [1968].

There is an alternative way of looking at the symmetry of the transition probabilities for JB-algebras. If $\mathfrak{A}_{\mathbb{R}}$ is a finite-dimensional JB-algebra, there exists a (real) inner product $(\, ,\,)$ on $\mathfrak{A}_{\mathbb{R}} \simeq \mathbb{R}^n$ such that the positive cone $\mathfrak{A}_{\mathbb{R}}^+$ is self-dual. Using the inner product to identify $\mathfrak{A}_{\mathbb{R}}$ with $\mathfrak{A}_{\mathbb{R}}^*$, this means simply that $\mathfrak{A}_{\mathbb{R}}^+ = (\mathfrak{A}_{\mathbb{R}}^*)^+$. The symmetry of the transition probabilities then eventually follows from the symmetry of the inner product. The close relationship between the self-duality of the positive cone and the symmetry of the transition probabilities in a finite-dimensional (formally real) Jordan algebra has been stressed by Haag [1996] in connection with the foundations of quantum mechanics. It turns out that finite-dimensional JB-algebras are characterized by the self-duality of the positive cone, plus the fact that the subgroup of $GL(\mathfrak{A}_{\mathbb{R}})$ that maps $\mathfrak{A}_{\mathbb{R}}^+$ into itself acts transitively on the interior of $\mathfrak{A}_{\mathbb{R}}^+$. See, for example, Faraut and Korányi [1994]. An infinite-dimensional analogue of this result is discussed in Iochum and Shultz [1983] and Iochum [1984]; the central property of facial homogeneity (originally due to Connes) occurring in their work is further analyzed in Ajupov et al. [1990], in which it is admitted that the physical relevance of this property is obscure.

Equation (2.67) appears in Roberts and Roepstorff [1969], who also prove the related result that the spaces $\mathbb{P}\mathcal{H}_\alpha$ in the decomposition $\mathcal{P}(\mathfrak{A}) = \cup_\alpha \mathbb{P}\mathcal{H}_\alpha$ (cf. 2.5.4) are precisely the components of $\mathcal{P}(\mathfrak{A})$ in the norm-topology. The result that $\|\rho - \sigma\| = 2$ for inequivalent pure states is due to Glimm and Kadison [1960]; the statement actually holds for arbitrary disjoint states (these are states whose GNS-representations have no equivalent subrepresentations), see Cor. 10.3.6 in Kadison and Ringrose [1986].

The transition probabilities (2.65) can be expressed in terms of the Fubini–Study metric g on $\mathbb{P}\mathcal{H}$. This metric can be normalized in such a way that

$$p(\rho, \sigma) = \tfrac{1}{2}(1 + \cos d(z, w)),$$

where d is the distance defined by g. For example, for $\mathcal{H} = \mathbb{C}^2$ the Fubini–Study distance $d(z, w)$ is just the angular distance measured along the (shortest) great circle connecting z and w (cf. 3.7.1). This expression has led to interesting connections with information theory, entropy, uncertainty, and statistical inference; cf., e.g., Hilgevoord and Uffink [1991], Petz [1994], and Brody and Hughston [1998]. In a different direction, the ensuing connection between quantum mechanics and Riemannian geometry has been exploited by Anandan [1991] and Ashtekar and Schilling [1998]. Another way to look at (2.65) relates this transition probability to the projective cross-ratio of algebraic geometry; see Hughston [1995].

Continuity properties of the transition probabilities (2.65) are studied in Archbold and Shultz [1989].

Although it is somewhat contrary to the spirit of the present work, one can define transition probabilities between mixed states (the physical relevance of such transition probabilities has been questioned by Roberts and Roepstorff [1969]). For general *-algebras this was done in Uhlmann [1976]; it was shown by Alberti [1983] that for unital C^*-algebras $\mathfrak{A}$ Uhlmann's general expression reduces to

$$p(\omega_1, \omega_2) = \inf \{\omega_1(A)\omega_2(A^{-1}) | A > 0,\, A \in \mathfrak{A},\, A^{-1} \in \mathfrak{A}\}.$$

For density matrices on a Hilbert space $\mathcal{H}$ (that is, states on $\mathfrak{B}_0(\mathcal{H})$) this is equivalent to a formula due to Araki [1972], namely

$$p(\rho_1, \rho_2) = \left(\mathrm{Tr}\sqrt{\rho_1^{1/2}\rho_2\rho_1^{1/2}}\right)^2.$$

Further information and references may be found in Uhlmann [1993]. Cantoni [1975] defines transition probabilities between arbitrary states in the context of lattice theory (cf. 3.6); see

Gudder [1979] for a review. The equivalence between Cantoni's and Uhlmann's transition probabilities on state spaces of unital C^*-algebras is shown in Raggio [1982] and Araki and Raggio [1982]. In more general situations the various approaches do not even coincide on the pure state spaces; see Pulmannová [1989].

I.3.1 This material is mainly taken from Landsman [1997]. Following a seminar the author gave in Göttingen, 1995, A. Uhlmann informed him that in his lectures on quantum mechanics $\mathfrak{A}_{\mathbb{R}}^{00}(\mathcal{P})$ had long been employed as the space of observables; also see Uhlmann [1996].

I.3.2 Theorem 3.2.1 originates in the following result of Shultz [1982]: If $A \in p\mathfrak{A}^{**}$ is such that A, A^*A, and AA^* are in $C_u(\mathcal{P}(\mathfrak{A}))$, then $A \in p\mathfrak{A}$. It was then shown by Brown [1992] that the hypothesis on A^*A and AA^* can be dropped. Our proof is based on that of Brown [1992], which also contains the corollary of the Stone–Weierstrass theorem that is used in the proof. The original Stone–Weierstrass theorem for C^*-algebras, due to Glimm (see Dixmier [1977] for a very detailed presentation), states that if $\mathfrak{A}$ is a unital C^*-subalgebra of a unital C^*-algebra $\mathfrak{B}$ which separates $\mathcal{P}(\mathfrak{B})^-$, then $\mathfrak{A} = \mathfrak{B}$. The usual Stone–Weierstrass theorem then follows by taking $\mathfrak{B}$ to be commutative.

Equation (3.5) follows from the property that the predual determines the order of $\mathfrak{M}$ (that is, $\omega(A) \geq 0$ for all pure normal states ω implies $A \geq 0$), and Lemma 3.4.1 in Dixmier [1977]. There is no such result for arbitrary von Neumann algebras, which may even have no pure normal states at all. In the given setting the pure normal states are abundant, because $\mathcal{N}(\mathfrak{B}^{**}) = \mathcal{S}(\mathfrak{B})$.

If $\mathfrak{A}$ has no unit, Theorem 3.2.1 can be adapted in two essentially equivalent ways. Firstly, one has $(\mathfrak{A}_{\mathbb{I}})_{\mathbb{R}} = \mathfrak{A}_{\mathbb{R}}(\mathcal{P}(\mathfrak{A})) \cap C_u(\mathcal{P}(\mathfrak{A}), \mathbb{R})$; this follows from the proof as given, plus Corollary 8 in Brown [1992]. Secondly, $\mathfrak{A}_{\mathbb{R}} = \mathfrak{A}_{\mathbb{R}}(\mathcal{P}(\mathfrak{A})) \cap C_u(\mathcal{P}(\mathfrak{A}) \cup 0, \mathbb{R})$ (cf. Theorem 6 of Brown [1992]). Indeed, the Stone–Weierstrass theorem for nonunital C^*-algebras is as in the unital case, but with $\mathcal{P}(\mathfrak{B})^-$ replaced by $\mathcal{P}(\mathfrak{B})^- \cup 0$.

Perfect C^*-algebras were introduced in Shultz [1982], and studied in detail in Akemann and Shultz [1985]. For nonunital algebras the definition is that $\mathfrak{A}$ is perfect if $\mathfrak{A}_{\mathbb{R}} = \mathfrak{A}_{\mathbb{R}}(\mathcal{P}(\mathfrak{A})) \cap C(\mathcal{P}(\mathfrak{A}) \cup 0, \mathbb{R})$; this means that $\mathfrak{A}$ is perfect iff its unitization is. The main motivation was that if $\mathfrak{A}$ is perfect, the Stone–Weierstrass theorem can be sharpened so as to state that $\mathfrak{A} = \mathfrak{B}$ if $\mathfrak{A}$ separates $\mathcal{P}(\mathfrak{B}) \cup 0$ (where "$\cup 0$" may be omitted in the unital case). The perfectness of $\mathfrak{B}_0(\mathcal{H})$ is a special case of the following result (Shultz [1982]): If $\mathcal{P}(\mathfrak{A})^-$ consists of multiples of normal states on $\pi_{\mathrm{ra}}(\mathfrak{A})''$, then $\mathfrak{A}$ is perfect. Also, arbitrary direct sums of perfect C^*-algebras are perfect.

The physical meaning of uniform structures on state spaces in quantum mechanics is discussed by Werner [1983].

I.3.3 This material is from Landsman [1997]; the second half of the proof of Proposition 3.3.3 is based on the proof of Thm. 12.12 in Alfsen and Shultz [1976]. For the proof of Lemma 3.3.4 see Alfsen et al. [1978], Shultz [1979], or Hanche-Olsen and Størmer [1984].

I.3.4 Proposition 3.4.1 is taken from Landsman [1997]. Exhaustive information on derivations and one-parameter automorphism groups on Banach spaces may be found in Bratteli and Robinson [1987].

Corollary 3.4.2 was inspired by Thm. 18 in Shultz [1982], but is phrased in different language and has an entirely different proof. In Shultz's result the condition that α^* be a Poisson map is replaced by the requirement that α^* preserve the orientation of $\mathcal{P}(\mathfrak{A})$.

There are similar results relating properties of an (auto)morphism α of a unital C^*-algebra $\mathfrak{A}$ to properties of its dual α^*, seen as a map on the entire state space $\mathcal{S}(\mathfrak{A})$. Kadison [1965] showed that α is a Jordan automorphism of $\mathfrak{A}$ iff α^* is an affine (w^*-) homeomorphism of

$\mathcal{S}(\mathfrak{A})$ (also cf. Bratteli and Robinson [1987]). Shultz [1981] extended this by proving that α is a morphism of $\mathfrak{A}$ iff α^* in addition preserves orientation.

Corollary 3.4.3 is a famous theorem due to Wigner [1931]. An **antiunitary operator** U is an antilinear bijection on $\mathcal{H}$ (i.e., $Uc\Psi = \overline{c}U\Psi$ for all $c \in \mathbb{C}$) that satisfies $(U\Psi, U\Phi) = (\Phi, \Psi)$ for all $\Psi, \Phi \in \mathcal{H}$. See Bargmann [1964], Roberts and Roepstorff [1969], Varadarajan [1985], Beltrametti and Cassinelli [1984], Shultz [1982], Cirelli et al. [1983], or Tuynman and Wiegerinck [1987] for various alternative approaches to this theorem.

Lemma 3.4.4 goes back to Kadison, but we here refer to Bratteli and Robinson [1987] (Example 3.2.14) for a detailed presentation. The following generalization is due to Alfsen et al. [1980] (Prop. 2.4): if $\pi_i : \mathfrak{A} \to \mathfrak{B}(\mathcal{H}_i)_{\mathbb{R}}$ $(i = 1, 2)$ are irreducible Jordan-equivalent representations of a JB-algebra $\mathfrak{A}$, then there exists a unitary or an antiunitary map from $\mathcal{H}_1$ to $\mathcal{H}_2$ that implements the equivalence (here Jordan equivalence means that there is a Jordan automorphism $\beta : \mathfrak{B}(\mathcal{H}_1)_{\mathbb{R}} \to \mathfrak{B}(\mathcal{H}_2)_{\mathbb{R}}$ such that $\pi_2 = \beta \circ \pi_1$).

I.3.5 A classic on lattice theory is Birkhoff [1967]. For orthomodular lattices see Maeda and Maeda [1970] or Kalmbach [1983] (which is highly readable and contains many attractive historical quotations and excursions). The connection between Hermitian forms and orthocomplementations is thoroughly discussed in Baer [1952], and Varadarajan [1985], as well as in Maeda and Maeda [1970]; recent reviews are Piziak [1991] and Holland [1995]. The fact that $\mathcal{L}(V)$ is not modular in infinite dimension follows, after an elementary but somewhat lengthy argument, from the existence of closed subspaces whose sum is not closed; cf. Kalmbach [1983] or Beltrametti and Cassinelli [1984].

A **division ring** $\mathbb{D}$ (sometimes called a skew-field) is a ring in which the equations $xa = b$ and $ay = b$ can be solved for x and y whenever $a \neq 0$. An **involution** of $\mathbb{D}$ is a linear bijection $\lambda \to \overline{\lambda}$ satisfying $\overline{\lambda\mu} = \overline{\mu}\overline{\lambda}$ and $\overline{\overline{\lambda}} = \lambda$. For general division rings, the definition of a sesquilinear form is the same as for $\mathbb{D} = \mathbb{C}$, namely a bilinear map $(\, , \,) : V \times V \to \mathbb{D}$ satisfying $(\lambda\Psi, \mu\Phi) = \overline{\lambda}(\Psi, \Phi)\mu$ and $(\Phi, \Psi) = \overline{(\Psi, \Phi)}$. Such a form is said to be nondegenerate if $(\Psi, \Phi) = 0$ for all Φ implies $\Psi = 0$.

A detailed proof of Proposition 3.5.7 is in Kalmbach [1983], Thm. 3.1; also cf. Birkhoff [1967], §8.

I.3.6 For a direct proof of Proposition 3.6.1 cf. Kalmbach [1983]. Detailed discussions of projections in von Neumann algebras are in Takesaki [1979] and in Kadison and Ringrose [1983]. It can be shown that $\mathcal{L}(\mathfrak{M})$ is irreducible iff $\mathfrak{M}$ is a **factor**, that is, $\mathfrak{M} \cap \mathfrak{M}' = \mathbb{C}\mathbb{I}$. This is because the center $C(\mathcal{L}(\mathfrak{M}))$ consists of the projections in $\mathfrak{M} \cap \mathfrak{M}'$. One should therefore be aware that $\mathfrak{M}$ may not act irreducibly on a Hilbert space (in the sense of 2.2.1), while $\mathcal{L}(\mathfrak{M})$ is nonetheless irreducible.

A modern reference on the lattices $\mathcal{L}(\mathfrak{A}^{**})$ and $\mathcal{F}(\mathcal{S}(\mathfrak{A}))$ is Akemann and Pedersen [1992], where Proposition 3.6.3 may be found; it goes back to Prosser [1963] and Effros [1963]. The "technical argument" on left ideals used in the proof is due to Effros [1963], and also appears in Pedersen [1979], Thm. 3.6.11. As shown by Topping [1967], these lattices have the property of **semimodularity** (also called M-symmetry), cf. Birkhoff [1967] or Maeda and Maeda [1970]. Call (y, z) a **modular pair** if (3.12) holds for all $x \leq z$. Semimodularity then means that (z, y) is a modular pair whenever (y, z) is. In an orthocomplemented atomic lattice, semimodularity is equivalent to the covering property defined in 3.7, cf. Thm. 30.2 in Maeda and Maeda [1970].

Proposition 3.6.3 can be generalized to a certain class of partially ordered Banach spaces (which includes unital JB-algebras and C^*-algebras); see Alfsen and Shultz [1976] and Edwards and Rüttimann [1985].

Proposition 3.6.4 is due to Zabey [1975] and Belinfante [1976]; also cf. Beltrametti and Cassinelli [1984] and Pulmannová [1986].

A third lattice associated with a unital C^*-algebra $\mathfrak{A}$, which is not isomorphic to either $\mathcal{L}(\mathfrak{A}^{**})$ or $\mathcal{L}(\mathfrak{A}_{\mathbb{R}}(\mathcal{P}))$, is the lattice $\tilde{\mathcal{F}}(\mathfrak{A})$ of w^*-closed faces of $\mathcal{S}(\mathfrak{A})$ (where $\leq$ is $\subseteq$). Since a w^*-closed face is norm-closed, one can associate a projection p to each such face by the construction in the proof of 3.6.3. Any projection thus associated to a w^*-closed face is called **closed**; if $\mathbb{I}-p$ is closed, then p is said to be **open**. There is a bijective correspondence between open projections p and norm-closed left ideals $\mathfrak{A}^{**}p \cap \mathfrak{A}$ in $\mathfrak{A}$. The lattice $\tilde{\mathcal{F}}(\mathfrak{A})$ is isomorphic to the lattice $\tilde{\mathcal{L}}(\mathfrak{A})$ of all closed projections, but note that the latter is not a sublattice of $\mathcal{L}(\mathfrak{A}^{**})$, since $x \vee y$ is not necessarily closed if x and y are. Hence $x \vee y$ has to be redefined as the smallest closed projection containing x and y. An intrinsic characterization of an open projection is that it is the ultraweak limit of an increasing net in $\mathfrak{A}$. It can be shown that all open and closed projections lie in $\mathfrak{A}_{\mathbb{R}}(\mathcal{P})$. The terminology comes from the special case where $\mathfrak{A}$ is abelian: In that case the open projections are precisely the characteristic functions of open sets in $\mathcal{P}(\mathfrak{A})$ (with the w^*-topology). It is clear from this example that the lattices $\tilde{\mathcal{L}}(\mathfrak{A})$ and $\tilde{\mathcal{F}}(\mathfrak{A})$ do not admit a (natural) orthocomplementation.

More information on this subject, sometimes called **noncommutative topology**, may be found in Akemann [1969], Giles [1970], Giles and Kummer [1971], Borceux and van den Bossche [1989], and Akemann and Pedersen [1992].

Yet another lattice (the fourth) associated with $\mathfrak{A}$ consists of all projections in $\sigma(\mathfrak{A})$, which is the so-called sequential completion of $\mathfrak{A}$. This is defined as the smallest σ-complete C^*-algebra on $\mathcal{H}_{\text{ra}}$ containing $\pi_{\text{ra}}(\mathfrak{A})$; here a C^*-algebra concretely acting on a Hilbert space $\mathcal{H}$ is called σ-complete if it contains the limits of all weakly convergent sequences in it. See Plymen [1968]. This lattice is generally neither atomic nor complete, though it is orthocomplemented. Also cf. Roberts and Roepstorff [1969] for the use of σ-complete C^*-algebras in algebraic quantum mechanics.

Proposition 3.6.5 and Theorem 3.6.6 are from Landsman [1997]. From the point of view of quantum logic, the first claim of Theorem 3.6.7 is that each $A \in \mathfrak{A}$ is an observable on the lattice $\mathcal{L}(\mathcal{P}(\mathfrak{A}))$ if the Borel sets B_i are mutually disjoint; an observable on a σ-complete orthocomplemented lattice $\mathcal{L}$ is defined as a lattice homomorphism $A : \mathcal{B}(\mathbb{R}) \to \mathcal{L}$ satisfying $A(\vee_{i=1}^{\infty} B_i) = \vee_{i=1}^{\infty} A(B_i)$, cf. Varadarajan [1985], §III.2. The study of observables on lattices is closely related to measure theory on lattices; see Varadarajan [1985], Rüttimann [1985], and Schindler [1990]. Yet another way of looking at this situation is that a given observable A defines a map $\rho \mapsto \mu_A^{\rho}$ from (pure) states into probability measures on $\mathbb{R}$ (supported on the spectrum of A).

Theorem 3.6.7 places our approach in the context of Mackey's [1963] axioms for quantum mechanics. See Plymen [1968], Roberts and Roepstorff [1969], Gudder [1979], and Holland [1995] for further development of Mackey's methodology.

I.3.7 The two-sphere property was inspired by Alfsen et al. [1980] and Shultz [1982]. Theorem 3.7.2 is from Landsman [1997]. The exclusion of $\dim(\mathcal{P}_\alpha) = 3$ is a consequence of the use of Theorem 3.7.4, which leads to the desired result for $\dim(\mathcal{P}_\alpha) \geq 4$ only (the case $\dim(\mathcal{P}_\alpha) = 2$ is covered directly by the axiom).

The covering property in atomistic lattices is equivalent to the **exchange property**, stating that if a and b are atoms and x is such that $x \wedge a = o$, then $a \leq x \vee b$ implies $b \leq x \vee a$. This, in turn, is equivalent to **Birkhoff's exchange axiom**; cf. Thm. 7.10 in Maeda and Maeda [1970] or Prop. 10.1 in Kalmbach [1983].

Lemma 3.7.3 is taken from Landsman [1997]. The final step in the proof is as follows. According to Ramsay [1965], a complete orthocomplemented lattice $\mathcal{L}$ with the covering property is a so-called dimension lattice (cf. Kalmbach [1983, 1986] for a detailed discussion). This implies that there is a function $d : \mathcal{L} \to \mathbb{R}^+$ with certain properties; in our case, $\mathcal{L} = \mathcal{L}(\mathcal{P})$, it is easily verified that d is proportional to the dimension defined in 2.7. More precisely, our proof of the finite-dimensional covering property (which was inspired by the proof of Prop. 6.15 in Alfsen et al. [1980]) implies that each interval $[o, Q]$, where $\dim(Q) < \infty$, is a dimension lattice.

By Thm. 13.2 in Kalmbach [1983] (or, equivalently, Prop. 8.2 in Kalmbach [1986]), a complete orthomodular lattice is modular iff it is a dimension lattice in which $d(l) < \infty$. Thus each $[0, Q]$ is modular as long as $\dim(Q) < \infty$, as is $I := \{Q \in \mathcal{L}(\mathcal{P}) | \dim(Q) < \infty\}$. The sublattice I is an ideal of $\mathcal{L}(\mathcal{P})$ (in the sense that $y \leq x$ and $x \in I$ imply $y \in I$), which is supremum-dense (this means that an arbitrary $x \in \mathcal{L}(\mathcal{P})$ may be written as $x = \vee_i x_i$ for some $x_i \in I$). The existence of a supremum-dense modular ideal means, by definition, that $\mathcal{L}(\mathcal{P})$ is locally modular. Thm. 8.17 in Kalmbach [1986] states that a complete orthomodular and locally modular lattice is a dimension lattice. Thm. 8.20 in Kalmbach [1986] says that a dimension lattice has the exchange property (we use only the implication (i)→(ii) of this theorem, since the converse, while true, has an incomplete proof). As remarked in the previous paragraph, in our context the exchange property is equivalent to the covering property. This completes the proof of Lemma 3.7.3.

Theorem 3.7.4 originated in projective geometry; the main contributions were by von Staudt, Hilbert, von Neumann, and Birkhoff (junior). Complete modern proofs may be found in Baer [1952], Freyer and Halperin [1956], and Varadarajan [1985], who also explain the connection between lattice theory and projective geometry (the connection between quantum mechanics and projective geometry clearly fascinated von Neumann; cf. Piron [1976] and Varadarajan [1985] for a full explanation of this connection). Summaries are in Maeda and Maeda [1970], Birkhoff [1967], Beltrametti and Cassinelli [1984], Kalmbach [1986], and Holland [1995]. The fact that length 3 is excluded is caused by the existence of so-called non-Desarguesian projective geometries in dimension 3; see Freyer and Halperin [1958] for a certain analogue of the coordinatization procedure in that case. Various other generalizations exist; for example, when $\mathcal{L}$ is not necessarily atomic, but modular, one can coordinatize $\mathcal{L}$ in terms of a so-called (von Neumann) regular ring (instead of a division ring). If $\mathcal{L}$ has no atoms at all, this leads to the subject of continuous geometry (cf. Maeda [1958] and von Neumann [1981]), created by von Neumann in connection with quantum mechanics and his work on rings of operators.

Lemma 3.7.5 is due to Kolmogorov [1932], and was used in exactly the same way in Zierler [1961] and in Cirelli and Cotta-Ramusino [1973].

The criteria setting out when a definition of convergence defines a topology are given in Kelley [1955]. They are almost trivially verified in Lemma 3.7.6, since our convergence is defined through convergence in $\mathbb{R}$.

Lemma 3.7.7 is taken from Landsman [1997]; the first argument in the proof is Lemma 3.3 in Cirelli and Cotta-Ramusino [1973].

The classification of topological division rings used in the proof of Lemma 3.7.8 is due to Pontrjagin [1946] (also cf. Weiss and Zierler [1958]). The classification of (continuous) involutions of $\mathbb{R}$, $\mathbb{C}$, and $\mathbb{H}$ is discussed in Varadarajan [1985] (§II.2 and Lemma IV.4.5); also cf. Wilbur [1977] for conditions guaranteeing the continuity of the involution.

Proposition 3.7.9 is due to Amemiya and Araki [1966] (it had previously been stated, with an incorrect proof, by Piron); also cf. Maeda and Maeda [1970] (Thm. 34.9), Varadarajan [1985] (Lemma 4.42), or Kalmbach [1986] (Thm. 11.9).

The generalization of Wigner's theorem used at the end of 3.7 is Theorem 4.29 in Varadarajan [1985]. For $\dim(\mathcal{H}) \geq 4$, it is equivalent to the fact that the group of lattice automorphisms of $\mathcal{P}(\mathcal{H})$ (for separable $\mathcal{H}$) is precisely the group of unitary and antiunitary operators on $\mathcal{H}$. More precisely, for any automorphism φ of $\mathcal{L}(\mathcal{H})$ there exists a unitary or antiunitary operator U on $\mathcal{H}$ such that $\varphi(\mathcal{K}) = U\mathcal{K}$, where $\mathcal{K}$ is a closed subspace of $\mathcal{H}$ (if we look at $\mathcal{L}(\mathcal{H})$ as the lattice $\mathcal{L}(\mathfrak{B}(\mathcal{H}))$ of projections p on $\mathcal{H}$, one would have $\varphi(p) = UpU^*$). This is also derived in Varadarajan [1985], §IV.3 (the results are stated for infinite-dimensional $\mathcal{H}$, but this restriction is not essential). The proof is not easy, relying on several steps in the proof of the coordinatization theorem 3.7.4. Wigner's generalized theorem (like its weaker counterpart) holds in any dimension, for one can embed a low-dimensional Hilbert space isometrically in a higher-dimensional one, and choose the bijection so that the embedded space is mapped into itself.

Section 3.7 should be seen in the light of a large body of work in which axioms on an orthocomplemented lattice $\mathcal{L}$ are given so as to make it isomorphic to $\mathcal{L}(\mathcal{H})$. This program goes back to Birkhoff and von Neumann [1936], and received considerable impetus from Mackey [1963]. See Zierler [1961], Wilbur [1977], Piron [1976], Gudder [1979], Beltrametti and Casinelli [1984], Kalmbach [1986], Piziak [1991], and Holland [1995]. The orthomodularity of $\mathcal{L}$ is somewhat justified from Mackey's layout of the logical structure of quantum mechanics. The covering property can to some extent be physically motivated in an operational framework (cf. Gunson [1967], Pool [1968], and Beltrametti and Casinelli [1984]), whereas irreducibility amounts to the absence of superselection rules. Completeness and atomicity seem more a matter of mathematical convenience.

Having arrived at a lattice of the type $\mathcal{L}(V)$, the main difficulty in the traditional approach lies in the determination of the division ring $\mathbb{D}$. An important mathematical breakthrough is the work of Solèr [1995] (reviewed in Holland [1995]), who gave surprisingly minimal conditions on $\mathcal{L}$ implying that V must be a Hilbert space over $\mathbb{R}$, $\mathbb{C}$, or $\mathbb{H}$. Her main condition on $\mathcal{L}$ is equivalent to the existence of an infinite orthogonal sequence in V, and therefore her theorem applies only to infinite-dimensional separable Hilbert spaces (moreover, her conditions are very hard to interpret physically).

Since the fields $\mathbb{R}$ and $\mathbb{H}$ are as irrelevant to quantum mechanics as other more exotic division rings, our approach has been to put in the choice of $\mathbb{C}$ as early as possible. Since it enters through an axiom on the transition probabilities, this has been done in a physically meaningful way (this was inspired by Schwinger [1970], who introduces $\mathbb{C}$ through the properties of filters). As an added bonus, the covering property did not have to be postulated separately, but could be derived.

I.3.8 This material, like that of the next section, is taken from Landsman [1997]. A simplified version appeared in Landsman [1998b].

I.3.9 Theorems 3.9.1 and 3.9.2 should be compared with the work of Alfsen et al. [1980], who characterized unital C^*-algebras in terms of their state spaces (cf. Alfsen [1977] and Asimow and Ellis [1980] for reviews); see Landsman [1997] for such a comparison. Araki [1980] provides a certain simplification of this characterization in the finite-dimensional case. The general program has been continued by Alfsen and Shultz [1998].

An interesting argument leading from JB-algebras to C^*-algebras is that only C^*-algebras admit a satisfactory notion of a tensor product, allowing one to combine physical systems; see Araki [1980] and Hanche-Olsen [1985].

The normal state space of a JBW-algebra or a von Neumann algebra has been characterized by Iochum and Shultz [1983]. The situation is qualitatively different from JB-algebras or C^*-algebras, since a normal state space may have no extreme points.

Chapter II

II.1.1 The quotation in the Introductory Overview is from Simon [1980]; it is sometimes displayed in papers and seminars, and R. Haag has repeatedly expressed similar sentiments.

For the early history of quantization and the classical limit, based on Bohr's correspondence principle, cf. Jammer [1974] and Mehra and Rechenberg [1982].

In condition **F** of sections III.1 and III.5 of von Neumann [1932], it is stated that if a self-adjoint operator R "corresponds" to a function $\mathfrak{R}$ on classical phase space, and $F : \mathbb{R} \to \mathbb{R}$ is an arbitrary function, then $F(R)$ (defined by the functional calculus) should correspond to $F \circ \mathfrak{R}$. Construing the correspondence $\mathfrak{R} \mapsto R$ as a quantization map $R = \mathcal{Q}_\hbar(\mathfrak{R})$, this stringent condition cannot hold in general, not even for $F(t) = t^2$. For this choice, in view of I.(3.10) the condition is equivalent to (1.2) without the limit, explaining our terminology. Dirac's condition (1.3), also without the limit, is proposed in §21 of his [1930] book; Dirac did recognize that his condition could not always be satisfied, and added the qualifying remark that the condition should be satisfied only by "the simpler" commutators.

The idea of deformation quantization, which appreciates the fact that the conditions of von Neumann and Dirac can hold only asymptotically, goes back to Berezin [1974, 1975a,b] (also cf. Vey [1975] and Bayen et al. [1978]). The mathematical framework was developed in a series of papers starting with Gerstenhaber [1964]. In the original setting one constructs a "deformed" associative product $\cdot_\hbar$ on a given Poisson algebra, in such a way that $f \cdot_\hbar g \to f \cdot g$ for $\hbar \to 0$ and $i(f \cdot_\hbar g - g \cdot_\hbar f)/\hbar$ converges to $\{f, g\}$ in the same limit. Here $f \cdot_\hbar g$ is defined by a formal power series expansion, and the $\hbar \to 0$ limit is handled accordingly.

This subject of "formal" deformation quantization reached a high point in the work of Fedosov [1994, 1996], who showed that every regular Poisson manifold P (or rather its associated Poisson algebra $C^\infty(P, \mathbb{R})$) is quantizable in the given sense (regularity here means that the rank of $B^\sharp$ is constant; cf. I.2.3). Also cf. Weinstein [1994]. The culmination of the subject is Kontsevich [1998], who proved that every finite-dimensional Poisson manifold can be quantized in the sense of formal deformation quantization.

In its current development, formal deformation quantization is remote from quantum mechanics and even from Hilbert space theory, using essentially different techniques from the ones described in this book. Moreover, no version of von Neumann's condition is imposed.

Strict deformation quantization was introduced by Rieffel [1989a, 1994], who in particular proposed what we call Rieffel's condition. In his approach the norm and the product in the C^*-algebras $\mathfrak{A}^\hbar$ depend on $\hbar$; in particular, the product in $\mathfrak{A}^\hbar$ is analogous to the product $\cdot_\hbar$ in formal deformation quantization. Further work in this setting, especially on Rieffel's condition, may be found in Nagy [1992, 1997, 1998a] and Blanchard [1996]. These papers contain applications to the theory of quantum groups (in the C^*-algebraic setting introduced by Woronowicz [1987, 1995]; also cf. Lance [1995]), as do Rieffel [1993b, 1995], Nagy [1993, 1998b], and Sheu [1996, 1997]. Related work may be found in Landstad [1994] and Landstad and Raeburn [1997].

Definition 1.1.1 is taken from Landsman [1993b]. The reformulation of strict (deformation) quantization in terms of the maps $\mathcal{Q}_\hbar$ simply adopts the perspective of a physicist, who looks at quantization in precisely this way. Mathematically, this reformulation is closely related to the concept of E-theory and its associated asymptotic morphisms (see Connes [1994]). The connection between E-theory, quantization, and operator K-theory is further developed by Nagy [1996, 1997] and Rosenberg [1996]; for the last two topics also see Rieffel [1993c].

We shall not discuss **geometric quantization** or prequantization in this book, referring the interested reader to Souriau [1969, 1997], Kostant [1970], Śniatycki [1980], Kirillov [1990], Woodhouse [1992], and Chernoff [1995]. The construction of the prequantization line bundle provided by this technique is often useful in the context of Berezin–Toeplitz quantization on Kähler manifolds (cf. the notes to 1.5). On the other hand, the prequantization of functions on phase space does not easily fit into a C^*-algebraic framework, because the prequantization of a bounded function is always an unbounded operator, a property that may persist even after the second step of quantization. Moreover, one works at a fixed value of $\hbar$.

II.1.2 See Fell [1962] or Dixmier [1977] for the traditional theory of continuous fields of C^*-algebras. Definition 1.2.1 is taken from Kirchberg and Wassermann [1995]; Lemma 1.2.2 and Proposition 1.2.3 show that their definition is equivalent to Dixmier's.

Blanchard [1996] defines a continuous field of C^*-algebras over a locally compact Hausdorff space X as a C^*-algebra $\mathfrak{C}$ equipped with a nondegenerate morphism from $C_0(X)$ to the center of the multiplier algebra of $\mathcal{G}$ (cf. IV.2.1), such that 1.2.1.1 holds with $\mathfrak{A}^x := \mathfrak{C}/C_0(X)^x\mathfrak{C}$ (where $C_0(X)^x$ is the ideal in $C_0(X)$ of functions vanishing at x, and φ_x the canonical projection. Condition 1.2.1.2 is then automatically satisfied, so that one obtains a continuous field in the sense of Definition 1.2.1. Conversely, given 1.2.1, one has $\ker(\varphi_x) = C_0(X)^x\mathfrak{C}$, so that the canonical isomorphism $\varphi_x(\mathfrak{C}) \simeq \mathfrak{C}/\ker(\varphi_x)$ leads to the equivalence between the two definitions in question.

The connection between strict (deformation) quantization and continuous fields of C^*-algebras was recognized by Rieffel [1989a]; it was initially thought that any such quantization would define a continuous field, but it was quickly realized (Rieffel [1993a]) that further assumptions were needed. Definition 1.2.5, which is a slight variation on a definition proposed in Rieffel [1998], seems a good compromise between Rieffel's earlier definitions and those in Landsman [1993b]. Results analogous to Theorem 1.2.4 are given in Nagy [1992, 1998a].

Somewhat against the spirit of the founding fathers, one could omit Dirac's condition from Definition 1.2.5. In some cases the ensuing continuous fields of C^*-algebras (in which $\mathfrak{A}^0$ is commutative) may nonetheless be seen as quantizations. For examples see Matsumoto [1991a,b], Matsumoto and Tomiyama [1992], Borthwick et al. [1993], and Exel [1994]. A unified approach to these examples is developed in Abadie and Exel [1997], where Dirac's condition reappears through the back door.

Partitions of unity are discussed in Pedersen [1989] and Jänich [1994]. Since these will often be used, we recall their definition. Let Q be a Hausdorff space, and let $\{\mathcal{N}_\alpha\}_{\alpha\in I}$ be a locally finite open cover of Q (i.e., each point of Q has a neighborhood that intersects only a finite number of the sets $\mathcal{N}_\alpha$). A **partition of unity** subordinate to the given cover is a collection of positive functions $\{u_\alpha\}_{\alpha\in I}$ such that $u_\alpha \in C_c(\mathcal{N}_\alpha, \mathbb{R})$ and $\sum_{\alpha\in I} u_\alpha = 1$. A partition of unity always exists when Q is paracompact; Hörmander [1983] proves that the u_α may be taken to be smooth when Q is a manifold.

II.1.3 Definition 1.3.1, anticipated by Emch [1984], Rieffel [1989b], and Landsman [1993a], is due to Nagy [1992, 1998a] and Blanchard [1996]. The last two authors look at $\mathfrak{C}$ as a $C_0(X)$ module, so that each set $\{\omega_x^\lambda\}_{x\in X}$ defines a $C_0(X)$-linear function $\varphi_\omega^\lambda : \mathfrak{C} \to C_0(X)$ by $\varphi_\omega^\lambda(A) : x \to \omega_x^\lambda(A_x)$.

Suppose one has a triple $(\mathfrak{C}, \{\mathfrak{A}^x, \varphi_x\}_{x\in X})$ as in Definition 1.2.1, that satisfies conditions 2 and 3. One may then still use Definition 1.3.1. Under the assumption that each $\mathfrak{A}^x$ is separable and nonzero, Blanchard [1996] proves that such a triple satisfies 1.2.1.1 (so that it is a continuous field of C^*-algebras) iff it admits a continuous field of states. The proof uses Kasparov's [1981] generalization of Stinespring's Theorem 1.4.2 (also see Lance [1995]). Analogous results are given in Nagy [1992, 1998a]. Previous applications of continuous fields of states by Rieffel [1989b] and Landsman [1993a] in proving Rieffel's condition (1.1) in certain models may be seen as embryonic versions of these results; Proposition 1.3.6 is a case in point.

Definition 1.3.3 is an abstraction of the notion of a coherent state, rewritten in the language of transition probabilities. For the standard theory of coherent states and their various generalizations, see Klauder and Skagerstam [1985], Perelomov [1986], Zhang et al. [1990], and Ali et al. [1995]. The usual definition stipulates as a minimal requirement that coherent states form a family $\{\Omega_\sigma | \sigma \in S\}$ for which the map $\sigma \mapsto \Omega_\sigma$ is (strongly) continuous, and $\int_S d\mu(\sigma)[\Omega_\sigma] = \mathbb{I}$ weakly, for some measure μ on S; various requirements may be added. For example, parametric dependence on $\hbar$ and good behavior for $\hbar \to 0$ were already studied in a special example by Schrödinger [1926]. Further work on the role of coherent states in the classical limit of quantum mechanics is cited in the notes to 2.7; also cf. Simon [1980], Yaffe [1982], as well as the first two books cited above. From our point of view, equation (1.11) is of central conceptual importance, for it shows that the transition probabilities on $q_\hbar(S)$ that are inherited from $\mathbb{P}\mathcal{H}_\hbar$ become classical when $\hbar \to 0$.

The main ideas of what is here called Berezin quantization go back to Davies and Lewis [1970], Holevo [1973] (cf. Davies [1976] and Holevo [1982] for a textbook presentation of the approach in these papers), and Berezin [1972, 1974, 1975a,b] (also cf. Perelomov [1986] for a summary of these four papers). Whereas the other authors concentrated on operational ideas and measurement theory, it was the specific contribution of Berezin to study operators of the type $\mathcal{Q}_\hbar^B(f)$ in connection with quantization theory and the classical limit, in particular analyzing their $\hbar$-dependence. In doing so he discovered, for example, the "quantization" of Planck's constant when one quantizes a compact phase space. In view of this, and of Berezin's premature death in a drowning accident (cf. Bogolyubov et al. [1981] and Dobrushin et al. [1996]), it seems reasonable to name the quantization method involving $\mathcal{Q}_\hbar^B$ after him.

Equation (1.16) appears in Ali and Doebner [1990] under the name **prime quantization** (with weaker conditions on the coherent states). Berezin [1972] calls f the **contravariant symbol** of $\mathcal{Q}_\hbar^B(f)$; for an arbitrary bounded operator on $\mathcal{H}$, the **covariant symbol** of an operator A is the function on S defined by $\sigma \mapsto (q_\hbar(\sigma))(A)$ (the terminology **lower** and **upper** symbol, respectively, is also found in the literature, e.g., Simon [1980]). Berezin actually looks for operators whose covariant symbol is well-behaved for $\hbar \to 0$, and regards such an operator as the quantization of the $\hbar \to 0$ limit of its covariant symbol. The **Berezin transform**

$$Bf(\rho) = \int_S d\mu_\hbar(\sigma)\, p(q_\hbar(\rho), q_\hbar(\sigma)) f(\sigma),$$

which is well-defined as map from $L^\infty(S)$ to itself, maps the contravariant symbol into the covariant one. Our condition (1.9) evidently states that the Berezin transform becomes the identity for $\hbar \to 0$. For a study of the Berezin transform on so-called **bounded symmetric domains** (these are certain bounded subspaces of $\mathbb{C}^N$; cf. Helgason [1978]) see Berezin [1975a], Peetre [1990], Unterberger and Upmeier [1994], and Engliš [1996].

A different approach to quantization theory based on coherent states is due to Klauder [1988, 1995]. For the connection between geometric quantization and coherent states see Rawnsley [1978], Tuynman [1987b], Odzijewicz [1988, 1992], and Rawnsley et al. [1990].

The coherent states constructed in these papers should satisfy Definition 1.3.3; on those Kähler manifolds that are not coadjoint orbits the limit in (1.10) is strictly necessary, and (1.13) does not hold. In fact, starting from Berezin–Toeplitz quantization on Kähler manifolds (see the notes to 1.5), the easiest way to find $\mu_\hbar$ is to use (1.18).

Bochner integrals, such as (1.16), will frequently occur in this chapter. The theory of such integrals may be found in Yosida [1980]. A function $f : S \to \mathcal{B}$ taking values in a Banach space $\mathcal{B}$ is Bochner-integrable with respect to a measure μ on S iff (i) f is weakly measurable (that is, for each functional $\omega \in \mathcal{B}^*$ the function $\sigma \to \omega(f(\sigma))$ is measurable), (ii) there is a null set $S_0 \subset S$ such that $\{f(\sigma)|\sigma \in S\backslash S_0\}$ is separable, and (iii) the function defined by $\sigma \mapsto \|f(\sigma)\|$ is integrable. It will always be directly clear from this whether a given operator- or vector-valued integral may be read as a Bochner integral; if not, it is understood as a weak integral, in a sense always obvious from the context. The Bochner integral $\int_S d\mu(\sigma) f(\sigma)$ may be manipulated as if it were an ordinary (Lebesgue) integral. For example, one has

$$\| \int_S d\mu(\sigma)\, f(\sigma)\| \leq \int_S d\mu(\sigma)\, \|f(\sigma)\|.$$

II.1.4 Theorem 1.4.2 is due to Stinespring [1955]; also cf. Paulsen [1986] (where the nonunital version may be found) and Kadison [1994]. Stinespring also proved 1.4.4.

Proposition 1.4.6 is equivalent to Theorem V.1.1 in Berezanskii [1968]. Positive-operator-valued measures are discussed abstractly by Riesz and Sz.-Nagy [1990], Appendix, and, in the context of quantum mechanics, by Davies [1976], Holevo [1982], Busch et al. [1995], and Schroeck [1996]. The measure theory in the proof of Proposition 1.4.8 is discussed in Pedersen [1989], §4.5. Corollary 1.4.9, due to Neumark, is actually valid for any space X with a σ-algebra; see Schroeck [1996], §II.11.F.

II.1.5 For coherent states see the notes to 1.3. The theory of Hilbert spaces with a reproducing kernel may be found in Aronszajn [1950] or Meschkowski [1962]; also see Ali [1985] for a summary. Schroeck [1996] contains a generalization to matrix-valued reproducing kernels. Definition 1.5.6 and the ensuing theory are due to Schwartz [1964], who develops a far-reaching generalization of the theory of reproducing kernels. Overviews of the connection between coherent states, reproducing kernels, and POV-measures are given by Davies [1976] and Ali and Doebner [1990].

If S is a complex manifold, (usually taken to be homogeneous and Kähler in this type of application), and $\tilde{\mathcal{H}}_\hbar$ consists of (anti-) holomorphic functions, operators of the type (1.44) are known as (generalized) **Toeplitz operators**. (Strictly speaking, the term "Toeplitz operator" refers to the case where $S = S^1$ and p projects onto the Hardy subspace of functions with positive Fourier coefficients only.) The reproducing kernel in $\tilde{\mathcal{H}}_\hbar$ is then known as a **Bergman kernel**, and p is sometimes called the **Szegö projection**; we refer to Meschkowski [1962] and Helgason [1978] for the first steps in the theory of this kernel.

See Boutet de Monvel and Guillemin [1981], Guillemin [1984], and Upmeier [1996] for the theory of generalized Toeplitz operators (the latter book is particularly relevant, since it describes the C^*-algebras generated by these operators in great detail). More generally, the projection p in (1.44) may project onto the space of (anti)holomorphic sections of a holomorphic line bundle over S. In either case, the quantization procedure defined by (1.44) is known as **Berezin–Toeplitz quantization.** In the context of homogeneous Kähler manifolds this quantization was introduced by Berezin [1974, 1975a]; also cf. Guillemin [1984], Berger and Coburn [1986], and Tuynman [1987a,b] for early work. More recent work on Berezin–Toeplitz quantization is cited in the notes to 2.4.

II.2.1 For exhaustive information on the Heisenberg group, and nilpotent Lie groups in general, see Corwin and Greenleaf [1989] or Leptin and Ludwig [1994]. Much useful information on $\tilde{H}_n$ and its irreducible representations is also contained in Folland [1989].

The representation theory of Lie algebras by unbounded operators on infinite-dimensional Hilbert spaces, which is relevant here as well as in the remainder of this chapter, may be found in Warner [1972] or Barut and Raçka [1977]; some relevant notions are also reviewed in III.1.5. For $G = \tilde{H}_n$ and $U = U_c^S$ (or U_c) on $\mathcal{H} = L^2(\mathbb{R}^n)$, it is not difficult to see that the space of smooth vectors is $\mathcal{S}(\mathbb{R}^n)$; see Howe [1980] or Corwin and Greenleaf [1989].

II.2.2 Proposition 2.2.1, as well as the construction of $\rho^\hbar$ in 2.2.2, go back to van Hove [1943]. In this context the **Groenewold–van Hove theorem** should be mentioned: This states, roughly speaking, that there exists no decent map $d\rho^\hbar$ from $C^\infty(T^*\mathbb{R}^n, \mathbb{R})$ to some Lie algebra of unbounded operators on a Hilbert space for which (2.38) can be extended beyond $\mathsf{P}^{\leq 2}$ and the restriction to $\mathsf{P}^{\leq 1}$ gives an irreducible representation of $\mathfrak{h}_n$. See Groenewold [1946], van Hove [1943], Guillemin and Sternberg [1984b], Abraham and Marsden [1985], and Gotay et al. [1996].

Equation (2.32) is a special case of the momentum map; see III.1.1.

For $\mathfrak{g} = \mathfrak{sp}(n, \mathbb{R})$ and $R = d\rho^\hbar$ on $\mathcal{H} = L^2(\mathbb{R}^n)$ a dense set of analytic vectors is given by the linear span of the Hermite polynomials. By the integrability conditions proved in the references cited in the notes to 2.1, there exists a unitary representation $\rho^\hbar$ of $\tilde{S}p(n, \mathbb{R})$ whose derivative in the sense explained above is indeed $d\rho^\hbar$. The explicit form of $\rho^\hbar$ and the metaplectic group $Mp(n, \mathbb{R})$ are discussed in, e.g., Segal [1959], Bargmann [1961], Shale [1962], Voros [1977], Guillemin and Sternberg [1984b], Littlejohn [1986], Folland [1989], and Kirillov [1990]. A different approach to the construction of $\rho^\hbar(Mp(n, \mathbb{R}))$ is based on the fact that $Sp(n, \mathbb{R})$ is contained in the automorphism group of $\tilde{H}_n$. Therefore, for each $M \in Sp(n, \mathbb{R})$ the map $h \mapsto U_{1/\hbar}(Mh)$ defines an irreducible representation of $\tilde{H}_n$, which in view of 2.1.4 is equivalent to $U_{1/\hbar}$. The unitary implementer is $\rho^\hbar(M)$; cf. (2.40). In any case, it turns out that $\rho^\hbar$ and $\rho^{\hbar'}$ are equivalent iff $\hbar$ and $\hbar'$ have the same sign.

II.2.3 The coherent states of 2.3.1 were discovered by Schrödinger [1926]; also see the notes to 1.3. In this case the Berezin transform becomes simply

$$Bf = e^{\frac{1}{2}\hbar\Delta_{2n}} f;$$

cf. (2.117), where the prefactor of $\hbar\Delta_{2n}$ is different.

Proposition 2.3.2 is due to Bargmann [1961]. He absorbs $\exp(-z\overline{z}/\hbar)$ into the measure on $\mathbb{C}^n$, so that the elements of $\tilde{\mathcal{H}}_\hbar$ are entire functions. Like many other authors, he actually uses holomorphic rather than antiholomorphic functions, but the latter choice is more natural in the context of coherent states (cf. Klauder and Skagerstam [1985]). The Hilbert space of entire analytic functions in $L^2(\mathbb{C}^n, \exp(-z\overline{z})dzd\overline{z}/(2\pi i))$ is often called the **Bargmann–Fock space**, since the use of this space in the context of the canonical commutation relations goes back to Fock [1928] (who defined the inner product directly in terms of the Taylor coefficients of a function, rather than through a Gaussian measure on $\mathbb{C}^n$). See Folland [1989] for a thorough discussion.

Exponential Hilbert spaces, called bosonic Fock spaces by physicists, go back to Segal [1956]; also see Guichardet [1972], who proves all claims we make on exponential vectors. The realization of the "canonical" coherent states in such spaces is due to Klauder [1970]. Also cf. Klauder and Skagerstam [1985]. A very detailed analysis of creation and annihilation operators is in Bratteli and Robinson [1981]. These operators derive their name from their action on the subspaces $\otimes_s^n \mathcal{K}$ of $\exp(\mathcal{K})$, which we do not need; it turns out that $a(z)$ maps $\otimes_s^n \mathcal{K}$ to $\otimes_s^{n-1}\mathcal{K}$ (for $n = 0$ one has $a(z)\Omega = 0$), whereas $a(z)^*$ maps it to $\otimes_s^{n+1}\mathcal{K}$.

The Riemannian geometry of $q_\hbar(\mathcal{K})$ as a submanifold of $\mathbb{P}\exp(\mathcal{K})$ is studied in Field [1996]; the most interesting result is that $q_\hbar(\mathcal{K})$ has zero intrinsic curvature with respect to the induced Fubini–Study metric (the extrinsic curvature is nonzero).

II.2.4 Theorem 2.4.1 is due to Coburn [1992] and Borthwick et al. [1993]; both acknowledge Klimek and Lesniewski [1992a] for the organization of the proof. Using less crude estimates they show that (2.76) is even $O(\hbar)$. Our proof of (1.1) is different from these references; the proof of nondegeneracy is taken from Berger and Coburn [1986]. By the nondegeneracy of $\mathcal{Q}_\hbar^B$ and the open mapping theorem, (2.73) implies that $\|f\|_\infty \leq C\|\mathcal{Q}_\hbar^B(f)\|$ for some $C > 0$, and $f \in C_0(S)$. See Berger and Coburn [1994] for a study of the constant C, and for deeper inequalities.

One can study $\mathcal{Q}_\hbar^B$ for function spaces larger than $C_0(\mathbb{C}^n)$, so that one leaves the compact operators. This is not particularly useful for physics (it introduces spurious superselection sectors), but leads to fascinating mathematical structures; see Guillemin [1984], Berger and Coburn [1986], Coburn and Xia [1995], and Upmeier [1996].

Theorem 2.4.1 has an analogue for the Berezin–Toeplitz quantization of bounded symmetric (Cartan) domains: See Borthwick et al. [1993], and Borthwick et al. [1995]. Riemann surfaces have been treated in Klimek and Lesniewski [1992a,b, 1994, 1996]. The analysis aspects of the proofs are similar to the one for $\mathbb{C}^n$, but one has to add detailed information about the structure of such domains (which are related to Jordan algebras; cf. Upmeier [1987, 1996]). Also see Cahen et al. [1994, 1995] for an approach through formal (rather than strict) deformation quantization. Bordemann et al. [1994] and Sheu [1996] apply Berezin–Toeplitz quantization to certain compact Kähler manifolds. Here $\hbar$ can assume only quantized values (cf. (1.12)), as we will confirm in certain special cases in III.1.11. See Cahen et al. [1993] for the same problem in formal deformation quantization. A strict Berezin quantization of the upper half-plane is given by Rădulescu [1998]. In all cases discussed so far, one does not obtain a strict deformation quantization from Berezin–Toeplitz quantization.

Theorem 2.4.3 is due to Berezin [1974, 1975a].

II.2.5 Weyl quantization is due to Weyl [1931], whose definition was (2.111). There is a huge mathematical literature on this subject in the context of the theory of pseudodifferential operators; principal sources are Grossmann et al. [1968], Voros [1977, 1978], Hörmander [1979, 1985a], Howe [1980], Robert [1987], Folland [1989], and Rieffel [1993a].

Identifying $T^*\mathbb{R}^n$ with $\mathbb{R}^{2n}$, one initially defines an isomorphism $\mathcal{Q}_\hbar^W : \mathcal{S}(\mathbb{R}^{2n}) \to \mathcal{S}(\mathbb{R}^{2n})$ by (2.107). By duality, one then immediately has $\mathcal{Q}_\hbar^W : \mathcal{S}'(\mathbb{R}^{2n}) \to \mathcal{S}'(\mathbb{R}^{2n})$. The Schwartz kernel theorem (cf. Reed and Simon [1975] or Hörmander [1983]) identifies $\mathcal{S}'(\mathbb{R}^{2n})$ with the space of continuous maps from $\mathcal{S}(\mathbb{R}^n)$ to $\mathcal{S}'(\mathbb{R}^n)$, so one eventually has a continuous map $\mathcal{Q}_\hbar^W(f) : \mathcal{S}(\mathbb{R}^n) \to \mathcal{S}'(\mathbb{R}^n)$ for each $f \in \mathcal{S}'(\mathbb{R}^{2n})$.

To get back into the realm of (possibly unbounded) operators on $L^2(\mathbb{R}^n)$, as well as other instances of good behavior (relevant to the theory of partial differential equations), one has to impose certain restrictions on f. One firstly assumes that $f \in C^\infty(\mathbb{R}^{2n})$, and secondly imposes conditions on the behavior of f and its derivatives at infinity. If these are satisfied, the expression (2.110), already meaningful as the Fourier transform of a distribution, makes direct sense as a so-called **oscillatory integral** (Hörmander [1983]), and defines the kernel (2.109) as an element of $\mathcal{S}'(\mathbb{R}^{2n})$. The corresponding operator $\mathcal{Q}_\hbar^W(f)$ then maps $\mathcal{S}(\mathbb{R}^n)$ into itself (rather than into $\mathcal{S}'(\mathbb{R}^n)$, as for general f). For example, $\mathcal{Q}_\hbar^W(f)$ thus defined lies in $\mathfrak{B}_0(L^2(\mathbb{R}^n))$ if $f \in C_0^\infty(T^*\mathbb{R}^n)$ (see Voros [1977]); as mentioned in the main text,

$\mathcal{Q}_\hbar^W : C_0^\infty(T^*\mathbb{R}^n) \to \mathfrak{B}_0(L^2(\mathbb{R}^n))$ is not continuous. Oscillatory integrals have routinely and correctly been computed by physicists since Dirac [1930].

In Rieffel [1993a] a similar procedure is used to define $\mathcal{Q}_\hbar^W(f)$ for functions f, all of whose derivatives (including the zeroth) are bounded. Such functions form a dense subspace of $C_b(T^*\mathbb{R}^n)$, and the image of this space under $\mathcal{Q}_\hbar^W$ is contained in the C^*-algebra of pseudodifferential operators of order 0 (see, e.g., Cordes [1987, 1995]). Rieffel [1993a, 1994] shows that this defines a strict deformation quantization.

The more traditional calculus of pseudodifferential operators, which goes back to Kohn and Nirenberg [1965], uses the "quantization"

$$\mathcal{Q}_\hbar^{\mathrm{KN}}(f)\Psi(x) := \int_{T^*\mathbb{R}^n} \frac{d^n p d^n y}{(2\pi\hbar)^n}\, e^{ip(x-y)/\hbar} f(p,x)\Psi(y),$$

rather than (2.107). This is not useful for quantum mechanics, because $\mathcal{Q}_\hbar^{\mathrm{KN}}$ does not preserve self-adjointness. It is, however, local in x, so that it may effortlessly be extended to manifolds (unlike $\mathcal{Q}_\hbar^W$); see Taylor [1984].

Early mathematical studies of Weyl quantization are Segal [1963] and Pool [1966]. Among other things, these authors showed that the map $f \mapsto \mathcal{Q}_\hbar^W(f)$ is unitary from $L^2(T^*\mathbb{R}^n)$ to $\mathfrak{B}_2(L^2(\mathbb{R}^n))$; this follows from a straightforward calculation. More recent work is, for example, Daubechies [1980, 1983] and Gracia-Bondía and Várilly [1988].

The deformed product defined by the Weyl quantization (which first occurred in von Neumann [1931]) may formally be written as (Groenewold [1946])

$$(f \cdot_\hbar g) = f \exp\left[\frac{\hbar}{2i}\left(\frac{\overleftarrow{\partial}}{\partial p}\frac{\overrightarrow{\partial}}{\partial q} - \frac{\overleftarrow{\partial}}{\partial q}\frac{\overrightarrow{\partial}}{\partial p}\right)\right] g.$$

This is (historically inaccurately) sometimes called the **Moyal product**, after Moyal [1949]. Though an attractive formal expansion, the Moyal product plays no role in our setting, but it can be given a precise meaning in various ways; see Voros [1977, 1978], Bayen et al. [1978], Folland [1989], Gracia-Bondía and Várilly [1988], and Estrada et al. [1989].

The expression (2.95), and the ensuing connection between Weyl quantization, the parity operator, and the Dirac delta function, are due to Grossmann [1976]. Equation (2.102) was first written down by Royer [1977]. The idea of inverting expressions of the type (2.95) by (2.97) goes back to Stratonovic [1957]. See Gadella [1995] for a review of Weyl quantization and Wigner functions from this perspective. Theorem 2.5.1 goes back to van Hove [1951]; also cf. Voros [1977], Hörmander [1979, 1985a], Folland [1989], Graffi and Parmeggiani [1990], and Borsari and Graffi [1994]. Our proof stresses the role of the parity operator.

Wigner functions were introduced in Wigner [1932]. Moyal [1949] was the first to recognize the connection between Weyl's quantization and Wigner's function. One may evidently define the Wigner function of a mixed state through (2.101) as well; in fact, in physics these functions are often used in quantum statistical mechanics. See Hillery et al. [1984] for a survey of Wigner functions in nonrelativistic physics (cf. de Groot et al. [1980] and Cariñena et al. [1990] for the relativistic case), and Folland [1989] for mathematical aspects. The physical interpretation of the Wigner function is that it is to some extent a probability distribution on phase space. For the right-hand side of (2.101) is of the form $\int_{T^*\mathbb{R}^n} \rho f$, which looks like the expectation value of f in a mixed state ρ in classical mechanics. However, in classical mechanics ρ is a probability measure, which the Wigner function fails to define because of its potential nonpositivity. This failure is a consequence of the uncertainty relations of quantum mechanics, which forbid sharp localization in phase space; see, e.g., Schroeck [1996].

Positivity of the Wigner function is actually quite rare: Hudson [1974] shows that for pure states $W_\hbar\lfloor\psi\rfloor$ is positive iff Ψ is a complex Gaussian; also cf. Folland [1989], and see Bröcker and Werner [1995] for the case of mixed states.

II.2.6 Theorem 2.6.1 is due to Rieffel [1993a, 1994], who actually proved it for the larger function space $\tilde{\mathfrak{A}}_{\mathbb{R}} = C_b^\infty(T^*\mathbb{R}^n, \mathbb{R})$ of smooth functions that together with all derivatives are bounded (cf. the notes to the previous section). Our proof is different from Rieffel's, and derives from Landsman [1993b]. A third approach to results of this type may be found in Elliott et al. [1996]. Equation (2.117) was found by Berezin [1974].

A sharper version of Lemma 2.6.2 was originally proved by Caldéron and Vaillancourt [1971] in the setting of the Kohn–Nirenberg calculus; a simple proof may be found in Hwang [1987]. See the references at the beginning of this section for the Weyl version. Both versions play a central role in the theory of pseudodifferential operators.

The smearing (2.116) goes back to Husimi [1940]; his motivation was that it leads to a positive phase space distribution function (which in our setting is the analogue of the Wigner function for Berezin quantization) $W_\hbar^{\mathrm{B}}\lfloor\psi\rfloor : (p,q) \to \hbar^{-n} p(\psi_\hbar^{(p,q)}, \psi)$, which is sometimes called the **Husimi function**; cf. Lee [1995] for a recent survey of its applications in physics. It is easily shown (cf. Prop. 1.99 in Folland [1989]) that replacing $\psi_\hbar^0$ in (2.116) by an arbitrary pure state ψ, where $\Psi \in L^2(\mathbb{R}^n)$, defines a collection of positive maps as well. These more general maps may not always correspond to a strict deformation quantization, though.

Proposition 2.6.3 is due to Helffer et al. [1987].

Convolution algebras of may be defined for any locally compact group; see III.1.7.

The unitary transformation W is the Plancherel transform for $\tilde{H}_n$, and the measure $d\hbar|\hbar|^n/(2\pi)^{2n}$ is the Plancherel measure on the unitary dual of $\tilde{H}_n$ (up to a set of Plancherel measure zero, namely the collection of one-dimensional representations of $\tilde{H}_n$). See Dixmier [1977] for this transform for locally compact unimodular groups in general, and Corwin and Greenleaf [1989] for the details for nilpotent Lie groups.

Proposition 2.6.4 has its roots in Dixmier [1960]; for a modern approach see, e.g., Packer and Raeburn [1992], Thm. 1.2 and Example 1.4.(2). Lemma 2.6.6 and Corollary 2.6.7 are stated without proof in Elliott et al. [1993].

II.2.7 Proposition 2.7.1 is the easiest version of a number of results in the literature that may be seen as adaptions of **Egorov's theorem** in the theory of pseudodifferential operators (cf. Taylor [1984] or Hörmander [1985a]) to the setting of quantization theory. Our approach follows Rieffel [1996], who considered the special case $\mathcal{Q}_\hbar = \mathcal{Q}_\hbar^W$, and proved the stronger version in which h and f are allowed to be in $C_b^\infty(T^*\mathbb{R}^n, \mathbb{R})$. The strongest result is Thm. IV-9 in Robert [1987] (also cf. Prop. 1.5 in Helffer et al. [1987]), who proves (2.131) for $\mathcal{Q}_\hbar = \mathcal{Q}_\hbar^W$ and $f \in C_b^\infty(T^*\mathbb{R}^n, \mathbb{R})$ under the following assumption on $H_\hbar$: There is a classical Hamiltonian $h \in C^\infty(T^*\mathbb{R}^n, \mathbb{R})$ satisfying

$$|\partial_p^\alpha \partial_q^\beta h(p,q)| \leq C_{\alpha\beta}(1 + |x| + |p|)^{\max\{2-|\alpha|-|\beta|,0\}}$$

for each multi-index (α, β) (the notation is explained below (2.114)), and $\| H_\hbar - \mathcal{Q}_\hbar^W(h)\| = O(\hbar)$. This includes all Hamiltonians considered by Rieffel [1996], as well as certain unbounded ones; cf. Theorem 2.5.1.

In all these cases, the convergence in (2.131) may actually be shown to be uniform in t; the same comment applies to (2.135) and (2.162). The completeness of the classical flow of h follows from Prop. 2.1.21 in Abraham and Marsden [1985].

Under similar assumptions, one may show that $\mathcal{Q}_\hbar(\alpha_t^0(f)) \sim \alpha_t^\hbar(\mathcal{Q}_\hbar(f)) + O(\hbar)$ as an asymptotic expansion in the sense of pseudodifferential operators; see Wang [1986], Robert [1987, 1998], and Paul and Uribe [1995].

The theory of Schrödinger operators may be found in Reed and Simon [1975, 1978], Cycon et al. [1987], and Hislop and Sigal [1996]. For a positive potential, self-adjointness in an external magnetic field is guaranteed by a theorem of Leinfelder and Simader [1981] (also cf. Cycon et al. [1987]), which states that (2.134) is essentially self-adjoint on $C_c^\infty(\mathbb{R}^n)$ when $V \geq 0$, $V \in L^2_{\text{loc}}(\mathbb{R}^n)$, $A_i \in L^4_{\text{loc}}(\mathbb{R}^n)$ for each i, and $\partial_i A_i \in L^2_{\text{loc}}(\mathbb{R}^n)$. Also cf. Combes et al. [1978] and Hogreve [1983].

Theorem 2.7.2 generalizes a result of Hepp [1974], who proved the special case $A = 0$, $Q_\hbar = \mathcal{Q}_\hbar^W$, and $f(p,q) = \exp i(uq - pv)$. For the inclusion of abelian as well as nonabelian (Yang–Mills) gauge fields see III.2.11 and its notes.

Hepp assumes only $V \in C^{2+\delta}(\mathbb{R}^n)$ for some $\delta > 0$; the proof then uses the Hölder continuity of $V^{(2)}$ rather than the second-order Taylor series of V. The meaning of the various steps in Hepp's proof needed some clarification. Much useful information on small fluctuations, coherent states, and their connection with the metaplectic representation may be found in Littlejohn [1986]. The theory of linearizing Hamiltonian equations of motion is in Marsden et al. [1991]; also cf. Marsden and Ratiu [1994] for a quick review. For arbitrary symplectic spaces S, the linearization in terms of h_1 proceeds exactly as in the case $S = T^*\mathbb{R}^n$. However, linearization along a classical trajectory with h_2 is more complicated in the general case, as one needs a so-called symplectic connection to identify the tangent spaces $T_{\sigma(t)}S$ with $T_\sigma S$, as well as to define the second derivative h''.

Proposition 2.7.3 is of interest in its own right, as it shows that the quantum fluctuations around the classical path are controlled by the quadratic term in the expansion of $H_\hbar$ around this path; see Hepp [1974] and Littlejohn [1986].

Further work in the direction of Theorem 2.7.2 is Yajima [1979], Hagedorn [1980, 1981, 1985], Robert [1987, 1998], Robinson [1988a, 1988b, 1993], Wang [1991], Combescure [1992], and Arai [1995]. Heuristic work includes Kurchan et al. [1989], Barnes et al. [1994], and Nauenberg et al. [1994], who review experiments on the semiclassical evolution of coherent states).

Hepp's approach differs fundamentally from the **time-dependent WKB method** (cf. the Introductory Overview), for which we refer to Truman [1976, 1977], DeWitt-Morette et al. [1979], Schulman [1981] (these two references explain what happens to the WKB approximation beyond caustics; also cf. DeWitt-Morette et al. [1983]), Maslov and Fedoriuk [1981], Saksena et al. [1991], Maslov [1994], and Robert [1998]. As in the time-independent case, rigorous work on the WKB approximation has been submerged into the theory of Fourier integral operators (see Hörmander [1985b]); the connection between the two is discussed in Guillemin and Sternberg [1977], Voros [1977, 1978], Robert [1987, 1992, 1998], and Paul and Uribe [1995]. Rather then follow any of these authors, a C^*-algebraist would rather introduce $\hbar$ into microlocal analysis through the theory of pseudodifferential operators on a tangent groupoid; cf. Nistor et al. [1997] and Monthubert and Pierrot [1997].

A sample of other approaches to the semiclassical behavior of nonstationary states is Marsden [1974], Albeverio and Høegh-Krohn [1977], Voros [1977, 1978], Berry and Balazs [1979], Jona-Lasinio et al. [1981], Blanchard and Sirugue [1985], Sławianowski [1991], Saksena et al. [1991], Omnès [1994, 1997a], Osborn and Molzahn [1995], Paul and Uribe [1995], Werner [1995], and Rezende [1996]. References to literature on the connection between the classical limit of quantum mechanics and the small-time limit of diffusion processes may be found in the notes to 3.7.

The semiclassical approximation of eigenvalues and eigenstates of Schrödinger operators is a topic with a different flavor, and a huge literature. The traditional theory of the **time-independent WKB method** is contained in Maslov and Fedoriuk [1981], and is summarized in Arnold [1989]. Geometric aspects of Maslov's version of the time-independent WKB method are discussed in Guillemin and Sternberg [1977], Arnold and Givental [1990], Woodhouse [1992], Littlejohn [1992], and Bates and Weinstein [1995]. An interesting heuristic overview may be found in Gutzwiller [1990]. Analytical aspects, closely related to microlocal analysis (cf. Hörmander [1985a,b]), are reviewed in Robert [1987, 1992, 1998] and Helffer [1988]. Different analytical techniques, closer to the usual theory of Schrödinger operators, are described in Hislop and Sigal [1996].

A given (normalized) energy eigenfunction $\Psi^{E_\hbar}$ (with eigenvalue $E_\hbar$) defines a pure state $\omega_\hbar^{E_\hbar}$ on $\mathfrak{A}^\hbar = \mathfrak{B}_0(L^2(\mathbb{R}^n))$ by I.(1.46). In the context of Definition 1.3.1, in which the continuous field of C^*-algebras is given by Theorem 2.6.5, one may try to choose the sequence $\{E_\hbar\}_\hbar$ in such a way that $\omega_\hbar^{E_\hbar}$ converges to a state ω_0^E on $\mathfrak{A}^0 = C_0(T^*\mathbb{R}^n)$, that is, to a probability measure μ_0^E on $T^*\mathbb{R}^n$. A necessary condition for this to happen is that the sequence $\{E_\hbar\}$ converges to some E; there is a formidable literature on the behavior of such sequences of eigenvalues, reviewed in Robert [1992, 1998]. It follows from Lemma 1.3.2, Theorem 2.6.5, and (2.52) that ω_0^E, if it exists, is the weak (vague) limit of the sequence $\{\mu^{E_\hbar}\}_\hbar$ of measures on $T^*\mathbb{R}^n$, defined by $d\mu^{E_\hbar}(p,q) := (2\pi\hbar)^{-n} p(\psi_\hbar^{(p,q)}, \psi^{E_\hbar}) d^n p d^n q$.

For example, if $H_\hbar = \mathcal{Q}_\hbar^W(h)$ and the flow of the classical Hamiltonian h on $h^{-1}(E) \subset T^*\mathbb{R}^n$ is ergodic, then under suitable assumptions almost all sequences of the above type converge to the Liouville measure on $h^{-1}(E)$; see Helffer et al. [1987]. WKB states $\Psi_\hbar \sim \exp(iS/\hbar)$, on the other hand, converge to measures supported on the so-called Lagrangian submanifold defined by S; cf. Colin de Verdière and Parisse [1994] and Werner [1995]. Further work in this direction may be found in Knauf [1989], Duclos and Hogreve [1993], and Paul and Uribe [1996]. It is quite remarkable that pure quantum states may well converge to mixed classical states.

II.3.1 The necessary background in affine and Riemannian geometry may be found in, for example, Helgason [1978], Klingenberg [1982], Gallot et al. [1990], Choquet-Bruhat et al. [1982], or Lang [1995].

For the group $\mathcal{G}_Q$ and its action on T^*Q cf. Guillemin and Sternberg [1984b] or Isham [1983]; also cf. the notes to 3.6. For the topology of Diff(Q), and some of its modifications, see Ebin and Marsden [1970], Marsden [1974], Ismagilov [1996], and Omori [1997].

II.3.2 In the Riemannian case, geodesics satisfying (3.4) are automatically **affinely parametrized**; this means that t is an affine function of the length of the geodesic. The equation satisfied by arbitrarily parametrized geodesics contains additional terms.

A proof of Proposition 3.2.2 may be found in most books on the subject; e.g., Helgason [1978] (Thm. I.6.2), Gallot et al. [1990] (Thm. 2.92), or Klingenberg [1982] (Thm. 1.9.7). Theorem 3.2.3 is named after Hopf and Rinow; a proof may be found in Klingenberg [1982], Thm. 2.1.3, or Gallot et al. [1990], Thm. 2.103 and Cor. 2.105. Theorem 3.2.5 is in Klingenberg [1982] (Thm. 2.1.14) and Gallot et al. [1990] (Prop. 2.113 and Scholium 3.78).

II.3.3 An extensive treatment of Riemannian geodesic and cogeodesic motion is in Klingenberg [1982]; also cf. Abraham and Marsden [1985]. Theorem 3.3.4 is closely related to Lemma 3.1.17 in Klingenberg [1982], and is suggested by comments in Marsden et al. [1991]. Linearization in the second sense explained in 2.7 (i.e., using the quadratic Hamiltonian (2.143)) is possible but awkward in the general Riemannian case, since the symplectic connection needed to identify the tangent spaces along a geodesic will usually not coincide with the Levi-Civita connection. This leads to cumbersome expressions.

One may equip TQ with a metric g such that the Hamiltonian flow on TQ is geodesic with respect to g. Using the decomposition (3.3), the metric g is simply defined by declaring the two copies of T_qQ to be orthogonal, and putting $\mathsf{g} = \mathbf{g}$ on each copy. It follows from the construction of g that the projection $\tau : TQ \to Q$ is an isometric submersion in the sense that τ^* is an isometry between the orthogonal complement of its kernel and its image. This implies that the horizontal lift of a curve is a geodesic in TQ iff the curve is a geodesic. It is then immediate from this property, combined with the definition of horizontal curves in TQ and the description of the flow in TQ in 3.3.3, that the Hamiltonian flow on TQ is geodesic with respect to g.

II.3.4 This section is mostly adapted from Landsman [1993b].

For the normal bundle see, e.g., Klingenberg [1982], 1.3.11–1.3.14 (which includes a detailed discussion of the diagonal map δ), and many other books on differential geometry. Theorem 3.4.2 is the **tubular neighborhood theorem**; see Lang [1995]. The proof uses a partition of unity argument; recall that our definition of a manifold includes paracompactness. Our choice (3.48) corresponds to the choice of $\{X\dot{+}-X\}$ as a complement to $T_{(q,q)}\delta(Q)$ in $T_{(q,q)}(Q \times Q)$. The more commonly used map $\nu'_\delta(X_q) = (\exp_q(X_q), q)$ corresponds to the complement $T_qQ \oplus 0 = \{X\dot{+}0\}$.

The geometric meaning of ν_δ^{-1} explained after Lemma 3.4.3 may be used to give an elegant direct proof of the tubular neighborhood theorem.

The abbreviation C^∞_{PW} in (3.49) stands for Paley–Wiener, and $\tilde{\mathfrak{A}}_0$ consists of the C^∞ functions f on T^*Q satisfying the following conditions: (i) the support of f, projected to Q, is compact; (ii) for each fixed $q \in Q$ the function $f(\cdot, q)$ on T^*_qQ has an extension to $T^*_qQ_{\mathbb{C}} \simeq \mathbb{C}^n$ as an entire analytic function f_q such that for every N there are constants C_N and H (independent of N) such that $|f_q(z)| \leq C_N(1+|z|)^{-N}\exp(H\,\mathrm{Im}\, z)$. These are simply the conditions of the Paley–Wiener theorem, which characterizes functions whose Fourier transform is in $C_c^\infty(\mathbb{R}^n)$; see Reed and Simon [1975] or Hörmander [1985a].

When Q is compact, one can show that the Wigner function (3.53) is continuous. The proof is analogous to the flat-space case (cf. the comment following (2.102)); see Landsman [1993b]. In the noncompact case one shows that $W_\hbar \in C_0(T^*Q, \mathbb{R})$ if the following condition is satisfied by the metric $\mathbf{g}$ on Q: The constant $\sup_{q_1,q_2}[d\mu(\rho(q_1;q_2))/d\mu(q_2)]$ should be finite. Here the supremum is taken over all pairs of points that can be connected by a unique geodesic, and $\rho(q_1;q_2)$ is the geodesic reflection of q_1 in q_2 (that is, $\rho(q_1;q_2) = \gamma(1)$, where γ is the affinely parametrized geodesic for which $\gamma(0) = q_1$ and $\gamma(1/2) = q_2$).

There are alternative attempts to generalize Weyl quantization to Riemannian manifolds. The proposal by Underhill [1978] (who only assumes the existence of an affine connection on TQ) corresponds to a Wigner function where, compared with (3.53), the factor J is absent. Liu and Qian [1992], on the other hand, have our factor J as well as an additional factor $\tilde{J}(\gamma(q, v, \frac{1}{2}), v; -\hbar)^{1/2}$. As mentioned in their paper, this quantization is not always self-adjoint, which can be traced back to this extra factor. Different from all these is the proposal of Emmrich [1993a], who did not compute the Wigner function (but cf. the notes to 3.7). For yet another Wigner function in curved space see Habib and Kandrup [1989].

A different approach is due to Upmeier [1991], who generalized the definition (2.91) of flat-space Weyl quantization (rather than (2.107)); also cf. Unterberger and Unterberger [1988]. This requires that the Riemannian structure on Q admits the analogue of the parity operator P, that is, Q must be a symmetric space (see Helgason [1978]). One may also attempt to generalize (2.99), (2.100); this turns out to be possible if the phase space is a homogeneous symplectic manifold, see Várilly et al. [1990], Figueroa et al. [1990], and Gracia-Bondía [1992]. Another generalization of Weyl quantization requiring a group structure may be found in Manchon [1993].

If one is interested merely in quantizing functions on T^*Q that are independent of, or linear in p, one does not even need an affine connection; the manifold structure of Q suffices. See Abraham and Marsden [1985]. The analogue of the (non-self-adjoint) Kohn-Nirenberg "quantization" (cf. the notes to 2.5) on a manifold with connection is developed by Bokobza-Haggiag [1969], Widom [1980], and Pflaum [1995].

II.3.5 The proof is from Landsman [1993b], with certain improvements added. Much useful computational information on Jacobi fields and their associated determinants is in Azencott et al. [1981] and in Molzahn et al. [1990]. The Jacobians (3.59) and (3.61) may alternatively be expressed through the derivative of the exponential map, which itself may be written in terms of Jacobi fields.

A different way to construct a positive quantization on a Riemannian manifold is given in Colin de Verdière [1985], who uses a so-called Friedrichs "quantization" (see Taylor [1984]). A general construction of coherent states on Riemannian manifolds may be found in Paul and Uribe [1995, 1996].

II.3.6 Proposition 3.6.2 goes back (at least) to Goldin [1971] and Goldin et al. [1980], who see quantization theory on Q as the problem of finding general unitary representations of $\mathcal{G}_Q$. A closely related approach is developed in Doebner and Tolar [1975] and Angermann et al. [1983]. See Isham [1983] and Ali and Goldin [1991] for reviews; also cf. Albertin [1991], Chernoff [1995], and Ismagilov [1996].

Theorem 3.6.3 is due to Landsman [1993b]. When φ is not an isometry, the classical and the quantum action of $\mathcal{G}_Q$ are related only in the limit $\hbar \to 0$; see the above reference, and Lemma 3.7.6 for a special case.

Proposition 3.6.4 is stated without proof in Abraham and Marsden [1985], which contains the lemma used in the proof as Lemma 2.6.13; also cf. Theorem VIII.10 in Reed and Simon [1972].

II.3.7 Quantum theory on Riemannian manifolds was discussed almost immediately after the birth of modern quantum mechanics; cf. Dowker [1974] for early references.

The coefficient $\frac{1}{3}$ in (3.93) is sensitive to the precise quantization scheme that is used; the scheme in Liu and Qian [1992] produces $\frac{1}{8}$, Underhill [1978] finds $\frac{1}{12}$, whereas the value $\frac{1}{6}$ has been found from geometric quantization (cf. Śniatycki [1980], Woodhouse [1992], and Wu [1998]). Emmrich [1993a] obtains the value zero. Our $\frac{1}{3}$ seems somewhat preferred by physicists: It is equivalent to having an extra term $-\frac{1}{6}\hbar^2\mathbf{R}$ in the classical Lagrangian, which is "naturally" induced by the measure in the path integral on curved space (see Dowker [1974]). The need to have the Ricci scalar in the quantum Hamiltonian was apparently first recognized by Pauli in 1950 (see Pauli [1973], pp. 161–174).

The proof of Theorem 3.7.3 is mainly based on Strichartz [1983]; also cf. Davies [1989]. Alternative proofs are in Chernoff [1973], Cheeger et al. [1982], and Cordes [1987]. These references, as well as Rosenberg [1997], should also be consulted for additional results in the analysis of the Laplace–Beltrami operator. Aspects of the theory of unbounded operators used in our proof may be found in Reed and Simon [1975]. The existence of Ψ_1 in case that $\overline{\Delta}$ is not self-adjoint follows from the Corollary to their Thm. X.1 on p. 137. For elliptic regularity see §IX.6 of this reference. Completeness is sufficient but not necessary for essential self-adjointness on $C_c^\infty(Q)$: For examples where $(Q, \mathbf{g})$ is incomplete but Δ is nonetheless essentially self-adjoint on $C_c^\infty(Q)$, see Horowitz and Marolf [1995]. On the other hand, it is easy to give examples where $(Q, \mathbf{g})$ is incomplete and Δ fails to be essentially self-adjoint on $C_c^\infty(Q)$. The simplest one is $Q = (0, 1)$ with flat metric (cf. Reed and Simon [1975], p. 178). More generally, boundary value problems always involve Laplacians whose different self-adjoint extensions describe the possible boundary conditions one imposes on the solutions. See, e.g., Berezanskii [1968] or Cordes [1987] for a modern treatment.

For the **Kato–Rellich theorem** see Reed and Simon [1975]. It states that $A+B$ (where A and B are densely defined linear operators on a Hilbert space) is self-adjoint on $D(A)$ (and essentially self-adjoint on any core of A) if $D(A) \subseteq D(B)$ and $\|B\Psi\| \leq a\|A\Psi\| + b\|\Psi\|$ for all $\Psi \in D(A)$ and some $a < 1$ (called the **relative bound**) and arbitrary b. In our application $A = \Delta$, and $B = \mathbf{R}$ is bounded, so its domain as a multiplication operator is all of $L^2(Q)$. Hence $a = 0$ and $b = \|\mathbf{R}\|_\infty = \sup_q |\mathbf{R}(q)|$.

Theorem 3.7.5 is analogous to a result of Hogreve [1983], which does not involve Weyl quantization and is a direct generalization of Hepp's version of Theorem 2.7.2 (cf. the notes to 2.7). His proof is somewhat different from ours; a third proof, using the time-dependent WKB approximation to the kernel of the propagator, is in Landsman [1993b]. A major advantage of the present proof along the lines of Hepp [1974] is that there are no difficulties with caustics (see below). As in the entire chapter, the metric is assumed to be smooth, though for the proof to go through this could be relaxed to be C^5. It would be interesting to generalize 3.7.5 to arbitrary, and particularly to incomplete Riemannian manifolds. See Paul and Uribe [1995] for a microlocal approach.

Elworthy and Truman [1981] give a certain analogue of Proposition 2.7.3 that for sufficiently small t is valid for arbitrary Riemannian manifolds, but holds for initial wave functions of the WKB type $\Psi = \rho \exp(iS/\hbar)$ rather than for coherent states; also cf. DeWitt-Morette et al. [1979], Schulman [1981], and Elworthy et al. [1985]. In general, the time-dependent WKB approximation in curved space that is used in these papers suffers from similar problems with caustics as in the flat case. In contrast to the flat case, the geodesic WKB-like approximation to the kernel of the propagator now has problems with caustics as well because of the cut locus. See Molzahn et al. [1990, 1992] for the approximation up to the cut locus, and DeWitt-Morette et al. [1979] and Azencott et al. [1981] for the situation beyond it.

The limit $\hbar \to 0$ in quantum mechanics is similar to the limit $t \to 0$ of a diffusion process; cf. Nagasawa [1993] for the general connection between the Schrödinger equation and diffusion theory. The relation between small-time diffusion and the classical limit of quantum mechanics on Riemannian manifolds is analyzed in DeWitt-Morette et al. [1979], Elworthy and Truman [1981], and Azencott and Doss [1985].

We have not studied the connection between eigenfunctions of the Laplace–Beltrami operator and geodesic flow; the enormous literature on this topic is reviewed in Robert [1992, 1998]. In fact, the ergodicity result of Helffer et al. [1987] quoted in II.2.7 was directly inspired by an analogous theorem concerning ergodic geodesic flow on a Riemannian manifold, whose final proof is due to Colin de Verdière [1985]. Further references may be found in the notes to III.2.11.

Chapter III

III.1.1 The Lie–Poisson structure, coadjoint orbits, and the momentum map (in a special case) are all in Lie [1890], but were rediscovered in the sixties in the work of Kirillov, Souriau, Kostant, Smale, and others. The momentum map, whose original definition was (1.8), plays a central role in modern symplectic geometry. For textbook accounts, which contain most results in this section, as well as in 1.2 and 1.4, see Guillemin and Sternberg [1984b], Abraham and Marsden [1985], Libermann and Marle [1987], Marsden and Ratiu [1994] (which contains historical notes, as well as a generalization of the momentum map to Poisson manifolds), and Souriau [1969, 1997]. These accounts all start from actions of G rather than $\mathfrak{g}$, but much of the theory depends only on the $\mathfrak{g}$-action, and is valid even when it is not integrable to a G-action.

General references for the cohomology of Lie groups and Lie algebras are Guichardet [1980] and de Azcárraga and Izquierdo [1995] (which contains many interesting applications to physics, particularly to the theory of anomalies). In the present context also cf. Guillemin and Sternberg [1984b] and Libermann and Marle [1987], as well as Varadarajan [1985] for central extensions of Lie algebras. The vanishing of $H^1(\mathfrak{g}, \mathbb{R})$ and $H^2(\mathfrak{g}, \mathbb{R})$ for semisimple Lie algebras is known as the Whitehead lemma(s).

III.1.2 Symplectic group actions are "canonical" in the terminology of classical mechanics, and are therefore a "classical" subject. See the notes to the preceding section for references. Theorem 1.2.1 is due to Palais [1957], which contains the full proof. A textbook account is in Hector and Hirsch [1986], §3.1.3. This result is particularly interesting in comparison with the well-known theorems on (essential) self-adjointness of unbounded symmetric operators on Hilbert spaces (for which see Reed and Simon [1972, 1975]).

The construction starting with (1.21) was introduced by Souriau [1969, 1997]. A good account is also in de Azcárraga and Izquierdo [1995].

Corollary 1.2.9 is due to Gotay and Tuynman [1991], whose proof is entirely different. Our proof relies on a theorem of van Est [1953] (also see Guichardet [1980]) to the effect that for a compact Lie group the cohomology with coefficients in any (linear) representation space is trivial.

The examples in 1.2.11 are "classical", but the modern formulation is due largely to Souriau [1969, 1997].

III.1.3 The analytic theory of multipliers for Lie groups and Lie algebras may be found in the book by Varadarajan [1985]. Our Theorem 1.3.3 is Thm. 7.21 of that book, and our Corollary 1.3.7 is a special case of Thm. 7.37. The last part of our Proposition 1.3.4 is Varadarajan's Cor. 7.30. Also cf. Guichardet [1980] and de Azcárraga and Izquierdo [1995].

The geometric approach to multipliers and central extensions may be found in the enlightening article of Tuynman and Wiegerinck [1987], who introduced the cohomology group $H^2(G, U(1))$ as defined in our main text (the analogous and better-known cohomology group defined on the basis of smooth cocycles is only relevant to topologically trivial central extensions). In particular, $H^2(G, U(1))$ is not necessarily isomorphic to $H^2(\mathfrak{g}, \mathbb{R})$, since not all $\Gamma \in Z^2(\mathfrak{g}, \mathbb{R})$ can be integrated to an element $c \in Z^2(G, U(1))$.

Proposition 1.3.6 is from Neeb [1996b], who reformulated a condition due to Tuynman and Wiegerinck [1987]. There are two alternative formulations of the condition stated in the second item. Firstly, regard Γ as a 2-form on G by left translation, and identify X with the corresponding right-invariant vector field ξ_X^R on G. A necessary and sufficient condition for Γ to define a $U(1)$-extension of G with Lie algebra $\mathfrak{g}_\Gamma$ is that for each $X \in \mathfrak{g}$ the 1-form $i_X\Gamma$ be exact. Secondly, consider the group $\tilde{G}_c$ appearing in the proof of 1.3.6. The adjoint action of $\mathbb{R} \subset \tilde{G}_c$ on the Lie algebra $\mathfrak{g}_c = \mathfrak{g}_\Gamma$ is trivial (since $\mathbb{R}$ is central in $\tilde{G}$), so that the adjoint action of $\tilde{G}_c$ quotients to a well-defined action of $\tilde{G} = \tilde{G}_c/\mathbb{R}$ on $\mathfrak{g}_c$. Since $G = \tilde{G}/\pi_1(G)$, we obtain an action of $\pi_1(G)$ on $\mathfrak{g}_c$ by restriction of the $\tilde{G}$-action. The second equivalent criterion is that this $\pi_1(G)$-action be trivial.

From the proof of 1.2.9 we therefore infer that $H^2(G, U(1)) = 0$ for compact simply connected G, and $H^2(\mathfrak{g}, \mathbb{R}) = 0$ for a Lie algebra $\mathfrak{g}$ whose associated simply connected Lie group is compact. Since this restriction implies that $\mathfrak{g}$ is semisimple (see, e.g., Thm. 3.8.2 in Barut and Raçka [1977]), the latter statement alternatively follows from the more general property stated after the proof of 1.1.12.

III.1.4 For the initial part of this section cf. Abraham and Marsden [1985], Libermann and Marle [1987], and Marsden and Ratiu [1994].

Theorem 1.4.4 is due to Kirillov [1976]. Corollary 1.4.8 has an alternative version, in which transitive symplectic G-spaces correspond to orbits in the space of 2-cocycles $Z^2(\mathfrak{g}, \mathbb{R})$; this has the advantage that only a single G-action on $Z^2(\mathfrak{g}, \mathbb{R})$ needs to be considered, and the disadvantage that $Z^2(\mathfrak{g}, \mathbb{R})$ is less intuitive than $\mathfrak{g}^*$. See Guillemin and Sternberg [1984b].

Martinez Alonso [1979] proved that given G, there exists a single central extension $\overline{G}$ (which was first constructed by Cariñena and Santander [1975] in the context of projective representations) such that any transitive symplectic G-space is a coadjoint orbit of $\overline{G}$ (or a covering space thereof).

III.1.5 Projective unitary group representations have a long history; the traditional theory may be found in Varadarajan [1985]. Our Proposition 1.5.1 is essentially his Thm. 7.5.

A modern presentation is given by Rieffel [1979], whose work, among other things, removes a number of separability assumptions in Varadarajan [1985]. In particular, the approach based on (1.64) is due to Rieffel.

For smooth vectors for U see Warner [1972] and Barut and Raçka [1977]. A seemingly alternative treatment of $dU(\mathfrak{g})$ is in terms of the **Gårding subspace** $\mathcal{H}_G \subset \mathcal{H}$, which consists of all vectors of the type $\int_G dx f(x)U(x)\Psi$, where f and Ψ run through $C_c^\infty(G)$ and $\mathcal{H}$, respectively. The density of $\mathcal{H}_U^\infty$ in $\mathcal{H}$ is then most easily proved by showing that $\mathcal{H}_G \subseteq \mathcal{H}_U^\infty$, and subsequently that $\mathcal{H}_G$ is dense in $\mathcal{H}$. However, it is shown in Dixmier and Malliavin [1978] that for connected G one actually has the equality $\mathcal{H}_U^\infty = \mathcal{H}_G$.

Sufficient conditions for the integrability of representations of Lie algebras by unbounded operators are reviewed in Barut and Raçka [1977].

III.1.6 The seminorms $p_{a_1\cdots a_n}\Psi := \|dU(T_{a_1})\cdots dU(T_{a_n})\Psi\|$ define a topology on $\mathcal{H}_U^\infty$, relative to which each $dU(X)$ is a continuous map on $\mathcal{H}_U^\infty$, and U (seen as a map from $G \times \mathcal{H}_U^\infty$ to $\mathcal{H}_U^\infty$) is separately continuous; see Corwin and Greenleaf [1989], Appendix.

The analysis of the momentum map on $\mathcal{H}_U^\infty$ is done in Michor [1990], who uses a particular notion of smoothness in infinite-dimensional manifolds due to Fröhlicher and Kriegl [1988]. He shows that the G-action on $\mathcal{H}_U^\infty$ (and hence on $\mathbb{P}\mathcal{H}_U^\infty$) is smooth, and that the momentum map (1.71) is smooth as well. The manifold $\mathbb{P}\mathcal{H}_U^\infty$ is weakly symplectic.

A great deal is known about the image of the momentum map for infinite-dimensional group representations; see Wildberger [1992], Arnal and Ludwig [1992], and Neeb [1995, 1996a]. Also cf. the notes to 1.10.

For the usual enveloping algebra see, e.g., Warner [1972] or Barut and Raçka [1977]. The author is indebted to A. Kent for the proof of Proposition 1.6.4.

Definition 1.6.5 is taken from Landsman [1993c]. Theorem 1.6.7 goes back to Berezin [1967]; also cf. Gutt [1983].

III.1.7 The theory in this section is usually discussed in the general setting of locally compact groups. For a discussion of unimodularity see, e.g., Gaal [1973]. A sufficient condition for a locally compact group G to be unimodular is that the identity e have a compact neighborhood that is invariant under inner automorphisms. It follows that all compact and all locally compact abelian groups are unimodular. Also, G is unimodular if it coincides with its commutator subgroup; this applies to all semisimple Lie groups.

A detailed discussion of group C^*-algebras may be found in Dixmier [1977] and in Pedersen [1979]. Twisted group C^*-algebras were introduced in Auslander and Moore [1966], and further studied in, e.g., Kleppner and Lipsman [1973], Green [1978], and Packer and Raeburn [1992]; also consult the review by Rosenberg [1994]. Theorem 1.7.3 is a special case of Thm. 3.3 in Busby and Smith [1970]. Twisted group C^*-algebras belong to the class of **twisted covariance algebras**; references on the structure of such algebras are Green [1978] and Packer and Raeburn [1989, 1990]. For (untwisted) covariance algebras, also called crossed products, see the notes to 3.4.

Amenability is discussed, e.g., in Pedersen [1979] and Paterson [1988]. The original definition (due to von Neumann) is different from the one above, and amounts to the existence of an invariant mean on the C^*-algebra $C_b(G)$. The fact that the amenability of G implies $C^*_r(G, c) = C^*(G, c)$ follows from Thm. 3.11 in Packer and Raeburn [1989]; see Packer [1994] for more general results in this direction.

III.1.8 For the Peter–Weyl theorem and the Plancherel transform in the unimodular case see Dixmier [1977], The nonunimodular case is treated in Kleppner and Lipsman [1972, 1973], which also contains various results equivalent to Theorem 1.8.1 and Corollary 1.8.3. For the theory of induced representations see the notes to 2.9.

Lemma 1.8.2 is a rather trivial case of the "Mackey machine", in which one constructs representations of a group extension by inducing from representations of the normal subgroup defining the extension; see Mackey [1958], Green [1978], Rieffel [1979], and the review by Rosenberg [1994].

Proposition 1.8.4 is well known, and contained (usually in practically unrecognizable generalizations) in all references for twisted group C^*-algebras given above. The equality $C^*(\mathbb{R}^{2n}, c) = C^*_r(\mathbb{R}^{2n}, c)$ alternatively follows from the fact that $\mathbb{R}^{2n}$ is amenable; see the notes to the preceding section.

Equation (1.121) has the following generalization to the case where c is degenerate. Define Γ by (1.34), and decompose $\mathbb{R}^{2n} = V_0 \oplus V_1$, such that Γ vanishes on V_0 and is nondegenerate on V_1 (hence V_0 and V_1 are even-dimensional). Then

$$C^*(\mathbb{R}^{2n}, c) \simeq C^*_r(\mathbb{R}^{2n}, c) \simeq C_0(\hat{V}_0) \otimes \mathfrak{B}_0\left(L^2\left(\mathbb{R}^{\frac{1}{2}\dim(V_1)}\right)\right).$$

Here we have written $\hat{V}_0$ in order to indicate that one has taken the Fourier transform in the variables in V_0. This degenerate case is investigated in Kastler and Mebkhout [1990].

III.1.9 The map $\mathcal{Q}^R_\hbar$ in 1.9.1 is due to Rieffel [1990a], who was the first to recognize that the group C^*-algebra $C^*(G)$ should be thought of as the quantization of the Lie–Poisson algebra $C^\infty(\mathfrak{g}^*_-)$. He also proved a version of Theorem 1.9.2 for exponential groups. Theorems 1.9.2 and 1.9.5 are due to Landsman [1998d]. For Lemma 1.9.3 see Helgason [1978], Ch. II.3, and Milnor [1976], §5. The general theory of (Riemannian and other) connections on Lie groups may be found in Kobayashi and Nomizu [1963, 1969].

III.1.10 The Cartan–Weyl theory is discussed in a large number of textbooks; good modern presentations are, e.g., Wallach [1973], Bröcker and tom Dieck [1985], and Knapp [1986]. The reformulation of the Cartan–Weyl theory in terms of coadjoint orbits is due to Kostant [1970]. An explicit construction of U_γ from $\mathcal{O}_\gamma$ is done through the Borel–Weil theory (see Wallach [1973], Knapp [1986], or Vogan [1987]), which in this application coincides with the approach through geometric quantization; see the notes to II.1.1 for references, and in the present context in particular cf. Hurt [1983].

Equation (1.147) is closely related to Prop. 4.12 in Knapp [1986].

In the context of 1.10.7 one may ask which properties single out the coadjoint orbit $\mathcal{O}_\gamma$ through $J(\psi_\gamma)$ among all coadjoint orbits contained in $J(\mathbb{P}\mathcal{H}_\gamma)$. This question was answered by Kostant [1973] and Atiyah [1982] in terms of convexity properties. A brief summary of the situation is as follows. One has a natural projection $\tau := \tau_{\mathfrak{g}^* \to \mathfrak{t}^*}$ given by restricting a $\theta \in \mathfrak{g}^*$ to $\mathfrak{t}$; the object $\tau \circ J$ is then evidently the momentum map for the T-action on $\mathbb{P}\mathcal{H}_\gamma$ given by restricting U_γ to T. A general theorem, due to Atiyah [1982] and Guillemin and Sternberg [1982, 1984a], states that the image of the momentum map of a torus action on a compact connected symplectic manifold is a convex polytope (see Kirwan [1984] for a generalization and Audin [1991] or Guillemin [1994] for reviews). Applying this to the situation at hand, it turns out that $\mathcal{O}_\gamma$ is singled out by the property that it contains the extreme points of $\tau \circ J(\mathbb{P}\mathcal{H}_\gamma)$; these extreme points are precisely the imgages of the highest weight state ψ_γ and its transforms under the Weyl group.

An overview of the role of coadjoint orbits in the representation theory of noncompact Lie groups is given in Guichardet [1985], Vogan [1987, 1992], and Kirillov [1990]; see the notes to II.2.1 for analogous references relevant to the nilpotent case.

III.1.11 Coherent states of the type studied in this section were introduced by Klauder [1963], and were rediscovered by Perelomov [1972], who added the perspective of the Cartan-Weyl theory. See Perelomov [1986] and Klauder and Skagerstam [1985] for more references, reprints of the original papers, and a general overview. A wealth of rigorous information is contained in Simon [1980]. The connection with the Borel-Weil theory is explained in Onofri [1975].

The fundamental idea of rescaling the label of an irreducible representation by multiplying with $1/\hbar$ (which accordingly has to be quantized in the compact case) is due to Berezin [1975a,b]. He in addition investigated certain noncompact Lie groups; also cf. Perelomov [1986]. In a more intuitive setting, this rescaling was explicit in the early years of quantum mechanics, and seems to comprise one of the faces of Bohr's correspondence principle; see Mehra and Rechenberg [1982].

One would expect that $\mathcal{Q}^B_\hbar(\tilde{X})$ equals $i\hbar dU_{\gamma/\hbar}(X)$, but after an arduous calculation (due to Simon [1980]) one actually obtains

$$\mathcal{Q}^B_{1/k}(\tilde{X}) = \frac{i}{k + c(\gamma)} dU_{k\gamma}(X),$$

where $c(\gamma) := 2(\gamma, \delta)/(\gamma, \gamma)$; recall that δ was defined after (1.164). While this result is expected to be true in general, Simon [1980] acknowledges that his proof is limited to the case that γ is a multiple of a fundamental weight.

Lemma 1.11.2 is from Gilmore [1979] (whose proof, as remarked in Simon [1980], is unnecessarily complicated). A detailed proof of (1.155) is in Duffield [1990], Prop. 4. Related results are in Berezin [1972], Lieb [1973], Simon [1980], and Hogreve et al. [1983]. For the Weyl dimension formula see, e.g., Wallach [1973], Bröcker and tom Dieck [1985], or Knapp [1986].

Theorem 1.11.4 is due to Landsman [1998c]. An entirely different proof of Dirac's condition, valid for arbitrary compact Kähler manifolds, is given by Bordemann et al. [1994]. Another relevant paper, which stresses the Kähler geometry behind Berezin quantization on coadjoint orbits of compact Lie groups, is Barmoshe and Marinov [1994].

The steepest descent method used in the proof of 1.11.4 is in Hörmander [1983].

III.2.1 A standard introductory reference for bundles and connections is Kobayashi and Nomizu [1963, 1969]; for a full meal see Greub et al. [1972, 1973]. Choquet-Bruhat et al. [1982] and de Azcárraga and Izquierdo [1995] are presentations directed at physicists.

Much of the theory of bundles (with the evident exception of the theory of connections) applies to general topological spaces and groups.

In the construction of principal bundles one may start with Q, H, a cover of Q by suitable open sets $\mathcal{N}_\alpha$, and a collection of transition functions satisfying the relation $h_{\alpha\beta}(q)h_{\beta\gamma}(q) = h_{\alpha\gamma}(q)$ whenever $q \in \mathcal{N}_\alpha \cap \mathcal{N}_\beta \cap \mathcal{N}_\gamma$. The bundle $\mathsf{P}(Q, H, \tau)$ can then be reconstructed from these data.

III.2.2 The insight that the physicists' gauge fields are the mathematicians' connections on a bundle goes back to Hermann [1975], Wu and Yang [1975], and Konopleva and Popov [1983] (relevant parts of which apparently date back to the sixties). The entire theory of elementary particle interactions is currently based on gauge fields; see Weinberg [1995, 1996].

III.2.3 The history of cotangent bundle reduction is described in Marsden [1993]; applications of this construction are surveyed in Marsden [1992].

A more common way of obtaining the reduced space $(\widetilde{T^*\mathsf{P}})^{\mathcal{O}}$ is via Marsden–Weinstein reduction; see IV.1.5.

Lemma 2.3.1 is "classical"; see, e.g., Cor. 4.2.11 in Abraham and Marsden [1985]. Connections are discussed from the cotangent bundle point of view in Guillemin and Sternberg [1984b].

Theorem 2.3.7, straightforward as it is in its final formulation, is the culmination of a development involving the work of Smale [1970] (who did the abelian case), Sternberg [1977], and Weinstein [1978]. The Poisson bracket (2.55) was first computed by Montgomery et al. [1984]. The symplectic form on $\mathsf{P}^0_V \times_H \mathcal{O}$ is discussed in Guillemin and Sternberg [1984b].

Writing $\mathcal{O} = H/H_\mu$, it is possible to embed $(\widetilde{T^*\mathsf{P}})^{\mathcal{O}}$ as a symplectic submanifold of $T^*(\mathsf{P}/H_\mu)$, equipped with a modified symplectic structure; see Abraham and Marsden [1985], Thm. 4.3.3, Kummer [1981], and Marsden [1992].

III.2.4 The exact sequences (2.60) and (2.64), along with the pertinent interpretation of connections, are due to Atiyah [1957]; cf. Mackenzie [1987a], App. A, for a detailed discussion. Propositions 2.4.2 and 2.4.3 are from Atiyah and Bott [1983]. In identifying $\mathfrak{aut}(\mathsf{P})$ as the Lie algebra of Aut(P) it is worth mentioning that an H-invariant vector field on P is complete iff its projection to Q is complete; see Kumpera and Spencer [1972], §33. Since the latter condition is satisfied as a consequence of the compact support on Q, elements of $\mathfrak{aut}(\mathsf{P})$ are automatically complete on P.

III.2.5 The construction of observables on $\mathsf{P}^0_V \times_H \mathcal{O}$ through the momentum map of the $\mathcal{G}^\tau_{\mathsf{P}}$-action on $T^*\mathsf{P}$ is taken from Landsman [1993b]; also see Meinrenken [1994] and Robson [1994, 1996]. This generalizes the approach of Isham [1983] and Guillemin and Sternberg [1984b] to nontrivial structure groups; cf. the notes to II.3.1.

The second term on the right-hand side of (2.89) can be understood from a Lagrangian point of view as the contribution to the Noether conserved charge due to the gauge field. In general, if the Lie derivative $L_{\xi^{\mathsf{P}}}\mathbf{A}$ vanishes for some vector field ξ^{P} on P, it does not follow that $L_{\xi^Q}s^*\mathbf{A}$ is zero, too. This leads to the above-mentioned contribution, see Jackiw and Manton [1980].

Cotangent reduction for more general symplectic structures on $T^*\mathsf{P}$ than the canonical one is discussed in Alekseevsy et al. [1994].

III.2.6 The Wong equations were first proposed by Wong [1970] on the basis of a heuristic study of the classical limit of the equation of a scalar quantum field coupled to a Yang–Mills field. The symplectic formulation is due to Sternberg [1977] and Weinstein [1978]; Montgomery [1984] related this to the construction of these equations due to Kerner [1968], and thereby proved Theorem 2.6.2.

The Wong equations are generalized to spinning particles in Künzle [1972] (which covers the abelian case), Arodz [1988], and Harnad and Pare [1991]; also cf. Linden et al. [1996]. Other aspects of the Wong equations are treated in, for example, Duval and Horvathy [1982], Balachandran et al. [1983, 1984], Chiang et al. [1985], Fehér [1986], and Chruściński and Kijowski [1996].

Proposition 2.6.1 is known in physics as the **Kaluza–Klein construction**. It is, in fact valid in the more general situation that $\mathbf{g}_H$ is merely right-invariant (in which case (2.100) is, of course, not valid). An amazing aspect of this construction is that the Einstein equations for $\mathbf{g}$ are equivalent to the coupled Einstein–Yang–Mills equations for $\mathbf{g}^Q$ and $\mathbf{A}$; this follows from (2.188), interpreting Q as space-time rather than space. This has led to the physical idea that all Yang–Mills fields as well the gravitational field on four-dimensional spacetime are shadows of the gravitational field in a higher-dimensional world, some of whose dimensions are compact, and so small as to be invisible. (The original version of Kaluza and Klein described the electromagnetic field in this way, assuming that the universe is 5-dimensional.) This idea is also fundamental to string theory, but there is no evidence that it is correct other than as a mathematical artifact. With certain restrictions, the construction is valid in the pseudo-Riemannian case as well; see Choquet-Bruhat and DeWitt-Morette [1989] for a precise statement. For an overview of "Kaluza–Klein physics" see the reprint volume Appelquist et al. [1985]; mathematical aspects are discussed in Coquereaux and Jadczyk [1988].

The correspondence between (2.112) and (2.111) is a special case of the passage between Lie–Poisson equations on $\mathfrak{g}^*$ and second-order equations on TG; see Marsden and Ratiu [1994].

III.2.7 For the H-connection see Kobayashi and Nomizu [1963], §II.11. The more general theory of connections invariant under some group action is in Kobayashi and Nomizu [1969], applications to physics being discussed in Forgács and Manton [1980], Jackiw and Manton [1980], Harnad et al. [1980], and Cant [1981].

The main reason for the popularity of the H-connection among physicists is that it solves the Yang–Mills equations; see, e.g., Laquer [1984]. In particular, many famous "topological" Yang–Mills configurations, such as instantons and monopoles, are special cases of the H-connection. See Bais and Batenburg [1985] and the notes to 2.12.

A different type of application of the H-connection is to the theory of the Berry phase; see Vinet [1988] and Giavarini and Onofri [1990]. For quantization theory on homogeneous spaces see the notes to IV.2.8.

III.2.8 This material is mainly adapted from Landsman [1993b] (which generalizes the treatment of the homogeneous case in Landsman [1990a, 1992]), but 2.8.2 is an elementary special case of Thm. 3.1 in Muhly et al. [1987], with a different proof. The special case $\mathsf{P} = G$ is Thm. 2.1 in Green [1980]. Corollary 2.8.3 is discussed in IV.2.7.

For the existence of the measure ν in (2.139) cf. Bourbaki [1963], Prop. VII.2.3 and 4. Equation (2.154) is Lemma 5.3 in Guillemin and Uribe [1986].

When $\mathcal{H}_\chi$ is separable, an equivalent construction of $\mathcal{H}^\chi$ starts from the vector space $\tilde{\mathcal{H}}^\chi$ of all μ-measurable functions $\tilde{\Psi}^\chi : \mathsf{P} \to \mathcal{H}_\chi$ that satisfy the properties that (2.145) holds, the function $x \to (\tilde{\Psi}^\chi(x), \tilde{\Psi}^\chi(x)_\chi)$ is locally integrable, and $(\tilde{\Psi}^\chi, \tilde{\Psi}^\chi) < \infty$. The Hilbert space $\mathcal{H}^\chi$ is then the quotient of $\tilde{\mathcal{H}}^\chi$ by the vectors of zero norm. This is proved in Moscovici [1969]; to apply his proof, note that the action of a structure group on a principal bundle is always proper.

III.2.9 An even more general construction of induced representations, which applies when P and H are merely locally compact, and the H-action on P is not necessarily free, is presented in Moscovici [1969] (though no bundle-theoretic interpretation is given). One then tries to define an induced representation of some group G that acts on P; the G- and H-actions must commute. For the construction (which is entirely analogous to the one in our main text) to apply, the H-action must be proper, and $Q = \mathsf{P}/H$ must possess a positive measure ν that is quasi-invariant under the natural G-action.

A detailed description of Mackey induction for locally compact groups may be found in Warner [1972], Gaal [1973], Barut and Raçka [1977], Varadarajan [1985], or Knapp [1986]. The formulation in terms of vector bundles appeared in Hermann [1966], following special cases in Bott [1957]. A discussion of harmonic analysis on vector bundles of the type $G \times_H \mathcal{H}_\chi$ may be found in Wallach [1973]. Attractive general overviews with applications and history, by the founder of the modern theory of induced group representations, are Mackey [1968, 1978, 1992]. Also cf. the notes to IV.2.8.

In the special case $\mathbf{A} = \mathbf{A}^H$, equation (2.180) is due to Doebner and Tolar [1990]. A different perspective on the quantum analogue of the classical momentum map for the $\mathcal{G}_{\mathsf{P}}^{\tau}$-action on $T^*\mathsf{P}^{\mathcal{O}}$ is offered by Meinrenken [1994].

III.2.10 For the Laplace–Bochner operator cf. Wallach [1973] (who simply calls it the Laplacian) and Kuwabara [1982]. Equation (2.187) is due to Berline and Vergne [1985] and Guillemin and Uribe [1986]. For (2.188) see the notes to 2.6. An intrinsic definition of $\mathbf{F}^2$ is presented in Atiyah and Bott [1983].

Lemma 2.10.3 is a special case of Thm. 3 in Nussbaum [1964].

An explicit expression for the Ricci scalar $\mathbf{R}_G$ on a Lie group G may be found, e.g., in Coquereaux and Jadczyk [1988], §2.5.

The identity (2.193) appears, e.g., in Strathdee [1983], Slebarski [1987], and in the present context of quantization theory also in Landsman [1992].

III.2.11 Theorem 2.11.1 is a reformulation of Theorem 6.1 of Hogreve et al. [1983]. Their derivation of their (6.33) provides a detailed proof of the claim preceding our (2.212), and their proof of their (6.35), in particular their Lemma 4.7, also proves our (2.214). The theorem and the given proof still apply if one includes a scalar potential $V(q)$ and a term $A_0(q)^i z_i$ in the Hamiltonian (as is done in the above reference). Our proof relates to that of Hogreve et al. [1983] in much the same way that our proof of Theorem II.2.7.2 stands to the proof of the corresponding result in Hepp [1974]; cf. the notes to II.2.7. Moschella [1989] claims to simplify the proof of the convergence of the quantum equations of motion to the classical Wong equations, but in fact his argument proves convergence in a certain class of mixed states, and with respect to different observables from the ones we are interested in. Accordingly, he does not obtain a quantization condition on $\hbar$.

The right-hand side of (2.209) may be written as $T \exp\left(-i \int_0^t H^{(2)}(s)/\hbar\right)$, where T stands for "time-ordering"; see Dollard and Friedman [1979], and cf. IV.(3.73).

There is an impressive body of mathematical literature on the semiclassical asymptotics of the eigenfunctions and eigenvalues of the Laplace–Bochner operator, and their connection with Hamiltonian trajectories given by solutions of the classical Wong equations. For example, Schrader and Taylor [1984, 1989] and Zelditch [1992] extend the work on classical limits of energy eigenstates cited in the notes II.2.7 to the case at hand. Eigenvalue asymptotics are investigated in Guillemin and Uribe [1985, 1986, 1989, 1990], Taylor and Uribe [1992], Brummelhuis and Uribe [1992], and Brummelhuis et al. [1995]. Other aspects of the motion of a quantum particle in a Yang–Mills field are discussed in Arodz [1983], Belov and Maslov [1990], and Oh [1996].

III.2.12 The quantum theory of magnetic monopoles started with the paper by Dirac [1931], who arrived at the quantization condition $eg = \frac{1}{2}n\hbar$. The factor $\frac{1}{2}$ corresponds to the use of the covering groups $SU(2)$ and $U(1)$ of $SO(3)$ and $SO(2)$. This leads to the Hopf fibration $SU(2)(S^2, U(1), \tau)$; see, e.g., any of the books cited in the notes to 2.1.

There exists a gigantic body of literature on monopoles. The modern understanding of the quantum case in terms of the line bundles H^n is due to Greub and Petry [1975] and Wu and Yang [1975, 1976]. The description in terms of induced representations is due to Langlands [1987]. Much information about both standard angular momentum and the role of $SO(3)$ in the theory of magnetic monopoles may be found in Biedenharn and Louck [1981a,b]. The fact that rotational symmetry forces the field to be a monopole configuration may be found in, e.g., Cant [1981] and Horvathy [1981].

One should actually start with the theory on $\mathbb{R}^3$, on which the monopole field potential is given by $A(r, \phi, \theta) = A(\phi, \theta)/r$. The magnetic field is $B = -gT_3 \otimes \mathbf{e}_r/r^3$, which is evidently singular at the origin 0. Hence one declares that the configuration space of a charged particle moving in a monopole field is $\mathbb{R}^3\backslash\{0\} \simeq S^2 \times \mathbb{R}^+$. The theory on S^2 contains all essential features of the situation on $\mathbb{R}^3\backslash\{0\}$.

Our treatment mainly follows Landsman [1990b], which includes a detailed discussion of the passage from S^2 to $\mathbb{R}^3\backslash\{0\}$. The proof of Proposition 2.12.1 is taken from Landsman and Linden [1991]; the ancillary result (2.224) may be found in Choquet-Bruhat et al. [1982], Problem III.5(8), in which we corrected a sign error.

The eigenfunctions of the operator $dU^n(C_2(SO(3)))$, which is usually taken as the quantum Hamiltonian, are so-called monopole harmonics; see Wu and Yang [1976], Biedenharn and Louck [1981b], Kuwabara [1982, 1984], and Landsman [1990b]. In the realization of the theory on the space $\mathcal{H}^n$ of $SO(2)$-equivariant functions on $SO(3)$, a monopole harmonic is simply a matrix element $U_j(x)^n_m$, where $U_j(SO(3))$ is the usual irreducible representation of spin j, and m runs from $-j$ to j. It may be checked that the corresponding functions in $\mathcal{H}^n_\pm$ are indeed in $\Gamma(\mathsf{H}^n)$.

As in the general case, the justification for using a Hilbert space of sections of a line bundle in quantum theory is that the smooth sections provide a domain of essential self-adjointness of the relevant operators (angular momentum and Hamiltonian). The topology of the line bundle enters the quantum-mechanical description in this way; the total Hilbert space $L^2(\mathsf{H}^n)$ is not sensitive to this topology.

III.3.1 For an overview of the theory of groupoids cf. Brown [1987] and Weinstein [1996a], as well as Renault [1980] and Mackenzie [1987a]. The shortest definition is that a groupoid is a small category with inverses. In the context of group representation theory a groupoid is sometimes called a virtual group; see Ramsay [1971] for an interesting account. Most of the theory that is not purely algebraic is done in the context of topological or measurable groupoids.

Lie groupoids were introduced by Ehresmann [1958]. The main modern sources are Mackenzie [1987a], Coste et al. [1987], and Albert and Dazord [1988]. Mackenzie refers to differentiable groupoids, reserving the name Lie groupoids for transitive differentiable groupoids. Weinstein [1996b] remarks that one can omit the smoothness of the inclusion from Definition 3.1.5. The theory of gauge groupoids is developed in Mackenzie [1987a, 1989].

III.3.2 For half-densities etc. see Guillemin and Sternberg [1977]; our treatment is somewhat different. For the Hilbert space of half-densities, due to Mackey, see Abraham and Marsden [1985]. The bundle $\sqrt{|\Lambda|}^{s\otimes t}\mathsf{G}$ is mentioned in Weinstein [1991] and Connes [1994], following a special case in Connes [1980] (involving the holonomy groupoid of a foliation).

III.3.3 Convolution of sections of $\sqrt{|\Lambda|}^{s\otimes t}$G was introduced by Connes [1980, 1994]. He symbolically writes convolution on a Lie groupoid as

$$f * g(\gamma) = \int_{\gamma_1\gamma_2=\gamma} f(\gamma_1)g(\gamma_2).$$

With an isomorphism similar to λ in (3.17) added, the equivalence between Connes's expression and (3.17) is easily established, using Prop. I.1.2 in Mackenzie [1987a]. An equivalent construction of the convolution algebra of a Lie groupoid appears in Bigonnet [1988], whereas a third approach is in unpublished lecture notes by Renault (see Ramazan [1998]).

For left Haar systems and the corresponding convolution see Hahn [1978b] (who acknowledges Westman [1968]) and Renault [1980], who work in the context of topological groupoids. As in the group case, one can introduce a twist (cocycle) into the convolution; see Renault [1980]. The first part of Proposition 3.3.3 is due to Ramazan [1998].

III.3.4 Crossed products were introduced by Doplicher et al. [1966]. A crossed product is usually defined as the C^*-completion of $C^*(G, \tilde{\mathfrak{A}})$, denoted by $C^*(G, \mathfrak{A})$; see Pedersen [1979] for the basic theory, and Green [1978] and Packer [1996] for advanced results. It is not necessary to assume that G is unimodular; the standard definition of a crossed product C^*-algebra contains a factor $\Delta(x)^{-1}$ on the right-hand side of (3.32), but our definition leads to an isomorphic algebra.

A complete proof of Theorem 3.4.4 may be found in Busby and Smith [1970], Thm. 3.3; a slightly different approach is in Pedersen [1979], Prop. 7.6.4. All these authors work in the L^1 rather than C_c^∞ context, allowing general locally compact groups G.

Action C^*-algebras are usually called transformation group algebras; see the notes to 3.7.

Systems of imprimitivity (for general locally compact groups) were introduced by Mackey in his study of induced group representations; see Barut and Raçka [1977] and Mackey [1968] (for the case $Q = G/H$), Varadarajan [1985], Mackey [1978], and Mackey [1992]. The definition in these books is stated in terms of projection-valued measures on Q; see IV.(2.118) and surrounding text. The equivalent approach we use goes back to Glimm [1962] (who, in the locally compact setting, of course worked with $C_0(Q)$ rather than $C_c^\infty(Q)$).

III.3.5 Representations of locally compact groupoids in the sense of Definition 3.5.1 are studied in Hahn [1978b] and Renault [1980]. For the more general case of an action of a groupoid on some space see 3.9.11 or Mackenzie [1987a]. Quasi-invariant measures are studied in Hahn [1978a]; also cf. Renault [1980]. The regular representation is developed in Hahn [1978b].

Our definition of a direct integral Hilbert space (originally due to von Neumann) follows Bratteli and Robinson [1987], §4.4.1. Also cf. Takesaki [1979] or Kadison and Ringrose [1986], among others.

III.3.6 The C^*-algebra of a locally compact groupoid first appeared in Connes [1980] and Renault [1980]. The structure of such algebras is beginning to be analyzed, cf. Muhly et al. [1987] and the series of papers by Muhly and Williams [1990, 1992, 1995].

Renault [1980], II.1.22, shows that every representation of C_c^∞(G) on a separable Hilbert space that is continuous with respect to the inductive limit topology on C_c^∞(G) and the weak topology on $\mathfrak{B}(\mathcal{H})$ is automatically bounded in the sense of (3.57); his proof is for locally compact groupoids, involving an additional condition that is automatically satisfied for Lie groupoids. A detailed proof of (3.59) is in Hahn [1978a], Thm. 3.8.

The groupoid analogue of the Banach algebra $L^1(G)$ is $L^I(\mathsf{G})$, defined as the space of measurable functions on G for which $q \mapsto \int_{\tau_{s,t}^{-1}(q)} d\mu_q^{s,t}(\gamma)|f(\gamma)|$ is essentially bounded with respect to a locally Lebesgue measure on Q. It is a Banach algebra under the norm $\|\cdot\|_I$ defined in (3.57) and the continuous extension of multiplication and involution in $C_c^\infty(\mathsf{G})$. It follows from Proposition 3.6.1 that every nondegenerate representation of $C_c^\infty(\mathsf{G})$ on a separable Hilbert space that is bounded (in the sense of (3.57)) extends to $L^I(\mathsf{G})$. See Hahn [1978b] for the proof of these claims. The appropriate generalization of Theorem 1.7.3 holds here, in that each such representation corresponds to a representation U of G as in 3.6.1. This result is due to Renault [1987] (following a special case in Renault [1980]).

The first two points of Theorem 3.6.2 are in Hahn [1978b] and Renault [1980]. The fourth point was inspired by Connes [1994], which contains a version of (3.65).

Parallel to the group case there is a concept of amenability of (locally compact) groupoids, which is expressed by the equality $C_r^*(\mathsf{G}) = C^*(\mathsf{G})$, either as a theorem or as a definition; see Renault [1980].

III.3.7 Theorem 3.7.1 is a special case of Thm. 3.1 in Muhly et al. [1987], with a different proof. Equation (3.76) is proved in Renault [1980] via the correspondence between representations of $C^*(\mathsf{G})$ and those of G, as mentioned above. Our proof is based on Rieffel [1972]. For $\mathsf{P} = G$, equation (3.78) is a special case of Thm. 2.13 in Green [1980] (who states the result for the transformation group C^*-algebra $C^*(G, G/H)$).

Starting with Effros and Hahn [1967], **transformation group C^*-algebras** (which we call action C^*-algebras) are much studied by C^*-algebraists; see the review by Packer [1994] for history and references. Corollary 3.7.4 is a special case of Thm. 7.6.6 in Pedersen [1979].

Corollary 3.7.6 is Mackey's (transitive) imprimitivity theorem; see the books listed in the notes to 3.4. Like Corollary 3.7.2, it is a special case of Rieffel's imprimitivity theorem, and will be rederived as such in IV.2.7.

The special case $C^*(G, G/H)$ of the C^*-algebra $C^*(G, Q)$ was introduced by Glimm [1962]. See the notes to IV.2.8 for applications to physics.

The proof of the Stone–von Neumann uniqueness theorem at the end of the section is due to Mackey [1963, 1968, 1978]. The technical details of Mackey's proof are slightly different, because he uses a different (equivalent) notion of systems of imprimitivity. In any case, the argument can be generalized to arbitrary locally compact abelian groups.

III.3.8 Lie algebroids were introduced by Pradines [1966], who also showed how they could be constructed from Lie groupoids. References and further development of the theory until 1987 may be found in Mackenzie [1987]; a more recent reference is Vaisman [1994]. Pradines's "grand scheme to generalise the standard construction of a simply connected Lie group from a Lie algebra to a corresponding construction of a Lie groupoid from a Lie algebroid" was completed by Mackenzie [1987] in the locally trivial case, and by Brown and Mucuk [1995, 1996] in general. Unlike the case of Lie algebras, there is a potential cohomological obstruction, so that not every Lie algebroid corresponds to a Lie groupoid. Apart from principal fiber bundles, the main context is the theory of foliations.

Equation (3.93) appears in Mackenzie [1987b], who attributes it to A. Weinstein (our derivation in the proof of Proposition 3.8.8 is different).

III.3.9 Propositions 3.9.1 and 3.9.2 are due to Courant [1990]. A different approach to the correspondence between Lie algebroids and linear Poisson structures, including a more intrinsic definition of the Poisson structure on $\mathfrak{G}^*$, may be found in Coste et al. [1987].

Definition 3.9.5, which includes the Poisson bracket (3.101) on $\mathfrak{g} \times Q$ as a special case, is due to Krishnaprasad and Marsden [1987], and was further studied by Weinstein [1987].

Theorem 3.9.6 is due to Xu [1992], as is Corollary 3.9.8 (which is a special case of his Cor. 4.1). We now sketch the remaining part of the proof of the latter; this depends on the concept of a symplectic groupoid discussed in the notes to IV.1.2. We have to show that the completeness of the map $\rho : S \to Q$ associated to $\tilde{\pi} : C^\infty(Q) \to C^\infty(S)$ by $\rho^* = \tilde{\pi}$ implies that $J = (J_{(1)}, \rho) : S \to \mathfrak{g}^*_- \times Q$ is complete.

Thinking of Q as a Poisson manifold with zero Poisson structure, it is integrable, with symplectic groupoid T^*Q. The existence of an equivariant momentum map for the pullback G-action on T^*Q (see Lemma 2.3.1) implies that one may equip $\mathsf{G} := T^*G \times T^*Q$ with the structure of a symplectic groupoid with base $\mathfrak{g}^*_- \times Q$; when the G-action on Q is trivial this would be the direct product with respect to the groupoid structures mentioned in the notes to IV.1.2. Applying Thm. 3.1 in Xu [1991b] in the direction "complete Poisson map $\to$ symplectic groupoid action" to ρ, there exists a symplectic T^*Q-action on S. The covariance condition (3.107) implies that ρ intertwines the G-actions on S and Q. By Thm. 4.1 in Xu [1992] this in turn entails that there exists a symplectic G-action on S associated to J. Applying Thm. 3.1 above in the opposite direction then leads to the desired conclusion that J is complete.

Corollary 3.9.10 is a special case of the classical transitive imprimitivity theorem IV.1.6.4.

Definition 3.9.11, originally due to Pradines, may be found in Mackenzie [1987]. However, Mackenzie's notion of a Lie algebroid action (which he calls a representation) is different from ours. A representation in his sense is a morphism of $\Gamma(\mathsf{V})$ into the Lie algebra of derivations on sections of some vector bundle. This concept is not appropriate in relationship with the representation theory of the Poisson algebra $C^\infty(\mathsf{V}^*, \mathbb{R})$.

A different line of research relating Lie groupoids to Poisson structures is discussed in the notes to IV.1.2. Here the generalized momentum map associated to an action of a Lie groupoid G on a symplectic manifold takes values in the base Q rather than in the dual of the Lie algebroid $\mathfrak{G}$, as in our approach.

III.3.10 Lemma 3.10.1 is Prop. III.3.3 in Mackenzie [1987]. The map Exp^L was introduced by Pradines [1968].

III.3.11 The role of Lie algebroids and groupoids in strict quantization as explained in this section originates with Landsman [1993b, 1996a].

In the context of formal deformation quantization, Dirac's condition has been proved for arbitrary Lie groupoids by Nistor et al. [1997], and by Ramazan [1998].

Theorem 3.11.3 was first mentioned by Rieffel [1989a], though not in the context of Lie algebroids and groupoids or Weyl quantization. This paper gives many interesting examples of strict deformation quantizations that do not fit into our scheme, such as the quantization of the Poisson algebra of functions on the 2-torus, equipped with a suitable Poisson structure, by the so-called **noncommutative torus.** As these examples show, a given C^*-algebra may be a strict deformation of more than one Poisson algebra.

An "unbounded" version of this deformation, generalizing the procedure in 1.6 from Lie algebras to Lie algebroids, may be found in Nistor et al. [1998].

III.3.12 The normal groupoid is constructed in Hilsum and Skandalis [1987], §3.1, and is further discussed in Weinstein [1989]. These authors use $I = [0, 1]$ instead of $I = \mathbb{R}$, and construct the manifold structure in a slightly different way. The special case of the tangent groupoid (again, for $I = [0, 1]$) is due to Connes [1994] (circulating in the eighties).

Lemma 3.12.4 is due to Lee [1976]. As explained in Elliott et al. [1993], it can be used to simplify the proof of Rieffel's condition in certain examples in Rieffel [1989b]. For the primitive spectrum and the Jacobson topology cf. Dixmier [1977], who gives our Lemma 3.12.5 as §§3.2.1,2.

The realization of $C_r^*(\mathrm{G}_N)$ as the C^*-algebra of a continuous field has the following generalization (G. Skandalis, private communication, June 1997). Let $\tilde{\mathrm{G}}$ be a Lie groupoid with base $\tilde{Q}$, and let p be a continuous and open map from $\tilde{Q}$ to some Hausdorff space X which is $\tilde{\mathrm{G}}$-invariant in the sense that $p \circ \tau_s = p \circ \tau_t$. Define $\tilde{\mathrm{G}}_x := (p \circ \tau_s)^{-1}(x)$ (this is a subgroupoid of $\tilde{\mathrm{G}}$ because of the $\tilde{\mathrm{G}}$-invariance of p), and $\mathfrak{A}^x := C^*(\tilde{\mathrm{G}}_x)$. Then the triple $(C_r^*(\tilde{G}), \{C_r^*(\tilde{\mathrm{G}}_x), \varphi_x\}_{x \in X})$, where $\varphi_x(f) := f \upharpoonright \tilde{\mathrm{G}}_x$, is a continuous field of C^*-algebras at those points x where $C^*(\tilde{\mathrm{G}}_x) = C_r^*(\tilde{\mathrm{G}}_x)$.

We apply this to our situation by taking $\tilde{\mathrm{G}} = \mathrm{G}_N$ and $X = \mathbb{R}$, whence $\tilde{Q} = \mathbb{R} \times Q$, and p is just projection onto the first variable. Continuity away from $\hbar = 0$ follows from the triviality of the field for $\hbar \neq 0$ (whether or not $C_r^*(\mathrm{G}) = C^*(\mathrm{G})$). The result above may be used to prove continuity at $\hbar = 0$ by noticing that $C_r^*(\mathfrak{G}) = C^*(\mathfrak{G})$; this follows because both sides are isomorphic to $C_0(\mathfrak{G}^*)$. In other words, from this point of view it is the amenability of $\mathfrak{G}$, regarded as a Lie groupoid as explained in 3.12.1, that lies behind Theorem 3.11.4.

An interesting application of Corollary 3.12.6 would lie in the development of generalizations of the Atiyah-Singer index theorem, noting that recent proofs of this theorem through deformation quantization (see Connes [1994], Fedosov [1996], and Elliott et al. [1996]) may be interpreted in the light of the special case of 3.12.6 in which G_N is the tangent groupoid of a manifold. The first step towards such generalizations, namely a good definition of the analytical index, has already been taken in Monthubert and Pierrot [1997].

An entirely different application of the normal groupoid to the classical limit of quantum mechanics is given Bellissard and Vittot [1990].

Chapter IV

IV.1.1 The theory of constraints and reduction has a venerable tradition; the modern era started with the work of Dirac [1950, 1964]. Efforts to put his approach on a geometric and rigorous footing were initiated by the Warszaw school of Tulczyjew and collaborators in the sixties and seventies; see, e.g., the books by Kijowski and Tulczyjew [1979] and Binz et al. [1988], and references therein. Further contributions were made by Lichnerowicz [1977] and Gotay et al. [1978] (also cf. Gotay and Nester [1980]). The approach of the latter starts from a presymplectic manifold S, and is therefore more general than the one presented in the main text. Also, the infinite-dimensional case is included.

In physics the constraints on the phase space S are usually derived from a Lagrangian; the so-called constraint algorithm then leads to the final constraint hypersurface C, on which the equations of motion defined by a Hamiltonian h are well-defined. Hence our C is supposed to be the endproduct of this algorithm. Apart from the references above, see Sundermeyer [1982] or Henneaux and Teitelboim [1992]; these books contain a wealth of information and examples from physics.

Theorem 1.1.2 goes back to Cartan [1958]; a heuristic version is implicit in Dirac [1950, 1964]. The given formulation may be found, e.g., in Libermann and Marle [1987], §14, which contains a proof that the null distribution is smooth in App. 4, Prop. 3.7.

The concept of a weak observable is due to Dirac and Bergmann (see Sundermeyer [1982] for an extensive list of references to the original literature).

The decomposition mentioned in the last paragraph of the section is well known to physicists; as a theorem it is proved, e.g., by Lichnerowicz [1977] (also cf. Thm. III.14.11 in Libermann and Marle [1987]).

IV.1.2 Theorem 1.2.2 is a reformulation of Prop. 2.1 in Xu [1991a], which in turn generalizes Thm. 3.12 in Mikami and Weinstein [1988]. In these papers special symplectic reduction is approached through the following theory, due to Karasev [1987], Weinstein [1987b], and Zakrzewski [1990a,b]. We give only a brief summary (following Weinstein [1991]); for more information we refer to Coste et al. [1987], Albert and Dazord [1990], Karasev and Maslov [1993], Vaisman [1994], and Weinstein [1998].

A **symplectic groupoid** (G, ω) is a Lie groupoid G $\overset{\leftarrow}{\Rightarrow}$ Q with a symplectic form ω, with the property that the graph $\{(\gamma, \gamma', \gamma\gamma')\}$ (where $\gamma, \gamma' \in$ G) of groupoid multiplication is a Lagrangian submanifold of G $\times$ G $\times$ G$^-$. This entails that the inversion I : G $\to$ G$^-$ is a Poisson map. The most important consequence of the definition is that there exists a Poisson structure on Q for which $\iota(Q)$ is a Lagrangian submanifold of G, and τ_s : G $\to Q$ and τ_t : G $\to Q^-$ are Poisson maps. Moreover, the Poisson subalgebras $\tau_s^* C^\infty(Q)$ and $\tau_s^* C^\infty(Q)$ of C^∞(G) commute; when the fibers of τ_s (and hence of τ_t) are connected, these subalgebras are even the Poisson commutants of one another.

The simplest example is G $= T^*Q$ (with its canonical symplectic form), where $\tau_s = \tau_t = \tau_{T^*Q\to Q}$, and groupoid "multiplication" is addition in a fiber of T^*Q. The associated Poisson structure on Q is the zero bracket. When Q is a Lie group G, still using the canonical symplectic form, one can put a different groupoid structure on T^*G by identifying T^*G with $G \times \mathfrak{g}^*$ in the right trivialization (cf. III.1.4), and regarding $G \times \mathfrak{g}^*$ as an action groupoid with respect to the coadjoint action (cf. III.3.1.4). Hence $\tau_s = -J^R$ and $\tau_t = J^L$; this assigns the (+) Lie–Poisson structure to $\mathfrak{g}^*$.

One may ask whether a given Poisson manifold P is **integrable**, in that there exists a symplectic groupoid whose base is P. As we just saw, any manifold P with the zero bracket and any dual Lie algebra $\mathfrak{g}^*$ with the Lie–Poisson structure are integrable; not every Poisson manifold is.

Recall Definition III.3.9.11 of a smooth action of a Lie groupoid G $\overset{\leftarrow}{\Rightarrow}$ Q on a manifold S. For later use, define the **orbit** Gσ of $\sigma \in S$ under a G-action on S in the obvious way, i.e., G$\sigma := \{\gamma\sigma \mid (\gamma, \sigma) \in$ G $*_Q S\}$. Mikami and Weinstein [1988] define such a Lie groupoid action to be **symplectic** when the graph $\{(\gamma, \sigma, \gamma\sigma)\}$ of the groupoid action is a Lagrangian submanifold of G $\times S \times S^-$. It easily follows that $J_\rho : S \to Q$ is a Poisson map. (Compare this with Proposition III.3.9.13, where as an alternative proposal we associate a generalized momentum map $J : S \to \mathfrak{G}^*$ to the G-action on S.) A deeper result is that J_ρ is necessarily complete, and that conversely, when the τ_s-fibers of G are connected, the completeness of a given Poisson map $J_\rho : S \to Q$ implies that there exists a symplectic G-action on S associated with J_ρ; see Thm. 3.1 in Xu [1991b]. The main example of such a symplectic groupoid action is derived from an ordinary group action with equivariant momentum map $J : S \to \mathfrak{g}^*_-$ by putting $J_\rho = -J$. The action of $(\theta, x)_R \in T^*G$ on $\sigma \in S$ is then defined when $\theta = J(x\sigma)$, and $(J(x\sigma), x)_R\sigma = x\sigma$.

Let now P be an integrable Poisson manifold, so that there exists a symplectic groupoid G $\overset{\leftarrow}{\Rightarrow}$ P. Xu's [1991a] formulation of special symplectic reduction starts from a pair of symplectic G-actions on S and S_ρ. Using the above definition of an orbit, one can form the quotient space $(S *_P S_\rho)/$G under the diagonal G-action on $S *_P S_\rho$. When the G-orbits are connected, Prop. 2.1 in Xu [1991a] then shows that this quotient coincides with the reduced space S_J^ρ as defined in (1.13). When there are disconnected orbits one has a situation similar to the one discussed around (1.26). As Mikami and Weinstein [1988] remark, this formulation of special symplectic reduction "turns out to be purely groupoid-theoretic, involving no symplectic geometry at all". (They work in the special situation that S_ρ is a symplectic leaf of P, with J_ρ the inclusion map.)

In the form given here, Definition 1.2.1 and Theorem 1.2.2 appear in Landsman [1995a]. Our formulation without symplectic groupoids is motivated by the clean analogy with the quantum situation; this analogy is obscure in the version cited above. For our use of transverse intersections see, e.g., exercise 1.6E in Abraham and Marsden [1985], or §27 in Guillemin and Sternberg [1984]. The dimension counting in the proof goes back to Kazhdan, Kostant, and Sternberg [1978].

Theorem 1.2.2 can actually be generalized to the case where S and S_ρ are Poisson manifolds. As in the symplectic case, the reduced space S^ρ is defined as the quotient of $S *_P S_\rho$ by the foliation defined by the vector fields $\hat{\xi}_f$, $f \in C^\infty(P, \mathbb{R})$; the alternative description in terms of the null foliation of ω_C is, of course, not available here. Rather than being symplectic, the manifold S^ρ is a Poisson space, which carries a reduced Poisson structure in the sense of Marsden and Ratiu [1986]. Exactly as in the proof of 1.2.2, one first shows that $\mathcal{N} \subset TC$. Secondly, one has $B^\sharp(\mathcal{N}_x^0) \subset T_x C$. To show this local property, take $\alpha = dg_1 + dg_2$, with $g_i = \tau_i^* h_i$, where the τ_i are the natural projections $\tau_1 : S \times S_\rho \to S$ and $\tau_2 : S \times S_\rho \to S_\rho$, and $h_1 \in C^\infty(S)$, $h_2 \in C^\infty(S_\rho)$. Then the property that $\alpha \in B^\sharp(\mathcal{N}_x^0)$ is equivalent to the equality $\{J^* f, h_1\} = \{\rho^* f, h_2\}$ for all $f \in C^\infty(P)$. Hence $J_* \xi_{g_1} = \rho_* \xi_{g_2}$, which proves the claim. The Poisson generalization of Theorem 1.2.2 now immediately follows from the "Poisson Reduction Theorem" in Section 2 of Marsden and Ratiu [1986].

IV.1.3 Symplectically complete foliations were first defined by Libermann [1983], which contains Proposition 1.3.2; also see Prop. III.9.7 in Libermann and Marle [1987]. See Dazord and Delzant [1987] for further developments.

Proposition 1.3.3 is given by Weinstein [1983] for the case where S/Φ is smooth. Our proof of the more general statement follows the proof of Thm. 1 in Karshon and Lerman [1997].

When conditions 1 and 2 in 1.3.4 are satisfied, one speaks of a **Weinstein dual pair**; the attribute "Weinstein" is sometimes omitted. When in addition 1.3.4.3 is met, one has a **full dual pair**. The theory of such dual pairs is due (independently) to Karasev [1989] (whose unpublished Russian original is from 1981) and Weinstein [1983] (the latter contains 1.3.2 as well). For a review also see Vaisman [1994] and Choquet-Bruhat and DeWitt-Morette [1989]. Weinstein [1990] introduces the concept of a **symplectic affinoid space**, which is a generalization of a full dual pair.

A symplectic groupoid $\mathsf{G} \rightrightarrows Q$ (cf. the notes to 1.2) provides an interesting example of a dual pair: The diagram in (1.20) then becomes $Q \xleftarrow{\tau_s} \mathsf{G} \xrightarrow{\tau_t} Q^-$.

Weinstein [1983] remarks that his dual pair is the classical analogue of a **Howe dual pair**; this is a pair of reductive subgroups of a symplectic group $Sp(2n, \mathbb{R})$, which are each other's centralizer. Such pairs were introduced by Howe [1989] (which had been around for a decade prior to publication). Howe dual pairs are studied from the perspective of constrained quantization in Landsman [1994] and Bowes and Hannabuss [1997]. In the author's opinion, the true quantum analogue of a classical dual pair is a quantum dual pair; cf. Definition 2.3.1. Corollary 1.3.6 is due to Weinstein [1983], who acknowledges Kazhdan et al. [1978].

Definition 1.3.7 and Proposition 1.3.9 are due to Xu [1991b] (which contains a great deal of additional information on Morita equivalence in the present context, as does Xu [1992]). The second example in 1.3.9 is attributed to Weinstein. The algebraic topology needed to complete the proof is the exact sequence (see Bott and Tu [1982], equation

(17.4)) $\pi_1(J_i^{-1}(\sigma_i)) \to \pi_1(S) \to \pi_1(S_i) \to \pi_0(J_i^{-1}(\sigma_i))$. By assumption, the first and the last entry are the trivial group, so that the claim follows.

IV.1.4 Theorem 1.4.1 is due to Xu [1991b]. Our proof, however, is taken from Landsman [1995a]; the relevant homotopy theory may be found in Jänich [1994]. The proof in Xu [1991b] is based on the theory of symplectic groupoids. Xu assumes that the Poisson manifolds in question are integrable; combining Thms. 4.18 and 5.2 in Weinstein [1990], one infers that this is always the case in the given situation. Xu's proof follows the lines of first showing that integrable Morita-equivalent Poisson manifolds have Morita-equivalent symplectic groupoids, which in turn have equivalent categories of complete symplectic realizations.

IV.1.5 Lemma 1.5.1 is due to Smale [1970]. For proper group actions and a proof of Proposition 1.5.3 see Abraham and Marsden [1985], particularly Prop. 4.1.23, or Cushman and Bates [1997], App. B.

Marsden–Weinstein reduction is due to Meyer [1973] and Marsden and Weinstein [1974]; it has a long pedigree in classical mechanics. Our formulation as a special case of special symplectic reduction coincides with the construction of Marsden–Weinstein quotients in Kazhdan et al. [1978]. There is a great deal more to say about this subject. For example, the construction may be carried out under less stringent conditions than the surjectivity of J_* (i.e., the regularity of θ). Extensive treatments may be found in Guillemin and Sternberg [1984], Abraham and Marsden [1985], and Libermann and Marle [1987]. Marsden [1992] gives an overview of applications of Marsden–Weinstein reduction in mechanics. A generalization to general Poisson manifolds is presented in Marsden and Ratiu [1986, 1994].

Theorem 1.5.5 is due to Marle; see Libermann and Marle [1987], prop. IV.6.8. It is also mentioned by Weinstein [1983].

Proposition 1.5.8 is implicit in, e.g., Weinstein [1983] and Xu [1992]. The completeness of J is explicitly proved in the latter paper using a different method based on Thm. 3.1 in Xu [1991b]. The procedure to (re)construct dynamics on S given the dynamics on S/H used in the proof of the completeness of τ is due to Marsden and Weinstein [1974]; also see Marsden et al. [1990] and Marsden [1992]. The existence of a complete solution of the equation $x(t)^{-1}\dot{x}(t) = X(t)$ is proved, e.g., in Dollard and Friedman [1979], Ch. I.

When H does not act freely on S one may still ask whether 1.3.3 holds in the context of Marsden–Weinstein reduction. For compact H this is analyzed by Karshon and Lerman [1997].

IV.1.6 In the special case $\mathsf{P} = G$ the reduced space $T^*\mathsf{P}^\rho$ of Figure 3 appeared in Kazhdan et al. [1978]; also see Guillemin and Sternberg [1984] and Zakrzewski [1986]. In this context one usually speaks of **symplectic induction.**

Theorem 1.6.1 is due to Duval et al. [1992]. Many interesting generalizations of Theorem 1.6.2 may be found in Xu [1992]. Theorem 1.6.4 is due to Ziegler [1996].

IV.1.7 Our proof of 1.6.4 is based on Theorem 1.4.1 (which was not used by Ziegler). The last part of the proof of Lemma 1.7.1, however, is based on Ziegler's [1996] proof of Theorem 1.6.4.

IV.1.8 Theorem 1.8.1 is due to Landsman [1995a]. It is motivated by Theorem 2.6.1 on Rieffel induction in stages. Theorem 1.8.2 is a straightforward generalization of Corollary 1.8.4, which appeared in Landsman [1995a]. Many special cases were known; see, for example, Marsden et al. [1984] and Guillemin et al. [1996]. Lemma 1.8.3 is Prop. A.4 in Weinstein [1987a].

Proposition 1.8.5 is taken from Sjamaar and Lerman [1991], who prove it for singular reduced spaces as well. Extensive information on strongly Hamiltonian product actions may be found in Libermann and Marle [1987].

IV.1.9 All unproved statements in this section may be found in Leptin and Ludwig [1994] or in Corwin and Greenleaf [1989]. The latter contains Lemma 1.9.2 as Thm. 1.3.3 or Prop. 3.1.18, whereas the "nontrivial fact" (due to Chevalley and Rosenlicht) used in the proof of 1.9.1 is Thm. 3.1.4. Finally, Theorem 1.9.3, which combines the work of Dixmier and Kirillov, is Thms. 2.2.2–4.

Theorem 1.9.1 is taken from Landsman [1995a]. Our use of the structure of the coadjoint orbits of nilpotent Lie groups should be distinguished from the "orbit philosophy" of Kirillov [1962, 1990], Souriau [1969], Kostant [1970], and others; our sole aim is the correspondence between symplectic reduction and the theory of induced representations, which is seen to be quite perfect in the case of connected, simply connected nilpotent Lie groups.

IV.1.10 Proposition 1.10.1 is due to Rawnsley [1975]. Theorem 1.10.2 appeared in Landsman [1995a] and in Guillemin et al. [1996]. Theorem 1.10.3 was first given by Wigner [1939] for the case that G is the Poincaré group. Barut and Raçka [1977] and Varadarajan [1985] are good sources for the theory of induced representations of semidirect products. Our proof is a straightforward C^*-algebraic reformulation of the proof given in these references.

Theorem 1.10.4 appeared in Marsden et al. [1984], which includes extensive references to related results, as well as applications; also cf. Guillemin and Sternberg [1984]. Both groups of authors use the equality in the opposite direction.

The quantization theory of Isham [1983] may be reconsidered in the light of 1.10.4. Given a homogeneous configuration space $Q = L/H$, he first looks for a vector space V with an L-action ρ, such that Q is diffeomorphic to some $\rho^*(L)$-orbit in V^*. He then accepts any irreducible representation of $G = L \ltimes_\rho V$ as a possible quantization of the cotangent bundle T^*Q. However, having found a $\tilde{p} \in V^*$ for which $H = L_{\tilde{p}}$, the space T^*Q is symplectomorphic to $\widetilde{T^*L}^0 = (J^R_{\restriction \mathfrak{l}_{\tilde{p}}})^{-1}(0)/L_{\tilde{p}}$, which by 1.10.4 is symplectomorphic to the coadjoint orbit $\mathcal{O}^G_{(0,\tilde{p})}$. Hence Isham's proposal amounts to accepting any irreducible representation of G as a possible quantization of this particular coadjoint orbit. Cf. Robson [1994, 1996] for a related discussion.

Corollary 1.10.5 is due to Leonard and Marsden [1997], who provide a very detailed proof, as well as giving applications to the motion of underwater vehicles.

Baguis [1998] gives a detailed study of the symplectic geometry of the coadjoint orbits of semidirect products.

IV.1.11 The mathematical theory of singular Marsden–Weinstein reduction started with Arms et al. [1981], who proved that the singularities in $J^{-1}(0)$ are conical. Arms et al. [1990] look at the singular case of general symplectic reduction, comparing various approaches, and include a good bibliography. A very detailed treatment is given in Cushman and Bates [1977], App. B.

Lemma 1.11.1 is Lemma 27.1 in Guillemin and Sternberg [1984]. Proposition 1.11.2 is taken from Cushman and Bates [1977], App. B.5.17 (our proof is a trifle different). Proposition 1.11.3, which in our presentation is of fundamental importance, is due to Sjamaar and Lerman [1991]. Definition 1.11.4 was first proposed by Arms et al. [1991], which contains the first half of Proposition 1.11.5; the second half is due to Sjamaar and Lerman [1991].

Theorem 1.11.6 summarizes results of Arms et al. [1981], Otto [1987], Arms et al. [1991], and Sjamaar and Lerman [1991]. Using more sophisticated techniques, the latter prove that the decomposition (1.85) is locally finite and satisfies the **condition of the frontier**; that is, the closure of each piece is the union of other pieces in the decomposition. Indeed, they show that the decomposition in question is a stratification in the sense of Goreski and MacPherson [1988]. Note that none of these properties is in general satisfied by the decomposition of a Poisson manifold into its symplectic leaves, which otherwise is somewhat comparable with the decomposition of a singular Marsden–Weinstein quotient into its symplectic pieces.

The $SO(2)$-example was given in Gotay and Bos [1986], who also generalize it to $SO(n)$. Lerman et al. [1993] further analyze this example (among many others), and refer to Schwarz [1975] for the proof of the claim on smooth functions on S^0. Proposition 1.11.7 is due to Sjamaar and Lerman [1991].

The second example is taken from Landsman [1998a], who also explains its relevance to cosmology. Similar examples appear in the literature; see, e.g., Śniatycki and Weinstein [1983] and Arms et al. [1990].

Singular Marsden–Weinstein reduction is of great importance to general relativity and Yang–Mills theories, where field configurations with symmetry project to singular points of the physical phase space (obtained by forming a Marsden–Weinstein quotient with respect to the gauge group; see IV.3). These applications are studied in Fischer et al. [1980], Isenberg and Marsden [1981], Arms [1981, 1986], Arms et al. [1981], and Emmrich and Römer [1990]. The two-dimensional case enables one to perform explicit calculations; see IV.3.6 and notes thereto.

Moduli spaces of flat connections on a compact Riemann surface provide closely related examples; from the large body of literature on this topic, starting with Atiyah and Bott [1983], we select Hitchin [1990], Weinstein [1995], Huebschmann [1996], and Jeffrey and Weitsman [1997] (and references therein to earlier work of these authors). For the moduli space of all Yang–Mills connections on a compact Riemann surface see the review by Sengupta [1997], and references therein.

IV.2.1 The theory of Hilbert C^*-modules over commutative C^*-algebras was initiated by Kaplansky [1953]. The generalization to the noncommutative case was studied by Paschke [1973], containing all results in this section except Corollary 2.1.4. Simultaneously, Rieffel [1974a] introduced pre-Hilbert C^*-modules. A recent textbook is Lance [1995], which contains 2.1.4.

For multiplier algebras in the present context see Wegge-Olsen [1993] or Lance [1995]. The advanced theory of Hilbert C^*-modules, which we do not cover, is mainly due to Kasparov [1980, 1981]. In his work, Hilbert C^*-modules are a basic tool in the K-theory of C^*-algebras, which is a noncommutative generalization of the theory of vector bundles. See Wegge-Olsen [1993] for a "friendly introduction" to this topic, and Connes [1994] for a high-level treatise. Blackadar [1986] and Skandalis [1991] review Kasparov's [1981] generalization of operator K-theory, known as KK-theory. Frank [1998] contains an exhaustive bibliography on all aspects of Hilbert C^*-modules. A detailed study of self-duality is in Frank [1990].

IV.2.2 Rieffel induction is due to Rieffel [1974a], which contains historical comments. He works entirely in the setting of pre-Hilbert C^*-modules on which (2.4) does not necessarily hold (which he refers to as pre-$\mathfrak{B}$-Hilbert spaces).

The construction revolving around (2.32) is taken from Hannabuss [1984]. Fell induction is due to Fell [1978]; also cf. Fell and Doran [1988] (which is an encyclopedic treatment of induction techniques in representation theory, including a vast bibliography). A related induction procedure is given by Bennett [1978].

IV.2.3 Operators of the type (2.47) appear in Rieffel [1974a], and the C^*-algebra $C_0^*(\mathcal{E}, \mathfrak{B})$ is defined in Paschke [1973]. Theorem 2.3.3 is a "completion" of Prop. 6.18 in Rieffel [1974a]; an essential step in the proof, namely the equality $\|\Psi\|_{\mathfrak{A}} = \|\Psi\|_{\mathfrak{B}}$, is equivalent to Prop. 3.1 in Rieffel [1979]. In the present form, Theorem 2.3.3 is a special case of Prop. 7.1 in Lance [1995].

IV.2.4 For the history of the concept of Morita equivalence see Morita's obituary by Arhangel'skii et al. [1997]; see, e.g., Bass [1968] for a textbook treatment. Morita's theorem in pure algebra states that two rings $\mathfrak{A}$ and $\mathfrak{B}$ have isomorphic categories of (left) modules iff $\mathfrak{A}$ is isomorphic to the endomorphism ring of a $\mathfrak{B}$-module $\mathcal{E}$, where $\mathcal{E}$ and $\mathfrak{B}$ are each a direct summand of some (possibly different) power of each other. In that case $\mathfrak{A}$ and $\mathfrak{B}$ are said to be Morita equivalent. An appropriate version of this concept is applied to C^*-algebras and W^*-algebras in Rieffel [1974b], and to general Banach algebras in Grønbæk [1995]. With an appropriate definition of modules and category equivalence, Rieffel [1974b] shows that two von Neumann algebras $\mathfrak{M}$ and $\mathfrak{N}$ have isomorphic categories of (left) modules iff $\mathfrak{M} = C^*(\mathcal{E}, \mathfrak{N})$ for some Hilbert C^*-module $\mathcal{E} \rightleftharpoons \mathfrak{N}$.

What we (following, e.g., Skandalis [1991] and [Lance [1995]) for simplicity call Morita equivalence in the main text should more properly be called **strong Morita equivalence**, which is indeed the terminology used in most of the literature. The original definition of this equivalence relation by Rieffel [1974a] consisted in the conditions of Proposition 2.4.4, on the basis of which he formulated and proved Theorem 2.4.5. Various generalizations of Rieffel's imprimitivity theorem are studied in Fell [1978] and Fell and Doran [1988].

The correspondence between the representations of $\mathfrak{A}$ and $\mathfrak{B}$ established in Theorem 2.4.5 can be shown to preserve weak containment, but not cyclicity.

An interesting result, due to Brown et al. [1977], is that two C^*-algebras $\mathfrak{A}$, $\mathfrak{B}$ with countable approximate identity (this is automatic when the algebras are separable) are strongly Morita equivalent iff they are stably isomorphic; that is, when $\mathfrak{A} \simeq \mathfrak{B} \otimes \mathfrak{B}_0(\mathcal{H})$ for separable $\mathcal{H}$. Also cf. Lance [1995]. The notion of **stable isomorphism** appearing here is a noncommutative generalization of the same concept for vector bundles.

IV.2.5 The idea of looking at Hilbert C^*-modules coming from a group representation first appeared, in a different context, in Rieffel [1988]. It was rediscovered in Landsman [1995a], which contains most results in this section.

It would be interesting to have a criterion on U or H guaranteeing that the dense subspace $\tilde{\mathcal{E}} \subset \mathcal{H}$ assumed in Theorem 2.5.4 exists. More generally, one could ask for conditions guaranteeing the existence of an $\tilde{\mathcal{E}}$ such that the function defined by (2.77) lies in $L^1(H)$ for all $\Psi, \Phi \in \tilde{\mathcal{E}}$. This question is well known when L^1 is replaced by L^2; see, e.g., Dixmier [1977].

The fact about amenable groups used in the proof of Theorem 2.5.4 may be found in §3.6 of Greenleaf [1969], or in §II.3 of Renault [1980] (where the existence of the U_j is even given as the definition of amenability).

One may generalize the construction to the case where U is a representation of a Lie groupoid as defined in III.3.5.1, with associated direct integral Hilbert space III.(3.53). The generalization of (2.76) is

$$(\pi_{\mathrm{R}}(f)\Psi)_q = \int_{\tau_t^{-1}(q)} d\mu_q^t(\gamma)\, f(\gamma^{-1})U(\gamma)\Psi_{\tau_s(\gamma)}.$$

Equation (2.77) becomes

$$\langle\Psi, \Phi\rangle_{C_c^\infty(\mathrm{G})} : \gamma \to (\Psi_{\tau_t(\gamma)}, U(\gamma)\Phi_{\tau_s(\gamma)})_{\tau_t(\gamma)}.$$

For $\Phi = \Psi$ this is positive when Γ is amenable as defined by Renault [1980]; the relevant part of the proof of Theorem 2.5.4 may simply be copied.

IV.2.6 Theorem 2.6.1 is due to Rieffel [1974a]; also cf. Fell and Doran [1988]. The special case Corollary 2.6.4 is due to Mackey, and holds for locally compact groups. For the original proof see the books cited in the notes to III.2.9. A somewhat different derivation of Theorem 2.6.1 may be found in Rieffel [1974a] and Fell and Doran [1988]. The intermediate case Theorem 2.6.2 incorporates the generalization of 2.6.4 given by Moscovici [1969].

IV.2.7 A different derivation of Mackey's transitive imprimitivity theorem from Rieffel's imprimitivity theorem may be found in Rieffel [1974a] and Fell and Doran [1988]. These authors do not use the groupoid C^*-algebra $C^*(\mathsf{P} \times_H \mathsf{P})$ and then specialize to $\mathsf{P} = G$, but directly construct an equivalence bimodule between $C^*(G, G/H)$ and $C^*(H)$. A very efficient proof of the transitive imprimitivity theorem is given by Ørsted [1979].

Theorem 2.7.3 is contained in Glimm [1961, 1962]. Our proof combines the easy part of the proof of Thm. 1 in Glimm [1961] (namely the implications $1 \to 2$ and $3 \to 4$) with §VI.5 of Varadarajan [1985] and some elements of the proof of Prop. 8.1 in Rieffel [1979]. A different approach to the proof is contained in the proof of Thm. 2.2 in Glimm [1962], which could be somewhat simplified by using Lemma 1.1 in Effros and Hahn [1967].

The fact that Gq_0 is homeomorphic to G/G_{q_0} under the regularity assumption is contained in Thm. 1 of Glimm [1961]. A simpler proof may be given by first noting that the orbit is Hausdorff (which is a trivial consequence of the regularity assumption), and then proceeding as in the proof of Prop. 7.1 in Rieffel [1979].

An interesting application of Theorem 2.7.3 to the theory of quantum groups is contained in Koornwinder and Muller [1997].

When the regularity assumption on (G, Q) does not hold, the classification of the irreducible representations of $C^*(G, Q)$ appears to be impossible, and the right object to study is the primitive ideal space of $C^*(G, Q)$. This study was initiated by Effros and Hahn [1967], whose main conjecture on the structure of this space was proved by Gootman and Rosenberg [1979].

IV.2.8 Definition 2.8.1 is usually given in terms of covariant POV-measures; see all books on POV-measures cited in the notes to II.1.4 also for the covariant case. Theorem 2.8.2 is due to Poulsen [1970], and was rediscovered in physics by Neumann [1972], with further contributions by Scutaru [1977], Cattaneo [1979], and Castrigiano and Henrichs [1980]. Applying this theorem to the covariant Berezin quantization of the coadjoint orbits of compact Lie groups studied in III.1.11, one is naturally led to the Borel–Weil theory (cf. the notes to III.1.10), in a way independent of geometric quantization.

The analysis of covariant localization on $\mathbb{R}^3$ was initiated by Newton and Wigner [1949], and was reformulated in terms of systems of imprimitivity by Wightman [1962]. He used the original definition of these systems, namely (2.118). The generalization to arbitrary homogeneous spaces is due to Mackey [1968, 1978]; see Varadarajan [1985] for a detailed technical account, and consult Mackey [1992, 1998] for historical comments.

The Mackey–Wightman approach was further developed in Doebner and Tolar [1975], who suggested that the natural quantum Hamiltonian on $\mathcal{H}^\chi$ is the middle term in III.(2.192); cf. the classical expression III.(2.135). Moreover, they remarked that one could state the theory in terms of the action C^*-algebra $C^*(G, G/H)$. This approach was further developed by Majid [1988, 1990] and Landsman [1990a,b, 1992]. Other aspects of quantization theory on homogeneous spaces are discussed in Emch [1982, 1983], Camporesi [1990], Landsman and Linden [1991], Marinov [1995], Robson [1994, 1996], and Wu [1998]. The last two references confirm the picture sketched at the end of the section via geometric quantization.

Unlike a massive particle, a photon cannot be localized in the configuration space $\mathbb{R}^3$ in an $E(3)$-covariant way. It may, however, be covariantly localized in the generalized sense normally applied to phase space localization; see, for example, Ali and Emch [1974], Kraus [1977], and Brooke and Schroeck [1996]. In the notation of (2.109) etc., the wave function Ψ of a photon belongs to the subspace of elements $\Psi \in \mathcal{H}^1$ that satisfy the transversality condition $\nabla \cdot \Psi(x) = 0$. The projection p onto this subspace (given by $p\Psi = \Psi - \Delta_L^{-1}\nabla\nabla \cdot \Psi$, where Δ_L is the Laplacian) commutes with $U^1(E(3))$, so that one is in the setting of 2.8.2, with $G = E(3)$, $H = SO(3)$, $U^\chi = U^1$, and $\tilde{\pi}^\chi = \tilde{\pi}^1$ as defined in (2.117).

By Proposition II.1.4.8, this leads to an $E(3)$-covariant POVM $\Delta \mapsto A(\Delta)$ on $\mathbb{R}^3$ in $p\mathcal{H}^1$. In line with Corollary II.1.4.9, this POVM is given by $A(\Delta) = pE(\Delta)p$, where $E(\Delta) = \chi_\Delta \otimes \mathbb{I}_1$ (cf. the main text). The position operators $Q_k = \int_{\mathbb{R}^3} dA(x)x_k$ (cf. II.(1.34) and the main text) do not commute with each other. The classical counterpart of this phenomenon is mentioned after (3.8). Indeed, Duval and Elhadad [1992] show that the geometric quantization of the canonical classical position variables precisely yields the quantum position operators Q_k just defined.

A different approach to photon localization, based on microlocal analysis, has been initiated by Omnès [1997b].

IV.2.9 The problem of quantizing constrained systems has been faced since the earliest days of quantum mechanics. A good historical overview of the treatment of gauge invariance in quantum electrodynamics is given in Weinberg [1995], which with Weinberg [1996] also contains an up-to-date treatment of heuristic techniques used by physicists to deal with gauge invariance and constraints.

Books more specifically dealing with constrained quantization include Dirac [1964] (which initiated the modern era), Sundermeyer [1982], Govaerts [1991], and Henneaux and Teitelboim [1992]. The technique of BRST quantization developed in the last two books (as well as in Weinberg [1996]) seems to perform well in quantum field theory and string theory, especially in their path-integral version. Applied to finite-dimensional systems, the operatorial BRST technique faces similar functional-analytic problems as the Dirac method; cf. Landsman and Linden [1992] for simple examples. Nonetheless, the BRST method remains the most highly developed and widely used method of constrained quantization to date. See Duval et al. [1991] for a "bosonic" reformulation of BRST.

There is an extensive literature on the geometric quantization of constrained systems; see, for example, Gotay [1986], Ashtekar and Stillerman [1986], Blau [1988], Tuynman [1990], Woodhouse [1992], and Robson [1994, 1996].

A C^*-algebraic approach to constrained quantization that is closer in spirit to the Dirac method than the technique described in the main text, has been developed by Grundling and Hurst [1985, 1987, 1988a,b]. Applications to quantum field theory are given in Grundling [1988]; similar techniques are used by Thirring and Narnhofer [1992], and Acerbi et al. [1993a,b]. This approach has the advantage of being able to handle second-class constraints (which in our method have to be brought into first class form by reformulating the classical situation), but lacks the connection with symplectic reduction and Hilbert C^*-modules.

Other mathematically sound attempts to rescue the Dirac method include Ashtekar and Tate [1994], Ashtekar et al. [1995], and Klauder [1997]. In Klauder's approach the projection $p_{\rm id}$ is replaced by approximate projections in the spirit of the $p^j_{\rm id}$ used in the proof of Theorem 2.5.4.

The use of Rieffel induction in constrained quantization started with Landsman [1995a]; Definition 2.9.1 appeared in Landsman [1998a].

The idea of constructing an inner product by group averaging as in (2.81) (with $\mathcal{H}_\chi = \mathbb{C}$) goes back at least to Nachtmann [1968]. In the context of constrained quantization see

Teitelboim [1982, 1984], Higuchi [1991], Halliwell and Hartle [1991], and Ashtekar et al. [1995].

IV.2.10 The analysis of the operator (2.127) is done with Weyl's method; see, for example, Thm. X.7 in Reed and Simon [1975]. Our operator is in the limit circle case at 0 and in the limit point case at ∞. The essential self-adjointness on $V_\chi C_c^\infty(\mathbb{R}^2)$ follows from Thm. 3 in Nussbaum [1964], or from a direct argument. Some abstract theory behind this example is developed in Wren [1997], who in addition discusses a profound generalization.

There is a great deal of literature on generalized eigenfunctions of the type $f_k^\pm$ and the corresponding expansions; see Berezanskii [1968] for an old but still adequate, and Poerschke et al. [1989] and Poerschke and Stolz [1993] for a more recent treatment. The group generated by the Lie algebra defined by (2.137) is the two-dimensional Poincaré group.

It is instructive to replace the classical constraint $\varphi = \frac{1}{2}(p_1^2 - p_2^2)$ by $\varphi_\pm = \varphi \pm \frac{1}{2}\exp(4q^1)$. Interestingly, the Hamiltonian flow of φ_- on $T^*\mathbb{R}^2$ is incomplete, so that the constraint fails to generate an action of $\mathbb{R}$.

The constraints are quantized on $L^2(\mathbb{R}^2)$ by $\mathcal{Q}(\varphi_\pm) = \mathcal{Q}(\varphi) \pm \frac{1}{2}\exp(4q^1)$, where the last term is a multiplication operator. The incompleteness of φ_- is reflected in the quantum theory, because $\mathcal{Q}(\varphi_-)$ is not essentially self-adjoint on $C_c^\infty(\mathbb{R}^2)$. What follows applies to any self-adjoint extension.

The most fundamental difference between φ and $\varphi_\pm$ is that the spectrum of (the closure of) $\mathcal{Q}(\varphi_+)$ and of $\mathcal{Q}(\varphi_-)$ is $\mathbb{R}$ with multiplicity one, whereas the spectrum of $\mathcal{Q}(\varphi)$ is $\mathbb{R}$ with multiplicity two. Consequently, for fixed k only one of the two generalized eigenfunctions of $\mathcal{Q}(\varphi_\pm)$ plays a role in the construction of the induced space (as opposed to the pair $f_k^\pm$ in the main text), which is naturally isomorphic to $L^2(\mathbb{R}, dk/2\pi|k|)$. These eigenfunctions may be deduced from the elementary theory of Bessel functions, but one needs more specific Hilbert space techniques to decide which one occurs in the spectral decomposition of $\mathcal{H}$. These techniques may be found, for example, in Picard [1989].

The details of the quantum treatment of φ and $\varphi_\pm$ may be found in Landsman [1998a]. These constraints are motivated by quantum cosmology and the question what the "wave function of the universe" should be; see Landsman [1995c] for this context. See Marolf [1997] and references therein for an analogous treatment of the constraints of quantum cosmology.

The above consideration on multiplicity is relevant to the constrained quantization proposal of Hájiček [1994], who suggests that all generalized solutions of the quantum constraints should be used in the construction of the physical Hilbert space of pure quantum states.

A different approach to the quantization of singular Marsden–Weinstein quotients is presented by Emmrich and Römer [1990], who face the problem of having to decide which self-adjoint extension of the reduced Hamiltonian to choose. See Śniatycki and Weinstein [1983] for yet another approach (further discussed in Wren [1997]). Meinrenken and Sjamaar [1998] look at the problem in the context of geometric quantization.

Despite the existence of a large number of papers on the quantization of the moduli space of flat connections on a compact Riemann surface (cf. the notes to 1.11), the effect of the singularities in this space on the quantum theory is not well understood (see Jeffrey and Weitsman [1992] for a careful treatment of the singular points in geometric quantization).

IV.3.1 For the Poincaré group and its use in physics see Barut and Raçka [1977], Varadarajan [1985], Woodhouse [1992], or Weinberg [1996]. The coadjoint orbits of the Poincaré group were first described by Souriau [1969, 1997] and Arens [1971a,b]. Further discussions of these orbits may be found in Guillemin and Sternberg [1984], Cariñena, Gracia-Bondía, and Várilly [1990], Woodhouse [1992], Duval and Elhadad [1992], and Schroeck [1996]. The noncommutativity of position coordinates of particles with nonzero spin or helicity is discussed by Bacry [1988].

IV.3.2 For $m > 0$, Theorem 3.2.2 is suggested in §I.20 of Guillemin and Sternberg [1984]. The general case is taken from Landsman and Wiedemann [1994], where the details of the proof may be found.

IV.3.3 For Proposition 3.3.1, originally due to Wigner [1939], see the first four references in the notes to IV.3.1 above. It is usually thought that the irreducible representations of P provide an adequate description of elementary particles in asymptotic states (and this was Wigner's motivation as well), but this description fails even for electrons (because of the photon cloud always surrounding them), and also, for different reasons, for quarks and gluons. See Buchholz [1996] for a promising new approach.

The trick involved in (3.20) is due to Carey et al. [1977, 1978]. Proposition 3.3.3 comes from Landsman [1995a]. For covariant representations see the notes to the next section.

One may wonder whether covariant Berezin quantizations of the coadjoint orbits of the Poincaré group exist; see the discussion following the proof of Theorem 2.8.2. It unfortunately turns out that for the physical orbits the condition in Corollary 2.8.3 cannot be satisfied; see Schroeck [1996] (who reaches this conclusion in a different way), and references therein on this issue. For attempts to construct nonetheless a covariant relativistic quantum mechanics of single particles, see Ali [1985], Ali et al. [1995], and Schroeck [1996]. As explained in Ali [1998], this leads to a hyperplane-dependent notion of localization in phase space. (For the analogous proposal of hyperplane-dependent localization in Minkowski space, see Butterfield and Fleming [1998].) At least for the massive orbits, covariant Weyl quantization turns out to be possible; see Cariñena et al. [1990]. This is reminiscent of the nonrelativistic theory, in which Weyl quantization has better covariance properties than Berezin quantization, too; cf. Theorems II.2.4.3 and II.2.5.1.

IV.3.4 Covariant representations of P and Proposition 3.4.1 have a long tradition (going back to Pauli and Wigner), culminating in the work of Weinberg [1995] (in which references to his original work in the early sixties may be found). For a more mathematical treatment of wave equations for massive fields see Barut and Raçka [1977] and Asorey et al. [1985].

The finite-dimensional representations of the Lorentz group are labeled by two positive integers j_1, j_2. The decomposition of $\mathcal{R}_{(j_1,j_2)}$ under the restriction to $SO(3)$ is given by the well-known Clebsch–Gordan series, so that all integral spins between $|j_1 + j_2|$ and $|j_1 - j_2|$ occur. In the restriction to $E(2)$ the only helicity that occurs as a proper subrepresentation is $j_1 - j_2$; see Weinberg [1995]. This result misses representations on reduced spaces, and therefore fails to explain the connection between masslessness and gauge invariance. It does explain why helicity 0 occurs in the proof of Proposition 3.4.2, since $\mathcal{R}_V = \mathcal{R}_{(1/2,1/2)}$.

The second half of 3.4.2 is taken from Landsman and Wiedemann [1994], as is Theorem 3.4.4. This paper also treats helicity ± 2, relating the masslessness of the graviton to the infinitesimal diffeomorphism invariance of linearized gravity.

The idea of gauge invariance lies at the basis of modern high-energy physics. The connection between masslessness and gauge invariance holds for most, but not all, interacting theories. The most famous exception is massless quantum electrodynamics in $d = 2$, as recognized by Schwinger [1962]; see Lowenstein and Swieca [1971] for the definitive treatment of the Schwinger model.

IV.3.5 A standard reference for the CCR-algebra is Bratteli and Robinson [1981], which, however, does not give Definition 3.5.1. The equivalence between our definition (which we learned from H. Grundling) and the standard one may be proved using a uniqueness theorem due to Slawny [1972]. As is clear from the fact that to define the CCR-algebra one quite unnaturally has to equip $\mathcal{K}$ with the discrete topology, it is an object best avoided.

Our definition of the Fock representation is also different from, but equivalent to, that in Bratteli and Robinson [1981]. In physics this representation is known as the **second quantization** of $\mathcal{K}$; the representation U_χ^F defined in Proposition 3.5.2 is usually called $\Gamma(U_\chi)$. See, for example, Reed and Simon [1975].

Proposition 3.5.3 is due to Grundling and Hurst [1987]; the present proof, due to H. Grundling, fills a gap in the proof in that reference. Araki [1963], Thm. 1(5), gave an arduous proof of the corresponding von Neumann algebra result $\pi(\mathfrak{W}(\mathcal{V}))'' = \pi(\mathfrak{W}(\mathcal{V}^\perp))'$ for any regular representation π. This result follows immediately from 3.5.3, so that the regularity assumption may evidently be dropped.

The Fock representation of $\mathfrak{W}(S^{0,\mathbb{R},V})$, with the conventions (3.20), was introduced by Carey et al. [1977, 1978], who regard it as a rigorous version of the **Fermi representation** of quantum electromagnetism.

The treatment of Radon measures based on his own theory of Hilbert subspaces (cf. the notes to II.1.5) may be found in Schwartz [1973]. Related approaches to measure theory on infinite-dimensional topological vector spaces are presented in Kuo [1975] and Guichardet [1972]. Malliavin [1997] is entirely concerned with Gaussian measures. Physicists will enjoy the discussion in DeWitt-Morette et al. [1979] and Choquet-Bruhat et al. [1982].

The Gaussian measure μ_γ on $\mathcal{V}$ defined by (3.53) is the image of a so-called cylinder measure μ_γ^c on $\mathcal{H}$, but this way of looking at things is complicated by the fact that in infinite dimension $\mu_\gamma(\mathcal{H}) = 0$, although $\mu_\gamma^c(\mathcal{H}) = 1$.

Theorem 3.5.5, generalizing the original result of Cameron and Martin [1944] for $\mathcal{H} = L^2([0,1], \mathbb{R}^n)$ and $\mathcal{V} = C([0,1], \mathbb{R}^n)_0$, is due to Thomas [1983]. Related results are in Kuo [1975] and Malliavin [1997]. A locally convex space is called quasi-complete when all closed and bounded sets are complete.

Theorem 3.5.7 is due to Landsman and Wiedemann [1994] (also see Wiedemann [1994]), who used cylinder measures on $\mathcal{G}$. The conventional treatment of the quantized free electromagnetic field may be found in Weinberg [1995].

There actually exists a construction of a group algebra for certain infinite-dimensional groups; see Grundling [1997]. It would be interesting to try to formulate Theorem 3.5.7 using Rieffel induction on this group algebra.

IV.3.6 The literature on two-dimensional Yang–Mills theories is formidable; much of it was triggered by the work of Rajeev [1988] and Witten [1991, 1992]. A review of the Euclidean theory is given by Sengupta [1997].

Definition 3.6.1 is due to Rajeev and Rossi [1995]. For Sobolev loop groups see Frenkel [1984] and Freed [1988]. (A basic reference on smooth loop groups is Pressley and Segal [1986].) Sobolev spaces of paths on Riemannian manifolds are studied in Klingenberg [1982]. In general dimensions, Sobolev gauge groups and spaces of connections are considered, e.g., in Mitter and Viallet [1981], Freed and Uhlenbeck [1984], and Kondracki and Sadowski [1986].

A proof of the smoothness claimed in Lemma 3.6.2 may be found in Freed and Uhlenbeck [1984], App. A, or in Rajeev and Rossi [1995]. The latter also contains the description of $\mathcal{G}$ as a semidirect product, as well as Lemma 3.6.4 (whose heuristic version first appeared in Rajeev [1988]).

Wilson loops come from physics, where one writes (3.73) as

$$\mathcal{W}(A) = P\,\mathrm{Exp}\left(-\int_0^1 d\alpha\, A(\alpha)\right).$$

Here P stands for path-ordering; see Dollard and Friedman [1979] for a rigorous discussion of such "path-ordered" or "product" integrals. Wilson loops are used in Yang–Mills theory as well as in gravity. An interesting monograph on this topic is Gambini and Pullin [1996], where further references may be found.

Proposition 3.6.3 and Theorem 3.6.5 are taken from Landsman and Wren [1997]. Gross [1993], Thm. 2.5, proves the statement preceding Theorem 3.6.5 for almost every A with respect to the measure $\mu_1^{W_e}$ defined in 3.7.

For Definition 3.6.6 see Bröcker and tom Dieck [1985]; Proposition 3.6.7 is due to Hall [1997b].

Langmann and Semenoff [1993] relate the appearance of the Weyl group W in the Stieffel chamber T/W to the so-called Gribov problem, which occurs when one tries to fix the gauge in this model. The geometry of Stieffel chambers is discussed in Bröcker and tom Dieck [1985]. The effect of the singularities in these chambers on constrained quantization is analyzed in Wren [1997, 1998b].

IV.3.7 Proposition 3.7.1 is due to Dimock [1996]. The representation (3.85) has been considered, in various realizations, by many authors, such as Albeverio and Hoegh-Krohn [1978], Frenkel [1984], and Ismagilov [1996]. It is a special case of a general class of "energy" representations of gauge groups in various dimensions introduced by Gelfand et al. [1977]. Further to these authors, Albeverio et al. [1981] and Wallach [1987] show that such representations are irreducible when the dimension of space is ≥ 3, and provide criteria for irreducibility in dimension 2. In our case of dimension 1, the representation U_γ is evidently reducible, but Driver and Hall [1998] prove that as in higher dimensions, it has no trivial subrepresentation.

A complete proof of Lemma 3.7.3 may be found in Guichardet [1972], Thm. 7.1. Equation (3.89) in Definition 3.7.4 is motivated by a construction of Dimock [1996], who uses stochastic calculus. The approach through (3.90) is due to Wren [1998a,b].

For Ito's map and all Wiener measures in this section see Frenkel [1984] and Malliavin and Malliavin [1990]. The Cameron–Martin formula (3.100) appears as Prop. (5.2.7) in Frenkel [1984], Thm. 1.3 in Malliavin and Malliavin [1990], and Thm. XI.1.4.3 in Malliavin [1997]. A further generalization to paths on Riemannian manifolds is given by Hsu [1995]; also cf. Malliavin [1997].

Different approaches to the quantization of Yang–Mills theory on a cylinder are presented by Hetrick [1994], Dimock [1996], and Hall and Driver [1998]. For the Euclidean theory see Witten [1991, 1992].

IV.3.8 For $H = U(1)$, Theorem 3.8.1 and its proof are due to Landsman and Wren [1997]; the general case was proved by Wren [1998a,b]. The connection between the Wiener measure and the heat kernel on $\mathbb{R}^n$ is classical; for loop groups see Frenkel [1984] and Malliavin and Malliavin [1990]. For smoothness and other properties of general heat kernels cf. Davies [1989]. The analyticity arguments used in the proof of 3.8.1.5 are developed in Hall [1994] (who establishes the unique analytic continuation of the heat kernel on H) and Wren [1998b].

Definition 3.8.3 is due to Hall [1994], who actually defined two inequivalent families of coherent states for compact Lie groups (a third family was added in Hall [1998]). The associated Segal–Bargmann transform is studied in Hall [1994, 1997a]. For a review cf.

Hall [1997c]. Hall's coherent states appear, in a different way, also in the quantization of Yang–Mills theory on a circle by Hall and Driver [1998].

We have not addressed the rather difficult issue of the Hamiltonian of the theory; see Dimock [1996] and Wren [1998a,b].

IV.3.9 The role of $\pi_1(Q)$ and the associated "θ-angles" in the quantization of a particle on a multiply connected configuration space was independently discovered by Schulman [1968, 1981] and Souriau [1969, 1997]. Another important early paper is Laidlaw and DeWitt [1970]. More recent treatments, all different from ours, are Śniatycki [1980], Isham [1983], Horvathy et al. [1989], Balachandran et al. [1991], and Giulini [1995].

In the works of all these authors the fundamental group $\pi_1(Q)$ plays a central role. To relate this to discrete reduction, we generalize the discussion of multiply connected Lie groups in the main text. Recall that a multiply connected space Q may be written as $Q = \tilde{Q}/\pi_1(Q)$, where $\tilde{Q}$ is the universal covering space of Q. As for $Q = G$, we have $T^*Q \simeq (T^*\tilde{Q})/\pi_1(Q)$. Hence the inequivalent quantizations of T^*Q are labeled by the unitary dual of $\pi_1(Q)$.

The emergence of θ-angles in quantum field theory was discovered by Lowenstein and Swieca [1971]. It later turned out that such angles are relevant to quantum chromodynamics; the physics literature is reviewed by Jackiw [1985] and Weinberg [1996]. For $U(1)$ gauge theory on the circle also see Manton [1985]. There is a fundamental difference between θ-angles in quantum field theories on a compact space, which are of a purely topological nature, as discussed in the main text, and also in Asorey [1981] and Jackiw [1985], and θ-angles in theories on a noncompact space. The latter are of a dynamical origin, and are closely related to the infrared behavior of the theory. See Acerbi et al. [1993b], and Löffelholz et al. [1996] for a rigorous discussion.

The treatment of quantum mechanics on the circle is taken from Landsman [1990b]; an alternative mathematical discussion may be found in Isham [1983] and in Asorey et al. [1983]. The Aharonov–Bohm effect was discovered by Aharonov and Bohm [1959]; for a rigorous discussion see Asorey [1982] and Ruijsenaars [1983]. The easiest way to prove (3.120) is to use the theorem of Dixmier and Malliavin [1978] quoted in the notes to III.1.5. This yields the boundary condition on Ψ; the precise domain then follows from Example 1 in Section X.1 of Reed and Simon [1975].

Our approach to θ-angles in constrained quantization, taken from Landsman and Wren [1997], is intended to explain the origin of θ-angles in quantum Marsden–Weinstein reduction by a disconnected gauge group. (The theory and applications of discrete reduction in classical mechanics may be found in Marsden [1992].) We hereby complement treatments based on Dirac's quantization method (such as the one of Jackiw [1985]).

References

Abadie, B. and R. Exel [1997] Deformation quantization via Fell bundles. *e-print funct-an/970600.*

Abbati, M.C., R. Cirelli, P. Lanzavecchia, and A. Mania [1984] Pure states of general quantum-mechanical systems as Kähler bundles. *Nuovo Cim.* **B83**, 43–59.

Abraham, R. and J.E. Marsden [1985] *Foundations of Mechanics*, 2nd ed. Addison Wesley, Redwood City.

Acerbi, F., G. Morchio, and F. Strocchi [1993a] Infrared singular fields and nonregular representations of canonical commutation relation algebras. *J. Math. Phys.* **34** (1993), 899–914.

Acerbi, F., G. Morchio, and F. Strocchi [1993b] Theta vacua, charge confinement and charged sectors from nonregular representations of CCR algebras. *Lett. Math. Phys.* **27**, 1–11.

Aharonov, Y. and D. Bohm [1959] Significance of electromagnetic potentials in quantum theory. *Phys. Rev.* **115**, 485–491.

Ajupov, S.A., B. Iochum, and N.D. Yadgorov [1990] Symmetry versus facial homogeneity for self-dual cones. *Lin. Alg. Appl.* **142**, 83–89.

Akemann, C.A. [1969] The general Stone–Weierstrass problem. *J. Funct. Anal.* **4**, 227–294.

Akemann, C.A. and G.K. Pedersen [1992] Facial structure in operator algebra theory. *Proc. London Math. Soc.* **64**, 418–448.

Akemann, C.A. and F.W. Shultz [1985] Perfect C^*-algebras. *Mem. Amer. Math. Soc.* **326**, 1–117.

Albert, C. and P. Dazord [1988] Théorie des groupoides symplectiques. I. Théorie générale des groupoides de Lie. *Publ. Dépt. Math. Univ. C. Bernard–Lyon I (nouv. sér.)* **B88–4**, 51–105.

Albert, C. and P. Dazord [1990] Théorie des groupoides symplectiques. II. Groupoides symplectiques. *Publ. Dépt. Math. Univ. C. Bernard–Lyon I (nouv. sér.)*, 27–99.

Alberti, P.M. [1983] A note on transition probabilityover C^*-algebras. *Lett. Math. Phys.* **7**, 25–32.

Albertin, U.K. [1991] The diffeomorphism group and flat principal bundles. *J. Math. Phys.* **32**, 1975–1980.

Albeverio, S. and R. Høegh-Krohn [1977] Oscillatory integrals and the method of stationary phase in infinitely many dimensions, with applications to the classical limit of quantum mechanics. *Inv. Math.* **40**, 59–106.

Albeverio, S. and R. Høegh-Krohn [1978]. The energy representation of Sobolev–Lie groups. *Compositio Math.* **36**, 37–52.

Albeverio, S., R. Høegh-Krohn, and D. Testard [1981] Irreducibility and reducibility for the energy representation of the group of mappings of a Riemannian manifold into a compact semisimple Lie group. *J. Funct. Anal.* **41**, 378–396.

Alekseevsy, D., J. Grabowski, G. Marmo, and P.W. Michor [1994] Poisson structures on the cotangent bundle of a Lie group or a principle bundle and their reductions. *J. Math. Phys.* **35**, 4909–4927.

Alfsen, E.M. [1970] *Compact Convex Sets and Boundary Integrals.* Springer, Berlin.

Alfsen, E.M. [1977] On the state spaces of Jordan and C^*-algebras. In: Connes, A. (ed.) *Algèbres d'opérateurs et leurs applications en physique mathématique.* Editions CNRS, Paris.

Alfsen, E.M., H. Hanche-Olsen, and F.W. Shultz [1980] State spaces of C^*-algebras. *Acta Math.* **144**, 267–305.

Alfsen, E.M. and F.W. Shultz [1976] Non-commutative spectral theory for affine function spaces on convex sets. *Mem. Amer. Math. Soc.* **172**, 1–120.

Alfsen, E.M. and F.W. Shultz [1978] State spaces of Jordan algebras. *Acta Math.* **140**, 155–190.

Alfsen, E.M. and F.W. Shultz [1979] On non-commutative spectral theory and Jordan algebras. *Proc. London Math. Soc.* **38**, 497–516.

Alfsen, E.M. and F.W. Shultz [1998] On orientation and dynamics in operator algebras. Part I. *Commun. Math. Phys.* **194**, 87–108.

Alfsen, E.M., F.W. Shultz, and E. Størmer [1978] A Gelfand–Neumark theorem for Jordan algebras. *Adv. Math.* **28**, 11–56.

Ali, S.T. [1985] Stochastic localisation, quantum mechanics on phase space and quantum space-time. *Riv. Nuovo Cim.* **8** (11), 1–128.

Ali, S.T. [1998] Systems of covariance in relativistic quantum mechanics. *Int. J. Theor. Phys.* **37**, 365–373.

Ali, S.T., J.-P. Antoine, J.-P. Gazeau, and U.A. Mueller [1995] Coherent states and their generalizations: a mathematical overview. *Rev. Math. Phys.* **7**, 1013–1104.

Ali, S.T. and H.-D. Doebner [1990] Ordering problem in quantum mechanics: prime quantization and a physical interpretation. *Phys. Rev.* **A41**, 1199–1210.

Ali, S.T. and G.G. Emch [1974] Fuzzy observables in quantum mechanics. *J. Math. Phys.* **15**, 176–182.

Ali, S.T. and G.A. Goldin [1991] Quantization, coherent states and diffeomorphism groups. In: Henning, J.D., W. Lücke, and J. Tolar (eds.) *Differential Geometry, Group Representations, and Quantization*, pp. 147–178. *Lecture Notes in Physics* **379**. Springer, Berlin.

Alvermann, K. [1985] The multiplicative triangle inequality in noncommutative JB and JB^*-algebras. *Abh. Math. Sem. Uni. Hamburg* **55**, 91–96.

Amemiya, I. and H. Araki [1966] A remark on Piron's paper. *Publ. RIMS (Kyoto)* **A2**, 423–427.

Anandan, J. [1991] A geometric approach to quantum mechanics. *Found. Phys.* **21**, 1265–1284.

Angermann, B., H.-D. Doebner, and J. Tolar [1983] Quantum kinematics on smooth manifolds. In: Andersson, S.I. and H.-D. Doebner (eds.) *Nonlinear Partial Differential Operators and Quantization Procedures*, pp. 171–208. *Lecture Notes in Mathematics* **1037**. Springer, Berlin.

Appelquist, T., A. Chodos, and P.G.O. Freund [1985] (eds.) *Modern Kaluza–Klein Theories*. Benjamin–Cummings, Menlo Park.

Arai, T. [1995] Some extensions of the semiclassical limit $\hbar \to 0$ for Wigner functions on phase space. *J. Math. Phys.* **36**, 622–630.

Araki, H. [1963] A lattice of von Neumann algebras associated with the quantum theory of a free Bose field. *J. Math. Phys.* **4**,1343–1362.

Araki, H. [1972] Bures distance function and a generalization of Sakai's non-commutative Radon–Nikodym theorem. *Publ. RIMS Kyoto* **8**, 335–362.

Araki, H. [1980] On a characterization of the state space of quantum mechanics. *Commun. Math. Phys.* **75**, 1–24.

Araki, H. and G.A. Elliott [1973] On the definition of C^*-algebras. *Publ. RIMS* **9**, 93–112.

Araki, H. and G.A. Raggio [1982] A remark on transition probability. *Lett. Math. Phys.* **6**, 237–240.

Archbold, R.J. and F.W. Shultz [1989] Characterization of C^*-algebras with continuous trace by properties of their pure states. *Pac. J. Math.* **136**, 1–13.

Arens, R. [1971a] Classical Lorentz invariant particles. *J. Math. Phys.* **12**, 2415–2422.

Arens, R. [1971a] Classical relativistic particles. *Commun. Math. Phys.* **21**, 139–149.

Arhangel'skii, A.V., K.R. Goodearl, and B. Huisgen-Zimmermann [1997] Kiiti Morita 1915–1995. *Notices Amer. Math. Soc.* **44**, 680–684.

Arms, J.M. [1981] The structure of the solution set for the Yang–Mills equations. *Math. Proc. Camb. Phil. Soc.* **90**, 361–372.

Arms, J.M. [1986] Symmetry and solution set singularities in Hamiltonian field theories. *Acta Phys. Polon.* **B17**, 499–523.

Arms, J. [1996] Reduction of Poisson algebras at nonzero momentum values. *J. Geom. Phys.* **21**, 81–95.

Arms, J.M., R.H. Cushman, and M.J. Gotay [1991] A universal reduction procedure for Hamiltonian group actions. In: Ratiu, T. (ed.) *The Geometry of Hamiltonian Systems*, pp. 33–52. Springer, New York.

Arms, J.M., M.J. Gotay, and G. Jennings [1990] Geometric and algebraic reduction for singular momentum maps. *Adv. Math.* **79**, 43–103.

Arms, J.M., J.E. Marsden, and V. Moncrief [1981] Symmetry and bifurcation of momentum mappings. *Commun. Math. Phys.* **78**, 455–478.

Arms, J.M., J.E. Marsden, and V. Moncrief [1982] The structure of the space of solutions of Einstein's equations II: Several Killing fields and the Einstein–Yang–Mills equations, *Ann. Phys. (N.Y.)* **144**, 81–106.

Arnal, D. and J. Ludwig [1992] Convexity of the moment mapping of a Lie group. *J. Funct. Anal.* **105**, 256–300.

Arnold, V.I. [1989] *Mathematical Methods of Classical Mechanics*, 2nd ed. Springer, New York.

Arnold, V.I. and A.B. Givental [1990] Symplectic Geometry. In: Arnold, V.I. and S.P. Novikov (eds.) *Dynamical Systems IV*, pp. 1–136. Springer, Berlin.

Arodz, H. [1983] Motion of a wave packet in an external Yang–Mills field. *Acta Phys. Polon.* **B14**, 757–773.

Arodz, H. [1988] Lagrangian and Hamiltonian formulations of dynamics of classical particles with spin and color. *Acta Phys. Polon.* **B19**, 697–708.

Aronszajn, N. [1950] Theory of reproducing kernels. *Trans. Amer. Math. Soc.* **68**, 337–404.

Ashtekar, A., J. Lewandowski, D. Marolf, J. Mourão, and T. Thiemann [1995] Quantization of diffeomorphism invariant theories of connections with local degrees of freedom. *J. Math. Phys.* **36**, 6456–6493.

Ashtekar, A. and T.A. Schilling [1998] Geometrical formulation of quantum mechanics. *e-print gr-qc/9706069*.

Ashtekar, A. and M. Stillerman [1986] Geometric quantization and constrained systems. *J. Math. Phys.* **27**, 1319–1330.

Ashtekar, A. and R.S. Tate [1994] An algebraic extension of Dirac quantization: examples. *J. Math. Phys.* **35**, 6434–6470.

Asimow, L. and A.J. Ellis [1980] *Convexity Theory and its Applications in Functional Analysis*. Academic Press, London.

Asorey, M. [1981] Some remarks on the classical vacuum structure of gauge field theories. *J. Math. Phys.* **22**, 179–184.

Asorey, M. [1982] Regularity of gauge equivalence in quantum mechanics and the Aharonov–Bohm effect. *Lett. Math. Phys.* **6**, 429–435.

Asorey, M., L.J. Boya, and J.F. Cariñena [1985] Covariant representations in a fibre bundle framework. *Rep. Math. Phys.* **21**, 391–404.

Asorey, M., J.G. Esteve, and A.F. Pacheco [1983] Planar rotor – The θ-vacuum structure, and some approximate methods in quantum mechanics. *Phys. Rev.* **D27**, 1852–1868.

Atiyah, M.F. [1957] Complex and analytic connections in fibre bundles. *Trans. Amer. Math. Soc.* **85**, 181–207.

Atiyah, M.F. [1982] Convexity and commuting Hamiltonians. *Bull. London Math. Soc.* **23**, 1–15.

Atiyah, M.F. and R. Bott [1983] The Yang–Mills equations over Riemann surfaces. *Phil. Trans. R. Soc. London* **A308**, 524–615.

Audin, M. [1991] *The Topology of Torus Actions on Symplectic Manifolds*. Birkhäuser, Basel.

Auslander, L. and C.C. Moore [1966] Unitary representations of solvable Lie groups. *Mem. Amer. Math. Soc.* **62**.

Ayupov, S.A., A. Rakhimov, and S. Usmanov [1997] *Jordan, Real and Lie Structures in Operator Algebra*. Kluwer, Dordrecht.

Azcárraga, A. de and J.M. Izquierdo [1995] *Lie Groups, Lie Algebras, Cohomology, and some Applications in Physics*. Cambridge University Press, Cambridge.

Azencott, R. et al. [1981] *Géodésiques et diffusions en temps petit. Séminaire de probabilités Université de Paris VII. Astérisque* **84–85**.

Azencott, R. and H. Doss [1985] L'équation de Schrödinger quand $\hbar \to 0$: une approche probabiliste. *Lecture Notes in Mathematics* **1109**, 1–17.

Bacry, H. [1988] *Localizability and Space in Quantum Physics. Lecture Notes in Physics* **308**. Springer, Heidelberg.

Baer, R. [1952] *Linear Algebra and Projective Geometry*. Academic Press, New York.

Baguis, P. [1998] Semidirect products and the Pukanszky condition. *J. Geom. Phys.* **25**, 245–270.

Bais, F.A. and P. Batenburg [1985] A new class of higher–dimensional Kaluza–Klein monopole and instanton solutions. *Nucl. Phys.* **B253**, 162–172.

Balachandran, A.P., G. Marmo, N. Mukunda, J.S. Nilsson, E.C.G. Sudarshan, and F. Zaccaria [1984] Non-abelian monopoles break color. 1. Classical Mechanics. *Phys. Rev.* **D29**, 2919–2935.

Balachandran, A.P., G. Marmo, B.S. Skagerstam, and A. Stern [1983] *Gauge Symmetries and Fiber Bundles – Applications to Particle Dynamics. Lecture Notes in Physics* **188**. Berlin, Springer.

Balachandran, A.P., G. Marmo, B.S. Skagerstam, A. Stern [1991] *Classical Topology and Quantum States*. World Scientific, Singapore.

Bargmann, V. [1961] On a Hilbert space of analytic functions and an associated integral transform. *Comm. Pure Appl. Math.* **14**, 187–214.

Bargmann, V. [1964] Note on Wigner's theorem on symmetry operations. *J. Math. Phys.* **5**, 862–868.

Barmoshe, D. and M.S. Marinov [1994] Realization of compact Lie algebras in Kähler manifolds. *J. Phys.* **A27**, 6287–6298.

Barnes, I.M.S., M. Nauenberg, M. Nockleby, and S. Tomsovic [1994] Classical orbits and semiclassical wavepacket propagation in the Coulomb potential. *J. Phys.* **A27**, 3299–3321.

Barut, A.O. and R. Raçka [1977] *Theory of Group Representations and Applications*. PWN, Warszawa.

Bass, H. [1968] *Algebraic K-Theory*. Benjamin, New York.

Bates, S. and A. Weinstein [1995] *Lectures on the Geometry of Quantization. Berkeley Mathematics Lecture Notes* **8**. University of California, Berkeley.

Bayen, F., M. Flato, C. Fronsdal, A. Lichnerowicz, and D. Sternheimer [1978] Deformation theory and quantization I, II. *Ann. Phys. (N.Y.)* **110**, 61–110, 111–151.

Belinfante, J.G.F. [1976] Transition probability spaces. *J. Math. Phys.* **17**, 285–290.

Bellissard, J. and M. Vittot [1990] Heisenberg's picture and non commutative geometry of the semi classical limit in quantum mechanics. *Ann. Inst. H. Poincaré* **A52**, 175–235.

Belov, V.V. and V.P. Maslov [1990] Quasi-classical trajectory coherent states in quantum mechanics with gauge fields. *Dokl. Akad. Nauk.* **311**, 849–854.

Beltrametti, E.G. and G. Cassinelli [1984] *The Logic of Quantum Mechanics*. Cambridge University Press, Cambridge.

Bennett, J.G. [1978] Induced representations of C^*-algebras and complete positivity. *Trans. Amer. Math. Soc.* **243**, 1–36.

Berezanskii, Ju.M. [1968] *Expansions in Eigenfunctions of Self-Adjoint Operators*. American Mathematical Society, Providence.

Berezin, F.A. [1967] Some remarks about the associative envelope of a Lie algebra. *Funct. Anal. Appl.* **1**, 91–102.

Berezin, F.A. [1972] Covariant and contravariant symbols of operators. *Math. USSR Izv.* **6**, 1117–1151.

Berezin, F.A. [1974] Quantization. *Math. USSR Izv.* **8**, 1109–1163.

Berezin, F.A. [1975a] Quantization in complex symmetric spaces. *Math. USSR Izv.* **9**, 341–379.

Berezin, F.A. [1975b] General concept of quantization. *Commun. Math. Phys.* **40**, 153–174.

Berger, C.A. and L.A. Coburn [1986] Toeplitz operators and quantum mechanics. *J. Funct. Anal.* **68**, 273–299.

Berger, C.A. and L.A. Coburn [1994] Heat flow and Berezin–Toeplitz estimates. *Amer. J. Math.* **116**, 563–590.

Berline, N. and M. Vergne [1985] A computation of the equivariant index of the Dirac operator. *Bull. Soc. Math. France* **113**, 305–345.

Berry, M.V. and N.L. Balazs [1979] Evolution of semiclassical quantum states in phase space. *J. Phys.* **A12**, 625–642.

Biedenharn, L.C. and J.D. Louck [1981a] *Angular Momentum in Quantum Physics. Encyclopedia of mathematics and its applications, Vol. 8.* Addison Wesley, Reading.

Biedenharn, L.C. and J.D. Louck [1981b] *The Racah–Wigner Algebra in Quantum Physics. Encyclopedia of mathematics and its applications, Vol. 9.* Addison Wesley, Reading.

Bigonnet, B. [1988] Construction d'une C^*-algèbre associée à certains feuilletages de Stefan. *C. R. Acad. Sci. Paris Sér. I Math.* **307**, 307–310.

Binz, E., J. Śniatycki, and H. Fischer [1988] *The Geometry of Classical Fields*. North–Holland, Amsterdam.

Birkhoff, G. [1967] *Lattice Theory* (3d ed.). *Amer. Math. Soc. Coll. Publ.* **25**. American Mathematical Society, Providence.

Birkhoff, G. and J. von Neumann [1936] The logic of quantum mechanics. *Ann. Math.* **37**, 823–843.

Blackadar, B. [1986] *K-Theory for Operator Algebras*. Springer, New York.

Blanchard, E. [1996] Deformations de C^*-algebras de Hopf. *Bull. Soc. math. France* **124**, 141–215.

Blanchard, Ph. and M. Sirugue [1985] Large deviations from classical paths. Hamiltonian flows as classical limits of quantum flows. *Commun. Math. Phys.* **101**, 173–185.

Blau, M. [1988] On the geometric quantisation of constrained systems. *Class. Quantum Grav.* **5**, 1033–1044.

Bogolyubov, N.N. et al. [1981] Berezin, Feliks Aleksandrovich – Obituary. *Russ. Math. Surv.* **36**, 209–216.

Bokobza–Haggiag, J. [1969] Opérateurs pseudo–différentiels sur une variété différentiable. *Ann. Inst. Fourier Grenoble* **19**, 125–177.

Borceux, F. and G. van den Bossche [1989] An essay on noncommutative topology. *Topology and Its Appl.* **31**, 203–223.

Bordemann, M., E. Meinrenken, and M. Schlichenmaier [1994] Toeplitz quantization of Kähler manifolds and $gl(N)$, $N \to \infty$ limits. *Commun. Math. Phys.* **165**, 281–296.

Born, M. [1926] Zur Quantenmechanik der Stossvorgänge. *Z. Phys.* **37**, 863–867.

Borsari, I. and S. Graffi [1994] Symbols of operators and quantum evolution. *J. Math. Phys.* **35**, 4439–4450.

Borthwick, D., A. Lesniewski, and M. Rinaldi [1995] Notes on the structure of quantized Hermitian symmetric spaces. *Rev. Math. Phys.* **7**, 871–891.

Borthwick, D., A. Lesniewski, and H. Upmeier [1993] Non-perturbative deformation quantization of Cartan domains. *J. Funct. Anal.* **113**, 153–176.

Bott, R. [1957] Homogeneous vector bundles. *Ann. Math.* **66**, 203–248.

Bott, R. and L. Tu [1982] *Differential Forms in Algebraic Topology*. Springer, New York.

Bourbaki, N. [1963] *Integration*. Hermann, Paris.

Boutet de Monvel, L. and V. Guillemin [1981] *The Spectral Theory of Toeplitz Operators. Annals of Mathematics Studies* **99**. Princeton University Press, Princeton.

Bowes, D. and K.C. Hannabuss [1997] Weyl quantization and star products. *J. Geom. Phys.* **22**, 319–348.

Bratteli, O. and D.W. Robinson [1987] *Operator Algebras and Quantum Statistical Mechanics, Vol. I: C^*- and W^*-Algebras, Symmetry Groups, Decomposition of States*, 2nd ed. Springer, Berlin.

Bratteli, O. and D.W. Robinson [1981] *Operator Algebras and Quantum Statistical Mechanics, Vol. II: Equilibrium States, Models in Statistical Mechanics*. Springer, Berlin.

Bredon, G. [1972] *Introduction to Compact Transformation Groups*. Academic Press, New York.

Bröcker, T. and T. tom Dieck [1985] Representations of compact Lie groups. Springer, Berlin.

Bröcker, T. and R.F. Werner [1995] Mixed states with positive Wigner function. *J. Math. Phys.* **36**, 62–75.

Brody, D.C. and L.P. Hughston [1998] Geometrisation of statistical mechanics. *e-print gr-qc/9708032*.

Brooke, J.A. and F.E. Schroeck [1996] Localization of the photon on phase space. *J. Math. Phys.* **37**, 5958–5986.

Brown, L.G. [1992] Complements to various Stone–Weierstrass theorems for C^*-algebras and a theorem of Shultz. *Commun. Math. Phys.* **143**, 405–413.

Brown, L.G., P. Green, and M.A. Rieffel [1977] Stable isomorphism and strong Morita equivalence of C^*-algebras. *Pac. J. Math.* **71**, 349–363.

Brown, R. [1987] From groups to groupoids: a brief survey. *Bull. London Math. Soc.* **19**, 113–134.

Brown, R. and O. Mucuk [1995] The monodromy groupoid of a Lie groupoid. *Cah. Top. Geom. Diff. Categ.* **36**, 345–369.

Brown, R. and O. Mucuk [1996] Foliations, locally Lie groupoids and holonomy. *Cah. Top. Geom. Diff. Categ.* **37**, 61–71.

Brummelhuis, R. and A. Uribe [1991] A semi-classical trace formula for Schrödinger operators. *Commun. Math. Phys.* **136**, 567–584.

Brummelhuis, R., T. Paul, and A. Uribe [1995] Spectral estimates around a critical level. *Duke Math. J.* **78**, 477–530.

Bub, J. [1981] Hidden variables and quantum mechanics – a sceptical review. *Erkenntnis* **16**, 275–293.

Buchholz, D. [1996] Quarks, gluons, color: facts or fiction? *Nucl. Phys.* **B469** (1996), 333–353.

Busby, R.C. and H.A. Smith [1970] Representations of twisted group algebras. *Trans. Amer. Math. Soc.* **149**, 503–537.

Busch, P., M. Grabowski and P.J. Lahti [1995] *Operational Quantum Physics*. Springer, Berlin.

Butterfield, J.N. and G. Fleming [1998] Strange positions. In: Butterfield, J.N. and C. Pagonis (eds.) *From Physics to Philosophy. A Festschrift for Michael Redhead*. Cambridge University Press, Cambridge.

Cahen, M., S. Gutt, and J. Rawnsley [1990] Quantization of Kähler manifolds I. *J. Geom. Phys.* **7**, 45–62.

Cahen, M., S. Gutt, and J. Rawnsley [1993] Quantization of Kähler manifolds II. *Trans. Amer. Math. Soc.* **337**, 73–98.

Cahen, M., S. Gutt, and J. Rawnsley [1994] Quantization of Kähler manifolds III. *Lett. Math. Phys.* **30**, 291–305.

Cahen, M., S. Gutt, and J. Rawnsley [1995] Quantization of Kähler manifolds IV. *Lett. Math. Phys.* **34**, 159–168.

Caldéron, A. and R. Vaillancourt [1971] On the boundedness of pseudo-differential operators. *J. Math. Soc. Japan* **23**, 374–378.

Cameron, R.H. and W.T. Martin [1944] Transformation of Wiener integrals under translations. *Ann. Math.* **45**, 386–396.

Camporesi, R. [1990] Harmonic analysis and propagators on homogeneous spaces. *Phys. Rep.* **196**, 1–134.

Cant, A. [1981] Invariant connections and magnetic monopoles. *J. Math. Phys.* **22**, 2283–2288.

Cantoni, V. [1975] Generalized "transition probability". *Commun. Math. Phys.* **44**, 125–128.

Carey, A.L., J.M. Gaffney, and C.A. Hurst [1977] A C^*-algebraic formulation of the quantization of the electromagnetic field. *J. Math. Phys.* **18**, 629–640.

Carey, A.L., J.M. Gaffney, and C.A. Hurst [1978] A C^*-algebraic formulation of gauge transformations of the second kind for the electromagnetic field. *Rep. Math. Phys.* **13**, 419–436.

Cariñena, J.F., J.M. Gracia-Bondía, and J.C. Várilly [1990] Relativistic quantum kinematics in the Moyal representation. *J. Phys.* **A23**, 901–933.

Cariñena, J.F. and M. Santander [1975] On the projective unitary representations of connected Lie groups. *J. Math. Phys.* **16**, 1416–1420.

Cartan, E. [1958] *Leçons sur les Invariants Intégraux*. 2nd ed. Hermann, Paris.

Castrigiano, D.P.L. and R.W. Henrichs [1980] Systems of covariance and subrepresentations of induced representations. *Lett. Math. Phys.* **4**, 169–175.

Cattaneo, U. [1979] On Mackey's imprimitivity theorem. *Comment. Math. Helvetici* **54**, 629–641.

Cheeger, J., M. Gromow, and M.E. Taylor [1982] Finite propagation speed, kernel estimates for functions of the Laplace operator, and the geometry of complete Riemannian manifolds. *J. Diff. Geom.* **17**, 15–53.

Chernoff, P.R. [1973] Essential self–adjointness of powers of generators of hyperbolic equations. *J. Funct. Anal.* **12**, 401–414.

Chernoff, P.R. [1995] Irreducible representations of infinite dimensional transformation groups and Lie algebras I. *J. Funct. Anal.* **130**, 255–282.

Chernoff, P.R. and J.E. Marsden [1974] *Properties of Infinite Dimensional Hamiltonian Systems* (LNM **425**). Springer, Berlin.

Chiang, C.C., S.C. Lee, and G. Marmo [1985] Lagrangian dynamics on higher–dimensional spaces with applications to Kaluza–Klein theories. *J. Math. Phys.* **26**, 1083–1092.

Choquet-Bruhat, Y., C. DeWitt-Morette, and M. Dillard-Bleick [1982] *Analysis, Manifolds, and Physics*, 2nd ed. North-Holland, Amsterdam.

Choquet-Bruhat, Y. and C. DeWitt-Morette [1989] *Analysis, Manifolds, and Physics. Part II: 92 Applications*. North-Holland, Amsterdam.

Chruściński, D. and J. Kijowski [1996] Equations of motion from field equations and a gauge–invariant variational principle for the motion of charged particles. *J. Geom. Phys.* **20**, 393–403.

Cirelli, R. and P. Cotta–Ramusino [1973] On the isomorphism of a 'quantum logic' with the logic of projections in a Hilbert space. *Int. J. Theor. Phys.* **8**, 11–29.

Cirelli, R., P. Lanzavecchia [1984] Hamiltonian vector fields in quantum mechanics. *Nuovo Cimento* **B79**, 271–283.

Cirelli, R., P. Lanzavecchia, and A. Mania [1983] Normal pure states of the von Neumann algebra of bounded operators as a Kähler manifold. *J. Phys.* **A16**, 3829–3835.

Cirelli, R., A. Mania, and L. Pizzochero [1990] Quantum mechanics as an infinite-dimensional Hamiltonian system with uncertainty structure. *J. Math. Phys.* **31**, 2891–2897.

Cirelli, R., A. Mania, and L. Pizzochero [1994] A functional representation for non-commutative C^*-algebras. *Rev. Math. Phys.* **6**, 675–697.

Coburn, L.A. [1992] Deformation estimates for the Berezin–Toeplitz quantization. *Commun. Math. Phys.* **149**, 415–424.

Coburn, L.A. and J. Xia [1995] Toeplitz algebras and Rieffel deformations. *Commun. Math. Phys.* **168**, 23–38.

Colin de Verdière, Y. [1985] Ergodicité et fonctions propres. *Commun. Math. Phys.* **102**, 497–502.

Colin de Verdière, Y. and B. Parisse [1994] Equilibre instable en regime semi-classique. I. Concentration microlocale. *Comm. Partial Differential Equations* **19**, 1535–1563.

Combes, J.M., R. Schrader, and R. Seiler [1978] Classical bounds and limits for energy distributions of Hamilton operators in electromagnetic fields. *Ann. Phys. (N.Y.)* **111**, 1–18.

Combescure, M. [1992] The squeezed state approach of the semiclassical limit of the time-dependent Schrödinger equation. *J. Math. Phys.* **33**, 3870–3880.

Connes, A. [1980] A survey of foliations and operator algebras. In: Kadison, R.V. (ed.) *Operator Algebras and Applications, Proc. Symp. Pure Math.* **38(1)**, pp. 521–628. American Mathematical Society, Providence.

Connes, A. [1994] *Noncommutative Geometry*. Academic Press, San Diego.

Coquereaux, R. and A. Jadczyk [1988] *Riemannian Geometry, Fibre Bundles, Kaluza–Klein Theories and all that.* World Scientific, Singapore.

Cordes, H.O. [1987] *Spectral Theory of Linear Differential Operators and Comparison Algebras.* (LMS Lecture Notes **76**). Cambridge University Press, Cambridge.

Cordes, H.O. [1995] *The Technique of Pseudodifferential Operators.* (LMS Lecture Notes **172**). Cambridge University Press, Cambridge.

Corwin, L. and F.P. Greenleaf [1989] *Representations of Nilpotent Lie Groups and Their Applications*, Part I. Cambridge University Press, Cambridge.

Coste, A., P. Dazord, and A. Weinstein [1987] Groupoides symplectiques. *Publ. Dépt. Math. Univ. C. Bernard–Lyon I* **2A**, 1–62.

Courant, T.J. [1990] Dirac Manifolds. *Trans. Amer. Math. Soc.* **319**, 631–661.

Cushman, R.H. and L.M. Bates [1997] *Global Aspects of Integrable Systems.* Birkhäuser, Basel.

Cycon, H.L., R.G. Froese, W. Kirsch, and B. Simon [1987] *Schrödinger Operators, with Applications to Quantum Mechanics and Global Geometry.* Springer, Berlin.

Daubechies, I. [1980] On the distributions corresponding to bounded operators in the Weyl quantization. *Commun. Math. Phys.* **75**, 229–238.

Daubechies, I. [1983] Continuity statements and counterintuitive examples in connection with Weyl quantization. *J. Math. Phys.* **24**, 1453–1461.

Davidson, K.R. [1996] *C^*-Algebras by Example. Fields Institute Monographs* **6**. American Mathematical Society, Providence (RI).

Davies, E.B. [1976] *Quantum Theory of Open Systems.* Academic Press, London.

Davies, E.B. [1989] *Heat Kernels and Spectral Theory.* Cambridge University Press, Cambridge.

Davies, E.B. and J.T. Lewis [1970] An operational approach to quantum probability. *Commun. Math. Phys.* **17**, 239–260.

Dazord, P. [1985] Feuilletages à singularités. *Indag. Math.* **47**, 21–39.

Dazord, P. and T. Delzant [1987] Le problème général des variables actions-angles. *J. Diff. Geom.* **26**, 223–251.

DeWitt–Morette, C., A. Maheshwari, and B. Nelson [1979] Path integration in non-relativistic quantum mechanics. *Phys. Rep.* **5**, 255–372.

DeWitt–Morette, C., B. Nelson, and T.-R. Zhang [1983] Caustic problems in quantum mechanics with applications in scattering theory. *Phys. Rev.* **D28**, 2526–2546.

Dimock, J. [1996] Canonical quantization of Yang–Mills on a circle. *Rev. Math. Phys.* **8**, 85–102.

Dirac, P.A.M. [1930] *The Principles of Quantum Mechanics.* Clarendon Press, Oxford.

Dirac, P.A.M. [1931] Quantized singularities in the electromagnetic field. *Proc. R. Soc. London* **A133**, 60–72.

Dirac, P.A.M. [1950] Generalized Hamiltonian Systems. *Can. J. Math.* **12**, 129–148.

Dirac, P.A.M. [1964] *Lectures on Quantum Mechanics*. Belfer School of Science, Yeshiva University, New York.

Dixmier, J. [1960] Sur les répresentations unitaires des groupes de Lie nilpotents, IV. *Can. J. Math.* **12**, 324–352.

Dixmier, J. [1977] *C^*-Algebras*. North–Holland, Amsterdam.

Dixmier, J. and P. Malliavin [1978] Factorisations de fonctions et de vecteurs indéfinéments différentiables. *Bull. Soc. Math. France* **102**, 305–330.

Dobrushin, R.L., R.A. Minlos, M.A. Shubin, and A.M. Vershik (eds.) [1996] *Contemporary Mathematical Physics. F.A. Berezin Memorial Volume. AMS Translations* **2–175**. American Mathematical Society, Providence.

Doebner, H.D. and J. Tolar [1975] Quantum mechanics on homogeneous spaces. *J. Math. Phys.* **16**, 975–984.

Doebner, H.D. and J. Tolar [1990] Mackey's quantization and invariant connections. In: Niederle, J. and J. Fischer, (eds.) *Selected Topics in Quantum Field Theory and Mathematical Physics*, pp. 234–238. World Scientific, Singapore.

Dollard, J.D. and C.N. Friedman [1979] *Product Integration*. Addison–Wesley, London.

Doplicher, S., D. Kastler, and D. W. Robinson [1966] Covariance algebras in field theory and statistical mechanics. *Commun. Math. Phys.* **3**, 1–28.

Doran, R.S. [1994] (ed.) *C^*-algebras: 1943–1993. Contemp. Math.* **167**. American Mathematical Society, Providence.

Doran, R.S. and V.A. Belfi [1986] *Characterizations of C^*-algebras*. M. Dekker, New York.

Dowker, J.S. [1974] Covariant Schrödinger equations. In: Arthurs, A.M. (ed.) *Functional Integration and its Applications*, pp. 34–52. Clarendon Press, Oxford.

Driver, B.K. and B.C. Hall [1998] A note on the one-dimensional energy representation. University of California at San Diego preprint.

Duclos, P. and H. Hogreve [1993] On the semiclassical localization of the quantum probability. *J. Math. Phys.* **34**, 1681–1691.

Duffield, N.G. [1990] Classical and thermodynamic limits for generalized quantum spin systems. *Commun. Math. Phys.* **127**, 27–39.

Duval, C., and J. Elhadad [1992] Geometric quantization and localization of relativistic spin systems. *Contemp. Math.* **132**, 317–330.

Duval, C., J. Elhadad, M.J. Gotay, J. Śniatycki, and G.M. Tuynman [1991] Quantization and bosonic BRST theory. *Ann. Phys. (NY)* **206**, 1–26.

Duval, C., J. Elhadad, and G.M. Tuynman [1992] Pukanszky's condition and symplectic induction. *J. Diff. Geom.* **36**, 331–348.

Duval, C. and P. Horvathy [1982] Particles with internal structure: the geometry of classical motions and conservation laws. *Ann. Phys. (N.Y.)* **142**, 10–33.

Ebin, D.G. and J.E. Marsden [1970] Groups of diffeomorphisms and the motion of an incompressible fluid. *Ann. Math.* **92**, 102–163.

Edwards, C.M. and G.T. Rüttimann [1985] On the facial structure of the unit balls in a GL-space and its dual. *Math. Proc. Camb. Phil. Soc.* **98**, 305–322.

Effros, E.G. [1963] Order ideals in a C^*-algebra and its dual. *Duke Math. J.* **30**, 391–412.

Effros, E. and F. Hahn [1967] Locally compact transformation groups and C^*-algebras. *Mem. Amer. Math. Soc.* **75**.

Ehresmann, C. [1958] Catégories topologiques et catégories differentiables. In: *Colloque de Géometrie Différentielle Globale*, pp. 137–150. Centre Belge des Recherches Mathematiques, Brussels.

Elliott, G.A., T. Natsume, and R. Nest [1993] The Heisenberg group and K-theory. *K-Theory* **7**, 409–428.

Elliott, G.A., T. Natsume, and R. Nest [1996] The Atiyah–Singer index theorem as passage to the classical limit in quantum mechanics. *Commun. Math. Phys.* **182**, 505–533.

Elworthy, D. and A. Truman [1981] Classical mechanics, the diffusion (heat) equation and the Schrödinger equation on a Riemannian manifold. *J. Math. Phys.* **22**, 2144–2166.

Elworthy, D., A. Truman, and K. Watling [1985] The semi-classical expansion for a charged particle on a curved space background. *J. Math. Phys.* **26**, 984–990.

Emch, G.G. [1972] *Algebraic Methods in Statistical Mechanics and Quantum Field Theory*. Wiley, New York.

Emch, G.G. [1982] Quantum and classical mechanics on homogeneous Riemannian manifolds. *J. Math. Phys.* **23**, 1785–1791.

Emch, G.G. [1983] Geometric dequantization and the correspondence problem. *Int. J. Theor. Phys.* **22**, 397–420.

Emch, G.G. [1984] *Mathematical and Conceptual Foundations of 20th Century Physics*. North Holland, Amsterdam.

Emmrich, C. [1993a] Equivalence of extrinsic and intrinsic quantization for observables not preserving the vertical polarization. *Commun. Math. Phys.* **151**, 515–530.

Emmrich, C. [1993b] Equivalence of Dirac and intrinsic quantization for non-free group actions. *Commun. Math. Phys.* **151**, 531–542.

Emmrich, C. and H. Römer [1990] Orbifolds as configuration spaces of systems with gauge symmetries. *Commun. Math. Phys.* **129**, 69–94.

Engliš, M. [1996] Berezin quantization and reproducing kernels on complex domains. *Trans. Amer. Math. Soc.* **348**, 411–479.

Est, W.T. van [1953] Group cohomology and Lie algebra cohomology in Lie groups. I, II. *Proc. Kon. Ned. Akad. Wet.* **A56**, 484–504.

Estrada, R., J.M. Gracia-Bondía, and J.C. Várilly [1989] On asymptotic expansions of twisted products. *J. Math. Phys.* **30**, 2789–2796.

Exel, R. [1994] The soft torus: a variational analysis of commutator norms. *J. Funct. Anal.* **126**, (1994), 259–273.

Falk, G. [1951] Überringe mit Poisson-Klammern. *Math. Ann.* **123**, 379–391.

Faraut, J. and A. Korányi [1994] *Analysis on Symmetric Cones*. Clarendon Press, Oxford.

Fedosov, B.V. [1994] A simple geometrical construction of deformation quantization. *J. Diff. Geom.* **40**, 213–238.

Fedosov, B.V. [1996] *Deformation Quantization and Index Theory*. Akademie-Verlag, Berlin.

Fehér, L.Gy. [1986] Classical motion of colored test particles along geodesics of a Kaluza–Klein spacetime. *Acta Phys. Hung.* **59**, 437–444.

Fell, J.M.G. [1962] The structure of algebras of operator fields. *Acta Math.* **106**, 237–268.

Fell, J.M.G. [1978] *Induced Representations and Banach *-algebra Bundles. Lecture Notes in Mathematics* **582**. Springer, Berlin.

Fell, J.M.G. and R.S. Doran [1988] *Representations of* -Algebras, Locally Compact Groups and Banach *-Algebraic Bundles, Vol. 2.* Academic Press, Boston.

Field, T.R. [1996] *The Quantum Complex Structure.* D. Phil. thesis, Oxford University.

Figueroa, H., J.M. Gracia-Bondía, and J.C. Várilly [1990] Moyal quantization with compact symmetry groups and noncommutative harmonic analysis. *J. Math. Phys.* **31**, 2664–2671.

Fillmore, P.A. [1996] *A User's Guide to Operator Algebras.* Wiley-Interscience, New York.

Fischer, A.E., J.E. Marsden, and V. Moncrief [1980] The structure of the space of solutions of Einstein's equations I: One Killing field. *Ann. Inst. H. Poincaré (Phys. Théor.)* **33**, 147–194.

Fock, V. [1928] Verallgemeinerung und Lösung der Diracschen statistischen Gleichung. *Z. Physik* **49**, 339–357.

Folland, G.B. [1989] *Harmonic Analysis on Phase Space.* Princeton University Press, Princeton.

Forgács, P. and N.S. Manton [1980] Space-time symmetries in gauge theories. *Commun. Math. Phys.* **72**, 15–35.

Frank, M. [1990] Self-duality and C^*-reflexivity of Hilbert C^*-modules. *Zeitschr. Anal. Anw.* **9**, 165–176.

Frank, M. [1998] Hilbert C^*-modules and related subjects – a guided reference overview. *http://www.mathematik.uni-leipzig.de/MI/frank/hilmod.html.*

Freed, D.S. [1988] The geometry of loop groups. *J. Diff. Geom.* **28**, 223–276.

Frenkel, I. B. [1984] Orbital theory for affine Lie algebras. *Inv. Math.* **77**, 301–352.

Freed, D.S. and K.K. Uhlenbeck [1984] *Instantons and Four-Manifolds.* Springer, New York.

Freyer, K.D. and I. Halperin [1956] The von Neumann coordinatization theorem for complemented modular lattices. *Acta Scient. Math.* **17**, 203–249.

Freyer, K.D. and I. Halperin [1958] On the construction of coordinates for non-Desarguesian complemented modular lattices. *Proc. Kon. Ned. Akad. Wet.* **A61**, 142–161.

Fröhlicher, A. and A. Kriegl [1988] *Linear Spaces and Differentiation Theory.* Wiley, Chichester.

Gaal, S.A. [1973] *Linear Analysis and Representation Theory.* Springer, Berlin.

Gadella, M. [1995] Moyal formulation of quantum mechanics. *Fortschr. Phys.* **43**, 229–264.

Gallot, S., D. Hulin and J. Lafontaine [1990] *Riemannian Geometry.* Springer, Berlin.

Gambini, R. and J. Pullin [1996] *Loops, Knots, Gauge Theories and Quantum Gravity.* Cambridge University Press, Cambridge.

Gelfand, I.M., M.I. Graev, and A.M. Vershik [1977] Representations of the group of smooth mappings of a manifold into a compact Lie group. *Compositio Math.* **35**, 299–334.

Gerstenhaber, M. [1964] On the deformation of rings and algebras. *Ann. Math.* **79**, 59–103.

Giavarini, G. and E. Onofri [1990] Vector coherent states and non-abelian gauge structures in quantum mechanics. *Int. J. Mod. Phys.* **A5**, 4311–4331.

Giles, R. [1970] Foundations for quantum mechanics. *J. Math. Phys.* **11**, 277–322.

Giles, R. and H. Kummer [1971] A non-commutative generalization of topology. *Indiana Univ. Math. J.* **21**, 91–102.

Gilmore, R. [1979] The classical limit of quantum nonspin systems. *J. Math. Phys.* **20**, 891–893.

Giulini, D. [1995] Quantum mechanics on spaces with finite fundamental group. *Helv. Phys. Acta* **68**, 438–469.

Glimm, J. [1962] Families of induced representations. *Pac. J. Math.* **12**, 885–911.

Glimm, J. and R.V. Kadison [1960] Unitary operators in C^*-algebras. *Pac. J. Math.* **10**, 547–556.

Goldin, G.A. [1971] Non-relativistic current algebra as unitary representation theory of groups. *J. Math. Phys.* **12**, 462–487.

Goldin, G.A., R. Menikoff and D.H. Sharp [1980] Particle statistics from induced representations of local current groups. *J. Math. Phys.* **21**, 650–664.

Goodearl, K. [1982] *Notes on Real and Complex C^*-algebras.* Shiva Publ., Nantwich.

Gootman, E.C. and J. Rosenberg [1979] The structure of crossed product C^*-algebras: a proof of the generalized Effros–Hahn conjecture. *Inv. Math.* **52**, 283–298.

Goreski, M. and R. MacPherson [1988] *Stratified Morse Theory*. Springer, New York.

Gotay, M. [1986] Constraints, reduction, and quantization. *J. Math. Phys.* **27**, 2051–2066.

Gotay, M.J. and L. Bos [1986] Singular angular momentum mappings. *J. Diff. Geom.* **24**, 181–203.

Gotay, M.J., H.B.G.S. Grundling, and G.M. Tuynman [1996] Obstruction results in quantization theory. *J. Nonlinear Sci.* **6**, 469–498.

Gotay, M.J. and J.M. Nester [1980] Generalized constraint algorithm and special presymplectic manifolds. In: Kaiser, G. and J.E. Marsden (eds.) *Geometric Methods in Mathematical Physics. Lecture Notes in Mathematics* **775**. Springer, Berlin.

Gotay, M.J., J.M. Nester, and G. Hinds [1978] Presymplectic manifolds and the Dirac–Bergmann theory of constraints. *J. Math. Phys.* **19**, 2388–2399.

Gotay, M.J. and G.M. Tuynman [1991] A symplectic analogue of the Mostow–Palais theorem. In: Dazord, P. and A. Weinstein (eds.) *Symplectic Geometry, Groupoids, and Integrable Systems*, pp. 173–181. Springer, New York.

Govaerts, J. [1991] *Hamiltonian Quantization and Constrained Dynamics*. Leuven University Press, Leuven.

Gracia-Bondía, J.M. [1992] Generalized Moyal quantization on homogeneous symplectic spaces. *Contemp. Math.* **134**, 93–114.

Gracia-Bondía, J.M. and J.C. Várilly [1988] Algebras of distributions suitable for phase-space quantum mechanics I, II. *J. Math. Phys.* **29**, 869–879, 880–887.

Graffi, S. and A. Parmeggiani [1990] Quantum evolution and classical flow in complex phase space. *Commun. Math. Phys.* **128**, 393–409.

Green, P. [1978] The local structure of twisted covariance algebras. *Acta Math.* **140**, 191–250.

Green, P. [1978] The structure of imprimitivity algebras. *J. Funct. Anal.* **36**, 88–104.

Greenleaf, F. [1969] *Invariant Means on Topological groups*. Van Nostrand Reinhold, New York.

Greub, W., S. Halperin, and R. Vanstone [1972] *Connections, Curvature, and Cohomology. Vol. I: de Rham Cohomology of Manifolds and Vector Bundles.* Academic Press, New York.

Greub, W., S. Halperin, and R. Vanstone [1973] *Connections, Curvature, and Cohomology. Vol. II: Lie Groups, Principal Bundles and Characteristic Classes.* Academic Press, New York.

Greub, W. and H.-R. Petry [1975] Minimal coupling and complex line bundles. *J. Math. Phys.* **16**, 1347–1351.

Grgin, E. and A. Petersen [1974] Duality of observables and generators in classical and quantum mechanics. *J. Math. Phys.* **15**, 764–769.

Griffiths, P. and J. Harris [1978] *Principles of Algebraic Geometry.* Wiley, New York.

Groenewold, H.J. [1946] On the principles of elementary quantum mechanics. *Physica* **12**, 405–460.

Grønbæk, N. [1995] Morita equivalence for Banach algebras. *J. Pure Appl. Alg.* **99**, 183–219.

Groot, S.R. de, W.A. van Leeuwen, and Ch.G. van Weert [1980] *Relativistic Kinetic Theory.* North-Holland, Amsterdam.

Gross, L.P. [1993] Uniqueness of ground states for Schrödinger operators over loop groups. *J. Funct. Anal.* **121**, 373–441.

Grossmann, A. [1976] Parity operator and quantization of delta-functions. *Commun. Math. Phys.* **48**, 191–194.

Grossmann, A., G. Loupias, and E.M. Stein [1968] An algebra of pseudodifferential operators and quantum mechanics in phase space. *Ann. Inst. Fourier (Grenoble)* **18**, 343–368.

Grundling, H.B.G.S. [1988] Systems with outer constraints. Gupta–Bleuler electromagnetism as an algebraic field theory. *Commun. Math. Phys.* **114**, 69–91.

Grundling, H.B.G.S. [1997] A group algebra for inductive limit groups. Continuity problems of the canonical commutation relations. *Acta Appl. Math.* **46**, 107–145.

Grundling, H.B.G.S. and C.A. Hurst [1985] Algebraic quantization of systems with a gauge degeneracy. *Commun. Math. Phys.* **98**, 369–390.

Grundling, H.B.G.S. and C.A. Hurst [1987] Algebraic structures of degenerate systems and the indefinite metric. *J. Math. Phys.* **28**, 559–572.

Grundling, H.B.G.S. and C.A. Hurst [1988a] A note on regular states and supplementary conditions. *Lett. Math. Phys.* **15**, 205–212; Err. *ibid.* **17**, 173–174.

Grundling, H.B.G.S. and C.A. Hurst [1988b] The quantum theory of second class constraints: kinematics. *Commun. Math. Phys.* **119**, 75–93; Err. *ibid.* **122**, 527–529.

Gudder, S.P. [1979] A survey of axiomatic quantum mechanics. In: Hooker, C.A. (ed.) *Logico-Algebraic Approach to Quantum Mechanics*, pp. 323–363. D. Reidel, Dordrecht.

Gudder, S.P. [1979] *Stochastic Methods in Quantum Mechanics* North-Holland, Amsterdam.

Guichardet, A. [1972] *Symmetric Hilbert Spaces and Related Topics. Lecture Notes in Mathematics* **261**. Springer, Berlin.

Guichardet, A. [1980] *Cohomologie des Groupes Topologiques et des Algèbres de Lie.* Nathan, Paris.

Guichardet, A. [1985] Théorie de Mackey et méthode des orbites selon M. Duflo. *Expo. Math.* **3**, 303–346.

Guillemin, V. [1984] Toeplitz operators in n dimensions. *Int. Eq. Op. Th.* **7**, 145–205.

Guillemin, V. [1994] *Moment Maps and Combinatorial Invariants of Hamiltonian T^n-Spaces*. Birkhäuser, Boston.

Guillemin, V., E. Lerman, and S. Sternberg [1996] *Symplectic fibrations and multiplicity diagrams*. Cambridge University Press, Cambridge, 1996.

Guillemin, V. and S. Sternberg [1977] *Geometric Asymptotics. Math. Surveys* **14**. American Mathematical Society, Providence.

Guillemin, V. and S. Sternberg [1982] Convexity properties of the moment mapping. I. *Inv. Math.* **67**, 491–513.

Guillemin, V. and S. Sternberg [1984a] Convexity properties of the moment mapping. II. *Inv. Math.* **77**, 533–546.

Guillemin, V. and S. Sternberg [1984b] *Symplectic Techniques in Physics*. Cambridge University Press, Cambridge.

Guillemin, V. and A. Uribe [1985] Band asymptotics on line bundles over S^2. *J. Diff. Geom.* **21**, 129–133.

Guillemin, V. and A. Uribe [1986] Clustering theorems with twisted spectra. *Math. Ann.* **272**, 479–506.

Guillemin, V. and A. Uribe [1989] Circular symmetry and the trace formula. *Inv. Math.* **96**, 385–423.

Guillemin, V. and A. Uribe [1990] Reduction and the trace formula. *J. Diff. Geom.* **32**, 315–347.

Gunson, J. [1967] On the algebraic structure of quantum mechanics. *Commun. Math. Phys.* **6**, 262–285.

Gutt, S. [1983] An explicit *-product on the cotangent bundle of a Lie group. *Lett. Math. Phys.* **7**, 249–258.

Gutzwiller, M.C. [1990] *Chaos in Classical and Quantum Mechanics*. Springer, New York.

Haag, R. [1996] *Local Quantum Physics*, 2nd ed. Springer, Berlin.

Haag, R., and D. Kastler [1964] An algebraic approach to quantum field theory. *J. Math. Phys.* **5**, 848–861.

Habib, S. and H.E. Kandrup [1989] Wigner functions and density-matrices in curved spaces as computational tools. *Ann. Phys. (N.Y.)* **191**, 335–362.

Hagedorn, G.A. [1980] Semiclassical quantum mechanics I: The $\hbar \to 0$ limit for coherent states. *Commun. Math. Phys.* **71**, 77–93.

Hagedorn, G.A. [1981] Semiclassical quantum mechanics III: The large order asymptotics and more general states. *Ann. Phys. (N.Y.)* **135**, 58–70.

Hagedorn, G.A. [1985] Semiclassical quantum mechanics IV: Large order asymptotics and more general states in more than one dimension. *Ann. Inst. H. Poincaré* **A42**, 363–374.

Hahn, P. [1978a] Haar measure for measure groupoids. *Trans. Amer. Math. Soc.* **242**, 1–33.

Hahn, P. [1978b] The regular representation of measure groupoids. *Trans. Amer. Math. Soc.* **242**, 35–72.

Hájiček, P. [1994] Quantization of systems with constraints. In: Ehlers, J. and H. Friedrich (eds.) *Canonical Gravity: From Classical to Quantum. Lecture Notes in Physics* **434**, pp. 113–149. Springer, Berlin.

Hall, B.C. [1994] The Segal–Bargmann "coherent state" transform for compact Lie groups. *J. Funct. Anal.* **122**, 103–151.

Hall, B.C. [1997a] The inverse Segal–Bargmann transform for compact Lie groups. *J. Funct. Anal.* **143**, 98–116.

Hall, B.C. [1997b] Phase space bounds for quantum mechanics on a compact Lie group. *Commun. Math. Phys.* **184**, 233–250.

Hall, B.C. [1997c] Quantum mechanics in phase space. *Contemp. Math.* **214**, 47–62.

Hall, B.C. and B.K. Driver [1998] Yang–Mills theory and the Segal–Bargmann transform. University of California at San Diego preprint.

Halliwell, J.J. and J.B. Hartle [1991] Wave functions constructed from an invariant sum over histories satisfy constraints. *Phys. Rev.* **D43**, 1170–1194.

Hanche-Olsen, H. [1985] JB-algebras with tensor product are C^*-algebras. *Lecture Notes in Mathematics* **1132**, 223–229.

Hanche-Olsen, H. and E. Størmer [1984] *Jordan Operator Algebras*, Pitman, Boston.

Hannabuss, K.C. [1984] Holomorphic and abstract inducing. *Math. Proc. Camb. Phil. Soc.* **96**, 453–468.

Harnad, J. and J.P. Pare [1991] Kaluza–Klein approach to the motion of non-abelian charged particles with spin. *Class. Quant. Grav.* **8**, 1427–1444.

Harnad, J., S. Shnider, and L. Vinet [1980] Group actions on principal bundles and invariance conditions for gauge fields. *J. Math. Phys.* **21**, 2719–2724.

Hartkämper, A. and H. Neumann [1974] (eds.) *Foundations of Quantum Mechanics and Ordered Linear Spaces. Lecture Notes in Physics* **29**. Springer, Berlin.

Hector, G. and U. Hirsch [1986] *Introduction to the Geometry of Foliations. Part A, 2nd ed.* Vieweg, Braunschweig.

Helffer, B. [1988] *Semi-classical Analysis for the Schrödinger Operator and Applications. Lecture Notes in Mathematics* **1336**. Springer, Berlin.

Helffer, B., A. Martinez, and D. Robert [1987] Ergodicité et limite semi-classique. *Commun. Math. Phys.* **109**, 313–326.

Helgason, S. [1978] *Differential Geometry, Lie Groups, and Symmetric Spaces*. Academic Press, New York.

Henneaux, M. and C. Teitelboim [1992] *Quantization of Gauge Systems*. Princeton University Press, Princeton.

Hepp, K. [1974] The classical limit of quantum mechanical correlation functions. *Commun. Math. Phys.* **35**, 265–277.

Hermann, R. [1966] *Lie Groups for Physicists*. Benjamin, New York.

Hermann, R. [1973] *Topics in the Mathematics of Quantum Theory*. MathSci Press, Brookline.

Hermann, R. [1975] *Gauge Fields and Cartan–Ehresmann Connections*. MathSci Press, Brookline.

Hetrick, J. E. [1994] Canonical quantization of two-dimensional gauge fields. *Int. J. Mod. Phys.* **A9**, 3153–3178.

Higuchi, A. [1991] Quantum linearization instabilities of de Sitter spacetime. I, II. *Class. Quantum Grav.* **8**, 1961–1981, 1983–2004.

Hilgevoord, J. and J. Uffink [1991] Uncertainty in prediction and in inference. *Found. Phys.* **21**, 323–341.

Hillery, M., R.F. O'Connel, M.O. Scully, and E.P. Wigner [1984] Distribution functions in physics – Fundamentals. *Phys. Rep.* **106**, 121–167.

Hilsum, M. and G. Skandalis [1987] Morphismes K-orientés d'espaces de feuilles et fonctorialité en théorie de Kasparov. *Ann. scient. Éc. Norm. Sup. (4ᵉ s.)* **20**, 325–390.

Hislop, P.D. and I.M. Sigal [1996] *Introduction to Spectral Theory with Applications to Schrödinger Operators*. Springer, New York.

Hitchin, N.J. [1990] Flat connections and geometric quantization. *Commun. Math. Phys.* **131**, 347–380.

Hogreve, H. [1983] *The Classical Limit of Quantum Theories: Particles in External Metrics and with Spin*. Ph.D. thesis, Freie Universität Berlin.

Hogreve, H., J. Potthoff, and R. Schrader [1983] Classical limits for quantum particles in external Yang–Mills potentials. *Commun. Math. Phys.* **91**, 573–598.

Holevo, A.S. [1973] Statistical decision theory for quantum systems. *J. Multivariate Anal.* **3**, 337–394.

Holevo, A.S. [1982] *Probabilistic and Statistical Aspects of Quantum Theory*. North–Holland, Amsterdam.

Holland, S.S. Jr. [1995] Orthomodularity in infinite dimensions; a theorem of M. Solèr. *Bull. Amer. Math. Soc.* **32**, 205–234.

Hörmander, L. [1979] The Weyl calculus of pseudo-differential operators. *Comm. Pure Appl. Math.* **32**, 359–443.

Hörmander, L. [1983] *The Analysis of Linear Partial Differential Operators, Vol. I*. Springer, Berlin.

Hörmander, L. [1985a] *The Analysis of Linear Partial Differential Operators, Vol. III*. Springer, Berlin.

Hörmander, L. [1985b] *The Analysis of Linear Partial Differential Operators, Vol. IV*. Springer, Berlin.

Horowitz, G.T. and Marolf, D. [1995] Quantum probes of spacetime singularities. *Phys. Rev.* **D52**, 5670–5675.

Horvathy, P.A. [1981] Rotational symmetry and Dirac's monopole. *Int. J. Theor. Phys.* **20**, 697–707.

Horvathy, P.A., G. Morandi, and E.C.G. Sudarshan [1989] Inequivalent quantizations in multiply connected spaces. *Nuovo Cim.* **D11**, 201–228.

Hove, L. van [1951] Sur certaines représentations unitaires d'un groupe infini de transformations. *Mem. Acad. Roy. de Belgique, Classe des Sci.* **26**, 61–102.

Howe, R. [1980] Quantum mechanics and partial differential equations. *J. Funct. Anal.* **38**, 188–254.

Howe, R. [1989] Remarks on classical invariant theory. *Trans. Amer. Math. Soc.* **313**, 539–570.

Hsu, E.P. [1995] Quasi-invariance of the Wiener measure on the path space over a compact Riemannian manifold. *J. Funct. Anal.* **134**, 417–450.

Hudson, R.L. [1974] When is the Wigner quasi-probability density non-negative? *Rep. Math. Phys.* **6**, 249–252.

Huebschmann, J. [1990] Poisson cohomology and quantization. *J. reine angew. Math.* **408**, 57–113.

Huebschmann, J. [1996] Poisson geometry of flat connections for $SU(2)$-bundles on surfaces. *Math. Z.* **221**, 243–259.

Hughston, L.P. [1995] Geometric aspects of quantum mechanics. *Lect. Notes Pure Appl. Math.* **169**, 59–79.

Hurt, N.E. [1983] *Geometric Quantization in Action*. Reidel, Dordrecht.

Husimi, K. [1940] Some formal properties of the density matrix. *Prog. Phys. Math. Soc. Japan* **22**, 264–314.

Hwang, I.L. [1987] The L^2-boundedness of pseudodifferential operators. *Trans. Amer. Math. Soc.* **302**, 55–76.

Iochum, B. [1984] *Cônes Autopolaires et Algèbres de Jordan. Lecture Notes in Mathematics* **1049**. Springer, Heidelberg.

Iochum, B. and F.W. Shultz [1983] Normal state spaces of Jordan and von Neumann algebras. *J. Funct. Anal.* **50**, 317–328.

Isenberg, J. and J.E. Marsden [1982] A slice theorem for the space of solutions of Einstein's equations. *Phys. Rep.* **89**, 179–222.

Isham, C.J. [1983] Topological and global aspects of quantum theory. In: DeWitt, B.S. and R. Stora (eds.) *Relativity, Groups and Topology 2*, pp. 1059–1290. North–Holland, Amsterdam.

Ismagilov, R.S. [1996] *Representations of Infinite-Dimensional Groups*. American Mathematical Society, Providence.

Jackiw, R. [1985] Topological investigations of quantized gauge theories. In: Treiman, S.B. et al. (eds.) *Current Algebra and Anomalies*, pp. 211–359. World Scientific, Singapore.

Jackiw, R. and N.S. Manton [1980] Symmetries and conservation laws in gauge theories. *Ann. Phys. (N.Y.)* **127**, 257–273.

Jammer, M. [1974] *The Philosophy of Quantum Mechanics*. Wiley, New York.

Jänich, K. [1994] *Topologie*. 4. Auflage. Springer, Berlin.

Jeffrey, L.C. and J. Weitsman [1992] Bohr–Sommerfeld orbits in the moduli space of flat connections and the Verlinde dimension formula. *Commun. Math. Phys.* **150**, 593–630.

Jeffrey, L.C. and J. Weitsman [1997] Toric structures on the moduli space of flat connections on a Riemann surface. II. Inductive decomposition of the moduli space. *Math. Ann.* **307**, 93–108.

Jona-Lasinio, G., F. Martinelli, and E. Scoppola [1981] The semi-classical limit of quantum mechanics: a qualitative theory via stochastic mechanics. *Phys. Rep.* **77**, 313–327.

Jordan, P. [1932] Über eine Klasse nichtassoziativer hyperkomlexen Algebren. *Nachr. Ges. Wiss. Göttingen*, 569–575.

Jordan, P., J. von Neumann, and E.P. Wigner [1934] On an algebraic generalization of the quantum mechanical formalism. *Ann. Math.* **36**, 29–64.

Kadison, R.V. [1958] Theory of operators, part II. Operator algebras. *Bull. Amer. Math. Soc.* **64**, 61–85.

Kadison, R.V. [1965] Transformations of states in operator theory and dynamics. *Topology* **3**, 177–198.

Kadison, R.V. [1982] Operator algebras – the first forty years. In: Kadison, R.V. (ed.) *Operator Algebras and Applications, Proc. Symp. Pure Math.* **38(1)**, pp. 1–18. American Mathematical Society, Providence.

Kadison, R.V. [1994] Notes on the Gelfand–Neumark theorem. In: Doran, R.S. (ed.) *C^*-algebras: 1943–1993. Contemp. Math.* **167**, pp. 21–53. American Mathematical Society, Providence.

Kadison, R.V. and J.R. Ringrose [1983] *Fundamentals of the Theory of Operator Algebras I. Elementary Theory*. Academic Press, New York.

Kadison, R.V. and J.R. Ringrose [1986] *Fundamentals of the Theory of Operator Algebras II. Advanced Theory*. Academic Press, New York.

Kalmbach, G. [1983] *Orthomodular Lattices*. Academic Press, London.

Kalmbach, G. [1986] *Measures and Hilbert Lattices*. World Scientific, Singapore.

Kaplansky, I. [1953] Modules over operator algebras. *Trans. Amer. Math. Soc.* **75**, 839–858.

Karasev, M. V. [1989] The Maslov quantization conditions in higher cohomology and analogs of notions developed in Lie theory for canonical fibre bundles of symplectic manifolds. I, II. *Selecta Math. Soviet.* **8**, (1989), 213–234, 235–258.

Karasev, M.V. and V.P. Maslov [1993] *Nonlinear Poisson Brackets: Geometry and Quantization*. American Mathematical Society, Providence.

Karshon, Y. and E. Lerman [1997] The centralizer of invariant functions and division properties of the moment map. *Ill. J. Math.* **41**, 462–487.

Kasparov, G.G. [1980] Hilbert C^*-modules: theorems of Stinespring and Voiculescu. *J. Operator Theory* **4**, 133–150.

Kasparov, G.G. [1981] The operator K-functor and extensions of C^*-algebras. *Math. USSR Izv.* **16**, 513–572.

Kastler, D. and M. Mebkhout [1990] Revisiting the Mackey–Stone–von Neumann theorem. The C^*-algebra of a presymplectic space. *Nucl. Phys. Proc. Suppl.* **B18**, 200–211.

Kazhdan, D., B. Kostant, and S. Sternberg [1978] Hamiltonian group actions and dynamical systems of Calogero type. *Commun. Pure Appl. Math.* **31**, 481–507.

Kelley, J.L. [1955] *General Topology*. Van Nostrand, London.

Kerner, R. [1968] Generalization of the Kaluza–Klein theory for an arbitrary non abelian gauge group. *Ann. Inst. H. Poincaré* **9**, 143–152.

Kijowski, J. and W. Tulczyjew [1979] *A Symplectic Framework for Field Theories. Lecture Notes in Physics* **107**. Springer, Berlin.

Kirchberg, E. and S. Wassermann [1995] Operations on continuous bundles of C^*-algebras. *Math. Ann.* **303**, 677–697.

Kirillov, A.A. [1962] Unitary representations of nilpotent Lie groups. *Russ. Math. Surv.* **17**, 53–104.

Kirillov, A.A. [1976] Local Lie algebras. *Russ. Math. Surv.* **31**, 55–75.

Kirillov, A.A. [1990] Geometric Quantization. In: V.I. Arnold and S.P. Novikov (eds.) *Dynamical Systems IV*, pp. 137–172. Springer, Berlin.

Kirwan, F. [1984] Convexity properties of the moment mapping. III. *Inv. Math.* **77**, 547–552.

Klauder, J.R. [1963] Continuous representation theory. II. Generalized relation between quantum and classical dynamics. *J. Math. Phys.* **4**, 1058–1073.

Klauder, J.R. [1970] Exponential Hilbert space: Fock space revisited. *J. Math. Phys.* **11**, 609–630.

Klauder, J.R. [1988] Quantization is geometry, after all. *Ann. Phys. (N.Y).* **188**, 120–141.

Klauder, J.R. [1995] Quantization without quantization. *Ann. Phys. (N.Y).* **237**, 147–160.

Klauder, J.R. [1997] Coherent state quantization of constrained systems. *Ann. Phys. (N.Y.)* **254**, 419–453.

Klauder, J.R. and B.-S. Skagerstam [1985] (eds.) *Coherent States*. World Scientific, Singapore.

Kleppner, A. and R.L. Lipsman [1972] The Plancherel formula for group extensions. I. *Ann. Scient. Ec. Norm. Sup.* **5**, 459–516.

Kleppner, A. and R.L. Lipsman [1973] The Plancherel formula for group extensions. II. *Ann. Scient. Ec. Norm. Sup.* **6**, 103–132.

Klimek, S. and A. Lesniewski [1992a] Quantum Riemann surfaces, I. The unit disc. *Commun. Math. Phys.* **146**, 103–122.

Klimek, S. and A. Lesniewski [1992b] Quantum Riemann surfaces, II. The discrete series. *Lett. Math. Phys.* **24**, 125–139.

Klimek, S. and A. Lesniewski [1994] Quantum Riemann surfaces, III. The exceptional case. *Lett. Math. Phys.* **32**, 45–61.

Klimek, S. and A. Lesniewski [1996] Quantum Riemann surfaces for arbitrary Planck's constant. *J. Math. Phys.* **37**, 2157–2165.

Klingenberg, W. [1982] *Riemannian Geometry*. de Gruyter, Berlin.

Knapp, A.W. [1986] *Representation Theory of Semisimple groups. An Overview Based on Examples*. Princeton University Press, Princeton.

Knauf, A. [1989] Coulombic periodic potentials: the quantum case. *Ann. Phys. (N.Y.)* **191**, 205–240.

Kobayashi, S. and K. Nomizu [1963] *Foundations of Differential Geometry. Vol. I.* Wiley, New York.

Kobayashi, S. and K. Nomizu [1969] *Foundations of Differential Geometry. Vol. II.* Wiley, New York.

Kohn, J. and L. Nirenberg [1965] An algebra of pseudo-differential operators. *Comm. Pure Appl. Math.* **18**, 269–305.

Kolmogorov, A. [1932] Zur Begründung der projektiven Geometrie. *Ann. Math.* **33**, 175–176.

Kondracki, W. and P. Sadowski [1986] Geometric structure on the orbit space of gauge connections. *J. Geom. Phys.* **3**, 421–434.

Konopleva, N.P. and V.N. Popov [1981] *Gauge Fields*. Chur, Switzerland.

Kontsevich, M. [1998] Deformation quantization of Poisson manifolds, I. *e-print q-alg/9709040*.

Koornwinder, T.H. and N.M. Muller [1997] The quantum double of a (locally) compact group. *J. Lie Theory* **7**, 101–120.

Kostant, B. [1970] Quantization and unitary representations. *Lecture Notes in Mathematics* **170**, 87–208.

Kostant, B. [1973] On convexity, the Weyl group and the Iwasawa decomposition. *Ann. scient. Éc. Norm. Sup.* **6**, 413–455.

Kraus, K. [1977] Position observables of the photon. In: Price, W.C. and S.S. Chissick (eds.) *The Uncertainty Principle and Foundations of Quantum Mechanics*, pp. 293–320. Wiley, New York.

Krishnaprasad, P.S. and J.E. Marsden [1987] Hamiltonian structure and stability for rigid bodies with flexible attachments. *Arch. Rat. Mech. An.* **98**, 137–158.

Kurchan, J., P. Leboeuf, and M. Saraceno [1989] Semiclassical approximation in the coherent-state representation. *Phys. Rev.* **A40**, 6800–6814.

Kummer, H. [1991] The foundation of quantum theory and noncommutative spectral theory. I, II. *Found. Phys.* **21**, 1021–1069, 1183–1236.

Kummer, M. [1981] On the construction of the reduced phase space of a Hamiltonian system with symmetry. *Indiana Univ. Math. J.* **30**, 281–291.

Kumpera, A. and D.C. Spencer [1972] *Lie Equations. Vol. 1.* Princeton University Press, Princeton.

Künzle, H.P. [1972] Canonical dynamics of spinning particles in gravitational and electromagnetic fields. *J. Math. Phys.* **13**, 739–744.

Kuo, H.H. [1975] *Gaussian Measures in Banach Spaces. Lecture Notes in Mathematics* **463**. Springer, Berlin.

Kuwabara, R. [1982] On spectra of the Laplacian on vector bundles. *J. Math. Tokushima Univ.* **16**, 1–23.

Kuwabara, R. [1984] Spectrum and holonomy of the line bundle over the sphere. *Math. Z.* **187**, 481–490.

Laidlaw, M.G.G. and C.M. DeWitt [1970] Feynman functional integrals for systems of indistinguishable particles. *Phys. Rev.* **D3**, 1375–1378.

Lahti, P.J. and S. Bugajski [1980] Fundamental principles of quantum theory. *Int. J. Theor. Phys.* **19**, 499–514.

Lahti, P.J. and S. Bugajski [1985] Fundamental principles of quantum theory 2. From a convexity scheme to the DHB theory. *Int. J. Theor. Phys.* **24**, 1051–1080.

Lance, E.C. [1995] *Hilbert C*-Modules. A Toolkit for Operator Algebraists. LMS Lecture Notes* **210**. Cambridge University Press, Cambridge.

Landau, L.J. [1996] Macroscopic observation of a quantum particle in a slowly varying potential – on the classical limit of quantum-theory. *Ann. Phys. (N.Y.)* **246**, 190–227.

Landi, G. [1997] *An Introduction to Noncommutative Spaces and their Geometries.* Springer, Berlin.

Landsman, N.P. [1990a] Quantization and superselection sectors I. Transformation group C^*-algebras. *Rev. Math. Phys.* **2**, 45–72.

Landsman, N.P. [1990b] Quantization and superselection sectors II. Dirac Monopole and Aharonov–Bohm effect. *Rev. Math. Phys.* **2**, 73–104.

Landsman, N.P. [1991] Algebraic theory of superselection sectors and the measurement problem in quantum mechanics. *Int. J. Mod. Phys.* **A6**, 5349–5372.

Landsman, N.P. [1992] Induced representations, gauge fields, and quantization on homogeneous spaces. *Rev. Math. Phys.* **4**, 503–528.

Landsman, N.P. [1993a] Deformations of algebras of observables and the classical limit of quantum mechanics. *Rev. Math. Phys.* **5**, 775–806.

Landsman, N.P. [1993b] Strict deformation quantization of a particle in external gravitational and Yang–Mills fields. *J. Geom. Phys.* **12**, 93–132.

Landsman, N.P. [1993c] Quantization and classicization: from Jordan–Lie algebras of observables to gauge fields. *Class. Quantum Grav.* **10**, S101–S108.

Landsman, N.P. [1994] The infinite unitary group, Howe dual pairs, and the quantization of constrained systems. *e-print hep-th/9411171.*

Landsman, N.P. [1995a] Rieffel induction as generalized quantum Marsden–Weinstein reduction. *J. Geom. Phys.* **15**, 285–319; Err. *ibid.* **17** (1995) 298.

Landsman, N.P. [1995b] Observation and superselection in quantum mechanics. *Stud. Hist. Phil. Mod. Phys.* **26**, 45–73.

Landsman, N.P. [1995c] Against the Wheeler–DeWitt equation. *Class. Quantum Grav.* **12**, L119–L123.

Landsman, N.P. [1996a] Classical and quantum representation theory. In: de Kerf, E. A. and H.G.J. Pijls (eds.) *Proceedings Seminar Mathematical Structures in Field Theory.CWI-syllabus* **39**, pp. 135–163. Mathematisch Centrum, CWI, Amsterdam.

Landsman, N.P. [1996b] Classical behaviour in quantum mechanics: a transition probability approach. *Int. J. Mod. Phys.* **B**, 1545–1554.

Landsman, N.P. [1997] Poisson spaces with a transition probability. *Rev. Math. Phys.* **9**, 29–57.

Landsman, N.P. [1998a] The quantization of constrained systems: from symplectic reduction to Rieffel induction. In: Strasburger, A., S.T. Ali, J.-P. Antoine, J.-P. Gazeau, and A. Odzijewicz (eds.) *Quantization, Coherent States and Poisson Structures. Proc. XIVth Workshop on Geometric Methods in Physics, Białowieża, 1995*, pp. 79–95. Polish Scientific Publishers, Warsaw.

Landsman, N.P. [1998b] Simple new axioms for quantum mechanics. *Int. J. Theor. Phys.* **37**, 343–348.

Landsman, N.P. [1998c] Strict quantization of coadjoint orbits. *J. Math. Phys.*, to appear.

Landsman, N.P. [1998d] Twisted Lie group C^*-algebras as strict quantizations. *Lett. Math. Phys.*, to appear.

Landsman, N.P. and N. Linden [1991] The geometry of inequivalent quantizations. *Nucl. Phys.* **B365**, 121–160.

Landsman, N.P. and N. Linden [1992] Superselection rules from Dirac and BRST quantization of constrained systems. *Nucl. Phys.* **B371**, 415–433.

Landsman, N.P. and U.A. Wiedemann [1995] Massless particles electromagnetism, and Rieffel induction. *Rev. Math. Phys.* **7**, 923–958.

Landsman, N.P. and K.K. Wren [1997] Constrained quantization and θ-angles. *Nucl. Phys.* **B502 [PM]**, 537–560.

Landstad, M.B. [1994] Quantizations arising from abelian subgroups. *Int. J. Math.* **5**, 897–936.

Landstad, M. B. and I. Raeburn [1997] Equivariant deformations of homogeneous spaces. *J. Funct. Anal.* **148**, 480–507.

Lang, S. [1995] *Differential and Riemannian Manifolds*, 3d ed. Springer, New York.

Langlands, R.P. [1987] The Dirac monopole and induced representations. *Pac. J. Math.* **126**, 145–151.

Langmann, E. and G.W. Semenoff [1993] Gribov ambiguity and non-trivial vacuum structure of gauge theories on a cylinder. *Phys. Lett.* **B303**, 303–307.

Laquer, H.T. [1984] Stability properties of the Yang–Mills functional near the canonical connection. *Michigan Math. J. 31*, 139–159.

Lee, H.-W. [1995] Theory and applications of the quantum phase space distribution functions. *Phys. Rep.* **259**, 147–211.

Lee, R.-Y. [1976] On the C^*-algebras of operator fields. *Indiana Univ. Math. J.* **25**, 303–314.

Lee, T. and P. Oh [1994] Non-abelian Chern–Simons quantum mechanics and non-abelian Aharonov–Bohm effect. *Ann. Phys. (N.Y.)* **235**, 413–434.

Leinfelder, H. and C. Simader [1981] Schrödinger operators with singular magnetic vector potentials. *Math. Z.* **176**, 1–19.

Leptin, H. and J. Ludwig [1994] *Unitary Representation Theory of Exponential Lie Groups*. De Gruyter, Berlin.

Leonrad, N.E. and J.E. Marsden [1997] Stability and drift of underwater vehicle dynamics: mechanical systems with rigid motion symmetry. *Physica* **D105**, 130–162.

Lerman, E., R. Montgomery, and R. Sjamaar [1993] Examples of singular reduction. In: Salomon, D. (ed.) *Symplectic Geometry, LMS Lecture Notes Series* **192**, pp. 127–155. Cambridge University Press, Cambridge.

Libermann, P. [1983] Problèmes d'equivalence et géométrie symplectique. *Asterisque* **107–108**, 43–68.

Libermann, P. and C.-M. Marle [1987] *Symplectic Geometry and Analytical Mechanics*. Reidel, Dordrecht.

Lichnerowicz, A. [1977] Les variétés de Poisson et leurs algèbres de Lie associées. *J. Diff. Geom.* **12**, 253–300.

Lie, S. [1890] *Theorie der Transformationsgruppen*. S. Teubner, Leipzig.

Lieb, E.H. [1973] The classical limit of quantum spin systems. *Commun. Math. Phys.* **62**, 327–340.

Linden, N., A.J. MacFarlane, and J.W. van Holten [1996] Particle motion in a Yang–Mills field – Wong's equations and spin 1/2 analogs. *Czech. J. Phys.* **46**, 209–215.

Littlejohn, R.G. [1986] The semiclassical evolution of wave packets. *Phys. Rep.* **138**, 193–291.

Littlejohn, R.G. [1992] The Van Vleck formula, Maslov theory, and phase space geometry. *J. Stat. Phys.* **68**, 7–50.

Liu, Z.-J. and M. Qian [1992] Gauge invariant quantization on Riemannian manifolds. *Trans. Amer. Math. Soc.* **331**, 321–333.

Löffelholz, J., G. Morchio, and F. Strocchi [1996] A quantum mechanical gauge model and a possible dynamical solution of the strong CP-problem. *Ann. Phys. (N.Y.)* **250**, 367–388.

Lowenstein, J.H. and J.A. Swieca [1971] Quantum electrodynamics in two dimensions. *Ann. Phys. (N.Y.)* **68**, 172–195.

Ludwig, G. [1985] *An Axiomatic Basis for Quantum Mechanics. Volume 1: Derivation of Hilbert Space Structure*. Springer, Berlin.

Mackenzie, K. [1987a] *Lie Groupoids and Lie Algebroids in Differential Geometry.* Cambridge University Press, Cambridge.

Mackenzie, K. [1987b] A note on Lie algebroids which arise from groupoid actions. *Cah. Top. Géom. Diff. Cat.* **28**, 283–302.

Mackenzie, K. [1989] Classification of principal bundles and Lie groupoids with prescribed gauge group bundle. *J. Pure Appl. Alg.* **58**, 181–208.

Mackey, G.W. [1958] Unitary representations of group extensions. I. *Acta Math.* **99**, 265–311.

Mackey, G.W. [1963] *The Mathematical Foundations of Quantum Mechanics.* New York, Benjamin.

Mackey, G.W. [1968] *Induced Representations.* Benjamin, New York.

Mackey, G.W. [1978] *Unitary Group Representations in Physics Probability and Number Theory.* Benjamin, New York.

Mackey, G.W. [1992] *The Scope and History of Commutative and Noncommutative Harmonic Analysis.* American Mathematical Society, Providence.

Mackey, G.W. [1998] The relationship between classical mechanics and quantum mechanics. *Contemp. Math.* **214**, 91–110.

Maeda, F. [1958] *Kontinuerliche Geometrien.* Springer, Berlin.

Maeda, F. and S. Maeda [1970] *Theory of Symmetric Lattices.* Springer, Berlin.

Majid, S. [1988] Hopf algebras for physics at the Planck scale. *Class. Quantum Grav.* **5**, 1587–1606.

Majid, S. [1990] Physics for algebraists: non-commutative and non-cocommutative Hopf algebras by a bicrossproduct construction. *J. Algebra* **130**, 17–64.

Malliavin, P. [1997] *Stochastic Analysis.* Springer, Berlin.

Malliavin, M.-P. and P. Malliavin [1990]. Integration on loop groups. I. Quasi invariant measures. *J. Funct. Anal.* **93**, 207–237.

Manchon, D. [1993] Weyl symbolic calculus on any Lie group. *Acta Appl. Math.* **30**, 159–186.

Manton, N. S. [1985] The Schwinger model and its axial anomaly. *Ann. Phys. (N.Y.)* **159**, 220–251.

Marinov, M.S. [1995] Path integrals on homogeneous manifolds. *J. Math. Phys.* **36**, 2458–2469.

Marolf, D. [1997] Refined algebraic quantization: systems with a single constraint. *Banach Center Publ.* **39**, 331–344.

Marsden, J.E. [1974] *Applications of Global Analysis in Mathematical Physics.* Publish or Perish, Boston.

Marsden, J.E. [1981] *Lectures on Geometric Methods in Mathematical Physics.* SIAM, Philadelphia.

Marsden, J.E. [1992] *Lectures on Mechanics.* Cambridge University Press, Cambridge.

Marsden, J.E. [1993] Steve Smale and geometric mechanics. In: Hirsch, M.W., J.E. Marsden, and M. Shub (eds.) *From Topology to Computation: Proceedings of the Smalefest*, pp. 499–516. Springer, New York.

Marsden, J.E., R. Montgomery, and T. Ratiu [1990] Reduction, symmetry, and phases in mechanics. *Mem. Amer. Math. Soc.* **436**, 1–110.

Marsden, J.E. and T.S. Ratiu [1986] Reduction of Poisson manifolds. *Lett. Math. Phys.* **11**, 161–170.

Marsden, J.E. and T.S. Ratiu [1994] *Introduction to Mechanics and Symmetry*. Springer, New York.

Marsden, J.E., T. Ratiu, and G. Raugel [1991] Symplectic connections and the linearization of Hamiltonian systems. *Proc. Royal Soc. Edinburgh* **177A**, 329–380.

Marsden, J.E., T. Ratiu, and A. Weinstein [1984a] Semidirect products and reduction in mechanics. *Trans. Amer. Math. Soc.* **281**, 147–177.

Marsden, J.E., T. Ratiu, and A. Weinstein [1984b] Reduction and Hamiltonian structures on duals of semidirect product Lie algebras. *Contemp. Math.* **28**, 55–100.

Martinez Alonso, L. Group-theoretical foundations of classical and quantum mechanics. II. Elementary systems. *J. Math. Phys.* **20**, 219–230.

Maslov, V.P. [1994] *The Complex WKB Method for Nonlinear Equations*. Birkhäuser, Basel.

Maslov, V.P. and M.V. Fedoriuk [1981] *Semi-Classical Approximation in Quantum Mechanics*. Reidel, Dordrecht.

Matsumoto, K. [1991a] Noncommutative three-dimensional spheres. *Japan. J. Math. (N.S.)* **17**, 333–356.

Matsumoto, K. [1991b] Noncommutative three-dimensional spheres. II. Noncommutative Hopf fibering. *Yokohama Math. J.* **38**, 103–111.

Matsumoto, K. and J. Tomiyama [1992] Noncommutative lens spaces. *J. Math. Soc. Japan* **44**, 13–41.

Mehra, J. and H. Rechenberg [1982] *The Historical Development of Quantum Theory. Vol. 1. The quantum theory of Planck, Einstein, Bohr, and Sommerfeld: its foundation and the rise of its difficulties, 1900–1925*. Springer, New York.

Meinrenken, E. [1994] Coherent states and classical limits. *J. Phys.* **A27**, 3257–3265.

Meinrenken, E. and R. Sjamaar [1998] Singular reduction and quantization. *e-print dg-ga/9707023*.

Meschkowski, H. [1962] *Hilbertsche Räume mit Kernfunktion*. Springer, Berlin.

Meyer, K. [1973] Symmetries and integrals in mechanics. In: Peixoto, M.M. (ed.) *Dynamical systems*, pp. 259–272. Academic Press, New York.

Michor, P.W. [1990] The moment mapping for unitary representations. *Ann. Global Anal. Geom.* **8**, 299–313.

Mielnik, B. [1968] Geometry of quantum states. *Commun. Math. Phys.* **9**, 55–80.

Mielnik, B. [1969] Theory of filters. *Commun. Math. Phys.* **15**, 1–46.

Mielnik, B. [1974] Generalized quantum mechanics. *Commun. Math. Phys.* **37**, 221–256.

Mikami, K. and A. Weinstein [1988] Moments and reduction for symplectic groupoids. *Publ. RIMS Kyoto Univ.* **24**, 121–140.

Milnor, J. [1976] Curvatures of left invariant metrics on Lie groups. *Adv. Math.* **21**, 293–329.

Mitter, P.K. and C.M. Viallet [1981] On the bundle of connections and the gauge orbit manifold in Yang–Mills theory. *Commun. Math. Phys.* **79**, 457–472.

Molzahn, F.H., T.A. Osborn, and S.A. Fulling [1990] Gauge invariant asymptotic expansions of Schrödinger propagators on manifolds. *Ann. Phys. (N.Y.)* **204**, 64–112.

Molzahn, F.H., T.A. Osborn, and S.A. Fulling [1992] Multi–scale semiclassical approximations for Schrödinger propagators on manifolds. *Ann. Phys. (N.Y.)* **214**, 102–141.

Montgomery, R. [1984] Canonical formulation of a classical particle in a Yang–Mills field and Wongs's equations. *Lett. Math. Phys.* **8**, 59 67.

Montgomery, R., J.E. Marsden, and T. Ratiu [1984] Gauged Lie–Poisson structures. *Contemp. Math.* **28**, 101–114. American Mathematical Society, Providence.

Monthubert, B. and F. Pierrot [1997] Indice analytique et groupoïdes de Lie. *C. R. Acad. Sci. Paris Sér. I Math.* **325**, 193–198.

Moschella, U. [1989] Classical limit of a quantum particle in an external Yang–Mills field. *Ann. Inst. H. Poincaré* **A51**, 351–370.

Moscovici, H. [1969] Generalized induced representations. *Rev. Roum. Math. Pures et Appl.* **14**, 1539–1551.

Moyal, J.E. [1949] Quantum mechanics as a statistical theory. *Proc. Camb. Phil. Soc.* **45**, 99–124.

Muhly, P.S., J.N. Renault, and D.P. Williams [1987] Equivalence and isomorphism for groupoid C^*-algebras. *J. Operator Th.* **17**, 3–22.

Muhly, P.S., and D.P. Williams [1990] Continuous trace groupoid C^*-algebras. *Math. Scand.* **66**, 231–241.

Muhly, P.S., and D.P. Williams [1992] Continuous trace groupoid C^*-algebras. 2. *Math. Scand.* **70**, 127–145.

Muhly, P.S., and D.P. Williams [1995] Groupoid cohomology and the Dixmier–Douady class. *Proc. London Math. Soc.* **71**, 109–134.

Nachtmann, O. [1968] Dynamische Stabilität im de-Sitter-Raum. *Sitz. Ber. Öst. Akad. Wiss. II* **176**, 363–379.

Nagasawa, M. [1993] *Schrödinger Equations and Diffusion Theory*. Birkhäuser, Basel.

Nagy, G. [1992] *A Framework for Deformation Quantization*. Ph.D. thesis, University of California at Berkeley.

Nagy, G. [1993] On the Haar measure of the quantum $SU(N)$ group. *Commun. Math. Phys.* **153**, 217–228.

Nagy. G [1996] E-theory with $*$-homomorphisms. *J. Funct. Anal.* **140**, 275–299.

Nagy, G. [1997] Deformation quantization and K-theory. *Contemp. Math.* **214**, 111–134.

Nagy, G. [1998a] A deformation quantization procedure for C^*-algebras. *J. Operator Th.*, to appear.

Nagy, G. [1998b] A rigidity property for quantum $SU(3)$ groups. In: Brylinski, J.-L., R. Brylinski, N. Handzy, and B. Tsygan (eds.) *Advances in Geometry and Mathematical Phyisics, Vol. I*. Birkhäuser, Basel.

Nauenberg, M., C. Stroud, and J. Yeazell [1994] The classical limit of an atom. *Sci. Amer.* **270(6)**, 44–49.

Neeb, K.-H. [1995] On the convexity of the moment mapping for unitary highest weight representations. *J. Funct. Anal.* **127**, 301–325.

Neeb, K.-H. [1996a] Coherent states, holomorphic extensions, and highest weight representations. *Pac. J. Math.* **174**, 497–542.

Neeb, K.-H. [1996b] A note on central extensions of Lie groups. *J. Lie Theory* **6**, 207–213.

Neumann, H. [1972] Transformation properties of observables. *Helv. Phys. Acta* **25**, 811–819.

Neumann, J. von [1931] Die Eindeutigkeit der Schrödingerschen Operatoren. *Math. Ann.* **104**, 570–578.

Neumann, J. von [1932] *Mathematische Grundlagen der Quantenmechanik.* Springer, Heidelberg.

Neumann, J. von [1936] On an algebraic generalization of the quantum mechanical formalism (Part I). *Math. Sb.* **1**, 415–484.

Neumann, J. von [1981] Continuous geometries with a transition probability. *Mem. Amer. Math. Soc.* **252**, 1–210 (edited by I.S. Halperin; MS from 1937).

Newton, T.D. and E.P. Wigner [1949] Localized states for elementary systems. *Rev. Mod. Phys.* **21**, 400–406.

Nistor, V., A. Weinstein, and P. Xu [1997] Pseudodifferential operators on differential groupoids. *e-print funct-an/970200.*

Nussbaum, A.E. [1964] Reduction theory for unbounded closed operators in Hilbert space. *Duke Math. J.* **31**, 33–44.

Odzijewicz, A. [1988] On reproducing kernels and quantization of states. *Commun. Math. Phys.* **114**, 577–579.

Odzijewicz, A. [1992] Coherent states and geometric quantization. *Commun. Math. Phys.* **150**, 385–413.

Oh, P. [1996] Classical and quantum mechanics of non-abelian Chern–Simons particles. *Nucl. Phys.* **462**, 551–570.

Omnès, R. [1994] *The Interpretation of Quantum Mechanics*. Princeton University Press, Princeton.

Omnès, R. [1997a] Quantum–classical correspondence using projection operators. *J. Math. Phys.* **38**, 697–707.

Omnès, R. [1997b] Localization of relativistic particles. *J. Math. Phys.* **38**, 708–715.

Omori, H. [1997] *Infinite-Dimensional Lie Groups*. American Mathematical Society, Providence.

Onofri, E. [1975] A note on coherent state representations of Lie groups. *J. Math. Phys.* **16**, 1087–1089.

Ørsted, B. [1979] Induced representations and a new proof of the imprimitivity theorem. *J. Funct. Anal.* **31**, 355–359.

Osborn, T.A. and F.H. Molzahn [1995]. Moyal quantum mechanics: the semiclassical Heisenberg dynamics. *Ann. Phys. (N.Y.)* **241**, 79–127.

Otto, M. [1987] A reduction scheme for phase spaces with almost Kähler symmetry. Regularity results for momentum level sets. *J. Geom. Phys.* **4**, 101–118.

Packer, J.A. [1994] Transformation group C^*-algebras: A selective survey. In: Doran [1994], pp. 183–217.

Packer, J.A. [1996] Crossed product C^*-algebras and algebraic topology. *Rev. Math. Phys.* **8**, 623–637.

Packer, J.A. and I. Raeburn [1989] Twisted crossed products of C^*-algebras. *Math. Proc. Cam. Phil. Soc.* **106**, 293–311.

Packer, J.A. and I. Raeburn [1990] Twisted crossed products of C^*-algebras. 2. *Math. Ann.* **287**, 595–612.

Packer, J.A. and I. Raeburn [1992] On the structure of twisted group C^*-algebras. *Trans. Amer. Math. Soc.* **334**, 685–718.

Palais, R.S. [1957] A global formulation of the Lie theory of transformation groups. *Mem. Amer. Math. Soc.* **22**, 1–123.

Palmer, T.W. [1994] *Banach Algebras and the General Theory of*-Algebras. Vol. I: Algebras and Banach Algebras.* Cambridge University Press, Cambridge.

Paschke, W.L. [1973] Inner product modules over B^*-algebras. *Trans. Amer. Math. Soc.* **182**, 443–468.

Paterson, A.L [1988] *Amenability.* American Mathematical Society, Providence.

Paul, T. and A. Uribe [1995] The semi-classical trace formula and propagation of wave packets. *J. Funct. Anal.* **132** (1995), 192–249.

Paul, T. and A. Uribe [1996] On the pointwise behavior of semi-classical measures. *Commun. Math. Phys.* **175**, 229–258.

Pauli, W. [1973] *Ausgewählte Kapitel aus der Feldquantisierung.* E.T.H. Zürich Lecture Notes 1950–51. In: Enz, C.P. (ed.) *Pauli Lectures in Physics, Vol. 6: Selected Topics in Field Quantization.* MIT Press, Cambridge (MA).

Paulsen, V.I. [1986] *Completely Bounded Maps and Dilations.* Longman, Harlow.

Pedersen, G.K. [1979] *C^*-Algebras and their Automorphism Groups.* Academic Press, London.

Pedersen, G.K. [1989] *Analysis Now.* Springer, New York.

Petz, D. [1994] Geometry of canonical correlation on the state space of a quantum system. *J. Math. Phys.* **35**, 780–795.

Peetre, J. [1990] The Berezin transform and Haplitz operators. *J. Operator Th.* **24**, 165–186.

Perelomov, A.M. [1972] Coherent states for arbitrary Lie groups. *Commun. Math. Phys.* **26**, 222–236.

Perelomov, A. [1986] *Generalized Coherent States and their Applications.* Springer, Berlin.

Pflaum, M.J. [1995] *Local Analysis of Deformation Quantization.* Ph.D. thesis, Ludwig-Maximilians-Universität München.

Picard, R. [1989] *Hilbert Space Approach to some Classical Transforms.* Longman, Harlow.

Piron, C. [1976] *Foundations of Quantum Physics.* Benjamin, Reading (Mass.).

Piziak, R. [1991] Orthomodular lattices and quadratic spaces: a survey. *Rocky Mount. J. Math.* **21**, 951–992.

Plymen, R. [1968]. C^*-algebras and Mackey's axioms. *Commun. Math. Phys.* **8**, 132–146.

Poerschke, T., G. Stolz, and J. Weidmann [1989]. Expansions in generalized eigenfunctions of self–adjoint operators. *Math. Z.* **202**, 397–408.

Poerschke, T. and G. Stolz [1993] On eigenfunction expansions and scattering theory. *Math. Z.* **212**, 397–357.

Pontrjagin, L. [1946] *Topological Groups.* Princeton University Press, Princeton.

Pool, J.C.T. [1966] Mathematical aspects of the Weyl correspondence. *J. Math. Phys.* **7**, 66–76.

Pool, J.C.T. [1968] Semimodularity and the logic of quantum mechanics. *Commun. Math. Phys.* **9**, 218–228.

Poulsen, N.S. [1970] *Regularity Aspects of the Theory of Infinite-Dimensional Representations of Lie Groups*. Ph.D. thesis, MIT.

Pradines, J. [1966] Théorie de Lie pour les groupoïdes différentiables. Relations entre propriétés locales et globales. *C. R. Acad. Sc. Paris* **A263**, 907–910.

Pradines, J. [1968] Géométrie différentielle au-dessus d'un groupoïde. *C. R. Acad. Sc. Paris* **A266**, 1194–1196.

Pressley, A. and G. Segal [1986] *Loop Groups*. Oxford University Press, Oxford.

Prosser, R.T. [1963] On the ideal structure of operator algebras. *Mem. Amer. Math. Soc.* **45**.

Pulmannová, S. [1986] Transition probability spaces. *J. Math. Phys.* **27**, 1791–1795.

Pulmannová, S. [1989] Mielnik and Cantoni transition probabilities. *Int. J. Theor. Phys.* **28**, 711–718.

Rădulescu, F. [1998] The Γ-equivariant form of the Berezin quantization of the upper half plane. *Mem. Amer. Math. Soc.* **630**.

Raeburn, I. [1988] Induced C^*-algebras and a symmetric imprimitivity theorem. *Math. Ann.* **280**, 369–387.

Raggio, G.A. [1982] Comparison of Uhlmann's transition probabilitywith one induced by the natural cone of von Neumann algebras in standard form. *Lett. Math. Phys.* **6**, 233–236.

Ramazan, B. [1998] *Deformation Quantization of Lie–Poisson Manifolds*. Ph.D. thesis, Université d'Orléans.

Ramsay, A. [1965] Dimension theory in complete orthocomplemented weakly modular lattices. *Trans. Amer. Math. Soc.* **116**, 9–31.

Ramsay, A. [1971] Virtual groups and group actions. *Adv. Math.* **6**, 253–322.

Rawnsley, J.H. [1975] Representations of a semi-direct product by quantization. *Math. Proc. Camb. Phil. Soc.* **78**, 345–350.

Rawnsley, J.H. [1977] Coherent states and Kähler manifolds. *Quart. J. Math. Oxford (2)* **28**, 403–415.

Rajeev, S.G. [1988] Yang–Mills theory on a cylinder. *Phys. Lett.* **B212**, 203–205.

Rajeev, S.G. and L. Rossi [1995] Some rigorous results for Yang–Mills theory on a cylinder. *J. Math. Phys.* **36**, 3308–3319.

Reed, M. and B. Simon [1972] *Methods of Modern Mathematical Physics. I: Functional Analysis*. Academic Press, New York.

Reed, M. and B. Simon [1975] *Methods of Modern Mathematical Physics. II: Fourier Analysis, Self-adjointness*. Academic Press, New York.

Reed, M. and B. Simon [1978] *Methods of Modern Mathematical Physics. IV: Analysis of Operators*. Academic Press, New York.

Renault, J. [1980] *A Groupoid Approach to C^*-algebras. Lecture Notes in Mathematics* **793**. Springer, Berlin.

Renault, J. [1987] Représentation des produits croisés d'algèbres de groupoïdes. *J. Operator Th.* **18**, 67–97.

Rezende, J. [1996] Stationary phase, quantum mechanics and semi-classical limit. *Rev. Math. Phys.* **8**, 1161–1185.

Rieffel, M.A. [1972] On the uniqueness of the Heisenberg commutation relations. *Duke Math. J.* **39**, 745–753.

Rieffel, M.A. [1974a] Induced representations of C^*-algebras. *Adv. Math.* **13**, 176–257.

Rieffel, M.A. [1974b] Morita equivalence for C^*-algebras and W^*-algebras. *J. Pure Appl. Alg.* **5**, 51–96.

Rieffel, M.A. [1979] Unitary representations of group extensions: an algebraic approach to the theory of Mackey and Blattner. *Adv. Math. Suppl. Stud.* **4**, 43–82.

Rieffel, M.A. [1988] Projective modules over higher-dimensional noncommutative tori. *Canad. J. Math.* **40**, 257–338.

Rieffel, M.A. [1989a] Deformation quantization of Heisenberg manifolds. *Commun. Math. Phys.* **122**, 531–562.

Rieffel, M.A. [1989b] Continuous fields of C^*-algebras coming from group cocycles and actions. *Math. Ann.* **283**, 631–643.

Rieffel, M.A. [1990a] Lie group convolution algebras as deformation quantizations of linear Poisson structures. *Am. J. Math.* **112**, 657–686.

Rieffel, M.A. [1990b] Proper actions of groups on C^*-algebras. In: Araki, H. and R.V. Kadison (eds.) *Mappings of Operator Algebras*, pp. 141–182. Birkhäuser, Boston.

Rieffel, M.A. [1993a] Deformation quantization for actions of $\mathbb{R}^d$. *Mem. Amer. Math. Soc.* **506**.

Rieffel, M.A. [1993b] Compact quantum groups associated with toral subgroups. *Contemp. Math.* **145**, 465–491.

Rieffel, M.A. [1993c] K-groups of C^*-algebras deformed by actions of $\mathbb{R}^d$. *J. Funct. Anal.* **116**, 199–214.

Rieffel, M.A. [1994] Quantization and C^*-algebras. In: Doran, R.S. (ed.) *C^*-algebras: 1943–1993. Contemp. Math.* **167**, pp. 67–97. American Mathematical Society, Providence.

Rieffel, M.A. [1995] Non-compact quantum groups associated with abelian subgroups. *Comm. Math. Phys.* **171**, 181–201.

Rieffel, M.A. [1996] The classical limit of dynamics for spaces quantized by an action of $\mathbb{R}^d$. *Can. J. Math.* **49**, 160–174.

Rieffel, M.A. [1998] Quantization and operator algebras. In: Bracken, A.J., D. De Wit, M. Gould, and P. Pearce, (eds.) *Meeting with the Platypus. Proc. XIIth Int. Congress of Mathematical Physics, Brisbane 1997*. International Press, Singapore.

Riesz, F. and B. Sz.-Nagy [1990] *Functional Analysis*. Dover, New York.

Robert, D. [1987] *Autour de l'Approximation Semi-Classique*. Birkhäuser, Basel.

Robert, D. [1992] (ed.) *Méthodes Semi-Classiques. Astérisque* **207**, 1–212, **210**, 1–384.

Robert, D. [1998] Semi-classical approximation in quantum mechanics. A survey of old and recent Mathematical results. *Helv. Phys. Acta* **71**, 44–116.

Roberts, J.E. and G. Roepstorff [1969] Some basic concepts of algebraic quantum theory. *Commun. Math. Phys.* **11**, 321–338.

Robinson, S.L. [1988a] The semiclassical limit of quantum mechanics. I. Time evolution. *J. Math. Phys.* **29**, 412–419.

Robinson, S.L. [1988b] The semiclassical limit of quantum mechanics. II. Scattering theory. *Ann. Inst. H. Poincaré* **A48**, 281–296.

Robinson, S.L. [1993] Semiclassical mechanics for time-dependent Wigner functions. *J. Math. Phys.* **34**, 2185–2205.

Robson, M.A. [1994] *The Geometric Quantization of Constrained Systems*. Ph.D. thesis, University of Cambridge.

Robson, M.A. [1996] Geometric quantization of reduced cotangent bundles. *J. Geom. Phys.* **19** (1996) 207–245.

Rodriguez Palacios, A. [1988] Jordan axioms for C^*-algebras. *Manuscripta Math.* **61**, 297–314.

Rosenberg, J. [1994] C^*-algebras and Mackey's theory of group representations. In: Doran [1994], pp. 151–181.

Rosenberg, J. [1996] Behavior of K-theory under quantization. In: Doplicher, S., R. Longo, J.E. Roberts, and L. Szido (eds.) *Operator Algebras and Quantum Field Theory*, pp. 404–415. International Press, Cambridge (MA).

Rosenberg, S. [1997] *The Laplacian on Riemannian manifolds*. LMS Student Texts **31**. Cambridge University Press, Cambridge.

Royer, A. [1977] Wigner function as the expectation of the parity operator. *Phys. Rev.* **A15**, 449–450.

Ruijsenaars, S.N.M. [1983] The Aharonov–Bohm effect and scattering theory. *Ann. Phys. (N.Y.)* **146**, 1–34.

Rüttimann, G.T. [1985] Facial sets of probability measures. *Prob. Math. Stat.* **6**, 187–215.

Saksena, A., T.A. Osborn, and F.H. Molzahn [1991] An asymptotic analysis of quantum evolution with electromagnetic fields. *J. Math. Phys.* **32**, 938–955.

Schindler, C. [1990] The unique Jordan–Hahn decomposition property. *Found. Phys.* **20**, 561–573.

Schrader, R. and M.E. Taylor [1984] Small $\hbar$ asymptotics for quantum partition functions associated to particles in external Yang–Mills potentials. *Commun. Math. Phys.* **92**, 555–594.

Schrader, R. and M.E. Taylor [1989] Semi-classical asymptotics, gauge fields and quantum chaos. *J. Funct. Anal.* **83**, 258–316.

Schrödinger. E. [1926] Der stetige Übergang von der Mikro- zur Makromechanik. *Naturwiss.* **14**, 664–666.

Schroeck, F.E. [1996] *Quantum Mechanics on Phase Space*. Kluwer, Dordrecht.

Schulman, L.S. [1968] A path integral for spin. *Phys. Rev.* **176**, 1558–1569.

Schulman, L.S. [1981] *Techniques and Applications of Path Integration*. Wiley, New York.

Schwartz, L. [1964] Sous-espaces hilbertiens d'espaces vectoriels topologiques et noyaux associés. (Noyaux réproduisants). *J. Anal. Math.* **13**, 115–256.

Schwartz, L. [1973] *Radon Measures on Arbitrary Topological Spaces and Cylindrical Measures*. Bombay, Oxford University Press.

Schwarz, G.W. [1975] Smooth functions invariant under the action of a compact Lie group. *Topology* **14**, 63–68.

Schwinger, J. [1962] Gauge invariance and mass. II. *Phys. Rev.* **128**, 2425–2429.

Schwinger, J. [1970] *Quantum Kinematics and Dynamics*. W.A. Benjamin, New York.

Scutaru, H. [1977] Coherent states and induced representations. *Lett. Math. Phys.* **2**, 101–107.

Segal, I.E. [1947] Postulates for general quantum mechanics. *Ann. Math.* **48**, 930–948.

Segal, I.E. [1956] Tensor algebras over Hilbert spaces. I. *Trans. Amer. Math. Soc.* **81**, 106–134.

Segal, I.E. [1959] Foundations of the theory of dynamical systems of infinitely many degrees of freedom I. *Mat.-Fys. Medd. Dansk. Vid. Selsk.* **31**, 1–38.

Segal, I.E. [1963] Transforms for operators and symplectic automorphisms over a locally compact abelian group. *Math. Scand.* **13**, 31–43.

Sengupta, A. [1997a] The moduli space of Yang–Mills connections over a compact surface. *Rev. Math. Phys.* **9**, 77–121.

Sengupta, A. [1997b] Yang–Mills on surfaces with boundary: Quantum theory and symplectic limit. *Commun. Math. Phys.* **183**, 661–705.

Sengupta, A. [1997c] Gauge theory on compact surfaces. *Mem. Amer. Math. Soc.* **600**.

Shale, D. [1962] Linear symmetries of free Boson fields. *Trans. Amer. Math. Soc.* **103**, 149–167.

Sheu, A.J.-L. [1996] The Weyl quantization of Poisson $SU(2)$. *Pac. J. Math.* **173**, 223–240.

Sheu, A.J.-L. [1997] Compact quantum groups and groupoid C^*-algebras. *J. Funct. Anal.* **144** (1997) 371–393.

Shultz, F.W. [1979] On normed Jordan algebras which are Banach dual spaces. *J. Funct. Anal.* **31**, 360–376.

Shultz, F.W. [1982] Pure states as dual objects for C^*-algebras. *Commun. Math. Phys.* **82**, 497–509.

Shultz, F.W. [1981] Dual maps of Jordan homomorphisms and *-homomorphisms between C^*-algebras. *Pac. J. Math.* **93**, 435–441.

Simon, B. [1980] The classical limit of quantum partition functions. *Commun. Math. Phys.* **71**, 247–276.

Sjamaar R. and E. Lerman [1991] Stratified symplectic spaces and reduction. *Ann. Math.* **134**, 375–422.

Skandalis, G. [1991] Kasparov's bivariant K-theory and applications. *Exp. Math.* **9**, 193–250.

Sławianowski, J. [1991] *Geometry of Phase Spaces*. PWN, Warszawa and Wiley, Chichester.

Slawny, J. [1972] On factor representations and the C^*-algebra of the canonical commutation relations. *Commun. Math. Phys.* **24**, 151–170.

Slebarski, S. [1987] The Dirac operator on homogeneous spaces and representations of reductive Lie groups I. *Amer. J. Math.* **109**, 283–301.

Smale, S. [1970] Topology and Mechanics. I. *Inv. Math.* **10**, 305–331.

Śniatycki, J. [1980] *Geometric Quantization and Quantum Mechanics*. Springer, Berlin.

Śniatycki, J. and A. Weinstein [1983] Reduction and quantization for singular momentum mappings. *Lett. Math. Phys.* **7**, 155–161.

Solèr, M.P. [1995] Characterization of Hilbert spaces with orthomodular spaces. *Comm. Algebra* **23**, 219–243.

Souriau, J.-M. [1969] *Structure des Systèmes Dynamiques*. Dunod, Paris.

Souriau, J.-M. [1997] *Structure of Dynamical Systems: a Symplectic View of Physics*. Birkhäuser, Basel.

Sternberg, S. [1977] On minimal coupling and the symplectic mechanics of a classical particle in the presence of a Yang–Mills field. *Proc. Nat. Acad. Sci.* **74**, 5253–5254.

Stinespring, W. [1955] Positive functions on C^*-algebras. *Proc. Amer. Math. Soc.* **6**, 211–216.

Strathdee, J. [1983] Symmetry aspects of Kaluza–Klein theories. In: Ferrara, S. et al. (eds.) *Supergravity 1982*, pp. 170–182. World Scientific, Singapore.

Stratonovic, R.L. [1957] On distributions in representation space. *Sov. Phys. JETP* **4**, 891–898.

Strichartz, R.S. [1983] Analysis of the Laplacian on a complete Riemannian manifold. *J. Funct. Anal.* **52**, 48–79.

Strocchi, F. [1966] Complex coordinates in quantum mechanics. *Rev. Mod. Phys.* **38**, 36–40.

Sundermeyer, K. [1982] *Constrained Dynamics. Lecture Notes in Physics* **169**. Springer, Berlin.

Sussmann, H.J. [1973] Orbits of families of vector fields and integrability of distributions. *Trans. Amer. Math. Soc.* **180**, 171–188.

Takesaki, M. [1979] *Theory of Operator Algebras I*. Springer, Heidelberg.

Taylor, M.E. [1984] *Pseudo Differential Operators*. Princeton University Press, Princeton.

Taylor, M.E. and A. Uribe [1992] Semi-classical spectra of gauge fields. *J. Funct. Anal.* **110**, 1–46.

Teitelboim, C. [1982] Quantum mechanics of the gravitational field. *Phys. Rev.* **D25**, 3159–3179.

Teitelboim, C. [1984] Explicit evaluation of group–invariant measure as by–product of path integration over Yang–Mills fields. *J. Math. Phys.* **25**, 1093–1101.

Thirring, W. [1981] *Quantum Mechanics of Atoms and Molecules*. Springer, Berlin.

Thirring, W. [1983] *Quantum Mechanics of Large Systems*. Springer, Berlin.

Thirring, W. and H. Narnhofer [1992] Covariant QED without indefinite metric. *Rev. Math. Phys.* **Special issue** (dedicated to R. Haag) 193–211.

Thomas, E.G.F. [1983] A simple proof of the Cameron–Martin theorem making use of Schwartz reproducing kernels. *Bol. Soc. Mat. Mexicana* **28**, 67–76.

Topping, D. [1967] Asymptoticity and semimodularity in projection lattices. *Pac. J. Math.* **20**, 317–325.

Truman, A. [1976] Feynman path integrals and quantum mechanics as $\hbar \to 0$. *J. Math. Phys.* **17**, 1852–1862.

Truman, A. [1977] The classical action in non-relativistic quantum mechanics. *J. Math. Phys.* **18**, 1499–1509.

Tuynman, G.M. [1987a] Generalized Bergman kernels and geometric quantization. *J. Math. Phys.* **28**, 573–583.

Tuynman, G.M. [1987b] Quantization: towards a comparison between methods. *J. Math. Phys.* **28**, 2829–2840.

Tuynman, G.M. [1990] Reduction, quantization, and nonunimodular groups. *J. Math. Phys.* **31**, 83–90.

Tuynman, G.M. and W.A.J.J. Wiegerinck [1987] Central extensions in physics. *J. Geom. Phys.* **4**, 207–258.

Uhlmann, A. [1976] The "Transition Probability" in the state space of a *-algebra. *Rep. Math. Phys.* **9**, 273–279.

Uhlmann, A. [1993] Density operators as an arena for differential geometry. *Rep. Math. Phys.* **33**, 253–263.

Uhlmann, A. [1996] Spheres and hemispheres as quantum state spaces. *J. Geom. Phys.* **18**, 76–92.

Underhill, J. [1978] Quantization on a manifold with connection. *J. Math. Phys.* **19**, 1932–1935.

Unterberger, A. and J. Unterberger [1988] Quantification et analyse pseudodifferentielle. *Ann. Scient. Ec. Norm. Sup.* **21**, 133–158.

Unterberger, A. and H. Upmeier [1994] The Berezin transform and invariant differential operators. *Commun. Math. Phys.* **164**, 563–579.

Upmeier, H. [1987] Jordan algebras in analysis, operator theory, and quantum mechanics. *CBMS Reg. Conf. Ser. Math.* **67**, 1–85.

Upmeier, H. [1991] Weyl quantization of symmetric spaces. I. Hyperbolic matrix domains. *J. Funct. Anal.* **96**, 297–330.

Upmeier, H. [1996] *Toeplitz Operators and Index Theory in Several Complex Variables*. Birkhäuser, Basel.

Vaisman, I. [1994] *Lectures on the Geometry of Poisson Manifolds*. Birkhäuser, Basel.

Vaisman, I. [1996] Reduction of the Poisson–Nijenhuis manifolds. *J. Geom. Phys.* **19**, 90–98.

Varadarajan, V.S. [1985] *Geometry of Quantum Theory* (2nd ed.). Springer, New York.

Várilly, J.C., J.M. Gracia-Bondía, and W. Schempp [1990] The Moyal representation of quantum mechanics and special function theory. *Acta Appl. Math.* **18**, 225–250.

Vey, J. [1975] Déformation du crochet de Poisson sur une variété symplectique. *Commun. Math. Helv.* **50**, 421–454.

Vinet, L. [1988] Invariant Berry connections. *Phys. Rev.* **D37**, 2369–2372.

Vogan, D.A. [1987] *Unitary Representations of Reductive Lie Groups*. Princeton University Press, Princeton.

Vogan, D.A. [1992] Unitary representations of reductive Lie groups and the orbit method. In: Tirao, J. and N. Wallach (eds.) *New Developments in Lie Theory and their Applications*, pp. 87–114. Birkhäuser, Basel.

Voros, A. [1977] Asymptotic $\hbar$-expansions of stationary quantum states. *Ann. Inst. H. Poincaré* **26**, 343–403.

Voros, A. [1978] An algebra of pseudodifferential operators and the asymptotics of quantum mechanics. *J. Funct. Anal.* **29**, 104–132.

Voros, A. [1989] Wentzel–Kramers–Brillouin method in the Bargmann representation. *Phys. Rev.* **A40**, 6814–6825.

Wallach, N. [1973] *Harmonic Analysis on Homogeneous Spaces*. Dekker, New York.

Wallach, N. [1987] On the irreducibility and inequivalence of unitary representations of gauge groups. *Compositio Math.* **64**, 3–29.

Wang, X.-P. [1986] Approximation semi-classique de l'equation de Heisenberg. *Commun. Math. Phys.* **104**, 77–86.

Wang, X.-P. [1991] Semiclassical resolvent estimates for N-body Schrödinger operators. *J. Funct. Anal.* **97**, 466–483.

Warner, G. [1972] *Harmonic Analysis on Semi-simple Lie Groups*, Vol. I. Springer, Berlin.

Wegge-Olsen, N.E. [1993] *K-theory and C^*-algebras*. Oxford University Press, Oxford.

Weinberg. S. [1995] *The Quantum Theory of Fields. Vol. I.* Cambridge University Press, Cambridge.

Weinberg. S. [1996] *The Quantum Theory of Fields. Vol. II.* Cambridge University Press, Cambridge.

Weinstein, A. [1978] A universal phase space for particles in a Yang–Mills field. *Lett. Math. Phys.* **2**, 417–420.

Weinstein, A. [1983] The local structure of Poisson manifolds. *J. Diff. Geom.* **18**, 523–557. Err. *ibid.* **22** (1985) 255.

Weinstein, A. [1987] Poisson geometry of the principal series and nonlinearizable structures. *J. Diff. Geom.* **25**, 55–73.

Weinstein, A. [1989] Blowing up realizations of Heisenberg–Poisson manifolds. *Bull. Sc. math.* **(2) 113**, 381–406.

Weinstein, A. [1990] Affine Poisson structures. *Int. J. Math.* **1**, 343–360.

Weinstein, A. [1991] Noncommutative geometry and geometric quantization. In: *Symplectic Geometry and Mathematical Physics. Progr. Math.* **99**, pp. 446–461. Birkhäuser, Basel.

Weinstein, A. [1995a] Deformation quantization. *Sém. Bourbaki* **789**. *Astérisque* **227**, 389–409.

Weinstein, A. [1995b] The symplectic structure on moduli space. In: *The Floer Memorial volume. Progr. Math.* **133**, pp. 627–635. Birkhäuser, Basel.

Weinstein, A. [1996a] Groupoids: unifying internal and external symmetry. *Notices Amer. Math. Soc.* **43**, 744–752.

Weinstein, A. [1996b] Lagrangian mechanics and groupoids. *Fields Inst. Commun.* **7**, 207–231.

Weinstein, A. [1997] The modular automorphism group of a Poisson manifold. *J. Geom. Phys.* **23**, 379–394.

Weinstein, A. [1998] Poisson geometry. *Diff. Geom. Appl.* **9**, 213–238.

Weinstein, A. and P. Xu [1991] Extensions of symplectic groupoids and quantization. *J. reine angew. Math.* **417**, 159–189.

Weiss, E. and N. Zierler [1958] Locally compact division rings. *Pac. J. Math.* **8**, 369–371.

Werner, R.F. [1983] Physical uniformities in the state space of nonrelativistic quantum mechanics. *Found. Phys.* **13**, 859–881.

Werner, R.F. [1995] The classical limit of quantum theory. *e-print quant-ph/9504016.*

Westman, J. [1968] Harmonic analysis on groupoids. *Pac. J. Math.* **27**, 621–632.

Weyl, H. [1931] *The Theory of Groups and Quantum Mechanics*. Dover, New York.

Widom, H. [1980] A complete symbolic calculus for pseudodifferential operators. *Bull. Sc. math., 2e série* **104**, 19–63.

Wiedemann, U.A. [1994] *Constraints and Spontaneous Symmetry Breaking in Quantum Field Theory*, Ph.D. thesis, University of Cambridge.

Wiedemann, U.A. and N.P. Landsman [1996] The Stueckelberg-Kibble model as an example of quantized symplectic reduction. *J. Math. Phys.* **37**, 2731–2747.

Wightman, A.S. [1962] On the localizability of quantum mechanical systems. *Rev. Mod. Phys.* **34**, 845–872.

Wigner, E.P. [1931] *Gruppentheorie und ihre Anwendung auf die Quantenmechanik der Atomspektren*. Vieweg, Braunschweig.

Wigner, E.P. [1932] On the quantum correction for thermodynamic equilibrium. *Phys. Rev.* **40**, 749–759.

Wigner, E.P. [1939] Unitary representations of the inhomogeneous Lorentz group. *Ann. Math.* **40**, 149–204.

Wilbur, W.J. [1977] On characterizing the standard quantum logics. *Trans. Amer. Math. Soc.* **233**, 265–282.

Wildberger, N.J. [1992] The moment map of a Lie group representation. *Trans. Amer. Math. Soc.* **330**, 257–268.

Witten, E. [1991] On quantum gauge theories in two dimensions. *Commun. Math. Phys.* **141**, 153–209.

Witten, E. [1992] Two dimensional gauge theories revisited. *J. Geom. Phys.* **9**, 303–368.

Wong, S.K. [1970] Field and particle equations for the classical Yang–Mills field and particles with isotopic spin. *Nuovo Cim.* **A65**, 689–694.

Wong, Y.-C. and K.-F. Ng [1973] *Partially Ordered Topological Vector Spaces*. Clarendon Press, Oxford.

Woodhouse, N.M.J. [1992] *Geometric Quantization*. 2nd ed. Clarendon Press, Oxford.

Woronowicz, S. L. [1987] Compact matrix pseudogroups. *Commun. Math. Phys.* **111**, 613–665.

Woronowicz, S. L. [1995] C^*-algebras generated by unbounded elements. *Rev. Math. Phys.* **7**, 481–521.

Wren, K.K. [1997] Quantization of constrained systems with singularities using Rieffel induction. *J. Geom. Phys.* **24**, 173–202.

Wren, K.K. [1998a] Constrained quantization and θ-angles. II. *Nucl. Phys.* **B521 [PM]**, 471–502.

Wren, K.K. [1998b] *Constrained Quantization of Yang–Mills Theory via Rieffel Induction*. Ph.D. thesis, University of Cambridge.

Wright, J.D.M. [1977] Jordan C^*-algebras. *Michigan Math. J.* **24**, 291–302.

Wu, T.T. and C.N. Yang [1975] Concept of non-integrable phase factors and global formulation of gauge fields. *Phys. Rev.* **D12**, 3845–3857.

Wu, T.T. and C.N. Yang [1976] Dirac monopoles without strings: monopole harmonics. *Nucl. Phys.* **B107**, 365–380.

Wu, Y. [1998] Quantization of a particle in a background Yang–Mills field. *J. Math. Phys.* **39**. 867–875.

Xu, P. [1991a] Morita equivalent symplectic groupoids. In: Dazord, P. and A. Weinstein (eds.) *Symplectic Geometry, Groupoids, and Integrable Systems*, pp. 291–311. Springer, New York.

Xu, P. [1991b] Morita equivalence of Poisson manifolds. *Commun. Math. Phys.* **142**, 493–509.

Xu, P. [1992] Morita equivalence and symplectic realizations of Poisson manifolds. *Ann. Sc. Ec. Norm. Sup.* **25**, 307–333.

Yaffe, L.G. [1982] Large *N* limits as classical mechanics. *Rev. Mod. Phys.* **54**, 407–428.

Yajima, K. [1979] The quasi-classical limit of quantum scattering theory. *Commun. Math. Phys.* **69**, 101–129.

Yosida, K. [1980] *Functional Analysis*, 6th ed. Springer, Berlin.

Zabey, P.C. [1975] Reconstruction theorems in quantum mechanics. *Found. Phys.* **5**, 323–342.

Zakrzewski, S. [1986] Induced representations and induced Hamiltonian actions. *J. Geom. Phys.* **3**, 211–219.

Zakrzewski, S. [1990a] Quantum and classical pseudogroups. I. Union pseudogroups and their quantization. *Commun. Math. Phys.* **134**, 347–370.

Zakrzewski, S. [1990b] Quantum and classical pseudogroups. II. Differential and symplectic pseudogroups. *Commun. Math. Phys.* **134**, 371–395.

Zelditch, S. [1992] On a "quantum chaos" theorem of R. Schrader and M. Taylor. *J. Funct. Anal.* **109**, 1–21.

Zhang, W.M., D.H. Feng, and R. Gilmore [1990] Coherent states: theory and some applications. *Rev. Mod. Phys.* **62**, 867–927.

Ziegler, F. [1996] *Méthode des Orbites et Représentations Quantiques*. Ph.D. thesis, Université de Provence.

Zierler, N. [1961] Axioms for non-relativistic quantum mechanics. *Pac. J. Math.* **11**, 1151–1169.

Index